Encyclopedic Reference of Immunotoxicology

Hans-Werner Vohr (Ed.)

Encyclopedic Reference of Immunotoxicology

With 144 Figures and 105 Tables

Professor Dr. Hans-Werner Vohr
PH-PD, Toxicology
Bayer HealthCare AG
Aprather Weg 18a
D-42096 Wuppertal
hans-werner.vohr@bayerhealthcare.com

3-540-44172-7 Springer Berlin Heidelberg New York

Library of Congress Control Number: 2004105677

Bibliographic information published by *Die Deutsche Bibliothek*.
Die Deutsche Bibliothek lists this publication in the Deutsche Nationalbibliografie;
detailed bibliographic data is available in the Internet at: http://dnb.ddb.de

This work is subject to copyright. All rights are reserved, whether the whole or part of the material is concerned, specifically the rights of translation, reprinting, reuse of illustrations, recitation, broadcasting, reproduction on microfilms or in other ways, and storage in data banks. Duplication of this publication or parts thereof is only permitted under the provisions of the German Copyright Law of September 9, 1965, in its current version, and permission for use must always be obtained from Springer-Verlag. Violations are liable for prosecution under the German Copyright Law.

Springer is part of Springer Science+Business Media

springeronline.com

© Springer-Verlag Berlin Heidelberg New York 2005
Printed in Germany

The use of registered names, trademarks, etc. in this publication does not imply, even in the absence of a specific statement, that such names are exempt from the relevant protective laws and regulations and therefore free for general use.

Product liability: The publishers cannot guarantee the accuracy of any information about the application of operative techniques and medications contained in this book. In every individual case the user must check such information by consulting the relevant literature.

Editor: Thomas Mager, Heidelberg
Development Editor: Andrew Spencer, Heidelberg
Production Editor: Frank Krabbes, Heidelberg
Cover design: Erich Kirchner, Heidelberg
Printed on acid free paper   SPIN: 10838455   14/2109 - 5 4 3 2 1 0

To Heide, Florian,
Hannah and Lucas

Preface

Over the last three decades, immunotoxicology has blossomed from a somewhat neglected subsidiary branch of immunology into an independent science. This process has been aided by incidents such as those involving dioxins in Italy, polybrominated biphenyls (PBBs) in Michigan and polychlorinated biphenyl (PCB) in Taiwan and Japan, which have drawn the attention of both the public and the authorities to the effects of substances on the immune system. Incidents such as these have resulted in generous financial support being made available for attempts to elucidate mechanisms potentially involved in the interaction of the immune system with chemicals. Intensified efforts to explain congenital human immunodeficiencies and the spread of acquired immunodeficiencies, for example after HIV infection, have naturally also increasingly brought the immune system to the attention of a wider public.

Many years' clinical experience of the use of immunomodulators, for example in the treatment of cancer or AIDS and in transplant recipients, have established a basis for documenting adverse effects on the immune system caused by drugs. On top of this, easy and reliable ways to identify the immunotoxicological potential of a substance were discussed for several years, aiming towards a reasonably small number of universally and easily applicable methods. Substantial information, however, was not gained until international ring trials and laboratories within the pharmacological and chemical industry started to investigate immunotoxicological aspects of new substances. Only then did the importance of specific substance requirements become clear.

Immunotoxicology therefore focuses on the undesirable effects of chemicals on the immune system. The exposure of humans or animals to such agents may be intentional (drugs) or unintentional (environment). Since the immune system is a very complex, well-regulated system, its interactions with substances, are as diverse as they are complicated. The side effects may lead to over-activation of the immune system, or equally to immunosuppression. The end points of dysregulation are therefore also varied: allergies, cancer, autoimmunity, poor resistance to infection, flu-like reactions, chronic inflammation and effects on hepatic metabolism. While some of these side-effects like allergies, flu-like reactions, or poor host resistance already appear after relatively short treatment periods, others are normally induced after longer treatments, e.g. cancer or autoimmunity. Therefore, showing proof that compounds are responsible for cancer or autoimmunity reactions by primary immunotoxicity is often difficult. On the other hand, the huge capacity of the immune system, normal immune reactions, the overall toxicological situation and beneficial therapeutic effects also have to be taken into account in the evaluation of "immunotoxic" findings.

All of these widely differing aspects are featured in this book. Without a basic knowledge of the key components of the immune system, the interactions of various organs with the immune system, and the normal reactions of the immune system to infections, it is easy to misinterpret data from immunotoxicological studies. These issues are therefore addressed in this book in their entirety. Each contribution has been written by an experienced expert in the field in question, with the result that a solid foundation of knowledge has been established.

This would not have been possible without the immense commitment of all involved. I should like to express my gratitude to them here, starting with the members of the Editorial Board: Jack Dean, Dori Germolec, Mike Holsapple, Robert House, Henk van Loveren, Mike Luster, Peter Ulrich and Kimber White jr. These people have been wonderfully supportive from the outset; they have given the project much impetus and put a lot of work into it themselves. Next, I should of course like to thank the many authors who have produced their contributions with great dedication and enthusiasm. All the reviewers' requested amendments and addenda were usually implemented in timely and very cooperative fashion. Lastly, my thanks are due to the staff of the publisher, Springer, in Heidelberg and above all to Dr. R. Lange (Editorial Director Medicine and Biomedicine) and Dr. T. Mager (Planning Drugs and Therapy), who introduced me to the book series and infected me with their enthusiasm for this encyclopaedia. It is intended not as a collection of loosely related papers but as a compact information source written by specialists in a way that is generally comprehensible – an intention that inspired me from the beginning and the objectives of which are borne out by this book. Special thanks are due to Mr. A. Spencer (Development Editor), who always kept tabs on all aspects of the book, never lost track of the progress of the many manuscripts and always managed to come up with constructive solutions to problems.

As the term itself suggests, immunotoxicology is a field that can be approached from the perspective of either immunology or toxicology. To generalize broadly, it might be said that the immunological arm deals mainly with fundamental mechanisms, while the toxicologists are concerned more with preclinical and clinical studies and so work in the field of applied immunotoxicology. It is to be hoped that the various contributions to this book manage to do justice to both groups. The book was naturally devised, structured, written and set in motion by human beings. It is therefore clear that the reader will here and there regret the absence of an entry, notice an error in a diagram or feel that certain articles are too short or too long. The prime concern of all those involved in producing this encyclopedia was to provide readers with entertaining but sound and scientifically correct information. They would therefore all welcome constructive criticism or offers of assistance with a future edition of the encyclopedia.

HANS-WERNER VOHR

Editorial Board

Prof. Dr. Jack Dean
Sanofi-Aventis
11 Great Valley Parkway
Malvern, PA 19355
USA

Dr. Dori Germolec
NIEHS, NIH
111 Alexander Dr, P.O. Box 12233
Research Triangle Park, NC 27709
USA

Prof. Dr. Michael Holsapple
Health and Environmental Sciences Institute
One Thomas Circle, NW, Ninth Floor
Washington, DC 20005-5802 USA

Prof. Dr. Robert V. House
DynPort Vaccine Company LLC
60 Thomas Johnson Drive
Frederick, MD 21702
USA

Prof. Dr. Henk Loveren
National Institute of Public Health and the Environment
Section Immunobiology and Haematology
PO Box 1 3720 BA Bilthoven Netherlands

Prof. Dr. Michael I. Luster
National Institute for Occupational Safety and Health Chief,
Toxicology and Molecular Biology Branch
1095 Willowdale Road
West Virginia, 26505-2888
USA

Dr. Peter Ulrich
Novartis Pharma AG
Preclinical Safety
MUT2881.329 Auhafenstrasse
CH-4132 Muttenz
Switzerland

Prof. Dr. Kimber L. White
Virginia Commonwealth University
Dept. of Pharmacology & Toxicology
PO Box 980613
Richmond, VA 23298
USA

List of Contributors

MARIO ASSENMACHER
Miltenyi Biotec GmbH
Friedrich-Ebert-Str. 68
D-51429 Bergisch Gladbach
Germany
Mario@miltenyibiotec.de

HAVA KARSENTY AVRAHAM
Division of Experimental Medicine, Beth Israel Deaconess
Medical Center
Harvard Institutes of Medicine
Boston, MA 02115
USA
havraham@caregroup.harvard.edu

SHALOM AVRAHAM
Division of Experimental Medicine, Beth Israel Deaconess
Medical Center
Harvard Institutes of Medicine
Boston, MA 02115
USA

SHUKAL BALA
Division Of Special Pathogen and Immunologic Drug
Products, Center For Drug Evaluation And Research
US Food and Drug Administration
USA

JOHN BARNETT
Dept of Microbiology and Immunology
West Virginia University, Health Sciences
Morgantown Center North
Morgantown, WV 26506-9177
USA
jbarnett@hsc.wvu.edu

DAVID BASKETTER
Applied Science and Technology Safety and Environmental
Assurance Centre
Unilever Colworth Laboratory
Sharnbrook
Bedford
MK44 1LQ
UK
david.basketter@unilever.com

YAACOV BEN-DAVID
Department of Medical Biophysics
University of Toronto, Canada and Sunnybrook and
Women's College Health Sciences Center
Toronto, Ontario,
Canada
bendavid@srcl.sunnybrook.utoronto.ca

CLAUDIA BEREK
Deutsches Rheuma Forschungszentrum
Schumannstr. 21–22
D-10117 Berlin
Germany
berek@drfz.de

JÖRG BLÜMEL
Merz Pharmaceuticals GmbH
Eckenheimer Landstrasse 100
D-60318 Frankfurt a. M.
Germany
joerg.bluemel@merz.de

ANNE PROVENCHER BOLLIGER
Rebbergstrasse 59
CH-4800 Zofingen
Switzerland
Acbolliger@bluewin.ch

BRAD BOLON
GEMpath Inc.
2540 N 400 W
Cedar City, UT 84720
USA
bradgempath@aol.com

S GAYLEN BRADLEY
College of Medicine/Research Affairs
Penn State University
500 University Drive
Hershey, PA 17033
USA
gbradley@psu.edu

KATHLEEN M BRUNDAGE
Dept of Microbiology, Immunology and Cell Biology
West Virginia University
Morgantown, PO Box 9177
USA
rmage@niaid.nih.gov

GEORG BRUNNER
Leiter der Tumorforschung
Fachklinik Hornheide an der Universität Münster
Dorbaumstrasse 300
D-48157 Münster
Germany
georg.brunner@fachklinik-hornheide.de

PETER J BUGELSKI
Experimental Pathology
Centocor, Inc.
R-4-2, 200 Great Valley, Parkway
Malvern, PA 19355
USA

SCOTT W BURCHIEL
College of Pharmacy Toxicology Program
The University of New Mexico
Albuquerque, NM 87131
USA
sburchiel@salud.unm.edu

LEIGH ANN BURNS-NAAS
Pfizer Global Research & Development
10777 Science Center Dr.
San Diego, CA 92121
USA
leighann.burns@pfizer.com

JEANINE L BUSSIERE
Amgen Inc.
One Amgen Center Drive
Thousand Oaks, CA 91320-1799
USA
bussierj@amgen.com

SCOTT B CAMERON
Division Infectious Diseases, Department of Medicine
University of British Columbia
Vancouver BC,
Canada

MICHELLE CAREY
NIEHS ND D2-01, Laboratory of Pulmonary Pathobiology
RTP, PO Box 12233
USA
carey1@niehs.nih.gov

KARIN CEDERBRANT
Safety Assessment
AstraZeneca
Södertalje
151 85
Sweden
karin.cederbrant@astrazeneca.com

ANTHONY W CHOW
Division Infectious Diseases, Department of Medicine
University of British Columbia
Vancouver BC,
Canada

MITCHELL D COHEN
Department of Environmental Medicine
New York University School of Medicine
57 Old Forge Road
Tuxedo, NY 10987
USA
cohenm@env.med.nyu.edu

DOROTHY COLAGIOVANNI
OSI Pharmaceuticals, Inc.
2860 Wilderness Place
Boulder, CO 80301
USA
dcolagiovanni@osip.com

MARCELA CONTRERAS
National Blood Service
Development and Research, North London Centre
Colindale Avenue
London
NW9 5BG
UK

JOEL B CORNACOFF
Centocor Inc.
200 Great Valley Parkway
Malvern, PA 19403
USA

EMANUELA CORSINI
Dapartment of Pharmacological Sciences
University of Milan
Via Balzaretti 9
Milan
20133
Italy
Emanuela.corsini@unimi.it

RENÉ CREVEL
Safety &, Environmental Assurance Centre
Unilever Colworth
Sharnbrook, Bedford
Bedfordshire
MK44 1LQ
UK
Rene.Crevel@unilever.com

CHRISTOPHER CUFF
Dept. Microbiology, Immunology, and Cell Biology
West Virginia University
Morgantown, Box 9177
USA

CHARLES J CZUPRYNSKI
Department of Pathological Sciences
University of Wisconsin
2015 Linden Drive W
Madison, WI 53706
USA
czuprync@svm.vetmed.wisc.edu

JAN GMC DAMOISEAUX
Dept Clinical and Experimental Immunology
University Hospital Maastricht
6202 AZ Maastricht
The Netherlands
jdam@limm.azm.nl

GEOFF DANIELS
Bristol Institute for Transfusion Sciences
National Blood Service
Soutmead Road
Bristol
BS10 5ND
UK

ANTHONY D DAYAN
Department of Toxicology
University of London
London
UK

REBECCA J DEARMAN
Syngenta Central Toxicology Laboratory
Alderley Park, Macclesfield
Cheshire
SK10 4TJ
UK

Sarah V M Dodson
Department of Microbiology, Immunology, and Cell Biology
West Virginia University Health Sciences Center
Morgantown, WV 26506
USA

Alan Ebringer
Division of Life Sciences
King's College, University of London
150 Stamford Street
London
SE1 8WA
UK
alan.ebringer@kcl.ac.uk

Andrea Engel
Biosciences Life Science Research
Becton-Dickinson
Tullastr. 8–12
69126 Heidelberg
Germany
Andrea_Engel@Europe.bd.com

Charlotte Esser
Inst. für Umweltmedizinische Forschung
Auf'm Hennekamp 50
40225 Düsseldorf
Germany
chesser@uni-duesseldorf.de

Kimberly J Fairley
National Institute for Occupational Safety and Health
1095 Willowdale Road
Morgantown, WV 26505
USA

Rafael Fernandez-Botran
Dept Pathology and Laboratory Medicine
University of Louisville
Louisville, KY 40292
USA

Rafael Fernandez-Botran
Dept Pathology and Laboratory medicine
University of Louisville
Louisville, KY 40292
USA

Dennis K Flaherty
Biology Department
Lamar University Beaumont
Beaumont, PO Box 10037
USA
flahertydk@hal.lamar.edu

Werner Frings
Covance Laboratories GmbH
Kesselfeld 29
D-48163 Münster
Germany

Shayne Cox Gad
Gad Consulting Service
102 Woodtrail Lane
Cary, NC 27511
USA
scgad@gadconsulting.com

Donald E Gardner
Inhalation Toxicology Associates Inc.
Raleigh, P.O. Box 97605
USA

Susan C Gardner
Inhalation Toxicology Associates Inc.
Raleigh, P.O. Box 97605
USA

Johan Garssen
Laboratory for Toxicology, Pathology and Genetics
National Institute for Public Health and the Environment
(RIVM)
3720 BA Bilthoven
The Netherlands

Anatoliy A Gashev
Department of Medical Physiology, College of Medicine,
Cardiovascular Research Institute Division of Lymphatic
Biology, Texas A&M University System Health Science
Center
336 Reynolds Medical Building
College Station, TX 77843
USA

Jorge Geffner
IIHEMA
Academia Nacional de Medicina
Pacheco de Melo 3081
Buenos Aires, 1425
Argentina
geffnerj@fibertel.com.ar

Gernot Geginat
Institut für Medizinische Mikrobiologie
Fakultät für klinische Medizin Mannheim der Universität
Heidelberg, Klinikum Mannheim
Theodor-Kutzer-Ufer 1–3
D-68167 Mannheim
Germany
geginat@rumms.uni-mannheim.de

Diethard Gemsa
Institute of Immunology
Philipps-University Marburg
Robert Koch-Strasse 17
D-35037 Marburg
Germany
gemsa@mailer.uni-marburg.de

Frank Gerberick
Human Safety Department
Procter & Gamble Company
Cincinnati, PO Box 538707
USA
Gerberick.gf@pg.com

Dori Germolec
SIEHS/NIH
111 Alexander Drive, PO Box 12233
Research Triangle Park, NC 27709
USA
germolec@niehs.nih.gov

Kathleen M Gilbert
Associate Professor, Department of Microbiology and Immunology
University of Arkansas for Medical Sciences, Arkansas Children's Hospital Research Institute
1120 Marshall Street
Little Rock
AR 72202

Jill Giles-Komar
Centocor Inc.
200 Great Valley Parkway
Malvern, PA 19403
USA

Elizabeth R Gore
Immunologic Toxicology Preclinical Safety Assessment
GlaxoSmithKline R&D
709 Swedeland Road
King of Prussia, PO Box 1539
USA

Peter Griem
Produktsicherheit/Toxikologie
Wella AG
Berliner Allee 65
D–64274 Darmstadt
Germany

Ina Hagelschuer
PH-R ZfV, Geb. 516
Bayer HealthCare AG
Aprather Weg 18
D-42096 Wuppertal
Germany
ina.hagelschuer.ih@@bayerhealthcare.com

Helen G Haggerty
Bristol-Myers Squibb Co.
6000 Thompson Road
EA Syracuse, NY 1305755
USA
helen.haggerty@bms.com

Andrew Hall
Dept. of Epidemiology & Population Health
London School of Hygiene & Tropical Medicine
Keppel Street
London
WC1E 7HT
UK
Andy.Hall@lshtm.ac.uk

S Hanneken
Department of Dermatology
University of Düsseldorf
Germany

Kenneth L Hastings
Division Of Special Pathogen and Immunologic Drug Products, Center For Drug Evaluation And Research
US Food and Drug Administration
USA
hastingsk@cder.fda.gov

Arie H Havelaar
Laboratory for Toxicology, Pathology and Genetics
National Institute for Public Health and the Environment (RIVM)
3720 BA Bilthoven
The Netherlands

Eckhart Heisler
PH-PD, Toxicology
Bayer HealthCare AG
Aprather Weg
42096 Wuppertal
Germany

Ricki M Helm
Arkansas Children's Hospital, Research Institute
University of Arkansas for Medical Sciences
1120 Marshall Street
Little Rock, AR 72202
USA
HelmRickiM@uams.edu

Reinhard Henschler
Institute for Transfusion Medicine und Immune Hematology, Department of Cell Production and Stem Cell Biology Group
German Red Cross Blood Donation Center
Sandhofstrasse 1
D-60528 Frankfurt a. M.
Germany
rhenschler@bsdhessen.de

Thomas Herrmann
Institute for Virology and Immunobiology
University of Würzburg
Versbacher Strasse 7
D-97078 Würzburg
Germany
herrmann-t@vim.uni-wuerzburg.de

Danuta J Herzyk
Immunologic Toxicology, Preclinical Safety Assessment
GlaxoSmithKline R&D
709 Swedeland Road
King of Prussia, PA 19406-0939
USA
Danuta.J.Herzyk@gsk.com

Rachel R Higgins
Department of Medical Biophysics
University of Toronto, Canada and Sunnybrook and Women's College Health Sciences Center
Toronto, Ontario,
Canada
rachel.higgins@qs-quote.com

Bettina Hitzfeld
Abt. Stoffe, Boden, Biotechnologie
BUWAL
3003 Bern
Switzerland
Bettina.Hitzfeld@buwal.admin.ch

Steven Holladay
Dept of Biomedical Sciences & Pathobiology
Virginia Tech
Southgate Drive
Blacksburg, A 24061-04422
USA
holladay@vt.edu

Michael Holsapple
Health and Environmental Sciences Institute
One Thomas Circle, NW, Ninth Floor
Washington, DC, 20005-5802
USA
mholsapple@ilsi.org

Robert V House
DynPort Vaccine Company LLC
64 Thomas Johnson Drive
MD 21702
Frederick,
USA
HouseR@dynport.com

Lucy Hughes
Division of Life Sciences
King's College, University of London
150 Stamford Street
London
SE1 8WA
UK

Tae Cheon Jeong
College of Pharmacy
Yeungnam University
214-1, Dae-dong
Kyungsan
712-749
Korea
taecheon@yu.ac.kr

Victor J Johnson
National Institute for Occupational Safety and Health
1095 Willowdale Road
Morgantown, WV 26505
USA

Wim H de Jong
Laboratory for Toxicology, Pathology and Genetics
National Institute for Public Health and the Environment (RIVM)
3720 BA Bilthoven
The Netherlands

Rob de Jonge
Laboratory for Toxicology, Pathology and Genetics
National Institute for Public Health and the Environment (RIVM)
3720 BA Bilthoven
The Netherlands

Arati Kamath
Brigham and Women's Hospital, Division of Rheumatology, Immunology, and Allergy
Harvard Medical School
Smith 518, 1 Jimmy Fund Way
Boston, MA 02115
USA
akamath@rics.bwh.harvard.edu

Norbert E Kaminski
Department of Pharmacology and Toxicology
Michigan State University
B440 Life Science Building
East Lansing, MI 48324
USA
kamins11@msu.edu

Ronald Kaminsky
Centre de Recherche Santé Animale
CH-1566 St-Aubin
Switzerland
ronald.kaminsky@ah.novartis.com

Meryl Karol
Department of Environmental and Occupational Health
University of Pittsburgh
130 Desoto Street
Pittsburgh, PA 15261
USA
mhk+@pitt.edu

Michael L Kashon
Biostatistics Branch
National Institute for Occupational Safety and Health
Morgantown, WV 26505
USA
MKashon@cdc.gov

Nancy I Kerkvliet
Oregon State University
Dept. Environmental and Molecular Toniology
Corvallis, OR 97331
USA

Ian Kimber
Syngenta Central Toxicology Laboratory
Alderley Park, Macclesfield
Cheshire
SK10 4TJ
UK
ian.kimber@syngenta.com

David M Knight
Centocor Inc.
200 Great Valley Parkway
Malvern, PA 19355
USA

AC Knulst
Afd. Dermatology/Allergology
University Medical Center
3508 SA Utrecht
The Netherlands
a.c.knulst@azu.nl

Eugen Koren
Clinical Immunology, Amgen Inc.
1 Amgen Center Drive
Thousand Oaks, CA,
USA
ekoren@amgen.com

Georg Kraal
Department of Molecular Cell Biology
Vrije Universiteit Medical Center
1007MB Amsterdam
The Netherlands
g.kraal@vumc.nl

Anke Kretz-Rommel
Principal Scientist
Alexion Antibody Technologies
Suite A, 3958 Sorronto Valley Rd
San Diego, CA 92121
USA
kretz-rommel@alxnsd.com

C Frieke Kuper
Toxicology and Applied Pharmacology
TNO Food and Nutrition Research Zeist
The Netherlands
kuper@voeding.tno.nl

Gregory Ladics
DuPont Haskell Laboratory
Newark, DE 19714
USA
gregory.s.ladics@usa.dupont.com

Michael Laiosa
NIAID/NIH
Building 4, Room 111
Bethesda, MD 20892
USA
mlaiosa@niaid.nih.gov

Kenneth S Landreth
Department of Microbiology, Immunology, and Cell Biology
West Virginia University Health Sciences Center
Morgantown, WV 26506
USA
klandret@wvu.edu

B Paige Lawrence
Department of Pharmaceutical Sciences
College of Pharmacy, Washington State University
Pullman, WA 99164 6534
USA

David A Lawrence
Laboratory of Clinical and Experimental Endocrinology and Immunology
Wadsworth Center
Albany, NY 12201-0509
USA

Byeong-Chel Lee
Division of Experimental Medicine, Beth Israel Deaconess Medical Center
Harvard Institutes of Medicine
Boston, MA 02115
USA

William Lee
Wadsworth Center
David Axelrod Institute for Public Health
Albany, PO Box 22002
USA
leew@wadsworth.org

Lasse Leino
Department of Clinical Chemistry
University of Turku
Turku
FI-20520
Finland
Lasse.leino@orionpharma.fi

Hilmar Lemke
Biochemisches Institut, Raum 116, 1. OG Altbau
Universität Kiel
Rudolf-Höber-Straße 1
D-24118 Kiel
Germany
hlemke@biochem.uni-kiel.de

JG Lewis
Department of Pathology
Duke University Medical Center
Durham, Box 3712
USA
lewis026@mc.duke.edu

Jutta Liebau
Fachklinik Hornheide
Dorbaumstrasse 300
D-48157 Münster
Germany
jutta.liebau@fachklinik-hornheide.de

Pier-Luigi Lollini
Cancer Research Section
Department of Experimental Pathalogy, University of Bologna
Viale Filopanti 22
Bologna
40126
Italy
pierluigi@lollini.dsnet.it

HENK VAN LOVEREN
Department of Toxicology
University of London
London
UK

BOB LUEBKE
Immunotoxicology Branch
Research Triangle ParkMail Drop B143-04, US EPA
USA
luebke.robert@epa.gov

MICHAEL I LUSTER
National Institute for Occupational Safety and Health
1095 Willowdale Rd
Morgantown, WV 26505
USA
mluster@cdc.gov

ROSE G MAGE
Molecular Immunogenetics Section Laboratory of Immunology
NIAID Building 10 11 N 311, MSC 1892 NIH, 10 Center Drive
Bethesda, MD
USA
rmage@niaid.nih.gov

CURTIS C MAIER
R&D, Toxicology Preclinical Safety Assessment
GlaxoSmithKline
709 Swedeland Road
King of Prussia, P.O.Box 1539
USA
Curtis.C.Maier@gsk.com

MICHAEL U MARTIN
Institute of Immunology
Justus-Liebig-University, Giessen
Winchesterstrasse 2
D-35394 Giessen
Germany
Michael.Martin@bio.uni-giessen.de

THOMAS MAURER
Toxicology
Swissmedic
Erlachstrasse 8
CH-3000 Bern 9
Switzerland
thomas.maurer@swissmedic.ch

SUSAN C MCKARNS
Laboratory of Cellular and Molecular Immunology
NIAID/NIH
Building 4, Room 111, MSC 0420, 4 Center Drive
Bethesda, MD 20892
USA
smckarns@niaid.nih.gov

B JEAN MEADE
National Institute for Occupational Safety and Health
1095 Willowdale Road
Morgantown, WV 26505
USA

BERNHARD MOSER
Theodor-Kocher Institute
University of Bern
Winchesterstrasse 2
CH-3012 Bern
Switzerland

SHIGEKAZU NAGATA
Osaka University Medical School
Osaka,
Japan
nagata@genetic.med.osaka-u.ac.jp

MARIANNE NAIN
Institute of Immunology
Philipps-University Marburg
Robert Koch-Strasse 17
D-35037 Marburg
Germany

NORBERT J NEUMANN
Department of Dermatology
University of Düsseldorf
Germany

DEBORAH L NOVICKI
Toxicology
Chiron Corp.
4560 Horton Street
Emeryville, CA 94608
USA
deborah_novicki@chiron.com

JOHN L OLSEN
Stony Brook University Medical School
327 Sheep Pasture Road
Setauket, NY 11733
USA
john.olsen@ravel.informatics.sunysb.edu

JÜRGEN PAULUHN
Toxikology
Bayer HealthCare AG
Aprather Weg
D-42096 Wuppertal
Germany
juergen.pauluhn.jp@@bayerhealthcare.com

WERNER PICHLER
Klinik für Rheumatologie & klinische Immunologie/Allergologie
Inselspital-Universtät Bern
3010 Bern
Switzerland
werner.pichler@insel.ch

RAYMOND PIETERS
Head Immunotoxicology
Institute for Risk Assessment Sciences (IRAS)
3508 TD Utrecht
The Netherlands
R.Pieters@iras.uu.nl

K Michael Pollard
Department of Molecular and Experimental Medicine
The Scripps Research Institute
10550 North Torrey Pines Road
La Jolla, CA 92037
USA
mpollard@scripps.edu

Klaus T Preissner
Biochemisches Institut
Universitätsklinikum der Justus-Liebig-Universität Gießen
Friedrichstrasse 24
D-35392 Giessen
Germany
Klaus.T.Preissner@biochemie.med.uni-giessen.de

Stephen B Pruett
Department of Cellular Biology and Anatomy
Louisiana State University
Health Sciences Center
Shreveport, Louisiana 71130
USA
spruet@lsuhsc.edu

Neil R Pumford
Adjunct Professor, POSC O-214
University of Arkansas
1260 W. Maple Street
Fayetteville, AR 72701
USA
npumford@uark.edu

Taha Rashid
Division of Life Sciences
King's College, University of London
150 Stamford Street
London
SE1 8WA
UK

Helen V Ratajczak
Boehringer Ingelheim Pharmaceuticals
900 Ridgebury Road
Ridgefield, CT 6877
USA
hratajcz@rdg.boehringer-ingelheim.com

Frank AM Redegeld
Dept Pharmacology + Patophysiology
University Utrecht Institute of Pharmaceutical Sciences
3508 TB Utrecht
The Netherlands
F.A.M.Redegeld@pharm.ruu.nl

Jean F Regal
School of Medicine
University of Minnesota
Duluth, MN 55812
USA

Klaus Resch
Institute of Pharmacology
Hannover Medical School
Carl-Neuberg-Str. 1
D-30625 Hannover
Germany
Resch.Klaus@MH-Hannover.de

Kathleen Rodgers
Livingston Research
University of Southern California
1321 N. Mission Road
Los Angeles, CA 90033
USA
krodgers@hsc.usc.edu

Danielle Roman
PCS Toxicology/Pathology
Novartis Pharma AG Muttenz
Switzerland
danielle.roman@pharma.novartis.com

Noel R Rose
Dept Pathology and Dept Molecular Microbiology and Immunology, The Johns Hopkins Medical Institutions
Bloomberg School of Public Health
615 North Wolfe Street
Baltimore, MD 21205
USA
nrrose@jhsph.edu

Gary J Rosenthal
Drug Development
RxKinetix Inc.
1172 Century Drive
Louisville, CO 80027
USA
rosenthal@rxkinetix.com

Tina Sali
NIEHS Mail Drop E4–09, Laboratory of Molecular Carcinogenesis
111 Alexander Drive
Research Triangle Park, PO Box 12233
USA
sali@niehs.nih.gov

Janneke N Samsom
Department of Molecular Cell Biology
Vrije Universiteit Medical Center
1007MB Amsterdam
The Netherlands

Huub FJ Savelkoul
Cell Biology and Immunology Group, Wageningen University Wageningen
The Netherlands
huub.savelkoul@wur.nl

Rosana Schafer
Dept. Microbiology, Immunology, and Cell Biology
West Virginia University
Morgantown, Box 9177
USA
rschafer@wvu.edu

Mark Schatz
Institute of Immunology
University of Mainz
Obere Zahlbacher 67
D-55131 Mainz
Germany
markschatz@gmx.de

Hansjoerg Schild
Institute of Immunology
University of Mainz
Obere Zahlbacher 67
D-55131 Mainz
Germany

David Shepherd
Center for Environmental Health Sciences, Dept of medical and Pharmaceutical Sciences
University of Montana
58 Skaggs Building, 32 Campus Drive
Missoula, MT 59812
USA
shepherd@selway.umt.edu

Tetsuo Shiohara
Department of Dermatology
Kyorin University School of Medicine
6-20-2 Shinkawa Mitaka
Tokyo, 181-8611
Japan
tpshio@kyorin-u.ac.jp

Allen Silverstone
Upstate Medical University
166 Irving Ave.
Bethesda, NY 13210
USA

Petia P Simeonova
Toxicology and Molecular Biology Branch
National Institute for Occupational Safety and Health
Morgantown, WV 26505
USA
PSimeonova@cdc.gov

Ralph J Smialowicz
National Health & Environmental Effects, Research Laboratory
US Environmental Protection Agency
Research Triangle ParkNC 27711
USA
Smialowicz.ralph@epa.gov

KGC Smith
Addenbrooke's Hospital, Cambridge Institute for Medical Research, Department of Medicine
Box 139
Cambridge
CB2 2XY
UK
kgcs2@cam.ac.uk

Jeanne M Soos
Immunologic Toxicology Preclinical Safety Assessment
GlaxoSmithKline R&D
709 Swedeland Road
King of Prussia, PO Box 97605
USA
Jeanne.M.Soos@gsk.com

Koert J Stittelaar
Institute for Virology
3000 DR Rotterdam
The Netherlands
k.stittelaar@erasmusmc.nl

Frank Straube
MUT-2881.330 Biomarker Development
Novartis Pharma AG
4002 Basel
Switzerland
Frank.straube@pharma.novartis.com

Courtney EW Sulentic
Department of Pharmacology and Toxicology
Michigan State University
B440 Life Science Building
East Lansing, MI 48824
USA
courtney.sulentic@wright.edu

B Swart
Child Health Research Institute, Women's and Children's Hospital
Adelaide
Australia

Katsuhisa Takumi
Laboratory for Toxicology, Pathology and Genetics
National Institute for Public Health and the Environment (RIVM)
3720 BA Bilthoven
The Netherlands

Maciej Tarkowski
Nofer Institute of Occupational Medicine
Lodz
Poland

Jan Willem Cohen Tervaert
Dept Clinical and Experimental Immunology
University Hospital Maastricht
6202 AZ Maastricht
The Netherlands

Peter T Thomas
Early Development
Covance Laboratories
Madison, Wisconsin
USA
werner.pichler@insel.ch

SALLY S TINKLE
Health Effects Laboratory Division
National Institutes for Occupational Safety and Health,
Centers for Disease Control and prevention
1095 Willowdale Road
Morgantown, WV 26505
USA

GEORGE TREACY
Centocor Inc.
200 Great Valley Parkway
Malvern, PA 19355
USA

KEVIN TROUBA
NIEHS Mail Drop C1-04, Envionmental Immunology Laboratory
111 Alexander Drive
Research Triangle Park, PO Box 12233
USA
trouba@niehs.nih.gov

HELEN TRYPHONAS
Toxicology Research Division
Food Directorate, Health Products and Food Branch
Frederick G Banting Research Center, Tunney's Pasture
Ottawa, Ont. K1A 0L2
Canada
HelenTryphonas@hc-sc.gc.ca

MARIAGRAZIA UGUCCIONI
Theodor-Kocher Institute
University of Bern
Winchesterstrasse 2
CH-3012 Bern
Switzerland

PETER ULRICH
Preclinical Safety, MUT2881.329
Novartis Pharma AG
Auhafenstrasse
CH-4132 Muttenz
Switzerland
peter.ulrich@pharma.novartis.com

MAURICE W VAN DER HEIJDEN
Dept Pharmacology + Patophysiology
University Utrecht Institute of Pharmaceutical Sciences
3508 TB Utrecht
The Netherlands

ROB J VANDEBRIEL
Laboratory for Toxicology, Pathology and Genetics
National Institute for Public Health and the Environment
3720 BA Bilthoven
The Netherlands
R.Vandebriel@rivm.nl

KRIS VLEMINCKX
Unit of Developmental Biology, Dept. Molecular Biomedical Research
Ghent University – VIB
Technologiepark 927
Ghent
9052
Belgium
kris.vleminckx@dmb.rug.ac.be

HANS-WERNER VOHR
PH-PD, Toxicology
Bayer HealthCare AG
Aprather Weg 18a
Wuppertal
42096
hans-werner.vohr@bayerhealthcare.com

GERHARD F WEINBAUER
Covance Laboratories GmbH
Kesselfeld 29
D-48163 Münster
Germany
gerhard.weinbauer@covance.com

I BERNARD WEINSTEIN
Columbia University
New York, NY
USA
weinstein@cuccfa.ccc.columbia.edu

HANS ULRICH WELTZIEN
Max-Planck-Institut für Immunbiologie
Stuebeweg 51
D-79108 Freiburg
Germany
weltzien@immunbio.mpg.de

AINSLEY WESTON
Division of Extramural Research and Training, National Institute of Environmental Health Sciences
National Institutes of Health
111 T. W. Alexander Drive
Research Triangle Park, NC 27709
USA

KIMBER L WHITE
Dept. of Pharmacology & Toxicology
Virginia Commonwealth University
Richmond, PO Box 980613
USA
klwhite@vcu.edu

CLYDE WILSON
Division of Life Sciences
King's College, University of London
150 Stamford Street
London
SE1 8WA
UK

MARK WING
Huntingdon Life Science Limited
Woolley Road
Alconbury, Huntingdon, Cambs
PE28 4HS
UK

ANNA MARIA WOLF
Department of Internal Medicine, Division of
Gastroenterology and Hepatology
Innsbruck University Hospital
Anichstrasse 35
AT-6020 Innsbruck
Austria
maria.wolf@uibk.ac.at

PARVEEN YAQOOB
School of Food Biosciences
The University of Reading Whiteknights
PO Box 226
Reading
RG6 6AP
UK

BERRAN YUCESOY
National Institute for Occupational Safety and Health
1095 Willowdale Road
Morgantown, WV 26505
USA

DAVID C ZAWIEJA
Department of Medical Physiology, College of Medicine,
Cardiovascular Research Institute Division of Lymphatic
Biology, Texas A&M University System Health Science
Center
336 Reynolds Medical Building
College Station, TX 77843
USA

JUDITH T ZELIKOFF
Institute of Environmental Medicine
New York University School of Medicine
57 Old Forge Road
Tuxedo, NY 10987
USA
judyz@env.med.nyu.edu

H ZOLA
Child Health Research Institute, Women's and Children's
Hospital
Adelaide
Australia

PA VAN ZWIETEN
Departments of Pharmacotherapy, Cardiology, Cardio-
Thoracic Surgery, Academic Medical Centre
University of Amsterdam
Meibergdreef 9
1105 AZ Amsterdam
The Netherlands

3'

Refers to the direction of the DNA strands which are double stranded with the top strand in the orientation of 5' to 3'. The bottom strand is complimentary to the top strand and is in the reverse orientation of 3' to 5'. For the top strand, 3' is also denoted by the term downstream.

▶ B Lymphocytes

5'

Refers to the direction of the DNA strands which are double stranded with the top strand in the orientation of 5' to 3'. The bottom strand is complimentary to the top strand and is in the reverse orientation of 3' to 5'. For the top strand, 5' is also denoted by the term upstream.

▶ B Lymphocytes

ABO Blood Group System

GEOFF DANIELS
Bristol Institute for Transfusion Sciences
National Blood Service
Soutmead Road
Bristol
BS10 5ND
UK

MARCELA CONTRERAS
National Blood Service
Development and Research, North London Centre
Colindale Avenue
London
NW9 5BG
UK

Synonyms

AB0 histo-blood group system; major human ▶ blood group system

Definition

The most important ▶ histocompatibility and blood group ▶ antigen system, consisting of two main antigens and four main phenotypes inherited in a Mendelian fashion.

Characteristics

The AB0 blood group system was discovered by Karl Landsteiner in 1901. By mixing the separated sera with suspensions of red cells obtained from the blood of different individuals, four patterns of agglutination were obtained. These patterns subdivide the population into four main blood groups (with approximate European Caucasian frequencies in parentheses): O (46.5%); A (42%); B (8.5%); and AB (3%). The frequencies of the four AB0 groups varies in different populations: native Americans are almost exclusively group O, while Asians have a proportionately higher incidence of group B. There are two ▶ antigens, A and B, though A is subdivided into A_1 and A_2. The O phenotype is the absence of A and B (Table 1).

Almost without exception, every person has ▶ antibodies in their serum to those A or B antigens they lack from their red cells and tissues. In addition to anti-A and anti-B, group O individuals have a cross-reacting antibody called anti-A,B. Testing of red cells with selected potent anti-A, anti-B and anti-A,B reagents, while simultaneously testing the sera of the same subjects with reagent red cells (group A_1, A_2, B and O), provides the basis for AB0 grouping.

The major subgroups of A are A_1 and A_2. A_2 is a weaker A antigen than A_1, but the difference between them is also qualitative. These subgroups can be distinguished with specific anti-A_1 reagents and are only significant clinically if the serum of an A_2 individual reacts with A_1 cells at 37° C and so may cause destruction of transfused group A_1 red cells. Anti-A_1 reagents can be a lectin prepared from ▶ Dolichos biflorus seeds, sera of group B subjects absorbed with group A_2 red cells, or mouse monoclonal antibodies. Naturally occurring anti-A_1 is present in the serum of 1%–8% group A_2 and 22%–35% group A_2B individuals, but is too weak to be used as a grouping reagent. Other variants of A (A_{int}, A_x, A_{end}, A_3, A_m, A_y, A_{el}) and B (B_3, B_x, B_m, B_{el}) are characterized by varying degrees of weakness of A or B antigens and by the absence of the appropriate AB0 antibodies from their plasma. For example, the red cells of A_x individuals fail to react with anti-A from group B individuals, although they react with strong anti-A in group O people and with some monoclonal anti-A reagents; A_x individuals do not have anti-A in their serum. A and B variants are rare and usually of little clinical significance in blood transfusion.

Structure of the ABO Antigens

A and B antigens are carbohydrate structures, synthesized by glycosylation of oligosaccharide precursors with ▶ H antigen activity. The H antigen is synthesized from its precursor by a glycosyltransferase, a fucosyltransferase that is encoded by a gene that is genetically independent of *AB0*. Carbohydrate chains carrying the A, B, and H antigens are present on (i) the highly branched *N*-linked polysaccharides of integral membrane proteins, (ii) the heavily branched polysaccharides that form the polyglycosyl moieties of either soluble glycoproteins present in secretions or of poly-

AB0 Blood Group System. Table 1 The AB0 blood group system

Phenotype	Antigens	Genotypes	Antibodies in Serum
A_1	A_1, A	A^1/A^1, A^1/A^2, A^1/O	Anti-B
A_2	A	A^2/A^2, A^2/O	Anti-B, (anti-A_1)*
B	B	B/B, B/O	Anti-A
O	None	O/O	Anti-A, -B, -A, B
A_1B	A_1, A, B	A^1/B	
A_2B	A, B	A^2/B	(Anti-A_1)*

* Present in the plasma of some A_2 and A_2B individuals.

glycosylceramides in the red cell membrane, and (iii) the short chain oligosaccharides of simple glycolipids in plasma. The immunodominant sugars of the A and B antigens are at the non-reducing ends of the various polysaccharide chains expressing A or B, and are invariably attached by an $\alpha1$–3 linkage to a fucosylated galactose residue with H antigen activity, such that the simplest A and B epitopes are trisaccharides with the structures given in Formula 1 (where R represents the remainder of the polysaccharide chain).

N-acetylgalactosamine (GalNAc) and galactose (Gal) are the immunodominant monosaccharides of the A and B ▶ epitopes, respectively. The presence of the fucose residue, the immunodominant sugar of the H antigen, is essential for A and B expression.

The β-Gal residue of the terminal trisaccharides can be attached to R in at least six different ways or types:

- Type 1 Galβ1→3GlcNAcβ1→R
- Type 2 Galβ1→4GlcNAcβ1→R
- Type 3 Galβ1→3GalNAcα1→R
- Type 4 Galβ1→3GalNAcβ1→R
- Type 5 Galβ1→3Galβ1→R
- Type 6 Galβ1→4Glcβ1→R

Of these peripheral core structures, type 2 is the most abundant on red cells; integral red cell membrane glycoproteins and glycolipids have almost exclusively type 2 sugars, though some glycolipids also have type 3 or type 4 structures. Red cells may also contain glycolipids, passively adsorbed from plasma, that have type 1 chains. The existence of these various epitopes on red cells probably explains the heterogeneity in reactivity of different A and B antibodies with group A and B variants. Types 1 and 2 are abundant in body secretions and endodermally derived tissues.

A GalNAc α 1→3 Gal —R
            ↑
         Fuc α1,2

B Gal α 1→3 Gal —R
            ↑
         Fuc α1,2

AB0 Blood Group System. Formula 1

Biosynthesis and Molecular Genetics

The genes controlling the expression of A and B antigens are codominant alleles at the *AB0* locus on chromosome 9q34. The products of the *A* and *B* genes are ▶ glycosyltransferases, which catalyze the biosynthesis of the A and B antigens. They comprise a 353 amino acid polypeptide organised into three domains: a short N-terminal domain; a hydrophobic domain that spans the Golgi membrane; and a large C-terminal domain containing a catalytic site. The *A* gene product is an *N*-acetylgalactosaminyltransferase that transfers GalNAc from a UDP-GalNAc donor to the C3 position of the fucosylated Gal residue of the H antigen, to produce an A-active structure (Figure 1). The *B* gene product is a galactosyltransferase that transfers Gal from UDP-Gal to the fucosylated Gal of H, to produce a B-active structure (Figure 1). The *O* allele produces no active enzyme, hence the H structure remains unconverted.

The *AB0* gene consists of seven exons. The two largest exons, exons 6 and 7, contain 77% of the coding sequence and are the most important in determining the substrate specificity of the gene products.

The *A* (or more specifically A^1) and *B* alleles differ at seven nucleotide positions, four of which (in exon 7) generate four amino acid differences. Two of these, at positions 266 (Leu from *A*, Met from *B*) and 268 (Gly from *A*, Ala from *B*), are responsible for determining whether the enzyme has predominantly GalNAc-transferase (A) or Gal-transferase (Gal) activity.

The most common *O* allele (O^1) has a nucleotide sequence almost identical to that of the A^1 allele, but with a single base deletion in exon 6, which generates a change in reading frame at amino acid position 87 and a new in-frame stop codon. Consequently, O^1 encodes a truncated polypeptide, which is only 116 amino acids long, lacks the catalytic domain and is enzymatically inactive. Another common *O* allele, (O^{1var}) has at least nine nucleotide differences from O^1, but still has the single base deletion and so is functionally identical to O^1. A third type of O (O^2)

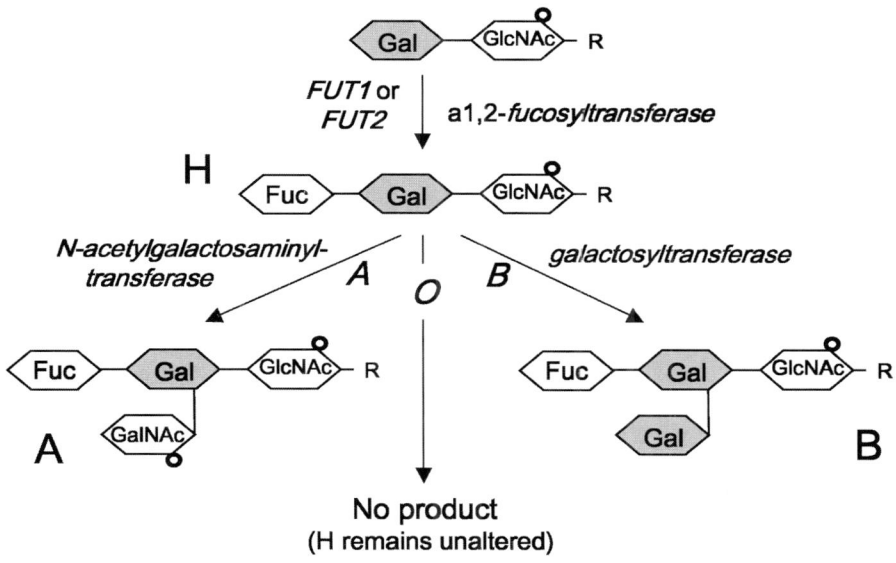

AB0 Blood Group System. Figure 1 Biosynthesis of H antigen from a common precursor and A and B antigens from H.

encodes a charged arginine, instead of neutral glycine (A) or alanine (B) at the vital 268 position, abolishing the enzymatic activity of the resultant protein.

The A^2 gene product is a GalNAc-transferase with different kinetics to those of the A^1-transferase, making it apparently less efficient. The A^2 allele closely resembles A^1, but has a single base deletion at the 3′ end of the gene, in the codon before the usual translation stop codon. The resultant reading-frame shift abolishes the stop codon, so the gene encodes an enzyme with an extraneous 21 amino acids at its C-terminus.

A variety of different mutations account for the rare AB0 subgroups and demonstrate that the molecular background to most of these variants is heterogeneous. These mutations include missense mutations, splice site mutations, nonsense mutations, and nucleotide insertions. In addition, there are many different hybrid genes in which exons 1–6 derive from one allele and exon 7 derives from another. For example, A^1–O^{1v}, B to O^{1v} and O^2–O^{1v} all give rise to an A_x phenotype, because exon 6 does not contain the single nucleotide deletion characteristic of O^1 and so produces an active enzyme and exon 7 has the O^{1v} sequence, which is similar to A, but encodes an important Phe216Ile substitution, accounting for the weak A activity.

Knowledge of the nucleotide sequences that distinguish the $AB0$ alleles has made it possible to devise polymerase chain reaction (PCR)-based methods for recognizing the presence of A^1, A^2, B, O^1, O^{1v} and O^2 alleles. Consequently, $AB0$ genotypes can be determined, though the methods are relatively complex because of the number of different sequence changes involved.

H antigen, the acceptor substrate for the A and B transferases, is synthesized by addition of fucose (Fuc) to the C2 position of the terminal Gal of a peripheral core structure (see above and Figure 1). This fucosylation is catalyzed by an α1,2-fucosyltransferase. Two genes on chromosome 19 encode α1,2-fucosyltransferases: *FUT1* is active in mesodermally derived tissues and is responsible for H expression on red cells; *FUT2* is active in endodermally derived tissues and is responsible for H expression in secretions, plasma and respiratory and digestive epithelia. Homozygosity for inactivating mutations in either of these genes results in absence of H in the appropriate tissues, and therefore absence of A or B antigens from those tissues, regardless of $AB0$ genotype. *FUT2* is polymorphic and inactive *FUT2* alleles are common. About 20% of Caucasians lack H, A, and B from their secretions and other endodermally derived tissues and are referred to as ABH non-secretors. They have normal ABH antigens on their red cells. Inactive *FUT1* alleles are rare and homozygosity results in very rare phenotypes in which the red cells lack H, A and B (regardless of $AB0$ genotype). Individuals who are homozygous for inactive alleles of both *FUT1* and *FUT2* have the extremely rare blood group known as the Bombay phenotype (red cell H-deficient non-secretors). They almost invariably make a potent anti-H, making it very difficult to provide compatible blood for transfusion.

Tissue Distribution and Ontogeny

The A and B transferases are abundant in intestinal and gastric mucosa, respiratory mucosa, salivary glands and epithelia of the urinary tract. H, A and B antigen expression in these tissues, however, is under the control of the *FUT2* locus, so the antigens are only expressed in those tissues in ABH secretors. The transferases are in free solution in plasma and secretions: mucin droplets, ovarian cyst fluid, milk and saliva. Molecules glycosylated by the transferases include membrane enzymes, membrane structural proteins and receptors, as well as secreted proteins, such as immunoglobulin A and coagulation factors.

During ontogeny ABH activity is at its highest in the early embryo from the fifth week post-fertilisation; ABH antigens are found in large amounts on endothelial cells and most epithelial primordia, and in practically all early organs, including blood islands of the yolk sac, erythropoietic foci of the liver, digestive tube epithelia, pharyngeal pouches, the thymus, the pituitary, thyroid glands, trachea and bronchi, hepatic and pancreatic diverticula, the cloaca, urachos and allantois, mesonephros and the ducts of the metanephros. The central nervous system, liver, adrenal glands and secretory tubules show no ABH activity at this stage. The number of A and B sites on the red cell is increased approximately fourfold in adults compared with neonates. There are $25–37 \times 10^4$ A sites per red cell in the newborn and $81–120 \times 10^4$ in the A_2 adult, and $20–32 \times 10^4$ B sites per red cell in the newborn and approx. 75×10^4 in adults.

Preclinical Relevance

The AB0 system is polymorphic (see ▶ Polymorphism) and the antigens are strongly immunogenic (see ▶ Antigen), capable of eliciting 'naturally occurring' and immune ▶ antibodies. These antibodies can give rise to acute intravascular ▶ hemolytic transfusion reactions and rejection of transplanted organs.

Relevance to Humans
Disease Associations

Many pathogenic microorganisms are capable of attachment to cell surface carbohydrate structures, so ABH antigens can be exploited as receptors for invasion of these cells. Secretor status may play an important role as it controls ABH expression in many tissues that are vulnerable to infection. Consequently, the degree of susceptibility to a variety of bacterial, viral, fungal and protozoan infections is associated with specific AB0 and secretor phenotypes.

Microorganisms that are reported to bind to ABH antigens include *Helicobacter pylori*, *Propionbacterium granulosum*, *Aeromonas hydrophila*, *Pseudomonas aeruginosa*, *Candida albicans*, *Streptomyces* and several strains of *Escherichia coli*. Heat-labile ▶ enterotoxin produced by *E. coli* isolated from humans preferentially binds to glycolipids isolated from A, B, and AB human red cells, compared with O cells.

Statistical associations between a multitude of diseases and AB0 and secretor phenotypes have been claimed. Though many may result from flawed statistics, the fact that these polymorphisms represent glycosylation changes on cell membranes and soluble glycoproteins makes almost any disease association feasible. To summarize the well-established associations, Samuelsson and Henry concluded that "There is tendency for bacterial infections to attack persons of group A, while virus infections tend, in a very general way, to be associated with group O. Cancers are mostly associated with group A, as are clotting diseases, while autoimmune diseases and bleeding are associated with O".

ABH activity is often absent from malignant tumours, despite being present on the surrounding epithelium. The prognostic value of this loss of ABH antigens is controversial. Another phenomenon associated with malignancy is the illegitimate A antigen, occasionally expressed on tumours of group O or B people. About 10% of colonic tumors from group O patients homozygous for the O^1 allele express A antigen and contain active A-transferase activity. This might result from loss of the product of exon 6 of *AB0* and the consequent absence of the nucleotide deletion characteristic of O^1, creating an A-active enzyme.

AB0 ▶ Antibodies

The clinical importance of the AB0 blood group system in blood transfusion derives from the high prevalence of its antibodies and their in vivo potency. The "naturally occurring" antibodies of the majority of group A or B individuals are mainly IgM and probably produced in response to environmental AB0 antigens, especially those of microbes in the gut and respiratory tract. Such IgM antibodies, although displaying optimal activity in the cold, are reactive at 37° C and can activate the complement cascade up to the C9 stage, leading to the immediate intravascular lysis of transfused incompatible red cells in vivo. Approximately one in every three random, ungrouped blood donations would be incompatible with a given recipient. Such incompatible transfusions can lead, in about 10% of cases, to renal failure, disseminated intravascular coagulation, and death. Severe haemolytic transfusion reactions occur mainly in group O people, who have stronger AB0 antibodies. The majority of the signs and symptoms of severe AB0 intravascular ▶ hemolytic transfusion reactions can be attributed to the generation of C3a and C5a fragments as a result of full complement activation, with the consequent release of vasoactive amines from mast cells and of cytokines such as interleukins IL-1, IL-6, IL-8 and tumor necrosis factor (TNF) from mononuclear cells. The release of thromboplastic substances from lysed red cells activates coagulation.

Group O adults and a small proportion of group A and B individuals have "naturally occurring" (usually weak) IgG in addition to stronger IgM AB0 antibodies. The IgG component can cross the placenta and bind to fetal red cells. Lysis of fetal red cells, however, is generally minimal and hemolytic disease of the newborn (HDN) caused by AB0 antibodies is usually mild or unapparent in Western Europe and North America. HDN due to AB0 antibodies only affects the offspring of group O mothers. In some parts of the world, AB0 HDN is more prevalent, though seldom severe, and this is attributed to environmental factors such as the greater stimulation of AB0 antibodies by microbes and parasites.

Some individuals possess plasma IgA AB0 antibodies, irrespective of immunization. AB0 antibodies of colostrum are often wholly IgA, although sometimes IgM antibodies can also be found.

Cord blood usually does not contain AB0 antibodies although maternally derived IgG anti-A or anti-B can sometimes be detected. Newborn infants do not produce AB0 antibodies until 3–6 months of age, reaching a maximal level at 5–10 years of age. The vast majority of healthy adults have easily detectable AB0 antibodies, except from those of AB phenotype. Weakening of AB0 antibodies can occur naturally in individuals aged over 50; a third of patients over 65 have low AB0 antibody levels. Very occasionally individuals lack the appropriate AB0 agglutinins, especially if hypogammaglobulinemic, or if their plasma IgM levels are low. Antibody levels can be substantially reduced by exhaustive plasma exchange (used therapeutically in AB0 incompatible bone marrow and organ transplantation) or by immunosuppression caused by therapy or by disease.

References

1. Chester MA, Olsson ML (2001) The AB0 blood group gene: a locus of considerable genetic diversity. Transfus Med Rev 15:177–200
2. Daniels G (2002) Human Blood Groups, 2nd ed. Blackwell, Oxford, pp 7–98
3. Mollison PL, Engelfriet CP, Contreras M (1997) Blood Transfusion in Clinical Medicine, 10th ed. Blackwell, Oxford, pp 116–131; 317–324; 358–367
4. Henry S, Samuelsson B (2000) AB0 polymorphisms and their putative biological relationships with disease. In: King MJ, ed. Human Blood Cells. Consequences of Genetic Polymorphism and Variations. Imperial College Press, London pp 1–103
5. Yamamoto F (2001) Cloning and regulation of the *AB0* genes. Transfus Med 11:281–294

ABO Histo-Blood Group System

▶ AB0 Blood Group System

Abscess

Accumulation of pus in a cavity originating after tissue colliquation.
▶ Dermatological Infections

Acquired Immunity

Requires stimulation of effector mechanisms following exposure to foreign materials (e.g. xenobiotics). Also known as adaptive immunity and exhibits antigen specificity, diversity, memory, and self/non-self recognition that is mediated by activated B and T cells. Therefore, acquired immunity can be subdivided into antibody-mediated immunity (AMI) and cell-mediated immunity (CMI).
▶ Humoral Immunity
▶ Immunotoxicology

Acrocyanosis

Arterial vasoconstriction with persistent cyanosis of hands and feet.
▶ Septic Shock

Activated Macrophages

Inflammatory macrophages exposed to both interferon-γ and lipopolysacchride (LPS), or primed macrophages exposed to LPS, or macrophages elicited with infectious agents such as mycobacteria that are the highest activated state for killing.
▶ Macrophage Activation

Activation-Induced Cell Death (AICD)

In the course of a proliferative T cell response, death-inducing molecules are being upregulated ultimately inducing cell death in the activated cells, thereby limiting the immune response.
▶ Tolerance

Activator Surface

A surface that allows massive activation of C3 and covalent binding of C3b. A nonactivator surface such as a host cell limits this activation using the normal control mechanisms of the complement system (e.g. factor H, CR1, presence of sialic acid).
▶ Complement, Classical Pathway/Alternative Pathway
▶ Complement System

Active Immunotherapy

Immunotherapy based on the stimulation of the immune system of the host. Therapeutic vaccination is a typical example of active immunotherapy. *See also* Passive immunotherapy.
▶ Tumor, Immune Response to

Active Lymph Pump

Also known as the "intrinsic lymph pump." Contractile activity of smooth muscle cells located in walls of lymphatic vessels. Lymphatic contractions cause a decrease in lymphatic diameter and generate an increase in intralymphatic pressure needed for lymph propulsion in the downstream direction.
▶ Lymph Transport and Lymphatic System

Acute Graft-Versus-Host Disease

▶ Graft-Versus-Host Reaction

Acute Inflammation

On contact with pathogens specialized sentinel cells of the immune system release cytokines and other proinflammatory mediators in order to initiate a local and acute response by activating surrounding tissue cells and recruiting leukocytes to the site of infection.
▶ Immune Response

Acute Lymphocytic Leukemia

▶ Leukemia

Acute Myelogenous Leukemia

▶ Leukemia

Adaptive Immune Response

The acquired arm of the immune system that produces a specific immune response to each infectious agent encountered and is capable of remembering the agent, thus protecting the host from future infection by the same pathogen. It is synonymous with acquired immune response.
As a first step of an adaptive immune response an antigen-presenting cell, such as a dendritic cell, traps an antigen in the periphery and migrates to the lymphoid tissues. Here it presents the antigen to T cells, evoking either a humoral response with the help of B cells, or a direct cytotoxic T cell response. Whereas the humoral responses are mainly directed against extracellular pathogens such as most bacteria, the cytotoxic T cell responses are in the case of infection with intracellular antigens such as by viruses.
▶ Assays for Antibody Production
▶ Aging and the Immune System
▶ Lymphocytes

Adaptive Immunity

The adaptive or specific arm of immunity comprises T and B lymphocytes that both express a discrete and individual antigen receptor which is created by genetic rearrangement of specific gene segments. This creates millions of individual lymphocytes each with discrete antigen specificity. T effector cells either help the innate and adaptive immune responses or they delete virus-infected cells. B cells produce antibodies as important reagents to provide immunological memory.
▶ Immune Response
▶ Graft-Versus-Host Reaction

Adaptors

Adaptors are molecular scaffolds that recruit other proteins. These proteins contain two or more domains (i.e. SH2 and SH3 domains) which bind other proteins. They mediate protein–protein interactions but usually have no intrinsic kinase activity. In lymphocytes, adaptors recruit other proteins to the activated receptor

where these proteins can be phosphorylated and activated.
▶ Signal Transduction During Lymphocyte Activation

ADCC

Antibody-dependent cellular cytotoxicity is a cytotoxic mechanism through which antibody-coated target cells are killed by different effector cells, such as polymorphonuclear leukocytes, mononuclear phagocytes, natural killer (NK) cells, dendritic cells, and platelets, which bear receptors for the Fc portion of antibodies.
▶ Antibody-Dependent Cellular Cytotoxicity

Adherens Junctions

An intercellular junctional structure, most prominent in epithelial cells. In the adherens junction, the cell–cell adhesion is mediated by Ca^{2+}-dependent adhesion molecules, the cadherins. The cytoplasmic tail of these cadherins is indirectly linked to the actin cytoskeleton.
▶ Cell Adhesion Molecules

Adhesion Molecules

Proteins expressed on the surface of cells that mediate binding of immune system cells to other cells. The system of adhesion molecules facilitates movement of immune system cells from the circulation to lymphoid tissues or to sites of immune system activity, e.g. infection or inflammation.
There are three major families of proteins including integrins, the immunoglobulin superfamily, and selectins.
▶ Cell Adhesion Molecules
▶ Glucocorticoids
▶ Leukocyte Culture: Considerations for In Vitro Culture of T cells in Immunotoxicological Studies

Adoptive Transfer PLNA

▶ Popliteal Lymph Node Assay, Secondary Reaction

Adrenocorticotropic Hormone (ACTH)

ACTH is secreted from the anterior pituitary gland in response to corticotropin-releasing hormone, enters the blood stream and is transported to the adrenal glands, stimulating the synthesis and release of glucocorticoids. Its production is increased in times of stress.
▶ Glucocorticoids
▶ Stress and the Immune System

Adult Respiratory Distress Syndrome (ARDS)

A descriptive term for diffuse infiltrative lung lesions of diverse etiologies which are accompanied by severe arterial hypoxemia.
▶ Septic Shock

Advanced or Extended Histopathology

▶ Histopathology of the Immune System, Enhanced

Afferent Lymphatics

Lymphatics are small vessels that contain clear fluid (lymph) that is collected from the tissues. The vessels that drain the tissues and transport fluid to lymph nodes are described as afferent lymphatics.
▶ Local Lymph Node Assay

Affinity Maturation of the Immune Response

▶ B Cell Maturation and Immunological Memory

Aflatoxins

Naturally occurring toxin metabolites produced from some strains of fungi. They act by combining with DNA, suppressing DNA and RNA synthesis and play a role in the etiology of cancer of the liver.
▶ Respiratory Infections

Agglutination

In principle agglutination is the clumping of particles. In the context of this encyclopedia these particles can be cells or erythrocytes agglutinated by antigen specific antibodies. The agglutination of red blood cells is called hemagglutination. This phenomenon is used as a diagnostic tool, e.g. for blood typing for transfusion, or for the Coombs Assay. Aggregation of erythrocytes in grapelike clusters are also seen on Romanofski stained peripheral blood smears of patients with IMHA.

▶ Antiglobulin (Coombs) Test

Aging and the Immune System

ANNA MARIA WOLF
Department of Internal Medicine, Division of Gastroenterology and Hepatology
Innsbruck University Hospital
Anichstrasse 35
AT-6020 Innsbruck
Austria

Synonym

immunosenescence

Definition

Aging is the process of growing older starting from birth, whereas senescence is referred to as the process of somatic deterioration at older age. Our body is constructed to function optimally until the age of reproduction. After this time point, increasing age-related alterations and changes affecting the organism on a whole as well as the immune-system can be observed. The deterioration of immune function in old age is termed "immunosenescence." The characteristics described here of the aging immune system are related to the post-reproduction period.

Characteristics

The thymus is the central lymphoid organ where bone-marrow derived T cells learn to distinguish between self and non-self. This organ is almost fully developed at birth, but its involution starts soon after puberty. At the age of 60 years thymic tissue is almost completely replaced by fat, resulting in a decreased thymic output of naive T cells in elderly persons. Aging is therefore accompanied by decreasing numbers of naive T cells. The loss of naive T cells is associated with a reduced IL-2 production, as observed in old age. Interestingly, the total count of T cells does not decrease with age, which is a consequence of proliferation of antigen-experienced memory cells which substitute for the decline of naive T cells. The increased number of memory/effector cells leads to altered cytokine production with a shift towards proinflammatory cytokines such as the interferon IFN-γ. The increased whole-body load of IFN-γ observed in the elderly may accelerate immune responses that lead to tissue injury. Elevated levels of proinflammatory cytokines are also associated with a number of age-related diseases (see Relevance to Humans).

A decreased T cell reactivity towards mitogens and antigens—which is probably due to increased membrane rigidity and decreased expression of costimulatory molecules such as $CD28^-$ has been reported. Another characteristic of the immune system in the elderly is a restriction in the T cell repertoire. While newborns show a diverse spectrum of antigen recognition, elderly persons are often affected by the dominance of huge expanded clones specific for only few antigens as a result of chronic infections with, for example, persistent viruses. The appearance of multiple $CD8^+$ T cells clonal expansions is one of the most dramatic qualitative changes in the memory cell population during aging. These clones often lack the costimulatory molecule CD28 and their ▶ telomeres are short, suggesting that they are end-stage cells. Concerning the humoral immunity, both the B cell mitogen response and absolute B cell number remain unaltered in old age. However the antibody response towards primary and secondary immunizations is lower compared with young subjects, probably due to a poorer cooperation between T and B cells.

Dendritic cells are the most professional antigen-presenting cells (APC) showing a unique ability to induce ▶ adaptive immune responses via the presentation of antigenic peptides to T cells. Dendritic cells generated in vitro from peripheral blood monocytes of elderly people are not impaired in their capacity to induce T cell responses and seem to persist unaltered in number, function, and surface marker expression during the aging process. In contrast, dendritic cells isolated directly ex vivo from old people are reduced in their functional capacity to stimulate immune responses. This may indicate a negative impact of an aged environment on the functional state of the dendritic cells, rather than an impaired cell function per se.

The innate immune system is not as dramatically affected as the specific immune system described above. Although natural killer (NK) cell lytic activity seems to be diminished in old age at the single-cell level, the overall cytotoxic activity remains intact as the numbers of NK cells have been reported to be higher in old than in young persons.

Investigations of the effect of aging on neutrophil bactericidal responses showed that neutrophils from elderly donors were able to generate superoxide and to opsonize *Escherichia coli* efficiently. In contrast, the phagocytic index was significantly decreased in neutrophils from the elderly, compared with young donors, proposing a contribution of aged neutrophils to immunosenescence. In summary, alterations of both specific and innate immunity result in an enhanced proinflammatory status which is characteristic of old age.

Preclinical Relevance
It is useful to distinguish between primary and secondary age-dependent alterations of immune reactivity. Primary age-related immune deficiencies occur also in healthy elderly persons due to an age-dependent intrinsic decline of immune function. Secondary age-related alterations result as a consequence of other environmental conditions such as malnutrition, insufficient blood supply, metabolic changes, and drugs.

Relevance to Humans
Infectious Diseases
It is well known that the frequency and severity of infections increases with advancing age. This can be attributed to a clear-cut decline of the immune function in the elderly. As explained, T cells in particular are affected by the aging process. Due to their declining helper function, the whole complex process of acquiring immunity following bacterial or viral infection or vaccination is disordered. Cohort studies showed declining antibody titers with ongoing age. This seems to be a problem, particularly when elderly persons are immunized with new antigens, such as tuberculin bacillin emulsion (TBE) or rabies.

Alzheimer's Disease
Alzheimer's disease is the most common form of dementia in the elderly. The critical step in the development of the disease is probably the deposition of amyloid leading to the formation of neuritic plaques and subsequently to cognitive impairment. As small amyloid deposits can also be found in the brain of healthy elderly persons and the aggregation and deposition of amyloid starts very early, probably 10–20 years before the onset of clinical symptoms, it is likely that further factors bias the outcome of the disease. Recently it has become evident that proinflammatory cytokines play a pivotal role in the pathogenesis of Alzheimer's. Large studies demonstrated that the disease was less frequent in patients treated regularly with antiinflammatory drugs compared to untreated control groups. Further, on combinations of the proinflammatory cytokines tumor necrosis factor α (TNF-α), or the interleukin-1α (IL-1α), and IFN-γ have been shown to trigger the production of amyloid. Amyloid aggregation per se also seems to induce a chronic inflammatory reaction in the brain. The increased production of proinflammatory cytokines in old age may therefore facilitate the development of dementia.

Atherosclerosis
For long it has been presumed that an autoimmune-inflammatory process forms the basis of the disease. According to a recent concept heat shock protein HSP 60 is a relevant antigen for this immune response. HSPs are highly conserved components of pro- and eukaryotic cells which are expressed upon exposure to stress. Antibodies and T cells reactive against HSP 60 seem to cause damage of arterial endothelial cells, especially in the areas of major hemodynamic stress. Moreover a cholesterol-rich diet showed additive effects in rabbits which were immunized with recombinant mycobacterial HSP 60, leading to more severe atherosclerosis than in normally fed animals. Hence, atherosclerosis may have its seeds in an immunologically mediated disease, starting early in life, and becoming increasingly evident with ongoing age and under the influence of additional risk factors such as smoking and high cholesterol intake.

Osteoporosis
The term osteoporosis describes a condition characterized by rarefaction of the bone mass that may be localized or involve the whole skeleton. Primary and secondary osteoporosis can be distinguished. Secondary osteoporosis may be the result of various underlying diseases such as rheumatoid disorders, malnutrition, malignancies, or side effects of drugs. Primary osteoporosis often occurs in terms of senile or postmenopausal osteoporosis after the age of 50 years and is associated with a loss of bone mass exceeding 1.5%–2% per year. Senile osteoporosis and postmenopausal osteoporosis are the most common primary forms of this condition. Low calcium intake, lack of physical activity, and low hormonal status are regarded as the main causes of age-dependent osteoporosis. Further the relative increase of proinflammatory cytokines in the elderly may disturb the balance between bone formation and resorption by activating and recruiting osteoclasts, and has therefore important effects in the development of osteoporosis.

Cancer
Malignant transformation is the end-point of multiple consecutive oncogenic damages leading to the final loss of cell-cycle control. In humans, the majority of cancer occurs in the final decades of life, culminating in a lifetime risk of 1 in 2 for men and 1 in 3 for women. The dramatic increase of malignant tumors in the elderly is probably due to a combination of several physiological changes throughout life, including telomere dysfunction, age-dependent deterioration

in genome maintenance and stability, epigenetic mechanisms promoting carcinogenesis, altered stromal milieu, and decreased control function of the immune system. As tumorigenesis—at least of certain malignancies—may be under control of the innate and the adaptive immunity a functional impairment of these defence mechanisms by immunosenescence may result in increased susceptibility to tumors.

Regulatory Environment

In the research on human immunosenescence only a limited number of animal models are available: mice live up to 2 years under germ-free laboratory conditions compared to humans with a lifespan of about 80 years in an unprotected environment; the nematode *Caenorhabditis elegans*, which is frequently used to study aging processes, lacks an immune system. So, further attempts have been made to standardise research guidelines in the human system. To exclude changes based on extrinsic factors such as illnesses, chronic diseases, or the use of medication, the SENIEUR protocol (from SENIor EURopean) was designed, defining "healthy elderly people." In this protocol, strict admission criteria for further immunogerontologic studies were specified. The SENIEUR protocol therefore helps to distinguish between any alterations caused by aging per se and those caused by diseases. However, the strict selection of admission criteria may limit the significance of the studies. Therefore, careful selection of a suitable model system is obligatory and different approaches may be used to compliment one another.

References

1. Globerson A, Effros RB (2000) Ageing of lymphocytes and lymphocytes in the aged. Immunol Tod 21:515–521
2. Grubeck-Loebenstein B, Wick G (2002) The aging of the immune system. Adv Immunol 80:243–284
3. Wick G, Jansen-Durr P, Berger P, Blasko I, Grubeck-Loebenstein B (2000) Diseases of aging. Vaccine 18:1567–1583
4. Miller RA (1999) Aging and Immune function. In: Fundamental immunology, 4th ed. Lippincott-Raven Publishers, Philadelphia, pp 947–966
5. Ligthart GH (2001) The SENIEUR protocol after 16 years. The next step is to study the interaction of ageing and disease. Mech Ageing Dev 122:136–140

Ah Receptor (AhR)

The endogenous receptor in mammalian cells for PAHs such as BaP and dioxin-like compounds that mediates signaling and gene transcription via the DRE.

▶ Polycyclic Aromatic Hydrocarbons (PAHs) and the Immune System

Air Pollution

▶ Respiratory Infections

Airborne Contagion

▶ Respiratory Infections

Alexin

▶ Complement, Classical Pathway/Alternative Pathway

Allelic Discrimination

A method to detect different forms of the same gene that differ by nucleotide substitution, insertion, or deletion. In a bi-allelic system, two different fluorochrome-labeled probes are designed to hybridize each to a specific allele and are included in a PCR amplification of sample material. An increase in fluorescence of both dyes indicates allelic heterozygosity while an increase in only one signal reflects allelic homozygosity.

▶ Polymerase Chain Reaction (PCR)

Allergen

Non-infectious antigens that induce hypersensitivity reactions, most commonly IgE-mediated type I reactions or cell-mediated type IV reactions.

▶ Flow Cytometry
▶ Food Allergy

Allergen Hypothesis

A relationship exists between the allergen concentrations experienced in infancy and the subsequent development of sensitization and asthma.

▶ Asthma

Allergic Contact Dermatitis

A delayed inflammatory reaction on the skin seen in type IV hypersensitivity, resulting from allergic sensitization.
▶ Contact Hypersensitivity
▶ Local Lymph Node Assay (IMDS), Modifications
▶ Skin, Contribution to Immunity

Allergic Reactions

▶ Hypersensitivity Reactions

Allergic Reactions to Drugs

▶ Drugs, Allergy to

Allergic Rhinitis (Hay Fever)

A typical immediate-type allergic reaction in the nasal mucosa. It is also known as hay fever, and causes runny nose, sneezing, tears.
▶ Hypersensitivity Reactions

Allergy

An immunological response to an allergen which may involve various organ systems.
▶ Food Allergy

Alloantigens

Alloantigens are surface molecules for example on erythrocytes (AB0 system) or lymphocytes (MHC molecules) which are expressed by an individual but not by others of the same species.
▶ Rodents, Inbred Strains

Allogeneic

This term describes the genetic relationship between individuals of the same species in an outbred population, i.e. it refers to the intraspecies genetic variations.

▶ Idiotype Network
▶ Graft-Versus-Host Reaction

Allogeneic Determinants

The part of the antigen molecule that binds to a receptor on T cells which have a genetic dissimilarity between the same species.
▶ Mixed Lymphocyte Reaction

Alloreaction

This describes the stimulation of T cells by non-self antigens and determines the recognition.
▶ Cyclosporin A

Alloreactive

Stimulation of T cells by MHC molecules other than those expressed on self.
▶ Mixed Lymphocyte Reaction

Allotransplantation

Transplantation of an allograft, that is a graft of tissue from an allogeneic or non-self donor of the same species.
▶ Mixed Lymphocyte Reaction

Allotype

Products of allelic genes encoding immunoglobulin heavy or light chains originally detected in rabbits by immunization of one rabbit with immunoglobulin from another (alloimmunization). Complex allotypes are due to multiple amino acid differences between alleles and lead to several allotypic determinants detectable with alloantisera. Simple allotypes result from single base changes in alleles that replace one amino acid with another.
The MHC locus is highly polymorphic, giving rise to a range of different allotypic MHC molecules.
▶ Antigen-Specific Cell Enrichment
▶ Rabbit Immune System

Allotypic Epitopes

Immunoglobulins isolated from one strain of a species and injected into another strain will induce a response of allotypic epitopes.
▶ Humanized Monoclonal Antibodies

Alternative Activation

▶ Macrophage Activation

Alternative Pathway

A pathway of the complement system that is activated by pattern recognition of foreign surfaces independent of antibody, and is initiated by the spontaneous hydrolysis of C3. This pathway includes the complement components C3, factor B and factor D, resulting in the formation of a C3 convertase to cleave C3.
▶ Complement, Classical Pathway/Alternative Pathway
▶ Complement and Allergy

Ambient Air

Air that is surrounding, encompassing an area; pertaining to the environment.
▶ Respiratory Infections

Amnestic (or Recall) Immune Response

▶ Memory, Immunological

ANA

▶ Antinuclear Antibodies

Anaphylactic Shock (Anaphylaxis)

A life-threatening acute immunological reaction to external allergens characterized mainly be appearance of cutaneous rashes, signs of respiratory distress, and circulatory failure.
▶ Molecular Mimicry

Anaphylatoxin

Small fragments of Complements are called anaphylatoxins. They are formed during Complement activation, and are able to bind to so-called "anchor residues" of MHC class I molecules. They are potent and effective chemoattractants and cell activators by inducing the release of a number of cytokines, chemokines, and inflammatory mediators. In synergy with other pro-inflammatory factors, such as lipopolysaccharide (LPS) or tumor necrose factor (TNF) they can cause severe effects, e.g. septic shock or the acute respiratory distress syndrome (ARDS). The most important anaphylatoxins are C5a and C3a. They are heat stable, 10 kD fragments of the amino terminus of the alpha chain of complement components C3 and C5, respectively. C3a and C5a interact with the C3a receptor or C5a receptor, respectively, to cause their biological effects. C4a is sometimes also included in the term anaphylatoxin, but is less potent than C3a and C5a.
▶ Anaphylatoxins
▶ Complement and Allergy

Anaphylaxis

Severe IgE-mediated allergy with involvement of different organs (urticaria, hypotension, cardiovascular collapse, bronchoconstriction). The prompt and severe reaction can be lethal (e.g. penicillin or bee venom).
▶ IgE-Mediated Allergies

Anaplastic Large Cell Lymphoma

▶ Lymphoma

Androgen

Androgens (testosterone) are steroid hormones produced in the testes. Biological activity of the androgens is conferred by interaction with the androgen receptor.
▶ Steroid Hormones and their Effect on the Immune System

Anemia

Anemia is a condition in which there is a decrease in the numbers of red blood cells in the blood, resulting in a decreased capacity of the blood to carry oxygen. Anemia may be associated with palor (paleness) of the skin, fatigue, palpitations of the heart, and shortness of breath on exertion.
▶ Leukemia

Anemia Associated with Immune Response

▶ Hemolytic Anemia, Autoimmune

Anergic

A form of immunologic tolerance and refers to lymphocytes that bind antigen but are functionally inactive.
▶ Humoral Immunity

Anergy

Anergy is a state of unresponsiveness of lymphocytes that occurs when immune cells encounter their specific antigen in the absence of necessary co-stimulatory molecules. These cells will subsequently be unresponsive to stimulation with the peptide even in the presence of co-stimulation.
▶ Autoimmunity, Autoimmune Diseases
▶ Tolerance

Angioedema

Angioedema is a type I reaction induced in deep dermal and subcutaneous tissues. Angioedema is often associated with urticaria. Drugs are among the most likely triggers for angioedema. Frequently affected sites include the eyelids, lips, and genitalia.
▶ Drugs, Allergy to
▶ Complement Deficiencies

Angiogenesis/Angiostasis

Angiogenesis is the process of vascularization of a tissue or tumor, involving the formation of new blood vessels induced by angiogenic factors (e.g. fibrinogen) or fibroblast growth factor (FGFα or FGFβ). Associated normally with wound healing, but also with chronic inflammatory diseases, tumor growth and metastasis. Angiostasis is the process of inhibition of angiogenesis.
▶ Chemokines
▶ Erythropoietin

Animal Models for Respiratory Hypersensitivity

JÜRGEN PAULUHN
Toxikology
Bayer HealthCare AG
Aprather Weg
D-42096 Wuppertal
Germany

Synonyms
Respiratory allergy assay, lung sensitization test, asthma models, respiratory hypersensitivity test

Short Description
The primary objective of respiratory allergy tests is to determine whether a low-molecular-weight chemical (hapten) of a high-molecular-weight compound (antigen) exhibits sensitizing properties to the respiratory tract. This may range from reactions occurring in the nose (allergic rhinitis), in the bronchial airways (allergic bronchitis or asthma) or alveoli (e.g. hypersensitivity pneumonitis). The clinical manifestations of response differ from site at which the response occurs. Asthma is defined as a chronic disease of the entire lung, and asthma attacks may either be immediate, delayed or dual in onset. The pathology of asthma is associated with reversible narrowing of airways, with prominent features that involve structural changes in the airway walls and extracellular matrix remodeling, including abnormalities of bronchial smooth muscle, eosinophilic inflammation of the bronchial wall, hyperplasia and hypertrophy of mucus glands. Current assays utilize two phases:
- an induction phase which includes multiple exposures to the test compound (sensitization) via the respiratory tract (e.g. by intranasal or intratracheal instillations, by inhalation exposures or by dermal contact.

- a challenge or elicitation phase in which the challenge can either be with the chemical (hapten), the homologous protein conjugate of the hapten or the antigen.

The choice depends both on the irritant potency and the physical form (vapor, aerosol) of the hapten. Endpoints to characterize a positive response range from the induction of immunoglobulins (e.g. total IgE), cytokines or lymphokines in serum, to (patho-)physiological reactions occurring in the lung (e.g. bronchoconstriction, influx of inflammatory cells). For the identification of chemical irritants nonirritating or mildly irritating concentrations must be selected for challenge exposures, as changes in breathing patterns caused by marked irritation may be clinically indistinguishable from an allergic response. None of the currently applied animal models duplicate all features of human asthma. Accordingly, the specific pros and cons of the selected animal model—including the induction regimen, animal species and strain selected—must be interpreted cautiously in order to arrive at a meaningful extrapolation for humans.

Characteristics

Most of the animal models used for studying specific respiratory tract hypersensitivity were developed using high-molecular-weight allergens, notably proteins. Fewer animal models have been developed as predictive tests for hazard identification and risk assessment in the area of chemical induced respiratory allergy. The models may differ from one class of chemicals to another, e.g. diisocyanates, organic acid anhydrides, reactive dyes. The majority of these models are based upon antibody-mediated events occurring as a result of induction. The models differ with regard to the following aspects:
- animal species utilized
- route of administration of the agent
- protocol for both induction and elicitation of responses
- type of response measured
- judgment of classifying a significant or a magnitude of response.

Guinea pig model

The guinea pig is known to respond vigorously to inhaled irritants by developing an asthmatic-like bronchial spasm. This species possesses a developed bronchial smooth muscle, which contracts intensively and rapidly in response to in vivo or in vitro exposure to antigen. This anatomical prerequisite is required for both the expression of bronchoconstriction of the immediate hypersensitivity reaction, which evolves in minutes, and for its late component, which evolves in hours. This anatomical feature renders this species especially susceptible to a nonspecific airway hyperreactivity bronchoconstrictive, as well as a specific hypersensitivity response. Therefore, this species has been used for decades for the study of protein-evoked anaphylactic shock and pulmonary hypersensitivity, and it can experience both immediate-onset and late-onset responses.

In comparison with other laboratory animal species, a high number of lymphocytes and eosinophils are detected in the bronchoalveolar lavage fluid of guinea pigs. It is particularly sensitive to airway resistance changes induced by aerosolized histamine—while rats are not. Airway hyperreactivity and eosinophil influx and inflammation can also be demonstrated in this animal species. For many years ovalbumin-induced pulmonary hypersensitivity in guinea pigs has been used as a model to study atopic asthma-like responses. However, mechanistic studies have been hampered by the lack of reagents needed to identify cells and mediators in respiratory allergy. Guinea-pig anaphylactic responses usually involve IgG_1 antibodies, even though the model can be tailored for the production of IgE. In this species no clear association of pulmonary hypersensitivity responses and elevated specific IgG_1 titers could be established. Thus, measurement of specific antibody formation provides ancillary evidence of an immunologically mediated response.

The key features of this animal model involve protocols using single or repeated inhalation or cutaneous exposures (or any other route) followed by a rest period until day 21. If sensitization is by inhalation, five consecutive exposures (3 hours/day) are commonly used for chemicals. Several groups of animals are needed to test concentrations from nonirritant to irritant. Generally, respiratory tract irritation is dose-limiting in inhalation studies. The advantage of topical induction regimens is, in turn, that substantially higher dosages can be used for induction. After the rest period, inhalation challenge with the hapten or antigen is performed. It focuses on identifying chemical sensitizers by measurement of the response, or the elicitation phase, of sensitization. Challenge by inhalation requires an exposure period of 15–30 min.

For irritant, volatile, reactive chemicals, which are preferentially scrubbed in the upper airways of nose breathing experimental animals, a conjugated hapten may be especially indispensable. When using the free chemical, the selection of adequate challenge concentrations of aerosol (e.g. trimellitic anhydride, diphenylmethane diisocyanate, reactive dyes) or vapor (e.g. hexamethylene diisocyanate, toluene diisocyanate) is critical. Changes in respiratory patterns may occur as a result of too-high concentrations used for challenge exposures and depend also on the location of the predominant deposition of the inciting and challenging agent within the respiratory tract.

Moreover, the interpretation of changes in respiratory pattern induced by irritant particulates is complicated further because of the size-dependency of the deposition of particles within the respiratory tract. Irritant aerosols that evoke bronchiolar or pulmonary irritation may produce a rapid, shallow breathing pattern (i.e. changes appear to be similar to those occurring following conjugate or antigen challenge). When sensitization is by inhalation to respiratory tract irritants, ensuing irritant-related inflammatory responses may render the respiratory tract more susceptible to subsequent challenge exposures. Thus, when sensitization is by inhalation, the similarity of the sites used both for induction and elicitation of respiratory allergy requires careful protocol considerations and selection of concentrations.

Respiratory patterns are often measured in volume displacement plethysmographs (see Figure 1). The analysis of response focuses on measurements during tidal breathing, that is peak expiratory flow, respiratory rate, respiratory minute volume, and inspiratory and expiratory times. Ideally, for each animal baseline data should be collected during a prechallenge adaptation period and during or following the subsequent challenge any response exceeding the mean ± 3 × standard deviations (SD) of this period might be classified as a positive response. This type of objective, quantitative analysis calculates the area exceeding the mean ± 3 × SD and can be used to express objectively the "intensity" of individual responses (see Figure 1). Nonspecific airway responsiveness to subsequently increased, stepped concentrations of aerosolized acetylcholine or methacholine can be measured in the same way. More recently developed methodologies use animals placed in a barometric, whole-body chamber, allowing a continuous measurement of the box pressure-time wave. Airway hyperresponsiveness is then expressed as changes in P_{enh} (enhanced end-expiratory pause), an indirect indicator of airflow obstruction and lung resistance.

A common pathologic accompaniment or cause of increased airway hyperresponsiveness is prolonged eosinophil-rich inflammatory leukocyte infiltration into the lungs of guinea-pigs after inhalation of specific antigen. It is suggested that this inflammation is responsible for the change in histamine or cholinergic agonist responsiveness. In contrast with the assays relying upon an induction of a specific set of characteristic endpoints, this model does not depend upon a preconceived mechanism of sensitization. Rather, it functions by reproducing the characteristics which typify the hypersensitivity reactions—the immediate-onset physiologic response of the airways (bronchoconstriction), and the ensuing inflammation quantified by lung lavage or histopathology. The characteristic features of such inflammation include an influx and activation of eosinophilic granulocytes.

Mouse IgE model

Two approaches to the identification of respiratory chemical allergens in mice have been described. The first focuses on the induction of total serum IgE, the second is cytokine fingerprinting. Both have as their theoretical foundation, the fact that chemical allergens of different types induce in *BALB/c* mice divergent immune responses characteristic of the selective activation of discrete T lymphocyte subpopulations. Respiratory allergens provoke T helper type 2 (Th2) responses. Contact allergens such as 2,4-dinitrochlorobenzene (DNCB) are considered not to cause sensitization of the respiratory tract; they stimulate in mice immune responses consistent with the preferential activation of Th1 cells. Such responses are associated with the production by draining lymph node cells (LNC) of interferon-γ (IFN-γ). The converse picture is seen with chemicals that have been shown to cause allergic respiratory hypersensitivity and occupational asthma in humans. Thus, chemical respiratory allergens such as trimellitic anhydride (TMA) elicit in mice Th2-type immune responses, associated with the production by draining LNC of high levels of interleukin (IL-4, IL-5, IL-10 and IL-13) and of other cytokine products of Th2 cells. IgE antibody responses are regulated by cytokines, the induction and maintenance of IgE responses being dependent upon the availability of IL-4, and being inhibited by IFN-γ. As a consequence it has been found that exposure of mice to TMA, but not to DNCB, results in the appearance of specific IgE antibody. In practice, assays are performed using three concentrations of the test material together with TMA and DNCB which serve, respectively, as positive and negative controls. To carry out the *BALB/c* mouse IgE test, chemical in vehicle is applied to the shaved flank of the mouse; 7 days later the chemical is applied to the dorsum of both ears; 14–

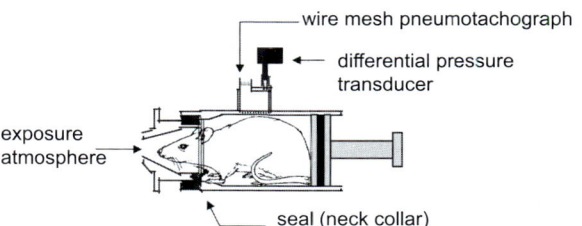

Animal Models for Respiratory Hypersensitivity. Figure 1 *Top panel*: Rat in a volume displacement plethysmograph for measurement of respiratory patterns during challenge.

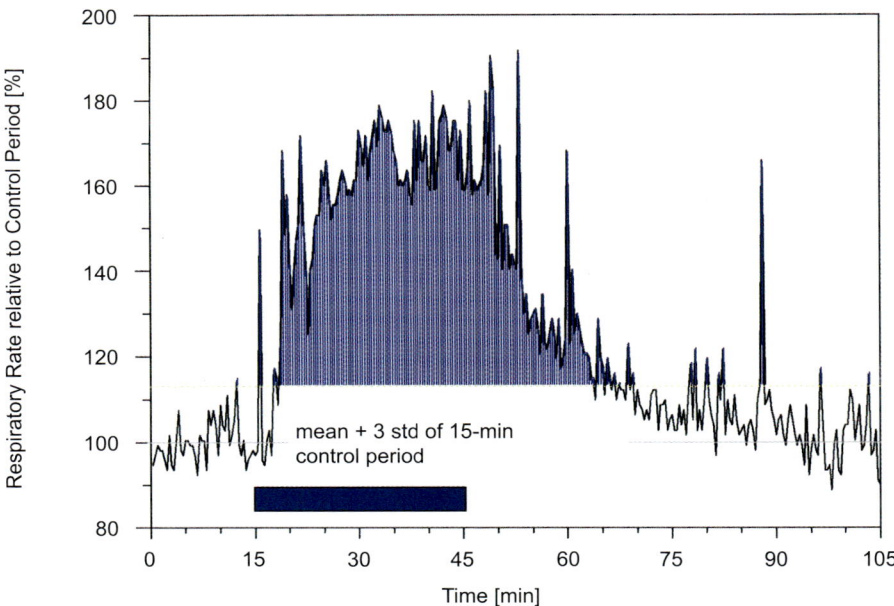

Animal Models for Respiratory Hypersensitivity. Figure 2 Analysis of the intensity of respiratory response of Brown Norway rats (8 rats per group) sensitized by epicutaneous administration of trimellitic anhydride (TMA) (1%, 5% or 25% in a vehicle) and challenge with TMA. Animals in the control groups (vehi) received vehicle only. Data represent the area under the curve exceeding ± 3 × SD of the animals' prechallenge period (top panel) or are presented as Tukey Box plot (lower panel). Boundaries of the box represent the 10th and 90th percentiles, the means and medians are displayed as dotted and solid lines, respectively.
MV: respiratory minute volume.

21 days later serum is drawn and total IgE assessed using an ELISA assay.

Rat model

Rat models of airway allergy are considered to demonstrate many features of allergic human asthma. In contrast to guinea pigs, which exhibit mast-cell-dependent bronchoconstriction to histamine, allergic bronchoconstriction in rats seems to be primarily mediated by serotonin. Similarities between responses in Brown Norway (BN) rats and humans include the production of IgE, a reasonable percentage of rats that have both immediate- and delayed-phase responses following aeroallergen challenge of sensitized animals, airway hyperreactivity to methacholine, acetylcholine or serotonin, and the accumulation of neutrophils, lymphocytes, and particularly activated eosinophils in lung tissue and bronchoalveolar lavage fluid. Elevations of the Th2 cytokines IL-4 and IL-5 and a reduction in the Th1 cytokine IFN-γ are also observed. However, some questions have been raised regarding the correlation between airway inflammation and airway hyperreactivity.

Further support for the BN rat is provided by comparison with other strains such as the Sprague-Dawley rat: the levels of eosinophilia and IgE parallel the airway responses. However, the rat is a weak bronchoconstrictor, and higher levels of agonist are required to induce the same level of response compared to guinea pigs. Thus, this animal model focuses on the induction of airway inflammation, which cause most of the characteristic features of asthma. In contrast to the BN rat, essentially no eosinophilic pulmonary inflammation is observed in Lewis or Fisher rats.

To induce an asthmatic state, Wistar rats are nose-only exposed for approximately 2 consecutive weeks by inhalation (5 hours/day, 5 days/week). To assess functional evidence (lung mechanics, forced expiratory maneuvers, diffusing capacity, acetylcholine bronchoprovocation, arterial blood gases), biochemical evidence (inflammatory parameters in bronchoalveolar lavage), and morphological evidence (influx of eosinophilic granulocytes into the tissue of the airways, secretory cell hyperplasia and metaplasia, smooth muscle hypertrophy and hyperplasia, epithelial desquamation, occlusion of the airway lumen with mucus and cellular debris), evidence of asthma-like lung disease and their regression during an observation period of approximately one to two months could be demonstrated.

More recently, the protocols used for guinea pigs were duplicated for BN rats. To probe respiratory hypersensitivity induced and elicited by chemical agents, such as TMA (either topically or by inhalation), the measurement of the respiratory minute volume and rate proved to be suitable to integrate (Figure 3) and quantify (Figure 4) the individual animal's response. To study the inflammatory component during disease development, the techniques commonly used include the same as those already presented for guinea pigs. When comparing topical and inhalation routes, it appears that important variables of this bioassay are related to both the route of induction and especially the total dose administered. It seems that the total dose required for topical sensitization in this animal model is rather high compared to the inhalation route.

Pros and Cons

Antigens entering through the skin and respiratory tract are recognized, processed, and carried by dendritic cells toward the respective draining lymph nodes. The local immune response may change as a result of the phenotype and function of local immune cells being altered by inhaled agents modifying the local microenvironment (e.g. inhaled irritants acting as adjuvant, and may not be representative of the systemic immune response). The immune response in the lung is compartmentalized (blood versus lung parenchyma) and observations in one compartment do not necessarily reflect the situation in another. This means that bioassays relying solely upon markers of response in serum do not necessarily mirror the response occurring in the critical organ—the lung.

The guinea pig bioassay offers advantages of integrating (patho-)physiological responses using relevant routes and procedures that can readily be compared to inhalation studies focusing on nonimmunological endpoints. This method seeks to identify chemicals that have the potential to elicit respiratory allergy, and attempts to define the respective threshold concentrations for induction and elicitation. Although costly and elaborate, judgment is based on several independent endpoints that include quantitative changes in breathing patterns following challenge, identification of the bronchial inflammatory response, and associated induction of specific IgG_1.

While the mouse IgE test potentially offers advantages of cost and speed, it is not without limitations. Not least of these is the fact that the method seeks to identify chemicals that have the potential to induce the quality of immune response required for sensitization of the respiratory tract. It is not necessarily the case that hazards identified in this way will translate into a risk of respiratory allergy in humans. Irrespective of the perceived benefits and drawbacks, the method must be considered as being as not yet validated.

The rat inhalation models used as adjunct to conventional repeated exposure inhalation studies to irritant asthmagens (e.g. diisocyanates) have the disadvantage that a high experimental sophistication is required, including the constraints on selecting effective test concentrations. Rat strains differ appreciably in their properties to demonstrate the essential features of allergic human asthma. Unlike guinea pigs, which exhibit mast-cell-dependent bronchoconstriction to histamine, the allergic bronchoconstriction in rats seems to be mediated primarily by serotonin. Endpoints related to pulmonary inflammation proved to be most sensitive in BN rats for demonstrating response. However, a breeder-specific background of spontaneously occurring pulmonary lesions has been reported in BN rats. This may hamper interlaboratory comparisons of studies. Also the rat model must be considered as incompletely validated.

Predictivity

Most bioassays define response in demonstrating some feature of allergic asthma using potentially relevant routes for induction and elicitation of response. However, no harmonized test guidelines are yet available, so none of the protocols currently applied can be viewed harmonized or validated with respect to the different classes of sensitizing or irritant—but nonsensitizing—chemicals. Predictivity is complicated by both the variability of the protocols used for induction in regard to the dose, route, and frequency of dosing, and by how to define positive response in using specific endpoints.

Relevance to Humans

Respiratory tract allergy and asthma in humans is characterized by a chronic type of pulmonary inflammation and increased responsiveness to specific and nonspecific stimuli. Different mechanisms are involved for low- and high-molecular-weight allergens, and they might be stimulated at lower levels compared to currently applied bioassays. The relevance of topical versus inhalation routes of induction, including adjuvant effects related to irritation and preexisting disease, appears to be yet unresolved. It seems, however, that in the currently employed bioassays the total dose required for successful sensitization seems to be appreciably higher when compared to that required to sensitize humans.

Regulatory Environment

In the regulatory arena there are several situations where data on the potential allergenicity of materials are required. The needs depend on the objectives of particular scopes. For example, premanufacturing notices (PMNs) required for the review and classification of new chemicals and setting of workplace concentrations are deemed to be safe.

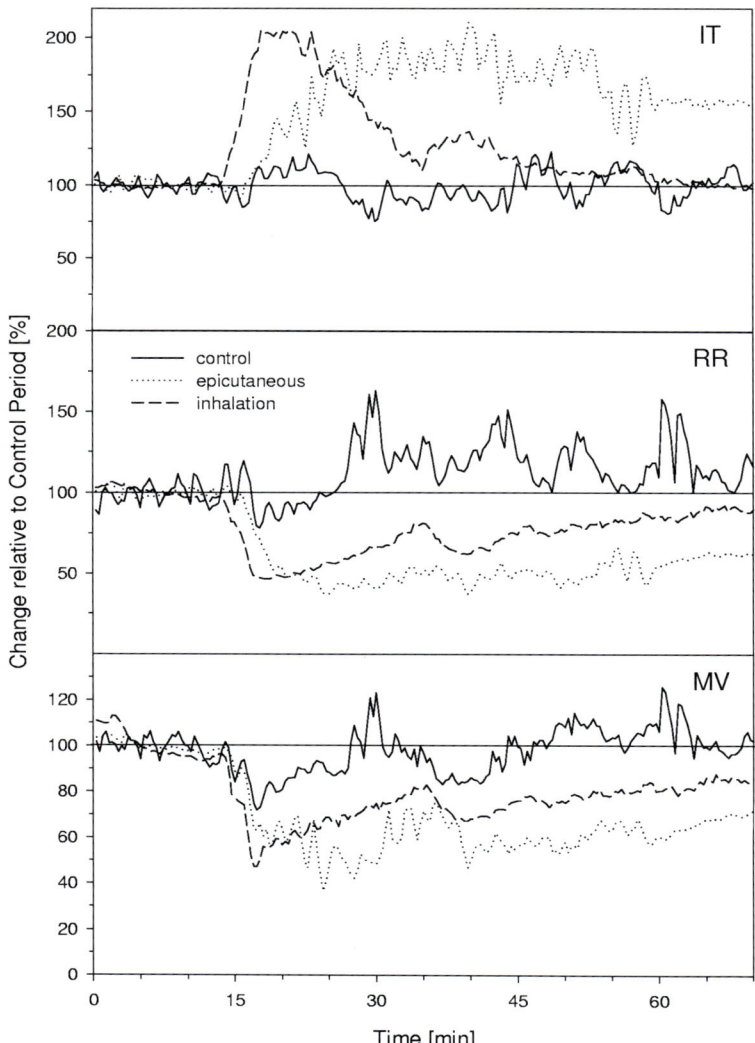

Animal Models for Respiratory Hypersensitivity. Figure 3 Change of respiratory patterns during a challenge with ≈ 23 mg TMA/m^3 (duration of challenge: 30-min). Brown Norway rats were either sensitized by epicutaneous administration of TMA in a vehicle or by 5 × 3-hrs/day inhalation exposures to 120 mg TMA/m^3. Respiratory response data were normalized to the mean of a 15-min pre-challenge exposure period (=100%). Before and after challenge the rats were exposed to conditioned air.
IT: inspiratory time, RR: respiratory rate, MV: respiratory minute volume.

Within the EU, regulatory status and implementation activities require identification and characterization of chemicals inducing respiratory tract irritation and sensitization. In accordance with the criteria given, the risk phrase R37 is assigned to chemicals acting as "irritants to the respiratory system". Conditions leading to classification with R37 are normally transient in nature and limited to the upper respiratory tract. The phrase R42 is assigned to chemicals that "may cause sensitization by inhalation". Classification is based on the chemical structure, human evidence, or positive results from appropriate animals tests.

Relevant guidelines

Harmonized testing guidelines are not yet available.

References

1. Briatico-Vangosa C, Braun CJL, Cookman G et al. (1994) Review: Respiratory allergy: Hazard identification and risk assessment. Fundam Appl Toxicol 23:145–158
2. Bice DE, Seagrave JC, Green FHY (2000) Animal models of asthma: Potential usefulness for studying

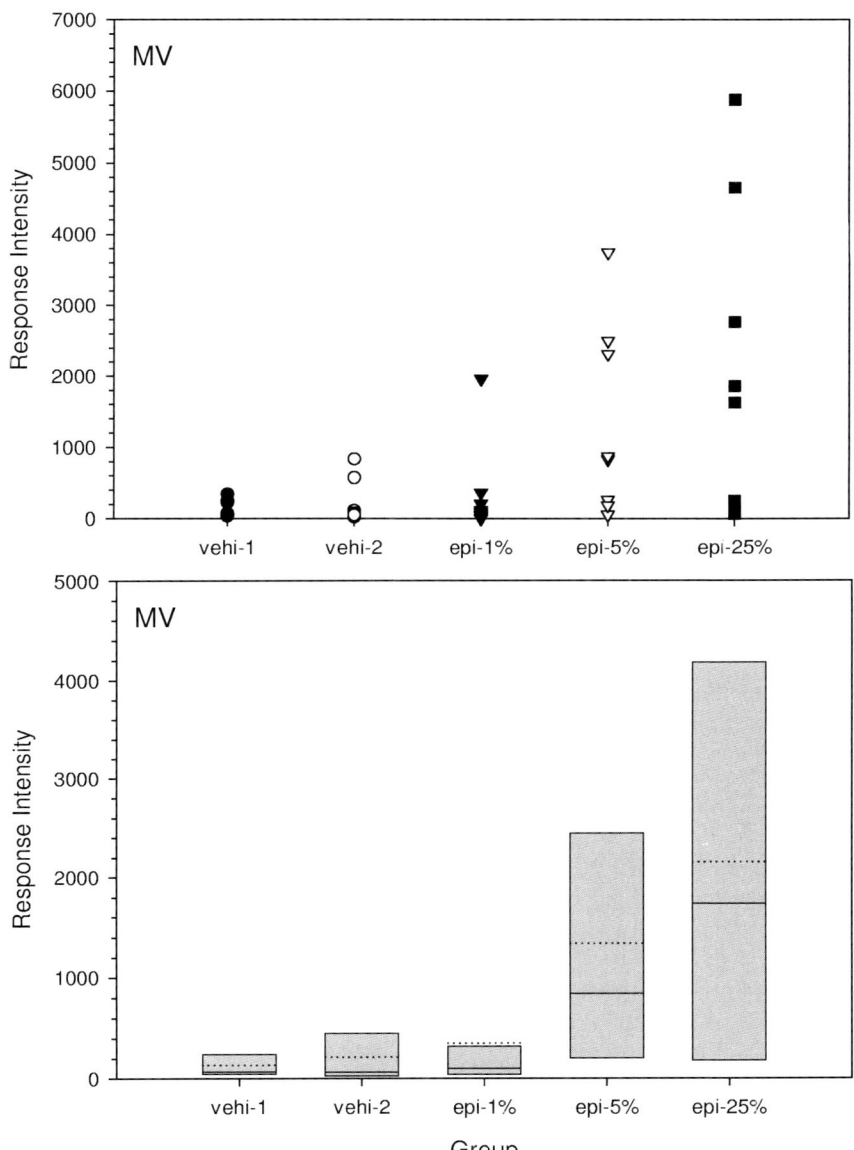

Animal Models for Respiratory Hypersensitivity. Figure 4 Analysis of the intensity of respiratory response of Brown Norway rats (eight rats per group) sensitized by epicutaneous administration of TMA (1%, 5% or 25% in a vehicle) and challenge with TMA. Animal of the control groups (vehi) received the vehicle only. Data represent the area under the curve exceeding ± 3 × SD of the animals3pre-challenge period (top panel) or are presented as Tukey Box plot (lower panel). Boundaries of the box represent the 10th and 90th percentiles, the means and medians are displayed as dotted and solid lines, respectively.
MV: respiratory minute volume.

health effects of inhaled particles. Inhal Toxicol 12:829–862
3. Kimber I, Dearman RJ (1997) Toxicology of Chemical Respiratory Hypersensitivity. Taylor and Francis, London
4. Karol MH, Thorne PS (1988) Respiratory hypersensitivity and hyperreactivity: implications for assessing allergic responses. In: Gardner DE, Crapo JD, Massaro EJ (eds) Toxicology of the Lung. Raven Press, New York, pp 427–448

Animal Models of Immunodeficiency

KENNETH L HASTINGS · SHUKAL BALA
Division Of Special Pathogen and Immunologic Drug Products, Center For Drug Evaluation And Research
US Food and Drug Administration

USA

Short Description

▶ Immunodeficient animals are those in which the immune system has been impaired to increase susceptibility to infections, tumors, or to similar diseases. Although immunodeficient animals are considered to be different from hyperresponsive models (that is, animals that develop immunopathies such as diabetes or lupus due to dysregulated immune function), they are often prone to develop autoimmune diseases. Immunodeficient animal models can be created using either physical, chemical, biological, or surgical methods, or by genetic manipulation.

The use of immunodeficent animals has been minimal in immunotoxicology research and has been limited, for the most part, to mechanistic studies. For example, immunodeficient mice have been used in host-resistance assays (due to susceptibility to infections of particular interest) or to study the role of particular endogenous molecules (such as cytokines) in immunomodulation by chemicals or drugs. Studies have been conducted to assess the effects of drugs/chemicals on morbidity and/or mortality due to experimental infection in immunocompromised animals (1). Gene knockout animal models have been used to indirectly assess the role of biological molecules such as cytokines in host resistance to infections and/or tumors (2). Animals can be made immunodeficient either by direct impairment using various techniques—chemical (e.g. dexamethasone), radiological (e.g. ionizing radiation exposure), biological (e.g. antilymphocyte antibody treatment), surgical (e.g. neonatal thymectomy)—or by genetic manipulation (e.g. continuous breeding of spontaneous mutants or by specific genetic alteration). Chemical, radiologic, or biological methods for inducing immunodeficiency have not been used commonly in immunotoxicology, primarily due to confounding adverse effects not related to the immune system. Surgical techniques are technically demanding and also have not been used to any great extent.

Genetically manipulated immunodeficient animal models have been used more commonly in immunotoxicology (3). Although monkey, dog, hamster, and rodent immunodeficiency models (including nude mice and rats) have been described, genetically altered mice are the most commonly used in immunotoxicology research. There are two general types of genetically-modified immunodeficient mice: those that were obtained by selective breeding of immunodeficent mutants (e.g. severe combined immunodeficient (SCID) mice), and those that were obtained by genetic engineering (e.g. transgenic mice; gene knockout mice). The actual impairment can be fairly broad (such as the *beige* mouse, which lacks genetic coding for natural killer (NK) cells and other immune cell functions), or can be relatively specific (as with knockout mice) where genes coding for specific immune system components (e.g. interleukin(IL)-2) have been removed. Two mutant strains, in addition to *beige* mice, have been used in immunotoxicology studies: the ▶ SCID mouse and the triple mutant ▶ *bg/nu/xid mouse*.

SCID mice are recombinase defective at the variable domain J (▶ VDJ) region, which results in a non-functional immunoglobulin heavy-chain gene rearrangement and a lack of functional T and B lymphocytes. Thus, they are unable to mount effective cellular or antibody-mediated immune responses. Triple mutant *bg/nu/xid* mice have reduced functions of NK cells, lymphokine-activated killer cells, and T and B cells. Immunodeficient animals often need to be housed under conditions that limit exposure to environmental pathogens. Thus, special containment facilities need to be used. These facilities should employ high-efficiency particulate air (HEPA) filtration, sterilization of animal bedding, food, and water, and use of fully protective laboratory outerwear by workers (including autoclaved gowns, sterile masks and gloves, and any other materials that could potentially come into contact with housed animals). These special procedures for animal housing and handling limit the usefulness of the models, but are not needed for all models. For example, these are necessary for SCID and beige mice, but are not necessary for some transgenic mice such as IL-4 knockout mice.

Characteristics

Immunodeficient animals are, in general, susceptible to diseases (especially autoimmune and infectious diseases) and have relatively short life spans. As an example, IL-10 knockout mice develop chronic enterocolitis, apparently due to aberrant immune responses to antigen exposure. This is not uniformly true, however, and some immunodeficient mice can appear to be relatively normal. It is thought that this is the result of redundant mechanisms of the immune system: indeed, much of what is known concerning redundant immune system mechanisms was derived from studies with knockout mice.

Pros and Cons

Immunodeficient animal models are useful in two particular situations: host-resistance studies and specific mechanism studies. These models have also been used for toxicology studies not specifically related to immunotoxicology.

Host-resistance assays in immunodeficient rodents (especially mice) take advantage of increased susceptibility to infectious agents as compared to wild-type animals. In most of these models, the experimental infection is likely to be fatal: thus time-to-death is taken as a convenient parameter to indicate immunotoxicity of an administered chemical. However, this is a relatively crude endpoint: combined percentage and time-to-death should be used as supporting indicators. Tissue microbial counts obtained from animals at predetermined time(s) of sacrifice should also be assessed and are probably better indicators of immunotoxicity. Also, immune parameters such as serum cytokine levels should be obtained on animals sacrificed on study. One advantage of genetically modified animals is that these can be used to replace models in which immune deficiency is induced using chemical treatment (such as steroid-induced immunodeficiency). This avoids unintended toxicities unrelated to the immune system which confound interpretation of study results.

Reconstitution of specific immune mechanisms by genetic manipulation or cell transfer in immunodeficient models can also be used. Animals thus reconstituted can be treated with a drug or chemical and assessed for impairment of the reconstituted function. For example, the immune systems of immunodeficient mice can be reconstituted with human immune cells, allowing for experimental *in vivo* evaluation of potential immunotoxicity of direct clinical relevance. Unfortunately, this methodology has not been exploited to any great extent, primarily due to inconsistencies in graft survival and immune function (4).

Predictivity

These models are probably not predictive of human health effects in any direct way. They are probably best viewed as hazard identification models. Knockout mice have proven to be valuable to the pharmaceutical industry, in pharmacological discovery studies. For example, knockout mice have been used to discover drugs with anti-inflammatory and immunomodulatory activity.

Relevance to Humans

In general, immunotoxicology studies conducted in immunodeficient mice are not directly relevant to human health. There are two potential exceptions to this, however: host resistance studies conducted in SCID mice and immunotoxicity studies conducted in SCID mice in which the immune system has been reconstituted with human immune cells (hu-SCID models).

In the first example, SCID could serve as a reasonably relevant model of human immunodeficiency diseases (such as human immunodeficiency virus (HIV) disease). The effect of administered xenobiotics (e.g. drugs) on host resistance in these models could be somewhat predictive of effects in humans with such diseases.

In the second example, hu-SCID models could be used to determine adverse effects of xenobiotics on discrete human immune cell types. However, in every example the state-of-the-art should be considered to support hazard identification versus risk assessment concerning potential human health effects.

Regulatory Environment

Immunodeficient animal models have value for hazard identification, but have not been widely accepted for risk assessment. Since these models have been used primarily to study underlying mechanisms in immunotoxicology, it is unlikely that they would be used as primary tools in regulatory toxicology. However, the models have been shown to be useful in discovery pharmacology and toxicology studies. For example, hu-CD4 transgenic mice have been used to assess the potential adverse effects of therapeutic monoclonal antibodies (5).

References

1. Bala S, Hastings KL, Kazempour K, Inglis S, Dempsey W (1998) Inhibition of tumor necrosis factor alpha alters resistance to *Mycobacterium avium* complex infection in mice. Antimicrob Agents Chemother 42:2336–2341
2. Nansen A, Pravsgaard Christensen J, Ropke C, Marker O, Scheynius A, Randrup Thomsen A (1998) Role of interferon-γ in the pathogenesis of LCMV-induced meningitis: unimpaired leukocyte recruitment, but deficient macrophage activation in interferon-γ knock-out mice. J Neuroimmunol 86:202–212
3. Lrvik M (1997) Mutant and transgenic mice in immunotoxicology: an introduction. Toxicology 119:65–76
4. Vallet V, Cherpillod J, Waridel F, Duchosal MA (2003) Fate and function of human adult lymphoid cells in immunodeficient mice. Histol Histopath 18:309–322
5. Herzyk DJ, Bugelski PJ, Hart TK, Wier PJ (2002) Practical aspects of including functional endpoints in developmental toxicity studies. Case study: immune function in HuCD4 transgenic mice exposed to anti-CD4 Mab in utero. Hum Exp Toxicol 21:507–512

Ankylosing Spondylitis

A chronic arthritic condition mainly affecting young adult males and characterized by progressive stiffness and fusions in the spinal joints, especially in individuals who are HLA-B27 positive.

▶ Molecular Mimicry

Anterior (Head) Kidney

This is the front part of fish kidney with immune functions comparable to mammalian bone marrow, i. e. it performs hematopoiesis.
▶ Fish Immune System

Anthracene

▶ Polycyclic Aromatic Hydrocarbons (PAHs) and the Immune System

Anti-Cancer Antibodies

Antibodies directed against tumor-specific antigens, used as therapeutic and/or diagnostic agents.
▶ Antibodies, Antigenicity of

Anti-DNA Antibodies

▶ Antinuclear Antibodies

Anti-Double Stranded (ds) DNA Antibodies

▶ Antinuclear Antibodies

Anti-Histone Antibodies

▶ Antinuclear Antibodies

Anti-Inflammatory Antibodies

Antibodies directed against pro-inflammatory molecules, such as cytokines, cell adhesion molecules and leukotrienes, used to suppress inflammation.
▶ Antibodies, Antigenicity of

Anti-Inflammatory Cytokine

A cytokine (e.g. interleukin-10) that downregulates an inflammatory process by reducing expression of proinflammatory cytokines.
▶ Cytokines

Anti-Inflammatory (Nonsteroidal) Drugs

PETIA P SIMEONOVA
Toxicology and Molecular Biology Branch
National Institute for Occupational Safety and Health
Morgantown, WV 26505
USA

Synonyms
Nonsteroidal antiinflammatory drugs, NSAIDs, aspirin, aspirin-like drugs, cyclooxygenase inhibitors, COX inhibitors.

Definition
NSAIDs are a group of chemically dissimilar agents, other than steroids, commonly used to treat a variety of conditions because of their analgesic, antiinflammatory, and antipyretic properties. They are widely used to reduce inflammation and pain in musculoskeletal disorders including osteoarthritis, rheumatoid arthritis, gout, tendonitis, and muscle strains. They are also used for treatment of headaches, fever, dental and other common painful conditions. Recently, the use of aspirin for prevention of diseases such as myocardial infarction, stroke and cancer has gained attention.

Characteristics
The mechanism of action of NSAIDs is primarily related to inhibition of prostaglandin synthesis (Figure 1).
Arachidonic acid, a 20-carbon fatty acid, is the precursor of the eicosanoids including prostaglandins. Enzymatic action of phospholipase A_2 on cell membrane phospholipids triggers the arachidonic acid cascade. There are two major pathways in the synthesis of eicosanoids. All eicosanoids with ring structures (including prostaglandins, thromboxanes, and prostacyclins) are synthesized through cyclooxygenase pathways. Cyclooxygenase exists in two isoforms, COX-1 and COX-2. Despite structural similarities, they are encoded by different genes and are distinct in their distribution and expression in various tissues. COX-1 is the constitutive isoform; its responsibilities include maintaining gastrointestinal mucosal integrity,

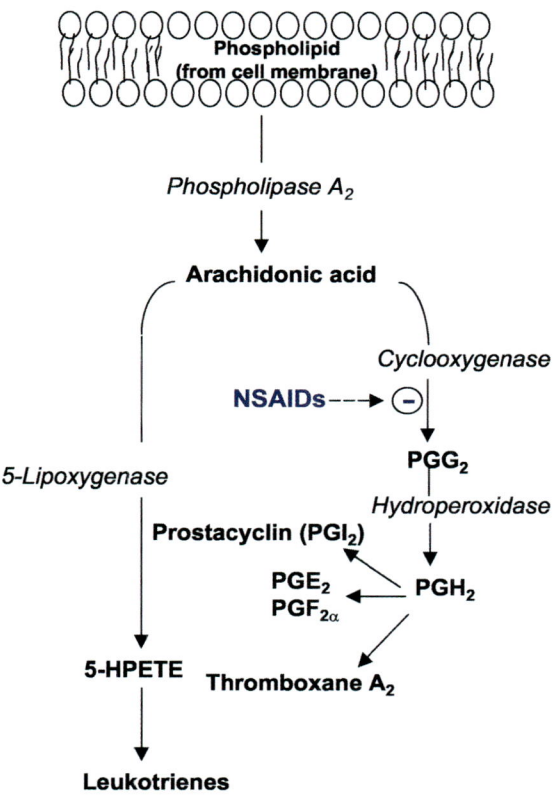

Anti-Inflammatory (Nonsteroidal) Drugs.
Figure 1 Synthesis of prostagandins (PGs) and leukotrienes.

platelet aggregation, and renal blood flow. COX-2 is the inducible isoform involved in inflammation, mitogenesis, and signaling pathways. Alternatively, several lipoxygenases act on arachidonic acid to form leukotrienes and related products. The NSAIDs act primarily by inhibiting the cyclooxygenase enzymes but not the lipoxygenase enzymes. Based on their specificity for the isoforms of the cyclooxygenase, the NSAIDs can be classified in four major groups: highly COX-1, equally COX-1/COX-2, relatively COX-2 and highly COX-2 selective (Table 1).

Aspirin is the most commonly used NSAID and the drug with which all other antiinflammatory drugs are compared. Aspirin is unique among the NSAIDs because it irreversibly and nonselectively inactivates (by acetylating) cyclooxygenase enzymes. The inhibition of cyclooxygenase activity diminishes the formation of prostaglandins at the peripheral target sites and at the thermoregulatory centers in the hypothalamus resulting in strong antiinflammatory and antipyretic effects. Furthermore, reduction of prostaglandin levels results in diminished sensitization of pain receptors to both mechanical and chemical stimuli. Low doses (60–80 mg daily) of aspirin used over many days can irreversibly inhibit thromboxane production in platelets, resulting in reduced platelet aggregation without markedly affecting the prostaglandin synthesis by most tissue.

The therapeutic use of aspirin and other NSAIDs is limited by their significant gastrointestinal and renal toxicity. Normally, prostacyclin (PGI_2) inhibits gastric acid secretion, whereas prostaglandins PGE_2 and $PGF_2\alpha$ stimulate synthesis of protective mucus in both the stomach and small intestine. Inhibition of COX-1 leads to increased gastric acid secretion and diminished mucus protection. The consequences might be epigastric distress, ulceration, and/or hemorrhage. Renal complications of NSAIDs are also related to inhibition of basal COX activity, particularly in the presence of vasoconstrictors such as angiotensin, norepinephrine, and vasopressin. The prostaglandins PGE_2 and PGI_2 are responsible for maintaining renal blood flow under these conditions. A reduction of these can result in retention of sodium and water or hyperkalemia in some patients.

The recently introduced selective inhibitors of COX-2 have strong antiinflammatory properties and reduced gastrointestinal toxicity. However, selective COX-2 inhibitors seem to have the same degree of renal toxicity and increased risk for thrombosis and myocardial infarction as the traditional NSAIDs. Selective COX-2 inhibition suppresses the synthesis of PGI_2 and has no effect on thromboxane$(Tx)A_2$, shifting the hemostatic balance toward the prothrombotic state. The original paradigm regarding COX-1 and COX-2 might be simplistic as they might share more complex physiologi-

Anti-Inflammatory (Nonsteroidal) Drugs. Table 1 Cyclooxygenase isoform selectivity of NSAIDs (adapted from Cryer and Dubois (1))

Selectivity			
Highly COX-1	**Equally COX-1/COX-2**	**Relatively COX-2**	**Highly COX-2**
Flurbiprofen	Aspirin	Meloxicam	Celecoxib
Ketoprofen	Ibuprofen	Nimesulide	Rofecoxib
Aspirin (low dose)	Indomethacin		
	Naproxen		

cal and pathophysiological roles. A new approach for reducing gastrointestinal and renal toxicity of NSAIDs is by the use of NSAIDs containing nitric oxide (NO) —for example NO-Aspirin. NO has a critical role in maintaining the integrity of the gastroduodenal mucosa. In theory, the NO-NSAIDs have the potential to provide the same or better therapeutic effects, including prophylaxis against myocardial and cerebrovascular ischemia, with lower level of toxicity.

Immunotoxicity of NSAIDs

Other adverse effects of NSAIDs are associated with sensitivity reactions. The prevalence of NSAID sensitivity ranges from 0.3% to 2.5% in the general population to around 10% in asthmatic patients. Two types of mechanism may account for the induction of NSAIDs sensitivity: allergic reactions (hypersensitivity) and pseudoallergic (idiosyncratic) reactions (Table 2). Allergic reactions are rare and could be cell-mediated or IgE-mediated. They can range from acute ▶ urticaria/angioedema to anaphylactic shock. Usually they are induced by a single drug and starting an alternative NSAID is helpful.

The most common NSAID-related sensitivity is a pseudoallergic (idiosyncratic) reaction, Such reactions mimic allergic reactions, but do not include immune recognition. They are associated with underlying allergic disease, for example asthma or urticaria and excessive production of leukotrienes. They are characterized by cross-reactions to different NSAIDs. Enhanced activity of key synthetic enzymes, perhaps genetically determined, has been implicated in affected people. Furthermore, a trigger event in the pathogenesis is NSAID-induced COX-1 inhibition. The subsequent decrease in the synthesis of prostaglandins, such as PGE_2, an inhibitor of 5-lipooxygenase, results in shifting the balance of the pathway in the direction of excessive leukotriene production. Most of the proinflammatory actions of the leukotrienes are mediated by binding to one of their high-affinity receptors, termed $CysLT_1$. Overexpression of this receptor on inflammatory cells has been proposed as an additional contributory mechanism in NSAID sensitivity. The leukotrienes can produce bronchospasm, increased bronchial hyperresponsiveness, mucus production, mucosal oedema, airway smooth-muscle cell proliferation, and eosinophil recruitment to the airways.

Aspirin-exacerbated respiratory disease (AERD), formerly referred to as aspirin-induced asthma or aspirin intolerant asthma, is the most well characterized and common example of NSAID-related pseudoallergic sensitivity. It is associated with progressive sinusitis, nasal polyposis, and asthma. Small single doses of aspirin, or other nonselective COX inhibitors, may cause rhinorrhea, bronchospasm, and shock symptoms. AERD is seen only in adulthood with higher prevalences in females.

People with NSAID sensitivity can be diagnosed definitively only through provocative tests. NSAID challenge can be by oral, bronchial, or nasal routes. The oral provocation test is one of the most commonly used methods and is the only one available in the USA. If 650 mg is administered without reaction, and the patient is not taking > 10 mg of prednisone or a leukotriene modifier drug, the challenge test is determined to be negative.

The treatment of NSAID sensitivity includes desensitization by repeated administration of increasing doses of the drug until all reactions have disappeared. NSAIDs share the phenomena of cross-desensitization. Acetaminophen (paracetamol), a weak peripheral COX-1 and COX-2 inhibitor (not included in the group of NSAIDs) can be used for analgesic and antipyretic treatment in patients with NSAID sensitivity. However, cross-reactions have been reported at high concentrations of this drug. The selective COX-2 inhibitors and new types of antiinflammatory drugs, including phospholipase A2 inhibitors (benzydamine) and leukotriene modifiers (zileuton and montelukast), might provide an alternative antiinflammatory approach in NSAID sensitive patients.

Aspirin given during viral infections, especially in children, has been associated with increased incidence of ▶ Reye's syndrome, an often fatal, fulminating hepatitis with cerebral edema. Although the pathogenesis of this fatal disease is not well understood, acetamin-

Anti-Inflammatory (Nonsteroidal) Drugs. Table 2 Classification of NSAID sensitivity (adapted from Stevenson et al (2))

Allergic reactions	Cross-reactions after first exposure
Single drug-induced urticaria/angioedema	No
Multiple drug-induced urticaria/angioedema	Yes
Single drug-induced anaphylaxis	No
Pseudoallergic reactions	
NSAID-induced rhinitis and asthma	Yes
NSAID-induced urticaria/angioedema	Yes

ophen is recommended instead of aspirin for children who need medication.

References

1. Cryer B, Dubois A (1998) The advent of highly selective inhibitors of cylooxygenase: a review. Prostagland Lipid Mediat 56:341–361
2. Stevenson D, Sachez-Borges M, Szczeklik A (2001) Classification of allergic and pseudoallergic reactions to drugs that inhibit cyclooxygenase enzymes. Ann Allergy 87:177–180
3. Namazy JA, Simon RA (2002) Sensitivity to nonsteroidal antiinflammatory drugs. Ann Allergy Asthma Immunol 89:542–550

Anti-Single Stranded (ss) DNA Antibodies

▶ Antinuclear Antibodies

Anti-Tumor Immunity

▶ Tumor, Immune Response to

Antibodies, Antigenicity of

EUGEN KOREN
Clinical Immunology, Amgen Inc.
1 Amgen Center Drive
Thousand Oaks, CA,
USA

Synonyms

Antibodies, monoclonal antibodies, recombinant antibodies, immunoglobulins.

Definition

Antibodies used for treatment and/or diagnosis of human disease are immunoglobulin molecules of variable origin and structure.

Characteristics

Currently, there are many different types of therapeutic and/or diagnostic antibodies in various stages of clinical use (1,2). Chimeric mouse to human antibodies are recombinant immunoglobulin molecules composed of murine variable and human constant domains. Humanized antibodies are recombinant molecules comprising murine complementarity determining regions (CDRs) grafted onto human immunoglobulin framework. Fully human antibodies produced by recombinant technology or by hybridoma technology in transgenic mice are also in clinical use. Additional variants of therapeutic and/or diagnostic antibodies are represented by recombinant constructs such as monovalent and multivalent antigen binding fragments, single-chain variable fragments, antibodies conjugated with toxins, enzymes, prodrugs and viruses, as well as radiolabeled antibodies or fragments thereof. There are also purified animal and human serum antibodies directed against specific targets or just plain ▶ immunoglobulins used for various clinical applications.

Major Indications and Efficacy

Thus far, the major therapeutic applications of recombinant antibodies are in oncology and inflammation. There has been successful treatment of breast cancer, colon cancer, and lymphoma by humanized and chimeric antibodies. Chimeric, humanized, and fully human antibodies with anti-inflammatory activity are effective in the treatment of rheumatoid arthritis and psoriasis. Most recently, a humanized anti-IgE antibody has been approved for allergic asthma. Transplantation is another important therapeutic area where a spectrum of antibodies, including mouse ▶ monoclonal antibodies, human serum antibodies, and recombinant humanized antibodies directed against lymphocytes are used to inhibit graft rejection. Animal antisera against venoms, as well as antibodies that provide passive immunity against microorganisms and/or their toxic products, have been in use for decades. Diagnostic applications of antibodies primarily include detection cancer by the use of radiolabeled ▶ recombinant antibodies directed against tumor antigens.

Antigenicity

Antigenicity of antibodies depends on several factors that apply to all therapeutic proteins. These include species differences, route of administration, dosing regimen, concomitant therapy, formulation, purity, presence of immunogenic epitope(s) as well as the overall complexity of the molecule (3).

Obviously, the greater the species divergence, the more antigenic are injected antibodies. For example, animal antibodies such as horse anti-venom immunoglobulins or mouse monoclonal antibodies invariably cause a high incidence of antibodies in humans. Conversely, human antibodies cause a strong antibody response in rodents and a less pronounced, but still present, antibody response in non-human primates. It should be pointed out that antigenicity of therapeutic antibodies can be diminished in a number of ways. Recently, horse anti-venom preparations have been replaced by ovine affinity-purified Fab anti-venom fragments, from which the most antigenic part of an immunoglobulin molecule—the Fc-fragment—is removed.

More importantly, recombinant technology allows for significant antigenicity reduction of murine monoclonal antibodies. Chimeric mouse–human antibodies containing approximately 30% of murine amino acid sequences are less antigenic than fully mouse antibodies, whereas the antigenicity of humanized mouse antibodies with only 3%–7% murine sequences is even further reduced.

Perhaps the most straightforward example for the significant reduction of antigenicity is the humanization of the mouse antibody directed against human T-cell receptor. The fully mouse version of this monoclonal antibody (OKT3) causes antibody response in virtually all treated (transplant) patients, but its humanized counterpart (OKT3-H1) elicits antibody response in less than 10% of patients. There is no published data that directly compares antigenicity of subcutaneously and intravenously injected antibodies; however, there is evidence for other therapeutic proteins that the subcutaneous route is more immunogenic than intramuscular or intravenous routes.

In addition, it is widely known that the subcutaneous injection of vaccines is the most effective way of inducing an antibody response. Frequent dosing, especially over a long period of time, is more than likely to elicit a greater antibody response, irrespective of the route of administration. This is again based on knowledge from the field of vaccination, and on direct experience with therapeutic antibodies and other therapeutic proteins given to both experimental animals and humans.

Concomitant therapy is an important factor also. Immunosuppressive treatments such as chemotherapy and/or radiation have been shown to attenuate antibody response to therapeutic proteins. It should be mentioned that an antibody (such as anti-CD20) could be immunosuppressive and could, therefore, diminish antibody response against itself.

Formulation can indirectly influence antigenicity if stability of the molecule is not properly maintained. For example, self-aggregation of a recombinant protein due to less than optimal formulation can lead to an increased uptake by antigen-presenting cells (APC) and augmented antigen presentation. Contaminating host-cell proteins and/or degradation products have been shown to increase antigenicity. Presence of immunodominant epitope(s) in a therapeutic protein could make even fully human molecule antigenic. Recently approved fully human antibody to TNFα has been shown to cause neutralizing antibody response in 10%–12% of patients with rheumatoid arthritis. This unexpected observation could be explained by the presence of immunogenic epitope(s) comprised within CDR(s).

Finally, it is conceivable that the immunoconjugates described above, together with various constructs composed of recombinant antibody fragments, could be viewed as non-self entities by the immune system, and could therefore stimulate antibody responses.

Preclinical Relevance

Antigenicity of antibodies in preclinical studies can have quite significant effects on antibody drug development. Pharmacokinetic and pharmacodynamic studies can be complicated because antibodies can enhance drug clearance or cause drug accumulation, alter bioavailability, and/or PK/PD relationship. Antibodies can adversely affect toxicology studies also, because clearing, sustaining, and neutralizing antibodies can all alter exposure to the drug, thus leading to inaccurate interpretation of toxicity data. In some cases, induced antibodies can form immune complexes with the antibody drug. This could lead to deposition of immune complex and/or complement activation, causing toxicity symptoms that the drug by itself may not generate. Antibodies can also cross the placenta to appear in the fetus or neonate, which further complicates toxicology studies.

Relevance to Humans

Antibody responses to therapeutic and diagnostic antibodies in humans raise somewhat different concerns. Obviously, patient safety is of primary importance. Potential allergic reactions may be a threat if an IgE response is present. Immune complex deposition and complement activation have often been observed with native animal antibodies, and also with recombinant antibody drugs especially in cases of pronounced antibody response. Reduced efficacy may become an issue if neutralizing or clearing antibodies occur (4).

Regulatory Environment

The regulatory environment for therapeutic and diagnostic antibodies has been generally favorable. Many of them, especially anti-cancer ones, received priority review status in the last decade. Antigenicity of therapeutic antibodies is routinely monitored, although it appears to be less of a concern than recombinant proteins that are likely to induce antibodies capable of cross-reacting with and neutralizing endogenous molecules (5).

It should be noted that regulatory agencies issue a number of guidelines that may be considered during development of therapeutic and/or diagnostic antibodies. They include:
- FDA Guidance for Industry and FDA Reviewers (Immunotoxicity Testing Guidance)

- Annex B of CPMP/SWP/1042/99 (Guidance on Immunotoxicity)
- ICH Guidelines for Preclinical Safety Evaluation of Biotechnology-Derived Pharmaceuticals.

References
1. Taylor PC (2003) Antibody therapy for rheumatoid arthritis. Curr Opin Pharmacol 3:323–328
2. Souriau C, Hudson PJ (2003) Recombinant antibodies for cancer diagnosis and therapy. Exp Opin Biol Ther 3:305–318
3. Koren E, Zuckerman LA, Mire-Sluis AR (2002) Immune responses to therapeutic proteins in humans—Clinical significance, assessment and prediction. Curr Pharm Biotechnol 3:349–360
4. Baert F, Noman M, Vermeire S et al. (2003) Influence of immunogenicity on the long-term efficacy of infliximab in Crohn's disease. N Engl J Med 348:601–608
5. Casadevall N, Nataf J, Viron B et al. (2002) Pure red-cell aplasia and anti-erythropoietin antibodies in patients treated with recombinant erythropoietin. N Engl J Med 346:469–475

Antibody

Antibodies are immunoglobulins, a family of structurally related glycoproteins produced by the immune system, which combine with antigen and then mediate various biological effects. The usual object of these reactions is to activate complement or the mononuclear phagocytic system with the purpose of inactivation, destruction, and removal of the antigen from the organism.

Antibodies are produced by the host immune cells (B lymphocytes, plasma cells) as a consequence of the immune response to an antigen. An antibody distinctively binds to its antigen, thus it is antigen-specific. There are five classes (isotypes) of immunoglobulins: IgM, IgG, IgE, IgA, IgD. Two dominant isotypes of antigen-specific antibodies produced and secreted into blood plasma are IgM and IgG. The IgM and/or IgG antigen-specific antibodies can be detected in plasma or serum samples using immunoassays.

▶ Immunoassays
▶ Autoantigens
▶ Cell Separation Techniques
▶ AB0 Blood Group System
▶ Antibodies, Antigenicity of
▶ Monoclonal Antibodies

Antibody Class

An antibody classification system based on structural and functional characteristics of the heavy chain. The five classes of antibodies are IgG, IgM, IgA, IgD, and IgE.

▶ Monoclonal Antibodies

Antibody-Dependent Cell-Mediated Cytotoxicity (ADCC)

Antigen-specific killing of target cells by cytotoxic cells of the natural or non-adaptive immune system (e.g. natural killer cells) that express receptors for the Fc region of immunoglobulins. Antibody-dependent cell-mediated cytotoxicity (ADCC) requires antibody binding via Fab to antigen on the surface of the target cells and via Fc to the Fc receptor on the surface of the cytotoxic cell.

▶ Tumor, Immune Response to
▶ Cell-Mediated Lysis
▶ Natural Killer Cells
▶ Antibody-Dependent Cellular Cytotoxicity

Antibody-Dependent Cellular Cytotoxic (ADCC) Cells

Once antibodies have bound to specific cell surface antigens of target cells or tissues (opsonization), they interact with a variety of Fc-receptor expressing effector cells. This initiates a cascade of biochemical events leading to target cell or tissue damage.

▶ Limiting Dilution Analysis

Antibody-Dependent Cellular Cytotoxicity

JORGE GEFFNER
IIHEMA
Academia Nacional de Medicina
Pacheco de Melo 3081
Buenos Aires, 1425
Argentina

Synonyms
antibody-dependent cell-mediated cytotoxicity, antibody-dependent cytotoxicity

Definition
Antibody-dependent cellular cytotoxicity (ADCC) is a mechanism mediated by different leukocyte populations bearing receptors for the Fc portion of immuno-

globulins (▶ FcRs) that enable them to kill a wide variety of antibody-coated target cells rapidly (1). ADCC is induced, in all cases, without major histocompatibility complex (MHC) restriction and therefore is a mechanism operative in syngeneic, allogeneic and xenogeneic systems. The specificity of ADCC is conferred by antibodies, which at extremely low concentrations, far below those required for complement-mediated lysis, are able to induce the destruction of targets by effector cells.

Characteristics

A variety of effector cell populations, using different FcRs, are able to mediate ADCC.

The ability of a given effector cell to mediate ADCC, however, is strongly dependent on several factors: the activation state of effector cells, the isotype and density of antibodies coating the target cell surface, the type of FcR involved, the nature of the target cells, and the intracellular domains of target antigens. In all cases, the cytotoxic response is triggered by the interaction of antibodies coating target cells with FcRs expressed by the effector cells without the participation of ▶ complement. It is also well established that cytotoxicity does not involve the participation of a diffusible factor—in fact, when mixtures of antibody-coated and uncoated target cells are cultured together with effector cells, only destruction of coated targets is observed.

Receptors for the Fc portion of immunoglobulins have been described in all cell types of the immune system. They are able to recognize the Fc portion of IgG (FcγR), IgA (FcαR) and IgE (FcεR) and trigger not only ADCC, but also a wide array of responses which range from effector functions (such as phagocytosis, pinocytosis and release of inflammatory mediators) to immunoregulatory signaling (such as modulation of antigen presentation, lymphocyte proliferation and antibody production).

Receptors for the Fc portion of IgG (FcγR) fall into two general classes:

- the activation receptors, which are characterized by association with the ▶ FcRγ chain, which bear an immunoreceptor tyrosine-base activation motif (▶ ITAM) in their cytoplasmic region, and are essential for the triggering of activation signals
- the inhibitory receptor FcγRIIb, which are characterized by the presence of an ▶ ITIM sequence.

Only activation FcγRs are able to mediate ADCC. These receptors fall into three main classes based on structural analysis of the genes and proteins: the high-affinity receptor FcγRI (CD64) and the low-affinity receptors FcγRIIa (CD32) and FcγRIII (CD16). They are expressed by monocytes, macrophages, dendritic cells, polymorphonuclear leukocytes (PMN), mast cells, and platelets. Most of these cells express different types of FcγRs. Using these receptors, mononuclear phagocytes, neutrophils, and NK cells mediate ADCC against a wide variety of target cells. Several lines of evidence support that IgG-mediated ADCC plays an important role in host-acquired immunity against different infectious agents (2). Interestingly, observations made in FcγR knockout mice showed that passive immunization with specific IgG antibodies directed to some infectious agents results in protection in wild-type, but not FcγR-deficient mice. Moreover, a role for IgG-mediated ADCC in antitumor immunity has also been proposed. Studies performed by Clynes and coworkers in a syngeneic model of metastatic melanoma demonstrate that antimelanocyte antibodies prevent tumor metastasis in wild-type animals but are ineffective in FcγR-deficient mice (3). Similarly, observations made in FcγR-deficient mice have shown that the therapeutic

Antibody-Dependent Cellular Cytotoxicity. Table 1 Antibody-dependent cellular cytotoxicity

Effector Cells	Antibody	Fc Receptor
Neutrophils	IgG	FcγRI, FcγRII, FcγRIIIb
	IgA	FcαRI
Eosinophils	IgG	FcγRII
	IgA	FcαRI
	IgE	FcεRI, FcεRI II
Monocytes and macrophages	IgG	FcγRI, FcγRII, FcγRIIIa
	IgA	FcαRI
	IgE	FcεRI, FcεRII
Natural killer cells	IgG	FcγRIIIa
Dendritic cells	IgG	FcγRII, FcRγIIIa
Platelets	IgE	FcεRI, FcεRII

efficacy of two of the most widely used ▶ monoclonal antibodies (mAbs) in oncology—rituximab and herceptin—are strongly dependent on FcγR expression, supporting a role for ADCC in the antitumor activity mediated by these mAbs. Rituximab is a chimeric anti-CD20 antibody which produces a response rate of approximately 50% in relapsed low-grade, B cell non-Hodgkin's lymphoma; herceptin is a humanized mAb directed at the product of the proto-oncogene Her-2, which is used to treat metastatic breast cancer. Besides expression on mast cells and basophils, the high-affinity receptor for the Fc portion of IgE, FcεRI, is also expressed on human Langerhans cells, monocytes, macrophages, eosinophils, and platelets. With the exception of Langerhans cells, all of these cells have been seen to perform IgE-mediated ADCC. Moreover, mononuclear phagocytes, eosinophils and platelets also express FcεRII (CD23), the low-affinity receptor for the Fc portion of IgE, which is the only FcR described so far that is not a member of the Ig superfamily. This receptor is also able to trigger ADCC. IgE-mediated ADCC appears to play an important role in immunity against helminthic parasites, as described by Capron and coworkers (4). Moreover, recent observations demonstrate that monocytes are also able to kill IgE-coated tumor cells efficiently, supporting the idea that that IgE antibodies may also be exploited for cancer immunotherapy.

The receptor for the Fc portion of IgA, FcαRI (CD89), is expressed in PMN, monocytes, and macrophages and enables them to mediate ADCC against IgA-coated target cells. Moreover, FcαRI appears to represent the most effective leukocyte FcR for initiation of CD20-targeted antibody therapy. Interestingly, the ability of PMN to kill IgA-coated tumor cells is abolished in mice that were deficient in the β_2 integrin Mac-1 (CR3, CD11b/CD18) as well as in human neutrophils blocked with anti-Mac-1 mAb. Recent studies have shown that in the absence of Mac-1 the interaction of IgA with FcαRI is not impaired, but neutrophils are completely unable to spread on IgA-opsonized targets—a phenomenon that appears to play a critical role in ADCC. A similar role for Mac-1 in IgG-mediated ADCC has also been described. Of note, recent observations performed in vivo have shown that Mac-1 is crucial for effective Fc receptor-mediated immunity to melanoma (5).

Which are the mediators responsible for target cell destruction in ADCC? The identity of these mediators appears to be different depending on the ADCC model analyzed. Reactive oxygen intermediates (ROI) such as superoxide anion, hydrogen peroxide, and hydroxyl radical, play a critical role in ADCC mediated by neutrophils, monocytes, and macrophages against IgG-coated red blood cells. By contrast, ADCC mediated by NK cells against IgG-coated tumor cells depends on the release of perforins and granzymes. Interestingly, using monocytes obtained from chronic granulomatous disease patients, which produce very limited amounts of ROI, it was found that ADCC against IgG-coated red blood cells, but not ADCC against IgG-coated lymphoblastoid cells is dependent on the action of ROI. These results support the notion that a single type of effector cell might mediate ADCC towards different targets through distinct mechanisms.

How is ADCC regulated? Different ▶ cytokines exert potent regulatory effects on ADCC. Interferon-γ increases both, the expression of FcγRI, and the production of ROI by neutrophils, monocytes and macrophages. Both effects appear to be responsible for the increase of IgG-mediated ADCC induced by interferon-γ treatment. On the other hand, while interleukins IL-4 and IL-13 inhibit the expression of all three FcγRs and ADCC mediated by mononuclear phagocytes, IL-10 stimulates both, FcγRI expression and ADCC. Tumor necrosis factor (TNF-α) stimulates ADCC mediated by neutrophils and eosinophils against tumor cells and schistosomes, respectively, while IL-5 enhances ADCC mediated by eosinophils but not neutrophils. Colony stimulating factors (CSFs) have also shown to exert powerful stimulating effects on ADCC mediated by phagocytic cells. Granulocyte macrophage-colony stimulating factor (GM-CSF) increases ADCC mediated by eosinophils, neutrophils and monocytes against antibody-coated tumor cells, as well as ADCC mediated by eosinophils against antibody-coated parasites. Granulocyte-CSF (G-CSF) is used to increase neutrophil counts in neutropenic patients, and induces the expression of FcγRI in neutrophils during the course of G-CSF therapy, acting on myeloid precursor cells. This effect results in a marked increase in neutrophil-mediated ADCC against tumor cells. Macrophage-CSF (M-CSF) on the other hand dramatically increases (by 10 to 100-fold) monocyte-mediated ADCC. Together, these results suggest that the ability of effector cells to perform ADCC in vivo may be strongly dependent on the presence of a number of cytokines at the inflammatory foci. Therefore, it is conceivable that the efficiency of unconjugated antibody-mediated tumor therapy, that partially depends on effector cell functions triggered through FcRs, may be improved by the simultaneous administration of specific cytokines, as supported by recent reports.

Bispecific antibodies are currently being developed as new agents for immunotherapy. By virtue of combining two specificities, they are able to bind a target cell directly to a triggering molecule expressed on the effector cell. Recent reports have shown that bispecific antibodies directed to *Candida albicans* and either FcαRI or FcγRI trigger in vitro potent antifungal responses. Moreover, bispecific antibodies directed to the FcαRI and tumor antigens efficiently promote cell-mediated cytotoxicity against tumor targets. Further studies are required to clarify the possible role of bispecific antibodies as anticancer agents.

Preclinical Relevance

Monoclonal antibodies were first described by Kohler and Milstein in 1975 (6). Because their immunogenicity, large molecular size, suboptimal biodistribution and side effects, the development of monoclonal antibodies as therapeutic agents was slow. Recent advances in antibody engineering open up exciting opportunities for antibody-based cancer therapies. In fact, over 150 different mAb are being developed to treat a variety of cancers. Some of them are designed to selectively target and destroy specific types of tumor cells by inducing ADCC (7,8).

Relevance to Humans

As mentioned above, several lines of evidence support that ADCC plays an important role in immunity against a wide variety of infectious agents. It also appears to participate in host-acquired antitumor immunity, as well as in the antitumor activity mediated by a number of mAb used in oncology. More than 70 mAbs are currently in commercial trials beyond Phase I and Phase II. Engineered mAb containing human constant regions are increasingly emerging as useful adjuncts to cancer therapy. Three major fields of research have emerged:
- unconjugated mAbs
- immunotoxin-conjugated mAbs
- radionuclide-conjugated mAbs.

Binding of unconjugated mAbs to antigens on tumor cells could exert antitumoral effects by three different mechanisms:
- complement activation
- ADCC
- crosslinking of membrane receptors on tumor cells which generates intracellular signals leading to apoptosis or growth arrest.

Unconjugated mAbs, such as those directed to CD20 (rituximab), Her-2/neu (herceptin), CD52 (CAM-PATH-1H), and EpCAM (17-1A, Panorex) have shown to be able to control the growth of certain neoplastic conditions. Since most unconjugated anticancer mAbs are ineffective when used as Fab' or F(ab')$_2$ fragments, or when tested in FcγR-deficient mice, it appears that ADCC is responsible, at least in part, for their therapeutic properties (7,8).

Regulatory Environment

Monoclonal antibody-based therapeutics are under regulation by special guidelines such as those from the US Food and Drug Administration (FDA), the European Agency for the Evaluation of Medicinal Products, and the Ministry of Health, Labour and Welfare in Japan.

References

1. Fanger MW, Shen L, Graziano RF, Guyre PM (1989) Cytotoxicity mediated by human Fc receptors for IgG. Immunol Today 10:92–99
2. Ahmad A, Menezes J (1996) Antibody-dependent cellular cytotoxicity in HIV infections. FASEB J 10:258–266
3. Clynes RA, Towers TL, Presta LG, Ravetch JV (2000) Inhibitory Fc receptors modulate in vivo cytotoxicity against tumor targets. Nat Med 6:443–446
4. Capron M, Capron A (1994) Immunoglobulin E and effector cells in schistosomiasis. Science 264:1876–1877
5. Van Spriel AB, van Ojik HH, Bakker A, Cansen MJ, van de Winkel JG (2003) Mac-1 (CD11b/CD18) is crucial for effective Fc receptor-mediated immunity to melanoma. Blood 101:253–258
6. Kohler G, Milstein C (1975) Continuous cultures of fused cells secreting antibody of predefined specificity. Nature 256:495–497
7. Cragg MS, French RR, Glennie MJ (1999) Signaling antibodies in cancer therapy. Curr Opin Immunol 11:541–547
8. Glennie MJ, Johnson WM (2000) Clinical trials of antibody therapy. Immunol. Today 21:403–409

Antibody-Dependent Cytotoxicity

▶ Antibody-Dependent Cellular Cytotoxicity

Antibody-Forming Cell

▶ B Lymphocytes
▶ Plaque-Forming Cell Assays

Antibody-Forming Cell Assay

▶ Plaque Versus ELISA Assays. Evaluation of Humoral Immune Responses to T-Dependent Antigens

Antibody Forming Cell Response

▶ Humoral Immunity

Antibody Fragments

Immunoglobulin fragments produced by enzymatic or chemical degradation or by recombinant methods, capable of binding to antigens (Fabs) or cell and complement receptors (Fc).

▶ Antibodies, Antigenicity of

Antibody Isotype

An antibody classification system within an antibody class based on structural and functional characteristics of the heavy chain.

▶ Monoclonal Antibodies

Antibody Response to Therapeutic and Diagnostic Antibodies

Production of antibodies directed against therapeutic and diagnostic antibodies due to presence of antigenic determinants such as idiotypes, allotypes or mismatching species fragments.

▶ Antibodies, Antigenicity of

Anticytokines

▶ Cytokine Inhibitors

Antigen

A substance that, when introduced to a host, is recognized by the host immune system as foreign and effectively elicits immune responses. Antigens are usually large molecules (macromolecules) consisting of peptides, proteins, glycoproteins or lipoproteins. Antigens can be detected in biological fluids using immunoassays, and are specifically recognized by an antibody or T cell receptor.

The term antigen is also used to describe a molecule that generates an immune response, but this is more accurately called an immunogen. The structure on an antigen molecule that interacts with the combining site of an antibody is an epitope.

▶ AB0 Blood Group System
▶ Graft-Versus-Host Reaction
▶ Autoantigens
▶ Immunoassays
▶ Monoclonal Antibodies
▶ Cell Separation Techniques
▶ B Lymphocytes
▶ Antinuclear Antibodies

Antigen-Antibody Binding Assay

▶ Immunoassays

Antigen-Dependent B Cell Development

▶ B Cell Maturation and Immunological Memory

Antigen Presentation via MHC Class II Molecules

Frank Straube
MUT-2881.330 Biomarker Development
Novartis Pharma AG
4002 Basel
Switzerland

Synonyms

Major histocompatibility complex class II antigen, HLA-DP, HLA-DQ, HLA-DR (human), H-2IA, H-2IE (mouse), RT1.B, RT1.D (rat)

Definition

T cells recognize protein-derived antigens, and T cell activation depends on antigen presenting cells (APC). APC degrade proteins into small peptides; if the peptides were derived from endogenously produced proteins they are complexed with MHC class I molecules. If the protein was taken up from outside the APC, the peptides bind to MHC class II molecules. The complexes of MHC molecule and antigenic peptide are then brought to the cell surface, where they can interact with T cell receptors (TCR). With the expression of adhesion molecules and costimulatory molecules an APC facilitates interactions with T cells, to find the few T cells expressing an antigen specific TCR in the T cell population. CD4 positive helper T cells only recognize antigenic peptides on MHC class II molecules (they are MHC class II restricted), while CD8 positive cytotoxic T cells recognize MHC class I peptide complexes (MHC class I restricted).

Characteristics

The term "major histocompatibility complex" (MHC) designates a genetic region with multiple gene loci that was found to be responsible for the acute rejection of transplanted allogeneic tissue. Because different regions within the MHC correlate with different types of transplant rejection, the MHC region was subdivided into class I, class II and class III regions. The proteins encoded in the MHC class I and II loci are generally called MHC molecules.

Immunologists learned that MHC molecules are of central importance in the immune system, because they present antigenic peptides to T cells. In contrast to the ubiquitously expressed MHC class I molecules, class II molecule expression is normally limited to DC, activated macrophages/monocytes and B cells (professional APC).

Thus, only a few cell types can present antigens to the CD4 positive helper T cells. However, helper T cells play a key role in the regulation of immune responses. Linked to helper T cells, antigen presentation via MHC class II molecules is therefore involved in nearly all physiologic and pathologic responses of the immune system.

Genetics of MHC class II molecules exhibit a number of peculiarities that explain, together with the individual spectrum for the TCR, why different individuals can induce different immune responses to the same stimulus. A MHC class II molecule is a heterodimer of α and β glycoprotein chains with a size of 33–35 kD and 28–30 kD, respectively. The expression of MHC class II genes is normally limited to DC, B cells, monocytes, macrophages and thymic epithelial cells. However, in an inflammatory environment MHC class II expression can be induced on fibroblasts, melanocytes, epithelial and endothelial cells. Also some other cells may express MHC class II molecules, e.g. subpopulations of human and rat, but not mouse T cells.

An important species difference is the number of MHC class II genes: while a human APC normally expresses three loci (HLA-DP, HLA-DQ, HLA-DR), mice and rats have only two genes (H-2IA, H-2IE; RT1.B, RT1.D). However, there is some level of evolutionary conservation; also birds and even amphibians seem to express two MHC class II gene loci. MHC genes, including class II genes are generally highly polymorphic in vertebrates (2). Paternal and maternal alleles are expressed co-dominantly and mixed dimers of the α and β chains inherited from father and mother therefore can be composed. Consequently, four different forms of each MHC class II gene are expressed on the same cell, what gives a total of eight different MHC class II molecules expressed in the mouse and even twelve expressed in humans. For most vertebrates, this polymorphism is sufficient to ensure that only closely related individuals may express the same combination of MHC class II alleles.

Expression of different alleles for MHC class II molecules lead to an individual pattern of antigens that can be presented. Differences normally affect sites important for antigen presentation, while the overall structure of MHC class II molecules is conserved. Both, the MHC class II α and the β chain have two extracellular domains, a transmembrane domain and a small cytoplasmic domain. The short cytoplasmic anchors seem to lack motives for cellular signalling, however, this function is taken over by associated protein chains. While the extracellular, membrane proximal $\alpha 2$ and $\beta 2$ domain chains show homology to the immunoglobulin-fold structure, the distal $\alpha 1$ and $\beta 1$ domains display unique folding. Together they form a surface of eight antiparallel β sheets and two antiparallel α helices that build a peptide-binding cleft. This groove-like structure binds an antigenic peptide, which requires a fitting central core of approximately 13 amino acids, that all may interact with the MHC structure. Because the binding groove of MHC class II molecules is open ended, longer peptides can bind, extending beyond either end of the cleft, but in general the peptides are 13 to 18 amino acids long. The MHC-peptide complex forms a relatively planar surface that can then interact with the surface of a TCR, which is also relatively planar. The complete complex of MHC class II molecule and peptide is recognized by the TCR. The interaction sites for CD4 are located laterally on the MHC class II molecule. MHC polymorphisms mainly affect the peptide binding site and regions on the α helices that interact with the TCR. Thus, individually expressed MHC class II alleles determine the possible spectrum of bound peptides, and they shape the individual T cell repertoire.

Peptides that are bound by MHC class II molecules result from cleavage of extracellular proteins and are loaded to MHC class II molecules via an own pathway (1). Using different mechanisms an APC takes up extracellular material and digests it via the lysosomal pathway. MHC class II molecules, after synthesis in the rough endoplasmic reticulum, are stabilized by binding to the so-called invariant chain Ii. They reach early lysosomes via the trans-Golgi network. With increasing proteolytic activity, Ii is degraded and a fragment called CLIP stays bound to the peptide binding groove. Catalysed by HLA-DM (human; mouse: H-2M; rat: RT1.M) and other chaperones, CLIP is then exchanged by peptides with a higher binding affinity before the peptide-MHC class II complex can move to the plasma membrane. By this mechanism a spectrum of different peptides is loaded to the MHC class II molecules sometimes with a few predominant peptides. As mentioned, MHC polymorphism broadens the range of bound peptides considerably because the co-dominant expression of allelic α and β chains generates four slightly different MHC class II molecules for each gene locus. Inbred strains of animals represent a relevant exception; they inherit the same ▶ haplotype from both parents. Although such animals have lost some variability in the MHC spectrum, they still induce adequate immune reactions against most challenges.

Neither the loading of MHC class II molecules, nor the generation of the TCR by nonprecise recombination of gene segments does contain a mechanism to discriminate between foreign antigens and endogenous structures. When MHC molecules are loaded with peptide, even in the presence of a pathogen, many peptides are derived from the organism's own proteins (▶ self antigens). Different mechanisms must therefore retain ▶ tolerance to self antigens. Most important is the tolerance induction during T cell maturation in the thymus. This organ firstly establishes MHC class II restriction of $CD4^+$ T cells, which guarantees that newly generated $CD4^+$ T cells only survive if their TCR has some affinity to the self peptide-loaded MHC class II molecules expressed in this organ. In a second step self tolerance is established by induction of apoptosis in all T cells that have a high affinity to the self peptide-MHC class II complex. When they leave the thymus, the TCR of surviving T cells has some affinity to MHC class II molecules but interactions with the spectrum of self peptides do not reach the threshold of affinity necessary for T cell activation. A T cell will be activated if new (normally foreign) peptides presented on MHC molecules can interact with its TCR with a high affinity. Only very few interactions between TCR and coreceptor CD4 the peptide MHC class II complex have the right affinity. Normally only in the order of one out of 10^5 T cells can get activated by a foreign peptide. However, with transplantation of a MHC missmatched organ, in the order of one in 1000 cells are activated by the foreign tissue because all MHC molecules display changed structures.

To ensure that the rare antigen specific T cells are activated be a foreign antigen, antigen presentation is a well ordered process involving a number of cells and specialised organs. The first activation of helper T cell in peripheral organs is regulated very tightly. Only activated cells of the dendritic cell system in ▶ secondary lymphoid organs express all necessary factors for the activation of these so called naive (never before activated) helper T cells. The second activation of effector T cells is much easier and does not depend on a DC only. But the activation of naive T cells follows a general scheme; DC of the peripheral organs take up extracellular proteins extremely efficiently. When they encounter "danger" signals they start a number of maturation steps by migration to the draining lymph nodes (during hours or days). There they express extended dendrites and encounter T cells. First, DC and T cells establish short (seconds to minutes) antigen-independent interactions via cell adhesion molecules that bring the cell membranes into close proximity. This enables interactions between the TCR and MHC class II-peptide complexes. If the TCR, supported by the coreceptor CD4 binds the MHC class II-antigen complex with sufficiently high affinity and with the right half life the cells will establish a longer lasting connection. They organize a structure that can last for hours and that is called immunological synapse (3). This is a dynamic structure where the signals necessary for T cell activation can be generated. In the center of the immunological synapse on the DC side the MHC class II/peptide complexes are found on the side of the T cell TCR and the coreceptor CD4. They are surrounded by a ring of costimulatory molecules like CD28 that give second signals for T cell activation. Outside the structure, large adhesion molecules like CD44 and CD45 are located. Finally the cells separate again and an activated helper T cell, expressing a polarised pattern of cytokines, leaves the lymph node.

Several factors influence the outcome of this encounter between DC and T cell. First of all: A certain number of MHC class II molecules must be loaded with the same peptide. However, in some cases less than 10 MHC molecules (out of about 50 000 in total) loaded with the identical peptide may be sufficient. Therefore even very low amounts of antigen can induce efficient immune reactions and there will be no pronounced dose-response curve for the induced T cell response. Still, to predict a pathological outcome of an immune response, it is not sufficient to know that T helper cells have been activated; the cytokine pattern of the response decides if an adverse immune reaction will occur. The induction of Th2 cytokines may for example lead to the induction of IgE mediated type 1 allergies. However, in case of an ▶ autoimmune reaction immune pathology may be prevented by a Th2 type reaction. It is still unresolved how the polarization of the T cell response is induced. The number of MHC peptide complexes, as well as the binding affinity for the TCR, can favour a certain helper T cell response. In addition other factors like the nature of secondary signals for T cell activation seem to be involved.

Antigen presentation via MHC class II molecules and resulting activation of helper T cells are central steps in many immune responses. However, relevant toxicological effects resulting from interference with this process are rare. Even the different viral strategies to interfere with nearly all steps involved in antigen presentation normally do not protect a virus from the immune attack (4). Coevolution with pathogens made this basic immunological step quite robust.

Preclinical Relevance

Antigen presentation via MHC class II molecules and activation of $CD4^+$ helper T cells are essential steps in cell mediated immune responses. In addition, the type of helper cell response elicited (Th1 or Th2) can lead to beneficial or pathologic conditions. However, the induction of an adequate immune response can be disturbed on many different steps of antigen presentation on MHC class II molecules. Therefore the EPAs "Health Effects Test Guidelines OPPTS 870.7800 Immunotoxicity" and the CDERs "Guidance for Industry, Immunotoxicology Evaluation of Investigational New Drugs" ask for testing immune functions like the production of T cell-dependent antibodies or the elicitation of a delayed-type hypersensitivity response. Both immune responses depend on T helper cell activation and therefore indirectly address antigen presentation via MHC class II.

Relevance to Humans

For some clinically relevant situations antigen presentation via MHC class II molecules is especially important; such as missmatched MHC molecules represent a major problem for organ transplantations. As described above, a MHC allele that is new for the organ recipient will be recognised by a high number of T cells. Because there was no tolerance induction in the thymus, many T cells will recognize peptides on MHC molecules that are derived from allelic variants of cellular proteins in the graft. Therefore pharmacological suppression of immune responses is crucial after an organ transplantation. Monoclonal antibodies that recognise the TCR block interaction with MHC molecules are in use. Nevertheless, they are normally used in combination with other immunosuppressive drugs. Currently, the most efficient drugs interfere with T cell signaling to achieve immune suppression. Very common diseases with a link to MHC class II presentation are allergies. Helper T cell activation is indispensable for the class switch from IgM to IgE (type I allergy) or IgG (type II) antibodies. However, the antigen presentation itself is not the problem, but the induction of the wrong helper T cell cytokine profile, which favors the class switch to allergy related antibody classes. During contact allergy (type IV) both MHC class I and class II molecules present hapten modified peptides; direct haptenization of MHC molecules seems to be of minor importance (5). The induced inflammation leads to the clinical manifestation of contact allergy.

A number of diseases with an autoimmune component shows genetic linkage to certain HLA alleles. However, the reasons for onset of the diseases are complex and influenced by multiple genes. Most individuals with these HLA molecules do not develop the disease and, with some exceptions, individuals with other HLA alleles can be affected (6).

Few data exist about the possible MHC class II linkage of adverse drug reactions like hypersensitivities; associations with certain genes are generally difficult to establish. Severe hypersensitivity reactions to carbamazepine may be associated with a linkage of particular TNF2, HLA-DR3 and -DQ2 alleles rather than with any single allele itself (7).

A special case are intoxications caused by superantigens. These proteins induce crosslinking of certain TCR chains with MHC class II. *Staphylococcus* species express several superantigens and infections can induce the so-called toxic shock syndrome that is in part caused by excessive T helper cell stimulation.

A very rare genetic disease is bare lymphocyte syndrome. Mutations in the MHC class II promoter regions prevent expression of the molecules; this leads to a severe combined immune deficiency syndrome.

Regulatory Environment

No immunotoxicology guideline directly addresses MHC presentation. However, guidelines do ask for quantification of T cell-dependent immune functions, like the production of helper T cell dependent antibodies, or the elicitation of a delayed-type hypersensitivity response.

Relevant guidelines

- Health Effects Test Guidelines OPPTS 870.7800 Immunotoxicity. EPA
- Guidance for Industry, Immunotoxicology Evaluation of Investigational New Drugs. CDER

References

1. Pieters J (2000) MHC class II-restricted antigen processing and presentation. Adv Immunol 75:159–208
2. Flajnik MF (1992) Amphibian immune system. In: Roith IM, Delves PJ (eds) Encyclopedia of Immunology, Volume 1. Academic Press, London, pp 58–61
3. Creusot RJ, Mitchison NA, Terazzini NM (2001) The immunological synapse. Molec Immunol 38:997–1002
4. Vossen MTM, Westerhout EM, Söderberg-Nauclér C, Wiertz EJHJ (2002) Viral immune evasion: a masterpiece of evolution. Immunogenetics 54:527–542
5. Weltzien HU, Moulon C, Martin S, Padovan E, Hartmann U, Kohler J (1995) T cell immune responses to haptens. Structural models for allergic and autoimmune reactions. Toxicology 107:141–151
6. Nepom GT (1995) Class II antigens and disease susceptibility. Ann Rev Med 46:17–25
7. Pirmobamed M, Park BK (2003) Adverse drug reactions: back to the future. Br J Clin Pharmocol 55:486–492

Antigen-Presenting Cell (APC)

Cells that can process protein antigens into peptides and express the peptides on their surfaces for stimulation of specific T cells. The peptides must be presented in as a complex with either class I or class II Major Histocompatibility Complex (MHC) molecules. The APC must also express the appropriate costimulatory ligands to promote T cell activation. Whereas all nucleated cells generally express MHC Class I molecules and are potential APCs for $CD8^+$ T cells, only specialized "professional" APCs express MHC class II molecules and can present antigen to $CD4^+$ cells. The antigen-specific receptor (TCR) of $CD4^+$ cells recognizes peptide associated with major histocompatability complex (MHC) class II molecules (mouse classes I-A, I-E; human leukocyte antigens (HLAs) DR, DQ, DP). The major professional APCs are dendritic cells, macrophages, and B cells.

▶ Memory, Immunological
▶ Helper T lymphocytes
▶ Autoantigens
▶ Leukocyte Culture: Considerations for In Vitro Culture of T cells in Immunotoxicological Studies
▶ Graft-Versus-Host Reaction
▶ Metals and Autoimmune Disease
▶ Assays for Antibody Production
▶ Antigen Presentation via MHC Class II Molecules
▶ Hapten and Carrier
▶ Maturation of the Immune Response

Antigen-Specific Cell Enrichment

AJ CUTLER · JR HAIR · KGC SMITH
Addenbrooke's Hospital
Cambridge Institute for Medical Research,
Department of Medicine
Box 139
Cambridge
CB2 2XY
UK

Synonyms
positive selection, fluorescence-activated cell sorting, paramagnetic cell selection

Short Description
The clonal selection theory of antibody formation put forward by Burnet proposed that the precursors of antibody-forming cells carry receptors of single specificity, with the vast majority of lymphocytes carrying receptors of unique antigenic specificity to enable recognition of the large number of antigens encountered during life. This diversity of receptor specificities is generated by rearrangement of the antigen receptor genes. The precursor frequency of lymphocytes with specificity for a given antigen is usually very low. Encounter with antigen induces the proliferation and expansion of individual clones with specificity for that antigen, but even in the event of clonal expansion, such antigen-specific cells may still exist at a very low frequency. For example, at the height of a B cell response to haptens such as (4-hydroxy-3-nitrophenyl) acetyl (NP), antigen-specific cells comprise less than 1% of the lymphoid population. Persisting (particularly viral) infections may maintain clonal populations at higher levels over time, however, most antigen-specific populations fall to very low levels in the long term (e.g. 0.01% for memory B cells specific for NP within a few months of immunization). Many of the problems involved in isolating rare antigen-specific cells have been overcome using transgenic mice, where the vast majority of T or B cells carry receptors with known specificity. Mouse lines have been created with receptors specific for autoantigens, model-protein or hapten antigens, viral and bacterial antigens.

Numerous strategies have been utilized to enrich antigen-specific cells. The techniques using nylon wool, panning, density gradients, complement mediated lysis and rosetting are not antigen-specific but can be still be used to pre-enrich antigen-specific cells by positive or negative selection. Current protocols used in the purification of rare antigen-specific T or B lymphocytes have taken advantage of the unique specificity of the antigen-binding receptors. B cells recognize antigen in native form and were the first lymphoid populations to be purified using specific antigen. T cells recognize peptide from antigen in the context of the major histocompatibility complex (MHC). Recent advances have enabled the isolation and enrichment of T cells based upon their receptor specificity. The cells are targeted as a function of their expression of exquisitely specific receptors using native or peptide antigens allowing the isolation of highly pure antigen-specific cells.

Characteristics
Pre-enrichment of T and B cells
The purification of highly pure antigen-specific B and T cells is largely carried out using multiparameter fluorescence-activated cell sorting (FACS). The efficacy of both techniques is greatly enhanced and less time consuming if the target cell population is enriched, either by positive selection of the desired lymphoid cell population, or by negative depletion of unwanted cells prior to antigen-specific selection.

Positive selection is achieved by purifying T or B cells using cell-specific surface molecules, such as CD19 or

CD20 on B cells and CD4, CD8 or CD3 on T cells. Paramagnetic beads coupled to monoclonal antibodies (mAb) with specificity for cell surface molecules can be used to select cell populations. The beads then attach cells to a column or to the side of a container when a magnetic field is applied, while the undesired cells pass through. Removal of the column or container from the magnet releases the cells, and an enriched population can be obtained. The efficacy of positive selection can depend upon the density of expression of the target antigen or the frequency of the target population. For example, murine B cells selected on the basis of CD19 can be selected to around 99% purity, whereas cells targeted on the basis of IgG1 are rare, so while 40-fold enrichment can be achieved, purity remains at less than 90%.

The negative selection of cells can be achieved in a number of ways. The physical properties of certain cell types can be used, thus macrophages will adhere to plastic, and T cells to nylon wool. Complement-mediated lysis is a commonly used and effective technique. Cocktails of antibodies specific for unwanted cell types are used to label such cell populations, and complement is added to lyse them. Paramagnetic beads can be used with cocktails of antibodies to deplete unwanted cells.

Cells labeled with beads will adhere to a magnetic column, whilst the desired population will remain untouched. This has the advantage of avoiding antibody binding and receptor cross-linking which may influence cell behavior in subsequent assays.

Purification of antigen-specific B cells: FACS sorting

Flow cytometric sorting provides the most accurate means to purify antigen-specific B cells. The antigen by which the subject has been immunized can be used as a probe to purify the desired cells. Hayakawa and colleagues used the fluorescent protein phycoerythrin (PE) to immunize mice, then used direct binding of the fluorochrome to identify PE-specific B cells by flow cytometry (1). Lalor et al used the hapten NP coupled to allophycocyanin (APC) to achieve a similar end, subsequently allowing identification and isolation of very rare memory B cells and bone-marrow plasma cells (2). This technique relies upon flow cytometry using multiple detection parameters. Cells not conforming to the size and granularity characteristics required are removed by forward and side-scatter gates respectively. Markers of contaminating unwanted cell types and dead cells can be placed into a single 'dump' channel. In the case of NP-specific B cells, IgM, IgD, Mac-1 and Gr-1, markers of naive B cells, monocytes and granulocytes respectively are used to remove cells that may bind NP non-specifically. Propidium iodide or 7-amino-actinomycin D (7-AAD) may be used to remove dead cells. The use of a dump channel can decrease the contamination by unwanted cells to less than 1 in 10^6. The remaining detection parameters can be used to fine tune the cell selection process. B220 or CD19 and IgG1 can be used to detect mature IgG1 B cells and NP-APC to detect antigen-binding cells (Fig. 1). The probability of non-specific cells being included using all six parameters is so low that cells as rare as 1 in 200 000 can be reliably and consistently sorted to more than 95% purity, even in the absence of pre-enrichment. Purity of cells sorted using these methods can be confirmed by single cell cloning and immunoglobulin analysis, enzyme-linked immuno-SPOT assay (ELISPOT), or single-cell polymerase chain reaction (PCR) (2).

Purification of antigen-specific B cells: Magnetic cell sorting

Antigen-specific B cells can be enriched using paramagnetic beads labeled with recombinant antigen. B cells are purified prior to using the antigen-specific beads. Autoreactive human B cells and B cells specific for tetanus toxoid antigen have been enriched using this approach.

Purification of antigen-specific T cells: Tetramer staining

Historically, research in the field of antigen-specific T cells has relied on techniques such as ELISPOT, cloning and limiting dilution analysis (LDA). Direct visualization of antigen-specific T cells has been hampered by the fast dissociation rate between the T cell receptor (TCR) and its monomeric MHC-peptide ligand. Stable binding is achieved only when multimeric MHC-peptide complexes are used to ligate more than one TCR on the cell surface. Whilst dimeric and trimeric multimers have been used, tetrameric MHC-peptide complexes display the lowest rate of dissociation and therefore the most stable labeling of antigen-specific T cells.

The methodology for creating tetramers has varied little since the first was constructed in 1996 (3). Altman et al. introduced a 15 amino acid BirA substrate peptide (BSP) into the carboxyl tail of an MHC class I heavy chain. After folding the MHC molecule in the presence of β_2 microglobulin and its specific peptide, the BirA enzyme was added to biotinylate a lysine residue in the BSP (Fig. 2a). Tetrameric MHC-peptide complexes were created by the addition of phycoerythrin (PE)-labeled avidin, which contains four individual biotin binding sites (Fig. 2b). The introduction of a fluorescent dye such as PE into the tetramer complex allows the flow cytometric analysis of T cell populations bearing the relevant TCR.

In combination with standard cell surface labeling, tetramers allow the identification of discrete populations of antigen-specific cells that may differ in differ-

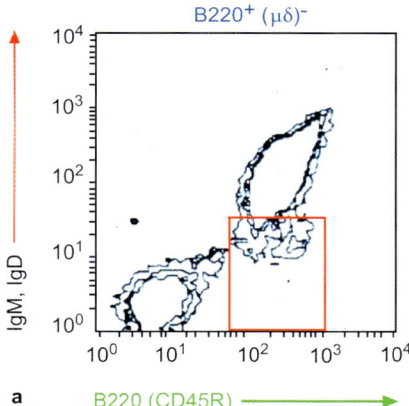
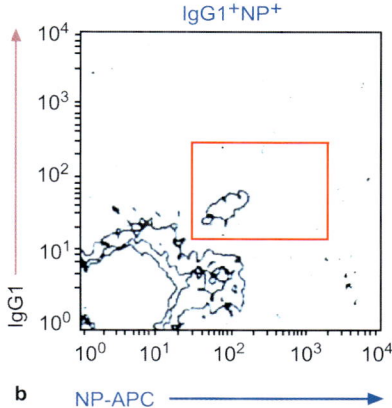

Antigen-Specific Cell Enrichment. Figure 1 Identification of NP-specific IgG1 positive B cells. [A] Cells staining for IgM, IgD, Mac-1, Gr-1 and Propidium iodide were excluded (y axis) and B cells identified with B220 (x axis). Cells thus identified (Box) were further analysed to detect those positive for IgG1 (y axis) and binding the original antigen NP coupled to APC (x axis) [B]. Thus IgG1 NP-specific B cells are shown in the box in [B], allowing further analysis with other FACS parameters, cell sorting etc.

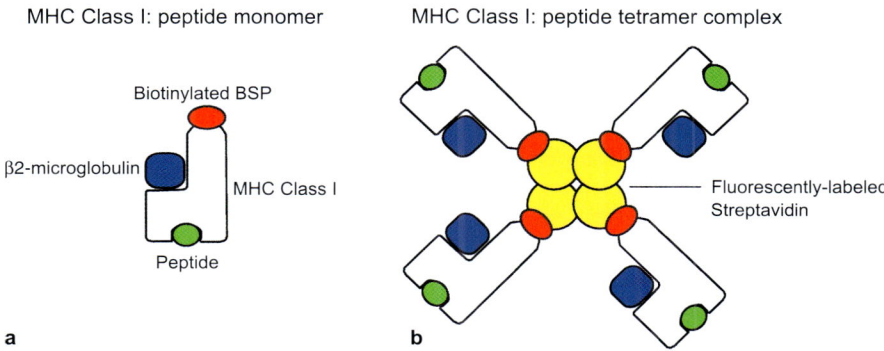

Antigen-Specific Cell Enrichment. Figure 2 Structure of MHC Class I:peptide tetramer complex. (A) Peptide is bound to the MHC Class I molecule and the BirA enzyme added to biotinylate the BirA Substrate Peptide (BSP) in the carboxyl tail. (B) Tetramers of MHC Class I:peptide molecules are created by the addition of streptavidin, which contains four biotin binding sites. Fluorescent labeling of the streptavidin molecule allows the detection of tetramer^{+ve} cells by FACS analysis.

entiation or activation markers. By applying FACS, assays of function such as ELISPOT, intracellular cytokine staining (ICS) and proliferation can be performed on each T cell subgroup. A number of different fluorochromes can be introduced into the tetramer complex and signal amplification using secondary and even tertiary antibodies has been successfully performed. Peptide-mismatched controls should be used to account for non-specific binding to the T cell surface. This helps to establish fluorescence staining limits. It is generally preferable to use whole blood when staining for CD8$^+$ T cells as repeated processing of the cells may reduce the labeling efficiency and cell recovery. Cryopreservation of samples may reduce the signal intensity. Binding of MHC-peptide complexes to the TCR does induce partial activation of the T cell and therefore functional assays performed using FACS-sorted populations require the inclusion of carefully considered controls. Staining using MHC class II tetramers requires other special considerations. There is a more marked temperature-dependence of labeling which may require optimization. The low in vivo frequency of antigen-specific CD4$^+$ T cells may necessitate pre-enrichment to reduce background staining 'noise'. One method is the use of clonotypic antibodies to isolate T cells with specific TCR variable regions that are known dominate in the particular immune response under investigation.

Transgenic mice

Enrichment of antigen-specific T and B cells in transgenic models uses similar procedures as for normal mice but of course is easier, as most or all of the lymphocytes may express the transgenic antigen receptor. Antigen-specific cells can be identified by anti-clonotypic or allotypic mAb that recognize the transgenic receptor, as well as by the methods described above.

Pros and Cons
Antigen-specific B cell enrichment

Multiparameter cell sorting enables the enrichment of antigen-specific B cells to an extremely high purity. This allows the analysis of individual antigen-specific cells at the most exquisite level. However, the rarity of antigen-specific B cells makes sorting a lengthy and expensive procedure. When sorting very rare cells, relatively small changes in experimental variables can lead to difficulties with reproducibility and purity.

Tetramer staining

The major advantage of tetramer staining is that it allows the highly specific and sensitive visualization and purification of antigen-specific T cells. Whilst the majority of studies have employed flow cytometry, in situ staining using tetramers has also been performed successfully. Very small percentages of tetramer-positive cells can be identified from samples, and cryopreservation of cells for later analysis (whilst not preferable), is possible. Rare T cell populations can be amplified in vitro for further study by applying either antigen-specific or non-antigen-specific mitogenic stimuli. Importantly, the isolation of antigen-specific cells by MHC-peptide complexes does not in any way rely on the functional capacity of the cell. This facet has been of particular value in longitudinal studies of certain viral infections in which tetramer-positive populations display differences in functional activity relative to the viral antigenic load.

The isolation of antigen-specific T cells using tetramers does require pre-existing knowledge of the peptide and particular MHC ▶ allotype to which the TCR binds. In addition, the highly specific MHC-peptide interaction makes it more difficult to study diseases in which the immune response is directed against a complex set of epitopes. The characterization of ▶ dominant peptides has recently been aided by the computer-based prediction of protein sequences that will bind strongly to MHC molecules. Advancements such as this will facilitate the production of tetramers to identify important T cell populations in less well-characterized systems.

Transgenic mice

Transgenic mouse lines have allowed insights into early activation events in immune responses where in normal instances, the precursor frequency of antigen-specific cells is extremely low. However, the artificially high incidence of antigen-specific cells can result in 'unphysiologic' immune responses. This can be overcome by adoptively transferring transgenic cells into non-transgenic syngeneic recipient mice so they make up only a minority of the lymphocyte populations.

Predictivity

A major point for consideration is that tetramer staining gives no information on the functional status of the T cells isolated. Tetramer-positive cells do not always show functional capacity, as has been detailed for many viral infections. The discrepancies that have arisen between tetramer staining and functional assays in the calculation of antigen-specific cell number are a function of the intrinsic differences in the read-out of these two techniques.

Relevance to Humans

Our understanding of the immune response to viruses during the acute and persistent phases of infection has been greatly enhanced by use of tetramers. Surprisingly large clonal expansions of antigen-specific $CD8^+$ T cells during acute viral infection have been identified for several human viruses including human immunodeficiency virus (HIV), Epstein-Barr virus (EBV) and human cytomegalovirus (HCMV). Analysis of the distribution and phenotype of antigen-specific T cells during infection has given insights into the development of T cell memory. By assaying the function of individual T cell populations ex vivo we have been able to infer their role during the progression of an infection in vivo. Such information is of particular importance for studying the development of immunity following vaccination. In a non-infectious context, tetramers have been valuable tools in the study of anti-tumor immunity mediated by antigen-specific $CD8^+$ T cells. However, the comparatively weak $CD8^+$ T cell response to tumor antigens in particular has hampered both detection and immunization strategies. Detection of antigen-specific B cells in humans is less advanced that that of T cells, but nonetheless enrichment of antigen-specific mouse and human B cells has allowed the dissection of immune responses from their initiation to the generation and maintenance of memory.

References

1. Hayakawa K, Ishii R, Yamasaki K, Kishimoto T, Hardy RR (1987) Isolation of high-affinity memory B cells: phycoerythrin as a probe for antigen-binding cells. PNAS 84:1379–1383

2. Smith KGC, Light A, Nossal GJV, Tarlinton DM (1997) The extent of affinity maturation differs between the memory and antibody-forming cell compartments in the primary immune response. EMBO J 16:2996–3006
3. Altman JD, Moss PA, Goulder PJ et al. (1996) Phenotypic analysis of antigen-specific T lymphocytes. Science 274:94–96
4. Klenerman P, Cerundolo V, Dunbar PR (2002) Tracking T cells with tetramers: new tales from new tools. Nat Rev Immunol 2:263–272

Antigenic Similarity

▶ Molecular Mimicry

Antigenic Variation

Various protozoan parasites (*Trypanosoma* spp., *Plasmodium* spp., *Pneumocystis carinii* etc.) have the ability to change regularly their antigen coat, and thus escape the immune response of the host. The main mechanism of antigenic variation in African trypanosomes is by gene replacement. African trypanosomes can exhibit up to 1000 different variant antigen types (VATs).

▶ Trypanosomes, Infection and Immunity

Antigenicity

Ability of a substance that stimulates a specific immune response resulting in antibodies or activated cells.

▶ Respiratory Infections

Antigens, T Dependent and Independent

Thymus-dependent antigens or TD antigens require the cooperation of both T and B cells. An antigen-specific immune response is exclusively dependent on the successive activation of APC, T and B cells. In contrast to this, some antigens are able directly to induce an activation of B cells and antibody production independent of T cell help. Antigens inducing such immune responses are called thymus-independent or TI antigens.

Antiglobulin (Coombs) Test

ANNE PROVENCHER BOLLIGER
Rebbergstrasse 59
CH-4800 Zofingen
Switzerland

Synonyms
direct Coombs test, DAT, direct antiglobulin test

Short Description
Direct Coombs test
Coombs test is a screening test used to demonstrate the presence of immunoglobulin and/or complement bound to a patient's (human or animal) red blood cells (RBCs) (1). It was introduced by Coombs et al in 1945 and remains the hallmark in the laboratory diagnosis of (auto)immune-mediated hemolytic anemia (IMHA or AIHA). The ▶ direct Coombs test can be performed in human and all common domestic and laboratory animals, due to the availability of a wide range of species-specific antiglobulin reagents (3).

Indirect Coombs test
The ▶ indirect Coombs test is a screening test used to demonstrate the presence of serum immunoglobulin and/or complement directed against a patient's RBCs (2).

Characteristics
Traditional tube method
Direct Coombs test, or DAT, demonstrates the presence of bound immunoglobulins or complement to the RBC membrane. It is performed traditionally by a test tube method in which antiglobulin reagent is added to washed erythrocytes, which are then centrifuged (1). The antiglobulin reagent is diluted and each dilution tested. The endpoint is the presence or absence of ▶ agglutination. Agglutination can be observed macroscopically and, when there is doubt, microscopically. Results are reported as the highest dilution in which agglutination still occurs (2).
The initial test is done with a polyspecific 'broad spectrum' antiglobulin reagent (containing anti-IgG and anti-C3d). The test should be performed at 37 °C, in order to avoid false-positive agglutination that can occur at lower temperatures (non-specific cold agglutinin). If the test is positive, further testing can be done with monospecific reagent when available (1).

Gel technique method
This newer technique can give higher agglutination scores than traditional tube testing. It can be used to identify causative antibodies other than IgGs (1).

Pros and Cons

Pros
Coombs test is easy to perform and it is fast. Agglutination is easy to observe and to interpret. It can be performed in many animal species and will detect most cases in most species (most cases are IgG alone, or IgG plus C3) (3).

Cons
There can be a significant proportion of false-negative results if low levels of clinically important IgGs or other immunoglobulins are involved (e.g. IgA, IgM). In those cases, it will be necessary to rely on other procedures and tests (e.g. ▶ regenerative anemia, presence of spherocytes, hemolysis) to diagnose IMHA (1). Indirect Coombs may also be useful.
In the gel technique, the detection of C3d-coated RBCs may not be optimal and false-negative results can occur (1). False positive results may occur if there is cross-reactivity between immunoglobulin reagent and the RBC membrane without causing hemolytic disease (4). False-positive agglutination can occur at lower temperatures or when the samples are stored at 4 °C (non-specific cold agglutinin) (1). False-positive results are frequent following blood transfusion. False-positive results are common in cats (3).

Predictivity
Direct Coombs test is the only diagnostic test for IMHA, in humans or animals. Although a newly developed gel method exists, the traditional tube testing remains the gold standard and is still the simplest, fastest, and most reliable (1,3).
In addition to the test, characteristic morphological features can be observed on the peripheral blood smear of people and animals with IMHA. These features include polychromasia, ▶ spherocytes, RBC fragments, nucleated RBCs, and occasionally erythrophagocytosis. Occasionally the patient may have leucopenia and ▶ neutropenia and possibly ▶ thrombocytopenia (▶ Evans syndrome) (4).

Relevance to Humans
The direct Coombs test is the same test, whether applied to humans or animals. However, the immunoglobulin reagents are species-specific and there should be used exclusively for the species tested.

Regulatory Environment
No official documentation was found with regard to the use of Coombs test in preclinical testing. A case-by-case approach should be used when there is a suspicion of IMHA or drug-induced hemolytic anemia in preclinical studies.

References
1. Manny N, Zelig O (2000) Laboratory diagnosis of autoimmune cytopenias. Curr Opin Hematol 7:414–419
2. Coles EH (1986) Veterinary clinical pathology, 4th ed. WB Saunders, Philadelphia, pp 437–438
3. Barker RN (2000) Anemia associated with immune responses. In: Feldman BV et al. (eds) Schalm's veterinary hematology, 5th ed. Lippincott, Williams & Wilkins, Philadelphia, pp 169–177
4. Packman CH (2001) Acquired hemolytic anemia due to warm-reacting autoantibodies. In: Beutler E et al. (eds) William's hematology, 6th ed. McGraw Hill, New York, pp 639–648

Antihistamines

Drugs that inhibit the major vasoactive substance histamine, which is released during an allergic reaction.
▶ Food Allergy

Antinuclear Antibodies

MICHAEL HOLSAPPLE
Health and Environmental Sciences Institute
1 Thomas Circle
Washington, DC 20005-5802
USA

Synonyms
ANA, anti-DNA antibodies, anti-double stranded (ds) DNA antibodies, anti-single stranded (ss) DNA antibodies, anti-histone antibodies

Definition
▶ Antinuclear antibodies (ANA) are antibodies directed against nuclear antigens, such as DNA or ▶ histones. As such, they are an important class of ▶ autoantibodies. The presence of ANA has been associated with a number of autoimmune syndromes, and represent the hallmark indicator of systemic lupus erythematosus (SLE).

Characteristics
SLE is an autoimmune disease of unknown etiology, characterized by the involvement of multiple organ systems (1). In fact, it was listed on a continuum of organ-specific to organ-nonspecific autoimmune diseases as the most organ-nonspecific disease (2). The importance of SLE in this section is that it is typified by the production of autoantibodies, especially ANA. In fact, a hallmark of SLE is the presence of ANA, and serum antibodies directed at nuclear constituents are

found in 95% of the patients. The central immunologic disturbance in SLE is autoantibody production and all clinical manifestations that have been elucidated in terms of pathogenetic mechanisms have been shown to be the direct or indirect result of autoantibodies. In fact, the heterogeneity of SLE appears to result from differences in the types of autoantibodies produced (1).

Because ANA are specific antibodies against components of the nucleus, it is appropriate to provide some background about ▶ autoimmunity, which is simply defined as immunological reactivity to self- or autoantigens. As such, autoimmunity is an inappropriate immune response to autoantigens that can result in the generation of autoantibodies and/or autoreactive T cells, and that can lead to tissue damage. An important distinction is that these manifestations of autoimmunity are not synonymous with pathogenicity or morbidity. A key point of this section will be to emphasize that although the presence of ANA can be associated with some types of autoimmune disease, the latter diagnosis absolutely requires the presence of clinical symptoms.

The mechanism(s) by which autoimmunity is triggered is (are) unclear and being studied. A number of possibilities are suggested by an understanding of the processes required to maintain T cell and B cell repertoires including: an influence on the development of the T cell repertoire in the thymus to either allow for the positive selection of autoreactive cells, or the deletion of a regulatory T cell specificity; a generalized failure to induce ▶ tolerance within either T cells (e.g. negative selection) or B cells (e.g. central tolerance); or a failure of peripheral tolerance mechanisms.

Consistent with the fact that the mechanisms by which autoimmunity is triggered are unclear in general, the specific mechanisms responsible for the onset and progression of SLE, are not known. As discussed above, the central immunologic disturbance in SLE is known to be the production of autoantibodies.

Importantly, B cells capable of secreting pathogenic autoantibodies characteristic of SLE appear to be absent from the functional repertoire of normal individuals, and they appear to arise during the process of developing the autoimmune disease. B cell activation in SLE has been studied and generalized hyperactivity ("polyclonal") has been demonstrated, even early in the disease. However, the majority of the results indicate that pathogenic autoantibody production in SLE is selective for only certain self antigens and is driven by ▶ antigen at the B cell level (1). There is no evidence that T cells play a direct role in the tissue damage associated with SLE. However, T cells are clearly involved in the production of the autoantibodies and evidence suggests that pathogenic B cells are driven by self antigens in a T-dependent process.

Studies have indicated three types of autoantibodies in SLE (1), reflecting different macromolecular antigenic targets:
1. nuclear targets: DNA, RNA, other protein/nucleic acid complexes
2. cytoplasmic targets: proteins associated with RNA
3. cell surface targets: molecules on surface of red blood cells (RBC), platelets or lymphocytes.

Anti-dsDNA antibodies, especially when present in high levels in serum, are only associated with SLE. The presence of anti-dsDNA antibodies also correlate best with the expression of glomerulonephritis, one of the hallmark clinical manifestations of SLE. As such, quantification of anti-dsDNA antibodies may be useful in diagnosis and management of patients. It is important to emphasize that the correlation with anti-dsDNA antibodies with renal damage is a general one. Some patients have high serum levels and yet no clinical evidence of renal disease, whereas others have severe disease and little detectable anti-DNA activity.

In contrast with dsDNA, anti-ssDNA antibodies have little diagnostic specificity and do not correlate with disease activity. Histones are small DNA-binding proteins and represent the largest protein component of the nucleus of eukaryotic cells. Histones and DNA associate to form nucleosomes, the basic units of chromatin. Anti-histone antibodies are among the most frequent autoantibodies seen in different rheumatic diseases, are not specific for SLE, and have not been associated with particular disease manifestations in the setting of SLE. One of the interesting aspects of anti-histone antibodies is their high level of production in drug-induced lupus, as discussed below. In general, the nuclear and cytoplasmic molecules targeted by autoantibodies in SLE are involved in critically important cell functions, including storage of genetic material, cell division, regulation of gene expression, RNA transcription and RNA processing. The mechanism by which most autoantibodies, especially those directed to intracellular structures, result in immunopathology remains unclear. There is little evidence that ANA can readily penetrate cellular membranes, and bind to their nuclear targets. There is therefore little information to suggest that disease results from their ability to inhibit intracellular processes dependent on these nuclear molecules. In contrast, autoantibodies to surface molecules can have clear clinical consequences—for example, immunoglobulin G (IgG) antibodies to RBC results in the autoimmune hemolytic anemia that is seen in some patients with SLE. The actual destruction of RBC is mediated by the reticuloendothelial system, especially by macrophages in the splenic sinusoids. In contrast to mechanisms associated with antibodies to cell surfaces, the severe renal disease in SLE appears to be mediated by the disposition or formation of immune complexes in glomeruli. Some other manifestations, such as arthralgia/arthritis, serositis and vasculitis, may be similar to renal disease associated with SLE and appear to be mediated by immune complexes. Most of the clinical associations of SLE are poorly understood.

Preclinical Relevance

Several murine models have contributed greatly to the elucidation of SLE pathogenesis and in so doing have increased our understanding of the role of ANA in autoimmunity (1). In particular, two models of lupus-like renal disease—New Zealand (NZ) hybrid mice and MRL mice (especially those that express the lymphoproliferation *lpr* defect)—are associated with the production of high levels of IgG autoantibodies to DNA and reveal the importance of heredity/genetics in the onset and progression of SLE.

NZ black mice (NZB) develop spontaneous autoimmune hemolytic anemia with autoantibodies to RBC (3). Although NZB mice also frequently produce IgM ANA, it is unusual for them to make IgG antibodies to dsDNA or histones, and lupus-like disease rarely occurs. NZ white mice (NZW) can make anti-ssDNA antibodies; but rarely demonstrate clinical evidence of autoimmune disease. In contrast, the F1 cross of NZB × NZW make high levels of IgG anti-dsDNA antibodies and develop a fatal immune-complex glomerulonephritis. It is clear that genes from both parental strains are necessary for manifestation of lupus-like disease and the role for heredity is obvious, albeit complex. The fact that 90% of females in some colonies die from their renal disease within the first year, and that, in contrast, nearly all males live longer than 1 year, indicates a role of sex hormones. Castration and hormone replacement studies clearly demonstrated in this model that estrogen accelerates and testosterone inhibits the formation of ANA and the expression of lupus-like renal disease. Although the mechanism for this hormonal effect is not clear, it is seen in humans, as described below.

A second lupus-like animal model is based on MRL mice and emphasizes the importance of the *lpr* gene (3), an autosomal recessive mutation that results in the massive accumulation of $CD4^-$ and $CD8^-$ T cells in lymphoid tissue (1). The *lpr* mutation has been shown to act as an accelerator of autoimmunity. When "bred" into normal mice, animals will produce ANA, including low titers of anti-DNA antibodies; but no pathology. However, when bred into an MRL background (*lpr / lpr* mice) high levels of anti-dsDNA antibodies are produced and a majority of animals develop a severe and fatal lupus-like glomerulonephritis. Interestingly, massive accumulation of double negative cells is not a clinical feature of SLE, which suggests that the MRL model is an imperfect surrogate for human SLE. However, the gene for *lpr* is a mutation of the gene that encodes for the FAS ligand, which is involved in control of apoptosis. Although the mechanism is unclear, it has been speculated that an apoptotic defect leads to the escape of autoreactive $CD4^+$ T cells that ultimately contribute to the production of autoantibodies in SLE.

Relevance to Humans

As discussed above, SLE is an autoimmune disease of unknown etiology, characterized by the involvement of multiple organ systems. SLE is typified by the production of autoantibodies, and a hallmark of the disease is the presence of ANA. Serum antibodies directed at nuclear constituents are found in 95% of patients with SLE. The clinical presentation and course of SLE is extremely variable. Women of childbearing age are primarily affected. It should be emphasized that although males develop SLE less frequently than do females, their illness is not milder. Similar results regarding the earlier onset and greater incidence in females were observed in the murine lupus models discussed above.

The animal models for lupus also clearly demonstrated an important role for heredity/genetics in the onset and progression of the disease. What has emerged in our understanding of clinical SLE pathogenesis is that one or more environmental triggers act on a genetically susceptible individual to create a lymphocyte defect (or defects) that result in IgG autoantibody production against a variety of targets. In terms of the environmental trigger, a number of candidates are worthy of discussion.

The role of sex hormones was clearly indicated in the murine lupus models. Although the mechanism for this hormonal effect is not clear, it is clearly also seen in humans. For example, there is evidence to show that the onset or exacerbation of SLE occurs during or shortly after pregnancy, and anecdotes of disease flares occurring coincident with estrogen replacement therapy are consistent with an important role by sex hormones.

Sun exposure is another environmental factor that can influence the expression of SLE. In susceptible individuals, exposure to UV light can result in exacerbations of skin disease, occasionally accompanied by flares of systemic disease. The mechanism is unclear; but there is some evidence that UV damage can result in cell surface expression and release of nuclear antigens, thereby allowing for interaction with autoreactive lymphocytes, ultimately leading to the production of ANA.

There is also some evidence that infections, both bacterial and viral, can influence the disease expression in humans or animals with SLE. Unfortunately, there remains little understanding of the underlying mechanisms.

Finally, studies in the mouse models of lupus have also indicated a role for diet. Caloric reduction, protein deprivation, zinc deficiency and alterations in the type of fat ingested have all been shown to reduce renal disease and prolong survival. The mechanisms of these effects of diet remain poorly understood. Although the role of diet in the onset and progression of clinical SLE is not known, there is no question that obesity is a frequently observed risk factor for a number of autoimmune conditions.

Although not included among the environmental triggers for SLE for reasons made clear below, some drugs have been implicated in inducing ANA production and causing a lupus-like disease in humans. Although a large number of drugs have been associated with a lupus-like condition, most of the cases have been attributed to treatment with hydralazine or procainamide. The knowledge of a drug-induced lupus-like condition may afford the best situation to study the induction of this disease in humans. Indeed, drug-induced lupus is associated with positive ANA test results and anti-histone antibodies are present in almost all patients. Other autoantibodies seen in SLE, such as anti-dsDNA antibodies, are usually absent. Unfortunately, the underlying immunological mechanisms in drug-induced lupus remain unclear. ANA and anti-histone antibodies are common in patients on procainamide and hydralazine even without clinical problems. For example, ANA have been detected in 30% of patients treated long-term with hydralazine, and most patients treated for 1 year with procainamide demonstrate a positive ANA test results. Fewer than a quarter of these serologically positive patients will actually manifest clinical evidence of disease. Moreover, drug-induced lupus in general differs from idiopathic lupus in that it tends to cause predominantly joint and pleural-pericardial involvement. Lupus-like renal disease and CNS manifestations are very unusual, and skin disease is also less common. Compared to SLE, drug-induced lupus occurs in older patients (coincident with the age of the patients receiving the medications) and the sex ratio is close to unity. The disease remits when the offending drug is discontinued; but the time to remission may be prolonged (by months) and residual manifestations may require treatment with nonsteroidal antiinflammatory drugs or steroids.

Regulatory Environment

Almost every review that has ever been written about immunotoxicology emphasizes that the consequences can include a number of possible outcomes such as immunosuppression, hypersensitivity/allergy and autoimmunity. However, in reality, the focus of scientific research has been heavily weighted with the following emphasis:

immunosuppression ≥ hypersensitivity/allergy >>> autoimmunity.

This focus can perhaps be most readily appreciated in the context of the regulatory environment. A number of regulatory agencies have issued guidelines and/or guidance documents to address immunosuppression and hypersensitivity. For example, the US EPA Office of Prevention, Pesticides and Toxic Substances (OPPTS) includes Health Effects Guidelines to address both Immunosuppression (OPPTS 870.8700) and Skin Sensitization (OPPTS 870.2600). These guidelines signal that immunotoxicology has evolved as a scientific discipline to the point where it can be used in risk assessment. However, consistent with the aforementioned focus, there has not been a lot of activity to address autoimmunity.

A workgroup from the US Agency for Toxic Substances and Disease Registry (ATSDR) recommended a strategy for evaluating the presence of autoimmunity or autoimmune diseases in communities located near hazardous waste sites (4). Importantly, the proposed strategy included measuring ANA, in addition to assessing levels of the following:

- C-reactive protein (an acute-phase reactive protein whose levels rise in response to tissue damage and infection)
- antithyroglobulin antibody (an autoantibody associated with a variety of thyroid disorders)
- rheumatoid factor (an autoantibody to immunoglobulin M)
- complete blood count including a five-part differential and a total lymphocyte count.

At the time this workgroup made their recommendations, no studies had been done to try to establish a causal relationship between autoimmunity or autoimmune disease and the kinds of exposures found in communities located near hazardous waste sites (4). Interestingly, there still has not been much progress. Over a hundred immunologists, clinicians, epidemiologists, molecular biologists and toxicologists came together in a workshop in September of 1998 to review the current knowledge about environmental links to autoimmune disease, and to identify data gaps and future research needs. That workshop (5) was sponsored by several branches of the US National Institutes of Health, by the US Environmental Protection Agency (EPA), by the American Autoimmune Related Diseases Association, and by the Juvenile Diabetes Foundation International. An entire volume of a journal was devoted to the results presented at the workshop. Unfortunately, the workshop did very little to shed light on the role of ANA in the onset or progression of autoimmune diseases.

References

1. Kotzin BL, O'Dell JR (1995) Systemic lupus erythematosus. In: Frank MM, Austen KF Claman, HN, Unanue ER (eds) Samter's Immunological Diseases, Volume II, 5th ed. Little, Brown and Co, Boston, pp 667–697
2. Roitt I (1989) Autoimmunity and autoimmune disease. In: Roitt IM, Brostoff J, Male DK (eds) Immunology, 2nd ed. JB Lippincott, Philadelphia, pp 23.1–23.12
3. Putterman C, Naparstek Y (1994) Murine models of spontaneous systemic lupus erythematosus. In: Cohen IR, Miller A (eds) Autoimmune Disease Models: A Guidebook. Academic Press, San Diego, pp 217–244
4. Ozonoff D, Tucker ES, Demers R et al. (1994) Test batteries to evaluate autoimmunity in environmental health field studies. In: Straight JM, Kipen HM, Vogt RF, Amler RW (eds) Immune Function Test Batteries for Use in Environmental Health Field Studies. US Department of Health and Human Services, Atlanta, GA, pp 45–54
5. Cooper GS, Germolec D, Heindel J, Selgrade M (1999) Linking environmental agents and autoimmune diseases. Environ Health Persp 107[Suppl 5]:659–660

Antioxidant (Levels)

Free radicals are atoms or groups of atoms with an odd (unpaired) number of electrons, which can be formed when oxygen interacts with certain molecules. They are generated during cellular metabolism, for example by activated neutrophils and macrophages as reactive oxygen species (oxidative stress). Once formed their chief danger comes from the damage they can do when they react with important cellular components such as DNA, or the cell membrane. Antioxidants can eliminate free radicals from the body. Also compounds other than vitamins C and E and carotenoids contribute a major portion of the increase in antioxidant capacity. Reduced level of antioxidants, or chronic activation of immune cells could contribute to destruction of normal tissue. Among the foods with the highest antioxidant capacity are oranges, cauliflower, and peas.

▶ Rodents, Inbred Strains

Antiprotease

Antiproteases contribute to the airway defense mechanisms. In the normal lung, proteinases are neutralized by antiproteinases secreted into the mucus. An imbalance in protease–antiprotease levels in the airways causes epithelial disruption, increased mucus secretion and reduced mucociliary clearance.

▶ Respiratory Infections

Aorto-Gonadomesonephros Region (AGM)

The aorto-gonadomesonephros, or AGM, region in the developing embryo originates from the mesodermal tissue. Organogenesis within this region results in formation of the heart, gonads, and kidneys. Hematopoietic stem cells appear in the AGM on day 7 of rodent gestation and constitute the population of stems cells that give rise to the postnatal blood-forming and immune systems.

▶ Rodent Immune System, Development of the

Ape

▶ Primate Immune System (Nonhuman) and Environmental Contaminants

Apoptosis

SHIGEKAZU NAGATA
Osaka University Medical School
Osaka,
Japan

Synonyms
Programmed cell death

Definition
Apoptosis is a cell death process which occurs during development and aging of animals. It is also induced by cytotoxic lymphocytes (CTL), anti-cancer drugs, c- or UV-irradiation, a group of cytokines called death factors and deprivation of survival factors.

Characteristics
Apoptosis was initially characterized by morphological changes of dying cells. During apoptosis cells shrink, and microvilli on the plasma membrane disappear. The nucleus is also condensed and fragmented. At the final stage of apoptosis the cells themselves are fragmented with all cellular contents inside. One of the biochemical hallmarks of apoptosis is the fragmentation of chromosomal DNA into nucleosome size units (180 bp).

Apoptotic cells can be recognized by staining of the condensed nuclei with fluorescence dyes Hoechst or DAPI. Apoptotic cells expose phosphatidyl-serine to the cell surface, which can be stained with fluoresently labelled annexin V. The fragmented DNA can be detected by TUNEL (terminal deoxynucleotidyltransferase-mediated UTP end labelling) procedure, or by electrophoresis of the isolated DNA on an agarose gel, which yields a ladder of DNA fragments with a unit size of 180 bp.

Cellular & Molecular Regulation

Apoptosis is mediated by a family of proteases called caspases that are activated by processing from its inactive precursor (zymogen). Thirteen members of the human caspase family have been identified. Some of the family members are involved in apoptosis, and these can be divided into two subgroups. The first group consists of caspase 8, caspase 9, and caspase 10, which contain a long prodomain at the N-terminus and function as initiators of the cell death process. The second group contains caspase 3, caspase 6, and caspase 7, which have a short prodomain and work as effectors, cleaving various death substrates that ultimately cause the morphological and biochemical changes seen in apoptotic cells. The other effector molecule in apoptosis is Apaf-1 (apoptotic protease activating factor), which, together with cytochrome C, recruits pro-caspase 9 in an ATP (or dATP)-dependent manner, and stimulates the processing of pro-caspase 9 to the mature enzyme.

The other regulators of apoptosis are the Bcl-2 family members. Eighteen members have been identified for the Bcl-2 family, and divided into three subgroups based on their structure. Members of the first subgroup, represented by Bcl-2 and Bcl-xL have an anti-apoptotic function. Members of the second sub-

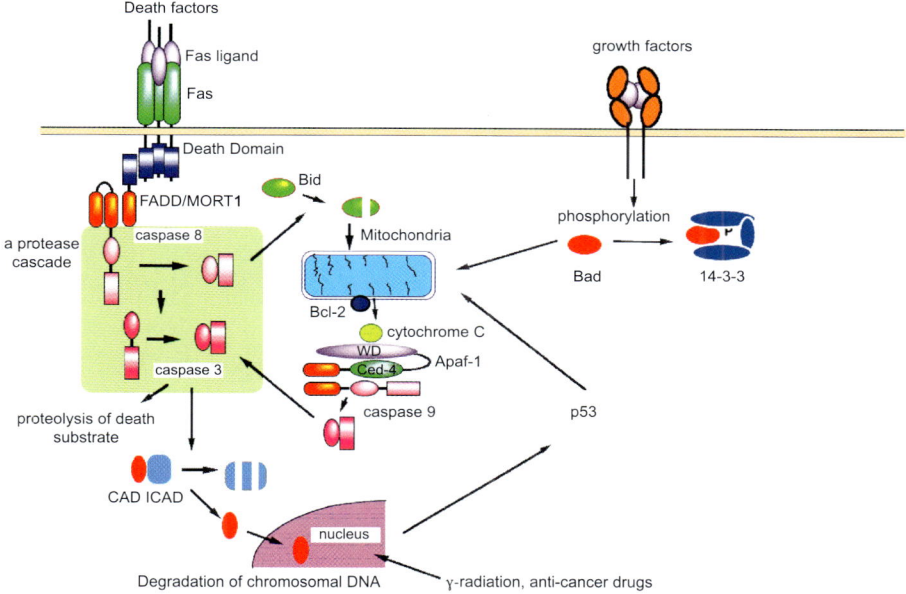

Apoptosis. Figure 1 Signal transduction for apoptosis. Inducers of apoptosis are categolized into three groups (death factors, genotoxic anti-cancer drugs, and factor deprivation). Fas ligand, a representative of death factors, binds to Fas receptor, and causes its trimerization. The trimerized death domain in the Fas cytoplasmic region recruits pro-caspase 8 through a FADD/MORT1 adaptor, and forms a DISC. The pro-caspase 8 is autoactivated at DISC, and becomes a mature active enzyme. Two routes have been identified to activate caspase 3 by caspase 8. In one route, caspase 8 directly processes pro-caspase 3 in the downstream, and caspase 3 cleaves various cellular proteins including ICAD. CAD is released from ICAD, and degrades chromosomal DNA. In another route, caspase 8 cleaves Bid, a pro-apoptotic member of Bcl-2, which translocates to mitochondria to release cytochrome C into the cytosol. Bcl-2 or Bcl-xL, anti-apoptotic members of the Bcl-2 family, inhibits the release of cytochrome C, the mechanism of which is not well understood. The cytochrome C then activates caspase 9 together with Apaf-1, and caspase 9 in turn activates caspase 3. The genotoxic anti-cancer drugs such as etoposide and c-radiation generate damage in chromosomal DNA. The signal seems to be transferred to mitochondria in a p53-dependent manner by as yet an identified mechanism. This releases cytochrome C from mitochondria, and activates caspase 9 as described above. The apoptosis induced by factor-deprivation is best studied with IL-3-dependent myeloid cell lines. In the presence of IL-3, the signal from the IL-3 receptor causes phosphorylation of Bad, a pro-apoptotic member of the Bcl-2 family. The phosphorylated Bad is trapped by an adaptor called 14-3-3. In the absence of IL-3, nonphosphorylated Bad is released from 14-3-3, and translocates to mitochondria to release cytochrome C to activate caspase 9.

Apoptosis. Table 1 The apoptosis factory.

worker	synonym	apoptosis job pro	apoptosis job contra	chromosome
Fas	CD95 Apo-1	+		10q24
FADD	MORT-1	+		11q13
granzyme B	GZBM	+		14q11
Apaf-1	CED4	+		12q23
Casp 2	ICH1 NEDD2	+		7q35
Casp 3	CPP32 Yama apopain	+		4q33
Casp 4	TX ICH-2 ICE-rel-II	+		11q22
Casp 6	MCH2	+		4q25
Casp 7	MCH3 ICE-LAP3	+		10q25
Casp 8	MACH MCH5 FLICE	+		2q33
Casp 9	APAF3 MCH6 ICE-LAP6	+		1q36.3-p36.1
Casp 10	MCH4	+		2q33
CAD	DFF40	+		1p36.3
Bak		+		6p21
Bax		+		19q13
Bcl-2			+	18q21
Bid		+		22q11
Bik		+		22q13.3
XIAP			+	Xq25
UBL1	SUMO-1 sentrin		+	2q32

group, represented by Bax and Bak [BAK1], as well as members of the third subgroup such as Bid and Bad are pro-apoptotic molecules.

The signal transduction pathway for a death factor (Fas ligand)-induced apoptosis has been well elucidated. Binding of Fas ligand to its receptor results in the formation of a complex (disc, death-inducing signaling comlex) consisting of Fas, FADD and pro-caspase 8. Pro-caspase 8 is processed to an active enzyme at the disc. There are two pathways downstream of caspase 8. In some cells, such as thymocytes and fibroblasts, caspase 8 directly activates 3. In type II cells such as hepatocytes, caspase 8 cleaves Bid, a member of the Bcl-2 family. The truncated Bid then translocates to mitochondria and stimulates release of cytochrome c, which activates caspase 9 together with Apaf-1. The activated caspase 9 causes processing of pro-caspase 3 to the mature enzyme. In addition to the death factors, anti-cancer drugs, c-irradiation or factor-depletion induce apoptotic cell death. Although cytochrome C is released from mitochondria during apoptosis induced by these stimuli, the molecular mechanism that triggers the release of cytochrome C from mitochondria is not known.

Caspase 3 activated downstream of the caspase cascade activates a specific DNase (CAD, caspase-activated DNase). CAD is complexed with its inhibitor, ICAD (inhibitor of CAD), in proliferating cells. When caspase 3 is activated in apoptotic cells, it cleaves ICAD to release CAD. CAD then causes DNA fragmentation in the nuclei.

Preclinical Relevance

Blocking of apoptosis by loss-of-function mutations of apoptosis-inducing molecules such as Fas, Fas ligand and caspases, or overexpression of apoptosis-inhibitory molecule such as Bcl-2, causes cellular hyperplasia. In some cases it leads to tumorigenesis, as evident in B-cell lymphomas, which over-express Bcl-2 due to the translocation of the Bcl-2 gene to the immunoglobulin gene locus. Some multiple myeloma and non-Hodgkin's lymphoma carry loss-of function mutations in the Fas gene. Somatic mutation in the Fas gene can also be found in patients of autoimmune diseases called Canale-Smith syndrome or autoimmune lymphoproliferative syndrome (ALPS).

Exaggeration of apoptosis causes tissue damage. For example, administration of Fas ligand, exposure to c-irradiation, or treatment with a high dose of glucocorticoid kill test animals by causing massive apoptosis in the liver or thymus. Hepatitis, insulitis, graft-versus-host disease, and allergic encephalitis are due to the excessive apoptosis by Fas ligand expressed on CTL. Apoptotic cells are detected in the brain of ischemia or Alzheimer patients, suggesting that apoptosis is at least in part responsible for the disease manifestation in these patients.

A proper dose of anti-cancer drugs or c-irradiation can kill cancer cells by activating the apoptotic death program in the target cells. Some cancer cells are resistant to these drugs by an unknown mechanism. It is hoped that elucidation of the molecular mechanism of apoptosis leads to development of an efficient cancer therapy.

References

1. Nagata S (1997) Apoptosis by death factor. Cell 88:355–365
2. Nagata S, Golstein P (1995) The Fas death factor. Science 267:1449–1456
3. Raff M (1998) Cell suicide for beginners. Nature 396:119–122
4. Vaux DL, Korsmeyer SJ (1999) Cell death in development. Cell 96:245–254

Arachidonic Acid

A 20-carbon polyunsaturated essential fatty acid.
▶ Prostaglandins

ARNT (Aryl Hydrocarbon Receptor Nuclear Translocator)

▶ Dioxins and the Immune System

Aroclor

▶ Polychlorinated Biphenyls (PCBs) and the Immune System

Arthritis Models

▶ Rheumatoid Arthritis and Related Autoimmune Diseases, Animal Models

Artificial Determinant

▶ Hapten and Carrier

Artificial Skin/Epidermis

▶ Three-Dimensional Human Skin/Epidermal Models and Organotypic Human and Murine Skin Explant Systems

Aryl Hydrocarbon Receptor

The aryl hydrocarbon receptor, also known as dioxin receptor, is an endogenous transcription factor of an evolutionarily highly conserved family of proteins, the PAS-bHLH family. Members of this family are involved in development and differentiation, xenobiotic metabolism, or rhythm. Some are transcription factors, some – like the AhR – are ligand activated transcription factors. The aryl hydrocarbon receptor becomes activated after binding to a ligand, whereupon it can translocate to the nucleus and transcriptionally activate a number of genes. Hundreds of genes are known as targets – directly or indirectly. Known ligands are plant products such as indole-3 carbinole or other indole derivatives, flavonoids and polyphenols, or anthropogenic substances like polycyclic aromatic hydrocarbons, such as dioxins or biphemyls as an endogenous ligand 2-(1'H-indole-3'-carbonyl)-thiazole-4-

carboxylic acid methyl ester was isolated from porcine lung.
▶ Dioxins and the Immune System

Aryl Hydrocarbon Receptor Nuclear Translocator (ARNT)

Aryl hydrocarbon receptor nuclear translocator (ARNT) is is a member of the PAS-bHLH family of DNA-binding proteins. ARNT dimerizes with other members of this family including the aryl hydrocarbon receptor. Dimerization allows binding to xenobiotic response elements (minimal sequence 5′ GCGTG-3′), half of which equals the E-box of the estrogen receptor. ARNT can function as a homodimer to activated transcritpion from E-box elements, although the physiological significance is unclear. Other partners of ARNT are the hypoxia-inducible factor HIF1-α, sim (a transcription factor involved in central nervous system development), and per (a protein involved in circadian rhythm). ARNT-deficient mice are embryonically lethal due to a defect in angiogenesis. The synonym of ARNT is HIF1-β.
▶ Dioxins and the Immune System

Aspirin

▶ Anti-inflammatory (Nonsteroidal) Drugs

Aspirin-Like Drugs

▶ Anti-inflammatory (Nonsteroidal) Drugs

Assays for Antibody Production

GREGORY LADICS
DuPont Haskell Laboratory
1090 Elkton Road
Newark, DE 19714
USA

Synonyms

humoral immune function, ▶ humoral immune response, primary antibody response, primary humoral immune response, secondary antibody response, secondary humoral immune response, T cell-dependent antibody response, T cell-independent antibody response

Definition

The acquired or ▶ adaptive immune response, which involves both specificity and memory, can be subdived into cell-mediated immunity and humoral immunity. Humoral immunity involves the production of antigen-specific antibody by B cells following a complex interaction between antigen presenting cells, T cells, cytokines (e.g. interleukins (IL)), and cell surface markers. There are five types of immunoglobulins or antibody that may be produced by B cells. These include immunoglobulin G (and various subtypes), IgM, IgD, IgA, and IgE. For immunotoxicity assessment, the focus has primarily been on assays to assess IgM, IgG, and IgE antibodies.

Characteristics

The production of antigen-specific antibodies represents a major defense mechanism of humoral immune responses. Following antigen exposure, the generation of an antigen-specific antibody response involves the cooperation and interaction of several immune cell types (Figure 1). These include antigen-presenting cells (APCs) such as macrophages or dendritic cells, T helper (Th) cells and B cells. The APCs uptake the antigen and subsequently process and present it in association with major histocompatibility complex (▶ MHC) class II molecules to antigen-specific Th cells. The Th cells produce a variety of cytokines (e. g. IL-2, IL-4, IL-6, interferon-γ) which then help B cells to proliferate and differentiate into antibody-producing plasma cells (terminally differentiated, antibody-secreting B cells). Thus, there are numerous targets that may be altered following chemical exposure, making assays that evaluate antigen-specific antibody production a relatively comprehensive and sensitive assessment of immune function.

Several assays have been developed to assess antibody production. These include radial immunodiffusion, hemagglutination, immunoprecipitation, immunoelectrophoresis, radioimmunoassay (RIA), radioallergosorbent test (RAST), passive cutaneous anaphylaxis (▶ PCA), plaque-forming cell (PFC) assay, enzyme-linked immunospot (ELISPOT) assay, and ▶ enzyme-linked immunosorbent assay (▶ ELISA). Of the available assays, the PFC and ELISA are currently the two most often used to assess immunotoxicity. In fact, the quantification of the PFC response (i.e. the specific-IgM antibody-forming cell response) was found to provide one of the best predictors of immunotoxicity in mice (1,2). These tests typically employ a T cell-dependent antigen, such as sheep red blood cells (SRBC). Additional T cell-dependent antigens that have been utilized to assess the primary humoral im-

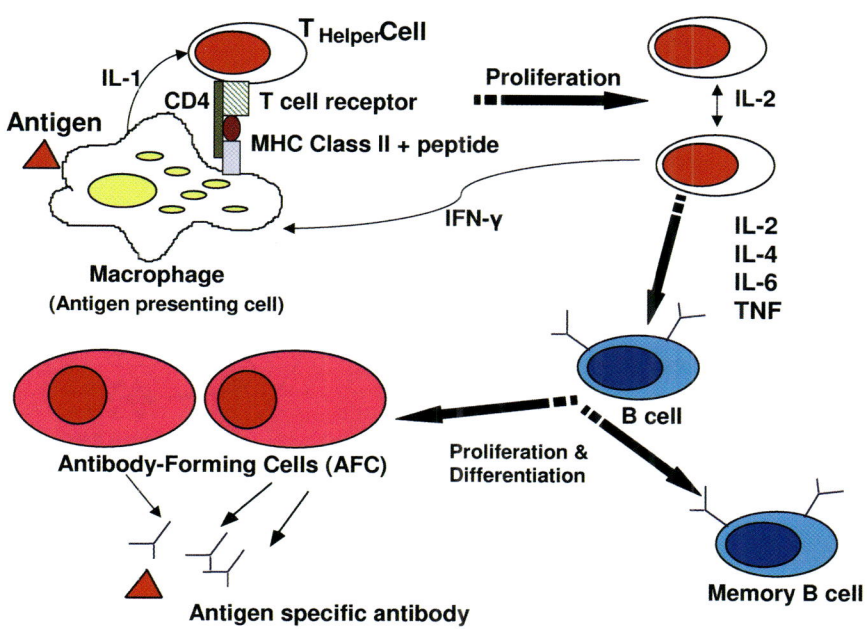

Assays for Antibody Production. Figure 1 Cellular Interactions Involved in Generating a Primary Antibody Response to a T Cell Dependent Antigen (e.g. SRBC).

mune response by ELISA include keyhole limpet hemocyanin or tetanus toxoid.

The PFC response to SRBC utilizes immunocompetent cells from lymphoid organs, primarily the spleen. Following immunization with SRBC, the spleens from immunized animals are removed (for rodents, typically 4 or 5 days later) and cells are incubated with SRBC and complement in a semisolid media (e.g. agar). Plasma cells produce IgM antibody specific for SRBC, which then bind to SRBC membrane antigens and cause complement-mediated lysis of the SRBC and the subsequent formation of plaques (clear areas of hemolysis around each antibody-forming cell (AFC)) that can then be counted visually. Data are usually expressed as IgM AFC (or PFC)/spleen or AFC (or PFC)/million spleen cells. Thus, the PFC does not quantitate the amount of antibody produced, but rather the number of specific antibody-producing plasma cells in a particular tissue (e.g. spleen), and therefore, does not account for antibody produced in other sites (e.g. bone marrow, lymph nodes). In addition, the PFC assay involves sacrificing the animal and is very labor intensive. The PFC assay can also be conducted entirely in vitro using immunocompetent cells obtained from either treated or naive animals (3). In the latter case, the immunocompetent cells are exposed to the test article and SRBC for the first time in tissue culture. This approach also allows for separation-reconstitution studies to identify the primary cell type(s) targeted by a test article, as well as a means to potentially distinguish between test article-induced direct or indirect effects (e.g. neuroendocrine alterations) on the immune system.

The ELISA quantitates antigen-specific antibody found in the serum of a subject generated from all antibody-producing tissues and can be designed to measure any class of antibody (Figure 2). The ELISA is more cost-effective and less time consuming in comparison to the PFC assay and can be automated, as ELISAs are typically performed in 96-well microtiter plates. In addition, the ELISA allows for a number of serum samples to be taken from the same animal and samples may be frozen for later assay. As a result, a time course of humoral immune function can be conducted, a recovery period following test article administration can be evaluated or, upon rechallenge with antigen, a secondary IgG-mediated immune response measured. One current limitation of the ELISA is the availability of reagents for species other than rats, mice, and primates.

The ELISPOT assay is a modification of the PFC assay that allows for the measurement of AFCs that produce antibody of different isotypes (IgM, IgG, IgE, or IgA). The ELISPOT assay is similar in methodology to the ELISA. Antigen is allowed to adhere to a solid support (e.g. plastic or nitrocellulose). Immunocompetent cells are then added and during an incubation period, AFCs secrete antibody that binds to sur-

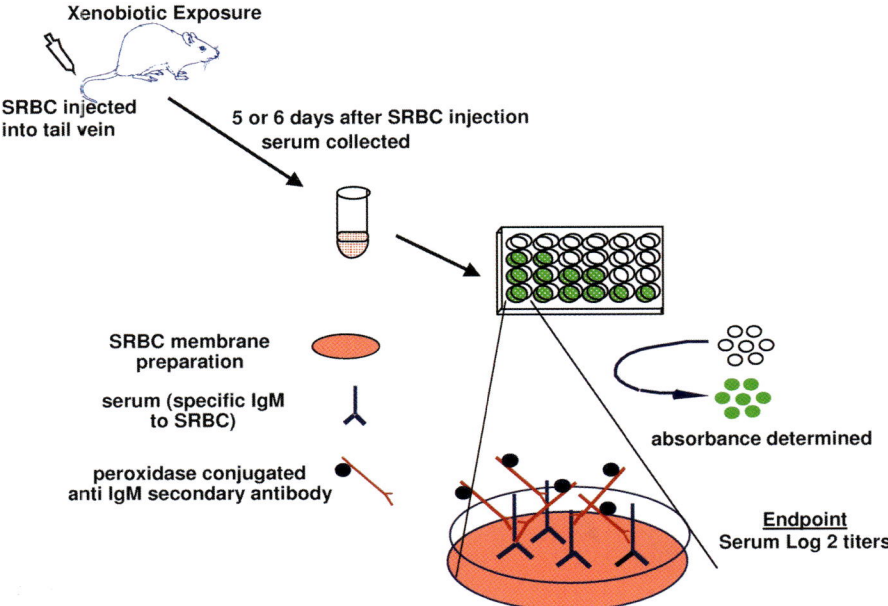

Assays for Antibody Production. Figure 2 Enzyme Linked Immunosorbent Assay (ELISA) to detect SRBC specific IgM Antibody in Sera.

rounding antigen. The AFCs detected by the ELISPOT assay are called spot-forming cells (SFC). To determine the isotype of SFC, an enzyme anti-immunoglobulin antibody conjugate specific for different heavy chains is used. A substrate is then added and an insoluble product produced in areas where antibody is bound to antigen. Each spot produced by the insoluble product represents an AFC.

The assessment of the allergenicity potential of proteins has become increasingly important, particularly in the biotechnology industry. There is a need to assess the safety of foods deriving from genetically modified (GM) crops, including the allergenic potential of novel gene products. Most food allergies are mediated by antigen-specific IgE and are characteristic of type-I reactions. Antigen-specific IgE can be measured by ELISA, ELISPOT, RAST, or PCA. In general, investigators utilize either a homologous or heterologous PCA to assess the presence and titer of antigen-specific, bioactive IgE following the exposure to a particular protein.

Preclinical relevance

As indicated below, there are guidelines that regulate the evaluation of humoral immune function with respect to chemical pesticides and pharmaceuticals.

Relevance to Humans

Due to the invasive nature (injection of antigen) that is required to assess humoral immunity, predictive testing in humans is rare. However, limited predictive testing can be performed on human peripheral blood. Serum concentrations of each of the major immunoglobulin classes (IgM, IgG, IgA, and IgE) can be measured, and natural immunity (antibody levels to ubiquitous antigens such as blood group A and B antigens, heterolysins, and antistreptolysin) can be assessed by ELISA. However, quantifying total immunoglobulin levels lacks the predictive value of assays that measure specific antibody responses following challenge with an antigen. Additionally, antibody responses following immunization to proteins (e.g. diphtheria, tetanus, poliomyelitis) and polysaccharides (e.g. pneumococcal, meningococcal) can be measured. For the most part, the tests available in humans for evaluating humoral immunity only assess the secondary recall response rather than a primary response to a new antigen. Primary immune responses, however, are a more sensitive measure of immune alteration compared to secondary responses (4). The clinical relevance of moderate or transient alterations in humoral immune function is also not known. Human data is limited to severe and long-lasting immunosuppression resulting from therapeutic drug treatments. Furthermore, what human data are available are difficult to interpret due to the idiosyncrasies of the immune system. The age, sex, or genetic background of an individual, and a number of other factors such as stress, malnutrition, chronic infections, or neoplasia can effect a "normal" immune response. Risk assessment is further complicated due to a lack of human exposure data to xenobiotics in general. Additionally, a biologically significant change in immune function does not necessarily produce a clinical health effect until the patient encounters a stress or insult. Further problems arise when evaluating dose-response relationships due to the immune systems reserve or redundant capacity.

Regulatory environment

It has been relatively recently that regulatory agencies have begun to require the evaluation of the primary antibody response. In the US for example, the ▶ Environmental Protection Agency (EPA) in 1998 published guidelines requiring chemicals used as pesticides to undergo an evaluation of the primary humoral immune response to a T-dependent antigen (i.e. SRBC) using either the PFC or ELISA following the administration of a test article to mice and/or rats for 28 days. Testing of pharmaceuticals for their ability to alter the antigen-specific antibody response is determined by a number of conditions. The US Food and Drug Administration (▶ FDA) suggests considering follow-up studies to investigate mechanism(s) of immunotoxicity that may include evaluating the antibody response to a T-dependent antigen among other endpoints if:

- there is evidence of immunotoxicity in repeat dose toxicology studies
- the test article or metabolites accumulate or are retained in reticuloendothelial tissues (i.e. there are pharmacokinetic effects)
- the test article is used for the treatment of HIV infection or related disease
- there are effects suggestive of immunosuppression that occur in clinical trials.

In Europe, conventional pharmaceuticals (not biotechnology derived or vaccines) under ▶ CPMP guidance must undergo an initial 28-day screening study in which the primary humoral immune response to a T-dependent antigen (e.g. SRBC) is conducted if an analysis of lymphocyte subsets and natural killer cell activity are unavailable. Additional studies are conducted on a case-by-case basis which consist of functional assays to further define immunological changes and may include a measure of the primary antibody response if not evaluated in the initial screening study. Testing for antigen-specific IgM or IgG antibody responses is regulated by various guidelines:

- FDA (CDER), Immunotoxicology Evaluation of Investigational New Drugs, 2002
- EPA OPPTS 870.7800, Immunotoxicity, 1998
- CPMP/SWP/2145/00, Note for Guidance on Non-Clinical Immunotoxicology Testing of Medicinal Products, 2001

There are no draft guidelines.

References

1. Luster MI, Munson AE, Thomas P et al. (1988) Development of a testing battery to assess chemical-induced immunotoxicity: National Toxicology Program's guidelines for immunotoxicity evaluation in mice. Fundam Appl Toxicol 10:2–19
2. Luster MI, Portier C, Pait DG et al. (1992) Risk assessment in immunotoxicology. I. Sensitivity and predictability of immune tests. Fundam Appl Toxicol 18:200–210
3. Kawabata TT, White KL Jr (1987) Suppression of the in vitro humoral immune response of mouse splenocytes by benzo(a)pyrene and inhibition of benzo(a)pyrene-induced immunosuppression by α-naphthoflavone. Cancer Res 47:2317–2322
4. National Research Council (1992) Biologic Markers in Immunotoxicology. National Academy Press, Washington DC, pp 1–206

Asthma

MERYL KAROL
Department of Environmental and Occupational Health
University of Pittsburgh
130 Desoto Street
Pittsburgh, PA 15261
USA

Definition

Asthma is a disease characterized by reversible airflow obstruction frequently accompanied by cough, chest tightness, wheeze, and breathlessness.

Characteristics

Asthma is diagnosed from characteristic symptoms combined with the demonstration of reversible airflow obstruction (1). Symptoms may occur spontaneously or as a result of airway hyperresponsiveness. The latter is an acute narrowing of the airways that is stimulated by either physical factors, such as exercise and cold air, or by non-specific agents, such as histamine or methacholine.

Most asthmatics have airway hyperresponsiveness. This can be measured by assessing airflow after inhalation of increasing concentrations of the provoking irritant, or of cold air or exercise. Hyperresponsiveness is determined from measurement of airflow, usually forced expiratory volume in the first second of exhalation (▶ FEV_1), or peak expiratory flow (▶ PEF). Measurement of PEF should be made serially over time, or in response to a bronchodilator (2). It should be noted that patients with demonstrable hyperresponsiveness may be asymptomatic.

The gender distribution of asthma changes with age. In early childhood the disease affects mainly boys; during the teenaged years there is equal occurrence of the disease in boys and girls.

The worldwide prevalence of asthma has increased with a high frequency of asthma recognized in the developed countries. Family size and birth order appear to have an influence on asthma occurrence. Children with few or no siblings have an increased risk of developing asthma, and it has been reported that asthma is usually seen in older siblings.

Preclinical Relevance

The occurrence of asthma varies geographically and temporally. This is suggestive of an environmental influence on the etiology of the disease. The frequent young age of onset of asthma suggests that early-life or prenatal factors may be of importance either in inducing the disease or provoking its symptoms in those with the disease (1).

Asthma may also arise in adulthood. An example is occupational asthma that occurs as a consequence of exposure to an airborne agent at work.

The role of allergen exposure is of critical importance in understanding disease onset. In developed countries, most asthmatic people report childhood allergy occurring to one or more common aeroallergens, such as pollens or dusts (3). Such individuals display symptoms upon exposure to the offending allergens and respond in a concentration-dependent manner, whereby they are more symptomatic in high-exposure environments. High levels of exposure to potent allergens such as the roach, house-dust mite or cat, are considered as risk factors for asthma. Indeed, the ▶ allergen hypothesis suggests a relationship between the allergen concentration experienced in infancy and subsequent development of sensitization and asthma. It is further proposed that a threshold concentration exists, such that exposure to subthreshold concentrations will not result in sensitization. However, it has been suggested that very high exposures to some allergens exert a tolerizing effect where sensitization is prevented from occurring.

The concentration dependence for sensitization and asthma to aeroallergens parallels that demonstrated for occupational allergy and asthma, and will be discussed later. It should be noted that resolution of symptoms and airway hyperresponsiveness may occur with avoidance of exposure (1).

Immunologic tests have been of considerable diagnostic importance. In the majority of asthmatic patients, immune sensitization to one or more allergens can be detected by skin prick testing or by the measurement of specific IgE antibodies in the serum. This state of IgE responsiveness to multiple antigens, usually demonstrating a hereditary tendency, is referred to as ▶ atopy. It should be noted that antibodies have been detected occasionally in asymptomatic antigen-exposed individuals.

Relevance to Humans

Asthma is difficult to study because of its pattern of remission and relapse. The frequency of asthma appears to have increased in more developed (westernized) countries.

The underlying pathology of asthma is airway inflammation. Frequently there is an inflammatory infiltrate in the bronchial walls, desquamated epithelium and mucus plugs within the airways, and goblet cell hyperplasia (2). The inflammation is characterized by increased numbers of T helper type 2 (Th2) lymphocytes and eosinophils. It is frequently accompanied by an IgE-associated allergy to inhaled allergens. Continual exposure to the causative allergen may result in a persistent inflammatory response.

Genetic influences are thought to influence the occurrence of asthma. However, the increased frequency of asthma in developed nations within the past 30 years has suggested environmental factors as causative elements. A focus of recent research has been early-life exposure to allergens and microbial agents, and the influence of such exposures on the development of the immune system in infants and children. The ▶ hygiene hypothesis has been proposed as an explanation for the recent increased occurrence of disease and disease distribution (1).

The most relevant allergens driving the increased prevalence of asthma appear to be those derived from domestic exposures, such as dust mites, cockroaches, and domestic pets (particularly cats). Outdoor agents that are associated with asthma include grass and tree pollens.

Outdoor air in some of the most polluted cities in Southeast Asia contains approximately 500 $\mu g/m^3$ particulate (3). However, the air inside houses in these regions contains 10 000 $\mu g/m^3$. Adverse health effects are seen at 50–100 $\mu g/m^3$ (4). Rural people in developing countries may receive as much as two-thirds of the global exposure to particulates. In developed countries, the quality of indoor air is frequently a problem in homes and commercial buildings because these structures were built to be airtight and energy efficient. Particulates from smoking, pets, fuels, microbial contaminants, and aerosols accumulate and may initiate or exacerbate an asthmatic condition. Having poor ventilation, airtight buildings may contain accumulations of molds, fungi, viruses and bacteria, as well as other biologic and chemical allergens.

The work place is considered to be the source of allergen exposure leading to occupational asthma.

About 300 agents have been associated with causation of workplace asthma including enzymes, proteins associated with laboratory animals, latex and chemicals such as diisocyanates and phthalates (5). An immunologic mechanism is assumed for sensitization and asthma to biological workplace agents, whereas the mechanism of asthma to chemical agents is much less clear (6). The risk of developing occupational asthma is related to the amount of exposure to causative allergens since reductions in airborne allergen exposures have been shown to lead to reductions in disease incidence. Exposure-response relationships have been used to suggest safe levels of exposure below which sensitization should not occur (2).

Regulatory Environment

According to the US Environmental Protection Agency (US EPA) asthma afflicts about 20 million Americans, including 6.3 million children. Since 1980, the biggest growth in asthma cases has been in children under 5 years of age. In 2000 there were nearly 2 million emergency room visits and nearly half a million hospitalizations due to asthma, at a cost of almost $2 billion, and causing 14 million school days missed each year.

In the USA, the Federal government has a long history of regulating outdoor air quality and the concentrations of airborne contaminants in industrial settings. It has established standards (regulations that limit allowable emissions or that do not permit degradation of air quality beyond a certain limit) for six outdoor pollutants (carbon monoxide, lead, nitrogen dioxide, ozone, particulate matter, and sulfur dioxide). Although indoor air has been identified as one of the foremost global environmental problems (3), and is associated with some of the most potent etiologic agents of asthma, the Federal government does not regulate ventilation in non-industrial settings.

Cost-benefit analyses of environmental regulations are increasingly mandated in the USA. Evaluations of ▶ criteria air pollutants have focused on benefits and costs associated with adverse health effects. Evidence of an association of ambient air pollution with provocation of asthma attacks has been obtained in multiple cities around the USA and internationally. Attempts have been made to consider the public health impacts of the criteria air pollutants—particulate matter (PM), ozone, carbon monoxide, sulfur dioxide, nitrogen dioxide, and lead. The ▶ US Clean Air Act (CAA) Amendments of 1990 (1990) included the provision (section 812) that the US EPA performs periodic analyses of the benefits and costs of the CAA. A retrospective analysis of the benefits and costs from 1970 to 1990 compared the costs of implementation of the CAA and its regulations with the health and welfare effects avoided because of decreases in criteria air pollutant concentrations and found that benefits outweighed costs between 11 and 95 times (4). A prospective analysis examining the benefits and costs of criteria air pollutant reductions (excluding lead) from 1990 to 2010 found that in 2010 benefits would outweigh costs by 4 to 1 (4).

The predicted health impacts of reduced air pollution have been calculated. Reductions in criteria air pollutants predicted to occur by 2010 because of CAA regulations have been estimated to produce the following impacts on child health (4):

- 200 fewer expected cases of postneonatal mortality
- 10 000 fewer asthma hospitalizations in children aged 1–16 years
- 40 000 fewer emergency department visits in children aged 1–16 years
- 20 million school absences avoided by children aged 6–11 years
- 10 000 fewer infants of low birth weight.

References

1. Cullinan P, Newman Taylor A (2003) Asthma: environmental and occupational factors. Br Med Bull 68:227–242
2. Redlich C, Karol MH (2002) Diisocyanate asthma: clinical aspects and immunopathogenesis. Int Immunopharm 2:213–224
3. World Development Report (1993) World Bank, Washington DC
4. Wong EY, Gohlke J, Griffith WC, Farrow S, Faustman EM (2004) Assessing the health benefits of air pollution reduction for children. Env Hlth Perspect 112:226–232
5. Yassi A, Kjellstrom T, de Kok T, Guidotti TL (2001) Basic environmental health. Oxford University Press, New York
6. Karol MH (2002) Respiratory allergy: what are the uncertainties? Toxicology 181–182:305–310
7. Prescott-Clarke P, Primatesta P (eds) (1998) Health Survey for England 1996. HMSO, Norwich

Asthma Models

▶ Animal Models for Respiratory Hypersensitivity

Asymptomatic

Absence of recognizable symptoms of an illness or condition.

▶ Mitogen-Stimulated Lymphocyte Response

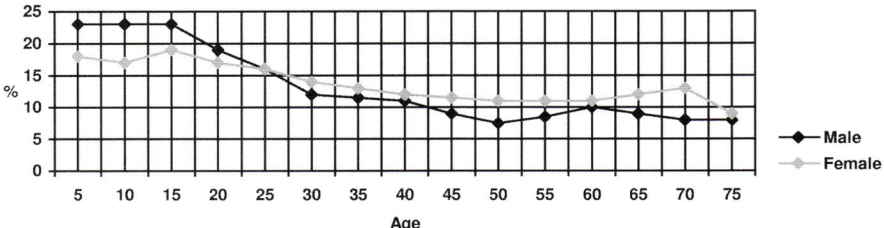

Asthma. Figure 1 Prevalence of asthma in England 1995–6 (7).

Atopic Allergy

▶ IgE-Mediated Allergies

Atopic Dermatitis

Chronic inflammatory disorder of the skin with a genetic disposition. Characteristically patients suffer from severe pruritus and show clinically acute, weeping or chronic, lichenified eczematous lesions. Often accompanied with a diathesis to develop allergic rhinitis or allergic asthma.
▶ Dermatological Infections
▶ Skin, Contribution to Immunity

Atopy

Atopy or atopic allergy is the clinical manifestation of immediate hypersensitivity due to a hereditary tendency for responsiveness to certain allergens. Typical atopic reactions include wheal and flare (skin), bronchoconstriction (airways), and diarrhea (ileum).
▶ Mast Cells
▶ Food Allergy
▶ Asthma

Attenuated Bacilli

Bacterial that have been killed or otherwise altered so as to no longer be infectious.
▶ Birth Defects, Immune Protection Against

Attenuated Organisms as Vaccines

Koert J Stittelaar
Institute for Virology, Erasmus MC
PO box 1738
3000 DR Rotterdam
The Netherlands

Synonyms

Immunization, prevention of infection, sterilizing immunity; or for therapeutic immunization. Weakened, less virulent—causing disease or mild disease—human or animal pathogens (i.e. bacteria, fungi, parasites or viruses) which are used for vaccination (*vacca* = cow (Latin) linked with cowpox matter; see below).

Definitions

There are many different live attenuated organisms used for vaccination approaches, for the greater part in experimental systems, but there are also a number of live attenuated vaccines licensed for human or veterinary use. The category of attenuated vaccines consists of live attenuated bacteria, fungi, parasites and viruses, either as homologous, heterologous, or recombinant live attenuated vaccines (recombinant or chimeric vectors for the expression or delivery of heterologous antigens). The properties, development and the historical and future application of attenuated organisms as vaccines will be explained below. Since the majority of attenuated organisms for vaccination purposes are viruses this chapter will focus on viruses, and vaccines based on other microorganisms will be dealt with briefly.

Characteristics
Properties of Attenuated Vaccines

The main points differentiating attenuated vaccines, or better, live attenuated vaccines (LAV) from other kinds of vaccines (such as killed vaccines, subunit vaccines, DNA vaccines, synthetic vaccines, conjugate vaccines) is that they are able to replicate in the vaccinee which in general results in the induction of strong humoral and cellular immune responses comparable with those induced by the pathogen. As a result, one single vaccine dose potentially provides lifelong protection against natural infection. In most cases the LAV is not only able to replicate, but its capacity even depends on the replication. In other words the inoculation dose initially does not harbor enough antigen to induce protective immune responses; rather the required antigenic load and longevity is regulated by replication of the vaccine. Furthermore, vaccine replication also generates an adequate cytokine milieu for promoting antigen-specific responses. The breadth of the immune response induced by live attenuated vaccine may account for its efficiency compared to (recombinant) subunit vaccines.

An important characteristic of attenuated viral vaccines is that they are able to infect host cells (enter the host cell). For viruses this is the first step of reproduction—once they have gained access into the cytosol, viral proteins will be translated and expressed. An accessory circumstance of this is that a part of the de novo synthesized viral protein is protealytically degraded by a multicatalytic protease of the host cell, the proteasome, resulting in the generation of antigenic peptides. These viral protein-derived peptides are translocated into the lumen of the endoplasmic reticulum by the transporter associated with antigen processing (TAP) where they are loaded onto major histocompatibility class I (MHC-I) molecules. MHC class I peptide complexes go to the surface of the infected host cell and can be recognized by cytotoxic T lymphocytes (CTL). As a result of this specific recognition the CTL will destroy the infected cell. Thus, CTL contribute substantially to protection because they can trace viruses even when they are hidden in host cells whereas antibodies specific to the virus are able to neutralize viruses before they can infect cells. Possible applications of attenuated vaccines are listed in Table 1. A prophylactic vaccine should induce complete protection against infection. In practice, even the most effective vaccines do not induce sterilizing immunity but rather prime the immune system which allows the host to contain the invading pathogen after subsequent wild-type infection. Consequently, the vaccine prevents disease and the host would be less contagious for others.

Effect of dose and strain

Several live attenuated vaccines are known for which the dose is quite particular for safety and efficacy. As mentioned below the currently used live attenuated vaccines against measles are not effective in very young children due to interference with maternally derived virus neutralizing antibody. In an attempt to overcome this problem the vaccine has been applied with doses that are 100-fold to 1000-fold higher. However, this apparently led to (poorly understood) increased mortality in girls in subsequent years as compared to infants vaccinated with standard titer measles virus vaccine.

The vaccine dose that can safely be administered is directly related to the degree of attenuation. Different live microorganisms used as vaccine display different levels and dependence of reproduction or replication. The measles virus vaccine is, for instance, dependent on replication, and a relative low dose (10^3 50% tissue culture infections dose TCID$_{50}$) is used for vaccination but, for instance, modified vaccinia virus Ankara (MVA) is completely replication-deficient, and high doses (10^8–10^9 pfu (plaque-forming units)) could be used warranted to be safe. Note that the doses/titers of different vaccines are expressed differently, and the titers—even when dealing with one kind of microorganisms like viruses—depend on the cell system and readout used to determine the titer. In order to evoke protective immunity a certain antigenic load must be reached. In the case of measles this will take several days, but for MVA the administered dose will determine its effectiveness. In addition, the immune responses induced by the measles virus vaccine will resemble those induced by natural measles virus infection more closely than responses induced by MVA resemble those induced by variola virus infection.

Dose-related effects have also been studied for the "classic" smallpox vaccine. However, these studies were done to see whether the existing supply of smallpox vaccine could be scaled up by dilution so that more doses would be available in the event of smallpox being used as a biological weapon. Diluting the vaccine reduced the rate of successful vaccination: 70% of vaccinees developed a vaccinia vesicle with the 10 × diluted vaccine compared to 95% with the undiluted vaccine.

Attenuated Organisms as Vaccines.
Table 1 Applications of attenuated vaccines

Prevention of bacterial infectious diseases
Prevention of viral infectious diseases
Prevention of fungal infectious diseases
Prevention of parasitic infectious diseases
Therapeutic vaccination against chronic infectious diseases
Prevention and therapy of cancers
Gene therapy
Preventive intervention in fertility

Development of Attenuated Vaccines

There are many different approaches to attenuating a microorganisms:
- selection of a pathogen from a patient showing a mild form of the disease (e.g. variolation)
- use of a related pathogen from another species (e.g. cowpox as a vaccine for smallpox)
- passaging the pathogen from one individual to the other and in cell cultures (e.g. the "classic smallpox vaccine" vaccinia virus
- passaging the pathogen numerous times in cell cultures (e.g. MVA; see below)
- selection for variants characterized by a small plaque phenotype
- chemical mutagenesis (viruses)
- construction of reassortant pathogens (live influenza vaccine)
- cold adaptation, in which the virus is grown at progressively lower temperatures (influenza); through this process viruses are produced which can thrive in the cooler distal ends of the respiratory tract (thereby eliciting mucosal immunity) and cannot survive in the lower respiratory tract and thus cannot cause a systemic/ pathological infection
- knocking out the gene(s) encoding for a known virulence factor—also referred to as targeted gene disruption (bacteria, fungi)
- targeting genes involved in a metabolic pathway whose function is crucial to in vivo survival or growth (bacteria)
- irradiation (parasites)

In some instances combinations of different approaches are used in order to obtain the desired grade of attenuation. In principle this may result in more stable vaccine strains with diminished risk for reversion to virulence.

Mechanisms of Attenuation

Although the treatment that caused the attenuation of the vaccine is often well documented (e.g. the number of passages and the identity of the cell cultures used) the actual underlying mechanisms of attenuation are not known. Despite the biotechnological revolution that has occurred since the first live attenuated vaccines were introduced, the molecular basis of attenuation remains elusive. The first attenuated vaccines (vaccinia, measles) were subjected to careful analysis for the molecular basis of attenuation long after they had been introduced as human vaccines. Understanding the molecular basis of attenuation may allow novel vaccine strains to be developed. For only a few attenuated viral vaccines has some of the "curtain" been lifted.

Live attenuated measles vaccine strains have only a few amino acid changes; these are found in the polymerase and accessory proteins but not in the glycoproteins. Thus the attenuation appears to be a transcriptional impediment rather than a change in receptor usage. Furthermore, the attenuated phenotype appears not to be restricted by the host immune responses.

MVA is one of the most serious alternative smallpox vaccine candidates, and one of the most serious vaccine vectors; it became avirulent by losing various genes, including host range factors, due to more than 570 passages in chicken embryo fibroblast (CEF) cells (1). MVA was completely sequenced and deletions in the genome were mapped and their influence on virulence was determined.

Attenuated flaviviruses, Japanese encephalitis virus, and Murray Valley encephalitis virus show an increased dependence glycosaminoglycans for cell attachment and entry associated with a lowered viremia and no spread from extraneural sites of replication into the brain.

Poliovirus vaccine strains exhibit attenuation determinants in a short conserved sequence in the 5' noncoding region, which may account for a weakened secondary structural element that is critical for translational initiation. As a result of this, binding of ribosomes to a site far downstream of the 5' end may be affected. For a chimeric yellow fever virus/Japanese encephalitis virus vaccine a point mutation has been defined to cause a significant increase in neurovirulence.

A certain point mutation on the poliovirus genome has been shown to be responsible for the attenuated phenotype of the vaccine virus.

The reason why bacille Calmette-Guérin (BCG) is attenuated is still largely unknown. Duplications of regulatory genes (*SigM*, *SigH*) observed in BCG may be responsible for the decreased virulence of the vaccine strain.

Analytical techniques have made it possible to select for targeted microbial components and genes known to be important for virulence. Attenuation by inactivating a metabolic gene has for instance the advantage that the resulting vaccine strain still expresses virulence determinants—something which may be crucial for the induction of protective immune responses. An attenuated typhoid fever vaccine was designed on basis of this principle, aimed at a conditional elimination of the bacterial O antigen which forms part of the cell surface lipopolysaccharide molecule.

Route of Administration

A live attenuated vaccine could be administered via the natural route of infection, which may be beneficial for inducing protective immune responses, both systemic and local. Vaccination strategies that would allow the induction of adequate mucosal immunity may have advantages in this respect. If it could be combined with the easy, cheap and safe administration of a stable vaccine, then efforts to control the infectious disease would considerably be facilitated. However this is not always done in practice for reasons that are unclear.

The currently used live attenuated measles virus vaccine, when parenterally administered, has proven to be quite successful. However, vaccine failures may (at least in part) be attributed to an inadequate vaccine-induced mucosal immunity: the current vaccine protects against measles but not necessarily against measles virus infection. Besides the development of new generations of measles virus vaccines, the question has been raised whether it would be feasible to apply the existing measles vaccine via mucosal routes instead of the currently used parenteral routes. This could lead to a better immune response at the site of virus entry and might allow a more effective vaccination in the presence of pre-existing measles virus neutralizing antibody. Actually, for measles vaccines this phenomenon was reported years ago by Okune 1965. They found that subcutaneously injected measles vaccine was neutralized in the presence of low levels of neutralizing antibody, whereas measles vaccine inhaled as aerosol was not. Since then, the concept of mucosal vaccination using the current measles vaccine has been studied frequently. Different routes of administration have been explored: inhalation of nebulized reconstituted vaccine, and inhalation of dry powder aerosols. Despite the fact that ample experience has been obtained with the current measles vaccine given via the subcutaneous route, if the same vaccine is administrated via an alternative route it will be considered as a new vaccine according to existing regulations. Thus, like for new vaccine formulations, measles vaccine administered via an alternative route would have to go through a complete process of registration and licensing. In addition, for vaccination against measles, oral administration using enteric-coated tablets has also been studied. In general, it is believed that a common mucosal immune system exists, although some reports argue against it. This implies that immune responses evoked at, for instance, the mucus of the intestine will also be active at other mucosal surfaces like those of the respiratory tract. *Shigella* and *Salmonella* strains are typically explored for the delivery of plasmid DNA or expression of heterologous antigens using the oral route.

To exploit the advantages of mucosal (particularly intranasal) vaccination, several live (intracellular) bacterial vectors have been developed and have been shown to elicit strong immune responses, including protective immunity against viruses, bacteria or parasites. Two main categories can be distinguished; those that are based on commensal bacteria, such as lactococci, lactobacilli or certain streptococci and staphylococci, and those that are based on attenuated pathogens, such as *Salmonella*, BCG and *Bordetella*. The quality of the immune responses may vary between the vector systems, but in most cases the immune responses obtained after intranasal administration are stronger than those obtained after oral administration of the same vaccines.

Pros and Cons

Virtually all vaccines cause adverse reactions that are usually self-limited, including fever, headache, fatigue, myalgia, chills, local skin reactions, nonspecific rashes, erythema multiforme, lymphadenopathy, and pain at the vaccination site. Below, examples of more specific vaccine-related problems associated with attenuated vaccines are given.

Oral poliovirus vaccine may typify the disadvantages of attenuated vaccines such as:
- genomic instability known to be able to result in increased neurovirulence
- vaccine-associated illness
- persistence of the vaccine virus in patience with reduced immune function
- transmission of vaccine virus to susceptible individuals who develop vaccine-associated illness
- shedding of vaccine virus into the environment, which could be a source of infection for humans in the future.

Each oral poliovirus vaccine lot contains attenuated and small amounts of wild-type viruses. Vaccines containing more than a certain limit of wild-type viruses may cause a vaccine-associated poliomyelitis. To provide safe vaccines for humans, each newly manufactured vaccine lot is tested in the monkey neurovirulence test. Vaccine strain derivatives that are designed to be more stable than the present vaccines, and therefore less likely to revert to virulence, are potentially useful in the strategically difficult final stages of poliomyelitis eradication.

The current live attenuated vaccine strains of poliovirus are genetically unstable and capable of rapid evolution in human hosts, resulting in reversion to neurovirulence and—in about 1 in a million vaccinees—paralysis, referred to as vaccine-associated paralytic poliomyelitis.

Rotavirus vaccine is not of human origin. The human vaccine, which covers four serotypes, is composed of an simian rotavirus and three reassortant viruses. It was licensed for human use in 1998. However, it was soon found to be a cause of intussusception in about 1 case per 10 000 infant vaccinees. Consequently, the vaccine has been withdrawn from the market although the etiology and pathogenesis of this vaccine-related illness is still largely unknown.

The Guillain-Barré syndrome occurs in about 1 in a million recipients of the influenza vaccine; there is a similar incidence of demyelinating encephalopathy after vaccination with live attenuated vaccine against measles.

The field of vaccinations exhibits a continuous tendency towards the development of combination vaccines. A preparation that contains all relevant vaccines, that provides long-lasting protection, and that can be administered without the use of needles, would be the so-called Holy Grail.

Combination vaccines help to overcome certain objections against vaccinations in general by limiting the number of injections. The mumps-measles-rubella (MMR) vaccine is a good example, illustrating that like inactivated vaccines (i.e. the *Corynebacterium diphtheriae*, *Bordetelia pertussis*, *Clostridium tetani*, Poliovirus, *Haemophilus influenzae B* vaccine cocktail) attenuated vaccines can be applied as safe and effective combination vaccines. The safety of administering measles vaccine (alone or as MMR) in children with allergy to egg protein has been addressed. A single dose showed only minor reactions at the site of injection, and cases of anaphylaxis after the administration of MMR could not be explained by allergy to eggs. There is some debate about a possible association between MMR vaccination and juvenile autism, Crohn disease and other forms of inflammatory bowel disease. However, solid proof based on epidemiological data has not been gathered and the most sensitive measles-specific polymerase chain reaction (PCR) system has found no sample to be positive. In some countries the MMR vaccine is given to HIV-infected children with or without symptoms as prophylaxis. There have been no reports of severe disease attributed to the attenuated mumps and rubella components of the cocktail vaccine in HIV-infected patients so far.

Since the efficacy of vaccination with attenuated vaccines seems (in particular cases like measles and immunodeficiency virus) to depend on the replication capacity of the vaccine virus, further attenuation for safety reasons might be limited. Therefore, attempts are being made to develop live attenuated vaccines that are as immunogenic as the parent pathogen. Expression of granulocyte-macrophage colony-stimulating factor (GM-CSF) by a live attenuated parainfluenza virus type 3 (PIV3) vaccine candidate eventuated in increased immunogenicity without increasing the level of virus replication. In contrast, an attempt to increase the immunogenicity of a live attenuated immunodeficiency virus vaccine candidate by expressing interleukin-2 (IL-2) resulted in a higher set point viral load and faster progression to AIDS after challenge. It appeared that a recombinant of the vaccine virus and the challenge could be detected, indicating that the emergence of more virulent viruses is an additional risk of live attenuated immunodeficiency virus vaccines. Generating recombinant viruses that express immune-stimulating factors should be done with great care. The difference between a biological weapon and a safe vaccine might be quite small. Recombinant vaccinia viruses expressing murine IL-4 were not cleared from immunodeficient mice, and the mice died. As they died more rapidly than immunodeficient mice inoculated with a control virus, it appeared that IL-4 contributed to their death and the IL-4-mediated toxicity was confirmed in normal immunocompetent mice.

Preclinical Relevance

Though it might seem a little disrespectful, there is talk of a tangle with regard to the availability of all kinds of recombinant live vaccines (see Table 3). Live-attenuated vaccines which are proven to be safe and effective are quite often subsequently explored as vaccine vectors for other systems. So far, there is no chimeric or recombinant vaccine licensed for human use; clinical trials are ongoing. However, at this moment such vaccine candidates are extremely helpful for understanding correlates of protective immunity against the targeted pathogen. It can be seen in Table 3 that poxviruses and BCG are the most widely explored vaccine vectors. This is probably due to their good efficacy and safety records for vaccination against smallpox and tuberculosis, respectively. Furthermore, it can be seen that for vaccination against HIV and measles the highest number of different approaches is being explored (regardless of the fact that other vaccine forms are not considered here).

Relevance to Humans
History

From time immemorial people have tried to protect themselves against diseases. Several major diseases of humans have been controlled through the use of attenuated vaccines (see Table 4). The history and specific aspects of vaccination against smallpox and tuberculosis are good illustrations of this. Vaccination against these diseases and against a selection of other viral, bacterial, fungal and parasitic diseases are described below.

Smallpox

Smallpox is a viral disease of humans caused by variola virus; it used to have a high mortality rate of 30%–50%. So far variola virus is the only pathogen globally eradicated by vaccination efforts. In the distant past people deliberately exposed themselves and their children to variola virus in the process of variolation. This involved taking virus from the pox lesion of an infected individual who manifested a mild form of the disease, or by being in close contact to a smallpox victim. Although this method involved the use of weakened causative agent, variolation was not safe,

Attenuated Organisms as Vaccines. Table 2 Pros and cons of live attenuated vaccines

Pros	Cons
Protective	Interference with pre-existing antibody
Broad immune responses	Dependent on cold chain
Balanced systemic and local immune responses	Potential risk in immunocompromised individuals
Balanced humoral and cell-mediated immune responses	Contraindication during pregnancy
Durable immune responses	Needles required
Easy to administer	Molecular basis of attenuation is often not known
Safe in immunocompetent individuals	Revertants not excluded
Combination vaccines possible	Need for neurovirulence test of vaccine lots
Low costs of production	Possible contamination introduced during production
Useful in gene therapy	Three components (vaccine, diluent and syringe)
Useful against infectious agents and cancers	Subclinical infections
	Spread to contacts
	Defective interfering particles
	Vaccine-related illness

with a mortality rate of about 0.2%. Edward Jenner (1749–1823) introduced a method based on observations that milkmaids who contracted cowpox had a lower incidence of smallpox infection. Cowpox primarily causes disease in rodents, and sporadically in cows. Jenner proved that cowpox material could be used to vaccinate humans against smallpox. Subsequently, several vaccine strains were derived from cowpox via many passages in different ways in various animals and from arm to arm of human vaccinees, now known as vaccinia viruses.

The WHO vaccine strain (Lister Elstree strain) used in the global eradication of smallpox was for the most part prepared on the skin of calves. Global eradication of variola virus, by a worldwide vaccination effort orchestrated by the WHO via massive vaccination campaigns and quarantine strategies, is one of the greatest achievements of modern medicine (1). The vaccinia viruses used in the smallpox eradication campaigns were highly efficacious and relatively safe. However they are contraindicated for persons with the following conditions or come into contact with someone with the following conditions:

- a history of atopic dermatitis (commonly referred to as eczema), irrespective of disease severity or activity
- active acute, chronic, or exfoliative skin conditions that disrupt the epidermis
- pregnant women or women who desire to become pregnant in the 28 days after vaccination
- persons who are immunocompromised as a result of human immunodeficiency virus or acquired immunodeficiency syndrome, autoimmune conditions, cancer, radiation treatment, immunosuppressive medication, or other immunodeficiency.

Adverse reactions that were associated with the classic vaccine include inadvertent inoculation, generalized vaccinia, eczema vaccinatum, progressive vaccinia, postvaccinial central nervous system disease, and fetal vaccinia. Inadvertent inoculation occurs when vaccinia virus is transferred from a vaccination site to a second location on the vaccine or to a close contact. The incidence of these serious adverse events (about 1 in 500 000 immunocompetent people and much higher in immunocompromised ones) rendered vaccination with these viruses less acceptable towards the completion of the eradication. This prompted the development of more attenuated vaccinia viruses. After eradication of variola virus, it was decided to discontinue vaccination. This may have created a niche in the human population for other orthopoxviruses, resulting in a spillover from animals to humans of monkey pox in Africa and cowpox in Europe in recent years. In addition, a major concern is the reintroduction of variola virus through bioterrorist acts.

Tuberculosis

Another classical example of the use of a live attenuated vaccine is the vaccination against human tuberculosis, caused by the bacterium *Mycobacterium tuberculosis*. As early as in 1886 Antonin Marfan observed that pulmonary tuberculosis occurred only rarely in individuals who had overcome lupus vulgaris—tuberculosis of the skin. Soon hereafter Robert Koch carried out experiments which showed that

Attenuated Organisms as Vaccines. Table 3 Overview of recombinant and/or chimeric microorganisms

Target agent	Vaccine vector							
	Measles	Poxviruses[1]	Adenovirus	PIV3	Rabies	HIV	Human rhinovirus	Semliki Forest virus
Measles	✓	✓						
HIV	✓	✓	✓			✓		✓
Dengue		✓						
Yellow fever		✓						
Japanese encephalitis		✓						
Rabies		✓	✓		✓			
Mycobacteriumtuberculosis		✓						
Leishmania		✓						
Listeriamonocytogenes								
Schistosomajaponicum								
Mumps	✓							
Lassa fever virus		✓						
Respiratory syncytial virus		✓		✓			✓	
Cytomegalovirus								
Parainfluenza virus		✓		✓				
Influenza virus		✓						
Herpes simplex virus		✓						
Human papillomavirus		✓						
Tumor[2]	✓	✓	✓					
Plasmodium[4]		✓						
Bordetellapertussis		✓						
Brucella abortus		✓						
Hepatitis C		✓						
Hepatitis B	✓							

Attenuated Organisms as Vaccines. Table 3 Overview of recombinant and/or chimeric microorganisms (Continued)

	Listeria monocytogenes	Influenza	Poliovirus	Dengue	Yellow fever	Vesicular stomatitis virus	Streptococcus gordonii	Shigella flexneri
Measles						✓		
HIV	✓	✓		✓		✓	✓	✓
Dengue					✓			
Yellow fever					✓			
Japanese encephalitis					✓			
Rabies								
Mycobacterium tuberculosis								
Leishmania								
Listeria monocytogenes								
Schistosoma japonicum								
Mumps								
Lassa fever virus								
Respiratory syncytial virus								
Cytomegalovirus								
Parainfluenza virus								
Influenza virus								
Herpes simplex virus								
Human papillomavirus								
Tumor[2]	✓				✓			
Plasmodium[4]		✓						
Bordetella pertussis								
Brucella abortus								
Hepatitis C								
Hepatitis B								

Attenuated Organisms as Vaccines. Table 3 Overview of recombinant and/or chimeric microorganisms (Continued)

	Leishmania enriettii	Salmonella[3]	Escherichia coli	BCG	Brucella abortus
Measles		✓		✓	
HIV		✓	✓	✓	✓
Dengue		✓			
Yellow fever					
Japanese encephalitis					
Rabies				✓	
Mycobacterium tuberculosis				✓	
Leishmania					
Listeria monocytogenes				✓	
Schistosoma japonicum				✓	
Mumps					
Lassa fever virus		✓			
Respiratory syncytial virus		✓		✓	
Cytomegalovirus					
Parainfluenza virus					
Influenza virus					
Herpes simplex virus					
Human papillomavirus					
Tumor[2]		✓			
Plasmodium[4]	✓	✓			
Bordetella pertussis				✓	
Brucella abortus					
Hepatitis C					
Hepatitis B					

[1] Includes fowlpox, MVA, ALVAC, NYVAC, vaccinia.
[2] Includes bladder cancer, colorectal cancer.
[3] Includes *Salmonella enterica, S. typhi, S. typhimurium*.
[4] Includes *Plasmodium falciparum, P. yoelli, P. berghei*.

Attenuated Organisms as Vaccines. Table 4 Milestones of important attenuated vaccines

	Infectious agent	Disease	First isolated by	Date of discovery	Founder of vaccine	Date of vaccine development / licensing
Bacteria	Bacillus anthracis	Anthrax	Koch, Davaine	1876	Pasteur	1881
	Pasteurella multocida	Rhinitis	Pasteur	1881		
	Mycobacterium tuberculosis	Tuberculosis	Koch	1881	Calmette	1921
	Yersinia pestis	Plague	Yersin	1894	Haffkine	1897
	Salmonella typhi	Typhoid fever			Leishman	1913
	Vibrio cholerae	Cholera				
Viruses	Variola virus	Smallpox			Jenner	1798
	Rabies virus	Rabies			Pasteur	1885
	Yellow fever virus	Yellow fever	Reed	1900	Theiler, Smith	1937
	Polio virus	Poliomyelitis	Heine, Medin	1840	Sabin	1957
	Measles virus	Measles	Enders, Peebles	1954	Enders	1961
	Rubella virus	Rubella	Parkman, Weller	1962	Parkman, Weller	1966
	Mumps virus	Mumps				1967
	Adenovirus	several[1]	Rowe	1953		1980
	Rotavirus	Diarrhea	Bishop	1973		1998
	Varicella-zoster virus	Chickenpox	Von Bokay	1892	Takahashi	1974
	Japanese encephalitis virus	Encephalitis				1992
	Hepatitis A	Liver cirrhosis				1995
	Influenzavirus A	Influenza				
Parasites	Leishmania	Kala-azar	Leishman, Donovan	1900, 1903		
Fungi	Blastomyces dermatitidis	Blastomycosis	Gilchrist	1894		

* Including acute febrile pharyngitis, pneumonia, gastroenteritis, hepatitis, acute hemorrhagic cystitis.

naive guinea pigs inoculated subcutaneously with live tubercle bacilli acquired protection against experimental infection. The site of infection showed a hard nodule which underwent ulceration and necrosis after 2–3 weeks. In contrast, when live or killed tubercle bacilli were inoculated into a tuberculous animal, ulceration occurred within 2–3 days (Koch phenomenon). This observation, which was the first description of the delayed-type hypersensitivity (DTH) response, resulted in the Mantoux test that is used for the diagnosis of tuberculous infection.

Louis Pasteur (1822–1895) showed for several pathogens that it was possible to weaken the pathogen in order to use it as a vaccine. His achievements prompted Albert Calmette and Camille Guerin to work on an attenuated vaccine strain of the tubercle bacillus. They passaged an isolate from a cow, *Mycobacterium bovis* (a close relative of the human tubercle bacillus) about 230 times in tissue cultures. The result-

ing strain, referred to as bacille Calmette-Guerin or BCG, proved not to revert to virulence in different animal models and conferred protection against wild-type challenge. From 1921 onwards BCG has been in use in humans and today it remains the most widely used vaccine in the world. Unfortunately, despite availability of BCG the incidence of tuberculosis is on the increase, which may be explained by different features of tuberculosis or the vaccine:
- safe and successful vaccination depends on the route of administration (subcutaneous, intradermal)
- coinfection with HIV increases the risk of developing tuberculosis 30-fold
- emergence of multidrug-resistant strains of *Mycobacterium tuberculosis*
- BCG protects children from meningeal and miliary tuberculosis but fails to provide complete protection against the most common form of the disease, namely pulmonary tuberculosis in adults.

In addition, BCG-related untoward effects were not seen in several studies of HIV-seropositive children. However, the safety of live attenuated BCG vaccine in HIV-positive adults or immunocompromised individuals remains unknown and is a matter of some concern.

At present, different alternative vaccine candidates for vaccination against tuberculosis are under development: further attenuated BCG vaccines, introduction of additional mutations, recombinant BCG that overexpress certain immunodominant secretory antigens of *M. tuberculosis*, subunit vaccines, and DNA vaccines. As a result of Pasteur's conclusions and data vaccinations for diphtheria, tetanus, anthrax, chicken cholera, silkworm disease, tuberculosis, and the dreaded plague were developed.

Following the previous eradication of smallpox, a handful of attenuated vaccines prevent illness or death for millions of individuals. They have dramatically reduced the burden of disease and death from polio, measles and tuberculosis. Although the first vaccines were in some respects crude, they have proved to be robust and efficient, and continue to be a source of inspiration for vaccine research and development. There are now several attenuated vaccines in the pipeline which are likely to be registered for human use. However, the future certainly belongs to multivalent vaccines where genes encoding vaccine antigens are inserted into nonpathogenic viruses or bacteria. The most promising model seems to be one that uses poxviruses.

Other examples of attenuated (candidate) vaccines
Viruses
Measles is—in theory—eradicable. Initially, in the 1960s, an inactivated vaccine adjuvanted with alum was used for vaccination against measles until it appeared that upon natural infection with measles virus, children vaccinated with this vaccine were predisposed for enhanced disease (atypical measles). In the 1970s vaccination against measles was pursued again, but now with the use of live attenuated vaccines which induced more balanced humoral and cell-mediated immune responses. Prior to the introduction of live attenuated measles vaccines, more than 130 million cases and about 8 million deaths occurred worldwide each year. These figures have been dramatically reduced to 45 million cases with 1 million deaths per year, with the majority of these in third-world countries. A major stumbling block of the measles virus vaccine used today is vaccine failure due to the presence of measles virus neutralizing antibody, of maternal origin or from previous vaccination. Among many different new vaccine candidates, a recombinant MVA construct has been shown to be able to induce protective immunity in the presence of passively transferred antibody.

Dengue viruses (serotype 1–4) belong to the most important emerging viruses. During the last four decades the incidence of dengue has increased dramatically, affecting more than 100 million people annually. In general, people recover from dengue without serious problems. However, each year about 25 000 people, mainly young children, die from severe forms of dengue: dengue hemorrhagic fever (DHF) and dengue shock syndrome (DSS). All four serotypes of dengue virus can cause dengue fever and DHF. After recovery from dengue virus infection caused by one of the four serotypes, the individual will be protected against a reinfection with this serotype, but usually not against infection with the other serotypes. Moreover, a heterologous secondary infection has been associated with severe disease.

At the time of writing, no dengue vaccine is on the market. The (tetravalent) live attenuated dengue virus vaccines belong to the most promising candidate vaccines at this moment. Several live attenuated dengue virus candidate vaccines have been or are being tested in humans. Optimally, a dengue virus vaccine should safely (without the risk of enhanced disease) induce protection against all four serotypes in the absence and presence of pre-existing immune responses against one or more dengue virus serotypes and other flaviviruses. The need to prepare a mixture of four attenuated dengue virus vaccine strains which induces comparable immune responses against the different serotypes complicates vaccine development.

Fungi
Blastomycosis is a systemic fungal infection of humans and animals caused by the dimorphic fungus *Blastomyces dermatitides*. Especially because this pathogen can cause opportunistic infection in the im-

munocompromised host (such as one with AIDS) a vaccine is desirable. It has been shown that targeted gene disruption or mutation of the gene *WI-1* which encodes for an adhesion-promoting protein causes substantial reduction of virulence indicating that *WI-1* knockout yeast can serve as an attenuated vaccine strain.

Candida albicans is the most common yeast pathogen in humans. Secreted aspartyl proteinases (SAPs) from *C. albicans* have been found to be important virulence factors. Targeted gene disruption of these factors revealed that deletion of one or more SAP isoenzymes causes attenuated virulence.

Parasites

Malaria, caused by the parasite *Plasmodium*, can not effectively be prevented by a vaccine at this moment. As for many pathogens there is poor understanding of the natural immune response to malaria—something that must be understood in order to know which type of immune response will be elicited by a vaccine. A major stumbling block in developing a vaccine against the malaria parasite is that the parasite harbors a high number of antigens that vary throughout its life-cycle. Therefore, the best option is probably for a multistage vaccine instead of a vaccine that only targets a specific stage of the malaria life-cycle. Malaria is mentioned here because one of the many vaccination strategies being explored is the use of irradiated sporozoites of *Plasmodium falciparum*. Irradiation can be considered to be a method of attenuation since it has been shown that the irradiation dose matters. Irradiated sporozoites enter hepatocytes and only partially develop within these cells. When sporozoites are over-irradiated, cytotoxic lymphocyte activity against infected hepatocytes is suppressed. Breakthrough infections, observed in some volunteers, may be addressed to insufficient attenuation. In these experimental vaccination studies volunteers were vaccinated via contact with irradiated mosquitoes. However, this vaccine candidate has not been widely pursued due to the difficulty in growing plasmodia in vitro. An example of a radiation-attenuated vaccine that has been successfully used for over three decades for the prevention of a parasitic disease—albeit a veterinary vaccine—is an irradiated larval vaccine against *Dictyocaulus viviparous* or lungworm.

Leishmaniasis is one of the major parasitic diseases targeted by the WHO. At this moment there is no effective vaccine available and vaccine development is complicated by the variety of different *Leishmania* species. A genetically attenuated vaccine candidate, which lacks an enzyme by gene replacement, has been shown to invade macrophages and to persist for several months without causing disease. However, this vaccine did not induce protective immunity and thus needs to be improved.

Bacteria

Tularemia, a disease caused by *Francisella tulerensis* in a range of vertebrates including humans, has an unknown worldwide incidence. Nevertheless, partly due to the threat of bioterrorism, attempts are being made to develop an effective vaccine against it. Attenuated strains have been generated by repeatedly subculturing fully virulent strains in the presence of antiserum or by drying the strain. These have proven to be effective after both subcutaneous and aerogenic vaccination.

Others

There has recently been the first ever success in experimental gene therapy treatment. A child suffering from the potentially life-threatening "bubble boy" disease, or severe combined immunodeficiency syndrome (SCID), caused by a mutation of the X chromosome, became ill with a leukemia-like disease. He was treated with a retrovirus construct. A retrovirus—from the same virus family as HIV—inserts the therapeutic DNA randomly into the cell chromosomes; this could have caused damage that would cause cells to proliferate wildly, leading to the leukemia. The decision to halt such trials is a major blow to genetic therapy medicine.

Regulatory Environment

National institutions govern the regulation of vaccines. Most countries have their own vaccination programs and surveillance systems. For example, in the US, the public health mission of the Center for Biologics Evaluation and Research (CBER) of the US Food and Drug Administration (FDA), governed by the Code of Federal Regulations, is to ensure the safety and efficacy of biological products and to facilitate their development and approval to the consumer to advance product development. Internationally, the WHO, through its Expert Committee on Biological Standardization (ECBS), plays a key role in reviewing scientific progress and in establishing International Reference Preparations and Recommendations on production and control of biological products. The ECBS provides information and guidance concerning the history, characteristics, production and control of attenuated vaccines, among other things, and facilitates progress towards the eventual international licensing of the vaccine. For example, the ECBS formulates guidelines to develop and standardize appropriate methods and criteria for certain tests, such as the neurovirulence test.

The regulations serve as the framework for product characterization, as well as preclinical and clinical testing strategies. Vaccines that protect against infection are typically developed through a series of different studies, or trials. After basic biological research and animal studies have been completed, progressive clinical trials in humans are conducted. Advancement from one phase of trials to the next depends on the successful completion of the previous set of trials.

- Phase I trial: this is the first setting for vaccine evaluation in humans in which the vaccine is tested in a small number (20–80) of healthy, low-risk, uninfected volunteers to determine the safety of the candidate vaccine and the optimal dosage and immunization schedule.
- Phase II trial: after it has been shown that the vaccine successfully produces the desired immune response and it is well tolerated, the vaccine is tested in larger numbers (up to a few hundred) of healthy, uninfected volunteers to further establish safety and to refine the dosage and immunization schedules.
- Phase III: a much larger trial involving thousands of uninfected, high-risk individuals to determine the protective efficacy of the vaccine. This is the last and most important step in the evaluation process before a vaccine is considered for licensing.
- Phase IV: includes any post-licensing vaccine clinical trial.

References

1. Behr MA (2002) BCG—different strains, different vaccines? Lancet Infect Dis 2:86–92
2. Breman JG, Arita I (1980) The confirmation and maintenance of smallpox eradication N Engl J Med 303:1263–1273
3. Henderson DA (1999) The looming threat of bioterrorism. Science 283:1279–1282
4. Henderson DA, Moss B (1999) Smallpox and vaccinia, In: Plotkin SA, Orenstein WA (eds) Vaccines, 3rd ed. WB Saunders, Philadelphia
5. Hull HF (2001) The future of polio eradication. Lancet Infect Dis 1:299–303
6. Mayr A, Stickl H, Muller HK, Danner K, Singer H (1978) The smallpox vaccination strain MVA: marker, genetic structure, experience gained with the parenteral vaccination and behavior in organisms with a debilitated defence mechanism. Zentralbl Bakteriol [B] 167:375–390
7. Stittelaar KJ, de Swart RL, Osterhaus AD (2002) Vaccination against measles: a neverending story. Exp Rev Vaccines 1:151–159
8. Young DB, Stewart GR (2002) Tuberculosis vaccines. Br Med Bull 62:73–86
9. Okune I et al. (1965) Studies on the combined use of killed and live measles vaccines. II. Advantages of the inhalation method. Biken J 8 (2):81–85

Autoantibodies, Tests for

Jan GMC Damoiseaux · Jan Willem Cohen Tervaert
Dept Clinical and Experimental Immunology
University Hospital Maastricht
P.O.Box 5800
6202 AZ Maastricht
The Netherlands

Synonyms

Humoral autoreactivity assays, autoantibody detection

Short Description

The purpose of these assays is to establish the presence of autoantibodies in the circulation, that is in serum or plasma. Autoantibodies may also be present in other body fluids, but most tests are not standardized for these preparations. There are many hundreds of reported autoantibodies (1), but demonstration of autoantibodies does not imply that there is also an autoimmune disease. Indeed, autoantibodies are relatively common in humans without autoimmune disease (especially in the elderly). Moreover, natural autoantibodies may be physiological and even protective. Basically, most commonly used tests for autoantibodies make use of solid-phase autoantigens to which the autoantibodies will bind (2). Next, the binding of autoantibodies is visualized by labeled secondary reagents. Other assays are based on the characteristics of antigen-antibody complexes to precipitate directly or indirectly upon addition of secondary reagents. This essay will focus on detection of human autoantibodies, but the techniques are, with minor adjustments, also applicable for animal autoantibodies.

Characteristics
Indirect Immunofluorescence Technique
For detection of autoantibodies by the indirect immunofluorescence technique (IFT) serum samples are incubated with antigen substrate to allow specific binding of autoantibodies. The antigen substrate may consist of tissue sections or cell suspensions on a glass slide that are either air-dried or prepared with a fixative to enable autoantibody binding. After washing to remove nonspecific antibodies, the substrate is incubated with an antihuman antibody reagent conjugated to fluorescein isothiocyanate (FITC). The final three-part complex, consisting of fluorescent secondary antibody, human autoantibody, and antigen can be visualized with the aid of fluorescence microscopy. Different types of autoantibody will give a distinctive reaction pattern with the tissue or cell suspension in which the respective antigen is present. The read-out of indi-

rect IFT may be hampered by autoreactivity to other autoantigens in the same tissue. Therefore, an experienced microscopist is required who is familiar with the relevant staining patterns. Indirect IFT on tissue sections and cell suspensions (typically Hep-2 cells or neutrophilic granulocytes) is widely used as a screening assay for the presence of autoantibodies in case of organ specific and systemic autoimmune diseases, respectively. Results obtained by indirect IFT can be confirmed in antigen-specific assays and quantitated (at best semiquantitatively) by testing serial, two-step dilutions. or perform quantitative image analysis of the serum samples that are positive in the screening assay. The latter technique, which is based on the principle of indirect IFT, quantitates fluorescence of a patient sample in comparison to the intensity of standardized calibrators and directly converts intensity into an antibody titer.

Counter-Immunoelectrophoresis

The principle of counter-immunoelectrophoresis (CIE) is based on the typical characteristics of immune complex formation, that is, insoluble immune complexes are only formed where an antibody encounters its antigen when present in an optimal concentration. A precipitation line will form at the point of equilibrium where autoantibodies are present in the serum sample. When antibody and antigen migrate only because of diffusion, the assay is called an Ouchterlony assay. When the migration is facilitated by applying a current across an electrolyte containing agarose gel the assay is called CIE. This technique is predominantly used for detection of autoantibodies to extractable nuclear antigens (ENA) that are negatively charged in the electrolyte solution and therefore migrate to the anode, whereas the autoantibodies are positively charged and migrate to the cathode. Appropriate standardization of these assays requires a reliable source of antigen or antibody of known specificity and purity. These preparations should be run in alternating wells from the patient sample, so as to allow evaluation of the degree of identity. In case of identity in the precipitating antigen-antibody combination the immunoprecipitation lines of two neighboring wells will fuse, whereas nonidentical combinations will result in crossing immunoprecipitation lines. Obviously, CIE is primarily a qualitative screening assay.

Hemagglutination

The principle of hemagglutination is that red blood cells (RBC) will aggregate when an antibody cross-links the antigens on their surfaces. Tests based on hemagglutination are fundamental and are widely used in blood group serology, but they may also be applied for detection of autoantibodies directed to a whole array of antigens. For detection of anti-RBC autoantibodies (Coombs test) the test may be either direct or indirect. The direct Coombs test is used for detecting autoantibodies (and/or complement) that have already bound to the surface of the erythrocytes in vivo. On addition of antihuman globulin reagent the RBC will agglutinate in case of bound autoantibodies (and/or complement). The indirect Coombs test is used to detect and/or to type circulating autoantibodies directed to RBC; the serum is first incubated with test RBC and next with antihuman globulin reagent. Again, agglutination indicates a positive test. Pretreatment of the RBC (addition of colloid, proteolytic enzyme treatment, or low ionic strength saline) is often required to increase the sensitivity of the indirect Coombs test and to detect IgG antibodies. Similar to anti-RBC antibody detection, this system can also be used to test other (auto)antibodies. For detection of autoantibodies directed to antigens not expressed by RBC the autoantigens are readily bound to the surface of RBC which have been treated with tannic acid. The erythrocytes originate from a different species as the serum to prevent false positive reactions to blood group antigens. Incubation of serum with the antigen-coated RBC will results in agglutination when the respective autoantibodies are present. This assay can be performed in serial dilutions in microtiter plates to obtain semiquantitative results when a reference reagent is included.

Enzyme-linked Immunosorbent Assay/Fluorescent Enzyme Immunoassay

The enzyme-linked immunosorbent assay (ELISA) is an immunoassay which employs an enzyme linked to either antiimmunoglobulin or antibody specific for antigen and detects either antibody or antigen. There is a whole array of distinct ELISAs: indirect vs capture ELISAs and competitive vs noncompetitive ELISAs. Competitive assays most commonly measure antigen instead of antibody and are therefore not further discussed here. In most cases the indirect, noncompetitive ELISA is the method of choice for detection of antigen specific autoantibodies. In this assay the antigen is solid-phase bound to the microtiter plates and unoccupied protein-binding sites on the carrier are blocked to prevent nonspecific binding of antibodies. In the first step antigen specific autoantibodies present in the serum sample bind to the antigen. After washing to remove nonspecific antibodies, the antihuman immunoglobulin is labeled by an enzyme (horseradish peroxidase or alkaline phosphatase); it binds to the antigen-antibody complex which leads to the formation of an enzyme-labeled three-part complex that converts the finally added substrate to form a colored solution. The rate of color formation is a function of the amount of autoantibody present in the serum sample. Therefore, the ELISA is considered a true quantitative assay when a reference standard is available, but the assay may also be used for qualitative purposes. When the enzyme labeled to the antihuman immunoglobulin is replaced by a fluorochrome, the assay is referred to as a fluorescent enzyme immunoassay (FEIA). Especially in cases where difficulties in purification of the autoantigen have to be met, a capture ELISA may be more appropriate. This assay utilizes a capturing monoclonal antibody specific for the respective antigen. The antibody will bind the autoantigen of choice and all other contaminants will be washed away. All further steps are essentially the same as in the indirect ELISA although it has to be taken into account that the antihuman immunoglobulin does not bind directly to the capturing monoclonal antibody. The free three-dimensional presentation of the autoantigen in the capture ELISA results in a high assay-sensitivity, as compared to the indirect ELISA, but the disadvantage is the blocking of the epitope recognized by the monoclonal antibody.

Radioimmunoassay
In contrast to the solid-phase ELISAs, the radioimmunoassay (RIA) is a liquid phase assay. The antigen is labeled with a radioactive marker. Alternatively, if the antigen is a receptor (acetylcholine receptor) with very high affinity for a ligand (the snake venom α-bungarotoxin), the ligand may be labeled instead of the antigen. Up on incubation with the serum sample, autoantibodies will attach to the labeled antigen and the resulting immune complexes are subsequently precipitated with antihuman immunoglobulin. The amount of radioactivity in the sediment is directly proportional to the concentration of autoantibodies in the sample. The use of standards of known concentration enables the report of quantitative results. Although this assay is hampered by the fact that pure antigen and special facilities for working with radioactive material are required, the advantage is that only high affinity antibodies are detected. This is particularly important for the clinical relevance of anti-double-stranded(ds)DNA antibodies as detected by the so-called Farr assay.

Immunoblotting
The procedure of immunoblotting is in principal analogous to the steps of the indirect IFT and ELISA. Proteins are immobilized on a membrane surface either by direct application in dots (dot-blot) or lines (line-blot), or by transfer from an electrophoresis gel (Western blot). In case of the dot-blot and line-blot pure antigen preparations are required, whereas in case of the Western blot crude antigen preparations present in body fluids or cell/tissue extracts may be used for separation by electrophoresis. Alternatively, a well-defined mixture of purified or recombinant antigens may by applied. The latter are often commercially available as prepared membranes. Next, free binding sites on the membrane are saturated by incubation with an irrelevant antigen to prevent nonspecific binding of antibodies. Then the blot is incubated with the serum sample and nonspecific antibodies are removed by washing. Bound autoantibodies are visualized by antihuman immunoglobulin conjugated to an enzyme and subsequent conversion of the appropriate substrate into insoluble, detectable products at the sites of protein immobilization. Similar to the CIE this assay will only reveal qualitative results, however, in case of the Western blot determination of the molecular weight of the recognized antigen will give an additional clue to the antigen specificity of the reaction.

Multiplex Analysis
Whereas immunoblotting enables the qualitative detection of many autoantibodies in one single test, new, fluorescent-based techniques are nowadays available that differentially quantitate the presence of multiple autoantibodies at the same time. The multiplex immunoassay employs a mixture of beads with an unique internal fluorescent signature emitted by the beads that are differentially labeled with antigen. Therefore, each bead population represents a separate immunoassay. Upon incubation with the serum sample the respective autoantibodies will bind to the beads and the nonspecific antibodies are removed by a washing step. Next the beads are incubated with fluorescent antihuman immunoglobulins and the presence of autoantibodies is evaluated by flow cytometry. Using known amounts of multiplexed protein, standard curves can be created from the signal intensity and the curves are then used to quantitate the autoantibodies in the samples.

Pros and Cons
The choice of the proper assay for detection of autoantibodies is determined by many variables:
- Are special equipment/facilities available (RIA, multiplex analysis)?
- Are technicians sufficiently trained (IFT, immunoblotting)?
- Are qualitative data sufficient for clinical interpretation (CIE, immunoblotting)?
- And, most important, are purified or recombinant antigens available (ELISA/FEIA, RIA, multiplex analysis)?

Autoantigens
The choice of autoantigen has a serious impact on assay characteristics like sensitivity and specificity (vide infra). The autoantigen may be purified or present in a whole mixture of antigens, native or recombinant, and of human or animal origin. For optimal

recognition of the autoantigen by the autoantibody the three-dimensional structure and interaction with associated molecules should be maintained. This is best achieved by indirect IFT although, if applied, fixation of the tissue/cells may affect the structure of the autoantigen. Also, human tissue for indirect IFT is not always readily available and therefore primate, or even rodent tissues are used. Obviously, interspecies differences may reduce the sensitivity of the assay. Other assays that utilize crude extracts containing the autoantigen include CIE and the Western blot assay. In both cases the three-dimensional structure of the autoantigen may be affected during electrophoresis by the electrolytes and the denaturing conditions, respectively.

The other assays make use of purified or recombinant autoantigens. The problem of contamination in purified autoantigen preparations may be overcome by using a capture ELISA or Western blot, but is a critical caveat in most other assays. The use of recombinant autoantigens is a good alternative because these antigens can be species specific and obtained in large and pure preparations (3). However, depending on the expression system used, the use of recombinant autoantigens is hampered by altered glycosylation and subsequent reduced sensitivity.

Antihuman Immunoglobulin Reagents

The choice of the antihuman immunoglobulin reagent is, analogous to the preparation of the autoantigen, especially critical in terms of outcome and clinical relevance of the test systems. Obviously, this only holds for assays were antihuman immunoglobulin reagents are required. Antihuman immunoglobulin reagents may react with all immunoglobulins or with specific isotypes. For diagnostic purposes IgG-specific antihuman immunoglobulins are most specific for the respective autoimmune diseases. Some exceptions include IgM rheumatoid factor and autoantigens that are expressed in mucosal tissues. Autoantibodies against mucosal antigens are preferentially of the IgA isotype. Overall, autoantibodies of the IgM isotype, especially in low titres, are less specific for autoimmune diseases because IgM autoantibodies may occur as natural autoantibodies and have only low affinity for the autoantigen. Moreover, detection of IgM autoantibodies may reveal false-positive results due to the presence of IgM rheumatoid factor in the serum sample. The isotype-specific antihuman immunoglobulin reagents are induced in domestic animals upon immunization with purified immunoglobulin isotypes. However, since these immunoglobulin preparations consist of heavy and light chains of the immunoglobulin molecule and because the light chains are common to all immunoglobulin isotypes, the collected sera should be absorbed with the other isotypes in order to purge for the antihumanimmunoglobulin antibodies that react with the light chains. Alternatively, the *Staphylococcus aureus* cell wall protein A or the group G *Streptococcus* cell wall constituent protein G may be used as antihuman immunoglobulin reagents. These reagents are specific for the IgG isotype, but while protein G recognizes all IgG subclasses, protein A does not recognize the IgG_3 subclass. Finally, if greater assay sensitivity is required the antihuman immunoglobulin reagent may not be directly coupled to the reporter molecule (the enzyme or the fluorochrome). Conjugation for instance to biotin and subsequent incubation with streptavidin conjugated with enzyme or fluorochrome will enhance the signal, obviously at the cost of increased background reactivity.

Predictivity

The extend to which detection of autoantibodies can be considered a diagnostic marker for a particular autoimmune disease can be expressed in a couple of statistical parameters. These include sensitivity and specificity. Sensitivity is defined as the probability of a positive test result in a patient with the disease under investigation. Specificity is the probability of a negative test result in a patient without the disease under investigation. Furthermore, it is of great practical concern to know the predictive value of positive and negative test results, that is the proportion of those with a positive test who actually have the disease, and the proportion of those with a negative test who actually do not have the disease, respectively. Whether a test result is positive or negative is dependent on the cut-off point of the assay. To pinpoint the cut-off point that results in optimal sensitivity and specificity a receiver operating characteristic (ROC) curve can be generated by plotting sensitivity against 1-specificity. Overall, the best cut-off maximizes the sum of sensitivity and specificity, which is the point nearest the top left-hand quadrant. It is of great practical concern to realize that the type of assay, autoantigen preparation, and antihuman immunoglobulin reagent will affect the statistical parameters of the test and that laboratories should therefore be recommended to determine their own test characteristics.

Relevance to Humans

The human relevance of autoantibody detection in terms of diagnosis and follow-up of autoimmune diseases is obvious. However, whether induction of autoantibodies up on exposure to chemicals results in autoimmune phenomena depends on several factors:
- the antibody may be pathogenic by itself or just a bystander effect of tissue damage
- the autoantibodies may be of the IgM isotype (low affinity, natural autoantibodies) or of the IgG and/or

IgA isotype (clinically most relevant autoantibodies)
- the genetic environment determines the balance between susceptibility and resistance genes.

Translation of results obtained in studies that expose animals to chemicals is hampered mainly by the homogenous genetic background of the animals, because there are typical autoimmune-prone and autoimmune-resistant inbred strains.

Regulatory Environment
To our knowledge there exist no specific regulations that impose the evaluation of autoantibody induction upon exposure to new chemical substances. Nevertheless, it should be realized that these chemicals, and especially intermediates of these chemicals, may conjugate to self antigens and thereby reveal neoepitopes that eventually may break tolerance to these self antigens. Several health organizations have proposed an autoantibody testing scheme for preliminary evaluation of individuals exposed to immunotoxicants (4). The World Health Organization (WHO) suggests testing for antinuclear antibodies, anti-dsDNA antibodies, antimitochondrial antibodies, and rheumatoid factor. The screening panel recommended by the US Centers for Disease Control and Agency for Toxic Substances and Disease Registry includes antinuclear antibodies, rheumatoid factor and antithyroglobulin antibodies, whereas the US National Academy of Sciences for studies of persons exposed to immunotoxicants recommends (besides the tests suggested by the WHO) antibodies to red blood cells.

References
1. Peter JB, Shoenfeld Y (1996) Autoantibodies. Elsevier science, Amsterdam
2. Rose NR, Conway de Macario E, Folds JD, Lane HC, Nakamura RM (1997) Manual of Clinical Laboratory Immunology. ASM press, Washington
3. Schmitt J, Papisch W (2002) Recombinant autoantigens. Autoimmun Rev 1:79–88
4. WHO (1996) Environmental Health Criteria 180: Principles and methods for assessing direct immunotoxicity associated with exposure to chemicals. World Health Organization, Vammala

Autoantibody

An antibody that reacts with at least one self or autoantigen. It is important to note that there is no distinction in the definition of autoantibodies that originate from diseased or from healthy individuals.

▶ Autoantigens
▶ Antinuclear Antibodies

▶ Autoimmune Disease, Animal Models
▶ Autoimmunity, Autoimmune Diseases

Autoantibody Detection

▶ Autoantibodies, Tests for

Autoantigens

MICHAEL HOLSAPPLE
Health and Environmental Sciences Institute
One Thomas Circle, NW, Ninth Floor
Washington, DC 20005-5802
USA

Synonyms
Self antigens, autoantibodies, natural antibodies.

Definition
Perhaps the best way to define the term "autoantigen" is to break it down into its components: the preface "auto" means self, which is why one of the synonyms is "self antigens"; the second part of the term ▶ antigen is defined as a substance capable of inducing a specific immunological response. An "antigen" really cannot be characterized without considering the qualities of the immune response against it. As discussed below, one of the critical roles of the thymus is "negative selection" to ensure that "specific immunological responses" are not elicited against "self". As such, one can easily appreciate that a specific immunological response against an autoantigen is an aberrant response by the immune system, and is suggested to be one of the mechanisms responsible for autoimmunity. When an ▶ antibody against an autoantigen can be quantified, it is generally referred to as an ▶ autoantibody, which is why this term is also offered as a synonym. Importantly, autoantibody refers to antibodies that react with at least one self antigen, whether the antibodies originate from diseased or from healthy individuals (1). It is important to distinguish autoantibodies from ▶ natural antibodies. The latter are defined as antibodies produced without apparent antigenic stimulation (2).

Characteristics
Because autoimmunity is an inappropriate immune response that is mediated against self, most of the mechanisms that have been proposed are centered on either a change in the expression of autoantigens, or a change in the way that the immune system recog-

nizes self. Very early studies in immunology led to the identification of a group of antigens in mice which—when matched between donor (*foreign*) and recipient (*self*) animals—markedly improved the ability of the graft to survive. Because they played such an important role in graft rejection, these antigens were named ▶ histocompatibility antigens.

It was also noted early on that these antigens were the products of one particular region of the genome, the major histocompatibility complex ▶ (MHC). Three major sets of molecules are encoded by the MHC: class I, class II and class III antigens. Class I molecules are composed of one MHC-encoded polypeptide and are expressed on all nucleated cells. Class II molecules are formed from two separate MHC-encoded polypeptides, have a much more restricted distribution and are only expressed on B lymphocytes, macrophages, monocytes and some types of epithelial cells (e.g. especially cells that can function as antigen-presenting cells (APC)). Although the MHC was originally identified by its role in transplantation, it is now understood and accepted that proteins encoded by this region, especially class I and class II molecules, are involved in many aspects of immunological recognition (*self* versus *foreign*), including interactions between different lymphoid cells and the interaction between lymphocytes and APC. Class III molecules are involved in the complement cascade and will not be further discussed here.

A major challenge of the immune system is the balance between generating a repertoire that is sufficiently diverse to ensure recognition of the myriad of potential foreign antigens, and yet avoids recognition of the vast array of self-antigenic determinants. This challenge is compounded by the facts that the relevant determinants of antigens—be they foreign or self—have similar structural components (e.g. amino acids and carbohydrates), and that the distinction between self and foreign determinants may be very subtle. The repertoire of T cells is selected in the thymus during early ontogeny and involves three steps (3).

The first step is the generation of immature T cells in an MHC-independent fashion with potential reactivity for all of the various MHC molecules expressed in the species as a whole.

The second step, positive selection, involves screening the immature T cell repertoire for cells expressing some degree of reactivity for the self MHC antigens displayed in the thymus. Positive selection involves the interaction of the T cell receptor (TcR) of double-positive (CD4$^+$/CD8$^+$) thymocytes with MHC molecules expressed on thymic epithelial cells. Only T cells displaying physiologic specificity for self MHC molecules survive and, depending on the nature of the interaction between the TcR and MHC, emerge as either CD4$^+$ or CD8$^+$ thymocytes. During positive selection, the vast majority of double-positive thymocytes die rapidly, presumably from programmed cell death.

The third step, negative selection, involves self-▶ tolerance induction and results in the destruction of potentially autoaggressive T cells expressing overt reactivity for self MHC molecules. The thymus has a huge task to accomplish tolerance to the enormous range of self-antigens. Many self components, such as serum proteins and surface molecules on various types of lymphoid or hematopoietic cells, readily enter the bloodstream. These circulating self antigens have ready access to the thymus where they are degraded by thymic APCs for presentation to newly formed T cells. A different situation applies to tissue-specific self antigens, which enter the circulation in negligible quantities. As a consequence, T cells generally display little or no tolerance to tissue-specific antigens, which is why these antigens are the main targets for autoimmune disease.

Although the selection of B cells does not occur in a process comparable to T cells, there are a number of mechanisms available to maintain the B cell repertoire (4). First, the principal means by which autoreactive B cells are purged from the repertoire is "central tolerance" that is induced by the stable cross-linking of sIg receptors by multivalent antigens in the absence of T cell help, and that results in either permanent inactivation or elimination of these cells. Although mature B cells can be inactivated by continuous exposure to high concentrations of certain antigens, under most circumstances central tolerance affects only newly developing B cells. This dichotomy in the susceptibility to tolerance induction by immature versus mature B cells has been speculated to explain the difference between self and nonself antigens (4). "Self" is defined by the spectrum of antigenic determinants that are present in the milieu of newly developing B cells—and other antigens, first encountered by mature B cells, would by default be recognized as "foreign". The attractiveness of this mechanism is that despite the diversity in the structure and concentration of self antigens, if they are present in the environment of newly developing B cells, autoreactive cells that recognize these determinants would be eliminated. However, the affinity threshold for tolerance induction of newly generated B cells is such that low-affinity antigen-activated cells, as well as those that recognize monovalent self determinants, are likely to escape without becoming tolerant and it is not surprising that reactivity to self antigens can be found among mature B cells (4). There are numerous peripheral mechanisms available to minimize the escape of B cells from tolerance induction and ensure against the generation of autoantibodies. Because B cell stimulation by most antigens occurs in a T-dependent fashion, B cells do not respond to some self antigens because of tolerance induced in T cells, as described above. Stimulation of B cells in the periphery can also be prohibited by either antibodies or T cells that recognize determinants on the surface immunoglobulin (sIg) receptors, so-called idiotypic or network suppression.

In recent years it has become increasingly evident that neither B nor T lymphocytes reactive with autoantigens are completely eliminated from the immune repertoire of normal individuals (5). In spite of the elaborate mechanisms in place to facilitate the elimination (e.g. clonal selection) or control (e.g. tolerance) of self-reactive clones, it is now recognized that these processes are not completely correct.

Preclinical Relevance

Much of what we know about the mechanisms of autoimmunity and the onset and progression of autoimmune disease has come from animal models, including the role played by autoantigens. Autoimmunity is an inappropriate immune response to autoantigens that can result in the generation of autoantibodies and/or autoreactive T cells, and that can lead to tissue damage. An important distinction is that these manifestations of autoimmunity are not synonymous with the pathogenicity or morbidity that characterizes autoimmune disease.

The mechanism (or mechanisms) by which autoimmunity is triggered is unclear and is being studied. A number of possibilities are suggested from the aforementioned discussion about processes to maintain T cell and B cell repertoires including:

- an influence on the development of the T cell repertoire in the thymus to either allow for the positive selection of autoreactive cells or the deletion of a regulatory T cell specificity
- a generalized failure to induce tolerance within either T cells (e.g. negative selection) or B cells (e.g. central tolerance)
- a failure of peripheral tolerance mechanisms.

Target autoantigens may be located on the cell surface (e.g. acetylcholine receptor and idiotypes of antigen receptors), inside the cell (e.g. DNA and ribosomal proteins), or may be extracellular molecules (e.g. insulin and intrinsinc factor).

The basis for how most autoantibodies result in immunopathology remains unclear. Antibodies against autoantigens located on the cell surface can have clear clinical consequences—e.g. immunoglobulin G antibodies to surface molecules on red blood cells (RBC) can result in autoimmune hemolytic anemia. The actual destruction of RBC is mediated by the reticuloendothelial system, especially by macrophages in the splenic sinusoids. In contrast, there is little evidence that autoantibodies can readily penetrate cellular membranes, and bind to intracellular or nuclear targets. Therefore, there is little information to suggest that disease results from the ability of autoantibodies to inhibit intracellular processes dependent on these molecules. Some manifestations of autoimmune diseases like lupus and rheumatoid arthritis, such as arthralgia/arthritis, serositis and vasculitis, appear to be mediated by immune complexes.

Ultimately, autoantibodies causative of pathology are rare. Stated another way, most immune responses to autoantigens are not associated with morbidity. As such, stringent criteria are needed to distinguish autoantibodies that are pathogenic from those that are not (5), including the following:

- autoantibodies isolated from an affected organ or tissue must react in vitro with the same autoantigen
- autoantibodies isolated from the diseased tissue, or produced in vitro with similar characteristics, should transfer an identical lesion in animals
- upon transfer it should be possible to reproduce the histopathological, functional, and biochemical abnormalities in the original disease.

As discussed below, there is at least one additional aspect to the presence of autoantibodies that must be considered. Some autoantibodies, especially so-called natural antibodies, may serve a regulatory function in some selected instances.

Relevance to Humans

Paradigms about autoreactivity have evolved considerably since Ehrlich proposed "horror autotoxicus" to illustrate the notion that antibody responses may not occur against self components (1). It is, of course, now recognized that autoantibodies do occur naturally and that their presence is not necessarily associated with pathobiology.

Although it is widely held that the presence of autoantibodies serves as one of the best markers for autoimmunity, this approach is not without confounders. The diagnosis of autoimmune disease must be based on the presence of both clinical and biological criteria in a given patient, and the relevance of identifying autoantibodies in asymptomatic patients is unknown. In light of the fact that titers of autoantibodies occur in asymptomatic, apparently healthy individuals and are known to increase with age, it is doubtful that immune responses to autoantigens detected in individuals without any clinical manifestations can be used reliably as early biomarkers of autoimmunity.

The failure to distinguish between the production of "natural" autoantibodies from pathogenic autoantibodies has contributed to the confusion associated with the mechanism of autoimmune diseases (5). Normal persons can be shown to have B cells that secrete autoantibodies (most of which are IgM), that bind to their target autoantigens with relatively low affinity and frequently cross-react with multiple antigens. In contrast, most pathogenic autoantibodies are IgG, bind to their antigens with high affinity and high specificity. Natural antibodies can usually be demonstrated after polyclonal stimulation of a normal B cell population. It has been suggested that natural autoantibodies may help in the clearance of senescent cells, cell constituents after cell death (e.g. DNA) or immune complexes. One example of such natural autoantibody production may be rheumatoid factor production in bacterial endocarditis or other chronic inflammatory conditions. Natural autoantibodies appear to be encoded by non-mutated (germline) Ig variable region genes; this is consistent with their IgM nature and their relatively low affinity. In contrast, most pathogenic autoantibodies are encoded by somatically mutated Ig genes, which is also consistent with their IgG nature and high affinity. Recent studies also suggest that a major proportion of natural autoantibodies are secreted by B cells that express the CD5 antigen. Although still somewhat controversial, natural autoantibodies appear to play little role in the pathogenesis of autoimmune diseases, and current evidence suggests that CD5+ B cells are not primarily involved in the production of pathogenic autoantibodies (2).

Regulatory Environment

Almost every review ever written about immunotoxicology emphasizes that the consequences can include a number of possible outcomes such as immunosuppression, hypersensitivity/allergy and autoimmunity. As noted throughout this section the importance of autoantigens is primarily in the context of autoimmunity. However, in reality, the focus of scientific research has been heavily weighted with the following emphasis:
immunosuppression ≥ hypersensitivity/allergy >>> autoimmunity.
This focus can perhaps be most readily appreciated in the context of the regulatory environment. A number of regulatory agencies have issued guidelines and/or guidance documents to address immunosuppression and hypersensitivity. For example, the US EPA Office of Prevention, Pesticides and Toxic Substances (OPPTS) includes Health Effects Guidelines to address both Immunosuppression (OPPTS 870.8700) and Skin Sensitization (OPPTS 870.2600). These guidelines signal that immunotoxicology has evolved as a scientific discipline to the point where it can be used in risk assessment. However, consistent with the aforementioned focus, there has not been much attention on autoimmunity within the regulatory environment.

References

1. Dietrich G, Kazatchkine MD (1994) Human natural self-reactive antibodies. In: Coutinho A, Kazatchkine MD (eds) Autoimmunity: Physiology and Disease. Wiley-Liss, New York, pp 107–128
2. Casali P, Kasaian MT, Haughton G (1994) B-1 (CD5 B) cells. In: Coutinho A, Kazatchkine MD (eds) Autoimmunity: Physiology and Disease. Wiley-Liss, New York, pp 57–88
3. Sprent J (1995) T cell biology and the thymus. In: Frank MM, Austen KF, Claman HN, Unanue ER (eds) Samter's Immunological Diseases, Volume I, 5th ed. Little, Brown & Company, Boston, pp 73–85
4. Klinman NR (1995) Biology of B cells. In: Frank MM, Austen KF, Claman HN, Unanue ER (eds) Samter's Immunological Diseases, Volume I, 5th ed. Little, Brown & Company, Boston, pp 61–72
5. Zanetti M (1994) Autoantibodies and autoimmune network: The evolving paradigm. In: Coutinho A, Kazatchkine MD (eds) Autoimmunity: Physiology and Disease. Wiley-Liss, New York, pp 129–141

Autoimmune Chronic Active Hepatitis

▶ Hepatitis, Autoimmune

Autoimmune Chronic Hepatitis

▶ Hepatitis, Autoimmune

Autoimmune Disease

A condition in which the immune system fails to recognize an organ or tissue as "self" and therefore attacks and harms the organ or tissue.
▶ Autoimmunity, Autoimmune Diseases
▶ Birth Defects, Immune Protection Against

Autoimmune Disease, Animal Models

Dori Germolec
NIEHS/NIH
111 Alexander Drive, PO Box 12233
Research Triangle Park, NC 27709
USA

Synonyms

Major histocompatability complex, MHC, systemic

lupus erythematosus, SLE, complete Freund's adjuvant, CFA.

Short Description

A number of animal models are currently used to evaluate etiologic agents, influences of chemical exposure and therapeutic efficacy for autoimmune diseases. While studies of autoimmune disease have been carried out in a wide variety of animals including primates, dogs, pigs, chickens and rabbits, rodents are most commonly used for these investigations and will be the focus of this essay. Rodent models fall into three broad categories:
- genetically predisposed animals
- animals in which disease is produced by immunization with specific antigens
- animals in which the disease is organic or chemical-induced.

In each type of model the development and severity of symptoms is dependent on multiple factors, including the genetic makeup of the animal, age, and hormonal and/or environmental factors.

Characteristics

A number of syndromes ranging from organ-specific to systemic diseases similar to that observed in humans can be mimicked in laboratory animals. As these models are relatively disease-specific, they are significantly different than screening tests such as the popliteal lymph node assay, which evaluate the immunostimulatory potential of chemical agents. The endpoints evaluated in rodent models of autoimmune disease are frequently similar to tests conducted in clinical practice. Measurement of serum or urinary parameters indicative of disease (i.e. autoantibodies, protein and glucose levels) can be conducted on rodent samples with considerable ease using commercially available products. Histological evaluation of tissue damage, measurement of cytokine levels and evaluation of tissue specific expression of adhesion molecules, cytokines and cell-surface receptors have provided significant mechanistic information.

The genetically predisposed models, whether naturally occurring, transgenic or knockout based, tend to be very reliable and therefore have been more commonly employed in ▶ autoimmunity research (1). In these models, mild to severe syndromes spontaneously develop, usually due to mutations in major histocompatability complex (MHC), cell surface receptor or cytokine genes, often inducing functional abnormalities of T helper (Th) cells. Transgenic and knockout strains have become critical tools to investigate the relative contributions of disease susceptibility genes and whether specific alleles may confer protection against the development of autoimmune reactions (1). As MHC class II (and less frequently class I) genes show the most consistent associations with disease susceptibility, mice which express human HLA molecules or mutated rodent genes have been used to study rheumatoid arthritis, type 1 diabetes, multiple sclerosis, myasthenia gravis, autoimmune thyroiditis and many other disorders (2). One of the most commonly used models of type 1 diabetes is the non-obese diabetic (NOD) mouse, in which disease has been linked to a variety of genes both within and outside the MHC. Using transgenic mice on the NOD genetic background, investigators have been able to evaluate the T cell autoimmune response, B cell antigen presentation, and the role of specific cytokines on disease progression (3). To investigate the role of infectious agents in autoimmune disease, transgenic animals that express viral or microbial proteins, often targeted to specific tissues have been developed. In some instances these antigens have significant sequence homology with self-proteins and stimulate an autoimmune response against these self-proteins. Alternatively, the viral or microbial proteins may stimulate a chronic inflammatory response that induces tissue destruction, release of self-antigens and activation of autoreactive lymphocytes leading to autoimmune disease. These types of models have been particularly important in investigating organ-specific autoimmune diseases.

The F1 cross between New Zealand black (NZB) and New Zealand white (NZW) mouse and the MRL/lpr mouse are likely the most frequently used rodent models of human systemic lupus erythematosus (SLE). A number of variants of the NZBxNZW F1, which differ in the genetic contribution of the parental strain and exhibit different disease phenotypes, have been used to map specific susceptibility loci and assess the importance of B cell hyperactivity and T cell involvement in ▶ autoantibody production in the development of disease (3). The genetic defect in the MRL/lpr mouse is a result of a mutation in the *Fas* gene leading to defective apoptosis. While both mouse strains develop high levels of serum immunoglobulins, antinuclear and anti-DNA antibodies and immune-mediated nephritis similar to that seen in human SLE, the MRL/lpr mice also exhibit rheumatoid factor autoantibodies and inflammatory joint disease characteristic of an arthritic response.

Immunization with purified antigens can elicit a specific autoimmune response, particularly when adjuvants are administered in conjunction with self-proteins or in the presence of polyclonal immune activation following infection. As in other types of autoimmune models, a permissive genetic background may be necessary for the development of disease. Thus while immunization of susceptible mouse strains (e. g. those containing $H-2^q$ or $H-2^r$ alleles) with type II

Autoimmune Disease, Animal Models. Table 1 Experimental models for autoimmune disease*.

Autoimmune Disease	CLASSIFICATION		
	Genetically Predisposed Strains	Immunization	Induction
Autoimmune Thyroiditis (Hashimoto's and Grave's)	MRL (m) BB (r) OS (ch) NOD H2h4 (m)	Thyroglobulin – (m,r) Thyrotropin receptor (m)	Dietary Iodine (m)
Autoimmune Uveitis		S-antigen (m) Rhodopsin (m) Phosducin (m) Interphotoreceptor retinoid binding protein (m)	
Inflammatory Bowel Disease (Colitis/Crohn's Disease)	TCR alpha/beta transgenic (m) MHC Class II transgenic (m)		
Insulin-Dependent Diabetes Mellitus	NOD (m) BB (r) DRBB (r) BN (r)		Streptozotocin (STZ: m)
Myasthenia Gravis	HLA-DR3 transgenic (m)	Acetylcholine receptor – (m,r)	Penicillamine (m, r)
Multiple Sclerosis	HLA-DR2 transgenic (m) MHC Class II transgenic (m) TCR transgenic (m)	Myelin Basic Protein – (m,r,c)	
Rheumatoid Arthritis	MRL/lpr (m) SCID (m) HLA B27 (r) IL-1 RA knockout (m)	CFA + Type II collagen (m,r,mo) CFA + Mycobacterium heat shock protein (m,r)	Streptococcal cell-wall (r)
Spondyloarthropathies	HLA-B27 transgenic (r) ank/ank (m)	Aggrecan (m) Versican (m)	
Systemic Lupus Erythematosus	MRL +/+ (m) MRL/lpr (m) MRL-mp-lpr/lpr (m) NZW 2410 (m) NZB/NZW (m)	CFA + antiDNA antibodies (m,r)	Mercury (m,r,mo) Penicillamine (m, r) Procainamide (m, r)
Systemic Sclerosis (Scleroderma)	TSK (m) UCD-L200 (c) Integrin 1alpha knockout (m)		Vinyl chloride (m) Bleomycin (m)

Complete Freund's Adjuvant (CFA), mouse (m), rat (r), chicken (ch), monkey (mo)
* This table is not meant to be a comprehensive review, but rather is a sampling of potential models for some of the most common autoimmune conditions. Adapted from refs 1–5.

collagen or cartilage glycoproteins, in the presence of adjuvant, can induce pathology remarkably similar to rheumatoid arthritis, in similarly immunized animals with a less permissive genetic background, the disease is self-limiting and severe pathology does not occur. Arthritis can also be induced in susceptible rat strains by immunization with ▶ complete Freund's adjuvant (CFA) containing killed *Mycobacterium tuberculosis* in oil. In collagen-induced models of arthritis, the immune response is directed against specific connective tissue antigens, while in the adjuvant-induced models the response is directed against a mycobacterium heat shock protein and pathology results from cross-reactive destruction of the proteoglycan found in joints (4). Another frequently used model of this type, experimental autoimmune encephalomyelitis, can be induced in a number of species by immunization with myelin basic protein and CFA. In rodents, the resulting pathology is a T helper-mediated autoimmune disease characterized by perivascular lymphocyte infiltration of the CNS and destruction of the myelin nerve sheath with resultant paralysis, similar to that observed in patients with multiple sclerosis. The disease can also be induced by adoptive transfer of myelin-protein specific T lymphocytes from immunized animals and via the injection of viral proteins that share homology with myelin basic protein (4). Organ-specific autoimmune diseases have been induced in a number of tissues including liver, kidney, thyroid, testis and ovary following immunization with tissue-specific antigens in the presence or absence of adjuvant. In addition, systemic diseases can also be induced by immunization with putative autoantigens. For example, mice immunized with anti-DNA antibodies or allogeneic leukocytes can develop lupus-like symptoms.

In models where autoimmunity is induced by exposure to chemical or biological agents, foreign substances are used to initiate the autoimmune disease state. These may include chemicals, drugs, or biological agents. As described above, injection of viral or microbial peptides that have sequence homology with self-antigens, particularly in the presence of adjuvant, may lead to the development of a number of autoimmune diseases. In some instances, infection with specific pathogens may directly result in disease in susceptible strains. For example, herpetic stromal keratitis, a T cell-mediated autoimmune disease of the eye, can be induced in some strains of mice following infection with herpes simplex virus type 1. Infection with coxsackie B3 virus has been associated with both viral myocarditis and type 1 diabetes (4). Evidence suggests that both viral and host factors may be important with regard to what tissue the virus targets and the specific immune pathways that are activated. Thus some mouse strains develop a response characterized by Th2-dependent production of antiviral antibodies that cross-react with cardiac myosin, which other strains generate a Th1-mediated cytotoxic T lymphocyte response (4). Neonatal infection with mouse T lymphotropic virus, which selectively depletes $CD4^+$ thymocytes, can result in the development of organ-specific autoimmune disease in susceptible strains.

One of the more commonly employed models of chemical-induced autoimmunity is the Brown Norway (BN) rat model, in which the animals are injected with non-toxic amounts of mercuric chloride ($HgCl_2$). The chemical exposure produces no overt signs of toxicity, yet the rats develop an immunologically mediated disease characterized by T cell-dependent polyclonal B cell activation, autoantibodies to laminin, collagen IV, and other components of the glomerular basement membrane similar to that observed in humans with autoimmune glomerulonephritis. The importance of cytokine regulatory networks and underlying genetic background has been elucidated through the use of this and other rat model, as the nature or the immunologic response and types of autoantibodies that develop following mercury exposure appear to be highly strain dependent. Numerous mouse strains have also been used to evaluate the development of autoantibodies following exposure to mercury, gold and cadmium (5).

A large number of drugs have been associated with risk of developing SLE and the diversity of compounds which elicit lupus-like disease indicate that multiple mechanisms may be involved. Procainamide, hydralazine, isoniazid, penicillamine and others have been shown to induce autoimmune manifestations similar to that observed in idiopathic SLE (4). Injection of procainamide hydroxylamine, a reactive metabolite of procainamide, into the thymus, but not the spleen or periphery, of normal mice results in the development of autoreactive T cells and autoantibodies against chromatin characteristic of procainamide-induced lupus. The establishment of self-▶ tolerance during positive selection appears to be altered by the presence of the drug metabolite in this model (5). As an example of drug-induced organ specific disease, streptozotocin has been shown to induce diabetes in a number of mammalian species. C57Bl/6 and NOD mice develop immune-mediated diabetes following multiple low-dose injections, while some strains are highly resistant. Treatment with high doses of streptozotocin are directly toxic to pancreatic β cells, however the mechanisms involved in generation of autoimmunity following low dose exposure are still in question. The roles of co-stimulatory molecules, nitric oxide production and free radical generation, as well as the identification of potential susceptibility genes have been investigated using this model.

Pros and Cons

The use of animal models has increased an understanding of the mechanisms of autoimmune diseases and provided an opportunity to identify susceptibility genes and investigate the role of environmental factors and therapeutic strategies. Animal models allow for the evaluation of exogenous factors on the onset or progression of disease. Animal models permit opportunities for prophylactic or preventative treatment that is often impossible in humans where disease is not diagnosed until clinical symptoms have manifested and tissue damage has occurred. However, while many of the animal models of autoimmune disease closely mimic aspects the pathophysiology found in humans, there are important differences that may limit their utility in predicting the development of human disease. Most laboratory studies are conducted in inbred strains that express specific genes in a controlled environment. Humans are highly outbred and are exposed to a variety of environmental factors that influence disease susceptibility and progression. Thus many animal models cannot adequately assess the multifactorial nature of human disease.

Predictivity and Relevance to Humans

There is little information available on the predictivity of the various animal models. While in many instances, the clinical manifestations of autoimmunity may be similar to those in humans, the complex interaction of genetics and environmental factors make it difficult to mimic many aspects of human disease. It has been shown that engineered genetic defects that result in immune dysfunction in rodent models do not necessarily mean that naturally occurring variants in those genes confer disease susceptibility in humans. In addition, there has been no systematic effort to use these models for the purposes of screening chemicals or drugs for their potential to induce autoimmunity.

Regulatory Environment

References

1. Boyton R, Altmann D (2002) Transgenic models of autoimmune disease. Clin Exp Immunol 127:4–11
2. Das P, Abraham R, David C (2000) HLA transgenic mice as models of human autoimmune diseases. Rev Immunogenet 2:105–114
3. Chernajovsky Y, Dreja H, Daly G et al. (2000) Immuno- and genetic therapy in autoimmune diseases. Genes Immunity 1:295–307
4. Lahita R (2002) Textbook of the Autoimmune Diseases. Lippincott Williams & Wilkins, Philadelphia
5. Selgrade MK, Cooper GS, Germolec DR, Heindel JJ (1999) Linking environmental agents to autoimmune diseases. Environ Health Perspect 107:659–813

Autoimmune Disorders

▶ Autoimmunity, Autoimmune Diseases

Autoimmune Heart Disease

▶ Cardiac Disease, Autoimmune

Autoimmune Models

▶ Rheumatoid Arthritis and Related Autoimmune Diseases, Animal Models

Autoimmune Reaction

Pathologic reaction of the immune system against self antigens.

▶ Antigen Presentation via MHC Class II Molecules

Autoimmunity

In general, the immune system does not mount an immune response against self-antigens. In order to achieve this, autoreactive T-cells generated in the thymus are eliminated by the so-called negative selection. Autoreactive T cells which have slipped through this process are kept in check in the periphery by lack of co-stimulation, and other mechanism. When these fail, autoimmunity can be the clinical outcome. Autoimmune diseases can be systemic (all tissues are targets of self-destruction) or tissue-specific. Some examples for autoimmune diseases are type I diabetes, systemic lupus erythemadodes, multiple sclerosis or myasthenia gravis.

▶ Autoimmunity, Autoimmune Diseases
▶ Molecular Mimicry
▶ Systemic Autoimmunity
▶ Dioxins and the Immune System
▶ Graft-Versus-Host Reaction
▶ Antinuclear Antibodies
▶ Autoantigens
▶ Autoimmune Disease, Animal Models

Autoimmunity, Autoimmune Diseases

NOEL R ROSE
Dept Pathology and Dept Molecular Microbiology and Immunology, The Johns Hopkins Medical Institutions
Bloomberg School of Public Health
615 North Wolfe Street
Baltimore, MD 21205
USA

Synonyms

Autoimmunity, autoimmune disease, autoimmune disorders.

Definitions

▶ Autoimmunity is defined as an immune response directed to antigens normally expressed in the host. This definition is not limited by any particular method of induction. An autoimmune response can be the consequence of immunization with a foreign antigen, such as an infectious microorganism or even a simple chemical; the response to an alloantigen, such as might occur following bone marrow transplantation and graft-vs-host disease; or a response to an antigen derived directly from the host itself. All possibilities are known to occur. Autoimmunity, in fact, is a common and even universal phenomenon, as demonstrated by the presence of naturally occurring ▶ autoantibodies in all normal individuals. Whether they result from an exogenous or endogenous stimulus, these natural autoantibodies are usually in the IgM class with relatively low antigen-binding affinities. They are widely cross-reactive and may represent an early general defense mechanism against invading pathogens. The presence of naturally occurring autoantibodies and their non-specific increase following inflammatory stimuli complicate research and clinical diagnosis of the autoimmune diseases ▶ autoimmune disease by blurring the distinction between harmless, natural autoimmunity and the pathogenic autoimmunity involved in autoimmune disease.

Autoimmune disease is defined as the pathologic consequence of an autoimmune response. Some autoimmune diseases are the direct result of an autoantibody. They include Graves' disease, where hyperthyroidism results from antibodies that stimulate the thyroid-stimulating hormone receptor; myasthenia gravis, where antibodies to cholinesterase block acetocholine transmission at the myoneural junction; the skin diseases, pemphigus and pemphigoid, where antibodies interfere with the essential intercellular connections and cause blistering; and autoimmune cytopenias, where autoantibodies destroy or limit the lifespan of circulating blood cells. In other autoimmune conditions, the antigen may not be normally accessible to antibody, but may be released following cell injury or death. A combination of liberated intracellular antigen and autoantibody produces immune complexes, such as those seen in systemic lupus erythematosus. The pathology is greatest in areas where the circulating immune complexes localize in venule or capillary beds, including the skin, lung, brain and, most significantly, the kidney. In many other autoimmune disorders, however, autoantibodies are present as signs or biomarkers of an autoimmune response, but their role in inducing pathology is uncertain. In these instances, autoreactive T lymphocytes are directly or indirectly responsible for pathology. Cytotoxic T lymphocytes may attack target cells directly to produce injury in the target organ or T cells may be stimulated to release cytotoxic cytokines and other mediators of injury. Macrophages and NK cells will also be attracted to areas of lymphocyte accumulation and these cells can release toxic products, such as reactive oxygen or nitrogen intermediates. It must be noted, in fact, that most of the cells accumulating at the site of autoimmune damage are not specific for the initiating cellular antigen, but rather localize non-specifically to join the inflammatory process. The pathologic outcome of an autoimmune response usually begins when T cells, during their normal migration, encounter their cognate self-antigen and then, ceasing their migration, are activated and stimulated to proliferate. Further cellular localization follows the up-regulation of adhesion molecules on the local vascular bed. Thus, in many autoimmune diseases, the ultimate organ damage and clinically detectable disease are mediated by a collection of immune and non-immune cells and their soluble products, including locally produced antibody.

Characteristics

The immunotoxic effect of xenobiotics on the immune system may manifest itself as immunodepression, hypersensitivity, or autoimmunity. Of the three phenomena, autoimmunity has been the least studied, but may represent a devastating and long-lasting result of exposure to environmental agents. A recent report by the National Institutes of Health indicates that more than 80 diseases are attributable to a self-directed immune response. Collectively, these diseases afflict an estimated 5% to 8% of the population of the United States. They impose a substantial medical challenge with respect to diagnosis and treatment as well as an enormous social and financial burden. Most of the autoimmune diseases are chronic debilitating disorders that lead to poor quality of life, high health care costs, and substantial loss of productivity. Autoimmune diseases can affect any site and impair the function of any organ of the body. They are highly varied in their clinical presentation, but some key features are common to all of them.

Regulation of Autoimmune Responses

The immune response has great potential for benefit or harm but must be carefully regulated. Although autoimmunity itself is not necessarily harmful, it must be kept within bounds, if the goal of good health is to be achieved. A first step in preventing harmful autoimmunity in the form of autoimmune disease is to eliminate the potential high-affinity, self-reactive lymphocytes. In T cells, this procedure is carried out in the thymus where negative selection follows on contact with self-antigen and self-MHC-presenting elements. The elimination of self-reactive T cells depends upon the presentation in the thymus of autologous antigens during the critical steps of negative selection. Self-antigens that are absent or poorly presented during this process are the ones most likely to allow self-reactive T cells to escape and later engender autoimmune responses in the periphery. One can actually envision an immunologic homunculus based on the effectiveness of self-antigen presentation during negative selection in the thymus. A similar process of ▶ clonal deletion occurs during the genesis of B cells in the bone marrow and, perhaps later, in or around germinal centers of lymph nodes. B cells have the additional mechanism of receptor editing, which allows them to trade a self-reactive receptor for one that does not recognize an autologous antigen.

Central clonal deletion by negative selection is usually successful in eliminating high-affinity, self-reactive T cells and B cells. The process is, however, incomplete, since not all self-antigens are adequately presented during the negative selection process. In addition, clonal deletion depends upon ▶ apoptosis of the self-reactive lymphocytes, so that genetic or induced impairment of apoptosis favors the production of an autoimmune response. Even under normal conditions, however, significant numbers of self-reactive T cells and B cells will escape the clonal deletion and be found in peripheral sites. Therefore, additional regulatory measures are required to maintain these cells in an unresponsive state and to prevent progression to autoimmune disease. Several such regulatory mechanisms have been investigated. They include ▶ anergy, a state of unresponsiveness of lymphocytes that occurs when the cell encounters its specific antigen in the absence of necessary co-stimulatory molecules. These anergic cells remain unresponsive for long periods or eventually die by apoptosis. A second mechanism of immunologic ignorance has been described as the phenomenon by which self-reactive lymphocytes ignore the autoantigen because of its inaccessibility or inappropriate presentation. Finally, self-reactive lymphocytes can be held in check by active suppression, imposed either by specialized populations of T cells, by macrophages, by NK T cells or by cytokine products of these cells. Removing such ▶ regulatory cells will sometimes lead to the spontaneous development of autoimmunity.

Genetic Predisposition

he notion of a ▶ genetic predisposition to autoimmune disease arose initially from clinical observations that particular autoimmune diseases, or autoimmune disease more generally, seemed to occur with an inordinate frequency in certain families. Considerable epidemiologic evidence shows that individuals with one autoimmune disease, such as Type 1 diabetes mellitus, have a markedly increased prevalence of another autoimmune disease, thyroiditis. It is often difficult from such observations to distinguish shared environmental risk factors from a shared genetic pool. More definitive support for a genetic predisposition to autoimmune disease has come from studies comparing the prevalence of autoimmune disease in genetically identical, monozygotic twins with the prevalence in dizygotic twins, based on the assumption that such twin pairs will share similar environmental factors. In instances where such twin studies have been carried out, the concurrence rate among monozygotic pairs has ranged from about 15% to 60% with a median value of about 30%. In contrast, the concurrence rate of dizygotic twins is generally 3% to 5%, a figure similar to that found in any other sibling pairs.

A more detailed genetic analysis of the inheritance of susceptibility to autoimmune disease has been carried out in experimental animals. The first clear association with heightened susceptibility to autoimmunity was with the class II major histocompatibility complex (MHC) genes and experimentally induced and spontaneous autoimmune thyroiditis. Since then, virtually every autoimmune disease has been shown to associate statistically with particular class II MHC alleles. Some alleles predict increased susceptibility, whereas others seem to determine protection. It seems likely that the class II MHC bias comes from differing affinities for the critical antigenic peptides involved in inducing autoimmune responses. MHC binding plays a role in the deletion of self-reactive T cells in the thymus as well as in the recognition by T cells of autologous antigen in the periphery. Even the change of a single amino acid in the class II MHC may alter binding affinity sufficiently to determine heightened or lessened susceptibility to autoimmunity. Class I MHC genes may also play an important role in autoimmune disease, since allelic differences may dictate the severity of disease itself, as has been shown in experimental autoimmune thyroiditis.

Based on many studies of the genetic susceptibility to autoimmune disease in experimental animals, it seems clear that only about half the inherited susceptibility can be attributed to MHC-related genes. The remaining half of genetic susceptibility is attributable to an incremental contribution of non-MHC genes that modulate the immune response or increase the vulnerability of the target organ. Identification of these non-MHC genes has been difficult, but is now progressing rapidly with the availability of methods for genome-wide screening. Some non-MHC-associated genes are directly involved in immunologic regulation. For example, allelic differences in CTL4 have been associated with increased susceptibility to Type 1 diabetes and autoimmune thyroid disease. Undoubtedly, many more non-MHC genes remain to be discovered. Their recognition will assist in recognizing individuals or populations with heightened genetic predisposition to autoimmune disease.

The evidence cited above suggests that about a third of the risk of developing autoimmune disease can be predicted by the genetic constitution of the host. That general statement naturally leads to the question of the remaining 70%. Part of the variability among genetically identical humans or rodents is undoubtedly related to the stochastic nature of the immune response itself. It is the mission of the immune system to generate sufficient diversity to respond to an almost limitless number of potential antigens. To do so, the immune system employs several devices that increase post-germline diversity. Such measures as B-cell receptor or T-cell receptor mini-gene recombination or hypermutation can explain a certain amount of the diversity between genetically identical twins and inbred mice. Beyond those stochastic events, however, it seems undeniable that environmental factors play an important role in the initiation of pathogenic autoimmune responses.

Relevance to Humans

The list of environmental factors implicated in precipitating autoimmune disease is lengthy, but firm evidence for their involvement is difficult to obtain. There are very few human diseases for which an environmental trigger has been definitively identified. The strongest evidence of an association in humans between xenobiotics and the development of an autoimmune disease is the tendency of certain drugs to precipitate autoimmune disease. A good example can be found in the association of procainamide with a form of systemic lupus erythematosus. Here, the association is based on the finding that the disease remits when the drug is discontinued and recurs when the drug is re-administered. Many other drug-associated autoimmune diseases have been described in the literature. A second likely environmental factor is infection. Autoantibodies often rise following an infectious process, but citing specific instances where disease is caused by autoimmunity has been difficult . Based on extensive epidemiologic evidence, the association of beta hemolytic streptococci with rheumatic heart disease is now well established and has led to the regular use of antimicrobial treatment to prevent recurrence of streptococcal infections in susceptible children. Although a great deal of circumstantial evidence suggests that childhood exposure to an infection may be important in the etiology of multiple sclerosis, no single organism has emerged as the likely culprit. Some cases of Type 1 diabetes seem to be a consequence of Coxsackievirus B4 infection, whereas myocarditis is often associated with a preceding Coxsackievirus B3 infection. EB virus profoundly affects immunoregulatory mechanisms and has often been cited as a probable infectious trigger for many autoimmune diseases, including lupus and rheumatoid arthritis.

Proving that a particular microorganism is the cause of a particular disease has turned out to be a difficult task, however, and usually requires that the disease be reproduced in an experimental animal in which appropriate adoptive transfer studies can be performed. In very few instances is the actual antigen responsible for the autoimmune disease identified and it is uncertain whether the antigen is derived from the microorganism representing molecular mimicry or is liberated as a consequence of the infection from the cells of the host itself.

Dietary factors have long been incriminated in the development of autoimmune disease. Another pertinent example is the epidemiologic and clinical evidence that increased iodine in the diet is associated with a rise in autoimmune thyroid disease. Celiac disease, an inflammatory bowel disease, is due to ingestion of glutein from wheat or oats. The disease can be controlled by eliminating glutein from the diet. Mercury, particularly organic mercury, has been shown to induce autoimmune disease in animals and is another likely dietary candidate. Environmental pollutants represent an important potential source of environmental triggers. Some studies have focused on the possible role of halogenated hydrocarbons as a contributing factor to lupus or other systemic autoimmune diseases. Epidemiologic evidence has associated silica with scleroderma and lupus. Finally, the fact that most autoimmune diseases are more prevalent in females suggests that endocrine disruptors from the environment may be important factors in precipitating autoimmune disease. The identification of environmental factors will be an important clue to developing strategies for preventing autoimmune disease in genetically susceptible individuals. It is likely that research on this issue will accelerate in the coming years.

References

1. Rose NR, Mackay IR (eds) (1998) The Autoimmune Diseases, 3rd ed. Academic Press, San Diego
2. Lahita RG, Chiorazzi N, Reeves W (eds) (2000) Textbook of the Autoimmune Diseases. Lippincott Williams & Wilkins, Philadelphia
3. Theofilopoulos AN, Bona CA (eds) (2002) The Molecular Pathology of Autoimmune Diseases, 2nd ed. Taylor & Francis, London pp 951–964

Autologous

This expression denotes a process in which organs, cells or molecules are removed and given back to the same individual.

▶ Idiotype Network
▶ Graft-Versus-Host Reaction

Autoreactive Cells

Cells that can react with autoantigens, sometimes leading to an autoimmune reaction.
▶ Flow Cytometry

Avidity

The number of epitope binding sites on an immunoglobulin molecule. IgM has an avidity of 10, since it has a pentameter configuration with two binding sites per subunit. IgG has an avidity of 2, two binding sites per each IgG molecule.
▶ Plaque Versus ELISA Assays. Evaluation of Humoral Immune Responses to T-Dependent Antigens

Azathioprine

A prodrug which is in vivo transformed into its active form, a purine analogue. It inhibits purine ring synthesis and nucleotide conversions, resulting in inhibition of lymphocyte proliferation.
▶ Cyclosporin A

Azo Dyes

Industrially used diazonium salts, derived from anilin and its derivatives by reaction with nitrous acid. They covalently modify aromatic amino acids in proteins, resulting in structure-dependent colors.
▶ Hapten and Carrier

AZT

Zidovudine (3'-azido-3'-deoxythymidine) antiviral drug used in treatment of acquired immunodefiency syndrome (AIDS).
▶ Plaque Versus ELISA Assays. Evaluation of Humoral Immune Responses to T-Dependent Antigens

B Cell

▶ B Lymphocytes

B Cell Antigen Receptor (BCR)

The B cell receptor for antigen (BCR) consists of membrane-bound antibodies of different classes.
▶ Idiotype Network

B Cell Maturation and Immunological Memory

Claudia Berek
Deutsches Rheuma Forschungszentrum
Schumannstr. 21–22
D-10117 Berlin
Germany

Synonyms

B lymphocyte, antigen dependent B cell development, affinity maturation of the immune response, germinal center reaction, B memory cell, B cell receptor modification.

Definition

Efficient protection from pathogens and toxins requires high affinity antibodies. These antibodies are generated by a process referred to as affinity maturation which takes place in the ▶ germinal centers. Within the micro-environment of the germinal center antigen activated B cells differentiate into plasma and memory cells. Plasma cells secrete large quantities of high affinity antibodies whereas memory B cells give long-term protection. On subsequent contact with the same antigen, memory cells can rapidly differentiate into plasma cells secreting high affinity protective antibodies.

Characteristics

B Cell Maturation

B cells are generated in the bone marrow from haematopoietic stem cells. By stepwise ▶ rearrangement (Fig. 1) of the ▶ Ig V-region genes (V, (D), J) these cells become competent to express a B cell receptor (BCR). The naive B cells express a broad repertoire of antigen-specific receptors, but for a given antigen there are only a few B cells available. In addition, the over all affinity of these B cells will be low.

The process of B cell affinity maturation is antigen dependent. It takes place in the peripheral lymphoid organs where the naive B cell is activated by antigen binding to its receptor. In order to induce a germinal center reaction the B cell needs help from T lymphocytes (T cell-dependent immune response) and if this is available then clonal expansion of the antigen specific B cell and differentiation into plasma and memory cells occurs (1).

Germinal Center Reaction

In order to induce a germinal center reaction, antigen-activated T and B cells—as well as antigen-presenting dendritic cells—have to come together in close vicinity. When this happens a micro-environment is set up which permits rapid proliferation of B cells so that within a few days a single antigen-activated B cell can form a large clone consisting of several thousand cells.

It takes about 1 week for the mature germinal center structure to be formed (2). Such a germinal center is composed of a dark and a light zone. In the dark zone B cells proliferate, whereas in the light zone they differentiate into plasma and memory cells (Fig. 2).

During proliferation in the germinal center a process of hypermutation is activated (1). This highly specific mechanism introduces single nucleotide exchanges into the genes coding for the V region of the ▶ H-chain and L-chain of the B cell receptor. In this way B cells with variant receptors are generated and these variants may have different affinities for the antigen. Only those B cells with high affinity receptors are selected to differentiate into plasma and memory B cells.

This selection process takes place in the light zone of the germinal center where the B cells are embedded in

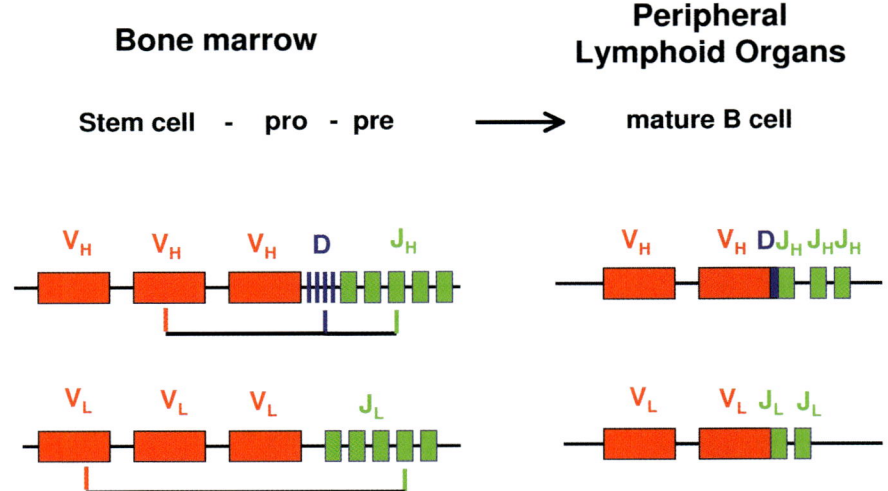

B Cell Maturation and Immunological Memory. Figure 1 Rearrangement.

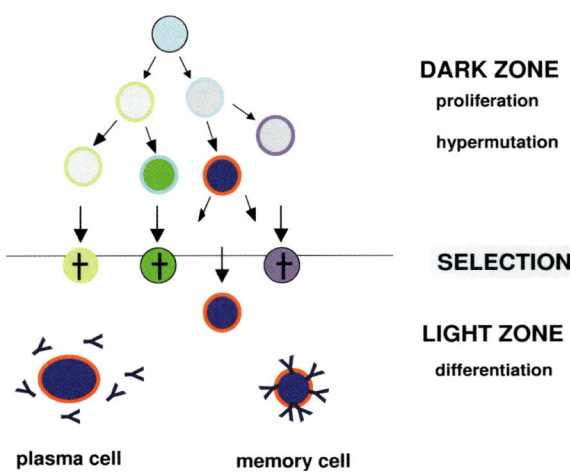

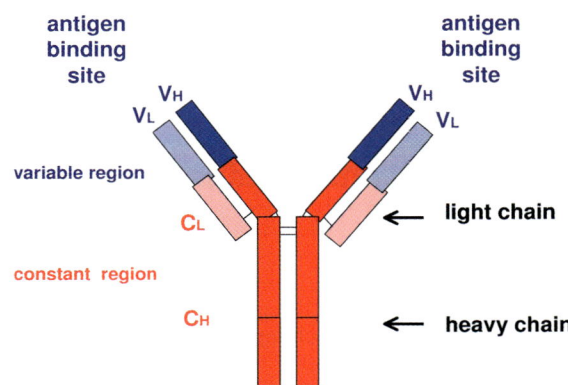

B Cell Maturation and Immunological Memory. Figure 3 Divalent antibody molecule.

B Cell Maturation and Immunological Memory. Figure 2 During B cell proliferation in the dark zone hypermutation is activated and variants of different affinity are generated (indicated by circles of different color). Only a few cells express high affinity receptors (blue/red circle) and only those are preferentially expanded. Antigen presented on FDC selects the few B cells with high affinity receptors to differentiate into memory and plasma cells. Low affinity variants die through apoptosis (+).

a network of follicular dendritic cells (1). These are specialised cells which do not process antigen, but rather present intact antigen in form of antigen-antibody complexes bound to their complement and Fc receptors. Here, the B cell variants compete for binding to the antigen. Those cells which—due to low affinity receptors—fail to interact with antigen die by apoptosis while those with high affinity receptors survive and differentiate. Thus the interplay of hypermutation and selection ensures that only B cells with high affinity receptors develop into plasma and memory cells.

Ig Class Switch and Hypermutation

The V region of the B cell receptor defines its antigen specificity. However, functions such as complement binding, secretion, opsonization depend on the constant region of the H-chain which defines the so-called Ig class of the B cell receptor (see Fig. 3). During B cell differentiation an Ig class switch may take place, so that the same V region gene may be combined with a different constant region gene. The naive B cell expresses Ig receptors of the IgM and the IgD class. A switch to IgG may take place in antigen-activated B cells during the germinal center reaction and it requires the activity of the enzyme AID (activated cytidine deaminase) which is also essential for hypermutation (3). Nevertheless, these two processes of B cell receptor modification are independently controlled. Somatic mutations are found in V region genes of memory B cells, independent of expression of IgM or IgG.

Immunological Memory

The process of affinity maturation is surprisingly efficient. However it is dependent on clonal expansion and clonal selection, both of which are time consuming processes. For immediate protection, for example against bacterial toxins, one needs a pool of memory B cells which have developed in a germinal center reaction and hence already have high affinity receptors for the antigen. This is the reason for vaccination. In contrast to the naive B cell, the memory cell responds immediately to an antigenic stimulus and rapidly differentiates into plasma cells secreting high affinity antibodies.

Plasma cells generated in a germinal center reaction have the capacity to migrate back into the bone marrow where they may survive for many months. In this way the level of antigen-specific antibodies stays elevated for a long time after antigenic contact and these antibodies provide protection from further infections. However, even long-lived plasma cells need finally to be replenished.

Little is known about the mechanisms controlling the long-term maintenance of immunological memory. It is still not clear whether it is the memory cell itself which is long-lived or whether it is the memory B cell clone that survives over the years. Low doses of antigen stored in the network of follicular dendritic cells may be sufficient to continuously stimulate memory B cell proliferation. There is also evidence that signals provided by the innate immune system (▶ innate immunity) might be involved in the ▶ homeostatic control (4). Since memory B cells—in striking contrast to naive B cells—express ▶ Toll-like receptors they may respond to non-specific environmental stimuli. Since the size of the memory B cell pool is limited, there must be strong competition for space and indeed the proliferative activity of these cells in the absence of antigen is low.

Preclinical Relevance

The germinal center reaction is a useful marker for the antigen-dependent activation of the immune system. In preclinical studies immunohistology is used to asses whether reagents have stimulating or immune suppressive activity. Specific staining of tissue section allows the visualisation of germinal centre formation. Germinal center evaluation belongs to the key parts of histopathological examinations in immunotoxicologic screening studies.

Relevance to Humans

The importance of B cell memory and affinity maturation is clearly shown in patients with immune deficiencies (5), such as in X-linked hyper-IgM syndrome. A genetic analysis revealed that these patients have a defect in the gene for the ▶ CD40 ligand, which prevents efficient communication between antigen-activated T and B cells. Without adequate T cell help no germinal center will develop and no memory B cells will be generated. In these patients B cells do differentiate into plasma cells, but there is no Ig class switch and the secreted antibodies are IgMs of low affinity. As a result patients suffer from recurrent infections from early in life.

References

1. MacLennan ICM (1994) Germinal centers. Annual Review in Immunology 12:117–139
2. Camacho SA, Kosco-Vilbois MH, Berek C (1998) The dynamic structure of the germinal center. Immunol Today 19:511–514
3. Muramatsu M, Kinoshita K, Faragasan F, Shinkai K, Yamada Y, Honjo T (2000) Class switch recombination and somatic hypermutation require activation-induced cytidine deaminase (AID), a member of RNA editing cytidine deaminase family. Cell 102:553–563
4. Bernasconi NL, Traggiai E, Lanzavecchia A (2002) Maintenance of serological memory by polyclonal activation of human memory B cells. Science 298:2199–2202
5. Fischer A (2002) Primary immunodeficiency diseases: natural mutant models for the study of the immune system. Eur J Immunol 32:1519–1523

B Cell Receptor Complex

The B cell receptor complex (BCR complex) is expressed on the surface of a B cell and is composed of six chains. There are two immunoglobulin (Ig) heavy chains and two Ig light chains that form the portion of the complex that is involved in antigen specificity. Associated with heavy and light immunoglobulin chains are two other chains, Igα and Igβ. These chains have long cytoplasmic regions and are involved in the signaling process.

▶ Signal Transduction During Lymphocyte Activation

B Cell Receptor Modification

▶ B Cell Maturation and Immunological Memory

B Lymphocytes

Norbert E Kaminski · Courtney EW Sulentic
Department of Pharmacology and Toxicology
Michigan State University
B440 Life Science Building
East Lansing, MI 48824
USA

Synonyms
B cell, plasma cell, antibody-forming cell

Definition
B lymphocytes (1–2) are a subset of white blood cells, also termed leukocytes, which are characterized by their ability to synthesize and secrete antibodies. B lymphocytes also express antibody molecules on their surface, which serve as antigen receptors allowing for the recognition of non-self. When terminally differentiated into antibody producing cells, B lymphocytes are termed plasma cells or antibody-forming cells. The primary function of B cells is to detect and tag foreign or non-self molecules (▶ antigens) through the secretion of antibodies that will specifically bind to the foreign antigen, allowing removal by other cells of the immune system, or activation of the complement cascade.

Characteristics
B lymphocytes develop from the pluripotent hematopoietic stem cells located in the bone marrow in mice, humans, and other mammals, or in the bursa of fabricius in birds, hence the denotation 'B'. Stem cell commitment to the lymphoid lineage, either B or T cell, is controlled by the *Ikaros* gene. The *Ikaros* gene codes for a DNA-binding protein which, in combination with other transcription factors, regulates the expression of genes that yield the phenotypic characteristics of lymphocytes including the immunoglobulin heavy chain, the light chain, the essential components of gene rearrangement machinery and members of the CD3 complex of the T cell antigen receptor. Once the stem cells have committed to the lymphoid lineage, the specific mechanism responsible for commitment to B lymphocytes is poorly understood but is thought to involve a progressive loss in the ability to differentiate into other lineages. However, B cell commitment has an essential requirement for an interaction with stromal cells. Stromal cells signal the progenitor cells to continue the B cell development program through cell-to-cell interactions that are mediated by adhesion proteins and through the release of soluble growth factors. Once stem cell commitment to the B cell lineage has been made, the B cell undergoes four stages of development. The first stage of development, termed the pro-B cell stage, occurs in the bone marrow and is identified by the expression of cell surface proteins associated with early B cell development and the capability of ▶ *VDJ* gene rearrangement and joining. The second stage is the pre-B cell stage, which is marked by the expression of small amounts of μ (immunoglobulin (Ig)M heavy chain) on the cell surface. It is notable that the expressed μ also possesses a surrogate light chain making up the pre-B cell receptor. In the third stage of B cell development, light-chain gene assembly takes place, which can be either the λ or κ form, resulting in the expression of surface IgM and is the hallmark of the immature B cell. B cell development up to the immature B cell stage takes place in the bone marrow. Immature B cells then undergo positive and negative selection in order to eliminate those cells that react to self antigens. B cells that survive the selection process and are present in the periphery are termed mature B cells. These mature B cells express IgM and IgD on their surface and are present in the circulation and in secondary lymphoid organs such as the spleen and lymph nodes. A single mature B cell can recognize a specific antigen through its B cell receptor (BCR) which is a membrane-bound Ig molecule. Binding of the antigen to the BCR with the help of secondary cellular and soluble mediators can result in activation of the B cell and proliferation or ▶ clonal expansion. Following clonal expansion, the B cells, all with specificity for the activating antigen, differentiate into antibody-forming cells (AFC; also termed plasma cells), or into ▶ memory cells. AFCs secrete antibodies, a soluble form of Ig, which coat a specific antigen and facilitate antigen clearance or activation of the complement cascade. The basic Ig molecule is composed of two identical heavy chains and two identical light chains, and has two antigen binding sites that are held together by disulfide bonds. There are two classes of light chains, κ and λ, adding to the diversity of the antigenic repertoire, and five heavy chain classes, IgM, IgG, IgA, IgE and IgD, that are encoded, respectively, by μ, γ, α, ε, and δ heavy chain genes. Each class of Ig appears to have unique biological properties. In addition, certain classes of antibodies can be joined together by additional disulfide bonds and by a polypeptide chain termed the J chain, which is also synthesized in B cells. For example, IgM is pentameric or hexameric resulting in high antigen valence (10 or 12 antigen-binding sites). IgM also has relatively low affinity for antigen, and is the major Ig involved in a ▶ primary antibody response. IgG is monomeric, has high affinity for antigen, can cross the placental barrier and is the hallmark of a ▶ secondary antibody response. IgE is also monomeric and is involved in allergic and an-

tiparasitic responses. IgA is monomeric or dimeric, is very efficient at bacterial lysis and is the main secretory antibody (found in saliva, mucus, sweat, gastric fluid, and tears). IgD is monomeric, is a major surface component on many B cells and has unknown biological properties. Regulation of Ig expression and isotype class switching is governed through a complex interaction of several regulatory elements whose activity is B cell specific and dependent on the state of B cell maturation. The most 5' regulatory element in the Ig heavy chain gene is the variable heavy chain (V_H) ▶ promoter, which lies immediately upstream of each variable region and contributes to B cell-specific activity of the Ig heavy chain. Located between the rearranged ▶ VDJ segments and the μ constant region is the intronic ▶ enhancer (Eμ) which contributes to B cell-specific activity and is involved early in B cell development where it regulates V to D-J joining and μ heavy-chain gene expression. However, processes late in B cell differentiation, such as upregulation of heavy chain expression and secretion, as well as class switching, occur normally in the absence of Eμ and appear to be regulated by another regulatory element(s) located ▶ 3' of the α constant region. Within this region, four separate enhancer domains, hs3A, hs1,2, hs3B and hs4 were identified and are collectively termed the 3'α enhancer. Activity of these enhancer domains is dependent on the developmental stage of the B cell with hs3A, hs1,2 and hs3B primarily active in activated B cells or plasma cells. In contrast, hs4 is active from a pre-B cell to the plasma cell stage. Expression of the light chain and J chain is also regulated through the activation of a 5' promoter and a 3' enhancer. Appropriate modulation of the above regulatory elements during B cell activation and differentiation results in Ig expression or class switch. However, exposure to therapeutic drugs, environmental compounds, or industrial chemicals may alter the activity of these regulatory elements, perhaps resulting in the suppression or enhancement of Ig expression and secretion.

Preclinical Relevance

Due to the similarity between the mouse and human immune systems and the availability of biological reagents, mouse models have primarily been utilized to understand the basic mechanisms of B cell function as well as the impact of potential therapeutic and toxic compounds on these mechanisms. Preclinical studies provide the necessary information to evaluate the potential hazard to humans from occupational, inadvertent, or therapeutic exposure to drugs, environmental compounds, or industrial chemicals.

Relevance to Humans

The B cell is a vital component of the immune system. Maintaining immunocompetence is essential to human survival and well-being. Therefore, immunotoxicologic studies are essential in identifying potential modifiers of B cell function which might be inhibited, leading to increased infections, or enhanced, leading to autoimmune and/or hypersensitivity reactions. Either situation could lead to morbidity and even mortality.

Regulatory Environment

Leukocyte phenotyping by flow cytometric analysis is routinely conducted in immunotoxicology testing to determine whether exposure to an agent produces a change in the number of cells in leukocyte-specific subpopulations. Changes in the number of B lymphocytes in circulation or in secondary lymphoid organs can be quantified by flow cytometry using fluorochrome-conjugated antibodies directed against B lymphocyte-specific surface peptides such as Ig. Similarly, stages of B cell differentiation can be monitored by measurements of cell surface expression of differentiation markers. Specifically, mature B cells express MHC class II, which is downregulated in plasma cells. Conversely, B cells that are differentiating into plasma cells upregulate syndecan.

References

1. Hardy RR (2003) B lymphocyte development and biology. In: Paul WE (ed) Fundamental Immunology, 5th ed. Lippincott, Williams & Wilkins, Philadelphia, pp 159–194
2. Goodnow CC, Rajewsky K, Alt F, Cooper M et al. (1999) The development of B lymphocytes. In: Janeway CA, Travers P, Walport M, Capra JD (eds) Immunobiology. The immune system in health and disease, 4th ed. Garland Publishing, New York, pp 195–226

B Memory Cell

▶ B Cell Maturation and Immunological Memory

B7.1 and B7.2

These are important co-stimulatory molecules on antigen-presenting cells. Their receptors on T lymphocytes are the CD28 molecule and CTL-4 (present only on activated cells). They co-stimulate T cell proliferation and cytokine secretion. Also known as CD80 (B7.1) and CD81 (B7.2).

▶ Interferon-γ

Bacteremia

The presence of bacteria in the blood.
▶ Streptococcus Infection and Immunity

Bacteremic Shock

▶ Septic Shock

Bactericidal

An agent or host defense mechanism capable of causing the death of bacteria.
▶ Respiratory Infections

B(a)P

Benzo(a)pyrene is the prototype of the polycyclic aromatic hydrocarbon class of compounds. It is an immunosuppressive compound which predominately effects humoral immunity.
▶ Plaque Versus ELISA Assays. Evaluation of Humoral Immune Responses to T-Dependent Antigens
▶ Polycyclic Aromatic Hydrocarbons (PAHs) and the Immune System

Bcl-2 Interacting Domain (Bid)

Bcl-2 interacting domain (Bid) is a novel member of the Bcl-2 family of proteins and is critical to the regulation of apoptosis induced by many stimuli including TNF-α. Bid belongs to the BH3-only family of pro-apoptotic regulators and can mediate apoptosis through two interacting pathways. Activation of caspase-8 at the death-inducing complex results in Bid cleavage and release of the truncated form tBid. tBid translocates to the mitochondrial membrane, where it facilitates the release of apoptogenic proteins like cytochrome C. cJun N-terminal kinase can also cleave Bid producing jBid. jBid binds and sequesters inhibitor of apoptosis proteins leading to increased activation of caspases and increased tBid formation and apoptosis.
▶ Tumor Necrosis Factor-α

Beige Mouse

A mouse that is deficient in NK cells and other cellular immune functions.
▶ Animal Models of Immunodeficiency

Benzo-e-pyrene

▶ Polycyclic Aromatic Hydrocarbons (PAHs) and the Immune System

Berylliosis

▶ Chronic Beryllium Disease

Beryllium Disease

▶ Chronic Beryllium Disease

Beryllium-Stimulated [or Beryllium-Specific Peripheral Blood?] Lymphocyte Proliferation Test (BeLPT)

This is an in vitro assay in which beryllium-stimulated cell proliferation is measured by tritiated thymidine incorporation. The test is performed on peripheral blood mononuclear cells to support a diagnosis of beryllium sensitization and on bronchoalveolar lavage cells to confirm chronic beryllium disease.
▶ Chronic Beryllium Disease

bg/nu/xid Mouse

A mouse with deficient function of NK cells, lymphokine-activated killer cells, and T and B cells.
▶ Animal Models of Immunodeficiency

Bioaerosols

Biological substances that are dispersed in air in the form of a fine mist intended for inhalation.

▶ Respiratory Infections

Biologic-Response Modifiers

▶ Immunotoxicology of Biotechnology-Derived Pharmaceuticals

Biologics

▶ Immunotoxicology of Biotechnology-Derived Pharmaceuticals

Biotherapeutics

▶ Immunotoxicology of Biotechnology-Derived Pharmaceuticals

Biotransformation

A process that converts liphophilic chemicals to water-soluble metabolites in general. The physical properties of the xenobiotics are generally changed from those favoring absorption to those favoring excretion. Sometimes, however, more reactive metabolites are produced by the biotransformation to cause toxicity.

▶ Metabolism, Role in Immunotoxicity

Biphenotypic Leukemia

▶ Leukemia

Birth Defects, Immune Protection Against

STEVEN HOLLADAY
Dept of Biomedical Sciences & Pathobiology
Virginia Tech
Southgate Drive
Blacksburg, A 24061-04422
USA

Synonyms
Immunoteratology (although this term may also be used for structural or functional defects of immune-related etiology)

Definition
Immune protection against birth defects refers to the ability of immune stimulation in mice to reduce the occurrence or severity of birth defects caused by diverse teratogenic exposures.

Characteristics
Female mice are exposed to any of a variety of agents that cause non-specific activation of the immune system, during or shortly before pregnancy. After the immune stimulation procedure, mice are also exposed to a ▶ teratogen. The immune-stimulated mice display reduced numbers of fetuses with birth defects, as compared to control mice that experience identical teratogen exposure, but without the immune stimulation. Immune stimulation procedures that have been used to cause reduced birth defects are diverse, and include intraperitoneal injection of ▶ attenuated bacilli or inert particles (pyran copolymer); intravascular, intrauterine or intraperitoneal injection of ▶ cytokines (i.e. interferon-γ or granulocyte macrophage-colony stimulating factor); footpad injection with ▶ Freund's complete adjuvant; and intrauterine or intravascular injection with splenocytes collected from rats. These immune stimulation procedures reduced several different birth defects caused by a variety of teratogens that included chemical agents, hyperthermia, x-rays, or metabolic disturbances (see Table 1).

The immune stimulation procedures in mice all cause increased production and release of cytokines. Some of these cytokines, including GM-CSF and transforming growth factor (TGF)-β cross the placenta where they may affect cellular proliferation, differentiation, or ▶ apoptosis to reduce birth defects (2). Presumably, these cytokines would cause these actions in the fetus by altering gene expression in target tissues of the teratogens. In this regard, immune stimulation in ethyl carbamate-exposed pregnant mice reduced fetal incidence of cleft palate and reversed affects of the teratogen on fetal palate genes that control cell cycle and cell death (3).

For unknown reasons, non-specific stimulation of the immune system in pregnant mice has a broad spectrum of efficacy for reducing birth defects. Immune stimulation procedures that are effective include footpad injection with Freund's complete adjuvant, intraperitoneal injection with inert particles (pyran) or attenuated bacillus Calmette-Guérin (BCG), intravascular, intrauterine, or intraperitoneal injection with the cytokines GM-CSF or IFN-γ, or intrauterine or intravascular injection with rat splenocytes. Birth defects that have been reduced include cleft palate, ▶ neural tube defects, digit defects, tail defects and craniofacial de-

Birth Defects, Immune Protection Against. Table 1 Immune protection against teratogenesis

Immune stimulant	Birth defect	Litter affected (%) with stimulation	Litter affected (%) without stimulation	Teratogen
Pyran	cleft palate	86	67	TCDD
	cleft palate + digit defects	25	6	ethyl carbamate
	cleft palate + digit defects	35	20	methyl nitrosourea
	digit defects	22	7	methyl nitrosourea
	digit defects tail defects	55	28	x-rays
Splenocytes (rat)	craniofacial + limb defects	81	49	cyclophosphamide
	exencephaly	28	13	hyperthermia
Granulocyte macrophage-colony stimulating factor (GM-CSF)	craniofacial + limb defects	78	50	cyclophosphamide
	neural tube defects	9	2	diabetes mellitus
Interferon-γ	cleft palate	70	48	ethyl carbamate
	neural tube defects	51	14	diabetes mellitus
Freund's adjuvant	cleft palate	70	26	ethyl carbamate
	neural tube defects	51	23	diabetes mellitus
	neural tube defects	53	0	valproic acid
BCG (bacillus Calmette–Guérin)	digit defects	19	0	ethyl carbamate

TCDD: 2,3,7,8-tetrachlorodibenzo-*p*-dioxin.
All data shown represent significant decreases in birth defects, $P \geq 0.05$.
Modified from Holladay et al. (1).

fects. Inducing factors for these defects include chemical teratogens, x-rays, hyperthermia, and diabetes mellitus.

Preclinical Relevance

Protection against birth defects as a result of maternal immune stimulation is a recently demonstrated phenomenon. Such protection has been demonstrated—and, for that matter, investigated—only in the mouse. The possibility that similar effects may occur in non-mouse rodent species, non-rodent species, or humans, remains uninvestigated.

Relevance to Humans

Women who work with certain chemicals during pregnancy are significantly more likely to deliver children with congenital malformations. Pesticide exposure in pregnant women working in agriculture-related occupations has been associated with ▶ orofacial clefts. More than 100 case reports link human birth defects with maternal exposure to toluene or trichloroethylene during pregnancy (4). Both hyperthermia and the antiepileptic drug valproic acid also increase risk of neural tube defects in humans. The incidence of malformed newborns in women with insulin-dependent diabetes mellitus is 6%–10%, approximately five times higher than among non-diabetic women (5). Relatively minor manipulations of maternal dietary conditions (supplementation with vitamins, retinoic acid, or nicotinamide) can reduce spontaneous or induced malformations in experimental animals. More recently, it has been demonstrated that folic acid supplementation during the periconception period reduces neural tube defects in both rodents and humans. The mouse has generally been a reliable predictor of immune responses in humans. Rodent data showing highly significant reduction in birth defects as a result of immune stimulation suggest the possibility of an immune-mediated beneficial effect on development in humans.

Regulatory Environment

At present no guidelines exist regulating maternal immune stimulation procedures that have been used in mice to reduce birth defects. However, it must be considered that immune stimulation in pregnant women may induce or exacerbate pathologic immune responses in genetically predisposed women, including autoimmune diseases ▶ autoimmune disease. Also, increased levels of some cytokines, such as IFN-γ, during early pregnancy may increase risk of pregnancy loss.

References

1. Holladay SD, Sharova LV, Punareewattana K et al. (2002) Maternal immune stimulation in mice decreases fetal malformations caused by teratogens. Internat Immunopharmacol 2:325–332
2. Sharova LV, Gogal RM Jr, Sharov AA, Crisman MV, Holladay SD (2002) Immune stimulation in urethane-exposed pregnant mice causes increased expression of genes for cytokines, including TGFß and GM-CSF, that have previously been suggested as possible mediators of reduced birth defects. Internat Immunopharmacol 2:1477–1489
3. Sharova LV, Sura P, Smith BJ et al. (2000) Non-specific stimulation of the maternal immune system. II. Effects on fetal gene expression. Teratology 62:420–428
4. Jones HE, Balster RL (1998) Inhalant use in pregnancy. Obs Gynecol Clin N Amer 25:153–167
5. Reece EA, Homko CJ, Wu YK (1996) Multifactorial basis of the syndrome of diabetic embryopathy. Teratology 54:171–183

Blastogenesis

Conversion of small lymphocytes into larger cells that are capable of undergoing mitosis.
▶ Mitogen-Stimulated Lymphocyte Response

Blood Cell Formation

▶ Bone Marrow and Hematopoiesis

Blood Clotting

▶ Blood Coagulation

Blood Coagulation

Klaus T Preissner
Biochemisches Institut
Universitätsklinikum der Justus-Liebig-Universität Gießen
Friedrichstrasse 24
D-35392 Giessen
Germany

Synonyms

Hemostasis, blood coagulation and fibrinolysis, blood clotting.

Definition

Upon vascular injury, the dynamic hemostasis system engages platelets and cell-derived ▶ microparticles, the stationary vessel wall, as well as humoral and cell-associated factors of blood coagulation and fibrinolysis to ensure a proper wound-healing response under the conditions of continuous and variable blood flow. These spatiotemporally regulated reactions prevent life-threatening bleeding and initiate wound healing and tissue repair mechanisms.

Characteristics
Definition of Components Involved

As the innermost monolayer of cells, the endothelium covers all blood vessels. Disturbance of its integrity initiates platelet adhesion and aggregation and the onset of blood clotting. Endothelial cells are actively and dynamically integrated in the control and regulation of hemostasis, since they express, bind and endocytose several of the factors involved.

Blood ▶ platelets are the smallest circulating cellular corpuscles (derived from megakaryocytes in the bone marrow) and serve to provisionally seal the wound in the initial phase of hemostasis. Platelets are devoid of a nucleus and are rich in different storage granules, which contain adhesive proteins, growth factors and low molecular weight agonists—indispensable for the vascular repair process.

Humoral factors, including coagulation and fibrinolytic proteins/proenzymes, circulate in their inactive form, and interactions with newly exposed surfaces at the site of vascular injury lead to their activation and accumulation into temporary ▶ multicomponent enzyme complexes. These are the backbone elements of the dynamic hemostasis system.

Initiation, Amplification, and Propagation of Blood Clotting

Following vascular injury or endothelial cell denudation, the exposed collagenous subendothelial extracel-

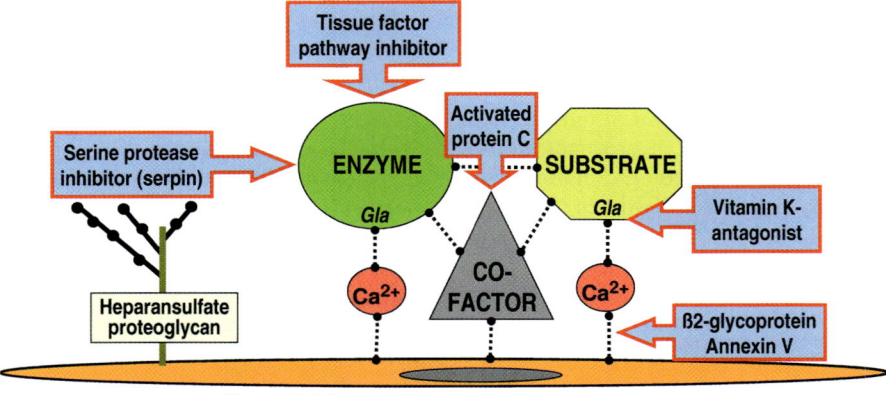

Blood Coagulation. Figure 1 Formation of multicomponent enzyme complexes on activated platelet membrane and the multiple control mechanisms provided by natural anticoagulants.

lular matrix containing von Willebrand factor and other adhesive proteins serves as a homing area for adhering platelets under conditions of varying blood flow. Deficiency in von Willebrand factor or its cognate receptor GPIb complex on platelets results in impaired platelet adherence at this stage and is associated with critical bleeding tendency. Upon platelet activation and fibrinogen-mediated platelet aggregation, the negatively charged phospholipid membrane areas of platelets become exposed. These serve as new recognition sites for circulating blood clotting factors. Together with the exposure of ▶ tissue factor (constitutively expressed in deeper cell layers of the vessel wall) towards plasma components in this initial phase, the assembly of surface-bound multicomponent enzyme complexes leads to initiation and propagation of the blood clotting cascade, culminating in the generation of initial amounts of ▶ thrombin.

At this stage, thrombin further amplifies the hemostasis system by enhancing platelet activation/aggregation, and by elevating further thrombin formation through activation of protein cofactors V and VIII, as well as by inducing activation of factor XI. These amplification reactions eventually lead to the generation of sufficient thrombin to induce fibrin formation in association with the temporary platelet plug.

Finally, stabilization of the fibrin clot by covalent cross-linking mediated by thrombin-activated factor XIII (transglutaminase) protects the wound site against unwanted bleeding, invasion of microbes or inflammatory reactions. During wound closure, previously secreted platelet components, such as growth factors and cytokines, promote proliferation and migration of vessel wall cells necessary for proper wound repair to regain a patent vessel wall.

Intrinsic Control of Blood Clotting

At the onset of blood clotting, both circulating and cell-associated tissue factor pathway inhibitor (TFPI) provide a stoichiometric threshold for tissue factor-dependent reactions. This is because TFPI reacts with both factors VIIa and Xa in order to prevent the initiation of blood clotting.

Protein cofactors (including tissue factor, factor V, factor VIII, and thrombomodulin) on different levels of the coagulation cascade are essential for triggering the enzymatic efficiency of each multicomponent enzyme complex in a spatiotemporal manner (Table 1). Specifically, diffusable thrombin loses its procoagulant activity by binding to the endothelial cell receptor thrombomodulin, and together with receptor-bound protein C this proenzyme becomes efficiently activated. Subsequently, activated protein C (APC) together with its cofactor protein S inactivates the procoagulant cofactors Va and VIIIa, thereby blocking further thrombin generation. In parallel, diffusable thrombin, and other serine proteases of the clotting cascade, are complexed and inactivated by circulating ▶ serine protease inhibitors (such as antithrombin) which thereby serves a low but progressive inhibitory control to prevent systemic thrombin action.

Fibrinolysis: Initiation, Amplification and Control

After the major events of wound sealing have occurred, the produced thrombus has to be removed by plasmin degradation in a controlled manner in order to regain the appropriate blood flow conditions of the patent vessel and to complete tissue regeneration. As soon as a fibrin clot surface is established, circulating fibrinolytic factors with affinity for fibrin, such as tissue plasminogen activator (t-PA), and plasminogen bind to the fibrin clot, and plasmin generation is induced. In order to ensure stabilization of the fibrin clot

Blood Coagulation. Table 1 Multicomponent enzyme complexes in hemostasis

Function (complex)	Enzyme	Substrate	Cofactor
Factor IXa generation (intrinsic)	XIa	IX	Kininogen
Factor IXa generation (extrinsic)	VIIa	IX	Tissue factor
Factor Xa generation (intrinsic tenase)	IXa	X	VIIIa
Factor Xa generation (extrinsic tenase)	VIIa	X	Tissue factor
Thrombin generation (prothrombinase)	Xa	Prothrombin	Va
Protein C activation	Thrombin	Protein C	Thrombomodulin
Factor Va/VIIIa inactivation	Protein Ca	Va/VIIIa	Protein S
Plasmin generation	t-PA	Plasminogen	Fibrin

and to prevent too early an onset of fibrinolysis, thrombin in complex with thrombomodulin activates a circulating procarboxypeptidase B known as TAFI (thrombin-activated fibrinolysis inhibitor). This regulates t-PA and plasminogen binding to the fibrin clot. Since fibrin itself serves as a promoting cofactor for t-PA-mediated plasmin formation, its subsequent degradation serves to limit fibrinolysis. Furthermore t-PA, in addition to plasmin, is controlled by serine protease inhibitors PAI-1 (plasminogen activator inhibitor-1) and α2-antiplasmin in order to prevent bleeding.

Preclinical Relevance

Hemostasis and Cell Functions

In addition to their "classical" functions, most of the cofactor proteins and enzymes of the hemostasis system exhibit activities that are related to cell proliferation, migration or differentiation. These are all mediated by unrelated receptors on a variety of cells in the body. For example, thrombin constitutes a potent mitogen for vascular smooth muscle cells and has been implicated in the pathogenesis of atherosclerosis. Although the entire functional repertoire of hemostatic factors in this regard has not been uncovered yet, essential functional links are apparent between hemostasis and angiogenesis, inflammation, vessel degeneration, tumor progression or neurological processes.

Mouse Models and Hemostasis

The genetic manipulation of mice resulting either in the overexpression of a particular gene for a hemostatic protein or its complete or partial knock-out lead to a variety of important insights into the biology of hemostatic factors and their receptors during embryonic development or during the challenge with pathologies in the adult phase. Here, almost any knock-out of a clotting factor or its respective receptor resulted in an embryonically lethal phenotype or the death of the affected mice perinatally. Based on these discoveries on the role of hemostasis "beyond" blood clotting, new therapeutic regimen for various vascular pathologies may become available in the future.

Relevance to Humans

Based on our understanding of the activation, amplification, progression and control of blood coagulation and fibrinolysis in vivo (also from knock-out and transgene animal experiments), the contribution of this system under pathological conditions for the risk of thrombotic, as well as bleeding complications and therapeutic consequences thereof, is obvious. Both acquired and hereditary deficiencies of blood clotting and fibrinolytic factors predispose the affected patients. Moreover, the diagnostic evaluation of hemostasis parameters that fall outside the normal physiological range are indicators and prognostic markers of disease conditions. These include the following factors:

- deficiency in vitamin K: reduction of active vitamin K-dependent clotting factors
- prothrombin F1/F2 fragment: increased production of thrombin
- fibrinopeptides A, B: increased production of fibrin
- fibrin degradation products: increased thrombus formation and dissolution
- fibrin D-dimer products: increased thrombus formation and dissolution
- plasmin/α2-antiplasmin complex: increased thrombus formation/fibrinolysis
- increased lipoprotein(a): less efficient fibrin-dependent thrombolysis
- prolonged clotting times of in vitro global clotting tests: deficiency or dysfunction of the blood coagulation cascade.

The following acquired or hereditary deficiencies will lead to or predispose for a significant disturbance of the blood coagulation and fibrinolysis systems in patients.

Defects in γ-Carboxylation, Defects in Biosynthesis, Isolated Deficiencies of Clotting Factor

Examples are functional deficiencies in vitamin K-dependent clotting factors, deficiency in protein cofactor VIII (haemophilia A) or factor IX (haemophilia B) which are associated with bleeding complications. Other defects include mutations in the prothrombin gene (thrombotic complications) or deficiency in plasminogen or t-PA (hypofibrinolysis), and mutations in fibrinogen (mostly asymptomatic but some associated with impaired wound healing).

Deficiency of Hemostasis Inhibitors

While antithrombin deficiency is associated with an impaired control of thrombin, $\alpha 2$-antiplasmin deficiency results in hyperfibrinolysis and bleeding complications. Increased PAI-1 levels are associated with an increased prothrombotic tendency, as well as a poor prognosis for atherothrombotic complications.

Gene Defects or Deficiencies of Protein C, Protein S, or Factor V

These defects are associated with an impaired intrinsic control of thrombin formation, whereby the Leiden mutation in factor V (known as APC resistance) is associated with the highest prevalence of thromboembolic complications in affected patients.

Therapeutic Interventions

Different therapeutic interventions exist in order to interfere with or prevent unwanted thrombotic or bleeding complications in patients.

Platelet aggregation—and to a certain extent platelet activation—is inhibited by antagonists of the glycoprotein IIb/IIIa integrin, inhibitors of ADP-receptors or by aspirin (an inhibitor of cyclooxygenase). Bleeding complications due to impaired platelet reactivity/function or deficiency/dysfunction of von Willebrand factor may be corrected by substitution therapy. Similarly, deficiency in factor VIII or other coagulation factors can be corrected by supplementing the respective (recombinant) factor. A decrease in ▶ thrombin formation and activity can be induced by oral vitamin K antagonists (such as warfarin), by heparin, or by substitution with natural inhibitors of the clotting system (such as TFPI, antithrombin or inactivated factor VIIa) as well as with hirudin (a natural anticoagulant from leech). Acute thrombolysis therapy with natural plasminogen activators (t-PA, urokinase) or streptokinase results in elevated (systemic) plasmin formation. In cases of hyperfibrinolysis, low molecular weight inhibitors that interfere with fibrin binding of t-PA and plasminogen can be applied.

Regulatory Environment

The potential toxicity in patients with hereditary or acquired disorders of hemostasis is brought about by the appearance of, for example, alloantibodies against mutated hemostatic factors or adhesion molecules, via drug-induced pathologies or interference with the inflammatory or immune systems. Conversely, in severe septic shock syndrome, bacterial infection followed by multiple cell activation and a massive consumption of hemostatic factors may lead to life-threatening situations.

References

1. Bertina RM (1999) Molecular risk factors for thrombosis. Thromb Haemost 82:601–609
2. Collen D (1999) The plasminogen (fibrinolytic) system. Thromb Haemost 82:259–270
3. Esmon CT (2001) Role of coagulation inhibitors in inflammation. Thromb Haemost 86:51–56
4. Mann KG (1999) Biochemistry and physiology of blood coagulation. Thromb Haemost 82:165–174

Blood Coagulation and Fibrinolysis

▶ Blood Coagulation

Blood Group System

A blood group is an inherited character of the surface of the red cell detected by a specific alloantibody. A blood group system consists of one or more blood group antigens encoded by a single gene or cluster of closely linked homologous genes.

▶ AB0 Blood Group System

Blotting

The transfer of protein, RNA, or DNA molecules from an acrylamide or agarose gel to a membrane (usually nylon or nitrocellulose) by capillarity or an electric field. Immobilized molecules can be detected by hybridization to a sequence-specific probe (DNA and RNA), or antibody labeling (protein).

▶ Southern and Northern Blotting

Blotting Membrane

The blotting membrane, usually consisting of nitrocellulose, polyvinylidene difluoride (PVDF), or nylon, is a membrane support for the electrophortic transfer of proteins out of polyacrylamide gels.

▶ Western Blot Analysis

Bone Marrow and Hematopoiesis

REINHARD HENSCHLER
Institute for Transfusion Medicine und Immune Hematology, Department of Cell Production and Stem Cell Biology Group
German Red Cross Blood Donation Center
Sandhofstrasse 1
D-60528 Frankfurt a. M.
Germany

Synonyms
haemopoiesis, hemopoiesis, blood cell formation

Definition
Hematopoiesis is the process of new blood cell formation. It is a continuous process, comprises the regeneration of all different blood cell lineages from a limited number of hematopoietic stem cells (HSC) and hematopoietic progenitor cells (HPC), and is capable of a fine-tuned adaptation to need.

Characteristics
Hematopoietic cells, as harvested from the bone marrow, include cells which belong to a continuum of different stages of a maturation hierarchy, starting from very primitive and undifferentiated, to fully mature and terminally differentiated cells (Table 1).
The most primitive hematopoietic cells, stem cells, are able to self-renew; that is, to undergo cell division resulting in at least one daughter cell which maintains the stem cell status. Primitive cells which are not capable of maintaining undifferentiated status, but still have the potential to undergo extensive (though finite) proliferation, are generally termed progenitor cells (Table 1). Towards increased ▶ differentiation status, the commitment to a particular cell lineage takes place.

Hematopoietic Growth Factors
Hematopoietic growth factors (HGF) are glycosylated polypeptides of a molecular weight between approximately 22 kD and 60 kD. They regulate the growth and differentiation of the various individual hematopoietic lineages. For example, granulocyte-macrophage colony stimulating factor (GM-CSF) stimulates growth and development of granulocytes and macrophages from precursor bone marrow cells in semisolid culture medium. Similarly, G-CSF stimulates the growth of granulocytic colonies, M-CSF those of macrophages, and multi-CSF of colonies containing multiple myeloid cell lineages (granulocytes, macrophages, erythrocytes and megakaryocytes). Erythropoietin (EPO) and thrombopoietin (TPO) were discovered by their ability to support erythrocytic or megakaryocytic development, respectively (1).

HGF are responsible for a regulated and adaptive response to need within the hematopoietic system. Principally, in this system, HGF-induced proliferation and differentiation of progenitor and immature hematopoietic cells are coupled, but HGFs can serve to increase the number of cell doublings and thus the number of mature cells produced from a precursor cell, thus providing a mechanism of fine-tuned regulation of mature cell production in the bone marrow. The commitment of undifferentiated progenitor cells to a single cell lineage is also ascribed to the effect of HGFs. It is irreversible and confines the further development of this cell. HGF withdrawal, on the other hand, results in apoptosis of progenitor cells which express the cognate receptor for a given HGF in a certain differentiation state. Apoptosis is continuously taking place to a certain degree in steady-state hematopoiesis, and inadvertent programmed cell death due to HGF withdrawal provides a negative regulating tool to demand-adapted blood cell maturation.

Bone Marrow and Hematopoiesis. Table 1 Characteristics of immature and mature hematopoietic cell populations

Cell type	Self-renewal	Characteristic morphology	Numbers/frequency	Lineage commitment	Proliferation potential
Stem cell	Yes	No	Very few (< 1 in 10 000)	No	> Life-long
Progenitor cell	No	No	Few (about 1 in 1000)	1–6	Extensive
Immature cell	No	Yes	Majority of bone marrow	1	Limited
Mature cell	No	Yes	Frequent	1	None

In addition to the HGFs which stimulate selective cell lineage cell development, additional cytokines such as the interleukin IL-1β or IL-6, stem cell factor/c-kit ligand (SCF) or FLT3 ligand (FL) were identified as synergistic molecules, which on their own cannot stimulate hematopoietic colony growth, but which strongly support the growth and development of progenitor cells initiated by CSFs (2). This is achieved both by shortening cell cycle times and by amplifying the numbers of cell divisions between an immature precursor and a finite differentiation stage. In particular when multiple factors are present, *HGF*s also regulate the survival of very primitive cells/stem cells.

Stromal Cell Regulation of Hematopoiesis

Important survival and differentiation-inducing but also cell adhesion signals for developing hematopoietic cells are provided by the hematopoietic microenvironment. It consists mainly of stromal cells and deposited extra-cellular matrix. The main known stromal cell types are bone marrow fibroblasts, adipocytes, osteoblasts, endothelial cells, and macrophages. These have been extensively characterized after the establishment of long-term bone marrow cultures (*LTBMC*) which allow the maintenance of HSC and a continuous in vitro hematopoiesis over a period of up to 8 weeks (3). Fibroblasts provide a mesh within the bone marrow cavity, and together with endothelial cells give hold to islands of developing hematopoietic cells which are located in islands. Macrophages play important roles in providing iron for erythropoietic cells in the process of hemoglobinization, and likely also nourish other maturing cell types ("nurse cells").

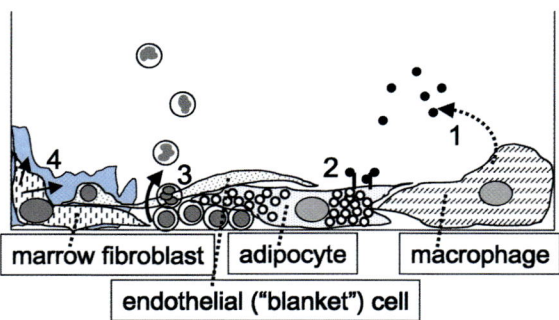

Bone Marrow and Hematopoiesis.
Figure 1 Schematic representation of stromal cells and their function in long-term bone marrow cultures. 1. Secretion of hematopoietic growth factors. 2. Topical binding of hematopoietic growth factors via stromal cell heparan sulfate proteoglycans. 3. Providing niches for development of primitive cells ("cobblestone areas") and allowing transmigration of maturing cells to the stromal surface. 4. Secretion of extracellular matrix molecules.

Adipocytes are the sign of well-proliferating long-term bone marrow cultures, and osteoblasts have been ascribed a role in the maintenance of quiescent HSC. Stromal cells provide the separation of areas of very primitive hematopoietic cells ("cobblestone areas") in LTBMC, and thus divide primitive cell development from islands of maturing cells (Figure 1).

Megakaryopoiesis is associated with sinusoidal endothelium, and release of platelets into the circulation can be seen to occur through egress via sinusoidal lining endothelial cells by electron microscopic preparation of bone marrow. A variety of extracellular matrix substances are produced by stromal cells, which can influence survival and development of hematopoietic cells in conjunction with HGFs; these included fibronectin, laminin, and collagen IV. Soluble HGFs have been shown to be bound to stromal cells by specific proteoglycans, such as GM-CSF and G-CSF, or are expressed as membrane-integral proteins in stromal cells (SCF). LTBMC have allowed detailed studies of the role of stromal components in hematopoiesis.

Ontogenetic Development and Sites of Hematopoiesis

Embryonal hematopoiesis arises from a small group of cells which emerge from the dorsal aorta in the aortogonado-mesonephros region. The earliest cells with hematopoietic capacity are termed hemangioblasts, and the endothelial cell differentiation potential remains associated with HSC during later stages of ontogenetic development. Following this, HSC are found in the yolk sac, and during the fetal period HSC immigrate into the liver. Before the fetal liver stage, a so-called primitive hematopoiesis prevails (as in mammals with nucleated erythrocytes and a macrophage-like population of leukocytes), whereas after this stage, in most higher organisms, definitive hematopoiesis develops and already bears the features of multilineage differentiation from HSC into lymphoid and myeloid precursors cells. During birth, the cord blood contains a substantial number of HSC in man, and therefore cord blood has been established as a transplant source which is especially well suited for children. In humans, adult hematopoiesis finds its place within the bone cavities, whereas in mice due to the relative restriction of caval bone, hematopoiesis often expands also to the spleen. In states of bone marrow fibrosis, hematopiesis re-locates to the liver and spleen also in humans.

Preclinical Relevance

Toxicity to human hematopoietic and hematopoietic-supportive cells can be assayed using several in vitro test systems. Colony-forming unit (CFU) assays for progenitor cells in semisolid medium give data on the direct effects of hematotoxic compounds on hematopoietic progenitor cells. They can be performed using murine or human progenitors. CD34 antigen-positive cells from human cord blood plated at 5000 cells per ml will give rise to approximately 100 hematopoietic colonies; dependent on the choice of added HGF, both granulocyte-macrophage and erythrocytic colonies are developing. Enriched progenitor cells populations are preferred over unselected cell populations, since especially mature macrophages display a source of metabolic activity towards many organic compounds. Stromal cells can be grown as cell lines which are of fibroblastic morphology, or as underlayers of LTBMC which then will include different stromal cell types. The stroma can be irradiated with 30 Gy to eliminate endogenous hematopoiesis, treated with immunotoxins, and then overlaid with HSC to assess toxic effects to the stromal cell compartment. In addition, of course, entire cultures can be treated to assess damage to the system in its entire complexity. The hematotoxic damage exerted by busulfan or cyclophosphamide is detected using exposure of LTBMC stromal layers to the substances. Interestingly, a concomitant depletion of the colony-forming stromal cell precursor cells (CFU-F) is detected.

In vivo models have been established for the detection of altered hematopoietic cell turnover by test compounds. The mouse bone marrow micronucleus test serves as a relatively sensitive and easy-to-handle test for detecting alterations in bone marrow cell turnover, and possible genotoxic damage to immature cells. Readout is confined to erythrocytic cells (reticulated and young polychromatic erythrocytes). Mouse models have also been validated to detect long-term hematopoietic stromal cell damage as observed after bone marrow irradiation. In the bone marrow of rats, stromal cells deteriorate and decrease in numbers about 3 months after bone marrow transplantation, indicating that stromal damage follows different kinetics as HSC damage, most likely resulting from a much slower turnover of stromal cells.

Relevance to Humans

Human HSC have been transplanted as a curative treatment for patients with a variety of hematologic malignancies for more than 20 years, since the possibility to test for histoincompatibility by anti-HLA antibodies, and the development of improved immunosuppressive medication (4). These patients have been observed closely, and found to have a number of specific alterations in their hematopoietic systems, most likely relating to the toxic effects of their intensive chemotherapy and/or irradiation. In the patient's bone marrow, the numbers of stromal cells and also stromal precursor cells are substantially reduced. Also, numbers of progenitor cells are reduced concomitantly, and the proportion of progenitors which are in cell cycle is highly elevated. Still, numbers of circulating blood cells are normal, as is the adaptive response of hematopoiesis, for example with increased production of neutrophils during states of infection. Also, development of leukemia as a consequence is increased during the first 5 years after transplantation, yet spontaneous rates of leukemia development return to normal levels thereafter in the transplanted patients. Therefore, it is not very likely that changes in numbers or behavior of human hematopoietic progenitor cells will reflect or predict bone marrow insufficiency states or malignant development from HSC. Changes in bone marrow CFC numbers and cellularity have been reported from workers heavily exposed to hematotoxic compounds, which parallel findings with the same compounds in animal or in vitro tests (5).

Regulatory Environment

So far, except for the bone marrow micronucleus test, standardized test systems have not been included in the routine investigation of potential HSC toxic compounds. However, the Declaration of Helsinki and national legislation for animal experimentation must be respected.

References

1. Metcalf D (1993) Hematopoietic regulators: redundancy or subtlety? Blood 82:3515–3523
2. Moore MAS (1991) Clinical implications of positive and negative hematopoietic stem cell regulators. Blood 78:1–19
3. Dexter TM, Allen TD, Lajtha LG (1977) Conditions controlling the proliferation of hematopoietic cells in vitro. J Cell Physiol 91:335–344
4. Thomas ED, Storb R, Clift RA et al. (1975) Bone marrow transplantation. N Engl J Med 292:832
5. Cody RP, Strawderman WW, Kipen HM (1993) Hematologic effects of benzene. Job-specific trends during the first year of employment among a cohort of benzene-exposed rubber workers. J Occupat Med 35:776–782

Bootstrap

A resampling technique in which multiple random samples (with replacement) are obtained from the empirical data and a test statistic is calculated on each new sample. The distribution of the test statistic is thought to reflect the characteristics of the underlying population from which the original sample was drawn.

▶ Statistics in Immunotoxicology

BPDE

▶ Polycyclic Aromatic Hydrocarbons (PAHs) and the Immune System

BP-7,8-diol

▶ Polycyclic Aromatic Hydrocarbons (PAHs) and the Immune System

BPQ

▶ Polycyclic Aromatic Hydrocarbons (PAHs) and the Immune System

Bronchitis

▶ Respiratory Infections
▶ Trace Metals and the Immune System

Bronchus-Associated Lymphoid Tissue

Bronchus-associated lymphoid tissue (BALT) refers to secondary lymphoid tissue in the respiratory tract.
▶ Mucosa-Associated Lymphoid Tissue

Buehler Test

▶ Guinea Pig Assays for Sensitization Testing

Buffy Coat

The thin, white, leukocyte-rich band that separates the separated serum from the mass of erythrocytes in a centrifuged whole blood sample.
▶ Lymphocytes

Burkitt's Lymphoma

▶ Lymphoma

C

C-Reactive Protein

C-reactive protein (CRP) is a plasma protein produced by the liver during acute inflammatory reactions.
▶ Fish Immune System

C3 Convertase

An enzyme that cleaves the complement component C3, converting it to an active state (C3a and C3b). The classical pathway C3 convertase is a complex of C4b2a. The alternative pathway C3 convertase is C3bBb.
▶ Complement and Allergy
▶ Complement, Classical Pathway/Alternative Pathway

C5 Convertase

An enzyme that cleaves the complement component C5, converting it to an active state (C5a and C5b). The classical pathway C5 convertase is a complex of C4b2a3b. The alternative pathway C5 convertase is C3bBb3b.
▶ Complement, Classical Pathway/Alternative Pathway
▶ Complement and Allergy

Cachectin

▶ Tumor Necrosis Factor-α

CAMs

▶ Cell Adhesion Molecules

Cancer and the Immune System

JÖRG BLÜMEL
Merz Pharmaceuticals GmbH
Eckenheimer Landstrasse 100
D-60318 Frankfurt a. M.
Germany

Synonyms
cancer immunosurveillance, cancer immunoediting, tumor immunology

Definition
This section briefly describes the role of the immune system in cancerogenesis, its capability of promoting either host resistance or tumor formation. The immune system has initially been hypothesized to play a tumor-suppressive action leading to the control of neoplastic diseases—cancer immunosurveillance. This hypothesis was recently extended to include the concept of cancer immunoediting encompassing both positive host-protecting properties and negative tumor-forming properties. The action of the immune system may provoke the complete elimination of the tumor, generate a non-protective immune phenotype, or favor the development of immunologic anergy or tolerance. Recent research focuses on the development of alternative immunotherapeutic strategies against human cancer as alternatives to conventional cancer therapy.

Characteristics
The immune system offers within both the innate and adaptive immune function a broad spectrum of mechanisms to recognize and eliminate foreign structures of various origin, e.g. bacteria or viruses, thereby protecting the host and maintaining the tissue homeostasis. The idea that the immune system may also play a protective role against tumor progression was introduced briefly early in the 20th century. This initial idea was developed further years later by Burnet [1] and Thomas [2] resulting in the hypothesis of cancer immunosurveillance. Both proposed independently an immunologic mechanism induced by non-physiological antigenic properties of nascent transformed cells. It

was speculated that this mechanism finally evokes an effective protective immune response followed by complete tumor regression.

Although initial experiments using mice with impaired immune function (e.g. athymic mice) failed to support the immunosurveillance concept in a physiological setting, modern science offers several explanation for these early failures. Firstly, important mechanisms and cell populations of the mature immune system and their functions were unknown (e.g. the function of natural killer cells as important cytotoxic effector cells of the innate immune system, or the extrathymic maturation of lymphocyte subpopulations like γδ T cells). Secondly, the limitations of the experimental models used were not fully known (e.g. the partial immunocompetence of athymic mice). Although these initial drawbacks led to a partial fading interest in the concept of immunosurveillance, advances in gene-targeting techniques—as well as our rapidly improving understanding of the processes involved in the host-protective function of the immune system—triggered further research, resulting finally in a revival of the initial concept. In particular, there was experimental verification that interferon-γ (IFN-γ) protects against tumor growth and the increased susceptibility of gene-targeted mice lacking the cytotoxic mediator ▶ perforin against chemically-induced tumor formation; these were milestone results strongly supporting the cancer immunosurveillance theory. However, the broad preclinical experimental evidence generated over the years by independent international laboratories supporting the existence of cancer immunosurveillance does not accurately match the pathophysiological situation of carcinogenesis in fully immunocompetent organisms. Therefore, recent research has focused on the impact of a functional immune response on tumor development and growth.

As a result, a modification of the original theory, the concept of cancer immunoediting (3) was introduced. This modified approach took into account both a host-protective and a tumor-surviving function of the immune system by distinguishing three stages:

- elimination
- equilibrium
- escape.

In the elimination phase, the immune system is able to control and eliminate the tumor completely, thereby reflecting the original cancer immunosurveillance concept. In the equilibrium stage, the interaction of the immune system with the tumor promotes the selection of tumor cell subpopulations with reduced immunogenicity, e.g. tumor cells that display MHC class I deficiencies or cells that present tumor antigens but do not deliver a costimulatory signal to specific $CD8^+$ cytotoxic T cells. The equilibrium stage is characterized by an increased ability of tumor cell subpopulations to survive even in the immunocompetent host. The terminal phase of escape describes the stage where tumors with modified immunogenicity are developed by the selective pressure of the immune system. These immunologically shaped tumors are capable of expanding uncontrolled in the host.

Striking scientific evidence supporting this modified approach is derived from various preclinical experiments. Using gene-targeted mice with a compromised immune system it was demonstrated that tumors derived of wild-type (intact) as well as gene-targeted (compromised immune function) mice grew similarly when transplanted in gene-targeted mice. In contrast, the incidence of tumor rejection increased significantly when tumors derived of gene-targeted mice were transplanted in wild-type mice. These data amongst others indicate that tumors developed without the selective pressure of an intact immune system exhibit a higher immunogenicity, thus providing strong evidence for the overall concept of immunoediting.

The molecular and cellular mechanisms underlying the immune response to a tumor seem to be extremely complex. Beside the cellular components of innate immunity like macrophages, NK and γδT cells, and the main effector cells of adaptive immunity, $CD4^+$ Th cells and $CD8^+$ cytotoxic T cells, other critical mediators in the interaction of the host's immune system with the tumor are IFN-γ and several chemokines. The role of IFN-γ during the process of cancer immunoediting was extensively studied in recent years. However, the detailed mechanisms by which IFN-γ achieves its protective effects remains unclear. It seems that non-immunologic mechanisms like antiproliferative, antimetabolic, angiostatic and pro-apoptotic effects, as well as immunologic mechanisms like chemoattraction of immunocompetent cells, induction of cytokine or chemokine secretion, enhancement of tumor immunogenicity, and directing the ▶ Th1/Th2 balance, are both essential for IFN-γ-induced tumor surveillance. The proposed pathway leading to tumor regression is most likely initiated by cells of the innate immunity such as NK cells or those T cell populations involved in innate immunity, like γδT cells. Initially, specific ▶ tumor-associated antigens (TAA), such as ▶ p53 tumor suppressor protein, HER/2neu, or CD20 are recognized by these cells. To date, a variety of TAA are known. These are either derivatives of one of the following:

- physiological self-antigens or tissue specific differentiation antigens that are dramatically overexpressed by tumor cells in comparison to other cells
- mutated self proteins or specific oncogenic antigens inappropriately expressed by tumor cells, or
- those derived from virally encoded antigens.

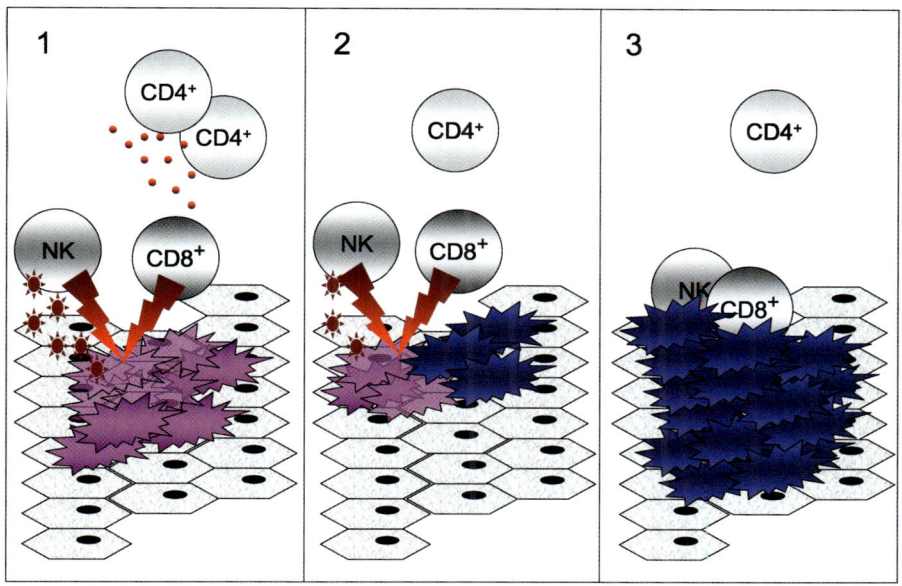

Cancer and the Immune System. Figure 1 Concept of cancer immunoediting. In the phase of elimination (1) the immune system is able to control and eliminate the tumor cells (purple). In the stage of equilibrium (2) the interaction of the immune system with the tumor promotes the selection of tumor cell subpopulations with reduced immunogenicity (blue). The terminal phase of escape (3) describes the stage where tumors with modified immunogenicity are developed and are capable of escaping the host-protective immune attack. Non-transformed cells (grey), lymphocyte subpopulations as marked. Red flashes illustrate cytotoxic action of the effector cells, [oval] red stars perforin-mediated cytolysis, and small red circles soluble mediators of the immune response. Modified after Dunn GP et al (3).

The recognition pattern induced by TAA triggers the secretion of IFN-γ by innate immune cells. This initial level of secreted IFN-γ induces the secretion of other angiostatic chemokines like CXCL10 (Interferon-γ-inducible protein-10, IP-10) and results in chemoattraction of further immune effector cells such as macrophages and more NK cells to the tumor site. These infiltrating effector cells themselves secret immunomodulatory cytokines such as IL-12, IL-18 and again IFN-γ, which in turn activate the cytotoxic properties of the infiltrated cells. Fragments of lysed tumor cells that are presented in the tumor site draining lymph nodes by professional antigen-presenting cells, e.g. dendritic cells, promote the immune response modulated by the adaptive immune system. Such a microenvironment characterized by high levels of IFN-γ and IL-12 promotes CD4$^+$ T cells (via MHC class II recognition) to trigger a Th1-like immune response with strong activation of CD8$^+$ cytotoxic T cells. This results, for instance, in perforin-mediated tumor cell lysis (via MHC class I recognition) and again further secretion of IFN-γ. These dramatically increased levels of IFN-γ enhance the expression of MHC class I molecules on the tumor cells, thereby increasing their immunogenicity. The angiostatic, antiproliferative and pro-apoptotic effects of IFN-γ are thereby amplified. Overall, IFN-γ is a key player orchestrating together with other important mediators the complex interaction of various elements of both the innate and adaptive immunity leading to an effective immune response against the tumor that may protect the immunocompetent host. Nevertheless, genetic alterations in tumor cells may also enable them to circumvent an effective immune response of the host.

Preclinical Relevance

A better understanding of the basic mechanisms and regulatory pathways involved in cancer immunoediting or surveillance derived from preclinical—especially gene-targeted—models may help to develop alternative immunotherapeutic strategies to conventional cancer therapy. Preclinical experiments in which either proinflammatory genes like those encoding for MHC class I or granulocyte-macrophage colony-stimulating factor (GM-CSF) were transferred to tumor cells, or antisense oligonucleotides were designed to inhibit the expression of immunosuppressive genes (e.g. genes encoding for TGF-β) showed an impressive stimulation of the antitumor response. Another area of research focuses on chemokines as an effective treatment against human cancer. It was shown that genes encoding for chemokines like CCL-3 (▶ macrophage inflammatory protein-1α, MIP-1α) or CCL-5 (▶ RANTES) expressed either in genetically modified cells or administered locally as a recombinant or fusion protein can induce protective immunity and tumor regression. Finally, gene-transfer studies using a combination of both genes encoding for chemokines like XCL-1 and cytokines like IL-2 expressed at the tumor site enhanced the lymphocyte infiltration and protected also from tumor growth.

Relevance to Humans

There is increasing evidence in support of the concept of immunosurveillance in both the experimental preclinical setting and the clinical human situation. In fact, there is accumulating epidemiologic evidence supporting the existence and physiological relevance of this concept in humans. Early data from patients with primary immunodeficiency syndromes or immunosuppressed transplant patients revealed an increased cancer risk in these populations. However, a significant number of the observed tumors in these individuals are of viral origin thereby reflecting more the impairment of the natural protective function of the immune system against infectious diseases rather than the loss of a specific tumor-suppressive function. Nevertheless, there is also broad evidence of an increased risk for the development of tumor types with no apparent viral etiology in these populations. A study analyzing the tumor incidences in 608 cardiac transplant patients showed a 25-fold increase in prevalence of lung cancer compared to the general population (4).

An assessment of 5692 patients receiving a renal transplant revealed an increased prevalence for the development of several tumor types of non-viral origin (5). In addition, a positive correlation was found between lymphocyte tumor infiltration, by $CD8^+$ T cells in particular, and patient survival. This was, for example, shown in a retrospective study investigating more than 500 patients with primary melanoma (6). Patients developed a significant infiltration of lymphocytes in the tumor during the vertical growth phase of cutaneous melanoma showed a significant increased survival time compared to the patients showing low or absent lymphocyte infiltration.

In conclusion, data obtained from both experimental preclinical studies and human epidemiology strongly support the important role of the immune system in tumor suppression, as well as the physiological relevance of the concept of cancer immunosurveillance in different species including man.

Recently, research has focused more on the interaction of the intact immune system with a developing tumor as a useful approach to overcome the resistance of several tumor subtypes to conventional therapeutic strategies. The principle of cancer immunogene therapy utilizes genetically modified human cells to stimulate the immune response against the tumor. However, clinical trials conducted with, for example, direct intratumoral MHC class I gene transfer, or vaccination with irradiated autologous plasma cells engineered to express IL-2 by adenoviral gene transfer, showed somewhat equivocal results and did not fully replicate the impressive preclinical results. The vaccination with DNA encoding for tumor-associated antigens induced also only a limited antitumor response. In contrast, another approach that showed impressive efficacy in clinical trials comprises the systemic administration of monoclonal antibodies directed against TAA. As a result, the first recombinant humanized antibodies directed against HER/2neu (breast cancer) or CD20 (non-Hodgkin's lymphoma) recently received marketing approval. Overall, it seems that the best strategy for the treatment of human cancer might be a combination of both conventional methods reducing the overall tumor burden and immunotherapeutic approaches effectively eliminating residual tumor cells.

References

1. Burnet FM (1970) The concept of immunological surveillance. Prog Exp Tumor Res 13:1–27
2. Thomas L (1982) On immunosurveillance in human cancer. Yale J Biol Med 55:329–333
3. Dunn GP et al. (2002) Cancer immunoediting: from immunosurveillance to tumor escape. Nature Immunol 3:991–998
4. Pham SM et al. (1995) Solid tumors after heart transplantation: lethality of lung cancer. Ann Thorac Surg 60:1623–1626
5. Birkeland SA et al. (1995) Cancer risk after renal transplantation in the Nordic countries, 1964–1986. Int J Cancer 60:183–189
6. Clark WH et al. (1989) Model predicting survival in stage I melanoma based on tumor progression. J Natl Cancer Inst 81:1893–1904

Cancer Immunoediting

▶ Cancer and the Immune System

Cancer Immunosurveillance

▶ Cancer and the Immune System

Cancer-Testis Antigens

A group of antigens originally discovered in human

melanoma that are also expressed in normal testis. Includes MAGE, GAGE, BAGE and NY-ESO-1 tumor antigens. Cancer-testis antigens are found in a variety of human tumors, but because of their historical association with melanoma are sometimes called "melanoma-testis antigens".

▶ Tumor, Immune Response to

Cancer Vaccine

Vaccine administered to cancer patients to elicit a therapeutic immune response against tumor cells. Antigen-based vaccines can be made of whole tumor cells, recombinant tumor antigens, synthetic peptides, or DNA-encoding tumor antigens. Dendritic cell-based vaccines consist of dendritic cells isolated from patients and exposed in vitro to a source of tumor antigens before re-injection in vivo.

▶ Tumor, Immune Response to

Canine Immune System

MARK WING
Huntingdon Life Science Limited
Woolley Road
Alconbury, Huntingdon, Cambs
PE28 4HS
UK

Synonyms
dog

Definition
The mammalian immune system consists of multiple cell types that circulate via the blood and lymphatics to specialized lymphoid and non-lymphoid tissues. Following exposure to pathogens which include viruses, bacteria, fungi, and parasites, the cells of the immune system interact at both the cell level and the molecular level in the lymphoid tissues, which drain the site of infection. Activated cells migrate out of the lymph nodes back to the site of infection to respond to the threat. As part of the first encounter with an infectious organism, resistance to repeated infection by the same organism is mediated through immunological memory.

Characteristics
Superficially, there is little to distinguish the development, structure, and function of immune system of one mammalian species from another. Lymphocyte development takes place in the primary lymphoid tissues (the thymus and bone marrow) with mature lymphocytes residing in the secondary lymphoid tissues (the lymph nodes, spleen and mucosa-associated lymphoid tissues (MALT)). However a more detailed review of the canine immune system reveal more subtle differences compared to other toxicology species and humans (for reviews see (1) and (2)).

The thymus is primarily responsible for T lymphocyte maturation and the deletion of autoreactive cells, but recent evidence suggests that lymphocyte follicles exist in the medulla of the canine thymus, a situation well documented in humans (Fig. 1).

▶ Immunohistochemical staining of these lymphoid aggregates revealed that when present, the ▶ germinal centers were predominantly composed of ▶ B lymphocytes, confirming the identity of these structures (3). The canine spleen differs to that of humans and other toxicology species in its greater capacity to store blood and sparse lymphoid tissue. The spleen is commonly used as a source of lymphocytes to perform ▶ ex vivo and in vitro functional assays (see below), however the lymph nodes and peripheral blood may provide a better source of canine cells.

Regarding the development of the immune system, the dog is more similar to humans in that the neonate is born with a largely intact immune system that matures postnatally—in contrast to the rodent immune system which is less well developed at birth. Other developmental similarities to humans include the age-related changes in serum ▶ immunoglobulins (Ig) with IgM reaching adult levels first, shortly followed by IgG, but with IgA levels lagging behind. Developmental changes in lymphocyte subsets also mirror those seen in humans. These include a decline in the percentage of peripheral blood B lymphocytes and an increase in

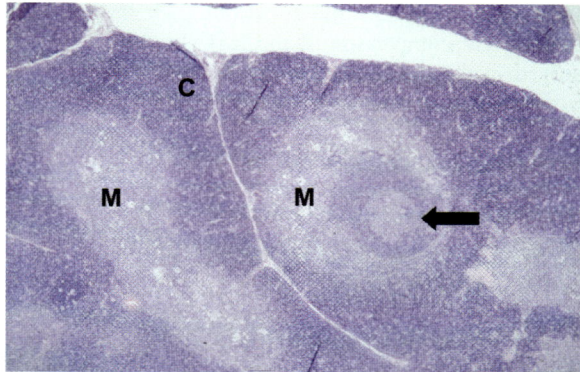

Canine Immune System. Figure 1 Canine thymus (original magnification × 120). Arrow denotes follicle containing active germinal centre.
C=cortex; M=medulla.
Kindly supplied by Dr Andrew Pilling of Huntingdon Life Sciences.

T lymphocytes observed in the weeks following birth, and considerably higher numbers of peripheral blood ▶ CD4$^+$ T lymphocytes are present at birth compared to CD8$^+$ cells. As observed with humans, the percentage of CD4$^+$ T lymphocytes decline after 10–12 weeks, with the proportion of CD8$^+$ cells reaching adult levels. Greater than 90% of the peripheral blood dog T lymphocytes at birth express the naive CD45RA isoform, which declines to 40–50% by 4 months, reflecting exposure to environmental antigens. Again a very similar pattern of CD45RA expression is seen in humans. One difference between humans and dogs relates to the degree of maternal IgG transfer, which in the case of the dog is largely achieved afater birth through colostrum, as opposed to placental transfer.

The proportion of peripheral blood neutrophils to lymphocytes is similar between adult dogs and humans, but is different when compared with rodents, which have a greater proportion of lymphocytes. Lymphocyte subset analysis reveals a qualitatively similar pattern of peripheral blood B lymphocytes and T lymphocyte ratios in the dog compared to other species including humans, whereby T lymphocytes > B lymphocytes, and CD4$^+$ T lymphocytes > CD8$^+$ T lymphocytes (Table 1).

The CD4 antigen is expressed on T lymphocytes in all mammalian species including the dog, however it is reported to be uniquely expressed at a high density on canine neutrophils.

Further evidence for an overall similarity between the immune system of the dog and humans is provided from studies involving the X-linked severe immunodeficiency syndrome (▶ XSCID). This syndrome, which is seen in humans, dogs, and mice, is caused by a mutation of the gamma chain common to the receptors of ▶ interleukins IL-2, IL-4, IL-7, IL-9, IL-15, and IL-21. These cytokines are important for the normal development and functioning of the immune system and comparison of the XSCID phenotype observed in the different species indicates that the biology of these cytokines is more similar between humans and dogs.

In short, although there are differences between the dog and human immune systems, this is probably less than those seen between humans and rodents.

Preclinical Relevance

The rodent is the most commonly employed species for safety assessment of all types of test compound on the immune system, with the exception of biopharmaceuticals where the need to use a pharmacologically active species and the issue of compound immunogenicity often requires the use of the non-human primate. As the most commonly used second species for safety assessments, again with the exception of biological drugs, the dog may be a suitable species for performing immunotoxicity assessments. Such situations arise where there is evidence suggesting that the dog is a more relevant species to humans, due perhaps to comparable metabolism, or to confirm immunototoxicity findings in the rodent. Finally, the dog would be the relevant species for efficacy studies where the dog is the intended population to be treated with an immunomodulatory drug or neutroceutical.

Relevance to Humans

As discussed above, the canine immune system displays many features common to humans but as with other toxicology species, exhibits some unique features. Providing these differences are understood, it should be possible to make a rational decision regarding the suitability of the dog, or any species for that matter, for an immunotoxicity study. At this time, the predictability of immunology and/or toxicology assays performed in animals is being evaluated, with a view to harmonizing the geographical differences that exist regarding regulatory requirements for immunotoxicity testing for small chemical drugs. On a case-by-case basis, in the absence of evidence to suggest that the dog is any more or less relevant to humans compared to rodents, animal welfare considerations should dictate that the rodent is the default species. Where the dog is justified, functional testing should be incorporated onto standard safety assessments where possible, consistent with the principles of the ▶ three Rs.

Canine Immune System. Table 1 Flow cytometry analysis of beagle dog peripheral blood lymphocyte populations

	Phenotype	Antibody clone	Cell number (per µl of blood)	% of lymphocytes
B lymphocytes	CD21$^+$	CA2.1D6	931.8	28.8
T lymphocytes	CD3$^+$	CA17.2A12	1911.9	61.5
T helper lymphocytes	CD3$^+$CD4$^+$	YKIX302.9	1353.3	47.2
Cytotoxic T lymphocytes	CD3$^+$CD8$^+$	YCATE55.9	433.7	16.5

Regulatory Environment

Repeated-dose immunotoxicity studies are increasingly being performed as part of regulatory submissions for test substances as diverse as food additives, industrial chemicals, and pharmaceuticals (see below for regulatory guidelines). Whilst the majority of these studies are performed on rodents, the regulatory authorities would accept data from canine studies where this second species, the dog, is considered to be more relevant.

Most of the assays employed to assess immune function can be performed in the dog using blood and surplus tissue obtained in-life or at necropsy respectively. Assays include the ▶ natural killer (NK) cell assay, ▶ flow cytometry, the ▶ primary antibody response, the mitogen assay, and the ▶ phagocytosis assay (4,5). Key differences include CTAC, a cell line derived from a canine thyroid adenocarcinoma, as the target of choice for the functional analysis of NK cells, and the use of peripheral blood or lymph nodes rather than the spleen as a source of cells (as discussed above).

Whilst reagents are increasingly becoming commercially available for performing leukocyte phenotyping by flow cytometry, few are canine specific, and the use of 'cross-reactive' antibodies should be undertaken with caution. Suitable reagents exist to perform a standard B lymphocyte and T lymphocyte subset panel. In addition to a limited range of antibodies to leukocyte surface antigens, there appear to be very few commercial reagents for the analysis of canine cytokines. This lack of standard reagents means that comparison of data from different laboratories is difficult, with limited historical data often cited as a significant disadvantage of the dog, a problem confounded by the outbreed nature of the species and the small group sizes typically employed in toxicology studies.

Regulatory guidelines requesting an assessment of the test substance on the immune system:
- Food additives. FDA 'Red Book' Draft 1993 and 2000
- Biochemical pesticides. EPA Biochemicals Test Guidelines 1996
- Agrochemicals and Industrial chemicals. EPA Health Effects Test Guidelines 1998
- Small chemical drugs. EMEA Repeat Dose Toxicity 2000
- Small chemical drugs. FDA (CDER) Immunotoxicology Evaluation of Investigational New Drugs 2002
- Small chemical drugs. MHLW/JPMA Draft Guidance for Immunotoxicity Testing 2003

References

1. Felsburg PJ (2002) Overview of immune system development in the dog: comparison with humans. Human Exp Toxicol 21:487–492
2. Hayley PJ (2003) Species differences in the structure and function of the immune system. Toxicology 188:49–71
3. Ploemen J-P, Raveskoot W, van Esch E (2003) The incidence of thymic B lymphoid follicles in healthy beagle dogs. Toxicol Pathol 31:214–219
4. Lanham DF, Bidgood J, Hunter EL, Wing MG (2002) Immunophenotyping and immune function assays in beagle dogs. Toxicol Lett 135:136
5. Finco-Kent DL, Kawabata TT (2003) Development and validation of an assay to evaluate the canine T-dependent antibody response. Toxicologist 72 [Suppl 1]:103

Carcinogenesis

I BERNARD WEINSTEIN
Columbia University
New York, NY
USA

Definition

Carcinogenesis is the process by which cancer develops in various tissues in the body.

Characteristics

In most cases carcinogenesis occurs via a stepwise process that can encompass a major fraction of the lifespan (multistep development). These progressive stages often include hyperplasia, dysplasia, metaplasia, benign tumors, and eventually malignant tumors. Malignant tumors can also undergo further progression to become more invasive and metastatic, autonomous of hormones and growth factors and resistant to chemotherapy or radiotherapy.

Causes

Known causes of carcinogenesis include various chemicals or mixture of chemicals present in several sources. This includes cigarette smoke, the diet, the workplace or the general environment, ultraviolet and ionizing radiation, specific viruses, bacteria and parasites and endogenous factors (oxidative DNA damage, DNA depurination, deamination).

According to the International Agency for Research on Cancer (IARC) 69 agents, mixtures, and exposure circumstances are known to be carcinogenic to humans (group 1), 57 are probably carcinogenic (group 2A) and 215 are possibly carcinogenic to humans. Some of these agents, or their metabolites, form covalent adducts to DNA and are mutagenic. Others act at the epigenetic level by altering pathways of signal transduction and gene expression. These include tumor promoters, growth factors and specific hormones.

Dietary factors also play an important role. Fruits and vegetables often have a protective effect. Excessive fat and/or calories may enhance carcinogenesis in certain organs. Hereditary factors can also play an important role in cancer causation. Indeed, human cancers are often caused by complex interactions between these multiple factors. An example is the interaction between the naturally occurring carcinogenaflatoxin and the chronic infection with hepatitis B virus in the causation of liver cancer in regions of China and Africa.

Molecular Genetics

Recent studies indicate that the stepwise process of carcinogenesis reflects the progressive acquisition of activating mutations in dominant acting oncogenes and inactivating recessive mutations in tumor suppressor genes. It is also apparent that epigenetic abnormalities in the expression of these genes also play an important role in carcinogenesis. Thus far over 100 oncogenes and at least 12 tumor suppressor genes have been identified. Tumor progression is enhanced by genomic instability due to defects in DNA repair and other factors. The heterogeneous nature of human cancers appears to reflect heterogeneity in the genes that are mutated and/or abnormally expressed. Individual variations in susceptibility to carcinogenesis are influenced by hereditary variations in enzymes that either activate or inactivate potential carcinogens, variations in the efficiency of DNA repair and other factors yet to be determined. Age, gender and nutritional factors also influence individual susceptibility.

Relevance to Humans

Cancer is a major cause of death throughout the world. Therefore, the prevention of carcinogenesis is a major goal of medicine and public health. The carcinogenic process can be prevented by avoidance of exposure to various carcinogenic factors such as cigarette smoking and excessive sunlight, dietary changes, early detection of precursor lesions and chemoprevention.

References

1. Kitchin KT (ed) (1999) Carcinogenicity, Testing, Predicting and Interpreting Chemical Effects. Marcel Dekker, New York
2. Weinstein IB, Santella RM, Perera FP (1995) Molecular biology and molecular epidemiology of cancer. In: Greenwald P, Kramer BS, Weed DL (eds) Cancer Prevention and Control. Marcel Dekker, New York, pp 83–110
3. Weinstein IB, Carothers AM, Santella RM, Perera FP (1995) Molecular mechanisms of mutagenesis and multistage carcinogenesis. In: Mendelsohn J, Howley PM, Israel MA, Liotta LA (eds) The Molecular Basis of Cancer. WB Saunders, Philadelphia, pp 59–85
4. Weinstein IB (2000) Disorders in cell circuitry during multistage carcinogenesis: the role of homeostasis. Carcinogenesis 22:857–864

Cardiac Disease, Autoimmune

NOEL R ROSE
Dept Pathology and Dept Molecular Microbiology and Immunology, The Johns Hopkins Medical Institutions
Bloomberg School of Public Health
615 North Wolfe Street
Baltimore, MD 21205
USA

Synonyms

Autoimmune heart disease, immune-mediated heart disease, inflammatory heart disease, myocarditis, cardiomyopathy

Definition

A number of inflammatory diseases of the heart have been associated with autoimmune or other immune-mediated pathogenic mechanisms. The disease can affect any portion of the heart: the pericardium (surface of the heart), the myocardium (heart muscle), or the endocardium (lining and valves of the heart). Inflammation may occur in the coronary vessels that supply blood to the heart itself and lead to atherosclerosis. These inflammatory processes are often accompanied by autoimmune responses in the form of antibodies to antigens found in the heart. There are, however, very few instances where one can clearly state that these autoimmune responses are the cause rather than the result of heart disease in humans. On the other hand, there are a number of well-defined animal models of autoimmune heart disease. They can be cited as indirect evidence supporting an autoimmune etiology of the comparable human disorder.

Characteristics

All of the autoimmune heart diseases are characterized by cardiac inflammation or by extensive fibrosis resulting from preceding inflammation. The clearest example of an autoimmune cardiac disease is ▶ myocarditis or inflammation of the heart muscle (1). The disease in humans sometimes follows viral infection.

In the laboratory this disease can be produced in experimental animals by infection with an appropriate virus that attacks the heart. A number of different viruses can induce autoimmune myocarditis in mice. They include ▶ Coxsackievirus B3 (a small RNA virus), encephalomyocarditis virus (a related small RNA virus), and murine cytomegalovirus (a large DNA herpesvirus). Following infection by any one of these three viruses, an inflammatory response occurs in the heart, consisting of large numbers of infiltrating mononuclear cells, such as macrophages, leukocytes, lymphocytes, and natural killer cells, distributed focally within the heart muscle. There may be evidence of cardiac cell death. Infectious virus can be isolated from the heart during this early stage of the disease, suggesting that the virus infection itself produces the pathology. After disappearance of the virus, the myocardial disease gradually resolves and, after a week or so, the heart appears perfectly normal in most strains of mice. In a few strains, however, the disease fails to resolve but rather changes in its character. The infiltration becomes largely lymphocytic and broadly distributed throughout the heart muscle surrounding the ventricles. Although there is little direct evidence of cardiac cell death, myocyte dropout suggests that many of the heart cells have died during previous stages of disease. In the mice that have developed this continuing phase of myocarditis, no infectious virus can be isolated. On the other hand, autoantibodies are evident. The most prominent population of autoantibodies is directed to cardiac ▶ myosin, a form of myosin that is uniquely produced by heart muscle cells. Thus it appears that, in certain genetically predisposed strains of mice, an autoimmune form of myocarditis has followed the earlier virus-mediated disease.

Direct evidence of an autoimmune basis of the later phase of disease comes from experiments in which mice are immunized with purified cardiac myosin or even with a short peptide sequence isolated from the large cardiac myosin molecule. Immunization with cardiac myosin reproduces the pathologic appearance of late-phase myocarditis in mice that are genetically susceptible to the late-phase disease following viral infection. Immunization with a closely related molecule, skeletal muscle myosin, produces no effect, illustrating the strict specificity of the autoimmune response. Other strains of mice do not respond to immunization with cardiac myosin, showing that the response is genetically restricted. Thus immunization with cardiac myosin or the myosin peptide is capable of reproducing autoimmune myocarditis even in the absence of virus.

Preclinical Relevance

Autoimmune myocarditis produced in mice or rats has proved to be a valuable model for studying the pathogenesis of human myocarditis (2). It has been shown, for example, that certain key mediators, called cytokines, are necessary for the progression from viral to autoimmune myocarditis. Among these critical cytokines are interleukin-1 (IL-1), tumor necrosis factor-alpha (TNF-α), and the third component of complement (C3). On the other hand, natural killer cells that are prominent in the early viral infection tend to diminish the later autoimmune disease. Nitric oxide (NO) is an important mediator of protection against the early viral infection which adds to the heart cell damage in the later autoimmune phase of myocarditis. These findings may prove to be valuable in designing new therapies for inflammatory heart disease in humans.

Relevance to Humans

Most cases of myocarditis occur in humans without warning. Many patients, however, report a recent viral infection and in about half of these cases serologic evidence of a recent Coxsackie B3 infection can be found [3]. A large number of other viruses, including adenoviruses, cytomegaloviruses, and even HIV, have been associated with myocarditis in humans. Although most humans appear to recover completely from a transient virus-induced myocarditis, a few may go on to chronic myocarditis with evidence of impaired cardiac function. Sometimes myocarditis evolves into ▶ dilated cardiomyopathy, a disease characterized primarily by extensive fibrotic changes in the heart muscle. This disease is the major cause of heart failure in young adults in industrialized countries. At this time, the only available treatment of dilated cardiomyopathy is cardiac transplantation.

In addition to viruses, a number of other microorganisms can cause myocarditis (4). The β-hemolytic streptococcus is associated with rheumatic fever and rheumatic heart disease, a condition that may affect all three portions of the heart, producing pericarditis, myocarditis, and endocarditis, with characteristic valvular lesions. Lyme disease can include myocarditis along with inflammation in other sites. In Central and South America, Chagas' disease due to infection by *Trypanosoma cruzi* is a common cause of myocarditis.

In addition to infectious agents, many chemicals have cardiotoxic effects and may cause inflammatory heart disease. They include ethanol, mercury, cobalt, anthracyclines, and the drug adriamycin. A severe dilated cardiomyopathy known as Keshan disease is associated with selenium deficiency.

In these instances, it is not clear whether the damage is due to direct, acute cardiotoxicity of the chemical or to indirect, immune-mediated cardiac inflammation triggered by the agent.

References

1. Afanasyeva M, Rose NR (2004) Viral infection and heart disease: Autoimmune mechanisms. In: Shoenfeld Y, Rose

NR (eds) Infection and Autoimmunity. Elsevier Science, Amsterdam
2. Rose NR, Baughman KL (1998) Immune-mediated cardiovascular disease. In: Rose NR, Mackay IR (eds) The Autoimmune Diseases, 3rd ed. Academic Press, San Diego, pp 623–636
3. Nugent AW, Daubeney PEF, Chondros P et al. (2003) The epidemiology of childhood cardiomyopathy in Australia. N Engl J Med 348:1639–1646
4. Abelmann WH (1988) Etiology, pathogenesis, and pathophysiology of dilated cardiomyopathy. In: Schultheiss HP (ed) New Concepts in Viral Heart Disease. Springer-Verlag, Heidelberg Berlin New York, pp 3–21

Cardiac Output (CO)

Measured in L/min/m^2, with a normal level of 2.6–4.2. Lower values indicate impaired myocardial function (heart insufficiency). For exact measurement, heart catheterization is necessary.
▶ Septic Shock

Cardiomyopathy

▶ Cardiac Disease, Autoimmune

Carrier

An immunogenic macromolecule (usually a protein) to which a hapten is attached, allowing the hapten to be immunogenic.
▶ Local Lymph Node Assay (IMDS), Modifications

CAS Number 17646-01-6-

▶ Dioxins and the Immune System

Caspase

A family of cysteine protease that cleave after an aspartate residue. The term caspase incorporates these elements (cysteine, aspartate, protease), which play important roles in the chain reactions that leads to apoptosis.
▶ Flow Cytometry

CD (Cluster of Differentiation)

Cell membrane molecules identified by monoclonal antibodies and used to differentiate leukocyte subpopulations.
▶ Humoral Immunity

CD Markers

H ZOLA · B SWART
Child Health Research Institute, Women's and Children's Hospital
Adelaide,
Australia

Synonyms
CD molecules, cluster of differentiation, (human) leukocyte differentiation antigens

Definition
CD markers are ▶ leukocyte cell surface molecules, as well as the respective ligands expressed by other tissues. CD markers are used to identify, count, study, purify, destroy, or in some other way work with leukocytes. The name originated from studies using antibodies against leukocytes, and the term CD marker is used in the context of antibody-based studies, although the CD number refers to the leukocyte molecule, not to the antibody used to detect it.
The CD nomenclature was devised to achieve standardization at a time when many new monoclonal antibodies were being described against leukocyte antigens. The number of published antibodies was increasing rapidly, and in many cases the corresponding antigens were unknown, and it was difficult to establish whether two antibodies with similar reactivities were in fact against the same antigen or not. The first International Workshop and Conference on Human Leukocyte Differentiation Antigens (HLDA) was organized to compare antibodies in a 'blind' manner through multilaboratory testing, to identify and characterize the corresponding antigens, and to develop an internationally agreed nomenclature (1). The first HLDA workshop has been followed by a series of Workshops, which are still continuing (www.hlda8.org).

Characteristics
CD molecules have a wide range of properties. Most of them are glycoproteins, but some are glycolipids. Many are integral membrane proteins, with at least one membrane-spanning hydrophobic sequence, but others

are extracellular proteins linked to the cell membrane through lipid anchors. Structures of a small selection of leukocyte markers are shown schematically in Figure 1. These molecules mediate a large number of diverse functions, as might be expected since leukocytes interact via their cell surface with other leukocytes, with endothelium, with foreign antigen, and with a large variety of signaling molecules.

Currently the CD nomenclature spans CD1–CD247, equating to some 300 molecules (because some molecules have been given designations such as CD1a and CD1b). The eighth HLDA workshop has tabulated approximately 200 additional leukocyte cell surface molecules which are candidate CD molecules (www.hlda8.org), and it has been estimated that the total number of distinct leukocyte cell surface molecules may be as high as 1000 (2).

Preclinical Relevance

CD markers, and the antibodies directed against them, are widely used in research studies on the human immune system and animal models of the immune system, and in studies of hemopoietic malignancies. For a review of the CD system for the laboratory mouse see Lai et al (3).

While the effects of exposure of humans to toxic substances on immunologic parameters is of major interest, in practice studies are better performed in laboratory animals in order to predict toxicity, rather than by evaluation of individuals who have suffered accidental exposure. Evaluation of immunotoxic effects in animals is an important aspect of safety evaluation of chemicals.

Relevance to Humans

CD markers are widely used in diagnostic immunology, hematology, and pathology, and have added enormous resolving power to these disciplines. For example, the number of circulating T cells is monitored using the CD3 marker in patients with HIV-AIDS, and treatment decisions are based on the result.

In healthy individuals the major cell types occur in proportions which lie within rather defined ranges—the 'normal range'. T lymphocytes constitute generally 70% of the lymphocyte fraction, with a range of 60% to 85%, while ▶ B lymphocytes generally comprise 3% to 20% of circulating lymphocytes in healthy individuals. The proportions in blood are very variable, for example changing transiently in response to exercise, and show a diurnal pattern of variation. Nevertheless, the numbers fall outside the normal range, in a number of disease situations, and these changes are measured using antibodies against appropriate CD markers. Figure 2 shows a typical analysis for two CD markers on blood cells from a control blood donor. Immunologic consequences, including changes in leukocyte populations as detected by CD markers, have been reported in individuals exposed to a wide range of toxic substances, ranging from lead and mercury to

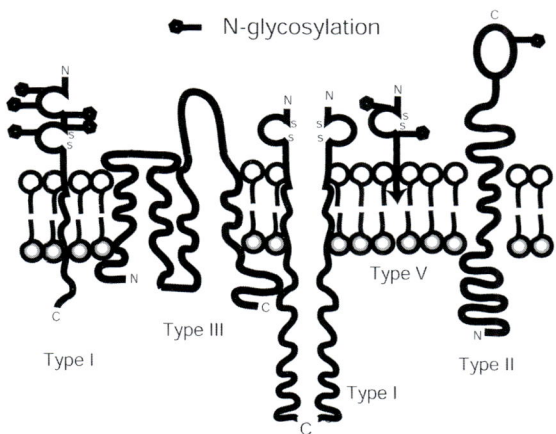

CD Markers. Figure 1 Schematic representation of a number of CD molecules. Type I membrane proteins have their C terminus inside the cell, a single membrane-spanning region, and an extracellular region which may consist of several domains with specific functions enabling interaction with other cells or extracellular signaling molecules. The intracellular sequence may contain sequences specialized to interact with intracellular signaling molecules. Type II membrane proteins are oriented the other way up, with their C terminus outside the cell, but are otherwise similar in structure. Type III molecules span the membrane more than once. They may have both N and C termini inside the cell, or only one terminus inside the cell, depending on whether they span the membrane an odd or even number of times. Type V membrane proteins do not span the membrane at all, but are linked to membrane lipid.

CD Markers. Table 1 The most widely used CD markers in diagnostic immunology

Cell type to be Identified	Most useful CD Marker
T lymphocyte	CD3
B lymphocyte	CD19
Monocyte	CD14
Natural killer cell	CD56, CD16
T helper lymphocyte	CD4
T suppressor/cytotoxic lymphocyte	CD8
Naive T lymphocyte	CD45RA
Memory T lymphocyte	CD45R0

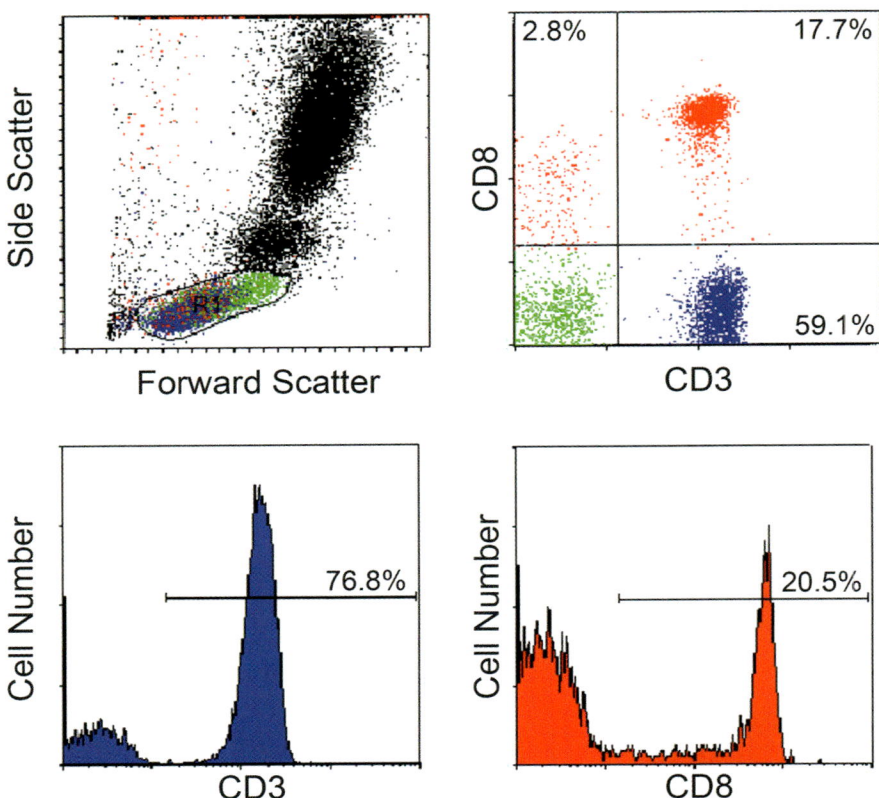

CD Markers. Figure 2 Typical flow cytometric analysis of leukocytes from blood using antibodies against CD markers. Analysis of light scatter in two directions (upper left panel) allows the lymphocyte fraction to be selected, while granulocytes, most monocytes, red cells, platelets, and dead cells fall outside the 'gated' region (outlined) and are excluded from analysis. In this experiment, antibodies against CD3 (which is on all T lymphocytes) and CD8 (which is on a subset of T lymphocytes that characteristically function to kill cells bearing foreign antigens or to suppress antibody responses) have been used, with different fluorochrome dyes. The lower left panel shows that CD3 divides the population of lymphocytes into two, with approximately 77% expressing CD3. The lower right panel shows that CD8 is expressed on 20% of lymphocytes. The upper right panel shows both dyes simultaneously, allowing enumeration of $CD8^+/CD3^+$ (cytotoxic/suppressor T cells, 17.7%), $CD8^+$ cells that do not express CD3 (2.8%, probably natural killer (NK) cells, which can be identified directly with additional CD markers), CD3 cells that do not express CD8 (59%, probably helper T cells, which can be identified directly with a CD4 antibody), and double-negative cells, which would include B cells, some monocytes which have not been excluded by the scatter gates, and some NK cells. Each of these cell types can be identified with other CD markers (see Table 1). Data from a Coulter Elite flow cytometer/cell sorter.

dioxins, and including cigarette smoking and air pollution. The fetus and newborn may be particularly susceptible.

Antibodies against CD markers are increasingly used therapeutically. Initially CD3, and more recently other antibodies against CD molecules, are used to treat or reverse organ graft rejection, while a number of antibodies against B cell molecules, especially CD20, are used increasingly in the treatment of lymphoma.

Regulatory Environment

Many monoclonal antibodies against CD markers have been accepted as reagents for diagnostic assays, and they are generally superior to the reagents they have replaced or they allow the use of superior assays. The environment for approval of CD monoclonal-antibody based diagnostics is thus favorable. By contrast, monoclonal antibodies as therapeutics are associated with a number of potential problems, leading to a very stringent regulatory environment. First, the specificity of a monoclonal antibody—while very high—does not rule out side effects due to reaction with other body components (for example, CD9 which might be useful in lymphoma treatment is ruled out because it is present on platelets and in the kidney). Second, antibodies may bind antigen and form complexes which can be deposited in the kidney and in other organs, causing disease. Even though CD markers are defined as cell surface molecules, many are shed from the surface and are therefore also present in the serum. Thirdly, the reaction triggered by the monoclonal antibody may have undesirable downstream effects. An example is the cytokine release reaction, which happens when antibodies against CD3 are administered to patients. The antibodies bind to the cells and induce the release of cytokines, which have powerful pharmacological effects. Finally, monoclonal antibodies are potentially immunogenic, leading to adverse reactions from the recipient's immune system.

The methodology of leukocyte marker determination is the subject of numerous recommendations and requirements bearing the authority of regulatory bodies such as the National Institutes of Health (NIH) and the Centers for Disease Control and Prevention (CDC) (4).

References

1. Bernard AR, Boumsell L, Dausset J, Schlossman SF (1984) Leukocyte typing I. Springer, Heidelberg Berlin NewYork
2. Zola H, Swart BW (2003) Human leukocyte differentiation antigens. Trends Immunol 24:353–354
3. Lai L, Alaverdi N, Maltais L, Morse HC 3rd (1998) Mouse cell surface antigens: nomenclature and immunophenotyping. J Immunol 160:3861–3868
4. Mandy FF, Nicholson JK, McDougal JS (2003) Guidelines for performing single-platform absolute $CD4^+$ T-cell determinations with CD45 gating for persons infected with human immunodeficiency virus. Centers for Disease Control and Prevention. MMWR Recomm Rep 52:1–13

CD Molecule

A marker expressed on the surface of leukocytes, which may be recognized by a monoclonal antibody and thus may be used to differentiate cell populations.
▶ CD Markers
▶ Canine Immune System

CD3

A five-chain molecular complex associated with the T cell receptor in the T cell plasma membrane. It occurs on all T cells, as well as on some subsets of natural killer cells, and anti-CD3 antibodies can thus be used as a marker for T cells. CD3 plays a key role in signal transduction, and in the formation of the immunological synapse, but many aspects of its function remain to be elucidated.
▶ Lymphocyte Proliferation

CD4

A single chain glycoprotein, also referred to as the T4 antigen, that has a molecular weight of 56 kD and is present on approximately two-thirds of circulating human T cells, including T helpers or T inducers. It is therefore a marker for T helper cells and functions as receptor for class II molecules of the major histocompatability complex (MHC).
▶ Trace Metals and the Immune System
▶ Idiotype Network

$CD4^+/CD8^-$

▶ Trace Metals and the Immune System

$CD4^+$ T Cells

▶ Helper T lymphocytes

CD8

An antigen, also referred to as the T8 antigen, that has a molecular weight of 32–34 kD. The CD8 antigen consists of two polypeptide chains, α and β, which may exist in combination as the α/α homodimer or the α/β heterodimer. This antigen binds to class I mixed histocompatibility cell molecules on antigen-presenting cells, and may stabilize interactions between antigen-presenting cells and class I cells.
▶ Trace Metals and the Immune System
▶ Idiotype Network
▶ Cytotoxic T Lymphocytes

CD28

A homodimeric molecule present on T cells. It acts as a receptor for CD80 and CD86 molecules. Cross-linking of CD28 with anti-CD28 antibody restores proliferation in the presence of suboptimal concentrations of anti-CD3, while ligation of CD28 with its natural li-

gands costimulates T cell effector and helper functions.
▶ Lymphocyte Proliferation

CD40 Ligand

Antigen-dependent T cell/B cell activation requires costimulatory signals. Immune deficiencies have shown that the interaction between CD40 on the B cell surface and CD40 ligand on the T cell surface is essential for the initiation of a germinal centre reaction. CD40 belongs to the family of tumor necrosis factor-like receptors.
▶ B Cell Maturation and Immunological Memory

CD45RO

Cell surface marker found on lymphocytes, activation dependent, and a developmental marker.
▶ Mucosa-Associated Lymphoid Tissue

Cell Adhesion Molecules

KRIS VLEMINCKX
Unit of Developmental Biology, Dept. Molecular Biomedical Research
Ghent University – VIB
Technologiepark 927
9052 Ghent
Belgium

Synonyms
cell adhesion receptors, adhesion molecules, CAMs

Definition
Cell adhesion molecules are transmembrane or membrane-linked glycoproteins that mediate the connections between cells or the attachment of cells to the substrate (such as stroma, basement membrane). Dynamic cell-cell and cell-substrate adhesion is a major morphogenetic factor in developing multicellular organisms. In adult animals, adhesive mechanisms sustain tissue architecture, allow the generation of force and movement, and guarantee the functionality of the organs (e.g. creating barriers in secreting organs, intestines and blood vessels) as well as generation and maintenance of neuronal connections. Cell adhesion is also an integrated component of the immune system and wound healing. At the cellular level, cell adhesion

Cell Adhesion Molecules. Figure 1 Different modes of cell-cell and cell-substrate adhesion and the mechanism of cytoskeletal strengthening. A: Three possible mechanisms by which cell adhesion molecules mediate intercellular adhesion. A cell-surface molecule can bind to an identical molecule (homophilic) on the opposing cell or interact with another adhesion receptor (heterophilic). Alternatively, cell adhesion molecules on two neighboring cells bind to the same multivalent secreted ligand (linker-mediated adhesion). Intercellular adhesion can take place between identical cell types (homotypic) or between cells of different origin (heterotypic) independent of the adhesion molecules involved. Cell-substrate adhesion molecules attach cells to specific compounds of the extracellular matrix (ECM). Cell-cell and cell-substrate adhesion can occur simultaneously. B: Intercellular and cell-substrate adhesion can be strengthened by indirect intracellular linkage of the cytoplasmic tail of the adhesion molecules to the cytoskeleton and by lateral clustering in the membrane.

molecules do not just function as molecular glue. Several signaling functions have been attributed to adhesion molecules, and cell adhesion is involved in processes such as ▶ contact inhibition, growth and ▶ apoptosis. Deficiencies in the function of cell adhesion molecules underlie a wide range of human diseases, including cancer, autoimmune diseases, and impaired wound healing.

Characteristics
At the molecular level, cell adhesion is mediated by molecules that are exposed on the external surface of the cell and are somehow physically linked to the cell membrane. In essence, there are three possible mechanisms by which such membrane-attached adhesion molecules link cells to each other (see Figure 1A). First, molecules on one cell bind directly to similar molecules on the other cell (homophilic binding). Secondly, adhesion molecules on one cell bind to other adhesion receptors on the other cell (heterophilic adhesion). Finally, two different adhesion molecules on two cells may both bind to a shared secreted multivalent ligand in the extracellular space. Also, cell-cell adhesion between two identical cells is called homotypic cell adhesion, while heterotypic cell adhesion takes place between two different cell types. In the case of cell-substrate adhesion the adhesion molecules bind to the extracellular matrix (ECM).

Cell Adhesion Molecules and the Cytoskeleton
Adhesion molecules can be associated with the cell membrane either by a glycosylphosphatidylinositol (GPI) anchor or by a membrane-spanning region. In the latter case the cytoplasmic part of the molecule

Cell-Cell Adhesion

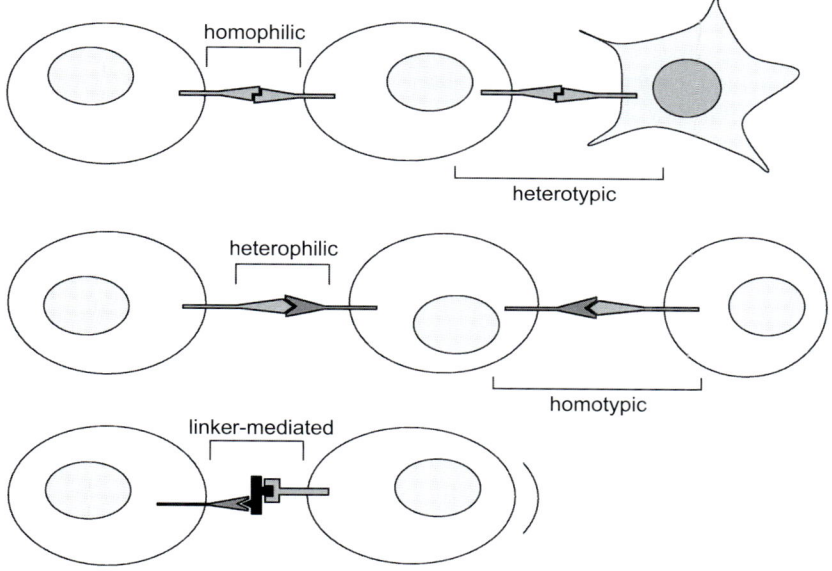

Cell-Substrate Adhesion

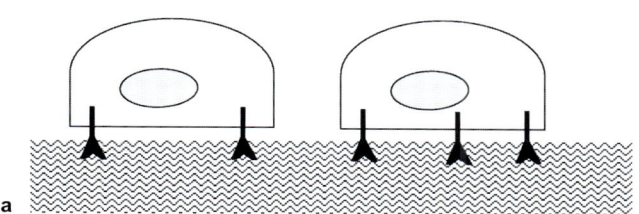

Cytoskeletal Strengthening

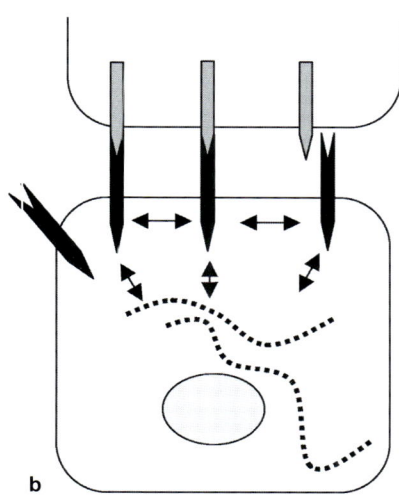
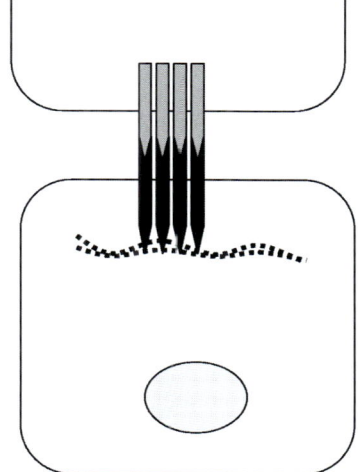

often associates indirectly with components of the ▶ cytoskeleton (e.g. actin, intermediate filaments, submembranous cortex). This implies that adhesion molecules, which by themselves establish extracellular contacts, can be structurally integrated with the intracellular cytoskeleton, and they are often clustered in specific restricted areas in the membrane—the so-called junctional complex (see Figure 1B). This combined behavior of linkage to the cytoskeleton and clustering considerably strengthens the adhesive force of the adhesion molecules. In some cases, exposed adhesion molecules can be in a conformational configuration that does not support binding to its adhesion receptor. A signal within the cell can induce a conformational change that activates the adhesion molecule (e.g. for integrins; see below). These mechanisms of regulation allow for a dynamic process of cell adhesion which, among others, is required for morphogenesis during development and for efficient immunological defence.

Classification of Cell Adhesion Molecules

Based on their molecular structure and mode of interaction, five classes of adhesion molecules are generally distinguished: the cadherins, the integrins, the immunoglobulin (Ig) superfamily, the selectins, and the proteoglycans (Figure 2).

Cadherins

Cadherins and proto-cadherins form a very large and diverse group of adhesion receptors. They are Ca^{2+}-dependent adhesion molecules, involved in a variety of adhesive interactions in both the embryo and the adult. Cadherins play a fundamental role in metazoan embryos, from the earliest gross morphogenetic events (e.g. separation of germ layers during gastrulation) to the most delicate tunings later in development (e.g. molecular wiring of the neural network). The extracellular part of vertebrate classical cadherins consists of a number of so-called cadherin repeats, whose conformation is highly dependent on the presence or absence of calcium ions. Only in the presence of calcium can homophilic interactions be realized, usually by the most distal cadherin repeat. Classical cadherins are generally exposed as homodimers and their cytoplasmic domain is tightly associated with the actin ▶ cytoskeleton. Cadherins are the major adhesion molecules in tissues that are subject to high mechanical stress, such as epithelia (E-cadherin) and endothelia (VE-cadherin). However, finer and more elegant intercellular interactions, such as synaptic contacts, also seem to involve cadherins.

Integrins

Integrins are another group of major players in the field of cell adhesion. They are involved in various processes such as morphogenesis and tissue integrity, hemostasis, immune response and inflammation. Integrins are a special class of adhesion molecules, not only because they mediate both cell-cell and cell-substrate interactions (with components in the ECM such as laminin, fibronectin and collagen), but also because they are functional as heterodimers consisting of an α subunit and β subunit. To date, at least 16 α subunits and 8 subunits have been identified. Of the theoretical

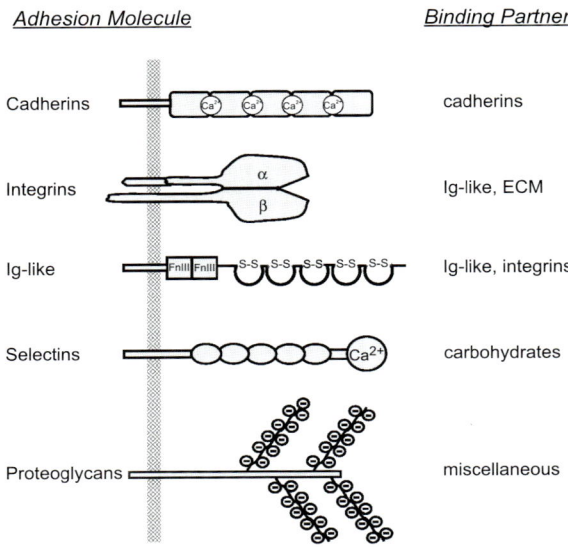

Cell Adhesion Molecules. Figure 2 The five major classes of cell adhesion molecules and their binding partners. Cadherins are Ca^{2+}-dependent adhesion molecules consisting of a varying number of cadherin repeats (five in the case of the classical cadherins). The conformation and activity of cadherins are highly dependent on the presence of calcium ions. In general, cadherin binding is homophilic. Integrins are functional as heterodimers consisting of an α subunit and a β subunit. They interact with members of the immunoglobulin superfamily or with compounds of the extracellular matrix (e.g. fibronectin, laminin). The immunoglobulin superfamily (immunoglobulin-like) is characterized by a various number of immunoglobulin-like domains (open circles) and more membrane-proximal often fibronectin type III repeats are observed (grey boxes). They can bind either homophilically, to other members of the immunoglobulin family, or to integrins. Selectins contain an N-terminal Ca^{2+}-dependent lectin domain (circle) that binds carbohydrates, a single EGF-like repeat (grey box) and a number of repeats related to those present in complement-binding proteins (ovals). Proteoglycans are huge molecules consisting of a relatively small protein core to which long side-chains of negatively charged glycosaminoglycans are covalently attached. They bind various molecules, including components of the extracellular matrix.

128 heterodimeric pairings, at least 21 are known to exist. While most integrin heterodimers bind to ECM components, some of them—more particularly those expressed on leukocytes—are heterophilic adhesion molecules binding to members of the Ig superfamily. The α subunit mostly contains a ligand-binding domain and requires the binding of divalent cations for its function (Mg^{2+}, Ca^{2+} and Mn^{2+}, depending on the integrin). Interestingly, integrins may be present on the cell surface in a nonfunctional and a functional configuration. Their cytoplasmic domain appears to be responsible for the conformational change that activates the integrin upon appropriate stimuli.

Ig Superfamily
Among the classes of adhesion molecules discussed here, the Ig superfamily is probably the most diverse. The main representatives are the neural CAMS (NCAMs) and vascular CAMs (VCAMS). As the name suggests, the members of this family all contain an extracellular domain consisting of different immunoglobulin-like domains. NCAMs sustain homophilic and heterophilic interactions that play a central role in regulation and organization of neural networks, specifically in neuron-target interactions and fasciculation. The basic extracellular structure consists of a number of Ig domains, which are responsible for homophilic interaction, followed by a discrete number of fibronectin type III repeats. This structure is then linked to the membrane by a GPI anchor or a transmembrane domain. The VCAM subgroup, including intercellular CAMs (ICAMs) and the mucosal vascular addressin adhesion molecule (MAdCAM), are involved in leukocyte trafficking (or homing) and extravasation. They consist of membrane-linked Ig domains that make heterophilic contacts with integrins. CD2 molecules are found on cytotoxic and T-helper cells and enhance their binding to antigen-presenting T cells. CD2 binding is pseudohomophilic, to highly homologous adhesion receptors. Other members of this family are LFA-3 (leukocyte function-associated antigen-3), carcinoembryonic antigen (CEA), "deleted in colon cancer" (DCC) and platelet endothelial (PE)CAM-1.

Selectins
These types of adhesion molecules depend on carbohydrate structures for their adhesive interactions. Selectins have a C-type ▶ lectin domain, which can specifically bind to discrete carbohydrate structures present on cell-surface proteins (often sialyl Lewis X). Intercellular interactions mediated by selectins are of particular interest in the immune system, where they have a fundamental function in trafficking and homing of leukocytes (e.g. E-, L- and P-selectin). They also play major roles in dendritic cells and Langerhans cells, both as antigen receptors and as adhesion molecules regulating migration of these antigen-presenting cells. An example of the latter is dendritic-cell specific ICAM-3 grabbing nonintegrin (DC-SIGN), important for both antigen uptake and migration.

Proteoglycans
Proteoglycans are very large extracellular proteins consisting of a relatively small protein core to which long chains of glycosaminoglycans are attached. Although poorly documented, proteoglycans may bind to each other or may be the attachment site for other adhesion molecules. An example is the homing receptor CD44 involved in the transmigration of lymphocytes.

Preclinical Relevance
Maintenance of Tissue Barriers
Adhesion molecules are crucial for maintaining functional and physical barriers in the organism. Barriers with the external world (e.g. in the intestine, the skin and oral mucosa) are important not only to avoid the penetration of chemical compounds and pathogens but also to prevent leaking out or evaporation of fluid. Also barriers within the body are vital, e.g. the blood-brain and blood-neuron barrier in the central and peripheral nervous system, respectively. Establishment and maintenance of these barriers is essential for normal body function.

Migratory Behavior of Leukocytes
In order to be able to fulfil their immense task, leukocytes migrate through the body and specifically traffic and home to the sites where they are needed. In general, three ways of migration are observed in the immune system.

- Dendritic antigen-presenting cells are found in peripheral organs like the skin and the intestinal epithelia where they make strong contacts with the surrounding cells and the ECM. When they capture and process antigens they become highly mobile and migrate to the lymphoid organs where they present their antigens to the lymphocytes. This migratory behavior requires a change in cell adhesion molecules, e.g. reduction of E-cadherin expression in the case of Langerhans cells.
- T and B lymphocytes survey the body, scanning for infectious pathogens, and for this purpose they constantly circulate in the vascular and lymphatic compartments. Lymphocytes can leave the blood vessels at the lymph nodes where they are confronted with the antigen-presenting cells, after which they proliferate and differentiate, traverse the lymphatic system, and then return to the vascular system.
- Granulocytes and monocytes circulate in the blood and extravasate into the surrounding tissue in response to inflammatory stimuli. Leukocyte migra-

tion involves a highly regulated adhesive mechanism, also known as the "multistep adhesion cascade". For simplicity, we will only discuss the homing and extravasation of neutrophils but the mechanism can be easily extrapolated to lymphocyte trafficking.

The Multistep Adhesion Cascade
The multistep adhesion cascade starts with the selective and local expression of selectins on the cells of the vessel wall in response to inflammatory stimuli. These selectins can bind to carbohydrates expressed on the neutrophils that pass by in the blood stream. This interaction, known as tethering, is of low affinity and transient and is easily disrupted by the continuous blood flow. As a result, the neutrophils roll along the surface of the endothelium. The neutrophils express integrins on their cell surface but these are in a nonfunctional state. However, chemokines released from the endothelial cells on which the neutrophils are rolling induce a G-protein-mediated conformational change in the integrins of the neutrophil. As a result these integrins are activated and can bind their targets, which are ICAMs, on the endothelial cells. Consequently, the neutrophils are arrested, attach firmly to the endothelium and migrate through it. Knowing that for lymphocytes also CD44 is involved in transmigration, and considering that to pass through the endothelial cell layer VE-cadherin mediated cell-cell contacts have to be disrupted, we can state that this "multistep adhesion cascade" of leukocytes involves all the major families of adhesion molecules discussed earlier.

Other Adhesive Interactions in the Immune System
Other adhesive interactions are involved in the immune system. It is for instance found that immature thymocytes require intimate interactions with the epithelial cells in the thymus. These contacts seem to be established by homophilic but heterotypic E-cadherin adhesion. Interestingly, E-cadherin can also form heterophilic interactions with a $\alpha_E \beta_7$ integrin on certain T lymphocytes. Aggregation of platelets also involves adhesion receptors, in this case members of the integrin family, and also here the integrins need to be activated by agonists like thrombin in order to induce effective adhesion. Finally, T cells interact with antigen-presenting cells through binding of the T-cell receptor with the antigen-major histocompatability complex (MHC). However, this binding is of very low affinity. Efficient interaction between the T cell and the antigen-presenting cell requires the formation of an immunological synapse with the cooperation of adhesion molecules, in most cases a heterophilic interaction between an LFA-1 and ICAM-1, or a pseudohomophilic interaction between CD2 adhesion molecules (see above).

Relevance to Humans
Inflammation: Impaired Immune Response
General Involvement

The central role of adhesion molecules in the immune response makes these molecules interesting therapeutic targets for controlling inflammatory diseases in humans (1). Several potential scenarios can be envisioned.

- The activity of specific adhesion molecules (e.g. integrins involved in extravasation of leukocytes) can be blocked with humanized monoclonal antibodies or with peptidomimetics.
- The influence of the cell on the activity of the adhesion molecules (inside-outside signaling) and, vice versa, the signaling properties of the adhesion receptors (outside-inside signaling) and their intimate and functionally important association with the cytoskeleton opens other roads for intervention.
- As the expression of adhesion molecules on leukocytes and endothelial cells is under strict control of cytokines and chemokines, also specific interference with the expression or the function of these secreted factors is therapeutically used to temper exaggerated immune responses.

Specific Genetic Diseases

Rare genetic leukocyte adhesion deficiency (LAD) diseases are associated with mutations in adhesion molecules. LAD type I is characterized by a β_2-integrin deficiency and is associated with the inability of leukocytes to emigrate from the vasculature (1). This results in recurrent pyogenic infections often leading to death from septicemia in the first or second decade. LAD type II is another genetic disease, associated with defects in the GDP-fucose transporter. This causes aberrant fucosylation of glycoproteins, including the specific ligands of the selectins. Aside from developmental abnormalities, the affected patients also suffer from diseases associated with impaired leukocyte trafficking.

Autoimmune Diseases

Autoimmunity is characterized by the reaction of the immune system against self-antigens. This has of course detrimental effects on the patients who have to rely on immunosuppressive drugs for their entire life. In some cases, the targets recognized by antibodies or T cells associated with autoimmunity are cell adhesion molecules. Paramount examples of the latter are pemphigus foliaceus and pemphigus vulgaris, where antibody-mediated autoimmune reactions are directed against desmoglein 1 and desmoglein 3, respectively (2). Desmogleins (and desmocollins) are members of the cadherin family that are localized at the ▶ desmosomes and are responsible for epidermal cell-cell adhesion in keratinocytes. The autoimmune diseases result in severe intraepidermal blistering.

Adhesion Molecules as Pathogen Receptors

The food-borne pathogen *Listeria monocytogenes* is able to enter nonphagocytic cells. Entry in target cells is mediated mainly by two proteins, internalin (InlA) and InlB. Interestingly, InlA binds specifically with E-cadherin, which is used as a receptor for internalization (3). This is achieved by cross-talk between the bacterium and the host cell through activation of specific signaling proteins inducing phagocytosis.

The human immunodeficiency virus (HIV-1) surface protein gp120 is able to bind the selectin DC-SIGN in dendritic antigen-presenting cells in peripheral tissues such as the mucosa or the skin (4). Binding of gp120 occurs without viral entry. Rather, the virus piggybacks on the dendritic cell when it travels to secondary lymphoid organs. In this compartment it is presented to T cells expressing CD4 and chemokine receptor, thus permitting infection.

T Cell Epitope Mimicry

In some specific cases, autoimmunity is thought to occur as a byproduct of the immune response against a microbial infection (5). An example is Lyme arthritis induced by the bacterium *Borrelia burgdorferi*. Lyme arthritis is an inflammatory joint disorder resembling rheumatoid arthritis. It occurs relatively late after infection and is frequently resistant to antibiotic treatment, suggesting it is actually an autoimmune-like disease. This appears to be the result of a T-cell response against an epitope in the outer surface protein A (OspA) of *B. burgdorferi*. The recognized peptide OpsA(165–173) is also present in the human LFA-1α protein (residues L326–345). Hence, the T-cell reaction against the OpsA(165–173) epitope automatically provokes an autoimmune response against the integrin LFA-1α.

Adhesion Molecules as Proteolytic Targets for Pathogens

Interestingly, some pathogens are disrupting intercellular adhesion through proteolytic activity. This is the case for *Staphylococcus aureus*, which secretes an exfoliating toxin that specifically binds and cleaves the desmosomal cadherin desmoglein 1 (6). This generates the staphylococcal scalded skin syndrome, a severe epidermal blistering disease that resembles the autoimmune disease pemphigus foliaceus (see above).

References

1. Marshall D, Haskard DO (2002) Clinical overview of leukocyte adhesion and migration: where are we now? Sem Immunol 14:133–140
2. Moll R, Moll I (1998) Epidermal adhesion molecules and basement membrane components as target structures of autoimmunity. Virchows Arch 432:487–504
3. Cossart P, Pizarro-Cerda J, Lecuit M (2003) Invasion of mammalian cells by Listeria monocytogenes: functional mimicry to subvert cellular functions. Trends Cell Biol 13:23–31
4. Benoist C, Mathis D (2001) Autoimmunity provoked by infection: how good is the case for T cell epitope mimicry? Nat Immunol 2:797–801
5. Figdor CG, van Kooyk Y, Adema GY (2002) C-type lectin receptors on dendritic cells and Langerhans cells. Nat Rev Immunol 2:77–84
6. Amagai M (2003) Desmoglein as a target in autoimmunity and infection. J Am Acad Dermatol 48:244–252

Cell Adhesion Receptors

▶ Cell Adhesion Molecules

Cell-Based Bioassays

▶ Cytokine Assays

Cell-Mediated Allergic

▶ Delayed-Type Hypersensitivity

Cell-Mediated Hypersensitivity, Type IV Immune Reaction

▶ Delayed-Type Hypersensitivity

Cell-Mediated Immunity

Cell-mediated immunity (CMI) is part of the acquired arm of immunity and is mediated by the direct actions of cells on foreign (non-self) agents such as viruses, bacteria, and other antigens. Cell-mediated immunity is more specifically defined as the T-cell-mediated responses such as delayed hypersensitivity or cytotoxic T cell activity, antibody-dependent cellular cytotoxicity mediated by natural killer cells, and soluble-factor-mediated macrophage cytotoxic responses. It protects against intracellular bacteria, viruses, and cancer, and is responsible for graft rejection.

▶ Leukemia

▶ Graft-Versus-Host Reaction
▶ Lymphoma
▶ Primate Immune System (Nonhuman) and Environmental Contaminants
▶ Rabbit Immune System
▶ Flow Cytometry

Cell-Mediated Lysis

B Paige Lawrence
Department of Pharmaceutical Sciences
College of Pharmacy, Washington State University
Pullman, WA 99164 6534
USA

Synonyms
CTL activity, NK cell killing, antibody-dependent cell-mediated cytotoxicity, ADCC, target cell killing.

Definition
Cell-mediated lysis describes the killing of specific cellular targets by cells of the immune system. The effector cells are most often differentiated CD8+ T cells (that is, cytotoxic T lymphocytes or CTL) or ▶ natural killer cells (NK cells). In some instances CD4+ T cells can acquire a cytolytic phenotype, but this is considered rare. With regard to host defense, cell-mediated lysis plays an important role in resistance to intracellular pathogens. Specifically, CTL primarily kill host cells that are infected with viruses, whereas NK cells kill virally-infected and bacterially-infected host cells. Additionally, both CTL and NK cells recognize and kill certain types of tumor cells, and on rare instances cells infected with bacteria are targets of CTL-mediated killing.

Characteristics
An obligatory feature of cell-mediated lysis is that it involves cell-cell contact between the effector and target cells. However, this interaction is directed by different molecules, leading to three distinct mechanisms of cell-mediated lysis.

MHC-Restricted Lysis
The acquisition of killing ability by CD8+ T cells requires the initial activation, clonal expansion and differentiation of naive CD8+ T cells that are specific for a particular peptide presented by an antigen-presenting cell (APC) in the context of major histocompatibility complex (MHC) class I molecules. This activation process requires co-stimulatory signals that are provided by the APC. Following this process, differentiated CD8+ T cells are now mature CTL that recognize and destroy target cells. CTL recognize their specific targets using T cell receptors (TCR) on the CTL, which bind to peptide-MHC class I molecules on the surface of target cells.

CTL can kill antigen-bearing target cells via three different mechanisms (1) which are described more fully in the section on cytotoxic T lymphocytes. One of these mechanisms involves the release of ▶ perforin and ▶ granzymes by the CTL in response to TCR engagement with peptide-MHC molecules on the target cell.

Upon secretion, perforin monomers self-assemble, creating membrane-spanning pores. These pores compromise the integrity of the target cell membrane, which permits granzymes access to the target cell through the holes created by perforin. Granzymes are serine proteases that stimulate programmed cell death machinery via the activation of caspase enzymes and other pro-apoptotic proteins.

In addition to the perforin-granzyme system, CTL can also destroy antigen-bearing targets using ▶ Fas ligand (FasL) and tumor necrosis factor (TNF)-α. Both Fas, the cognate co-receptor for FasL, and type I TNF receptors contain "death domains" in their cytoplasmic tails. Binding of receptors with death domains by their respective ligands activates an intracellular signaling cascade, leading to apoptotic cell death.

NK Cell-Mediated Lysis
While NK cells are lymphocytes, they do not require prior exposure to antigen in order to acquire cytolytic activity, nor do they exist as naive precursors. Instead, mature NK cells constitutively synthesize and store perforin and granzymes, and the powerful cytolytic capacity of NK cells has been recognized for many years.

In contrast to T cells—which have antigen-specific receptors and undergo a maturation process that deletes cells with self-reactive TCR—NK cells lack antigen-specific receptors and do not undergo clonal selection or screening. Instead, target cell recognition and the cytolytic activity of NK cells are controlled by the expression of a large family of cell surface receptors referred to as the NK gene complex (NKC). Members of this family act as either inhibitory or stimulatory receptors (2,3).

Inhibitory Receptors
While other mechanisms likely also exist, one well-established pathway by which NK cells recognize target cells as "self" is the interaction of inhibitory receptors on NK cells and MHC class I molecules on another cell. This concept is often referred to as the "missing self hypothesis" and is predicated on the idea that NK cells require a constant "off" signal, delivered through cell surface inhibitory receptors. In the ab-

gently removed. After washing immobilised cells might also be recovered by vigorous pipetting or with a cell scraper.

Affinity Chromatography
In principle similar to the panning technique, antibodies are also immobilised on a solid support. For affinity chromatography antibodies are conjugated to different kinds of beads (e.g. polyacrylamide), which are then used to prepare immunoaffinity columns. Cells are applied to the immunoaffinity column. While cells recognised by the antibodies bind to and are retained on the column, unbound cells flow through and are easily recovered. Recovery of the cells bound to the column is also possible, but can be difficult and inefficient. Therefore affinity chromatography is mainly used for cell depletion approaches.

Particle Sedimentation
Antibodies are attached to large, high-density particles (e.g. 10 μm diameter nickel beads) for example by passive adsorption. Cells are mixed with the antibody-coated high-density particles. Then particles are allowed to settle. Settled particles and the bound cells form a pellet, thus unbound cells stay in suspension and can be easily recovered. Unbound cells might be subjected to several rounds of processing to efficiently remove antigen-expressing cells. Because recovery of the cells bound to the high-density particles is difficult and inefficient, particle sedimentation is mainly used for depletion approaches.

Magnetic Cell Sorting (MACS)
For immunomagnetic cell separation antibodies are attached to super-paramagnetic beads consisting of iron oxide and polymer. Significantly different types of magnetic cell separation methods have been evolved.

In one technique cells are labeled with antibodies attached to large polystyrene beads (2–5 μm diameter) under continuous mixing. Magnetically labelled cells are then separated from unlabelled cells in a separation vessel (e.g. tube) by exposing the cell mixture to a magnetic field using a permanent magnet. Unlabelled cells stay in suspension and can easily be recovered. Labelled cells and free beads are attracted to the magnet and retained on the side of the separation vessel. They can be recovered after removal of the magnetic field. Because these types of beads interfere with some subsequent applications of the cells (e.g. immunofluorescent analysis or cell culture) it might be necessary to release the beads from the labelled cells (which can be achieved in a number of different ways).

In the MACS technology cells are labelled with antibodies covalently coupled to very small super-paramagnetic MicroBeads (20–100 nm diameter) within minutes. The cell suspension is then passed over a column containing a ferromagnetic matrix (e.g. iron spheres) placed in a strong external magnetic field, which creates a high-gradient magnetic field in the column. The magnetically labelled cells are retained in the column, while unlabelled cells pass through. After removal of the column from the magnetic field, the magnetically retained cells can be eluted. Both labelled and unlabelled cells can be recovered and directly used for analysis as well as cell culture (including in vivo transfer). Due to their size and composition (iron oxide and polysaccharide) MicroBeads do not affect light scattering or fluorescent properties of labelled cells, and they are biodegradable.

Fluorescence-Activated Cell Sorting (FACS)
For immunofluorescence analysis and separation by flow cytometry cells are labelled with specific probes, especially antibodies conjugated to fluorochromes. A flow cytometer is an instrument to measure the optical properties of individual stained cells. Cells focused in a liquid stream are sequentially passing a laser beam. Scatter and fluorescence signals of illuminated cells are detected. A fluorescence-activated cell sorter is a flow cytometer equipped to separate individual cells based on their detected properties.

Use of different antibodies with different fluorochromes, that is with distinct fluorescence colours (i.e. emission spectra), allows analysis and separation of cells for multiple parameters at the same time. Because cells are processed sequentially (one at a time) flow sorting allows sorting of single cells, e.g. for single cell analysis or cloning. However therefore sorting of large cell numbers can take substantial time. Flow cytometry allows the quantitative analysis of the density of antigen expression on single cells over a large dynamic range and thereby allows very precise defining of the target cells for separation.

Pros and Cons
Specific methods for the identification and separation of cells from the immune system have, in the last 30–40 years, tremendously improved our understanding of the nature and the function of many different cell types of the immune system. However, especially cell functions observed in vitro and their relevance need to be confirmed in vivo—with adoptive transfer experiments or in vivo manipulation.

Beside particular advantages and limitations of each of the cell separation methods there are a number of "general" separation parameters, which are important for the decision of which method(s) to choose for the separation of a specific target cell population from a given starting population:

- purity and starting frequency of target cells (enrichment/depletion rate)

- recovery (or yield)
- viability and functionality of target cells
- total starting cell number
- total number of target cells
- reproducibility, complexity, time and costs.

Several methods can result in reasonable purity if the starting frequency of target cells before separation is high (e.g. for $CD3^+$ T cells among PBMC). However only a few methods result in a good purity, if the starting frequency of target cells before enrichment is low (e.g. for $CD34^+$ cells from PBMC). Methods leading to comparable purities of target cells might differ significantly with respect to the recovery of target cells or the time required for separation. The latter is especially relevant for FACS, the only "serial" separation method (compared to the other "parallel" separation methods). In flow cytometric sorting the separation time is strongly influenced by the total cell number to be processed. A conventional flow sorter can separate about 10,000 cells per second (separation of 10^7 cells takes 17 minutes) but separation of 10^{10} cells (e. g. from leukapheresis harvest) takes 278 hours (compared to minutes with some "parallel" separation methods).

Some of the old methods are not frequently used anymore, e.g. methods of group II, because they give lower purity and often also lower recovery and they are less versatile than modern immuno-separation methods of group III. Among methods of group III MACS and FACS are state of the art and are commonly used nowadays.

However especially methods of group I are still widely used, because they allow processing of large cell numbers in a short time with low costs. They are primarily used for "general" cell preparation (like PBMC from peripheral blood). In addition they can significantly improve quality, time and/or costs of other subsequent cell separation methods, e.g. as "pre-enrichment" steps before MACS or FACS.

Relevance to Humans

Beside the impact of cell separation techniques on biomedical research, some techniques have also particular relevance to humans in the field of cellular therapies.

The large-scale isolation of haematopoietic stem cells from leukapheresis or bone marrow harvests, e.g. with automated magnetic cell selection devices is used for autologous as well as allogeneic transplantation.

Furthermore there are many ongoing attempts to use different cell types of the immune system for the treatment of infectious diseases, tumours or autoimmunity. This includes, for example, dendritic cells (either derived from monocytes or hematopoietic stem cells, or directly isolated from leukapheresis harvests), NK cells, T cells, or particular T cell subsets (CD4 cells, CD8 cells, antigen-specific T cells or regulatory T cells). In addition clinical research is investigating the potential use of (adult) stem cells for tissue engineering and regeneration.

References

1. Radbruch A (ed) (1999) Flow Cytometry and Cell Sorting, 2nd ed. Springer-Verlag, Berlin
2. Recktenwald D (ed) (1997) Cell Separation Methods and Applications. Marcel Dekker, New York

Cell Sorting

▶ Cell Separation Techniques

Cellular Immune Reactions

An adaptive immune reaction mediated by antigen-specific (allergen-specific) effector T cells. Such an immune response is directed to cell-bound antigens, and cannot be transferred to naive recipients by injection of antibodies.

▶ Hypersensitivity Reactions

Central Tolerance

Tolerance mechanisms induced in the thymus for T cells and in the bone marrow for B cells.

▶ Tolerance

Centroblast

Intermediately differentiated B lymphocyte present in germinal centres; a medium to large cell with a round to ovoid nucleus.

▶ Germinal Center

Centrocyte

Intermediately differentiated B lymphocyte present in germinal center; medium-sized cell with irregular nucleus.

▶ Germinal Center

CFA

▶ Autoimmune Disease, Animal Models

CFU-Meg

▶ Colony-Forming Unit Assay: Methods and Implications

Chagas Disease

Chagas disease is caused by the protozoan parasite *Trypanosoma cruzi* which is transmitted by various *Reduviidae* (bloodsucking bugs), and occurs in south and central America. In the chronic phase of the disease, trypanosomes live intracellularly as amastigotes in various tissues including muscle cells.

▶ Trypanosomes, Infection and Immunity

Chemical Allergen

▶ Hapten and Carrier

Chemical Structure and the Generation of an Allergic Reaction

DAVID BASKETTER
Applied Science and Technology Safety and Environmental Assurance Centre
Unilever Colworth Laboratory
Sharnbrook
Bedford
MK44 1LQ
UK

Synonyms
Structure activity relationships, quantitative structure activity relationships.

Definition
The structure of a chemical and its associated ▶ physicochemical properties are fundamental determinants of the nature, type and degree of allergic reaction(s) that it may cause. Typically, chemicals which cause allergic reactions have a molecular weight in excess of 1000 Daltons, allowing ready penetration through epithelial surfaces, including the skin. However, exceptions do occur.

Characteristics
Chemicals must bind firmly with proteins in order to behave as ▶ haptens. For organic chemicals, this means the formation of covalent bonds; for the few allergenic metals (like nickel) the formation of coordination complexes is key. Extremely rarely—for example, for some drug allergens—tight interactions of organic chemical with protein, such as those encountered in ligand-receptor interactions, have been proposed. Thus the primary elements examined in relating structure to allergy are those chemical substructures which are—or can lead to—reactive moieties capable of covalently binding with proteins. Theoretical predictions on this topic were made more than 20 years ago by Dupuis and Benezra. However, the broadest identification of chemical substructures associated with allergic sensitization has been compiled into a computer-based expert system DEREK (Deductive Estimation of Risk from Existing Knowledge). The original skin sensitization rule-base contained around 40 rules, which were derived from an historical database containing data from guinea-pig maximization tests on 135 chemicals that had been classified as skin sensitizers according to European Union criteria, as well as a similarly sized group of non-sensitizers. As a result of development of the system over subsequent years, the number of ▶ structural alert rules for skin sensitization currently stands at 61 in version 6 of the programme.

Other expert system approaches to the prediction of skin sensitization include TOPKAT (TOxicity Prediction by Komputer Assisted Technology) and CASE (Computer Automated Structure Evaluation) systems. These computer-based systems are built upon varying approaches, but all employ physicochemical descriptors of chemical sensitizers and non-sensitizers as a means to provide a more general characterization of potential allergens. None of the computer systems has undergone any formal validation, indeed there has only been one significant independent assessment of a system, DEREK, undertaken by representatives of the German regulatory authority. The review was positive, but indicated much work to be done.

Allergy is not an all-or-none phenomenon, so an important question is whether any information on allergen potency can be derived from a consideration of its structure. Initial attempts to derive quantitative structure-activity relationships (QSARs) for skin sensitizers by Roberts and Williams in 1982 tried to relate the elements of dose, skin penetration, and electrophilic reactivity. This proved successful for several families of chemical allergens. Work has continued with successful QSARs recently being reported for groups of aldehydes. However, QSARs cover only a tiny fraction of the world of chemistry and, to date, more general rules relating reactivity to skin sensitization have not been elucidated. The only exception to this state of affairs has come from the work of Dave Roberts, a pioneer of QSAR, who with his colleague Grace Patlewicz from Unilever succeeded in developing a QSAR which appeared to cover both aldehydes and ketones—a small but perhaps significant, step.

A vital component in the appreciation of how chemical structure relates to sensitization may arise from the development of a fuller knowledge of how chemical allergens actually react with protein. This challenge has come under renewed scrutiny in recent years. A body of work has been published by Lepoittevin and colleagues describing in details the reactions of selected haptens with protein nucleophiles. Unfortunately (though importantly) these studies suggest that protein hapten interactions may be more complex than previously thought. For example, it is suggested that reaction specificity, rather than rate of reaction, might be a key determinant of whether a chemical can behave as a hapten. Such thoughts are echoed in the work of others who have also presented views on the role of derivatization of specific amino acids, notably cysteine, as the primary driver of the allergic response. Currently however what is most evident is that more data is needed before general conclusions can be drawn.

Finally, it is important to remember that the structure of the chemical applied to skin may differ from that which actually reacts with endogenous protein. Two primary routes exist which can convert otherwise unreactive chemicals into reactive species—air oxidation and metabolic conversion in the skin. The first of these has been extensively studies by the group of Karlberg and colleagues, with particular focus on colophony and on limonene. In contrast, skin metabolism of chemicals in relation to allergy is largely a theoretical science. The probable important routes of metabolism, as well as the current state of knowledge, have been condensed into a book by Smith and Hotchkiss. At present however, it is still very hard to predict to what, if any, extent either oxidation and/or metabolism may impact upon skin sensitization.

It is worth noting that in contrast to contact allergy very little progress has been made on understanding the relationship between chemical structure and respiratory allergy. It is a paradox that, although (it is assumed that) chemical respiratory allergens must also bind covalently with protein, the great majority of contact allergens appear not to present significant respiratory allergenic potential. This implies that there exists a level of complexity in the relationship between chemical structure and the generation of allergenic reactions which we have still to understand.

Preclinical Relevance

There are special guidelines regulating the investigation of contact allergy for chemicals, agrochemicals, pharmaceuticals and cosmetics, and some attention is paid to the ability of chemicals to cause respiratory allergy. For contact allergy, SARs may be taken into account in the general consideration of whether a chemical may be a sensitizer, although the information is not sufficient to classify a substance. For respiratory allergy, SAR essentially is not considered; however, isocyanate chemicals are assumed to be respiratory allergens unless proven otherwise.

Relevance to Humans

The predictivity of (Q)SARs for humans has not been assessed in a formal manner. Much of what is known concerning how chemistry drives allergy has been based on data from guinea-pig and mouse predictive tests. Although these may have up to 90% correlation with humans, important gaps remain in our knowledge —not least the impact of intraspecies differences, notably in metabolism, and how this impacts not only specificity but also the intensity of the allergic response.

Regulatory Environment

Both contact and respiratory allergy represent important human health risks and so are addressed in regulatory toxicology. Currently, the impact of (Q)SAR on these regulations is limited.

References

1. Barratt MD, Basketter DA, Roberts DW (1997) Quantitative structure activity relationships. In: LePoittevin J-P, Basketter DA, Dooms-Goossens A, Karlberg A-T (eds) The Molecular Basis of Allergic Contact Dermatitis. Springer-Verlag, Heidelberg Berlin New York, pp 129–154
2. Dupuis G, Benezra C (1982) Allergic Contact Dermatitis to Simple Chemicals: A Molecular Approach. Marcel Dekker, New York Basel
3. Patlewicz GY, Wright ZM, Basketter DA, Pease CK, Lepoittevin J-P, Gimenez, Arnau E (2002) Structure activity relationships for selected fragrance allergens. Cont Derm 47:219–226
4. Roberts DW, Williams DL (1982) The derivation of quantitative correlations between skin sensitization and physico-chemical parameters for alkylating agents and their application to experimental data for sultones. J Theor Biol 99:807–825
5. Rodford R, Patlewicz G, Walker JD, Payne MP (2003) Quantitative structure–activity relationships for predicting skin and respiratory sensitization. Environ Toxicol Chem 22:1855–1861
6. Smith CK, Hotchkiss SAM (2001) Allergic Contact Dermatitis: Chemical and Metabolic Mechanisms. Taylor & Francis, London

Chemoattractants

▶ Inflammatory Reactions, Acute Versus Chronic

Chemokine

A cytokine that causes chemotaxis in leukocytes by inducing a directional movement up the concentration agent. Most of the around 50 chemokines are rather preferential chemoattractants for certain leukocyte subclasses. Chemokines also induce adhesiveness and leukocyte activation.

▶ Cytokines

Chemokine Receptor Antagonists

These are chemokine receptor blocking agents that do not induce receptor-mediated signal transduction. Numerous antagonists have been generated by means of chemokine structure modifications. For potential use in the treatment of inflammatory diseases small molecular weight (non-peptide) compounds are favored over chemokine-derived inhibitors.

▶ Immune Cells, Recruitment and Localization of

Chemokine Receptors

Seven-transmembrane-domain receptors responsible for transducing signals generated upon binding of chemokines. They belong to a subset of the G protein-coupled receptor (GPCR) superfamily.

▶ Chemokines

Chemokines

RAFAEL FERNANDEZ-BOTRAN
Dept Pathology and Laboratory Medicine
University of Louisville
Louisville, KY 40292
USA

Synonyms
Chemotactic cytokines, small secreted cytokines.

Definition
Family of small cytokines with chemotactic properties that signal through seven transmembrane receptors coupled to G_I proteins. Besides regulating leukocyte recruitment and migration, chemokines have many other physiologic functions, including angiogenesis/angiostasis, lymphocyte development, immunoregulation, and metastasis (1–3).

Characteristics

Although the first chemokines were originally identified and characterized based on their biological activity, many members of this family have been recently identified through the application of bioinformatics and expressed sequence tag (EST) databases (2).

Approximately 40 different chemokines have been identified to date. Chemokines have been classified into four subfamilies based on the motif displayed by the first two cysteine residues located near their N-terminal end: CC-, CXC-, C- and CX_3C (3,4). The CXC subfamily is further divided into two groups depending on the expression of an ▶ ELR motif before the first cysteine (that is, ELR-CXC and non-ELR-CXC chemokines). Traditionally, the names given to chemokines were based primarily on their cellular sources or functional properties, creating a great deal of ambiguity and confusion. A new systematic nomenclature based on the different subfamilies has been proposed for the chemokines (ligands) and their receptors (4). The human chemokines, with their systematic and classical nomenclature, are listed in Table 1.

Structurally, chemokines are small proteins, usually 60–90 amino acids in length. With exception of the C- chemokines—they have four highly conserved cysteine residues that form two disulfide bonds, which function to stabilize their structure and limit their configuration. Because of this, most cytokines have similar three-dimensional structures (5). Although some chemokines form dimers at high concentrations, the biologically active molecule is the monomeric form (5). While the genes encoding most human CXC and CC chemokines are tightly clustered in chromosomes 4 and 17, respectively, at least five other chromosomes are known to contain chemokine genes, indicating considerable complexity in terms of chromosomal organization (5).

The biological effects of chemokines are exerted through interaction with specific receptors (chemokine receptors) located on the membrane of target cells. These are seven-transmembrane-domain receptors that belong to a subset of the ▶ G protein-coupled receptor (GPCR) superfamily (2,3). Like the chemokines, the genes encoding ▶ chemokine receptors are also tightly clustered, most of them located in chromosomes 2 and 3 (5). At least 17 different chemokine receptors have been identified (see Table 1) in addition to several orphan GPCR with high similarities to che-

Chemokines. Table 1 List of human chemokines, showing their systematic and classical names and the main receptors with which they interact

Systematic Name	Classical Name	Chemokine Receptors
CXC Family		
CXCL1	GROα/MGSA-α	CXCR2, CXCR1
CXCL2	GROβ/MGSA-β	CXCR2
CXCL3	GROγ/MGSA-γ	CXCR2
CXCL4	PF4	?
CXCL5	ENA-78	CXCR2
CXCL6	GCP-2	CXCR1, CXCR2
CXCL7	NAP-2	CXCR2
CXCL8	IL-8	CXCR1, CXCR2
CXCL9	Mig	CXCR3
CXCL10	IP-10	CXCR3
CXCL11	I-TAC	CXCR3
CXCL12	SDF-1α/β	CXCR4
CXCL13	BLC/BCA-1	CXCR5
CXCL14	BRAK/bolekine	?
CXCL15	? (lungkine in mouse)	?
CC Family		
CCL1	I-309	CCR8
CCL2	MCP-1/MCAF	CCR2
CCL3	MIP-1α/LD78α	CCR1, CCR5
CCL4	MIP-1β	CCR5
CCL5	RANTES	CCR1, CCR3, CCR5
CCL6	? (MRP-1 in mouse)	?
CCL7	MCP-3	CCR1, CCR2, CCR3
CCL8	MCP-2	CCR3
CCL9/10	? (MRP-2, MIP-1γ in mouse)	?
CCL11	Eotaxin	CCR3
CCL12	? (MCP-5 in mouse)	CCR2
CCL13	MCP-4	CCR2, CCR3
CCL14	HCC-1	CCR1
CCL15	HCC-2/Lkn-1/MIP-1δ	CCR1, CCR3
CCL16	HCC-4/LEC	CCR1
CCL17	TARC	CCR4
CCL18	DC-CK1/PARC AMAC-1	?
CCL19	MIP-3β/ELC/exodus-3	CCR7
CCL20	MIP-3α/LARC/exodus-1	CCR6
CCL21	6Ckine/SLC/exodus-2	CCR7
CCL22	MDC/STCP-1	CCR4
CCL23	MPIF-1	CCR1
CCL24	MPIF-2/Eotaxin-2	CCR3
CCL25	TECK	CCR9
CCL26	Eotaxin-3	CCR3
CCL27	CTACK/ILC	CCR10

Chemokines. Table 1 List of human chemokines, showing their systematic and classical names and the main receptors with which they interact (Continued)

Systematic Name	Classical Name	Chemokine Receptors
C Family		
XCL1	Lymphotactin/SCM-1α	XCR1
XCL2	SCM-1β	XCR1
CX3C Family		

mokine receptors (presumably their ligands are still unknown) (2,5).

A salient characteristic of chemokines and chemokine receptors is the high degree of redundancy and binding promiscuity. For example, a particular chemokine can bind to more than one receptor and a particular receptor can interact and transduce signals for several chemokines (4). There are, however, certain receptors that are specific for a given chemokine. On the other hand, the ▶ Duffy antigen, a molecule related to chemokine receptors, can bind several chemokines but does not have any signal transducing function (5).

Interactions with the Immune System

The chemotactic function of chemokines was the first of their activities to be discovered and characterized (1). Initially, chemokines were thought to be involved in directing the ▶ extravasation and migration of neutrophils and other cells of the innate immune system. It is now well established, however, that chemokines function to attract different types and subsets of lymphocytes, and many other types of cells (2–4). Moreover, it is now recognized that chemokines play many important roles besides ▶ chemotaxis, including regulation of ▶ angiogenesis/angiostasis, lymphoid traffic, lymphoid organ development, differentiation of Th1/Th2 T helper cells, wound healing, and malignant cell metastasis (1–4). Some chemokine receptors also play an important role in viral infections, acting as co-receptors for different viruses, notably ▶ HIV co-receptors for HIV-1 (3,5). Some of the most important functions of chemokines are discussed below.

Leukocyte Migration and Activation

One of the primary functions of chemokines and chemokine receptors is the recruitment of different subsets of leukocytes to different tissues. Chemokine-induced cell migration takes place, not only in the classical context of inflammation, but also in the context of normal leukocyte homing, and lymphoid homeostasis (3). These functions are mediated by "inflammatory" and "lymphoid" chemokines, respectively.

- "Inflammatory" chemokines mediate the recruitment and activation of neutrophils, monocytes and other cells of the innate immune system. In addition, effector and memory lymphocytes are also attracted. They are produced by endothelial cells, epithelial cells and leukocytes in response to inflammatory stimuli, such as endotoxin, tumor necrosis factor(TNF)α and interleukin-1 (2–4). Examples are CXCL8 (IL-8), CXCL10 (interferon-inducible protein-10 (IP-10)), CCL2 (monocyte chemoattractant protein-1 (MCP-1)), CCL5 (▶ RANTES) and CCL11 (eotaxin).
- "Lymphoid" chemokines are constitutively produced within lymphoid tissues and function to regulate normal leukocyte homing, cellular compartmentalization and lymphoid organ development. These are also known as "homeostatic" chemokines (3). Examples are CXCL13 (BCA-1), CCL14 (HCC-1), CCL21 (SLC) and CCL25 (TECK).

Angiogenesis/Angiostasis

ELR-CXC chemokines have potent angiogenic activity and can induce chemotaxis on endothelial cells (e.g. CXCL1–3 (GROα/β/γ), CXCL6 (GCP-2) and CXCL8 (IL-8)). In contrast, non-ELR-CXC chemokines possess angiostatic activity and inhibit chemotaxis of endothelial cells (e.g. CXCL9 (Mig), CXCL10 (IP-10) and CXCL11 (I-TAC)) (3).

Hemopoiesis

Several chemokines have been shown to have inhibitory activities on hemopoietic progenitor cell proliferation. For example, CCL23 (MPIF-1) and CCL24 (MPIF-2) inhibit committed progenitors of granulocyte and monocyte lineages (CFU-GM) but not erythroid or megakaryocytic precursors (5).

Metastasis

Chemokines have the potential to affect tumor growth and metastasis through their effects on angiogenesis and cell migration (3). Alterations in the expression of angiogenic chemokines by tumor cells may promote an angiogenic environment, contributing to tumor growth, while expression of chemokine receptors on tumor cells may determine differential migration/metastasis to certain tissues. For example, the role of CXCR4 expression by breast cancer cells in their metastasis to tissues expressing their ligand CXCL12 (▶ stromal cell-derived factor-1 (SDF-1)) such as the bone marrow, has been demonstrated (3).

Lymphocyte Development
Chemokines are involved in the development of both B and T cells, and in regulating the migration of mature and immature lymphocytes to lymphoid organs (3,4). For example, several chemokines, such as CCL17 (TARC), CCL21 (6Ckine/SLC), CCL22 (MDC) and CCL25 (TECK) are highly expressed in the thymus, where they function to promote migration of immature thymocytes. Indeed, thymocytes at different stages of differentiation express different chemokine receptors (4). Moreover, the expression of the chemokine CXCL12 (SDF-1) and its receptor CXCR4 in the bone marrow are essential for B cell development (3,4). Finally, there is evidence that the homing of different lymphocyte subsets to secondary lymphoid organs (e.g. lymph nodes, spleen) and the cellular architecture of those organs are regulated by different chemokines (3,4).

Immunoregulatory Activities
Besides regulating lymphocyte development and homeostasis, chemokines play important roles in several aspects of immunologic responses. For example, dendritic cell-derived chemokines (e.g. CCL17 (TARC), CCL22 (MDC)) may play a role in the initiation of immune responses by recruiting different subsets of T cells and may even have adjuvant effects (4). Chemokines and chemokine receptors (e.g. CCR7) also regulate the migration of dendritic cells to lymph nodes, promoting antigen presentation (3,4). Furthermore, the differential expression of chemokine receptors on subsets of Th1 and Th2 cells, coupled to differential chemokine production by these subsets, suggest that chemokines may play a considerable influence on the control of Th1/Th2 responses (2,4).

Role in HIV Pathology
Chemokine receptors play a significant role in the pathogenesis of human immunodeficiency virus (HIV) infection. Two receptors, CXCR4 and CCR5, have been reported to act as the primary co-receptors for the entry of HIV-1 into cells (5). Moreover, the ligands of these receptors (CXCL12 (SDF-1), CCL3 (▶ macrophage inflammatory protein-1α (MIP-1α)), CCL4 (MIP-1β) and CCL5 (RANTES)) act as HIV-1-suppressive factors by interfering with viral entry (5).

Relevance to Humans
From the brief discussion of their multiple activities above, it is clear that chemokines are extremely important molecules in human physiology. The traditional view of chemokines as chemoattractants is definitely an oversimplification, as exemplified by their roles both in maintaining homeostasis in the lymphoid system and in the regulation of inflammatory and immune responses. Given their important regulatory roles, it is not surprising then that alterations in their expression or production often result in pathological consequences. Indeed, the involvement of different chemokines in a wide variety of inflammatory, infectious and autoimmune diseases has been established (2–5). Moreover, the recently recognized roles of chemokine receptors, both in promoting tumor metastasis and in acting as co-receptors for the entry of HIV-1 into cells, underscore the great importance of the chemokines and their receptors in human disease.

Because of their important physiological roles and their involvement in different types of disease, including HIV infection and cancer, a great amount of interest has been generated in the therapeutic uses and/or targeting of chemokines. Fortunately, the nature of the chemokine receptors, as seven-transmembrane G protein-coupled receptors, makes them likely candidates for the development of small drug therapeutic agonists or antagonists.

Among the main potential uses of chemokines as therapeutic agents are applications as adjuvants in antitumor therapies. This is based on the abilities of some chemokines, particularly those of the inflammatory type (e.g. CXCL9 (Mig), CXCL10 (IP-10), CCL2 (MCP-1)) to recruit T lymphocytes, natural killer cells, macrophages and dendritic cells to the tumor site. Moreover, some of these chemokines also have angiostatic effects (3).

Chemokine receptor antagonists may have important applications in modulating deleterious immune responses, allergies and/or excessive inflammation. In this regard, chemokine antagonists could be used as a more specific type of "anti-inflammatory" agents, in comparison to corticosteroids, cyclosporin A and similar drugs. Several types of cytokine antagonists have been generated, including monoclonal antibodies against chemokines or their receptors, modified or truncated chemokines, and small molecule receptor inhibitors.

A very important and promising therapeutic application of chemokines or chemokine receptor inhibitors is as HIV-suppressive factors (3,5). This is based on the observations that some chemokines block the entry of HIV into cells by interfering with the interactions between the virus envelope and co-receptors (CXCR4, CCR5) on the membrane of cells.

References

1. Oppenheim JJ, Zachariae COC, Mukaida N, Matsushima K (1991) Properties of the novel supergene "intercrine" cytokine family. Ann Rev Immunol 9:617–648
2. Sallusto F, Mackay CR, Lanzavecchia A (2000) The role of chemokine receptors in primary, effector, and memory immune responses. Ann Rev Immunol 18:593–620
3. Rossi D, Zlotnik A (2000) The biology of chemokines and their receptors. Ann Rev Immunol 18:217–243
4. Zlotnik A, Yoshie O (2000) Chemokines: a new classification system and their role in immunity. Immunity 12:121–127
5. Schweickart VL, Raport CJ, Chantry D, Gray PW (1999) The chemokine gene family. Similar structures, diverse functions. In: Hébert CA (ed) Chemokines in Disease. Biology and Clinical Research. Humana Press, Totowa NJ, pp 3–18

Chemotactic Cytokines

▶ Chemokines

Chemotaxis

Movement of cells along a concentration gradient of a diffusible, chemotactic substance. The chemotaxis of leukocytes is mediated by gradients of chemokines.
▶ Chemokines

Chemotaxis of Neutrophils

LASSE LEINO
Department of Clinical Chemistry
University of Turku
FI-20520 Turku
Finland

Synonyms
Migration of neutrophils

Definition
Chemotaxis is the reaction by which the direction of neutrophil locomotion is determined by chemical substances in the environment of the cell. This is distinct from chemokinesis where the speed, or frequency, of locomotion is determined by chemical substances while lacking a directional component. Typically, a chemical substance, often called a chemoattractant in this connection, determines both cell speed and direction; thus a neutrophil moving directionally and also accelerating is showing both chemotaxis and chemokinesis. Transmigration is a special form of chemotaxis which defines neutrophil locomotion through a cell barrier, e.g. endothelial cell monolayer, along an increasing concentration of chemoattractant.

Characteristics
Neutrophils are the first cells of the body's defence system to be recruited at sites of inflammation and infection. The circulating neutrophils are selectively accumulated into inflamed tissues by proinflammatory molecules within minutes of the generation of these mediators. The initial recruitment of neutrophils is prompted by the expression of specific adhesion receptors on leukocytes and their counterparts on activated endothelial cell surface, which interact with each other to promote neutrophil tethering and adherence on endothelium. After transmigration (see above), neutrophils progress towards a gradient of increasing levels of endogenous (e.g. ▶ complement fragments, interleukin-8, chemokines) and/or pathogen-derived (e.g. formylated ▶ peptides) chemoattractants released at the site of inflammation and/or infection.

Chemoattractants bind to specific plasma membrane receptors on the neutrophil cell surface. In seconds this binding initiates intracellular reactions that lead to alterations in cellular metabolism, facilitating and orchestrating neutrophil locomotion. Typically, very low concentrations of chemoattractants (e.g. 10^{-9} mol/l) are needed for chemotaxis compared to other functional responses (e.g. ▶ respiratory burst, degranulation) mediated by the same receptors in neutrophils.

Upon stimulation with chemoattractant, neutrophils undergo rapid morphological changes from rounded and relatively smooth cells to elongated and ruffled cells with pseudopodia. Pseudopodia are formed very quickly, within minutes, and they form broad, thin lamellipodia that are extended anteriorly in the direction of an increasing chemoattractant concentration gradient. Also, a contractile uropod is formed posteriorly, which results in a polarized cell. Neutrophil chemotaxis is a complex event in which the cells repetitively extend lamellipodia in the direction of the chemoattractant gradient and retract uropodia toward the cell body. Several comprehensive reviews dealing with the cellular and molecular events in leukocyte chemotaxis have been published and readers wishing to obtain more information about this area are encouraged to read this literature (1,2).

The assessment of neutrophil chemotaxis relays mainly on in vitro assays of the net migration of large cell populations. Of these assays, the filter assay is the most popular and widely used. This assay was first introduced by Boyden in 1962, and since then it has been used successfully as such, or with some variations, to study the mechanisms of chemotaxis and identify new chemoattractants.

Briefly, in this method neutrophils are separated from a solution of chemoattractant by a porous filter or membrane. The assay is typically run in a chamber, or more recently in multiple-well microtiter plates with cells on the top of the separating filter and the chemotactic substance below. A chemoattractant concentration gradient is formed through the filter, inducing the migration of neutrophils into the pores of the filter. Eventually the cells will reach the surface of the chemoattractant side. The quantification of the neutrophil chemotactic activity is based on counting the cells found on the filter surface at the lower side. Alternatively, the cell count can be determined in the solution under the filter, or a biochemical marker (such as ▶ myeloperoxidase activity) can be use as an quantitative indicator of neutrophil presence in the solution. In a leading front-filter assay, first described by Zigmond and Hirsch in 1973, cells were not allowed to penetrate the whole depth of the filter but the distance migrated by the front of the leukocyte population was scored.

Also other forms of chemotaxis assays have been developed. For instance, there are assays where neutrophils are loaded onto the upper surface of an attractant-containing collagen or fibrin gel, and they are allowed to migrate into the gel matrix. Finally, many visual techniques have been developed to study chemotaxis at a single-cell level. These range from a simple microscopic visualisation of cell movement to computer-assisted time-lapse videorecording systems. A common feature of all these methods is that they provide very important information about cellular locomotion in single cells, but they are usually laborious and time-consuming to perform and thus not as suitable, for instance, for a screening-assay. A reader interested in methodological aspects of chemotaxis assays should see Wilkinson (3).

In practise, chemotaxis measurements are made only with neutrophils obtained from peripheral blood. This should be considered when carrying out in vivo exposure tests of exogenous substances in test animals or human subjects: only agents with systemic access either directly or indirectly (e.g. via a mediator) are expected to show alterations in the chemotaxis of circulating neutrophils. A negative result with a topically applied substance does not rule out the possibility of chemotaxis modulation locally. Therefore, an in vitro exposure test with isolated cells should always be done in parallel with in vivo testing.

Preclinical Relevance

Due to the lack of guidance in the investigation of neutrophil chemotaxis, its relevance is discussed below.

Relevance to Humans

As neutrophil chemotaxis is an essential part of host defence against infection, a chemotaxis deficiency, either innate or acquired, may lead to increased susceptibility to infections. Several human conditions are known where a defect in the chemotaxis of neutrophils has been noted. For example, chemotaxis disorders have been reported in patients with diabetes, cirrhosis, hairy cell leukemia, ▶ Hodgkin's disease, Job's syndrome of hyper-IgE, systemic lupus erythematosus, ▶ juvenile periodontitis, cancer, and leukocyte adhesion deficiency. As is evident from this nonexhaustive list of diseases with a broad range of pathophysiological mechanisms, a chemotaxis deficiency in neutrophils may be a disorder with multiple origins. Therefore the human relevance is obvious for exogenous chemicals with potential systemic exposure. In particular, pharmaceuticals targeting cells of the immune system should be evaluated for their effect on neutrophil chemotaxis.

In addition to causing functional suppression, chemical substances may act also as immunostimulators. As discussed above, chemotaxis is usually the first neutrophil response responding to low levels of a potential activator. When the activator concentration is increased, the response repertoire is accompanied with other forms of responses, such as respiratory burst and degranulation, which bear the potential for self-destructive tissue damage, that is they may promote inflammation. Therefore, a chemical substance showing chemoattractant-like properties should be tested in other neutrophil functional assays at a broad concentration range.

Regulatory Environment

There are no direct guidelines determining the testing of neutrophil chemotaxis. However, assays for testing drug effects on neutrophil functions are recommended in several immunotoxicity guidance documents for pharmaceutical industry.

- Guidance for Industry and FDA Reviewers: Immunotoxicology Testing Guidance, FDA-CDRH, 1999
- Guidance for Industry: Immunotoxicology Evaluation of Investigational New Drugs FDA-CDER, 2002

References

1. Cicchetti G, Allen PG, Glogauer M (2002) Chemotactic signaling pathways in neutrophils: from receptor to actin assembly. Crit Rev Oral Biol Med 13:220–228
2. Worthylake RA, Burridge K (2001) Leukocyte transendothelial migration: orchestrating the underlying molecular machinery. Curr Opin Cell Biol 13:569–577

3. Wilkinson PC (1998) Assays of leukocyte locomotion and chemotaxis. J Immunol Methods 216 (1–2):139–153

Chimera

An organism composed of two genetically distinct types of cells. In case of knockouts a mouse in which some cells and tissues are derived from embryonic stem (ES) cells bearing an engineered mutation, while the others originate from stem cells present in the blastocyst into which the gene-targeted ES cells were introduced.
▶ Knockout, Genetic
▶ Transgenic Animals

Chip Array

Chip arrays are a molecular gadget used to identify the relative abundance of RNAs in a given cell type. Small unique sequences of thousand of different genes are spotted on a small chip, which can be hybridized with suitably labelled RNA (or cDNA thereof). RNA-types present in the sample will bind and give a signal (radioactive of fluorescent). Comparison of different samples, e.g. TCDD treated hepatocytes versus untreated hepatocytes, then allows identification of upregulated or downregulated genes.
▶ Dioxins and the Immune System

Chlorobiphenyl

▶ Polychlorinated Biphenyls (PCBs) and the Immune System

$C_{62}H_{111}N_{11}O_{12}$

▶ Cyclosporin A

Chromate

The hexavalent monomeric base salt or ester of chromic acid (H_2CrO_4).
▶ Chromium and the Immune System

Chromium and the Immune System

MITCHELL D COHEN
Department of Environmental Medicine
New York University School of Medicine
57 Old Forge Road
Tuxedo, NY 10987
USA

Definition

Pallas was the first to discover chromium in 1765; the element itself was not isolated until 1797 by Nicolas Vauquelin. Fourcroy and Hauy (1798) were first to suggest the name from *chroma* (Greek for color) due to the many colored compounds found to contain the element. By 1816, manufacture of chromium chemicals had begun; by 1820, mordant dyeing with chromium agents was initiated; by the mid-1800s, leather tanning with chromic acid became a common practice; by 1879, chromite was routinely used in manufacturing refractory furnaces; by 1910, use of chromium for metallurgy became common; and in the 1920s chromium agents became important for manufacture of metal alloys and found extensive application in growing markets for automobiles and home appliances.

Molecular Characteristics

Chromium (Cr) is a Group VI B ▶ transition element that can exist in oxidation states from −2 to +6. Only Cr^0, Cr^{2+}, Cr^{3+}, and Cr^{6+} forms are routinely encountered; Cr^{5+} and Cr^{4+} complexes have been seen, but are not stable, with a half-life of minutes. Valence factors into:
- the types of ligands that complex chromium: ▶ hexavalent chromium compounds (Cr^{6+}) exist almost exclusively as oxides/oxocomplexes while Cr^{2+} and Cr^{3+} are primarily halide, sulfide, oxide, or amine complexes
- complex ▶ stereochemistry▶ : Cr^{2+} complexes exist mostly as high-spin distorted octahedral (d-O_h) or tetrahedral (d-T_h) structures. Nearly all Cr^{3+} complexes are ▶ hexacoordinate low-spin O_h, and Cr^{6+} compounds exist uniformly as T_h structures
- potential chromium interactions with (bio)ligands: while ligand displacement reactions in some ▶ trivalent chromium compounds (Cr^{3+})—such as those mediated by hydroxo, amino, or oxo groups—gives rise to polynucleated bridge complexes, Cr^{6+} compounds do not readily form bridges because of the greater extert of multiple bonds between chromium and O atoms, making it chemically difficult to displace the latter.

Ultimately, it seems that by being a factor in chromium

agent composition and structure, ▶ valency also has a role in how each might interact with cells. As Cr^{6+} compounds at neutral pH are predominantly T_h ▶ chromate anions, these can enter cells using a general anion transport system—a relatively nonselective anion channel used by other T_h physiological anions (SO_4^{2-} and PO_4^{3-}). Conversely, as the majority of Cr^{3+} complexes exist as O_h cations (though a few anionic compounds can be found) these agents only can cross membranes very slowly by simple diffusion or, with less soluble forms, via endocytic or ▶ pinocytic uptake after adherence to membrane cationic binding sites.

Putative Interaction with the Immune System
Contact Dermatitis

Allergic ▶ contact dermatitis due to chromium is most commonly observed during occupational contact with low to moderate levels of chromates. This hypersensitivity usually occurs in the presence of other metal allergens, like nickel or cobalt; however, the coexisting hypersensitivities are not due to immune cross-reactivities, but rather to concomitant host sensitization. Most reports of severe chromium dermatitis are limited to the hands and lower forearms, but it is not uncommon for the condition to develop at other body sites with frequent contact with chromium. Dermatitis on the neck, ear, and ankle is known to arise from contact with jewelry containing chromium; dermatitis on the foot is associated with wearing shoes made with chromium-tanned leather; and dermatitis on the arms and thighs is related to wearing military uniforms dyed with water-soluble chromium dyes. Contact hypersensitivity reactions in conjunction with stainless-steel prosthetic devices have been observed near the site of implantation which most often involves sites located near the arm, leg, or hip.

The occurrence of allergic contact dermatitis due to chromium exposure is peculiar because several factors need to be overcome for a response to manifest. Such factors include:
- a lack of universal contact sensitivity despite widespread chromium distribution in the environment
- relatively weak allergenic potency for chromium
- variations in skin penetrability by different chromium compounds
- long periods of exposure are required for clinical manifestations to become evident.

While the levels of chromium needed to induce sensitization are often only slightly greater than physiologic ones, chromium at very low or high concentrations is known to induce immunologic unresponsiveness. Under conditions of repeated exposure, immunological tolerance rather than increased incidences of allergic reactions occurs.

The elicited contact sensitivity is a four-stage hypersensitivity response that depends on T-cell activation rather than on formation of antibodies against a chromium-containing allergen. In the first phase (the refractory period) following initial contact with the chromium, Cr^{6+} ions penetrate cell membranes and undergo intracellular reduction; resulting Cr^{3+} ions then bind cellular proteins to form chromium-protein complexes. Precisely which protein is conjugated is uncertain, but serum albumin, heparin, and glycosaminoglycans have been suggested as the potential allergens (though some researchers question the importance of specificity of the carrier conjugate protein). If a level of damage sufficient enough to cause cell death occurs, the damaged/dead cell is engulfed and processed by resident antigen-presenting cells (APC); similarly, APC can engulf chromium-protein complexes if chromium-induced cell lysis occurs and chromium-protein complexes are released into the tissue microenvironment. The APC then present the chromium-modified proteins to naive T-cell and initiate an expansion and proliferation of effector and memory cells specific for individual chromium-bearing protein/peptide complexes. Any subsequent exposure of the individual to chromium will then induce a type IV hypersensitivity response characterized by both induction and elicitation. Induction occurs as a result of the APC presentation of chromium-protein/peptide complexes to memory T-cell. The elicitation phase arises from subsequent release from activated T-cell of lymphokines which stimulate chemotaxis, inflammation, and edema. This cascade of cellular events also enhances further chromium-peptide/protein-specific effector T-cell proliferation. The final phase, persistence, is achieved through the continuous renewal of memory T-cell specific for each of the APC-expressed chromium-protein/peptide complexes.

Hosts that display chromium-dependent allergic contact dermatitis also display increased serum immunoglobulin IgM and IgA levels, increased chromium-induced lymphocyte transformation and proliferation, increased formation of immediate (E) rosettes, and decreased suppressor index values reflective of changes in relative numbers of CD^{4+} T-helper (Th) and CD^{8+} T-suppressor (Ts) cells. An overall reduction in Ts cell activity (either through a decrease in cell number or chromium-mediated alteration in function) is thought to be responsible, at least in part, for increases in levels of circulating antibodies and immune complexes. While the chromium-induced lymphocyte proliferation was found to be monocyte dependent, it is not clear whether monocytes (or mature macrophages) themselves, or even inflammation-associated polymorphonuclear leukocytes, are affected by chromium in ways that might contribute to onset/development of the allergic response.

Asthma

Asthma, a chronic illness characterized by persistent bronchial hyperactivity, is an immune-mediated response that has been historically linked with exposure to common allergens. Over 200 industrial compounds have been associated with high incidences of occupational asthma, including chromium. Cases of occupational asthma have been reported in conjunction with exposure to dichromates, ammonium bichromate, chromic acid, chromite ore, chromate pigments, and to welding fumes. Evidence suggesting that chromium exposure is a cause of occupational asthma is limited to a small number of case reports and case series. In most instances, a consistent pattern between chromium exposure and the onset of asthma is noted. In some cases, hypersensitivity to chromium is confirmed either by patch testing or by challenge a chromium bronchodilator. Both immediate and delayed asthmatic reactions have been reported suggesting that chromium-induced asthma may be mediated by both immune and nonimmune mechanisms.

Effects on Immune System Cells

As inhalation is the most likely means of maximal chromium exposure by humans, studies have examined the impact of chromium on the function of the cells most essential to maintaining lung immunocompetence - lung macrophages. Morphologically, macrophages recovered from experimental animals after inhalation of Cr^{6+} or Cr^{3+} agents display increased numbers of chromium-filled cytoplasmic inclusions, enlarged lysosomes, surface smoothing, and decreased membrane blebs for cell mobility and target contact. Functionally, macrophages display reduced phagocytic activity, oxygen consumption rates following zymosan stimulation, and production of reactive oxygen intermediates used for killing. Other studies have, however, reported opposite effects on macrophage numbers and function—as well as no effect on morphology—after inhalation of chromium. Clearly, choice of agent, exposure regimen, dose, and cumulative length of exposure are critical for the type of immunomodulation that may manifest.

The majority of these effects of chromium on macrophage structure and function have also been reproduced in vitro using alveolar macrophages form a variety of hosts. However, unlike in in vivo studies, Cr^{3+} compounds are mainly ineffective. Treatment of U937 monocytes or isolated human monocytes and macrophages with chromium enhanced release of interleukin (IL)-1 and tumor necrosis factor(TNF)-α, suppressed release of transforming growth factor(TGF)-β, and induced proliferation of the monocytes/macrophages.

The involvement of oxidative stress in the toxicities in macrophages has been suggested based on several interrelated findings. These include:
- ex vivo exposure of primary rat lung macrophages with Cr^{3+} or Cr^{6+} reduced cell chemiluminescence/oxygen consumption
- exposure of J774A.1 macrophages resulted in increased nitric oxide (NO) and superoxide ($\cdot O_2^-$) production
- Cr^{3+} treatment induced DNA single strandbreaks in J774A cells while Cr^{6+} had no effect.

Immunotoxic effects arising from chromium exposure also occur in lymphocytes. Lymphocytes exposed to Cr^{6+} in vivo or in vitro display an increased incidence of chromosomal aberrations (including strandbreaks, gaps, interchanges) and increased levels of DNA-protein complex formation. Though the implications from the defects are uncertain, it has been suggested that genetic alterations/damage to DNA integrity might result in changes in lymphocyte proliferation. At the immunologic level, lymphocytes recovered from chromium-exposed hosts display alterations in mitogenic responsiveness. At low concentrations, soluble Cr^{6+} was slightly stimulatory, but became overtly inhibitory with increasing levels; Cr^{3+} was ineffective at all doses tested. Effects similar to those in cells from chromium-exposed hosts were observed in vitro using cultured human and rat lymphocytes. Rat splenocytes in mixed lymphocyte cultures or in combination with B-cell-specific or T-cell-specific mitogens (LPS or PHA) again demonstrated a very narrow concentration-dependent biphasic (stimulatory, then inhibitory) effect from Cr^{6+}. However, when using peripheral blood lymphocytes from chromium-treated rats, mitogenic responsiveness was enhanced overall, with an even greater effect when exogenous chromium was added. Analysis of cell chromium content found that altered proliferation (and immunoglobulin production) was related to total amounts of chromium in the cells. One reason for the discrepancies between the in vitro and vivo outcomes is that chromium added to naive splenocyte cultures may have reacted with cell surface proteins (the surface mitogen receptors) to block proliferative effects, while extensive periods of exposure to chromium (via ingestion) may have resulted in host sensitization and, ultimately, selection of lymphocytes that would proliferate in the presence of chromium ions/chromium-conjugated haptens.

Lastly, other chromium-induced alterations in macrophages and lymphocytes include changes in production and/or release of proteins required for proper immune cell function and induction of cell activation in an immune response. These include:
- alterations in the levels of circulating antibody in response to viral antigens

- reduced formation of interferons in response to viruses or antigenic stimulation
- decreased production of IL-2 for proliferation and differentiation of B-lymphocytes during the onset of humoral immunity.

These chromium-induced disturbances of immune cell intercommunication likely serve as a basis for the reductions in cell-mediated and humoral immunity observed *in vivo*, and for the subsequent increases in incidence/severity of infectious diseases, and (possibly) for the cancers manifest in animals and humans exposed to chromium compounds over extended periods of time.

Relevance to Humans

Chromium is ubiquitous in nature and found at levels ranging from < 0.1 µg/m^3 in ambient air to > 4 ppt chromium in contaminated soils. Global input of chromium into the environment is only 30% from natural sources (volcanic emissions, rock weathering); 70% is due to manmade emissions (fossil fuel combustion, coal burning, welding, general metal-use processes). Almost all naturally occurring chromium is in the Cr^{3+} form; environmental Cr^{6+} is almost uniformly derived from human activities.

Ambient chromium levels in most nonindustrialized areas are found at < 0.1 µg/m^3 and average rural air levels have been estimated at < 0.01 µg/m^3; average levels in most American cities are around 0.1 µg/m^3. Overall distribution of Cr^{3+} to Cr^{6+} is about 2 : 1 in atmospheric chromium; this is because most Cr^{6+} that enters air is reduced by common environmental constituents and/or other pollutants. In general, intake from air is estimated at < 1 µg/day by average human populations.

Levels of chromium in water are variable and dependent on salinity. Average chromium levels in American rivers and lakes range from 1–30 mg/l, and 0.1–5 mg/l in seawater. Drinking water often contains higher chromium levels than river water. International and national drinking water standards have been established that allow municipal drinking water to not contain more than > 50 µg/l (as Cr^{6+}). Most municipal supplies (except in regions of naturally high chromium levels or intensive industrialization) rarely have burdens > 0.5 µg/l. As a result, daily intake of chromium solely from water ingestion is estimated at 4 µg/day. Significant chromium exposure is via ingestion of food; high chromium sources are vegetables, unrefined sugars, and meats, while fruit, fish, and vegetable oils contain fairly low levels. Estimated adult intakes are 100–200 mg/day, though large interindividual variations occur. The highest bioavailability of chromium is from glucose tolerance factor (GTF) predominantly found in Baker's yeast, liver, and meat. Regardless of source, ingested Cr^{3+} is less well-absorbed than Cr^{6+}, although uptake of either species is still a function of solubility, and most ingested Cr^{6+} is converted to Cr^{3+} as a result of stomach acids and other gastrointestinal components, Overall, human intestinal absorption efficiencies for chromates average 2%–6% but only 0.4%–1.0% for Cr^{3+} agents. Because chromium deficiency results in impaired glucose and lipid metabolism, a recommended dietary intake of 50–200 mg/day has been established for adults. To assure chromium sufficiency, direct supplementation with various chromium compounds was initially a common medical practice. However, when it was shown that optimal uptake of Cr^{3+} occurred after chelation to GTF, inclusion of products containing Baker's or Brewer's yeast was accepted as a safer approach. Today, Cr^{3+} picolinate is increasingly used because this form greatly enhances cell absorption of Cr^{3+}.

The other major means for introducing chromium into the body is dermal contact. Overall, because Cr^{3+} binds readily to skin constituents, it is not dermally absorbed to any significant extent—soluble Cr^{6+} passes more readily through the epidermal barrier. There are cases where contact surfaces (like walls) have elevated chromium levels and the majority is as Cr^{6+}. This differs from a scenario involving skin contact with soil in that walls are conducive to permeation by solubilized Cr^{6+} agents, while excluding Cr^{3+} agents. Significant amounts of dermal exposure can also arise from daily contact with many household materials (e.g. bleaches, detergents). Since sweat is a primary vehicle for liberating chromium and for concentrating it on skin during evaporation, clothing (particularly tanned leathers) are another major chromium source; lesser sources include match heads, magnetic tape, gaming-table felt, and tattoo dyes.

References

1. Polak L (1983) Immunology of chromium. In: Burrows D (ed) Chromium Metabolism and Toxicity. CRC Press, Boca Raton, pp 51–136
2. ATSDR (1992) Toxicological Profile for Chromium and Compounds. US Public Health Service, Atlanta
3. Zelikoff J, Cohen MD (1997) Metal immunotoxicology. In: Massaro EJ (ed) CRC Handbook on Human Toxicology. CRC Press, Boca Raton, pp 811–852
4. Arfsten DP, Aylward LL, Karch NJ (1998) Chromium. In: Zelikoff J, Thomas P (eds) Immunotoxicology of Environmental and Occupational Metals. Taylor and Francis, London, pp 63–92
5. Burns-Naas L (2000) Arsenic, cadmium, chromium, and nickel. In: Cohen MD, Zelikoff J, Schlesinger RB (eds) Pulmonary Immunotoxicology. Kluwer Academic Publishers, Norwell, pp 241–266
6. Cohnen MD (2004) Pulmonary immunotoxicology of select metals: Aluminum, arsenic, cadmium, chromium, copper, manganese, nickel, vanadium, and zink. Journal of Immunotoxicology I (I):39–69

Chromium Release Assay

Cytotoxicity assay that utilizes the radioisotope, 51chromium, as a marker to assess lytic capacity of effector cells such as natural killer cells and cytotoxic T cells.
▶ Cytotoxicity Assays

Chromophore

Light-absorbing molecule.
▶ Photoreactive Compounds

Chronic Beryllium Disease

SALLY S TINKLE
Division of Extramural Research and Training,
National Institute of Environmental Health Sciences
National Institutes of Health
111 T. W. Alexander Drive
Research Triangle Park, NC 27709
USA

AINSLEY WESTON
Health Effects Laboratory Division
National Institute for Occupational Safety and Health,
Centers for Disease Control and Prevention
1095 Willowdale Road
Morgantown, WV 26505
USA

Synonyms
beryllium disease, berylliosis

Definition
Beryllium, an alkaline earth metal, is the fourth element of the periodic table, with an atomic weight of 9.012. It is the 44th most abundant element, dull grey in color and occurs naturally in the environment. Beryllium was discovered by Vanquelin in 1797, but commercial manufacture of beryllium products did not begin until after beryllium-copper-aluminium alloy was patented in 1926 (1).
A chemical pneumonitis was described in beryllium workers in the early 1940s, but it was not attributed to their beryllium exposure. Beryllium toxicity was first recognized in workers engaged in the manufacture of fluorescent lamps and was termed acute beryllium disease (ABD). In 1949 the Atomic Energy Commission implemented an 8-hour exposure limit of 2 $\mu g/m^3$, and ABD became rare. ABD is commonly associated with very high exposures to soluble salts whereas chronic beryllium disease (CBD), a debilitating, persistent granulomatous lung disease, is associated with exposure to insoluble beryllium materials. Although involvement of liver, spleen, skin, lymph node, kidney and muscle may accompany CBD, the lung is the primary site of disease.

CBD begins as a delayed type IV hypersensitivity that, over time and in the presence of persistent pulmonary beryllium, progresses to a non-caseating granulomatous lung disease (2). The HLA-beryllium-TCR interaction is critical in the initiation of beryllium sensitization. Given the highly polymorphic nature of HLA (human leukocyte antigen) and TCR (T cell receptor) and the small subset (about 20%) of exposed workers who become sensitized, the genetic structure (and, hence, the protein structure) of these molecules will significantly impact upon—if not determine—susceptibility to beryllium disease.

Beryllium's adverse effects, although primarily immunotoxicologic, also include disruption of homeostatic metabolism in liver and vascular endothelium. Using mammalian cell lines, researchers determined that beryllium binds covalently to a serine residue in the active site of alkaline phosphatase, phosphoglucomutase, and nuclear protein kinases and inhibits substrate phosphorylation. Additionally, beryllium causes arrest of cell division in lung and skin fibroblasts through a blockade of the G_1 restriction point.

Characteristics
Immunopathology of Beryllium Sensitization
The beryllium-stimulated type IV hypersensitivity is a major histocompatibility complex (MHC) class II-restricted, T helper 1, CD4$^+$ T cell response (2). Consistent with the model of a cell-mediated immune response, beryllium—probably in association with a peptide and in the context of human leucocyte antigen (HLA)—is presented by an antigen-presenting cell to the T cell. Antigen presentation requires direct contact between the HLA molecule, beryllium and the T cell receptor (TCR; Fig. 1). Activation of a T cell causes clonal expansion and the generation of beryllium-specific memory T cells. This cellular process is driven by ▶ proinflammatory cytokines, including tumor necrosis factor (TNF-α) and the interleukin (IL) -6, and culminates in the development of a T helper type 1 lymphocytic response. This T helper type 1 cytokine response has been demonstrated in vitro with beryllium-stimulated bronchoalveolar lavage cell production of interferon γ (IFN-γ) and IL-2.

Beryllium sensitization (BeS) is asymptomatic, however diagnosis is supported by a positive beryllium-specific peripheral blood lymphocyte proliferation test

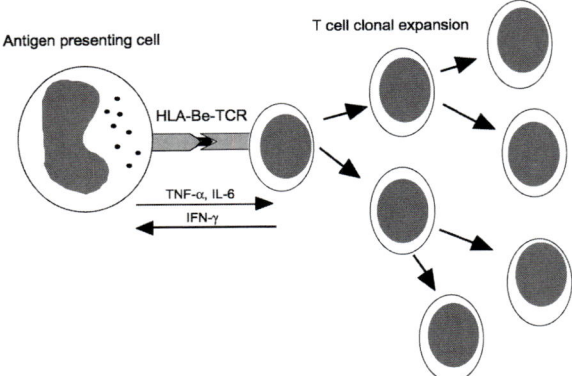

Chronic Beryllium Disease. Figure 1 The HLA-beryllium-TCR interaction is pivotal in the development of beryllium sensitization and disease.

(BeLPT). This test exploits the antigen-specific, cell-mediated immune response and measures increased cell proliferation in response to beryllium in vitro.

Immunopathology of Chronic Beryllium Disease

In its initial stages, CBD is characterized by chronic pulmonary inflammation and T cell alveolitis. Recent studies have identified the $CD4^+/CD28^-$ effector memory phenotype of the beryllium-specific pulmonary T cells (3). Effector memory T cells are primed for immediate cytokine release when antigen is detected in the microenvironment but have diminished proliferative capacity and an increased rate of apoptosis. As disease progresses, beryllium stimulates phagocyte accumulation, granuloma formation, calcific inclusion formation and deposition of fibrotic materials. The pathogenesis of CBD is similar to other pulmonary granulomatous diseases, and the pulmonary lesions in beryllium-, aluminum- and titanium-induced granulomas are virtually indistinguishable from each other and from sarcoidosis (2).

▶ Non-caseating granuloma formation is the hallmark pathobiologic response to persistent beryllium lung burden. In the response to antigen, the $CD4^+$ T cells release proinflammatory molecules that attract macrophages and other inflammatory cells to the site of antigen deposition. Antigen persistence maintains the cell-mediated immune response and, over time, inflammatory cells—and the $CD4^+$ T-cells in particular—accumulate in the lung. This proinflammatory microenvironment promotes macrophage differentiation into multinucleated giant cells. The granuloma forms as a discrete nodule containing a cluster of multinucleated giant cells surrounded by a ring of densely packed lymphocytes. Laminated calcific densities can be observed in the multinucleated giant cells.
Little is known about the mechanism that causes pulmonary fibrosis in CBD. The pathogenesis of dust-induced fibrosis relates chronic activation of macrophages to activation and proliferation of fibroblasts. Increasing numbers of fibroblasts in the pulmonary interstitium in CBD patients and the close association of the profibrotic, basic fibroblast growth factor (bFGF)-positive mast cells with beryllium granulomas has been shown.

Clinical symptoms of beryllium disease generally correlate with the development of pulmonary granulomas and fibrosis. Individuals with a low granuloma and fibrosis lung burden frequently have no respiratory impairment. As the disease progresses, individuals experience shortness of breath, especially on exertion, conditions that are confirmed by abnormal lung function tests and chest x-ray, and fatigue and night sweats. At present, treatment options are limited to palliative measures such as antiinflammatory steroids.

Preclinical Relevance
Genetic Susceptibility to Beryllium

The molecular specificity of the HLA-beryllium-TCR interaction, coupled to the 2%–10% sensitization rate in the exposed worker population, suggests ▶ genetic susceptibility to beryllium (4,5). Several laboratories confirmed the requirement for this interaction by blocking beryllium-stimulated CBD bronchoalveolar lavage lymphocyte proliferation with monoclonal antibodies directed against the HLA antigen-presenting moiety. The HLA-DP antibody provided the strongest inhibition of the BeLPT.

Genes coding for HLA-DP, -DR and -DQ are all located on chromosome 6p12.3, and several laboratories have evaluated their association with risk of CBD and beryllium sensitization. These studies have shown that among beryllium exposed workers, inheritance of a HLA-DPB1 gene coding for a glutamic acid residue in the 69th position of the mature protein (Glu69) carries a high risk of BeS and CBD (Fig. 2).

There are 108 known polymorphic variants at the HLA-DPB1 locus, 36 coding for glutamic acid at codon 69, 67 coding for lysine and 5 coding for arginine. Among the 36 coding for glutamic acid it has been suggested that some alleles carry a higher risk of CBD in exposed workers than others.

The mode of interaction between beryllium and HLA-DPB1 is speculative. Several laboratories have suggested that the negatively charged glutamate at position 69 of HLA-DPB1 may be the binding site for the positively charged beryllium ion. This would place beryllium directly in the antigen-binding groove of the Glu69-containing HLA isomers. In addition to Glu69, a comparison of the electrostatic potential of nine isoforms of HLA-DPB1 Glu69 (HLA-DPB1*0201, *0401, *0601, *0901, *1001, *1301, *1601, *1701, *1901) suggests a possible role for

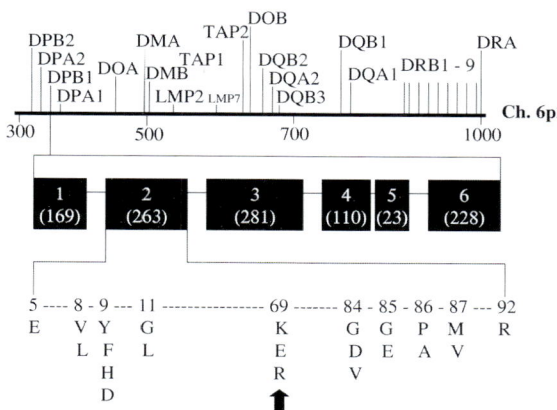

Chronic Beryllium Disease. Figure 2 Chromosome 6p, indicating the relative positions of **MHC Class II** loci. The *HLA-DP* locus is found at 6p12.3.

the negatively charged amino acids, at positions 55, 56, 84 and 85 on the HLA beta chain. These data are consistent with a previous molecular epidemiologic analysis.

A critical component of the T cell activation paradigm, the TCR, is a heterodimeric transmembrane protein formed by significant genetic recombination with the highest degree of recombination occuring in ▶ complementarity-determining region (CDR) 3 (CDR3). The amino acid sequence of CDR3 and its central location in the TCR make it a critical region in HLA-antigen-CR recognition and binding. Studies of beryllium-stimulated T cell oligoclonality have identified a homologous motif in the CDR3 region of the TCR Vβ chain in 11 of 28 CBD subjects studied. Of the approximately 100 possible TCRA V combinations, eight clones from two CBD subjects expressing VB3 co-expressed the same a chain, TCRA V22S1. Direct involvement of this TCR motif and HLA-DPB1*0201in beryllium-specific T cell clonal expansion has been confirmed in vitro.

TNF-α is a potent proinflammatory cytokine in the beryllium-stimulated bronchoalveolar lavage cell-mediated immune response. At least nine ▶ genetic polymorphisms in the *TNF-α* promoter region have been described and linked to variability in TNF-α production and to autoimmune and infectious disease severity and disease progression. Published reports link the *TNF-α*02* allelotype to the magnitude of beryllium-stimulated bronchoalveolar lavage TNF-α production and to disease severity, however, differences in the *TNF-α* promoter genotype were not linked to the magnitude of TNF-α production.

These data on genetic susceptibility to occupational exposure to beryllium provide a framework for understanding beryllium sensitization under a defined set of molecular interactions. Continued analysis of the immunologic mechanism for sensitization and determination of the relationship of these findings to larger patient populations will add a third component, response, to gene-exposure studies.

Relevance to Humans

Beryllium has unique physicochemical properties; it is light and strong (stiffer than steel), highly conductive (of heat and electricity), a neutron moderator, x-ray transparent, springy, non-sparking, non-corrosive, wear resistant, and has a melting point of 1560° K (1). These properties make it ideal for a large number of specialized technological applications. Beryllium, as a metal, an oxide, or an alloy, has become indispensable to multiple industries: aerospace (satellites, gyroscopes, tools, landing gear), automotive (springs, brake terminals, air-bag triggers), telecommunications (cell phones, transistor mountings, undersea housings), electronics (switches, microwaves, nuclear reactors, heat exchangers), fire prevention (tools, sprinklers), biomedical (x-ray windows, lasers, dental prostheses, electron microscopes), defense (mirrors, sights, springs, missile guidance, nuclear triggers), and others (camera shutters, molds, bellows, phonographic equipment).

Regulatory Environment

Environmental exposure to beryllium is well below the EPA standard in the absence of a polluting event. For example, the permissible concentration of beryllium in drinking water is 4 μg/L, and the average measurable concentration is 0.19 μg/L with a range of 0.1–1.22 μg/L (1). Immunological sensitization to beryllium is generally associated with beryllium workers. On average, 2%–10% of exposed workers become sensitized to beryllium. After a variable latency period of months to decades, many sensitized individuals develop pulmonary, non-caseating granulomas.

Epidemiologic studies suggest that disease prevalence varies by exposure type and intensity (4). The exposure limit for most work environments was set in 1971 and remains 2 μg/m³ (OSHA PEL, 8-hour time-weighted average); however, occurrence of CBD among workers with low exposure levels prompted the Department of Energy to set an action level of 0.2 μg/m³ for their facilities. Although most industrial exposures are measured as mass per unit area, other physicochemical characteristics, rather than mass, may be associated with increased risk of disease. This is consistent with the observation that tasks that generate beryllium dust (machining and lapping) convey a higher risk of disease. Although pulmonary exposure to beryllium has been considered to be the primary route for sensitization, recent evidence also suggests that sensitization may be related to skin exposure or to systemic burdens of beryllium.

Beryllium was classified as a potential human carcinogen by the International Agency for Research on Cancer (IARC) in 1993, but the carcinogenic potential of beryllium remains controversial. Evidence to support beryllium carcinogenicity includes an increased risk of lung tumors following beryllium inhalation in rodents, rabbits and monkeys, and in vitro, beryllium-induced sister chromatid exchange, chromosomal aberrations, HPRT gene mutations and other morphological cell transformations that are potentially tumorigenic. Epidemiological evidence supports the carcinogenic potential of beryllium, however crude exposure classifications and tobacco smoking may be confounders. In contrast to these published studies, a recent meta analysis of several beryllium worker cohort mortality studies found no statistically significant standard mortality ratio that supports causal association between occupational beryllium exposure and lung cancer.

References

1. Kolanz ME (2001) Introduction to beryllium: uses, regulatory history, and disease. Appl Occup Environ Hyg 16:559–567
2. Rossman MD (2001) Chronic beryllium disease: a hypersensitivity disorder. Appl Occup Environ Hyg 16:615–618
3. Fontenot AP, Gharavi L, Bennett SR, Canavera SJ, Newman LS, Kotzin (2003) CD28 costimulation independence of target organ versus circulating memory antigen-specific CD4$^+$ T cells. J Clin Invest 112:776–784
4. Kreiss K, Mroz MM, Zhen B, Wiedman H, Barna B (1997) Risks of beryllium disease related to work processes at a metal, alloy, and oxide production plant. Occ Env Med 54:605–612
5. McCanlies EC, Kreiss K, Andrew M, Weston A (2003) HLA-DPB1 and chronic beryllium disease: A huge review. Am J Epidem 157:388–398

Chronic Graft-Versus-Host Disease

▶ Graft-Versus-Host Reaction

Chronic Inflammation

Chronic inflammation consists of a sustained inflammatory reaction, perpetuated by persistence of the causative agent and far more often by an autoimmune response. Elevated levels of proinflammatory cytokines are a major part of its pathophysiology.

▶ Cytokine Inhibitors

Chronic Inflammatory Disease

Chronic inflammation is the pathophysiological cause of many diseases, some of which are of great economical importance, including rheumatoid arthritis, osteoarthritis, and inflammatory bowel disease. It also contributes to diseases such a arteriosclerosis, chronic lung diseases such as chronic obstructive pulmonary disease and asthma, or neurological disorders such as Parkinson's Disease.

▶ Cytokine Inhibitors

Chronic Lymphocytic Leukemia

▶ Leukemia

Chronic Myelogenous Leukemia

▶ Leukemia

Chronic Obstructive Pulmonary Disease

▶ Trace Metals and the Immune System

Classical Pathway

A pathway of the complement system that is activated by antigen-antibody complexes. This pathway includes the complement components C1, C4, and C2, resulting in the formation of a C3 convertase to cleave C3.

▶ Complement and Allergy
▶ Complement, Classical Pathway/Alternative Pathway

Clonal Deletion

Negative selection which eliminates high-affinity, self-reactive T cells and B cells.

▶ Autoimmunity, Autoimmune Diseases

Clonal Expansion

The lymphocyte receptors for antigen are uniquely expressed on individual lymphocyte clones. While individual lymphocytes are mono-specific, within secondary lymphoid organs a broad repertoire of antigen reactivity exists. Engagement of the B cell receptor (BCR, surface immunoglobulin) or T cell receptor (TCR) by specific antigen induces the selective activation of the receptor-bearing lymphocyte clone. Consequently, the activated clone undergoes multiple, successive rounds of cell division. This proliferative phase leads to a large population of daughter cells capable of responding to the same antigenic determinant as the progenitor cell.
▶ Memory, Immunological
▶ B Lymphocytes
▶ Humoral Immunity

Clonotypic Antibodies, T Cell

The clonal expansion of T cells from a single precursor yields a population in which all cells bear the same unique T cell receptor. Antibodies raised against this specific receptor are said to be 'clonotypic'.
▶ Antigen-Specific Cell Enrichment

Clophen

▶ Polychlorinated Biphenyls (PCBs) and the Immune System

Cluster Determinant (CD)

This abbreviation, followed by a number, indicates the cell surface receptor or ligand.
▶ Leukocyte Culture: Considerations for In Vitro Culture of T cells in Immunotoxicological Studies

Cluster of Differentiation (CD)

Nomenclature for leucocyte surface antigens. The cluster designation refers to groups (clusters) of monoclonal antibodies, which bind specifically to the particular antigens. Antigens are assigned a CD number by the International Workshop on Human Leucocyte Differentiation Antigens.

▶ CD Markers
▶ Mucosa-Associated Lymphoid Tissue
▶ Cell Separation Techniques

Co-Cultured

The in vitro cultivation of two distinct cell types or populations.
▶ Mixed Lymphocyte Reaction

Co-Stimulation

In order to be activated, T and B cells require additional activation through specific receptors such as B7 family members for T cells.
▶ Tolerance

Cold Agglutinins Disease

Cold agglutinins disease (cryopathic hemolytic disease) is IMHA caused by autoantibodies that bind optimally to erythrocytes at temperatures below body temperature.
▶ Hemolytic Anemia, Autoimmune

Cold Autoantibodies

Autoantibodies that react as well, or more strongly, at 4°C than at lower temperatures.
▶ Hemolytic Anemia, Autoimmune

Colony Forming Unit (CFU)

Colony forming unit, or CFU, assays are in vitro limiting dilution assays of hematopoietic cells. When bone marrow cells are seeded in semisolid culture medium in the presence of hematopoietic growth factors, colonies develop from a small subfraction of precursors cells which have a high proliferation potential. According to the hematopoietic growth factors that are added and the differentiation potential of the progenitors present, development may be limited to cells of one lineage, or include multilineage differentiation. CFU are termed accordingly, e.g. CFU-GM for granulocytic-macrophage colonies, or CFU-GEMM for co-

lonies containing megakaryocytic, erythrocytic, granulocytic, and macrophage cells.
- Bone Marrow and Hematopoiesis
- Colony-Forming Unit Assay: Methods and Implications
- Rodent Immune System, Development of the

Colony-Forming Unit Assay: Methods and Implications

HAVA KARSENTY AVRAHAM · BYEONG-CHEL LEE · SHALOM AVRAHAM
Division of Experimental Medicine, Beth Israel Deaconess Medical Center
Harvard Institutes of Medicine
Boston, MA 02115
USA

This work was supported in part by National Institutes of Health Grants HL55445 (SA), HL51456 (HA), CA096805 (HA), CA 76226 (HA) and K18 PAR-02-069 (HA). This work was done during the term of an established investigatorship from the American Heart Association (HA).

Synonyms
fetal bovine serum, FBS, committed precursor of megakaryocytes, CFU-Meg

Short Description
Hematopoiesis is sustained by uncommitted ▶ stem cells that give rise to progenitors capable of producing mature blood cells (1). This process is governed by numerous factors that regulate the differentiation of these progenitors into specified lineages (2). Distinct subclasses of primitive cells defined by functional assays have led to the identification and hierarchical arrangement of stem cells comprising the human hematopoietic system.

In the late 1970s, Dexter et al (3) developed culture conditions that permitted the growth of bone marrow cells ex vivo. Over the subsequent years, culture conditions have improved the expansion of immature hematopoietic cells. In the early 1990s, a closed clinical-scale bioreactor system with continuous medium perfusion was developed for the ex vivo expansion of cells.

Mature blood cells have a limited lifespan and are continuously replaced by the proliferation and differentiation of a very small population of ▶ pluripotent hematopoietic stem cells (HSCs) found primarily in the bone marrow of healthy adults. HSCs have the ability to replenish themselves because of their self-renewal capabilities, and to differentiate into progenitor cells and mature blood cells of all hematopoietic lineages.

Hematopoietic culture assays are used to detect the proliferation and differentiation ability of hematopoietic cells and their distinct, successive stages of differentiation, as well as to measure the frequency of these cells in hematopoietic tissues and purified cell populations. The most common approaches to quantify multi- or single lineage-committed hematopoietic progenitors, called colony-forming cells (CFCs) or colony-forming units (CFUs), utilize viscous or semisolid matrices and culture supplements. These methods promote cell proliferation and differentiation and allow the clonal progeny of a single progenitor cell to form a colony of more mature cells.

In vitro clonal assays provide an environment for hematopoietic cells to proliferate, and the number of colonies formed is proportional to the number of viable progenitors. In these assays, the progenitor cell populations are detected by in vitro clonogenic soft gel systems where the clonogenic cells proliferate and form colonies of recognizable mature cells. The cells giving rise to colonies are identified in standard assays for GM-CFU (granulocyte/macrophage colony forming unit), BFU-E (burst forming unit-erythroid, where "burst" describes the appearance of the colony), mixed colony forming units (CFU-mix) and granulocyte, erythrocyte, macrophage and megakaryocyte colony forming units (CFU-GEMM).

The development of longterm cultures has allowed investigators to study cell-cell interactions in vitro with a system that incorporates an order of complexity close to that exhibited by hematopoiesis in-vivo. In addition, the availability of these hematopoietic cultures has led to the isolation and characterization of at least 50 ▶ cytokines that interact in a complex network involved in the control of hematopoiesis, including inflammatory and immune responses.

Characteristics

Culture media have been optimized for the outgrowth of erythroid, monocyte/macrophage, granulocytic, megakaryocytic and multipotent progenitor cells. Colony assays are used to quantitate and characterize hematopoietic progenitors from different sources (e.g. bone marrow, umbilical cord blood, mobilized peripheral blood, fetal tissues, patient samples) and for ensuring quality control of clinical stem cell collections, processing and cryopreservation.

Two key elements of hematopoietic colony assays are:
- the use of a culture medium that maximizes the growth and differentiation of progenitors of interest, and

- the use of a gelling agent which increases the viscosity of the medium without converting it to a solid; this allows the clonal progeny of a single progenitor cell to stay together.

These properties greatly facilitate the recognition and enumeration of distinct colonies.
The growth efficiency of colonies in these assays is increased by using methylcellulose in the semisolid phase instead of agar. This results in tighter colonies that are easier to evaluate and count (see also (4)). For now, we will focus on the human clonal assays.

Bone Marrow Cell Collection

Human bone marrow (BM) is aspirated from the iliac crest of normal volunteers under a protocol approved by the Committee on FDA Research Involving Human Subjects. Mononuclear cells (MNC) are collected after gradient centrifugation (density ≤ 1.077 g/ml) on Ficoll-PaquePLUS (Pharmacia, Piscataway, NJ), then washed and suspended in Iscove's modified Dulbecco's medium (IMDM; GIBCO, Grand Island, NY). Enrichment of $CD34^+$ cells is accomplished by positive selection using a hapten-conjugated $CD34^+$ antibody (QBEND/10, mouse IgG1), which is then magnetically labeled and selected over a column in a magnetic field (MACS separator; Miltenyi Biotec, Sunnyvale, CA). Average purity of the resultant $CD34^+$ population is then calculated by flow cytometry analysis. MNC or $CD34^+$ cells are counted with a hemocytometer and cell viability determined by phase microscopy.

In Vivo Hematopoietic Stem Cell Assay

The severe combined immunodeficiency (SCID) repopulating cell (SRC) xenotransplant system (SRC assay) provides a powerful tool for analyzing human hematopoietic stem cells. SCID mice-engrafting cells are more primitive than most cells that can be assayed in vitro. This in vivo model allows us to monitor whether HSCs or gene-modified primitive HSCs can sustain their self-renewal capacity or maintain the potential to differentiate into different blood cell lineages. However, the frequency and/or homing efficiency of SRC in bone marrow is inefficient. In addition, transduced SRCs are often associated with the loss of reconstituting activity, posing a problem for the development of clinical HSC gene therapy.
6–8 week old mice are sublethally irradiated with 275 cGy (dose rate 50 cGy/min) in a cesium irradiator. Twenty-four hours after irradiation, human HSCs are injected via the tail vein with normal carrier cells. After 6–8 weeks, mouse bone marrow cells or peripheral blood cells can be analyzed using antihuman CD45 (Becton Dickinson Immunocytometry Systems, San Jose, CA) or southern blot analysis with specific human cDNA probes.

Hematopoietic Clonal Assays

Methylcellulose is widely used as a gelling agent because it is inert, with no ionic charge and is stable over a wide pH range. In addition, it permits better growth of erythroid colonies than other types of semisolid support systems, while allowing optimal granulocyte/macrophage colony formation. Committed progenitors for both erythroid and granulocyte/macrophage lineages, as well as multipotential progenitors, can be assayed simultaneously in the same culture dish. Briefly, there are four main steps.
- 1. Prepare the cells.
- 2. Add cells to methylcellulose-based media with the appropriate growth factors.
- 3. Plate and incubate human cells for 14–16 days and murine cells for 7–14 days in a humidified incubator at 37°C and 5% CO_2.
- 4. Count colonies and evaluate colony types using an inverted microscope and gridded scoring dishes. Alternatively, individual colonies may be used for staining, PCR analysis, or cytogenetic analysis.

Briefly, methylcellulose (0.9%) is used as the semisolid matrix for the clonal assays supplemented with fetal bovine serum (FBS), 1% bovine serum albumin, 0.1 mM 2-mercaptoethanol and growth factors as detailed in Table 1 (StemCell Technologies).
Cells are plated in the culture mixture in 35 mm Petri dishes (three per group) and incubated at 37°C in a humidified atmosphere with 5% CO_2 for 13–16 days (Table 1). All colony countings are performed according to standard methodology under an inverted 3-D microscope (5). CFU-G colonies are comprised of small cells, whereas CFU-GM colonies contain both small and large cells in tight formation.
Both erythroid and myeloid elements are found in CFU-GEMM colonies giving them a reddish tint in culture. Large clusters of myeloid cells greater than 2 mm in diameter are indicative of HPP-CFC colonies (high proliferative potential colony forming cell).

Data Analysis

Colony counts are expressed as plating efficiency (PE), that is by the number of colonies per 10^5 nucleated marrow cells. Percent colony inhibition is determined by comparing the PE in the treated groups (T) to the PE in the negative controls (C), thus:
% Inhibition = $(C - T) / (C) \times 100$

Pros and Cons

These batteries of clonal assays for myeloid progenitor cells are highly powerful tools for evaluating the effects of drugs, compounds, radiation and therapeutic

Colony-Forming Unit Assay: Methods and Implications. Table 1 Myeloid clonal assay components

Component	CFU—G	CFU—GM	CFU—GEMM	HPP—CFC
rhG—CSF	20 ng/ml	—	—	20 ng/ml
rhGM—CSF	—	20 ng/ml	10 ng/ml	20 ng/ml
rhIL—3	—	—	10 ng/ml	20 ng/ml
rhIL—6	—	—	—	20 ng/ml
rhSCF	—	—	50 ng/ml	50 ng/ml
rhEPO	—	—	3 U/ml	—
Cell type	MNC	MNC	CD34$^+$	CD34$^+$
Cells/dish	100 000	100 000	1000	300
Incubation	13 days	13 days	16 days	15 days

CSF, colony stimulating factor; EPO, erythropoietin; IL, interleukin; rh, recombinant human; SCF, stem cell factor.

treatment on hematopoiesis. The recent advances in recombinant growth factors, media, sera, the purity of reagents and standardization of these assays have led to *less* heterogeneity of results. While myeloid colony assay systems have been well established, efficient T- and B-lymphoid progenitor assay systems still need to be developed.

Several applications of clonal assays include:
- quantitation and characterization of hematopoietic progenitors
- assessment of cells from patients with myeloproliferative disorders
- screening for new growth factors and/or inhibitors
- quantitation of progenitor numbers following ex vivo expansion
- testing in vitro sensitivity of hematopoietic progenitors.

However, there is still a lack of easy, accessible and optimal in vitro assays for human stem cells, which rely on ▶ stroma cell based methods to detect primitive cells by their ability to give longterm hematopoietic reconstitution in vivo. Nevertheless, the CFU assays are more established in the literature and the cytokines required are well defined.

Predictivity

The establishment of cell culture conditions, growth factors, media and serum has led to the standardization and predictivity of these clonal assays. Of note, Stem-Cell Technologies (Vancouver, Canada) has developed a variety of reagents and kits for hematopoietic cells, which include negative and positive controls. These kits include detailed information on experimental procedures and technical suggestions which are very useful to the investigators.

Relevance to Humans

Clonal assays can be used to investigate progenitor responses to growth factors, inhibitors and drugs, as read outs for LTC-IC assays, to quantitate progenitor cell numbers after ex vivo expansion, and to assess gene transfer efficiencies to stem cells and progenitor cells.

In addition, myelosuppression is the most common dose-limiting toxicity in cancer treatment. Conventional antineoplastics, high-dose therapies and drug combinations lead to severe bone marrow suppression with the potential for infection and hemorrhagic complications. Therefore, it is highly advantageous to predict during drug development whether a new antineoplastic agent will be clinically myelosuppressive, including severity (grade of cytopenia), onset, and duration (time of recovery).

The in vitro clonal assays provide an environment for hematopoietic cells to proliferate, and the number of colonies formed is proportional to the number of viable progenitors. The degree of inhibition of colony formation resulting from in vivo or in vitro drug exposure can be used to evaluate the cytotoxicity of a variety of compounds. Using these clonal assays, a compound can be classified as toxic or not and erythroid-myeloid progenitor cells can be qualitatively predictive of clinical cytopenia and utilized to determine the hematologic effects of chemotherapeutics. Thus, the CFU assays for in vitro studies can be recommended for their correlations to drug-induced neutropenia, predictive value and technical simplicity. The battery of a primitive and highly proliferative progenitor population (HPP-CFU), a multilineage myeloid-erythroid-megakaryocytic progenitor (CFU-GEMM), a bipotent progenitor for granulocytes and monocytes (CFU-GM) and a mature progenitor restricted to neutrophil production (CFU-G), will allow analysis of the most sensitive precursor cells.

References

1. Morrison SJ, Wandycz AM, Hemmati HD, Wright DE, Weissman IL (1997) Identification of a lineage of multipotent hematopoietic progenitors. Development 124:1929–1939
2. Domen J, Weissman IL (1999) Self-renewal, differentiation or death: regulation and manipulation of hematopoietic stem cell fate. Mol Med Today 5:201–208
3. Dexter TM, Allen TD, Lajtha LG (1977) Conditions controlling the proliferation of haemopoietic stem cells in vitro. J Cell Physiol 91:335–344
4. Freshney RI, Pragnell IB, Freshney MG (eds) (1994) Culture of Hematopoietic Cells, Ch 16. Wiley-Liss, New York
5. Eaves CJ, Lambie K (1995) Atlas of Human Hematopoietic Colonies. StemCell Technologies, Vancouver

Committed Precursor of Megakaryocytes

▶ Colony-Forming Unit Assay: Methods and Implications

Common Chain

In some families of cytokine receptors discrete α-chains specifically bind their corresponding ligand and then associate with one or more common signaling subunits which are shared between members of related cytokines and cytokine receptors (e.g. the common γ-chain in the Interleukin-2 receptor family).
▶ Cytokine Receptors

Common Cold

▶ Respiratory Infections

Competitive PCR

▶ Polymerase Chain Reaction (PCR)

Complement

Complement is a collective term for a large family of proteins with enzymatic activity that interact in a classic or alternate pathway to lyse target cells, or stimulate immune adherence, chemotaxis or opsonization. The cascade of the classic pathway is initiated when an antigen and antibody react, leading to the binding of C1, C4 and C2 to activate C3. The alternate pathway does not involve the binding of C1, C2 or C4 by the antigen-antibody complex.

▶ Complement, Classical Pathway/Alternative Pathway
▶ Streptococcus Infection and Immunity
▶ Antibody-Dependent Cellular Cytotoxicity
▶ Opsonization and Phagocytosis
▶ Humoral Immunity

Complement and Allergy

JEAN F REGAL
School of Medicine
University of Minnesota
Duluth, MN 55812
USA

Definition

▶ Anaphylatoxin was originally described as a toxin produced in blood after incubation with immune precipitates. This toxin caused adverse symptoms resembling severe systemic anaphylaxis and allergy when injected into animals. As the biochemistry of the complement system was revealed, anaphylatoxin activity was found to originate from the complement components C3 and C5. These initial observations formed the basis for the longstanding idea that the complement system is important in allergy. The anaphylatoxins C3a and C5a are approximately 10 kD fragments of the amino terminus of the alpha chain of C3 and C5, respectively. C4a is also an anaphylatoxin, but is less potent than C3a and C5a and its relevance in vivo is less clear. When injected into animals, C3a and C5a cause symptoms resembling immune-mediated hypersensitivity. This provides the rationale for investigations of the complement system in allergy and hypersensitivity. Increasing information on the biochemistry of the complement system, as well as the availability of animal models of complement deficiencies, has recently allowed more rigorous investigation of the role of the complement system in allergy and hypersensitivity.

Characteristics
Complement and Immune-Mediated Disease

Gell and Coombs in the 1960s proposed a method of classifying immune-mediated disease, with the terms type I, type II, type III and type IV hypersensitivity. Various modifications of this classification scheme have evolved. Classically, complement system participation in immune-mediated diseases or hypersensitivity states is relegated to type II and III hypersensitivity, where antigen-antibody complexes or antibody fixed to cells results in classical pathway complement activation and all of its consequences. Both type I and type IV hypersensitivity are classically considered complement-independent mechanisms of immunopathology. However, it is increasingly recognized that representing any disease as a distinct type I, II, III or IV hypersensitivity is too confining. Current evidence suggests that the complement system is important in asthma and contact dermatitis, which have historically been regarded as complement-independent type I and IV disease.

Because of the known links between complement system activation and adaptive immunity, interfering with complement function can have multiple effects in hypersensitivity. It is important to differentiate effects in the induction (or sensitization) vs the elicitation (challenge or effector) phase of the allergic response. For example, in the absence of C3, the complement system via complement receptor 2 cannot augment the signal to the B cell and the induction phase of asthma may be affected. Also, in the absence of C3, C3a cannot be generated so that any contribution of C3a to the effector phase of asthma will be eliminated. Using animals genetically deficient in complement will affect both the induction and elicitation phases, whereas use of enzyme inhibitors or antagonists can selectively target induction or elicitation separately.

Redundancy of inflammation and immunity is a consideration in mechanistic investigations of hypersensitivity. In asthma, it is not clear that any single mediator is responsible for the bronchoconstriction or eosinophilia. Numerous substances can cause bronchoconstriction and cell infiltration, including C3a and C5a. Many endogenous substances likely contribute, and the extent of the contribution may differ with the allergen, the dose, the induction versus elicitation phase, and the acute or chronic state of the disease.

Complement in Respiratory Hypersensitivity

Evidence for a role for the complement system in respiratory allergy in humans is circumstantial. Information is not available regarding allergy or asthma incidence in individuals genetically deficient in complement components. In segmental allergen challenge of asthmatics, studies demonstrated increased levels of C3a and C5a in the asthmatic lungs at time of the allergic response (in the elicitation phase of the disease). Asthma is a chronic disease, but therapeutics aimed at inhibiting complement activation long term have not been vigorously pursued because of the essential role of complement in host defense. However, more targeted therapy at effector molecules such as C3a and C5a may prove beneficial, without compromising host defense.

In animal models of asthma, where specific antibody to the antigen is present, the evidence indicates that the importance of an intact complement system in respiratory allergy differs depending on the allergen, animal model and endpoint being examined (1). Most studies have examined the allergic response to ovalbumin as a prototype antigen. Whether the results would be similar if examining food allergens or various occupational allergens is unknown. Studies have concentrated on the role of C3, C5, C3a and C5a, because anaphylatoxins can mimic the symptoms of respiratory allergy. Asthma in animal models is characterized by antigen-induced bronchoconstriction, airway hyperresponsiveness to methacholine, and lung eosinophilia. Animals deficient in C3 and C5 provide information on the role of complement in both induction and elicitation of the asthmatic response. Mice deficient in C3 have reduced airway hyperresponsiveness and lung eosinophilia in the elicitation phase of the asthmatic response, using ovalbumin in combination with *Aspergillus* as the antigen. Consistent with what we know about the links between complement and adaptive immunity, these C3-deficient mice had a reduced number of IL-4 producing cells and attenuated antigen specific IgE and IgG1 responses, suggesting a critical role for C3 in induction of respiratory allergy and a Th2 immune response. It is not clear whether the reduced airway hyperresponsiveness and eosinophilia observed in the elicitation phase was caused by lack of C3 during the induction phase. A number of inbred mouse strains show spontaneous deficiencies in complement component C5. Using these mice, C5 has been shown to be a susceptibility locus for airway hyperresponsiveness in the mouse asthma model. C5-deficient mice are more responsive to methacholine after allergen sensitization and challenge when compared to mice with normal C5 levels. Whether sensitization was equivalent in C5 deficient versus C5 sufficient mice was not reported. From studies to date, it appears that the complement components C3 and C5 have opposing influences on the asthmatic response.

Using mice deficient in the receptors for C3a and C5a, the role of complement activation products C3a and C5a can be determined. Mice deficient in the C3a receptor appear to have a normal immune response to ovalbumin, that is with no difference in levels of ovalbumin-specific antibody in C3a receptor-deficient animals compared to controls. This suggests that the induction phase of asthma is unaffected by C3a receptor deficiency. In the elicitation phase of asthma, antigen induced airway hyperresponsiveness was completely inhibited in the absence of the C3a receptor, but lung eosinophilia was unaffected. The asthmatic response in C5a receptor deficient mice has not been reported. The existence of a second C5a binding receptor, C5L2, complicates interpretation of experiments in mice lacking only the originally described C5a receptor CD88.

A strain of guinea pigs deficient in the C3a receptor has been described. Anaphylaxis induced by intravenous allergen exposure in these animals is partially reduced; the hypotensive response is inhibited but the bronchoconstriction is minimally affected (2). In this same strain of C3a receptor deficient guinea pigs, the bronchoconstrictor response to aerosol allergen challenge was reduced by about 30%, suggesting a minor role for C3a in this event. Lung eosinophilia was not affected. Similar to the mouse, the induction phase of asthma was not affected by C3a receptor deficiency in the guinea pig.

Studies to examine the importance of entire complement pathways in the allergic response have used either cobra venom factor or soluble complement receptor 1 (sCR1). Cobra venom factor intraperitoneally is used to activate the ▶ alternative pathway of complement and deplete C3 and C5, as well as C6–9 from the circulation, thus eliminating the influence of all terminal complement components. CR1 is a normal membrane-associated controller of complement activation that acts by causing decay of the ▶ C3 convertase and ▶ C5 convertase. sCR1 was created by eliminating the membrane-spanning region of the molecule so that it could limit complement activation in the fluid phase. sCR1 inhibits complement activation from C3 through C9, but does not activate the pathway and deplete the components like CVF. CVF and sCR1 treatment are effective for a limited time period because an immune response to the foreign protein ensues, limiting its complement depleting activity.

A clear drawback of the use of CVF is massive activation of the complement system with associated effects, including temporary neutrophil sequestration in the lungs, and release of mast cell mediators and cytokines. In fact intravenous injection of CVF results in an enhanced response to intravenous antigen, most likely due to the temporary sequestration of neutrophils in the lung. An advantage of CVF and sCR1 is that complement depletion can be targeted to a specific phase of the asthma—the elicitation phase. Using CVF in the guinea pig, complement depletion significantly inhibited antigen-induced lung injury and eosinophil infiltration into the airspace using the occupational allergen trimellitic anhydride (3).

In contrast, with ovalbumin as the antigen, complement depletion using CVF did not significantly inhibit cell infiltration in a similar guinea pig asthma model. With intravenous antigen challenge, sCR1 significantly reduced both bronchoconstriction and hypotension in response to ovalbumin. In a rat model of asthma using ovalbumin, inhibition of complement activation with sCR1 significantly attenuated both the early and late airway responses to intratracheal antigen. Thus, the effectiveness of complement inhibition differs with species and antigens, and generalizations regarding the role of the complement system in respiratory hypersensitivity cannot be made.

Complement in Contact Dermatitis

Contact dermatitis is often cited as the classical example of type IV or delayed-type hypersensitivity that lacks involvement of antibody and complement. However, studies have shown involvement of C5a and IgM anti-hapten antibody produced by B-1 cells in the mechanism of induction or initiation of contact sensitivity. Activation of the classical complement pathway by antigen binding to natural IgM antibody results in the generation of C5a. C5a appears to be crucial for local early T cell recruitment in contact sensitivity, making sure that the effector T cells get to the skin site (4). Other studies suggest that C3 fragments through stimulation of the CR3 receptor inhibit induction of contact sensitivity via IL-12. Nickel and cobalt are two metals that cause contact dermatitis. Recent evidence indicates that nickel and cobalt stimulate alternative pathway activation up to four times faster than magnesium, the endogenous metal required (5). The authors suggest that the ability of nickel and cobalt to act as contact sensitizers may be related to the increased stimulation of complement activation. Again, the role of the complement system in contact dermatitis varies depending on the complement component or pathway investigated.

Activation of the Complement System by Allergens and Environmental Pollutants

As indicated previously, allergen exposure of asthmatics results in increased levels of C3a and C5a in the lung washings. Also, ragweed allergen extracts activate the complement system in the serum of ragweed allergic patients, with the degree of complement activation correlating with the nasal allergy symptoms. Both asthma and nasal allergy are generally associated with increased IgE which does not activate complement in vitro. Any involvement of IgG in complement activation in these allergic patients was not investigated. Clearly complement activation occurs in sensitized individuals, and complement activation products can produce symptoms resembling asthma and nasal allergy. At this point, the mechanism or pathway leading to complement activation is unknown.

In occupational asthma, a clear relationship between specific antibody and symptoms is not always evident. The suggestion is that, in some instances, asthma symptoms may be due to a mechanism other than antigen interaction with specific antibody (i.e. pseudoallergic reaction). Various triggers of asthma are also known to activate complement in the absence of antibody. Plicatic acid, the occupational allergen from Western red cedar, activates complement via the ▶ classical pathway in the absence of antibody. A number of allergens have serine protease type activity (e.g. dust mites, *Bacillus subtilis* protease) and can cleave C3 and C5, generating active complement fragments. Besides allergens, various toxic substances and environmental pollutants also can activate the complement system. Asbestos fibers and crystalline silica, along with many biomaterials, activate the complement system, generating C5a. Evidence also indicates that diesel exhaust particles, cigarette smoke, endotoxin in dust, house dust, and particulate matter activate the complement system. Recent studies in a C3 deficient mouse demonstrate the critical importance of C3 in airway hyperresponsiveness induced by particulate matter in a mouse model (6).

Preclinical relevance

Currently, screening for potential allergenicity of compounds does not include evaluating their ability to activate the complement system. Screening however does look for IgE induction as an indicator of airway or respiratory hypersensitivity. As mechanisms are clarified it may be advantageous to examine the ability of a suspected allergen or pollutant to activate complement.

Relevance to Humans

Circumstantial evidence indicates that C3a and C5a are putative mediators of allergy and asthmatic responses. Because of the known link between complement and adaptive immunity, a compound's ability to activate C3 could significantly affect its ability to be immunogenic. However, the role of the complement system in the induction and elicitation of the allergic response is still under active investigation.

Regulatory Environment

The 2002 Guidance for Industry from the Food and Drug Administration entitled "Immunotoxicology Evaluation of Investigational New Drugs" (http://www.fda.gov/cder/guidance/index.htm) does not directly recommend evaluation of effects of new drugs on the complement system. However, in instances where the administration of a substance results in an anaphylactic-like reaction, complement activation and generation of ▶ anaphylatoxins could be suspect. The complement system could be causing a ▶ pseudoallergic reaction—a reaction where no specific antibody is involved. Alternatively, the substance could be reacting with specific antibody to cause complement activation and a hypersensitivity reaction. Whether with or without antibody, participation of the complement system needs to be considered if symptoms of allergy occur with administration of a substance.

References

1. Gerard NP, Gerard C (2002) Complement in allergy and asthma. Curr Opin Immunol 14:705–708
2. Regal JF, Klos A (2000) Minor role of the C3a receptor in systemic anaphylaxis in the guinea pig. Immunopharmacology 46:15–28
3. Regal JF (1997) Hypersensitivity reactions in the lung. In: Sipes IG, McQueen CA, Gandolfi AJ, Lawrence DE (eds) Comprehensive Toxicology, Volume 5: Toxicology of the Immune system. Pergamon Press, New York, pp 339–352
4. Tsuji RF, Szczpanik M, Kawikova I et al. (2002) B cell-dependent T cell responses: IgM antibodies are required to elicit contact sensitivity. J Exp Med 196:1277–1290
5. Acevedo F, Vesterberg O (2003) Nickel and cobalt activate complement factor C3 faster than magnesium. Toxicology 185:9–16
6. lters DM, Breysse PN, Schofield B, Wills-Karp M (2002) Complement factor 3 mediates particulate matter-induced airway hyperresponsiveness. Am J Respir Cell Mol Biol 27:413–418

Complement Cascade

The consecutive activation of complement serum proteins by enzymatic activity. This activation can be induced by interaction with antibodies (classical pathway) or bacteria (alternative pathway). Activated components are important for the stimulation of phagocytes, and result in lytic attack on cell membranes.
▶ Hypersensitivity Reactions

Complement Deficiencies

MICHELLE CAREY
NIEHS ND D2-01, Laboratory of Pulmonary Pathobiology
P.O.Box 12233
RTP, NC 27709
USA

Synonyms

Immunodeficiency.

Definition

A group of inherited and acquired disorders characterized by reduced levels of specific proteins—complement—necessary for proper functioning of the innate and adaptive immune systems.

Characteristics

The complement system is a group of at least 30 different serum proteins, produced primarily in the liver, which circulate in their inactive forms. These proteins, when activated, produce various complexes that play critical roles in immunity. Examples of such roles include ▶ opsonization, chemotaxis and activation of leukocytes, lysis of bacteria and cells, augmentation of antibody responses, and clearance of immune complexes and apoptotic cells.

Complement activation can occur via three different pathways:

- the classical pathway
- the alternative pathway
- the mannose-binding lectin pathway.

There are deficiencies associated with each pathway and with complement regulatory proteins and receptors (Table 1).

Complement deficiency can be acquired or inherited. Acquired deficiency can be caused by ailments that involve a lot of protein loss, such as serious burns, liver and kidney disease, by acute infection, or in conjunction with chronic autoimmune disorders. Exposure to certain chemicals (e.g. tetrachloroethylene) can alter serum complement levels. Most complement deficiencies are inherited as autosomal recessive conditions with the exception of properdin deficiency, which is sex-linked. Inherited deficiency is rare in the general population with an estimated frequency in the order of 0.03%.

Relevance to Humans
Diagnosis

The patient's history and clinical presentation are im-

Complement Deficiencies. Table 1 Complement deficiencies and disease associations

Proteins and Pathways	Consequences of Deficiency	Clinical Manifestations
Classical pathway C1q C1r C1s C4 C2	Failure to activate classical pathway Defective immune complex clearance Impaired immunoregulation	Immune complex disease (e.g. systemic lupus erythematosus) Recurrent infections (*Streptococcus pneumoniae*, *Neisseria meningitides*, *Haemophilus influenzae*)
MBL pathway MBL MASP	Impaired first-line defence against microbes	Recurrent infections Accelerated course of systemic lupus erythematosus and rheumatoid arthritis
Alternative pathway Properdin Factor B Factor D	Defective clearance of immune complex Impaired first-line defence against microbes	Severe fulminant pyogenic neisserial infections with high mortality rate
C3	Major opsonin critical to all pathways Failure to activate membrane attack complex	Severe infections Severe autoimmune disease
C5–9	Critical for lysis of cells and bacteria C5 important for chemotaxis	Neisserial Infections Autoimmune disease
Plasma regulatory proteins C1 inhibitor Factor H and Factor I	Loss of regulation of C1 and failure to activate kallikrein Failure to regulate activation of c3 Severe secondary C3 deficiency	Hereditary angioedema Hemolytic-uremic syndrome Membranoproliferative glomerulonephritis
Cell membrane regulators PI-linked proteins (DAF; CD59) CR3	Important in degrading C4, C3 and regulating lysis Important in phagocytosis	Paroxysmal nocturnal hemoglobinuria Leukocyte adhesion defect Infection

portant in making the correct diagnosis. Complement deficiencies should be suspected, for example, when there is:
- a patient history or family history of recurrent systemic meningococcal infection
- meningococcal disease (especially non-group B meningococci) in patients older than 10 years
- a family history of systemic lupus erythematosus
- a family history of meningococcal disease in males which might be suggestive of properdin deficiency.

The two most common screening tests are the complement hemolytic activity (CH_{50}) and the alternative pathway hemolytic activity (AP_{50}) which identify the group of complement components that have a defect. These tests measure the complement activity in dilutions of patient plasma on sheep erythrocytes that have been coated with anti-sheep erythrocyte antibody. The antibody-coated sheep erythrocyte immune complex activates the complement cascade. If all components of the complement system are present and functioning, the erythrocytes are lysed and the hemolysis can be measured and compared to a reference range.

More specific blood tests to identify components are then performed. Additional tests, such as C1 esterase level, Ham's (acidified serum) test, and white blood cell count, may also be useful.

Treatment

There are no specific treatments available for genetic complement deficiencies. However, antibiotics are used to treat infections, and vaccinations are given to reduce the risk of disease. In some cases (e.g. ▶ Hemoglobinuria, Paroxysmal Nocturnal) a bone marrow transplant may be recommended. Acute attacks of hereditary ▶ angioedema have been successfully treated with infusion of vapor heated C1 esterase inhibitor and androgen therapy may be used to prevent such attacks. Genes have been cloned for individual complement deficiencies and thus gene therapy may be a potential therapeutic option in the future.

Morbidity/Mortality

The prognoses for individuals with complement deficiency varies widely due to the range of disorders associated with such deficiencies. Some deficiencies are associated with a high mortality rate (e.g. paroxysmal nocturnal hemoglobinuria). Other patients are hospitalized frequently due to infections which are sometimes life-threatening. Patients with autoimmune disorders may have a normal life expectancy, and other patients remain healthy and asymptomatic throughout their lives.

References

1. Folds JD, Schmitz JL (2003) 24. Clinical and laboratory assessment of immunity. J Allergy Clin Immunol 111 [Suppl 2]:S702–711
2. Frank MM (2000) Complement deficiencies. Pediatr Clin N Amer 47 (6):1339–1354
3. Ross SC, Densen P (1984) Complement deficiency states and infection: epidemiology, pathogenesis and consequences of neisserial and other infections in an immune deficiency. Medicine (Baltimore) 63 (5):243–273
4. Walport MJ (2001) Complement. First of two parts. New Engl J Med 344 (14):1058–1066

Complement Fixation Test

The Complement Fixation (CF) test has traditionally been used for the screening of antibodies against a variety of possible pathogenic microbes (viruses, bacteria, parasites, fungi). The detection system is complement consumption by antibodies of a patient. The test is performed in two steps: first the serum is mixed with a standard amount of the antigen and complement (usually guinea pig complement is used); second antibody-sensitized indicator cells (rabbit antibody-coated sheep erythrocytes) are added to detect unconsumed complement.

▶ Complement Fixation Test

Complement Fragments

The complement system is made up of a series of about 25 proteins that work to "complement" the activity of antibodies in destroying bacteria, either by facilitating phagocytosis or by puncturing the bacterial cell membrane. Complement also helps to eliminate antigen–antibody complexes in the body. Complement proteins circulate in the blood in an inactive form. When the first of the complement substances is triggered, usually by antibody interlocked with an antigen, it starts a sequence of biochemical events where each component is activated in turn—the "complement cascade". Simultaneously, various fragments of complement proteins are cleaved during the course of the cascade, which then can produce other consequences. For instance, one fragment, C5a, is able to stimulate and attract neutrophils.

▶ Chemotaxis of Neutrophils

Complement System

JEAN F REGAL
School of Medicine
University of Minnesota
Duluth, MN 55812
USA

Synonyms

Alexin, complement

Definition

When originally discovered, the complement system was defined as a heat labile substance responsible for the bactericidal activity of blood. Over time, it has come to be recognized as a group of proteins functioning as a humoral immune amplification system, both in innate immunity as well as in amplification of the adaptive immune response. The major function of the complement system is elimination of foreign organisms and immune complexes. The complement system consists of three activation pathways (classical, alternative, and mannose-binding lectin), a terminal lytic pathway, plasma and membrane associated control proteins, and cell surface receptors for the effector molecules generated on activation of the system. Activation of the complement proteins occurs in the plasma and extracellular space, utilizing cell associated proteins as regulators and membrane receptors.

Characteristics

General Considerations

A combination of more than 30 plasma and membrane-associated proteins make up the complement system (1). By the use of three different activation pathways (Figure 1) the complement system can respond to varied stimuli. Both conformational changes and enzymatic cleavage of complement proteins lead to the sequential formation of the enzymes, ▶ C3 convertase and ▶ C5 convertase. As their names imply, the enzymes cleave C3 and C5, respectively. Cleavage of C3 in the presence of an ▶ activator surface leads to covalent attachment of the fragment C3b to the surface. C4 activation via either the classical or ▶ mannose-binding lectin pathway leads to C4 cleavage and covalent binding of C4b to the ▶ activator surface. C3 activation is central to each of the activation pathways, and in conjunction with activation of C5, leads to generation of important effector molecules, including molecules for opsonization and clearance of immune complexes, the proinflammatory ▶ anaphylatoxins (C3a and C5a), and the membrane attack complex. The ultimate goal of activation of the pathways is elimination of foreign organisms, by lysis or opsonization, and clearance of immune complexes. In addition, activation of the complement pathway leads to production of inflammatory mediators and amplification of the adaptive immune response for continued protection of the host with specific immunity.

Effect of Complement Deficiencies

Individuals genetically deficient in complement system proteins have provided insight into the critical functions carried out by the pathways (2,3). Consequences of defects in the complement system in humans include uncontrolled edema, gaps in host defense, and the appearance of immune complex diseases such as systemic lupus erythematosus (SLE) and glomerulonephritis. In the absence of C1 inhibitor (C1INH), hereditary angioedema occurs. In the absence of components of the ▶ classical pathway, an increased incidence of infection and immune complex disease has been noted. Defects in C3 and the alternative pathway are associated with more severe infection, including *Neisseria*. Deficiencies in any of the components of the terminal lytic pathway (C5 through C9) also contribute to an increased risk of *Neisseria* infection.

Complement is also recognized for its key role in the solubilization and elimination of immune complexes. In deficiencies of classical pathway components the risk of immune complex disease is increased. Because of redundancies in the innate and adaptive immune response, deficiencies in complement components do not always result in disease expression. Clinically relevant disease may only emerge after the immune system is overwhelmed by multiple insults. Thus, complement deficiencies increase the risk of infection and immune complex disease, but do not necessitate that disease will occur.

Activators

As depicted in Figure 1, activation of the complement system can be initiated via three different pathways, each resulting in the formation of a ▶ C3 convertase. Each pathway is activated differently, with a major distinction being participation of adaptive immunity (Table 1). The classical pathway relies on specific immunity and antibody, whereas activation of either the mannose-binding lectin pathway or the ▶ alternative pathway occurs in the absence of any specific antibody. Classical pathway activation is primarily initiated by antigen-antibody complexes, though C1q can also recognize a variety of pathogens in the absence of specific antibody. The mannose-binding lectin pathway recognizes foreign carbohydrates and is initiated by sugar residues such as *N*-acetylglucosamine and terminal mannose groups. The alternative pathway is activated by pattern recognition of foreign surfaces, including cells and negatively charged

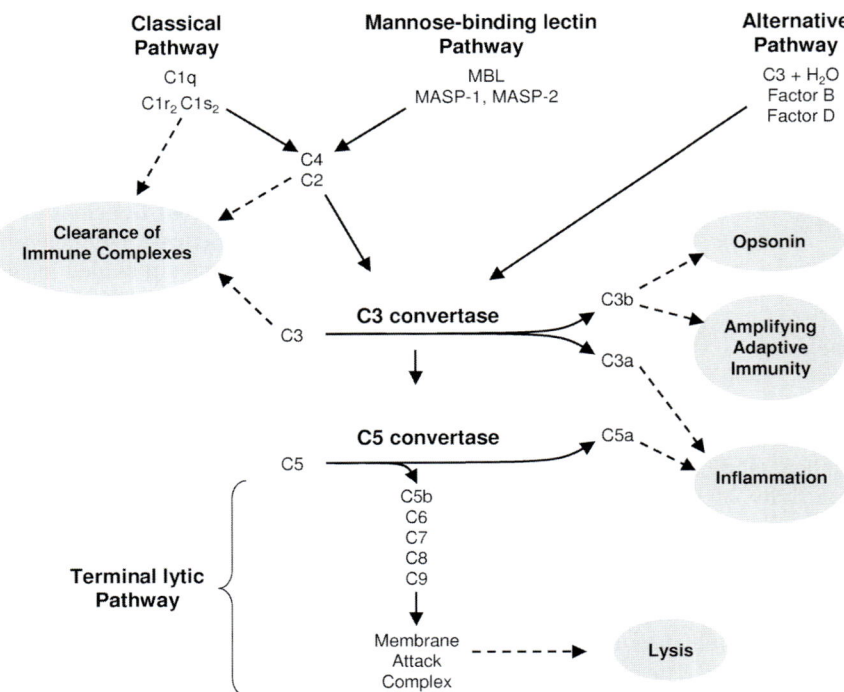

Complement System. Figure 1 Major functions associated with complement system pathways. Dashed arrows depict functions of portions of the pathways. MBL, mannose-binding lectin; MASP-1, MASP-2, mannose-binding lectin associated serine protease 1 and 2.

surfaces, and is closely controlled in order to prevent damage to the host. In the alternative pathway, low-grade hydrolysis of C3 continually occurs. If proteins controlling complement system activation are absent or overwhelmed, this low-grade hydrolysis is allowed to amplify and alternative pathway activation ensues.

Complement System. Table 1 Select activators of the complement system

Classical pathway
Antigen-antibody complexes
Mannose-binding lectin pathway
Sugar residues such as *N*-acetylglucosamine and mannose
Alternative pathway
Lipopolysaccharide (LPS) from gram-negative bacteria
Teichoic acid from gram-positive cell walls
Fungal cell walls (zymosan)
Aggregated immunoglobulins
Cobra venom factor
Anionic polymers (dextran sulfate)
Pure carbohydrates (agarose, inulin)

Controlling Complement Activation

Control of complement activation is essential to prevent damage to host cells. The system is regulated partially by the short half-lives of the many participating enzymes. This confines activation to a local area. The critical control, however, is the presence of numerous proteins in the host, both soluble and membrane associated, that prevents complement activation on the host and limits activation to foreign surfaces (Table 2).

Just as with other plasma enzyme systems, continued activation is dependent on the laws of mass action, a continuous competition between activation and decay of the proteolytic enzymes. Fluid phase inhibitors prevent widespread systemic activation at multiple steps in the activation pathways. For example, C1INH in binding to C1 will inhibit C4 activation. Control of activation is also critical at the level of C3 and C5 activation, as well as the formation of the membrane attack complex.

Host cells are protected from ▶ alternative pathway activation in a number of ways. Membrane-associated control proteins on autologous cells limit action of the ▶ C3 convertase formed, and fluid-phase inhibitors cause decay of the convertase so activation does not continue. Pathogens in general lack the membrane-associated control proteins that limit complement activa-

Complement System. Table 2 Proteins controlling complement activation

	Synonym or abbreviation	Inhibition of		
		C4 activation	C3 & C5 activation	MAC formation
Protein soluble in plasma				
C1 inhibitor	C1INH	+		
C4b binding protein	C4bp	+	+	
Factor H	FH		+	
Factor I	C3b inactivator		+	
S protein	Vitronectin			+
Apolipoprotein J	Clusterin, SP-40,40			+
C3a/C5a inactivator	C3a/C5a INA Carboxypeptidase N			
Membrane-associated protein				
Complement receptor 1	CR1, CD35		+	
Decay accelerating factor	DAF, CD55		+	
Membrane cofactor protein	MCP, CD46		+	
CD59	HRF, homologous restriction factor			+

tion, so activation continues unabated on the foreign surface. In addition, host cell sialic acid limits activation of the system by binding factor H. Factor H then inactivates the ▶ C3 convertase before amplification and further cleavage of C3 and C5 can occur. The surface of many pathogens is low in sialic acid so factor H does not bind and limit convertase activity on the pathogen. Identification of a surface as an "activator" or "nonactivator" surface involves the participation of plasma proteins as well as membrane bound regulators of complement activation. In the absence of sufficient control, complement activation will continue on an ▶ activator surface until the membrane attack complex is inserted in the lipid bilayer, resulting in lysis of the foreign organism.

Complement System in Health
Host Defense
The complement system functions in host defense by lysis of a variety of microorganisms, viruses and nucleated cells. Regardless of the activation pathway, if activation exceeds control, the terminal lytic pathway is initiated and the membrane attack complex formed. As its name implies, the membrane attack complex is a pore-like structure assembled in the membrane of the foreign organism resulting in lysis. The complement system also functions in viral neutralization by facilitating removal of virus, as well as interfering with viral infectivity by blocking attachment to host cells. Depending on the situation, if activation of the complement system does not continue on to the terminal lytic pathway, the complement system can still participate in host defense by opsonization. That is, if activation proceeds through C4 and/or C3 cleavage, then the foreign organism is tagged with C4b, C3b or the degradation products C3d or iC3b. These tags on the surface of the foreign organism are ligands for the corresponding receptors on phagocytes (CR1, CR2 and CR3). This receptor ligand interaction promotes phagocytosis (Table 3). Tagging of the foreign surface by deposition of the C3 and C4 fragments leads to cell adhesion and the facilitation of uptake of particulate antigen by phagocytes. Molecules such as C3b and C4b that bind to the antigen and the phagocyte to enhance phagocytosis are called ▶ opsonins. Even in the absence of the terminal lytic pathway, the complement system can be very effective in host defense.

Clearance of Immune Complexes
Besides functioning in host defense, the complement system plays a critical role in solubilization and clearance of immune complexes. In the presence of excess antigen, antigen is bound by antibody in the circulation to form immune complexes that are small and soluble. Continued binding of antibody forms large insoluble complexes. However, if complement is present, ▶ classical pathway activation occurs, and C3b is covalently attached to the complexes, preventing formation of large insoluble complexes. CR1 receptors on red blood cells then interact with C3b on the im-

Complement System. Table 3 Receptors of the complement system

Receptor	Synonyms	Major ligand	Selected functions
C1qRp	C1qR	C1q, mannose-binding lectin	Enhanced phagocytosis
C3aR	C3a Receptor	C3a	Inflammation Smooth muscle contraction
C5aR	CD88	C5a	Inflammation Smooth muscle contraction
CR1	CD35	C3b, C4b	Immune adherence Phagocytosis
CR2	CD21	C3d, iC3b	Augments B cell stimulation
CR3	CD11b/18, Mac-1	iC3b	Phagocytosis, cell adhesion
CR4	CD11c/18	iC3b	Phagocytosis, cell adhesion

mune complexes, carrying the immune complexes to the reticuloendothelial system for clearance. The complexes are transferred from red blood cells to CR1 receptors on macrophages. The macrophages then internalize and degrade the immune complexes. If classical pathway function is impaired or absent, immune complexes are not cleared, and an increased risk of immune complex disease results.

Proinflammatory Effects of Complement Activation

Cleavage of C3 results in covalent attachment of C3b to the ▶ activator surface with subsequent liberation of a small-molecular-weight fragment of C3 (C3a) into the fluid phase. An analogous situation is when C5 is activated and the small fragment C5a is released into the fluid phase. C3a and C5a are classically termed ▶ anaphylatoxins. This terminology came from the observation by Friedberger in 1910 that incubation of immune precipitates with serum resulted in the formation of a toxin. When injected into a guinea pig, the toxin caused adverse symptoms resembling severe systemic anaphylaxis. To cause these effects, C3a and C5a interact with their respective receptors present on numerous cell types. The biological effects associated with C3a and C5a receptor occupation include —but are not limited to—smooth muscle contraction, histamine release from mast cells and basophils, chemotaxis of white blood cells, increased vascular permeability, vasodilation, and increased arachidonate metabolism. The C3a and C5a receptors are seven transmembrane G protein-coupled receptors. Evidence for a second C5a receptor designated C5L2 is also emerging. C3a and C5a activity are controlled by cleavage of the carboxy-terminal arginine by the plasma enzyme carboxypeptidase N (Table 2). Loss of the terminal arginine generally reduces the potency of C3a and C5a.

Relationship Between Adaptive Immune Response and Complement

The complement system is generally considered part of innate immunity, but it is increasingly clear that it plays a very important role as a link to adaptive immunity. The two primary links are:

- activation of the classical pathway by antigen and specific antibody
- interaction of C3b degradation products (C3d and iC3b) with CR2 on B lymphocytes to augment stimulation of the humoral immune response

CR2 associates with CD19 and CD81 in the B cell membrane. An antigen coated with C3d binds to CR2 and crosslinks the CR2/CD19/CD81 receptor complex with the B cell receptor to significantly augment immunoglobulin production. Thus, by activation of C3, humoral immunity is amplified.

Complement System in Disease

Simplistically, problems with the complement system can result from either too little complement or too much activation of complement. As seen in the natural experiments with humans, the normal functions of host defense and clearance of immune complexes are impaired if complement deficiencies occur. The more common problem, however, is too much complement activation. Complement activation contributes to pathology in many situations including ischemia reperfusion injury, cardiopulmonary bypass, renal dialysis and reactions to implanted biomaterials. In addition, the mechanism of ▶ pseudoallergic reactions can involve misdirected complement activation. In hypersensitivity reactions or immune-mediated disease with aberrant responses to antigen, complement activation can lead to systemic injury (e.g. sepsis or anaphylaxis) or local tissue injury (e.g. glomerulonephritis).

Complement has traditionally been implicated in mechanisms of type II and type III hypersensitivity, leading to deposition of complement fragments and tissue injury in organs such as the kidney and lung. However, studies have also pointed to the importance of the complement system in mediating type I (asthma) and type IV (delayed-type) hypersensitivity. Thus, excessive complement activation or chronic uncontrolled activation in response to antigen can have serious consequences to the host.

Direct activation of the complement system by therapeutic agents, in the absence of specific antibody, is also a clinically important issue. This has been termed complement activation-related pseudoallergy. Infusion of radiographic contrast media, heparin or heparin substitutes, antibodies as therapeutics, phosphorothioate oligonucleotides, or anticancer drugs in liposomal formulations can lead to complement activation systemically with elaboration of C3 fragments, C3a and C5a, and the membrane attack complex, all with pathological effects. Availability of inhibitors of complement activation can minimize such adverse immunotoxic reactions if the mechanism is identified.

Preclinical Relevance

As is evident from the preceding discussion, the complement system is an important player in host defense and clearance of immune complexes. Exposures that reduce complement system activity for an extended period of time may have immunotoxic effects. These effects may be apparent in decreased resistance to infection or increased risk of immune complex disease. Compromising the complement system in the short term is probably not an issue. But just as with any immunotoxic insult, long-term suppression of inflammation and immunity can have deleterious effects. In addition, in the assessment of risk, determining which complement component or activation pathway is compromised is important. If a key element such as C3 is affected, the consequences can be severe.

Generally too much complement activation is of greater concern than deficiencies of complement. Once an individual is sensitized to a substance, an allergic response involving complement activation via the ▶ classical pathway could ensue, leading to all the adverse effects associated with products of complement system activation, including anaphylaxis and death. The route of administration needs to be considered, with intravenous exposures more likely to cause adverse effects from complement activation compared to other routes. Studies in the guinea pig have demonstrated that severe hypotension in anaphylactic shock in the guinea pig requires complement activation and is partially mediated via the C3a receptor (4,5). Substances may also directly activate the complement system without antibody, leading to the symptoms of allergy. Clinically relevant examples of pseudoallergy include adverse reactions to radiographic contrast media or infusions of liposomal preparations of taxol. In addition, exposure to particulate matter leads to a complement dependent airway hyperresponsiveness in animal models.

Relevance to Humans

Studies involving individuals genetically deficient in certain complement components have provided definitive information in the human regarding the consequences of reduced complement activity. Thus, any exposure which compromises the complement system for an extended time period can also be expected to have similar deleterious effects. However, as also seen from the natural experiments, because of redundancy of systems, the impairments in the complement system may not be revealed unless the insult is of sufficient intensity and duration to also compromise some of the backup systems of host defense.

Antibody to C5 is being tested as a therapeutic for the treatment of various immune-mediated diseases. This provides an example where immunotoxicity involving complement should clearly be considered. Administration of antibody to C5 leads to a significant reduction in C5 concentrations, the intended therapeutic endpoint. However, because of our knowledge of the functions of the complement system, it is very important to monitor these individuals for the immunotoxic effect of increased risk of infection with *Neisseria*.

Regulatory Environment

The 2002 Guidance for Industry from the Food and Drug Administration entitled "Immunotoxicology Evaluation of Investigational New Drugs" (http://www.fda.gov/cder/guidance/index.htm) does not directly recommend evaluation of the effects of new drugs on the complement system. However, in instances where the administration of a substance results in an anaphylactic-like reaction, complement activation and generation of ▶ anaphylatoxins could be suspect. If the administration of a substance results in increased meningococcal infections, a breach in the terminal lytic pathway of complement could be suspected. If autoimmunity is a side effect of a substance, then the ability of the complement system to solubilize immune complexes may be hindered. Participation of the complement system needs to be considered if host defense is compromised by exposure to a toxic substance, if an allergic like reaction occurs with administration of the substance, or if the incidence of immune complex disease increases.

References

1. Morley BJ, Walport MJ (2000) The Complement Facts Book. Academic Press, San Diego

2. Ross GD (1986) Immunobiology of the Complement System: An Introduction to Research and Clinical Medicine. Academic Press, Orlando
3. Morgan BP (1990) Complement: Clinical Aspects and Relevance to Disease. Academic Press, London
4. Regal JF, Fraser DG, Toth CA (1993) Role of the complement system in antigen-induced bronchoconstriction and changes in blood pressure in the guinea pig. J Pharmacol Exp Ther 267:979–988
5. Regal JF, Klos A (2000) Minor role of the C3a receptor in systemic anaphylaxis in the guinea pig. Immunopharmacology 46:15–28

Complementarity-Determining Region (CDR)

The complementarity-determining regions (CDR) are those parts of the variable regions of BCR and TCR which participate in the binding of epitopes and peptide fragments, respectively. The three CDRs of each of the two receptor chains (H/L chains of BCR and α/ß chains of TCR) together form the epitope-binding paratope. A synonymous expression for CDR is hypervariable (hv) region. CDR1 and 2 are encoded in the germ line and positioned over the residues at either end of an antigenic peptide. CDR3 is positioned over the central peptidic residues and generated by genetic recombination. The variability of the CRD3 amino acid sequence and its central location make it a critical region in HLA–antigen–TCR recognition and binding.
▶ Idiotype Network
▶ Rabbit Immune System
▶ Chronic Beryllium Disease
▶ Superantigens
▶ Humanized Monoclonal Antibodies

Complementary DNA (cDNA)

Single-stranded DNA that is synthesized from an RNA template or complex sample of mRNAs using reverse transcriptase. It can be used to deduce the amino acid sequence of the protein that it encodes.
▶ Polymerase Chain Reaction (PCR)

Complete Freund's Adjuvant

Oil emulsion containing heat killed Mycobacterium tuberculosis given to augment T- and B- cell mediated immune responses.
▶ Autoimmune Disease, Animal Models

Completely Randomized Design

An experimental design structure where individual subjects are randomly selected and randomly assigned to treatment conditions.
▶ Statistics in Immunotoxicology

ConA

▶ Polyclonal Activators

Concanavalin A (ConA)

A lectin (glycoprotein) extracted from the jack bean (*Canavalia ensiformis*). In immunology it is used to stimulate the division of T lymphocytes, and can therefore be used to test their proliferative capacity. See also Phytohemagglutinin.
▶ Lymphocyte Proliferation
▶ Polyclonal Activators

Concordance

Statistical term used in this case to determine the probability of correctly identifying a compound which is either immunosuppressive or not immunosuppressive.
▶ Plaque Versus ELISA Assays. Evaluation of Humoral Immune Responses to T-Dependent Antigens

Conditional Gene Expression

Gene expression under the control of an outside physical or chemical influence.
▶ Transgenic Animals

Conditioning

The pretreatment of a recipient organism before transplantation of hematopoietic stem cells. In leukemia, the conditioning regimen has the purpose to eliminate all malignant hematopoietic clones, in order to allow the permanent replacement with normal, nonmalignant transplanted hematopoietic stem cells and the ensuing entire hematopoietic system.
▶ Hematopoietic Stem Cells

Connective Tissue Diseases

▶ Systemic Autoimmunity

Connective Tissue Mast Cells

▶ Mast Cells

Constant Regions (C Regions)

The segments of immunoglobulin heavy and light chains that have identical sequences in chains of the same allotype and isotype.
▶ Rabbit Immune System
▶ B Lymphocytes

Consumption

▶ Respiratory Infections

Contact Dermatitis

Skin dermatitis reactions instigated by allergen or irritant. Allergic dermatitis (contact allergy) is immunologically-mediated, involves T cells sensitized to allergen, and is characterized by delayed reaction 12–48 h after contact. Irritant contact dermatitis is a nonimmunologic inflammatory response to contact irritant characterized by redness and sharply demarcated vesicles. Contact sensitivity/hypersensitivity is an adaptive immune response to contact exposure in an exaggerated form. Type I, II, and III reactions are antibody-mediated; type IV is a result of an activated T cell response to antigen and is manifest in four phases: refractory, induction, elicitation, and persistence.
▶ Chromium and the Immune System
▶ Contact Hypersensitivity

Contact Hypersensitivity

SHAYNE COX GAD
Gad Consulting Service
102 Woodtrail Lane
Cary, NC 27511
USA

Synonyms

Contact dermatitis, allergic contact dermatitis, local skin immune hyperreaction

Definition

Contact hypersensitivity is a form of ▶ delayed-type hypersensitivity (DTH) or cell-mediated immune reaction that is expressed in the epidermis. It is the most frequent form of DTH reaction and the most prominent clinical manifestation of a type IV immune reaction. But it is also part of immediate contact dermatitis, according to type I immune reactions. Although contact hypersensitivity may also be directed towards parasites, bacteria or fungi, the issue of immunotoxicity evaluation is focused on the interaction with foreign proteins and small chemicals acting as haptens.

Characteristics

Dermatitis responses remain the most common industrial hygiene issue in the work place. Of these, contact hypersensitivity responses are the most difficult to deal with. Contact hypersensitivity covers the entire range of nonirritant dermal responses arising from contact with foreign substances. Contact (or dermal) hypersensitivity reactions are inflammatory reactions of the skin that can either defend the host against pathologic agents, or damage host tissue and cause disease. The protective effects of hypersensitivity are a desirable part of host "immunity", while the detrimental effects arise from immune-mediated lesions, defined as immunopathologic disease. The terms "allergy" and "hypersensitivity" commonly denote deleterious immune reactions, which involve the pathophysiologic interaction of antigens (substances that induce an immune response) with specific antibodies (gamma globulin proteins) or with sensitized T lymphocytes. The term "allergy" generally designates immediate or humoral antibody reactions, while "hypersensitivity" usually signifies delayed cellular immune reactivity. According to the classification of Gell and Coombs (1), hypersensitivity reactions can be thought of as being of one of four major types. Of these, contact hypersensitivities are primarily types I and IV.

Type I

Contact urticaria syndrome is characterized by an im-

Contact Hypersensitivity. Table 1 Classification of hypersensitivity reactions based on immunophathologic mechanisms

Reaction type	Immunologic mechanism	Reaction time	Predominant immunocyte	Gammaglobulin	Primary cytokine	Tissue injury
Type I	Anaphylaxis asthma	Immediate (10–20 min)	Mast cell, basophil	IgE, (IgG)	Histamine, SRS, kinins	Smooth muscle contraction
Type II	Ig-dependent cytotoxicity	Variable	K cell (p complement)	IgG, IgM	None	Cell destruction
Type III complex	Immune (6–8 hours)	Intermediate (complement)	Polymorphs, IgM	IgG, enzymes	Lysosomal	Basement membrane damage
Type IV	Cell-mediated	Delayed	T_{DTH} lymphocyte	None	Lymphokines	Granuloma, dermatitis

mediate contact dermatitis reaction of normal or eczematous skin within minutes to an hour or so after agents capable of producing this type of reaction have been in contact with the skin. The skin reactions disappear within 24 hours, usually within a few hours. Local wheal-and-flare is the prototype reaction of contact urticaria. Generalized urticaria after a single local contact is uncommon. Minute vesicles may rapidly appear on the fingers in protein contact dermatitis. Apart from the dermatitis, effects may also appear in other organs in cases of strong hypersensitivity, leading to the use of the term contact urticaria syndrome. In some cases, immediate contact reaction can be demonstrated only on slightly or previously affected skin and it can be part of the mechanism responsible for maintenance of chronic eczemas.

Confusion has arisen in the use of terms such as contact urticaria, immediate contact reactions, atopic contact dermatitis, and protein contact dermatitis (Table 2). Immediate contact urticaria includes both urticarial and other reactions, whereas protein contact dermatitis means allergic or nonallergic eczematous dermatitis caused by proteins or proteinaceous materials.

There are two main mechanisms underlying contact reactions: immunologic (immunoglobin (Ig)E-mediated) and nonimmunologic immediate contact reactions. However, there are substances causing immediate contact reactions whose mechanism (immunologic or not) remains unknown.

Agents that have been reported to cause immediate contact reactions include chemicals in medications, industrial chemicals, latex, components of cosmetic products, and of foods and drinks, and chemically undefined environmental agents. The pathogenetic classification (nonimmunologic versus immunologic) is also given but in many instances it is arbitrary, because the mechanisms of various contact reactions are unclear, or (mainly) because a pathogenic evaluation was not performed.

Increasing awareness of immediate contact reactions will expand the list of etiologic agents, and more thorough understanding of pathophysiologic mechanisms will lead to a better and more rational classification of these reactions than at present. The international epidemic of latex-protein contact urticaria has led to an awareness of the syndrome among surgeons, anesthesiologists, pediatricians, and gynecologists. Tests used to identify such agents include human single application prick, scratch, scratch chamber, or open tests, and are described in Marzulli and Maibach (2).

Type IV

The most publicly recognized variety of hypersensitivity reactions are the antigen-specific T-cell mediated variety which occur 24–48 hours after exposure—

Contact Hypersensitivity. Table 2 Terminology of contact urticaria syndrome

Term	Remarks
Immediate contact reaction	Includes urticarial, eczematous, and other immediate reactions
Contact urticaria	Allergic (type I) and nonallergic (type II) contact urticaria reactions
Protein contact dermatitis	Allergic or nonallergic eczematous reactions caused by proteins or proteinaceous material

either epicutaneously (as in tuberculin testing) or dermally (as is the major source of concern occupationally and environmentally). In these, T lymphocytes that have previously been specifically sensitized to an antigen migrate to the region of exposure "recruiting" macrophages leading to the accumulation of basophils. This is accompanied by mediator release. The result is erythema and edema in the region of contact. The erythema is the basis of the traditional guinea pig-based sensitization tests. The edema portion of the sensitization response can be detected when the site exposure to an antigen is the skin, but it is not easily distinguished from background, and therefore is not used as an endpoint marker. Activated T-cell migration is the basis for the mouse-based test systems (LLNA, MEST, and PLNA). The preferred term for delayed dermal hypersensitivity to contact allergens is "allergic contact dermatitis", clinically known as dermatitis venenata.

Acute irritant contact dermatitis arises on first contact with an adequate concentration of a direct-acting cytotoxic chemical. On the other hand, allergic contact dermatitis (ACD) usually arises following more than one skin contact (induction and elicitation) with an allergenic chemical. The skin response of ACD is delayed, immunologically mediated, and consists of varying degrees of erythema, edema, and vesiculation. In the Gell and Coombs system (1) it is classified as a type IV allergy.

The best known example of ACD is the skin response that is often seen hours after contact with poison ivy (*Toxicodendron radicans*), at which time itching is a prominent symptomatic feature. Allergenic chemicals penetrate the skin as small molecules (usually < 400 MW) and they are incompletely allergenic (haptens) until they bind to protein and form a complete allergen. In ACD, the first significant exposure to a haptenic chemical activates the immune system (induction) and sensitizes the skin. After sensitization (which takes a few days to a few weeks), subsequent antigenic exposures result in the evocation of an altered (allergic") skin response (elicitation) that is, one that is more pronounced than the original response. In order for sensitization to take place, the allergenic chemical must first penetrate the skin, so that it can reach and interact with key elements of the underlying immune system. A certain level of allergen entry must be achieved that represents a threshold for triggering the immune system. The threshold can be reached following a single skin exposure to a sufficiently high amount or concentration of allergenic chemical, or after contact with a large area of skin, or as a consequence of repeated skin applications. Thereafter, there is a threshold (lower than the inducing level) of exposure which is required to evoke a response from the sensitizing entity of a related molecule that the T lymphocytes respond to as "close enough" (cross-sensitization reactions).

Once the allergenic chemical has transited the horny barrier layer of skin and entered the viable layer of the epidermis, it makes contact and binds with Langerhans cells. These dendritic cells direct the allergen to a regional lymph node where interaction with T lymphocytes is followed by replication of sensitized T lymphocyte population, completing the induction phase of the sensitization process. In the sensitized individual the next contact with the allergenic chemical results in the elicitation of a hypersensitive skin response that is due to a reaction between circulating sensitized lymphocytes and allergen at the skin site where allergen has entered the living epidermis.

The development of predictive and diagnostic human and guinea pig tests for skin sensitization focused further attention on ACD (3), as did regulatory and legal requirements for evaluating drug and cosmetic safety (4). These tests are employed for identifying (or at least screening for) an immense variety of potential commercial concerns—from the occupational setting, to cosmetics, personal care products, medical devices, and topical drugs.

Early in the study of ACD humans were the primary investigative test species. Later, guinea pigs were added as the animal model of choice. More recently the mouse has been used extensively. A wide range of study designs are currently employed to evaluate the potential for inducing contact dermatitis. The more common ones, classified by test species, are listed in Table 3.

Careful consideration and correlation of developments in genetics and molecular biology, based on mouse studies, with the entry of the mouse as a test species for ACD potential led ultimately to the finding that cytokines play a role in both irritant dermatitis and ACD. A detailed interpretation of cellular and molecular events of ACD is given in Sauder and Aastore (5). Entrance of an irritant or allergenic chemical into the epidermis signals the release of a cascade of cytokines from affected keratinocytes, suggesting a key role for keratinocytes in these inflammatory processes. It is not yet clear how cytokines differ qualitatively and quantitatively during ACD and irritant reactions, but this is likely to be an important area of future research.

While foundations for the overall picture of events of ACD appear secure, the future may unhinge some present interpretations of the details.

Keratinocytes comprise the main cellular composition of the human epidermis. They are involved in synthesis of various cytokines during both normal and abnormal cell functions. Cytokines are regulatory proteins that mediate cell communication, and include interleukins, growth factors, colony-stimulating factors, and interferons. When keratinocytes are damaged

Contact Hypersensitivity. Table 3 Test systems for delay contact hypersensitivity potential

Guinea pig	Mouse	Human
Guinea pig maximization test (GMPT)	Local lymph node assay (LLNA)	Repeat insult patch tests (RIPT)
Buehler	Mouse ear swelling test (MEST)	
Open epicutaneous	Popliteal lymph node assay (PLNA)	

during contact with irritant or allergenic compounds, various inflammatory elements, including cytokines, adhesion molecules, and chemotactic factors, are released (6,7). Current research interest in this area has sparked a continuously expanding literature.

It is virtually impossible to distinguish irritant dermatitis from ACD with precision, on gross and even microscopic inspection. Recently, Brasch et al. (8) reported an attempt to idetnti a chemical, which was administered to seven sensitized subjects along with the allergenic chemical, on two separate test skin sites, in order to produce experimental irritant and allergic contact dermatitis on a small scale. Both skin sites responded similarly in clinical appearance, histology, and immunohistology. A large battery of monoclonal antibodies directed against numerous antigens (surface, intracellular, and nuclear) failed to uncover and differentiate ACD from irritant dermatitis (9,10).

A special form of allergic or toxic contact dermatitis can be induced by photoreactive chemicals, i.e. photoirritation or photoallergy. These undesired side effects are observed after both dermal and oral administration and are a concern with pharmaceuticals, agrochemicals, and industrial chemicals.

Preclinical Relevance

As described below, there are special guidelines regulating the investigation of contact hypersensitivity with agrochemicals, pharmaceuticals, and cosmetics. With respect to industrial chemicals, requests for such tests depend on a number of factors (see below).

Relevance to Humans

The human relevance is obvious for topically applied pharmaceuticals and cosmetics. People handling agrochemicals or industrial chemicals also need to know if a compound is a skin sensitizer or not. However, the predictivity or accuracy of the preclinical test systems presents a problem. Up to the end of the last decade contact sensitivity was tested exclusively in guinea pigs (▶ guinea pig assays for sensitization testing). Overall, this animal model has a positive predictivity of about 100% and an accuracy of about 73%. It had been thought that the introduction of more objective parameters into such tests—like the measurement of cell proliferation or skin reaction instead of assessment of skin reddening—would improve the accuracy (e.g. the (▶ MEST), the ▶ local lymph node assay (LLNA), the (IMDS). However, in spite of the huge advantages, these mouse tests still present some problems. It may turn out that some mouse tests are not more predictive than guinea pig tests.

In comparison to these mouse tests the ▶ popliteal lymph node assay (▶ PLNA) or the ▶ RA-PLNA are still far away from a sound validation status.

Regulatory Environment

As described above, testing of industrial chemicals for their skin-sensitizing properties is determined by a number of conditions. For example, in Europe new chemicals have to be tested in relation to the amount produced per year. Also short-lived intermediates produced during synthesis of any new chemical have to be tested if they appear as "isolated" intermediates.

Compounds have been tested for decades in guinea pigs for contact hypersensitivity (allergic contact dermatitis). However, in the past few years the local lymph node assay (LLNA) or modifications of it (such as the IMDS) have been increasingly accepted by the authorities as stand-alone alternatives.

Testing for contact hypersensitivity is regulated by various guidelines:

- OECD 406 Guideline for Testing Chemicals, Skin Sensitisation 1992
- CPMP/SWP/2145/00 Note for Guidance on Non-Clinical Local Tolerance Testing of Medicinal Products 2001
- OECD 429 Guideline for Testing Chemicals, Skin Sensitisation: Local Lymph Node Assay, 2002
- CPMP/SWP/398/01 Note for Guidance on Photosafety Testing 2002 (as modified lymph node assay)
- FDA (CDER) Guidance for Industry. Immunotoxicology Evaluation of Investigational New Drugs 2002
- EPA OPPTS 870.2600, Skin Sensitization 2003

References

1. Gell PGH, Coombs RRA (1968) Clinical aspects of immunology, 2nd ed. Blackwell, Oxford
2. Marzulli FN, Maibac HI (1996) Dermatotoxicology, 5th ed. Taylor & Francis: Philadelphia
3. Schwartz L (1941) Dermatitis from new synthetic resin fabric finishes. J Contact Dermatol 4:459–470
4. Draize J (1959) Dermal toxicity. In: Appraisal of the safety of chemicals in foods, drugs, and cosmetics.

Association of Food and Drug Officials of the United States, Texas State Department of Health, Austin, pp 46–49
5. Sauder DN, Pastore S (1993) Cytokines in contact dermatitis. Am J Contact Dermatol 4:215–224
6. Barker JNWN, Mitra RS, Griffiths CEM, Dixit VM, Nickoloff BJ (1991) Keratinocytes as initiators of inflammation. The Lancet 337:211–215
7. Kupper TS (1989) Mechanisms of cutaneous inflammation. Arch Dermatol 125:1406–1412
8. Brasch J, Burgard J, Sterry W (1992) Common pathogenetic pathways in allergic and irritant contact dermatitis. J Invest Dermatol 998:166–170
9. Enk AH, Katz S (1992) Early events in the induction phase of contact sensitivity. Soc Invest Dermatol 99:39s–41s
10. Paludan K, Thestrup-Pedersen K (1992) Use of the polymerase chain reaction in quantification of interleukin 8 mRNA in minute epidermal samples. Soc Invest Dermatol 99:830–835

Contact Inhibition

Contact inhibition occurs when cells are grown in a monolayer their growth is arrested when they contact each other and reach confluency. Under these circumstances, cancer cells usually continue growth and pile up on top of one another.

▶ Cell Adhesion Molecules

Contact Photoallergy

▶ Photoreactive Compounds

Coombs Test

An immunological agglutination test used to help in the diagnosis of hemolytic anaemia. It is either direct, to detect the presence of antibodies that have coated the surface of red cells in vivo—or indirect, to detect the presence of antibodies that has not coated the surface of red cells in vivo.

▶ Molecular Mimicry
▶ Hemolytic Anemia, Autoimmune

COPD

▶ Trace Metals and the Immune System

Coreceptor Competition

In some cytokine receptor families decoy receptors exist which have a modulatory function. They serve as ligand sinks for the cytokine and like in the interleukin-1 receptor system may recruit the coreceptor molecule into a nonsignaling complex thus depriving the signaling receptor from its indispensable coreceptor.

▶ Cytokine Receptors

Coreceptor of the TCR

CD4 or CD8 stabilize the TCR interaction with the MHC–peptide complex and through the intracellularly bound kinase $p56^{lck}$ they contribute to CD3 signaling.

▶ Antigen Presentation via MHC Class II Molecules

Cornifying/Cornification

During their differentiation process, keratinocytes undergo dramatic changes in cell morphology. Cornification of the cells starts in the upper spinous layer of the epidermis and predominantly depends on the intracellular calcium ion level. Upon release of free fatty acids, ceramides and cholesterol from the laminar bodies of keratinocytes to the extracellular space, epidermal cells form the water-insoluble barrier of the epidermis. The cornification process of keratinocytes results in the formation of cornified cell envelopes (CCE), and this is accompanied by degradation of cell organelles. After the transition of keratinocytes from the *stratum granulosum* to the *stratum corneum* the life-cycle of the cells (now called corneocytes) ends.

▶ Three-Dimensional Human Skin/Epidermal Models and Organotypic Human and Murine Skin Explant Systems

Coronavirus

A virus that causes bronchitis, pneumonia, and possibly gastroenteritis.

▶ Respiratory Infections

Corrosive

A chemical that causes visible destruction of, or irreversible alterations in, living tissue by chemical action at the site of contact. For example, a chemical is considered to be corrosive if, when tested on the intact skin of albino rabbits by the method described by the US Department of Transportation in appendix A to 49 CFR part 173, it destroys or changes irreversibly the structure of the tissue at the site of contact following an exposure period of 4 hours. This term shall not refer to action on inanimate surfaces.
▶ Three-Dimensional Human Skin/Epidermal Models and Organotypic Human and Murine Skin Explant Systems

Cortex

The outer layer of the thymus (in mammals) with a high density of (immature) T lymphocytes.
▶ Thymus

Corticosteroid-binding globulin (CBG)

Sometimes referred to as transcortin, this binds to corticosteroids with high avidity. The CBG–corticosteroid complex is too large to enter cells and is thus biologically inactive, since the glucocorticoid receptor is located in the within the cytosol. Circulating levels of CBG are decreased by stressful events, thus increasing the fraction of biologically active glucocorticoids.
▶ Glucocorticoids

Corticotrophin-Releasing Hormone

A neuropeptide released by the hypothalamus that stimulates the release of adrenocorticotropic hormone by the anterior pituitary gland.
▶ Stress and the Immune System

Corticotropin-Releasing Hormone/Factor

This is released in response to stressors. It is produced in the hypothalamus and released into a local venous plexus that communicates with the anterior pituitary gland (adenohypophysis).

▶ Glucocorticoids

Covariance, Analysis of

A set of procedures similar to the analysis of variance, but where statistical methods are used to adjust for the effects of a variable which is correlated with the dependent variable.
▶ Statistics in Immunotoxicology

COX

▶ Prostaglandins

COX Inhibitors

▶ Anti-inflammatory (Nonsteroidal) Drugs

Coxsackievirus

A heterogeneous group of small RNA viruses (picornaviruses) that cause a variety of diseases in humans.
▶ Cardiac Disease, Autoimmune

CpG Motifs

▶ DNA Vaccines

CPMP

This is the committee for proprietary medical products ot the European Agency for the Evaluation of Medical Products (EMEA) which is responsible for the evaluation of medicines for human use.
▶ Assays for Antibody Production

CREST

An acronym for a variant of scleroderma characterized by **c**alcinosis, **R**aynaud phenomenon, **e**sophageal involvement, **s**clerodactyly, and **t**elangiectasia.
▶ Systemic Autoimmunity

Criteria Air Pollutants

Six common air pollutants for which EPA has set national air quality standards—ozone, carbon monoxide, nitrogen dioxide, particulate matter, sulfur dioxide, and lead.
▶ Asthma

Cross-Reactivity

▶ Molecular Mimicry

Crossover Design

An experimental design in which each subject will receive each treatment combination, and the order in which treatments are received is randomized. This design allows for the estimation of carry-over effects related to each treatment.
▶ Statistics in Immunotoxicology

Crossreactivity

Development of an allergic response to structurally similar allergens.
▶ Food Allergy

Cryopreservation of Immune Cells

In 1949 Polge used glycerol successfully to cryopreserve both animal and human spermatozoa at $-80°C$. In recent decades several protocols have been adapted to optimize the preservation of all kinds of immune competent cells like bone marrow, lymphocytes and hybridomas. Today cryopreservation techniques in liquid nitrogen ($-196°C$) permits conservation of large numbers of cells of different types for several months or years.
▶ Cryopreservation of Immune Cells

CTL

▶ Cytotoxic T Lymphocytes

CTL Activity

▶ Cell-Mediated Lysis

CTMC

▶ Mast Cells

Cutaneous Anaphylaxis, Passive (PCA)

In 1964 Ovary published this technique as a highly sensitive method for the determination of antigen specific IgE antibodies. Different dilutions of the serum to be examined for IgE antibodies are injected intra-cutaneously into guinea pigs followed by an intravenous injection of antigen and dye three hours later. Evaluation is done 30 minutes after this last treatment by analysis of colored areas in the skin.
▶ Cutaneous Anaphylaxis, Passive (PCA)
▶ Assays for Antibody Production

Cyclooxygenase (COX)

Cyclooxygenase COX-1 and COX-2 enzymes are required for the synthesis of prostaglandins and related compounds from arachidonic acid and can be inhibited by NSAIDs.
▶ Anti-inflammatory (Nonsteroidal) Drugs
▶ Prostaglandins

Cyclosporin A

P ULRICH
Preclinical Safety
Novartis Pharma AG
MUT2881.329
CH 4002 Basel
Switzerland

Synonyms
$C_{62}H_{111}N_{11}O_{12}$

Definition
Cyclosporin A (CsA), an undecapeptide, was isolated as a product of a fungus classified as *Tolypocladium inflatum Gams* and was shown to inhibit activation by

specifically blocking the respective signal transduction pathway (1). These pharmacological properties have made CsA a widely used immunosuppressant for prevention of transplant rejection and for therapy of rheumatoid arthritis and diseases with autoimmune features like ▶ psoriasis. Rejection of transplanted organs is mediated by effector T cells, which have been activated via distinct pathways

- the direct pathway, in which alloreactive T cells recognize intact major histocompatability complex (MHC) alloantigens expressed by donor antigen-presenting cells (APC) (*see also*▶ alloreaction)
- the indirect pathway, which involves allopeptides derived from processing and presentation of allogeneic MHC molecules by recipient APC.

The role of both pathways in allograft rejection is currently discussed, however it appears that direct activation of T cells plays a predominant role during acute rejection, whereas the indirect pathway of T cell activation contributes to chronic or late rejection (2,3). Administration of CsA to transplant recipients inhibits the activation of T cells by direct interference with signal transduction and thus, the subsequent clonal expansion of T cells.

Characteristics
Molecular Characteristics and Interaction with the Immune System

Cyclosporin A was the first immunosuppressive drug to inhibit specifically T cell activation and proliferation, which play a central role in immune responses. Upon activation by antigen-presenting cells, which interact with T cells through binding of their MHC-peptide complex with matching T cell receptors, a cascade of signaling events is induced in the T cell leading to the transcription of the interleukin IL-2 gene. IL-2 or T cell growth factor induces proliferation of T cells in an autocrine and a paracrine way. The initial step is the activation of phospholipase Cγ1, which activates protein kinase C and leads to an increased Ca^{2+} influx. Protein kinase C activates transcriptions factors AP1 and NFκB, while Ca^{2+} influx turns calcineurin into its active, catalytic form. Activated calcineurin dephosphorylates NFATp, a subunit of nuclear factor of activated T cells (NFAT), which then can move from the cytoplasm into the nucleus.

In the nucleus the formation of NFAT by binding of AP1 proteins Fos and Jun with dephosphorylated NFATp leads to the activation of the IL-2 promotor. In parallel, calcineurin inactivates the NFκB inhibitor IκB, thus facilitating nuclear localization of NFκB, which together with NFAT is required for maximal transcription of the IL-2 gene. The molecular targets of CsA are proteins with *cis*-trans peptidyl-prolyl isomerase (PPIase) activity, cyclophilin A and B. These "immunophilins" lose their PPIase activity upon binding to CsA and acquire a new affinity to the calcineurin B binding domain on calcineurin. It is noteworthy that PPIase activity is not related to the signal transduction events leading to IL-2 gene transcription. Binding of the CsA-cyclophilin complex inhibits the phosphatase activity of calcineurin and prevents nuclear localization of NFAT and NFκB and, in turn, the initiation of IL-2 gene expression. FK506, which is structurally unrelated to CsA, binds to other immunophilins, FK506 binding proteins FKBP12 and FKBP12.6 with the same inhibitory consequences as stated for CsA. Rapamycin, a structural analogue of FK506, binds also to FKBP12, but does not block calcineurin phosphatase activity. The target of the rapamycin-FKBP12 complex is a cellular protein called FRAP/RAFT, which displays kinase activity to phosphorylate the protein kinase $p70^{S6K}$. The complex leads to dephosphorylation and inactivation of $p70^{S6K}$ and, as a result, inhibits IL-2-mediated T cell proliferation and cell cycle progression rather than the expression of IL-2. Thus, CsA and FK506 block the progression of T cells from the G0 to the G1 phase, whereas rapamycin interferes with G1/S phase transition, which is regulated by IL-2 in T cells (4,5).

Relevance to Humans

CsA is used widely used to prevent rejection of transplanted organs. CsA is often applied in combination with ▶ azathioprine and ▶ prednisone to reduce the dose and thus the toxicity of each single drug. CsA, azathioprine, prednisone and OKT3 (anti-CD3 monoclonal antibody) or antilymphocyte globulin (ALG) is sometimes given to manage acute, severe rejection periods. CsA is also applied in patients suffering from rheumatoid arthritis and psoriasis. Side effects

Cyclosporin A. Figure 1 Chemical structure of Cyclosporin A.

are opportunistic infections and tumours of the skin and the lymphoid system, which are related to the permanent suppression of immune function.

The development of post-transplant lymphoproliferative disorders (PTLD) in immunosuppressed patients is mainly induced by the failure of immune surveillance of B cells infected with Epstein-Barr virus (EBV), a human gamma herpesvirus with a tropism for B cells. More than 95% of the adult population carries EBV as a lifelong asymptomatic infection, since the growth-transforming capacity of EBV is usually controlled by the host's immune system. The virus defence is mainly mediated by cytotoxic T lymphocytes, but there is also evidence that CD4 T cells can respond to virus antigens and therefore contribute to the control of a latent EBV infection.

The main clinical side effect due to CsA treatment is kidney toxicity, which can be subdivided into functional toxicity (increase in serum creatinine, hyperuricemia, increase in serum potassium and magnesium, metabolic acidosis) and structural toxicity (tubulopathy, microangiopathy), where the first can occur without structural signs of toxicity (6). This kidney toxicity can be associated with decreased levels of calcium-binding protein (calbindin-D 28 kDa) in the kidney found in most of the human kidney biopsy sections (7). However, the introduction of low-dose regimens together with new galenic formulations resulted in a "safer" but still immunosuppressive exposure to CsA, making the incidence of kidney toxicity rare.

Regulatory Environment

The European Agency for the Evaluation of Medicinal Products (EMEA) has issued a guidance (CPMP/SWP/1042/99), which asks for immune function testing in repeated-dose rodent toxicity studies. The interlaboratory validation immunotoxicity studies with CsA as a standard conducted by different groups (8,9) and the report of an international workshop (10) certainly provided the foundation for the decision of the CPMP to require immune function testing as regular preclinical screening in drug development.

References

1. Borel JF, Feurer C, Gubler HU (1976) Biological effects of cyclosporin A: A new antilympholytic agent. Agents Action 6:468–475
2. Auchincloss H Jr, Sultan H (1996) Antigen processing and presentation in transplantation. Curr Opin Immunol 8:681–687
3. Suciu-Foca NP, Harris E, Cortesini R (1998) Intramolecular and intermolecular spreading during the course of organ allograft rejection. Immunol Rev 164:241–246
4. Thomson AW, Starzl TE (1993) New immunosuppressive drugs: mechanistic insights and potential therapeutic advances. Immunol Rev 136:71–98
5. Cai W, Hu L, Foulkes JG (1996) Transcription-modulating drugs: mechanism and selectivity. Curr Opin Biotechnol 7:608–615
6. Mihatsch MJ, Kyo M, Morozumi K, Yamaguchi Y, Nickeleit V, Ryffel B (1998) The side-effects of ciclosporin A and tacrolimus. Clin Nephrol 49:356–363
7. Aicher LD, Wahl A, Arce O, Grenet, Steiner S (1998) New insights into Cyclosporin A nephrotoxicity by proteome analysis. Electrophoresis 19:1998–2003
8. ICICIS (1998) Report of validation study of assessment of direct immunotoxicity in the rat. The ICICIS Group Investigators. International Collaborative Immunotoxicity Study. Toxicology 125:183–201
9. Richter-Reichhelm HB, Schulte AE (1998) Results of a cyclosporin A ring study. Toxicology 129:91–94
10. Richter-Reichhelm HB, Stahlmann R, Smith E et al. (2001) Approaches to risk assessment of immunotoxic effects of chemicals. Toxicology 161:213–228

Cytochrome P450s

Heme-containing proteins that absorb light maximally at 450 nm when carbon monoxide binds to the proteins. Among the phase I biotransforming enzymes, the cytochrome P450 system ranks first not only for its catalytic versatility, but also for the number of xenobiotics it detoxifies or activates to reactive intermediates.

▶ Metabolism, Role in Immunotoxicity

Cytogenetics

Cytogenetics is a diagnostic process that analyzes both the number and the shape of the chromosomes within cells. In addition to identifying chromosome alterations, such as those characteristic of specific disease states, sometimes specific genes affected can be identified.

▶ Leukemia
▶ Lymphoma

Cytokeratin

Intermediate filament found in epithelial cells (molecular mass ~ 40–68 kDa) as supportive filament for cell shape. Highly sulfated.

▶ Immunotoxic Agents into the Body, Entry of

Cytokine Assays

Curtis C. Maier
R&D, Toxicology Preclinical Safety Assessment
GlaxoSmithKline
709 Swedeland Road
P.O.Box 1539
King of Prussia, PA 19406-0939
USA

Synonyms
Methods used to detect cytokines include the enzyme-linked immunosorbent assay (e.g., ELISA and ELISPOT), intracellular staining by flow cytometry, microsphere-based multiplex assays, gene expression analysis (e.g; RT-PCR and microarray), and cell-based bioassays.

Short Description
Cytokines are immunomodulatory proteins that mediate proinflammatory and anti-inflammatory responses as well as hematopoietic development. Assays that measure cytokines use specific reagents capable of detecting production and secretion of cytokines at the protein or mRNA level. Antibody-based assays, or immunoassays, are used to measure cytokines at the protein level that have been secreted into supernatants or found in body fluids and tissue extracts (biofluids). A variation of the immunoassay, intracellular staining, uses fluorescently-labelled cytokine-specific antibodies to detect cytokines directly in the cells that manufacture them. Cell-based bioassays, which utilize cells dependent on a specific ▶ cytokine for a given functional endpoint, are used to measure the presence of cytokines specifically in their bioactive form. Since many cytokines are regulated at the transcriptional level, assays have also been developed to detect cytokine mRNAs, including real-time reverse transcriptase-polymerase chain reaction (RT-PCR) and multiplexed microarray assays. Detailed methods for selected and broadly used assays can be found elsewhere (1).

Characteristics
Antibody-Based Assays (Immunoassays)
The most common and well established ▶ immunoassay used for detecting cytokines in supernatants and biofluids is the sandwich ELISA. In this assay, one antibody is coated onto a solid support, such as a plastic microwell plate, and used to capture the cytokine of interest from supernatant or biofluid. The plate is then washed and a second antibody, directed against a different site on the cytokine, is added to detect the captured cytokine. The second antibody is labelled with an enzyme (e.g., alkaline phosphatase or peroxidase), and a color reaction develops upon addition of substrate. The intensity of the color reaction, read on a spectrophotometer, is correlated with the amount of cytokine captured from the biofluid. The amount of cytokine is quantified against a standard curve of titrated recombinant cytokine run on the same plate. The sensitivity of the assay, usually in the low picogram/mL range, can be enhanced by using a biotinylated detection antibody and adding the enzyme conjugated to avidin (or streptavidin) in a tertiary step. Sandwich immunoassays that use fluorescently labelled antibodies in place of enzyme-linked antibodies are also commercially available. The largest limitation to the ELISA is that only a single cytokine can be measured at once, while limitations to immunoassays in general are that the antibody reagents may cross react with structurally homologs or biologically inactive forms of the cytokines.

Microsphere-based fluorescent immunoassays, like the sandwich immunoassays, also measure the amount of cytokine in biofluids (2). In this assay fluorescently-labelled microspheres (e.g., a mix of orange and red fluorophores) are covalently coupled with capture antibodies specific for a given cytokine. After incubating the microspheres with biofluid samples, a biotinylated detection antibody and tertiary streptavidin conjugated with a different fluorophore (e.g., phycoerythrin or PE) is used to detect any captured cytokine. The microspheres are read on a flow cytometer and the intensity of the PE signal is correlated with the amount of cytokine captured. The key advantage of this assay is that it is readily multiplexed, or used to simultaneously measure several cytokines in a single sample. In multiplexed assays, a panel of microspheres are combined, where each subgroup in the panel has its own unique fluorescence wavelength (by using different proportions of orange and red fluorophores) and is conjugated to a unique cytokine-specific capture antibody. The flow cytometer distinguishes the different microsphere subgroups (and hence different cytokines) based on their emission wavelength and reads the intensity of the PE signal for each microsphere subgroup. Panels that measure 12 or more cytokines in a single sample are routinely assembled. This assay is as sensitive as the ELISA, requires less sample, and is quantitative over a larger dynamic range. One limitation of the assay is the need for a specialized flow cytometer that can read and analyze the multiplexed samples.

The ELISPOT is another sandwich-based enzyme-linked immunoassay used to detect the frequency of individual cells secreting a cytokine of interest. In this assay, the capture antibody is coated onto a tissue culture microwell plate and cells are cultured in these plates with a stimulant to induce activation and secretion of the cytokine. To improve sensitivity and background, tissue culture plates with PVDF membranes rather than plastic bottoms are available. Cytokine is captured as it is secreted and concentrates at the location of the cell. The cells are then washed away and any captured cytokine is detected with a secondary enzyme-conjugated antibody. When substrate is added, colored spots develop at the site where the cells were located that secreted the cytokine. Enumerating the spots gives the frequency of cells secreting the cytokine of interest. Instruments are available that automate spot counting. The ELISPOT assay is sensitive enough to detect one cytokine producing cell in approximately 10^5 cells.

The frequency of cells elaborating a cytokine of interest can also be determined in a fluorescent-based intracellular staining method, that when analyzed on the flow cytometer, also allows for the simultaneous identification of the cell type (3). In this assay, cells are activated in vitro in the presence of a protein transport inhibitor (e.g., brefeldin A) that prevents the secretion of newly synthesized cytokines. The cells are then fixed and permeabilized and cytokines detected with fluorescently labelled antibodies. While sensitivity is sacrificed in the flow-based assay compared to the ELISPOT, it has other advantages including the capability to measure multiple cytokines and cell surface markers in a single sample by multicolor analysis.

mRNA Detection Assays

Induction of cytokine production is also routinely measured at the mRNA level. There are two general approaches to mRNA analysis, either rational-based targeted analysis of specific genes, usually done by RT-PCR, and non-targeted explorations, usually done by microarray analysis. The RT-PCR method recently underwent a dramatic transformation where the generation of the PCR product can now be measured during the amplification reactions (real-time RT-PCR) (4). This done by adding a fluorescently labelled probes specific for the cytokine PCR product to the reaction and measuring the fluorescence signal at end of each amplification cycle. The earliest cycle that the PCR product is first detected is tightly correlated with the number of copies of cytokine mRNAs that were present in the sample. When the appropriate standard curve is run on the same plate, the data can be quantitative and reported as copy number, but more frequently the copy number is normalized to an endogenous reference gene signal, such as rRNA. Real time RT-PCR is extremely sensitive, detecting approximately 5–10 copies of template per reaction, and linearly quantitative over a broad dynamic range (up to 10^8 copies).

Microarray assays, either in the microchip format or more limited membrane format, can simultaneously measure the upregulation and downregulation of many genes, including cytokines and the pathways mediated by the cytokines. In this assay, mRNA is converted to cDNA and amplified, labelled and hybridized to gene-specific oligos fixed on a chip or membrane. The signal for each gene is generally normalized to the geometric mean of several housekeeping genes. Microarray analysis is neither as quantitative nor as sensitive as real-time RT-PCR, however because thousands of genes are represented in the microarray, genes or pathways can be identified that were not previously known to be involved or thought significant in a particular biological or immunopathological process. Changes in targets identified by microarray are generally confirmed in more accurate assays such as real-time RT-PCR.

Cell-Based Bioassays

Cell-based bioassays detect the presence of cytokines by measuring bioactivity of biofluids or supernatants. In these assays, a cell line dependent on a particular cytokine for growth or specific cell functional response, is cultured with the biofluid samples. The magnitude of the functional response (e.g., proliferation of CTLL-2 cells to IL-2 or cytotoxicity of WEHI 164 cells to TNF-α) is correlated with the amount of cytokine present in the biofluid. Standard curves of titrated recombinant cytokine are run simultaneously to quantify the amount of cytokine in the sample. In general, bioassays should also be run with anti-cytokine neutralizing antibodies to demonstrate that the functional response is specific to the cytokine of interest and not a result of a other cytokines in the sample that have redundant biological activity. While bioassays tend to be the more sensitive than immunoassays and detect only bioactive forms of the cytokines, they are cumbersome and lengthy, requiring specific cell lines and are therefore generally run when confirmation of immunoassays is necessary.

Pros and Cons

Each of the cytokine assays have advantages and disadvantages, as detailed above, and in the critical analysis of cytokine expression patterns, more than one assay is usually employed. The most important consideration in selection of an assay is whether the endpoint measured satisfactorily addresses the needs of the investigation. For example, activity of TGF-β is not necessarily correlated with its protein or mRNA levels due to latent forms being secreted, so a bioassay would be more appropriate than an immunoassay or RT-PCR when measuring this cytokine. In other situations, such as evaluating drug hypersensitivity, knowing the frequency of sensitized T cells may be more important in risk assessment than the precise pattern of cytokines elaborated by those cells (5). Other factors to consider are ease of use, throughput and cost, and in the regulated environment, standardization of operating procedures.

Predicitivity

Cytokines usually act in a local microenvironment, either in a paracrine or autocrine manner, making it difficult to measure or interpret measurements of cytokines directly from sera or plasma samples. Therefore, cytokines are generally measured after ex vivo stimulation of tissues or peripheral blood cells. Furthermore, because cytokines act in concert with others, measuring the expression of several cytokines to determine a profile is usually more beneficial in understanding biological significance than measuring a single cytokine in isolation. It is not always clear whether the patterns of cytokines induced ex vivo actually recapitulate the in vivo microenvironment, however when the inducible cytokine patterns are compared between drug-treated and control groups, differences could indicate potentially adverse effects on the immune system. For example, decreased IL-2, IFN-γ and IL-4 production suggests immunosuppression, while polarization towards ▶ type 1 or type 2 T cell responses suggest propensity for inflammatory or allergic diseases (6).

Relevance to Humans

All of the assays and reagents available for non-clinical assessment of cytokine production are also available for human samples. The most common source of cells for evaluating cytokine induction in humans is peripheral blood. When there is a risk of immunotoxicity, such as immunosuppression, hypersensitivity, autoimmunity or inflammation, monitoring changes in inducible cytokine profiles can be useful in understanding the progression of the immunopathology.

Regulatory Environment

The FDA and EMEA guidance papers on nonclinical immunotoxicity evaluation of investigational new drugs do not directly request measuring cytokines. However, when drugs are suspected to have potential adverse effects on the immune system, cytokine measurements can be incorporated into follow up functional studies to better characterize the nature of the adverse response.

Relevant Guidance Documents

CPMP/SWP/1042/99, Note for guidance on repeated dose toxicity, 2000
FDA (CDER) Immunotoxicology Evaluation of Investigational New Drugs, 2002

References

1. Coligan JE, Kruisbeek AM, Margulies DH, Shevach EM, Stober W (eds) (1991) Current protocols in immunology. John Wiley and Sons Inc., New York
2. Vignali DAA (2000) Multiplexed particle-based flow cytometric assays. J Immunologic Methods 243:243–255
3. Pala P, Hussell T, Openshaw PJM (2000) Flow cytometric measurement of intracellular cytokines. J Immunologic Methods 243:107–124
4. Heid CA, Stevens J, Livak KJ, William PM (1996) Real time quantitative PCR. Genome Res 6:986–994
5. Lebrec H, Kerdine S, Gaspard I, Pallardy M (2001) Th1/Th2 responses to drugs. Toxicol 158:25–29
6. Romagnani S (1997) The Th1/Th2 paradigm. Immunol Today 18:263–266

Cytokine Inhibitors

KLAUS RESCH
Institute of Pharmacology
Hannover Medical School
Carl-Neuberg-Strasse 1
D–30625 Hannover
Germany

Synonyms

Anticytokines.

Definition

Cytokines are small (glyco)proteins (with molecular weights of 8–75 kDa) which play a vital role in hematopoiesis, immune reactions and inflammation (see Cytokines). In disease situations where immune and inflammatory reactions are a prevailing, cytokines offer a useful therapeutic target. Cytokine inhibitors consist of a heterogeneous group of drugs which
- decrease the synthesis of cytokines
- decrease their concentration in free active form
- block their interaction with specific receptors, or
- interfere with the signaling of cytokine receptors (see Figure 1).

Characteristics
Inhibitors of Cytokine Synthesis

All drugs which decrease the number of producing cells implicitly also inhibit cytokine synthesis. Examples are general cytostatic drugs, some of which—such as azathioprine or methotrexate—are approved as immunosuppressants or antiinflammatory agents. A cytostatic drug with a higher selectivity for immune cells is mycophenolate. Antibodies directed against structures of T lymphocytes, as the monoclonal antibody against parts of the antigen receptor of T lymphocytes, muromonab CD3, decreases selectively the circulating pool of this lymphocyte subpo-

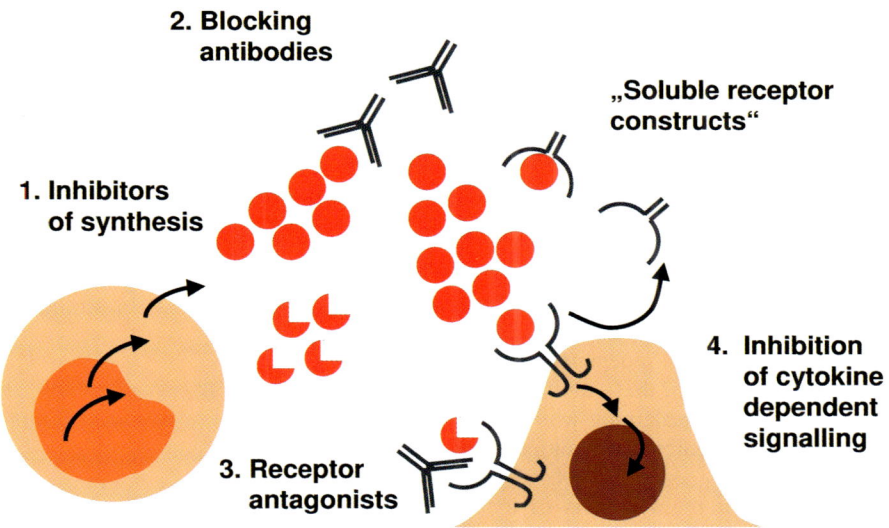

Cytokine Inhibitors. Figure 1 Suppression of immune and inflammatory reactions: Inhibition of cytokines.

pulation—predominantly by complement lysis—and in this manner is immunosuppressive.

Cytokine synthesis can be inhibited without affecting viability of cells. Such a mode of action is exerted by the glucocorticoids (e.g. prednisone), the anti-inflammatory or immunosuppressive effects of which were in clinical use long before it became known that inhibition of the synthesis of cytokines such as interleukin-1, tumor necrosis factor or interleukin-2, is the major mechanism of their action. Glucocorticoids bind to the glucocorticoid receptor in the cytoplasm of cells, releasing it from binding to the heat shock protein HSP 90. The receptor then translocates to the nucleus and binds to a "glucocorticoid responsive element" (GRE) that is present in the promoter of more than 100 genes. It inhibits cytokine gene expression by several mechanisms.

- The GRE interacts negatively with other promoter elements implicated in its activation such as the NFκB site.

Cytokine Inhibitors. Table 1 Cytokine inhibitors

Modes of action		Drug examples
Inhibition of synthesis		
Reduction of the number of cytokine producing cells:	Cytostatic drugs	Azathioprine
	Monoclonal antibodies to cells	Muromonab CD3
Regulation of cell activity:	Regulatory cytokines	Interleukin-4, Interleukin-10
Inhibition of signal transduction:	Cyclosporine and related drugs	Cyclosporin, Tacrolimus
Regulation of gene expression:	Glucocorticoids	Prednisone
Decrease of the concentration in active (free) form		
Monoclonal antibodies against: (Soluble cytokine receptors)	Cytokines	Infliximab Etanercept
Receptor blockade		
Monoclonal antibodies against: (Cytokine antagonists)	Cytokine receptors	Basiliximab Anakinra
Inhibition of cytokine dependent signaling		
Protein kinase inhibitors		Sirolimus

- The bound glucocorticoid receptor protein negatively interacts with transcription factors (e.g. NFκB).
- Proteins are induced, which interfere with signaling pathways such as the inhibitor of NFκB, IκB.
- The glucocorticoid receptor in its ligated form directly binds to signaling elements of receptors which induce cytokine synthesis (e.g. NFκB or the p38 kinase) and blocks their action.

Note that the first three of these mechanisms depend on the binding of the glucocorticoid receptor to its promoter site, while the fourth is independent of it. Most likely the combined action of these multiple mechanisms provide the basis for explaining why glucocorticoids are so effective in inhibiting cytokine synthesis.

While glucocorticoids inhibit the synthesis of many cytokines involved in immune responses and inflammation, some modern immunosuppressive drugs predominantly affect the T lymphocyte growth factor interleukin-2 and thus prevent expansion of these cells which is a prerequisite for an effective cellular immune reaction. This includes cyclosporine and tacrolimus. Both drugs, after binding to different intracellular receptors, as their predominant way of action inhibit the activity of calcineurin, a protein phosphate phosphatase. This phosphatase is required so that the transcription factor NFAT (nuclear factor of activated T cells) can shuttle from the cytoplasm into the nucleus and bind to the promoter of the interleukin-2 gene. The exclusive occurrence of NFAT in T lymphocytes explains the high selectivity of cyclosporine or tacrolimus for these cells.

In some cases cytokines behave antagonistically to the effects of others. Thus, interleukin-4 and interleukin-10 exert strong antiinflammatory actions. Especially interleukin-10 is tested—also in some clinical studies for this indication.

Blocker of Cytokines in Free Active Form

A very specific way, in principle, to block cytokines is to generate ▶ monoclonal antibodies which combine with the cytokine and thus prevent it from binding to its receptor. Although this has been probed for many different cytokines only one so far has proven to be of sufficient clinical value—infliximab, which is a humanized monoclonal antibody directed against tumor necrosis factor. This monoclonal antibody originally was raised in mice; the antigen-binding parts then were exchanged by gene technology in a human immunoglobulin IgG molecule ("humanized") which drastically increases the half-life in human beings and decreases the chance of eliciting antibodies (to mouse protein!) which limit the activity.

Cytokines exert their biological activities by binding to membrane receptors of target cells. As a physiological regulatory mechanism the extracellular—cytokine binding—parts can be released, which curtails the biological action of the respective cytokine. This principle of "soluble receptors" has been successfully exploited. To enhance effectiveness and half-life in vivo, soluble receptors were transferred (by gene technology) to non-antigen-binding parts of the IgG molecule. Such a construct is etanercept containing the soluble receptor for tumor necrosis factor. Both infliximab and etanercept have a strong antiinflammatory efficiency.

Cytokine Receptor Blockers

Similar to cytokine inhibitors, monoclonal antibodies can also be effective against their receptors by preventing the function of the cytokine in question. Basiliximab and daclizumab are ▶ humanized monoclonal antibodies against the interleukin-2 receptor. By blocking the receptor for this central T lymphocyte growth factor they suppress cellular immune reactions.

Interleukin-1 (together with tumor necrosis factor) is a very efficacious proinflammatory cytokine. Provided it is singular, nature has created an antagonist which when secreted dampens the action of interleukin-1. Anakinra represents this antagonist which is produced by gene technology and displays anti-inflammatory properties.

With the potential of modern molecular biological methods, cytokines (being proteins) can be modified with the aim to change their biological properties. Thus a mutated protein (mutein) of interleukin-4 has been generated which behaves as an IL-4 receptor antagonist (and also as an IL-13 receptor antagonist as both two chain receptors share one common receptor chain). This mutein is at present tested in clinical studies for its effectiveness to treat allergic diseases (predominantly asthma).

Similarly, chemokines mutated to chemokine receptor antagonists are evaluated in human immunodeficiency virus (HIV) infections as this virus uses some chemokine receptors such as CxCR5, or CCR3, and CCR5 as co-receptors for its entry into cells.

Blockers of Cytokine Receptor-Dependent Signaling

After the binding of a cytokine to its receptors the biological response is transduced by metabolic signals, generally involving multiple signaling cascades. Sirolimus (rapamycin) by selectively inhibiting the protein kinase mTOR (target of rapamycin) is an effective immunosuppressive drug as mTOR is an essential component of the signaling of the receptor for interleukin-2. mTOR plays a central role in several biological processes, including cell cycle control, and thus its inhibition prevents proliferation of activated T lymphocytes.

From the great number of efforts to elucidate the signal transducing mechanism of receptors for proinflammatory cytokines, as well as the availability of some lead compounds interfering with signaling components, it can be anticipated that a plethora of new drugs which inhibit protein kinases will emerge in the near future.

Preclinical Relevance

The possibility of inhibiting the specific function of a cytokine—together with generation of mouse mutants lacking the cytokine in question—is vital for elucidating its role in disease models. Anti-inflammatory or immunosuppressive drugs, acting at the level of signal transduction, can help to elucidate specific signaling cascades, as in the case of cyclosporine with the antigen receptor of T lymphocytes.

Relevance to Humans

Inflammation as part of the body's defense system against infection is indispensable for survival in a natural habitat. When it becomes dysregulated, as occurs in ▶ chronic inflammatory diseases where the inflammatory reaction is perpetuated by autoimmune reactions, or mounts an inadequate and exaggerated response to an (often rather bland) causative agent as in allergy, it turns into a disease itself. Thus chronic inflammatory diseases, such as rheumatoid arthritis or inflammatory bowel disease, or allergic diseases including asthma, make up a high percentage of diseases requiring adequate medication. Moreover it becomes increasingly clear that in rather important diseases, such as arteriosclerosis, or even neurologic "degenerative" diseases such as Parkinson's disease, inflammation plays a large role in the pathophysiology.

Cytokine inhibitors have developed to core antiinflammatory or immunosuppressive drugs. Glucocorticoids such a prednisone, though interfering with several processes relevant for inflammation or immune reactions, owe their effectiveness predominantly to inhibition of the synthesis of many cytokines. Thus they are still the most effective drugs for treatment of diseases involving ▶ chronic inflammation like rheumatoid arthritis or inflammatory bowel disease, or allergic diseases like hay fever, asthma or anaphylactic shock, as well as allergic skin diseases. Their usefulness, however, is limited by various side effects. Thus more selectively acting cytokine inhibiting drugs have been developed, the first of which have been approved for several diseases.

Infliximab, etanercept, and anakinra are indicated in rheumatoid arthritis, osteoarthritis or inflammatory bowel disease. Cyclosporine, tacrolimus and sirolimus (together with glucocorticoids) are components of standard regimens in the life-long prevention of rejection in organ transplantation (often in combination). They are also used to suppress graft versus host reactions after bone marrow or stem cell transplantation. They are also effective in autoimmune diseases which require immunosuppression. Muromomab CD3, basiliximab or daclizumab are used in the initial phase of organ transplantation, as well as to combat acute rejection episodes. Further cytokine inhibitors—regulatory cytokines, monoclonal antibodies to various cytokines or their receptors, ▶ muteins, with antagonistic properties or inhibitors of signal transduction—are in the pipeline as new drugs for treating, with increasing selectivity, inflammatory or allergic diseases.

Regulatory Environment

As for all drugs to be used in humans or animals, cytokine inhibitors are regulated by specific laws—in Germany the "Arzneimittelgesetz". Government agencies approve drugs for specific indications—in Germany either the "Bundesinstitut für Pharmaka and Medizinprodukte" (BfArM) or for biologicals, such as antibodies or cytokines, the Paul-Ehrlich-Institut (PEI). Such biotech products are internationally regulated by a ICH Harmonised Tripartite Guideline: Preclinical Safety Evaluation of Biotechnology-Derived Pharmaceuticals, July 16, 1997.

References

1. Ciliberto G, Savino R (2001) Cytokine Inhibitors. Marcel Dekker, New York
2. Fitzgerald KA, O'Neill LAJ, Gearing A, Callard RE (2001) The Cytokine Facts Book, 2nd ed. San Diego Academic Press, San Diego
3. Gemsa D, Kalden JR, Resch K (1997) Immunologie, 4th ed. Thieme Verlag, Stuttgart
4. Janeway CA, Travers P, Walport M, Shlomchik M (2002) Immunologie, 5th ed. Spektrum Verlag, Heidelberg
5. Mantovani A, Dinarello CA, Ghezzi P (2001) Pharmacology of Cytokines. Oxford University Press, Oxford

Cytokine Network

The interaction of various cytokines with their target cells whereby one cytokine may induce the production of another cytokine which either enhances the overall reactivity of a cell or leads to a downregulation of a cell response. A cascade-like production of cytokines, often in a certain hierarchical order.
▶ Cytokines

Cytokine Polymorphisms and Immunotoxicology

BERRAN YUCESOY · VICTOR J JOHNSON ·
MICHAEL I LUSTER
National Institute for Occupational Safety and Health
1095 Willowdale Road
Morgantown, WV 26505
USA

Synonyms
Single amino acid polymorphisms (SAPS), single base transition, single nucleotide polymorphisms (SNPs)

Definition
A polymorphism is a variation in DNA sequence that has an allele frequency of at least 1% in the population. There are several types of polymorphisms in the genome: single nucleotide polymorphisms, repeat polymorphisms, and insertions or deletions, ranging from a single base pair to thousands of base pairs in size. Most of the DNA sequence variation in the human genome is in the form of single nucleotide polymorphisms, or SNPs, which result from single-base changes that substitute one nucleotide for another. The vast majority of the polymorphisms in cytokine genes are of the SNP variety. Polymorphisms in cytokine genes can have profound influences on their expression and thus impact inflammatory and immune-mediated diseases. Research over the past several years has implicated that, similar to polymorphism in drug metabolizing enzymes respond differently to toxins, cytokine polymorphisms modify xenobiotic-induced immunotoxicity and immune-mediated disease processes.

Characteristics
The frequency of SNPs across the human genome is higher than for any other type of polymorphism. SNPs are stable from an evolutionarily standpoint, making them easier to follow in population studies. The SNPs that are most likely to have a direct impact on the protein product of a gene are either located in the coding region (cSNPs) and change the amino acid sequence, or in the regulatory region (rSNP), and affect expression levels. Tumor necrosis factor (TNF)α (−308), interleukin(IL)-1RA (+2018), IL-1β (+3953) and transforming growth factor(TGF)-β1(+915) polymorphisms represent examples of functional SNPs in cytokine genes in that they effect expression levels. If all variants in all individuals were identified, it has been estimated that, there may be as many as 10 to 30 million SNPs in the human genome. However, only about 1% of identified SNPs alter an amino acid in a protein encoded by a gene and fewer than that affect expression (functional).

Some SNPs affect expression indirectly, by altering the function of regulatory sequences that control gene expression or the rate of synthesis or degradation of the mRNA transcript. There are potential advantages of employing SNPs to identify the genetic components of complex human diseases. SNPs are highly abundant, stable and distributed throughout the genome, and thus represent biomarkers in dense positional cloning investigations. SNPs also exhibit patterns of linkage disequilibrium and haplotypic diversity that can be used to enhance gene mapping. Linkage disequilibrium mapping using SNPs is an important strategy for mapping genes involved in many common diseases. The approach relies on the expectation that a candidate region for a disease can be refined by examining allelic associations between the disease and a series of SNPs genotyped at high density. Furthermore, interpopulation differences in SNP frequencies can be used in population-based genetic studies. It is not practical or possible to genotype and test all SNPs in the genome for association with phenotypes. Therefore, it is important to select a limited number of representative SNPs to genotype and consider their frequencies in control and experimental populations. SNPs, with frequencies of at least 5% or greater in the general population, are more likely to be useful in candidate gene studies.

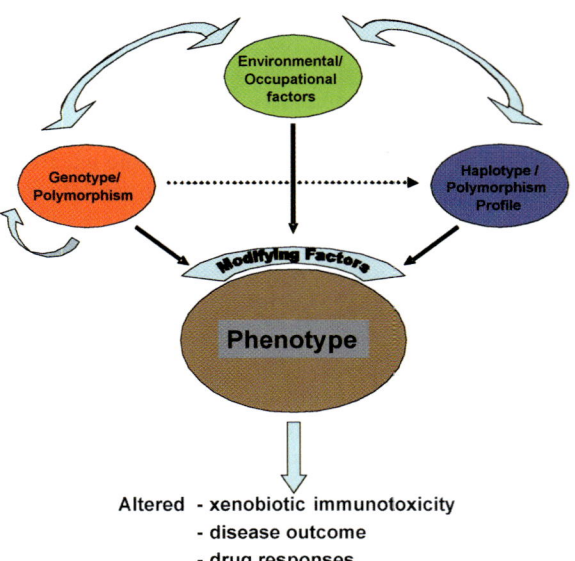

Cytokine Polymorphisms and Immunotoxicology. Figure 1 A model showing gene-gene and gene-environment interactions.

Disease and Environmental Interactions

SNPs have been used increasingly as a method for investigating the genetic etiology of complex human diseases. The most common human diseases are thought to be multigenic and multifactorial. Thus, by assessing the role of allelic variants in disease susceptibility and progression, the influence of gene-gene as well as gene-environment interactions in disease etiology can be determined (Figure 1).

Asthma is a good example of a complex, multifactorial disease with strong environmental/occupational influences. Candidate genes with known variants involved in asthma include, but are not limited to, IL-4, IL-4 receptor α (IL-4Rα), IL-13, IL-10, TGFβ1, interferon (IFN)-γ, IL-12β, TNF-α and IL-1β. This suggests an immune component (IL-4, IL-4Rα, IL-13, IL-10, TGFβ1) is likely involved in the initiation of asthma and an inflammatory component (IFN-γ, IL-12β, TNF-α and IL-1β) likely modifies disease severity. The potential for gene-gene interaction in disease initiation and severity also exists for polygenic diseases such as asthma. For example, a significant gene-gene interaction exists between IL-4Rα (+478) and IL-13 (−1111), such that individuals both homozygous for the IL-4Rα variant and heterozygous or homozygous for the IL-13 variant are five times more likely to develop asthma than the general population. Functionally, individuals with this haplotype display elevated serum IgE levels and bronchial hyperresponsiveness. Identification of interactions between SNPs can expose biological interactions that are vital to our understanding of the disease pathogenesis and treatment.

The potential areas of application for SNP technology are much broader than just cytokines and include gene discovery and mapping, pharmacogenetics, association-based candidate polymorphism testing, diagnostics and risk profiling, homogeneity testing, the prediction of response to environmental stimuli, and epidemiological study design. Genetic polymorphisms are also important variables for future biomonitoring studies with the aim of risk assessment.

Understanding how genetic variations effect responses to chemicals has major therapeutic relevance in the tailoring of an appropriate treatment (pharmacogenetics). One of the major limitations to the use of SNPs in investigations of complex disease genetics is related to the statistical issues about the rate of false-positive tests and the level of statistical significance and power. Some of these may be related to technical issues in SNP genotyping and ethnical differences.

Relevance to Humans

The use of SNPs associated with drug responses can be judged in terms of how well they predict drug response in patients and the proportion of patients who will benefit from the test. The issue of differences in response, or susceptibility, concerns pre-existing factors independent of exposure that affect the severity of toxic effects. The application of SNPs to investigations of complex human diseases requires the stratification of populations into responders, nonresponders and those with adverse side effects due to genetic determinants. The goal of such stratification would be to expedite targeted drug discovery and development and to improve the efficacy of drug-based interventions. In addition to immune-mediated diseases there are several reports showing examples of the influence of cytokine SNPs on drug responses. The IL-10 (−1082) and TGF-β (+29) genotypes are reported to be associated with resistance to combined antiviral therapy. The polymorphic TNF-a2 microsatellite and TNF-α (−308) variant are associated with a risk of chemotherapy-induced pulmonary fibrosis and severe carbamazepine hypersensitivity reactions, respectively. However, the frequency of a gene affecting responsiveness to a particular drug and potential interactions with other genetic (gene-gene interactions) and environmental factors (gene-environment interactions) must be assessed in genetically heterogeneous populations using case-control or cohort association study designs. Figure 1 depicts a model of gene-gene and gene-environment interactions as modifying factors in immunotoxicity. This is particularly important for extrapolation from specific clinical trials to general clinical use in the highly admixed, heterogeneous industrialized populations where the diseases are most common. Examples of potentially important environmental factors that might interact with underlying genetic susceptibilities include exposure to cigarette smoke, exposure and sensitization to common inhalant allergens, exposure to viral infections, housing and lifestyle factors, in utero factors acting during pregnancy, UV radiation and diet. Specific examples of occupational and environmental diseases where cytokine SNPs have been shown to contribute to disease processes include chronic obstructive pulmonary disease (COPD), silicosis, chronic beryllium disease (CBD), occupational asthma, coal worker's pneumoconiosis (CWP), alcohol and chemical-induced hepatitis, and allergic and contact dermatitis (Table 1). Some polymorphisms are effect modifiers in that they influence disease severity, and disease appearance is independent of the genotype. A strong association between silicosis and the TNF-α (−238) variant revealed that individuals with the variant are predisposed to more rapid development of severe silicosis. IL-1α (−889) polymorphism was found to be a disease modifier in myasthenia gravis, whereas IL-4B1 allele is associated with late onset of multiple sclerosis and therefore might represent a modifier of age of onset.

Cytokine Polymorphisms and Immunotoxicology. Table 1 Examples of associations between cytokine polymorphisms and occupational and environmental diseases

Disease	Cytokine Polymorphism
Alcohol and chemical-induced hepatitis	TNF-α (−308), (−238); IL-1β (+3953), (−511)
Allergic and irritant contact dermatitis	TNF-α (−308)
Asthma	IL-4 intron 2, IL-4RA, IL-13 (−1111); TGF-β (−509); TNF-α (−308)
Chronic beryllium disease (CBD)	TNF-α (−308)
Chronic obstructive pulmonary disease (COPD)	TNF-α (−308), (+489)
Coal worker's pneumoconiosis (CWP)	TNF-α (−308)
Farmer's lung disease	TNF-α (−308)
Sarcoidosis	TNF-α (−308)
Silicosis	TNF-α (−238); IL-1RA (+2018)

IL, interleukin; TGF, transforming growth factor; TNF, tumor necrosis factor.

Techniques to Genotype SNPs

A variety of molecular strategies are used for SNP analysis. In-silico SNP mapping is a cost-efficient way for new polymorphism identification. Gel electrophoresis-based genotyping methods for known polymorphisms include polymerase chain reaction (PCR) coupled with restriction fragment length polymorphism (RFLP) analysis, multiplex PCR, oligonucleotide ligation assay (OLA) and mini sequencing. Fluorescent dye-based genotyping technologies are emerging as high-throughput genotyping platforms, including OLA, pyrosequencing, single-base extension with fluorescence detection, homogenous solution hybridization such as Taq-Man®, and molecular beacons. The invader assay is able to genotype directly from genomic DNA without PCR amplification. DNA chip-based microarray and mass spectrometry genotyping technologies are the latest development in genotyping techniques and offer rapid and reliable performance.

References

1. Shi MM (2001) Enabling large-scale pharmacogenetic studies by high-throughput mutation detection and genotyping technologies. Clin Chem 47:164–172
2. Landegren U, Nilsson M, Kwok P (1998) Reading bits of genetic information: Methods for single-nucleotide polymorphism analysis. Genome Res 8:769–776
3. Palmer LJ, Cookson W (2001) Using single nucleotide polymorphisms as a means to understanding the pathophysiology of asthma. Resp Res 2:102–112
4. Tabor HK, Risch NJ, Myers RM (2002) Candidate-gene approaches for studying complex genetic traits: practical considerations. Nature 3:1–7
5. Yucesoy B, Kashon ML, Luster MI (2003) Cytokine polymorphisms in chronic inflammatory diseases with reference to occupational diseases. Curr Mol Med 3:485–505

Cytokine Receptor

Cytokine receptors are highly specific binding proteins for cytokines, which bind their ligand with a high affinity. Trans-membrane cytokine receptors bind the ligand on the outside of the cell and translate this binding into intracellular signaling events. Signaling cytokine receptors usually are complexes of two or more identical or different subunits, of which the ligand binding protein is termed α-chain. Additionally, cytokine receptors exist in soluble forms, which are capable of binding cytokines.

▶ Cytokine Receptors

Cytokine Receptor Complexes

Maybe with the exception of the class of receptors with seven transmembrane spanning regions (e.g. chemokine receptors), all signaling cytokine receptors are complexes of two or more transmembrane molecules. The basic principle of translating the docking of the cytokine on the outside of the cell to an event inside the target cell, is the ligand-mediated association of two or more identical or different transmembrane molecules with the result that the intracellular domains of the receptor subunits form a novel scaffold for the association of adapter molecules or enzymes such as protein kinases.

▶ Cytokine Receptors

Cytokine Receptors

MICHAEL U. MARTIN
Institute of Immunology
Justus-Liebig-University, Giessen
Winchesterstrasse 2
D-35394 Giessen
Germany

Synonyms
Receptors for mediators of the immune system

Definition
▶ Cytokine receptors are specific binding proteins for mediators of the immune system. Usually binding of a cytokine to its specific transmembrane cytokine receptor initiates signal transduction and results in cell activation. However, some forms of cytokine receptors function as decoy receptors and modulate cytokine responsiveness. Soluble cytokine receptors may activate cells in a process known as trans-signaling, but more often, soluble cytokine receptors specifically bind and neutralize their respective cytokine. Soluble cytokine receptors have proven to be effective as highly specific ligand scavengers in anticytokine therapies.
In this context the term cytokine is solely used for mediators of the immune system, and includes interleukins, interferons, and chemokines, but excludes hematopoietic growth factors (although there is a clear cross-talk between these different mediators).

Characteristics
▶ Cytokines are mediators of the innate and adaptive arms of the immune response. All cytokines are proteins or glycoproteins. They may be produced constitutively regulating differentiation and homeostasis of naive leukocytes and the formation of secondary lymphoid organs such as lymph nodes. Or they may be synthesized de novo in the course of an immune reaction with the aim of organizing the communication between leukocytes themselves or between leukocytes and tissue cells. As cytokines are relatively large proteins, ranging from 8 kDa to 100 kDa, they cannot pass the plasma membrane and therefore require specific receptors on the surface of target cells. Cytokine receptors translate the docking of the respective ligand on the outside of the cell into intracellular signals which result in an altered behaviour of the target cell, very often due to gene induction and de novo protein biosynthesis. Cells may express an individual panel of discrete receptors for different cytokines, thus responses to cytokines are cell-type specific. Cytokine responses include induction of proliferation and/or differentiation, or the production and release of effector molecules which often are ▶ secondary cytokines (schematically summarized in Figure 1).

Cytokine Receptors are Often Complexes of Transmembrane Molecules

Some receptors for cytokines are monomers. More often, signaling cytokine receptors form complexes of two or more subunits. In many cases the subunit which binds the ligand first or exclusively, is designated the α-chain. The α-chain binds its ligand with an intermediate or high affinity, which may be increased by further subunits. Normally, the subunits which form a signaling cytokine receptor complex are different but may also be identical. They may exist in preformed complexes in the plasma membrane awaiting their ligand, or they form after ligand binding to the α-chain by recruitment of the coreceptor(s). Four examples of prototypic ▶ cytokine receptor complexes are depicted in Figure 2.
Typically, cytokine receptors are highly specific for their individual ligand, but some cytokine receptors may bind different ligands. In some cytokine receptor families a certain degree of pleiotropy exists at the receptor level, so that one cytokine may bind to two different cytokine receptors and initiate (partially) different signals. A further feature of related cytokine receptors of one family is the sharing of one or more ▶ common chains. This may cause redundancy on the receptor level, as cytokines from one cytokine family may initiate partially overlapping signals due to usage of common signaling subunits in their specific receptor complexes.

Signaling Through Cytokine Receptors
Signal transduction is mainly achieved through the association of previously separated transmembrane receptor components. This results in the close positioning of either signal transducing moieties or protein interaction domains, thereby creating scaffolds for pro-

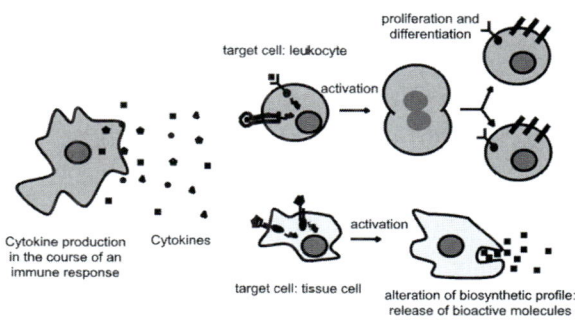

Cytokine Receptors. Figure 1 Principles of cytokine action and cytokine signaling.

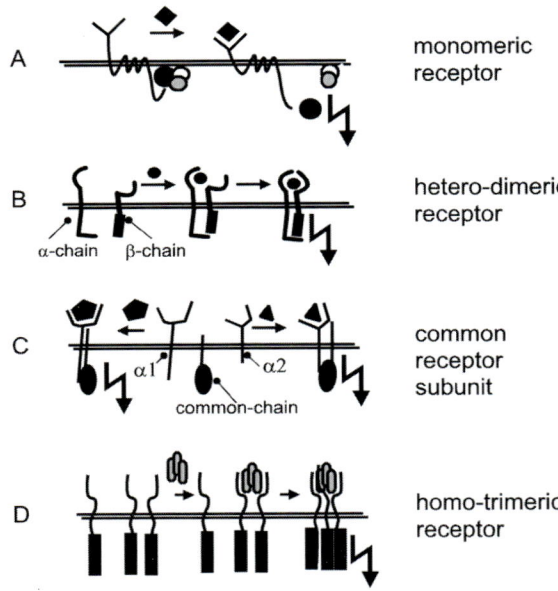

Cytokine Receptors. Figure 2 Prototypic cytokine receptor complexes.

tein interactions. Some cytokine receptors are already associated with kinases in the quiescent state, whereas other cytokine receptors recruit kinases either directly or by the means of adapter proteins after ligand binding. These kinases may be protein tyrosine-kinases, however, some cytokine receptors utilize serine/threonine-specific kinases. Chemokine receptors typically employ trimeric G proteins to initiate intracellular signaling cascades. Different classes of cytokine receptor families employ different signaling mechanisms, which may in part overlap. The consequence of signal transduction and cell activation is an altered behaviour of the target cell, which is cell-type specific. This includes rapid changes, by remodeling of the cytoskeleton or activation of existing enzymes. Frequently, kinase cascades are activated, which result in the activation of transcription factors, leading to induction of gene transcription, and subsequently to de novo protein synthesis. Although each cytokine induces its individual set of signaling pathways, an abundance of cross-talk exists between different signaling pathways. This explains the complexity of the bioresponse of a given cell type which faces a cocktail of cytokines in an ongoing immune response.

Regulation of Cytokine Receptors

Availability of cytokine receptors on the surface of a cell can be regulated on different cellular levels.

- **Regulation at the gene level:** Many cytokine receptors are constitutively expressed, but some are synthesized de novo only after appropriate stimulation. Frequently, the expression of cytokine receptors is enhanced after activation of a cell in the course of an immune response. In contrast, at a later stage in the immune response, the gene transcription of some receptors for proinflammatory cytokines may be downregulated by antiinflammatory cytokines.
- **Regulation at mRNA level:** It remains to be shown whether the stability of mRNA for cytokine receptors is modulated as has been demonstrated for some cytokine mRNAs. However, mRNA for some cytokine receptors is differentially spliced resulting in membrane-inserted or soluble cytokine receptors.
- **Regulation at protein level:** Once cytokine receptors appear on the surface of the cell, regulation (removal from the surface) may occur by two mechanisms. First, cytokine receptors can be internalized after ligand binding. In some cases internalization seems to be required for efficient signaling. After internalization ligand and receptor become either degraded or are recycled. Alternatively, or in parallel, cytokine receptors are cleaved from the cell surface by ektoproteases. Both mechanisms lead to a transient disappearance of these cytokine receptors, resulting in a transient nonresponsiveness of the target cell to the respective cytokine, before newly synthesized receptors or recycled receptors reach the surface.

Decoy Receptors

Decoy cytokine receptors are highly specific receptors, which bind their ligand with a high affinity but do not initiate and transduce a signal. They act as molecular traps for the cytokine and in some cases for coreceptor molecules required for efficient signal transduction. This concept was first identified in the interleukin IL-1 receptor system, where the type II IL-1 receptor (which is incapable of signaling due to the lack of a cytoplasmic signaling domain) is able to bind the ligand and compete with the signaling type I receptor for the essential coreceptor (▶ coreceptor competition;

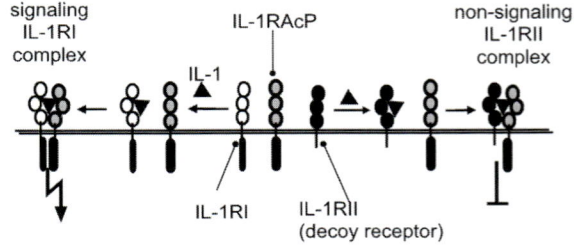

Cytokine Receptors. Figure 3 Decoy receptor in the Interleukin-1 receptor system.

Figure 3). The decoy receptor sets the threshold at which a given cell type responds to IL-1. In the meantime ▶ decoy receptors have been identified in several cytokine receptor families as additional levels of regulating cytokine action (1).

Soluble Cytokine Receptors

A series of cytokine receptors exist in a soluble form. Intact soluble cytokine receptors are capable of binding their respective ligand. Soluble cytokine receptors can be generated either by differential mRNA splicing or by ▶ receptor shedding from the plasma membrane. Different functions of soluble cytokine receptors are conceivable. They may bind and protect the ligand from proteolysis (chaperon function), serve as a carrier, or even pass the ligand (▶ ligand passing) to the plasma membrane-inserted receptor. Agonistic soluble cytokine receptors may associate with a transmembrane subunit and induce signal transduction in a process termed trans-signaling in the presence of ligand. ▶ Trans-signaling renders a cell responsive to a cytokine for which it lacks a receptor. However, most soluble cytokine receptors behave as ligand scavengers, which sequester and neutralize the cytokine in a fashion comparable to a neutralizing anticytokine antibody (2,3).

Preclinical Relevance

Cytokines are centrally involved in many acute and chronic inflammatory diseases or in autoimmune disorders. Almost always is the cytokine production disregulated, that is, the cytokine levels are too high and/or persist for too long. Therefore, the specific and effective sequestration of these surplus cytokine molecules is highly desirable.

This is in principal possible by targeting the cytokine-cytokine receptor system at several levels (induction of cytokine production, release of cytokine, neutralization of circulating cytokine, blocking of cytokine receptor, inhibition of cytokine-specific signaling pathways; for details see Cytokine inhibitors).

Cytokine Receptors as Target for Pharmaceutical Intervention

In principal three options are available:
- blocking of cytokine receptors by natural or recombinant receptor antagonists
- blocking the ligand binding site of the receptor with blocking antibodies
- using soluble coreceptor molecules as competitive antagonists (summarized in Figure 4).

In animal models all three strategies work, however, to

A : Natural antagonist (example IL-1Receptor Antagonist)

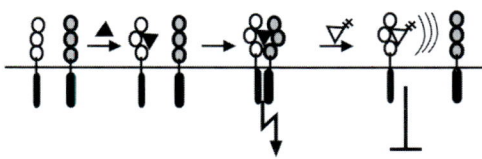

B : Blocking antibodies (example sIL-1RII)

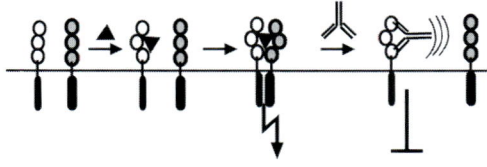

C : Recombinant soluble coreceptor molecules as competitive antagonists

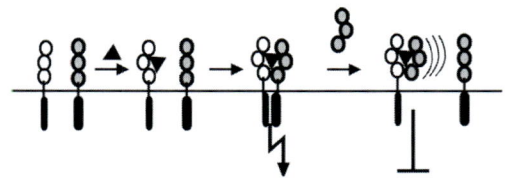

Cytokine Receptors. Figure 4 Cytokine receptors as targets for drugs in the interleukin-1 receptor system.

A : Homodimeric Fc fusion protein (example sIL-1RII-Fc)

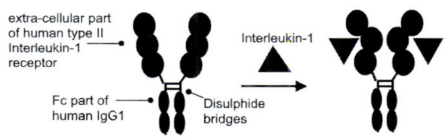

B : Heterodimeric Fc fusion protein (example sIL-1RII / sIL-1RAcP-Fc)

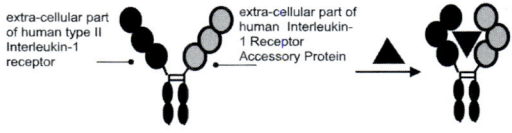

C : "inline" heterodimeric Fc fusion protein (example sIL-1RI/sIL-1RAcP-Fc)

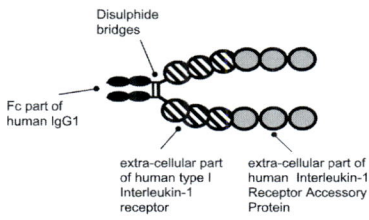

Cytokine Receptors. Figure 5 Recombinant soluble cytokine receptors as cytokine scavengers in the interleukin-1 receptor system.

date only the IL-1 receptor antagonist (Anakinra) is approved for clinical use.

Soluble Cytokine Receptors as Drugs to Neutralize Cytokines
Soluble cytokine receptors ideally fulfil the requirement of highly specific drugs, which combine high affinity binding with very low ▶ immunogenicity. As the natural soluble receptors have a rather short protein half-life in vivo genetically modified molecules have been generated, which show greater stability. Most of these recombinant molecules are ▶ fusion proteins consisting of the constant part of a human immunoglobulin chain and the soluble part of the cytokine receptor. In order to increase the affinity of soluble receptors, ▶ ligand traps were designed in which the relevant parts of the α-chains and β-chains of high-affinity cytokine receptor complexes were fused individually to one Fc part or together "in line"(4). These heterodimeric ligand traps show an enhanced affinity compare to those consisting of a homodimeric cytokine receptor fusion protein (5) (summarized in Figure 5).

In the IL-6 receptor system, a fusion protein of soluble IL-6 receptor α-chain and IL-6 (hyper IL-6) was very potent in activating cells that express only the transmembrane signaling subunit (gp130) in a process called trans-signaling. This type of agonistic cytokine receptor fusion protein may develop into a very potent reagent in situations where IL-6 effects are desirable on certain cell types (e.g. in hematopoiesis) (2).

Relevance to Humans
Soluble cytokine receptors are very attractive and potent reagents to neutralize cytokines. Since its approval by the FDA in 1998 the soluble tumor necrosis factor (TNF) receptor (as the fusion protein Etanercept) is used successfully in the treatment of rheumatoid arthritis and other chronic inflammatory diseases. Etanercept is a disulfide-linked homodimeric fusion protein consisting of two constant regions of the Fc-part of the human immunoglobulin G (IgG1) chain and the soluble part of the human p75 receptor for tumor necrosis factor containing the TNF binding site. Its size equals that of an antibody. It is produced by recombinant DNA technology in a mammalian expression system. A similarly constructed fusion protein of the soluble type II IL-1 receptor is in clinical trials also for the treatment of chronic inflammatory diseases. Novel strategies including the cytokine-specific ligand traps (discussed above) are presently in clinical trials (4,5).

Regulatory Environment
Recombinant proteins designed as therapeutic drugs, are regulated by specific laws, in Germany the "Arzneimittelgesetz". Governmental agencies approve drugs for specific indications: in Germany generally the Bundesinstitut für Pharmaka und Medizinprodukte (BfArM), for biologicals such as antibodies and cytokines the Paul-Ehrlich-Institut (PEI). To date only one soluble cytokine receptor (Etanercept) has been approved by the FDA in the USA, and subsequently by a series of other countries.

References
1. Mantovani A, Locati M, Vecchi, A, Sozzani S, Allavena P (2001) Decoy receptors: a strategy to regulate inflammatory cytokines and chemokines Trends Immun 22:328–336
2. Jones SA, Rose-John S (2002) The role of soluble receptors in cytokine biology: the agonistic properties of the sIL-6R/IL-6 complex. Biochim Biophys Acta 1592:251–256
3. Schooltink H, Rose-John S (2002) Cytokines as therapeutic drugs. J Interferon Cytokine Res 22:505–526
4. Dinarello CA (2003) Setting the cytokine trap for autoimmunity. Nat Med 9:20–22
5. Ecconomides AN, et al. (2003) Cytokine traps: multicomponent, high-affinity blockers of cytokine action. Nat Med 9:47–52

Cytokines

MARIANNE NAIN · DIETHARD GEMSA
Institute of Immunology
Philipps-University Marburg
Robert Koch-Strasse 17
D-35037 Marburg
Germany

Synonyms
Non-approved terms: peptide regulatory factors, immunotransmitters.
Subgroup cytokines: ▶ interleukins, ▶ lymphokines, ▶ monokines, chemokines.

Definition
Cytokines are low-molecular-weight proteins or glycoproteins of usually less than 30–40 kDa. They are mainly released from leukocytes, to a lesser degree from other cell types in response to exogenous stimuli. At picomolecular concentrations, they are high-affinity ligands for specific membrane receptors on other cells (occasionally their own cells) and alter or regulate their function.

Characteristics
Originally discovered as biologically active factors in the supernatants of in vitro cultured leukocytes, they were defined and received their name because of their specific activities on other leukocytes. Usually they

are soluble components that do not transfer antigen specificity, as soluble antibodies do, but efficiently alter cell functions.

Classification of Cytokines

With the advent of gene-cloning techniques, the identification, purification and production of cytokines from cloned genes was facilitated. Meanwhile, around 200 cytokines have been identified. Based on structural studies, many of them belong to one of four defined groups (see Figure 1):
- hematopoietin family
- interferon family
- chemokine family
- tumor necrosis factor family.

Similarly, cytokine receptors may be structurally allocated to five families (see Figure 1):
- class I cytokine hematopoietin receptors
- class II cytokine interferon receptors
- chemokine receptors
- tumor necrosis factor receptors
- immunoglobulin superfamily receptors.

Particularly in class I and class II receptors, multiple subunits with different ligand binding affinities and a common signal transduction chain significantly modulate the cellular response to a given cytokine (see ▶ Cytokine Receptors).

Biological Effects

Although initially thought to be produced exclusively by and to act within the immune system, it was later found that other cell systems, such as endothelial or epithelial cells, may also secrete and respond to cytokines. Thus, some cytokines create a link between the immune, neuronal, endocrine and other systems in which one compartment mutually affects the other.
Most cytokines are not stored inside the cells but are produced and released upon appropriate stimulation. Depending on the property of the released cytokine and its target cell receptor, either growth, differentiation, antimicrobial activation, cytotoxicity or another cell function is induced. A classical example is the de novo synthesis and release of the ▶ tumor necrosis factor TNF-α and the interleukin IL-1 from macrophages after stimulation with a bacterial antigen. Both cytokines are proinflammatory, aid in antigen stimulation of lymphocytes, lead to IL-2 and further successive production of other cytokines, which when taken together constructs an entire network of cytokines with the intention to enforce the interaction of cells. Once the inciting agent has been eliminated, the ▶ cytokine network is downregulated and finally cytokine production is abolished.

Signaling by Cytokines

In some aspects, cytokines resemble classic hormones in that the soluble product of one cell affects the function of another cell. The notable and distinguishing difference is that cytokines are not produced by special glands, their action is locally restricted, a systemic distribution is usually prevented, and a broader spectrum of target cells exist.
Cytokines act as "first messengers" which, upon binding to its receptors, initiate a signal transduction cascade ("second messengers") that leads to gene activation and expression of a specific cell function. For most class I and class II cytokine receptors, the highly specific binding of a cytokine to its receptor induces dimerization of receptor subunits which activates kinases of the Janus family (JAK) by phosphorylation, thereby generating docking sites for one of several transcription factors STAT (signal transducers and activators of transcription). After translocation of a STAT into the nucleus, the phosphorylated and dimerized STAT activates transcription of a distinct subset of those genes that are permitted to be expressed in a given cell type.

Cytokine Network

The large number of cytokines, its pleiotropy, synergy or antagonism, and its induction in a cascade manner, establishes a cytokine network which is essential for innate and adaptive immunity, but also for cell growth, differentiation and lastly controlled cell death. Despite the immunological nonspecifity of cytokines, a high degree of selectivity is achieved by strongly regulated cytokine receptor expression and close interaction of cytokine producer and cytokine responder cells. The most prominent example of a cytokine network is the cytokine secretion pattern of the two subgroups of activated CD4$^+$ T helper (Th) cells. Th1 cells release the ▶ interferon IFN-γ, TNF-β and IL-2, which are required for cell-mediated immunity. Th2 cells secrete interleukins IL-3, IL-4, IL-5, IL-10 and IL-13, which are required for antibody production. The critical balance is provided on the one hand by IFN-γ antagonizing the activity of Th2, and on the other hand by IL-10 antagonizing the activity of Th1. It represents an efficient cross-regulation that determines the outcome of many diseases.

Chemokines

A prominent group of cytokines, the ▶ chemokines, has attracted special attention as key mediators of inflammation. Two subgroups have been identified, the C-C subgroup in which the conserved cysteines are contiguous and the C-X-C subgroup in which the cysteines are interrupted by an amino acid. More than 50 chemokines and 15 C-C or C-X-C receptors have been characterized. Chemokines are usually induced by in-

RECEPTOR FAMILY	LIGANDS	
a) Immmunoglobulin superfamily receptors 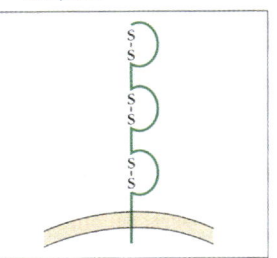	IL-1 M-CSF C-Kit	
b) Class I cytokine receptors (hematopoietin) 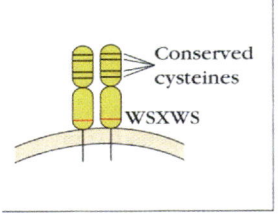	IL-2 IL-3 IL-4 IL-5 IL-6 IL-7 IL-9 IL-11 IL-12	IL-13 IL-15 GM-CSF G-CSF OSM LIF CNTF Growth hormone Prolactin
c) Class II cytokine receptors (interferon) 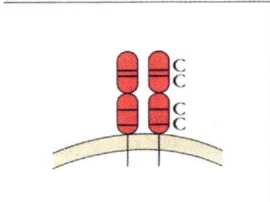	IFN-α IFN-β IFN-γ IL-10	
d) TNF receptors 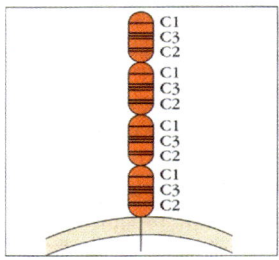	TNF-α TNF-β CD40 Nerve growth factor (NGF) FAS	
e) Chemokine receptors 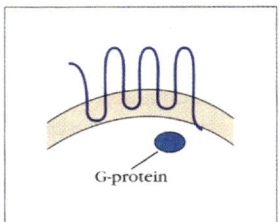	IL-8 RANTES MIP-1 PF4 MCAF NAP-2	

Cytokines. Figure 1 Schematic structural features of the five types of cytokine receptors and a selection of cytokine ligands binding to them. C, conserved cysteines; CNTF, ciliary neutrophilic factor; G-CSF, granulocyte colony stimulating factor; GM-CSF, granulocyte-macrophage colony stimulating factor; LIF, leukocyte inhibitory factor/leukocytosis-inducing factor; MCAF, monocyte chemotactic and activating factor; M-CSF, macrophage colony stimulating factor; MIP, macrophage inflammatory protein; NAP; neutrophil activating peptide; OSM, ovine submaxillary mucin; PF, platelet factor; WSXWS, tryptophan-serine-any amino acid (x)-tryptophan-serine.

fectious diseases and cause leukocyte immigration by building a gradient to attract leukocytes towards the highest chemokine concentration at the inflammatory site. In addition, chemokines promote leukocyte adhesion to endothelial cells and induce antimicrobial activation. The seven-transmembrane chemokine receptors bind more than one type of chemokine, and signal transduction proceeds via heterotrimeric G proteins and second messengers such as cyclic AMP and inositol triphosphate IP_3. Of particular human relevance are the chemokine receptors C-X-CR4 and C-C R5 as coreceptors for HIV infection.

Antagonistic Principles and Viral Mimics

Cytokine-suppressive cytokines such as IL-10 adjust the cytokine network to a well-balanced immune response. Additional, antiinflammatory and antagonizing mechanisms play a role. The IL-1 receptor antagonist (IL-1Ra) binds without own activity to the IL-1 receptor and thereby blocks access of IL-1α and IL-1β. Further cytokine inhibitors are derived from cleavage of membrane-bound receptors, thus generating soluble receptors that already bind cytokines in the circulation or tissue before they can reach their target cells. Prominent examples are the soluble receptor for IL-2 (sIL-2R), often useful as diagnostic marker for T cell activation in immune diseases, and the soluble p75 TNF receptor which already finds clinical application.

The importance of cytokines in "running" the immune response is also underlined by the fact that viruses have developed a strategy during evolution to capture or mimic cytokines or cytokine receptors, with the goal to undermine the antiviral response. Epstein-Barr virus and cytomegalovirus produce IL-10-like molecules capable of downregulating Th1-dependent cell-mediated immunity. Poxviruses, myxoma viruses and herpes viruses release cytokine receptor homologs or binding proteins directed against proinflammatory or antiviral cytokines, such as interferons, chemokines and TNF-α. More than 50 viral homologs of cytokines, cytokine receptors or cytokine binding proteins have been described which all, in one way or the other, manipulate antiviral immunity to the advantage of virus survival.

Preclinical Relevance

Under- or overexpression of cytokines and cytokine receptor defects in several disease states manifests the central importance of cytokines. In X-linked severe combined immunodeficiency (XSCID) a defect in the common γ-chain gene of the IL-2 subfamily receptors is present which results in loss of all functions mediated by interleukins IL-2, IL-4, IL-7, IL-9, and IL15. Classical examples of cytokine overexpression are the bacterial septic shock and bacterial toxic shock syndromes. In the first, Gram-negative bacterial lipopolysaccharides, ▸ endotoxins, are liberated during Gram-negative sepsis and act as exogenous ▸ pyrogens to release endogenous pyrogens from leukocytes. Responsible endogenous pyrogens are mainly TNF-α and IL-1 that are overproduced by monocytes and macrophages. When they reach too high concentrations in the circulation, they cause high fever, widespread blood clotting, diarrhea, and cardiac shock. A rather similar mechanism underlies bacterial shock syndrome. Superantigens such as staphylococcal enterotoxins bind simultaneously outside to major histocompatability complex (MHC) class II molecules and the Vβ domain of the T cell receptor, eventually activating 5% to 10% of all T cells present, which results in massive cytokine production and potential shock symptoms.

On the one hand cytokine-deficient mice may be very interesting models for the development of new drugs. On the other hand, however, the interaction of the cytokine network with other organ systems, as described above, may lead to misinterpretation of the impact of a drug on the immune system by results obtained in toxicologic studies. Therefore, a knowledge about this network and different dependencies is important for a sound hazard and risk assessment.

Relevance to Humans

As outlined above, a dysregulated cytokine production requires an adequate therapy. Measurements to curtail cytokine overproduction, such as capture of overproduced TNF-α by recombinant soluble TNF receptors, antibodies against the TNF receptor or against TNF-α, are described in the entries Cytokine Inhibitors and Cytokine Receptors. Here, some cytokine-based therapies in clinical use will be listed. Cloning of cytokine genes and biotechnological production provides sufficient cytokine amounts for clinical practice. Application of the following cytokines have been officially approved:

- interferon-α2a for the treatment of hairy cell leukemia, hepatitis B and C, and Kaposi's sarcoma
- interferon-α2b for hairy cell leukemia, hepatitis B and C, Kaposi's sarcoma, melanoma, and follicular lymphoma
- interferon-β1b for multiple sclerosis
- interferon-γ for chronic granulomatous disease and osteopetrosis
- interleukin-11 for thrombocytopenia during tumor therapy
- erythropoietin for anemia
- G-CSF and GM-CSF to stimulate neutrophil production in chemotherapeutically treated tumor patients and after bone marrow transplantation
- interleukin-2 for the treatment of kidney carcinoma and melanoma.

Further cytokines will be expected to be in clinical use, except those of the proinflammatory type because of their undesirable side effects.

Regulatory Environment

Cytokines as therapeutic agents are regulated by specific laws, in Germany by the "Arzneimittelgesetz." The responsible agency is the Paul Ehrlich Institut (PEI). All cytokines listed under section "Relevance for Humans" have been approved in the US by the Food and Drug Administration (FDA) and the PEI. As most of these therapeutics are biotech products they are internationally regulated by a ICH Harmonised Tripartite Guideline: Preclinical Safety Evaluation of Biotechnology-Derived Pharmaceuticals, July 16, 1997.

References

1. Gemsa D, Kalden JR, Resch K (1997) Immunologie, 4th ed. Thieme-Verlag, Stuttgart
2. Mantovani A, Dinarello CA, Ghezzi P (2001) Pharmacology of Cytokines. Oxford University Press, Oxford
3. Schooltink H, Rose-John S (2002) Cytokines as therapeutic drugs. J Interferon Cytokine Res 22:505–516
4. Fitzgerald KA, O'Neill LAJ, Gearing AJH, Callard RE (2001) The Cytokine Facts Book, 2nd edn. Academic Press, San Diego
5. Goldsby RA, Kindt TJ, Osborne BA, Kuby J (2003) Immunology, 5th ed. WH Freeman & Co, New York

Cytomegalovirus

Cytomegalovirus causes a systemic infection.
▸ Host Resistance Assays

Cytoskeleton

The cytoskeleton is an intracellular network of structural protein filaments (including actin filaments, microtubules, and intermediate filaments like keratins, vimentin, desmin, and neurofilaments). The cytoskeleton directs cell shape, mediates the anchoring and movement of cell organelles and is necessary for cell division. The cytoskeleton is linked to cell-cell and cell-substrate adhesion molecules and increases the structural integrity of tissues and organs.
▶ Cell Adhesion Molecules

Cytostatic

Inhibition or suppression of cellular growth and replication.
▶ Mixed Lymphocyte Reaction

Cytotoxic Activity

Several cell types of the immune system have the ability to kill specific target cells via distinct mechanisms. Examples are cytotoxic T lymphocytes (CTL), natural killer (NK) cells, lymphocyte-activated killer (LAK) cells, and antibody-dependent cellular cytotoxic (ADCC) cells.
▶ Limiting Dilution Analysis

Cytotoxic T Cell

A thymus-derived lymphocyte that is directly involved in cell-mediated immunity. These cells circulate in search of target cells to interact with and to neutralize at the site.
▶ Respiratory Infections

Cytotoxic T Lymphocyte (CTL) Assay

▶ Cytotoxicity Assays

Cytotoxic T Lymphocytes

B Paige Lawrence
Department of Pharmaceutical Sciences
College of Pharmacy, Washington State University
Pullman, WA 99164 6534
USA

Synonyms
CTL, $CD8^+$, ▶ effector cells.

Definition
Cytotoxic T lymphocytes (CTL) are T cells that, upon contact with antigen-bearing cells, directly kill the target cell. Generally, CTL-mediated killing is major histocompatability complex (MHC) class I-restricted, thus CTL are most often differentiated $CD8^+$ T lymphocytes. However, in some instances, $CD4^+$ T cells can acquire a cytolytic phenotype (1). With regard to host defense, CTL play an important role in resistance to intracellular pathogens. Specifically, CTL primarily kill virus-infected cells and, during a primary infection, are generally considered the most important cellular component of antiviral immunity. Additionally, they recognize and kill certain types of tumor cells, and occasionally bacterially-infected cells are targets of CTL-mediated killing.

Characteristics
CTL can be identified and studied using phenotypic and functional criteria. Using flow cytometry, CTL are most commonly defined phenotypically as those $CD8^+$ T cells that express high levels of the activation and adhesion molecule CD44, and which have down-regulated expression of the leukocyte adhesion molecule CD62L (2). Cells bearing a $CD8^+CD44^{hi}CD62L^{lo}$ phenotype are typically referred to as effector CTL (CTLe). Sorting CTLe using a flow cytometer has consistently affirmed that cells which express this phenotype are armed, antigen-specific cells capable of killing antigen-bearing target cells.
CTL synthesize and release several types of molecules (1,3). Therefore, they can be characterized functionally by the production of specific immunoregulatory and cytotoxic molecules. Whereas the various cytotoxic molecules released by CTL are directly involved in the destruction of antigen-bearing targets, the immunoregulatory molecules secreted by CTL are not exclusively involved in cell killing. They also drive the activation of other immunoregulatory cells, thereby facilitating multiple mechanisms for the destruction of antigen-bearing targets. The primary immunoregulatory molecules made by CTL are the cytokines interferon (IFN)-γ and tumor necrosis factor (TNF)-α.

IFN-γ is almost always produced by CTL and is a potent activator of macrophages and natural killer cells. Furthermore, IFN-γ stimulates MHC class I and class II gene expression, inhibits the growth of type 2 helper T cells and drives immunoglobulin class switching to IgG_{2a}. TNF-α is produced by some CTL and, like IFN-γ, sends a powerful activation signal to macrophages. In particular, it induces nitric oxide synthase-2, leading to the production of nitric oxide. TNF-α is a pleiotropic cytokine, and some of its other effects include activation of the vascular endothelium and recruitment of inflammatory cells. As described below, via binding to TNF receptors, TNF-α can also cause the death of type I TNF receptor-bearing cells. Thus, TNF-α serves dual roles as both an immunoregulatory cytokine and a mediator of cell death.

Once armed, effector CTL induce apoptosis in antigen-bearing target cells via three different—but not mutually exclusive—pathways (4). The primary mechanism for CTL-directed killing involves the calcium-dependent release of perforin-containing and granzyme-containing vesicles. The release of ▶ perforin and granzyme occurs immediately following cytotoxic T lymphocyte (TCR) engagement with peptide-MHC molecules on the target cell. Upon secretion, perforin monomers self-assemble, creating membrane-spanning pores. These pores compromise the integrity of the target cell membrane, which by itself may kill the cell. However, cell death is more likely driven by ▶ granzymes, which enter the target cell through the pores created by perforin. Granzymes A and B are a family of serine proteases that stimulate programmed cell death machinery via the activation of caspases and other pro-apoptotic proteins.

Experiments using perforin-deficient mice demonstrate that most CTL killing is mediated by perforin and granzymes (1,4). However, two perforin-independent mechanisms also exist. Many CTL express two TNF-family members: ▶ Fas ligand (FasL) and TNF-α. Both Fas, the cognate co-receptor for FasL, and type I TNF receptors contain "death domains" in their cytoplasmic tails. Binding of receptors with death domains by their respective ligands activates an intracellular signaling cascade, leading to apoptotic cell death. In contrast to perforin and granzymes, which are stored by CTL, FasL is not. Therefore, killing of target cells in a FasL-dependent manner is delayed because it requires the synthesis and secretion of FasL. The overall contribution of TNF-α-mediated target cell death is not clear, as it is difficult to distinguish the immunoregulatory and cytolytic roles of the TNF-α produced by CTL.

In summary, via the secretion of granules that contain perforin and granzymes, and by the expression of membrane-bound and soluble FasL and TNF-α, CTL kill target cells using multiple cytolytic mechanisms. In addition to the high degree of specificity with which CTL recognize their targets, another important feature of CTL is their ability to serially kill multiple antigen-bearing target cells in a short amount of time. In contrast to naive $CD8^+$ T lymphocytes, CTL contain perforin and granzymes in preformed cytoplasmic granules. This permits immediate release of granule contents upon TCR engagement. Binding of the TCR by peptide-MHC molecules also stimulates the synthesis of more perforin and granzymes by the cell, a process that actively replenishes intracellular stores of these cytolytic molecules. Likewise, TCR engagement stimulates CTL to rapidly synthesize additional FasL. In other words, CTL are able to rapidly restock their weaponry; a feature that enables them to kill multiple targets in rapid-fire succession until the target cells have been successfully eliminated. While most CTL die after the resolution of infection, a few antigen-specific $CD8^+$ T cells survive, generating a small population of memory CTL. Upon re-exposure to the same antigen, memory CTL are activated immediately, resulting in a CTL response that is more rapid and of greater magnitude than the initial response to that antigen.

Thus, the effects of a test agent on CTL are often screened using a tier approach and host resistance assays, such as viral infection or tumor challenge (5). Using either type of antigen, performing a CTL assay (i.e. measuring antigen-specific cytolytic activity) is considered the hallmark test of CTL function. However examining the effects of exposure to a test substance on the expansion of $CD8^+$ T cells, IFN-γ production, and acquisition of an effector CTL phenotype should not be overlooked as informative endpoints. In fact, due to the creation of fluorochrome-labeled MHC class I-restricted tetrameric reagents, it is now increasingly possible examine the effects of a test substance on the activation, clonal expansion, and death of antigen-specific CTL in vivo. In contrast, simply determining the effects of exposure to a test chemical on the percentage or number of bulk $CD8^+$ T cells in a test animal may not be predictive of toxicity. In other words, immunophenotypic analysis alone generally is not sufficient and the data can not stand alone as an endpoint when one assesses the potential immunotoxicity of a test agent.

Preclinical Relevance

As described above, CTL play an important role in host defense against pathogens, detect and destroy tumor cells, and contribute to immune-mediated pathology in hypersensitivity reactions and autoimmune disease. Thus, it is important to consider toxicity of environmental contaminants and pharmaceuticals toward $CD8^+$ T cells and CTL.

Regulatory Environment

Given that CTL function is a critical aspect of host resistance to intracellular viral infection, testing for effects of a chemical on $CD8^+$ T cell function should be considered during immunotoxicity screening. However there is not currently agreement on exactly how it should be tested and whether assessment of a CTL response should *always* be performed (5–7).

Testing for CTL activity is discussed in the following guidelines:

- EPA OPPTS 880.3550 Immuntoxicity. Environmental Protection Agency (1996) Biochemicals Test Guidelines: OPPTS 880.3550 Immunotoxicity. Prevention, Pesticides and Toxic Substances. EPA 712-C-96-280.
- CDRH Immunotoxicity Testing Guidance. Center for Devices and Radiological Health (1999) US Department of Health and Human Services, Food and Drug Administration. Guidance for Industry and FDA Reviewers: Immunotoxicity Testing Guidance.
  http://www.fda.gov/cdrh/ost/ostggp/immunotox.pdf
- CPMP/SWP/1042/99cor (2001) CPMP/SWP/1042/99cor (2001). European Agency for the Evaluation of Medicinal Products Committee for Proprietary Medicinal Products (CPMP) Safety Working Party (SWP) Note for Guidance on Repeated Dose Toxicity 1042/99.
- CDER Guidance for Industry: Immunotoxicology Evaluation of Investigational New Drugs. Center for Drug Evaluation and Research (2002) US Department of Health and Human Services, Food and Drug Administration. Guidance for Industry: Immunotoxicology Evaluation of Investigational New Drugs.
  http://www.fda.gov/cder/guidance/index/htm

Relevance to Humans

Insufficient, excessive and inappropriately directed CTL responses contribute to human disease. Suppressed CTL function leads to increased susceptibility to pathogens, particularly those that replicate inside host cells. They also play an important role in tumor surveillance, thus individuals with impaired CTL function are more susceptible to certain types of tumors. The importance of CTL in host resistance and tumor destruction has been known for quite some time. More recently, these cells are gaining a larger role in prophylactic immunotherapy. Vaccination with live, attenuated pathogens drives a strong, host-protective memory response leading to highly effective vaccines. In contrast, vaccination with killed pathogens or pathogen-derived constituents generally leads to a less robust immune response, and often does not provoke a strong cell-mediated response. In particular, very few vaccines that employ killed pathogens or pathogen constituents provoke an effective CTL response. Therefore protective immunity in humans relies predominantly on the generation of memory B cells and antibodies. The concern with this paradigm is that, for some pathogens, memory CTL are more important than antibodies for protective immunity. Therefore, the design of vaccines that promulgate cell-mediated immunity is an active area of basic and clinical immunology research.

Given their powerful cytolytic machinery, which can be directed very specifically at antigen-bearing target cells, CTL are also an attractive candidate for tumor therapy. While this is an area that is presently more in the realm of research and development rather than clinical application, it is important to keep this therapeutic exploitation of CTL in mind when one considers the possible therapeutic benefits of harnessing the antigen-specific activity of CTL to kill tumor cells.

In contrast to the beneficial activity of pathogen- and tumor-specific CTL, there are also CTL responses that are detrimental to the host. CTL mediate some forms of delayed-type (type IV) hypersensitivity reactions, such as contact dermatitis. Likewise, autoreactive $CD8^+$ T cells underlie the destruction of healthy cells in several prevalent autoimmune diseases, such as insulin-dependent diabetes mellitus. Therefore, diminution of CTL activity is often a therapeutic goal for the treatment of hypersensitivity and autoimmune diseases.

In addition to causing tissue damage in hypersensitivity and autoimmune diseases, CTL play a role in the rejection of transplanted tissue. In fact, about 10% of T cells in the body recognize allogeneic MHC molecules. Furthermore, even in a good human leukocyte antigen (HLA) match, transplant rejection often occurs via the recognition of minor histocompatability molecules, of which most are associated with MHC class I. Therefore rejection due to incompatible type I HLA and minor histocompatability molecules is most often driven by CTL.

References

1. Kagi D, Ledermann B, Burki K, Zinkernagel RM, Hengartner H (1996) Molecular mechanisms of lymphocyte-mediated cytotoxicity and their role in immunological protection and pathogenesis in vivo. Ann Rev Immunol 14:207–232
2. Doherty PC, Topham DJ, Tripp RA (1996) Establishment and persistence of virus-specific CD4+ and CD8+ T cell memory. Immunol Rev 150:23–44
3. Berke G (1994) The binding and lysis of target cells by cytotoxic lymphocytes: Molecular and cellular aspects. Ann Rev Immunol 12:735–773

4. Russell JH, Ley TJ (2002) Lymphocyte-mediated cytotoxicity. Ann Rev Immunol 20:323–370
5. Luster MI, Munson AE, Thomas PT et al. (1988) Development of a testing battery to assess chemical-induced immunotoxicity: National Toxicology Program's guidelines for immunotoxicity evaluation in mice. Fund Appl Toxicol 10:2–19

Cytotoxic T Lymphocytes (CTL)

A type of T lymphocyte whose major function is to recognize and kill host cells infected with viruses or other intracellular microbes. CTLs, expressing CD8, recognize antigenic peptides that are displayed by major histocompatibility class I (MHC-1) molecules on the surface of infected cells. In certain instances also $CD4^+$ T cells acquire the ability to kill other cells.
▶ Cytotoxicity Assays
▶ Flow Cytometry
▶ MHC Class I Antigen Presentation
▶ Cell-Mediated Lysis

Cytotoxicity

The ability of an immune cell to kill or eliminate a target cell, usually by apoptotic or granulocytic mechanisms.
▶ Natural Killer Cells

Cytotoxicity Assays

ELIZABETH R GORE
Immunologic Toxicology Preclinical Safety Assessment
GlaxoSmithKline R&D
709 Swedeland Road
P.O.Box 1539
King of Prussia, PA 19406-0939
USA

Synonyms

cytotoxicity assay, ▶ natural killer (NK) cell assay, cytotoxic T lymphocyte (CTL) assay, ▶ chromium release assay

Short Description

Cytolytic activity is a property of specialized immune cells, such as natural killer (NK) cells and ▶ cytotoxic T lymphocytes, enabling surveillance and ultimate destruction of tumor cells and virus-infected cells. NK cells, components of the innate immune response, display direct lytic activity that is regulated by cell surface stimulatory and inhibitory receptors. Cytotoxic T lymphocytes (CTLs) kill target cells on the basis of recognizing specific antigens (viral, tumor, or alloantigens) expressed in the context of self-major histocompatibility complex (MHC) molecules.

Characteristics

A procedure for measuring cell mediated cytotoxicity by ^{51}chromium (^{51}Cr) release assay has been well established (1). Measurement of NK cytotoxicity is performed by incorporating radiolabel into tumor-derived target cells (e.g. ▶ YAC cells and ▶ K-562 cells) followed by coincubation with NK cells at various effector : target ratios. NK cells can be derived from spleen or peripheral blood mononuclear cell (PBMC) populations; purification of NK cells from heterogenous cell populations is not required but can enhance sensitivity of the assay. In the presence of NK cells the labeled target cells are lysed and subsequent release of ^{51}Cr is measured quantitatively. Similarly, CTLs can be coincubated with ^{51}Cr-labeled syngenic cells expressing a specific antigen. The "primed" CTLs recognize and lyse the labeled cells causing release of ^{51}Cr. Cytotoxicity is expressed as a ratio relative to total release of ^{51}Cr, which can be induced by addition of detergent (e.g. triton X-100) to target cells. Spontaneous release of label should not exceed 15% of the total release. In some instances (e.g. comparing lytic activity of different effector cells over a range of effector : target ratios), it is advantageous to express cytotoxicity in lytic units (LU). Calculation of a lytic unit has been described previously in detail (2). In brief, it represents the quantity of effector cells needed to kill a certain percentage of a predetermined number of target cells.

Modifications to the conventional ^{51}Cr release assay have been evaluated, particularly utilizing a non-isotopic label as a marker for target cell lysis (3). In addition, entirely new formats have been assessed, namely flow cytometric assay (FCA) for determination of cytotoxicity (4). Fluorescent and flow cytometric-based cytotoxicity assays, conducted in parallel with ^{51}Cr release assays, have demonstrated similarity to the conventional procedure and therefore support use of alternative, non-radioactive approaches to cytotoxicity assessment.

The fluorescent-based approach, similar in format to the ^{51}Cr release assay, substitutes the lanthanide, ▶ europium, for ^{51}Cr. Europium, with its chelate diethylenetriaminopentaacetate (Eu^{3+}-DTPA), is incorporated into target cells, and upon its dissociation and release into solution forms a highly fluorescent chelate that can be measured rapidly and with high level of sensitivity by ▶ time-resolve fluorometry. A comparison between ^{51}Cr and europium release assays in determining cytolytic activity of monkey peripheral NK cells, cultured with or without recombinant human IL-2 (a known inducer of NK activity), against K-562 target cells is shown in Table 1.

A novel approach to cytotoxicity assessment by flow cytometry is described by Piriou et al (2000). The assay is based on the use of two fluorescent dyes: green fluorescent ▶ DIOC18$_3$ used to label target cells, and the red fluorescent dye, propidium iodine, for determination of dead cells. Thus discrimination between effector and target cell populations can be made, and the dead target cells identified based on uptake of propidium iodine. Cytolysis of K-562 cells by human NK cells has been demonstrated by flow cytometric methods with parallel comparison to ^{51}Cr release assay. Published results, shown in Table 2, indicate significant concordance ($R = 0.96$; $p < 0.001$) between the two methodologies.

Pros and Cons

Conventional ^{51}Cr release assay is well established, readily adaptable and does not require highly trained or specialized staff. Moreover, the application of this method has been evaluated not only with various sources of NK cells (e.g. whole blood, PBMCs, splenocytes, purified cell populations) but also for assessing CTL responses and antibody-dependent cellular cytotoxicity. A drawback to the conventional format is radioisotope usage. The europium release assay by time-resolve fluorometry is a good alternative to the ^{51}Cr release assay in that it eliminates radioisotope usage, consists of experimental set-up and data collection/analysis procedures similar to the conventional assay, and does not require highly trained technical staff. In contrast, the FCA methodology requires specialized scientists, particularly for collecting and analyzing data. Unless flow cytometry exists as a core capability, adapting this methodology may be time and cost prohibitive.

Predictivity

Much effort has been made to establish the predictability of various immune function tests in identifying immunotoxicity of new compounds (5). Relationships between NK function and altered host resistance were made with over 30 compounds utilizing mouse models, and the outcome indicated a 70% concordance between effect on NK function and altered host resistance. When NK cytotoxicity tests were combined with surface marker expression analysis the concordance with host resistance was increased to 90%. These findings support use of NK functional tests for preclinical assessment of immunotoxic potential of development compounds.

Relevance to Humans

While the clinical significance of altered cytolytic activity in humans has not been clearly established, there is obvious potential for biological consequences due to altered NK and CTL activity that include decreased resistance to infectious disease and certain tumor diseases. Preclinical evaluations on new chemicals for potential adverse effects on NK cell and CTL function may provide useful information for clinical monitoring, especially as it pertains to increased incidence of infections.

Regulatory Environment

Based on a high concordance between NK function

Cytotoxicity Assays. Table 1 Comparison of monkey natural killer cell activity against K-562 target cells by ^{51}chromium and Europium^{3+}-diethylenetriaminopentaacetate (DTPA) release assays

| | | % Cytotoxicity | | | |
| | rHuIL-2 | Effector : Target Ratio | | | |
Format	U/mL	5 : 1	10 : 1	20 : 1	40 : 1
^{51}Chromium	0	25.9	35.8	41.1	48.7
Europium^{3+}-DTPA	0	24.0	26.0	32.0	56.0
^{51}Chromium	50	40.9	43.9	53.5	58.8
Europium^{3+}-DTPA	50	30.0	40.0	57.0	62.0
^{51}Chromium	500	51.4	56.9	70.3	78.1
Europium^{3+}-DTPA	500	43.0	51.0	63.0	80.0
^{51}Chromium	5000	50.3	63.9	71.7	87.4
Europium^{3+}-DTPA	5000	57.0	67.0	83.0	100.0

From Gore ER unpublished results.

Cytotoxicity Assays. Table 2 Comparison of human natural killer cell cytotoxicity by ^{51}chromium release assay and flow cytometric assay (FCA)

Format	% Cytotoxicity Effector : Target Ratio				
	1.5 : 1	5 : 1	17 : 1	50 : 1	150 : 1
Flow cytometric assay					
Donor 1	2	10	25	34	52
Donor 2	5	23	58	74	65
^{51}Chromium					
Donor 1	2	11	14	25	38
Donor 2	8	12	20	36	53

From Piriou et al. (4).

and surface marker analysis with altered host resistance (5), European regulatory guidelines now recommend incorporation of these two parameters in the initial screening for immunotoxicity of new chemicals. In the US, both the FDA (Food and Drug Administration) and EPA (Environmental Protection Agency) guidelines on immunotoxicity testing recommend conducting the NK cell assay on a case-by-case basis depending on the outcome of the routine, standard toxicity evaluations indicated in the respective guidelines. Relevant guidelines are:

- CPMP/SWP/1042/99 Note for Guidance on Repeated Dose Toxicity 2000
- FDA (CDER) Immunotoxicology Evaluation of Investigational New Drugs, 2002
- EPA (OPPTS 870.7800) Health Effects Test Guidelines: Immunotoxicity 1998.

References

1. Brunner KT, Engers HD, Cerottini JC (1976) The ^{51}Cr-release assay as used for the quantitative measurement of cell mediated cytolysis in vitro. In: Bloom BR, David JR (eds) In vitro methods of cell mediated and tumor immunity. Academic Press, New York, pp 423–428
2. Bryant J, Day R, Whiteside TL, Heberman RB (1992) Calculation of lytic units for the expression of cell mediated cytotoxicity. J Immunol Meth 146:91–103
3. Blomberg K, Granberg C, Hemmila I, Lovgren T (1986) Europium-labelled target cells in an assay of natural killer cell activity. J Immunol Meth 92:117–123
4. Piriou L, Chilmonczyk S, Genetet N, Albina E (2000) Design of a flow cytometric assay for the determination of natural killer and cytotoxic T lymphocyte activity in human and in different animal species. Cytometry 41:289–297
5. Luster MI, Portier C, Pait DG, White KL Jr, Gennings C, Munson AE, Rosenthal GJ (1992) Risk assessment in immunotoxicology I. Sensitivity and predictability of immune tests. Fund Appl Toxicol 18:200–210

D

2-D Gel Electrophoresis

In two-dimensional (2-D) gel electrophoresis, proteins are first separated by isoelectric focusing according to their isoelectric point followed by separation by SDS-PAGE based on their molecular size.
▶ Western Blot Analysis

3-D Human Skin/Epidermal Equivalents

▶ Three-Dimensional Human Skin/Epidermal Models and Organotypic Human and Murine Skin Explant Systems

DAT

▶ Antiglobulin (Coombs) Test

DBPCFC

Double-blind placebo-controlled food challenge, the gold standard for the diagnosis of food allergy. A suspected food is given to a patient, in which the allergen is hidden.
▶ Food Allergy

Death-Inducing Signaling Complex

Binding of TNF-α to its receptors results in trimerization via intracellular death domains (DD). This leads to the docking of adapter proteins also containing DD. Adapter proteins include TNF-R1-associated death-domain-containing factor (TRADD), Fas-associated death-domain-containing protein (FADD), and TNFR-associated factor-2 (TRAF2). This complex serves to recruit and activate other accessory proteins with catalytic activity including various kinases and proteases. These enzymes function to activate the various signaling pathways culminating in cell death and/or cell survival.
▶ Tumor Necrosis Factor-α

Decavanadate

Decavanadate is a polymeric form of vanadium consisting of ten vanadium atoms linked primarily through oxygen atom bridges.
▶ Vanadium and the Immune System

Decoy Receptor

Decoy receptors are highly specific cytokine receptors, which bind their ligand with high affinity but do not initiate and transduce signals. Usually they lack the signaling domain. A decoy receptor may serve as a ligand sink and may compete for a coreceptor thus depriving the signaling cytokine receptor of its indispensable second signaling subunit (e.g. type II interleukin-1 receptor binds IL-1 and associates with interleukin-1 receptor accessory protein, but does not signal as it lacks the signaling domain). Decoy receptors are an additional level of regulating cytokine action.
▶ Cytokine Receptors

Defensin

Defensins are a family of 3.5–4.5 kD cationic peptides which are characterized by three internal disulfide bonds. They are synthesized in many cells, predominantly in phagocytic cells such as polymorphonuclear neutrophils. They exhibit a broad spectrum of antimicrobial activity by inserting into target membranes and forming pores. For example human-β-defensins 2 and 3 represent peptides with antimicrobial effects, which are produced by keratinocytes after contact with either microbial peptids or bacteria.

▶ Immune Response
▶ Dermatological Infections

Delayed-Type Hypersensitivity

HANS-WERNER VOHR
PH-PD, Toxikology
Bayer HealthCare AG
42096 Wuppertal
Germany

Synonyms
cell-mediated hypersensitivity, type IV immune reaction, contact hypersensitivity, cell-mediated allergicreaction, DTH

Definition
Delayed-type hypersensitivity (DTH) is an allergic immune reaction (▶ hypersensitivity reaction which may be transferred by lymphocytes of sensitized animals instead of serum (type I–III reactions). This type of reaction is, therefore, called cell-mediated hypersensitivity. DTH is characterized by inflammatory responses starting macroscopically 6–12 hours after antigen administration in sensitized individuals. Reactivity peaks at 24–72 hours after antigen contact with numerous exceptions.
Historically, three types of DTH reactions have been described:
- contact hypersensitivity
- ▶ tuberculin-type reaction (the first DTH described by R. Koch)
- granulomatous hypersensitivity (see Table 1).

Delayed-type hypersensitivity must be distinguished from (delayed) cell-mediated immune reactions, such as rejection of transplanted tissue (host-versus-graft response), which are beyond the scope of this encyclopedia.

Characteristics
Allergens processed by antigen-presenting cells (APC) are presented to T-helper cells (CD4$^+$ T cells) on MHC class II molecules in local tissue. Such antigens comprise bacteria, viruses, parasites, foreign proteins and chemicals acting as haptens. Upon renewed antigen contact (challenge phase) interaction of sensitized lymphocytes leads to a characteristic inflammatory reaction. During this effector phase T lymphocyte subsets release different chemokines and cytokines, such as INFγ, TNFβ, IL-3 and GM-CSF. In many DTH reactions, involvement of CD8$^+$ T cells (cytotoxic T lymphocytes) has also been described. With respect to the Th1/Th2 dichotomy the above-mentioned lymphokine pattern would favor Th1 cells as the central effector lymphocyte for the mediation of DTH reactions. However, there is evidence that several subsets of T cells are involved.

Allergic contact dermatitis (contact hypersensitivity) is the most frequent type of DTH reaction and the most prominent clinical manifestation of a type IV immune reaction. Normally, induction of DTH is caused by local (epidermal, intradermal) administration. Systemic forms of administration such as intravenous injections are usually not effective in inducing DTH reactions.

After penetration through the stratum corneum, the antigen reaches the epidermis and is taken up as a foreign protein or hapten-carrier complex by the professional antigen-presenting cells in the skin, the Langerhans cells (LC). After interaction with surrounding cells (especially ▶ keratinocytes) Langerhans cells are activated and migrate via the afferent lymph into the draining lymph nodes. Here they stimulate antigen-specific (mainly CD4$^+$ but also CD8$^+$ cells). Upon activation, these T cells will proliferate, differentiate and infiltrate the skin as memory T cells (CD4$^+$) or as effector cells (CD8$^+$). Reexposure to the relevant antigen will reactivate these T cells (mainly CD4$^+$) and cause a local inflammatory response. Clinically this reaction manifests as allergic eczema characterized by skin reddening, swelling and edema formation accompanied by severe itching. After long and intensive antigen contact the clinical picture may change to hyperkeratosis, lichenification and rhagades (lines).

This form of allergic contact dermatitis must be distinguished from nonspecific toxic contact dermatitis. Macroscopically these two forms are not easily differentiated in patients. However, while the toxic (cytotoxic, irritating) reactions are always restricted to the contact area and independent of the genetic background of the individual, allergic contact dermatitis is genetically restricted and may spread over the whole body. This is due to a battery of mediators released during the onset and effector phase of an allergic skin reaction.

During the onset of contact hypersensitivity, mediators (tumor necrosis factor(TNF)α, interleukins IL-1α, IL-8, GM-CSF, IL-1β and IL-6) released from keratinocytes and LC, play a crucial role in the further control of the reaction. The effector phase, however, is dominated and regulated by the release of different chemokines and cytokines from T lymphocyte subsets: interferon(INF)γ, TNFβ, IL-2, IL-3, GM-CSF, IL-4 and IL-10. In many cases CD8$^+$ T cells (cytotoxic T lymphocytes) are also involved in the inflammatory reaction, and may induce a primary late skin response.

Delayed-Type Hypersensitivity. Table 1 Causes and reactions of delayed-type hypersensitivity

Delayed-type hypersensitivity reaction	Allergens	Consequences	Peak
Contact hypersensitivity	Haptens (e.g. pentadecacatechols, dinitrofluorobenzene, oxazolone) Metal ions (e.g. nickel, chromate) Poison ivy Poison oak Rubber	Eczema (skin reddening, swelling, edema), itching, hyperkeratosis, lichenification	48–72 hours
Tuberculin-type reaction	Beryllium Soluble antigen of *Mycobacteria tuberculosis*, *M. leprae* or *Leishmania tropica* Tuberculin (intradermal) Zirconium	Local induration, swelling	48–72 hours
Granulomatous hypersensitivity	Beryllium Immune complexes Silica Talc zirconium	Hardening (skin, lung)	21–28 days

Hence, there is much evidence that several subsets of T cells are involved, and that local immune reactions do not follow a simple Th1/Th2 dichotomy.
As described above, the regulation of the responses is controlled by a whole network of cytokines, chemokines, inflammatory proteins produced by immunecompetent lymphocytes and tissue-specific cells such as Langerhans cells and keratinocytes. In particular, LC involved in this reaction differentiate and change their morphology, their mediator-expression pattern and surface molecules in the course of activation. A simplified model of this network interaction during the onset of contact dermatitis is illustrated by the scheme in Figure 1.

Preclinical Relevance

There are special guidelines regulating the investigation of contact hypersensitivity (as type IV reaction) with chemicals, agrochemicals, pharmaceuticals and cosmetics, but none of the other types of hyperreactions. This is understandable by the fact that allergic contact dermatitis is the most frequent type of DTH.

Relevance to Humans

Regarding the contact hypersensitivity the human relevance is obvious for topically applied pharmaceuticals and cosmetics. People handling agrochemicals or industrial chemicals also need to know if a compound is a skin sensitizer or not. However, with a sensitivity of only 71% the predictivity or accuracy of the preclinical test systems presents a problem. Up to the end of the last decade contact sensitivity was tested exclusively in guinea pigs (▶ guinea pig assays for sensitization testing) by the evaluation of subjective parameters like skin reddening. The introduction of more objective parameters based on the immunological understanding of the mechanism into such tests (MEST, LLNA or IMDS) has been thought to improve the accuracy and/or sensitivity. However, in spite of the huge advantages, these mouse tests still present some problems. It may turn out that some mouse tests are not more predictive than guinea pig tests.
Local intracutaneous injection of microorganisms like mycobacteria, or fungi, or their protein extract (e.g. tuberculin, lepromin) into sensitized individuals has some diagnostic relevance in clinics (tuberculin-type reaction). Especially in the light of growing numbers of HIV-infected patients, rapid and reliable diagnostic tools like enzyme-linked immunoassay (ELISA) or chemiluminescent DNA probes are needed.
A similar diagnostic problem exists for fungal infections of increasing numbers of immune compromised patients.

Regulatory Environment

Although granulomatous hypersensitivity is clinically the most important form of DTH, investigation of contact hypersensitivity (as type IV reaction) is the only area regulated by special guidelines (cf. related section). Nevertheless, granulomatous-like hyperreactions may be taken into consideration if macrophages or granulocytes are confronted with an antigen which they cannot destroy.

References

1. Friedmann PS (1989) Contact hypersensitivity. Curr Opin Immunol 1:690–693

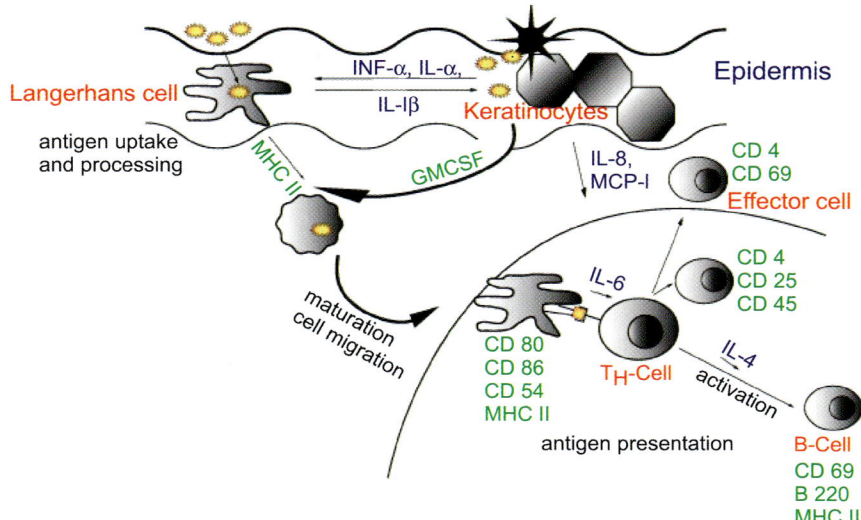

Delayed-Type Hypersensitivity. Figure 1 Proposed cellular and molecular events during the onset of allergic (yellow star = allergen) or toxic (black star = cytolytic, corrosive event) contact hypersensitivity. (Modified by E. Heisler).

2. Godfrey H, Gell PH (1978) Cellular and molecular events in the delayed onset hypersensitivities. Rev Physiol Biochem Exp Pharmacol 84:1–32
3. Polak L, Barnes JM, Turk JL (1968) The genetic control of contact sensitisation to inorganic metal compounds in guinea pigs. Immunology 14:707–711
4. Poulter LW, Seymour GJ, Duke O, Janossy G, Panay G (1982) Immunohistochemical analysis of delayed-type hypersensitivity in man. Cell Immunol 74:358–369
5. Salerno A, Dieli F, Sireci G et al. (1997) Three cell subsets are required for the transfer of delayed-type hypersensitivity reaction by antigen-specific T cell lines. Cell Immunol 175:157–163

► Contact Hypersensitivity

Delayed Type IV Hypersensitivity

This is a cell-mediated immune response in which antigen-specific T cells are activated to produce cytokines that will resolve an exposure and re-establish homeostasis. Type IV hypersensitivity requires repeated exposure to antigen.
► Chronic Beryllium Disease

Delayed-Type Hypersensitivity (DTH)

Delayed-type hypersensitivity is a form of cell mediated immune reaction in which the effector cells are activated mononuclear phagocytes. It is an increased response to an antigen or allergen that does not appear until 24–48 hours after exposure to the antigen. It can often be separated into an early and a late response. During the early response, mast cells and basophils release compounds that attract other inflammatory cells and enable them to migrate into the tissue. Activation of these cells causes the delayed-type hypersensitivity response.
► Helper T lymphocytes
► Mast Cells
► Immunotoxicology

Dendritic Cell

Dendritic cells are a specialised cell type of the immune system. They are the most potent antigen-presenting cells and stimulate T cells. Immature dendritic cells take up antigen in the periphery and become activated themselves. They then mature and migrate to the lymph nodes and spleen, where they present antigen to T cells. Dendritic cells are derived from the bone-marrow, they have a distinct morphology (thus the name) and express many co-stimulatory molecules. Dendritic cells in the skin are called Langerhans cells.
► Dioxins and the Immune System
► Lymphocytes
► Skin, Contribution to Immunity

Deoxyribonucleic Acid (DNA)

An informational biomolecule, originally described by Crick and Watson, found in the nucleus of a cell. It consists of deoxyribose units joined in the 3' and 5' positions through a phosphodiester linkage with a purine or pyrimidine base attached to the 1' position.

▶ Southern and Northern Blotting

Depletion

▶ Cell Separation Techniques

Dermatological Infections

SANDRA HANNEKEN · NORBERT J. NEUMANN
Department of Dermatology
University of Düsseldorf
Germany

Definition

Infections of the skin (Table 1) comprise a very large field of different causative pathogens, as well as a variety of clinical findings. Several microorganisms including fungi, bacteria, and viruses (as well as helminths, epizoons and protozoons) possess pathogenetic relevance for dermatological infections. They are responsible for skin conditions very commonly encountered by dermatologists. Microorganisms have developed several mechanisms to overcome the host defense and to grow in the host organism.

Pathogenetic bacteria possess the ability to adhere, grow on and invade the host. Obligate pathogenetic microorganisms include, for example, *Streptococcus pyogenes* and *Staphylococcus aureus*. Infections with these microorganisms almost always cause clinical symptoms. Well known clinical pictures represent the pyodermas, such as ▶ impetigo contagiosa, erysipelas or furuncles. Human herpes viridae are involved in the etiopathology of well known dermatological infections like herpes simplex (herpes simplex virus), shingles (varicella zoster virus), chickenpox (varicella zoster virus) and multiple types of verrucae (human papilloma virus). Infections with dermatophytes are causative for frequent dermatoses like onychomycosis, tinea capitis or tinea corporis. *Candida albicans* is associated with a variety of opportunistic infections including mucocutaneous candidiasis.

Characteristics

Important factors for the development of infections are
- virulence of the microorganism
- local factors
- the immunological situation of the host.

Dermatological infections are often related to trauma of the skin. For the development of many skin infections, only a very small trauma which represents the entry for pathogenetic microorganisms is sufficient. In addition, patients with underlying diseases like diabetes, or with abnormalities of the immune system, are prone to develop several skin infections. Acquired disposition to develop infections can be related to the systemic intake of immunosuppressive medications like antibiotics, cytostatics, or glucocorticoids. Furthermore preexisting diseases of the skin might facilitate the manifestation of dermatological infections.

Patients suffering from ▶ atopic dermatitis exhibit immunological abnormalities of the skin, as well as a disturbed skin barrier which is associated with a disposition to develop skin infections. Common infections in those patients are pyodermas caused by infections with pathogenic *S. aureus* or *S. pyogenes*. Those patients are also prone to viral infections of the skin (e.g. herpes simplex virus, molluscum contagiosum virus, human papilloma virus). Additionally, dermatological infections in patients with atopic dermatitis often show a more severe course, as in ▶ eczema herpeticum, and eczema molluscatum.

Furthermore dermatological infections can be a sign of an underlying systemic infection, commonly seen in patients with the acquired immunodeficiency syndrome (AIDS) or, less frequently, in tuberculosis. Infections of the skin can also occur due to the production of bacterial toxins or immune complexes, for example exotoxin-producing bacteria like *Staphylococcus aureus* can lead to potentially life-threatening conditions like toxic shock syndrome.

With its structural barrier, the skin delivers protection against microorganisms. However, the skin is physiologically colonized by several microorganisms including *Staphylococcus epidermidis*, *Micrococcus*, *Corynebacterium* species and yeasts, which constitute the resident flora. The microbes of the resident flora are commonly nonpathogenic and live in synergy with the host. They usually do not possess virulence factors which are a characteristic feature of pathogenic bacteria. In rare cases, colonization with nonpathogenic microbes can result in a clinical infection. *S. epidermidis* is a coagulase-negative facultative pathogenic bacterium which can be causative for severe infections (sepsis, endocarditis) under certain conditions. Pathogenic relevance of *S. epidermidis* is related to immunosuppression or to the presence of

Dermatological Infections. Table 1 Infections of the skin

Type of infection	Relevant skin disorder
Bacterial infections	
Actinomyces israelii	Actinomycosis
Bacillus anthracis	Anthrax
Bartonella henselae	Bacillary angiomatosis, cat scratch disease
Borrelia burgdorferi	Lyme disease
Calymmatobacterium granulomatis	Granuloma inguinale
Chlamydia trachomatis	Urethritis, pelvic inflammatory disease, Reiter syndrome
Clostridium perfringens	Gas gangrene
Corynebacterium diphtheriae	Diphtheria
Corynebacterium minutissimum	Erythrasma
Corynebacterium tenuis	Trichobacteriosis palmellina
Erysipelothrix rhusiopathiae	Erysipeloid
Francisella tularensis	Tularemia
Haemophilus ducreyi	Chancroid
Mycobacterium leprae	Leprosy
Mycobacterium marinum	Swimming-pool granuloma
Mycobacterium tuberculosis	Tuberculosis
Mycoplasma hominis *Mycoplasma genitalium*	Urinary and genital tract infection
Neisseria gonorrhoeae	Urethritis, cervicitis, proctitis, urinary tract infection, septicemia
Neisseria meningitidis	Septicemia, Waterhouse-Friderichsen syndrome
Propionibacterium acnes	Acne
Pseudomonas aeruginosa	Folliculitis, whirlpool dermatitis, ecthyma, wound infection
Salmonella paratyphi *Salmonella typhi*	Typhoid
Staphylococcus aureus	Folliculitis, furuncle, carbuncle, abscess, toxic shock syndrome, staphylococcal scalded skin-syndrome
Streptococcus pyogenes	Erysipelas, phlegmon, impetigo, ecthyma, scarlet fever, necrotizing fasciitis
Treponema carateum	Pinta (carate)
Treponema pallidum	Syphilis
Treponema pertenue	Yaws
Viral infections	
Cytomegalovirus	Exanthema, mucosal ulcer
Coxsackievirus	Herpangina, hand-foot-mouth disease
Epstein-Barr virus	Oral hairy leukoplacia, Burkitt's lymphoma
Hepatitis viruses	Gianotti-Crosti syndrome, lichen planus, hepatic-cutaneous porphyria
Herpes simplex virus Herpes labialis Herpes genitalis	Herpetic gingivostomatitis, eczema herpeticatum, herpes neonatorum
Human herpesvirus 6	Roseola infantum
Human herpesvirus 7	Pityriasis rosea (possibly)
Human herpesvirus 8	Kaposi sarcoma

Dermatological Infections. Table 1 Infections of the skin (Continued)

Type of infection	Relevant skin disorder
Human immunodeficiency virus	Acquired immunodeficiency syndrome
Human papillomavirus	Verrucae vulgares, plantares, planae, condylomata acuminata
Measles virus	Measles
Molluscum contagiosum virus	Mollusca contagiosa
Orf virus	Orf (ecthyma contagiosum)
Parvovius B 19	Erythema infectiosum
Rubella virus	Rubella
Smallpox virus	Smallpox
Varizcella zoster virus	Shingles, chickenpox
Fungal infections	
Candida albicans	Onychomycosis, intertrigo, folliculitis, mucocutaneous and genital candidiasis
Cryptococcus neoformans	Cryptococcosis of the skin
Histoplasma capsulatum	Histoplasmosis of the skin
Malassezia species	Pityriasis versicolor, seborrhoic dermatitis
Trichphyton rubrum and others	Tinea capitis, tinea corporis, dermatophytes, onychomycosis
Sporothrix schenkii	Sporotrichosis
Helmintic infections	
Dracunculis medinensis	Dracunculosis
Loa loa	Calabar swelling
Wuchereria bancrofti	Filariasis
Epizoal infections	
Demodex folliculorum	Demodicosis
Pediculus humanus Pthirus pubis	Pediculosis capitis, pediculosis pubis
Sarcoptes scabiei	Scabies

plastic foreign bodies in the host organism (e.g. intravascular catheter or artificial heart valves). The resident flora plays an important role in the defense of several pathogenic microorganisms.

Microorganisms which are not part of the resident flora are inactivated by means of fatty acids which contribute to the skin pH of 5.5. The adherence of microorganisms to the surface of skin or mucus membrane is essential for the colonization of microbes. The adherence is related to the interaction of bacterial adhesive factors and receptors on the host cell. Invasion of the microorganisms requires the presence of invasive factors. Some bacteria penetrate the skin barrier by the production of hyaluronidase (spreading factor), collagenase, elastase, lipase or ▶ streptokinase. These enzymes degrade the hyaluronic acid of connective tissue and thus diminish the connection of the tissue structure.

Some bacteria possess polysaccharide capsules which prevent phagocytosis. Others, like *Streptococcus pyogenes* contain surface protein (substance M) which delivers protection against phagocytosis. Leucocidine of *S. aureus* and lecithinase of *Clostridium perfringens* are capable of killing phagocytes. Some microorganisms are immunosuppressive for the host. *Neisseria gonorrhoea* degrades IgA antibodies and therefore disturbs the mucosal defence of the host. Proteins of *S. aureus* bind to IgG antibodies and thus hinder the interaction between antibodies and phagocytes.

Microorganisms such as *Chlamydia* and several

viruses can damage the host cell by growing intracellularly. Some microorganisms produce exotoxins. These toxins can lead to the damage of the host membrane cell. Examples are lecithinase of *C. perfringens*, α-toxin of *S. aureus* and streptolysine of *S. pyogenes*. After pathogenic microorganisms have penetrated the horny layer, the early and nonspecific defence of the host is activated. Langerhans cells of the epidermis produce cytokines and neutrophilic granulocytes which reach the site of infection and activate the complement system. Special relevance is related to the production of phagocytes, acute-phase proteins, interferons, and ▶ natural killer cells. The interaction between phagocytes and microorganisms results in the production of several enzymes (such as proteases and hydrolases) and proteins (▶ defensins), and thus in the elimination of the microorganism. Interferons inhibit the spreading of viruses in the organism. Natural killer cells also possess antiviral activity. The acute-phase reaction leads to the expression of adhesion molecules on endothelial cells and consecutively to the adherence of granulocytes to blood vessels. Thus granulocytes are able to reach the site of infection.

The late immune response is associated with the production of antibodies and the activation of the specific cellular immunity. This results in the elimination of infectious microorganisms and in the existence of a permanent immunity. In various infections this permanent immunity prevents the occurrence of clinical symptoms after a second or repetitive contact with a microorganism related with a prior skin disease, or its toxins.

Treatment Modalities

Bacterial infections are commonly treated with antibiotics. This antimicrobial metabolic products of bacteria and fungi have bacteriostatic or bacteriocidal properties. Depending on the spectrum of bacteria they comprise, some are effective in the treatment of very few bacteria (e.g. penicillin) whilst others can be used against a broad spectrum of different bacteria (e.g. tetracycline). The bactericidal penicillins (β-lactam antibiotics) bind to so-called penicillin-binding proteins in the cytoplasm membrane, thus disturbing the synthesis of the bacterial cell membrane and leading to the elimination of penicillin-sensitive bacteria such as the streptococci. Fluorochinolones (e.g. ciprofloxacin) inhibit bacterial DNA-gyrase and thus have bactericidal activity by preventing the supercoiling of DNA. Tetracyclines have bacteriostatic properties by inhibiting the interaction of tRNS and ribosomes.

Replication of viruses is completely integrated into the host cellular metabolism. Therefore development of antiviral agents is generally complicated by the fact that suppression of viral synthesis can also affect the host cellular metabolism. Dermatological relevant antiviral substances like acyclovir are specifically effective against herpesviruses by phosphorylation of virus-specific kinases, and thereby inhibiting virus-specific DNA-polymerase which is not present in human cells. A common antifungal mechanism is the inhibition of the biosynthesis of ergosterol in the infecting fungal cell. Ergosterol is essential for the fungal metabolism. A topical or systemic application is effective depending on the infected part of the skin.

Recently developed immunomodulatory treatment with, for example, imidazoquinolines, or inhibitors of calcineurin, either stimulate or suppress the innate or acquired immune system. They have immunomodulatory properties by influencing the cellular cytokine profile, the action of T cells and dendritic cells. Their benefit for treatment of dermatological infections will be more obvious after more extensive experience has been gained.

Relevance to Humans
Bacterial Infections of the Skin

Dermatologically relevant bacteria are the gram-positive streptococci and staphylococci which are causative for pyodermas. Important for the virulence of *S. aureus* is the production of several enzymes (nucleases, proteases, lipases, catalases, hyaluronidases, lysozymes, β-lactamase, coagulases, collagenases, proteinases) and several exotoxins (hemolysins and leucocidin). After penetrating the skin *S. aureus* usually remains at the site of infection and produces a localized infection. The extracellular products lead to the formation of a pus-filled cavity. Common dermatological infections caused by *S. aureus* are mainly bound to hair follicles: folliculitis, furuncles (boils), carbuncles, and abscesses. The toxic shock syndrome (TSS) and the staphylococcal scalded skin-syndrome (SSSS) are toxin-related syndromes due to the production of an exotoxin by *S. aureus*. SSSS usually occurs in children and is characterized by the formation of big blistering lesions and the loss of large areas of the superficial layers of the epidermis. TSS is a severe disorder with fever, exanthema and multiple organ involvement.

Infections with pathogenic *S. pyogenes* (β-hemolytic streptococci) are usually not bound to hair follicles. The virulence is based on the production of extracellular enzymes and toxins like streptolysin, hyaluronidases and streptokinases. *S. pyogenes* reaches the skin through minor trauma, grows and extends into the tissue. It causes common dermatological infections like erysipelas, phlegmons, impetigo contagiosa, ecthyma, and scarlet fever. For the development of erysipelas a very small trauma which represents the entry for (in most cases) pathogenic *S. pyogenes* is usually sufficient. In contrast to erysipelas, which are localized in the corium and extend lymphogeneously, phlegmons are usually localized in the subcutis and show a characteristic diffuse spreading and abscess-forming extension. Phlegmons extend diffuse and build abscesses. Affected patients are usually severely ill.

Some infections of the skin, like phlegmons and abscesses, can represent mixed infections with streptococci and staphylococci. In addition, skin diseases like impetigo contagiosa or erysipelas can be caused by either *S. pyogenes* or *S. aureus*. Clinically, dermatological infections caused by the latter often show blistering skin lesions.

Infections of the skin caused by *S. pyogenes* can bear the risk of late complications like acute glomerulonephritis and, less frequently, ▶ rheumatic fever. Erysipelas can be complicated by the development of bacteremia and sepsis.

Other gram-positive pathogenic bacteria for the skin are *Corynebacterium tenuis* and *C. minutissimum* which are involved in the formation of several common and usually harmless skin diseases like ▶ erythrasma, trichobacteriosis palmellina, and keratoma sulcatum. Those bacteria constitute the resident flora of the skin and prefer lipid-containing and moist areas. They live superficially in the stratum corneum. The development of skin symptoms is related to a lack of hygiene and to hyperhidrosis. *Propionibacterium acnes* is involved in the etiology of acne.

Gram-negative spirochetae represent a group of bacteria involved in the pathogenesis of several well known dermatological infections: infection with *Treponema pallidum* causes syphilis and occurs through direct contact with microtraumatized skin or mucus membrane. *T. pertenue* and *T. carateum* cause dermatological infections in tropical areas (e.g. yaws and pinta). Skin manifestations of an infection with *Borrelia burgdorferi* are erythema chronicum migrans, acrodermatitis chronica atrophicans, and lymphadenosis cutis benigna. *B. burgdorferi* is transmitted by the bite of an infected tick.

Mycobacteria are causative for the development of tuberculosis (*Mycobacterium tuberculosis*) or leprosy (*M. leprae*). Cutaneous tuberculosis can occur as a manifestation of systemic tuberculosis or through direct contact with infected material. Direct contact is important for the transmission of *M. marinum* in swimming-pool granuloma.

Neisseria gonorrhoeae and *Chlamydia trachomatis* are examples of important bacteria for sexually transmitted diseases. *N. gonorrhoeae* possesses pili and surface proteins which are responsible for the adherence to mucosal tissue. It usually results in a localized tissue infiltration and a purulent infection. Clinical symptoms range from urogenital symptoms, skin manifestations, arthritis and—in rare cases—sepsis. Infection with *N. gonorrhoeae* can lead to strictures in the urogenital tract.

Fungal Infections of the Skin

Fungal infections are common in dermatological patients. Dermatophytes like *Trichophyton rubrum* infect tissues containing a lot of keratin (epidermis, hair and nail). They possess several enzymes, including keratinase, elastase and collagenase, which are able to invade and lyze the horny substance of skin and skin appendages. They are transmitted between humans and from animals (e.g cats, dogs and hamsters) to humans. The site of contact is usually the primary localization of the infection. *Candida albicans* is physiologically part of the resident flora. *C. albicans* causes opportunistic infections of skin and mucus membranes in immunosuppressed patients with impaired cellular immunity. Colonization with *Candida* is also associated with a disturbed mechanical skin barrier (e.g. intertrigo). Further common manifestations of skin infections with *C. albicans* are onychomycosis and folliculitis, as well as mucocutaneous and genital candidiasis.

An underlying fungal infection may represent a possible entry for pathogenic *S. pyogenes* and thus for the development of erysipelas.

Subcutaneous mycoses become apparent in skin and subcutaneous tissue (sporotrichosis, chromomycosis, and madura mycosis). They occur mainly in tropical areas. These pathogenic fungi reach the subcutaneous tissue through injured skin and result in chronic granulomatous infections.

Viral Infections of the Skin

The human herpesviridae (HHV-1 to HHV-8) with their dermatotropic and neurotropic properties, possess strong relevance for dermatological infections. The ability to persist lifelong in an infected host organism is a common characteristic of human herpersviridae, and is decisive for their pathogenicity. Infection depends on the host immune system and can be latent (clinically asymptomatic), or endogenously reactivated. The severity of an infection with human herpesvirus typically correlates with the severity of immune deficiency.

Characteristically, an infection with herpes simplex virus can be differentiated into primary infection, latent, or reactivated infection. Herpetiform blistering lesions are a clinical hallmark of this infection. The primary infection with herpes simples virus type I can either be asymptomatic or can, for example, in children show painful vesicles and ulcers in the oral cavity accompanied by general malaise, fever and enlargement of the lymph nodes. Herpes simplex virus type II is associated with sexually transmitted genital infection. Reactivation of latent herpes simplex virus infection usually occurs at sites related to the primary infection. Patients with severe immunodeficiency show a generalized and more severe course sometimes associated with complications like encephalitis or pneumonia. In pregnancy, infection with HHV can result in severe congenital infection in relation to the type of virus and to the point of time when infection occurs. Further infections of the skin caused by HHV are chickenpox, ▶ shingles, infectious mononucleosis, and various exanthems. Oncogenic potential has been described for HHV-8 and it is associated with the development of Kaposi sarcoma.

Poxviridae are epitheliotrop and have toxic effects on the host cells during their replication. Transmission is either aerogenically or by direct contact. Poxviridae are usually very contagious. Clinical examples are smallpox (now eradicated according to WHO), ecthyma contagiosum (orf), and infection by molluscum contagiosum virus. Generally, viral infections are responsible for various forms of exanthems (e.g. paramyxovirus causes clinical measles (rubeola). It is assumed that this virus grows in lymphatic tissue prior to hematogenous distribution.

Very common skin infections are related to human papilloma viruses (HPV). They infect exclusively epithelial cells of skin and mucosal tissue. Infection may occur after microtrauma of the skin. The viruses reach the basal layer of the epidermis and after a period of latency cause enhanced cell growth. The infectious potential is related to the immune status of the infected person, the quantity of transmitted virus, as well as the way and intensity of transmission. HPV are responsible for multiple types of verrucae, condylomata and papilloma of the skin and mucus membranes. In addition, they exhibit a potential risk factor for the development of neoplasm.

Infection with the human immunodeficiency virus (HIV) can be associated with several dermatoses which occur in relation to immune competence. Particularly in the advanced stages of immunosuppression, AIDS patients with impaired cellular immunity are prone to develop viral and fungal infections of the skin. Because of the impaired cellular immunity these infections show a severe course and frequently abnormal clinical findings.

There is a well known association between several bacteria and viruses and the initiation of various dermatoses, e.g. psoriasis and *S. pyogenes*, erythema exsudativum multiforme and herpes simplex virus or erythema nodosum and mycobacterium tuberculosis or streptococci.

A complete overview concerning relevant dermatological infections (including those without mention in the text) is shown in Table 1.

References

1. Bisno AL (1984) Cutaneous infections: microbiologic and epidemiologic considerations. Amer J Med 76:172–179
2. Roth RR, James WD (1989) Microbiology of the skin: resident flora, ecology, infection. J Amer Acad Dermatol 20:367–390
3. Sharma S, Verma KK (2001) Skin and soft tissue infection. Indian J Pediatr 68: S46–S50
4. Odds FC (1994) Pathogenesis of *Candida* infections. J Amer Acad Dermatol 31:S2–S5

Dermis

The dermis has important functions of thermoregulation and supports the vascular network to supply the avascular epidermis with nutrients. The dermis is subdivided into two zones, a papillary dermis and a reticular layer. The dermis contains mainly fibroblasts, but immune cells are also present that are involved in defense against foreign invaders passing through the epidermis.

▶ Skin, Contribution to Immunity

Desaturation

Desaturation describes the process whereby an enzyme termed a desaturase introduces a double bond between a pair of adjacent carbon atoms in a fatty acid.

▶ Fatty Acids and the Immune System

Desmosomes

These are spot-like dense plaques mostly found in epithelial cells. Intercellular adhesion in the desmosomes is mediated by desmocollins and desmogleins (adhesion molecules, members of the cadherin family) which are cytoplasmically linked to the intermediate filaments. Desmosomes are crucial to withstand the tensile and shearing forces to which tissues like epithelia are subjected.

▶ Cell Adhesion Molecules

Developmental Immunotoxicology

JOHN BARNETT
Dept of Microbiology and Immunology
West Virginia University, Health Sciences
Morgantown Center North
Morgantown, WV 26506-9177
USA

Synonyms
Prenatal immunotoxicology, in utero immunotoxicology.

Definition
Developmental immunotoxicology is defined as the study of the consequences of exposing the immune system to toxic agents during embryonic and fetal development. Thus, the qualifying term "developmental" refers directly to that period of embryonic and fetal differentiation and maturation of the immune system. This includes the ▶ organogenesis of the various anatomical structures that are required for the normal function of the immune system, such as the thymus, spleen, lymph nodes, the ▶ microenvironment of the bone marrow, Peyer's patches, lymph and blood distribution systems during prenatal development. In addition, the functionality of the immune system depends directly on the normal development of the hematopoietic systems which provides the various cellular elements of the immune system (such as lymphocytes and myeloid cells).

Although there is maturation of the immune system postnatally In the human, organogenesis and differentiation of the immune system is completed early during gestation—certainly prior to ▶ parturition. However, in mice (a common experimental animal for developmental toxicology) there is evidence that the immune system continues to develop for a short time after birth (1). Developmental immunotoxicity refers to the effect(s) of a toxicant on the immunocompetence of the fetus or offspring and has not traditionally considered the effect that various toxic agents have on implantation and placental development—these topics are usually considered as factors of reproductive toxicology, although there is evidence that failure of placental development has an immunological basis (2).

Characteristics
The basic operational premise of developmental immunotoxicity is that exposing a subject to a toxic agent during the time of organogenesis and maturation of the anatomical structures and cells associated with the immune response may lead to a detrimental effect on the functionality of the immune system. The final outcome of such an exposure depends on a number of factors that are not a consideration in adult-exposure immunotoxicity. There are at least three major factors to consider:
- the timing of exposure during gestation (or for rodents, especially the mouse, this may also include the immediate neonatal period)
- the ability of the agent to pass through the placental tissues so that the fetus is exposed to the agent
- differences in the immunological milieu of the fetus.

For each of these factors, there are species-specific considerations that will be illustrated below. The final effect of this toxic insult may (or may not) differ from the effect measured after adult exposure.

The time required for full development of the immune system varies from species to species. For humans the various elements of the immune system are fully formed early in the second trimester (approximately 13–20 weeks of gestation) (1). It is generally accepted that exposure to xenobiotics during the formative stages of immune development (prior to 20 weeks gestation) will be more detrimental; however, there is little data to support this conclusion for humans. The milestones for immune system development in animal models differ substantially, thus, requiring cautious extrapolation of animal data to humans. For example, as previously mentioned, mice do not complete their immune system development until shortly after birth. Similarly, mice do not transfer ▶ hematopoiesis from fetal liver to the bone marrow until much later in gestation—approximately 1 day before birth in the mouse compared to 15–20 weeks of gestation in the human. An intent to extrapolate the conclusions to humans is often the underlying purpose of laboratory studies using a variety of laboratory mammalian species. However, there is growing concern about the effects of chronic (perhaps constant) environmental toxicant exposure on feral animal populations. Interpretation of these experiments can be hampered by lack of information on the chronology of immune system development, as well as by comprehensive understanding of the normal immune response in many wildlife species. Developmental immunotoxicology (or developmental toxicology, in general) using mammalian species must consider the anatomical structure of the placenta, which is quite variable among species. The terminology used to describe different types of placenta varies depending on whether it is classified according to origin, shape, internal structures, relation to maternal tissues, or composition of the placental membrane (3). Humans have a hemochorial placenta consisting of no maternal membrane and two fetal membranes, a chorion and endothelium. While apes and old-world monkeys have an identical placental structure, the mouse placenta is classified as hemoendothelial (no maternal membrane and only an endothelial membrane). Other common laboratory species, such as dogs and cats, have more complex placental structures. Placental structure becomes important when considering whether maternal exposure translates into fetal exposure as well. Although organic compounds, which are often very lipophilic, pass through the placenta and into the fetus regardless of the placental structure, hydrophilic compounds may or may not, depending on the placental structure. This is well illustrated by the differences in passage of immunoglobulins from maternal circulation to the fetus. Humans actively transport several IgG subtypes through the placenta, affording high concentrations of protective antibody to the newborn at birth (4). Thus, it is possible that xenobiotics which bind to proteins may be carried into the fetus via this mechanism.

The fetal hematopoietic milieu differs substantially from the adult milieu. Helper T cells have been classified into two functional categories defined as type 1 (Th1) and type 2 (Th2) based on their cytokine production (4). Th1 cells exclusively produce interferon-γ (IFN-γ) and tumor necrosis factor-β (lymphotoxin) whereas Th2 cells exclusively produce interleukins IL-4, IL-5 and IL-10. The fetus has predominantly a Th2 cell cytokine response which is deemed important for the survival of the pregnancy because IFN-γ (produced by Th1 but not Th2 cells) activates cytotoxic T cells that could destroy the placenta (2). IL-4 (produced by Th2 but not Th1 cells) functions as a "switching" factor, that can cause B cells to switch from IgG production to IgE production. IgE is the antibody type that is responsible for immediate hypersensitivity reactions, such as atopy and systemic hypersensitivity (anaphylaxis) (4). It has been hypothesized that this skewed T-helper response indicates that the fetus may be more susceptible to developing hypersensitivity or autoimmune responses to toxicants which are also antigenic or haptenogenic, capable of passing through the placental barrier into the fetus. It is well established that the human fetus is capable of developing an IgE antibody response prior to birth (2). Some immunotoxic compounds are capable of inducing an immediate hypersensitivity response and if they are also capable of passing through the placental barrier, they may also induce fetal immediate hypersensitivity.

Thus, developmental immunotoxicology must consider several factors that are unique to the fetus or neonate. These include:

- possible interruption of normal organogenesis or maturation of the cells during fetal life
- alteration of the signals associated with organogenesis and/or maturation of ▶ immunocompetent cells
- potential for a toxicant to reach the fetus
- the uniquely skewed T cells populations present during gestation.

The immunotoxic response to a number of toxicants has been demonstrated to differ depending on whether the exposure occurs during development (as defined herein) or as an adult. These differences have been effectively reviewed by Holladay & Smailowicz (1).

Preclinical Relevance

Guidelines for developmental immunotoxicology testing have recently been proposed (5). They provide a summary of the recommended immunotoxicology assays as well as a schematic diagram of proposed experiments. Because of the extensive use of the rat for preclinical developmental toxicology testing, particular emphasis was placed on developing experimental designs that use the rat as the test animal, so that immunotoxicology testing could be incorporated into existing protocols.

Relevance to Humans

Obviously no-one would deliberately expose a human fetus to an environmental toxicant to ascertain its developmental immunotoxicity. Therefore, all human studies demonstrating a detrimental effect of prenatal exposure to environmental toxicants on immune function are the result of some inadvertent or accidental exposure. These studies are often hampered by a multitude of factors; for example, a wide range of ages of the exposed population at the time of assessment, or uncertainty about the actual concentrations of the toxicant to which the population has been exposed making dose-effect conclusions tenuous at best. Despite these difficulties several studies have attempted to correlate the exposure to an environmental toxicant during human fetal development with immunotoxicity. As a result of these studies, there is highly suggestive evidence that humans are as susceptible to developmental immunotoxicity as animal models. These studies have been reviewed by Typhonas (6). This review does not include the outcome of incidental prenatal fetal exposures that result from therapeutic treatment of pregnant women (e.g. for neoplasia, psychiatric disorders, drug abuse) although there is extensive literature on these topics.

References

1. Holladay SD, Smialowicz RJ (2000) Development of the murine and human immune system: differential effects of immunotoxicants depend on time of exposure. Environ Health Perspect 108 [Suppl 3]:463–473
2. Bjorksten B (1999) The intrauterine and postnatal environments. J Allergy Clin Immunol 104:1119–1127
3. Ramsey EM (1982) The Placenta: Human and Animal. Praeger, New York

4. Janeway CAJ et al. (2001) Immunobiology; The Immune System in Health and Disease. Garland, New York
5. Luster MI, Dean JH, Germolec DR (2003) Consensus workshop on methods to evaluate developmental immunotoxicity. Environ Health Persp 111:579–583
6. Tryphonas H (2001) Approaches to detecting immunotoxic effects of environmental contaminants in humans. Environ Health Persp 109:877–884

Dexamethasone

A potent synthetic glucocorticosteroid with a relatively long biological half-life. Because it is not bound by corticosteroid-binding globulin, its greater bioavailability accounts for greater in vivo potency than natural glucocorticosteroid. Dexamethasone crosses the placenta and is not inactivated by fetal enzymes, and is therefore biological activity in the fetus.
▶ Glucocorticoids

DHA

▶ Fatty Acids and the Immune System

Diabetes and Diabetes Combined with Hypertension, Experimental Models for

PA van Zwieten
Departments of Pharmacotherapy, Cardiology, Cardio-Thoracic Surgery, Academic Medical Centre
University of Amsterdam
Meibergdreef 9
1105 AZ Amsterdam
The Netherlands

Synonyms
Obese Zucker rat,Zucker rat; streptozotocin-induced diabetes mellitus, STZ-induced diabetic rat; streptozotocin-induced spontaneously hypertensive rat,STZ-SHR rat.

Diabetes Mellitus: Introduction
Diabetes mellitus (in particular, type 2) and hypertension are both frequently occurring diseases. Obesity and increasing age are well known risk factors for both diseases. Furthermore, both diabetes mellitus and hypertension are highly relevant risk factors for well known cardiovascular and renal complications, such as coronary heart disease, congestive heart failure, stroke, and renal damage (nephropathy).
Hypertension and diabetes mellitus (in particular type 2) frequently occur simultaneously, particularly in obese individuals. Since several of the aforementioned sequence of both diseases can be influenced favorably by drug treatment, the necessity to develop animal models for both diseases and their combination has been felt since several decades.
In the present survey we will briefly review a few relevant animal models for diabetes mellitus (types 1 and 2), hypertension, and the combination of both diseases in one and the same model.
Without attempting completeness we will emphasize a few models that are well known to us and suitable for pharmacological and pathophysiological investigations.

Diabetic Animals: General Aspects
Diabetic animals are now readily available for research purposes. Experimental diabetes resembling insulin-dependent diabetes mellitus (IDDM, type 1 diabetes) can be induced in rodents, for instance by means of alloxan or (preferably) streptozotocin (STZ). An IDDM-like syndrome is observed in animal strains which spontaneously develop diabetes, such as non-obese diabetic (NOD) mice, BioBreeding (BB) rats, Long-Evans Tokushima-Lean (LETL) rats, Chinese hamsters, and Keeshond dogs. non-insulin-dependent diabetes mellitus (NIDDM) (type 2) syndromes in animals occur predominantly on a genetic basis. A few examples of such animals are:
- db/db mouse (obese, severe diabetes)
- ob/ob mouse (obese, mild diabetes)
- New Zealand obese mouse (obese, mild diabetes)
- obese Zucker rat (obese, mild diabetes)
- WKY fatty rat (obese, mild diabetes)
- Swiss-Hauschka mouse (diabetes developed by selective inbreeding)
- Cohen diabetic rat (diabetes developed by selective inbreeding)
- Koletsky rat (SHROB) (obese, diabetic, hypertensive)

An NIDDM-like syndrome can also be induced by nutrition, such as in the sand rat, the spiny mouse, and the Mongolian gerbil. The literature contains several reviews on animal models of diabetes (1–4).
We will discuss the streptozotocin-induced diabetic rat and the obese Zucker rat in some more detail because these models are widely used in experimental pharmacology and medicine.

Streptozotocin-Induced Diabetes Mellitus: Short Description
Streptozotocin (STZ) is an antibiotic cytostatic agent,

extracted from *Streptomyces achromogenes*, that has been used since 1963 to produce diabetes in experimental animals, particularly in rats. STZ is cytotoxic with respect to pancreatic islet β cells. STZ alkylates DNA in the β cells, thus leading to functional deterioration of these cells, and then to diabetes, which resembles human type 1 diabetes (IDDM), although insulin treatment is not necessary for the survival of the animals.

Characteristics
The degree of β cell deterioration is proportional to the dosage of STZ. According to our own experience, optimal results are obtained by the intravenous injection (into a lateral tail vein) of 55 mg/kg STZ in rats aged 12 weeks (body weight 260–280 g). Hyperglycemia develops in 1–2 days which is definitely stable at a level around 18 mmol/l or more after 8 weeks. The increase in body weight over these 8 weeks is significantly retarded when compared with non-treated control animals. Treatment with insulin is not necessary for survival (5,6).

Pros and Cons
The STZ model has been used widely and under well controlled conditions. Accordingly, much experience and reference data are available. The model is reproducible and easy to handle for pharmacological and biochemical investigations (5,6).

The STZ model is preferred to the formerly used model of alloxan-induced diabetes since STZ displays a more selective toxicity to the pancreatic β cells than does alloxan. The animals usually develop increased diuresis and diarrhea leading to changes in plasma volume and the loss of sodium and potassium ions.

Predictivity
The STZ model resembles human type 1 diabetes (IDDM) although not in all aspects, as there appears to be no need for insulin treatment in order to keep the animals alive.

Relevance to Humans
See Predictivity.

Regulatory Environment
The STZ model is also widely accepted by ethical committees for experimental animals.

The Obese Zucker Rat: Short Description
The obese Zucker rat (7–9) was first described by Zucker in 1965. It is a result of cross-breeding between Sherman and Merck Stock M rats. Obesity is inherited as an autosomal recessive mutation with the name *fa*. Homozygous rats develop hyperphagia and extreme obesity. Obesity is evident at 4 weeks of age, but many of the biochemical abnormalities and the size and metabolism of the rat tissue are already altered in the pre-weaning period. Hyperphagia, low physical activity, and increased efficiency of energy utilization all lead to a positive energy balance in these animals. The excess energy is stored in the adipose tissue which reflects abnormal metabolism. The mass of the fat cells increases as a result of both cellular hypertrophy and hyperplasia. The obese Zucker rats develop insulin resistance, hyperinsulinemia, glucose intolerance, hypertension, and hyperlipidemia along with their obesity. They continue to be hyperinsulinemic and demonstrate only mild hyperglycemia without progression to overt diabetes. Anatomically, the pancreatic islets are large because of both hyperplasia and hypertrophy of the β cells. The hypothesis that abnormal insulin secretion is the primary defect in the obesity syndrome is supported by the finding of insulin hypersecretion in the young fat rats. Insulin secreting β cells, glucagon secreting α cells, and somatostatin secreting δ cells all react abnormally to secretagogues. Insulin has an anorectic effect in lean but not in obese rats. Thus, in the obese Zucker rat, appetite is not suppressed despite the hyperinsulinemia, a condition leading to continuous and uninhibited food intake. It has been submitted that the primary defect is located in the central nervous system. A central defect, transmitted via the vagus nerve, would thus lead to insulin secretion and hyperinsulinemia, causing obesity and insulin resistance.

Lean Zucker rats are the genetic counterpart of the obese Zucker rats. The lean Zucker rats do not develop hyperglycemia or obesity. As such they are useful control animals for studies in obese Zucker rats.

Characteristics
Obese *(fa/fa)* and lean *(fa/)* Zucker rats can be obtained commercially (for example, from Harlan Olac, Bicester, UK) usually at an age of approximately 9 weeks. The time course of body weight and plasma glucose values of both categories of Zucker rats are listed in Table 1.

Our own experiences (6,10) have proved that at 9 weeks of age the body weights of the obese Zucker rats are already significantly higher than those of their age-matched controls, whereas the plasma glucose values are the same. The obese Zucker rats rapidly increase body weight, and this remains significantly higher than that of the lean controls (observations up to 22 weeks of age). Plasma glucose values in obese rats obtained at the age of 14, 18, and 22 weeks progressively increase and are significantly higher than in lean control animals.

In these observations, the obese animals gained body weight due to an increase of adipose tissue, in parti-

Diabetes and Diabetes Combined with Hypertension, Experimental Models for. Table 1 Time course of body weight (gram) and plasma glucose values (mmol/l) in lean and obese Zucker rats

	Age (weeks)	Lean Zucker rat	Obese Zucker rat
Body weight (g)	9	237.5 ± 4.2	338.1 ± 6.8*
	14	331.5 ± 5.1	490.3 ± 8.0*
	18	369.4 ± 6.3	531.5 ± 11.0*
	22	407.0 ± 5.2	587.0 ± 12.3*
Plasma glucose (mmol/l)	9	7.2 ± 0.1	7.7 ± 0.2
	14	7.1 ± 0.2	8.4 ± 0.3*
	18	7.6 ± 0.2	9.4 ± 0.6*
	22	6.2 ± 0.2	11.0 ± 0.8*

Data presented as means ± standard error mean (SEM), n = 35.
Data modified from Kam et al (10).
* $P < 0.05$ vs lean Zucker rat.

cular in the abdominal region. There was no loss of animals due to premature death.

Plasma insulin and serum lipid levels in obese Zucker rats were significantly higher than in lean Zucker rats. The blood pressure parameters of the obese Zucker rats were moderately but significantly higher than in the control rats. However, heart rates proved significantly lower in the obese rats than in the lean animals.

Pros and Cons

The obese Zucker rat model approaches the syndrome of diabetes mellitus type 2, combined with moderate hypertension. An advantage is the availability of highly relevant control animals which are devoid of obesity and hyperglycemia (lean Zucker rat). The model is reproducible and easy to handle. The somewhat high cost would be a disadvantage of large-scale use.

Predictivity

In brief, the obese Zucker rat develops the following symptoms:
- obesity, as a result of a positive balance induced by both hyperphagia and low physical activity
- insulin resistance, hyperinsulinemia, and glucose intolerance
- hyperlipidemia
- arterial hypertension (mild).

This combination of symptoms closely resembles the profile of human type 2 diabetes combined with obesity and (mild-to-moderate) hypertension.

Relevance to Humans

See Predictivity.

Regulatory Environment

The obese Zucker rat is widely accepted as a relevant model by ethical committees for experimental animals.

Description

Hypertensive Animals

Hypertensive disease can be acquired on the basis of DOCA-salt exposure, whereby one kidney is removed, deoxycorticosterone acetate (DOCA) is injected and 1% saline supplied as drinking water to cause an impairment to excrete a sodium load. Other experimental models are the Goldblatt 2 kidney/1 clip hypertension model and the induced aortic stenosis rat (ASR) model. In the spontaneously hypertensive rat (SHR) and the spontaneously hypertensive stroke-prone rat (SHR-SP), hypertension develops on a genetic basis. Other inbred genetic models of hypertension include the Dahl salt-sensitive rat, the New Zealand genetically hypertensive rat, the Lyon hypertensive rat, the Milan hypertensive rat, the Sabre hypertensive rat, and the spontaneously hypertensive mouse.

A vast body of literature concerns the very well known animal model of the SHR. For a review on hypertension models the reader is referred to the monograph by de Jong (11).

Hypertensive Diabetic Animals: General Aspects

As outlined in the Introduction there exists a need for experimental models where diabetes and hypertension occur simultaneously in the same animal. As discussed above (Hypertensive Animals) the obese Zucker rat is obese, diabetic, and usually mildly hypertensive.

The obese spontaneously hypertensive rat (SHROB, Koletsky rat) is not to be discussed here in detail; the condition resembles the human syndrome X, with obvious hypertension, obesity, hyperinsulinemia, and glucose intolerance (12).

An animal model that has been recently introduced is the spontaneously hypertensive rat made diabetic by the administration of streptozotocin (SHR).

The Streptozotocin-Induced Spontaneously Hypertensive Rat: Short Description

Spontaneously hypertensive rats (SHR) can be made diabetic by the administration of streptozotocin (STZ), in a manner similar to that described for normotensive rats. The STZ-SHR model developed for fundamental pharmacological and biochemical studies is easy to handle and is reproducible.

Characteristics

Spontaneously hypertensive rats (SHR) aged 12 weeks (body weight 260–280 g) are treated with an intravenous injection of 55 mg/kg STZ. As described for normotensive STZ rats, the animals are kept for 8 weeks. Their stable plasma level of blood glucose then amounts to 18 mmol/l or more. The increase in body weight of the STZ-SHR is then significantly less than that of normotensive control rats (see Figure 1). STZ-WKY (normotensive) rats are used as controls (5,6,12,13).

Pros and Cons

In contrast to the genetic models of hypertension and diabetes (obese Zucker rat, Koletsky rats) the STZ-SHR is very easy to obtain under conventional laboratory conditions and is also much cheaper than the genetic models. STZ-WKY rats are suitable controls. The STZ-SHR model has proved reproducible and suitable for pharmacological studies.

A possible disadvantage is the fact that STZ diabetes resembles type 1 more than type 2, whereas in humans the simultaneous occurrence of hypertension is more frequently associated with type 2 diabetes.

Predictivity

The STZ-SHR model resembles human type 1 diabetes together with hypertension, although not in all aspects. As in normotensive STZ animals, there appears to be no need for insulin treatment in order to keep the animals alive.

Relevance to Humans

See Predictivity.

Regulatory Environment

The STZ-SHR model is widely accepted also by ethical committees for experimental animals.

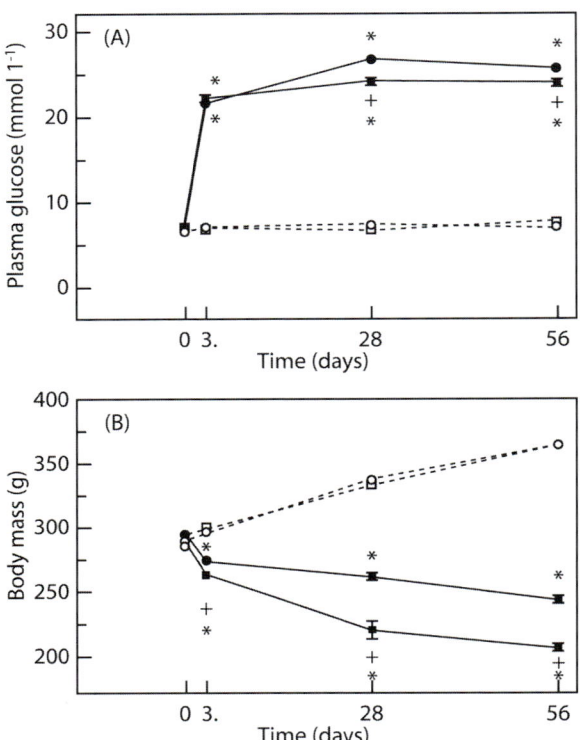

Diabetes and Diabetes Combined with Hypertension, Experimental Models for. Figure 1
(A) Mean plasma glucose levels (mmol/l) and (B) body mass (g) established at several times (days) just before and after the injection of streptozotocin (STZ) (day 0) or vehicle in the four groups of rats: STZ-induced diabetic normotensive (STZ–WKY) and hypertensive (STZ–SHR) rats and age-matched control rats (WKY and SHR), respectively. Data represent means ± SEM, $n \geq 28$, $P < 0.05$ vs control (WKY or SHR), + $P < 0.05$ vs WKY (STZ or control); ○ control WKY; ● STZ–WKY, □ control SHR, ■ STZ–SHR. Data from the author's department.

References

1. Fischer LJ (1985) Drugs and chemicals that produce diabetes. Trends Pharmacol Sci 6:72–75
2. Chappel CI, Chappel WR (1983) The discovery and development of the BB rat colony: an animal model of spontaneous diabetes mellitus. Metabolism 32:8–10
3. Factor SM, Bhan RAJ, Minase T, Wolinsky H, Sonnenblick EH (1981) Hypertensive diabetic cardiomyopathy in the rat, an experimental model of human disease. Am J Pathol 102:219–228
4. Karasik A, Hattori M (1994) Use of animal models in the study of diabetes. In: Kahn CR, Weir GC (eds) Joslin's Diabetes Mellitus, 13th edn. Lea & Febiger, Philadelphia, pp 217–350
5. Kam KL, Hendriks MGC, Pijl AJ et al. (1996) Contractile responses to various stimuli in isolated resistance vessels from simultaneously hypertensive and streptozotocin-diabetes rats. J Cardiovasc Pharmacol 27:167–175

6. van Zwieten PA, Kam KL, Pijl AJ et al. (1996) Hypertensive rats in pharmacological studies. Pharmacol Res 33:95–105
7. Bray GA (1977) The Zucker-fatty rat: a review. Fed Proc 36:148–153
8. Clark JB, Palmer J, Shaw WN (1983) The diabetic Zucker fatty rat (41611). Proc Soc Exp Biol Med 183:68–75
9. Kurtz TW, Morris RC, Parshadsingh HA (1989) The Zucker fatty rat as a genetic model of obesity and hypertension. Hypertension 13:896–901
10. Kam KL, Pfaffendorf M, van Zwieten PA (1996) Pharmacodynamic behaviour of isolated resistance vessels obtained from hypertensive-diabetes rats. Fund Clin Pharmacol 10:329–336
11. de Jong W (1984) Experimental and genetic models of hypertension. In: Birkenhäger WH, Reid JL (eds) Handbook of Hypertension, Volume 4. Elsevier, Amsterdam
12. Beenen OHM, Pfaffendorf M, van Zwieten PA (1995) Contractible responses to various inotropic agents in isolated hearts obtained from hypertensive diabetic rats. Blood Press 4:372–378
13. van Zwieten PA (1999) Diabetes and hypertension: experimental models for pharmacological studies. Clin Exp Hypertensi 21:1–16

Diethylstilbestrol (DES)

A model endocrine disruptor, originally administered to pregnant women in THE hope that it would prevent miscarriage. However DES exposure has been shown to have negative health effects in both the mothers and prenatally exposed children. DES has been associated with an increased risk to clear cell adenocarcinoma (a rare vaginal/cervical cancer) as well as significant increases in autoimmune diseases.
▶ Steroid Hormones and their Effect on the Immune System

Differentiation

The process of primitive cells (e.g. hematopoietic stem cells) to acquire characteristics such as cell surface markers and intracellular proteins which are specific for a mature cell type.
▶ Bone Marrow and Hematopoiesis
▶ Hematopoietic Stem Cells

DiGeorge Syndrome

A genetic syndrome in which patients are born with a very small or even no thymus. Depending on the severity, patients lack T cells and have very poor or no T-cell dependent responses to antigens.
▶ Thymus: A Mediator of T Cell Development and Potential Target of Toxicclogical Agents

Dilated Cardiomyopathy

Enlargement of the chambers of the heart.
▶ Cardiac Disease, Autoimmune

7,12-dimethylbenz-a-anthracene

▶ Polycyclic Aromatic Hycrocarbons (PAHs) and the Immune System

DIOC18$_3$

A lipophilic carbocyanine membrane dye that can be used to label either effector or target cells for flow cytometric-based cytotoxicity assays.
▶ Cytotoxicity Assays

Dioxin Response Elements (DRE)

The AhR-dependent gene promoter regions which bind xenobiotics and initiate the transcription of genes responsible for their metabolism and biologic activity.
▶ Polycyclic Aromatic Hydrocarbons (PAHs) and the Immune System

Dioxins

▶ Dioxins and the Immune System

Dioxins and the Immune System

Charlotte Esser
Inst. für Umweltmedizinische Forschung
Auf'm Hennekamp 50
40225 Düsseldorf
Germany

Synonyms

Aryl hydrocarbon receptor nuclear translocator, ARNT, hypoxia-inducible factor HIF1-β, polychlorinated dibenzodioxins, PCDDs, dioxins, 2,3,7,8-tetrachlorodibenzo-p-dioxin, TCDD, Seveso-Dioxin, Seveso-Poison, CAS number 17646-01-6-

Definition

Dioxins belong to the chemical class of poylchlorinated aromatic hydrocarbons, along with furans and biphenyls. Seventy-five dioxins exist, differing in the number and position of chlor atoms on the backbone of the molecule. According to this, various dioxins have different shapes (e.g. planar) and different toxicities. The term dioxin is often used to refer to the most widely studied and most toxic dioxin: 2,3,7,8-tetrachlorodibenzo-*p*-dioxin (TCDD). Dioxins are usually found in mixtures, so complicating risk evaluation. Therefore, the concept of toxic equivalents (TEQ) has been developed for dioxins, furans, and biphenyls. As the most toxic compound, 2,3,7,8-TCDD has been assigned a toxic equivalency factor of 1.0, all other compounds are related to this; thus the total toxic equivalence to TCDD of a mixture can be calculated. The concept is based on the assumption that the mechanism of action is the same for all of these compounds, albeit the strength differs. The common mechanism of action underlying dioxin toxicity is the dioxin-mediated activation of an endogenous transcription factor, the ▶ aryl hydrocarbon receptor (AhR).

Dioxins can be produced by natural processes such as forest fires, but the major source appears to be human activities. Dioxins can form during combustion processes, e.g. by waste incineration or burning fuels. They are also generated in certain types of chemical manufacturing and metallurgy, they are present in car exhausts and even in cigarette smoke. Dioxin emission can be (and has been in some countries) reduced significantly by appropriate technical measures.

Acute and chronic toxic effects of dioxins include carcinogenicity, teratogenicity, interference with lipid metabolism, chloracne, neurobehavioral effects, endocrine disruption, and immunotoxicity (as described here). A number of biochemical and molecular effects—which may or may not lead to adverse health effects—are hallmarks of dioxin exposure. These include induction of many genes and oxidative stress.

Molecular Characteristics

Colorless crystals
Molecular weight: 322
Vapor pressure: 2×10^{-7} Pa (at 25°C)
Melting point: 305°C
Water solubility: 8–200 ng/L (25°C)
High thermal stability: destruction starts above 800°C
Biological half-life in humans: 5.8 years

2,3,7,8-TCDD is a symmetric and planar molecule of two unsaturated rings connected by a third ring with two oxygens (oxin bridges). Because of its size and shape 2,3,7,8-TCDD fits well into the $1.4 \times 1.2 \times 0.5$ nm ligand receiving groove of the ▶ aryl hydrocarbon receptor, an endogenous transcription factor. The affinity of the binding is very high, its K_d is 10^{-9} to 10^{-12} M, depending on the genetic isoform of the aryl hydrocarbon receptor. Binding to the aryl hydrocarbon receptor leads to conformational changes of the protein-ligand complex, nuclear translocation, heterodimerization with the partner protein ARNT (aryl hydrocarbon receptor nuclear translocator), and eventually to specific DNA-binding and gene transcription of target genes.

Dioxins are lipophilic. As a consequence they accumulate preferentially in adipose tissue, and in the milk of lactating animals (also in nursing women). Dioxins easily cross cell membranes and diffuse into cells without the need of active transport. The tissue distribution of dioxins within the body depends on the level of exposure. However, no strict correlation between tissue concentration and adverse effects on that tissue can be made. Apart from adipose tissue, TCDD is found in blood, liver, skin, muscle, thyroid and adrenal gland, and pancreas—but not in the brain. The aryl hydrocarbon receptor is also differentially abundant: high in liver, lung, and lymphoid tissue, low in muscle or brain tissue. Interaction of TCDD with the aryl hydrocarbon receptor to achieve transcriptional gene activation is controlled at many levels, e.g. at cell stage, or competition for additional DNA-binding factors. This has to be considered in any interpretation of molecular and cellular events by TCDD.

Putative Interaction with the Immune System

In all animals studied so far dioxins lead to a number of adverse biological effects. Notable systemic effects include teratogenesis, tumor promotion, thymus atrophy, immunosuppression, cachexia, dermatological problems, cardiac problems, endocrine disruption, cancer, or changes in cell proliferation. Not all effects are found in all animals though, and not to equal extents, or at similar doses. Acute lethal toxicity varies from 1 μg/kg body weight for guinea pigs to more than 1000 μg/kg for gold hamsters. For humans the acute lethal dose is not known. The immune system is a highly sensitive target of TCDD, and ▶ immunosuppression is a hallmark of even low doses of dioxin exposure in all laboratory species. A dose of only 10 ng TCDD/kg body weight suffices in mice to induce ▶ thymus atrophy. The prime underlying basic mechanism of dioxin toxicity is its binding to and activating of the aryl hydrocarbon receptor. Mice genetically engineered to lack this receptor were relatively unaffected in liver and thymus by 2000 μg/kg body weight TCDD, a dose 20 times as high as the the normal median lethal dose (LD_{50}). However, some limited effects of high-dose TCDD were seen, suggesting other, additional pathways of TCDD action. AhR-independent effects of TCDD have been inferred also from other observations, but their extent and importance relating to TCDD-toxicity is still debated.

Immunosuppression

Immunusuppression by dioxins has been measured in several ways, analyzing different cell types and response types. In general, dioxin causes immunosuppression, although results have been negative in some particular assays. For instance exposed mice are more susceptible to bacterial and viral infections, and are positive in a classical test for immunosuppression— the contact hypersensitivity assay. Studies in rodents and some nonhuman primates have shown that over a wide range of doses the function of various cell types tested was impaired, especially T cells and B cells. The ▶ innate immune response (phagocytic activity and natural killer cell activity) seems to be unaffected by TCDD. Hemopoietic progenitor cells have also been identified as targets of TCDD with consequences for the T-cell compartment. In vivo the functions of T cells are affected, yet in vitro no direct effects of TCDD were observed.

Thymus Atrophy and TCDD-Induced Changes in the Thymus

Thymic atrophy (measured as a reduction of thymus weight and cellularity) is a particularly sensitive marker of TCDD exposure, and is found in all species analyzed so far. Several mechanisms contribute to TCDD-induced thymus atrophy of about 80%; one is enhanced apoptosis of cortical thymocytes. Whether or not apoptosis is the primary reason for thymus atrophy is still under debate. The other possible cause is the reduction of prothymocytes, and the proliferation rates of the immature thymocyte precursor pool; yet another an observed block in differentiation. Thymus stroma cells, which are necessary for thymocyte development, are also targets of TCDD. Moreover, TCDD accelerates maturation and skews thymocyte differentiation towards mature $CD4^-CD8^+$ cells, the future cytotoxic T cells.

Conceivably, such changes in T cell differentiation in the thymus can affect the functional capacity of T cells. Thus, they might become either incapable of mounting a normal immune response, or the selection processes against potential autoreactive T cells are disturbed in the thymus. Despite the manifold interference in thymocyte maturation and differentiation by TCDD, no evidence has been found for the latter possibility in mice so far. Moreover, an important mechanism to maintain tolerance outside of the thymus— the so-called activation-induced cell death—is enhanced rather than decreased. However, in mice from a strain prone to autoimmunity, which were exposed to TCDD during gestation, postnatal ▶ autoimmunity was exacerbated. TCDD can cross the placenta, thus it is conceivable that TCDD interferes with the developing immune system of the fetus.

T cells as Targets

As early as 1981, T cells were recognized as major target cells of immunosuppression by TCDD. T cells integrate the ▶ humoral immune response by providing help to B cells, they are immune reactive as killer cells, and they act as T-suppressor cells. An immunotoxic agent like TCDD may kill T cells directly or change the antigen-specific immune responses of the T cells, for instance through removal or introduction of inhibitory signals on T cells. For both scenarios evidence exists, that is for preferentially enhanced apoptosis of T cells by TCDD exposure, and for hyporesponsiveness of of cytotoxic T cells. Expression of costimulatory molecules and of cytokines was decreased in the latter experimental situation, completing the immunosuppressive picture.

B cells as Targets

B cells are sensitive target of TCDD action. T cell-independent as well as T cell-dependent antibody responses are suppressed by TCDD. For instance, the AhR is directly involved in the inhibition of IgM production through Xenobiotic response element (XRE elements) in the 3′μ-heavy chain enhancer. Again, the effects of TCDD are stage-specific, differing for precursor B cells and effector B cells. Fewer B-cell precursors are generated in TCDD-treated mice. On a systemic level, in a study in rats TCDD decreased allergic responses.

Modulation of Gene Expression by Dioxin

Due to its capacity to activate aberrantly the aryl hydrocarbon receptor, TCDD leads to changes in gene expression of genes with the AhR-binding sequence XRE in their promoters. Indeed, hundreds of genes can be targets of the AhR as shown on a one-by-one basis and now also by ▶ chip arrays. As stressed above, TCDD action is specific for cell type and cell stage. In immune cells, many TCDD-inducible genes have been identified now by using various methods. Genes responsive to TCDD include, for instance, cytokine genes (interleukins IL-1β and IL-2, and tumor necrosis factor TNFα), apoptosis genes (FasL, TRAIL), differentiation markers (Notch-1, CD44, CD69), costimulatory signals (CD40, CD80), depending on the cell type and cell differentiation stage.

Relevance to Humans

Exposure

The major route of exposure to dioxins is oral uptake with food. Exposure to dioxins in the environment has led to an average body burden of 4–20 ng/kg TCDD in adults. The current daily intake of toxic equivalents is 1–3 pg TEQ/kg body weight, for breast-fed babies it can be 100 pg TEQ/kg. Suggested tolerable daily intake doses range from 1–4 pg TEQ/kg/day (WHO) to 0.006 pg/kg/day (US Environmental Protection Agency) depending on the risk assessment approach employed. The endpoint in these models is carcinogenicity but there is growing concern over noncancer effects. Whereas immunological endpoints are easily detected in animals, in humans the effects of dioxin exposure on the immune system are much less clear. Effects of dioxins on the immune system of humans have been studied after occupational exposure, after major accidents, and environmental exposure (see Table 1). Thus, data on short-term and comparatively high acute exposure and medium level to low-level chronic exposure are available. Unborn children exposed in utero, breast-fed infants and adults have been considered separately. Results are not always conclusive due to small numbers of subjects, the problem of meaningful diagnostic parameters for "immune effects", and the lack of exposure data needed for correlating effects.

In vitro Studies with Human Cells

Controlled exposure studies in vitro with lymphoid cells have been done, and show that dioxins inhibit the functional differentiation of ▶ dendritic cells, and alter calcium homeostasis in B cells, T cells and monocytes. TCDD suppressed mitogen-induced proliferation and IgM secretion by low-density tonsillar lymphocytes. However, proliferation of peripheral blood lymphocytes was not affected in all studies. Contradictory results on changes in blood lymphocyte subsets exposed to dioxins in vitro have also been reported. Which in vitro parameters are suitable as biomarkers of exposure, or as indicators of immunosuppressive action of TCDD, remains unresolved.

Epidemiological Studies and in vivo Parameters

Many groups of dioxin-exposed people have been studied, even though only a few data on immune functions are available. Table 1 lists parameters studied in various groups where TEQ levels are known and can be correlated to observed effects. In some cases, phenotypic effects were detectable even decades after exposure. Immunosuppression by TCDD has been evaluated in epidemiological studies and, indeed, in exposed children a higher incidence of infections was reported. Autoimmune reactions or allergies have also been suggested as a consequence of dioxin exposure. Human epidemiological data—looking at adults only—are not conclusive in this respect, and rather suggest that ▶ autoimmunity (e.g. an increase in autoimmune diabetes) is not among the effects of TCDD exposure. Again, comparability and power of the studies are hampered by the complex nature of the immune system, and by the lack of a prominent and simple marker of impaired immune status. Also, many studies were carried out years after exposure, when immunotoxic effects might have been no longer measurable. Note that in none of the studies the important role of the aryl hydrocarbon receptor phenotype (i.e. the genetic variation) for dioxin toxicity could be considered. Further, exposure in some incidents included polychlorinated biphenyls and dibenzofurans. These substances may have AhR-independent toxic effects other than those of TCDD.

Risk to Infants

The developing immune system is the most sensitive target in animal studies. The available epidemiological evidence indicates that infectious disease is increased in childhood, especially when exposure to contaminants takes place in the period of pregnancy and nursing. An increased rate of infections is a sign of immunosuppression. Common infections acquired early in life may prevent the development of allergies, and this might explain the reduced prevalence of allergies in dioxin-exposed children. Comparative and follow-up epidemiological studies and clinical examination of infants and children at risk during upbringing are still needed, as safety factors used in risk assessment should consider children as the most sensitive part of the population.

References

1. Holsapple M et al. (1991) A review of 2,3,7,8-tetrachlorodibenzo-*p*-dioxin-induced changes in immunocompetence: 1991 update. Toxicology 69:219–255
2. Kerkvliet N (1995) Immunological effects of chlorinated dibenzo-*p*-dioxins. Environ Health Persp 103 [Suppl 9]:47–53
3. Holladay SD (1999) Prenatal immunotoxicant exposure and postnatal autoimmune disease. Environ Health Persp 107 [Suppl. 5]:687–691

Dioxins and the Immune System. Table 1 Reports on effects on the immune system in subjects with increased body burden of dioxins, furans, or biphenyls

Group	Parameter analysed	Effects
Vietnam Veterans (exposed 1962–71)	Delayed-type hypersensitivity Lymphocyte subsets Serum IgA Autoantibodies	None None Increased None
Occupational exposure (Coalite Oils, Great Britain)[1]	Lymphocyte subsets Natural killer cells Antinuclear antibodies	Normal Higher Higher
Occupational exposure (1951–72) at two plants	Lymphocyte counts Mitogen response Serum immunoglobulin Complement C3	Fewer $CD26^+$ T cells Decreased spontaneous proliferation Normal Normal
Occupational exposure (Bayer Ürdingen, Germany)	T-helper cell function	Decreased
Occupational exposure (Japan, Nose Bike Center)	Natural killer activity Mitogen stimulation History of allergy	Increased Decreased Increased
Environmental exposure (Missouri)*	$CD8^+$ T cell count Mitogen proliferation Immunoglobulin IgA/IgG levels	Increased Normal Increased/normal
Japanese babies, breast-fed	CD4/CD8 ratio in blood lymphocytes	Changed
Yucheng/Taiwan children, prenatal exposure	Incidence of infection Lymphocyte subsets Serum immunoglobulins	Higher incidence Normal Normal
Yucho, Japan	Serum immunoglobulins Autoantibodies Lymphocyte subsets	Normal Possible increase in rheumatoid factor Normal
Two Dutch cohorts of children (prenatal/postnatal exposure)	Incidence of infection Incidence of allergy Specific antibodies in serum Lymphocyte subsets	Higher incidence Lower incidence None None
Seveso children	Lymphocyte subsets Serum immunoglobulins Complement Mitogen stimulation	Increased Normal/increased Normal Normal/increased
Flemish adolescents	Blood cell counts Lymphocyte phenotypes Serum immunoglobulins	Eosinophil changes Natural killer cell changes IgA increase

Note of caution: For a detailed interpretation, and any valid comparison between studies the original data must be consulted; the number of exposed persons in these studies varied from 11 to 281.
* In these studies body burden of toxic equivalents (TEQ) was not measured.

4. van den Berg M, Peterson RE, Schrenk D (2000) Human risk assessment and TEFs. Food Addit Contam 17:347–58

5. Poellinger L (2000) Mechanistic aspects—the dioxin (aryl hydrocarbon) receptor. Food Addit Contam 17:261–266

Direct Antiglobulin Test

▶ Antiglobulin (Coombs) Test

Direct Coombs test

The direct Coombs test (or the direct antiglobulin test, DAT) is used to detect the presence of antibodies or complement on erythrocytes.

▶ Antiglobulin (Coombs) Test

Disseminated Intravascular Coagulation (DIC)

A life-threatening bleeding disorder presenting with skin and mucous membrane bleeding, hemorrhage, and peripheral microembolism. It is triggered by necrotic material, trauma, tumors, and other conditions. Patients have thrombocytopenia, prolonged coagulation parameters, and reduced fibrinogen.

▶ Septic Shock

DMBA

▶ Polycyclic Aromatic Hydrocarbons (PAHs) and the Immune System

DNA and RNA Blotting

▶ Southern and Northern Blotting

DNA-Derived Products

▶ Immunotoxicology of Biotechnology-Derived Pharmaceuticals

DNA Fingerprinting

A technique for identifying specific organisms based upon the uniqueness of their DNA pattern.

▶ Polymerase Chain Reaction (PCR)

DNA Vaccines

DEBORAH L NOVICKI
Toxicology
Chiron Corp.
4560 Horton Street
Emeryville, CA 94608
USA

Synonyms

gene vacc., genetic vaccines, polynucleotide vaccines, nucleic acid vaccines

Definition

The National Institute of Allergy and Infectious Diseases (NIAID) defines a DNA vaccine as follows: "direct injection of a gene (s) coding for a specific antigenic protein (s), resulting in direct production of such antigen (s) within the vaccine recipient in order to trigger an appropriate immune response".

DNA vaccines are considered third-generation vaccines. The first- and second-generation vaccines were live attenuated or killed whole cell vaccines and purified component or genetically engineered recombinant protein vaccines, respectively. The first- and second-generation vaccines contain the actual antigens to which the immune response is desired.

In contrast, DNA vaccines are based on the introduction of a DNA construct that contains a gene or genes coding for the antigen (s) of interest into the animal or human. Cellular transfection and subsequent RNA and protein synthesis results in the intracellular production of the protein antigen (s) of interest, which are then presented by the host's cells, leading to the induction of antigen-specific humoral and cellular immune responses. In addition to antigen-specific immune responses, the ▶ innate immune system is also engaged due to the presence of bacterial DNA in the vaccine.

Characteristics

In the early 1990s several groups of investigators, almost concurrently, demonstrated that rodents and primates given DNA vaccines developed B cell, cytotoxic T cell and helper T cell responses to various pathogens and cancers. Subsequently, it has been shown that dendritic cells (1) and other cells of the innate immune system play important roles in the immune responses to DNA vaccines.

In addition to skin or muscle cells at the injection sites, resident dendritic cells (professional antigen-presenting cells) could be transfected. The ability of a small number of transfected dendritic cells to express, process, and present antigen was responsible, in part, for the immune responses to DNA vaccines.

The role of the innate immune system in the immune responses to DNA vaccines is being defined. The cells of the innate immune system, macrophages, dendritic cells, and B lymphocytes, possess well-conserved pattern-recognition receptors (PRR) or Toll-like receptors (TLRs) that are capable of recognizing specific pathogen-associated molecular patterns (PAMPS) and triggering adaptive immune responses that are "custom-designed" to deal with the pathogen in question (2). For example, the production of interleukins (IL-6 and IL-12), tumor necrosis factor (TNF)-α, and interferon (IFN)-γ can be triggered in response to CpG motifs in bacterial plasmid DNA (consisting of unmethylated cytosine-phosphate-guanosine dinucleotides) which are recognized by TLR9.

Each component of the immune response to a DNA vaccine is an opportunity for tailoring those responses. A major advantage of DNA vaccines is the ability to rationally design the DNA to contain desired elements. To increase the efficacy of DNA vaccination, manipulation of the vector backbone DNA sequences, optimization of antigen-coding sequences, inclusion of immunostimulatory DNA (CpG) sequences or coexpressed proteins, and the use of delivery systems and targeting formulations have all been explored (3). The innate stability and ease of manufacturing DNA allows for the development of creative formulations and delivery systems.

The most frequently used approach to designing DNA vaccines has been the use of bacteria-derived plasmid vectors. The fundamental requirements are a eukaryotic promoter that controls the expression of the antigen-coding gene(s) of interest, the antigen-coding region(s) followed by a transcription-termination signal, polyadenylation sequences, and an antibiotic-resistance gene with a bacterial origin of replication to allow for the selection of plasmid-containing bacteria during manufacturing.

DNA vaccines can be administered by needle injection, injection with electroporation, using fluid-jet injection, and via air or gene guns. Routes of administration have included intramuscular, intradermal, subcutaneous, intravenous, intraperitoneal, oral, vaginal, intranasal, and non-invasive dermal delivery. Animal models of viral, bacterial and parasitic diseases, cancer, allergy, and autoimmunity have been studied. Generally, DNA vaccines have been more effective in smaller than larger mammals. Doses of 10–100 μg per animal are effective compared to doses of up to 2500 μg in man. However, based on success seen in non-clinical studies, a number of clinical trials have been and continue to be performed using this promising technology. Clinical trials aimed toward prophylaxis against HIV, influenza, hepatitis B, malaria, and Ebola infection, as well as immunotherapeutic trials in cancer and HIV, have been performed or are ongoing.

Preclinical Relevance

Because DNA vaccines have been more effective in laboratory animals than in humans, much effort has been directed toward enhancing immunogenicity to optimize the likelihood of success in man. Several approaches have been investigated including the use of adjuvants (traditional and genetic, such as granulocyte-macrophage colony-stimulating factor, GM-CSF), cytokines and chemokines (as recombinant proteins or as cotransfected genes), adding CpG motifs, and delivery and targeting strategies.

Optimizing the delivery of DNA to cells or tissues has been investigated by formulating DNA in/on alum, liposomes, polyvinylpyrrolidine/polylysine, poly-lactide coglycolide microspheres, chitosan, gold particles, poly(ortho ester) microspheres, cochleates, and other biocompatible materials. These approaches have enhanced the immunogenicity compared to naked DNA. Auxotrophic mutant intracellular bacteria are also under investigation as in vivo gene delivery systems for mucosal vaccines.

Because there are so many possibilities in creating a DNA vaccine, the ▶ preclinical safety assessment issues are dependent upon the specific vaccine under consideration. There are the general questions to be answered regarding the systemic toxicity of the DNA, any of its impurities, the adjuvant, or formulation or delivery components, and the expressed protein(s). More specific to DNA vaccines is the potential for chromosomal integration of the exogenous DNA into either somatic or germ cell genomes of the animal or human vaccinee. The safety concern was an increase in the risk of malignancy or reproductive effects. Although integration frequencies higher than the spontaneous mutation rate have not been a problem to date, it continues to be a concern as DNA vaccine formulations and delivery systems become more sophisticated. Other theoretical safety issues include:

- the development of tolerance due to DNA persistence and longer-term expression of antigen following DNA vaccination
- the development of autoimmunity resulting from the induction of antinucleic acid antibodies (ANA) or from immune responses against transfected cells
- the stimulation of cytokine production that alters the ability of the vaccinee to respond to other vaccines or pathogens.

For novel plasmid vectors, formulations, or delivery systems, to address the question of integration, single-dose studies are performed in which selected tissues are collected at various time points post-injection. Genomic DNA is purified and PCR performed to detect the presence of any plasmid sequences. Biodistribution studies are performed to define the tissue distribution and persistence of the exogenous DNA. Defining the persistence of the DNA in tissues and the time course of antigen expression potentially aids in determining dosing regimens as well as the risk of adverse effects or developing tolerance. The induction of ANA is evaluated using commercially available kits, and if ANA is negative, no further testing is required. If ANA is positive, it may be necessary to develop specific anti-plasmid antibody assays and test sera to define the antibody-eliciting stimulus.

Aside from the integration, biodistribution, and ANA studies described above, the non-clinical safety assessment of DNA vaccines is comparable to the requirements for non-DNA vaccines. Local reactogenicity (inflammation at the administration site), pyrogenicity (fever), immunotoxicity (either humoral or cell-mediated), induction of autoimmunity due to the similarity to self-epitopes, exacerbation of subclinical autoimmunity, hypersensitivity reactions, and systemic adverse effects are all potential toxicities of any vaccine. There are currently no validated experimental models available for assessing the potential of vaccines to induce autoimmune diseases.

In addition to the standard toxicology study parameters, immunotoxicity should be assessed using a tiered approach, as for any biologic or small molecule, and DNA vaccine toxicology studies should minimally include the following immunotoxicity-related endpoints:

- general appearance and body weight
- hematologic tests (blood cell counts and differentials)
- blood chemistry (albumin to globulin ratio (A/G))
- organ weights (weights of spleen, thymus and adrenals)
- histopathologic examinations (spleen, thymus, lymph nodes, bone marrow, intestines including Peyer's patches, liver, kidneys, and adrenals).

If any of these parameters are impacted by treatment, then additional assessments may be warranted ("tier I or II" tests as appropriate). It is recommended that immunogenicity be monitored during the course of repeated-dose toxicity studies to demonstrate exposure to antigen-specific antibodies.

Relevance to Humans

As mentioned above, clinical trials with DNA vaccines for HIV, cytomegalovirus, hepatitis B, malaria, Ebola, and cancer (melanoma, prostate, multiple myeloma) have been performed or are ongoing. To date, DNA vaccines appear to be safe and well tolerated in humans at intramuscular doses up to 2500 μg per dose for three doses. In earlier trials, cellular and humoral responses were generated, but DNA vaccines have generally been less potent in man than in laboratory animals. Based on these findings, non-clinical experiments aimed at increasing potency will drive the next generations of DNA vaccines. Combining a DNA vaccine prime (for ▶ cell-mediated immunity) with a protein antigen boost (for ▶ humoral immunity) is an approach that is currently being explored for AIDS, malaria, tuberculosis, and cancer. The first clinical trial using this approach in malaria has yielded positive immunogenicity results and exhibited an acceptable safety profile.

Regulatory Environment

Several global regulatory documents provide guidance for the development of vaccines in general, and for DNA vaccines in particular. The following are the most pertinent:

- Center for Biologic Evaluation and Research (CBER) Points to Consider on Plasmid DNA Vaccines (1996)
- EMEA guidelines for preclinical toxicological testing of vaccines CPMP/SWP/465/95 (1997)
- Committee for Proprietary Medicinal Products (CPMP) Note for guidance on preclinical pharmacological and toxicological testing of vaccines 6/1998
- Note for Guidance on the Quality, Preclinical, and Clinical Aspects of Gene Transfer Medicinal Products CPMP/BWP/3088/99 Draft
- US Food and Drug Administration. FDA Draft Guidance for Industry: Considerations for Developmental Toxicity Studies for Preventive and Therapeutic Vaccines for Infectious Disease Indications (2000) (revised)
- Proceedings of the CBER/SOT meeting: December 2–3, 2002
- EMEA Note for Guidance on Repeated Dose Toxicity CPMP/SWP/1042/99
- FDA Guidance for Industry: Immunotoxicology Evaluation of Investigational New Drugs, October 2002.

References

1. Mohamadzadeh M, Luftig R (2004) Dendritic cells: in the forefront of immunopathogenesis and vaccine development—a review. J Immune Based Ther Vaccines 2 (1):1
2. Armant MA, Fenton MJ (2002) Toll-like receptors: a family of pattern-recognition receptors in mammals. Genome Biol 29; 3 (8):3011.1–3011.6
3. Gurunathan S, Klinman DM, Seder RA (2000) DNA vaccines: immunology, application, and optimization. Ann Rev Immunol 18:927–974
4. Donnelly JJ (2003) DNA vaccines. In: Ellis R, Brodeur B (eds) New Bacterial Vaccines. Plenum Publishing Corp, New York, pp 30–44

5. Ertl H (2003) DNA vaccines. Kluwer Academic Publishers, New York

Docosahexanoic Acid

A 20-carbon fatty acid containing six double bonds, with the first double bond on the third carbon atom from the methyl end. This fatty acid can be obtained from oily fish or fish oil supplements.
▶ Fatty Acids and the Immune System

Dog

▶ Canine Immune System

Dolichos Biflorus

An extract from the seeds of *Dolichos biflorus* (horse gram) is a lectin used in blood grouping. It recognises N-acetylgalactosamine and when used at an appropriate dilution is a reagent for distinguishing A_1 and A_1B red cells (which are agglutinated) from A_2 and A_2B cells (which are not).
▶ AB0 Blood Group System

Dominant Negative

A state in which an engineered gene encodes a protein that reduces or cancels the normal activity of an endogenous protein.
▶ Knockout, Genetic

Dominant Peptides

The immune response is directed against specific pathogen proteins. A hierarchy of peptide dominance exists with the most dominant eliciting large clonal expansions of peptide-specific T cells.
▶ Antigen-Specific Cell Enrichment

Draining Lymph Node

Lymph nodes are knots in the net of lymphatic vessels that serve as filtering stations through which lymph percolates on its way to the blood stream. Lymph nodes are numerous and are found everywhere in the body. Lymph nodes filtering lymph from a particular region or organ are often referred to as regional or draining lymph nodes because all material and cells transported out of a certain region have to pass through a certain lymph node packed with B lymphocytes, T lymphocytes and dendritic cells.
▶ Popliteal Lymph Node Assay, Secondary Reaction

Drug Allergies

▶ Drugs, Allergy to

Drug Hypersensitivity Syndrome

A severe clinical drug allergy involving the skin (exanthema), the blood system (massive eosinophilia, > 1.5 G/l), and internal organs (frequently, hepatitis). It is often elicited by anti-epileptic drugs, and occasionally by others such as diltiazem, and sulfonamides. It is mediated by T cells.
▶ Lymphocyte Transformation Test

Drug-Induced Hypersensitivity

▶ Drugs, Allergy to

Drug-Induced Hypersensitivity syndrome (DIHS)

Drug-induced hypersensitivity syndrome (DIHS) is a potentially fatal drug eruption, usually accompanied by abnormalities of multiple organ systems (e.g hepatitis and nephritis). This syndrome typically develops 3 weeks to 6 months after starting treatment with anticonvulsants, beginning with a fever which is shortly followed by a generalized maculopapular eruption and variable degrees of lymphadenopathy. Withdrawal of the causative drug is often followed by exacerbations, rather than rapid clearing of the reaction. Reactivation of human herpesvirus 6 has been implicated in the pathogenesis.
▶ Drugs, Allergy to

Drugs, Allergy to

TETSUO SHIOHARA
Department of Dermatology
Kyorin University School of Medicine
6-20-2 Shinkawa Mitaka
Tokyo 181-8611,
Japan

Synonyms

Drug allergies, allergic reactions to drugs, hypersensitivity reactions to drugs, idiosyncratic drug reactions, drug-induced hypersensitivity.

Definition

Adverse drug reactions are defined as "an appreciably harmful or unpleasant reaction resulting from an intervention related to the use of a medical product" (1). Although the mechanism for most adverse drug reactions remains uncertain, immunologic reactions appear to explain the minority of these reactions. Adverse drug reactions that can be explained by immune mechanisms are called drug allergies. Thus, the term "allergy to drugs" is defined as "immunologically mediated, unpredictable adverse drug reactions occurring only in susceptible patients". Although reactions of this type represent only part of a broad spectrum of adverse drug reactions, their unpredictable and serious nature makes them an important clinical problem: this is because they are important causes of morbidity. Although patients who have experienced reactions occurring after administration of drugs commonly state that they have "drug allergies" to many drugs, few reactions that they have experienced are truly allergic. Although they are often referred to as rare, with a typical incidence of from 1 in 100 to 1 in 100 000, they may be more common than generally thought. When we describe an immunological idiosyncratic response to a drug that is not related to its pharmacological actions, the term "drug allergies" is preferable to other terms such as "idiosyncratic drug reactions" and "hypersensitivity reactions". The terms "allergic reactions" or "hypersensitivity reactions" can be used interchangeably to describe reactions resulting from the involvement of immunological mechanisms.

Characteristics

A broad spectrum of allergic drug reactions cannot be easily organized under a simple classification scheme. However, several practical classification systems have been proposed. The most practical one is a classification system based on the combination of the predominant clinical features of reactions, the predominant organs involved in the reactions, and the types of hypersensitivity reactions.

Most practically, allergic drug reactions can be classified into systemic and organ-specific reactions. Systemic allergic reactions include anaphylaxis, serum sickness, drug fever, and drug-induced autoimmune diseases. Anaphylaxis is generally believed to be caused by the release of a variety of mast cell-derived mediators in either an IgE-dependent or IgE-independent fashion. Characteristically, anaphylaxis begins immediately (usually within 1–2 hours) after administration of drug, and ranges in severity from mild urticaria to laryngeal, cardiovascular and pulmonary collapse, and death. Serum sickness and serum-sickness-like reactions occur 7–21 days after administration of the biological agents such as heterologous antisera, vaccines and drugs, respectively. This delayed onset reflects the time required for the formation of antigen-antibody complexes and subsequent activation of complement. Although fever as a allergic drug reaction was first recognized more than 100 years ago, the mechanism remains poorly understood. Drugs are often capable of producing a clinical syndrome indistinguishable from idiopathic autoimmune diseases: they include systemic lupus erythematosus (SLE), dermatomyositis, and pemphigus. These drug-induced autoimmune diseases are usually milder than the idiopathic forms and they improve with the discontinuation of the causative drug; however, some symptoms that are indistinguishable from the idiopathic form may remain or develop. There is, therefore, debate as to whether some drug-induced autoimmune diseases represent a predisposition to subsequent development of the idiopathic form of the disease. Drugs that have been implicated in the development of drug-induced autoimmune diseases include procainamide, hydralazine, isoniazid, chlorpromazine, hydantoin, D-penicillamine, quinidine, and minocycline.

Organ-specific allergic drug reactions occur in any organ, but the skin is the most commonly affected. Clinical symptoms observed in drug-induced cutaneous reactions are variable and include ▶ erythroderma, urticaria/angioedema, ▶ morbilliform eruptions, erythema multiforme, fixed drug eruption, Stevens-Johnson syndrome, toxic epidermal necrolysis, and photoallergic and phototoxic reactions.

▶ Morbilliform eruptions (or maculopapular eruptions) are the cutaneous reactions most frequently seen. Because this type of eruption is often observed in association with a viral infection, it is difficult to differentiate it from a viral exanthem on clinical and laboratory grounds in some cases. The fact that drug allergies occur far more often during viral infections makes it more difficult to diagnose this type of reaction. Morbilliform reactions usually occur within 1–2 weeks after therapy has begun. Urticaria/angioedema may appear as part of anaphylaxis and in some cases with morbilliform eruptions. Erythema multiforme (EM) is a cutaneous reaction to viruses or drugs, which is characterized by a typical "target" or "iris" lesion and histopathologic findings. Although EM was initially described as a self-limited mild skin disease, EM often evolves to a more severe disorder characterized by bullae, widespread skin detachment, fever, and erosive mucus membrane lesions. These syndromes include Stevens–Johnson syndrome (SJS) and toxic epidermal necrolysis (TEN). Recently, severe EM, SJS, and TEN have been regarded as variants within a continuous spectrum. Severe EM is at the mild end of a spectrum, while TEN represents the most severe extreme of the spectrum. One current classification scheme defines SJS and TEN according to the amount of epidermal detachment (< 10%, 10%–30%, > 30%), and according to whether pre-existing focal lesions or a diffuse redness are present (2). The incidence of TEN is estimate at 0.4–1.2 cases per million person-years, and of SJS at 1–6 cases per million person-years. The most common offending drugs are sulfamethoxazole-trimethoprim, other sulfonamide antibiotics, anticonvulsant agents (barbiturates and carbamazepine), allopurinol, non-steroidal anti-inflammatory drugs, and non-sulfonamide antibiotics. SJS and TEN are more common among people with HIV infection, patients with SLE, and bone-marrow transplant recipients. The mortality rate for patients with TEN is as high as 30%. These cutaneous drug reactions present a common histopathology characterized by epidermotropic infiltration of T cells, despite a broad spectrum of clinical features, except for urticaria/angioedema. In recent years, it has been widely accepted that drug-specific T cells that can recognize drug–hapten-modified major histocompatability complex (MHC) class II or I molecules play a central role in the pathogenesis of these cutaneous drug reactions. Cytotoxic granule (perforin/granzyme-B) exocytosis—rather than the Fas/Fas L pathway—represents the main pathway of cytotoxicity mediated by $CD4^+$ as well as $CD8^+$ T cells in these reactions (3). In TEN, Fas/Fas L interactions are shown to be directly involved in keratinocyte destruction, although the initial step may be mediated by $CD8^+$ T cells with natural killer(NK)-like receptors.

Fixed drug reactions (FDE) are a localized variant of cutaneous drug reactions characterized by their relapse in the same site with administration of the same drug, which may occur on any part of the body. Recent studies have demonstrated that intraepidermal $CD8^+$ T cells residing in the resting FDE lesions are responsible for localized epidermal injury seen in FDE via the release of interferon-γ and exocytosis of perforin/granzyme-B. These intraepidermal $CD8^+$ T cells have several unique properties similar to those of cells belonging to the innate immune system, such as NK cells.

Cutaneous drug reactions—often referred to as drug-induced hypersensitivity syndrome (DIHS)—or drug reactions with eosinophilia and systemic symptoms (DRESS) are life-threatening, with reactions in multiple organ systems characterized by EM-like rash, fever, tender lymphadenopathy, hepatitis, and leukocytosis with eosinophilia. The drugs most commonly causing DIHS are anticonvulsant agents (carbamazepine, phenytoin, and phenobarbital), allopurinol, and minocycline. The hallmark of the syndrome is the long interval between introduction of the drug and onset of symptoms: this syndrome typically appears after prolonged exposure to the offending drug (which can be 4 weeks or more). This syndrome also often has peculiar features that cannot be solely explained by drug-specific T cells: it sometimes shows persistent intolerance to many other chemically distinct drugs and some patients continue to deteriorate or show flare-ups for weeks or months after cessation of the causative drug. These findings suggest that this reaction should not be simply regarded as representing a reaction primarily mediated by immune responses to drugs or their toxic metabolites. In this regard, recent studies have provided evidence to indicate that reactivation of human herpesvirus 6 (HHV 6), a typical β-herpesvirus that has been implicated in the pathogenesis of certain immunologic disorders such as graft-versus-host disease (GVHD), contributes to the development of DIHS (4). Suzuki et al (4) noted a marked increase in anti-HHV-6 immunoglobulin G titer; this was seen 3–4 weeks after the onset of the syndrome, and viral DNA was detected just prior to the increase in HHV-6 IgG titers in many patients. Thus, this syndrome offers a unique opportunity to link drug-specific T cells with anti-viral immune responses: the occurrence of this syndrome is likely to be determined by the interplay of multiple factors, and viral infections may constitute one of the most important factors.

Recently, Matzinger (5) proposed an alternative mechanism by which drug allergies associated with viral infections are more likely explained. According to this hypothesis, referred to as the "danger hypothesis", a drug does not evoke an immune response under normal conditions (in which "danger" signals usually generated by "stressed" or "injured" cells are not provided). However, once danger signals are provided by inflammatory cytokines (produced in response to a viral or bacterial pathogen) they directly or indirectly activate drug-specific T cells and result in the development of drug allergies (see Table 1).

Drugs also cause hematologic disorders by immunologic mechanisms. They include hemolytic anemia, thrombocytopenia, and agranulocytosis. This reactions results from the binding of a drug or drug metabolite

Drugs, Allergy to. Table 1 Classification and pathogenesis of allergic drug reactions

Drug Reactions	Mediator(s)
Systemic	
Anaphylaxis	Mast-cell-derived mediator
Serum sickness	Immune complex, complement
Drug fever	Cytokines
Autoimmune diseases	Immune complex, complement
Organ-specific	
Cutaneous reactions	Cytotoxic T cells, cytokines, mast cell-derived mediators
Hematologic reactions	Immune complex
Pulmonary reactions	Cytotoxic T cells?
Hepatic reactions	Cytotoxic T cells?
Renal reactions	Cytotoxic T cells?

to the surface of blood cells such as red blood cells, platelets, and granulocytes.

Pulmonary hypersensitivity reactions to drugs are associated with eosinophilia, bilateral interstitial infiltrates, and finally pulmonary fibrosis.

Immune-mediated drug-induced hepatic disorders result from two major forms of injury: (hepatocellular damage often mimicking) viral hepatitis and cholestasis. The most common drug-induced renal disorder is interstitial nephritis. Although the mechanisms of these organ-specific drug reactions are unclear, drug haptens might undergo antigen processing and be presented to T cells, and antigen-specific cytotoxic T cell responses might be responsible for the clinical pathology observed in various organ-specific drug reactions.

Preclinical Relevance

Pharmaceutical companies try to detect any potential adverse effect of a drug before it is marketed. Unfortunately it is almost impossible to predict the complete range of adverse effects before a drug is marketed to a broad and diverse population, because the drug can only be exposed to a limited number of individuals (perhaps a few 1000) during its development. In particular, allergic reactions to a drug are notoriously unpredictable based on pre-marketing surveillance before licensing. Furthermore, there is no gold standard in-vitro test to predict how often a drug will cause allergic reactions in susceptible individuals.

Relevance to Humans

Although some drugs cause characteristic clinical reactions, it is difficult to predict which ones elicit allergic reactions in patients solely on clinical grounds: this is because one drug can be responsible for causing a range of allergic reactions. The most important step for the diagnosis of drug allergy is to obtain an accurate medical history. It is important to document all possible culprits, including information on dates of administration and discontinuation, dosages, and any previous drug exposure history. The physician also should determine whether the patient is taking over-the-counter formulations and products that may not be thought of as traditional medicines (i.e. herbal remedies). The temporal relationship between introducing the drug and the occurrence of the reaction is also important for assessment of suspected drug allergy: allergic reactions to drugs usually develop 1–2 weeks after starting therapy. However, a long interval does not necessarily rule out an allergic reaction. Almost all drugs are capable of causing a wide range of allergic reactions in humans. To date, there are no animal models in which a wide range of allergic reactions to drugs observed in humans can be easily reproduced.

Although there are several objective tests to confirm drug allergy, overall they still have limited practical value for the physician. In-vivo skin testing, such as intradermal and patch tests, are available for assessing drug allergies but the clinical value is unclear largely because it is unknown whether drug metabolites or degradation products—rather than native drug—are responsible for the allergic reactions in many allergic reactions. Even if the skin test is negative, the presence of drug-specific antibodies or T cells cannot be ruled out. This is because the relevant products may not be used as the skin test reagent. However, in some drug-induced cutaneous reactions (such as morbilliform eruptions, FDE, EM, SJS, and TEN) patch testing is helpful for assessing drug-specific T cell responses. There are also no generally available and reliable in-vitro tests for the diagnostic assessment of drug allergies, although some in-vitro testing methods may provide valuable information for investigators. They include in-vitro tests for drug-specific antibodies to IgE, IgG or IgM and lymphocyte-transformation testing in response to a specific drug.

Lack of knowledge of the clinically relevant antigenic determinants also makes it difficult to interpret a negative result. However, in some cutaneous reactions, certain drugs have been shown to induce the proliferation of drug-specific T cells and the production of cytokines. Although drug-specific T cells can be present in a patient with drug allergy, their presence does not necessarily mean that they are responsible for the development of drug allergy. It is difficult to exclude the possibility that their presence may represent an epiphenomenon. The most reliable tests to confirm a drug allergy is to rechallenge with the drug. Such rechallenge should be done at a far lower dose in patients with severe drug allergy. For SJS or TEN, rechallenge with the putative drug should be done only with great caution, by physicians who are well experienced in the method and its contraindications.

Regulatory Environment

The simplest approach is to avoid the offending drug if an alternative (clinically effective) therapeutic approach is available for the patient. However, if alternative agents are not available and if the patient cannot manage symptoms without a medicine that has caused allergic reactions, drug desensitization may be considered. Desensitization represents a process by which a drug-allergic person is converted from a highly sensitive state to a non-sensitive state. Nevertheless, desensitization should not be considered in cases where previous reactions were severe (like SJS and TEN). Information about which drug has caused allergic reactions in patients and how the diagnosis has been made should be forwarded to the manufacturer and to a national center with responsibility for the safety of medicines. National centers send this information to the WHO Collaborating Center for International Drug Monitoring Center (the Uppsala Monitoring Center), by whom this information is analyzed (1).

References
1. Edwards IR, Aronson JK (2000) Adverse drug reactions: definitions, diagnosis, and management. Lancet 356:1256–1259
2. Betsuji-Garin S, Rzany B, Stern RS, Shear NH, Naldi L, Roujeau JC (1993) Clinical classification of cases of toxic epidermal necrolysis, Stevens–Johnson syndrome, and erythema multiforme. Arch Dermatol 129:92–96
3. Pichler WJ, Yawalkar N, Britschgi M et al. (2002) Cellular and molecular pathophysiology of cutaneous drug reactions. Am J Clin Dermatol 3:229–238
4. Suzuki Y, Inagi R, Ando T, Yamanishi K, Shiohara T (1998) Human herpesvirus 6 infection as a risk factor for the development of severe drug-induced hypersensitivity syndrome. Arch Dermatol 134:1108–1112
5. Matzinger P (1994) Tolerance, danger, and the extended family. Ann Rev Immunol 12:991–1045

DTH

▶ Delayed-Type Hypersensitivity

Duffy Antigen

Membrane protein related to chemokine receptors expressed on erythrocytes, endothelial cells, and several types of epithelial cells. It binds IL-8 and several other chemokines without transmitting a signal, acting thus as a "cytokine sink". Contains also antigenic determinants for the Duffy blood group system and functions as the erythrocyte receptor for *Plasmodium vivax*. Also known as the Duffy antigen receptor for chemokines (DARC).

▶ Chemokines

E

ECSA

▶ Erythropoietin

Ectopic Lymphoid Tissue

A loosely organized, poorly circumscribed aggregate of immune response cells (chiefly T lymphocytes) that forms at or near the site of a chronic inflammatory stimulus. Due to their proximity, the cells in such ectopic foci direct and regulate the adaptive immune response to the antigen(s) that is inciting the reaction more efficiently than can the more distant secondary lymphoid organs.
▶ Lymphocytes

Eczema Herpeticum

An acute and disseminating infection with herpes simplex virus. Affects patients with underlying immunosuppression or with severe atopic dermatitis.
▶ Dermatological Infections

Edema

The swelling caused by entry of fluid and cells from the blood into tissues. This represents a central manifestation of inflammation.
▶ Inflammatory Reactions, Acute Versus Chronic

Effector Cells

Activation of antigen-specific lymphocytes, leads to clonal expansion and cell differentiation into effector and memory cells. Effector cells are short-lived cells involved in resolving the current antigenic challenge. Plasma cells are B cell effectors, secreting antibodies at a high rate. "Effector cell" is most commonly applied to activated CD4 T cells that secrete high levels of cytokines in response to antigen. In a more general sense it is used for a lymphocyte that is able to directly or indirectly drive the destruction of a pathogen, pathogen-infected cell, tumor cell, or other foreign body without the need of differentiation or proliferation.
▶ Memory, Immunological
▶ Cytotoxic T Lymphocytes

Eicosanoids

A family of hormone-like compounds formed by hydroxylation of 20-carbon polyunsaturated fatty acids, chiefly arachidonic acid including the prostaglandins, thromboxanes, hydroxy-eicosatetraeonic acids (HETEs), and leukotrienes. Eicosanoids act rapidly and locally to regulate inflammation, lymphocyte function, chemotaxis, blood pressure, pain, and blood clotting.
▶ Fatty Acids and the Immune System
▶ Prostaglandins

Eicosapentanoic Acid

Eicosapentaenoic acid (EPA) is a 20-carbon fatty acid containing five double bonds, with the first double bond on the third carbon atom from the methyl end. This fatty acid can be formed from its precursor, α-linolenic acid, but in humans this conversion is inefficient. It can alternatively be obtained from oily fish or fish oil supplements.
▶ Fatty Acids and the Immune System

Electroblotting

Electroblotting is a synonym for western blot analysis.
▶ Western Blot Analysis

Electrochemoluminescent Immunoassay (ECLIA)

An immunoassay in which a light-generating chemical reaction occurs during binding between a ruthenium tri-bispyridine (TAG)-labeled analyte—captured in an antigen–antibody–bead complex—and tripropylamine (TPA). The chemical reaction is registered on reduction by an applied voltage.
▶ Immunoassays

Electrophoresis

The separation of molecules based on their mobility in an electric field.
▶ Southern and Northern Blotting

Electrophoretic Protein Transfer

Electrophoretic protein transfer is part of western blot analysis and represents the elution of proteins from polyacrylamide gels onto a membrane support for further analysis.
▶ Western Blot Analysis

Elicitation Dose

The dose of a chemical or other antigen that suffices for eliciting a secondary immune response is called the elicitation dose. It is usually lower than the dose that is required for a primary immune response that leads to sensitisation against the chemical or other antigen.
▶ Popliteal Lymph Node Assay, Secondary Reaction

ELISA

Enzyme-linked immunosorbent assay.
▶ Maturation of the Immune Response
▶ Plaque-Forming Cell Assays

ELISPOT

Immunological assay used to quantitate number of antigen-secreting cells. Cells are cultured in an ELISPOT plate. Secreted analyte is captured by immobilized antibody and identified by detection reagent. Spots formed are proportional to the number of antigen-secreting cells present in the culture.
▶ Plaque Versus ELISA Assays. Evaluation of Humoral Immune Responses to T-Dependent Antigens
▶ Enzyme-Linked Immunospot Assay
▶ Reporter Antigen Popliteal Lymph Node Assay

ELISPOT Assay

Cells are placed in a plastic dish over an immobilized antigen or antibody. The cells' secreted product is trapped by the antigen or antibody and can be detected by adding a separate antibody that cleaves a colorless substrate to produce a localized colored spot.
▶ Enzyme-Linked Immunospot Assay
▶ Antigen-Specific Cell Enrichment

ELR

Three amino acid motif (Glu-Leu-Arg) found before the first cysteine residue in the NH_2-terminus of some CXC chemokines. The presence of the ELR motif correlates with angiogenic activity. In contrast, non-ELR CXC chemokines are angiostatic.
▶ Chemokines

Embryonic Stem (ES) cell

Totipotent cells from early embryos that, when introduced onto the inner cell mass (the embryo precursor region) of a blastocyst, are capable of contributing to the formation of all adult tissues.
▶ Knockout, Genetic
▶ Transgenic Animals

Emphysema

▶ Trace Metals and the Immune System

Encapsulatus

▶ Klebsiella, Infection and Immunity

Encephalopathy

A complex neuropsychiatric syndrome with disturbances in consciousness and behavior, personality changes, and fluctuating neurologic signs.
▶ Septic Shock

Endocrine Disrupters

Endocrine disruptors, also known as endocrine disrupting chemicals (EDCs), are chemicals found throughout the environment that have estrogenic activity. Endocrine disruptors may affect normal biological processes because of their ability to bind to steroid receptors, to mimic hormones, antagonize hormones, alter hormonal binding to receptors, and/or alter metabolism of natural or endogenous hormones.
▶ Steroid Hormones and their Effect on the Immune System

Endocrine Disrupting Chemicals

▶ Steroid Hormones and their Effect on the Immune System

Endocytosis

Internalization of extracellular macromolecules by enclosing them in the plasma membrane, which internalizes and becomes an intracellular vesicle containing the macromolecules.
▶ Humoral Immunity

Endoplasmic Reticulum

An organelle consisting of interconnected membrane-bound cavities, which is the major site of lipid and protein synthesis. The endoplasmic reticulum of various cell types is used to store calcium ions. In immune cells this calcium plays a role in cell signaling upon activation.
▶ Mast Cells

Endotoxin

The biological term to describe the toxic but also stimulating effects of the cell wall lipopolysaccharides (LPS) of Gram-negative bacteria on the immune system.
▶ Cytokines

Endotoxin Shock

▶ Septic Shock

Enhanced, Extended or Advanced Histology/Histopathology

Enhanced histology of lymphoid organs/tissues is the application of a structured, semiquantitative assessment of histological changes in the principal compartments of these organs/tissues, in addition to the assessment of 'true' histopathological changes like necrosis and granulomata.
▶ Histopathology of the Immune System, Enhanced

Enhanced Histological Assessment

▶ Histopathology of the Immune System, Enhanced

Enhancer

A regulatory region usually located within close proximity of the promoter that enhances promoteractivity and gene expression.
▶ B Lymphocytes

Enrichment

▶ Cell Separation Techniques

Enteramine

▶ Serotonin

Enterocytes

Cells of the intestinal epithelium. Through mucus production they form a protective barrier and in addition they play an important role in absorption of nutrients. Interconnected by tight junctions molecules—and pathogens—have to pass through the enterocytes from intestinal lumen to underlying connective tissue.
▶ Immunotoxic Agents into the Body, Entry of

Enterotoxin

Enterotoxins are substances that are toxic to the intestinal tract. They are usually produced by bacteria and often cause vomiting and diarrhoea.
▶ AB0 Blood Group System

Environmental Estrogens

▶ Steroid Hormones and their Effect on the Immune System

Environmental Protection Agency (EPA)

The United States governmental agency responsible for regulating and licensing commodity chemicals, herbicides, rodenticides, fungicides and insecticides.
▶ Immunotoxicology
▶ Plaque-forming Cell Assay

Enzyme-Linked Immunosorbent Assay

▶ Cytokine Assays

Enzyme-Linked Immunosorbent Assay (ELISA)

An immunoassay that uses enzyme-labeled reagents to detect binding between an antigen and its antibody bound in a complex and immobilized on a solid surface (i.e. plastic 96-well microtiter plate). The ELISA method is based on measurement of optical density (OD) generated in the chromogenic reaction between an enzyme substrate (e.g. *para*-nitrophenyl phosphate) and an enzyme (e.g. alkaline phosphatase) conjugated with a protein that binds the analyte (antigen or antibody of interest).
▶ Immunoassays
▶ Assays for Antibody Production

Enzyme-Linked Immunospot

▶ Plaque-Forming Cell Assays

Enzyme-Linked Immunospot Assay (ELISPOT)

GERNOT GEGINAT
Institut für Medizinische Mikrobiologie
Fakultät für klinische Medizin Mannheim der
Universität Heidelberg, Klinikum Mannheim
Theodor-Kutzer-Ufer 1–3
68167 Mannheim
Germany

Synonyms
enzyme-linked immunospot assay, ELISPOT; reverse enzyme-linked immunospot assay, RELISPOT

Short Description
The ELISPOT assay is a versatile test system for the quantitative assessment of the function of B cells, T cells and monocytes. Both drug-related immunotoxicity due to immunosuppression and the inherent immunogenicity of compounds can be monitored by ELISPOT. The assay visualizes proteins (e.g. antibody, cytokines, enzymes) secreted by individual cells. The resulting spots represent secretory footprints of individual B or T cells from which the frequency of cells secreting a specific cytokine or antibody can be estimated.

Characteristics
The enzyme-linked immunospot assay (ELISPOT) was originally described and used for the enumeration of cells secreting antibodies specific for a defined antigen (1,2) and was later adopted for the detection of cytokine-secreting cells. Because the latter assay detects the secretion of defined antigens with specific antibodies (and not specific antibodies with defined antigens as the original assay) it was also termed reversed ELISPOT assay (▶ RELISPOT). Table 1 summarizes some of the cell populations and the cellular

Enzyme-Linked Immunospot Assay (ELISPOT). Table 1 Selected applications of the ELISPOT assay

Molecule detected	Cell type	Function
Antibody	B cells	B cell-mediated immunity
IFN-γ	CD8 T cells, natural killer cells	T cell effector function
TNF-α	CD8 T cells	T cell effector function
IL-2	Th1 cells	T cell help for T cells
IL-4	Th2 cells	B cell development
IL-5	T helper cells	T cell help
IL-6	Monocytes, macrophages	T cell development (Th1/Th2)
IL-10	Th2 cells	Suppression of Th1 cells
IL-12	Monocytes, macrophages	T cell development (Th1/Th2)
Trx and Trx reductase	Monocytes, granulocytes, others	Oxidative stress

ELISPOT, enzyme-linked immunospot; IFN, interferon; IL, interleukin; Th, T helper cell; TNF, tumor necrosis factor; Trx, thioredoxin.

functions that can be assessed with the ELISPOT assay.

Test Principle

The ELISPOT assay is used for the quantification of cells which secrete specific antibodies or cytokines. Thus, the assay involves the handling of viable cells that have to be kept in a functional state during the assay and proper cell culture technique is essential. In many cases cytokine secretion is not spontaneous but must be triggered by specific stimuli, e.g. in the case of T cells the cognate antigenic peptides must be present. The sensitivity of the ex vivo analysis of rare cell populations can be further enhanced if responder cells undergo an additional in vitro expansion before ELISPOT analysis (3). Subsequently, the ELISPOT assay has to be performed in the time window when maximum cytokine secretion occurs.

After development of the ELISPOT, stained spots represent the footprints of individual cells, and the number of cells secreting a specific cytokine or antibody can be counted. Similar to the common enzyme-linked immunosorbent assay (▶ ELISA), the solid phase is coated with either antigen or antibody. Bound antibodies or antigens (e.g. cytokines) are detected with a biotin-linked secondary antibody with specificity for the molecule to be detected, followed by phosphatase or horseradish peroxidase-labeled streptavidin. Binding of the enzyme-linked secondary antibody is visualized by incubation with an appropriate substrate that by enzymatic activity is converted into a permanent stain. Generally, aminoethylcarbazole (AEC) is used with horseradish peroxidase, and the 5-bromo-4-chloro-3-indolyl-phosphate (BCIP)/4-nitroblue-tetrazolium chloride (NBT) system with phosphatase yielding red-dish brown and blue spots, respectively. Figure 1 summarizes the steps involved in a typical ELISPOT assay. In order to obtain spots that represent footprints of individual cells, diffusion of the stain has to be prevented. Originally, this was achieved by performing the substrate reaction in an agar overlay on the solid phase. More conveniently, the dye reaction is performed directly on nitrocellulose, nylon, polyvinylidene difluoride (PVDF), or other membrane supports. Even direct staining in conventional polystyrene 96-well ELISA-type microtiter plates is possible. The membrane stain is permanent, and after development of the assay the ELISPOT plates can be stored for later analysis and documentation.

As with a common ELISA the optimal antibody or antigen concentrations for coating and the concentration of the secondary antibody have to be identified by titration. Generally the first antibody is applied in a concentration 5–10 times that used for coating common ELISA plates. Internal controls should include nonactivated negative cells and activated cells as a positive control. The latter should be titrated to yield between 10 and 200 spots per well. With proper technique, recovery of seeded activated cells is above 90%.

The ELISPOT assay can be performed in a multiplex format that allows the simultaneous detection of different cytokines secreted by single cells. The multiplex ELISPOT assay requires the use of secondary antibodies linked to different enzymes and a dual or even multicolor staining procedure. Multiplex ELISPOT assays are not frequently employed.

Enumeration of Spots

Spots originating from single cells are circular and show an intense staining in the centre of the spot

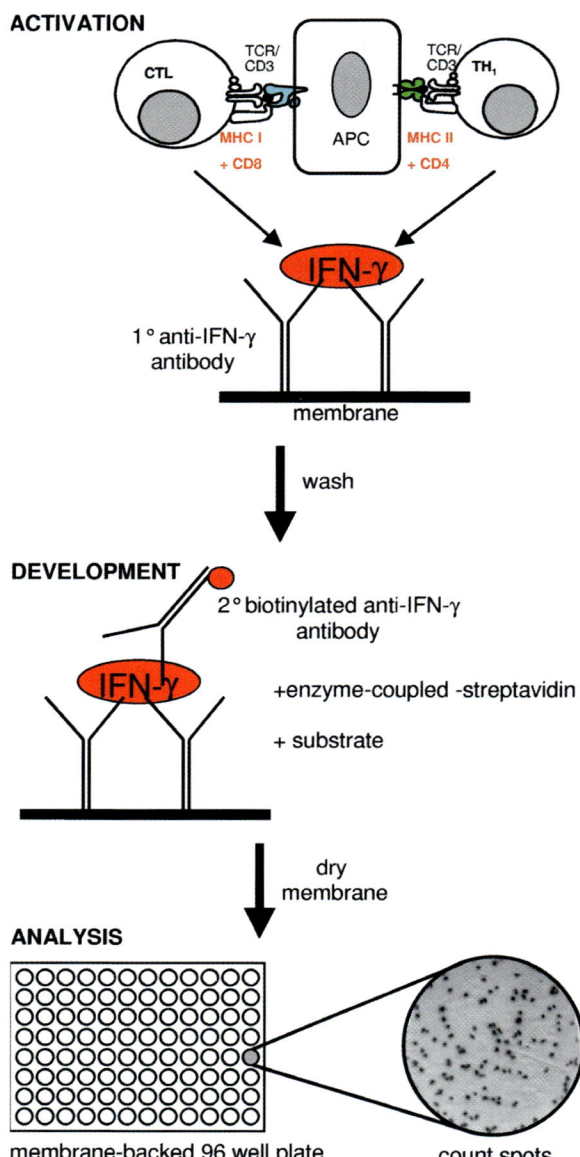

SPOT plates with detachable membranes is that detached membranes can be stored without the plastic frames and can be scanned using a conventional flat bed scanner.

Sensitivity and Dynamic Range

Compared to a conventional ELISA, the ELISPOT assay has an enhanced sensitivity because cytokine or antibody secretion is measured locally and less dilution occurs compared to the measurement of cytokines secreted into cell culture supernatants. The ELISPOT assay is generally considered to be approximately 10–100 times more sensitive than a comparable ELISA. Under optimal conditions as few as 10 positive cells per 1×10^6 cells can be detected.

A problem of the ELISPOT assay is its rather short dynamic range. By visual inspection with a dissection microscope up to approximately 200–300 spots per well can be counted reliably. This can be increased to more than 1000 spots per well by image analysis software. To further extend the dynamic range of the assay, responder cells have to be seeded in different concentrations. However, the concentration of cells also influences the efficiency of activation that may be required to induce cellular cytokine secretion. Thus, results obtained by dilution of cells are not strictly linear, and if possible only frequencies obtained at identical cell concentrations should be compared directly.

Currently, no external standards are available for the ELISPOT assay. Controls, for example B cell or T cell lines, have to be prepared freshly and individually for each assay.

Pros and Cons

The ELISPOT assay allows the quantitative ex vivo analysis of B and T cell-mediated immune responses. Compared to other assays that measure cell-mediated immunity, e.g. intracutaneous skin testing to measure delayed type hypersensitivity reactions, it allows a more specific, objective and quantitative assessment of defined responder cell populations.

As the ELISPOT assay relies on the detection of antibodies or cytokines secreted by living cells, proper cell culture techniques are important for optimal and reproducible results.

As external standards are not available, it is difficult to compare data obtained in different settings using different stimulation conditions and data analysis methods.

As the ELISPOT assay is generally performed in 96-well microtiter plates, simultaneous testing of hundreds of samples is possible, e.g. for screening purposes (4). Even higher sample numbers would be possible with automated liquid handling. Due to the high sensitivity of the assay, rare cell populations can also

that fades out towards the periphery. Artifacts generated by endogenous enzymatic activity of cells are generally smaller and show a more homogenous staining.

Counting of spots can be done by inspection with a dissection microscope or with a dedicated ELISPOT reader with pattern recognition software that is also best for the counting of a large number of samples. Software-based image analysis further enables the quantitative analysis of the spot size. The quality of image analysis if further improved if the 96-well ELISPOT plates are produced from a nonreflective and nontranslucent material that prevents artifacts due to mirror images in the wells. One advantage of ELI-

◀ **Enzyme-Linked Immunospot Assay (ELISPOT).**
Figure 1 Steps involved in a typical ELISPOT assay. The diagram shows the detection of activated T helper 1 (Th1) or cytotoxic T cells (CTL) by the detection of interferon-γ (IFN-γ) secreted by individual activated T cells. Activation: the activation step involves coincubation of responder cells with antigen-presenting cells expressing matching major histocompatibility complex (MHC) antigens and loaded with an antigenic peptide, that is recognized by the T cells. Recognition of the peptide–MHC complex results in T cell activation and the secretion of IFN-γ, that binds to the solid phase coated with anti-IFN-γ antibodies. Development: the ELISPOT assay is subsequently developed by incubation with a second enzyme-linked anti-IFN-γ antibody followed by enzyme-linked streptavidin and an appropriate substrate. Analysis: after drying spots are either counted visually using a dissection microscope or are quantified with a dedicated ELISPOT reader and image analysis software.

be analysed, which is especially important if cells are to be analysed directly ex vivo.

In comparison to other assays analysing cytokine secretion on a single-cell level (cytofluorometric detection of intracellular cytokines), the main advantages of the ELISPOT assay are its scalability and the economic use of responder cells; the main disadvantage is the limited dynamic range of the assay. The cytofluorometric detection of intracellular cytokines allows the analysis of crude cell populations that can be analysed after electronic gating of the population of interest. The analysis of IFN-γ-specific T cells with both methods yielded similar T frequencies in different infectious disease models. A problem of the ELISPOT assay and also of the cytofluorometric analysis of intracellular cytokines is the background activity due to spontaneously activated cells. If T cells of defined specificity are to be enumerated, staining with fluorescence-labeled major histocompatability complex tetramers (▶ MHC tetramers) is an alternative that enables the function-independent analysis of T cell frequencies. This is because it avoids those artefacts due to improper stimulation of responder cells or those due to spontaneously activated responder cells.

Predictivity

The ELISPOT can be applied to monitor various immune-related parameters in humans and rodents. The ELISPOT is generally used for the assessment of systemic immunity. It is therefore of potential value for the assessment of drug-mediated immunosuppression, drug-inherent immunogenicity, and drug-related autoimmunity. Very little experience exists in the application of the ELISPOT for the analysis of general and localized drug-mediated hypersensitivity reactions. As external standards for the ELISPOT are not available, a principal problem that is not yet solved is how to compare data obtained from different laboratories cooperating in multicenter studies.

Relevance to Humans

The ELISPOT assay is a reliable tool for the quantitative functional analysis of B and T cells. During the monitoring of immune responses in phase I trials of investigational new drugs the ELISPOT assay can be applied for the monitoring of drug-mediated immunosuppression, drug-inherent immunogenicity, and drug-related autoimmunity.

The presence of antigen-specific B and T cells correlates with humoral and cellular immunity respectively, against microorganisms expressing the tested antigen. Therefore, the monitoring of B or T cell frequencies by ELISPOT analysis is currently part of many clinical vaccination trials.

Regulatory Environment

Application of the ELISPOT assay for immunotoxicity assessment in the evaluation of investigational new drugs is suggested by a FDA guidance for industry (5).

References

1. Czerkinsky CC, Nilsson LA, Nygren H, Ouchterlony O, Tarkowski A (1983) A solid-phase enzyme-linked immunospot (ELISPOT) assay for enumeration of specific antibody-secreting cells. J Immunol Meth 65:109–121
2. Sedgwick JD, Holt PG (1983) A solid-phase immunoenzymatic technique for the enumeration of specific antibody-secreting cells. J Immunol Meth 57:301–309
3. McCutcheon M, Wehner N, Wensky A et al. (1997) A sensitive ELISPOT assay to detect low-frequency human T lymphocytes. J Immunol Meth 210:149–166
4. Geginat G, Schenk S, Skoberne M, Goebel W, Hof H (2001) A novel approach of direct ex vivo epitope mapping identifies dominant and subdominant CD4 and CD8 T cell epitopes from *Listeria monocytogenes*. J Immunol 166:1877–1884
5. FDA (2002) Guidance for industry. Immunotoxicology evaluation of investigational new drugs. US Department of Health and Human Services. Food and Drug Administration. Centre for Drug Elevation and Research http://www.fda.gov/cder/guidance/4945fnl.doc

Eosinophilia

Elevated amounts of eosinophilic leukocytes in the circulation (> 0.4 G/l). Elevated levels can be found in many diseases. Frequent causes might be drug hypersensitivity and parasite infestations and atopic allergy (asthma, hay fever, atopic dermatitis). It is as-

sumed that 100-fold more eosinophils are found in the tissue than in the blood.
▶ Lymphocyte Transformation Test

Eosinophilia–Myalgia Syndrome (EMS)

An illness resulting from ingestion of impure L-tryptophan which occurred in the USA in 1989. It is characterized by myalgia, eosinophilia, fatigue, muscle cramps, and scleroderma-like skin changes.
▶ Systemic Autoimmunity

Eotaxin

A CC chemokine important in the migration of eosinophils. Also known as CCL11.
▶ Chemokines

EPA

▶ Fatty Acids and the Immune System

Epidemics

▶ Respiratory Infections

Epidemiological Investigations

ANDREW HALL
Dept. of Epidemiology & Population Health
London School of Hygiene & Tropical Medicine
Keppel Street
London
WC1E 7HT
UK

Synonyms
Population studies, field studies.

Definition
Epidemiology is the study of the distribution and determinants of health-related states or events in specified populations, and the application of this study to control of health problems.

Characteristics
In the context of immunotoxicology this definition can be applied to exposure to the toxin, the immune effects of that toxin and the diseases that result from such exposure—all measured at a population level. Epidemiology has three major types of design. Two of these are observational, the third is an intervention.

Descriptive Studies
The first observational type is the descriptive study. These studies involved measuring the frequency of the health-related event in the population. Thus a study that estimated the proportion of people exposed to pesticides in a population is descriptive. The number of people developing lymphoma per year would be descriptive. The two essential components for a descriptive study are the number of events and a count of the population at risk of the event. These are then used to calculate prevalence (a proportion) or incidence (a rate or risk). These simple measures may be broken down into age groups, sex, geographical areas, calendar time, or any other variable that has been collected on both events and population denominators.

Analytical Observational Studies
Analytical observational studies attempt to relate a putative cause of disease to the disease itself. These studies take one of four major designs:
- ecological
- cross sectional
- case-control
- cohort.

In each design there are three types of variable. The exposure—in immunotoxicology this would be a measure of exposure to the toxin—either behavioural (e.g. occupation) or biological (a blood or tissue measurement). The outcome is typically a disease or death from a specific disease, but an intermediate outcome could be used such as an immunological measure of toxicity (e.g. CD4 count).
Finally there are confounding variables. These are other "exposures" which are associated with the exposure of interest (toxin exposure) and influence the outcome of interest (immune parameter or disease). Age is a typical "confounding variable". The level of exposure to a toxin often varies by age of the individual and most diseases show a particular pattern of occurrence by age. However in any one situation thought must be given to what the likely confounding variables are, and an effort made to measure them so that they can be controlled in the analysis.
Quality control of information on exposure, confounding variables and outcome measurements is critical to all epidemiological studies. Repeatability of the meth-

ods used, both within and between observers is crucial for questionnaire and clinical data. In the laboratory strict use of reference reagents with positive and negative controls is required. Misclassification of variables will reduce the chance of finding a positive association.

Ecological studies compare the average exposure in a population to the population risk of disease (measured as a risk or a rate). This comparison may be geographical, for example as in studies of rates of congenital malformations around waste dumps contaminated by chemicals. Alternatively, it may be over time; how does the rate of disease change in relation to changes in exposure over time?

Cross-sectional studies examine a population at one point in time. They simultaneously examine exposure and outcome (as a point prevalence).

Case-control studies take a group of people with the outcome (cases) and compare them to people without the outcome (controls) in relation to an estimate of previous exposure.

Cohort studies start from measurements of exposure and follow people up over time to measure who develops the outcome of interest. This follow up may be historical—as is often the situation with occupational studies. Thus a group of workers with varying levels of exposure are identified in the past and by obtaining death certificates or by reexamining them the proportion of the exposed and unexposed who have the outcome is determined.

The advantages and disadvantages of different designs rests primarily with the time taken versus their interpretation. Thus ecological studies may be very rapid—particularly where they use routine data sources (mortality or morbidity statistics and chemical sales by country for example). However it is very difficult to control confounding in such studies. Many variables change over time or vary between geographical areas as well as the variable of interest. Cross-sectional studies and case-control studies can be carried out moderately quickly, in just 1–2 years. However, interpretation of the causal direction can be difficult—did the exposure precede the effect? Cohort and intervention studies overcome this problem in that the exposure is observed, or applied, before disease is present. However the duration of such studies may be very long, particularly when there is a long interval from exposure to the toxin and the effect. The use of intermediary variables, such as immunological parameters, can overcome this to some extent. However caution is needed in extrapolating from an intermediate laboratory variable to disease.

In all of these designs the measurement and adjustment for confounding variables is critical. This is the main problem with observation studies—confounding can only be controlled as well as it is measured. Appropriate statistical analysis is crucial to the interpretation of analytical epidemiology. A qualified biostatistician, usually specialized in epidemiology, is an essential member of any investigative team.

Intervention Studies

Intervention studies remove confounding factors by randomization. The random allocation to two or more groups prior to intervention means that the level of the confounding variables in each group is identical. Therefore any difference found between the groups must be a result of the intervention. In these studies it is ideal (though not always possible) that neither the investigator nor the subjects know which group they are in. This is frequently possible where drugs are involved since one group can take a pseudo-medicine (placebo) which is identical in every way to the intervention substance apart from the active ingredient. In toxicology it is not usually feasible to carry out intervention studies where one group is exposed to a potential toxin. Therefore intervention studies involved removal, or reduction, in exposure. Even in this situation ethical issues arise as to whether a group should be observed who are exposed when a potentially effective intervention is available.

More complex intervention designs are increasingly popular in public health. Since the unit of intervention is usually the population rather than the individual designs in which populations (villages or clusters) are randomized are particularly instructive. However since relatively few populations are randomized confounding is not necessarily controlled. This means that more sophisticated statistical techniques are required in their design (sample size calculation for example) and analysis than is the case for simple individual randomization.

Preclinical Relevance

None. Epidemiological investigations always involve human subjects.

Relevance to Humans

Epidemiological studies are the critical studies to determine whether effects observed in animals are seen in humans. In addition they are a key method of determining whether unexpected adverse reactions to chemical occur and to allow the balance of risk and benefit to be measured. In general, studies in humans have three key problems.

- First, they are time consuming; even comparatively rapid studies such as the cross sectional design will take a year or more to plan, conduct and analyze. Some cohort studies take decades.
- Second, the ethics of the study require careful attention. Participants need to give informed consent to be part of the investigation. Their safety and

well-being must be at the forefront of investigators minds.
- Third, epidemiological studies are expensive. Recruitment of subjects, laboratory investigations and the employment of personnel over these long periods of study all add up.

References

1. Beaglehole R, Bonita R, Kjellstrom T (1993) Basic Epidemiology. WHO, Geneva
2. Last JE (ed) (1989) A Dictionary of Epidemiology, 2nd ed. Oxford University Press, Oxford
3. Rothman KJ, Greenland S (1998) Modern Epidemiology. Lippincott Williams & Wilkins, Baltimore
4. MacMahon B, Trichopoulos D (1996) Epidemiology Principles and Methods, 2nd ed. Little Brown, New York

Epidermal Cells

Cells present in the epidermis, including keratinocytes, Langerhans cells, melanocytes and Merkel cells.
▶ Skin, Contribution to Immunity

Epidermis

The most superficial layer of the skin, providing the first barrier of protection from the invasion of foreign substances into the body. The epidermis is subdivided into five layers or strata, the stratum germinativum, the stratum spinosum, the stratum granulosum, the stratum lucidum, and the stratum corneum in which a keratinocyte gradually migrates to the surface and is sloughed off in a process called desquamation.
▶ Skin, Contribution to Immunity

Epinephrine (Adrenaline)

A catecholamine hormone produced and secreted by the adrenal medulla and a neurotransmitter released by some neurons. It is released in response to exercise, stress, hypoglycemia, and other stimuli. It is a potent stimulator of the sympathetic nervous system acting on α- and β-adrenergic receptors. It is a powerful cardiac stimulant that increases heart rate and cardiac output, causes vasodilatation of small arteries, and promotes glycogenolysis and other metabolic effects. Also called adrenaline.
▶ Stress and the Immune System

Epitope

A molecular region, usually an amino acid sequence, on the surface of an antigen that is capable of eliciting a specific immune response. Synonymous with "antigenic determinant". Epitopes are recognized by complementary antigen receptors on B cells or T cells. Epitopes for T cells combine amino acids from MHC and associated peptide.
▶ Hapten and Carrier
▶ AB0 Blood Group System
▶ Therapeutic Cytokines, Immunotoxicological Evaluation of

9,10-epoxide

▶ Polycyclic Aromatic Hydrocarbons (PAHs) and the Immune System

Epstein–Barr Virus (EBV)

A gamma herpesvirus which infects selectively human B lymphocytes by binding to the complement receptor 2 (CD21). The virus causes infectious mononucleosis (Peiffer's glandular fever or disease) and leads to a lifelong infection of B cells, which is controlled by T lymphocytes.
▶ Cyclosporin A

Erythema Multiforme (EM)

Erythema multiforme (EM) is a benign, self-limited rash of the skin and mucus membranes characterized by symmetrical target-shaped or iris-shaped lesions with a tendency for recurrence. EM can be induced not only by drugs but also by infectious agents, such as herpes simplex virus and *Mycoplasma*. Depending on the severity of the target lesions and the mucus membrane lesions, EM can be subdivided into two forms: EM major and EM minor.
▶ Drugs, Allergy to

Erythrasma

Infection of the skin with *Corynebacterium minutissimum*. Clinical findings are reddish brownish and discreetly scaling lesions in intertriginous areas.

▶ Dermatological Infections

Erythroderma

Erythroderma, or exfoliative dermatitis, is characterized by diffusely inflamed red skin with varying degrees of scaling. It represents the end stage of cutaneous inflammation induced by various agents including drugs. Drug eruptions that appear first as morbilliform eruptions usually coalesce into erythroderma over time.

▶ Drugs, Allergy to

Erythroid Colony Stimulating Activity

▶ Erythropoietin

Erythropoiesis Stimulating Factor

▶ Erythropoietin

Erythropoietin

RACHEL R. HIGGINS · YAACOV BEN-DAVID
Department of Medical Biophysics
University of Toronto, Canada and Sunnybrook and Women's College Health Sciences Center
Toronto, Ontario
Canada

Synonyms
ESF (erythropoiesis stimulating factor); ECSA (erythroid colony stimulating activity)

Definition
Erythropoietin (from Greek erythro for red, and poietin to make) is a small glycoprotein hormone that is essential for the production of red blood cells. Erytropoietin (Epo) promotes the survival, proliferation and differentiation of erythroid progenitor cells (CFU-E, BFU-E) to mature erythrocytes and initiates hemoglobin synthesis. Cells responsive to Epo have been identified in adult bone marrow, fetal liver and adult spleen.

Characteristics
Epo is produced and secreted primarily in adult kidney and fetal liver cells. It is an acidic hormone with a molecular weight of 34–37 kD. Epo is synthesized as a 193 amino acid precursor that is cleaved to yield an active protein of 166 amino acids. It is relatively heat- and pH-stable (pI=4.5), is N-glycosylated at asparagine residues 24, 36 and 83 and O-glycosylated at serine 126. Epo is also sialylated and contains two disulfide bonds at positions 7/161 and 29/33. Glycosylation, which comprises approximately 40% of the molecular mass of Epo, is important to the pharmacokinetic behavior of the protein in vivo; non-glycosylated Epo has a very short biological half-life. The DNA sequence of monkey and mouse shows identity at about 90% and 80% to human Epo, respectively. Cellular and Molecular Regulation The Epo gene, which contains at least five exons, resides on chromosome 7q21-q22 in humans and chromosome 5 in mice. The synthesis of Epo in liver and kidney is induced by anemia, decrease in arterial oxygen tension (hypoxia), hepatic viral infection and exposure to hepatotoxic substances. Transcriptional response of the Epo gene to hypoxia is mediated partly by promoter sequences but mainly by a 24 base pair hypoxia-response element located 3' to STET, the human Epo gene. The biological activity of Epo is mediated by specific receptors present at 300 to 3000 copies per cell that undergo phosphorylation in response to Epo. Although pluripotent embryonic stem cells and multipotent hematopoietic cells express the Epo receptor, cells that are committed to non-erythroid lineage cease to express the receptor. The mouse Epo receptor consists of 507 amino acids with an extracellular domain, a single hydrophobic transmembrane domain and a cytoplasmic domain. The human Epo receptor is a 66 kD protein comprised of 508 amino acids. It consists of 8 exons spanning some 6 kb on human chromosome 19p13.3. A point mutation at position 129 of the mouse Epo receptor gene results in constitutive activation of the receptor without stimulation with Epo. Mice infected with a retrovirus expressing this aberrant receptor develop erythroleukemia and splenomegaly. The interaction of Epo with its receptor, a member of the cytokine super-family of receptors, results in the formation of a homodimer and its subsequent internalization (Figure 1). Dimerization of the receptor results in the autophosphorylation of Janus kinase 2 (JAK2), a protein kinase that is tightly associated with the Epo-receptor. Once activated, JAK2 phosphorylates eight tyrosine residues located in the cytoplasmic domain of the Epo receptor. Phosphorylation of the Epo-receptor leads to the recruitment and phosphorylation of a number of signal transduction proteins. One such protein is STAT5, a transcription factor that binds to tyrosine 343 and 401 of the Epo-receptor. Once phophorylated, STAT5 translocates to the nucleus to activate the expression of several genes.

Binding of Epo to its receptor triggers the activation of several signaling cascades. Examples include phosphatidylinositol 3-kinase (PI3K) that binds to tyrosine 479 and is involved in erythroblast survival, and Grb2 that binds to tyrosine 464 and is involved in the activation of the ras pathway. The Epo-receptor mediated activation of phospholipase A2 and C also leads to the release of membrane phospholipids, the synthesis of diacylglycerol and increase in intracellular calcium levels and pH. Since phosphorylation of the Epo-receptor by Epo is diminished after 30 minutes of stimulation, a number of tyrosine phosphatases have been identified that are involved in attenuating the signal. The tyrosine phosphatase SHP-2 binds to tyrosine 401 of the Epo-receptor and stimulates erythroid proliferation, while SHP-1 binds to tyrosine 429 and inhibits proliferation. Abnormalities of the Epo receptor do not appear to play a role in the pathogenesis of hematological diseases. However, prolonged activation of STAT5 has been observed in cells transfected with mutant (tyrosine 429) Epo receptor suggesting that STAT5 DNA binding activity may play a role in the pathogenesis of erythrocytosis.

Preclinical relevance

The synthesis of Epo is subject to a complex circuit that links bone marrow and kidney in a feedback loop. Epo synthesis is increased under hypoxic conditions, where the oxygen sensors in the kidney are believed to be a heme protein. The production of Epo is also influenced by other factors such as testosterone, thyroid hormone and growth hormone. The pathophysiological excess of Epo leads to erythrocytosis that is accompanied by increased blood viscosity and may cause heart failure and pulmonary hypertension. Chronic kidney disease causes the destruction of Epo-producing cells resulting in hyporegenerative normochrome normocytic anemias. Epo is therefore clinically used for the treatment of patients with severe kidney insufficiency (hematocrit below 0.3) occurring in approximately 50% of dialysis patients. In uremic patients, prolonged bleeding times have been shown to be improved by Epo treatments, and hemodialysis treatment with recombinant Epo also improves platelet adhesion and aggregation. Hypertony is an important complication in the treatment of renal anemia with Epo. The main reason for insufficient response to recombinant Epo therapy is iron deficiency, which can be overcome by concomitant intravenous iron administration. An important application of Epo is the pre-surgical activation of erythropoiesis allowing for the collection of autologous donor blood. Epo is also used to treat non-renal forms of anemia caused by chronic infections, inflammation, radiation therapy and cytostatic drug treatment.

Until recently Epo was thought to act solely on the erythroid compartment. However, it is now becoming increasingly clear that Epo has pleiotropic effects. Therefore, anemia on the one hand or tumor growth on the other detected in preclinical studies could well be based on interaction of the test compound with the Epo regulation. Epo deficiency is associated with reduction in red blood cell number while high levels of Epo are known to induce proliferation, chemotaxis and angiogenesis, and to inhibit apoptosis.

Regulatory Environment

The only regulation for Epo is under clinical practice guidelines set out by the American Society of Clinical Oncology and the American Society of Hematology for the treatment of cancer-related anemia. But Epo is not mentioned in any of the relevant guidelines or guideline drafts of immunotoxicity testing.

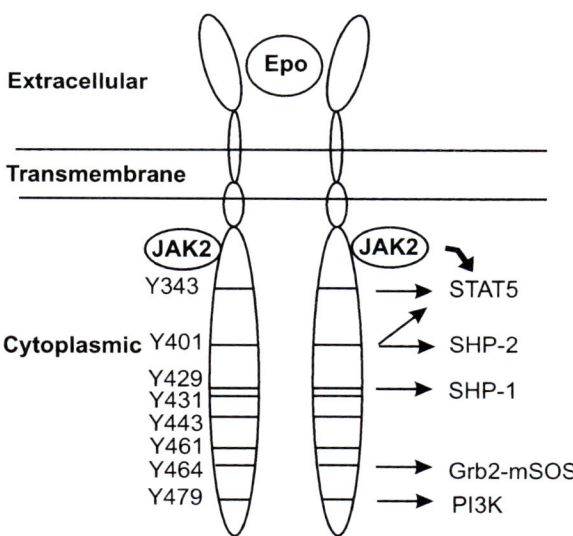

Erythropoietin. Figure 1 Schematic diagram of the Epo Receptor depicting the positions of tyrosine (Y) residues in the cytoplasmic domain and attachment sites of signal transduction proteins such as STAT5, SHP-1 and SHP-2. Binding of Epo to its receptor results in the autophosphorylation and activation of JAK2, which in turn phosphorylates eight tyrosine residues in the cytoplasmic domain of the Epo receptor.

References

1. Krantz SB (1992) Erythropoietin. Blood 77:419–434
2. Jelkman W (1992) Erythropoietin, structure, control of production and function. Physiol Rev 72:449–489
3. Wojchowski DM, Gregory RC, Miller CP, Pandit AK, Pircher TJ (1999) Signal transduction in the erythropoietin receptor system. Exp Cell Res 253:143–156
4. McMullin MF, Percy JM (1999) Erythropoietin Receptor and Hematological Disease. Am J Hematol 60:55–60

ESF

▶ Erythropoietin

Essential Elements

Those transition metals (occupying the middle portions of the periodic table and in most cases characterized by variable oxidation states) essential for life due to their ability to control metabolic and signaling functions. These elements include (but are not limited to) iron, cobalt, zinc, manganese, magnesium, and copper.
▶ Trace Metals and the Immune System

Essential Fatty Acids

Essential fatty acids cannot be synthesized *de novo* in mammalian cells, because they lack the necessary desaturases, and therefore have to be obtained through the diet. There are two essential fatty acids, linoleic acid (an n-6 polyunsaturated fatty acid) and α-linolenic acid (an n-3 polyunsaturated fatty acid).
▶ Fatty Acids and the Immune System

Essential Trace Nutrients

▶ Vitamins

Esterases

Esterases are any of a group of enzymes that hydrolyze esters into alcohols and acids.
▶ Viability, Cell

Estrogen

Estrogen is a steroid hormone produced in the ovaries. Biological activity of estrogen is mediated by interaction with steroid hormone receptors estrogen receptor alpha and beta (ERα and ERβ).
▶ Steroid Hormones and their Effect on the Immune System

Europium

A lanthanide with fluorescent properties that can be utilized as a marker in numerous protein and cell-based immunoassays.
▶ Cytotoxicity Assays

Evaluation of Humoral Immunity

▶ Plaque Versus ELISA Assays. Evaluation of Humoral Immune Responses to T-Dependent Antigens

Evans Syndrome

Co-occurrence of immune-mediated hemolytic anemia and immune-mediated thrombocytopenia.
▶ Antiglobulin (Coombs) Test

Ex vivo

An in vitro assessment made using cells or tissues obtained from animals that have undergone an in-life procedure, e.g. dosing with compound.
▶ Canine Immune System

Experimental Design

▶ Statistics in Immunotoxicology

Exposure Route and Respiratory Hypersensitivity

B Jean Meade · Kimberly J Fairley
National Institute for Occupational Safety and Health
1095 Willowdale Road
Morgantown, WV 26505
USA

Synonyms
pulmonary hypersensitivity

Definition
Respiratory hypersensitivity results first from the induction of humoral- or cellular-mediated sensitization

in the lung. This may be a local or systemic immunological response. Once sensitization develops, subsequent exposure to the sensitizing agent may elicit a clinical response. To help understand the mechanisms underlying the response, Coombs and Gell classified hypersensitivity reactions as types I–IV, with type I responses being mediated by IgE antibody and type IV resulting from T cell activation. Type II and type III responses are mediated by IgG and/or IgM antibodies often coupled with the complement system. Although these classifications are useful, often xenobiotic exposure leads to the development of more than one response.

Characteristics

IgE is a common mechanism of pulmonary hypersensitivity. Both proteins and low-molecular-weight chemicals have the potential to induce IgE-mediated responses. Examples include latex proteins, ovalbumin, and animal proteins, and the low-molecular-weight agents, trimellitic anhydride (TMA), and isocyanates. Regardless of the effector site, following chemical or protein exposure, IgE-mediated disease is initiated by the uptake of allergen by antigen-presenting cells (APCs). APCs and T cells interact, which then allows T cells to differentiate into T helper type 2 (Th2) cells, producing cytokines that promote IgE production. Cytokines involved in driving the IgE response include IL-4 and IL-13. Another Th2 cytokine, interleukin IL-5, is involved in eosinophil chemotaxis and maturation, impacting the inflammatory response associated with IgE-mediated disease. The mast cell growth and differentiation factor, IL-9, is also upregulated following IgE inducing chemical exposure. IL-9 and IL-13 both upregulate mucin production at epithelial surfaces, contributing to the pathophysiology of asthma. Once IgE is produced and enters the circulation, it binds to tissue mast cells, which are abundant in the airways, and to circulating basophils throughout the body. Upon subsequent exposure to allergen, cell-bound IgE is cross-linked resulting in the activation of signal transduction pathways, leading to both early- and late-phase hyperreactivity responses. The early phase response, occurring 5–30 minutes post challenge, results from mast cell and basophil degranulation and the release of preformed mediators including histamine, carboxypeptidase A, and serine and neutral proteases. The late phase occurs approximately 4–24 hours post challenge, and is the effect of the initiation of the arachidonic acid cascade with subsequent production of leukotrienes, prostaglandins, and cytokines, and the recruitment of eosinophils, T cells, and PMNs to the site, all of which mediate the inflammatory response.

Although less frequent, pulmonary responses have also been attributed to hypersensitivity types I–IV. In addition to IgE-mediated disease, there is evidence for type II and type III pulmonary responses following exposure to low-molecular-weight chemicals including TMA and toluene diisocyanate (TDI). Examples of chemicals initiating pulmonary sensitization through a T cell-mediated mechanism include beryllium and picryl chloride.

It is well established that sensitization of the respiratory tract can lead to an increase in airway hyperreactivity and asthma and classically, pulmonary sensitization has been thought to be induced by respiratory tract exposure. Research using murine models has shown that exposure to the upper airways (intranasal) or lower airways (intratracheal) can lead to pulmonary hypersensitivity. When ovalbumin exposure during both sensitization and challenge phases was limited to the upper airways, McCusker and colleagues demonstrated an elevation in serum antigen-specific IgE and an increase in both upper and lower airway inflammation (1). Upper airway exposure to latex proteins has also been shown to induce sensitization, as Thakker et al (2) demonstrated airway hyperreactivity following intravenous challenge in mice previously sensitized to latex proteins via the intranasal route. In addition, Woolhiser et al demonstrated an increase in latex-specific airway hyperreactivity following sensitization by either the intranasal or intratracheal routes of exposure (3).

Recent data has demonstrated that dermal exposure may play a significant role in the development of respiratory tract sensitization. Murine studies have demonstrated the local production of Th2 cytokines and IgE in the lymph nodes draining the site of dermal chemical exposure (4,5). In addition to local IgE production, elevations in systemic levels of IgE have been demonstrated following dermal chemical and protein exposure. Further studies have demonstrated dermal exposure to protein and chemical allergens leading to subsequent airway hyperactivity upon respiratory tract challenge. Dermal exposure of BALB/c mice to ovalbumin has been shown to induce elevated levels of serum IgE and subsequent AHR following a single inhalation dose of ovalbumin and intravenous challenge with methacholine (6). Howell et al (7) demonstrated that exposure of BALB/c mice to latex proteins by the dermal or respiratory routes using a similar dosing protocol resulted in the development of similar levels of IgE and the elicitation of both non-specific and specific airway hyperreactivity following inhalation challenge with methacholine or latex proteins, respectively. Similar results have been found following exposure to low-molecular-weight chemicals. Dermal exposure, as compared to inhalation exposure, to diphenylmethane-4,4'-diisocyanate (MDI) was shown to be more effective in inducing respiratory tract sensitization in guinea pigs (8). Zhang and colleagues demonstrated that dermal exposure to TMA either in vehicle or by dry-powder patch application, induced elevated levels of antigen specific IgE and resulted in the development of both an early phase and late phase pulmonary response following inhalation challenge in Brown Norway rats (9).

Although assessing the route of exposure leading to human sensitization can be difficult, there is human data supporting the potential for dermal exposure to lead to respiratory tract sensitization. New onset asthma-like symptoms were reported from workers at a newly constructed manufacturing plant using methylene diphenyl diisocyanate (MDI). The plant had been designed with engineering controls to reduce worker exposure and a personal protection program was in place including respiratory protection. The prevalence of cases was found to be increased progressively in work areas with low to high exposure to liquid MDI. Additionally, the prevalence of asthma-like symptoms was greater in the group of workers who reported MDI staining on their skin (10). Historically, while efforts to reduce beryllium sensitization focused primarily on respiratory protection, individuals were still becoming sensitized. Recent engineering controls have been implemented to reduce the amount of dermal contact with beryllium. Although only preliminary data has been collected, the data suggest that dermal and respiratory avoidance together may decrease the incidence of beryllium sensitization (11).

In animal models, subcutaneous exposure to low-molecular-weight chemicals and proteins have also been associated with the induction of airway hyperreactivity. Matheson and colleagues demonstrated in C57BL/6 mice that subcutaneous injection followed by inhalation challenge with TDI resulted in increases in IL-4 mRNA, total serum IgE, and non-specific airway hyperreactivity following methacholine challenge (12). Although human exposure to allergens is generally via the dermal, inhalation, or oral routes, the potential for subcutaneous exposure exists. The prevalence of latex allergy among healthcare workers (who are presumably exposed primarily by the dermal and inhalation routes) has been estimated to be as high as 17% (13), but among spina bifida patients (who are potentially additionally exposed to latex proteins via the mucosal route and through surgical procedures and implants) prevalence as high as 65% has been reported (14). Although subcutaneous exposure cannot be directly correlated with the increase prevalence in these children, multiple surgeries were shown to be significantly associated with latex sensitization.

Although uncommon, the potential for the development of respiratory hypersensitivity exists following ingestion of food allergens. Typical foods that induce respiratory reactions include egg, milk, peanuts, soy, fish, shellfish, and tree nuts. People who experience respiratory symptoms to a food allergen generally co-exhibit the more common symptoms of food allergy, including cutaneous and gastrointestinal symptoms (15).

Preclinical Relevance

Animal and computer models have been developed as screening tools for the evaluation of the potential for chemicals to induce respiratory hypersensitivity responses. However, unlike the mouse local lymph node assay (LLNA) and assays for contact hypersensitivity in guinea pigs, none of these assays have been validated or widely accepted as stand alone methods. Most models of respiratory hypersensitivity have focused on the potential of a chemical to induce IgE, or in the case of the guinea pig IgG_1–mediated responses. Early assays focused on the potential of a chemical to induce cutaneous or systemic anaphylaxis. In the passive cutaneous anaphylaxis assay, serum from a previously sensitized animal is injected intradermally into the test animal, which is injected intravenously with antigen and Evans Blue dye 24–48 hours later. The site of intradermal injection is then evaluated for the presence of dye resulting from local vasodilatation and the extravasation of serum indicating an urticarial response. The active cutaneous anaphylaxis assay is conducted similarly, with the difference that sensitization and elicitation occurs in the same animal. In the active systemic anaphylaxis assay, the animal is sensitized with the test article and then injected intravenously with the test article and evaluated for signs of systemic anaphylaxis. Plethysmography has also been used as a tool to evaluate the elicitation of a respiratory response following challenge in previously sensitized animals. Although these models work well for proteins and other species with molecular weights greater than 1000 kD the predictive potential of these assays for low-molecular-weight agents is considered to be poor (16).

Methods have also been developed to evaluate xenobiotics based on the sensitization phase of the response. Robinson and colleagues developed a murine intranasal model for the identification of respiratory sensitizers by evaluating the production of IgG_1 antibodies induced by detergent enzymes (17). As IgE is produced locally in lymph nodes draining the site of chemical exposure, the quantitation of IgE+B220+ cells (B cells binding soluble IgE through the low-affinity IgE receptor CD23) in draining lymph nodes by flow cytometry has been proposed as an additional method for the evaluation of chemical sensitizers (5). The evaluation of serum IgE or IgG_1 levels is frequently used in conjunction with plethysmography. Cytokine profiling both at the message and protein levels is being widely investigated as a screen to differentiate T cell-mediated sentizers from IgE-inducing sensitizers (4). Cytokine profiling is frequently undertaken following the demonstration of the sensitization potential of the chemical in the LLNA. Differentiation has been based on the observation of increased levels of the Th2 cytokines, primarily IL-4 and IL-10 following exposure to IgE-inducing sensitizers versus an increase in interferon-γ in animals exposed to chemicals with the potential to induce delayed-type hypersensitivity responses. Although these assays show promise, issues still remain regarding optimization of the assays as related to exposure protocols and the timing of evaluation as well as the appropriate endpoint, RNA versus protein levels.

Gene arrays are also being investigated as a potential tool to identify sensitizing chemicals. Using cluster analysis, He et al demonstrated differential gene expression in murine cells from lymph nodes draining the site of exposure to IgE inducing chemicals, irritants, and chemicals with the potential to induce a T cell-mediated response (18). Pennie and Kimber have proposed the evaluation of gene expression by dendritic cells, antigen-presenting cells that play a key role in the initiation of sensitization, as a means to identify and differentiate chemical sensitizers (19). These studies can possibly lead to the development of custom gene chips to identify and differentiate sensitizers based on their gene expression profile.

With efforts to reduce the use of animal testing, numerous quantitative structure activity relationship (QSAR) models are being developed. Although the majority of these models have focused on the identification of irritants and contact sensitizers, attempts have been made to model respiratory sensitizers (20). The MultiCASE model gives users the ability to self-define the database. Using this database, Karol and colleagues developed a model for respiratory sensitizers based on a learning set of 39 respiratory sensitizers (selected from the literature including both human and animal asthmagens) and 39 non-sensitizers. If QSAR models are to be effective in identifying the potential for chemicals to induce sensitization by differing routes of exposure, the bioavailability of the chemicals by each route must be factored into the model.

In an attempt to integrate in vitro and in vivo testing for the identification of low-molecular-weight respiratory sensitizers, Sarlo and Clark developed a 4-tier identification system. In this model, tier 1 includes evaluation of structure activity followed by in vitro haptenization studies in tier 2. When positive results are obtained in tiers 1 and 2, further in vivo studies are performed. Tiers 3 and 4 evaluate the immunogenicity and allergeneticity in guinea pigs following injection or inhalation exposures, respectively (21). Although none of these methods have been accepted as stand-alone tests, significant progress is being made in the area of preclinical identification of respiratory sensitizers.

Regulatory Environment

In relation to respiratory tract sensitization, the US FDA provides regulatory guidance only for drugs that are to be administered via inhalation. The recommendation is for the use of the guinea pig inhalation induction and challenge assay; however alternative assays can be used if appropriateness is demonstrated (16). At this time the US EPA does not require testing for respiratory hypersensitivity-inducing potential.

Relevance to Humans

Classically, respiratory sensitization has been considered to occur following inhalation exposure to chemicals and proteins. Recent studies have demonstrated that alternative routes of exposure may be involved in the sensitization phase of the immune response. Consequently, dermal as well as respiratory protection may be required to prevent respiratory tract sensitization.

References

1. McCusker C, Chicoine M, Hamid Q, Mazer B (2002) Site-specific sensitization in a murine model of allergic rhinitis: Role of the upper airway in lower airway disease. J Allergy Clin Immunol 110:891–898
2. Thakker JC, Xia JQ, Rickaby DA et al. (1999) A murine model of latex allergy-induced airway hyperreactivity. Lung 177:89–100
3. Woolhiser MR, Munson AE, Meade BJ (2000) Immunological responses of mice following administration of natural rubber latex proteins by different routes of exposure. Toxicol Sci 55:343–351
4. Dearman RJ, Betts CJ, Humphreys N et al. (2003) Chemical allergy: Considerations for the practical application of cytokine profiling. Toxicol Sci 71:137–145
5. Manetz TS, Meade BJ (1999) Development of a flow cytometry assay for the identification and differentiation of chemicals with the potential to elicit irritation, IgE-mediated, or T cell-mediated hypersensitivity responses. Toxicol Sci 48:206–217
6. Spergel JM, Mizoguchi E, Brewer JP, Martin TR, Bhan AK, Geha RS (1998) Epicutaneous sensitization with protein antigen induces localized allergic dermatitis and hyperresponsiveness to methacholine after single exposure to aerosolized antigen in mice. J Clin Investigat 101:1614–1622
7. Howell MD, Weissman DN, Meade BJ (2002) Latex sensitization by dermal exposure can lean to airway hyperreactivity. Int Arch Allergy Immunol 128:204–211
8. Rattray NJ, Botham PA, Hext PM et al. (1994) Induction of respiratory hypersensitivity to diphenylmethane-4,4'-diisocyanate (MDI) in guinea pigs. Influence of route of exposure. Toxicology 88:15–30
9. Zhang XD, Fedan JS, Lewis DM, Siegel PD (2002) Airway responses after specific challenge of rats sensi-

tized via skin exposure to trimellitic anhydride (TMA) The Toxicologist 66S:1184
10. Petsonk EL, Wang ML, Lewis DM, Siegel PD, Husberg BJ (2000) Asthma-like symptoms in wood product plant workers exposed to methylene diphenyl diisocyanate. Chest 118:1183–1193
11. Deubner D (2003) Inclusion of skin exposure reduction in a total hygiene program to reduce exposure to beryllium: Background and results. The Toxicologist 72:S94
12. Matheson JM, Lange RW, Lemus R, Karol MH, Luster MI (2001) Importance of inflammatory and immune components in a mouse model of airway reactivity to toluene diisocyanate (TDI). Clin Exper Allergy 31:1067–1076
13. Yassin MS, Lierl MB, Fischer TJ, O'Brien K, Cross J, Steinmetz C (1994) Latex allergy in hospital employees. Ann Allergy 72:245–249
14. Meade BJ, Weissman DN, Beezhold D (2002) Latex allergy: Past and present. Int Immunopharmacol 2:225–238
15. James JM (2003) Respiratory manifestations of food allergy. Pediatrics 111:1625–1630
16. DHHS, US FDA, Center for Drug Evaluation and Research (2002) Guidance for industry: Immunotoxicology evaluation of investigational new drugs. FNL 4945, Rockville MD
17. Robinson MK, Babcock LS, Horn PA, Kawabata TT (1996) Specific antibody responses to subtilisin carlsberg (alcalase) in mice: Development of an intranasal exposure model. Fund Appl Toxicol 34:15–24
18. He B, Munson AE, Meade BJ (2001) Analysis of gene expression induced by irritant and sensitizing chemicals using oligonucleotide arrays. Int Immunopharmacol 1:867–887
19. Pennie WD, Kimber I (2002) Toxicogenomics; transcript profiling and potential application to chemical allergy. Toxicol in Vitro 16:319–326
20. Karol MH, Graham C, Gealy R, Macina OT, Sussman N, Rosenkranz HS (1996) Structure-activity relationships and computer-assisted analysis of respiratory sensitization potential. Toxicol Lett 86:187–191
21. Sarlo K, Clark ED (1992) A tier approach for evaluating the respiratory allergenicity of low-molecular-weight chemicals. Fund Appl Toxicol 18:107–114

Expression Profiling

▶ Toxicogenomics (Microarray Technology)

Extracellular Matrix

Extracellular matrix (ECM) is the connective tissue filling up the space between cells, consisting of a network of protein fibers in a polysaccharide matrix. The compounds making up the extracellular matrix are mainly secreted by fibroblasts. The extracellular matrix serves as structural element of tissues to which cells attach through cell–substrate adhesion molecules.
▶ Cell Adhesion Molecules

Extravasation

Escape of fluids or cells from the vessels into the tissues. The extravasation of leukocytes is dependent on their interactions with endothelial cells, mediated by adhesion molecules, and chemotactic signals provided by chemokines.
▶ Chemokines

Extravascular Hemolysis

Red blood cell destruction occurring inside and outside the blood vasculature.
▶ Hemolytic Anemia, Autoimmune

Extrinsic Control, Neural and Humoral

Complicated, and still not completely understood, system of different neural and humoral mechanisms which influence lymphatic contractility to match it to the current body conditions.
▶ Lymph Transport and Lymphatic System

F

FACS

Fluorescent-activated cell sorter.
▶ Maturation of the Immune Response
▶ Flow Cytometry

Facultative Anaerobes

Bacteria that can grow in air using oxygen as a terminal electron acceptor or anaerobically using fermentation reactions to obtain energy.
▶ Streptococcus Infection and Immunity

Fas

A member of the tumor necrosis factor (TNF) receptor family that is expressed on the surface of many different types of cells. Expression of Fas renders a cell susceptible to killing by cells that express or secrete Fas ligand.
▶ Cell-Mediated Lysis
▶ Cytotoxic T Lymphocytes

Fats

▶ Fatty Acids and the Immune System

Fatty Acids and the Immune System

PARVEEN YAQOOB
School of Food Biosciences
The University of Reading Whiteknights
P.O.Box 226
Reading
RG6 6AP
UK

Synonyms

Fats, lipids, PUFA, polyunsaturated fatty acids, eicosapentanoic acid, EPA, docosahexanoic acid, DHA.

Definition

Fatty acids are hydrocarbon chains, which can be saturated, monounsaturated, or polyunsaturated (PUFA). There are two ▶ essential fatty acids, linoleic and α-linolenic acid, which cannot be synthesized de novo in animal cells. Linoleic acid is an n-6 PUFA present in large quantities in many vegetable oils; it can be described by its shorthand notation of 18:2n-6, which refers to an 18-carbon fatty acid with two double bonds, the first of which is on carbon atom 6 from the methyl end. α-Linolenic acid is an n-3 PUFA present in green leafy vegetables and some seed oils (eg linseed oil); its shorthand notation is 18:3n-3, describing an 18-carbon fatty acid with three double bonds, the first being positioned at carbon atom 3 from the methyl end. Both essential fatty acids can be further elongated and desaturated in animal cells forming the n-6 and n-3 families of PUFA (see Figure 1); the major endproduct of the n-6 pathway is arachidonic acid and the major endproducts of the n-3 pathway are ▶ eicosapentanoic acid (EPA) and ▶ docosahexanoic acid (DHA).

Since plankton is very rich in α-linolenic acid, fish contain lipid which contains a high proportion of the long chain n-3 PUFA, EPA, and DHA. ▶ Oily fish, such as salmon, herring and mackerel, store this lipid throughout their flesh, whereas lean fish such as cod store it in their livers.

In animal cells, there is competition between linoleic acid and α−linolenic acid for the Δ6 desaturase. The

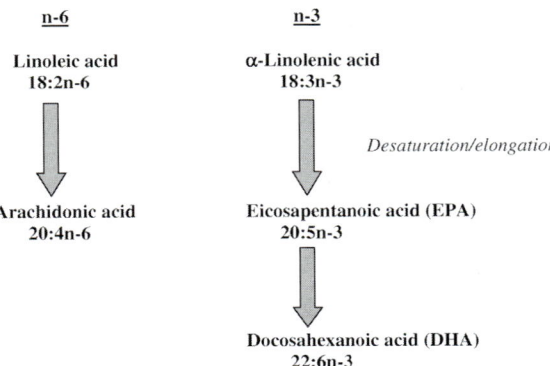

Fatty Acids and the Immune System. Figure 1 Key metabolites in essential fatty acid metabolism.

conversion of linoleic acid to arachidonic acid dominates in humans, since most diets contain relatively greater quantities of linoleic acid compared with α-linolenic acid. Furthermore, in humans the conversion of α-linolenic acid to EPA is very inefficient and virtually no α-linolenic acid is converted to DHA. For this reason, EPA and DHA are chiefly derived from consumption of oily fish. Consumption of oily fish or supplements containing concentrated fish oils results in enrichment of cell membranes with EPA and DHA and this occurs due to partial replacement of arachidonic acid in cell membranes.

The PUFA linoleic acid, arachidonic acid, EPA and DHA perform important roles in cell structure, neural function, growth and development and immune function. In addition to the n-6 and n-3 families of fatty acids, an n-9 family derived from the metabolism of oleic acid, which is monounsaturated, also exists. However, saturated and ▶ monounsaturated fatty acids are not essential, since they can be synthesized de novo in mammalian cells.

Characteristics

The fatty acid composition of leukocytes is sensitive to changes in the fatty acid composition of the diet. The n-6 PUFA, arachidonic acid, is a precursor for the synthesis of a family of hormone-like compounds termed ▶ eicosanoids (including prostaglandins, leukotrienes, lipoxins, and thromboxanes), which have important roles in the regulation of inflammation and immunity. Consumption of n-3 PUFA can have a marked impact on the synthesis of eicosanoids, since these fatty acids can partially replace arachidonic acid in cell membranes. Thus, the synthesis of arachidonic acid-derived eicosanoids can be decreased by up to 75% following consumption of fish oil, depending on the dose and duration. Furthermore, EPA can itself act as an alternative precursor for the synthesis of a separate and distinct family of eicosanoids, which often have different biological actions or potencies than those formed from arachidonic acid. Consumption of large quantities of fish oil results in suppression of some lymphocyte and macrophage functions in animals and may have similar, albeit more conservative, effects in humans (1,2). The n-3 PUFA are considered to have anti-inflammatory properties, which may be achieved by selective suppression of Th1 responses (1,2). Although some of the effects of n-3 PUFA may be brought about by the modulation of the amount and type of eicosanoids produced, it is evident that some of the effects of fatty acids on immune function are elicited through eicosanoid-independent mechanisms (1,2). These mechanisms are not well understood.

Preclinical Relevance

The current ratio of intake of n-6 PUFA to n-3 PUFA is approximately 6:1. This ratio is believed to have shifted considerably over the past 50 years, in favor of the n-6 PUFA. The dietary intake of n-3 PUFA can be supplemented by fish oil capsules, which provide varying doses. These products are widely available from health food retailers and selected products are licensed for use as prescribed medication in the ▶ secondary prevention of ▶ myocardial infarction and also for lowering blood ▶ triglycerides.

Relevance to Humans

Consumption of n-3 PUFA may have a role in the prevention and/or therapy of chronic inflammatory diseases, particularly those characterized by a dominant Th1-type response and excessive production of arachidonic acid-derived eicosanoids. Evidence for therapeutic effects of fish oil in rheumatoid arthritis exists, but claims for therapeutic effects in inflammatory bowel disease, asthma and inflammatory skin conditions remain controversial. Although it might be anticipated that consumption of large amounts of n-6 PUFA (abundant in vegetable oils) could predispose an individual to chronic inflammatory disorders, there is no evidence to suggest this is the case. Arachidonic acid is not present in the diet in large amounts and most of the arachidonic acid in cells is derived from dietary linoleic acid. Increased consumption of linoleic acid does not significantly increase the proportion of arachidonic acid in tissues and therefore does not increase eicosanoid synthesis.

Consumption of n-3 PUFA is recognized to be associated with ▶ secondary prevention of cardiovascular disease and may also have a role in primary prevention of cardiovascular disease. These fatty acids have multifaceted actions, including effects on cardiac arrhythmias, blood lipids and blood clotting. However, cardiovascular disease is increasingly recognized to have a chronic inflammatory component and inflammation within atherosclerotic lesions is at least partly responsible for their rupture, triggering an acute (and potentially fatal) cardiovascular event. It is therefore possible that the antiinflammatory actions of n-3 PUFA contribute to their protective effects in cardiovascular disease.

Patients who have suffered a myocardial infarction and patients with rheumatoid arthritis are routinely encouraged to increase their consumption of oily fish or to supplement their diet with fish oil. Patients who have suffered a myocardial infarction can, in some countries, be prescribed selected fish oil supplements, which are patented for this application.

A potential concern regarding the effects of n-3 PUFA on aspects of immune function is that they may inadvertently impair host defense. This has been demonstrated in some animal experiments, but it is unlikely that this would be the case in humans because the quantities of n-3 PUFA required are too high to be achievable in a normal diet or at recommended levels of supplementation with fish oil.

Regulatory Environment

A number of countries provide specific recommendations for the optimum intake of n-3 PUFA, but these can vary considerably between different countries and even between different bodies within the same country. For example, in the UK, the Department of Health in 1994 recommended an intake of EPA and DHA of 1.5 g per week, equivalent to one small serving of oily fish. This has remained the same since 1994, although the Scientific Advisory Committee on Nutrition concluded in 2002 that the scientific evidence for the recommendations was now stronger and that there may be a case for the recommended intakes to be increased. In contrast, the British Nutrition Foundation recommends intakes of 8 g per week for women and 10 g per week for men, which is equivalent to 2–3 medium portions of oily fish per week (3,4). An international workshop held in the US in 1999 recommended intakes of long-chain n-3 PUFA of 4.55 g per week. These higher recommendations are loosely based on the doses reported to be required for secondary prevention of cardiovascular disease (0.5–1 g/day). There are no recommended upper limits of intake of n-3 PUFA. However, it is recognized that n-3 PUFA are particularly susceptible to oxidation and for this reason fish oil supplements are normally fortified with vitamin E, a lipid-soluble antioxidant.

References

1. Calder PC, Yaqoob P, Thies F, Wallace F, Miles EA (2002) Fatty acids and lymphocyte functions. Br J Nutr 87:S31–S48
2. Yaqoob P (2003) Lipids and the immune response—from molecular mechanisms to clinical applications. Curr Opin Clin Nutr Metab Care 6:133–150
3. British Nutrition Foundation (1992) Unsaturated Fatty Acids: Nutritional and Physiological Significance. Report of the British Nutrition Foundation's Task Force. Chapman and Hall, London
4. British Nutrition Foundation (1999) N-3 Fatty Acids and Health Briefing Paper. The British Nutrition Foundation, London

FBS

▶ Colony-Forming Unit Assay: Methods and Implications

FC

▶ Flow Cytometry

Fc Region

Originally defined as a crystalizable fragment of an antibody after digestion with the proteolytic enzyme papain, the Fc region consists of the CH2 and CH3 domains of an antibody molecule.

▶ Monoclonal Antibodies

FcR

Receptors for the Fc portion of immunoglobulins (FcR) recognize the Fc portion of IgG (FcγR), IgA (FcαR) or IgE (FcεR) and are able to trigger a number of immune effector and regulatory functions.

▶ Antibody-Dependent Cellular Cytotoxicity

Fcγ Chain

The Fcγ chain plays an essential role in the expression and signaling of FcεRI, FcγRI, FcγRIII and FcαRI.

▶ Antibody-Dependent Cellular Cytotoxicity

FDA

The United States governmantal agency responsible for regulating and licensing pharmaceuticals, food additives, medicial devices, biologics, cosmetics and animal food and drugs.

▶ Assays for Antibody Production

Fetal Bovine Serum (FBS)

FBS is added to the culture media of hematopoietic cells.

▶ Colony-Forming Unit Assay: Methods and Implications

FEV_1

Forced expiratory volume in the first second of exhalation.

▶ Asthma

Fibronectin

High-molecular-weight multifunctional glycoprotein found on cell surface membranes in body fluids. They function as adhesive ligand-like molecules that play a role in contact inhibition.

▶ Respiratory Infections

Fibrosis

The formation of fibrous tissue, which is tissue at a wound site that is initially vascularized but later becomes avascular and dominated by collagen.

▶ Systemic Autoimmunity

Field Studies

▶ Epidemiological Investigations

Fish Immune System

BETTINA HITZFELD
Substances, Soil, Biotechnology Division Swiss Agency for the Environment, Forests and Landscape
CH-3003 Bern
Switzerland

Definition

In principle the immune system of the fish equals that of other vertebrates. However, some smaller differences, especially in histopathology and location of important immune competent cells, are to be mentioned.

Characteristics

The functions of the fish immune system are equivalent to those in other vertebrates: resistance against disease and protection against neoplastic cells (1). Fish comprise the largest vertebrate class, are very diverse in evolutionary terms and may be divided into jawless fish (such as the lampreys) and jawed fish. The latter can be further subdivided into cartilaginous fish (e.g. sharks) and bony fish (e.g. teleosts). The following discussion will be restricted to the modern teleosts such as the trout, since their immune system is the most intensively studied.

Immune Organs

The immune organs in teleost fish differ somewhat from those known in other vertebrates such as mammals. They comprise the thymus, kidney, intestinal tract, liver, and spleen (2).

Thymus

The thymus is located under the upper half of the gill operculum and may be distinguished by its white tissue surrounded by a thin layer of skin (1). It is far less developed than in mammals, but also seems to consist of a medulla containing immature T lymphocytes, and a cortex containing smaller, more developed lymphocytes. These T lymphocytes seem to be involved in allograft rejection, increased macrophage function, and B-cell stimulation (2). The size of the thymus is very much dependent on age, season (spawning) and stress: it has been shown to become involuted with age or after long periods of stress.

Kidney

The kidney in teleosts is a dark red organ located along the ventral surface of the vertebrae (2). It is the anterior part of the kidney in particular, the so-called anterior or head kidney, which performs important immune functions. The anterior kidney is the functional equivalent of the bone marrow in mammals and is involved in ▶ hematopoiesis. Mitotic and large immature lymphocytes can be seen in the anterior kidney (3). But immune cells located in the kidney also perform other immune functions, such as ▶ phagocytosis and antigen processing. The kidney contains antibody-producing B lymphocytes, which are important in the development of humoral immunity and immunological memory. Furthermore, aging blood cells and particulate matter are filtered by the reticular endothelial system of circulating and tissue macrophages (3).

Intestinal Tract
The intestinal tract has been shown to be an important area of antigen uptake and processing (2). In trout it is located along the lowest portion of the posterior peritoneal cavity. In contrast to mammals, the fish intestine does not contain aggregates of lymphocytes. The granular cells of the stratum granulosum have, however, been implicated in mucosal immunity (3). This gut-associated lymphoid tissue includes macrophages, B lymphocytes, immunoglobulin-negative T lymphocytes and natural cytotoxic cells.

Liver
The liver is a large organ located in the anterior portion of the peritoneal cavity and also performs many of the functions known in mammals (2). The liver is furthermore involved in presentation of particulate antigens; evidence for this comes from the presence of phagocytic mononuclear cells in the liver sinusoids. They are considered to be equivalent to mammalian ▶ Kupffer cells (3). Rainbow trout liver is also capable of producing ▶ C-reactive protein (CRP) (3).

Spleen
The spleen, as in other vertebrates, is the major filter of blood-borne antigens but it also performs immunopoietic functions (1). In teleost fish it is involved in hematopoiesis and may have immune functions that are comparable to lymph nodes in mammals (fish do not have lymph nodes) (3). It is situated in the lower posterior abdominal cavity, has a smooth texture and a dark red color. The spleen has an outer capsule consisting of connective tissue and a pulp matrix. The pulp contains both haematopoietic red pulp and lymphopoietic white pulp. The spleen may retain a large number of mature erythrocytes, which can be released into the circulation when needed, and it is the major site of thrombocyte production (4).

Blood Cells
The blood of teleost fish contains many of the cells known from mammalian blood. In contrast to mammalian blood one may, however, also find precursor cells, which in mammals can only be found in bone marrow. Fish erythrocytes contain basophilic nuclei and are larger (13–16 μm long) than mammalian erythrocytes (3). In a rainbow trout (*Oncorhynchus mykiss*) blood smear the immature, mature and degenerating red blood cells can be distinguished (see Figures 1 and 2). The leukocyte subpopulation is mainly made up of lymphocytes (> 90%) followed by neutrophilic granulocytes, thrombocytes, monocytes and natural cytotoxic cells (NCC) which are the equivalent to the natural killer cells (NK) of the mammalian immune system (Figure 2).

The presence of B lymphocytes has been shown in many teleost species by using monoclonal antibodies against fish immunoglobulin (Ig, sIgM$^+$) and through the identification of heavy and light chain genes (4). In contrast to B cells, fish T cells have been described only by the absence of surface Ig (sIgM$^-$) (5) and through functional assays (4). In rainbow trout, neutrophilic granulocytes can be recognised by their multilobed nuclei, while basophilic or eosinophilic granulocytes are rarely seen. In carp (*Cyprinus carpio*), on the other hand, all three types may be found (4). The leukocyte functions are largely equivalent to their mammalian counterparts. Teleost monocytes and neutrophilic granulocytes are chemotactic as well as phagocytic and bactericidal. Macrophages have also been described in the tissues of fish (4). Like their mammalian counterpart, teleost macrophage stain nonspecific esterase, acid phosphatase, periodic acid-Schiff reagent, and peroxidase positive. A unique feature of fish is the presence of so-called ▶ melanomacrophage aggregates which are found in many lymphoid tissues, such as spleen, liver, and kidney, and also the gonads (4). Thrombocytes have been shown to release clotting factors in response to collagen exposed by wounds (3). Natural cytotoxic cells, small agranular lymphocytes,

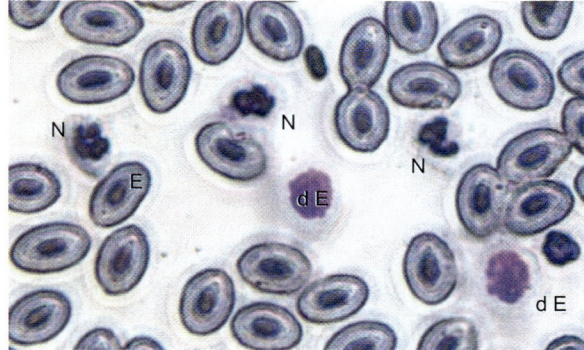

Fish Immune System. Figure 1 Typical blood smear of Rainbow trout (*Oncorhynchus mykiss*) peripheral blood. Nucleated erythrocytes (E), degenerate erythrocytes (d E), and mature neutrophilic granulocytes (N) can be seen.

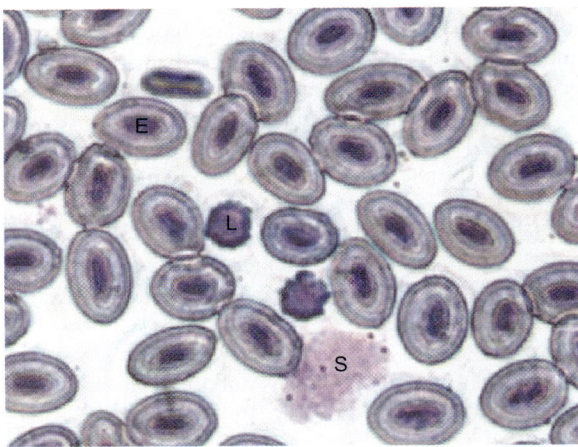

Fish Immune System. Figure 2 Typical blood smear of Rainbow trout (*Oncorhynchus mykiss*) peripheral blood. Nucleated erythrocytes (E), smudge cells (S; degradation form of erythrocytes), and lymphocytes (L) can be seen.

are functionally similar to mammalian natural killer (NK) cells.

Nonspecific Immune System

Fish live in an aquatic environment, which is very conducive to the transmission of disease-bearing organisms (1). Fish therefore possess a well-developed nonspecific immune system. The skin, ▶ lateral line, and gills are the first line of defence against pathogens (2). The skin of fish is coated by mucus, which is continuously secreted by goblet cells and contains antibodies as well as lysozyme. The cells of the epidermis are the next barrier followed by the scales (1). Fish are often more vulnerable in areas not covered by mucus or scales, such as the gills; macrophages may therefore be found on gill surfaces. It has furthermore been shown that antigens originating from pathogens such as *Yersinia ruckeri* bacteria, given as suspension in water, are taken up very effectively by rainbow trout and trigger strong immune responses (2). Other pathogens such as *Aeromonas salmonicida* do not elicit such a strong response. The nonspecific immune response in rainbow trout relies on the activation of mononuclear phagocytes (i.e. macrophages) and on polymorphonuclear granulocytes (PMN), which can be stimulated by opsonic antibodies and complement (1). In fish, complement activation can arise both from the classical and the alternative pathway (4). Both pathways lead to lysis of pathogens, but teleosts also possess other lytic activities in their serum, whose identity has not been elucidated yet. Fish are capable of inflammatory reactions (1,4) which involve migration of neutrophils, eosinophils, basophils (not in all species), macrophages, and lymphocytes to the site of infection; even slight increases in temperature have been detected (1). Chemotaxis to the site of inflammation is stimulated by soluble factors such as cytokines, eicosanoids, and complement components. Not many cytokines have unequivocally been shown in teleost fish. Interleukin(IL)-1, transforming growth factor (TGF)-β, and tumor necrosis factor(TNF)-α show biological activity in fish, suggesting the presence of receptors for these cytokines (4). Molecular cloning and sequencing has furthermore suggested the presence of an IL-2-like, IFN-like, IL-1β, and an TGF-β gene. It has also been found that carp macrophages and neutrophilic granulocytes may secrete an IL-1-like factor and supernatants from carp leukocyte cultures stimulated with mitogen or alloantigen were found to contain IL-2-like lymphocyte growth-promoting activity.

Specific Immune System

As mentioned before, fish are not only able to take up antigens in the water through their skin, the gills, the lateral line but also orally (1). After antigen presentation via major histocompatibility complex (MHC) I or II, T and B cells are activated, produce cytokines and thus induce plasma cells to produce antibodies (5). Up to the present time, only IgM antibodies have been identified in fish. Furthermore, in contrast to mammalian IgM, which is pentameric, fish IgM is a tetrameric molecule. After immunization, fish produce specific antibodies that have properties such as agglutination, precipitation, complement fixation, ▶ opsonization and skin sensitization (4). Isotype switching, rapid titer increases, affinity, or maturation have not been described in fish species. The production of antibodies in fish is largely temperature dependent and immunizations as well as cultivation of leukocytes should be carried out at the optimum temperature for the species, e.g. 15–20°C for trout (5). As mentioned above, fish have been shown to express MHC class I and II molecules. Using molecular cloning, full-length MHC class I α-chain cDNA and full-length MHC class II β-chain have been described in, for example, Atlantic salmon (*Salmo salar*), carp (*C. carpio*) and zebrafish (*Danio rerio*) and in rainbow trout (*O. mykiss*) (4). MHC class II β-chain has been found to be expressed in thymus, head kidney, spleen and peripheral blood of both carp and Atlantic salmon, albeit at differing expression levels.

Relevance to Humans

Apart from the scientific interest in the immune system of fish from a functional or an evolutionary perspective, scientists are also studying the fish immune system from a toxicological viewpoint. Fish are an integral part of the aquatic environment and are thus exposed to a wide variety of pollutants such as heavy metals, pesticides (1) and/or pharmaceuticals. Fish are therefore indicators of the quality of rivers, lakes or oceans and as such are very useful in ecotoxicological studies. A variety of fish species is furthermore an important food source and their safe "production" in aquaculture systems is gaining increasing significance (5). In order to obtain healthy and contaminant-free food for an increasing human population it is thus important to understand not only basic toxicological characteristics of potential pollutants, such as bioconcentration in various tissues but also to guarantee sustainable fish populations. The observed susceptibility of farmed fish to infectious diseases may be due to immunosuppression as a result of high population densities and thus to stress but also as a result of environmental contaminants (5). It is thus important to develop test systems that are able to detect immunosuppressive or immunostimulatory changes to immune functions in fish. These test system are not yet as developed as the ones for mammals, but due to the increasing interest in and understanding of the immune system of fish, these assays will become available in due course.

Regulatory Environment
There is currently no regulatory requirement for testing chemicals on fish immune parameters.

References
1. Anderson D, Zeeman M (1995) Immunotoxicology in fish. In: Rand G (ed) Fundamentals of Aquatic Toxicology: Effects, Environmental Fate, and Risk Assessment, 2rd ed. Taylor and Francis, Washington DC, pp 371–404
2. Powell D (2000) Gross functional anatomy: immune system. In: Ostrander G (ed) The Laboratory Fish. Academic Press, San Diego, pp 219–223
3. Powell D (2000) Microscopic functional anatomy: immune system. In: Ostrander G (ed) The Laboratory Fish. Academic Press, San Diego, pp 441–449
4. Pastoret P-P, Griebel P, Bazin H, Govaerts A (eds) (1998) Handbook of Vertebrate Immunology. Academic Press, San Diego
5. Köllner B, Wasserrab B, Kotterba G, Fischer U (2002) Evaluation of immune functions of rainbow trout (*Oncorhynchus mykiss*)—how can environmental influences be detected? Toxicol Lett 131:83–95

Fixed Drug Eruptions (FDE)

Fixed drug eruptions (FDE) are drug-induced cutaneous reactions that can occur on any part of the body, but characteristically recur at the same site when the causative drug is given. Although typical lesions are easily identified because of their hyperpigmentation, it is difficult to recognize non-pigmented variants of FDE as being "fixed", where the lesions are symmetrical and not followed by pigmentation. Unusual forms of FDE are likely to be misdiagnosed as erythema multiforme and other skin diseases.

▶ Drugs, Allergy to

Flow Cytometry

A technique used to analyze cell suspensions on the basis of fluorescence. For instance, cocktails of fluorescently conjugated monoclonal antibodies which recognize different CD antigens can be used to measure simultaneously different leukocyte populations from samples obtained from toxicology and pharmacology studies.

▶ Canine Immune System

Flow Cytometry Technique

DANIELLE ROMAN
PCS Toxicology/Pathology
Novartis Pharma AG
Muttenz
Switzerland

Synonyms
FACS, fluorescence activated cell sorter, FC

Short Description
Flow cytometry is a general method for rapidly analyzing large numbers of cells (usually 10,000 or more) individually using light-scattering, fluorescence, and absorbance measurements. The power of this method lies both in the wide range of cellular parameters that can be determined, and in the ability to obtain information on how these parameters are distributed in the cell population.

Information on the immune system is acquired using numerous established assays and procedures in animal or in vitro models which are generally referred to as tier testing. Tier I assays constitute a broad stroke analysis of chemical effects on humoral-mediated and ▶ cell-mediated immunity in animals (usually mice). Tier II tests are designed to assess specific target cells and responses affected by chemicals and to give insights into mechanisms of immunotoxicity. Flow cytometry is routinely used in tier II tests to assess effects of chemicals on subsets of lymphoid cells. During recent years, the uses of multiparameter flow cytometry assays have been expanded to give valuable information on biochemical changes occurring in subsets of lymphoid cells and insights on cellular and molecular mechanisms of immunotoxicity (1).

The purpose of clinical investigators is the characterization of parameters that can be assessed by flow cytometry and that serve as surrogates for measures of immune function. Conversely, preclinical investigators are focused on characterising sensitive models of immune function (i.e. ▶ primary immune responses) and have sought to apply flow cytometry to animal studies to more fully understand the mechanisms of immunotoxicity or to explore its potential ability to predict results likely to occur in humans (2).

Characteristics

Flow cytometry can be used to identify, quantify and isolate cells within a heterogeneous population. Analysis of cell surface markers can aid in the detection of direct immune toxicants affecting particular cell populations. Rapid and objective measurements can be made on single cells with high accuracy and reproducibility. The advent of clinical flow cytometers and the availability of a multitude of fluorescent probes have rendered this technology appropriate for the use in the toxicologic clinical pathology laboratory (3).

Cell Surface Markers and Cell Activation Indicators

Analysis of cell surface markers can aid in the detection of direct immune toxicants affecting particular cell populations.

Lymphocyte ▶ immunophenotyping can be performed on the blood, bone marrow, spleen, thymus, and lymph nodes. The appropriate tissue to use for immunophenotyping of laboratory animals may need to be determined on a case-by-case basis. Bone marrow, spleen, thymus, and lymph node cell sampling involves a fatal surgical procedure, and the small blood volume of mice limits the potential for repeated blood sampling. In addition, the cellular composition of lymphoid organs varies from tissue to tissue and from species to species.

Immunophenotyping of the spleen rather than the peripheral blood should be done when the goal is hazard identification and the prediction of immunotoxic effects.

There are numerous cell surface markers of cell activation that can be detected in different species with fluorescent antibody systems. The T cell activation complex, which is a component of the inteleukin-2 receptor system, has long been used as a marker for T cell activation. B cell activation can be measured by examining the major histocompatibility complex class II antigen expression. There has been also interest in the use of cell adhesion molecules, such as CD62L and CD44, as a function of cell activation and sensitization (3). Finally, intracellular staining of gene products such as cytokines has recently become a routine technique in flow cytometry (3).

Cell Cycle, Viability and Apoptosis Indicators

There has been considerable interest from immunotoxicologists about the use of DNA probes for cell cycle analysis, because many xenobiotics affect cell proliferation or cause alteration in DNA (3).

▶ Apoptosis is a form of cell death, distinct from necrosis, that has been observed in many tissues and that plays a critical role in both development and homeostasis of the immune system: thymic selection, cytotoxicity, deletion of ▶ autoreactive cells, and regulation of the size of the lymphoid compartment by activation-induced cell death. Numerous protocols have been developed and use antibodies to Bcl-2 family proteins, ▶ caspases, and p53 for the detection of early molecular events associated with the induction of apoptosis (3).

Assessment of Cytotoxic T Lymphocyte Function and Natural Killer Cell Activity Assays

▶ Cytotoxic T lymphocytes (CTL) represent one of several types of cells of the immune system that have the capacity to directly kill other cells. They also play a major role in host defense against infections. The ▶ natural killer (NK) cell assay is used to evaluate the host resistance system responsible for the elimination of virally infected cells and certain tumor cells including the regulation of various specific and non-specific host defense parameters.

Assessment of the CTL and NK cells function by phenotypic analysis represents an alternative to the conventional ^{51}chromium release assay. Standardization and validation studies are ongoing. The flow cytometric assay can be also completed with additional studies that further define the time course of the changes in the expression of surface adhesion and activation molecules expression during CTL activation and differentiation (1). This method facilitates also the incorporation of the NK cell activity assays into repeated-dose toxicity studies as fewer cells are needed and it circumvents the need to use radioactive material.

Cell Sorting

In cell sorting, positive or negative selection methods can be used to identify target cells with defined parameters. Cells can be physically separated into populations that can be assessed for immunotoxicity, or cell separation-reconstitution studies can be performed (3).

Immediate- or Delayed-Types of Hypersensitivity

Immediate-type or delayed-type of hypersensitivity reflect preferential activation of different helper T lymphocyte subpopulations (T helper type 1 (Th1) and T helper type 2 (Th2) cells) which can be identified by their cytokine secretion patterns. Single cell analysis of cytokine production is possible using flow cytometry and intracellular staining of cytokine using labeled monoclonal antibodies.

In addition, adaptations of the local lymph node assay have been reported in which lymph node cell phenotypes determined by flow cytometry have been shown to distinguish irritants from ▶ allergens (1).

Other studies suggest that flow cytometry analysis of suction blister-derived epidermal cells may be a useful approach for evaluating human allergic and irritant patch test responses (1).

Pros and Cons
Pros
Flow cytometry often offers a more rapid, sensitive, accurate, and quantitative means of analyzing a particular cell population in a heterogeneous cell suspension as compared to more traditional microscopic methods.

There are numerous advantages to using flow cytometry for assessing the effects of chemicals on cells and tissues, especially for an assessment of immunotoxicity. Since circulating lymphoid cells exist as single cell suspensions, they are readily adapted to flow cytometry analysis. There are numerous reagents available for the identification of surface markers on lymphoid cells that can be used alone or in combination for biochemical studies or subset analyses. In addition, the application of the method in the clinical situation is allowed as human blood is generally available and readily accessible, and can be obtained through a minimally invasive procedure and multiple samples can be collected over time. Because of the high sensitivity of laser excitation, and electronic detection methods, quantitative information on single cells can be achieved. In addition, since flow cytometric analyses can be performed at rapid flow rates (500–800 cells/sec), it is possible to perform analyses on extremely rare cells. Perhaps the most unique aspect of flow cytometry is the ability to utilise more than one parameter for simultaneous analysis and to reanalyse data that has been stored on computers in "list mode." Finally, the ability of flow cytometry to physically sort cells into phenotypically defined subpopulations of cells allows for further characterization of the effects of chemicals on cells collected under defined conditions (1).

Cons
Further efforts are needed in the validation of many of these assays for routine use in immunotoxicologic testing.

Peripheral blood immunophenotyping only provides information on the types, activity, and/or the number of cells in the periphery, and thus reflects the trafficking of immune cells at the time of sampling. The types, numbers and other characteristics of peripheral blood lymphocytes are unlikely to reflect lymphocyte populations in the lymphoid tissues as the immune response occurs principally in the primary and ▶ secondary lymphoid organs. In addition, patterns of cellular traffic in the periphery can be affected by glucocorticoids, epinephrine (adrenaline), and other stress factors, which may confound the distinction between treatment-induced changes and changes resulting from stress (e.g. blood sampling) (2).

Another issue related to the application of flow cytometry to immunotoxicology testing is the ability to detect dose-dependent changes in immunophenotypic markers. Actually, doses need to be sufficiently high to detect such changes (2). In addition, although bone-marrow cell differentials can be evaluated by flow cytometry, this method cannot reliably assess cell morphology or identify stages of maturation (4).

Predictivity
There is evidence that conventional toxicology testing will not always pick up immunotoxicity in susceptible populations, such as neonates and juveniles. Additionally, there is a lack of human data investigating the immune function of patients and people exposed to immunotoxic molecules. However, the tiered testing strategies that have been developed were successfully used to identify immunotoxic compounds (4).

Clinicians are sceptical of using changes in the distribution of various phenotypic markers to predict the risk of infections in patients. Even flow cytometric monitoring of the recovery of the immune system in transplantation patients provides little insights into their risk for developing infections (2). Information on the effects of immunotoxicants on peripheral blood leukocytes is limited and more studies in various species are needed to determine whether data from peripheral blood correlates with data from the spleen. It is suggested that conducting immunophenotyping studies in animals should evaluate both peripheral blood and spleen cells to generate the requisite data. If a response in the peripheral blood of mice can be related to a response in the mouse spleen, then changes in human peripheral blood might be considered indicative of changes occurring in the human spleen (2).

Additionally, as already mentioned, the location of blood sampling for flow cytometric evaluation may hamper the interpretation and the correlation of rodent versus human data. The location of blood sampling giving the best correlation with human data should be identified first.

Relevance to Humans
There are limited data available that offer insight how

much change in blood lymphocyte subsets is sufficient to cause clinical concern. Previous studies have shown that changes in immune cells phenotypes as determined by flow cytometry are 64%–83% concordant with the classification of immunotoxicity. However this is based on findings for a small number (< 59) of chemicals that are often studied at high doses and with a limited number of phenotypic markers (2). Therefore, blood lymphocyte phenotyping in patients is probably more useful for understanding the mechanisms associated with ▶ immune competence.

Immunophenotyping alone is not considered useful for assessing a population of people unintentionally exposed to poorly defined amounts of a xenobiotic. Although changes in phenotypic markers in humans can be measured, there is no basis for establishing how much of a change is important or even how to interpret observations of statistically significant changes (2).

Because most immunotoxicologic testing in people has to be performed with cells from peripheral blood, one will intuitively assume that results obtained with animal blood lymphocytes offer a better basis for comparison with results observed in man than results obtained with spleen lymphocytes. However, this has to be proven, and there is an open question about which changes in what cells correspond best with disease-causing malfunction of the immune system in man.

Nevertheless, immune cell phenotype changes have demonstrated one of the best single correlations with host resistance against pathogens or tumors (5) and the method can be used to monitor adverse effects in clinical trials.

Regulatory Environment

Europe has already taken a position in the revised Note for Guidance on Repeated Dose Toxicity Testing (referred to as the CPMP note for guidance). The draft Food and Drug Administration documents advocates a case-by-case approach to the need for functional assays. The Japanese draft guidance recommends routine lymphocyte subset phenotyping or spleen immunohistochemistry.

Currently, only Europe requests functional assays in routine screening. In Europe, the routine screening of every compound should include the primary antibody response to a T cell-dependent antigen. As an alternative, CPMP allows the sponsor to satisfy the requirement for immunotoxicity testing by determination of drug exposure on lymphocyte subsets (by phenotyping) and natural killer cell activity.

Immunotoxicity data should be evaluated and reviewed like other toxicity data. This implies that dose-response relationship, other types of toxicities, and the general health status of the animal should be taken into consideration in making the risk-benefit analysis. Generally preclinical toxicity data support the design of clinical trials and may trigger enhanced screening for certain effects in clinical trials.

One of the goals of preclinical immunotoxicity information is to trigger the inclusion of immunotoxic parameters in clinical studies. Consequently, testing should be performed early in development, although requirements regarding timing of testing are not in force in the CPMP document. Most companies indicate that they prefer to perform the histopathological evaluation early in development and defer the functional evaluation to Phase 2 unless the histopathology study reveal an aberration (4).

References

1. Burchiel SW, Kerkvliet NL, Gerberick GF, Lawrence DA, Ladics GS (1997) Assessment of immunotoxicity by multiparameter flow cytometry. Fund Appl Toxicol 38:38–54
2. Immunotoxicology Technical Committee (2001) Application of flow cytometry to immunotoxicity testing: summary of a workshop. Toxicology 163:39–48
3. Burchiel SW, Lauer FT, Gurulé D, Mounho BJ, Salas VM (1999) Uses and future applications of flow cytometry in immunotoxicity testing. Methods 19:28–35
4. Putman E, van Loveren H, Bode G et al. (2002) Assessment of the immunotoxic potential of human pharmaceuticals: a workshop report. Drug Info J 36:417–427
5. Luster MI, Portier C, Pait DG et al. (1993) Risk assessment in immunotoxicology. II. Relationships between immune and host resistance tests. Fund Applied Toxicol 21:71–82

Fluorescence Activated Cell Sorter

▶ Flow Cytometry

Fluorescence-Activated Cell Sorting

▶ Antigen-Specific Cell Enrichment

Fluoroimmunoassay (FIA)

An immunoassay that is based on the application of chemicals with fluorescent properties like lanthanide chelates (e.g. europium diethylenetriaminopentaacetate, Eu^{3+}–DTPA) to conjugate with a protein that binds the analyte (antigen or antibody of interest). FIAs are sometimes referred to as dissociation enhanced lanthanide fluoroimmunoassays (DELFIA) or time-resolved fluoroimmunassays (TRFIA).

▶ Immunoassays

Follicular Dendritic Cell (FDC)

In the lymphoid follicles B cells home into a network of follicular dendritic cells. These cells sequester antigen and present it to the B cell in form of antigen to antibody complexes. The complexes are bound to the surface of the follicular dendritic cell via the high affinity receptor for the constant region of IgG (FcRIIB) or via the complement receptor.
▶ B Cell Maturation and Immunological Memory
▶ Germinal Center

Food Allergies

Allergic reactions to foods, especially peanut, tree nuts, seafood, cow's milk, and hens' egg.
▶ Food Allergy

Food Allergy

A C KNULST
Afd. Dermatology/Allergology
University Medical Center
P.O.Box 85 500
3508 SA Utrecht
The Netherlands

Definition
The term food allergy is usually used to describe an immune-mediated disease which is ▶ IgE-mediated and results in type I acute allergic reactions (according to the definition of Gell and Coombs), mostly within half an hour to 2 hours after ingestion. Sometimes the term food allergy is also used for non IgE mediated disease e.g. celiac disease, in which IgA antibodies play a role. However its symptoms diverge considerably from type I IgE-mediated food allergy and the underlying pathophysiological mechanism is completely different. This paper will therefore focus on IgE-mediated food allergy.

Characteristics
IgE-mediated food allergy is characterized by the development of acute symptoms, usually within minutes, but sometimes after an interval of 1–2 hours after eating the culprit food. Often symptoms start with itching in the mouth and (a feeling of) swelling which might extend to the ears and the throat. Usually this so-called oral allergy syndrome is mild, but it can be severe and life-threatening, especially when swelling of the throat occurs. It's the only specific symptom of food allergy, since all other symptoms (see below) can be found in other diseases as well. In the case of severe food-allergic reactions, it may be the first symptom to manifest itself and is therefore cause for alarm.

Other target organs are the skin, the gastrointestinal tract, the upper and lower airways, and the cardiovascular system. The various symptoms that can occur during the course of an allergic reaction to food are summarized in Table 1. In more severe reactions often more than one organ system is involved. Near-fatal and fatal reactions may occur. The incidence of fatal reactions is not precisely known, but is estimated at around 0.53 per 1 000 000. Most at risk are adolescents with underlying asthma. Concomitant exercise or intake of alcohol or aspirin (and aspirin-like drugs) may result in more severe reactions, possibly by facilitation of absorption of the allergen, but also by causing direct release of vasoactive substances such as histamine (1).

Adverse reactions to food comprise a wide variety of disorders for which various, often confusing, terms are employed. For this reason the European Academy of Allergy and Clinical Immunology (EAACI) has proposed a classification based on the underlying mechanism, which is helpful in describing the various diseases related to food (2).

According to this classification, adverse reactions to food can be divided into toxic and nontoxic reactions. Toxic reactions are caused by the ingestion of bacterial toxins, e.g. from *Salmonella* species or histamine in fish poisoning, and may occur in any person exposed to a sufficient amount of the toxin. Among non-toxic

Food Allergy. Table 1 Symptoms of food allergy.

Type	Syndrome
Skin and mucosa	Oral allergy syndrome Conjunctivitis Exanthema Urticaria Angioedema Eczema
Airways	Rhinitis Asthma
Gastrointestinal	Nausea Abdominal cramps Vomiting Diarrhea
Cardiovascular	Drop of blood pressure Shock

reactions the immune- and the non immune-mediated are distinguished. Immune-mediated reactions to food are known as food allergy and depend on the individual sensitivity of the patient to the food in question and probably also on allergen-related factors, such as the stability of the ▶ allergen to heat and proteolytic enzymes. For nonimmune mediated reactions to food the term intolerance is used. This group is very heterogeneous and so far not very well defined. Pharmacological and metabolic factors, as well as idiosyncratic factors, may be involved. Within this group there also seems to be individual variability in susceptibility to the respective foods involved.

Preclinical Relevance

Food ▶ allergy is a disease seen exclusively in humans. It has proven difficult to establish animal models. However, animal studies are especially important when it comes to addressing the issue of allergic sensitization, thus the factors that lead to the induction of an allergic immune response to an essentially harmless, although foreign, substance such as food. The current hypothesis is that food allergy, like inhalant allergy, is the result of a combination of genetic and environmental factors. The genetic factors involved have not yet been further elucidated. Environmental factors that have been implicated are exposure to smoke and inhalant allergens, breastfeeding, vaccination, and contact with bacterial products. In food-allergic patients the normal balance in the immune system between so-called T helper type 1 (Th1) and type 2 (Th2) cells seems to be shifted in favour of the Th2 cells that are predominant in allergy. From studies dedicated to the development of spontaneous tolerance it has become clear that the immune system of the newborn is allergy-prone due to a preponderance of the Th2 cells necessary in utero to prevent rejection of what is, for the mother, a partly foreign foetus. This imbalance is normally corrected after birth, but may persist, resulting in ▶ atopy.

Relevance to Humans
Epidemiology

Although epidemiological data are scarce, it is generally accepted that the prevalence of allergic diseases (eczema, rhinitis, asthma, and food allergy) has been increasing over the past decade. Environmental factors are probably responsible for this phenomenon. The current prevalence of food allergy in children is estimated at 5%–8% and in adults at 2%–3%. The foods involved vary due to different eating habits between children and adults, and also between population groups. In childhood, cow's milk and hen's egg are the most important allergens, followed by peanut, whereas in adults peanut, tree nuts, seafood, fruits, and vegetables are most often implicated.

Diagnosis

The diagnosis of food allergy is based on a thorough history, with specific attention to the interval between the occurrence of symptoms and the ingestion of the food, the specific symptoms for each culprit food and possible reactions to hidden and crossreactive food allergens. Additionally, sensitization is detected by skin tests using commercial extracts or the native food and determination of specific IgE. In cases of doubt both techniques should be used, because results are not completely concordant. In rare cases even the combination of both tests may result in a false-negative outcome. However, an even greater problem is the high frequency of false-positive test results. The reason for this is not fully understood, but may well be related to the presence of IgE antibodies to irrelevant "allergens," such as the so-called crossreactive carbohydrate determinants (CCD) on various plant food allergens. Formal proof of the diagnosis can only be obtained by elimination of the suspect food, followed by double-blind placebo-controlled food challenge (▶ DBPCFC), which is the gold standard. Although this method is labour intensive and therefore not widely used, it provides an invaluable method to avoid radical elimination diets, with all their nutritional and psychosocial ramifications.

Allergens

It is unclear what makes an allergen an allergen. However, progress has been made in further characterization of the structures involved. Plant-derived proteins responsible for allergy include various families of pathogenesis-related (PR) proteins, protease and α-amylase inhibitors, peroxidases, profilins, seed-storage proteins, thiol proteases, and lectins. The large group of PR proteins consists of proteins that are induced by pathogens, wounding, or environmental stress. These proteins have been classified into 14 families. Among these is the large PR 10 family of proteins homologous to the major birch pollen allergen Bet v 1, which are present in many fruits, vegetables, and nuts (3). Aller-

Food Allergy. Table 2 Important food allergens.

Peanut
Tree nuts
Cow's milk
Hen's egg
Fish
Crustaceans and molluscs
Wheat
Soy

genic animal proteins belong to the muscle proteins, enzymes, and serum proteins.

Worldwide, peanut is the most important food allergen, both because of its prevalence, and because it is responsible for the most severe reactions. Very minute amounts are able to induce severe symptoms in strongly sensitive patients. A complicating factor is that peanut is used ubiquitously in prepackaged foods, often without labelling, which poses a threat for many patients.

The most important food allergens are listed in Table 2.

Despite the fact that wheat and soy are included in the list of the "big eight," it must be stated that clinically relevant allergy to ingested wheat and soy seems to be relatively rare, despite a high prevalence of allergic ▶ sensitization to these allergens, which is probably due to ▶ crossreactivity with grass pollen and peanut, respectively. Because of changing dietary customs, this list may have to be revised in future.

Crossreactivity

This is a major issue in ▶ food allergy. Crossreactivity occurs at the level of sensitization, that is, the presence of specific IgE, but also at the level of clinical reactivity. Sensitization may occur primarily by another food via the oral route, but also by environmental allergens via the inhalant route, such as by pollen allergens, which share epitopes with food allergens. The reason for crossreactivity lies in the structural homology between allergens. Therefore crossreactivity occurs frequently within plant families (legumes such as peanut, soy, and green pea) but is certainly not restricted to these. Especially in the case of primary sensitization to inhalant pollen allergens extensive crossreactivity may occur. The major birch pollen allergen Bet v 1 cross reacts with structures in fruits of the Rosaceae family, but also with vegetables from the Apiaceae family, with tree nuts and even with latex. Crossreactivity does not always result in clinically relevant allergy. In the case of peanut allergy, there is only a 5% risk of developing an allergy to other legumes, whereas for tree-nut allergy this percentage is 37%. On the other hand, the percentage of cow's milk-allergic patients who develop an allergy to goat's milk is as high as 92% (2).

Potential Allergenicity of Genetically Modified (GM) Foods

GM foods are currently developed for a number of reasons. Firstly, the development of GM foods might increase resistance to diseases, thereby increasing productivity, and decreasing the need to use pesticides. Secondly, GM foods might have improved nutritional value, due to enhanced protein composition or vitamin content. Thirdly, genetic modification might increase the taste, color, or storage life of foods.

In future genetic modification may be of great help in developing hypoallergenic allergy vaccines.

Despite these advantages, there is increasing public awareness of the possible disadvantages, including the introduction of new allergens to the food chain. Therefore, a joint FAO/WHO expert consultation (2001) has proposed guidelines for the evaluation of allergenicity of GM foods (4). These include assessment of sequence homology to known allergens, assessment of the protein with sera of patients allergic to the source material, and a so-called targeted serum screening with sera from patients allergic to allergens broadly related to the source material. When these evaluations do not show potential allergenicity, the protein is further checked for immunogenicity and allergenicity in animal models and for stability to pepsin digestion in vitro. A protein that gets successfully through these tests is ready for testing in patients. In particular, ▶ skin prick tests and oral challenges have to be considered. As even the best test procedures do not assure a 100% safety, post-marketing surveillance is strongly recommended.

Treatment

At present no curative therapy is available. Patients have to avoid the offending foods. This can be relatively simple in case of a food allergy that is limited in severity and as regards the number of foods involved, but can be difficult in case of multiple and severe allergies that can result in severe reactions when encountered unexpectedly as hidden allergens. In the latter case supervision by an experienced dietician is obligatory.

In addition, patients need information on how to deal with allergic reactions when they occur. If symptoms are mild, prescription of oral ▶ antihistamines is sufficient, but when angioedema or severe gastrointestinal, airway, or cardiovascular reactions can be expected, prescription of rescue medication is required. For this purpose an adrenaline autoinjector is on the market, which can be life saving. Patients and parents of allergic children, and teachers and other caretakers have to be instructed how and when to use it. Corticosteroids are invaluable in preventing late reactions, and may also have a beneficial effect in the early stages of an allergic reaction.

Regulatory Environment

The present regulations regarding food labelling are quite insufficient when it comes to protection of the allergic patient. Present law does not even require that the most important food allergens (the so-called "big eight," see Table 2) are listed as ingredients. Furthermore, at least in European countries food manufacturers are not obliged to specify the ingredients of composite foods that comprise less than 25% by weight of a certain product. It is therefore not surprising that unwanted and unexpected allergic reactions to food occur frequently. These may be due to unclear or absent labelling, but also to unexpected contamination. The latter is a major problem, which can occur during the whole line from harvest and storage to transport, packing, and labelling. To control this production chain, easy detection methods for small residual amounts of allergens have to be developed. At present, these are available for only a limited number of allergens.

In addition for risk assessment, knowledge of the sensitivity of allergic patients to a given food is required. Only the combination of both of these types of information makes it possible to determine acceptable levels of contamination and subsequent labelling (5). This is even more important since it is technically not possible to avoid contamination completely. Together these efforts will hopefully lead on the one hand to a larger number of safe foods available to the allergic consumer and on the other hand to patients' increased ability to take precautionary measures with respect to offending or suspect foods.

References

1. Sicherer SH (2001) Clinical implications of cross-reactive food allergens. J Allergy Clin Immunol 108:881–890
2. Bruijnzeel-Koomen C, Ortolani C, Aas K et al. (1995) Adverse reactions to food. Allergy 50:623–635
3. Breiteneder H, Ebner C (2000) Molecular and biochemical classification of plant-derived food allergens. J Allergy Clin Immunol 106:27–36
4. FAO/WHO (2001) Evaluation of allergenicity of genetically modified foods. Report of a joint FAO/WHO expert consultation on allergenicity of foods derived from biotechnology. Rome
5. Taylor SL, Hefle SL, Bindslev-Jensen et al. (2002) Factors affecting the determination of threshold doses for allergenic foods: How much is too much? J Allergy Clin Immunol 109:24–30

Food and Drug Administration (FDA)

The United States governmantal agency responsible for regulating and licensing pharmaceuticals, food additives, medicial devices, biologics, cosmetics and animal feed and drugs.

▶ Immunotoxicology
▶ Plaque-Forming Cell Assay

Formylated Peptides

N-Formylated peptides are uniquely used by bacteria to initiate protein synthesis. Because formylated peptides are specific for bacteria, they signal the presence of invading microbes. Most N-formylated peptides are chemotactic for leukocytes.

▶ Chemotaxis of Neutrophils

Framework Regions (FR)

Four segments (FR1, FR2, FR3, FR4) of the rearranged variable regions of antibody heavy and light chains that are located in the structurally conserved beta-pleated sheets of the immunoglobulin.

▶ Rabbit Immune System

Freund's Complete Adjuvant

Bacterial fragments in an oil immersion, which can be injected to cause stimulation of immune cells.

▶ Birth Defects, Immune Protection Against

Fusion Proteins

Fusion proteins in a pharmacological sense are proteins generated by gene technology which consist of two or more functional domains derived from different proteins. A typical example is the homodimeric fusion protein Etanercept which is used to neutralize tumor necrosis factor in rheumatoid arthritis. It consists of disulfide-linked parts of the constant regions of human immunoglobulin G and the ligand binding domain of the p75 receptor for tumor necrosis factor.

▶ Cytokine Receptors

G

G Proteins

Heterotrimeric proteins involved in transmembrane receptor signaling. Their name is based on their ability to bind the guanine nucleotides, GDP and GTP. G proteins are located in the inner surface of the plasma membrane, where they associate with transmembrane receptors of a variety of hormones, chemokines and other mediators. These are called G protein-coupled receptors (GPCRs).
▶ Chemokines

Gamma Interferon Activation Sites (GAS)

These are DNA sequences in the promoter regions of interferon-γ-activated genes that interact with activated Stat proteins and modulate gene transcription.
▶ Interferon-γ

Gastric Mucosa

Mucous membrane of the stomach.
▶ Serotonin

Gastroenteritis

▶ Salmonella, Assessment of Infection Risk

Gene Conversion (GC)

Non-reciprocal transfer of information from one DNA duplex to another first observed in fungi and later found to contribute to the process of diversification of rearranged antibody heavy and light chain variable region genes in chickens and rabbits. In rabbits, gene conversion occurs in young appendix where it may contribute to primary repertoire development and in sites of secondary immune responses, such as germinal centers of lymph nodes and spleen.
▶ Rabbit Immune System

Gene Expression

The full use of the information in a gene via transcription and translation leading to production of a protein and the appearance of the phenotype determined by that gene.
▶ Southern and Northern Blotting
▶ Transgenic Animals

Gene Expression Analysis

▶ Cytokine Assays

Gene Profiling

▶ Toxicogenomics (Microarray Technology)

Gene-Targeted Mouse

▶ Knockout, Genetic

Genetic Polymorphism

Genetic polymorphism is the phenomenon of multiple forms of the same gene. This can be detected at the level of the DNA sequence, which may or may not change the amino acid sequence of the protein for which the gene codes.
▶ Chronic Beryllium Disease

Genetic Predisposition

The association of specific genotypes, particularly in the major histocompatibility complex, with heightened susceptibility to autoimmunity.

▶ Autoimmunity, Autoimmune Diseases

Genetic Susceptibility

This explains interindividual variation in adverse health effects suffered due to exposure to environmental and occupational toxins. People who inherit certain forms (variants) of a specific gene or genes may be at greater risk of a toxin-induced disease than others who have inherited different forms of the same genes.

▶ Chronic Beryllium Disease

Genetic Vaccines

▶ DNA Vaccines

Genetically Defined Rodents

▶ Rodents, Inbred Strains

Genetically Engineered Mouse

▶ Knockout, Genetic

Genetically Modified

▶ Transgenic Animals

Genomic DNA

Molecular analysis using the total set of genes (DNA) carried by an individual or cell.

▶ Southern and Northern Blotting

Germ Center

▶ Germinal Center

Germinal Center

C FRIEKE KUPER
Toxicology and Applied Pharmacology
TNO Food and Nutrition Research
Zeist
The Netherlands

Synonyms
Germ center.

Definition
Germinal centers are areas, predominantly found in secondary lymphoid tissues, were B cell proliferation and differentiation, memory generation, antibody isotype switching, and affinity maturation occurs upon stimulation by an antigen. These processes result in the generation of memory cells and plasma precursor cells.

Characteristics
General
B cell follicles are normally found in secondary lymphoid organs (spleen, lymph nodes, mucosa-associated lymphoid tissues) and in the thymic medulla in some animal species. Under pathological conditions, they also can occur in the thymic medulla and in nonlymphoid organs, such as in the human thyroid gland in Hashimoto thyroiditis. Follicles in which no (auto)antigen-driven processes take place consist of recirculating, small resting virgin B lymphocytes in a network of follicular dendritic cells (FDCs). Such follicles are designated as primary follicles. During stimulation with antigen, germinal centers develop within B cell follicles and the follicles change from primary into secondary follicles. The small recirculating B cells are excluded from the follicular center and form the follicular mantle that surrounds the germinal center (Figure 1).

The mantle zone contains small resting B cells, similar to those in the primary follicles, and is broadest at the side nearest the capsule of lymph nodes or nearest to the apical light zone of the germinal center. The germinal center contains B lymphoblasts—▶ centrocytes and ▶ centroblasts. Histologically, the mantle stains densely basophilic while the germinal center is paler. The germinal center is generally divided into a dark and a light zone, the light zone being subdivided into a

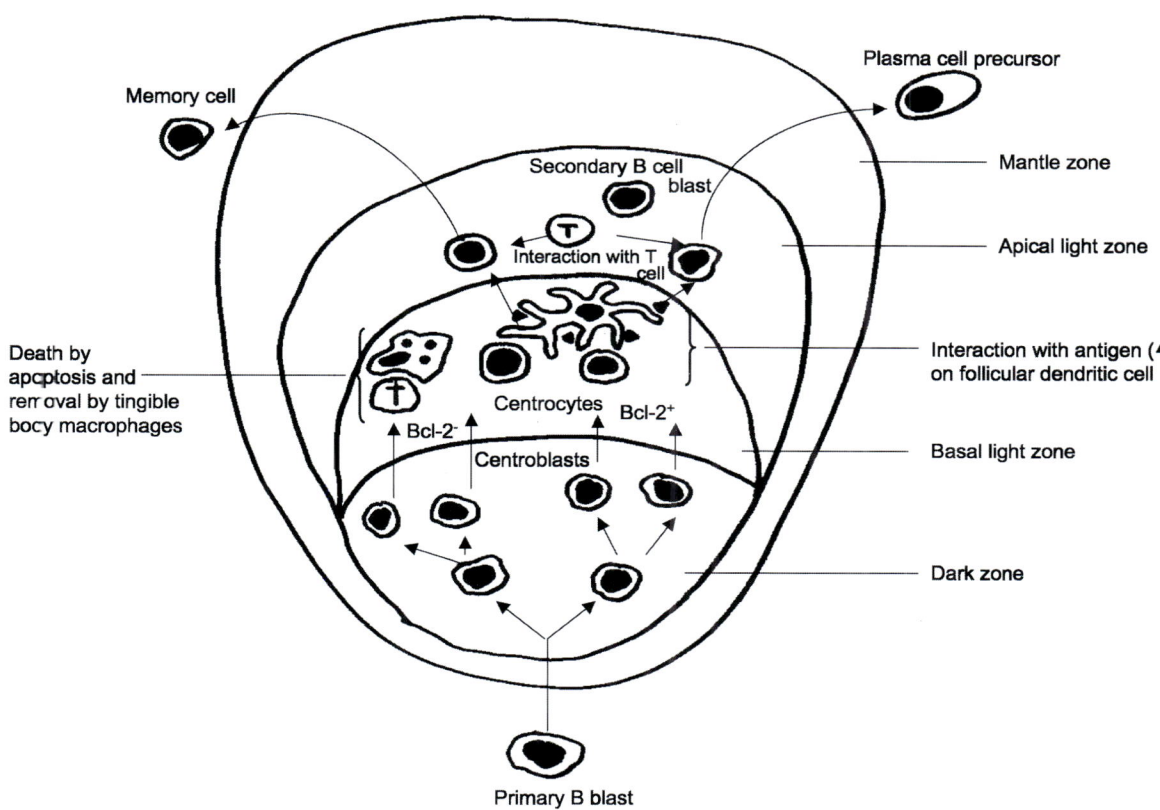

Germinal Center. Figure 1 Schematic presentation of a germinal center and B cell maturation.

basal and an apical light zone (1). In the germinal center, antigens are presented to B cells in the form of immune complexes trapped in cytoplasmic extensions of FDCs. CD4$^+$ T cells, normally present at relatively high density in the germinal center, assist in B-cell activation and are thought critical for germinal center development, at least at the later stages (2). The low numbers of secondary follicles in congenitally athymic rodents reflects the T cell dependency of most follicular responses.

Primary follicles have been found in various secondary lymphoid organs of mice within the first week of birth, although in mesenteric lymph nodes they were detected first at the age of 12 days (3). Germinal centers appeared at day 28, and were preceded by clusters of mature FDCs at day 14. From then on, T-dependent antibody responses could be generated. Germinal center activity may be diminished in old mice, as suggested by a reduction in the numbers of high-affinity antibody-forming cells.

Germinal Center Development and B Cell Maturation

The trapping of antigen by FDCs and the migration of primary B blasts into the follicle are regarded as the first events in germinal center formation. Most probably, these primary B blasts have encountered antigen already in the T-cell-dependent areas of lymphoid organs, before migration, during the first few days of primary T-cell-dependent antibody responses. Upon entering the follicle, the primary B blasts proliferate exponentially and alter phenotypically, e.g. by losing their surface immunoglobulin (sIg); in this state they are called centroblasts. Centroblasts move apically and at that time the dark and light zone of the germinal center become apparent. Centroblasts then give rise to centrocytes that are nondividing and again express sIg. Centrocytes are selected in the basal light zone on the basis of their capacity to be activated by antigen, which is held on FDCs. Centrocytes with high affinity for antigen express the cell survival gene *bcl-2* and survive. Centrocytes with low or no affinity for antigen held on FDCs do not express *bcl-2* and are lost by apoptosis. There is a high death rate among centrocytes as shown by the presence of several tingible body macrophages which contain nuclear debris from the centrocytes. Surviving centrocytes subsequently enter the apical light zone where antibody isotype switching by DNA rearrangements occurs, most probably with help of T lymphocytes. Antibody isotype switching comprehends the switch from the production of immunoglobulin predominantly of the IgM class early in the antibody response to one of the IgG subclasses (or IgA or IgE in germinal centers of mucosa-associated lymphoid tissues) later in the response. The differentiated B blasts leave the follicle as memory B cell, which maintain the antibody response, or as ▶ plasma cell precursor (also named antibody-forming cell or AFC) or stay as secondary B blast. In lymph nodes, plasma cell precursors migrate to the medullary cords where they complete their differentiation into plasma cells capable of secreting large amounts of antibody.

The antibodies produced in a secondary response and in the late stage of the primary response to a T-cell-dependent antigen have a higher affinity than those produced (early) in the primary response. This is done by slight changes in the structure of antibodies (somatic mutation), produced by B cell progeny, leading to refinement of antibody specificity. The refinement process is called affinity maturation.

The network in the germinal center is predominantly formed by FDCs. FDCs are long-lived cells which seldom divide and which change their morphology and phenotype during the generation of antibody responses. Most studies indicate that these cells are of fibroblastic or mesenchymal origin. They produce adhesion molecules, receptors, and FDC-specific surface antigens, and they exert a chemotactic effect on lymphocytes. Antigen is taken up by FDCs in the form of immune complexes and can be held there in a nondegraded form for many months.

Preclinical Relevance

Germinal centers appear within a few days after antigen administration in spleen and lymph nodes. In secondary lymphoid tissues which have low background germinal center activity, germinal center reactions last about 3 weeks after antigen administration. The presence, size and number of germinal centers in secondary lymphoid organs like spleen and lymph nodes can thus be used as a parameter of immune activation. In lymphoid organs like the thymus, which seldom exhibits germinal centers in most species, and in non-lymphoid organs like the thyroid, the formation of germinal centers may indicate the existence of pathologic autoimmune processes.

Relevance to Humans

There are no indications that the formation of the germinal center and its related processes show distinct interspecies-related differences. Therefore, it might be expected that results from animal studies will extrapolate quite well to humans.

Regulatory Environment

Several guidelines require examination of spleen, mucosa-associated lymphoid tissues and lymph nodes that drain the exposure site as well as nondraining lymph nodes. These tissues are the prime sites for the induction of germinal center formation, and thus can be used to study potential immune activating action of xenobiotic agents. Further, potential immune inactivation by an agent can be studied by decreased numbers of germinal centers in, for example, the continuously activated mesenteric and mandibular/cervical lymph nodes.

▶ Lymphocytes
▶ Canine Immune System
▶ Humoral Immunity

References

1. Liu Y-J, Johnson GD, Gordon J, MacLennan ICM (1992) Germinal centers in T-cell-dependent antibody responses. Immunol Today 13:17–21
2. De Vinuesa CG, Cook MC, Ball J et al. (2000) Germinal centers without T cells. J Exp Med 191:485–494
3. Hoshi H, Horie K, Tanaka K et al. (2001) Patterns of age-dependent changes in the numbers of lymph follicles and germinal centers in somatic and mesenteric lymph nodes in growing C57Bl/6 mice. J Anat 198:189–205

Germinal Center Reaction

▶ B Cell Maturation and Immunological Memory

Germinal Centers

Germinal centers develop in the B cell follicles of secondary lymphoid tissues during T cell-dependent antibody responses. Memory B cells and plasma cells that exit the germinal center frequently have undergone isotype switching and have diversified the sequences of the variable regions of their rearranged heavy and light chain genes. In rabbits, as in other species, affinity maturation takes place in the germinal centers. These structures provide a specialized microenvironment within which B cells proliferate rapidly in the dark zones (centroblasts) undergo somatic hyper-mutation and (in rabbits and chickens) gene conversion. The B cells undergo selection for expression of surface B cell receptors with high affinity for the immunizing antigen. Centrocytes in the light zones interact with antigen–antibody complexes bound through Fc receptors on follicular dendritic cells. The B cells present antigen to T cells which produce survival signals. Cells not receiving such signals may

die by apoptosis or undergo further rounds of mutation/conversion.
▶ Rabbit Immune System
▶ B Cell Maturation and Immunological Memory

Global Gene Expression Analysis

▶ Toxicogenomics (Microarray Technology)

Glomerulonephritis

A disease induced by destruction of glomerular or tubular membranes in the kidney, resulting in glomerular sclerosis and renal failure. This destruction occurs during autoimmune diseases, and is then mediated by immune complexes deposited in the glomeruli, or the glomerular or tubular basement membranes themselves are the targets for the autoimmune attack.
▶ Hypersensitivity Reactions

Glucocorticoid Receptors

These are cytosolic (GR) or membrane-bound (mGR) receptors for glucocorticoids. Glucocorticoid binding to the GR initiates a series of signaling events that lead to gene transcription and, ultimately, biological activity. Some effects of glucocorticoids occur more rapidly than is possible if gene transcription must precede the biological response to glucocorticoids. Some believe that rapid responses to glucocorticoids are mediated via a mGR, although the concept is still quite controversial.
▶ Glucocorticoids

Glucocorticoids

Bob Luebke
Immunotoxicology Branch
Mail Drop B143-04, US EPA
Research Triangle Park, NC 27709
USA

Synonyms
Corticotropin, adrenocorticotropin, glucocorticosteroids, corticotropin-releasing hormone, corticotropin-releasing factor, transcortin, corticosteroid-binding globulin.

Note that "corticotropin" and "adrenocorticotropin" are not synonyms for "glucocorticoid". "Corticotrophin" and "adrenocorticotropin" are trade names for a pituitary extract, typically of bovine origin, of ACTH. Corticotropin releasing hormone and corticotrophin releasing factor are products of the hypothalamus, rather than the adrenal glands. Also, transcortin (CBG) and albumin are synthesized primarily in the liver, not in the adrenals. These terms are certainly related to the topic, and would be appropriate in a section of keywords, but should not be listed as synonyms.
In addition to "glucocorticosteroids", other synonyms for glucocorticoids include "adrenocortical steroids" and "adrenocortical hormones".

Definition
▶ Glucocorticoids are steroid hormones that are synthesized in the cortex of the adrenal glands. They are required for normal immune system development and homeostasis; however, the adaptive response to stressful physical (e.g. heat, cold, vigorous exercise, infection), chemical, and psychological events causes the release of much higher levels, in quantities that are sufficient to alter both innate and antigen-driven immune system responses. Glucocorticoids increase blood glucose levels (hence the name "glucocorticoids") by stimulating synthesis of glucose in the liver and decreasing the utilization of glucose by cells, and they promote mobilization of factors required for increased energy production. In humans and most other species, cortisol (hydrocortisone) is the predominant glucocorticoid, while corticosterone is the most plentiful in mice and rats.

Molecular Characteristics
Synthesis, Release and Elimination
Glucocorticoids are synthesized from cholesterol, the precursor to all steroids (androgens, estrogens, progestins and corticoids) and are released following stimulation of the hypothalamic-pituitary-adrenal (HPA) axis. Stressors trigger release of corticotropin-releasing hormone (CRH or CRF, factor) from the hypothalamus into a local venous plexus that communicates with the anterior pituitary gland (adenohypophysis). Adrenocorticotropic hormone (ACTH) is secreted from the anterior pituitary gland in response to CRH, enters the blood stream and is transported to the adrenal glands, stimulating the synthesis and release of glucocorticoids. The adrenals do not store glucocorticoids to any great extent, thus plasma levels reflect synthesis, rather than release.
As with many other hormones, glucocorticoid levels are controlled by a feedback mechanism, in which high circulating levels of glucocorticoids suppress ACTH release, and low levels stimulate ACTH re-

lease. In addition, the intensity of the immune system response to stimulation (e.g. immunization or inflammation) appears to be under the homeostatic control of certain ▶ cytokines, including ▶ interleukins IL-1 and IL-6 and tumor necrosis factor(TNF)-α, that appear to act by stimulating release of CRH; IL-6 can also act directly (not through CRH) to stimulate glucocorticoid synthesis and release. Cytokine control of glucocorticoid secretion is at least in part under local control-receptors and mRNA for TNF-α and IL-6 are present in cortical cells, and cytokine release is stimulated by both ACTH and IL-1.

Under normal conditions, human males produce 15–20 mg of cortisol and approximately 4 mg of corticosterone per day; females produce approximately 10% less. All glucocorticoids are present in the circulation both as free hormone (< 10% of total) and bound to corticosteroid-binding globulin (CBG, or transcortin) or albumin. However, the glucocorticoid-protein complex is too large to cross the plasma membrane of target cells and therefore does not engage the cytoplasmic glucocorticoid receptor (GR); thus, only the unbound form is biologically active. In general, stressful events decrease circulating levels of CBG, effectively increasing the quantity of unbound glucocorticoids available to mediate biological effects (although gender- and stressor-related exceptions have been described).

Recent evidence suggests that CBG may serve to increase glucocorticoid levels locally (e.g. at sites of inflammation) or may perhaps have a role in active transport of "glucocorticoid" into certain cells via a putative CBG receptor, where CBG is cleaved, freeing glucocorticoids to bind to cytosolic receptors. Cortisol is cleared quickly at physiologic concentrations, with a half-life of just over 1 hour, although its half-life can double at higher concentrations. Glucocorticoids are primarily metabolized in the liver and approximately 90% of secreted glucocorticoids are eliminated in the urine.

The synthetic glucocorticoids, betamethasone and ▶ dexamethasone (DEX), cross the placenta and are pharmacologically active in the fetus; their use during pregnancy is restricted to cases where delivery of active glucocorticoids to the fetus is desired. In contrast, hydrocortisone crosses the placenta but is inactivated by fetal enzymes; similarly, prednisone, prednisolone and methylprednisolone are enzymatically inactivated in the placenta. Ratios of maternal : fetal synthetic glucocorticoid concentrations range from 10 : 1 (prednisone) to 2 : 1 (dexamethasone) with therapeutic use. Little if any of systemically administered glucocorticoids enter the breast milk.

Glucocorticoid Receptors and Biologic Effects

▶ Glucocorticoid receptors (GRs) are found in the cytoplasm of all cells that are sensitive to glucocorticoids, as a complex with heat shock proteins (HSP 70 and HSP 90) and immunophilins. Following glucocorticoid binding, the activated ligand-receptor complex is transported to the nucleus and binds to glucocorticoid responsive elements (GREs), specific DNA sequences that promote or inhibit gene transcription. GRs may also directly interact with the ▶ transcription factors AP-1 and NF-κB, blocking their activity and ultimately decreasing gene expression. There are, however, some biological effects of glucocorticoids that occur too rapidly to be mediated by RNA transcription and de novo protein synthesis, or that do not require protein synthesis (e.g. release of pre-formed enzymes, opening of ion channels). These events are apparently initiated by a membrane-bound glucocorticoid receptor (mGR); this form of the receptor differs from the classical cytosolic receptor structurally, in binding affinity and in ligand/receptor responses. The mGR may have an active role in apoptotic events associated with the negative selection of immature lymphocytes. It has also been suggested that mGR and a membrane form of the estrogen receptor may be involved in autoimmunity secondary to exposure to endocrine system—disrupting environmental contaminants.

Synthetic Glucocorticoids

A variety of synthetic glucocorticoids are used as antiinflammatory drugs. The synthetic glucocorticoids have 5–30 times the antiinflammatory activity of an equal dose of cortisol or corticosterone. DEX is one of the most potent synthetics; it does not bind to CBG or albumin (in contrast to prednisone and its active metabolite, prednisolone) and has a serum half-life of 2.5 hours. DEX is often used in animal studies, both to investigate the effects of glucocorticoids on immune function, and as a positive control for immunosuppression.

Putative Interaction with the Immune System

Immune system sensitivity to glucocorticoids is traditionally considered to be species-dependent. Humans, monkeys, guinea pigs and ferrets are classified as glucocorticoid-resistant, while mice, rats, rabbits and hamsters are glucocorticoid-sensitive, following either in vitro or in vivo exposure. A dose that causes atrophy of the thymic cortex and lymphocytopenia in a sensitive species has a minor or inconsistent effects in a resistant host. These differences may be more apparent than real, however; lysis of human lymphocytes may simply take longer at concentrations that produce lysis in "sensitive" species.

T Cells and T Cell Maturation

Glucocorticoids cause apoptosis, particularly in

CD4⁺CD8⁺ thymocytes; mature, single-positive (CD4⁺ or CD8⁺) cells in the thymic medulla and resting lymphocytes are resistant to in vitro or in vivo glucocorticoid exposure. Although the mechanism of glucocorticoid-induced apoptosis remains elusive, indirect evidence suggests that the mitochondrial pathway of programmed cell death is involved. Glucocorticoids may have a role in thymocyte maturation, contributing to the death of negatively selected cells and protecting positively selected cells from activation-induced apoptosis. Thymic epithelial cells serve as an important source of glucocorticoids during the critical fetal and neonatal phases of immune system maturation, and produce approximately twice as much glucocorticoid in the neonate as in 1-month-old animals. Recent studies have shown that glucocorticoid-receptor knockout mice experience normal T cell development, however, calling into question a pivotal role for glucocorticoids in T cell maturation.

Cytokine Responses

Exogenous administration of glucocorticoids decreases production of many cytokines, including those associated with both inflammation and antibody synthesis, in most cases by interfering with gene expression. In some cases (e.g. IL-1, IL-2, IL-6, IL-8, GM-CSF and TNF-α) the stability of RNA message for cytokines is also reduced by glucocorticoids. The nature of the T cell response may also be influenced by the presence of glucocorticoids during activation of T cells by antigen. DEX has been shown to increase production of IL-4, and to interfere with transcription factors for IL-12 production in rodents, thus favoring development of a T helper type 2(Th2)-dominated response. In contrast, human data suggest that both Th1 and Th2 cytokine production are inhibited by glucocorticoids.

Cytokine production is also influenced by endogenous glucocorticoids. Adrenalectomized (ADX) animals produce more inflammatory cytokines (e.g. TNF-α, IL-6, IL-12) than controls, and overproduction is reversed by glucocorticoid replacement. ADX animals are also more susceptible to certain infections (e.g. cytomegalovirus) because they are unable to control infection-related, cytokine-driven inflammation. Exogenous replacement of glucocorticoid reverses increased susceptibility.

Antibody Responses

Glucocorticoids may suppress antibody synthesis in a variety of ways. Expression of surface Ia antigen on B cells, critical for interaction with T helper cells during induction of the T cell-dependent antibody response, is downregulated by DEX. Failure to express Ia causes profound suppression of antibody responses. Shipping stress is sufficient to increase corticosteroid concentrations and to suppress the antibody response to the T-dependent antigen sheep red blood cells (SRBC) in mice (1). Other events generally considered to be stressful do not always cause immunosuppression, however. For example, minipump implantation increases circulating corticosterone above control levels, but the antibody response to SRBC is not affected (2). In this study the increase was brief and not of the magnitude induced by shipping stress. Thus, caution must be exercised when interpreting results of circulating glucocorticoid levels; a simple increase does not necessarily equate to immunosuppression.

DEX injection causes a dose-related decrease in the number of spleen cells producing antibody in response to bacterial endotoxin, a T-independent antigen. However, DEX causes a redistribution of B cells to the bone marrow, and numbers of bone marrow antibody producing cells and the titer of antiendotoxin antibody is actually slightly increased by high doses of DEX (3). This example illustrates the importance of evaluating immune function in multiple compartments. Evaluating only the splenic response to immunization would lead to the conclusion that the host's ability to mount a T-independent response was suppressed, when in reality circulating antibody titers were not suppressed. It is also important to note that there are strain differences in the sensitivity of rats to DEX. Peers et al. (4) reported that the IgG response of Lewis rats was unaffected by 0.01 mg DEX/kg, a dose that suppressed the response of Wistar and Brown Norway rats by about 40%. Suppression was achieved in both strains at 0.1 mg/kg.

Other Effects

Corticosteroids may also affect immunocompetence by altering patterns of lymphocyte circulation. For example, localization of both T and B cells to the bone marrow is increased (as illustrated above) and expression of ▶ adhesion molecules (e.g. ICAM-1 and E-selectin) are decreased following glucocorticoid exposure. Changes in homing patterns may limit the influx of cells to an inflammatory site and to draining lymph nodes, thus decreasing the intensity of the immune response.

Unlike lymphocytes, neutrophils (polymorphonuclear neutrophils, PMNs) are generally resistant to glucocorticoids. PMN survival is prolonged, possibly as a result of glucocorticoids-induced elevation of leukotriene B_4 production. Glucocorticoids also stabilize and alter the content of lysosomes, thus decreasing the inflammatory response. The relative resistance of PMN to glucocorticoids is believed to be the basis for resistance of DEX-exposed mice to infection with the bacterium *Listeriamonocytogenes*, even though T cell functions are dramatically suppressed by DEX. While

T cells have a critical role in recovery from infection, it appears that at less than lethal doses of DEX, PMN function alone is sufficient to protect against infection. Dendritic cells (DC) are critical in antigen presentation to and clonal expansion of T cells. In most DC model systems, DEX has been reported to suppress expression of costimulatory or major histocompatability complex (MHC) class II molecules, to alter the maturation pathway or to induce apoptosis in DC. These effects all lead to a decreased ability to stimulate T cell proliferation.

Glucocorticoids and Immunotoxicity

In addition to physical and psychological stressors, chemicals have been reported to elevate, or in fewer cases, suppress glucocorticoid levels. These include certain therapeutic and recreational drugs, pesticides, solvents, and other organic pollutants (5). In some cases, coadministration of a GR antagonist, or adrenalectomy done before chemical exposure, blocks immunosuppression, suggesting that glucocorticoid release is responsible for the chemical's immunotoxic properties. However, it is also clear that other products of the HPA axis (e.g. ACTH and CRH) may directly suppress immune function, as may the host of neuroendocrine system products released as part of the stress response. Drawing on the published literature, Pruett et al. (5) estimated that immunotoxicity is associated with circulating corticosterone concentrations ≥ 200 ng/ml, if immune function is evaluated at least 6–12 hours after the animal is exposed.

Glucocorticoids are given to premature infants with respiratory distress syndrome (RDS) to reduce the acute inflammation of RDS, thus preventing chronic lung disease. However, animal studies suggest that early-life exposure to glucocorticoids may predispose the host to later development of autoimmune disease: in rat pups exposed to exogenous glucocorticoids during the neonatal period (when glucocorticoid levels are low and unresponsive to HPA stimulation) there is life-long hyporesponsiveness of the HPA and increased susceptibility to certain types of experimentally-induced autoimmune disease. It has also been suggested that neonatal exposure may permanently bias the profile of cytokine production towards a Th2-type response, possibly increasing the risk of allergy.

Relevance to Humans

Glucocorticoids are widely used in humans to suppress inflammation and tissue damage, particularly in rheumatic and autoimmune diseases, and in combination with other immunosuppressive drugs to suppress rejection of organ transplants. Use during pregnancy is generally considered safe, since large population studies have found no increased risk of terata at normal therapeutic doses. However, DEX is reported to suppresses cellular and humoral immunity in mice exposed during the second and third trimesters of gestation, although immunosuppression occurred at doses that caused low birth weight. As noted above, animal models also suggest that therapeutic use of glucocorticoids in neonates may predispose to allergy and possibly autoimmunity later in life.

At lower doses (~ 50 mg) glucocorticoids do not generally increase the risk of infection in humans, particularly if treatment is relatively short term, or if doses are < 10 mg/day. Glucocorticoids are given to patients with certain infections (e.g. bacterial meningitis) to lessen the damage caused by the immune response to infection. Glucocorticoids are also used to reduce postoperative pain and to reduce recovery time after some types of surgery, with no evidence of increased rates of infection. However, exacerbation or reactivation of certain infections (e.g. tuberculosis, or *Pneumocystis* or *Herpes simplex* infections) may occur. Immunization with live vaccines should also be avoided while taking glucocorticoids.

References

1. del Rey A, Besedovsky H, Sorkin E (1984) Endogenous blood levels of corticosterone control the immunologic cell mass and B cell activity in mice. J Immunol 133:572–575
2. Rowland RRR, Reyes E, Chukuwuocha R, Tokuda S (1990) Corticosteroid and immune responses of mice following mini-osmotic pump implantation. Immunopharmacology 20:187–190
3. Benner R, van Oudenaren A (1979) Corticosteroids and the humoral immune response of mice. II. Enhancement of bone marrow antibody formation to lipopolysaccharide by high doses of corticosteroids. Cell Immunol 48:267–275
4. Peers SH, Duncan GS, Flower RJ (1993) Development of specific antibody and in vivo response to antigen in different rat strains: effect of dexamethasone and importance of endogenous corticosteroids. Agents Actions 39:174–181
5. Pruett SB, Ensley DK, Crittenden PL (1993) The role of chemical-induced stress responses in immunosuppression: A review of quantitative associations and cause-effect relationships between chemical-induced stress responses and immunosuppression. J Toxicol Environ Health 39:163–192

Glucocorticoids and Stress

A group of corticosteroids produced by the adrenal cortex that provides for the response to stress. It regulates carbohydrate, lipid, and protein metabolism, has antiinflammatory and immunosuppressive activity, and inhibits the release of adrenocorticotropic hormone. In rats and mice, corticosterone; in humans,

cortisol. Glucocorticoids are commonly used as anti-inflammatory agents.
▶ Stress and the Immune System
▶ Steroid Hormones and their Effect on the Immune System

Glucocorticosteroids (Glucocorticoids)

Steroid hormones with immunomodulatory and antiinflammatory properties.
▶ Glucocorticoids

Glucortisol

▶ Steroid Hormones and their Effect on the Immune System

Glucose Tolerance Factor (GTF)

A chromium-containing organic complex present in Brewer's (*Saccharomyces carlsbergensis*) or Baker's yeast (*Saccharomyces cerevisiae*) found to be an essential dietary component for maintenance of normal glucose uptake and metabolism.
▶ Chromium and the Immune System

Glutathione Redox Cycle

This is a multicomponent system that utilizes three enzymes—glucose-6-phosphate dehydrogenase (G6PDH), glutathione reductase (GSHRX), and glutathione peroxidase (GSHPX), to help in the removal of cellular organic peroxides and to maintain an adequate intracellular supply of reduced glutathione (without the need for *de novo* synthesis).
▶ Vanadium and the Immune System

Glycosyltransferase

Enzymes that catalyze the transfer of a specific monosaccharide from a nucleotide donor substrate, and its attachment, in a specific glycosidic linkage, to its acceptor substrate. Glycosyltransferases are responsible for the sequential addition of monosaccharides to generate a carbohydrate chain and for the branching of that chain.
▶ AB0 Blood Group System

Goblet Cells

Specialized epithelial cells found in intestines and respiratory epithelia. Unicellular glands of mucins, a mixture of glycoproteins and proteoglycans, functioning as lubricant and barrier.
▶ Immunotoxic Agents into the Body, Entry of

Gonadal Hormones

▶ Steroid Hormones and their Effect on the Immune System

Goodpasture's Syndrome

Goodpasture's syndrome is characterized by auto-antibodies directed against renal or pulmonary basement membranes. Type IV collagen was identified as an auto-antigenic structure. Pathogenesis is caused by a switch from T-independent B cell activation (low affinity antibodies) to a T-dependent induction of high affinity antibody production.
▶ Goodpasture's Syndrome

Graft-Versus-Host Disease

▶ Graft-Versus-Host Reaction

Graft-Versus-Host Reaction

MICHAEL HOLSAPPLE
Health and Environmental Sciences Institute
One Thomas Circle, NW, Ninth Floor
Washington, DC 20005-5802
USA

Synonyms

graft-versus-host reactions, graft-versus-host disease, acute graft-versus-host disease, chronic graft-versus-host disease

Definition

Graft-versus-host reaction (GVHR) is a cell-mediated immune (CMI) reaction mounted by donor (graft) T cells against ▶ histocompatibility antigens of the host (recipient). Graft-versus-host disease (GVDH) is a condition caused by ▶ allogeneic donor lymphocytes reacting against host tissue in an immunologically compromised recipient. The immunosuppressed host is unable to reject the donor cells. Depending on the extent of histo*in*compatibility, the donor lymphocytes can recognize the foreign tissue ▶ antigens of the host. The donor cells can divide, react against the recipient tissue or cells and recruit large numbers of host cells to the resulting inflammatory sites. This cycle can lead to the death of the host/recipient.

Because the basis for a GVHR is the ability of cells from the donor within the graft to react to the recipient's tissues, this condition is characterized by the transfer of immunocompetent cells. As such, the problem of GVHD is particularly associated with bone marrow transplantation, a treatment for a variety of hematological disorders. In bone marrow transplantation, total obliteration of the host immune competence is necessary to enable the graft to take. The immunosuppression in the host contributes to the onset and progression of a GVHR, as noted above. While death is a potential outcome, the primary target organs of acute GVHD in both animals and humans are the skin, gastrointestinal tract and liver (1). Therefore, the manifestations of GVHD include skin rash, liver function abnormalities, abdominal pain, and diarrhea, in addition to the increased incidence and severity of infections and profound deficiencies of immune function—anticipated consequences of obliterating the host immune system.

Characteristics

As noted above, GVHD is a possible outcome of bone marrow transplantation. As such, a GVHR (for example, where the immune response of the donor/graft predominates) is essentially the antithesis of a graft rejection (for example, where the immune response of the recipient/host predominates). Therefore, it is generally accepted that the immunologic and genetic conditions that contribute to graft rejection also play an integral role in the basis for a GVHR.

By way of background, it is important to emphasize that graft rejection displays the two key features of ▶ adaptive immunity: memory and specificity, and that these characteristics can be demonstrated by grafting skin from one animal to another. Memory is indicated when a second allogeneic skin graft from one donor is rejected more quickly by the recipient than the first graft from the same donor. Extensive evidence has accumulated which indicates that specific T cells and cell-mediated immunity are mainly responsible for rejection of solid grafts. For example, nude mice do not reject foreign skin grafts, and even ▶ xenogeneic grafts are accepted.

Similarly, depleting donor marrow of T lymphocytes by immunologic or mechanical means has proven to be an effective way to reduce acute GVHD. Thus, there is no doubt that specific T cells play a role in the manifestation of a GVHR. The relationship of immunologic memory in a GVHR is far more difficult to study. While the risk factors for the development of chronic GVHD include previous acute GVHD (1), no evidence could be found that would be consistent with a critical role for immunological memory in the progression from acute to chronic GVHD. In fact, the most typical manifestation of an animal model for chronic GVHD does not include acute GVHD and is dependent on the expression of antibodies (2). As such, this model of chronic GVHD is an expression of ▶ humoral immunity and is not an example of CMI, as described above for a GVHR. The trigger for this chronic model of GVHD is dependent on the depletion of $CD8^+$ cytotoxic T cells in the donor inoculum, and it is characterized by the development of autoimmune features reminiscent of systemic lupus erythematosus (SLE) (2).

Very early studies in immunology led to the identification of a group of antigens in mice which, when matched between donor and recipient animals, markedly improved the ability of the graft to survive. Because they played such an important role in graft rejection, these antigens were named "histocompatibility antigens". It was also noted early on that these antigens were the products of one particular region of the genome, the major histocompatibility complex ▶ (MHC), which was also logically demonstrated to play a critical role in the graft rejection process. Three major sets of molecules are encoded by the MHC: class I, II and III antigens. Class I molecules are composed of one MHC-encoded polypeptide and are expressed on all nucleated cells. Class II molecules are formed from two separate MHC-encoded polypeptides, have a much more restricted distribution, and are only expressed on B lymphocytes, macrophages, monocytes, and some types of epithelial cells (especially cells that can function as antigen-presenting cells(APC)). Although the MHC was originally identified by its role in graft/transplant rejection, it is now understood and accepted that proteins encoded by this region, especially class I and class II molecules, are involved in many aspects of immunological recognition including interactions between different lymphoid cells and the interaction between lymphocytes and APC. Because class I and class II molecules are involved in immunological recognition, they play a critical role in understanding the basis for a GVHR. Class III molecules are involved in the complement cascade and will not be further discussed in this section.

Tissue graft rejection has been identified as one of three broad areas (the others being hypersensitivity and ▶ autoimmunity) in which the immune system acts detrimentally (3). As discussed above, it was discovered in the early part of the 20th century that successful transplantation depended on the donor and recipient sharing a number of independently segregating alleles, which were eventually characterized as the histocompatibility genes. Although the phenomenon of graft rejection was recognized before 1920, the role of adaptive immunity in this process was not identified until the 1950s.

The important similarities between the processes that are responsible for graft rejection and those that provide the basis for a GVHR have already been noted. What has not been emphasized is that the rejection of foreign tissue (the basis for both the host response against the graft and the donor response against the tissues of an immunocompromised host) really has no normal physiologic function. In view of the unphysiologic nature of tissue transplantation, it may seem surprising that the immune system provides such a formidable barrier to success. Although the rejection of foreign tissue has no normal physiologic function, the processes involved in recognition of foreign determinants on cell surfaces (so-called "allogeneic recognition") (3) are accepted as being similar to those involved in the recognition of viral antigens on infected cells. Most virally infected cells display viral antigens on the surface of their plasma membranes which can be recognized by cytotoxic T cells in association with the class I MHC molecules on the surface of the infected cells.

Allogeneic MHC molecules, the primary drivers for both graft rejection and the GVHR, are highly immunogenic. The precise way in which allogeneic recognition by T cells occurs is still debated (3). One hypothesis proposes that fragments of allogeneic MHC molecules are processed in the same way as conventional ▶ antigens and are then presented in association with "self molecules" by APC. In the context of a GVHR and GVHD, this hypothesis would indicate that the graft is not only heavily populated by reactive T cells but also by APC. A second hypothesis proposes that the T cells react directly with the foreign MHC molecules. Some studies have suggested that T cells can see allogeneic MHC in the same way as self MHC plus antigen. These studies indicate that these cells are not exclusively alloreactive, but contribute to the T cell repertoire that recognizes foreign antigens. It has been estimated that the number of T cells that can recognize a particular allogeneic MHC in unprimed animals is relatively high and may be as great as 0.1% of the total T cell pool (3).

In the context of GVHR and GVHD, this hypothesis would be consistent with the critical role played by T cells in the donor inoculum and would not require the presence of APC in the graft.

Preclinical Relevance

There is little doubt that the preclinical relevance of GVHR and GVHD is in the context of contributing to our understanding of the conditions needed for successful transplantation. As noted above, a GVHR is dependent on a fully functional CMI component, and the GVHR has occasionally been used in immunotoxicology studies as an indicator of CMI. However, the assessment of the competence of CMI has been far more frequently based on other parameters, as discussed below. In the interest of completeness, there is one last point from a preclinical perspective about the GVHR that needs to be emphasized. The popliteal lymph node assay (PLNA) has been suggested to be an animal model capable of predicting the autoimmunogenic potential of some drugs and other small-molecular-weight chemicals (4). Interestingly, it has been suggested that the PLNA manifests some characteristics reminiscent of a GVHR (4).

Relevance to Humans

As discussed above, in clinical transplantation, the major obstacle to successful transplantation is the ▶ MHC. Studies have clearly indicated the important value of matching MHC molecules between donor and recipient and it appears that matching class II molecules is particularly important for graft survival. Consistent with the parallel relationship between graft rejection and GVHR, the severity of acute GVHD also increases as the degree of disparity for major histocompatibility antigens increases. Theoretically, some form of acute GVHD after human marrow transplantation would be expected in all cases except those with ▶ syngeneic and ▶ autologous transplants. This result can be anticipated because of potential differences in polymorphic minor histocompatibility antigens even between genetically identical siblings (1).

The long-term success of marrow grafting for the treatment of hematologic malignancies depends on avoiding three interrelated problems: GVHD, graft failure/rejection, and the recurrence of the underlying leukemia. Prevention or management of acute GVHD has most often involved the use of immunosuppressive drugs after transplant. For most patients, immunosuppressive therapy can be stopped after 3–6 months, when a state of graft-host ▶ tolerance has been achieved. Consistent with the fact that a GVHR is dependent on CMI, removal of T cells from donor marrow can also prevent—or at least dramatically reduce—the manifestation of acute GVHD. However, this approach can have two important consequences:

first, the percentage of successful marrow transplants is lower; and second, there is a higher recurrence rate for tumors in the host, presumably because GVH cytotoxicity is an effective mechanism for suppressing tumor cell expansion. For complete eradication of the tumor to occur, a graft-versus-leukemia effect may be needed. Whether the graft-versus-leukemia effect can be truly distinguished from GVHD is not clear (1). There are studies suggesting that the T-cell reactivity for the graft-versus-leukemia and GVHD can be distinguished. Certain cells in the donor inoculum may react specifically with the unique antigens expressed on leukemia cells, while others may react more broadly with histocompatibility antigens on host tissue cells. Obviously, an ongoing treatment manipulation is to expand the population of T cells in the donor inoculum that can react specifically with the tumor antigens.

Chronic GVHD develops in 25%–50% of recipients of allogeneic marrow at 3–15 months after transplant. The clinical manifestations resemble those seen with systemic collagen vascular diseases, including skin lesions, keratoconjunctivitis, buccal mucositis, esophageal and vaginal strictures, intestinal involvement, chronic liver disease, generalized wasting, and pulmonary insufficiency(1). Risk factors for the development of chronic GVHD include previous acute GVHD, as noted above, and older age. If left untreated, patients with chronic GVHD become disabled or die of infection. Prednisone, with or without cyclosporine, is the most effective treatment for chronic GVHD. In 50% of patients therapy can be discontinued after 9–12 months. Importantly, the management of chronic GVHD remains unsatisfactory as 25% of affected patients die from infections that result from their immunocompromised status.

Regulatory Environment

Because of concerns that repeated exposure to drug and non-drug chemicals may affect immune status, global regulatory bodies have established guidance and test guidelines to assess for immunotoxicity potential. In regard to the regulatory environment, it is important to emphasize that a GVHR is an example of a cell-mediated immune response, and that few immunotoxicology guidance documents or guidelines require the evaluation of cell-mediated immunity. There are a number of models that are appropriate to assess the status of cell-mediated immunity in addition to the GVHR, including the delayed hypersensitivity response, the mixed lymphocyte reaction, cell-mediated lympholysis and a variety of host resistance models. In actuality, in the annals of immunotoxicology, the GVHR has been a relatively infrequent choice as a model to assess cell-mediated immunity.

References

1. Storb R, Thomas ED (1995) Transplantation of bone marrow. In: Frank MM, Austen KF Claman, HN, Unanue ER (eds) Samter's Immunological Diseases, Volume II, 5th ed. Little, Brown and Company, Boston, pp 1471–93
2. Florquin S, Goldman M (1994) Allogeneic diseases. In: Cohen IR, Miller A (eds) Autoimmune Disease Models: A Guidebook. Academic Press, San Diego, pp 291–301
3. Welsh K, Male D (1989) Transplantation and rejection. In: Roitt IM, Brostoff J, Male DK (eds) Immunology, 2nd ed. Lippincott, Philadelphia, pp 24.1–24.10
4. Descotes J, Verdier F (1995) Popliteal lymph node assay. In: Burleson GR, Dean JH, Munson AE (eds) Methods in Immunotoxicology, Volume 1. Wiley-Liss, New York, pp 189–96

Gram-Negative Bacteria

Non-spore-forming, rod-shaped unicellular vegetable microorganisms which, when stained by Gram's method (using aniline water-Gentian violet or carbolic Gentian violet, iodine, alcohol, and a counterstain, eosin) are decolorized and take on the contrast stain.
▶ Klebsiella, Infection and Immunity

Granulocyte

Cell lineages (basophils, eosinophils, neutrophils) that have densely staining cytoplasmic granules. Also called polymorphonuclear leukocytes due to their oddly shaped nuclei.
▶ Lymphocytes
▶ Neutrophil

Granulocyte-Macrophage Colony-Stimulating Factor (GM-CSF)

Granulocyte-macrophage colony-stimulating factor (GM-CSF) is a cytokine inducing the proliferation and differentiation of hematopoietic progenitor cells to macrophages and/or granulocytes and activates the differentiated cells. It is produced in response to a number of inflammatory mediators by mesenchymal cells present in the hemopoietic environment and at peripheral sites of inflammation.
▶ Cancer and the Immune System

Granuloma

Persistent infections, e.g. by mycobacteria, or non-degradable foreign bodies, can trigger a local chronic inflammation. Typically, a center of macrophages and giant cells surrounded by T lymphocytes form a granuloma.
▶ Granuloma

Granulopoiesis

The process in the bone marrow for formation of mature polymorphonuclear leukocytes with granule-containing cytoplasm. Cell types in this class include neutrophils, eosinophils, and basophils.
▶ Therapeutic Cytokines, Immunotoxicological Evaluation of

Granzymes

A family of serine proteases that stimulate programmed cell death machinery via the activation of caspases and other pro-apoptotic pathways.
▶ Cytotoxic T Lymphocytes
▶ Cell-Mediated Lysis

Green Fluorescent Protein (GFP)

A protein expressed as transgene after transfection, which results in a fluorescent response in live cells. The cDNA has been engineered to express more stable forms, and the expression can be directed to specific tissues by using selective promoters.
▶ Hematopoietic Stem Cells

Guanine-Exchange Factor

Guanine-exchange factors (GEFs) are regulatory proteins that act as molecular switches. They regulate the activation state of small G proteins by inducing the exchange of bound GDP for GTP.
▶ Signal Transduction During Lymphocyte Activation

Guidelines in Immunotoxicology

▶ Regulatory Guidance in Immunotoxicology

Guinea Pig Assays for Sensitization Testing

HANS-WERNER VOHR
PH-PD, Toxikologie
Bayer HealthCare AG
D-42096 Wuppertal
Germany

Synonyms

Skin sensitization assay, Buehler test, Magnusson-Kligman maximization test

Short Description

The aim of such tests is to determine whether a test compound exhibits skin-sensitizing properties, that is whether the chemical is likely to induce a delayed-type hypersensitivity reaction in the skin. These assays can be split into two phases: the induction phase, which includes multiple skin contacts with the test compound (sensitization), and a challenge or provocation phase, which evaluates the pathological endpoint of this local immune reaction after applying the test compound in a low and nonirritating concentration.

Characteristics

More than 30 years ago, Buehler published a procedure for the screening of delayed contact hypersensitivity in guinea pigs using an occluded topical patch technique on the pre-shaven animal's back (1). He recommended the use of 20 animals in the sensitized group, 10 naive animals for the challenge, and 10 naive control animals for rechallenge. The concentration of the test compound was to be determined by establishing the minimal irritative concentration in a preliminary irritation study. Like other guinea pig assays, interpretation of the Buehler test is based on a subjective evaluation of the reactions graded for erythema 24 h and 48 h after patch removal.
Many modifications have been made to the basic principle of the Buehler test that aim to increase its sensitivity:
- pre-treatment of the skin with moderate irritants for nonirritating test substances
- additional intradermal injection of the test compound during the induction phase
- use of Freund's adjuvant to enhance the immune reaction to the test material.

Several of these modifications are currently accepted by regulatory agencies around the world (Table 1), although the tests described by Buehler (1), and Magnusson and Kligman (2) (known as the maximization test) are the ones most frequently employed.

Pros and Cons

Guinea pigs are relatively large animals. Several doses of antigen can therefore be tested on the same animal. Although guinea pig assays have never been validated —as new tests (like the local lymph node assay, LLNA) must be nowadays—the fact that they have been carried out for several years under standardized good laboratory practice (GLP) conditions has produced a lot of experience and "historical" reference data in laboratories that perform such tests. However, despite many years of experience the evaluation of skin-reddening and the reliability of such evaluations depend heavily on the availability of well-trained technicians. The irritant potential and/or the color of the test material interferes with a subjective evaluation.

Guinea pig assays are expensive and time-consuming. It is expensive to purchase and maintain the animals and the whole procedure, including pre-testing, can easily take 6–8 weeks. From the immunological point of view it is important to realize that there are only a few inbred strains and guinea pig-specific reagents available. This hampers mechanistic investigations in animals of this species.

Predictivity

There is always some debate about the predictivity of guinea pig skin testing. The reason for this could partly be the subjective evaluation involved. Well-trained individuals with some experience are required to judge skin-reddening and to distinguish a positive reaction from effects caused by shaving, stripping or other mechanical irritation of the skin.

In addition, a positive reaction depends heavily on the ability of the test chemical to penetrate into the skin through the stratum corneum. This penetration will be influenced dramatically by the carrier used for the test. Thus, guinea pig tests are at best able to predict the potential of moderately to strongly sensitizing compounds. It is clear that the uncertainty surrounding the identification of weak to mild sensitizers in animal tests derives from the relatively low number of animals used in these tests. Depending, of course, on the data included in the statistical evaluation, the accuracy of tests in guinea pig versus tests in humans is about 73%.

Relevance to Humans

One common error is to equate potential with risk. However, risk is absolutely dependent on:
- the genetic background, age and sex of the individual
- the duration of skin contact
- the concentrations used
- the condition of the skin (normal versus atopic).

With respect to these variables, even human predictive patch tests (such as the Draize or human maximization tests) are generally not designed to give a sound basis for risk assessment. On the other hand, many clinical "epidemiological" reports on patch testing do not take into account the variables mentioned above and are therefore not of great value for an overall risk assessment.

Regulatory Environment

In recent decades the Buehler test and the maximization assay have been considered the "gold standards"

Guinea Pig Assays for Sensitization Testing. Table 1 Guinea pig assays for sensitization testing

Test	Induction	Challenge
Buehler patch test*	Epidermal (occlusive)	Epidermal (occlusive)
Open epidermal test	Epidermal	Epidermal
Split adjuvant test	Epidermal (occlusive) Adjuvant	Epidermal (occlusive)
Freund's complete adjuvant test	Intradermal (+/− adjuvant)	Epidermal
Maximization test*	Intradermal (+/− adjuvant) Epidermal (occlusive)	Epidermal (occlusive)
Optimization test	Intradermal (+/− adjuvant) Epidermal (occlusive)	Intradermal
Draize test	Intradermal	Intradermal

* Favored by OECD 406 Guideline for Testing Chemicals: Skin Sensitisation, 1992.

for skin sensitization testing. However, the aforementioned problems have increasingly drawn attention to alternative assays that are based on objective endpoints.

In the past few years it has become more evident that measurable immunological endpoints, such as the proliferation of immune-competent cells, need to be employed to investigate immunological reactions—rather than the simple evaluation of skin erythema and edema (i.e. measuring immune reactions in the draining lymph nodes (LLNA/IMDS)).

Relevant Guidelines

- OECD 406 Guideline for Testing Chemicals: Skin Sensitisation, 1992
- EPA OPPTS 870.2600 Skin Sensitization, 1998 (according to these guidelines the LLNA was recommended for assessment of skin sensitization as a first-stage screening study). This passage has been eliminated in the revised guideline of 2003.
- CPMP/SWP/2145/00 Note for Guidance on Non-Clinical Local Tolerance Testing of Medicinal Products, 2001
- OECD 429 Guideline for Testing Chemicals, Skin Sensitisation: Local Lymph Node Assay, 2002

Draft Guidelines

- CPMP/SWP/398/01 Note for Guidance on Photosafety Testing: Draft, 2001 (as modified lymph node assay)

References

1. Buehler EV (1965) Delayed contact hypersensitivity in the guinea pig. Arch Dermatol 91:171–175
2. Magnusson B, Kligman AM (1969) The identification of contact allergens by animal assay. The guinea pig maximization test. J Invest Dermatol 52:268–276
3. ICCVAM (1999) The murine local lymph node assay: A test for assessing the allergic contact dermatitis potential of chemicals/compounds. The results of an independent peer review evaluation coordinated by the ICCVAM. NIH Publication No 99-4494. National Institutes of Health, Bethesda. Website: http:/iccvam.niehs.nih.gov

Gut-Associated Lymphoid Tissue (GALT)

In rabbits, the organized follicles of the GALT are B lymphocyte-rich sites where repertoire development and diversification of antibody heavy and light chain sequences occurs after birth. Lymphocytes and other accessory cells of the immune system are located in mucosal sites extending from the oropharynx through the gastrointestinal tract. Some important organized tissue sites in rabbits include appendix, the sacculus rotundus, and Peyer's patches.

▶ Rabbit Immune System
▶ Mucosa-Associated Lymphoid Tissue

H

H Antigen

H antigen is the precursor of the A and B antigens of the AB0 blood group system and represents the acceptor substrate of the *A* and *B* gene-specified glycosyltransferases. It is expressed abundantly on red cells and other tissues of group O people, but is present at much lower levels in group A and B people. The immunodominant sugar of H is L-fucose and the biosynthesis of H is catalyzed by 1,2-fucosyltransferases encoded by *FUT1* in mesodermally derived tissues (including red cells) and by *FUT2* in endodermally derived tissues.

▶ AB0 Blood Group System

H-Chain and L-Chain

The immunoglobulins consist of two types of polypeptide chains. The larger one is called heavy (molecular weight between 50 and 65 kDa), the smaller one is the light chain (molecular weight about 25 kDa). The basic unit of the antibody molecule consists of two identical heavy chains and two identical light chains. They form a heterodimer with two identical antigen binding sites (divalent).

▶ B Cell Maturation and Immunological Memory

H-2IA

▶ Antigen Presentation via MHC Class II Molecules

H-2IE (Mouse)

▶ Antigen Presentation via MHC Class II Molecules

HAART

An acronym for highly active antiretroviral therapy. A combination of drugs used to control replication of the HIV virus in infected patients.

▶ Thymus: A Mediator of T Cell Development and Potential Target of Toxicological Agents

Haemopoiesis

▶ Bone Marrow and Hematopoiesis

Hairy Cell Leukemia

▶ Leukemia

Haplotype

The linked set of alleles for all MHC gene loci, that is present on the chromatid of one parental chromosome.

▶ Antigen Presentation via MHC Class II Molecules

Hapten

A non-immunogenic compound of low relative molecular mass (usually below 1000 Dalton), which becomes immunogenic after conjugation with a carrier protein or cell and so induces an immune response. Binding of a hapten to a protein allows presentation of the hapten in the peptide-binding groove of MHC molecules so that it can be "seen" by T lymphocytes and binding of several identical haptens to the same carrier molecule allows direct B cell activation by B-cell receptor cross-linking.

▶ Local Lymph Node Assay (IMDS), Modifications
▶ Popliteal Lymph Node Assay, Secondary Reaction

▶ Chemical Structure and the Generation of an Allergic Reaction
▶ Idiotype Network

Hapten and Carrier

HANS ULRICH WELTZIEN
Max-Planck-Institut für Immunbiologie
Stuebeweg 51
D-79108 Freiburg
Germany

Synonyms
Protein-modifying compound, chemical allergen, artificial determinant

Definition
The term hapten (half antigen) was introduced in the 1920s by Karl Landsteiner to define low molecular weight chemical reagents which, as such, will not induce immune responses. However, immune responses to haptenated macromolecular carriers—that is, to proteins—lead to the production of hapten-specific antibodies and T lymphocytes. In the case of autologous (self) proteins these responses will be specifically directed against the artificial and chemically well-defined hapten determinants. Carriers that are foreign to the immunized individuals result in antibodies and T cells directed against the attached hapten as well as the carrier protein (1).

Even though free hapten fails to induce antibody production, hapten-reactive antibodies bind to free hapten as well as to carrier-coupled hapten. The free hapten molecules, therefore, compete for antibody bound to hapten-carrier conjugates and may be employed to elute hapten-specific antibodies from ▶ immunoabsorbents under nondenaturing conditions. Structurally defined hapten determinants have greatly advanced the elucidation of the molecular details of antibody-antigen interactions.

There are practically no limitations to the chemical nature of potential haptens, as long as they contain reactive groups to form covalent bonds with proteins. Acid chlorides or anhydrides, active esters, aldehydes, aryl halogenides, ▶ azo dyes or reactive radicals may attach to carrier proteins via ε-amino groups of lysine, sulfhydryl groups of cysteine, hydroxy groups of serine or threonine, or via aromatic rings of tyrosine or tryptophan. Typical model haptens in immunological research are 2,4-dinitrofluoro (or dinitrochloro) benzene (DNFB or DNCB), 2,4,6-trinitrochloro benzene (TNCB), 4-hydroxy-3-nitrophenyl (NP) derivatives, or oxazolone. Many drugs, like penicillins and other beta-lactam antibiotics, industrial chemicals, and additives in cosmetics, as well as metal ions, may act as haptens. Ions of metals such as nickel (Ni^{2+}), cobalt (Co^{2+}), copper (Cu^{2+}), beryllium (Be^{2+}), chromium (Cr^{3+}), or aluminum (Al^{3+}) differ from classical haptens in that they do not form covalent conjugates, but rather form reversible coordination complexes with proteins. Still other chemicals or drugs do not react with proteins per se. However, some of their intermediates formed upon intra or extracellular metabolism may become protein-reactive. Such compounds have been termed prohaptens. A typical example is urushiol, the toxic and allergenic principle of the American poison ivy. Others can be formed from prohaptens by UV light activation. In such cases the resulting metabolite may act as hapten and induce photoallergic reactions or may cause direct cytotoxicity, i. e. photoirritation. A typical example of the first is olaquindox while most of the fluoroquinolones (antibiotics) are exclusively photoirritating.

Characteristics
Once bound to protein carriers, haptens form antigenic determinants that may activate the antigen receptors of B lymphocytes as well as of T lymphocytes (1–3). Controlling the reaction conditions during carrier modification allows for the production of defined multiplicity of hapten ▶ epitopes. B cells require repetitive modification of individual protein molecules to be effectively activated. This has favored the widely accepted hypothesis that activation of immunoglobulin receptors on B cells requires their aggregation by multivalent antigen. It is only very recently that this aggregation model has seriously been questioned.

T cells, in contrast to B cells, recognize hapten neither as free hapten nor as a hapten-carrier conjugate. They require the presentation of hapten in conjunction with ▶ MHC molecules (major histocompatibility complex determined proteins) on the surface of antigen-presenting cells (APC; MHC restriction). The T-cell antigen receptor appears to require hapten modification of the MHC-associated peptides rather than of the MHC structure itself. However, exceptions in the case of nickel-reactive T cells have also been reported. In vivo, hapten determinants for T cells may be formed either by direct modification of MHC-associated peptides, or via modification of extra- or intracellular proteins which are subsequently degraded to form hapten-modified peptides that are inserted into MHC molecules.

For many hapten-specific T cells the carrier peptides mainly serve to anchor and position the hapten moiety on the MHC surface, whereas the amino acid sequence of the peptide contributes only little to specificity. Such T cells recognize the same hapten on different peptides as long as those expose the hapten in a defined position and contain the correct anchoring amino acid motifs for the restricting MHC molecule. This usually applies to situations where the peptide is modified in a rather central position. If the hapten modification is located closer to the edge of the MHC binding groove, T cell receptors tend to contact peptide epitopes as well as the hapten, resulting in specificity for hapten on only one particular peptide sequence. In such instances cross-reactivities with nonmodified peptides have been observed. If the carrier is derived from a self protein this situation might ultimately result in a hapten- (drug-)induced autoimmune reaction. Hapten-specific T cells include helper ($CD4^+$) as well as cytotoxic T cells ($CD8^-$). Both can be isolated from experimental animals following immunization either by skin painting with reactive haptens, or by injection of hapten modified proteins or of MHC-binding hapten-peptide conjugates. Human hapten-reactive T cells can be expanded and cloned by in vitro stimulation of blood-derived or skin lesion-derived T cells from hapten-allergic individuals. In the case of penicillin allergy human leukocyte antigen (HLA)-restricted, drug-specific T cells have been identified that reacted to autologous APC in the presence of penicillin-modified HLA-associating peptides (4).

Preclinical Relevance

Synthetic model haptens such as DNFB, TNCB and others in murine systems, as well as metal salts, drugs and other allergens in humans are widely used to study their effects on the immune system in vitro and in vivo. As inducers of adverse immune reactions, in particular of allergic hyperreactivities such as ▶ contact hypersensitivity or ▶ delayed-type hypersensitivity, model haptens are representative for a vast variety of chemical hapten ▶ allergens in drugs, cosmetics, food additives, and agrochemicals. These studies have led to a detailed molecular view of type IV and type I allergies in particular. The relevance of the above mentioned MHC-associating hapten-peptides as allergenic determinants has been best demonstrated by ▶ sensitization of animals via injection of such peptides for elicitation of contact hypersensitivity symptoms following exposure to the free hapten, e.g. TNCB.
Predictive screening for potential contact allergens by skin tests in guinea pigs or ▶ local lymph node assays (LLNA) in mice involves large numbers of experimental animals. Great efforts are therefore made to define in vitro test methods which differentiate between allergen and irritant properties of individual chemicals. Most promising results are generated from three-dimensional organ cultures like 3D-skin models.

Relevance to Humans

The relevance to humans is apparent if one considers the enormous numbers of potentially hapten-like or prohapten-like chemical compounds among drugs, cosmetics, or industrial and agricultural chemicals. While ▶ contact hypersensitivity (type IV allergy) is the most prominent hapten-related disease, type I allergies and a variety of less well defined adverse reactions to drugs or other chemicals also have to be considered. $CD4^+$ as well as $CD8^+$ T cells reactive to the hapten allergens can be isolated from hapten-allergic patients, including $CD25^+$ regulatory T cells. Hapten-specific T cells are considered to be responsible for the induction of skin pathology in contact hypersensitivity as well as for the signals that cause B cells to switch from IgG to IgE production in type I allergies. Selective silencing or elimination of these cell populations are prime targets in the development of new regiments in the treatment of allergy.
As mentioned above, the hapten-like or prohapten-like nature of drugs may further be involved in the phenomena of drug-induced autoimmune diseases. On the other hand, hapten-reactive T cells are being employed in a completely different setup to therapeutically control tumor metastases. Vaccination of metastatic melanoma patients with dinitrophenyl-modified autologous tumor cells has been reported to result in occasional regression of metastases (5). Though the mechanisms underlying these rejections are not yet resolved, inflammation and infiltration of $CD8^+$ T cells into the metastatic cell mass have been observed. It seems reasonable to speculate that hapten-modification of tumor-specific, MHC-associated peptides may result in the activation of hapten-carrier-specific T cells which exhibit cross-reactivity towards unmodified tumor peptides. Such effector cells may contribute to tumor cell destruction either by direct cytotoxicity or, more likely, by attraction of other cells via the secretion of soluble factors.

Regulatory Environment

For regulations concerning the testing and screening for potentially allergenic chemicals see ▶ contact hypersensitivity, ▶ delayed-type hypersensitivity, or ▶ autoimmunity.

References

1. Boak JL, Mitchison NA, Pattison PH (1971) The carrier effect in the secondary response to hapten-protein conjugates. 3. The anatomical distribution of helper cells and antibody-forming cell precursors. Eur J Immunol 1:63–65
2. Weltzien HU, Moulon C, Martin S, Padovan E, Hartmann U, Kohler J (1996) T cell immune responses to haptens. Structural models for allergic and autoimmune reactions. Toxicology 107:141–151
3. Budinger L, Hertl M (2000) Immunologic mechanisms in hypersensitivity reactions to metal ions: an overview. Allergy 55:108–15

4. Weltzien HU, Padovan E (1998) Molecular features of penicillin allergy. J Invest Dermatol 110:203–206
5. Berd D (2001) Autologous, hapten-modified vaccine as a treatment for human cancers. Vaccine 19:2565–2570

Hapten Theory

The basic paradigm for the induction of skin sensitization postulates chemical reactivity of a low-molecular-weight chemical, which leads to the formation of hapten–carrier conjugates composed of self-proteins covalently linked with the chemical. It is assumed that the immune system can specifically react to a small chemical in connection to a larger structure only, which is called the carrier. In almost all cases of skin sensitization this paradigm is valid, with the exception of heavy metals which can form stable co-ordinative bonds with proteins and thereby change their quaternary structure. In some cases a prohapten can be bioactivated and transformed into a hapten. This paradigm allows a search for structural alerts based on this chemical reactivity leading to hapten–carrier conjugates. The hapten concept can also be applied to some drug allergies, which are induced after oral administration of the drug like immune-mediated anemias and leukocytopenias. However, the same diseases can be induced by drugs via a different, non-hapten mechanism.

▶ Local Lymph Node Assay (IMDS), Modifications

Hassall's Bodies

Concentrically lamellated structure formed by, often keratinised, epithelial cells in the thymic medulla.
▶ Thymus

Helper T Cells (Th Cells)

A subset of cells that express the CD4 surface marker and are essential for initiating numerous immune functions, including antibody and cytokine production and activating cytotoxic T cells. Th1 cells produce mainly interferon (IFN)-γ, which activates macrophages so that they become able to destroy intracellular parasites like viruses, fungi, *Mycobacterium*, *Listeria*, etc. Th2 cells produce interleukins IL-4 and IL-5 and stimulate B lymphocytes to produce immunoglobulins IgG, IgA, IgM, and IgE. Th3 cells produces primarily TGF-β1 and IL-10.

▶ Transforming Growth Factor β1; Control of T cell Responses to Antigens

Helper T Lymphocytes

ARATI B KAMATH
Brigham and Women's Hospital, Division of Rheumatology, Immunology, and Allergy
Harvard Medical School
Smith 518, 1 Jimmy Fund Way
Boston, MA 02115
USA

Synonyms
CD4$^+$ T cells, T helper 1 cells, Th1 cells, T helper 2 cells, Th2 cells.

Definition
T cells that express the CD4 molecule on their surface upon maturation in the thymus. The CD4+ T cells that are classified as T helper type 1 cells (Th1), characterized by the secretion of interleukin-2 (IL-2) and interferon-γ (IFN-γ) but not IL-4 and IL-5, and T helper type 2 cells (Th2), are characterized by the secretion of IL-4 and IL-5 but not IL-2 and IFN-γ.

Characteristics
Stem cells that are first committed to becoming T cells arise in the bone marrow and migrate to the thymus during fetal and adult life. The cells that first appear in the fetal thymus do not express T cell receptor (TCR) molecules, CD3, CD4 or CD8, and are incapable of recognizing or responding to antigens. T cells in the thymus are also called thymocytes and immature thymocytes reside in the cortex. These immature thymocytes or T cell precursors migrate through the thymus where they undergo a maturation process which goes through the stages of CD4−CD8− (double negative), CD4+CD8+ (double positive) and then mature lineage committed CD4+ or CD8+ (single positive) T lymphocytes. The mature T lymphocytes are detected in the medulla from where they leave the thymus to peripheral lymphoid tissues.

In 1986 Mosmann and colleagues provided evidence that murine CD4+ T clones identical to each other in terms of surface antigen expression could be assigned into two subsets on the basis of distinct, non-overlapping cytokine secretion patterns in response to stimulation with antigens (I). Later this was also shown in human T cells. The T cells were classified as T helper type 1 cells (Th1), characterized by the secretion of IL-2 and IFN-γ but not IL-4 and IL-5, and T helper type 2 cells (Th2), which were characterized by the secretion of IL-4 and IL-5 but not IL-2 and IFN-γ (1). Th1 cells are involved in cell-mediated immune responses, namely, cytotoxic and pro-inflammatory reactions. These cells are also proficient at macrophage activation and have been shown in numerous disease models to activate the immune responses to intracellular pathogens, including viruses, bacteria, yeast and protozoa. Th2 cells are involved in promoting the secretion of IgG1 and IgE, in immediate hypersensitivity reactions and thereby lead to the development of humoral immunity. They also mediate extra-cellular immunity. Uncontrolled Th1 and Th2 responses can lead to chronic inflammatory autoimmune diseases and allergies, respectively. A third subset of CD4+ T cells has also been identified as Th0 cells and these cells produce a mixture of the two cytokine patterns. Thus, the Th cell populations differ with respect to the roles they play in the expression of a variety of immune responses, such as protection against infectious and neoplastic diseases, in the mediation of autoimmune diseases or allergic reactions. Furthermore, host factors such as concurrent infections, stress, age, hormones, ultraviolet radiation, and exposure to toxic substances also affect the balance between the Th1 and Th2 responses (1,2).

Factors Controlling the Differentiation of CD4+ T helper Cells into Th1 or Th2 Phenotypes

Although Th1 and Th2 cells arise from the common progenitor CD4+ T cells they can be skewed towards either phenotype. This selective differentiation of either subset is established during priming and can be significantly influenced by a variety of factors such as: cytokine environment, priming by different doses of antigens and others such as altered peptide ligands and altered co-stimulatory molecules (3–5).

Cytokines

The control of differentiation of uncommitted T cell precursors can be distinguished from the regulation of the resulting effector cells since some but not all of the cytokines influencing these two processes are the same (2,3). This differentiation process is initiated by the ligation of the TCR and directed by cytokines present at the priming of a T cell response which results in highly polarized immune responses in the case of chronic infections, such as parasitic infections. IL-4 promotes Th2 development, whereas, IL-12 plays a key role in controlling the development of Th1 cells. The key cytokines involved in the development of Th1 cells are IL-12 and IFN-γ.

- IL-12 is the primary cytokine responsible for directing the development of Th1 cells from antigen-stimulated naive CD4+ T helper cells. It is a 75 kDa heterodimer that activates STAT3 and STAT4 in Th1 cells and is produced by macrophages and dendritic cells upon their exposure to microbial products.
- IFN-γ plays a key role in the Th1 development leading to host defense against intracellular pathogens. It promotes Th1 development by its action on macrophages to upregulate the production of IL-12 and by maintaining the expression of functional IL-12 receptors on Th1 cells.

The key cytokines in the development of Th2 cells are IL-4 and IL-10:
- IL-4 plays an important role in polarizing the CD4+ T cells towards the Th2 phenotype, both in vivo and in vitro. Naive CD4+ cells in the presence of IL-4, develop into Th2 cells capable of producing IL-4, IL-5 and IL-13.
- IL-10 was initially described as cytokine secretion inhibition factor. It plays an important role in down-regulating the expression of activating and co-stimulatory molecules as well as the production of pro-inflammatory cytokines by macrophages, and also downregulating the Th1 responses.

Recently, other cytokines such as IL-18 and IL-21 have been shown to play a role in Th1/Th2 development.

Antigen Dose

The dose of the antigen may play a major role in determining the choice of effector functions (4). In vitro and in vivo studies have implicated increasing antigen doses may switch the immune response from Th1 to Th2 or vice versa. The term altered peptide ligand (APL) was originally used by Allen's group to describe analogs of immunogenic peptides in which the TCR contact sites have been manipulated. APL can also skew the Th1/Th2 development. This was shown by Constant and colleagues who provided evidence for the first time that the Th1/Th2 differentiation could be influenced by the affinity of an agonist peptide for a major histocompatability complex (MHC) molecule and suggested provision of a strong versus weak ligating signal to the TCR as an alternate mechanism whereby immune responses might be skewed. Early in vivo studies have shown that both low and high amounts of antigen primed for ▶ delayed-type hypersensitivity (DTH), whereas moderate levels stimulated antibody production. Similar effects are also seen in infectious disease models such as BALB/c mice infected with *Leishmania*.

Transcription Factors

A wide variety of transcription factors are known to be

involved in determining the fate of naïve CD4+ Th cells (3).

- T-bet is a recently discovered member of the T-box family, expressed selectively in thymocytes and Th1 cells, which controls the expression of IFN-γ.
- STAT-1 is the signal transducer and activator of transcription is activated by cytokines. It mediates the expression of T-bet that is induced readily in naive T cells by IFN-γ signaling. Both T-bet and STAT-1 are involved in Th1 cell development.
- C-MAF is a transcription factor expressed selectively by Th2 cells which regulates IL-4 expression.
- GATA-3 is another Th2 specific transcription factor whose expression is rapidly induced by IL-4, through STAT6 signaling. It is present at low levels on naive CD4+ T cells and its expression increases during Th2 differentiation and decreases during Th1 differentiation. GATA-3 is also essential for normal thymocyte development and embryonic survival.

Co-stimulatory Molecules

The antigen-presenting cell (APC) plays an important role in shaping the differentiation pathway of naïve Th cells because it provides the precursor T cell with its first activation signals. APCs undergo wide-ranging reprogramming of their surface and cytokine phenotype. The APC express co-stimulatory molecules, such as B7 molecules, which bind to its receptors, CD28 or CTLA-4 on T cells. CD28 is constitutively expressed by the T cells whereas CTLA-4 is expressed only upon activation. The role for CD28 in T helper regulation was first determined by using clones, where it was shown that blocking CD28/B7 interactions reduced the IL-2 production and proliferation in Th1 clones whereas the Th2 cells were unaffected. Other studies in mice and humans have supported the role of CD28 in early differentiation of T helper subsets. On the other hand, the presence of one of the B7 molecules on APC is essential for naive CD4 T cells to produce IL-2 and to proliferate as shown by blocking studies. Variable effects of anti-B7 antibodies in different disease states could reflect the relative importance of different APC and the concentration of antigen available at different stages of infection.

CD4+ T helper differentiation is also regulated by structural changes in the chromatin surrounding the T helper key cytokine genes, IL-4 and IFN-γ. Chromatin structure has been studied in vitro by digestion with DNAse I enzyme which cleaves decondensed DNA to which it has access—also known as DNAse I hypersensitivity. Naive CD4+ T cells acquire distinct patterns of DNAse I hypersensitivity at the IFN-γ and IL-4 loci. Th1 cells show DNAse I hypersensitivity at the IFN-γ locus and not at the IL-4 locus and the reverse is true for Th2 cells. Currently, the molecular mechanisms that regulate histone acetylation, methylation, and phosphorylation during T helper cell differentiation are an active area of research.

Preclinical Relevance

Recognition of antigenic targets by the CD4+ T helper cells holds promise for immune therapy, especially in cancers. The cytokines produced by the T cells or adoptive transfer of the T cells themselves have a tremendous potential in immunotherapy. Understanding cellular and molecular mechanisms that distinguish the Th1 and Th2 subsets can lead to better strategies for controlling the immune responses for therapy.

Relevance to Humans
Diseases

There are many well-documented Th1 and Th2 responses for many disease conditions. Many parasitic organisms stimulate highly polarized CD4+ T cell responses characterized by dominant Th1 or Th2 cytokine production profiles depending on the nature of the pathogen and/or the genetic status of the host. It has also been suggested that antigen-presenting cells (APCs) play a key role in determining the differential pattern of cytokine production which leads to the Th1/Th2 phenotype. APCs can get conditioned for Th1 priming by upregulating their surface co-stimulatory molecules and changing the MHC expression, thereby altering the cytokines they produce. Mostly, intracellular parasites induce Th1-dominated responses, whereas extracellular helminths preferentially trigger a Th2-type response. Several protozoans such as *Leishmania major, Trypanosoma cruzi*, and *Toxoplasma gondii* (all intracellular parasites) have been well studied with respect to the type of response generated. In the case of *L. major* or *T. gondii* infections in mice, it has been reported that IL-12 is important for maintaining the protective Th1 responses. Furthermore, IL-12 is also required for IFN-γ production in T cells that mediate resistance to *T. gondii* infection. In contrast, extracellular parasites such as *Heligmosonoides polygyrus, Schistosoma mansoni*, and *Trichuris murs* favor a Th2 bias by the GATA-3-mediated activation of IL-4 gene expression. Activated Th2 cells then become dependent on the co-stimulatory molecule signaling to provide for sustained Th2 cytokine production.

Chemical Toxicity

Many factors including exposure to xenobiotic compounds can modulate the balance between Th1 and Th2 cell responses. It has been demonstrated that both lead and mercury inhibit host resistance to a variety of viral and bacterial infections which are controlled by cell-mediated immunity. Furthermore, a skewing of the Th1 response towards the Th2 increases the chances of infectious or autoimmune disease. It has been previously demonstrated that lead can enhance the in vitro proliferation of Th2 clones and inhibit Th1 clones. Regardless of the route of exposure to lead, it causes a transient rise in IgE followed by autoimmune glomerulonephritis with increased humoral immunity due to the preferential activation of the Th2 cells in susceptible animals. Lead is more prominent in terms of its effects on the immune system by modifying IgE, IL-4 and IFN-γ synthesis in vivo and in vitro when compared to mercury. The development of APCs, the production of cytokines that differentially influence the activation of the T cells, the expression of co-stimulatory molecules and the activation of macrophages can all be influenced by toxic metals. By inducing a strong Th2 response, these compounds suppress the Th1 response and thereby resulting in immunopathology.

Asthma and allergies caused by exposure to chemicals in the environment lead to a Th1 or Th2 type CD4+ T cell response. Airway hypersensitivity is of the Th2 phenotype whereas contact hypersensitivity is Th1 associated. Since distinct patterns of cytokine production are created, these cells can be used to distinguish between chemicals that cause airway and contact hypersensitivity. Several studies have shown a good correlation between increased air pollution and increased symptoms of asthma.

Radiation

Exposure to ultraviolet radiation is the primary cause of non-melanoma skin cancer in humans. Ultraviolet (UV) radiation also alters the immune functions by suppressing the rejection of UV-induced skin tumors and the generation of delayed-type hypersensitivity reactions, both of which are cell mediated immune responses. Increase in the levels of Th2 type cytokines such as IL-4 and IL-10 and a subsequent downregulation of the cell-mediated Th1 response may result in the suppression of the immune response and tolerance induction seen in UV-irradiated animals.

Regulatory Environment

The use of recombinant gene technology to produce commercially available amounts of cytokines indicates an era of clinical applications of immunotherapy. Ideally, a T-cell assay not only needs to be sensitive, specific, reliable, reproducible, simple, and quick to perform but also should display a close correlation with clinical outcome. Cytokines produced by the CD4+ T cells are being used actively in immune therapy in humans. Tumor necrosis factor (TNF) has been approved for clinical use in Europe to treat sarcomas and melanomas. Also, antibodies to TNF are also being used to treat Crohn's disease. Anti-TNF immune therapy using TNF-receptor 2 has also been approved for treating rheumatoid arthritis.

Recently, several investigators reported successful clinical treatment with IFN-γ or IL-12 in patients with intractable tuberculous and non-tuberculous mycobacteriosis and IL-2 in treatment of renal cell cancers. Understanding cellular and molecular mechanisms that distinguish the Th1 and Th2 subsets can lead to better strategies for controlling the immune responses for therapy.

The development of all such biotech proteins are internationally regulated by an ICH Harmonised Tripartite Guideline: Preclinical Safety Evaluation of Biotechnology-Derived Pharmaceuticals, July 16, 1997.

References

1. Mosmann TR, Sad S (1996) The expanding universe of T-cell subsets: Th1, Th2 and more. Immunol Today 17:138–146
2. Murphy KM, Reiner SL (2002) The lineage decisions of helper T cells. Nature Rev Immunol 2:933–944
3. Agnello D, Lankford CSR, Bream J et al. (2003) Cytokines and transcription factors that regulate T helper cell differentiation: New players and new insights. J Clin Immunol 23:147–161
4. Constant SL, Bottomly K (1997) Induction of the Th1 and Th2 CD4+ T cell responses: The alternate approaches. Ann Rev Immunol 15:297–322
5. Selgrade MJK, Lawrence DA, Ullrich SE, Gilmour MI, Schuyler MR, Kimber I (1997) Modulation of T-helper cell populations: Potential mechanisms of respiratory hypersensitivity and immune suppression. Toxicol App Pharmacol 145:218–29

Hematopoiesis

Synthesis and maturation of blood cells including red cells, white cells and platelets from pluripotent haematopoietic stem cell giving rise to committed myleoid and lymphoid stem cells.

▶ Developmental Immunotoxicology
▶ Fish Immune System

Hematopoietic Growth Factors (HGF)

Polypeptide factors elaborated by bone-marrow stromal cells or activated immune cells which stimulate the proliferation and differentiation of hematopoietic stem cells and hematopoietic progenitor cells. Some HGF have initially been found to act in a lineage-selective and lineage-specific way; however the concept nowadays is that in primitive cells, several HGFs synergistically stimulate self-renewal and differentiation and act in concert with other (non-HGF) stimuli, whereas in more mature cells, stimulation of lineage-specific development by individual lineage-specific HGF prevails.

▶ Bone Marrow and Hematopoiesis
▶ Hematopoietic Stem Cells

Hematopoietic Progenitor Cell (HPC)

A precursor cell which has proliferative potential and can give rise to mature cells. Some authors imply hematopoietic stem cells to be part of the progenitor cell pool, whereas many others use the term to name the pool of primitive precursor cells which are not hematopoietic stem cells but still show a high proliferation potential.

▶ Bone Marrow and Hematopoiesis
▶ Hematopoietic Stem Cells

Hematopoietic Stem Cells

REINHARD HENSCHLER
Institute for Transfusion Medicine and Immune Hematology, Department of Cell Production and Stem Cell Biology Group
German Red Cross Blood Donation Center
Sandhofstrasse 1
D-60528 Frankfurt a. M.
Germany

Synonyms
HSC

Definition
Hematopoietic stem cells (HSC) are defined by their ability to undergo ▶ self-renewal during cell division, whereby at least one daughter cell is an HSC, whereas other daughter cells are able to undergo ▶ differentiation into all blood cell lineages. Via hematopoietic progenitor cell intermediates, HSC eventually give rise to fully mature differentiated blood cells, and contribute to all cell types of the hematopoietic system in a host organism after transplantation. HSC are able to permanently sustain the entire hematopoietic system of a recipient.

Characteristics
Morphology and Phenotype
HSC are small lymphocyte-like cells, typically spherical, with little cytoplasm and few cell organelles, which lack differentiated cell lineage markers (granulocytic, monocytic, erythrocytic, or lymphocyte-specific cell surface antigens; see Table 1) (1). They can be enriched by depletion of more mature cells which express cell differentiation markers after labelling with lineage-specific antibodies or, in addition, by positive selection for cells which express, e.g. Sca-1, c-kit (the receptor for stem cell factor), or Thy-1/CD90. Alternatively, efflux of fluorescent membrane vital dyes such as Hoechst 33342 or rhodamine and subsequent cell sorting of HSC by their ability to rapidly exclude these dyes has been used to highly enrich HSC. In the human system, the cell surface antigens CD34 and CD133 have been used as positive selection markers to enrich in a one-step procedure a proportion of approximately 1–3% of primitive precursor cells from bone marrow, yielding a cell population which will comprise almost all detectable HSC in man. HSC enrichment has also been successfully reached by counterflow centrifugal elutriation, yielding an approximately 20–50-fold enrichment of HSC in the low-density or "rotor-off" fractions (Table 1).

Frequency and Location
HSC are rare cells. Generally less than 1 in 10 000 (in mice) or 1 in 100 000 (in humans) of bone marrow cells are HSC. Other locations where HSC are found (at even lower frequencies) are the spleen, liver, muscle, or fat. In normal blood, small numbers of HSC are normally present. However, during states of bone marrow regeneration where hematopoiesis is increased (such as infection, status after chemotherapy or irradiation) HSC numbers in the blood are transiently elevated. A strong increase in numbers of circulating HSC is however observed after treatment with hematopoietic growth factors. This phenomenon, termed

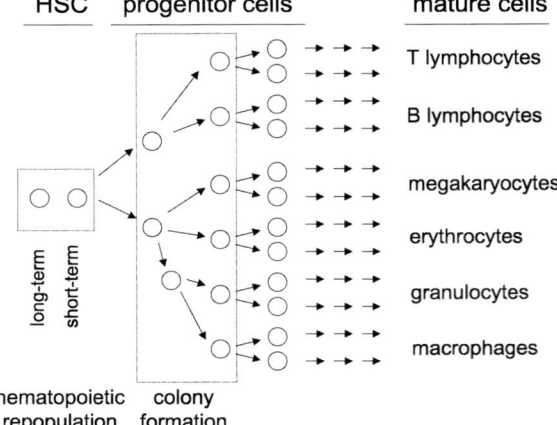

Hematopoietic Stem Cells. Figure 1 Schematic representation of hematopoietic stem cell differentiation. Arrows represent cell divisions; for reasons of simplicity, some cell lineages have not been represented.

Hematopoietic Stem Cells. Table 1 Phenotypes of hematopoietic stem cells and use for purification/enrichment

Marker Type	Examples	Mode of HSC Enrichment	Technology
Lineage differentiation	Gr-1, Mac-1, TER119, B220	Negative selection	Flow cytometric sorting; magnetic sorting
Expressed on stem cells	Sca-1, c-kit, Thy-1 (mouse) CD34, CD133 (human)	Positive selection	Flow cytometric sorting; magnetic sorting
Membrane efflux pump	Hoechst 33342, Rhodamine	High and early efflux selectively in HSC	Flow cytometric sorting
Cell density	–	Gradient sedimentation	Counterflow centrifugal elutriation

HSC ▶ mobilization, has been used clinically and nowadays provides the basis for the efficient donation of most human HSC used as transplants for patients. Within in the bone marrow, HSC locate preferentially to endosteal areas. Interaction of HSC with osteoblasts has been shown to induce the production of growth factors within osteoblasts which support HSC survival. Furthermore, the extracellular matrix environment of osteoblasts interacts with HSC and may also adapt specifically.

Cell Cycle Status

Most HSC are quiescent cells and reside within the G_0 phase of the cell cycle. However, HSC still display the highest sensitivity to ionizing radiation among all tissue types with a lethal dose of approximately 10 Gy. The pretreatment of recipient mice before transplantation (termed ▶ conditioning) for a mouse typically consists of 6–11 Gy of gamma irradiation; doses of more than 10 Gy have to be split and are completely myeloablative, whereas lower doses may result in partial regeneration from endogenous residual HSC, but are easier to apply and better tolerated.

Transplantation of HSC

The function of HSC can be assessed after transplantation by their ability to repopulate all different cell lineages of the hematopoietic system in a recipient organism. Mostly, mice (C57BL and other strains) are used. The function of human HSC can be assessed in animal models after transplantation using immunodeficient mice. Typically, non-obese severe combined immunodeficiency (NOD/Scid) strains are used, yet Scid or interleukin-2 receptor common/γ chain$^{-/-}$-0 or β_2-microglobulin NOD/Scid$^{-/-}$ mice have also been shown suitable recipients for transplantation of human HSC. One advantage of the latter two strains is their ability to develop T lymphocyte reconstitution after transplantation, in addition to the regeneration of human B lymphocytes, natural killer cells (NK cells), macrophages, neutrophils, erythroid cells, or megakaryocytes. In addition to their hematopoietic differentiation and repopulation potential, the ability of HSC to self-renew after transplantation can be tested by their ability to repopulate another preconditioned host organism after sequential transplantation (secondary transplantation).

Regeneration and Hematopoietic Chimerism

The cell dose used to repopulate a pre-irradiated mouse and thus save it from the lethal effect of the conditioning regimen (e.g. whole body irradiation) is usually between 1×10^6 and 5×10^6 whole bone marrow cells. The regeneration of granulocytes and macrophages will typically reach pretransplant levels between days 7 and 10 after transplantation, whereas platelet levels may regenerate up to 1 week later. Due to the longer survival times of erythrocytes in mice (> 10 days) than platelets (1 day) or neutrophils (6 hours), hemoglobin levels will usually only fall only by approximately 20–30%, and not decrease below critical levels.

Interestingly—as a general rule—after hematopoietic transplantation either all hematopoietic lineages will be regenerated, or none will be. A selective regeneration of single lineages is not known in normal organisms. This clearly demonstrates the ability of HSC—and only of HSC—to regenerate the hematopoietic system. Therefore, transplantation assays are selective for the detection of HSC. Experiments using enriched primitive hematopoietic precursor cells which do not include HSC have also shown the inability of any non-stem cell hematopoietic progenitor cell populations to significantly contribute to hematopoietic regeneration. Systems used to recognize the donor origin of the regenerated hematopoietic system in a recipient of transplanted HSC are depicted in Table 2. They all include genetic markers; and if possible the expression of easily detectable proteins in all daughter cell types regenerated from the HSC. Whereas GFP or lac^z transgenes are also expressed on nonhematopoietic cells, expression of Ly 5.1/5.2 (the CD45 antigen) is restricted to leukocytes, progenitors and HSC, and hemoglobin variants are used to selectively detect red cell repopulation.

Generally, the transplantation models are syngeneic (congenic) in order to exclude any side effects from immune reactions such as graft rejection, or graft-versus-host disease (GVHD). However, allogeneic transplantation is feasible in certain genetic constellations (histocompatibility antigen types) between donor and recipient animals, and can be specifically selected to investigate mechanisms of transplant-related immune reactions. As a control for successful conditioning, a survival curve in a control group of animals is established.

The time point at which post-transplant hematopoietic regeneration is determined is of importance. So far, short-term repopulating HSC have been described which contribute to hematopoietic regeneration at early intervals after transplant, generally between 2 weeks and 3 months. Regeneration of hematopoiesis by long-term repopulating HSC requires a follow-up period of between 3 months and up to 2 years; usually 6 months are used.

HSC Niches and Frequency Determination

HSC have been shown to engraft in recipient mice even without irradiation or other preconditioning. Yet, the degree of donor/recipient chimerism reaches only about 40% after five repetitive high doses of HSC. These findings resulted in the concept of a number of HSC "niches" which can be re-seeded better after previous damage or stress to residing HSC.

Models for HSC purification have shown that very few HSC can reconstitute the entire hematopoietic system in a mouse (2). These models also show that HSC cannot be purified to homogeneity as yet, but can be enriched by a factor of about 1000- to 10 000-fold. Transplantation of HSC at limiting dilution into mice has been used to determine the incidence of HSC in the transplanted sample by calculation of the percentage of negative non-engrafted mice in different groups of recipient mice which received different HSC doses and using Poisson statistics. Frequency analysis of transplanted HSC may also be used to estimate the degree of HSC self-renewal before, and at a given time point after transplantation (the latter by secondary transplantation) (3).

Homing of HSC

In part, the impossibility of purifying stem cells to homogeneity can be explained by the fact that intravenous transplantation of HSC does not result in seeding of every transplanted HSC into a site where it will initiate active hematopoiesis. The seeding efficiencies of HSC and of primitive progenitor populations have been calculated to be 2–40% using different transplant models. The more primitive the assayed progenitor or stem cell population is, the higher is the observed seeding efficiency. Cell surface molecules on HSC which play predominant roles in their homing to the hematopoietic organs are the β_1 integrin VLA-4 and the chemokine receptor CXCR4. However, tissue-specific molecules which determine HSC homing into the bone marrow have not been found so far (4).

Transformation

Leukemia can be induced in mouse bone marrow cells by retroviral insertion of oncogenes isolated from

Hematopoietic Stem Cells. Table 2 Markers used to detect donor cells after transplantation of hematopoietic stem cells in mice

Marker	Characteristics	Expression	Detection
Ly5.1/Ly 5.2	Natural mutation; polymorphism of CD45 antigen	All lymphocytes, HSC, progenitor cells	Antibody
Green fluorescent protein (GFP)	Transgene; also enhanced GFP (eGFP)	As regulated by promotor, e.g. ubiquitous	Fluorescence (microscopically, flow cytometrically), or antibody
β-galactosidase (lac^z)	Transgene	As regulated by promotor, e.g. ubiquitous (Rosa-26 mice); several tissue-specific mice available	Histochemical stain using color indicating substrate or antibody
Y chromosome	Sex specific	Ubiquitous	Polymerase chain reaction (PCR); fluorescence in situ hybridisation (FISH)
Hemoglobin variant	Protein	Erythrocyte specific	Gel electrophoresis

human leukemias or by infection with leukemogenic viruses. Cell populations enriched for HSC (most likely HSC themselves) are the target of the transfected oncogene. Importantly, only a rare subpopulation of the leukemic cells gives rise to another leukemia after secondary transplantation. Thus, the concept of HSC as the only cells capable of self-renewal, and thus of the continuous support of hematopoietic proliferation, also translates to leukemia, and the cells which initiated leukemias after transplant have therefore been termed leukemia-initiating cells (L-IC). Human leukemias have been successfully transplanted in NOD/Scid mice. So far, however, normal human stem cells have not been successfully transformed by any type of deliberate manipulation.

Pre-Clinical Relevance

In Vitro Substitute Assays for HSC

Enriched HSC populations have been demonstrated to give rise to colonies using clonogenic assays in semisolid medium in the presence of hematopoietic growth factors (HGF) to varying degrees. However, these assays cannot indicate the presence of HSC, since they are generally not able to indicate growth and differentiation of all hematopoietic lineages in one assay, and cannot measure permanent contribution to hematopoiesis. Also, progenitor cell populations which are devoid of HSC activity also display high ▶ plating efficiencies in these assays. More complex cell culture systems which include bone marrow stromal cells, such as the cobblestone area-forming assay or the long-term bone marrow culture initiating cell assay, have therefore been developed (5). Although they were able to better indicate, or even very well indicate the presence of bone-marrow reconstituting HSC in the murine system, they did not prove to be reliable measures of human HSC.

Mouse transplantation models have therefore been established as the best systems to measure damage to HSC which are exposed to chemicals in vitro (Table 2). Transplantation with HSC can be performed as competitive HSC repopulation, i.e. two HSC populations are injected in a given ratio, of which one has been exposed to an experimental variable (e.g. a chemical). After hematopoietic regeneration, the ratio of the progeny of the two HSC types is compared, and gives a measure of the potency of e.g. a drug or a chemical to damage (or stimulate) HSC. Typically, in these models cell-cycle specific cytotoxic drugs (e.g. 5-fluorouracil or cytosine arabinoside) will spare most HSC because these are not in cell cycle. Conversely, these assays will pick up the cell cycle-independent stem cell toxic effects of the cytotoxic drug, busulfan, or of ionizing radiation.

In Vivo Assays for Toxic Effects to HSC

Animals, mostly mice, have been exposed to a variety of hematotoxic compounds. For example, the hematopoiesis-suppressive effects of Pb, benzene or its metabolites, and of polycyclic aromatic hydrocarbons, become visible by the induction of an anemia (Pb), or a decrease of colony-forming progenitor cells in the bone marrow of mice. These are also referred to in the ▶ Hematopoiesis entry.

However, murine leukemia induction models by immunochemicals have in recent years received higher attention through the use of transgenic mouse strains. Bcl-2 transgenic and p53-deficient mice show reliable increases of leukemia after treatment with benzene or critical metabolites of benzene such as t,t-mucondialdehyde or hydroquinone. Also, in mice, a hypersensitivity of progenitor cells to hematopoietic growth factors (HGF) granulocyte-macrophage colony stimulating hormone (GM-CSF) or IL-3 has been seen, which is similar to the hypersensitivity of human bone marrow progenitors observed during preleukemia or bone marrow hyperproliferative syndromes which are also associated with leukemia induction.

Stromal Regulation of Hematopoiesis

For all in vivo leukemogenesis models, the involvement of the entire organism in the process has to be taken into account. Very likely, damage to stromal cells which support hematopoiesis in the bone marrow plays an important role in leukemogenesis. Indications for this are altered ability of stromal cells to elaborate HGFs in states of bone marrow insufficiency such as aplastic anemia, or the suppression of colony-forming stromal cell precursors in preleukemic states or states of bone marrow insufficiency.

Relevance to Humans

A problem exists regarding the induction of leukemia by chemicals in mice and the relevance to the situation in humans. Since it is generally believed that human leukemogenesis is a multistep process which requires transit through phases of preleukemia and which may take several years to occur, the 2-year lifespan of mice may not be suitable to replay these phenomena in an adequate fashion. Also, murine leukemias are likely different in nature since leukemic transformation can be reached by the expression of only one (human leukemia-derived) oncogene, e.g. a chromosomal translocation product. Thirdly, spontaneous murine leukemias (and possibly also drug-induced leukemias and leukemias by insertional oncogenesis) have been shown to activate endogenous retroviruses in mice. Lastly, a major aspect, which has been noted to play a predominant role in human tumors (the deregulation of telomere length and telomerase function) has not been seen to be involved to a significant extent in murine tumors. Thus, murine leukemogenesis will likely diverge from leukemogenesis in humans in critical aspects.

With regard to normal human HSC, very little validation has so far been reported for known hematopoiesis-suppressive agents and their effects on human HSC as assessed through transplantation models. Therefore, no definitive conclusions can drawn up to now.

Regulatory Environment

Assays for HSC toxicity and leukemogenicity have so far been more or less solely confined to the basic research field, and regulation has so far not specified the use of quantitative or qualitative assays for HSC.

References

1. Weissman IL, Anderson DJ, Gage F (2001) Stem and progenitor cells: origins, phenotypes, lineage commitments, and transdifferentiations. Annu Rev Cell Dev Biol 17:387–403
2. Spangrude GJ, Heimfeld S, Weissman IL (1988) Purification and characterization of mouse hematopoietic stem cells. Science 241:58–62
3. Szilvassy SJ, Humphries RK, Lansdorp PM, Eaves AC, Eaves CJ (1990) Quantitative assay for totipotent reconstituting hematopoietic stem cells by a competitive repopulation strategy. Proc Natl Acad Sci USA 87:8736–8740
4. Papayannopoulou T (2003) Current mechanistic scenarios in hemopoietic stem/progenitor cell mobilization. Blood 103:1580–1585
5. Dexter TM, Allen TD, LG Lajtha (1977) Conditions controlling the proliferation of hematopoietic cells in vitro. J Cell Physiol 91:335–344

Hematopoietic Stem Cells (HSC)

Hematopoietic stem cells are a population of self-renewing cells committed to blood cell formation. Stem cells have the capacity to repopulate the blood forming and immune systems in irradiated hosts (bone marrow transplantation). HSC differentiate and generate hematopoietic progenitor cells restricted to myeloid and lymphoid cell lineages.

▶ Rodent Immune System, Development of the

Hemoflagellates

▶ Trypanosomes, Infection and Immunity

Hemoglobinuria, Paroxysmal Nocturnal

Hemoglobinuria is the presence of free hemoglobin in the urine that may makes the urine look dark. Paroxysmal nocturnal hemoglobinuria is the presence dark urine in the morning but the urine lightens during the day.

▶ Complement Deficiencies

Hemolytic Anemia, Autoimmune

ANNE PROVENCHER BOLLIGER
Rebbergstrasse 59
CH-4800 Zofingen
Switzerland

Synonyms

immune-mediated hemolytic anemia, IMHA, anemia associated with immune response, hemolytic anemia due to ▶ warm autoantibodies or ▶ cold autoantibodies

Short Description

Autoimmune hemolytic anemia (AIHA) is part of a group of diseases called the immune-mediated anemias. These diseases form an important group of anemias in humans and animals. They are caused by the binding of immune proteins to red blood cells (RBCs) or their precursors (1). This immune fixation will create a series of reactions ending by direct lysis of RBCs in circulation (▶ intravascular hemolysis) or by the phagocytosis of RBCs by macrophages (▶ extravascular hemolysis) (2).

Classification

Different classification scheme exist. Usually AIHA are classified in two large categories:
- AIHA due to warm reactive autoantibodies
- AIHA due to cold reactive autoantibodies (i.e. ▶ cold agglutinins disease, cryopathic hemolytic disease).

Autoimmune hemolytic anemias are sometimes differentiated by mechanism of disease and classified as:
- autoimmune hemolytic where RBC destruction is mediated by autoantibodies (warm or cold)
- neonatal isoerythrolysis where RBC destruction is due to an uptake of isologous blood group specific antibodies.

Whatever the classification's scheme used, the anemia

is further classified by the presence or absence of an underlying disease:

- primary or idiopathic AIHA, where there is no recognizable disease
- secondary AIHA where the anemia is the manifestation or complication of an underlying disease.

Characteristics
Etiology and pathogenesis
Etiology of AIHA is unknown. The autoantibodies that mediate RBC destruction are predominantly (but not exclusively) IgG globulins with high binding affinity at body temperature.

In most patients with primary AIHA, erythrocyte autoantibodies are the only recognizable immunologic aberration. It is not due to a defect in immune regulation but it represents an aberrant immune response to a self antigen or an immunogen that mimics self antigen.

In secondary AIHA, the disease can be associated with an immune system disturbance (e.g. systemic lupus) (3).

Cold agglutinin disease or AIHA due to cold agglutinins is caused by autoantibodies that bind optimally to RBCs at temperatures below body temperature. The ability of these autoantibodies to injure RBCs is directly related to their ability to fix complement. Two types of cold reactive autoantibodies to RBCs are recognized: cold agglutinins and cold hemolysins. Cold agglutinins are typically IgM, although occasionally they may be globulins of other isotypes. Cold hemolysins are usually IgG isotypes (4).

Laboratory features
By definition, patients and animals with AIHA have anemia, ranging from very mild to severe (life threatening). Routine hematological evaluations reveal low RBC count, decreased packed cell volume (PCV) and hematocrit (Ht), and decreased hemoglobin (Hg).

The reticulocyte count is usually elevated, but early in the disease some patients and animals may have transient reticulocytopenia despite having an erythroid hyperplasia in the bone marrow. The mechanism for this phenomenon is unknown, although it has been speculated that autoantibodies reacting against antigens expressed on reticulocytes may lead to their destruction, or that an underlying bone-marrow disease may exist in patients and animals showing this finding.

Evaluation of blood smears is very important and can reveal anisocytosis, polychromasia (reticulocytosis), and the presence of ▶ spherocytes (small, dense, darker staining RBCs). The presence of spherocytes is strongly suggestive of AIHA. RBC fragments, nucleated RBC, occasional erythrophagocytosis and ▶ agglutination can also be observed.

Bone marrow evaluation usually reveals erythroid hyperplasia. Hyperbilirubinemia (unconjugated) is highly suggestive of hemolysis and hemolytic anemia but its absence does not exclude the disease. Increased urobilinogen, serum lactate dehydrogenase (LDH) and decreased serum haptoglobin are variably present in patients with AIHA (1,3).

Most of the hematological changes of AIHA are not unique to the disease, so RBC-specific autoantibodies should be identified with ▶ Coombs test (direct antibody test). Direct Coombs test, or DAT, demonstrates the presence of bound immunoglobulin or complement to the RBC membrane. It is performed traditionally by a test tube method in which antiglobulin reagent is added to washed erythrocytes, which are then centrifuged (5). Agglutination can be observed macroscopically and, if in doubt, microscopically. Results are reported as the highest dilution in which agglutination still occurs (6). Serum autoantibodies can be measured in serum by ▶ indirect Coombs test (2,5).

Preclinical Relevance
AIHA is the most frequent cause of hemolytic anemia in humans and animals. It has been reported also in horses, cows, cats, and rabbits. It is the most common manifestation of autoimmunity in dogs. Experimental animal models exist in mice and can also occur spontaneously in this species (1).

Relevance to Humans
The frequency of primary AIHA is estimated to be close to 50% of all AIHA (primary and secondary). Non-autoimmune disease can also result in spherocytic anemia (such as ▶ hereditary spherocytosis in man and animals). Alloimmune hemolytic anemia of transplant recipients resembles AIHA and shows similar laboratory features. Patients with idiopathic AIHA have unpredictable clinical courses, with variable relapses and remissions. Children and adults can be affected, from those in the first months of life to those in their 80s. There is a tendency to see more women affected than men (2,3).

References
1. Barker RN (2000) Anemia associated with immune responses. In: Feldman BV et al. (eds) Schalm's Veterinary hematology, 5th ed. Lippincott, Williams & Wilkins, Philadelphia, pp 169–177
2. Rochant H (2001) Auto immune hemolytic anemias. Rev Prat 51:1534–1541
3. Packman CH (2001) Acquired hemolytic anemia due to warm-reacting autoantibodies. In: Beutler E et al. (eds) William's Hematology, 6th ed. McGraw Hill, New York, pp 639–648
4. Packman CH (2001) Cryopathic hemolytic syndromes. In: Beutler E et al. (eds) William's Hematology, 6th ed. McGraw Hill, New York, pp 649–656
5. Coles EH (1986) Veterinary clinical pathology, 4th ed. WB Saunders, Philadelphia, pp 437–438

6. Manny N, Zelig O (2000) Laboratory diagnosis of autoimmune cytopenias. Curr Opin Hematol 7:414–419

Hemolytic Disease of the Newborn (HDN)

Hemolytic disease of the newborn (HDN) should more precisely be called hemolytic disease of the fetus and newborn (HDFN) or alloimmune haemolytic disease of the fetus and newborn. It is a condition in which the lifespan of red cells or erythroid precursors of a fetus or newborn infant is shortened by the action of specific antibody derived from the mother by placental transfer. The outcome may be severe, leading to death in utero as early as about week 18, or it may be mild, causing neonatal anaemia and/or jaundice.

▶ AB0 Blood Group System

Hemolytic Plaque Assay

▶ Plaque-Forming Cell Assays

Hemolytic Transfusion Reaction

A reaction in which signs of increased red cell destruction are produced by transfusion.

▶ AB0 Blood Group System

Hemopoiesis

▶ Bone Marrow and Hematopoiesis

Hemostasis

▶ Blood Coagulation

HEPA Filtration

High-efficiency particulate air filtration.
▶ Animal Models of Immunodeficiency

Hepatitis, Autoimmune

NEIL R PUMFORD
POSC O-214
University of Arkansas
1260 W. Maple Street
Fayetteville, AR 72701
USA

KATHLEEN M GILBERT
Department of Microbiology and Immunology
University of Arkansas for Medical Sciences,
Arkansas Children's Hospital Research Institute
1120 Marshall Street
Little Rock, AR 72202
USA

Synonyms

Autoimmune chronic hepatitis, lupoid hepatitis, idiopathic chronic active hepatitis, autoimmune chronic active hepatitis.

Definition

Autoimmune hepatitis (AIH) is a progressive inflammatory disease where an abnormal immune system damages liver cells (1). The idiopathic AIH is associated with hypergammaglobulinemia and autoantibodies. Histologically AIH is characterized by portal mononuclear cell infiltration with periportal lesions or piecemeal necrosis. Fibrosis is generally present. Similar histopathology can be found in other autoimmune liver diseases such as primary biliary cirrhosis, primary sclerosing cholangitis, and chronic viral hepatitis. The disease is found predominantly in females (70%). The condition is responsive to immunosuppressive therapy. Patients show a good response with corticosteroids with or without azathioprine. Many patients with AIH have other autoimmune disorders such as thyroiditis, vitiligo, diabetes mellitus, ulcerative colitis, or Sjögren syndrome.

Characteristics

There are three categories of AIH.
- Type 1 is characterized by the production of antinuclear (ANA) and/or antismooth muscle autoantibodies, and type 1 represent the majority of AIH cases (80%–85%).
- Type 2 patients have liver/kidney microsomal type 1 (LKM-1) autoantibodies. The target of anti-LKM1 is the cytochrome P450 2D6.
- Type 3 AIH patients have autoantibodies to soluble liver antigen or liver pancreas antigen.

There is a massive inflammatory cell infiltration in AIH. CD4$^+$ T cells have been show to be activated in patients with AIH. In addition, hepatocytes appear to express human leukocyte antigen (HLA) class II receptors that could present peptides to activate CD4$^+$ T cells. CD4$^+$ T cells have been shown to produce both T helper (Th) 2 and Th1 cytokines in AIH patients (2). Therefore, CD4$^+$ T cells probable play an important role in the liver damage in AIH.

There appears to be both a genetic as well as an environmental component to AIH. The HLA DR3 and HLA DR4 phenotypes of the major histocompatibility complex (MHC) class II are risk factors for the disease. Environmental factors or triggers may be required to initiate the disease in susceptible individuals. The environmental triggering agent for AIH is unknown. Potential environmental activators are viral infections from measles, hepatitis, or Epstein-Barr virus.

The female preponderance of AIH suggests that hormonal regulation may also be a precipitating event.

Other possible contributors to disease etiology are xenobiotics that could exacerbate or trigger autoimmune disease in susceptible individuals. No chemical has been identified to cause AIH, but several drugs produce clinical and pathological symptoms similar to AIH, and may precipitate the disease in susceptible people. Several drugs have been associated with AIH-like diseases, including halothane, tienilic acid, minocycline, methyldopa, dihydrolazine, and nitrofurantoin (3).

It seems unlikely that the immune pathology caused by xenobiotics is initiated against the drugs or chemicals themselves because they are too small to be immunogenic. However, the chemical may be metabolized to a reactive intermediate that binds to self proteins and then it can be recognized by the immune system as foreign. Hypersensitivity reactions directed against the chemical (hapten) will occur, but there can also be an immune reaction against the hapten-carrier complex. This response may cross-react with the carrier, which in many cases is the enzyme that produced the reactive metabolite. Halothane and tienilic acid are good examples of drugs that cause hepatitis with antibodies directed against not only the hapten (drug) but also against the carrier proteins (Figure 1).

This type of hepatitis is generally a hypersensitivity reaction with an autoimmune component of autoantibodies directed against self proteins and thus resembles AIH. The histological damage in the liver from halothane or tienilic acid is primarily localized in the centrilobular region with infiltration of mononuclear cells (Table 1 and Figure 1).

This is different from idiopathic AIH. There are exceptions: about 4% of the halothane hepatitis cases are chronic active hepatitis and are clinically similar to AIH. Yet, the majority of halothane hepatitis cases appear to be primarily hypersensitivity reactions. Autoantibodies include antinuclear antibodies—similar to idiopathic AIH—but other autoantibody targets (primarily the enzyme that metabolizes the chemical to a reactive metabolite such as cytochrome P450) are also found with drug-induced hepatitis. For example the primary autoantibody associated with tienilic acid hepatitis is anti-LKM-2 (cytochrome P450 2C9, CYP 2C9).

Additional drugs associated with chronic hepatitis that clinically resembles AIH include methyldopa, minocycline, and nitrofurantoin (Table 1). Most cases of minocycline-induced AIH are histologically identical to idiopathic AIH except that minocycline rarely induces fibrosis or cirrhosis (4). The average duration of therapy is over 2 years before the development of autoimmune disease. The mechanism for xenobiotic-associated AIH is unknown. Autoantibodies to liver microsomal proteins, such as cytochrome P450, are common for this group although the pathological relevance is not known.

Recently, an animal model for xenobiotic-associated AIH has been developed using the environmental contaminant trichloroethylene (5). The autoimmune-prone mice MRL$^{+/+}$ were treated with trichloroethylene in drinking water for 32 weeks. ANA and increased levels of total immunoglobulin were detected as early, after 4 weeks of trichloroethylene treatment. There was also dose-dependent activation of the CD4$^+$ T cells in both the spleens and lymph nodes of mice treated with trichloroethylene. Following chronic trichloroethylene treatment, there was a significant increase in hepatic mononuclear infiltration localized to the portal region with hepatotoxicity consistent with AIH.

Preclinical Relevance

There are no standard methods for determining the autoimmune potential of experimental drugs or chemicals. Potential screening assays are still in the development stage. Potential assays may include the popliteal lymph node assay (PLNA) and expression of T cell activation markers. The relevance of autoantibody detection is questionable. In order to develop predictive models more work is needed to identify the mechanisms involved in chemical-associated autoimmune disease.

Regulatory Environment

A better understanding of the role of xenobiotics in autoimmunity is needed—from the standpoint of mechanisms, epidemiology, chemical surveillance, or postmarketing—before there can be any attempt to regulate chemicals that induce autoimmunity.

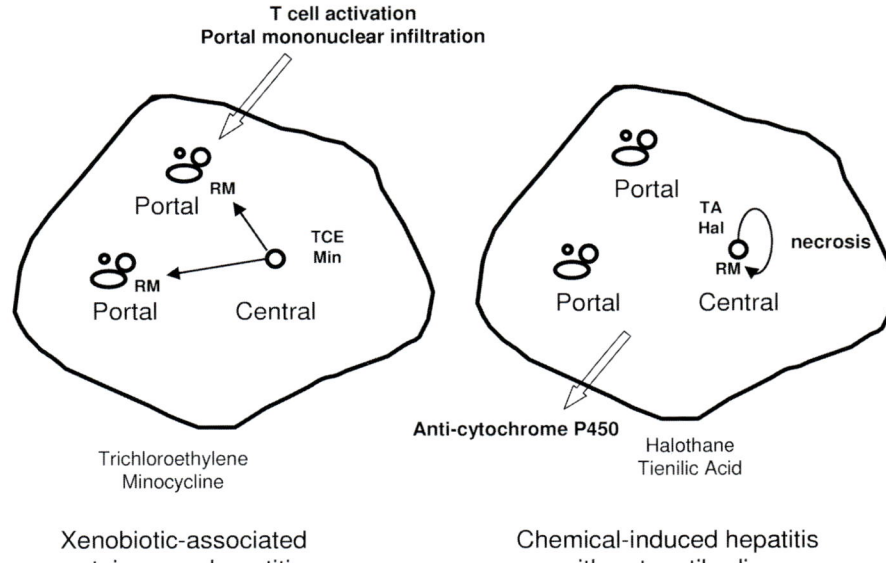

Xenobiotic-associated autoimmune hepatitis

Chemical-induced hepatitis with autoantibodies

Hepatitis, Autoimmune. Figure 1 Mechanisms by which chemicals can trigger autoimmunity. With xenobiotics that cause an autoimmune hepatitis (AIH)-like disease, the chemical is metabolized to a reactive metabolite that somehow migrates or causes T cells to migrate to the periportal region. In contrast, chemicals that cause hepatitis with an autoimmune component primarily cause hypersensitivity reaction to the chemical-modified protein, but can also stimulate the immune system to produce autoantibodies that recognize the native enzyme. RM, reactive metabolite.

Hepatitis, Autoimmune. Table 1 Characteristics of autoimmune hepatitis

Type	Histology	IgG	Autoantibodies
Idiopathic autoimmune hepatitis	Portal mononuclear infiltration with piecemeal necrosis	+++	Type I: ANA +/- antismooth muscle antibodies Type II anti LKM1 Type III antisoluble liver antigen
Xenobiotic-associated autoimmune hepatitis Minocycline Methyldopa Nitrofurantoin Trichloroethylene	Portal mononuclear infiltration with piecemeal necrosis	++	ANA Cytochrome P450 autoantibodies
Chemical-induced hepatitis with Autoantibodies Halothane Tienilic Acid	Centrilobular to midzonal necrosis with mononuclear infiltration	+	Antimicrosomal autoantibodies including cytochrome P450 autoantibodies (anti-CYP 2E1, LKM-2 or CYP 2C9)

ANA, antinuclear antibodies.

References

1. Krawitt EL (1996) Autoimmune hepatitis. N Engl J Med 334:897–903
2. Vergani D, Choudhuri K, Bogdanos DP, Mieli-Vergani G (2002) Pathogenesis of autoimmune hepatitis. Clin Liver Dis 6:439–449
3. Liu ZX, Kaplowitz N (2002) Immune-mediated drug-induced liver disease. Clin Liver Dis 6:467–486
4. Goldstein NS, Bayati N, Silverman AL, Gordon SC (2000) Minocycline as a cause of drug-induced autoimmune hepatitis. Report of four cases and comparison with autoimmune hepatitis. Am J Clin Pathol 114:591–598

5. Griffin JM, Gilbert KM, Lamps LW, Pumford NR (2000) CD4+ T-cell activation and induction of autoimmune hepatitis following trichloroethylene treatment in MRL+/+ mice. Toxicol Sci 57:345–352

Hereditary Spherocytosis

Hereditary disease caused by erythrocyte membrane disorder and characterized by hemolytic anemia of varying severity, presence of spherocytes, increased red cell osmotic fragility, and splenomegaly.
▶ Hemolytic Anemia, Autoimmune

Heterophilic Adhesion

Adhesion between two undefined cell types mediated by different adhesion molecules.
▶ Cell Adhesion Molecules

Heterotypic Adhesion

Adhesion mediated by undefined adhesion molecules between two different cell types.
▶ Cell Adhesion Molecules

Hexacoordinate

Descriptive term for a metal that bears or coordinates with six ligand groups (although the compound can then display one of several geometries—planar hexagonal, trigonal prismatic, or octahedral). Although six positions about the metal core are filled, the number of ligands can vary depending on whether ligands are unidentate, bidentate, or tridentate.
▶ Chromium and the Immune System

Hexavalent Chromium

The ionic form of chromium when the maximal number (six) of outer shell electrons (the one from 4s and all five from 3d) have been shed, thereby giving the atom an overall charge of +6.
▶ Chromium and the Immune System

Hexose–Monophosphate Shunt

The hexose–monophosphate shunt is also known as the phosphogluconate pathway and the pentose phosphate pathway. It is the secondary pathway in glucose metabolism (primary is glycolysis). In the context of immunomodulation, a target (primarily the enzyme G6PDH) for effects on production of reducing equivalents (NAD(P)H) required for optimal function of glutathione redox cycle and several energy-requiring functions in phagocytes.
▶ Vanadium and the Immune System

Histocompatibility

Histocompatibility is the ability to accept grafts between individuals. With relevance to transplantation, the genetic identity between donor and tissues required for acceptance of donor grafts.
▶ AB0 Blood Group System
▶ Autoantigens
▶ Graft-Versus-Host Reaction

Histoincompatible

Differences in histocompatibility antigens between donor and recipient which would lead to rejection of a graft.
▶ Mixed Lymphocyte Reaction

Histones

Small DNA-binding proteins and represent the largest protein component of the nucleus of eukaryotic cells. Histones and DNA associate to form nucleosomes, the basic units of chromatin.
▶ Antinuclear Antibodies

Histopathology of the Immune System, Enhanced

C Frieke Kuper
Toxicology and Applied Pharmacology
TNO Food and Nutrition Research
Zeist
The Netherlands

Synonyms
immunopathology, advanced or extended histopathology, enhanced histological assessment

Short Description
Enhanced pathology in the context of immunotoxicology was originally applied to additional measurements in the 28-day subacute oral toxicity test (OECD guideline 407) (1), namely: (i) weight determination and examination of lymphoid organs in addition to the ones already required, and (ii) application of structured assessment of changes in the principal ▶ lymphoid organ compartments and tissues (2). Presently, enhanced pathology of lymphoid organs applies primarily to the latter, although it is feasible that still more lymphoid organs or tissues may be added in the future to the list of required organs in guideline-driven toxicity studies. The semiquantitative, structured histological assessment of lymphoid organs should not be regarded as a standalone test because its results should be evaluated within the general pathological evaluation of the body.

Characteristics
Necropsy
Weight and gross morphology of lymphoid organs are the first parameters studied in toxicity assessment. Response to injury is often expressed as a change in tissue size and weight. Hyperplastic lymphoid tissue may be characterized by white nodules and increased organ pallor. The white areas represent lymphoid follicles with lymphocyte proliferation. Involution of lymphoid tissue is characterized by decreased organ size and increased similarity of the tissue color to surrounding tissues. The thymus in aged or immunosuppressed or -deficient animals may hardly be visible grossly and is therefore difficult to clean from surrounding mediastinal tissue and parathymical lymph nodes. Lymph nodes can be divided into those that drain the site of exposure and those that do not. It is possible that non-draining lymph nodes are stimulated, for instance, when they drain the mucosae, which are continuously exposed to substances from the diet or inhaled air. Bone marrow is either fixed by immersion of the bone with the marrow in fixative, for the preparation of sections, or expelled from the bone for preparation of cell suspensions. The lack of reported effects on Peyer's patches (PP) may be due to the selection of PP for microscopic examination. Possibly not all PP are equally sensitive to immunomodulating substances. Counting of all grossly visible PP per animal may be helpful in detection of, especially immunosuppressive, effects. In inhalation toxicity studies, nasal-associated lymphoid tissue (NALT) should be collected instead of, or together with, PP as representative of the mucosal immune system. NALT can be collected as part of the nasal passages, and be included in nasal cross sections, although this procedure does not enable the detection of subtle effects.

Structured Histological Assessment of Changes in the Principal Compartments of Lymphoid Organs/Tissue
Initial histologic examination is done mostly on formalin-fixed, paraffin-embedded and hematoxylin and eosin-stained sections of lymphoid organs. For immunohistochemistry, frozen tissue sections are often needed, with mild fixation using, for instance, acetone. A number of antibodies have become available that react to antigens in formalin-fixed, paraffin-embedded tissue, provided that the fixation period has been short. Especially cytoplasmic substances like immunoglobulins can be identified on short-term formalin-fixed tissue. In most cases the identification of membrane-bound antigens, such as those used to identify lymphocyte subpopulations, requires frozen tissue. These requirements preclude routine use of immunohistochemistry in guideline-driven toxicity studies.

Quantitative microscopy may help to objectify the microscopic observations. Sometimes it can help to identify changes in compartment size or cell density, which probably would go unnoticed upon (semiquantitative) histologic examination. For instance, when the sizes of all compartments are affected more or less according to ratio, and when there is no accompanying histopathology. Haematoxylin and eosin-stained sections may not always suffice to achieve optimal contrast between compartments, especially when automated image-analysis equipment is used. Immunohistochemical staining is often helpful for the identification of the compartments.

There are several critical factors in the process of quantitative histology: the number of animals needed, the number of sections examined, the way samples are selected, and the prevention of errors due to investigator bias. Quantification requires an understanding of the distribution of specific cells and compartments, and the characteristics of the changes. For instance, splenic white pulp tissue is distributed more or less homogeneously throughout the spleen and thus two samples or sections might be sufficient. The needs for proper quantitative examination are not exclusively applicable for this technique. In general, the quality of histology should be such that it can be used for histologic examination as well as an overall quantitative measurement.

Indisputable histopathological lesions, like necrosis, fibrosis, and granulomatous inflammation, are easily recognized in lymphoid organs, and their relationship with treatment generally does not pose problems. With these lesions, the main concern is the use of precise nomenclature.

Morphological reflections of functional disturbances may be described in quantitative terms like altered cellularity instead of atrophy, involution, hyperplasia, etc. When a semiquantitative term like decreased cellularity is used the following points should be considered. "Decreased cellularity" may refer to a decrease in cell density or to decreased organ or compartment size, because the size of the compartment may change rather than the cell density, when the number of lymphocytes is decreased due to the flexibility of the stroma in most compartments. Moreover, there may be a decrease in one type of cells with a concomitant increase in another type of cells, e.g. a decrease in lymphocytes and an increase in tingible body macrophages in corticoid-like lymphoid depletion in the thymus. Therefore, the terminology of the relevant microscopic changes needs to be defined in the report.

The dynamics of the immune system should be carefully considered in histological assessment of immunotoxicity. The histology of the thymus depends on factors like nutritional and hormonal (e.g. pregnancy and lactation) state of the animal (3). The histology of lymph nodes is highly dependent on (local) antigenic stimulation. In untreated control animals mesenteric and cervical lymph nodes are in a state of chronic stimulation (draining respectively the gastrointestinal tract and the oronasopharynx). In the non-immunized animal, they contain well-developed paracortex, considerable numbers of macrophages in the sinuses and paracortex (mesenteric lymph nodes), and prominent secondary follicles or germinal centers and medullary cord plasma cells (cervical lymph nodes).

The range of what is the "normal" or "control" histology needs to be established first. This can be done by considering the variability in the entire concurrent control group as the range of "normal", although this may lead to a rather broad "normal" range. One may also consider at least half of the concurrent control group as normal, leaving out the outliers. This option can be strengthened by the use of historical control references and/or background histology and the experience of the pathologist. The need to establish the range of "normal" histology implies that the initial examination cannot be performed without knowledge of the treatment—the identification of control and test groups. In a later phase, so-called "blind scoring" can be a valuable tool to ascertain histologic changes and to set grades.

Morphologic alterations are reported per compartment, because immunotoxic compounds may have an effect on one compartment and leave others unaffected. Information on the affected compartment is of interest for the evaluation of the mode of action of a compound because distinct compartments within a lymphoid organ feature one or more specific functions, and each houses lymphoid and non-lymphoid cells of different lineages and in different ratios (Table 1). Such an approach has been used in several interlaboratory studies (4,2,5) A compartment-driven approach is not possible in bone marrow because compartments have not been identified in hematoxylin and eosin-stained sections of bone marrow from a specific bone, which is unexpected because the bone marrow is both a primary (antigen-independent) and a secondary (antigen-dependent) organ. At present it is even unknown if there are regional differences in bone marrow from different bones, for instance between femur and sternum (the two sources of bone marrow that have been studied most in rodents).

Some Examples of Immunotoxicants

Although precursor (immature) T lymphocytes find a protected environment in the thymic cortex, e.g. the thymus cortex is relatively free of environmental antigens and separated from the blood circulation by the "blood-thymus barrier", the cortical thymocytes are more often affected than the medullary thymocytes. For instance, glucocorticosteroid hormones have their main effect on cortical, CD4/CD8 double-positive thymocytes, apparently because of the abundance of glucocorticosteroid receptors in these cortical thymocytes. Some organotin compounds affect the lymphoblast population in the outer cortex, by inhibition of cell division. With both compounds, the cortical depletion can be such that a reversed pattern of lymphocyte density temporarily exists, in which a higher lymphocyte density is observed in the medulla. Cytostatic agents like azathioprine affect DNA synthesis and hence cell proliferation, which mainly occurs in the outer cortex. All these compounds, after prolonged exposure, induce a decrease in the numbers of peripheral T cells, observable for example in blood (lymphopenia), the periarteriolar lymphocyte sheet (PALS) in spleen and paracortex of lymph nodes, and as reduced lymphoid organ weights. Alteration of the size of a thymic compartment without a major change in the size or weight of the whole organ has occurred, for example, after exposure to lower doses of the immunosuppressive drug cyclosporin, which induces an increase in the cortex to medulla ratio, or the food additive 2-acetyl-4(5)-(1,2,3,4-tetrahydroxybutyl) imidazole, which induces an expansion of the medulla at

Histopathology of the Immune System, Enhanced. Table 1 Compartments and cells of lymphoid tissues (adapted from (5))

Organs	Compartments	Cells
Bone marrow		Hematopoietic cells, mature leukocytes, plasma cells
Thymus	Cortex	Fine reticular epithelium, macrophages, immature T lymphocytes
	Medulla	Plump reticular epithelium, macrophages, dendritic cells, T lymphocytes
	Corticomedullary zone	Immature and mature lymphocytes
	Epithelium-free areas	Immature T lymphocytes, macrophages
Lymph node and spleen	Paracortex (lymph node), PALS (spleen)	Interdigitating cells, T lymphocytes
	Primary follicles, follicle mantle of secondary follicles	Dendritic cells, macrophages, B lymphocytes, low number of T lymphocytes
	Germinal center	Follicular dendritic cells, macrophages, B and T lymphocytes
	Medulla (lymph node), red pulp (spleen)	Plasma cells, T lymphocytes, reticular cells, granulocytes
	Marginal zone (spleen)	Macrophages, marginal zone B lymphocytes
Organized MALT	Lymphoepithelium	M (microfold) epithelial cells, lymphocytes
	Follicles	Dendritic cells, macrophages, B lymphocytes, low number of T lymphocytes
	Interfollicular areas	Interdigitating cells, T lymphocytes

MALT, mucosa-associated lymphoid tissue; PALS, periarteriolar lymphocyte sheet.

the cost of the cortex compartment. Decreased cellularity in the medulla, if it occurs, mainly follows lymphodepletion of the cortex, depending on the experimental conditions. There are no reports showing preferential lymphodepletion of the medulla with maintenance of the architecture and cellular composition in the cortex. Interestingly, in conditions where the thymus becomes extensively depleted of lymphocytes in cortex and medulla, perivascular spaces can be quite prominent and can be filled with leukocytes to a higher density than in the thymic epithelial microenvironment.

Besides agents like vaccines, which intentionally stimulate lymph nodes and spleen, only a few compounds have been reported with morphologically observed increased cellularity, hyperplasia and/or germinal center development. The limited data available indicate mostly (increased) development of germinal centers in the thymus and spleen, lymph node enlargement with germinal center development, increased size of paracortex and/or increased numbers of macrophages, macrophage accumulations or granulomata. Hyperplastic responses as an expression of toxicity are exemplified by the effects of hexachlorobenzene on spleen and mesenteric lymph nodes. Hyperplasia of the B lymphocyte system is observed as an increase in follicles and in the marginal zone of the spleen. Hexachlorobenzene also induces an increase in the number of high endothelial venules in T-dependent areas of the lymph nodes and Peyer's patches. Stimulated lymph nodes (with increased cellularity in paracortex and/or well-developed germinal centers) can be observed at the draining sites of skin applied with compounds with allergenic properties like dinitrochlorobenzene and trimellitic anhydride, and at the draining sites of the footpad injected subcutaneously with a number of drugs with immunomodulating potential. The increase in lymph-node weight during such reactions formed the basis of the auricular and popliteal lymph node assay.

Evaluation

Specific components of the lymphoid organs may be decreased in number (suppressed—involuted) or increased (stimulated—expanded), but that does not necessarily reflect the overall effects on the immune system. For example, allergic and autoimmune-related diseases are considered most often to be the consequence of deranged immunostimulation. Indeed, allergic and autoimmune inflammation in organs like lungs, thyroid, kidneys, skin and liver have been found in company with and following thymic follicular hyperplasia and lymphoid hyperplasia in lymph nodes and spleen. However, thymic atrophy, thymic epithelial defects and blood lymphopenia are observed as well. For instance, thymic atrophy may accompany autoimmune disease in the autoimmune-prone mastomys and (NZBxNZW) F1 mouse. Thymectomy and cyclosporin-induced partial thymus depletion in young animals can result in autoimmune(-like) diseases. Therefore, certain forms of thymus depletion may be considered as possible early indicators of autoimmune disease. Moreover, particular substances can display immunosuppressive properties in certain animal species or strains under certain conditions, but autoimmune diseases in other, sensitive animals. For instance, mercuric chloride induces in BN rats autoimmune nephropathy and Sjögren syndrome-like adenitis in the lacrimal and salivary glands, while it has immunosuppressive properties in Lewis rats. In addition, oral exposure to vomitoxin can both suppress and stimulate immunity, depending on the dose, gender, and animal species tested. Thus, substance-related depletion in lymphoid organs may be used as an indication for further research into deranged autoimmunity. The most appropriate way for interpretation of immunotoxicity and establishing the severity of immunotoxicity is to perform enhanced pathology in conjunction with functional testing.

High doses may disturb body homeostatis in such a way that non-specific stress is induced which clearly involves the immune system. Discrimination between effects due to non-specific stress and direct effects of the compound should be made, whenever possible.

Pros and Cons

The structured assessment of changes in the principal compartments of lymphoid organs and tissues is not aimed primarily on the detection of lymphoid organ diseases, but of disturbances in the immune system that may lead to disease. It is therefore also denoted as enhanced histological assessment of lymphoid organs/tissues. The structured assessment increases the awareness of the complexity of the lymphoid organs and of the broad spectrum of normal morphologic appearances of these dynamic organs, thereby increasing the sensitivity of the microscopic examination. For the experienced pathologist, it may not be necessary to use elaborate forms or schemes to detect induced changes in lymphoid organs. Rather, such forms help to gain experience and may help in selected cases. The enhanced histopathology requires good-quality sections. Conventional histotechnology enables the evaluation of the effects of xenobiotics on main cell subsets by assessing their distinct cytomorphology or tissue location. In this way, the effects on lymphocytes of T and B lineages or on components of the supporting stroma can be evaluated. Staining for lymphocyte and macrophage subpopulations (immunohistochemistry) and quantitative microscopy and flow cytometry are of great use but are not easily incorporated in routine toxicity studies in each laboratory.

Present knowledge regarding the relationship between structure and function of lymphoid organs is such that often a preliminary hypothesis of the possible mechanisms of toxicity to the lymphoid system can be formulated following histologic examination. This in turn may help in elucidation of normal lymphoid organ function and in the discovery of new drugs. However, different mechanisms of tissue injury can yield similar histopathological features and thus additional studies and/or techniques may be required.

The systemic immune system is covered adequately by the selection of lymphoid organs and tissues examined in routine toxicity studies. Examination of the local immune system in toxicology is generally focussed on Peyer's patches, bronchus-associated lymphoid tissue (BALT), NALT, and the regional lymph nodes. However, the enormous bulk of lymphocytes in and underneath the skin, intestinal and airway epithelium differs markedly from the T cell in the organized lymphoid tissues in terms of phenotype, origin, and function. It is therefore questionable whether effects of toxic compounds on organized lymphoid tissues are representative of effects on the single T cell pool. Unfortunately, the diffuse localization of this lymphocyte pool hampers the sampling and examination.

The enhanced histopathology should not be used as a standalone test, out of the context of a toxicity test with haematology, clinical chemistry, and the examination of non-lymphoid organs—especially the endocrine organs. Moreover, the results of the enhanced histological assessment should be evaluated together with indisputable histopathological changes like all kinds of inflammation and tumours in lymphoid and non-lymphoid organs.

Predictivity

Histology of lymphoid organs has been found quite sensitive in a number of studies, including interlaboratory validation studies, although it did not flag every model immunotoxic substance (4,2,5,6). Sensitivity of lymphoid organ histology can be increased by using a standardized semiquantitative approach (enhanced histology) that takes into account the various lymphoid organ compartments, and by training of pathologists in using this approach. The thymus has been found particularly sensitive to immunotoxicants (rat 10; mouse 4). In addition, cellularity and germinal center development in splenic and lymph node follicular areas were found to be sensitive parameters (4).

Relevance to Humans

The use of immunosuppressive drugs, cytotoxic anticancer drugs, high doses of glucocorticosteroids, and cytokine drugs has shown that humans are susceptible to immunosuppressive and immunostimulating agents in much the same way as laboratory animals. Differences between individual responses and between responses in humans and laboratory animals may be more dependent on differences in metabolism and toxicokinetics than on differences in immune system responses. For proper hazard identification, pathologists should acknowledge interspecies-dependent and intra-strain-dependent differences in the histophysiology of lymphoid organs, but even more so the enormous changes in lymphoid organs—especially the thymus—during life. This intraindividual age-related variability is possibly greater than the interspecies and inter-strain variability. The high sensitivity at young age for a number of immunotoxicants has received specific attention in immunotoxicity testing (7,8).

Enhanced histopathology has been found to increase the sensitivity of immunotoxicity studies in laboratory animals and may thus add to a proper hazard identification and risk assessment in man.

Regulatory Environment

Histopathological assessment of tissues is an important component of preclinical studies. For practical purposes the histologic examination is restricted to some key lymphoid organs and tissues. The Organisation for Economic Cooperation and Development or OECD 407 guideline (28-day oral toxicity) includes weighing of thymus and spleen and examination of these organs, bone marrow, draining and distant lymph nodes and Peyer's patches as representative of mucosa-associated lymphoid tissues (MALT).

A similar selection of lymphoid organs and tissues has been made by the Food and Drug Administration (FDA) and the European Agency for the Evaluation of Medicinal Products (EMEA). Blood is examined generally through differential counting of leukocytes on May-Grunwald Giemsa-stained blood smears. Issues which need special attention at present are validation of examination of bone marrow, approaches to studying mucosa-associated lymphocytes and macrophages, either single or in organized lymphoid tissues, the study of lymphoid organ development, and the application of molecular pathology.

The pathologist should indicate the likely pathogenesis of any treatment-induced effects, to discriminate between direct adverse effects and exacerbation of spontaneous disease unique to a particular laboratory animal species, because these processes may have quite different meanings for the human safety of a chemical or drug. This implies that the effects should be described very precisely, especially with regard to type of cells involved, the compartment of tissue involved, and criteria on terminology and grading.

References

1. OECD (1996) Report of the OECD Immunology Work Group Meeting, Research Triangle Park, NC, 11–12 December. Paris: Organisation for Economic Cooperation and Development
2. International Collaborative Immunotoxicity Study (ICICIS) Group Investigators (1998) Report of a validation study of assessment of direct immunotoxicity in the rat. Toxicology 125:183–210
3. Kendall MD, Atkinson BA, Munoz FJ, de la Riva C, Clarke AG, von Gaudecker B (1994) The noradrenergic innervation of the rat thymus during pregnancy and in the post partum period. J Anat 185:617–625
4. Germolec DR, Nyska A, Kashon M et al. (2004) Extended histopathology in immunotoxicity testing: Interlaboratory validation studies. Toxicol Sci 78(1):107–115
5. Richter-Reichhelm HB, Dasenbrock CA, Descotes G et al. (1995) Validation of a modified 28-day rat study to evidence effects of test compounds on the immune system. Regul Toxicol Pharmacol 22:54–56
6. Vos JG, Krajnc EI (1993) Immunotoxicity of pesticides. In: Hayes AW, Schnell RC, Miya TS (eds). Developments in the science and practice of toxicology. Elsevier, Amsterdam, pp 229–240
7. Holsapple MP (2003) Developmental immunotoxicity testing: a review. Toxicology 185:193–203
8. Van Loveren H, Vos J, Putman E, Piersma A (2003) Immunotoxicological conseque nces of perinatal chemical exposure: A plea for inclusion of immune parameters in reproduction studies. Toxicology 185:185–191
9. Kuper CF, Harleman JH, Richter-Reichhelm HB, Vos JG (2000) Histopathological approaches to detect changes indicative of immunotoxicity. Toxicol Pathol 28:454–466
10. Schuurman H-J, Kuper CF (1995) Pathology of the thymus: Changes induced by xenobiotics and gene targeting. APMIS 103:481–500

HIV Co-Receptors

Cell membrane molecules used by human immunodeficiency virus HIV-1 to gain entry into cells, in addition to the CD4 molecule. The CXCR4 and CCR5 chemokine receptors have been identified as HIV co-receptors.

▶ Chemokines

HLA-DP

▶ Antigen Presentation via MHC Class II Molecules

HLA-DQ

▶ Antigen Presentation via MHC Class II Molecules

HLA-DR (Muman)

▶ Antigen Presentation via MHC Class II Molecules

Hodgkin's Disease

A malignant disease, a form of lymphoma, characterized by enlargement of the lymph nodes, spleen, and liver.
▶ Chemotaxis of Neutrophils
▶ Lymphoma

Hodgkin's Lymphoma

▶ Lymphoma

Homeostatic Control

New lymphocytes are continuously generated in primary and secondary lymphoid organs. In the different immune compartments they compete with resident cells for survival. Homeostatic control ensures that the number and distribution of lymphocyte populations is stable.
▶ B Cell Maturation and Immunological Memory

Homogeneity/Heterogeneity of Variance

Situations which describe whether groups of observations come from populations with equal (homogeneity) or unequal (heterogeneity) variances.
▶ Statistics in Immunotoxicology

Homologous Recombination

Recomination involving exchange of homologous genetic loci. Important technique in the generation of null alleles in knockout transgenic animals.

▶ Transgenic Animals

Homophilic Adhesion

This is adhesion between two undefined cell types mediated by identical adhesion molecules.
▶ Cell Adhesion Molecules

Homotypic Adhesion

This is adhesion mediated by undefined adhesion molecules between identical cell types.
▶ Cell Adhesion Molecules

Homozygous

The state in which both copies of a given gene are identical. In knockout mice, this condition occurs in animals with the engineered null mutation (homozygous knockout or −/−) as well as in wild type animals (homozygous normal or +/+).
▶ Knockout, Genetic

Hormone

A chemical substance, formed in one organ or part of the body and carried in the blood to another organ or part which it stimulates to functional activity.
▶ Serotonin

HOSEC

▶ Three-Dimensional Human Skin/Epidermal Models and Organotypic Human and Murine Skin Explant Systems

Host Defense Systems

▶ Respiratory Infections

Host Resistance

The function of the innate and acquired immune system to protect against infections.
▶ Host Resistance Assays

Host Resistance Assays

H VAN LOVEREN
National Institute of Public Health and Environmental Protection
Bilthoven
The Netherlands

H-W VOHR
Bayer HealthCare AG
Wuppertal
Germany

Synonyms
infectious agents models

Short Description
Although different kinds of functional assays could be subsumed to the expression host resistance assays, it is used in the field of immunotoxicology as the resistance to parasites, bacteria, viruses, and tumors. Investigations of reactions of a host to such pathogens have been introduced into an immunotoxicity testing battery very early on. The topic of this chapter is ▶ host resistance in a more narrow sense, that is immune function tests using infectious agents.

Characteristics
As the defense mechanisms that are put into action differ for different pathogens, different host resistance models need to be employed to study the consequences of immunotoxicity. Models that are used include bacterial, viral, and parasitic challenges. The most commonly used ▶ infection models are described below.

Listeria monocytogenes
Relevant mechanisms of defence against ▶ Listeria monocytogenes include phagocytosis by macrophages, and T cell-dependent lymphokine production that enhances phagocytosis. Humoral immunity, in contrast to T cell-dependent immunity, is not relevant in terms of protection against the infection in this Listeria model. Mortality due to infection is a common endpoint monitored; however, clearance and organ bacterial colony counts can also be conducted. Clearance of Listeria after infection, for example, via the intravenous or the intratracheal route, can be assessed at various times after infection by determining the numbers of colony-forming units in the spleen or lungs respectively.

Differences in the numbers of bacteria, retrieved from the organs, are an indication for the clearance of the bacteria, which is the rate at which the host disposes of the bacteria after infection. Besides clearance of bacteria from different organs after infection, histopathology due to a Listeria infection can also be a valuable parameter. For instance, ozone exposure prior to intratracheal infection with Listeria has an effect on pathologic lesions due to the infection. Pulmonary infection with Listeria induces histopathologic lesions that are characterized by foci of inflammatory cells such as lymphoid and histiocytic cells, accompanied by local cell degeneration and influx of granulocytes. If rats are exposed to ozone for 1 week before infection, the lesions found are much more severe than in non-exposed animals and still present at time points that both ozone-associated as well as infection-associated effects alone would have resolved. Besides the severity and duration of the histopathology after infection, the quality of the lesions is also influenced by prior exposure to ozone. Mature granulomas were found in Listeria-infected rats that were also exposed to ozone. The Listeria model is the host resistance assay most often used in immunotoxicologic assessment of compounds and numerous examples of its use can be found in the immunotoxicologic literature (1).

Streptococcus pneumoniae
▶ Streptococcus pneumoniae is a Gram-positive cocci whose host resistance is multifaceted, This bacterial model has proved to be one of the most sensitive indicators for detecting changes in host resistance in the

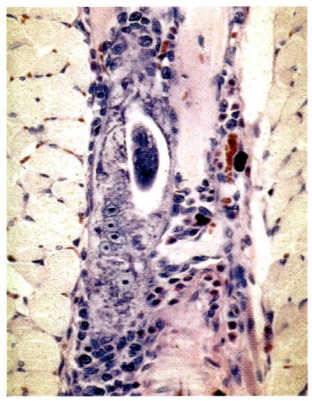

Host Resistance Assays. Figure 1 Liserteria monocytogenes, phagocytosed by macrophages.

mouse. This may be due to the fact that several immune defense mechanisms participate, to varying degrees, in the protection from this organism. The first line of defense against this organism is the complement system. Activation of the complement system can result in direct lysis of certain strains of *Streptococcus pneumoniae*. However, due to the nature of their cell wall, some strains are resistant to lysis by complement. Complement can still participate directly in the removal of these bacteria as a result of complement component C3 being deposited on their cell surface which facilitates phagocytosis by polymorphonuclear leukocytes (PMNs) and macrophages. In the later stages of the infection, antigen-specific antibody plays a major role in controlling the infection. Three compounds affecting complement, PMNs, B cell maturation and proliferation or the production of antibody can be evaluated with this model system. *Streptococcus pneumoniae* represents an excellent model for evaluating immunotoxicity as it elicits multiple immune components which participate in the host resistance, each of which can be a potential target site for an adverse effect of a xenobiotic (2). To date this model showed limited success in the rat.

Cytomegalovirus

▶ *Cytomegalovirus* (CMV) infections are widely distributed throughout the human population. Around 60% to 90% of the population is infected. The vast majority of the naturally occurring primary CMV infections are clinically asymptomatic. More severe disease may occur especially in immunocompromized hosts, such as transplantation patients. Immunosuppression may be an important factor necessary for development of CMV-associated diseases by reactivation of latent virus. For this reason a CMV model is of interest especially for testing potentially immunotoxic compounds. The role of several aspects of the immune system in the resistance against CMV has been extensively studied in mice. Antibodies play a role in neutralization of murine CMV and in antibody-dependent cell-mediated cytotoxicity. CMV-specific cytotoxic T cells can be detected in CMV-infected mice. Natural killer (NK) cell activity appeared to be most effective especially during the initial stages of CMV infections. A role for macrophages in resistance against CMV is doubtful. However recovery of CMV from macrophages several years after infection suggested that murine macrophages became latently infected and served as a reservoir for future CMV infections.

The various strains of CMV are species specific. CMV concentration in tissue can be determined using a plaque assay. In the rat, rat cytomegalovirus (RCMV) is already detectable 8 days after infection of PVG rats although the virus load at day 15 up to day 20 after infection is much higher. The virus load in the salivary gland is the highest compared to the other organs such as spleen, lungs, kidney, and liver. Immunosuppression induced by cyclophosphamide, antilymphocytic sera, or total-body irradiation results in inhibition of immunological resistance of mice to CMV (3,4) Also in the rat exposure to immunotoxic agents, such as organotin compounds, leads to altered resistance to rat CMV.

Influenza virus

Influenza virus A2/Taiwan 22 has been used as a viral challenge for evaluating alterations in host resistance in mice following exposure to various compounds. Among the compounds tested which decreased host resistance to the virus are dimethylnitrosamine and 2,3, 7,8-tetrachlorodibenzo-*p*-dioxin. Mortality is the endpoint routinely used to evaluate decreased host resistance to ▶ influenza which is usually instilled intranasally (5). Host resistance to this virus has been reported to be mediated by antibody and interferon. This model had been suggested for use in evaluating compounds that effect humoral immunity; however, the lack of effect with compounds that affect the humoral immune response raises concern that this model is suited for evaluating compounds that decrease antibody production. A possible explanation for this discrepancy is that because the virus is administered by intranasal instillation, it may involve local immune mechanisms of the lung and not adequately reflect systemic immunocompetence.

The influenza virus has been used in evaluating immunotoxicity in the rat. Use of this virus in the rat has required adaptation. It was shown that rats exposed to phosgene had significantly decreased host resistance. As a human pathogen, when using the influenza virus, appropriate safety precautions need to be followed.

Trichinella spiralis

A parasitic host resistance model that can be used both in rats and mice that uses the helminth ▶ *Trichinella spiralis*. Its life cycle is as follows: after ingestion of infected meat, the encysted larvae are excysted in the acid pepsin environment of the stomach, and are then passed down to the jejunum, where they mature within 3–4 days. After copulation the adult females penetrate the intestinal mucosa. This is associated with an inflammatory response, comprising mast cells and eosinophils. The penetrated worms produce viviparous larvae that emigrate through the lymph and blood flow towards striated muscle tissue 1–3 weeks after infection, where they are encapsulated by a host-derived structure. In these capsules that are surrounded by inflammatory cells of the host, larvae can stay alive for a long time. In contrast to these larvae, the adult worms that reside in the gut do not stay in the host for very long; they are expelled from the gut within 2–3 weeks.

Expulsion of the worms from the gut is an immunological phenomenon, in which T cell-dependent immunity plays a crucial role. Also, antibody responses to *T. spiralis* play a role in resistance to this parasite; antibodies of IgM, IgC, IgA, and IgE classes are all induced during infection, largely dependent on T cell-dependent immunity.

In immunosuppressed animals the numbers of muscle larvae that can be encountered is considerably higher than in fully immunocompetent animals. The increased larval burden of the muscle tissues may be a reflection of decreased resistance, but obviously also depends on the increased numbers of adult worms and the fact that the worms reside in the gut of such immunosuppressed animals for longer periods of time. Inflammatory responses in the gut as well as around encysted muscle larvae is very scarce compared to immunocompetent animals, and antibody responses are diminished. Thus, endpoints in this model are numbers of worms in the gut, and numbers of *Trichinella* muscle larvae. These are determined using standard parasitological methods. Another endpoint is determination of the inflammatory infiltrate around encysted muscle larvae, evaluated by histological procedures. Serum antibody titers of IgM, IgG, IgF, and IgA are endpoints that are evaluated using enzyme-linked immunosorbent assays (ELISA), with antigen prepared from muscle larvae (6). Infection of animals with muscle larvae is executed by gastric intubation. In animal facilities there is no danger of spreading of the infection in the colony, as animals can only be infected by consumption of infected tissue. For animal-care personnel, standard hygienic procedures are sufficient to preclude accidental infection.

As *T. spiralis* infection can be experimentally approached in rats and mice, studies with immunosuppressed animals show differences in immunocompetence with the different assessable parameters in this model, and as man can be a target for *Trichinella* infection the model is well suited to assess immunotoxic risk caused by chemicals. An example is the organotin compound TBTO, or the virustatic agent acyclovir. In an experiment in which rats were exposed to (lower concentrations of) TBTO for a period of more than 1 year, the functional deficiencies shown by the various tests appeared less pronounced than those with shorter exposure to higher doses. However, the infection models used were still able to show depressed reactivity. The *T. spiralis* test was in fact able to show depressed resistance, and thus a more severe state of disease at a concentration of TBTO (5 mg/kg food) that was unable to cause alterations in any of the other immune parameters tested. These data indicate that a reserve capacity of the immune system that undoubtedly exists is, with respect to resistance to *T. spiralis*, rather modest to say the least.

Plasmodium

Two strains of ▶ *Plasmodium* have been found useful for evaluating the potential immunotoxicity of compounds. The first strain *Plasmodium yoelii* (I7XNL) is a non-lethal strain which produces a self-limiting parasitemia in mice. Host resistance to this organism is multifaceted, including specific antibody, macrophage involvement, and T-cell mediated functions. In this assay animals are injected with 10^6 parasitized erythrocytes, and the degree of parasitemia is monitored over the course of the infection by taking blood samples. In control animals peak response usually occurs between days 10 and 14 after injection. The degree of parasitemia can be evaluated manually by blood smears or flow cytometry, enumerating the number of infected cells based on their size. Another method that easily lends itself to the automated features of flow cytometry is measurement of acridine orange incorporation into infected cells containing *Plasmodium*. This procedure is easy to perform and allows a large number of cells to be counted, thus giving a more precise measure of parasitemia. Host resistance to *P. yoelii* has been used in immunotoxicologic assessment of compounds that include: benzidine, diphenylhydantoin, TCDD, pyran copolymer, gallium arsenide, and 2'-deoxyco-formycin.

Plasmodium berghei is a strain of *Plasmodium* that is lethal in both mice and rats. This protozoan model has been used in the immunotoxic assessment of compounds and in the development of potential antimalarial drugs (6). Host resistance includes specific antibody production and ingestion and destruction of antibody-coated *Plasmodium* by phagocytic cells such as macrophages. T lymphocytes may also be involved in host resistance to the organism. Mortality is the end-

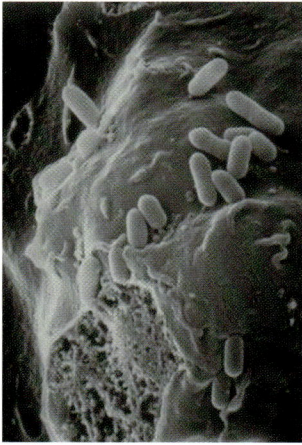

Host Resistance Assays. Figure 2 Muscle larve of Trichinella spiralis in striated muscle tissue.

point evaluated following the injection of 106 *Plasmodium*-infected erythrocytes. Compounds which have been evaluated for immunotoxicity using this system include 4,4'-thiobis(G-t-butyl-m-cresol, dietary fish oil supplement, and styrene. Neither *P. yoelii* nor *P. berghei* are infectious in humans and infection of other mice can only occur through contact with contaminated blood.

Pros and Cons

Host resistance models are valuable for studying consequences of exposure to immunotoxicants in terms of resistance to infections. Mortality due to the infection is often used as endpoint. It should be noted that obtaining statistically significant increases in mortality either requires substantial immunosuppression, or larger numbers of experimental animals per test group; sometimes, in addition to different dose groups with respect to the chemical, different dose groups with respect to the infectious agent are also necessary. Hence, this endpoint may not always be very sensitive. Burdens of the infectious agent in different target organs in certain models provide more detailed information.

One drawback of such parameters can be the more time consuming nature of the analysis. These special assays are normally not part of the routine of an immunotoxicity screening (tier I), therefore most laboratories have little or no experience with such tests. In addition, human pathogens may be used which require specific prevention and official approval. Therefore, the number of laboratories able to conduct such critical tests is limited.

Predictivity

In the 1980s the American National Toxicology Program (NTP) developed a tiered battery to assess chemical-induced immunotoxicity. The predictivity of different immunologic parameters to pick up immunosuppressive properties of test compounds were analyzed and statistically evaluated. Immunosuppressive effects were determined under immune modulating conditions in mice suffering from infectious agents. A good correlation was found between changes in T-dependent functional assays like the plaque-forming cell assay (PFCA) and immunosuppressive effects induced by different chemicals. By this the host resistance to infectious agents became a 'golden standard' for the evaluation of risk assessment in immunotoxicology, and was recommended as a tier II investigation in several guidelines and guideline drafts.

Relevance to Humans

The defense mechanisms that are put into action differ for different pathogens, so different host resistance models have been developed and are employed to study the relevance of immunotoxicity. The models that have been developed and are currently in use either use human pathogens, or are adequate models for human pathogens, in that mechanisms of resistance to the infectious agent in humans and animals are comparable. Hence, these models offer endpoints that are clinically relevant, and are therefore useful for estimating risk to humans. There is difficulty extrapolating the results of immunotoxicity studies using infection models to humans, whereby in experimental studies the infection with an infectious agent is controlled, unlike the case in the general population. Even strong immunosuppression may not become clinically evident in terms of development of an infection, if an encounter with that particular infectious agent does not take place. Exposure to infectious agents is influenced highly by hygienic, socioeconomic, and climatologic conditions.

Regulatory Environment

Assessment of the potential direct immunotoxicity of chemicals can be carried out successfully in experimental animals. Histopathology of lymphoid organs, in addition to differential cell counting, determination of lymphocyte subpopulations in the spleen, and measurement of total serum immunoglobulin levels have proved valuable screening parameters for potential immunotoxicity. An array of in vivo assays of the immune system, in addition to ex vivo/in vitro assays is available for studying the functional consequences of such toxic effects. Parameters should address the different aspects of the immune system, the non-specific responses (macrophage and natural killer activity), humoral responses (antibody responses to antigens, mitogen responsiveness), and cellular responses (mitogen responsiveness, mixed lymphocyte reactions, cytotoxic T cell responses, delayed-type hypersensitivity). Host resistance models are particularly valuable in this respect, because these models offer endpoints that are clinically relevant, and therefore useful for estimation of risk to man.

In several immunotoxicity guidelines or guideline drafts, the host resistance models discussed in this chapter are recommended for confirming possible immunotoxic effects observed in screening studies (tier I). The choice of the relevant host resistance model should be discussed with the responsible authority in a case-by-case manner.

▶ Streptococcus Infection and Immunity

References

1. Van Loveren H, Rombout PJA, Wagenaar SjSc, Walvoort HG, Vos JG (1988) Effects of ozone on the defense to a respiratory *Listeria monocytogenes* infection in the rat. Suppression of macrophage function and cellular immu-

nity and aggravation of histopathology in lung and liver during infection. Toxicol Appl Pharmacol 94:374–393
2. Winkelstein JA (1981) The role of complement in the host defense against *Streptococcus pneumoniae*. Rev Infect Dis 3:289–298
3. Bruggeman GA, Debie WMH, Grauls G, Majoor G, Boven GPA (1983) Infection of laboratory rats with a new cytomegalo-like virus. Arch Virol 76:189–199
4. Bruggeman GA, Meijer H, Bosman F, Boven GPA (1985) Biology of rat cytomegalovirus. Intervirology 24:1–9
5. Thomas F, Fugmann R, Aranyi G, Barbera F, Gibbons R, Fenters J (1985) Effect of dimethylnitrosamine on host resistance and immunity. Toxicol Appl Pharmacol 77:219–229
6. Van Loveren H, Luebke RL, Vos JG (1995) Assessment of immunotoxicity using the host infectivity model *Trichinella spiralis*. In: Methods in immunotoxicology, 2nd edn. Burleson G (ed). Wiley, New York, pp 243–271
7. Luster MI, Tucker AN, Hayes HT et al. (1985) Immunosuppressive effects of benzidine in mice: Evidence of alterations in arachidonic acid metabolism. J Immunol 135:2754–2761
8. Holsapple MP, White KL, McGay JA, Bradley SG, Munson AE (1988) An immunotoxicological evaluation of 4,4'-thiobis-[6-t-bntyl-m-cresol] in female B6G3F1 mice. Fund Appl Toxicol 10:701–716

HSC

▶ Hematopoietic Stem Cells

HSE

▶ Three-Dimensional Human Skin/Epidermal Models and Organotypic Human and Murine Skin Explant Systems

5-HT

5-hydroxytryptamine; serotonin.
▶ Serotonin

HuCD4 Mouse

Immunodeficient mice in which the immune system has been partially reconstituted with human CD4 cells by genetic insertion.
▶ Animal Models of Immunodeficiency

Human Leukocyte Antigens (HLA)

The human leucocyte antigen (HLA) summarizes human MHC or transplantation antigens. Human MHC class I = HLA-A, HLA-B, HLA-C. MHC class II = HLA-DP, HLA-DQ, HLA-DR, and are highly polymorphic transmembrane proteins. Genes are clustered in the HLA locus on human chromosome 6. The antigen-binding groove in the extracellular domain is composed of a platform of eight antiparallel beta sheets and two antiparallel alpha helices. The structural and electrical configuration of this site is determined by the sequence of amino acids lining the groove which, in turn, is dictated by genetic recombination during transcription.
▶ Chronic Beryllium Disease
▶ Hapten and Carrier

Human Pathogen

Any virus, microorganism, or other substance causing disease to *Homo sapiens*.
▶ Klebsiella, Infection and Immunity

Human Skin Recombinants

▶ Three-Dimensional Human Skin/Epidermal Models and Organotypic Human and Murine Skin Explant Systems

Humanized Monoclonal Antibodies

Joel B Cornacoff · Jill Giles-Komar
Centocor Inc.
200 Great Valley Parkway
Malvern, PA 19403
USA

Synonyms
Hyperchimeric

Short Description
Humanized antibodies are those ▶ monoclonal antibodies (mAbs) that are mostly, but less than fully, human. They may originate from one or more different techniques:
- nonhuman (i.e. murine) monoclonal antibodies that have then been manipulated by molecular biologi-

cal techniques such as: CDR (complementary-determining region) grafting, resurfacing and hyperchimerization to replace some or most of the animal sequences with human sequences
- fully human monoclonal antibodies that have subsequently been made less than fully human by having the genes encoding for them transfected into animal cells which then synthesize the Mab which acquires animal glycosylation patterns
- fully human antibodies that have been manipulated by molecular biological techniques to modify the amino acid sequence in order to alter specificity, affinity or biological functions, thereby possibly acquiring sequences that are not part of the human repertoire.

Characteristics

Since each clonally derived antibody is unique, each one will have unique characteristics. Either its binding specificity or affinity for antigen will be different from other mAbs with the same nominal specificity or its ability to cause biological effects through Fc or complement-binding will vary depending on glycosylation patterns or variations in the constant regions of the molecule. The prime—if not sole—interest in making human or humanized monoclonal antibodies concerns the perception (backed up by some data) that non-human antibodies are immunogenic in man. The corollary of this is the belief that human or humanized antibodies will not be immunogenic in man. It should be noted however, that other factors contribute to immunogenicity. These factors include:
- whether the recognized antigen is a soluble molecule or is on the surface of a cell
- the form of administered antibody (micro- or macroaggregation will increase immunogenicity)
- the route of injection
- **allotypic** and idiotypic variants which will be immunogenic for some members of the population
- the presence of immunogenic amino acid sequences or immunogenic tertiary structures.

Additionally, it is known that natural human proteins (e.g. Factor VIII) can generate an immune response in hemophiliacs fairly often and for unknown reasons, thus demonstrating the complexity of the problem.

Preclinical Relevance

In vivo assays for safety and efficacy using humanized antibodies in test animals usually will not be readily transferable to humans. Specificity, affinity and circulating half-lives are frequently quite different in test animals from those in humans for any given Mab that was obtained using human antigen.

One challenge in evaluating the safety of humanized mAbs occurs when there is limited interspecies cross-reactivity. This is the case with certain molecules that only target receptors on humans and chimpanzees. This is compounded by the fact that there is a moratorium on the breeding of chimpanzees and the available animals have been exposed to numerous biotherapeutic molecules. This has been circumvented in limited cases by using a homologous or surrogate antibody in rodents. If surrogate antibodies are used, specificity will be much improved along with affinity, but they may not be predictive of half-life in humans. Additionally, the relevant target antigen may not have the same functions in the test species as in the human patient (1). Additionally, there is an increasing role for the incorporation of transgenic rodents, knock-out, and knock-in models, where interspecies cross-reactivity is an issue.

Relevance to Humans

Since the advent of murine mAbs as described by Kohler and Milstein (2) in 1975 it has been recognized, first theoretically and later experimentally, that the immunogenicity of these proteins in a foreign, that is a human host would be a serious problem which might be overcome either by using fully human monoclonal antibodies or by humanizing murine mAbs. With the exception of the etiologic agents of infectious diseases, it is ethically difficult to carry out programmed immunization of humans to obtain a source for making truly human mAbs. Therefore, we are compelled to generate mAbs by molecular techniques or in animal species, usually mice, and then deal with the problem of high immunogenicity of the resulting therapeutic agent. Thus, humanized mAbs are becoming the accepted mode for therapeutic antibodies.

Although humanization decreases immunogenicity, it is highly dependent upon dose and number of repeated injections. Anti-idiotypic responses may still be generated with repeated injections, even in patients who share an Ig allotype with the humanized antibody. Ultimately the problem of immunogenicity may be more readily solved through the use of immune tolerance strategies or immunosuppressive therapy, in addition to strategies employed to reduce immunogenicity (3). Between 1997 and 2004, eight humanized mAbs have been approved for marketing (Table 1) (4) and at present there are a significant number of humanized mAbs either in discovery or in various stages of human clinical trials. An increasing number of these agent have either a direct or indirect effect on the immune system (Humira, Zenapax). One must be aware that even in the absence of a humoral immune response, cellular immunoregulatory pathways can be activated, resulting in serious clinical complications.

Regulatory Environment

The development of humanized mAbs is on a case-by-

Humanized Monoclonal Antibodies. Table 1 Marketed humanized monoclonal antibodies

Trade name	Generic name	Manufacturer	Approved	Indication
Avastin	bevacizumab	Genentech	2004	Colon cancer
Campath	alemtuzumab	Ilex Pharmaceuticals	2001	Chronic lymphocytic leukemia
Herceptin	trastuzumab	Genentech	1998	Her2-positive breast cancer
Mylotarg	gemtuzumab	Wyeth	2000	Acute myeloid leukemia
Raptiva	efalizumab	Genentech	2003	Psoriasis
Synagis	palivizumab	MedImmune	1998	Respiratory syncytial virus
Xolair	omalizumab	Genentech/Novartis	2003	Asthma
Zenapax	daclizzumab	Roche Protein Design Labs	1997	Organ rejection prophylaxis

case basis. The design and nature of preclinical and non-clinical studies depends on numerous factors including: target, route of administration, species, indication, pharmacology, pharmacokinetics, and clinical plan. The documents listed below are "Guidelines" and "Points to Consider" but should not be considered regulatory requirements. Some relevant guidelines exist:

- Guidance for Industry. S6 Preclinical Safety Evaluation of Biotechnology-Derived Pharmaceuticals (http://www.fda.gov/cder/guidance/1859fnl.pdf)
- Points to Consider in the Manufacture and Testing of Monoclonal Antibody Products for Human Use. FDA/CBER 1997
- Annex 11 to the EC CPMP Guideline 111/5271/94 Draft 5: Production and Quality Control of Monoclonal Antibodies

References

1. Green JD, Black LE (2000) Status of preclinical safety assessment for immunomodulatory biopharmaceuticals. Hum Exp Toxicol 19:208–212
2. Kohler G, Milstein C (1975) Continuous cultures of fused cells secreting antibody of predefined specificity. Nature 256:495–497
3. Kuus-Reichel K, Grauer LS, Karavodin LM, Knott C, Krusemeier M, Kay NE (1994) Will immunogenicity limit the use, efficacy, and future development of therapeutic monoclonal antibodies? Clin Diag Lab Immunol 11:365–372
4. The Physician's Desk Reference (2003) Thomson Healthcare, Montvale

Humanized Monoclonal Antibody

Most monoclonal antibodies are generated in mice. When used as drugs in humans they elicit an antibody response which limits their function. Therefore the antigen-binding parts are exchanged (by gene technology) in human immunoglobulin molecules, generally IgG, which diminishes antibody formation and increases half life.

▶ Cytokine Inhibitors

Humanized Mouse

A genetically engineered mouse into which a human gene or genes has been introduced. In humanized knockout animals, the human gene often is made so that it replaces the corresponding mouse gene.

▶ Knockout, Genetic

Humoral Autoreactivity Assays

▶ Autoantibodies, Tests for

Humoral Immune Assay

▶ Plaque-Forming Cell Assays

Humoral Immune Function

▶ Assays for Antibody Production

Humoral Immune Response

The immune response, which is mediated by antibodies, is called humoral immune response. Depending on several factors, like the type of pathogen, rout of exposure, or genetic background of responder, a given immune response might be dominated by antibodies. The production of antigen-specific antibodies by

B cells following a complex interaction between antigen presenting cells, T cells, cytokines, and cell surface markers.

▶ Dioxins and the Immune System
▶ Plaque-Forming Cell Assays
▶ Assays for Antibody Production

Humoral Immune System

The cells and molecules that lead to immunity mediated by antibodies (immunoglobulins) in body fluids. The group of serum proteins collectively known as complement also make important contributions to humoral immune mechanisms and hence the biological effectiveness of antibodies.

▶ Rabbit Immune System

Humoral Immunity

COURTNEY EW SULENTIC · NORBERT
E KAMINSKI
Department of Pharmacology and Toxicology
Wright State University
206 Health Sciences Building, 3640 Colonel Glenn Hwy
Dayton, OH 45435
USA

Synonyms
antibody forming cell response

Definition
Humoral immunity (1,2), a component of ▶ acquired immunity, defends the host against extracellular bacteria, parasites, and foreign macromolecules and is mediated by soluble antibodies which are secreted from terminally differentiated B cells (plasma cells or antibody-forming cells). Secreted antibodies coat a specific antigen and facilitate antigen clearance by recruiting ▶ complement components and non-specific immune cells which destroy and remove the antigen.

Characteristics
A single mature B cell can recognize a specific antigen through its B cell receptor (BCR) which is a membrane-bound immunoglobulin (Ig) molecule. The BCR serves two primary functions: to transduce activation signals into the B cell, and to mediate ▶ endocytosis of antigens. In the latter case, once endocytosed, the protein antigen is delivered to intracellular sites for degradation and association with MHC class II molecules followed by exportation to the cell surface in association with MHC class II. With the exception of certain microbial antigens, the ability of mature resting B cells to respond to most protein antigens requires T cell help. Those antigens requiring T cell help in order to initiate a humoral immune response are termed T cell-dependent or thymus-dependent antigens, where as those antigens that can trigger a humoral immune response in the absence of T cell help are referred to as being T cell-independent or thymus-independent. Thymus dependency and independency, when referring to antigens, derives from studies of humoral immune responses originally conducted in athymic mice (those without T cells).

T cell help in the activation of B cells occurs through cell-to-cell interactions and through the release of soluble regulatory factors called interleukins. Once T cell-dependent antigens have been processed and expressed in association with MHC class II molecules on the B cell surface, the B cell is now capable of interacting with antigen specific ▶ $CD4^+$ T cells. The cell-to-cell interactions occur through proteins that are either constitutively expressed or induced on the surface of both cell types. Specifically, the presentation of MHC class II with antigen by the B cell to $CD4^+$ T cells via the T cell antigen receptor, provides the first activation signal to the T cell. This interaction results in the upregulation of CD40 ligand (CD40L) on the T cell surface, which in turn binds with CD40 on the B cell surface to deliver the second activation signal to the B cell. As in the case of the T cell, the first activation signal for the B cell is through antigen binding to the BCR. Interaction between CD40 and CD40L in turn induces the upregulation of the B7 protein on the B cell surface, which serves as the ligand for the CD28 receptor on the surface of T cells. The binding of upregulated B7 and CD28, provides the T cell with its second activation signal. It is important to emphasize that for effective activation of T cells and B cells, both cell types require co-stimulation through co-receptors, CD40 in the case of the B cell and CD28 in the case of the T cell. In the absence of a second activation signal through co-stimulation via CD40 (B cell) and CD28 (T cell), the antigen-stimulated lymphocytes either become ▶ anergic or are induced to undergo apoptosis. The second phase of T cell help in humoral immune responses occurs through the production of ▶ cytokines, including IL-4, IL-5, and IL-6, by the activated $CD4^+$ T cells which act as B cell growth and differentiation factors. In the case of a T cell-independent response, the antigens are capable of providing both activation stimuli to the B cell.

Binding of the antigen to the BCR followed by the aforementioned cell-to-cell interactions and production of soluble mediators can result in B cell proliferation and ▶ clonal expansion. Following clonal expansion, the B cells, all with specificity for the activating antigen, differentiate into antibody-forming cells (AFC), also termed plasma cells, or into memory cells. AFCs synthesize and secrete antibodies, a soluble form of Ig, which coat a specific antigen and facilitate antigen neutralization or clearance by cells of innate and acquired immunity or activation of the complement cascade. The complement cascade refers to the activation of a series of enzymes that lead to the formation of transmembrane channels that disrupt the cellular integrity of the antigen resulting in cellular lysis and enhanced ▶ phagocytosis.

The basic Ig molecule is composed of two identical heavy chains and two identical light chains and has two antigen binding sites, which are held together by disulfide bonds. Two or more Ig molecules can be joined together by additional disulfide bonds and by a polypeptide chain termed the J chain, which is also synthesized in B cells. There are two classes of light chains, κ or λ, which add to the diversity of the antigenic repertoire, and five classes of heavy chains, IgM, IgG, IgA, IgE and IgD, each possessing unique biological properties. For example, IgM is pentameric or hexameric resulting in high antigen valence (with 10 or 12 antigen-binding sites), has relatively low affinity for antigen, and is the major Ig involved in a primary antibody response. IgG is monomeric, has high affinity for antigen, can cross the placental barrier, and is the hallmark of a secondary antibody response. IgE is also monomeric and is involved in allergic and antiparasitic responses. IgA is monomeric or dimeric, is very efficient at bacterial lysis, and is the main secretory antibody (found in saliva, mucus, sweat, gastric fluid, and tears). IgD is monomeric, is a major surface component on many B cells, and has unknown biological properties. During a secondary humoral immune response, class switching to different Ig isotypes is governed primarily by the profile of cytokines that are produced by $CD4^+$ T cells. For example, in the mouse, IL-4 primarily induces switching to IgG1 and IgE.

An initial response or primary response to a T-dependent antigen results in a burst of IgM production at the site of B cell activation; however, some B cells, prior to differentiation, migrate into the follicular region of secondary lymphoid organs and initiate germinal centers. Activated B cells forming germinal centers undergo rapid proliferation along with rapid somatic mutation of their Ig genes. Following somatic mutation, ▶ germinal center B cells experience positive selection for high affinity membrane-bound Ig. Ig class switching, which is independent from affinity maturation (somatic mutation and selection), also occurs in germinal center B cells. Positively selected B cells will either terminally differentiate to plasma cells or to quiescent memory cells. Most plasma cells live for only a few days before programmed cell death; however, these cells may migrate to the gut or bone marrow, where they secrete antibody and may live for more than 20 days. Memory cells are maintained for months to years and mediate second and subsequent responses to a particular antigen which are more rapid and of a longer duration and greater intensity than a primary antibody response. In addition, the antibody titer is greater in a secondary response and consists primarily of IgG as opposed to IgM from the primary response. This enhanced secondary antibody response is due to immunologic memory.

Preclinical Relevance

Due to the similarity between the mouse and human immune systems and the availability of biological reagents, mouse models have primarily been utilized to understand the basic mechanisms of humoral immunity as well as the impact of potential therapeutic and toxic compounds on these mechanisms. Preclinical studies provide the necessary information to evaluate the potential hazard to humans from occupational, inadvertent, or therapeutic exposure to drugs, environmental compounds, or industrial chemicals.

Relevance to Humans

The medical practice of vaccinations and immunizations has arisen to capitalize on the immunologic memory of acquired immunity. As children, we are vaccinated to several infectious organisms, including measles, mumps, diphtheria, tuberculosis, rubella, and poliomyelitis. In addition, many adults and children receive yearly influenza vaccinations. Non-pathogenic (inactivated or attenuated) forms of infectious microbes used in vaccinations and immunizations induce the generation and expansion of antigen-specific memory cells, thus boosting our immunity to these organisms. In the case of humoral immunity, the second and subsequent responses to a particular antigen are more rapid and of a longer duration and greater intensity than a primary antibody response. In addition, the antibody titer is greater in a secondary response and consists primarily of IgG antibodies, which have a higher affinity for antigen as opposed to IgM. Therefore, humoral immunity is a vital component of acquired immunity and maintains protection against a myriad of extracellular bacteria, parasites, and foreign macromolecules.

Regulatory Environment

The ability of xenobiotics to alter the primary IgM antibody response to a T cell-dependent antigen such as the sheep erythrocyte has been, and continues to be, one of the 'gold' standard assays routinely employed for identifying immunotoxicants. The assay system most commonly employed in immunotoxicology testing to assess adverse effects on humoral immune responses is performed by using variations of the Jerne plaque assay, also often called the antibody-forming cell assay, or simply the plaque assay. This assay quantifies the number of antigen-specific B cells that have been driven to terminally differentiate into plasma cells, also termed antibody-forming cells. The assay has been found to be exquisitely sensitive and reproducible. The sensitivity of the assay is attributable in large part to the fact that the response is dependent on antigen recognition, processing, and presentation by antigen-presenting cells as well as the activation, proliferation, and differentiation of effector B and T lymphocytes.

▶ Hypersensitivity Reactions
▶ Graft-Versus-Host Reaction
▶ Lymphoma
▶ DNA Vaccines

References

1. McHeyzer-Williams M (2003) B cell signaling mechanisms and action. In: Paul WE (ed) Fundamental immunology, 5th ed. Lippincott, Williams & Wilkins, Philadelphia, pp 195–225
2. Picker L, Frank MM, SGordan S et al. (1999) The humoral immune response. In: Janeway CA, Travers P, Walport M, Capra JD (eds) Immunobiology. The immune system in health and disease, 4th ed. Garland, New York, pp 307–362

Humoral-Mediated Immunity

Host defense that is mediated by antibody present in the plasma, lymph, and tissue fluids. It protects against extracellular bacteria and foreign macromolecules.

▶ Flow Cytometry

Hu-SCID Mouse

SCID mice in which the immune system has been partially reconstituted with human immune cells.

▶ Animal Models of Immunodeficiency

Hybridization

Technique in which single-stranded nucleic acids are allowed to interact so that complexes, or hybrids, are formed by molecules with similar complementary sequences.

▶ Southern and Northern Blotting

Hybridoma

An immortal antibody-producing cell line formed by the fusion of a single antibody-producing B cell and an immortal cell line.

▶ Monoclonal Antibodies

Hydralazine

Hydralazine (1(2H)-phthalazinone) is a peripheral vasodilator that is used as an antihypertensive.

▶ Systemic Autoimmunity

5-hydroxytryptamine

▶ Serotonin

Hygiene Hypothesis

Hypothesis that exposure to viral infections or endotoxins in early childhood is protective against the development of asthma.

▶ Asthma

Hyperchimeric

▶ Humanized Monoclonal Antibodies

Hypergammaglobulinemia

Unusually high titer of immunoglobulin resulting from overproduction by plasma cells. There are two types of hypergammaglobulinemia, the polyclonal, and the monoclonal type. An increase in polyclonal antibody titer are induced by extreme antigenic stimuli such as infection, inflammation or neoplasm. An increase in monoclonal antibodies is most often due to neoplastic transformation of a single clone of plasma cells.

▶ Hypergammaglobulinemia

Hyperplasia

Also numeric or quantitative hypertrophy. Describes

organ or tissue enlargement by increasing cell counts. In contrast to neoplasia, hyperplasia is reversible.

▶ Local Lymph Node Assay (IMDS), Modifications

Hyperreactions

▶ Hypersensitivity Reactions

Hypersensitivity Reactions

HANS-WERNER VOHR
PH-PD, Toxikologie
Bayer HealthCare AG
D-42096 Wuppertal
Germany

Synonyms
allergic reactions, type I–IV reactions, allergy, hyperreactions

Definition
Hypersensitivity reactions are inappropriate or excessive responses of the adaptive immune system to various foreign substances (allergens). Allergic reactions usually occur when an already sensitized person is re-exposed to the same allergen. This kind of hyperreaction can result in a wide range of individual inflammatory reactions and tissue damage.
Coombs and Gell classified these reactions into four categories (types I–IV) according to the effector mechanisms involved. Although this classification may be too simple for the underlying mechanisms, it is useful for understanding the relation between effector mechanisms and the pathologic picture of a hypersensitivity reaction.

Characteristics
As mentioned above, hypersensitivity reactions can be classified into four categories, i.e. types I–IV (see Table 1). This categorization is based on the different mechanisms and effector cells involved. The induction (sensitization) and challenge (re-exposure) phases are the main elements of hypersensitivity reactions, and all four categories can be distinguished primarily in the challenge phase, i.e. by the cells and components mediating the pathologic hypersensitivity reaction. Type I–III reactions are based on pathologic activation of antibodies (▶ humoral immunity), while type IV reactions are due to ▶ cellular immune reactions. However, these limits are not strictly delineated. For example, antigen-specific antibodies are involved in type IV hypersensitivity as well, but the cellular reaction is much more prominent.

Type I
This type represents what are commonly known as "allergic reactions," including allergic rhinitis (hay fever), ▶ asthma, or ▶ systemic anaphylaxis, which are known as "immediate hypersensitivity reactions." They are dominated by reactions mediated by IgE antibodies binding on mast cells via an IgE-specific receptor (high-affinity receptor FcεRI). Cross-linking of these IgE-receptor complexes by the antigen (allergen) results in degranulation of mast cells. The release of mediators (histamine, chemokines, cytokines, prostaglandins) will cause all the pathologic findings known to occur in allergic reactions.
Similar to tissue-specific macrophages, tissue-specific mast cells also exist. Mast cells in connective tissue and mast cells in mucosa can be distinguished by their ability to release different patterns of inflammatory mediators, and they may have varying sensitivity to external stimulation (IgE, complement proteins C3a/C5a or neuropeptides). In addition, several cells are able to bind IgE by other receptors, i.e. FcεRII (CD23) or εBP. In this way, cells such as basophils or neutrophils also contribute to the overall "allergic reaction."
Altogether the mechanisms involved in induction and manifestation of allergic reactions are complex and are still only partly understood. For many years it has been clear that genetic background has an influence. These genetic factors among others determine whether an individual reacts to an allergen with a "normal" specific immune reaction, or whether this allergen will lead to hyperimmune reactions (see ▶ MHC Class I Antigen Presentation; ▶ Antigen Presentation via MHC Class II Molecules). In addition to this genetic predisposition, some factor "X" is also important in establishing an allergic reaction.
This factor is thought to interfere with the immune system during the host reaction to the allergen. "X" could be an infection, temporary immune suppression, or a cross-reaction with other chemicals which have triggered an immune reaction. These interactions can enhance each other and finally cause deregulation which ends in the manifestation of a hyperimmune reaction.
It is important to know that two different circumstances are also known to have an impact on the development of hyperreactivity: the duration of contact, and the concentration of the allergen. It seems that both prolonged contact with low concentrations of an allergen and brief contact with relatively high concentrations of an allergen could increase the probability of hypersensitivity.

Hypersensitivity Reactions. Table 1 The four categories of hypersensitivity reactions, and their reactants and mediators

	Type I	Type II	Type III	Type IV
Allergic reaction	Anaphylaxis Asthma Hay fever Allergic urticaria	Immunoglobulin-dependent cytotoxicity Some drug allergies	Serum sickness Arthus reaction	Allergic contact dermatitis Chronic asthma Chronic rhinitis Tuberculin reaction
Reactant	IgE (IgG)	IgG (IgM)	IgG (IgM)	T helper type 1 T helper type 2 Cytotoxic T lymphocytes
Mediators	Mast cells (histamine, leukotrienes, cytokines, platelet activating factor)	FcR$^+$ cells (natural killer cells, phagocytes) Complement	FcR$^+$ cells Complement	Cytotoxic T lymphocytes (perforin, fas, granzymes) Macrophages, eosinophils (cytokines, chemokines, cytotoxins)

Type II

Reactions to red blood cells are the most common examples of type II hypersensitivity. The mediators of these adverse immune reactions are IgG and IgM antibodies. One classic example of a type II reaction is hemolytic disease of the newborn (HDNB) which is caused by the mismatch of a rhesus factor negative (Rh$^-$) mother carrying a rhesus factor positive (Rh$^+$) infant. Sensitization (the induction phase) of the mother is usually caused by fetal red cells crossing into the mother's blood stream during delivery. In the event of subsequent pregnancies, antibodies to the Rhesus antigen can then cross the placental barrier and result in destruction of the fetal red blood cells.

A second example is incompatibility of red blood cells according to the AB0 system. Destruction of blood cells induced by this mechanism is known as the transfusion reaction. It was first described in the early 20th century by Landsteiner. In regard to immunotoxicity, induction of hyperreactions in modified host (patient) cells—in particular red blood cells—is of importance. If a substance or its metabolite binds to the surface proteins of red blood cells, these cells will be recognized as foreign and attacked by the immune system. Unfavorable circumstances (see Type I above) ultimately result in the development of hypersensitivity to the modified red cells, i.e. allergic anemia.

The adverse effect of this type of hypersensitivity is mediated by substance-specific IgM and IgG antibodies. Enhanced amounts of such antibodies bind to the modified self-antigens (altered-self). Activation of the classic ▸ complement cascade either leads to direct cell destruction or facilitates the attraction and activation of neutrophils and macrophages/monocytes by which opsonization and phagocytosis of the target cells occurs. The overall complex mechanism is described in the relevant entries.

Type III

Formation of immune complexes is not an adverse effect per se but a normal event during the humoral immune response. Equal amounts of soluble antigen and binding antibodies give rise to complex formation. Such complexes are usually removed by macrophages in the liver. Transportation to the liver is managed by erythrocytes which bind complexes via their complement receptors (C3b). As can be seen in Table 2, three situations can lead to pathologic complex formation. As can be seen in Table 1, it is impossible to distinguish between hypersensitivity reactions and autoimmune disease. Any formation of immune complexes after application of a substance should therefore serve as an alert to immunotoxicologists that potential induction of autoimmune reactions must be clarified.

Hypersensitivity Reactions. Table 2 Causes and consequences of pathologic complex formation

Cause	Site of Deposition	Consequences
Persistent bacterial or viral infection	Skin Kidney Infects organs	Local organ damage Glomerulonephritis
Persistent inhalation of environmental antigen	Lungs	(Allergic) alveolitis "Farmer's lung" Arthus reaction Serum sickness
Formation of high-affinity autoantibodies	Skin Kidney Joints Arteries	Organ-specific and non-specific autoimmune reactions Glomerulonephritis Systemic lupus erythematosus Rheumatoid arthritis

Such signals may, for example, come from findings of unusual, inexplicable inflammatory reactions in preclinical studies.

In principle, there are two "experimental" models which have contributed to present knowledge about immune complex formation. These are the Arthus reaction and serum sickness. The Arthus reaction is a localized form of immune complex-mediated ▸ vasculitis. After cutaneous injection, the antigen is bound by specific antibodies (of the sensitized animal or patient), and complexes deposit in the walls of small arteries. On activation of the complement, the inflammatory cascade starts as described above.

The underlying mechanism causing serum sickness is similar to that of the Arthus reaction. The difference is that serum sickness is a pathologic systemic reaction instead of a local reaction. Deposits of immune complexes in blood vessel walls can lead to inflammatory disease such as ▸ glomerulonephritis and rheumatoid arthritis. For example, serum sickness can be experimentally induced in rabbits by intravenous injection of a foreign soluble protein (bovine serum albumin). About 1 week after injection, hemolytic complement titers decline as a result. Granulocytes and platelets migrate to the cellular basement membranes and into small vessels.

Although there are several animal models now available for autoimmune disease, the classic model for such investigations involves hybrid F1 NZB/NZW mice. At 2–3 months after birth, female animals of this hybrid spontaneously develop hemolytic anemia, antinuclear antibodies, and circulating immune complexes.

Similar lupus-like reactions can also be induced with several subcutaneous injections of heavy metal salts such as $HgCl_2$ into susceptible mice.

More detailed information about the underlying mechanism, animal models or different autoimmune diseases are available under the relevant entires.

Type IV

In contrast to types I–III, this type of hypersensitivity is mediated by a cellular immune reaction. As the peak reaction appears at a late stage (> 12 hours) after contact to the antigen, these reactions are also known as "delayed-type hypersensitivity" (DTH). Unlike other forms of hypersensitivity, the type IV allergy cannot be transferred from one animal to the other by serum but rather by T cells. During the initial contact (sensitization phase) via skin (and possibly lungs), antigen-specific T memory cells develop which cause an enhanced reaction after re-exposure. The most prominent and important form of DTH is what is known as (allergic) contact hypersensitivity (ACH), which often causes skin reddening and eczema.

For further information on these two forms, refer to ▸ Hypersensitivity Reactions▸ Delayed-type Hypersensitivity.

Preclinical Relevance

In regard to hypersensitivity reactions, only contact hypersensitivity is measured routinely for all chemicals which are expected to come into contact with human or animal (domestic or working animals) skin. Established animal models are available, such as the guinea pig maximization assay (GPMT), the Bühler patch test or the local lymph node assay (LLNA or IMDS).

For determination of respiratory tract hypersensitivity, several models involving guinea pigs, rats and mice are available, but none of these have been really validated, although most of these investigations are accepted by authorities if they are conducted under well-controlled conditions. A special entry about animal models for respiratory sensitization can be found in this book.

Special testing guidelines from national or international authorities are not yet available for respiratory hypersensitivity.

Other forms of humoral-mediated hypersensitivity reactions are not included in the routine toxicology package. These investigations are only conducted if specific evidence emerges from ongoing toxicological investigations.

Relevance to Humans

While the predictivity for the development of skin sensitization based on the results of existing animal models is relatively good, it is worse for hypersensitivity of the respiratory tract. This is due to the complex mechanisms and interaction of immunocompetent cells involved in the induction and elicitation phase of this kind of hypersensitivity. But it is also due to the absence of an internationally accepted and thoroughly validated animal model.

Regulatory Environment

As mentioned above, contact hypersensitivity (type IV reaction) is the only area regulated by special guidelines (see related sections). Nevertheless, other forms of hyperreactions have to be clarified on a case-by-case basis. In addition, the structural relationship, knowledge of related compounds, and of course the route of application, should be taken into consideration by the investigator for incorporation into special studies.

References

1. Janeway CA Jr, Travers P, Walport M, Shlomchik M (2001) Immunobiology, the immune system in health and disease, 5th edn. Garland Publishing, New York

2. Coombs RRA, Gell PGH (1975) Classification of allergic reactions for clinical hypersensitivity and disease. In: Gell PGH, Coombs RRA, Lachmann PJ (eds) Clinical aspects of immunology. Blackwell Scientific, Oxford, pp 761–781
3. Gleichmann E, Vohr H-W, Stringer C, Nuyens J, Gleichmann H (1989) Testing the sensitization of T cells to chemicals. From murine graft-versus-host (GvH) reactions to chemical-induced GvH-like immunological diseases. In: Kammüller ME, Bloksma N, Seinen W (eds) Autoimmunity and toxicology. Elsevier, Amsterdam, p 363

Hypersensitivity Reactions to Drugs

▶ Drugs, Allergy to

Hypothalamic-Pituitary-Adrenal (HPA) Axis

The neuroendocrine machinery involved in the production of glucocorticoids.
▶ Glucocorticoids

Hypothalamus

A portion of the brain that lies beneath the thalamus and secretes substances that control metabolism by exerting influences on the pituitary gland function. It is also involved in the regulation of body temperature, blood sugar, and fat metabolism, and regulates other glands such as the ovaries, parathyroid, and thyroid.
▶ Stress and the Immune System

Hypovolemia

Inadequate intake of fluids due to elevated demand, loss of fluid or both. Severe forms result in decreased ventricular filling with hypovolemic shock.
▶ Septic Shock

Hypoxia-Inducible Factor HIF1-β

▶ Dioxins and the Immune System

ICH

The International Conference on Harmonization of Technical Requirements for Registration of Pharmaceuticals for Human Use
▶ Immunotoxicology

ICICIS

ICICIS stands for the International Collaborative Immunotoxicity CEC-IPCS Study. This study was set up in the late 1980s to find out whether conventional toxicologic studies could be enhanced in order to detect potential immunotoxicants. The outcome of this study, which encompassed laboratories from several continents, demonstrated that endpoints to detect immunotoxicity could be integrated into conventional toxicity studies. It provided a highly influential input into the revision of the OECD testing guideline 407 (28-day repeat-dose rodent study).
▶ Lymphocyte Proliferation

ICS

Intracellular cytokine staining.
▶ Maturation of the Immune Response

Idiopathic Chronic Active Hepatitis

▶ Hepatitis, Autoimmune

Idiopeptides

Idiopeptides are peptide fragments of the variable regions of BCR H chains and L chains and of TCR α and β chains. Idiopeptides are generated by intrinsic digestion or by professional antigen-presenting cells. They can be recognized by idiotype-specific T cells in the conext of class I and II molecules of the major histocompatability complex (MHC).
▶ Idiotype Network

Idiosyncratic Drug Reactions

▶ Drugs, Allergy to

Idiosyncratic Reaction

Pseudoallergic reaction.
▶ Anti-inflammatory (Nonsteroidal) Drugs

Idiotype

The term idiotype indicates the individuality of a particular BCR or TCR. It consists of a collection of conformational antigenic determinants (idiotopes) which are located within the combined variable domains of antibody H chains and L chains (F_V fragment) *or* the α and β chains of the TCR.
▶ Idiotype Network

Idiotype Network

HILMAR LEMKE
Biochemisches Institut in der Medizinischen Fakultät
Christian-Albrechts-Universität zu Kiel
Olshausenstrasse 40
D-24098 Kiel
Germany

Definition

Antigen receptors of B (the B cell antigen receptor, or BCR) and T lymphocytes (the T cell antigen receptor, or TCR) may recognize each other as ▶ idiotypes and anti-idiotypes and thereby form a web of interacting clones designated as the idiotype/idiotypic network. The term idiotype indicates the individuality of a particular BCR/TCR. It consists of a collection of conformational antigenic determinants (▶ idiotopes) which are located within the combined variable domains of antibody heavy (H) chains and light (L) chains (F_V fragment) or the α and β chains of the TCR.

Conformational idiotopes can only be recognized by anti-idiotypic antibodies. In contrast, anti-idiotypic T cells recognize idiotype specific peptides of BCR and TCR (▶ idiopeptides) in an MHC class I- or II-restricted fashion when such peptides are presented by the lymphocytic clone itself or by professional antigen-presenting cells (APC) like e.g. macrophages, Langerhans cells or the various forms of dendritic cells.

Characteristics
Idiotypic Network Theory

In 1963 it was discovered that the immune system is not only able to mount antibody responses to external antigens, but also to antigenic determinants located in the variable regions of immunoglobulin H and L chains. Such idiotype-specific antisera could not only be raised in a different species (▶ xenogeneic host), but also in the same species when genetically different (▶ allogeneic) or identical inbred animals (▶ syngeneic) were immunized, and even in the ▶ autologous host. An autologous anti-idiotypic response could also be observed when mice were immunized with an antigen—that is, the antigen stimulated the corresponding antigen-specific antibodies (idiotypes or antibody 1, or Ab1) which thereafter induced the autologous anti-idiotypic response (anti-idiotype, or Ab2). It could be shown formally that Ab2 were able to stimulate anti-anti-idiotypic (Ab3) responses and these again could activate anti-anti-anti-idiotypic, or Ab4, antibody responses. These various internal responses forced the idea that B cells within a given immune system can be interconnected by idiotype-specific mutual recognition and that such clonal interactions exert important regulatory functions. As formulated in the idiotypic network theory (1), these multiple idiotypic interactions were supposed to enable an equilibrium among the involved cell clones which, after disturbance by stimulation with external antigens, can be re-established by idiotypic-anti-idiotypic reactions.

Since then huge numbers of investigations have been performed to elucidate the idiotypic specificities and the functions of the idiotypic network (2) and to enable an application of idiotypic internal responses for the treatment of infectious and autoimmune diseases as well as cancer (3).

Idiotypic Characterization of Antibodies

It has been estimated that an antigen-binding variable fragment (Fv) of antibodies can express up to 20 different idiotopes which can be recognized by anti-idiotypes (aId). Hence, even when an antibody is *mono*-specific for a particular epitope it is *multi*-reactive with internal network aId antibodies. Therefore, a particular Id can either be defined by polyclonal antisera through recognition of different sets of idiotopes or by monoclonal antibodies through recognition of specific single idiotopes.

Since the binding of an aId to its corresponding Id can be variably sensitive to inhibition by macromolecular antigen or ▶ hapten, it was concluded that idiotopes may be located at different distances from or may overlap with the paratope. This led to two assumptions: that aId recognizing such idiotopes may carry an internal image of the corresponding antigenic determinant; and, therefore, the immune system probably contains such internal images of all epitopes of external antigens. Indeed, the subsequent proposal that internal image aId could be useful as surrogate antigens for therapeutic vaccinations has been proven in numerous investigations—that is, that certain aId have been successfully employed to induce immunity to infectious diseases or against tumor cell growth.

However, it has been observed that xenogeneic and allogeneic anti-idiotypic antisera may not only react with the Id used for immunization, but in addition can detect other antibodies of the same and probably of different antigen specificities. Hence, the reactivity of allogeneic and xenogeneic aId demonstrated the expression of cross-reactive idiotopes (CRI or IdX). In contrast, several investigations have proven that syngeneic and autologous anti-idiotypic sera or monoclonal antibodies were never induced against such CRI, but exclusively against individual idiotopes (IdI). Thus, the greater the genetic difference between Id and aId donors, the broader the cross-reactivity of the corresponding aId will be, and the opposite seems to hold true when aId are raised in a syngeneic host—or even the autologous.

These observations relate to the question of whether internal image aId really exist. In the syngeneic/autologous host, only IdI are able to activate an internal anti-idiotypic response which, consequently, will *not* contain real internal image aId. Hence, such antibodies may only be inducible in allogeneic or xenogeneic hosts. This corresponds to earlier observations that an aId can not be defined as an internal image antibody by in vitro tests, but only by its ability to induce antigen-reactive antibodies in vivo since this therapeutic effectiveness also depends on pre-existing conditions in the network. Furthermore, it has been shown that even aId which are directed at individual idiotopes, and therefore certainly do *not* represent internal image antibodies, may nevertheless be able to function as surrogate antigen and induce neutralizing antiviral antibodies. Thus it can be concluded that antigen-specific Ab1 responses can even be induced by IdI-specific anti-idiotypic antibodies when at least a fraction of those Ab1 expresses relevant IdI. This type of activation depends on co-expression of the epitope-specific paratope and an Ab2-reactive individual idiotope, and not on the expression of an internal image of antigenic determinants by such anti-idiotypic antibodies. Further investigations on the specificity of syngeneic and autologous anti-idiotypic antibodies proved an exceedingly strong association of individual or private idiotopes with the third complementarity-determining region (CDR) of the heavy chain (CDR3) which is encoded by D gene segments. In principle, this has been demonstrated with two different experimental approaches:

- The reactivity of syngeneic and autologous Ab2 was found to be strictly dependent on the amino acid sequence of the V_HCDR3.
- Antibodies raised against V_HCDR3 peptides either in the same or a different species qualified as IdI-specific anti-idiotypic antibodies.

These reactivities allow us to conclude that in the syngeneic or autologous host, only private/individual antibody idiotopes are immunogenic and that these private idiotopes are at least mainly determined by and are probably located within or close to the V_HCDR3. However, as conformational epitopes they may also be influenced by other parts of the variable and even the constant domains, as demonstrated in a few reports. Moreover, individual idiotopes can be generated by the ▶ somatic hypermutation process during thymus-dependent immune responses. The reactivities of syngeneic/autologous anti-idiotypic antibodies with a particular idiotype are shown in Fig. 1.
In some cases, antibody-induced aId have been found to react with T cells and such Ab2 could be used to isolate the antigen receptor from these T cell with the correct α and β chains. Hence, B and T cell antigen receptors show ▶ idiotypic cross-reactivity.

Idiotypic Recognition by T cells

Since anti-idiotypic responses depend—irrespective of the immunized host (autologous, syngeneic, allogeneic, or xenogeneic)—on the activation of idiotype-specific T helper cells they are mainly of the immunoglobulin G class.
Hence, the in vivo existence of idiotype-specific T cells has clearly been demonstrated, and corresponding T cells lines have been established.
However, T cells do not recognize conformational determinants of external antigens as B cells do. Therefore, T cells are also neither able to detect antibody-associated idiotopes (BCR idiotopes) nor those of antigen receptors of other T cells (TCR idiotopes), but recognize BCR as well as TCR idiotypic peptides (idiopeptides) which are generated in different ways:

- When a protein antigen is bound to the cell surface of a B cell via its antigen receptor, this ▶ thymus-dependent antigen (TD) is endocytosed, processed and peptide fragments are presented on the cell surface via major histocompatability complex (MHC) class II molecules to antigen-specific T helper cells. At the same time, however, the intrinsic BCR is processed too, and its various peptide fragments are also presented on the cell surface via MHC class II. It was shown that these processed peptides of the intrinsic BCR can be recognized by allotype-specific or idiotype-specific normal T cells or T cell lines, respectively.
- Such a presentation of BCR idiopeptides has also been observed in resting B cells, but engagement of the BCR enhanced this process.
- In addition, evidence has been presented that autologous antibodies can be taken up by professional APCs (as immune complexes?). They are then processed and peptide fragments are presented to T cells, including those with anti-idiotypic specificity.
- Furthermore, T cells can present idiopeptides of their intrinsic TCR on MHC class I molecules which then can be recognized by anti-idiotypic T cells in an MHC-restricted fashion. Hence, a mutual idiotype-specific interaction of T cells is also possible.

Interestingly, when the specificity of anti-idiotypic T cells was tested in detail it became clear that the primarily detected BCR as well as TCR idiotopes were located in the third complementarity-determining region of the immunoglobulin H chain or the TCR β chain which is like the former encoded by D gene segments. This recognition is extremely specific in that not even minimal sequence deviations from the original V_HCDR3 idiopeptide are allowed.
Taken together, these investigations have shown that autologous and syngeneic idiotope-specific B as well as T cells mainly recognize non-germline-encoded V_HCDR3- or V_βCDR3-associated variability or somatically generated mutations within the variable regions of the H/L and α/β chains of BCR and TCR, respectively. The collective idiotypic interactions within the adaptive immune system are depicted in Fig. 2.

Idiotypically Mediated Immunological Imprinting During Early Ontogeny

Early ontogeny is a decisive developmental period in which idiotypic clonal interactions exert obvious and important regulatory functions (4).

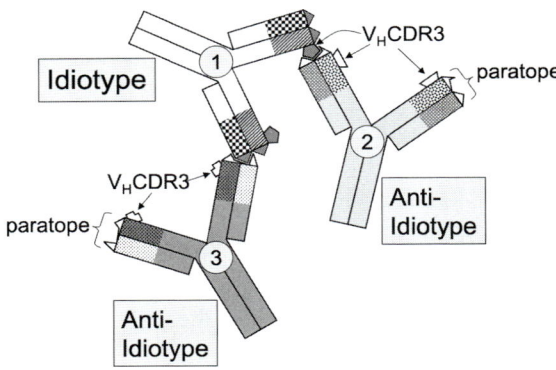

Idiotype Network. Figure 1 Localization of individual idiotopes (IdI) on immunoglobulin variable domains detected by syngeneic or autologous anti-idiotypic antibodies (Ab2). In the autologous or syngeneic host, an anti-idiotypic immune response to a particular idiotype (1) can only be induced against individual immunogenic idiotopes. Such IdI are generated by imprecisions in the VDJ recombination process (P nucleoties and N insertions) and are therefore mainly associated or located at or near the complementarity-determining region 3 of the H chain (V_HCDR3 or D region) as depicted for the binding of anti-idiotype (2). However, IdIs are also generated during thymus-dependent immune responses by the somatic hypermutation process and can then be associated the other CDRs of the H and L chains. This situation is shown for the binding of anti-idiotype (3) which recognizes a paratope-associated idiotope. In this figure, paratope and D region are shown as rather separated entities. However, there is no agreement on this point. While several reports have claimed an extraordinary importance of the D region for antigen binding, numerous others have documented that a given set of antibodies may have identical antigen specificities despite huge D region variability. Moreover, this figure shows that anti-idiotypes (Ab2) can function as a surrogate antigen if the corresponding idiotype (Ab1) co-expresses D region-associated idiotopes plus the antigen-reactive paratope. This view can explain the observation that even individual anti-idiotypes can induce antigen-reactive immune responses and thus function as if they were internal image antibodies. The shaded areas indicate the variable domains of H chains and L chains.

Whereas idiotypic suppression by anti-idiotypic antibodies in the adult is only of short duration, idiotypic responses in the neonatal period induce long-term effects which can even persist until adulthood or the reactive idiotype may even be permanently lost. Such anti-idiotypic suppression can not only be induced experimentally by exogenous antibodies, but also by maternally derived anti-idiotypic antibodies. Maternal antibodies induce T cell-dependent idiotypic responses which can be recognized as suppression of particular clones, induction of normally silent ones, or even a distortion of the whole *adult* repertoire.

These functions, however, can only be induced during defined *developmental windows* which are thus important for a neonatal *immunological imprinting* period that lasts, in the mouse, for about the first 3 weeks of life. The corresponding time period in humans may be the first 5–6 years.

In experimental animals it has shown that these stimulatory properties of maternal antibodies likewise improve the fight against microbial infections via:

- an enhancement of immune responses and even conversion of a primary into a secondary response (also in F2 animals)
- induction of antigen-reactive IgM antibodies in *non*-immunized F1 animals after maternal immunization with antigen or *anti*-idiotype
- diversification of the adult repertoire
- selection of primary antibodies with strongly enhanced affinities
- transfer of carrier sensitivity from dams to offspring
- most strikingly, however, even *non*-antigen-reactive antibodies—namely anti-idiotypes—can transfer anti-microbial protection
- maternally transferred immune or monoclonal IgG antibodies induce an allergen-specific suppression of IgE responsiveness.

Relevance to Humans

The involvement of idiotype-specific T cells in physiological cellular interactions has been particularly well documented in animal experimental systems of various diseases, such as experimental systemic lupus erythematosus (SLE), experimental autoimmune encephalomyelitis (EAE), autoimmune diabetes, or in tumor models like B cell lymphoma or myeloma. EAE is an animal model for multiple sclerosis (MS) that is induced by ▶ CD4+ T cells which react with myelin basic protein (MBP) (2). Consequently, attenuated MBP-reactive CD4+ T cell clones can be employed for a T cell vaccination (TCV) to induce an anti-EAE immunity, which depends on cytolytic T cells of the ▶ CD8+ phenotype. These are idiotypically specific for the inducing encephalitogenic CD4+ T cells. Because the CD8+ T cells are specific for Vβ chain idiotopes of the encephalitogenic T cells, immunizations with TCR idiotypic peptides of the Vβ chain, either of the CDR2 or the VDJ joining sequences, were also able to prevent EAE.

- Identical cellular interactions seem to exist in humans (2,5,6), e.g. a vaccination of MS patients with irradiated MBP-reactive T cells depleted the MBP-reactive cells from the circulation, and cytolytic CD8+ T cell clones could be isolated from these patients. Such an idiotype-specific T cell vaccination was also possible with peptides of the MBP-

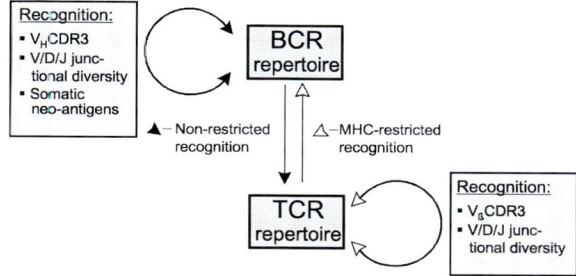

Idiotype Network. Figure 2 Idiotypic interactions within the adaptive immune system. Physiological clonal interactions in the idiotype network occur at different levels. Within the repertoire of B cell antigen receptors (BCR), the mutual recognition of antibodies as idiotype and anti-idiotype is mainly directed at (a) the CDR3 of the VH region for which the junctional diversity between V, D and J genes is most important and (b) somatic neo-antigens generated through the somatic hypermutation process. In addition, anti-idiotypic antibodies may recognize idiotopes of T cell antigen receptors (TCR) in a non-MHC-restricted way, thus indicating a BCR-TCR cross-reactivity. Both reactivities are indicated by black arrows. T cells are able to recognize BCR idiopeptides which are presented by B cells themselves or other antigen-presenting cells. This recognition is MHC restricted. T cells themselves process and present their TCR idiopeptides via MHC molecules and these presented idiopeptides can be recognized by other T cells. This idiotypic T-T interaction is initiated when the activation of one of the cell clones leads to an enhanced presentation of TCR idiopeptides. The MHC-restricted recognition is indicated by white arrows.

reactive clones and the resulting T cells were cytolytic for the auto (MBP)-reactive T cells expressed the CD8 membrane marker and preferentially recognized VβCDR3-related idiotopes. Anti-idiotypic CD4+ T cells of the T_H2-type were also induced, which only reacted with activated target T cells and contributed to the inhibition of the MBP-reactive clones. Furthermore, such an anti-idiotypic T cell vaccination also induced an anti-idiotypic antibody response that contributed to the suppression of the MBP-reactive T cells in those patients.

- In other clinical studies it has been shown that autologous anti-TCR-idiotypic antibodies may be generated and T cells have been observed which react with autologous anti-rabies virus antibodies in an Id-specific manner. Similarly, joint-derived T cells from patients suffering from rheumatoid arthritis may react with self-immunoglobulin or anti-immunoglobulin antibody. Moreover, autoantibody-reactive Id-specific T cells develop in patients with myasthenia gravis and SLE. In tumor patients, Id-specific T cells have been observed which may even confer protection against B cell lymphoma or may be directed against myeloma proteins.

Various autoimmune diseases in humans have successfully been treated with intravenous immunoglobulin (IVIg). It is assumed that anti-idiotypic antibodies which are contained in the IVIg preparations suppress the autoantibody secreting cell clones. Moreover, the efficient suppression of active EAE (in animals) by an intravenous administration of IVIg proves that aId influence autoaggressive T cell clones. Hence, the therapeutic efficacy of IVIg in the treatment of a set of autoimmune diseases in experimental animals—as well as humans—underscores the functionality of the idiotype network.

Conclusion

Clonal interactions of B and T cells by idiotope-specific mutual recognition of their antigen receptors form a web of unknown density, referred to as the idiotype network.
The idiotype network is not directly genetically determined, but develops during ontogeny through (i) maternal influences, (ii) during generation and maintenance of autotolerance and (iii) through multiple immune responses to external antigens. Therefore, idiotype networks are very complex and they are unique even among genetically identical individuals. At present, this uniqueness of idiotype networks prevents an easy and wide therapeutic application for the treatment of infectious, autoimmune and malignant diseases.

References

1. Jerne NK (1974) Towards a network theory of the immune system. Ann Immunol (Paris) 125C:373–389
2. Lemke H, Lange H (2002) Generalization of single immunological experiences by idiotypically mediated clonal connections. Adv Immunol 80:203–241
3. Shoenfeld Y, Kennedy RC, Ferrone S (1997) Idiotypes in Medicine: Autoimmunity, Infection and Cancer. Amsterdam: Elsevier Science BV
4. Lemke H, Continho A, Lange H (2004) Lamarekian inheritance by somatically acquired maternal IgG phenotypes. Trends Immunol 25:180–186
5. Stinissen P, Zhang J, Medaer R, Vandevyver C, Raus J (1996) Vaccination with autoreactive T cell clones in multiple sclerosis: overview of immunological and clinical data. J Neurosci Res 45:500–511
6. Jiang H, Chess L (2000) The specific regulation of immune responses by CD8+ T cells restricted by the MHC class Ib molecule, Qa-1. Annu Rev Immunol 18:185–216

Idiotypic Cross-Reactivity

Different antibodies may idiotypically cross-reactive when they express identical idiotopes which are detected by a particular anti-idiotypic antibody.
▶ Idiotype Network

Idiotypic Epitopes

When immunoglobulins are isolated from one inbred strain and injected into the same strain, the epitopes that are recognized are known as idioptypic epitopes.
▶ Humanized Monoclonal Antibodies

IFN-γ

Interferon-gamma.
▶ Maturation of the Immune Response

Ig V-Region

The binding site of the B cell receptor is formed by the variable regions of the immunoglobulin molecule. The variable region of the heavy chain (V_H) together with the variable region of the light chain (V_L) determines the specificity of the immunoglobulin.
▶ B Cell Maturation and Immunological Memory

IgE

Immunoglobulin E, a subclass of antibodies that mediates acute allergic reactions.
▶ Food Allergy

IgE-Mediated Allergies

WERNER PICHLER
Klinik für Rheumatologie & klinische Immunologie/Allergologie
Inselspital-Universtät Bern
3010 Bern
Switzerland

Synonyms
type 1 reactions according to Gell and Coombs, atopic allergy

Definition
An immune response to a soluble foreign protein (e.g. pollen protein) or a hapten–modified autologous protein can involve the formation of IgE-antibodies directed to the antigen (1). Thereby, the protein is immunogenic for B and T cells. Under certain circumstances the B-cell immune response results in the formation of IgE antibodies. Most studies indicate that IgE-mediated drug allergies are not related to an atopic predisposition, but that side effects, when they appear, might be more severe in atopic patients.

Characteristics
The IgE-mediated type of immune reaction has some peculiarities: It can be induced by minute amounts of the antigen (one estimates that during the pollen season only 1 μg of the relevant protein is inhaled). Moreover, very small amounts of the allergen can also elicit symptoms, which appear rapidly (within minutes) after contact with the allergen.
IgE is present in very low quantities in the serum (in

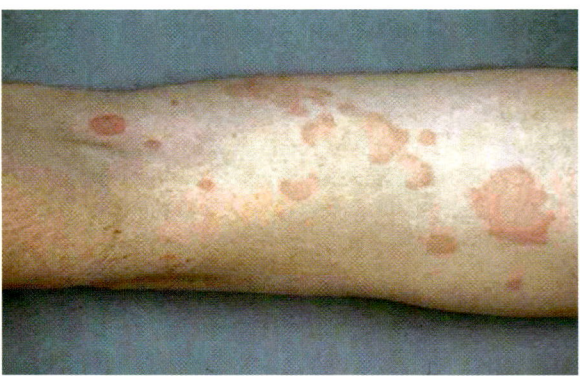

IgE-Mediated Allergies. Figure 1 Urticaria. Typical wheal reaction, often with surrounding erythema, which normally lasts < 12 hours. Massive generalized urticaria can be associated with circulatory collapse due to hypovolemia. Urticaria is also a frequent sign of anaphylaxis.

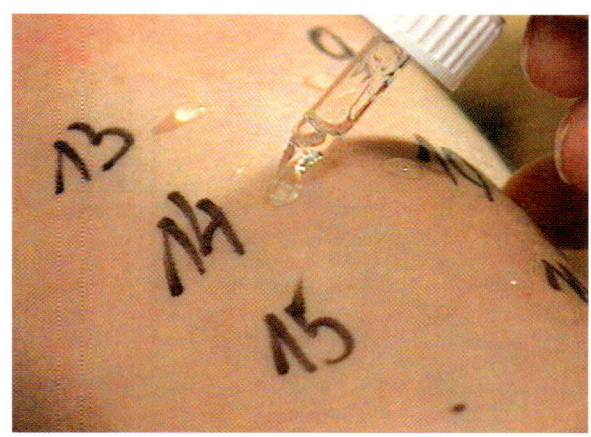

IgE-Mediated Allergies. Figure 2 IgE skin prick test. Typical wheal and flare reaction, normally with a peak 15–20 minutes after application.

terms of nanograms, about 10,000 lower then IgG1). Nevertheless, it is distributed all over the body due to very high affinity for Fc-IgE receptors type I (10^{-9} kD), which are present on tissue mast cells and basophilic leukocytes, and in a modified form on Langerhans cells of the skin and some monocytes (2). This cell-bound IgE serves as a very efficient amplification mechanism of the reaction. Indeed, it is astonishing to observe how sensitive this system is, as in certain highly sensitized patients, skin testing alone (intradermal application of penicillin-derivatives) may cause generalized ▶ urticaria or even ▶ anaphylaxis.

After a sensitization phase, the allergen (e.g. a penicillin-modified protein) is recognized by specific IgE on these cells. If cross-linking of Fc-IgE receptors occurs, mediators like histamine, leukotrienes, prostaglandins and cytokines are released and clinical symptoms develop. One differentiates between the immediate symptoms, which appear within minutes and are transient. In addition, such reactions are followed by delayed reactions: these are due to the immigration of inflammatory cells, in particular eosinophils and basophils, and lead, if it occurs repeatedly (chronic allergen exposure), to an eosinophilic inflammation (e.g. asthma bronchiale).

The typical symptoms can be reproduced in an immediate skin test with the substance. In sensitized individuals a wheal and flare reaction develops within 15–20 minutes. Clinically, type I allergies manifest as urticaria, bronchospasm, rhinitis, conjunctivitis, and, in the most severe form, as anaphylactic shock, which involves all of these symptoms together with circulatory collapse (a drop in blood pressure) due to a volume shift to the extra vascular space. If the reactions occur repeatedly due to persistent allergen exposure, a chronic inflammation develops, like asthma.

Anaphylaxis, asthma, and laryngeal oedema can be lethal, which explains the great fear of these reactions. Besides penicillins, cephalosporins, muscle relaxants, diclofenac, and pyrazolone-derivatives are common elicitors of IgE-mediated reactions.

Preclinical Relevance

IgE preferentially forms against soluble proteins or protein-hapten complexes.

Any hapten (e.g. cephalosporins, penicillins) are potential inducers of such reactions as they can elicit an immune response to the hapten-carrier construct (hapten-modified protein). The immune response can involve the formation of antibodies (IgE, IgG). IgE-mediated allergies to penicillins do still occur, but are surely less frequent than in the 1950s and 1960s. This is probably due to the wider use of orally given preparations. IgE-mediated allergies are sometimes side-chain specific, that is that structurally different antibiotics bearing the same side chain can elicit symptoms in sensitized individuals (3).

In the early 20th century IgE antibodies to foreign proteins, such as horse proteins, were frequent causes of anaphylaxis. More recently the generation of recombinant proteins, like chimeric antibodies consisting of mouse-derived antibody binding sites and human constant parts are at risk of inducing such reactions. They manifest themselves initially as local wheal and flare reactions, later in generalized urticaria and possibly anaphylaxis. The immunogenicity of these recombinant proteins causes a major concern for the future therapeutic use of these proteins. One tries to make the recombinant proteins as similar to the autologous protein as possible (e.g. humanized antibodies), which surely reduces the immunogenicity. But sometimes even a single amino acid difference to the original substance may cause an immune response (well documented with different insulins). In addition, the route of application is important as clearly the intravenous route is less sensitizing than subcutaneous application. Often only a rather harmless IgG response develops.

It should be emphasized that at the moment there is no generally accepted preclinical testing strategy for the allergenic and autoimmune potential of drugs; standardized means are only used for testing for the potential of haptens for skin sensitization and photoallergy. Nevertheless, with the increased use of recombinant allergens in the clinic, the immunogenicity of these proteins might become interesting. However, species differences determine the immunogenicity and therefore animal experiments might be of limited value (e. g. a mouse protein is in most instances less immunogenic in rats than in humans). Only clinical trials may reveal the sensitizing potential in humans.

In addition, haptens could be evaluated for immunogenicity in animals because these small molecules bind covalently to autologous proteins and make them immunogenic. But the clinical meaning of a positive test remains often unclear, thus amoxicillin is a well-known hapten and is potentially immunogenic, but in reality (in clinical practice) the frequency of a sensitization and allergy is just 4% to 8%, and the majority of treated people do not react (6). Moreover, unknown factors determine whether a (dangerous) IgE-mediated allergy or a rather harmless exanthema develops.

Relevance to Humans

IgE-mediated allergies have enormous relevance to humans because allergic diseases are some of the most common diseases affecting mankind. However, most of these allergic diseases are not severe, due to limited inhalation of the causative allergens. Clinical symptoms consist of rhinitis, conjunctivitis, and possible asthma, but severe reactions also occur—especially after food intake (especially peanuts), *Hymenoptera* stings, and drugs. In the USA about 150 people die annually due to food-induced allergic shock alone. The precise number of deaths resulting from drug-induced allergies is unknown—as many deaths might be due to ▶ pseudoallergy reactions, as in aspirin intolerance, or anaphylaxis due to radiocontrast media, where symptoms are identical, but drug-specific IgE is rarely detectable.

Skin testing for penicillin allergy is routine in allergy clinics, and many other allergies to drugs (pyrazolones, quinolones recombinant proteins) can be detected by both prick tests and intradermal tests, as well as serology (4,5).

Regulatory Environment

Up to now, only the cell-mediated delayed-type hypersensitivity is regulated by specific guidelines (see ▶ Delayed-Type Hypersensitivity). Other allergic reactions—local or systemic—are only mentioned in principle in different guidelines:
- the Japanese immunotoxicity draft guideline: Draft Guidance for Immunotoxicity Testing, 2003
- the CDER's Guidance for Industry, Immunotoxicology Evaluation of Investigational New Drugs (FDA, 2002)
- the ICH harmonised tripartite guideline: Preclinical Safety Evaluation of Biotechnology-Derived Pharmaceuticals (1997).

The fact that no specific guideline exists for the investigation of systemic allergic reactions is obviously because there is a lack of accepted and validated test models for all antibody-driven hyperimmune reactions.

References

1. Coombs PRA, Gell PGH (1968) Classification of allergic reactions responsible for clinical hypersensitivity and disease. In: Gell RRA (ed) Clinical Aspects of Immunology. Oxford University Press, Oxford, pp 575–596
2. Janeway CA, Travers P, Walport M, Shlochik M (2001) Immunobiology. Garland Publishing, New York
3. Blanca M (1994) The contribution of the side chain of penicillins in the induction of allergic reactions. J Allergy Clin Immunol 94:562–563
4. Himly M, Jahn-Schmid B, Pittertschatscher K et al. (2003) IgE-mediated immediate-type hypersensitivity to the pyrazolone drug propyphenazone. J Allergy Clin Immunol 111:882–8
5. Manfredi M, Severino M, Testi S et al. (2004) Specific IgE to quinolones J Allergy Clin Immunol 113 (1):155–60
6. Pichler WJ (2001) Predictive drug allergy testing: an alternative viewpoint. Toxicology 158:31–41

IMHA

▶ Hemolytic Anemia, Autoimmune

Immediate-Type Hypersensitivity

Immediate-type hypersensitivity (type I hypersensitivity) is an IgE-mediated response that occur within 20 minutes. Reactions are to environmental allergens, especially allergens to the skin, inhaled allergens, and food allergens. Characteristics immediate reactions are hay fever, asthma or urticaria.

▶ Mast Cells
▶ Immunotoxicology

Immediate-Type or Delayed-Type Hypersensitivity

Exaggerated immune responses that cause damage to the individual. Immediate hypersensitivity (types I, II, III) is mediated by antibody or immune complexes, and occurs within seconds or minutes after the second exposure to the allergen. Delayed-type hypersensitivity (type IV) is mediated by T_{DTH} cells and occurs 24–72 hours after the challenge of the sensitized individual.

▶ Flow Cytometry

Immune Cells, Recruitment and Localization of

BERNHARD MOSER
Theodor-Kocher Institute
University of Bern
CH-3012 Bern
Switzerland

MARIAGRAZIA UGUCCIONI
Institute for Research in Biomedicine
Via Vincenzo Vela 6
CH-6500 Bellinzona
Switzerland

Definition
Motility is a hallmark of leukocytes. This property is of crucial importance for all aspects of immunity and forms the basis for hematopoiesis and immune defense. Breakdown in the control of leukocyte mobilization contributes to chronic inflammatory diseases and, consequently, small molecular weight compounds that selectively interfere with leukocyte recruitment represent a promising approach for the development of novel anti-inflammatory medication. Chemotactic migration of leukocytes largely depends on adhesive interaction with the substratum and recognition of a chemoattractant gradient. ▸ Chemokines are secreted proteins of 67 to 127 amino acids and have emerged as key controllers of integrin function and cell locomotion. Numerous distinct chemokines exist, which target all types of leukocytes, including hematopoietic precursors, mature leukocytes of the innate immune system, as well as naive, memory, and effector lymphocytes. The combinatorial diversity in responsiveness to chemokines ensures the proper tissue distribution of distinct leukocyte subsets under normal and inflammatory/pathological conditions. Besides leukocyte chemoattraction, some unrelated processes—such as angiogenesis, tissue remodeling and tumor metastasis—are also influenced by chemokines.

Characteristics
Chemokines: The Largest Family of Cytokines
Interleukin-8 (IL-8/CXCL8), the first chemokine, was discovered 15 years ago on the basis of its neutrophil chemoattractant properties. Two of the four NH_2-terminal conserved Cys residues are separated by one amino acid, which typifies IL-8 as a CXC chemokine (Table 1). The monocyte chemoattractant protein MCP-1/CCL2 with the two NH_2-terminal Cys residues in adjacent positions is a CC chemokine. It was discovered shortly after IL-8. The large majority of approximately 50 human chemokines fall into either the group of CXC or the group of CC chemokines. In addition, there are two C chemokines, Ltn-α/XCL1 and Ltn-β/XCL2, in which two out of the four conserved Cys residues are missing, and a single CX3C chemokine, called fractalkine/CX3CL1, with three amino acids separating the two NH_2-terminal Cys residues. Two systems of nomenclature are used in the current literature, the traditional abbreviations dating back to the time of chemokine discovery (such as IL-8 and MCP-1), and a systematic nomenclature that combines structural motifs (CXC, CC, XC, CX3C) with L for ligand and the number of the respective gene (for access to recent updates see http://cytokine.medic.kumamoto-u.ac.jp) (Table 1) (1,2). In this section, traditional abbreviations are appended by their systematic nomenclature at the first instance and then used on their own throughout the text. Chemokine receptors are designated according to the type of chemokine(s) that they bind (CXC, CC, XC, CX3C), followed by R for receptor and a number indicating the order of discovery.

Chemokine Receptors
Chemokine receptors belong to the large family of seven-transmembrane domain receptors which couple to heterotrimeric GTP-binding proteins (G proteins) (1–3). Experiments with *Bordetella pertussis* toxin indicated that these receptors typically require G proteins of the G_I type for signal transduction. Biochemical and functional analysis with IL-8 and related chemokines demonstrated that neutrophils express two types of IL-8 receptors, CXCR1 with selectivity for IL-8 and GCP-2/CXCL6, and CXCR2 with promiscuous binding of IL-8 and numerous other related CXC chemokines. The subsequent cloning of the corresponding receptor cDNAs confirmed these early observations (Table 1). There are considerable species differences, as demonstrated by the presence of a single IL-8 receptor in rabbits and mice but not in rats or humans, and the lack of a IL-8 homologue in mice but not in rabbits or humans. Sequence information about the first chemokine receptors quickly led to new discoveries, giving rise to a total of 18 currently known human chemokine receptors, including six CXCRs, ten CCRs, one XCR, and one CX3CR (Table 1).

Cellular responses to chemokines are typically rapid in onset and transient in duration. The negative control mechanism elicited during chemokine receptor signaling is termed cellular desensitization and is by itself also transient, i.e. cells regain responsiveness to a given chemokine after a short period of culturing in chemokine-depleted medium. For more detailed information see also ▸ chemokines.

Rapid internalization allows the continuous redistribu-

Immune Cells, Recruitment and Localization of. Table 1 The human chemokine system

Systematic[a]	Common[b]		Receptor(s)[c]
CXC chemokines			
CXCL1	GROα	Growth-related protein α	CXCR2
CXCL2	GROβ	Growth-related protein β	CXCR2
CXCL3	GROγ	Growth-related protein γ	CXCR2
CXCL5	ENA-78	Epithelial cell-derived neutrophil activating peptide 78	CXCR2
CXCL6	GCP-2	Granulocyte chemotactic protein 2	CXCR1,-2
CXCL7	NAP-2	Neutrophil-activating peptide 2	CXCR2
CXCL8	IL-8	Interleukin-8	CXCR1,-2
CXCL9	Mig	Monocyte/macrophage-activating, IFNγ-inducible protein	CXCR3
CXCL10	IP10	IFNγ-inducible 10 kDa protein	CXCR3
CXCL11	I-TAC	IFN-inducible T cell alpha chemoattractant	CXCR3
CXCL12	SDF-1	Stromal cell-derived factor 1	CXCR4
CXCL13	BCA-1	B cell attracting chemokine 1	CXCR5
CXCL14	BRAK	Breast and kidney chemokine	n.d.
CXCL16	—	—	CXCR6
CC chemokines			
CCL1	I-309	Intercrine-β glycoprotein 309	CCR8
CCL2	MCP-1	Monocyte chemoattractant protein 1	CCR2
CCL3	MIP-1α	Macrophage inflammatory protein α	CCR1,-5
CCL4	MIP-1β	Macrophage inflammatory protein β	CCR5
CCL5	RANTES	Regulated on activation normal T cell expressed and secreted	CCR1,-3,-5
CCL7	MCP-3	Monocyte chemoattractant protein 3	CCR1,-2,-3
CCL8	MCP-2	Monocyte chemoattractant protein 2	CCR1,-2,-3,-5
CCL11	Eotaxin	Eosinophil chemoattractant protein	CCR3
CCL13	MCP-4	Monocyte chemoattractant protein 4	CCR2,-3
CCL14	HCC-1	Hemofiltrate CC chemokine 1	CCR1
CCL15	HCC-2	Hemofiltrate CC chemokine 2	CCR1,-3
CCL16	HCC-4	Hemofiltrate CC chemokine 4	n.d.
CCL17	TARC	Thymus- and activation-regulated chemokine	CCR4
CCL18	DC-CK1	Dendritic cell chemokine 1	n.d.
CCL19	ELC	EBV-induced receptor (EBI1) ligand chemokine	CCR7
CCL20	LARC	Liver- and activation-regulated chemokine	CCR6
CCL21	SLC	Secondary lymphoid tissue chemokine	CCR7
CCL22	MDC	Macrophage-derived chemokine	CCR4
CCL23	MIP-3	Macrophage inflammatory protein 3	CCR1
CCL24	Eotaxin-2	Eosinophil chemoattractant protein 2	CCR3
CCL25	TECK	Thymus-expressed chemokine	CCR9
CCL26	Eotaxin-3	Eosinophil chemoattractant protein 3	CCR3
CCL27	CTACK	Cutaneous T-cell attracting chemokine	CCR10
CCL28	MEC	Mucosae-associated epithelial chemokine	CCR3,-10
C chemokines			
XCL1	Ltn-α	Lymphotactin α	XCR1
XCL2	Ltn-β	Lymphotactin β	XCR1
CX$_3$C chemokine			
CX$_3$CL1	Fractalkine	—	CX$_3$CR1

[a] Systematic nomenclature as defined in http://cytokine.medic.kumamoto-u.ac.jp. Numerical gaps designate positions for orphan chemokines (such as CXCL4/platelet factor 4) or mammalian chemokines with unknown human homologue.
[b] One of several traditional names that is frequently used in the literature.
[c] Agonistic receptor selectivity.
n.d., not determined.

tion of chemokine receptors on leukocytes for maintaining polarized chemokine sensing and directed cell migration along a chemokine gradient (Figure 1).
In addition, certain chemokine receptors with "decoy" function, i.e. chemokine-binding receptors that are unable to signal, control chemokine responsiveness. Such decoy receptors include the Duffy antigen-related chemokine receptor (DARC) on red blood cells and the tissue-expressed D6 receptor, which are both characterized by broad chemokine-binding activity. Finally, a chemokine "decoy" function may also be induced in normal chemokine receptors, as shown in IL-10-treated monocytes and dendritic cells (DCs).

Leukocyte Mobilization (and other Chemokine Functions)

Chemoattractant activity and the "four Cys-residue" fingerprint arrangement prompted the term chemokines to designate this novel group of chemotactic cytokines. The "classical" chemoattractant agonists lacking the chemokine-typical structural motifs—such as complement component C5a, lipid derivatives leukotrienes and platelet-activating factor, and bacterial N-formylmethionyl peptides—do not belong to the chemokine family.
In addition to structural classification, chemokines are also grouped into functional subsets (3,4). Inflammatory chemokines control the recruitment of effector leukocytes in infection, inflammation, tissue injury and tumors. Many of the inflammatory chemokines have broad target-cell selectivity and act on cells of the innate as well as the adaptive immune system. Homeostatic chemokines, by contrast, navigate leukocytes during hematopoiesis in the bone marrow and thymus, during initiation of adaptive immune responses in the spleen and lymph nodes, and in ▶ immune surveillance of healthy peripheral tissues. Recent findings indicate that several chemokines cannot be assigned unambiguously to either one of the two functional categories and, therefore, may be referred to as "dual-function" chemokines.

Recruitment of circulating leukocytes to sites of pathogen entry or inflammation involves two separate migration processes, termed extravasation and chemotaxis (Figure 1). Adhesion to the luminal side of blood vessels, ▶ transendothelial migration, and subsequent chemotaxis of leukocytes are highly complex processes, which are controlled by "outside-in" and "inside-out" signaling events during cellular interactions with chemokines and adhesion ligands. Triggering of chemokine receptors in leukocytes by endothelia-associated chemokines is a requirement for extravasation and induces a rapid increase in integrin affinity/avidity, which results in firm but transient leukocyte adhesion (5). Subsequently, the adherent leukocytes move across the endothelial cell layer and the underlying basement membrane and are released into the tissue. Of note, only those types of leukocytes are able to transmigrate at a given vascular site, which are capable of responding to the chemokines present on the local endothelium. In other words, chemokines and their receptors largely determine the selectivity in leukocyte extravasation (3,4). For instance, T and B cells with receptors for chemokines present in secondary lymphoid tissues will not be recruited to peripheral tissues but instead are destined to recirculate through spleen, lymph nodes and Peyer's patches. By contrast, effector leukocytes bearing receptors for inflammatory chemokines are efficiently recruited to sites of inflammation and disease. This leukocyte traffic control function forms the basis for the development of chemokine-based anti-inflammatory therapy (see below). Chemokines are also thought to contribute to hematopoiesis by controlling the localization leukocyte precursors within distinct microenvironments in the bone marrow and thymus. Of note, SDF-1/CXCL12 and its receptor CXCR4 were shown to be critically involved in myelo- and B lymphopoiesis, whereas mostly dual-function chemokines and their receptors (including CXCR3, CCR4, CCR8, CCR9 as well as CXCR4) control thymocyte development.

Finally, recent reports demonstrated the involvement of chemokines in tumor metastasis, a finding that seems to be related to the chemokine-typical control of leukocyte traffic (6). Possibly, chemokine-mediated tissue cell relocation or retention may also contribute

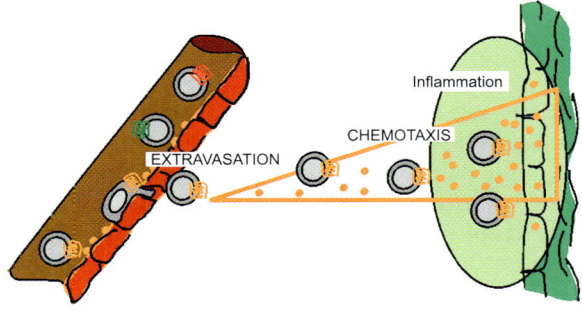

Immune Cells, Recruitment and Localization of. Figure 1 Leukocyte extravasation and chemotaxis: Inflammation induces a cascade of events that includes the production of chemokines in the affected tissue and microvasculature. Chemokines present on the luminal side of blood vessels trigger firm adhesion only in those leukocytes that express the appropriate chemokine receptors. After transendothelial migration and crossing of the basement membrane, leukocytes reach the site of inflammation via chemotaxis along increasing concentrations of chemokine(s). Leukocytes stop moving at the site of highest chemokine concentration.

to organogenesis, angiogenesis and tissue remodeling (see below).

Migration-unrelated chemokine activities include the modulation of leukocyte effector functions (cytotoxicity, mediator release, cytokine production), tissue cell growth, differentiation, and survival (organogenesis, angiogenesis), and human immunodeficiency virus (HIV) infection. HIV-1 requires the co-receptors CCR5 and CXCR4 for entry into $CD4^+$ target cells, and the chemokines—which bind to these co-receptors (RANTES/CCL5, MIP-1α/CCL3 and MIP-1β/CCL4 for CCR5, and SDF-1 for CXCR4)—interfere with this process (4). In infected individuals, the level of these HIV suppressor chemokines correlates indirectly with disease progression, suggesting that inhibitors for CCR5 and CXCR4 may be beneficial in the treatment of this disease. Migration-unrelated functions are major topics in current chemokine research.

Leukocyte Relocation During Immune Responses

Both inflammatory and homeostatic chemokines are involved in leukocyte traffic control during initiation of immune responses (3,4). Cells of the innate immune system (granulocytes, natural killer cells, monocytes/macrophages, immature DCs) are equipped with numerous receptors for inflammatory chemokines, which enable their "first line of defense" function at the site of pathogen entry. Of interest is the activation-induced switch in migratory behavior in DCs (Figure 2).

As sentinels in peripheral tissues, notably in the skin, lung and gastrointestinal tract, immature DCs quickly localize via receptors for inflammatory chemokines at the site of pathogen entry or tissue damage. Antigen uptake and processing, as well as maturation into potent T cell stimulators, is rapidly induced by the local inflammatory conditions. Importantly, this DC maturation results in the secretion of high levels of inflammatory chemokines, which leads to cellular desensitization towards these chemokines, and expression of CCR7 (see below), which enables their exit from the site of infection and their co-localization with T cells in reactive lymph nodes.

Naive T cells continuously recirculate through secondary lymphoid tissues (spleen, lymph nodes, Peyer's patches) and are largely excluded from peripheral tissues (skin, lungs and gastrointestinal tract) (Figure 2). Their lymph node homing phenotype is largely defined by the expression of CCR7, which recognizes ELC/CCL19 and SLC/CCL21, as well as by the expression of a set of adhesion molecules (CD62L, LFA-1 or $\alpha 4\beta 7$), which interact with the corresponding vascular ligands present at these sites. During contact with newly recruited antigen-presenting (mature) DCs, $CD4^+$ T helper (Th) cells rapidly acquire follicular homing properties by expression of CXCR5, which selectively responds to follicular chemokine BCA-1/CXCL13 (Figure 2). These newly primed T cells, referred to as follicular B helper T (T_{FH}) cells, are recruited to the B cell compartment by means of BCA-1, co-localize with B cells and support antibody responses (4).

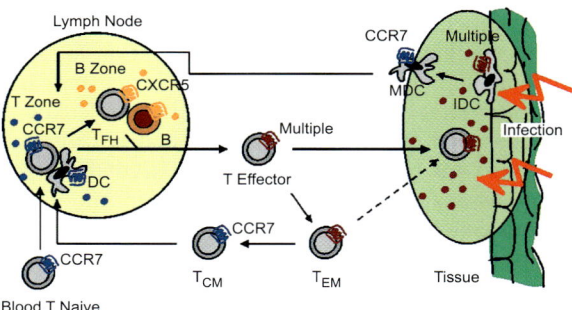

Immune Cells, Recruitment and Localization of. Figure 2 Initiation of adaptive immune responses: Dendritic cells (DCs) fulfil important functions as sentinel cells in peripheral tissues and as initiator of T cell responses in local lymph nodes. Stimulation of T cells within the T zone results in the transitory CXCR5-expressing follicular B helper T cells (T_{FH}) cells, which are the precursors of effector/memory T cells. By contrast to naive and T_{FH} cells, effector T cells and certain memory T cells (T_{EM} cells) express numerous receptors for inflammatory chemokines for their efficient relocation to inflammatory sites. During the primary immune response, some of the effector T cells become T_{EM} cells and—eventually—long-lived central memory T cells (T_{CM}) cells, which express CCR7 for recirculation through secondary lymphoid tissues (splenic T zone, lymph nodes, Peyer's patches). (Multiple, multiple chemokine receptor expression; IDC, immature DCs; MDC, mature DCs)

Two types of antigen-experienced T cell subsets are produced during adaptive immune responses, namely large numbers of short lived effector T cells, which express the full compliment of receptors for inflammatory chemokines for direct participation in ongoing immune responses, and long-lived ▶ memory T cells for rapid involvement in recall (vaccination) responses (3,4). Peripheral blood memory T cells are further divided into two major subsets characterized by different migratory and functional properties (Figure 2). Central memory T (T_{CM}) cells express CCR7 (and CD62L) for continuous recirculation through spleen, lymph nodes and Peyer's patches whereas effector memory T (T_{EM}) cells lack these lymph node homing receptors but instead express a variety of inflammatory receptors for immediate recruitment to inflammatory sites.

Recent evidence in mice suggests the following order in T cell differentiation during primary immune responses:

naive (antigen-inexperienced) T cells → effector T cells → T_{EM} cells → T_{CM} cells.

Helper T cells of the Th1 and Th2 phenotype express a characteristic set of chemokine receptors which mainly bind inflammatory chemokines, and some of these appear to contribute to immune response polarization (see below). CXCR3, CXCR6 and CCR5 are frequent on Th1 cells. Of note, CXCR6 binds the membrane-anchored chemokine CXCL16 and, thus, may primarily control Th1 cell adhesion and transendothelial migration. By contrast, Th2 cells frequently express CCR3, CCR4 and CCR8. CCR3 is highly selective for Th2 cells whereas CCR4 and CCR8 are also found on Th1 cells, skin-homing T cells and peripheral blood regulatory ($CD4^+CD25^+$) T cells. Collectively, the "polarized" expression of the chemokine receptors is not strict and a simple non-overlapping assignment of inflammatory chemokine receptors to distinct effector/memory T cell populations cannot be made. This may explain in part the marginal "phenotypes" seen in mice with single deficiencies in genes for inflammatory chemokines or chemokine receptors and further suggests that multiple chemokine receptors need to be targeted for effective anti-inflammatory intervention (see below).

Chemokines in Pathology

A vast literature documents the expression of various chemokines in inflamed tissue (7). Nevertheless, the real significance of multiple chemokine expression in inflammation is a controversial issue. Studies on chemokine expression in human tissue samples from different diseases have revealed that resident and infiltrating cells can produce a variety of inflammatory chemokines that do not entirely account for the type of leukocyte infiltration and for the disease characteristics. The information about the role of chemokines in human pathology is largely limited to studies of their occurrence and distribution in samples of disease tissues, and their concentrations in plasma, exudates and others body fluid. Most of the chemokines are produced and act locally rather then systemically, and the levels in body fluid are mostly a pale reflection of the disease process, which should be monitored in situ. Still, chemokine research has dramatically changed our understanding of leukocyte traffic in immune pathology and has opened attractive perspectives for new therapeutic approaches in the treatment of chronic inflammation and infectious diseases (8).

Antagonists and Chemokines

Also, recent studies have revealed natural chemokines with antagonistic activity, further underscoring the importance of chemokines as controllers of leukocyte navigation. On the one hand, multiple NH_2-terminally truncated forms of chemokines were identified in natural sources, which are thought to be the product of inflammation-derived or tissue cell-derived proteases. Many of these natural chemokine truncation variants have antagonistic activity and, therefore, chemokine-selective proteases are proposed to be important factors in the control of leukocyte traffic. On the other hand, several intact and active chemokines were demonstrated to act also as antagonists. For instance, Mig/CXCL9, IP-10/CXCL10 and I-TAC/CXCL11, the ligands for CXCR3, are antagonists for CCR3, the receptor for eotaxin/CCL11 and several other CC chemokines. Since CXCR3 and CCR3 are differentially expressed in Th1 and Th2 cells, these findings suggest that the CXCR3 ligands contribute to Th1-type immune response polarization by blocking the migration of $CCR3^+$ (Th2) cells. Other natural chemokines with agonistic and antagonistic activities are eotaxin/CCL11 and eotaxin-3/CCL26, which attracts eosinophils, basophils and Th2 cells *via* CCR3, while blocking $CCR2^+$ cells, and MCP-3/CCL7, a potent agonist for CCR1, CCR2 and CCR3 that blocks $CCR5^+$ cells. The combination of stimulatory and inhibitory properties represents yet another level of control that influences cellular traffic during immune responses.

Preclinical Relevance

Numerous chemokines have been associated with diverse pathological states in humans. Notably, the local profile of chemokines is thought to determine the cellular composition of the inflammatory infiltrates and aberrant regulation of chemokine expression may contribute to chronicity of inflammatory conditions. In support, work in many mouse models replicating human disease states, such as neuroinflammation, cardiovascular diseases, arthritis and allergy, have provided evidence for direct involvement of mostly inflammatory chemokines in inflammatory diseases (7). It needs to be emphasized, however, that the cell recruitment control systems show redundancy, i.e. that many chemokines with overlapping receptor selectivities are produced locally under inflammatory conditions. This probably explains the weak to undetectable phenotypes seen in mice with single-chemokine deletions. By contrast, prominent phenotypes were frequently observed in mice lacking homeostatic chemokines, which show greater selectivity for single chemokine receptors. Finally, translation of mouse data to human conditions should be done with caution since the chemokine systems differ between mouse and human in many respects, including presence/absence of certain chemokines and altered chemokine receptor selectivity profiles.

Relevance to Humans

There are clear indications for a role of chemokines in tumor biology, but the study of this area is still in its beginning. High-levels of multiple chemokines are expressed in tumor cells, tumor tissues and transformed cell lines. It has been suggested that chemokines can act as growth factors and may have angiogenic or angiostatic properties. Tumors could also express chemokines in order to weaken immune defense. More recently it has also been shown that chemokines may mediate tumor cell migration and metastasis, suggesting that chemokine receptors expressed on tumor cells are novel targets in antitumor treatment (6,8).

Characteristic for chemokines are two functionally conserved regions that are essential for induction of leukocyte migration, one mediates receptor binding and the other one enables cell surface or extracellular matrix fixation (3). The core structure and COOH-terminal α helix in chemokines are equipped with glycosaminoglycan-binding domains for their immobilization on extracellular matrices and cell surfaces (Figure 3).

Glycosaminoglycans are negatively charged carbohydrate moieties on glycoproteins and are portrayed as essential "chemokine-presenting and - clustering" sites that enable chemokine gradient formation within tissues. Receptor binding and triggering are mediated by distinct epitopes in the NH_2-terminal region of chemokines, which are in opposite position to the glycosaminoglycan-binding sites. Structure-activity studies in chemokines revealed the importance of the NH_2-terminal domain preceding the first Cys residue and the "N-loop" defined by the second and third Cys residues, for receptor interactions and led to a two-step chemokine-binding model (Figure 3). In a first "docking" step, an epitope in the N-loop of the chemokine interacts with the chemokine receptor, which is thought to lead to conformational changes in the receptor (and possibly also in the chemokine). The subsequent "triggering" step involves the binding of the NH_2-terminal region of the chemokine to the (now accessible) receptor-binding pocket. The two binding sites in chemokines are small (consisting of few amino acids) and are kept in close proximity by the two highly conserved disulfide bonds.

Modifications in the NH_2-terminal regions yielded numerous chemokine variants with antagonistic properties, demonstrating the potential therapeutic application (8). Obviously, the therapeutic potential of peptides is limited. Instead, the global effort of the pharmaceutical industry is focused non-peptide inhibitors of low molecular weight, which target chemokine receptors involved in the control of leukocyte recruitment to inflammatory sites or those functioning as HIV-1 co-receptors. Thus far, current drug development programs are concentrated on single chemokine receptors and ongoing clinical trials will tell us more about their strength as novel anti-inflammatory medication. However, given the functional redundancy in the chemokine receptors controlling inflammatory infiltrates, compounds with multiple receptor-blocking activities need to be considered as more efficient anti-inflammatory agents.

Regulatory Environment

There are no special remarks or recommendations, so far, in international guidelines in immunotoxicology with respect to cell traffic or chemokines.

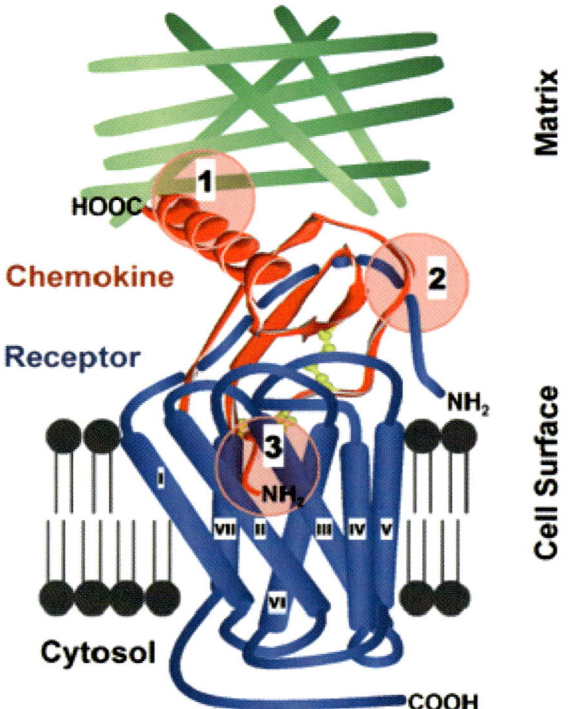

Immune Cells, Recruitment and Localization of. Figure 3 Structural segregation of chemokine anchorage and receptor triggering. (1) Matrix fixation site in the COOH-terminal α-helix or core structure; (2) N-loop region defined by second and third Cys residues; and (3) NH2-terminus, which interacts with the chemokine-binding pocket present in the transmembrane region of the chemokine receptor.

References

1. Murphy PM, Baggiolini M, Charo IF, Hebert CA, Horuk R, Matsushima K et al. (2000) International union of pharmacology. XXII. Nomenclature for chemokine receptors. Pharmacol Rev 52:145–176
2. Murphy PM (2002) International Union of Pharmacology. XXX. Update on chemokine receptor nomenclature. Pharmacol Rev 54:227–229
3. Loetscher P, Moser B, Baggiolini B (2000) Chemokines and their receptors in lymphocyte traffic and HIV infection. Adv Immunol 74:127–180

4. Moser B, Loetscher P (2001) Lymphocyte traffic control by chemokines. Nat Immunol 2:123–128
5. Laudanna C, Kim JY, Constantin G, Butcher E (2002) Rapid leukocyte integrin activation by chemokines. Immunol Rev 186:37–46
6. Murphy PM (2001) Chemokines and the molecular basis of cancer metastasis. N Engl J Med 345:833–835
7. Gerard C, Rollins BJ (2001) Chemokines and disease. Nat Immunol 2:108–115
8. Onuffer JJ, Horuk R (2002) Chemokines, chemokine receptors and small-molecule antagonists: recent developments. Trends Pharmacol Sci 23:459–467

Immune Competence

Denotes mature lymphocytes that are capable of recognizing a specific antigen and mediating an immune response.
▶ Flow Cytometry

Immune Complexes and Complement Activation

Soluble or insoluble antibody–antigen complexes capable of activating classical complement pathway and causing cell injury.
▶ Antibodies, Antigenicity of

Immune Interferon

▶ Interferon-γ

Immune-Mediated Heart Disease

▶ Cardiac Disease, Autoimmune

Immune-Mediated Hemolytic Anemia (IMHA)

Anemia caused by binding of autoantibodies to red blood cells and/or their precursors. The two major diseases are autoimmune-mediated hemolytic anemia (AIHA), in which erythrocyte destruction is mediated by autoantibodies, and neonatal isoerythrolysis (NI), which is caused by uptake of isologous, blood-group-specific autoantibodies.
▶ Antiglobulin (Coombs) Test

▶ Hemolytic Anemia, Autoimmune

Immune Response

MICHAEL U MARTIN
Institute of Immunology
Justus-Liebig University Giessen
Winchesterstrasse 2
D-35394 Giessen
Germany

KLAUS RESCH
Institute of Pharmacology
Hannover Medical School
Carl-Neuberg-Strasse 1
D-30625 Hannover
Germany

Synonyms
Immunity, innate immunity, adaptive immunity, immunological memory.

Definitions
Immune response is the process of recognition of potentially harmful agents by specialized cells of the immune system, initiated by a rapid activation of the innate arm of immunity in a process known as ▶ acute inflammation. Subsequently, the adaptive immune response is triggered in order to provide means of amplifying innate mechanisms and to develop an immunological memory. Harmful agents may be microorganisms such as viruses, bacteria, fungi, protozoa and helminths, and also tumour cells. Both arms of ▶ immunity—the innate and adaptive responses—involve soluble factors (humoral immunity) and immune competent cells (cellular immunity) as summarized in Table 1.

Characteristics
Sentinel Cells Alarm the Immune System and Orchestrate Immediate Responses
Recognition of pathogens is achieved by specialized sentinel cells (comprising dendritic cells, tissue macrophages, and mast cells) which are located at sites of possible pathogen entry like skin and mucosa. Upon contact or infection with pathogens these cells become activated and release endogenous alarm mediators including prostaglandins, leukotrienes, and proinflammatory ▶ cytokines, and—in the case of mast cells—preformed histamine. If possible, pathogens are phagocytosed and killed to avoid expansion and systemic spreading. ▶ Opsonization, phagocytosis, and killing

Immune Response. Table 1 Components of the innate and adaptive immune response

Cells	Molecules	Arm of immunity
Monocytes/macrophages Dendritic cells Mast cells Granulocytes (neutrophilic; eosinophilic; basophilic) Natural killer-cells	Complement factors Defensins Cytokines Enzymes Oxygen radicals	Innate (nonspecific immune response): rapid with preformed factors and existing cells, pathogen recognition receptors in germline configuration
T lymphocytes B lymphocytes	T-cell cytokines Antibodies	Adaptive (specific immune response): delayed, clonal expansion, antigen receptors generated by genetic rearrangement
Cellular immunity	Humoral immunity	

of viruses and bacteria is enhanced by humoral components of ▸ innate immunity such as the complement system, ▸ defensins, and antibacterial enzymes. Cellular components of innate immunity comprise dendritic cells, monocytes and macrophages, mast cells, neutrophilic, eosinophilic, and basophilic granulocytes, as well as natural killer cells.

The proinflammatory products of activated dendritic cells and macrophages recruit and activate neutrophilic granulocytes which preferably mount an antibacterial response, whereas the products of activated mast cells provide an environment enabling attraction and activation of eosinophils aimed at fighting large intruders like helminths. Infiltrating leukocytes are activated at the site of inflammation and support the resident phagocytic cells by effectively removing and/or killing the pathogens (neutrophilic granulocytes and monocytes/macrophages) or by killing (virus-)infected host cells (NK cells). During ongoing inflammation, the mechanisms of innate and ▸ adaptive immunity closely interact. Thus effector mechanisms of innate immunity are greatly enforced by cytokines like interferon-γ (which potentiates macrophage function) or by antibodies (which enhance the effect of complement factors and greatly increase ▸ phagocytosis).

Professional Phagocytes are Antigen-Presenting Cells which Link Innate and Adaptive Immune Responses

After having phagocytosed pathogens, or after having been infected by viruses, professional antigen-presenting cells (APC) like dendritic cells or macrophages leave the site of inflammation in the tissue and migrate via the lymphatics to the nearest draining lymph node, where they present peptides on major histocompatibility (MHC) antigens. MHC class II molecules sample endosomal compartments in which phagocytosed particles are digested into peptides; whereas viral proteins which are synthesized in the cytoplasm of the infected cells are presented on MHC class I molecules.

In the protected environment of the lymph node, APC express costimulatory molecules and produce cytokines which provide optimal conditions to stimulate naive T cells if their T cell antigen receptor recognizes the presented peptide in the MHC molecule and an ▸ immunological synapse can be formed. Thus professional APC actively link the innate immune response to adaptive immunity.

Natural killer cells recognize cells that are devoid of MHC class I molecules, which are normally present on all nucleated cells and may by downregulated due to viral infection or neoplastic transformation (effector functions of leukocytes are summarized in Table 2).

T and B Lymphocytes Mount the Specific Immune Response

Each individual T or B lymphocyte expresses an antigen receptor with a defined specificity. Antigen receptors are generated randomly by sequential genetic rearrangement of the gene segments. In T cells these genes code for the β- and α-chain of the T cell antigen receptor. The antigen receptor of B cells is a membrane-inserted immunoglobulin (antibody) and consists of heavy and light chains. Genetic rearrangement generates millions of individual mature naive T cells and B cells, each with a discrete antigen specificity, thus creating a huge repertoire capable of recognizing all potential pathogens. Upon encountering its specific antigen one lymphocyte proliferates and expands to become an effector cell (clonal selection). After antigen recognition T and B cells require some time to proliferate and generate a pool of antigen-specific progeny or effector cells thus explaining the delay of the adaptive immune response.

The expansion of antigen-specific lymphocytes is regulated by T helper lymphocytes, which recognize

Immune Response. Table 2 Leukocytes and their characteristic features

Cells	Characteristics
Monocytes/macrophages	Sentinel cells Excellent phagocytes Produces of inflammatory cytokines Antigen presentation
Dendritic cells	Sentinel cells Phagocytes Produces of inflammatory cytokines Excellent antigen presentation after activation
Mast cells	Sentinel cells in tissue and mucosa Mount an IgE-mediated anti-parasite response Initiate acute inflammation

Granulocytes
  Neutrophilic
  Eosinophilic
  Basophilic
	Effector cells of innate immunity
	  Excellent phagocytes and killers of bacteria and fungi
	  Specialists in fighting larger pathogens, e.g. helminths
	  Producers of IL-4

Natural killer-cells	Specialists in recognizing and killing cells with atypically low or missing MHC class I expression after virus infection or neoplastic transformation

T lymphocytes
  Th0
  Th1
  Th2
  Treg
  Tcytotox
	Thymus-derived T cell antigen receptor positive lymphocytes
	  Naive cells
	  Helper cells producing IL-2 and interferon-γ cellular immunity
	  Helper cells producing IL-4 and IL-5, humoral immunity
	  Regulatory T cells producing mainly IL-10 and transforming growth factor-β
	  Kill virus-infected cells

B lymphocytes	Bone marrow-derived lymphocytes producing antibodies

their specific antigen when it is presented on MHC class II molecules. T helper cells derive from a Th0 cell and may differentiate either to Th1 cells, which provide the central T cell growth factor, interleukin-2. Th1 cells also secrete interferon-γ which represents the strongest activator of macrophages. Alternatively, Th0 cells become Th2 cells, which promote B cell responses and antibody production by releasing a distinct set of cytokines including interleukins IL-4, IL-5, IL-6, IL-10, IL-13. Distorting the subtle balance of Th1 versus Th2 lymphocytes in immune responses may result in chronic inflammation (Th1 bias) or allergic diseases (Th2 bias).

In most cases B lymphocyte activation and clonal expansion rely on T cell help. Upon binding of an antigen to its immunoglobulin receptor a B cell becomes activated and starts to proliferate and differentiate into plasma cells, if the appropriate cytokines and costimulatory molecules are provided by T helper cells. Plasma cells secrete antibodies with exactly the specificity of the B cell antigen receptor of the B lymphocyte originally activated by the pathogenic structure. Initially, pentameric immunoglobulin M is secreted, but upon continued or repeated antigen binding the B lymphocyte switches to other isotypes such as immunoglobulins IgG, IgA, or IgE, always preserving the original antigen specificity. The different antibody classes fulfill different effector functions.

Generation of Immunological Memory

Innate immune cells efficiently combat pathogens and are usually killed in this process by their own aggressive antimicrobial effector molecules. Thus, they are lost during the immune response and cannot contribute to ▶ immunological memory. Lymphocytes, however, can develop an immunological memory which allows a quicker and more effective immune response upon rechallenge with an antigen which had been dealt with before. Immunological memory which may last for life consists of T memory cells, B memory cells, and antibodies. The usefulness of immunological memory for protection of reinfection is impressively demonstrated by the success of vaccination.

Preclinical Relevance

As an effective immune response is of vital importance to protect from infections, testing of influences on the immune response is a standard procedure in preclinical protocols of drug or pesticide development. These test batteries recommended by the authorities follow a tiered approach—thus the first step is identification of a possible hazard. After a positive hazard identification (flagging) additional tests have to be performed to gain insight into the mechanism for the interaction of the test compound with the immune system. These so-called tier II tests performed on a case-by-case basis are considered to be reasonable data for a risk assessment. The immunological capacity of immune cells may also be tested in vitro by whole blood assays in which cytokine synthesis or surface expression of immune-relevant molecules may be ascertained.

Relevance to Humans

Deviations of the immune system from a physiological, regulated response are associated with disease. Many congenital immunodeficiency syndromes occurring with low frequency are known, ranging from rather mild clinically inapparent defects, to conditions where a life-long struggle with infections exists, or even to conditions whereby there is nearly a complete absence of specific lymphocytes or nonspecific leucocytes, which eventually leads to death in early childhood.

Immunodeficiencies are also serious sequelae of cytotoxic therapies, for example against tumors. Overreaction of the immune system also causes disease, such as allergies, where symptoms include mild affection such as itching and rhinitis ("hay fever") but also chronic illnesses such as asthma, or acute life-threatening conditions such as anaphylactic shock (see Allergy). The immune system may also react against the normally hidden structures of the individual itself. Autoimmune disorders of different severities result, and these are also the cause of several chronic inflammatory diseases, such as rheumatoid arthritis.

Regulatory Environment

Drugs that interfere with the immune system, or components of the immune system that are designed as therapeutic drugs, are regulated by specific laws—in Germany the "Arzneimittelgesetz." Government agencies approve drugs for specific indications: in Germany for drugs in general it is the Bundesinstitut für Pharmaka und Medizinprodukte, and for biologicals such as antibodies and cytokines it is the Paul-Ehrlich-Institut. Similar regulations exist in other European countries, in the United States, and in Japan.

Besides these special guidelines there are general guidelines for immunotoxicologic screening of drugs and pesticides that cover different aspects of the possible influence of chemicals on the immune response. Such guidelines are mentioned under the appropriate entries.

▶ Immunoassays

References

1. Janeway CA, Travers P, Walport M, Shlomchik M (2001) Immunobiology, 5th ed. Churchill Livingstone, St Louis
2. Abbas AK, Lichtmann AH, Pober JS (1997) Cellular and Molecular Immunology, 4th ed. WB Saunders, Philadelphia
3. Gemsa D, Kalden JR, Resch K (1997) Immunologie. Georg Thieme Verlag, Stuttgart

Immune Response to Cancer

▶ Tumor, Immune Response to

Immune Surveillance

A hypothesis (sometimes called theory) originally proposed in 1959 by Thomas and then formalized by Burnet, whereby the immune system constantly operates to eliminate arising tumors. Tumors that grow and progress should be poorly immunogenic to escape immune surveillance. This term is, hoverer, often used in a more general sense. In these cases it refers to immune protection of peripheral tissues (e.g. skin) by a special subset of memory T cells. Skin-selective immune surveillance T cells are present in normal healthy skin. They are produced during a previous immune response against skin-associated antigens, are long-lived, and provide immediate (vaccination) responses during pathogen re-exposure.

▶ Tumor, Immune Response to
▶ Immune Cells, Recruitment and Localization of

Immunity

Immunity describes the ability of higher organisms to deal with pathogens by developing an appropriate protective response using soluble factors and specialized effector cell types (leukocytes).

▶ Immune Response

Immunization

▶ Attenuated Organisms as Vaccines

Immunoabsorbent

Defined antigen or antigenic determinant bound to inert carrier material (beads) such as agarose, sepharose, or latex etc. May be used to specifically absorb and isolate antibodies or cells of defined specificity from heterogenous mixtures.
▶ Hapten and Carrier

Immunoassay

Bioanalytical methods based on the formation of specific antigen-antibody complexes used to detect or quantify an analyte of interest (e.g., cytokine) captured in the complex. Either monoclonal or polyclonal antibodies can be used as reagents.
▶ Cytokine Assays

Immunoassays

Danuta J Herzyk
Immunologic Toxicology, Preclinical Safety Assessment
GlaxoSmithKline R&D
709 Swedeland Road
King of Prussia, PA 19406-0939
USA

Synonyms

antigen-antibody binding assay, immunochemical-based assays

Definition

Immunoassays (IA) are methods applied in analyses of proteins (also known as macromolecules). Proteins can act as ▶ antigens—substances that effectively elicit an ▶ immune response when introduced to an animal or human in vivo. The immune response to an antigen results in the production of a specific ▶ antibody—an immunoglobulin that distinctively binds its antigen. Immunoassays as bioanalytical methods are based on formation of the specific antigen-antibody complex for detecting the presence and/or quantitation of the analyte (either antigen or antibody) captured in the complex. Immunoassays require generation of specific reagents:(a) an antigen of interest and (b) an antibody against this antigen.

Characteristics

Immunoassays can be developed in various formats.

The principle of any format is the same and utilizes the binding between antigen and antibody. Two mechanisms have been postulated for antigen and antibody interactions:
- the "lock and key" model (1), in which there is no change of the structure of the antibody when it binds the antigen
- the "induced fit" model (2), in which both antigen and antibody change their confirmation upon complex formation.

Antibodies are Y-shaped modular glycoproteins made up of four polypeptide chains linked by disulfide bridges. The two heavy chains (H) are identical to each other, as are the two light chains (L). Both L and H chains consist of a number of subunits or domains, some of which are similar (constant domains, C) whereas others are variable (variable domains, V). The variable domains on light and heavy chains (V_L and V_H) on each arm of the Y-shaped immunoglobulin are clustered in space to form the antigen-combining site that contain hypervariable loops called complementarity-determining regions (CDRs). The CDRs determine the specific chemistry, nature and shape of the antigenic determinant, that is the contact surface or epitope on the antigen that binds to this site.

In order to be detectable, the formed antigen-antibody complexes need to be immobilized on a solid surface, either a plastic plate or bead with physical properties enabling covalent binding of the proteins. The signals generated as a consequence of an antibody-antigen reaction can be registered by adding a "label" attached to either the antigen or antibody (analyte) or a secondary reagent (e.g. another antibody) serving as a carrier of the label. Based on different label systems (e.g. radioisotopic, fluorescent, enzymatic, electroluminescent), immunoassays can be divided into four categories:
- radioimmunoassays (RIA)
- fluoroimmunoassays (FIA)
- ▶ enzyme-linked immunosorbent assays (ELISA)
- electrochemoluminescent immunoassays (ECLIA).

The application of RIA, an early method, is most often based on the use of ^{125}I-labeled antigens to generate counts per minute of the radiolabeled antigen-antibody complexes. In recent years RIAs are being used less often than the other immunoassay techniques. Both fluorescence-based and enzyme-based immunoassays generally utilize the well-characterized high-affinity binding between biotin and avidin for detection of complexed proteins. FIAs are based on application of lanthanide chelate, europium diethylenetriaminopentaacetate (Eu^{3+}–DTPA) as a fluorescent label. FIAs are referred to as dissociation enhanced lanthanide fluoroimmunoassays (DELFIA) or time-resolved fluoroimmunassays (TRFIA).

In these methods, the response is generated by a fluorescent agent (i.e. europium) conjugated to streptavidin reacted with the bound biotin antibody-antigen complex. In ELISA methods the chromogenic response, measured by a spectrophotometer, is generated by enzyme substrates (e.g. *para*-nitrophenyl phosphate or *ortho*-phenylenediamine) added to enzyme (alkaline phosphatase or horseradish peroxidase, respectively) which is conjugated with secondary antibodies or streptavidin-biotin that binds the complexed analyte (see Figure 1). The ECLIA technology is based on a light-generating chemical reaction during binding between a ruthenium tribispyridine (TAG)-labeled analyte captured in an antigen-antibody-bead complex, and tripropylamine (TPA) that luminesces upon reduction by an applied voltage.

In addition to immunochemical-based assays, new biophysical methods such as surface plasmon resonance (SPR) technology have emerged. The SPR method is a real-time biospecific analysis for detecting binding of ligands, such as antibodies, to immobilized antigens on a sensor chip surface. In SPR-based assays the signals (resonance units) are generated when binding of a molecule to the biosensor surface causes a change in refractive index (i.e. light reflection from a conducting film at the interface between two media). This technology has the advantage of direct detection of binding interactions without adding secondary label-conjugated reagents.

In all immunoassays, antigen-antibody reactions generate signals proportional (directly or inversely) to the amount of the complexed analyte. Immunoassays are capable of quantifying the analyte of interest. The concentration of analyte in a test sample can be determined by interpolation from a calibration (standard) curve of the control analyte (e.g. purified antigen or antibody of known concentration).

ELISA is the most frequently employed immunoassay because commercial reagents and ELISA kits are available for determination of a variety of proteins. Also, the method requires relatively common and inexpensive laboratory instruments (i.e. spectrophotometer) to conduct signal measurements. Alternatively, methods based on FIA, ECLIA and SPR, while requiring more sophisticated instruments, may offer higher throughput of testing.

Development of immunoassays of a particular format may be dictated by the availability of reagents and instruments in a given laboratory. Critical factors for the validation of any immunoassay method include the demonstration of the assay sensitivity, specificity, reproducibility, controls for positive and negative responses, and potential presence of interfering substances. For quantitative immunoassays, determination of dynamic ranges (linear response between the lowest and the highest signal) and the lower and upper limits of quantification (LLOQ and ULOQ, respectively) should also be established (3).

Immunoassays are designed to detect and measure multiple biologic molecules, most notably, cytokines, growth factors, and hormones as well as antibodies to cytokines, growth factors, and hormones, in biological fluids (plasma, serum, tissue extracts, urine, cultured cell supernatants). These molecules can either be natural or derived by recombinant DNA technologies.

Preclinical Relevance

Immunoassays are broadly used in research labora-

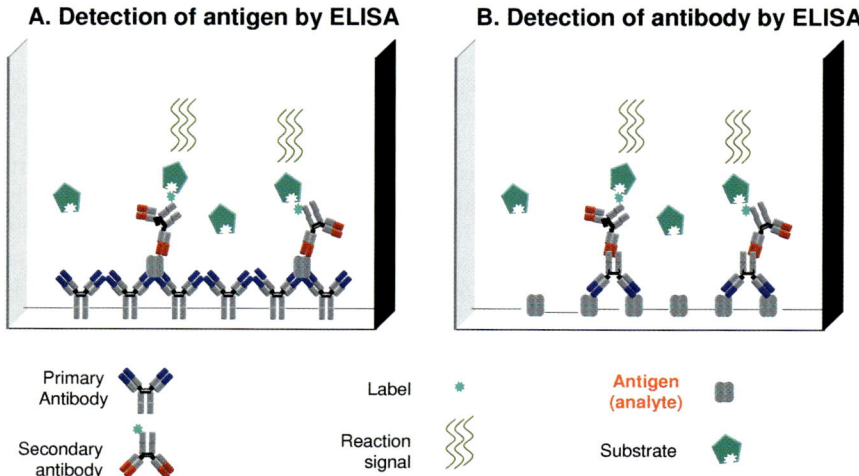

Immunoassays. Figure 1 Diagram of antigen-antibody reactions in immunoassays.

tories in many scientific fields: biochemical, medical, pharmaceutical, and environmental.

During drug development in the pharmaceutical industry, immunoassays are frequently used in bioanalysis of protein molecules and in preclinical animal evaluation of biotechnology-derived drugs. Such applications require development of novel and unique (drug-specific) immunoassay methods and their validation (4).

One of the immunoassays, ELISA, is frequently used in the preclinical testing of chemicals for immunotoxicity potential in the environmental, agricultural and pharmaceutical industries. In immunotoxicology evaluations, antibodies to a T cell-dependent antigen, e.g. sheep red blood cells (SRBC) or keyhole limpet hemocyanin (KLH) are measured by ELISA in serum samples of mice or rats exposed to the antigen.

Relevance to Humans

Immunoassays are frequently applied in diagnostic and therapeutic medicine. They are used for quantitation of biomarker molecules (e.g. thyroid hormones, insulin, prolactin, cardiac enzymes, viral serology, tumor antigens) to indicate disease progression or regression, and antibodies elicited in response to treatment with therapeutic proteins (biopharmaceuticals) to monitor safety. These applications require development of multiple immunoassay methods, validated and accepted for clinical use. Several manufacturers commercially develop diagnostic immunoassays (mainly ELISA kits) for clinical use. In biopharmaceutical drug development, drug-specific immunoassays for use in clinical trials have to be designed, developed and validated for each individual molecule (drug) of interest (5).

Regulatory Environment

Available guidance documents addressing the use and validation of immunoassays are very limited. Some aspects of analytical method validation have been discussed at the International Conference on Harmonization (FDA Federal Register 62 (1997) 27464–27467). More recently, a draft Guidance for Industry on Bioanalytical Methods Validation for Human Studies publication has become available from the Food and Drug Administration (FDA 64 (1999) 517).

References

1. Amit AG, Mariuzza RA, Phillips SEV, Poljak RJ (1986) Three-dimensional structure of an antigen-antibody complex at 2.8 A resolution. Science 233:747–753
2. Rini JM, Schulze-Gahmen U, Wilson IA (1992) Structural evidence for induced fit as a mechanism for antibody-antigen recognition. Science 255:959–965
3. Findlay JWA, Smith WC, Lee JW et al. (2000) Validation of immunoassays for bioanalysis: a pharmaceutical industry perspective. J Pharmaceut Biomed Anal 21:1249–1273
4. Shah VP, Midha KK, Dighe S et al. (1992) Analytical methods validation: Bioavailability, bioequivalence and pharmacokinetic studies. Pharmaceut Res 9:588–592
5. Koren E, Zuckerman LA, Mire-Sluis AR (2002) Immune response to therapeutic proteins in humans—Clinical significance, assessment and prediction. Curr Pharmaceut Biotechnol 3:349–360

Immunoblotting

Immunoblotting is a synonym for western blot analysis.

▶ Western Blot Analysis

Immunochemical-Based Assays

▶ Immunoassays

Immunocompetence

Immunocompetence refers to the capacity of the immune system to mount a specific immune.

▶ Neonatal Immune Response

Immunocompetent

Immunocompetent or immunocompetence refers to the ability of an organism generate a normal (robust) immune response or its individual cells to react normally in response to an immune stimulus. The antonym of this term is immunocompromised.

▶ Developmental Immunotoxicology

Immunodeficiency

A defect or absence of some component of the immune system that may result in inability of the host to eliminate or neutralize foreign substances.

▶ Respiratory Infections
▶ Complement Deficiencies

Immunodeficient Animal

An animal with an impaired immune system.
▶ Animal Models of Immunodeficiency

Immunogenicity

Synonym for antigenicity. The ability of a substance to be immunogenic, that is, to induce an immune response in a host. Immunogenically active compounds tend to be high-molecular-weight proteins that are recognized as "non-self" or are chemicals that are covalently bound to protein carriers as "haptens". The extent of immunogenicity of a molecule is a function of its size, its source, and the route of application. Recombinant proteins or monoclonal antibodies produced in animals or animal cells have a high immunogenicity when applied to human beings and may cause the generation of neutralizing antibodies.
▶ Therapeutic Cytokines, Immunotoxicological Evaluation of
▶ Cytokine Receptors
▶ Respiratory Infections
▶ Monoclonal Antibodies

Immunoglobulin

Plasma proteins produced by activated B lymphocytes that bind specifically within their hypervariable regions to antigenic "non-self" epitopes that elicited their production. Their constant domains bind to receptors (Fc receptors) on leukocytes to mediate opsonization and other effector functions. T lymphocyte 'help' is required for a T-dependent antigen.
▶ Opsonization and Phagocytosis
▶ Canine Immune System
▶ Antibodies, Antigenicity of

Immunoglobulin Class Switching

This is a process whereby the constant region (isotype) of an antibody is changed without alterations to its variable region, hence allowing the effector response to change while retaining the same antigen-binding specificity. It involves rearrangement of the heavy chain immunoglobulin genes (in the B lymphocyte) by DNA recombination. In some cases RNA processing is also involved. Some cytokines act as "switch factors", inducing immunoglobulin class switch (e.g. interluekin IL-4 induces switch from IgM to IgG1 and IgE, whereas interferon (IFN) γ induces switch to IgG2a in the mouse).
▶ Interferon-γ
▶ B Cell Maturation and Immunological Memory

Immunoglobulin. Subclasses and Functions

DAVID SHEPHERD
Center for Environmental Health Sciences
Department of Pharmaceutical Sciences, University of Montana
32 Campus Drive, 154 Skaggs Building
Missoula, MT 59812
USA

Synonyms

Immunoglobulin (Ig), Immunoglobulin A (IgA), Immunoglobulin D (IgD), Immunoglobulin E (IgE), Immunoglobulin G (IgG), Immunoglobulin M (IgM), antibody, B cell antigen receptor

Definition

Immunoglobulins (Ig) are a family of proteins produced exclusively by B lymphocytes that mediate humoral immune responses. Human Ig are comprised of nine subclasses that have varying effector functions in the adaptive immune response to specific antigens.

Characteristics

Humoral immunity is mediated by a variety of B lymphocyte-derived immunoglobulins that invoke multiple effector functions. Based on the antigenic uniqueness of their heavy chains, five human immunoglobulin (Ig) classes or isotypes have been identified and include IgA, IgD, IgE, IgG, and IgM. In addition, IgA and IgG can be further subdivided into closely related subclasses, IgA1 and IgA2, and IgG1, IgG2, IgG3, and IgG4, respectively. Some immunoglobulins are secreted into the plasma and are called antibodies whereas others are membrane-bound and function as the specific antigen receptor on B cells. Many of the physical characteristics of the different immunoglobulins are summarized in Table 1. One common feature of all immunoglobulins is that they are protein structures that share a common core consisting of two identical light chains and two identical heavy chains (1). The generation of each immunoglobulin subclass is influenced by B cell development, activation and differentiation. During the maturation of naïve B lymphocytes, IgM is the first isotype expressed following

Immunoglobulin. Subclasses and Functions. Table 1 Physical characteristics of human immunoglobulins

	Immunoglobulin subclass								
	IgM	IgD	IgG1	IgG2	IgG3	IgG4	IgA1	IgA2	IgE
Molecular weight (kDa)	970	184	146	146	165	146	160	160	188
Serum level (mean adult mg/mL)	1.5	0.03	9	3	1	0.5	3	0.5	5×10^5
Serum half-life (days)	10	3	21	20	7	21	6	6	2

successful germline rearrangement of the immunoglobulin genes in immature B cells and serves as the primary antigen specific receptor. On the surface of almost all mature B cells, IgM is co-expressed with IgD, although the function of this specific isotype is unknown. Following B cell activation via antigen receptor cross-linking or mitogenic stimulation, copious amounts of IgM are produced and secreted into the plasma. Furthermore, upon receipt of differentiation signals from cognate (activated T helper cells) or soluble (cytokine) sources, B cells may be induced to undergo immunoglobulin class switching and produce "downstream" isotypes such as IgG, IgA, or IgE (2). Immunoglobulins of all isotypes can be produced either in the secreted form or as a membrane-bound receptor. Ultimately, however, the determination of the secreted versus membrane-bound form of Ig is achieved by alternative RNA splicing and influenced by activation signals (1).

The isotype of the immunoglobulin which is determined by the constant region of the heavy chain (C_H) confers functional specialization. Immunoglobulins of different isotypes have distinct properties for recruiting the help of other immune cells and molecules to destroy and dispose of pathogens to which the antibody has bound. The constant (Fc, or Fragment crystallizable) regions of antibodies have three main effector functions which include (1) recognition by specific surface receptors on phagocytic cells, (2) binding to complement and initiating the complement cascade, and (3) facilitating the delivery of antibodies to normally unreachable tissues and bodily fluids via active transport mechanisms (2). Not all immunoglobulin isotypes have the same capacity to enlist each of the three effector functions. The differential capabilities of each isotype are summarized in Table 2 and also briefly described below.

IgA is secreted primarily by mucosal lymphoid tissues and functions to neutralize injurious agents in mucosal secretions. In a normal individual, more IgA is synthesized than any other isotype but comprises only a small fraction of the overall Ig in the plasma because it is transported so efficiently into the mucosal lumen (2). IgA is characterized by α heavy chains and is most commonly found as a homodimer.

IgD exists primarily as surface immunoglobulin on mature, naïve B cells but its function is unknown. It may play a role in the control of B cell activation and the development of immunological memory, or may be simply an evolutionary relic (1). IgD is characterized by the expression of δ heavy chains.

IgE is the immunoglobulin isotype that is involved in allergic reactions via the sensitization of mast cells that are distributed in the skin and mucosa, and along blood vessels in connective tissue. Activation of mast cells occurs following antigen binding of IgE and results in the release of powerful chemical mediators that induce coughing, sneezing, and vomiting. IgE is found at only very low concentrations in the blood and extracellular fluid (1). It is characterized by ε heavy chains and is secreted exclusively as a monomer.

IgG is the most abundant class of immunoglobulin found in the plasma and extracellular spaces of the internal tissues. The primary function of IgG is to eliminate or inactivate invading pathogens and their products and this is accomplished by binding to complement and/or specific receptors (Fcγ receptors) on various phagocytic cells (2). IgG is transferred across the placenta to provide protection to the human infant in neonatal life. This class of immunoglobulin is characterized by γ heavy chains and is monomeric in nature.

IgM is the first class of immunoglobulin to appear on the surface of B cells and the first to be secreted during the primary humoral response because it can be expressed without isotype switching. It acts as the initial B cell antigen receptor in a monomeric, membrane-bound form and subsequently can be secreted in a pentameric form by activated B cells (2). IgM is characterized by μ heavy chains and the secreted form is very effective in the activation of the complement system.

Preclinical Relevance

Immunoglobulin levels can be affected by a number of drugs and environmental chemicals. Although chemi-

Immunoglobulin. Subclasses and Functions. Table 2 Functional characteristics of human immunoglobulins

	Immunoglobulin subclass								
	IgM	IgD	IgG1	IgG2	IgG3	IgG4	IgA1	IgA2	IgE
Binding to macrophages and other phagocytes	–	–	+	–	+	–/+	+	+	+
High-affinity binding to mast cells and basophils	–	–	–	–	–	–	–	–	+++
Reactivity with staphylococcal Protein A	–	–	+	+	–/+	+	–	–	–
Activation of classical complement pathway	+++	–	++	+	+++	–	–	–	–
Activation of alternative complement pathway	–	–	–	–	–	–	+	–	–
Placental transfer	–	–	+++	+	++	–/+	–	–	–

cally-induced changes in the profile of the various subclasses of immunoglobulins have been demonstrated to possess a high concordance with altered host resistance, it remains to be determined if these alterations in immune function are physiologically relevant (3). The investigation of serum Ig levels and shifts in the relative composition of Ig subclasses following exposure to drugs and environmental chemicals is regulated by special guidelines (3). Evaluation of serum Ig levels and the ability of PFC to secrete IgM and IgG are commonly employed as an initial screening test for use in non-clinical, repeated dose immunotoxicity studies.

Relevance to Humans

Because of the various effector functions that are ascribed to the different isotypes of immunoglobulin, loss or deficiency of a particular Ig subclass may have significance in human health. The importance of atypical Ig production is best exemplified in individuals with inherited immunodeficiencies but also may apply to people exposed to certain drugs or environmental chemicals that modulate Ig production. For example, individuals who suffer from X-linked agammaglobulinemia possess defects preventing normal B lymphocyte development and resulting in an inability to generate any immunoglobulin subclass (1). As a result, infants with this disease are highly susceptible to developing recurrent infections by pyogenic bacteria such as Streptococcus pneumoniae. Another example of an inherited immunoglobulin-related immunodeficiency is selective IgA deficiency. Although no obvious disease susceptibility is associated with selective IgA defects, a lack of IgA may predispose individuals to lung infections by various pathogens since chronic lung disease is more common in these individuals than in the general population (1). This disease may also be related to another syndrome called common variable immunodeficiency that involves a deficiency of both IgA and IgG. The predominant manifestation of this disease is hypogammaglobulinemia and it is characterized by recurrent bacterial infections, especially of the upper and lower respiratory airways, and is also associated with an increased incidence of autoimmune and neoplastic disorders (4).

Although several clinical immunoglobulin immunodeficiencies have been characterized, there exists controversy relating to the overall significance of these diseases, in part due to the well-established redundancy within the immune system. For example, it is questionable if a low serum concentration of one or more of the IgG subclasses represents a disease state. As in common variable immunodeficiency, low concentrations of IgG1 can be found in primary and secondary immunodefiencies but not in isolation (5). Low IgG2 concentration in some individuals but not others is associated with an increased risk of bacterial infections (5). Deficiencies in IgG3 and IgG4 have not been convincingly demonstrated to date (5). Thus, identification of deficiencies in one or more IgG subclasses does not convincingly identify individuals at risk (6). However, low concentrations of antibodies to specific bacterial antigens such as tetanus, diptheria, and pneumococcus does identify individuals at risk and has been recommended as the preferred immunoglobulin measurement (5).

On the other hand, the overproduction of immunoglobulins has been well-characterized and can be strongly linked to disease. The primary example for excessive Ig production leading to increased morbidity is found in atopic individuals that generate abnormally high levels of allergen-specific IgE. Increased secretion of this class of immunoglobulin to a wide variety of common environmental allergens contributes to the development of allergic diseases such as hay fever and asthma. It has been reported that as many as 40% of people in Western populations show an exaggerated tendency to mount unwanted allergic IgE responses (1). Moreover, a common strategy to clinically treat atopic individuals is to inhibit their production of IgE via desensitization (ultimately shifting the antibody response from allergy-inducing IgE to lesser reactive IgG).

In addition to the overproduction of IgE, aberrant production of IgG can also lead to disease. This is illustrated in the generation of hypersensitivity responses mediated by IgG that specifically binds to drugs or innocuous soluble antigens. For example, antibody-mediated destruction of erythrocytes (hemolytic anemia) or platelets (thrombocytopenia) is an unwanted adverse effect often associated with the intake of drugs such as penicillin, quinidine, or methyldopa (1). These are examples of type II hypersensitivity responses. In the case of a type III hypersensitivity response, however, pathology can arise from immune complex formation following the binding of IgG to soluble antigen. This form of Ig-mediated disease is induced in serum sickness and farmer's lung (1). Also, the generation of antibodies that recognize self-antigens can contribute to the generation of autoimmune disease. Although autoantibodies (usually of IgM or IgG isotype) can invoke similar manifestations of disease as seen in type II or III hypersensitivity responses, they are usually specific for a diverse collection of self-antigens such as blood group proteins, DNA, histones, pancreatic β-cell antigen, synovial joint antigen, or myelin basic protein instead of innocuous antigens (1). Thus, the overproduction of specific immunoglobulin subclasses can often be associated with detrimental outcomes with regard to human health.

Regulatory Environment

Immunoglobulin production, especially IgM and IgG, is a standard readout in the evaluation of the immunotoxic potential of new drugs or chemicals. Tier testing as prescribed by the National Toxicology Program includes this measurement along with testing of other non-functional and functional parameters of the immune system (3). In addition, both the U.S. Food and Drug Administration and the European Medicines Evaluation Agency recommend the analysis of serum immunoglobulin levels as well as IgM and IgG production in the PFC assay as part of routine pre-clinical evaluations on all new chemicals (7).

References

1. Janeway CA, Travers P, Walport M, Shlomchik M (2001) Immunobiology: The immune system in health and disease (5th ed). The distribution and functions of immunoglobulin isotypes. Garland Publishing, New York, pp 82–85
2. Abbas AK, Lichtman AH, Pober JS (1994) Cellular and molecular immunology (2nd ed). Effector functions of antibodies. WB Saunders, Philadelphia, pp 51–55
3. Burns-Naas LA, Meade BJ, Munson AE (2001) Casarett and Doull's Toxicology: The basic science of poisons (6th ed). Toxic responses of the immune system. McGraw-Hill, New York, pp 419–470
4. Di Renzo M, Pasqui AL, Auteri A (2004) Common variable immunodeficiency: a review. Clin Exp Med 3:211–217
5. Maguire GA, Kumararatne DS, Joyce HJ (2002) Are there any clinical indications for measuring IgG subclasses? Ann Clin Biochem 39:374–377
6. Jefferis R, Kumararatne DS (1990) Selective IgG subclass deficiency: quantification and clinical relevance. Clin Exp Immunol 81:357–367
7. Gore ER, Gower J, Kurali E, Sui JL, Bynum J, Ennulat D, Herzyk DJ (2004) Primary antibody response to keyhole limpet hemocyanin in rat as a model for immunotoxicity evaluation. Toxicology 197:23–35

Immunohistochemical Staining

A technique which utilizes monoclonal antibodies to detect antigens on tissue sections allowing, for instance, the identification of lymphocyte subtypes.

▶ Canine Immune System

Immunological Memory

The immune system reacts more quickly and more efficiently to a challenge with a pathogen it has encountered previously due to the generation of memory T cells and B lymphocytes. These memory cells reside for a long time in the body. Upon re-entry of the specific pathogen these memory cells can be activated rapidly and mount a fast pathogen-specific cellular and humoral response. Circulating antibodies provide a long-lasting protection against pathogens.

▶ Immune Response

Immunological Synapse

The cellular contact site which is formed when specific lymphocytes interact with antigen-bearing target cells (such as antigen presenting cells) is—in analogy to the synapses of neurons—termed an immunological synapse. It is formed on the side of the lymphocyte by the antigen receptor, accessory coreceptors, adhesion

molecules, and components which initiate signal transduction.
▶ Immune Response

Immunological Unresponsiveness

▶ Tolerance

Immunomodulation

▶ Immunotoxicology

Immunonutrition

▶ Nutrition and the Immune System

Immunopathology

▶ Histopathology of the Immune System, Enhanced

Immunopharmacology

A branch of pharmacology that is concerned with the application of immunological techniques and theory to the study of the effects of drugs and other experimental substances on the immune system.
▶ Therapeutic Cytokines, Immunotoxicological Evaluation of
▶ Immunotoxicology

Immunophenotyping

Immunophenotyping refers to the process of identification of specific cell types using tagged antibodies to specific cell-surface antigens. It may be used as a clinical diagnostic tool or in basic research. The tags used are often fluorescent materials that can be identified by standard laboratory instrumentation such as a flow cytometer.
▶ Leukemia
▶ Flow Cytometry

Immunopoiesis

Synthesis and maturation of immune-competent cells such as plasma (B) cells and mature T cells from pre-B and pre-T cells.
▶ Fish Immune System

Immunopotentiation

▶ Immunotoxicology

Immunoreceptor Tyrosine-Based Activation Motif

An immunoreceptor tyrosine-based activation motif (ITAM) is a protein motif found in some signaling molecules. This motif is composed of two tyrosine (Y) residues separated by 9–12 amino acids. The canonical sequence is YXX[L/V]X_{6-9}YXX[L/V] where X is any amino acid, L is leucine, and V is valine. Phosphorylation of the Y (tyrosine) residues recruits other proteins (with SH2 domains) to the signaling complex.
▶ Signal Transduction During Lymphocyte Activation

Immunosenescence

▶ Aging and the Immune System

Immunosuppression

Immunosuppression is an operational term which describes the fact, that due to circumstances the immune system is less efficient than normally. Immune responses may start later, they might be weaker, or less efficient. Immunosuppression might be general, or restricted to certain pathogens. Whether caused intentionally by pharmacotherapy or unintentionally by environmental agents or genetic influences, immunosuppressed individuals are more susceptible to infections and the development of spontaneous cancers. Epidemiologically, immunosuppression can be measured by comparing the average healthy population with a particular subgroup, e.g. workers exposed to a toxic substance. However, no easily accessible and universally accepted markers for immunosuppression have been identified in vivo. Loss of immune cells (unless very large scale like in AIDS), or a shift in proportion of cells in the blood or in lymphoid organs is a difficult parameter, and has been proven of limited value in humans. Even if a shift in cell subpopulations is observed, its medical significance or biological relevance, and any causal relationship are largely unknown.

▶ Dioxins and the Immune System
▶ Leukemia
▶ Lymphoma
▶ Immunotoxicology

Immunosuppressive

This is a descriptive term for any agent that gives rise to alterations which reduce the optimal immune response of an exposed host.

▶ Vanadium and the Immune System

Immunoteratology

Teratology is the study of birth defects. Immunoteratology is the study the relation of the immune system to increased or decreased rates of birth defects.

▶ Birth Defects, Immune Protection Against

Immunotox

▶ Immunotoxicology

Immunotoxic Agents into the Body, Entry of

GEORG KRAAL · JANNEKE N SAMSOM
Department of Molecular Cell Biology
Vrije Universiteit Medical Center
P.O. Box 1057
1007 MB Amsterdam
The Netherlands

Synonyms

Definition
Entry of immunotoxic agents into the body is a result of breaching the natural barriers of skin and mucosal epithelia. In addition to being physical barriers, epithelia have evolved specific adaptations to prevent uptake of material or to remove or present material to the immune system via specialized cell types.

Characteristics
The epithelia of the skin and the mucus membranes of the respiratory, gastrointestinal and urogenital tract represent an extensive surface area. They are constantly exposed to a great variety of micro-organisms and antigenic and toxic substances. The skin constitutes a strong mechanical barrier through its keratin layer and through secreted gland products, in particular sebum. In the mucus membranes, which are only covered by a thin specialized epithelium, other mechanisms such as ciliary function and nonspecific and antimicrobial factors (e.g. mucin and lysozyme) are involved in preventing or combating infections. In addition, specific immune reactions play a part and in both types of epithelia elaborate antigen handling (uptake, processing and presentation) systems can be found.

The hallmark of specific mucosal immunity is the production and action of ▶ secretory immunoglobulin A (sIgA), which displays its functions at the mucosal surface ("outside" the body) and of effector T cells. Thus, the epithelia of the skin and the mucus membranes are primarily engaged in exclusion of substances, and only under pathological conditions (e.g. wounding, infections) is there a "free" exchange of antigens between the external and internal environment.

Skin as Target for Immunotoxicologic Agents
Several types of nonepithelial cells can be found in the skin, such as lymphocytes and macrophages in the dermis, and Langerhans cells in the epidermis. Langerhans cells are involved in the processing and presentation of substances that enter the skin. They form an extensive network in the basal layers of the epidermis and are efficient antigen-processing cells, capable of taking up exogenous antigens by virtue of several specialized surface receptors. These antigens may be derived from outside the body but may also be constituents of surrounding cells or matrix components that have been altered under the influence of external factors. In the case of nickel toxicology, for example, it is likely that nickel ions bind covalently to self-proteins, thus altering the antigenic structure.

The resulting display of antigenic peptides by major histocompatibility complex (MHC) molecules on the surface of the cell permits presentation to passing T cells. This process can take place in the epidermis, but Langerhans cells can also leave their sentinel position in the basal layers of the epidermis under the influence of factors produced in the skin as a result of stimulation (1). As shown in kinetic studies using fluoresceinated dyes, they can be found in lymphatic vessels and migrate into the draining lymph nodes.

During this migration they show characteristic, extended membrane processes, for which they are termed veiled cells. Once they have arrived in the lymph node they localize in the paracortical areas where, as dendritic, interdigitating cells, they play a pivotal role in the primary activation of T lymphocytes. Thus Langerhans cells, veiled cells and interdigitating cells are different stages of the dendritic cell lineage (Figure 1). While Langerhans cells are extremely efficient at processing antigen, once they arrive in the lymph node they differentiate into potent presenting cells at the expense of their processing capacity. This is reflected in the expression of MHC molecules and accessory molecules, which increases during the differentiation into interdigitating cells. It is only under pathological conditions that cells in the skin other than Langerhans cells are involved in antigen handling. In such cases keratinocytes can be induced to express MHC class II molecules and are able to present antigen.

Studies on squamous (mucosal) epithelia have shown that antigen-processing and antigen-presenting dendritic cells also occur in the nonkeratinizing epithelia of the oral and nasal mucosa, the trachea and bronchi, the esophagus and the tonsils, and can form comparable networks at these sites. In simple epithelial membranes, such as the epithelium lining the lung, the stomach and the gut, dendritic cells reside in large numbers in the underlying ▶ lamina propria where they exert similar functions (Figure 1).

Antigen Uptake by Mucosal Surfaces
Antigen Uptake by the Gut

Intestinal epithelial cells express many proteins that are, directly or indirectly, involved in forming or maintaining a protective epithelial barrier. Secretory, gel-forming mucins are produced in specialized mucus cells of glandular tissues and ▶ goblet cells of the gastrointestinal tract. As the primary constituents of extracellular mucus and the cellular barrier they are essential for a proper epithelial barrier function but both inflammatory mediators and bacterial factors are identified that can influence the expression of mucins. Moreover, mucins are very important in the contact of many microorganisms with the intestinal mucosa. Therefore, a primary defect in mucins could breach the epithelial barrier or lead to altered mucosal-bacterial interactions. On the other hand, the changing effects of immunological or bacterial factors during initial or ongoing inflammation could also influence the mucin production, and may influence the uptake and effects of potentially toxic compounds.

In addition to the mucus layer, intestinal epithelium is characterized by the presence of tight junctions, so that macromolecules have to transgress this barrier by endocytosis. Certain diseases or altered states (malnutrition, vitamin A deficiency, decreased gastric acidity, allergy) may allow for increased amounts of macromolecular transport. It is thought, for example, that the tight junction complexes present between the mucosal epithelial cells are broken down during malnutrition and vitamin A deficiency.

In the epithelium itself numerous lymphoid cells, the intraepithelial lymphocytes (IELs) can be found, mainly localized in the basal area of the epithelium. They bear either cytotoxic T cell or natural killer cell markers, but it is unclear whether these cells influence the barrier or uptake function of the gut epithelium. In the lamina propria underneath the epithelium granulocytes, macrophages and plasma cells may interact with the absorbed antigenic molecules, thus forming a second barrier of defense against penetration of antigens into the circulation.

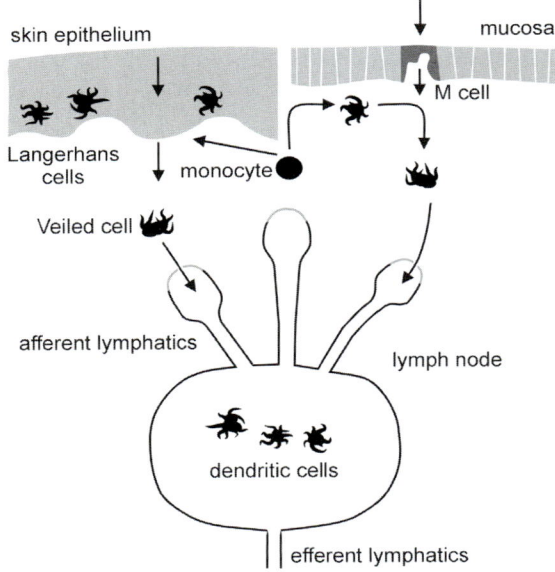

Immunotoxic Agents into the Body, Entry of. Figure 1 Migration of antigen presenting Langerhans cells from the skin epithelium into the draining lymph node. Damage to the skin induced by allergens, or damage, activates keratinocytes to produce multiple factors. Langerhans cells in the basal layers of the epidermis process allergens and are stimulated by the keratinocyte factors to leave the epidermins and migrate to the draining lymph nodes, where they interact with T cells after differentiating into antigen-presenting dendritic cells. Langerhans cells are replaced by monocyte-derived precursors. In mucosae (right of the scheme), antigen processing cells may acquire antigenic substances after uptake via M cells located in the epithelium of the mucosae, after wich transportation to the draining lymph nodes takes place. In the lymph nodes actual presentation to naive T cells will occur.

At specific sites in the gastrointestinal and respiratory tracts a specialized antigen transport system exists which allows for the efficient transport of macromolecules from the lumen to underlying organized lymphoid tissue. The epithelium covering these lymphoid tissues (gut-associated, nose-associated and bronchus-associated lymphoid tissue; GALT, NALT and BALT, respectively) contains specialized antigen-transporting membraneous (M) cells.

In the gut M cells are found in the epithelia covering the Peyer's patches (PPs), the appendix and scattered lymphoid follicles. M cells are commonly interspersed between ▶ enterocytes; their numbers vary between species. M cells are physically attached to enterocytes via conventional junctional complexes. M cells differ from the rest of the epithelial cells in histochemical and ultrastructural respects, but although the intermediate filament composition and glycosylation state of M cells is slightly different from that of the surrounding enterocytes, the junctional complexes and the presence of epithelial-specific ▶ vimentin and ▶ cytokeratin expression in M cells provide evidence for an epithelial origin of M cells, in contrast to all other types of antigen-handling cells, which are of hematopoietic origin. At the microscope level these cells can be recognized by their close association with lymphocytes. M cells have evolved special mechanisms to transport macromolecules and particles across the epithelium. The basolateral surface of M cells is deeply invaginated whereby large intraepithelial pockets are formed, that serve as an initial collection site for transcytosed material. Together with extensive arrays of intermediate filaments this pocket forms the cardinal characteristic of the M cell. Inside the pocket lymphocytes as well as macrophages can be found, in close association with the M cell membrane, suggesting some sort of interaction. Furthermore, M cells form long basal processes that are found to extend deeper into the lymphoid tissue underneath the epithelium, where they can associate with other cells such as antigen-presenting cells. M cells seem to be specialized in transcytosis (2). Bound particles such as bacteria and viruses induce changes in the apical membrane and cytoskeleton, after which phagocytosis takes place. Reorganization of the cell membrane and extension of cellular processes around the particles have been described, and the particles are transported through the cell in a phagocytic vesicle. Smaller particles and macromolecules are taken up through clathrin-coated pits or by receptor-mediated phagocytosis. All material that has been taken up is released again in the basolateral pocket, but it is unclear whether any degradation of the transported material or particles has taken place during the transit and whether the cells play an active role in antigen presentation.

Nevertheless there seems to be an active interaction between the M cells and underlying lymphocytes because after depletion of lymphocytes by irradiation the intraepithelial pocket of the M cells in ▶ Peyer's patches is lost. In conventionally reared animals the number of M cells increased with age and their morphology altered, leading to extensive cytoplasmic interfoliation and wide intercellular spaces filled with amorphous material.

In Peyer's patches the majority of cells that are localized in the M cell pockets are $CD4^+$ T cells. They can have characteristics of naive T cells, but also are found to show activation markers such as CD69 and CD45RO. B cells are also found in the pockets, and both naive B cells, based on CD45RA expression, as well as more activated cells have been described. The presence of B cells has led to the speculation that these cells are involved in the actual antigen presentation to T cells in these special epithelial niches.

It is suggested that part of the intestinal antigens—in particular soluble, low molecular weight antigens—are absorbed by the enterocytes of the villous epithelium. Enterocytes, in particular those of the small intestine, constitutively express MHC class II molecules and it has been shown that these cells are able to present antigen. There are indications that antigen presentation by enterocytes leads to the generation of T suppressor cells which are involved in systemic suppression and tolerance induction. These findings have significant implications in local gut immune responses and may partially explain the poorly understood phenomena of oral tolerance induction and systemic suppression.

Antigen Uptake by the Upper and Lower Respiratory Tract

The nasal and oral cavities are covered by a (pseudo) stratified epithelium, that acts, due to surface protection by mucus and sIgA, as an adequate barrier. This epithelium possesses antigen-presenting Langerhans-like (dendritic) cells. Painting of this area with fluoresceinated dyes in rodents has indicated that dendritic cells can bind antigen and transport it to the draining lymph nodes for an adequate immune response. The epithelium of the mucosal surface of the oral and upper respiratory tract, except for the epithelium above lymphoid structures (Waldeyer's lymphoid ring), is involved in antigen exclusion rather than in antigen uptake. The epithelium above the ring of Waldeyer—in the nose the lymphoid tissue has been described under the name NALT—contains antigen-transporting M cells by which the local mucosal immune system is allowed to react adequately. Antigen-presenting dendritic cells which constitutively express MHC class I and II molecules have also been described in the pseudostratified epithelium of the trachea and the bronchi, and also in the alveoli. The majority of them are localized in the mucosa, either above or below the basement membrane, and in the lamina propria. Although there are no clear data about the migration route of these cells, there are indications that after antigen uptake dendritic cells migrate to the draining lymph nodes for antigen presentation (3). The lower respiratory tract, in particular the lung, is mainly involved in antigen exclusion. Clearance of antigens from this part of the respiratory tract depends on mucociliary activity and alveolar macrophages. The alveolar macrophage population, however, appears to be heterogeneous, both morphologically and functionally. Some macrophages act as real scavenger cells (they phagocytose and digest antigen, and supposedly leave the lungs via the mucociliary trap), some are apparently inhibitory, while others display a stimulating effect on the local immune response by presenting antigen to lymphoid cells. Most probably this latter effect is brought about by migration of alveolar macrophages from the alveoli into the lung tissue. It has been shown that antigen-laden alveolar macrophages cross the lining of the alveoli and migrate to and present antigen in the draining lymph node. In this way alveolar macrophages act as dendritic cells forming a specialized antigen handling system in the lungs. Antigens can reach the draining lymph nodes via diverse routes and there are indications that the site where antigen is taken up, processed and presented determines the outcome of the immune response.

References

1. Romani N, Ratzinger G, Pfaller K et al. (2001) Migration of dendritic cells into lymphatics—the Langerhans cell example: routes, regulation, and relevance. Int Rev Cytol 207:237–270
2. Neutra MR, Mantis NJ, Kraehenbuhl JP (2001) Collaboration of epithelial cells with organized mucosal lymphoid tissues. Nat Immunol 2:1004–1009
3. Luster MI, Simeonova P, Gallucci R, Matheson J, Yucesoy B, Sugawara T (2000) Overview of immunotoxicology and current applications to respiratory diseases. Immunopharmacology 48:311–313

Immunotoxic Intermediates

▶ Metabolism, Role in Immunotoxicity

Immunotoxicant

Any agent which causes a decrease in immune cell development or function.

▶ Thymus: A Mediator of T Cell Development and Potential Target of Toxicological Agents

Immunotoxicity

Immunotoxicity refers to the alteration of any immune component that increases the risk of an adverse effect caused by a drug, chemical or physical agent. Immunotoxicity encompasses immunosuppression, impairment of normal host defenses such as phagocytosis and complement, and antigenicity of the agent including induction of hypersensitivity.

▶ Streptococcus Infection and Immunity
▶ Therapeutic Cytokines, Immunotoxicological Evaluation of

Immunotoxicological Evaluation of Therapeutic Cytokines

PETER T THOMAS
Early Development
Covance Laboratories
Madison, Wisconsin
USA

Definition

Cytokines are regulatory proteins secreted by white blood cells as well as other cells whose actions include a variety of pleiotropic effects on both hematopoietic and non-hematopoietic cells, as well as modulation of inflammatory responses (1). These physiologic activities include apoptosis, natural and adoptive immunity, response to neoplasia, and cellular differentiation, among other things. Over the last decade, efforts to leverage the unique characteristics of these molecules for the treatment of human diseases has lead to the approval of several as therapeutic drugs (Table 1).

Nomenclature

The beginnings of the understanding of cytokines came from early descriptive studies (2) identifying a substance produced by sensitized lymphocytes that inhibited the migration of non-immune macrophages from the site of inflammation. This putative factor was coined macrophage ▶ migration inhibitory factor (MIF). Since that time, over the last 25 years, additional classes of immunoregulatory substances have been described as having antiviral properties (the interferons), growth and functional control of various leukocytes, (the interleukins such as IL-1, IL-2 and more recently IL-23). In addition other hematopoietic and non-hematopoietic growth factors that control nerve and epidermal cell growth and differentiation, among others, have been characterized.

Systematic evolution of cytokine nomenclature is due,

Immunotoxicological Evaluation of Therapeutic Cytokines. Table 1 Cytokines and growth factors approved for human use

Product Type	Clinical Indication
Erythropoietin	Anemia associated with chronic renal failure
	Anemia associated with chemotherapy
Granulocyte-colony stimulating factor (G-CSF)	Bone marrow transplantation
Granulocyte macrophage-colony stimulating factor (GM-CSF)	Neutropenia associated with transplantation
Interferon-α2b	AIDS-related Kaposi's sarcoma
Interferon-γ1	Chronic granulomatous disease
Interferon-β	Multiple sclerosis
Interleukin-2	Renal cell carcinoma

Adapted from Talmadge 1998. (5)

in part, to the disparate origins of the basic research that lead to their description. The product of immunologically sensitized lymphocytes was first described as a lymphokine. The term "cytokine" was coined in the early 1970s to acknowledge that regulatory proteins could be produced by non-immune cells. The term "interleukin" was later proposed to describe a group of distinct molecules that regulated communication between leukocytes and other non-hematopoietic and somatic cell types. Presently, many cytokines fall within the interleukin nomenclature. There are, however, several additional molecules that have retained their original functional names (e.g. the interferon IFN-γ, transforming growth factor TGF-β, leukocyte inhibitory factor LIF, and others).

Characteristics

Many properties of cytokines are also shared by hormones and growth factors. However, there are important differences. Most cytokines are small polypeptide molecules with a molecular weight of 30 000 Daltons, or less. However, some cytokines form higher-molecular-weight oligomers or heterodimers. For example, the recently described composite cytokine IL-23 is comprised of the IL-12 p40 subunit and a novel p19 protein. Not surprisingly, it exhibits biological activities similar to, and unique from, IL-12 (3). Unlike hormones which often act at sites distant to the site of production, most cytokines act locally. Production of cytokines is not constitutive and is usually in response to various stimuli, primarily by those resulting from an immune response. Activity is usually at the level of mRNA transcription or translation. With the exception of several of the proinflammatory cytokines (e.g. IL-1, TNF, IL-6, IFN-γ), their effective pharmacodynamic range is usually localized. However unlike most hormones that have limited action and target cell specificity, physiologic effects displayed by cytokines are diverse with some targeting hematopoietic cells. Differences aside, it appears that cytokines, polypeptide hormones, and growth factors all serve to facilitate extracellular signaling pathways. Furthermore, many have common structural features. Once bound to a receptor, the signal transduction pathways appear to be the same. Research into the structure and function of cytokines and their receptors has resulted in classification of these molecules into loose families (1). Grouping of cytokines and their receptors is largely based upon primary or higher order structural homologies among the molecules.

Preclinical Relevance

The purpose of ▶ preclinical safety assessment studies is to characterize potential toxicity of new drug candidates, to relate any toxicity to ▶ pharmacokinetics and pharmacodynamic action with the intention of selecting relevant dose levels for follow-on human clinical studies. Unlike small molecular weight compounds, where standard approaches to safety assessment usually apply, special considerations apply when designing approaches to safety assessment approaches of biologics, including cytokines. These include limited species specificity, the ability to differentiate toxicity from exaggerated immunopharmacology, unusual dose-response relationships, potential immunogenicity, and ▶ immunotoxicity (4).

Many candidate cytokines demonstrate limited species specificity with respect to pharmacologic action. Moreover, the ability of animal models to consistently predict immunotoxicity has been, at time, of limited value. Species specific ▶ immunopharmacology has been demonstrated in studies with granulocyte macrophage colony stimulating factor (GM-CSF) and IFN-γ (human vs rodent), recombinant IL-1 (▶ granulopoiesis and ▶ neutropenia in the dog vs mouse models), and IL-6 (lymphoproliferation and nephritis in the mouse vs human). When examined in the appropriate species, the pleiotropic action of many of these molecules makes differentiation between toxicity and exaggerated immunopharmacology difficult. For example, production of numerous cytokines including TNF-α, IL-1, IL-15 and IL-18 are important in the pathogenesis of cytokine-induced shock in humans. Mechanistic studies have suggested that natural killer cells are important effector cells responsible for this human response.

Due to the unique properties of cytokines and other immunoregulatory molecules, traditional dose-response relationships seen with traditional small-molecular-weight new chemical entities, often do not apply. Talmadge (5) suggested that this is due to many factors, including biodistribution of the molecules in the body, circadian processes, specific receptor-mediated events, and indirect, downstream effects from the site(s) of action. Bell shaped or biphasic dose-response curves have been seen in studies with IFN-γ (murine tumor metastasis models) as well as TGF-β (wound healing models), among others.

An important factor in safety assessment of cytokines, like other biologics, is the potential of antibody production to the therapeutic or to endogenous product in the test species. The nature and duration of the antibody response is dependent on such factors as homology with the endogenous molecule, presence of protein aggregates and fragments formed during production, and route of exposure. Subcutaneous and intradermal treatments are the most immunogenic, followed by inhalation, intravenous and oral exposures. Strategies to modify proteins to make them less immunogenic include glycosylation, and elimination of antigenic epitopes by site-specific mutagenesis, exon shuffling or by "humanization" of the molecule. If it occurs, the antibody response is typically first detected within 2 weeks of treatment, making long-term chronic testing problematic. The immunopharmacologic effect of the antibody is dependent upon the particular ▶ isotype that is produced. Of concern to the toxicologist is appearance of antibody that alters the pharmacology, half-life, and distribution of the cytokine. Despite these concerns, appearance of antibody does not necessarily mean the same effect will occur in humans. In fact, these data are valuable for the design of the clinical pharmacology studies.

Aside from antibody effects, direct or indirect immunotoxicity following treatment with therapeutic cytokines is thought to be the cause of many potential adverse events seen during clinical treatment. These range from interferon or TNF-induced flu-like symptoms, to more serious and less well understood effects including thyroid disorders, systemic lupus erythematosis, and diabetes (Table 2). The animal models currently used for safety assessment of therapeutic cytokines do not have a strong history of predicting these adverse effects. Therefore, the challenge for the practicing toxicologist is to be able to differentiate true immunotoxicity from exaggerated immunopharmacology due to treatment.

Unless there is a specific reason to do so, measurement of cytokine levels during preclinical safety studies is not particularly useful for screening. However, understanding the role these molecules directly and indirectly play in immune regulation provides important mechanistic information that can aid in safety assessment. As reviewed by House (6), if measurement of cytokine levels is undertaken, many factors must be considered, including the biological source, effect of sample processing on activity, and—most importantly—whether or not to measure function or simply presence of the molecule(s) in question.

Relevance to Humans

Therapeutic human cytokines currently in clinical use include the interferons (IFN-α, IFN-β and IFN-γ), interleukins (IL-2 and IL-12) and hematopoietic factors such as granulocyte colony-stimulating factor (G-CSF), GM-CSF, thrombopoietin, and erythropoietin. While these drugs have proven and significant clinical benefit, progress made to develop cytokines as drug therapies has been hampered somewhat by a number of factors, including an incomplete understanding of mechanism(s) of action. Furthermore, the unique nature of these molecules has made development of suitable and predictive animal models for safety testing challenging. Nevertheless, over the past decade, cytokines that modulate the immune responses have proven efficacious in the treatment of human tumors and certain viral and bacterial infections. In addition, cytokines in the growth factor category have found utility in wound healing, in the restoration of cellular function following chemotherapy and radiotherapy, as well as in support of bone marrow transplantation.

It is safe to say that these biotherapeutics, given either as a monotherapy or in combination with other drugs, will continue to produce significant clinical benefit in a wide variety of diseases. As our understanding increases of the role these molecules play in modulating the immune response, and in regulating cell growth and differentiation, our ability to develop and deliver safer and more effective doses of these substances will improve.

Regulatory Environment

While there are no specific guidances that regulate development of therapeutic cytokines, guidelines that influence the approach for evaluating the impact on the immune system of drug and biological products have been published. The International Conference on Harmonization of Technical Requirements for Reg-

Immunotoxicological Evaluation of Therapeutic Cytokines. Table 2 Immunologically mediated clinical side effects of cytokines in humans

Cytokine	Effect
IFN-α	Thyroid disorders, autoimmune disease
IL-2	Insulin dependent diabetes mellitus
IL-2, IL-4, granulocyte macrophage colony stimulating factor (GM-CSF)	Vascular leak syndrome
IFN-β	Allergic contact dermatitis?
IFN-γ	Multiple autoantibodies in the absence of clinical disease
IL-2 ± lymphokine-activated killer cells (LAK) cells	Thyroid dysfunction, hypothyroidism Antithyroid antibodies
Colony stimulating factors	Autoimmune thyroiditis? Neutrophil/eosinophil inflammatory diseases

Adapted from Vial and Descotes 1995 (10)

istration of Pharmaceuticals for Human Use published guidelines for preclinical safety evaluation of biotechnology-derived pharmaceuticals (7). This document forms the basis for safety evaluation of biotherapeutics and stresses the importance of choosing the most relevant species for safety evaluation as many molecules, including cytokines, are highly species-specific. In contrast to traditional drugs, standard immunotoxicity tier tests are deemed not relevant for initial safety evaluation. Furthermore, measurement and characterization of any antibody response as it effects the host through indirect mechanism(s) or the ▶ pharmacodynamics of the test compound is stressed.
The US FDA published its guidance document on ▶ immunotoxicology evaluation for Investigational New Drugs (8). According to the FDA, all investigational new drugs should be evaluated for immunosuppression, through the use of standard clinical and anatomic pathology measures, rather than immune function tests. With respect to ▶ immunogenicity, while acknowledging that determining allergic potential is difficult in non-clinical toxicology, approaches and methods have been developed to do so. In addition to specific endpoints, the FDA provides guidance on the scope of nonclinical immunotoxicity safety testing, depending upon the disease and targeted patient population(s). Prior to June of 2003, the Center for Drug Evaluation and Research was the leading center at the FDA for the review and approval of chemically synthesized peptides, and the Center for Biologics Evaluation and Research was the leading center for cell-expressed therapeutic large molecules. In an effort to further harmonize processes and procedures within FDA, CDER now is the leading center for the review and approval of all large molecules intended for therapeutic use, including cytokines.

As far as the European Union is concerned, the EMEA provides more specific recommendations for immunotoxicity testing and mandates it in at least one repeat-dose toxicity study (9). When it comes to biotechnology products, this guidance refers to the ICH publication (7). At the present time, the ICH is poised to begin the harmonization process for immunotoxicity testing across all the three global regions.

References

1. Vilcek J (1998) The cytokines: An overview. In: Thompson A (ed). The cytokine handbook, 3rd edn. San Diego: Academic Press, 1–20
2. Bloom B, Bennett B (1966) Mechanism of a reaction in vitro associated with delayed-type hypersensitivity. Science 153:80–82
3. Oppmann B, Lesley R, Blom B et al. (2000) Novel p19 protein engages IL-12p40 to form a cytokine, IL-23 with biological activities similar as well as distinct from IL-12. Immunity 13:715–725
4. Thomas PT (2002) Nonclinical evaluation of therapeutic cytokines: Immunotoxicologic Issues. Toxicology 174:27–35
5. Talmadge J (1998) Pharmacodynamic aspects of peptide administration of biological response modifiers. Adv Drug Deliv Rev 33:241–252
6. House R (1999) Theory and practice of cytokine assessment in immunotoxicology. Methods 19:17–27
7. ICH (1997) Guidance S6. Preclinical safety evaluation of biotechnology-derived pharmaceuticals. July 1997. International Conference on Harmonization
8. US FDA (2002) CDER Guidance for Industry. Immunotoxicology Evaluation of Investigational New Drugs. October, 2002. Food and Drug Administration
9. EMEA (2000) Committee for Proprietary Medicinal Products. Note for Guidance on Repeat Dose Toxicity.

Appendix B. Guidance on Immunotoxicity. July 2000. CPMP/SWP/1042/99
10. Vial T, Descotes J (1995) Immune-mediated side-effects of cytokines in humans. Toxicology 105:31–57

Immunotoxicology

DENNIS K FLAHERTY
Biology Department
Lamar University Beaumont
P.O.Box 10037
Beaumont, TX 7710
USA

Synonyms

immunotox, immunopharmacology, immunomodulation, immunopotentiation

Definitions

There are several definitions of immunotoxicology. This subdiscipline of toxicology can be informally defined as the science that deals with purposeful or inadvertent changes or effects on the immune system induced by drugs, foods or environmental chemicals. A change refers to any adverse effect on the structure or function of the immune system, or on other systems as a result of immune system dysfunction. An effect is considered adverse or immunotoxic if it:

- decreases humoral or cellular immunity needed by the host to defend itself against infectious agents or cancers (▶ immunosuppression)
- causes tissue damage (autoimmunity or chronic inflammation)
- increases the frequency of immediate allergic reactions and/or decreases the frequency or intensity delayed hypersensitivity reactions (immunomodulation).

Regulatory agencies use a more global definition of immunotoxicity which embraces the concept that there is a complex balance between the immune system and other body systems (e.g. nervous and endocrine) that may utilize or be affected by the same biological mediators (e.g. neuropeptide and steroid hormones). The US Food and Drug Administration (FDA) considers an immunotoxic event to be "any change in the structure or function of the immune system that is permanent or reversible". As defined by the World Health Organization (▶ WHO) immunotoxicology is that specialty that is concerned with the study of adverse events resulting from the interaction of xenobiotics and the immune system. These adverse events may result in (i) a consequence of an activity of substances and/or their biotransformation products on the immune system, and (ii) an immunological response to a substance or its biotransformation products.

Characteristics

The immune system is distinguished by its functional and biological complexity. The system is composed of different organs and tissues that can act autonomously or in concert with peripheral blood, multiple lymphocyte effector cells (whose function can be modulated by hormones, growth factors and small protein messengers such as interleukins and cytokines), and redundant effector mechanisms. Also, a functional reserve must be exceeded before there is an increased susceptibility to infections or cancers. Moreover, immunotoxicity can be induced by prenatal, perinatal or adult exposure. Prenatal or perinatal exposure often yields more dramatic immunotoxicity when compared to adult exposured.

In general, immunotoxicology is the study of either innate or ▶ acquired immunity. ▶ Innate immunity is generally non-specific and passive. Anatomic and physiological barriers, antimicrobial peptides, inducible enzymes and pluripotential molecules facilitate innate immunity. In contrast, acquired immunity requires stimulation of effector mechanisms following exposure to foreign materials (e.g. xenobiotics). Acquired immunity can be subdivided into antibody-mediated (AMI) and cell-mediated immunity (CMI). The former plays the major role in response to extracellular bacteria and some viruses, while the latter is involved in defense against intracellular viruses, parasites and tumors.

Preclinical Relevance

Some immunotoxicants may modulate immune function without inducing overt toxicity or histological changes in organs, tissues or effector cells. Since standard toxicity testing primarily relies on histological evaluations, immunotoxicity may not always be detected in standard preclinical studies designed to determine potential toxicity or safety considerations. Specialized static and functional measurements of the immune system are necessary to access decrements in immune function.

Since a functional immune system is necessary for human health and survival of the species, preclinical assessment of immunotoxicity should be an essential part of safety assessment. Assuming that the absorption, distribution, metabolism and excretion (ADME) of xenobiotics are similar in rodents and humans, it is possible to assess potential human risks using data from preclinical animal studies. Translation of data to humans is possible because the immune systems of humans and rodents are almost identical in structure and function. A decision on whether a material or device is immunotoxic should rely on the weight of the evidence from both preclinical test results and clinical evaluation.

Relevance to Humans

Interest in immunotoxicology was heightened by the development of analytical methods with detection limits measured in parts per billion. Increased analytical capabilities allowed detection of trace materials such as 2,3,7,8-tetrachlorodibenzo-*p*-dioxin (TCDD) present in a range of industrial and agricultural products. The possible relationship between exposure to trace amounts of TCDD and human immunotoxicity constituted a vigorous scientific debate for two decades. When an increased rate of cancer was demonstrable in patients receiving long-term immunosuppressive therapy to avoid transplant rejection, the medical community became concerned about direct and inadvertent toxic effects on the immune system. Finally, emergence of the AIDS virus and the subsequent AIDS pandemic demonstrated that significant immunosuppression could cause life-threatening, opportunistic infections in people.

In response to the scientific issues, toxicologists attempted to determine whether the standard toxicology studies used for the last 40 years for product safety assessment could be used to assess immunotoxicity. In 1977, Vos compiled information concerning the effects of xenobiotics on the immune system and concluded that standard toxicity testing in animals underestimated chemical effects on the immune system (1). Effects on the immune system were not noted because of limited methodology, lack of information on the mode of action, and inadequate risk assessment paradigms. This sentinel publication triggered a number of different workshops and symposia in North America and Europe between 1981 and 1983. The Toxicology and Pathology Section of the National Institute of Environmental Sciences (NIEHS), the University of Surrey, and the Commission of European Communities all convened immunotoxicology symposia. The focus of these meetings was to critically examine the possible toxic effects of xenobiotics on the immune system and the ramifications of toxicity on drug design, immunosuppression or susceptibility to cancers and infectious disease. Other investigators called attention to the fact that chemicals could cause hypersensitivity reactions that were either immediate (within a few minutes) and immunoglobulin E-mediated or delayed (24–48 hours) and cell mediated. Also, it was suggested that some xenobiotics could cause an autoimmune response.

As a consequence of public pressures, governmental regulatory agencies began to consider implementing immunotoxicity testing into guidelines and regulations. The NIEHS was charged with developing and validating testing batteries useful for standard immunotoxicity testing. Of particular interest were tests that detect immunomodulation (immune suppression or potentiation), hypersensitivity (that is, allergy), and autoimmunity.

Immunotoxicologists were able to successfully modify in vitro assays for use with rodent splenic lymphocytes. In addition, they adapted ex vivo assays (e.g. the plaque forming cell assay and the lymphocyte transformation assay) that were designed to detect decrements in antibody or cell-mediated immunity. In a series of published studies, the accuracy, sensitivity and reproducibility of the assays were determined. At the same time, rodent host defense models for infectious diseases and cancers were developed and validated (2).

Using the validated ex vivo assays, Luster et al. (3) developed a battery of tests composed of various immune function, immunopathology and host resistance tests, the results of which could help establish the potential of chemical and biological agents to cause immunosuppression in animals. The authors determined which individual tests or testing configurations could accurately identify immunotoxic compounds. Only one or two tests (e.g. lymphocyte phenotyping and the B cell plaque forming culture assay) were needed to detect biologically relevant changes that correlated with increased susceptibility to infectious disease and cancers as measured in host defense models.

The "tier" approach assumes that any statistically significant, dose-related changes in immune function are biologically relevant. Because of redundant effector mechanisms and the functional reserve in the immune system, it is very difficult to determine the magnitude of the response necessary to induce biologically relevant immunosuppression.

While the NIEHS testing paradigm provided valuable information of immunosuppression related to infection and tumors, it yielded no information on a xenobiotics potential for inducing immediate allergic or autoimmune reactions. Development of new tests to predict allergic reactions in man is extremely important. It has been estimated that 30–35 million Americans suffer from allergic disease, of which 2%–5% is from occupational exposure. Allergic reactions often can be severe and, in some cases, result in death from anaphylactic shock. At the present time, there are no validated tests for determining immediate hypersensitivity or autoimmunity in animal models.

There are several validated methods for assessing both the frequency and intensity of delayed hypersensitivity reactions in animals. Delayed hypersensitivity reactions mediated by cellular components of the immune system are usually determined in the guinea pig maximization test (GPMT) or the Buehler test (BT). The local lymph node assay (LLNA) is a preferred alternative method to the traditional guinea pig test. It demonstrates an equivalent prediction of human allergic contact dermatitis as compared to the other sensitization tests, provides quantitative data and an assessment of dose response, gives consideration to animal welfare concerns, and is suitable for testing colored substances. All three assays are acceptable to governmental regulatory agencies as part of safety assessment studies.

Regulatory Environment

The regulatory environment for immunotoxicology differs with geography and the type of product (e.g. pharmaceuticals, agrochemicals or commodity chemicals). In the USA, the FDA regulates pharmaceuticals. Within the FDA, the requirements for immunotoxicity testing are fragmented. The Center for Food Safety and Applied Nutrition (CFSAN) has defined immunotoxicity testing protocols. Other FDA centers, such as the Center for Devices and Radiological Health and the Center for Drug Evaluation and Research (CDER), have recently issued guidance documents on immunotoxicity testing. Guidance is intended to provide a coherent strategy for assessing potential immunotoxic effects involving devices and pharmaceuticals. The European Agency for the Evaluation of Medicinal Products has also issued guidance documents suggesting that all new medicinal products should be tested for immunotoxic potential in preclinical testing. It is anticipated that an immunotoxicity testing guidance document will be incorporated into protocols being developed by The International Conference on Harmonization of Technical Requirements for Registration of Pharmaceuticals for Human Use (▶ ICH). The mission of the ICH is to make recommendations on ways to achieve greater harmonization in the interpretation and application of technical guidelines and requirements for pharmaceutical product registration in the US, Europe and Japan.

In the USA, regulation of commodity and agricultural chemicals fall under the purview of the Environmental Protection Agency (EPA). The agency has issued Harmonized Immunotoxicity Testing Guidelines for all chemicals registered under the Federal Insecticide, Fungicide and Rodenticide (FIFRA) and Toxic Substances Control (TSCA) acts. It is anticipated that the European Organization for Economic Cooperation and Development (▶ OECD) will modify existing 28-day rodent bioassays to include immunotoxicity endpoints. Because of the lack of validated testing methods, regulatory agencies have not expressed interest in testing for immediate allergic and autoimmune reactions induced by xenobiotics.

References

1. Vos JG (1977) Immune suppression related to toxicology. CRC Crit Rev Toxicol 5:67–97
2. Luster MI, Pait DG, Portier et al. (1992) Qualitative and quantitative experimental models to aid in risk assessment for immunotoxicology. Toxicol Lett 64–65:71–78
3. Luster MI, Portier C, Pait DG et al. (1992) Risk assessment in immunotoxicology. I. Sensitivity and predictability of immune tests. Fundam Appl Toxicol 18:200–210

Immunotoxicology, Definition of

A branch of toxicology that is concerned with identifying the adverse effects of drugs, chemicals or environmental agents on the structure and/or function of the immune system and elucidating the mechanisms responsible for such effects.

▶ Therapeutic Cytokines, Immunotoxicological Evaluation of

Immunotoxicology of Biotechnology-Derived Pharmaceuticals

GARY J ROSENTHAL
Drug Development
RxKinetix Inc.
1172 Century Drive
Louisville, CO 80027
USA

Synonyms

biotherapeutics, biologics, biologic-response modifiers, peptides, proteins, DNA-derived products, monoclonal antibodies

Definition

Biotechnology-derived pharmaceuticals represent therapeutic agents produced from characterized cells or other defined systems, often through the use of a variety of microorganisms or expression systems including bacteria, yeast, insect, plant, mammalian cells and whole animals. Active pharmaceutical ingredients represented by this class of therapeutics include ▶ proteins and ▶ peptides that includes ▶ cytokines, ▶ interleukins, and growth factors, plasminogen activators, recombinant plasma factors, fusion proteins, enzymes, receptors, antisense sequences, ribozymes, hormones and ▶ monoclonal antibodies.

Characteristics:

Biotechnology-derived therapeutics can be as simple as a 3-amino acid peptide or as complex as a humanized monoclonal antibody with remarkable specificity (Table 1).

Many biotechnology-derived therapeutics have been shown to be functionally pleiotropic, resulting in pharmacologic activity in addition to inadvertent modulation of cells and tissues other sites. Structurally, biotherapeutic peptides have two or more amino acids coupled by an amide link between the carboxylic acid group of one amino acid and the α-amino group of the other. Proteins are large macromolecular entities composed of one or more polypeptides. The complexity of biotherapeutic polypeptides is often amplified with secondary and tertiary structural features that can be intimately associated with pharmacologic activity. Regardless of the approach taken to produce the biotherapeutic, it is fundamental that the identity, purity, impurities, potency and quantity can be measured and controlled in a reproducible fashion because consistent and predictable manufacturing is crucial to clinical performance.

Preclinical Relevance

Alongside immunopharmacology, preclinical analysis of the immunotoxicity potential of biotherapeutic agents can be a key component in drug development. While regulatory guidance has been recently promulgated for drugs, the decision to commence nonclinical immunotoxicology studies during the course of biotherapeutic development is generally made on a case-by-case basis (1), depending on the targeted disease, the class of compound, and results of nonclinical studies.

When the decision is made to assess immunotoxicologic potential, a wide range of methods and approaches are available (Table 2) (2–5). While predictive methodology exists in assessing unintended immunosuppression, challenges still remain in assessing wayward immune stimulation such as immunogenicity, allergic/anaphylactoid reactions or autoimmunity. Evidence of immunotoxicity often results in supplementary mechanistic investigation, depending on the intended use of the therapeutic. Additional factors given consideration include whether the effect is an exaggerated pharmacologic response, or if the target disease is likely to be uniquely influenced by the immunomodulation (e.g. HIV).

Relevance to Humans

Biotherapeutics have emerged as a tremendously important class of drugs that have positively influenced a wide range of human diseases including cancer, infectious disease and diseases of the blood. Unfortunately, a number of immune-mediated side effects have been seen in man with this class of drug, from contact dermatitis to vascular leak syndrome. A rapidly expanding class of biopharmaceuticals, such as monoclonal antibodies, often target immune-related systems (Table 1). The immunologic basis of such therapeutics, along with the immune-related targets, suggests that human immunotoxicology will continue to be an important area of drug development. A current need in the area is a sufficient understanding and if possible prediction of immunogenicity in humans, which continues to be a hurdle to the development of protein-based biotherapeutics. Further research is needed to prospectively identify therapeutic candidates predisposed to this response, particularly as generic biotherapeutics begin to emerge.

Regulatory Environment

The regulatory environment relating to biotechnology-derived pharmaceuticals is evolving, as a greater un-

Immunotoxicology of Biotechnology-Derived Pharmaceuticals. Table 1 Examples of therapeutic antibodies, targets and indications

Therapeutic Antibody	Target	Indication
Orthoclone OKT3 (muromonab CD3)	CD3	Transplant rejection
ReoPro (abciximab)	GPIIb/IIIa	Blood clots
Zenapax (daclizumab)	IL-2R (CD25)	Transplantation
Remicade (infliximab)	TNF-α	Crohn's disease, rheumatoid arthritis
Simulect (basiliximab)	IL-2R (CD25)	Transplant rejection
Humira (adalimumab)	TNF-α	Rheumatoid arthritis
Rituxan (rituximab)	CD20	Non-Hodgkin's lymphoma
Herceptin (trastuzumab)	HER-2/neu	Breast cancer
Mylotarg (gemtuzumab ozogamicin)	CD33	Relapsed acute myeloid leukemia
Campath 1H (alemtuzumab)	CD52	Chronic lymphocytic leukemia
Zevalin (ibritumomab tiuxetan)	CD20	Rituximab-failed non-Hodgkin's lymphoma

Immunotoxicology of Biotechnology-Derived Pharmaceuticals. Table 2 Examples of endpoints assessed in immunotoxicology assessment

Immunotoxicology Endpoint
Hematology
Immunopathology: lymphoid organ histology, lymphoid organ weights and cellularity
Immune cell phenotyping
Non-specific immune assays (natural killer cells, macrophage functional activity)
T cell-dependent antibody response
Hypersensitivity: delayed-type, guinea pig maximization
Lymphoproliferation
Cytotoxic T cell response
Cytokine transcription, translation, production
Host resistance: infectious, tumor challenge
Immunogenicity: popliteal lymph node

derstanding of the clinical utility and safety of these agents is obtained. As noted above, the immunotoxicologic assessment of biotechnology-derived therapeutic has followed a scientifically based, case-by-case approach, depending on a variety of factors. Immunotoxicologic assessment of biotechnology-derived pharmaceuticals is directly or indirectly covered under a variety of regulatory guidance documents noted below:
- ICH S6 Document: Safety Studies for Biotechnological Products
- ICHS5a Document: Detection of Toxicity to Reproduction for Medicinal Products
- ICH S7A Document: Safety Pharmacology Studies for Human Pharmaceuticals (section 2.8.2.4)
- Guidance for Industry: Content and Format of Investigational New Drug Applications (INDs) for Phase 1 Studies of Drugs, Including Well-Characterized, Therapeutic, Biotechnology-derived Products, 11/1995
- Guidance for Industry: Immunotoxicology Evaluation of Investigational New Drugs, 10/2002 (specifically for drugs—not biotherapeutics—but useful foundation for a case-by-case approach to biotherapeutics)
- Guidance for Industry: Drugs, Biologics, and Medical Devices Derived from Bioengineered Plants for Use in Humans and Animals (draft)
- DRAFT Guidance for Industry: Medical Imaging Drug and Biological Products
- Part 1: Conducting Safety Assessments, 5/2003
- EMEA/CPMP Note for Guidance on Comparability of Medicinal Products
- Containing Biotechnology Derived Proteins as a Drug Substance, 7/2002

References
1. Hincks DJ, Remandet B (1998) Immunotoxicology assessment in the pharmaceutical industry. Toxicol Lett 102–103:247–255
2. Talmadge JE (1998) Pharmacodynamic aspects of peptide administration biological response modifiers. Adv Drug Deliv Rev 33:241–252
3. Hastings KL (2002) Implications of the new FDA/CDER immunotoxicology guidance for drugs. Internat Immunopharmacol 2:1613–1618
4. Thomas PT (2002) Nonclinical evaluation of therapeutic cytokines: immunotoxicologic issues. Toxicology 174:27–35
5. Luster MI, Munson AE, Thomas PT et al. (1988) Development of a testing battery to assess chemical induced immunotoxicity: National Toxicology Program's guidelines for immunotoxicology for immunotoxicology evaluation in mice. Fundam Appl Toxicol 10:2–19

Immunotransmitters

▶ Cytokines

Impetigo Contagiosa

Infection of the epidermis with *Streptococcus pyogenes* and less frequently with *Staphylococcus aureus*. Characteristically thin-walled vesicles which convert to honey-colored and serous crusting lesions are found predominantly on the faces of children.
▶ Dermatological Infections

In utero Immunotoxicology

▶ Developmental Immunotoxicology

In vitro Culture

▶ Maturation of the Immune Response

In vitro Engineered Skin/Epidermal Substitutes

▶ Three-Dimensional Human Skin/Epidermal Models

and Organotypic Human and Murine Skin Explant Systems

In vivo Immunotoxicology Testing

▶ Streptococcus Infection and Immunity

Inbred Strains

▶ Rodents, Inbred Strains

Inbreds

▶ Rodents, Inbred Strains

Indirect Coombs Test

This test is used to detect the presence of antibodies to erythrocytes in serum.
▶ Antiglobulin (Coombs) Test
▶ Hemolytic Anemia, Autoimmune

Indirect Immunotoxicity

▶ Metabolism, Role in Immunotoxicity

Infection Models

Experimental models in which effects of exposure to immunotoxicants is investigated by following the course of an experimental infection in exposed animals.
▶ Host Resistance Assays

Infectious Agents Models

▶ Host Resistance Assays

Inflammation

▶ Inflammatory Reactions, Acute Versus Chronic

Inflammatory Cytokines

▶ Inflammatory Reactions, Acute Versus Chronic

Inflammatory Heart Disease

▶ Cardiac Disease, Autoimmune

Inflammatory Macrophages

Monocytes that migrate into areas of inflammation upregulate certain functions associated with killing such as phagocytosis and secretion of oxidants.
▶ Macrophage Activation

Inflammatory Reactions, Acute Versus Chronic

MICHAEL I LUSTER
National Institute for Occupational Safety and Health
1095 Willowdale Rd
Morgantown, WV 26505
USA

Synonyms
Inflammation, chemoattractants, mononuclear cell function, inflammatory cytokines

Definition
Inflammation represents the consequence of capillary dilation followed by edema and emigration of phagocytic leukocytes (Figure 1). It is induced in response to injury from microbial agents, physical agents (burns, radiation, trauma), neoplasia, immune disorders, or toxic agents and the process is characterized by pain, heat, redness, swelling, and loss of function.

Characteristics
Inflammation can be of an acute or chronic nature. Acute inflammation is of relatively short duration and, independent of the agent responsible, is relatively stereospecific. It is associated with edema and neutro-

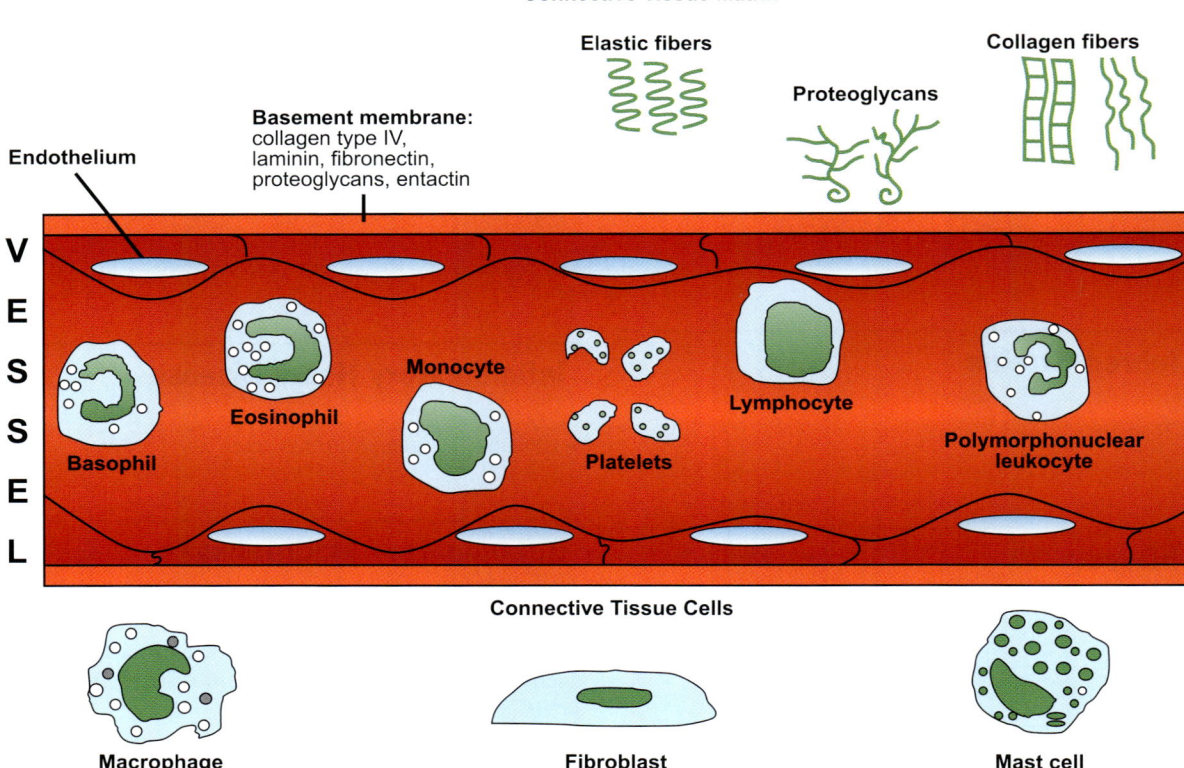

Inflammatory Reactions, Acute Versus Chronic. Figure 1 Cells and their localization in inflammation (from Cotran et al (1) with permission from Elsevier Science).

phil infiltration. Resolution of acute inflammation involves the destruction, dilution or segregation of the injurious agent. A hallmark of acute inflammation are changes in vascular flow and caliber which occur in an orderly sequence of events and eventually results in increased vascular permeability and exudation of fluid containing plasma proteins (▶ edema). Initially, transient vasoconstriction of arterioles occurs followed by vasodilation in the arteriole resulting in new microvascular beds in the affected area. The vasodilation is important because it causes increased blood flow, a hallmark of early hemodynamic changes in acute inflammation. Increased permeability of the microvasculature induces slowing of the circulation and stasis. Once stasis begins leukocytes, primarily neutrophils, adhere to the vascular endothelium (leukocytic margination and adhesion). Some of the adhered leukocytes eventually migrate through the vascular wall into the interstitial tissue and emigrate toward chemoattractants produce by injured tissue (emigration). These leukocytes phagocytize and degrade bacteria, immune complexes and cell debris through oxygen-dependent and oxygen-independent mechanisms. The generation of oxygen metabolites is attributed to the activation of NADPH oxidase which oxidizes NADPH and in the process reduces oxygen to superoxide ion (O^{2-}) which is converted to hydrogen peroxide (H_2O_2). Active substances stored in leukocyte granules, such as lysozyme, are responsible for oxygen-independent mechanisms.

In contrast to acute inflammation, chronic inflammation is less uniform and of longer duration, ultimately resulting in additional tissue destruction and increased connective tissue (fibrosis). Chronic inflammation can result from repeated bouts of acute inflammation or when the inciting stimulus is persistent, causing a low-grade smoldering response. The latter is seen with certain intracellular microorganisms (e.g. tubercle bacilli), autoimmune diseases (e.g. rheumatoid arthritis) or toxic pulmonary fibers or crystals. Histologically, chronic inflammation is associated with the infiltration of mononuclear nuclear cells, (i.e. macrophages and lymphocytes rather than neutrophils) and proliferation of endothelial cells and fibroblasts which leads to the formation of neovascularization. The activation of local macrophages and the continuous recruitment and activation of mononuclear cells from the periphery play a major role in chronic inflammation through the release of secretory products. Some agents induce a characteristic pattern of chronic inflammation referred to as granulomas resulting in granulomatous disease. Granulomas consist of a collection of modified macrophages, referred to as epithelioid cells, surrounded by a rim of lymphocytes. The fusion of these epithelioid cells forms multinucleated giant cells and eventually granulomas.

Mediators of Inflammation

Both acute and chronic inflammatory responses are initiated, propagated and arrested by mediators originating from the plasma, damaged cells, bystander cells or infiltrating cells. Although inflammatory mediators are released to some extent in a sequential manner, it is more appropriate to consider them as acting in concert. Mediators synthesized by the liver and found in plasma include proteins composing the complement, kinin and clotting/fibrinolytic systems. Preformed products released from cells include vasoactive amines, such as histamine and serotonin as well as lysosomal enzymes. Examples of newly synthesized substances released from cells include prostaglandins and leukotrienes. Products secreted from leukocytes, in particular macrophages, play a major role in chronic inflammation. These products include reactive oxygen and nitrogen species, enzymes such as proteases, collagenases and elastases, bioactive lipids, and cytokines. Cytokines have numerous roles in inflammation including cell growth and chemotaxis and, when locally produced, affect the hypothalamic-pituitary-adrenal axis, hepatocyte growth, hepatic metabolic responses, cardiovascular system, muscle and adipocyte catabolism, and hematopoietic-lymphoid reactions.

Preclinical Relevance

With respect to the evaluation of data obtained in immunotoxicity studies acute inflammatory reactions have to be taken into considerations. Such activation of immune competent cells by unspecific pathways could mimic or cover specific immunomodulating properties of the test compounds.

Relevance to Humans

Most forms of acute and chronic inflammation are amplified and propagated as a result of recruitment of humoral and cellular components of the immune system. Thus, drug-induced autoimmune reactions would by definition represent an inflammatory response. In addition, it is generally believed that allergic diseases, including allergic contact dermatitis and respiratory hypersensitivity, require induction of an inflammatory response as a prerequisite for sensitization. There are numerous occupationally and environmentally related diseases that represent chronic inflammatory processes (Table 1). Most of these are organ-specific responses many of which occur in the lung as a result of deposition and persistence of the toxic agent, although others (such as solvent induced hepatitis) may result from multiple exposures causing repeated bouts of acute inflammation. Some of the agents do not produce inflammation by themselves, but rather are found in the environment associated with inflammatory agents, the classical example being the association of cotton dust with endotoxin.

References

1. Cotran R, Kumar V, Robbins S (1989) Inflammation and repair. In: Staff W (ed) Robbin's Pathologic Basis of Disease, 4th ed. WB Saunders, Philadelphia, pp 39–86
2. Luster M, Simeonova P, Gallucci R, Matheson J (1999) Tumor necrosis factor alpha and toxicology. Crit Rev Toxicol 29:491–511
3. Schook L, Laskin D (eds) (1994) Xenobiotics and Inflammation. Academic Press, San Diego
4. Gallin J, Snyderman R (eds) (1999) Inflammation: Basic Principles and Clinical Correlates, 3rd ed. Lippincott Williams & Wilkins, Philadelphia

Influenza

▶ Respiratory Infections

Influenza Virus

A virus that causes an infection of the upper respiratory tract.
▶ Host Resistance Assays

Ingestion

▶ Opsonization and Phagocytosis

Innate Immune Response

The inborn, antigen-independent immune response that represents the initial effort to contain a disease-causing agent prior to the onset of the antigen-specific adaptive immune response. Innate immunity is present in all individuals at all times, and in contrast to the adaptive immune response mediated by B-cells and T-cells it does not improve or adapt to a pathogen, or provide protection against future infection of the same pathogen. Players of the innate immune response are macrophages and neutrophils bearing receptors for

Inflammatory Reactions, Acute Versus Chronic. Table 1 Examples of toxic substances associated with inflammatory-mediated diseases

Organ	Substance	Disease
Lung	asbestos	asbestosis; mesothelioma
silica	silicosis	
coal	pneumoconiosis	
cotton dust	byssinosis	
beryllium	chronic beryllium disease (granulomas)	
mercury	hypersensitivity pneumonitis	
cigarette smoke	chronic obstructive pulmonary disease	
ozone	pulmonary edema, asthma exacerbation	
bleomycin	fibrosis	
iron dust	siderosis	
tin dust	stannosis	
Liver	acetaminophen	hepatotoxicity
	carbon tetrachloride	hepatotoxicity
	organic solvents	hepatitis
Skin	ultraviolet radiation	contact dermatitis, immunosuppression
	sodium lauryl sulfate	contact dermatitis

common bacterial components, NK-cells, and also complement and cytokines. The early innate immune response influences also the adaptive response.
▶ Dioxins and the Immune System
▶ Lymphocytes

Innate Immune System

The non-specific arm of the immune system. The immune responses generated by innate immune cells recognize conserved motifs and do not increase with repeated exposure.
▶ Natural Killer Cells
▶ DNA Vaccines

Innate Immunity

Non-specific and passive immunity mediated by anatomic and physiological barriers, antimicrobial peptides, inducible enzymes, and pluripotential molecules. It is a phylogenetically ancient means of defense which involves phagocytes such as macrophages and dendritic cells to detect an infection. They do so using receptors (e.g. Toll-like receptors, mannose receptors, scavenger receptors) directed against conserved essential structures present on microorganisms. Innate immunity provides a means of combating an infection over the initial few days during which the specific immune system is being activated.
▶ B Cell Maturation and Immunological Memory
▶ Immune Response
▶ Immunotoxicology
▶ Humoral Immunity

Intercellular Adhesion Molecule-1 (ICAM-1)

Intercellular adhesion molecule-1 (ICAM-1) is also known as CD54. It is an interferon-γ-inducible cell membrane glycoprotein involved in leukocyte adhesion and migration, and is of importance in immune and inflammatory responses. It is the ligand of lymphocyte function-associated antigen-1 (LFA-1).
▶ Interferon-γ

Interferon-γ

Rafael Fernandez-Botran
Dept Pathology and Laboratory Medicine
University of Louisville
Louisville, KY 40292
USA

Synonyms

Immune interferon, type II interferon

Definition

Interferon-γ (IFNγ) is a cytokine which is produced mainly by activated natural killer (NK) cells, CD4$^+$ T lymphocytes of the T helper type 1 (Th1) subset, and CD8$^+$ T lymphocytes of the TC1 phenotype (1). IFNγ is a member of the interferon family, a group of cytokines with the ability to inhibit viral replication (1). Interferons have been divided into two main groups: type I or viral interferon, including IFNα (leukocyte derived), IFNβ (fibroblast derived) and IFNω; and type II or immune interferon, including IFNγ. Whereas type I interferons have similar activities and bind to the same membrane receptors, IFNγ displays a different set of immunoregulatory activities and binds to a different receptor, IFNγR (2). IFNγ plays a central role in the regulation of immune responses and is a key factor in the resistance to a variety of infectious pathogens (1,3).

The production of IFNγ is inducible and regulated at the level of transcription (1,3). In NK cells, macrophage-derived cytokines such as TNFα and interleukins IL-12 and IL-18, are the main inducing stimuli. In T cells, IFNγ production is triggered mainly by antigenic or ▸ mitogenic stimulation; however, IL-12 and IL-18 are involved in the differentiation of naive CD4$^+$ T cells into IFNγ-producing Th1 effector cells (1,4). IFNγ is the product of a single-copy gene located on chromosome 12 in the human and chromosome 10 in the mouse (1,4). The IFNγ gene has four exons and encodes a single 1.2 kb mRNA species and a 166-amino acid residue polypeptide containing a signal peptide and two potential *N*-glycosylation sites. Biologically active IFNγ is a noncovalent homodimer with a molecular weight of 34 kDa (4).

IFNγ exerts its effects through interaction with a specific cell-surface receptor (▸ interferon-γ receptor (IFNγR)), which is ubiquitously present on most nucleated cells (5). The IFNγR consists of a heterodimer consisting of a 90-kDa α-chain, involved in high-affinity binding, and a β-chain, required primarily for signaling (5). The α and β chains are associated with the Janus kinases, JAK1 and JAK2, respectively. Binding of a IFNγ homodimer to two IFNγRα-chains induces receptor dimerization and the activation of ▸ JAKS, with subsequent phosphorylation of the cytoplasmic transcription factor STAT1α. Upon phosphorylation, STAT1α dimerizes and translocates into the nucleus, where it binds to specific DNA sequences (▸ gamma interferon activation sites (GAS)) and initiates transcription (5). The list of IFNγ-regulated genes includes more than 200 different genes, and for many of these the functional role is still unclear (1).

Characteristics
Interactions with the Immune System

IFNγ was first described on the basis of its antiviral activity (1,3,4). It is now well established that IFNγ is an extremely pleiotropic cytokine with multiple activities, including immunoregulatory, antiviral, antibacterial, antiproliferative, antitumor and gene-modulatory effects (1,3). It needs to be pointed out that despite the complexity of the activities of IFNγ, its main role in the immune system appears to be the protection against infectious agents, particularly intracellular microorganisms (1). Indeed, the main phenotype of mice with disrupted genes for IFNγ, IFNγRα, or STAT1, is an increased susceptibility to a variety of intracellular pathogens (5). The effects of IFNγ are briefly discussed below.

Immunoregulatory Activities

IFNγ plays a major immunoregulatory role as one of the key cytokines involved in Th1-dominated responses (1,3,4). Generally, these types of responses are associated with cell-mediated immunity and resistance to intracellular pathogens. In this regard, IFNγ acts on different types of cells to promote resistance at multiple levels, including both innate and specific immunity (1). Indeed, interference with the production or activity of IFNγ leads to increased susceptibility to a variety of intracellular microorganisms (e.g. *Leishmania, Toxoplasma, Listeria*). The main immunomodulatory effects of IFNγ are described below.

1. Promotion of Th1 differentiation: IFNγ plays a role not only as an effector molecule but in promoting the differentiation of Th1 cells as well. IFNγ primes macrophages for production of IL-12, a key cytokine involved in the generation of Th1 responses and promotion of IFNγ production by NK cells (1).
2. Regulation of ▸ immunoglobulin class switching: IFNγ acts on B lymphocytes as the main switch factor for IgG2a and promotes that of IgG3 in the mouse. It also antagonizes IL-4-mediated class switch to IgG1 and IgE (1). IgG2a is involved in mediating antibody-dependent cell cytotoxicity (ADCC).
3. Enhancement of antigen presentation: IFNγ upregulates the expression of major histocompatibility complex (MHC) class I and class II molecules and

promotes antigen presentation by both pathways (1,3,4). IFNγ also induces the expression of other molecules involved in antigen processing, such as transporter associated with antigen processing (TAP), proteasomes, and ▶ molecular chaperones (1). Furthermore, IFNγ induces the expression of a number of co-stimulatory molecules on leukocytes, including intercellular adhesion molecule-1 (ICAM-1), leukocyte function-associated antigen-3 (LFA-3), B7.1 (CD80) and B7.2 (CD86) (1).

4. Enhancement of leukocyte-endothelial cell interactions: IFNγ upregulates expression of adhesion molecules on leukocytes and endothelial cells and induces secretion of chemokines (such as ▶ IP-10, macrophage inflammatory protein-1α (MIP-1α), and ▶ RANTES) that mediate lymphocyte and monocyte recruitment (1).

Antiviral Activity

IFNγ has both direct and indirect antiviral activities (1,3,4). Although multiple mechanisms are responsible for the direct antiviral effects of IFNγ, the most important are the induction of a double-stranded RNA activated protein kinase (PKR), 2'-5'oligoadenylate synthetase (2-5A synthetase), and a double-stranded RNA-specific adenosine deaminase (dsRAD). The indirect antiviral effects of IFNγ are related to the activation of cytotoxic T cells and NK cells (1,3,4).

Antibacterial Effects

IFNγ activates macrophages, increasing their capability to destroy intracellular parasites, including bacteria, fungi and protozoa. The antibacterial effects are related to the increased production of reactive oxygen intermediates (superoxide anion, hydrogen peroxide) and nitric oxide (NO) through induction of respiratory burst and inducible nitric oxide synthase (iNOS), respectively (1,3). Similar effects are also observed in neutrophils. In addition, IFNγ stimulates expression of indoleamine 2,3-dioxygenase (IDO), an enzyme involved in tryptophan metabolism and in the inhibition of parasite growth through intracellular tryptophan depletion (3).

Antiproliferative Effects

IFNγ slows down the growth of a number of normal and malignant cells, for example keratinocytes, endometrial cells, and vascular smooth muscle cells (3). Particularly in malignant cells, the antiproliferative effects of IFNγ are related to a differentiating effect or induction of apoptosis. Many mechanisms take part in the antiproliferative effects of IFNγ, including suppression of oncogene expression, inhibition of growth factor production, decreased expression of ▶ transferrin receptors, inhibition of DNA synthesis, tryptophan starvation (through induction of IDO), and blocking of the cell cycle (3).

Antitumor Effects

The effects of IFNγ on tumors are a combination of their direct antiproliferative activities and their stimulatory activities on immune effector mechanisms, including macrophage and NK cell activation, enhanced antigen presentation, and effects on T lymphocytes. Exposure of macrophages to IFNγ leads to an enhancement of their tumoricidal activity. In some systems, IFNγ has been shown to have angiostatic activity as well (3,4).

Modulation of Gene Activity

By modulating the expression of a variety of genes, IFNγ exerts many of the immunostimulatory, antiproliferative, antitumor and bactericidal effects discussed above. IFNγ also induces the expression of additional cytokines and chemokines, influencing inflammation and immune responses. It should be pointed out that the responses of cells to IFNγ are dependent on the cell type and other factors (3).

Relevance to Humans

The central role of IFNγ in mediating resistance to infectious pathogens, both in humans and animals, cannot be overemphasized. Thus, alterations in the production and/or activity of IFNγ often result in increased susceptibility to infectious diseases, particularly by intracellular microorganisms. Moreover, alterations in the normal production of IFNγ (e.g. autoimmune diseases) often contribute to the pathophysiology of a variety of human diseases, for example multiple sclerosis, graft-versus-host disease, and systemic lupus erythematosus (SLE).

IFNγ has sparked great interest because of its potential therapeutic applications. These have been investigated in experimental studies and many applications are currently being evaluated in clinical trials. The clinical applications of IFNγ are based on its antiviral, antibacterial, antitumoral and immunostimulatory activities, and thus IFNγ therapy has targeted a variety of infectious, immune and malignant diseases (Table 1). Because of its many gene-regulating activities, therapy with IFNγ has also many contraindications, including its use during pregnancy (to avoid risk of fetal damage) or in individuals with diseases in which it plays a pathophysiologic role, such as multiple sclerosis, Kaposi's sarcoma, ▶ thrombocytopenic purpura, some lymphatic leukemias and hemoglobinopathies (because of suppression of hemopoiesis). Side effects of IFNγ use have been observed mostly with the parenteral route of administration and include moderate fever, fatigue, headache, myalgia, nausea, anorexia, diarrhea and leukopenia. Rarely, more severe side ef-

Interferon-γ. Table 1 Clinical applications of interferon (IFN) γ and diseases in which IFNγ has demonstrated curative effects

Disease Type	Examples	Likely Mechanism of Action
Infectious disease		Antiviral effects, promotion of Th1-type responses, NK and CTL cell stimulation, increased antigen presentation, macrophage activation
Viral disease	Herpes simplex (types 1 and 2), human papillomaviruses HPV (condyloma acuminata), respiratory viruses (influenza)	
Bacterial disease	Mycobacteria (tuberculosis)	
Protozoal disease	Leishmania	
Fibroproliferative disorder	Systemic sclerosis, scleroderma, idiopathic pulmonary fibrosis	Antifibrogenic, anti-proliferative effects
Other disease	Osteopetrosis, allergic conditions	Metabolic effects, enhancement of Th1 responses
Malignant disease	Renal cell carcinoma, melanoma, sarcoma	NK and CTL activation, enhanced antigen presentation, antiproliferative effects, enhanced tumoricidal activity

CTL, cytotoxic T lymphocyte; NK, natural killer cells; Th1, T helper cell.

fects have been reported, including ▶ lymphadenopathy, cardiovascular, neurophysiological and neuropsychological disturbances, kidney and liver insufficiency and bone marrow hypoplasia (3).

References

1. Boehm U, Klamp T, Groot M, Howard JC (1997) Cellular responses to interferon-γ. Annu Rev Immunol 15:749–795
2. Aguet M, Dembic Z, Merlin G (1988) Molecular cloning and expression of the human interferon γ receptor. Cell 55:273–280
3. Tsanev RG, Ivanov IG (2002) Immune Interferon. Properties and Clinical Applications. CRC Press, Boca Raton, pp 5–174
4. de Maeyer E, de Maeyer-Guignard J (1991) Interferons. In: Thomson A (ed) The Cytokine Handbook. Academic Press, London, pp 215–239
5. Bach EA, Aguet M, Schreiber RD (1997) The IFNγ receptor: a paradigm for cytokine receptor signaling. Annu Rev Immunol 15:563–5 91

Interferon (IFN)

Originally found to be released from virus-infected cells and to induce an antiviral state in other cells, interferons were later additionally characterized as cytokines that participate in immune regulation. Three groups exist: IFN-α, IFN-β, and IFN-γ. Among interferons, the T lymphocyte and NK cell product IFN-γ has been found to be the main macrophage activating cytokine.

▶ Cytokines
▶ Respiratory Infections

Interferon-Inducible Protein 10 (IP-10)

Interferon inducible protein 10 (kD) belongs to the CXC(α) chemokines and binds to the chemokine receptor CXCR-3. It is also known as CXCL10.

▶ Three-Dimensional Human Skin/Epidermal Models and Organotypic Human and Murine Skin Explant Systems
▶ Chemokines

Interferon-γ Receptor (IFNγR)

This is a membrane molecule involved in the responses of cells to IFNγ. The IFNγR consists of a heterodimer consisting of a 90-kDa α chain, involved in high-affinity binding, and a β chain, required pri-

marily for signaling. The IFNγ-mediated signal transduction event involves IFNγ-binding, activation of IFNγ-R-associated janus kinases (JAK1 and JAK2), activation of STAT1 with subsequent translocation into the nucleus and interaction with specific DNA sequences (gamma-regulated sites, or GAS).
► Interferon-γ

Interferons

Proteins that are formed in the presence of a virus, that prevent viral reproduction, and that can induce resistance to a variety of viruses.

Interleukin (IL)

A group of proteins (cytokines) referred to by number (e.g. IL-1) that are produced primarily by leukocytes. Various combinations of interleukins and other cytokines control the initiation, maintenance and homeostasis of specific and nonspecific immune responses. They induce growth, differentiation, activation or cooperation between leukocytes.
► Cytokines
► Primate Immune System (Nonhuman) and Environmental Contaminants
► Glucocorticoids
► Canine Immune System

Interleukin-1 Receptor Accessory Protein

IL-1 receptor accessory protein (RacP) is protein involved in the binding of IL-1 to its receptor responsible for modifying the bond from low affinity to high affinity.
► Interleukin-1β (IL-1β)

Interleukin-1 Receptor Antagonist

Interleukin-1 receptor antagonist (IL-1ra) is an endogenous receptor antagonist for IL-1.
► Interleukin-1β (IL-1β)

Interleukin-1 Receptor Associated Kinase I

IL-1 receptor associated kinase I (IRAK-1) is involved in the signal transduction cascade of IL-1. It is shown to play a role in susceptibility to organ-specific autoimmunity.
► Interleukin-1β (IL-1β)

Interleukin-1 Receptor Associated Kinase II

IL-1 receptor associated kinase II (IRAK-2) is involved in the signal transduction cascade of IL-1.
► Interleukin-1β (IL-1β)

Interleukin-1β (IL-1β)

DOROTHY B COLAGIOVANNI
OSI Pharmaceuticals, Inc.
2860 Wilderness Place
Boulder, CO 80301
USA

Synonyms
Interleukin-1F2, lymphocyte-activating factor

Definition
Interleukin-1β (IL-1β) is a proinflammatory cytokine found in the serum of both healthy and diseased humans. One of the main roles of IL-1 is to increase antigen-induced T-cell activation. Interleukin-1β is produced by a variety of cell types, including endothelial cells, macrophages, osteoblastic cells, hepatocytes, fibroblasts, dendritic cells, and T lymphocytes. While not produced by every cell type, IL-1β production influences nearly every cell type in the body (1). Interleukin-1β is secreted by immune cells in response to foreign pathogens and in many autoimmune diseases. Following exposure to lipopolysaccharide (LPS), for example, IL-1β is secreted from mononuclear cells within 90 minutes and initiates activation of inflammatory cascades (2). There are both local and systemic effects of IL-1β. This is in contrast to IL-1α which is locally active and associated with the surface of cells involved with antigen presentation. A local effect of IL-1β is activation of vascular endothelium or lymphocytes in response to a bacterial infection. Systemically, fever and IL-6 induction occur following IL-1β secretion

from cells. Interleukin-1β is one of the body's key mediators in response to microbial invasion, inflammation, and immunologic reactions (2). While acting as a critical regulator of immune function, this cytokine is also an important cause of tissue injury. Polymorphonuclear leukocytes (PMNs) and mast cells release cytokines including IL-1β and ▶ tumor necrosis factor-α (TNF-α) at sites of inflammation. Aberrant production of these cytokines can result in deleterious consequences, as in the case of rheumatoid arthritis where excess production of IL-1β and TNF-α mediates cartilage and bone destruction associated with increased disease severity.

Molecular Characteristics

Interleukin-1β is a polypeptide with a molecular weight of approximately 17 kD. It is a member of the IL-1 family that includes IL-1α, IL-1β, IL-1 receptor antagonist (IL-1ra) and IL-18. Among the cytokines, IL-1β is unique in having an endogenous competitive inhibitor for IL-1 binding in the form of IL-1ra. IL-1ra is able to bind to the IL-1 receptor subtype I (IL-1RI) and prohibit IL-1β from initiating potent inflammatory activity. During infection, high concentrations of IL-1ra are produced. Even in the non-disease state the levels of IL-1ra exceed IL-1β by ten times in the systemic circulation, yet IL-1β binds far more tightly to the IL-1RI than IL-1ra, enabling IL-1β to modulate activity at lower concentrations (1).

Both IL-1β and IL-1α can bind to the IL-1RI or subtype II (IL-1RII) receptor. Soluble portions of IL-1RI and IL-1RII circulate normally in healthy humans and may serve to buffer concentrations of IL-1α, IL-1β and IL-1ra. The IL-1RI receptor is found on endothelial cells, hepatocytes, fibroblasts, and T lymphocytes. The IL-1RII receptor is located on B lymphocytes, monocytes and PMNs. Some cells express both subtypes of receptor and compete for IL-1 binding. The competition between signaling and non-signaling receptors for the same ligand is another unique characteristic of the IL-1 family. The IL-1RII receptor is considered to be a "decoy receptor" as it does not transduce a signal when IL-1β binds to it. IL-1β can tightly bind to the IL-1RII to prevent binding to IL-1RI. This avid bond (100 pmol/l) with a dissociation rate in excess of two hours may act to ameliorate IL-1 mediated effects (1). In contrast, IL-1β has a lower affinity for cell bound IL-1RI (500 pmol/l to 1 nmol/l) (1).

The benefits of IL-1RII and IL-1ra modulating IL-1β bonding to IL-1RI are two-fold: (i) excessive circulating concentrations of IL-1β can be diminished to prevent a runaway inflammatory cascade; and (ii) natural antagonism may prevent levels of IL-1β from fluctuating too far out of a given range by acting as a systemic regulatory mechanism. The balance of the agonist, IL-1, versus the antagonist, IL-1ra, may be critical to disease outcome.

The number of IL-1RI receptors required for IL-1β to initiate a signal appears to be very small. As few as 10 receptors expressed per cell appear to be adequate. Interleukin-1β must attach at two IL-1RI receptors for activation. At the cell membrane, signal transduction is initiated by complex formation between extracellular IL-1 and the transmembrane IL-1RI and IL-1R accessory protein (IL-1RacP). The initial binding of IL-1β to IL-1RI is a low affinity bond, however once the IL-1RacP joins the complex there is a shift to a high affinity bond. This high affinity bond may occur because IL-1β bonding to two receptors causes a structural/conformational change. This change may allow IL-1RacP to attach to a different portion of the molecule not previously accessible. The complex is then translocated intracellularly and associates with MyD88, the myeloid differentiation protein 88. Next, MyD88 associates with and recruits interleukin-1 receptor-associated kinase-1 and interleukin-1 receptor-associated kinase-2 (IRAK-I and IRAK-2) to the IL-1 receptor complex in response to IL-1β binding. Activation of these kinases then induces ▶ tumor necrosis factor-receptor associated factor-6 (TRAF-6) that can transduce multiple signals from TOLL-like receptors (4). As part of the signaling, TRAF-6 activation from IL-1β stimulation induces P38 kinase and ▶ Jun NH2-terminal kinase (JNK). The culmination of the cascade of intracellular signaling events is the translocation of NF-κB to the nucleus and the nuclear transcription of inflammatory genes including ▶ macrophage inflammatory protein-1β (MIP-1β), signal transducer and activation of transcription(STAT)-1α/β and ▶ RANTES (regulated on activation, T cell expressed and secreted).

Unlike the binding of IL-1β or IL-1α to IL-1RI with signal transduction initiation, IL-1ra does not possess agonist activities. While it readily binds to IL-1RI, no signal occurs. IL-1ra is unable to form the high affinity complex with IL-1RacP, possibly explaining why IL-1ra binding to IL-1RI does not induce signal transduction. Again, this may serve to modulate the *in vivo* activity of IL-1β.

Putative Interaction with the Immune System

To be the first recognized interleukin is a testament to the importance of IL-1. There are four critical areas of IL-1 interaction with the immune system: pathogen defense, T cell activation and response, autoimmune disease manifestation, and tumor suppression.

Interleukin-1β is critical in combating infections from numerous foreign pathogens, including the species *Listeria*, *Pneumocystis*, and *Klebsiella* (for a comprehensive review of the role of IL-1 in infections see Dinarello (3)). Interleukin-1β acts as a non-specific

Interleukin-1β (IL-1β). Table 1 In vivo activity of interleukin-1β of interest to immunotoxicology

Stimulates production of arachidonic acid metabolites
Increased antibody production (adjuvant effect)
Increased synthesis of acute phase proteins
Enhanced spleen mitogenic response to lipopolysaccharide
Activation of inducible nitric oxide synthase
Fever induction
Suppression of total cytochrome P450
Stimulation of the hypothalamic–pituitary–adrenal axis
Induction of gene expression for type 2 phospholipase A2 and cyclooxygenase-2
Increase in gene expression of c-Kit on bone marrow cells
Increased expression of ELAM-1, VCAM-1 and ICAM-1
Proliferation of fibroblasts, smooth muscle cells and mesangial cells
Hematopoietic stem cell activation

ELAM, endothelial leukocyte adhesion molecules; ICAM, intercellular adhesion molecule; VCAM, vascular cell adhesion molecule.

antigen mediator functioning as an activating signal in T-cell-dependent antigen specific immune responses. IL-1β and IL-1α act in concert to combat foreign invaders by modulating fever, upregulating receptor expression of intracellular adhesion molecule (ICAM)-1 and recruiting neutrophils to sites of infection. Upregulation of additional cytokines is also pivotal to the inflammatory cascade. IL-1β induces IL-2 and IL-6 synthesis and secretion. Three particular genes, inducible nitric oxide synthase (iNOS), type 2 cyclooxygenase (COX-2) and phospholipase A2 (PLA-2) are exquisitely sensitive to IL-1. Once activated via IL-1β in infection, these genes cause production of nitric oxides, prostaglandins, leukotrienes, and platelet activating factor that are proinflammatory mediators.

Interleukin-1 is known to act as a maturation signal for T lymphocytes, preparing them to respond to antigen. In the signal transduction cascade of IL-1β, two components, MyD88 and IRAK-1 may be critical for the generation of a Th1 type response. Studies in IL-1RI- or IRAK-1-deficient mice demonstrate a complete resistance to experimental autoimmune encephalomyelitis—a model with pathogenesis reminiscent of multiple sclerosis. Experimental autoimmune encephalomyelitis is a Th1-mediated autoimmune disease. IRAK-1 deficient T cells exhibit impaired priming, failure to proliferate, and failure to secrete IFN-γ in response to antigen (4). Results suggest that the IL-1β signaling pathway plays an essential role in T cell priming and demonstrate that development of autoimmunity may be mediated via innate immunity (4).

In light of the beneficial role IL-1 plays in fighting invading pathogens and priming T lymphocytes, it is ironic that IL-1β is also a major cause of cellular damage and tissue injury. Its involvement in autoimmune diseases such as rheumatoid arthritis and multiple sclerosis is well proven. The extracellular domain of the IL-1RI and IL-1RII receptors are found as "soluble" molecules in the circulation of patients with sepsis and in the synovial fluid of those with arthritis where high circulating levels of IL-1β are detectable (1).

IL-1 regulates the production of metalloproteinases such as collagenase and stromolysin by synovial cells and chondrocytes (5). Chronic exposure to inflammatory mediators like IL-1 causes joint tissue to thicken. This thickened tissue invades cartilage and bone and is responsible for the characteristic bone erosions associated with rheumatoid arthritis. Some data suggest that immune-mediated diseases are caused by exposure to foreign pathogens and it may be the actual host's response to infection influencing the development of autoimmunity. The body's response to an infective agent may be causing the overproduction of IL-1β that results in tissue damage.

Lastly, IL-1β has demonstrated roles in tumor suppression and development. The administration of IL-1 in tumor-bearing animals results in regression of tumors by immune-mediated mechanisms. IL-1β treatment has also been shown to increase the generation of lymphokine activated killer T cells in models of adoptive immunotherapy that might be beneficial in targeting tumor cells. In addition to direct effects on tumor cells, others have shown a benefit to combining IL-1β with G-CSF to enhance recovery of myelopoeitic cells following chemotherapy- or radiation-induced myelosuppression (6). While this suggests that IL-1β treatment may be beneficial in tumor therapy, some tumors are known to secrete IL-1. Blockade of IL-1β secretion has been evaluated as a therapeutic strategy for tumor suppression. Because of the heterogeneous nature of tumors, in general, it is difficult to predict the benefit or adverse effect of IL-1β therapy. Tumors might be screened for IL-1 production and gene expression to assess the value of antagonist or agonist therapy.

Relevance to Humans
Members of the IL-1 family are found on the long arm of chromosome 2. In a 430-kb region at 2q13–14, the genes for IL-1α, IL-1β and IL-1ra and encoded by *IL-1A*, *IL-1B* and *IL-1RN*, respectively (7). Some diseases appear to be linked, in part, to genetic polymorphisms of *IL-1B* and *IL-1RN*. These conditions may include periodontal disease, *Helicobacter pylori*-related gastric carcinogenesis, rheumatoid arthritis, ankylosing spondylitis, and inflammatory bowel disease. Studies suggest a relationship with IL-1 polymorphisms and susceptibility and pathogenesis of diseases.

In healthy individuals, IL-1β concentrations do not appear to change with age, although in elderly people IL-1ra increases 10-fold compared to younger individuals (1). The production of IL-1β in women varies in association with the stages of the menstrual cycle. In studies assessing effects of IL-1β administration on healthy adults, 1 ng/kg of IL-1β by intravenous administration caused fever and hypotension, mimicking sepsis (1). Additionally, there is a general enhancement of hematopoiesis following IL-1β administration indicated by increased mononuclear cell and platelet counts.

In disease settings, heightened concentrations of circulating IL-1β have been reported following viral, bacterial, fungal, and parasitic infections. Additionally, elevated IL-1β levels have been measured in patients with HIV, solid tumors, surgical trauma, graft-versus-host disease and following exposure to UV radiation. Significant elevations from normal circulating concentrations can be measured in conditions such as sepsis (> 30 pmol/l) or following severe burns (1).

Modulation of inflammatory diseases through regulation of cytokine production and local cytokine levels has tremendous therapeutic potential. Some drug therapies have demonstrated systemic reductions in circulating IL-1β levels. These therapies include corticosteroids and COX-2 inhibitors. These substances suppress IL-1β gene expression and secretion and increase IL-1RII expression (1). Cytokine inhibitors have been shown to reduce disease severity in experimental animal models. Additionally, clinical trials have yielded successful results with monoclonal antibodies to inflammatory cytokines by inhibiting disease pathology. Currently a recombinant form of hu-IL-1ra is available for rheumatoid arthritis patients. Trials have demonstrated clinical benefit to disease status and quality of life.

If IL-1β levels in diseased patients are modulated by drug therapy, a logical question would be: are reductions of proinflammatory cytokine responses altering the normal protective mechanisms against invading pathogens? Inhibition of these cytokines has been shown to adversely influence host immune function. IL-1β blockade, with specific antibodies or murine gene knockout strains, demonstrates increases in mortality in animal models of infection (8). Additionally, some recent clinical trial data suggest that blockade of IL-1 and TNF-α simultaneously may lead to an increased risk of infection. While suppressing excessive cytokine concentrations may be advantageous to the patient, modulation of IL-1β should be carefully monitored to maintain effective immune system function.

References
1. Dinarello CA (1996) Biologic basis for interleukin-1 in disease. Blood 87:2095–2147
2. Cohen J (2002) The immunopathogenesis of sepsis. Nature 420:885–891
3. Dinarello CA (1992) The role of interleukin-1 in infectious diseases. Immunol Rev 127:119–146
4. Deng C, Radu C, Diab A et al. (2003) IRAK-1 regulated susceptibility to organ-specific autoimmunity. J Immunol 170:2833–2842
5. Tiku K, Thakker-Varia S, Ramachandrula A, Tiku ML (1992) Articular chondrocytes secrete IL-1, express membrane IL-1 and have IL-1 inhibitory activity. Cell Immunol 140:1–20
6. Moore MA, Warren DJ (1987) Synergy of IL-1 and G-CSF: in vivo stimulation of stem cell recovery and hematopoietic regeneration following 5-fluorouracil treatment of mice. Proc Natl Acad Sci 84:7134–7138
7. van der Paardt M, Crusius JBA, Garcia-Gonzalez MA et al. (2002) Interleukin-1β and interleukin-1 receptor antagonist gene polymorphisms in alkylosing spondylitis. Rheumatology 41:1419–1423
8. Havell EA, Moldawer LL, Helfgott D, Kilian P, Sehgal B (1992) Type I IL-1 receptor blockade exacerbates murine listeriosis. J Immunol 148:1486–1492

Interleukin-1F2 (IL-1F2)

▶ Interleukin-1β (IL-1β)

Interleukin-4 (IL-4)

▶ Maturation of the Immune Response

Interleukin-5 (IL-5)

A cytokine mainly produced by T cells. It stimulates differentiation and activation of eosinophilic leukocytes. Drug-specific T cells produce it in high amounts, which might explain why eosinophilia is a common feature of drug hypersensitivity.
▶ Lymphocyte Transformation Test

Interleukin-8 (IL-8)

Interleukin-8 (IL-8) is an ELR-CXC chemokine also known as CCL8. It is an important mediator of the migration of neutrophils during inflammation.
▶ Chemokines

Interleukin-12 (IL-12)

Interleukin-12 (IL-12) is a heterodimeric 70 kDa glycoprotein (IL-12p70) consisting of a 40 kDa subunit (IL-12p40) and a 35 kDa subunit (IL-12p35) linked by disulfide bonds that are essential for its biological activity. It is secreted by peripheral lymphocytes, mainly by B cells, and to a lesser extent by T cells, after induction. Strong stimuli for the induction of IL-12 production are bacteria, their products and parasites, whereas IL-10 inhibits its production. IL-12 exhibits different effects on T Helper (Th) cell subpopulations. In Th1 cells, it induces the synthesis of IL-2, interferon (IFN)-γ or tumor necrosis factor (TNF), whilst in Th2 cells the synthesis of IL-4, IL-5 and IL-10 is inhibited by IL-12. IL-12 stimulates the proliferation of lymphoblasts, activates CD56$^+$ natural killer cells and promotes allogenic cytotoxic T cell reactions. IL-12 enhances myelopoiesis of bone marrow progenitor cells and synergizes with colony stimulating factors to induce proliferation.
▶ Cancer and the Immune System

Interleukin-18 (IL-18)

Interleukin-18 (IL-18) is one of the proinflammatory cytokines. It is produced during the acute immune response by macrophages and immature dendritic cells. It lacks a classical signal sequence necessary for secretion. IL-18 is synthesized as a precursor protein with limited biological activity and is processed by caspase-1, to generate the fully active molecule. An important function of IL-18 is the regulation of functionally distinct subsets of T helper (Th) cells required for cell-mediated immune responses. IL-18 is a pleiotropic cytokine. It induces activated B cells to produce interferon (IFN)-γ and has been shown to strongly augment the production of by IFN-γ by T cells and natural killer cells.
▶ Cancer and the Immune System

Internalization

▶ Opsonization and Phagocytosis

International Collaborative Studies on the Detection of Immunotoxicity

ANTHONY D DAYAN
Department of Toxicology, Queen Mary and Westfield College
University of London
London
UK

HENK VAN LOVEREN
RIVM
Bilthoven
The Netherlands

RALPH J SMIALOWICZ
EPA
Research Triangle Park,
USA

IAN KIMBER
Central Toxicology Laboratory
Syngenta plc
Macclesfield
UK

This report has been reviewed by the Office of Research and Development of the Environmental Protection Agency and has been approved for publication. Approval does not signify that the contents necessarily reflect the views and policies of the Agency, nor does mention of trade names of commercial products constitute endorsement or recommendation of use.

Definition
Regulatory Toxicology and Immunotoxicology

The immunotoxicologist, toxicologist and clinician potentially have an interest in detecting immunotoxicity. One is as a means to explore basic mechanisms of the immune system and its role in health and disease. The other, far more pragmatic interest lies in detecting substances that carry a risk of immunotoxicity, so that exposures can be controlled, rational preventive and therapeutic measures can be considered and appropriate warnings given. The second type of interest is part of practical toxicology, which is normally closely linked to the development and regulation of chemicals by industry and official agencies.

The objectives and often the techniques of the two processes are quite different as the former academic, investigative research must be free to exploit any experimental method, any model system and any substance that meets its needs in a specific investigation. Reproducibility and consistency are needed only to the extent that any truly scientific investigation may justify confirmation by repetition. Specificity and sensitivity are of limited importance because there is flexibility in the way in which experiments can be done, and "accuracy" has little relevance because the objective of the research is to discover some new information more than to display reproducibility and comparability if an investigation is repeated.

In regulatory toxicology, however, important practical, legal, and industrial decisions will be based on the results of the experiments ("tests"), so reproducibility, sensitivity, specificity and accuracy are essential if methods are to gain acceptance. The genesis and course of the international collaborative studies of immunotoxicity were heavily influenced by the use to be made of the results, which focused on their potential regulatory value. The aim of the work was to examine the adequacy and feasibility of various laboratory procedures for detecting and characterizing the potential of substances to affect the immune system, notably considerations of dosage, duration of exposure and test system, so that any *hazard* of immunotoxicity might be detected (i.e. potential of the substance to affect that system). It was considered important to be able to extrapolate from that information to the possible *risk* of immunotoxicity in humans or other animals (i.e. the likelihood of occurrence, the nature and the severity of the toxic effect after a given exposure of the target species). That prediction could be the basis for suggesting restrictions on use and exposure to a substance, warning labels to be applied to it and constraints on its application in industry, the home, and the environment. The collaborative work was deliberately directed at methods that might give reproducible results appropriate to support regulating exposure to a chemical. The importance and legal standing of decisions to be based on a finding or exclusion of immunotoxic potential were such that methods were to be robust, readily transferable between laboratories, reproducible and specific, and as sensitive as possible. Further constraints on the development of such methods and procedures were the ethical and economic needs. These include control of any use of experimental animals and other costly human and financial resources, and the importance of demonstrating which methods were likely to be most readily accepted internationally because of the growing importance of global regulation of chemical hazards.

The distinction between "scientific" and "regulatory" methods is far from absolute, because initially the only available knowledge about a new subject area will probably have come from academic research techniques, which are likely to require modification and development to make them suitable for regulatory purposes. Also, as new substances are submitted for "immunotoxicity" testing it is probable that anomalous results will emerge that suggest new paths for basic investigation.

Characteristics
Background to International Collaborative Studies of Immunotoxicity (ICIS)

At the scientific level, by the early 1980s there was extensiv understanding of the immune system, awareness that changes in it could be induced by chemical and physical agents, and experience of diseases of man and animals due to abnormal functioning of the immune system. This suggested the potential value to governments and industries of being able to predict whether a substance might affect the immune system of humans and animals before there was wide exposure to it. There had not been wide consideration of possible test methods, of how to interpret and employ their results for regulatory purposes; laboratory skills were concentrated in few institutions and were mostly directed at the particular interests of forward-looking investigators rather than efficient and effective means of generally detecting "immunotoxicity". Furthermore, there was no experience of the validation and use for the purpose of controlling chemicals (including drugs) of results intended to show such a broad class of effects on this diverse physiological system (1–4).

The International Programme on Chemical Safety, a joint activity of WHO, the International Labour Office (ILO) and the United Nations Environmental Programme (UNEP) took the initiative in 1984 to arrange a series of international discussions on immunotoxicity testing. Subsequently, an international collaborative study between a large number of laboratories was developed with the intent to show the value and reproducibility of various experimental methods to reveal the occurrence of "immunotoxicity", focused on direct toxic effects of chemical exposure on components of the immune system.

The guiding principles behind the establishment and subsequent conduct of the International Collaborative Immunotoxicity Study (ICIS) from 1984 to its successful completion in 1998 (5,6) were agreed at its first discussion meeting (2). These guiding principles are listed below.

- The first requirement was to show what convenient experimental techniques could be recommended as capable of demonstrating "immunotoxicity" due to a chemical without necessarily demonstrating its nature and pathogenetic mechanism, i.e. which were useful as a screening method to reveal an effect—the hazard of immunotoxicity—that might subsequently deserve more detailed investigation if it were necessary to characterize the risk.
- Studies should be conducted in vivo because no in vitro system offered the possibility of detecting the necessary variety of actions and the important dose-response relationship.
- The rat was preferred to the mouse as the species of which there was most experience among toxicologists, and it offered the possibility of economizing on use of animals if immunotoxicity investigations could be done on animals already under study in formal, regulatory toxicity tests.
- As far as possible, use would be made of specimens from animals in conventional regulatory tests. The 28-day oral toxicity test is the most common study done for regulatory purposes on most chemicals, and this, plus awareness from other investigations of the time required for some effects on the immune system to appear, led to selection of the standard OECD 28-day oral toxicity test as the basic type of experimental approach.
- It was realized that this meant that the value of the test indicating immunotoxicity would come from a more or less conventional pattern of laboratory investigations applied at the end of a toxicity test, and that it would be biased towards static pathological results rather than dynamic measures of immune system function. The latter were very likely to require additional animals, use of procedures that were unfamiliar to many possible participants and their adaptation to the rat model of techniques and reagents so far available only for the mouse.
- As far as possible use would be made of results obtained from the conventional OECD-type test. It was considered necessary to extend the then current protocol by what was termed "enhanced pathology investigations", that is, examination of histological sections from additional lymphoid tissues not then in the protocol for that test (5), namely mesenteric and popliteal lymph nodes, bone marrow, and gut-associated lymphoid tissue. The microscopic examination of those tissues was extended to semiquantitative grading of the state and changes in the T- and B-dependent areas of the principal lymphoid tissues and organs, as well as of the peripheral blood white cell and differential counts.
- Laboratories able to do so would conduct in parallel tests of immune system function. Their techniques and reagents were standardized and arrangements were made for experienced investigators to help to train those unfamiliar with particular techniques.
- All experimental procedures were standardized as far as possible, i.e. the studies were performed under GLP-like control, conventional methods were used for conventional investigations, e.g. weighing organs and making white cell counts, and standardized instructions were circulated for novel methods. For example, the histopathologists circulated instructions and specimen photographs of tissue sections to show the areas graded and the appearance of changes of varying degrees of severity. A workshop was held at which histopathologists compared slides and assessments.
- All qualitative and quantitative data were collected centrally and analyzed and reported according to preset guidelines and after evaluation and agreement by all the participants.

The first discussion meeting of ICIS agreed that it would be necessary to conduct separate collaborative experiments on immunosuppressants and immunostimulants. However, there was initial uncertainty about suitable compounds for elicitation of both types of immune response. Also, individual laboratories and investigators had only a limited capacity to cope with the very large amount of work, all to be done at their own expense. It was also agreed that no general method for detecting the likelihood of causing autoimmune disorders had yet been proven to have sufficient general promise to justify its incorporation into ICIS.

The first compound tested was azathioprine, chosen as a well-characterized immunosuppressant. Subsequently cyclosporin was studied in the same way as an example of an immunosuppressant that did not act by causing cytotoxicity.

There was much discussion and some uncertainty about a possible immunostimulant compound. Eventually there was agreement that hexachlorobenzene would probably be the best example of a substance with this type of effect, but its study was left for later as dealing with the two immunosuppressants took up all the available time and resources from 1988 to 1991–92.

Results of ICIS

The results and analyses of the "enhanced OECD-type" test and the functional studies on azathioprine and cyclosporin were reported in detail by IPCS (1). They showed that addition of the "enhanced pathology" component to the then conventional OECD Guideline 407 test protocol permitted the detection and evaluation of the effects of both azathioprine and cyclosporin on the immune system. Of the function tests, the standard sheep red blood cell (SRBC) antibody response was equally effective (i.e. level of circulating antibody or antibody-producing cells in the spleen after intraperitoneal injection of SRBCs during dosing with the test compound).

The basic investigations in the OECD test without the enhanced pathology and the many alternative immune function methods either did not demonstrate the anticipated effects or proved too capricious for recommendation for general use.

Subsequently, the OECD adopted the enhanced pathology as part of its revision of the 28-day oral toxicity test in the rat (7). They have strongly urged routine deployment of the SRBC antibody response in addition. CPMP, in their guideline for testing drugs, and the US EPA in their "Immunotoxicity" guideline, have made functional testing mandatory, with analysis of antibody to SRBC being suggested as such a functional test.

Further International Studies—the BGVV Ring Tests

The German Federal Institute for Health Protection of Consumers and Veterinary Medicine (Bundes Institut für Gesundheitlichen Verbraucherschutz und Veterinärmedizin; BGVV) coordinated two further multilaboratory studies into the detection of immunotoxicity, to familiarize itself and the scientists with the techniques and results and to examine the capability of the ICIS-type protocol to reveal the action of a putative immunostimulant.

The first BGVV study was based on examination of cyclosporin in the rat by five participating institutes of the BGVV. It employed a similar protocol to that of ICIS and added analysis of a number of phenotypic surface markers of circulating white cells and several functional tests, including the SRBC antibody response.

The importance of the enhanced pathology component in demonstrating specific effects on the immune system was confirmed and so were phenotypic analysis of leucocytes and several functional tests. Greater standardization of the histopathological grading was recommended, as was better control of the function tests.

It was concluded that studies done according to the old OECD Test Guideline 407 would not have revealed "immunotoxicity" whereas work that followed the new "enhanced" OECD Guideline 407 would do so, at least as far as an immunosuppressant compound like cyclosporin was concerned.

Subsequently the BGVV arranged a further comparative test in the rat, this time of hexachlorobenzene as an example of an immunostimulant compound (7).

The second comparative test in particular revealed the value of combined evaluation of enhanced histopathology and additional immunological methods such as flow cytometric analysis. Histopathological evaluation of immune organs, the analysis of subpopulations of lymphocytes by flow cytometry, and the incorporation of one functional test in combination with the overall evaluation of general toxicity was best able to demonstrate whether the the immune system was a primary target of a test compound (8). These additional parameters proved to be of decisive value in hazard identification in immunotoxicity.

The value of the "enhanced" OECD Test Guideline 407 was confirmed and so were several functional tests, but the particular range of actions of this compound suggested that additional or alternative methods might also have been worth employing.

Conclusions

The work done in ICIS for WHO and other official bodies and for the BGVV shows the value of multi-laboratory comparative studies as a feasible and effective means of familiarizing laboratories with novel test methods and simultaneously showing how robust and transferable those methods are whilst generating results on which to begin to judge the practical value of the techniques. These points and other more general and very important factors in the regulatory detection and evaluation of the immunotoxicity of chemicals have been summarized by many of the participants in the ICIS and BGVV projects (7).

This report has been reviewed by the Office of Research and Development of the Environmental Protection Agency and has been approved for publication. Approval does not signify that the contents necessarily reflect the views and policies of the Agency, nor does mention of trade names of commercial products constitute endorsement or recommendation of use.

Regulatory Environment

Relevant guidelines mentioned in the text are as following:
- OECD Guidelines for the Testing of Chemicals Number 407: 28-Day Oral Toxicity Test in Rodents. Adopted 12 May 1981. OECD, Paris, 1981
- OECD Guidelines for the Testing of Chemicals Number 407: Repeated Dose 28-Day Oral Toxicity Test in Rodents. Adopted 27 July 1995. OECD, Paris 1995]

- CPMP CPMP/SWP/1042/99 Note for Guidance on Repeated Dose Toxicity, 27 July 2000
- US-EPA Health Effects Test Guidelines, Immunotoxicity. OPPTS 870.7800, August 1998

References

1. International Programme on Chemical Safety (IPCS) and Commission of the European Communities (CEC) (1987) Immunotoxicology. Proceedings of the International Seminar on the Immunological System as a Target for Toxic Damage: Present status, open problems and future perspectives. Martinus Nijhoff, Dordrecht
2. International Programme on Chemical safety (1984) Development of predictive testing for determining the immunotoxic potential of chemicals. Report of a Technical Review Meeting. WHO, Geneva
3. Luster MI, Munson AE, Thomas PT et al. (1988) Development of a test battery to assess chemical-induced immunotoxicity: National Toxicology Programme's guidelines for immunotoxicity evaluation in mice. Fund App Toxicol 10:2–19
4. Van Loveren H, Vos GJ (1989) Immunotoxicological considerations: a practical approach to immunotoxicity testing in the rat. In: Dayan AD, Payne AJ (eds) Advances in Applied Toxicology. Taylor and Francis, London, pp 143–163
5. Dayan AD, Kuper F, Madsen C et al. (1988) Report of validation study of direct immunotoxicity in the rat. International Collaborative Immunotoxicity Study. Toxicology 125:183–201
6. Dayan AD, Smith EMB, Sundaram S, Wing M (1995) International Workshop on Environmental Immunotoxicology and Human Health. International Programme on Chemical Safety and UK Department of Health. Proceedings of a workshop held in Oxford, 22–25 March 1994. IPCS Joint Series 17. Hum Exp Toxicol 14(79)
7. IPCS (1996a) International Collaborative Immunotoxicity Study (Phase I). Report of the interlaboratory comparison of the effects of azathioprine on the pathology and functional tests of immunotoxicity in the rat. WHO, Geneva
8. Vohr HW, Rühl-Fehlert C (2001) Industry experience in the identification of the immunotoxic potential of agrochemicals. Science Total Environ 270:123–134

Intracellular Adhesion Molecule

ICAM-1 is a molecule expressed on activated vascular endothelium that binds LFA-1 receptors on effector T cells.
▶ Interleukin-1β (IL-1β)

Intracellular Cytokine Staining (ICS)

Cells which are producing specific cytokines can be visualized by permeabilizing the cells and adding fluorescently-labeled anticytokine antibodies.
▶ Antigen-Specific Cell Enrichment
▶ Maturation of the Immune Response

Intracellular Parasites

These parasites are incapable of replicating except when inhabiting the inside of a cell. All viruses are obligate intracellular parasites, and a number of bacteria are obligate intracellular parasites (for example, the chlamydias and the rickettsias).
▶ Viability, Cell

Intracellular Staining by Flow Cytometry

▶ Cytokine Assays

Intraepithelial Lymphocytes

Intraepithelial lymphocytes (IELs) are T lymphocytes interdigitated between epithelial cells in many mucosal sites.
▶ Mucosa-Associated Lymphoid Tissue

Intraluminal Pressure and Flow

Two main factors of lymphodynamics which modulate lymphatic contractility, adjusting it to the current levels of lymph formation and flow.
▶ Lymph Transport and Lymphatic System

Intravascular Hemolysis

Red blood cell destruction occurring inside the blood vasculature (intravascular).
▶ Hemolytic Anemia, Autoimmune

Ionophores

Chemicals capable of forming a complex with an ion and transporting it across a biological membrane.

▶ Mitogen-Stimulated Lymphocyte Response

IP-10

IP-10 is an interferon-inducible protein-10. It is a non-ELR-CXC chemokine, which is also known as CXCL10.
▶ Interferon-γ

Irritant

A chemical which is not corrosive but which causes a reversible inflammatory effect on living tissue by chemical action at the site of contact. A chemical is a skin irritant if, when tested on the intact skin of albino rabbits by the methods of 16 CFR 1500.41 for 4 hours exposure or by other appropriate techniques, it results in an empirical score of 5 or more. A chemical is an eye irritant if so determined under the procedure listed in 16 CFR 1500.42 or other appropriate techniques.
▶ Three-Dimensional Human Skin/Epidermal Models and Organotypic Human and Murine Skin Explant Systems

Irritant Contact Dermatitis

A form of skin inflammation induced by primary contact with chemicals, thought not to be mediated by lymphocytes, showing redness, swelling, infiltration, scaling, and sometimes vesicles and blisters.
▶ Skin, Contribution to Immunity

Isolation

▶ Cell Separation Techniques

Isotype

Any of the categories of antibodies determined by their physicochemical properties (such as molecular weight) and antigenic characteristics that are generally present in all individuals of a given species. For example, IgG, IgA and IgM are individual antibody isotypes that can be produced in response to an immunogen. However, their molecular weights differ significantly from each other.
▶ Therapeutic Cytokines, Immunotoxicological Evaluation of
▶ Rabbit Immune System

ITAM

Immunoreceptor tyrosine-based activation motifs (ITAMs) are motifs composed of two tyrosine residues separated by 9–12 amino acids, which are present in the cytoplasmic tails of Igα and Igβ, in the accesory chains involved in signaling from the T cell receptor, and in receptors for the Fc portion of IgG. Phosphorylation of the tyrosines in ITAMs leads ultimately to the activation of a number of effector functions.
▶ Antibody-Dependent Cellular Cytotoxicity

ITIM

Immunoreceptor tyrosine-based inhibitory motifs (ITIMs) are found in several receptors which exert and inhibitory effect on activation signals.
▶ Antibody-Dependent Cellular Cytotoxicity

J

JAKS

JAKS are janus kinases, which are cytoplasmic tyrosine kinases that mediate signal transduction by cytokine receptors. JAKS physically interact with cytokine receptors and are activated by cytokine binding. There are four known mammalian JAKS: JAK1, JAK2, JAK3, and Tyk2.
▶ Interferon-γ

Janus Kinase (JAK)

A family of non-receptor protein tyrosin kinases of approximately 130 kDa, consisting of JAK1 (Janus kinase-1), JAK2 (Janus kinase-2), JAK3 (Janus kinase-3), and TyK2 (tyrosine kinase-2). They play a crucial role in the initial steps of cytokine signaling, in cooperation with STATs (signal transducers and activators of transcription).
▶ Cytokines

Jun NH2-Terminal Kinase

Jun NH2-terminal kinase (JNK) is mitogenic activated protein (MAP) kinase involved with *c-Jun* phosphorylation and activation of NF-κB.
▶ Interleukin-1β (IL-1β)

Juvenile Periodontitis

Juvenile periodontitis is an aggressive and destructive inflammatory disease of the connective tissue of gums affecting adolescents and young adults.
▶ Chemotaxis of Neutrophils

K

K-562

The human erythroleukemia cell line. It is highly sensitive to human and non-human primate NK-mediated lysis.
▶ Cytotoxicity Assays

Kanechlor

▶ Polychlorinated Biphenyls (PCBs) and the Immune System

Keratinocytes

The major epidermal cells, which undergo a program of terminal differentiation to the production of the stratum corneum. They act as signal transducers, converting nonspecific exogenous stimuli into the production of cytokines, adhesion molecules and other inflammatory mediators.
▶ Skin, Contribution to Immunity
▶ Delayed-Type Hypersensitivity

Keratins

Keratins belong to the superfamily of intermediate filament proteins. They are the most abundant proteins in epithelial cells and are known to be responsible for the formation of cytoskeletal filaments by copolymerization. More than 20 type I keratins and about 15 type II keratins are described. At least one of the type I and one of the type II keratins are expressed by every epithelial cell. For post-mitotic cornifying cells in the epidermis, for example, a characteristic coexpression of K1 and K10 keratin is observed.
▶ Three-Dimensional Human Skin/Epidermal Models and Organotypic Human and Murine Skin Explant Systems

Keyhole Limpet Hemocyanin (KLH)

Keyhole-Limpet Hemocyanin, a T-dependent protein antigen derived from the hemocyanin of the mollusk *M. crenulata*. One of the primarily T-dependent antigens used in ELISA studies.
▶ Plaque Versus ELISA Assays. Evaluation of Humoral Immune Responses to T-Dependent Antigens
▶ Immunoassays

Klebsiella, Infection and Immunity

HELEN V RATAJCZAK
Boehringer Ingelheim Pharmaceuticals
900 Ridgebury Road
Ridgefield, CT 06877
USA

Synonyms

Encapsulatus

Definition

Klebsiella, also called *Encapsulatus*, is a genus of bacteria, of the tribe *Escherichieae* and family *Enterobacteriaceae*, containing several species causing infections primarily of the respiratory tract in man and some of the lower animals (1).

Klebsiella is among the enteric bacilli included in the coliform group, characterized as fermentative Gram-negative rods that inhabit the intestinal tract and nasopharynx of man and other animals without causing disease. However, when the organisms get outside these sites they cause serious disease. The Center for Disease Control (CDC) in Atlanta lists the percentage of endemic hospital infections caused by *Klebsiella* at 8% and of epidemic outbreaks at 3% of all pathogens. *Klebsiella pneumoniae* (Friedländer's bacillus) has been considered a significant respiratory pathogen since 1882. *Klebsiella* is a Gram-negative bacterium related to *Enterobacter* (formerly *Aerobacter*) and *Serratia* organisms which cause serious pulmonary and
▶ urinary tract infections in hospitalized patients and

people with underlying diseases such as alcoholism, diabetes, and chronic lung disease. *Klebsiella* is second to *Escherichia coli* as cause of Gram-negative bacteremia but is usually more virulent than *Escherichia coli* in urinary tract infections. It is important to differentiate among the bacteria because they have wide differences in antibiotic susceptibility and pathogenicity. Biochemical tests are used to differentiate the organisms (2).

Characteristics

The Gram-negative, non-spore-forming rods included in the family *Enterobacteriaceae* are relatively small (2–3 by 0.4–0.6 microns). The rods occur singly or in pairs and are non-motile, lacking flagella. The *Enterobacteriaceae* grow readily on ordinary media under aerobic or anaerobic conditions. They utilize glucose fermentatively with the formation of acid or of acid and gas, reduce nitrates to nitrites, and give a negative oxidase reaction. Lactose fermentation, recognized by the formation of colored colonies on solid media containing lactose and an appropriate indicator (e.g. neutral red) delineates the coliform organisms including *Klebsiella*.

The genus *Klebsiella* is ubiquitous in nature: In the environment it is in surface water, sewage, soil, and on plants. In animals it is on mucosal surfaces of mammals such as humans, horses, or swine, which they colonize. In humans the nasopharynx and the intestinal tract are the most common habitant sites.

The genus *Klebsiella* consists of five species: *K. pneumoniae* (subspecies *pneumoniae*, *ozaenae*, and *rhinoscleromatis*), *K. oxytoca*, *K. terrigena*, *K. planticola*, *K. ornithinolytica*. The most medically important is *K. pneumoniae* followed, to a much lesser degree, by *K. oxytoca*. *K. terrigena* and *K. planticola* were originally considered to have no clinical significance and to be restricted to water, plants, and soil. However, recent reports do describe them as occurring in human clinical specimens. Klebsiella pneumoniae is the most important ▶ human pathogen of the *Klebsiella* group.

Pathogenicity Factors

K. pneumoniae produces multiple adhesins, which help the microorganisms to adhere to host cells, a critical step in the infectious process. Some of the adhesins are fimbrial (pili) and others are non-fimbrial, each with distinct receptor specificity. In *K. pneumoniae* there are five adhesin types of which two types (1 and 3) of pili predominate and play a role in mediating adhesion to various epithelial cells. Type 1 pili agglutinate guinea pig erythrocytes. The agglutination is inhibited by mannose. This mannose sensitive type 1 fimbriae is common in many members of enterobacteria and plays a role in mediating adhesion to the upper respiratory tract. The type 3 fimbriae is characterized by its ability to agglutinate tannin-treated erythrocytes and is designated mannose-resistant, *Klebsiella*-like (MR/K-HA) fimbriae. This type of pili is made by many enteric genera and is capable of binding to various human cells such as endothelial cells and epithelial cells of the respiratory and urinary tracts. Recent studies have shown *K. pneumoniae* can internalize into epithelial cells. Three new types of *K. pneumonia* adhesins have been reported:

- the R-plasmid-encoded non-fimbrial CF29K adhesin shown to mediate adherence to human intestinal cell lines
- a new capsule-like extracellular adhesin that seems to confer an aggregative pattern of adhesion to intestinal cell lines
- another fimbria-like adhesin designated KPF-28, suggested to mediate adhesion to and colonization of the human gut.

All members of the species produce complex acidic polysaccharide capsules and large, moist, often very mucoid colonies. The capsules are antiphagocytic and are responsible for the organism's invasive properties. The capsules determine the pathogenicity of the bacteria and structurally form the basis for classification into 77 capsular serotypes. The serotypes differ in their pathogenicity and epidemiological relevance, with serotypes K1 and K2 considered especially likely to be virulent. (3)

Lipopolysaccharides (LPS) are used to divide *K. pneumoniae* into eight different serotypes (LPS, O-antigen). The O1 serotype is the most common O-antigen found among clinical isolates. LPS is a major factor in the ability of the bacterium to resist the host serum bactericidal activity. LPS is able to activate complement, with the deposition of C3b onto the LPS molecule. However, the location of the C3b prevents the formation of the lytic membrane attack complex (see below) (3).

Two types of siderophores (high-affinity, low-molecular-weight iron chelators) are secreted by *K. pneumoniae*. The siderophores compete effectively for iron bound to host proteins. The siderophores provide the bacteria with iron, taking it from intracellular (hemoglobin, ferritin, hemosiderin, myoglobin) and extracellular (lactoferrin and transferrin) proteins. Aerobactin, one of the siderophores, is considered to have a virulence enhancing effect (3).

Preclinical Relevance

K. pneumoniae is found in the respiratory tract and feces of 5 to 10% healthy subjects and is frequently present as a secondary invader in the lungs of patients with chronic pulmonary disease. It causes about 3% of all acute bacterial pneumonias (2).

The most important predisposing factors to infection from *Klebsiella* are granulocytopenia and qualitative phagocyte defects, cellular immune dysfunction, humoral immune dysfunction, and splenectomy (3).

Relevance to Humans
Infection
Klebsiella is a pathogen causing severe pyogenic community-acquired pneumonia, which mainly affects immune-compromised people and has a high fatality rate if untreated. *Klebsiella* is among the top forms of pathogens causing infection in neonatal intensive care units and is the second most common causative agent of Gram-negative neonatal bacteremia.

▶ Pneumonia caused by *Klebsiella pneumoniae* is characterized by the production of thick gelatinous sputum and a high bacterial population density in the edema zones of the active lesions. The destructive action of the unphagocytized organism on the pulmonary tissue interferes with antimicrobial therapy and often results in chronic lung abscesses requiring surgical resection (2).

Klebsiella is an opportunistic pathogen which can give rise to severe infections such as septicemia, pneumonia, urinary tract infections, and soft tissue infections. *Klebsiella* species have been implicated in chronic inflammatory upper respiratory tract infections: *K. ozenae* in ozena, a progressive fetid atrophy of the nasal mucosa; and *K. rhinoscleromatis* in rhinoscleroma, a destructive granuloma of the nose and pharynx. The main targets of *Klebsiella* are hospitalized immune-compromised hosts, particularly those with serious granulocytopenia, with severe underlying diseases. *Klebsiella* is the causative agent for 5%–7% of all hospital-acquired infections and is among the most important nosocomial pathogens. *Klebsiella* is among the eight most important pathogens in hospitals, second only to *E.coli* as the most common cause of Gram-negative sepsis. *Klebsiella* infections are observed in almost any body site, although infections of the urinary and respiratory tracts predominate. *Klebsiella* infections are associated with reactive arthritis in some individuals and may cause cutaneous infection (2).

Treatment
Kanamycin, gentamycin, the polymyxins, chloramphenicol, cephalothin, and streptomycin are commonly used in treatment. In urinary tract infections nalidixic acid and nitrofurantoin are effective. Although some *Klebsiella* strains are resistant, most strains of *Klebsiella* are susceptible to cephalothin, distinguishing this bacterium from *Enterobacter* and *Serratia*, which produce a cephalosporinase (2).

Strains of *Klebsiella* have emerged which are antibiotic-resistant including strains which produce extended spectrum β-lactamase and a new *Klebsiella* species (*K. planticola* and *K. terrigens,* respectively).

Immunity
Symptomatic *K. pneumoniae* infections exhibit a severe inflammatory process. Innate immune mechanisms include phagocytosis by polymorphonuclear leukocytes, deprivation of the bacteria of iron, and activation of complement. Two pathways of complement activation have been described:
- in the classical pathway the so-called natural *Klebsiella*-specific antibodies react with the *Klebsiella* to activate complement
- in the alternative pathway, activation of complement is achieved by antigens on the bacterial surface via the properdin system.

Both pathways lead to the activation of C3 and the formation of C3b on the bacterial surface, mediating phagocytosis and helping form the terminal C5b–C9 complex which lyses the bacteria. The alternative pathway of complement activation is considered the major innate immunity against *K. pneumoniae*.

Several of the pathogenicity factors described more fully above help the bacteria avoid these innate immune mechanisms. Although the lipopolysaccharide (LPS) present in the capsule which surrounds the bacterium activates complement, and C3b is deposited, the location of the C3b is on the longest O-polysaccharide side, far away from the bacterial cell membrane. Therefore the formation of the lytic membrane attack complex (C5b–C9) is prevented and bacterial cell death does not take place. Also the bacteria secrete low-molecular-weight iron chelators, called siderophores, that compete effectively for iron bound to host proteins. Perhaps the most important protective mechanism of the bacteria is their capsules which are composed of complex acidic polysaccharides. The capsule protects the bacteria from phagocytosis and inhibits the activation of or uptake of complement components.

Other non-specific innate immunity mechanisms which protect against *Klebsiella* infection include nitric oxide and the T helper 1-type cytokines: tumor necrosis factor-α (TNFα), interferon-γ, macrophage inflammatory protein-2, lipopolysaccharide-binding protein (LPB), CD-14, interleukin-12, γ-interferon, and nitric oxide. In contrast, Th2-driven immune responses (e.g., IL-4 and IL-10) appear to be detrimental to the host (4).

Several different approaches are being taken to provide protection against *Klebsiella*. Ribosomal immunotherapy combine ribosomes from different bacterial strains and provides increased innate as well as specific immunity. Vaccines include lysates and proteoglycans. Newer approaches include genetic inactivation of bacteria, e.g. the formation of bacterial ghosts created by expression of a cloned PhiX174 gene *E* which results in lysis of the bacteria. The latter has been shown to induce specific humoral and cellular immune responses and to confer protective immunity (5).

For immunotoxicity studies, a host resistance model is used in mice to monitor pulmonary host defense mechanisms (6).

Regulatory Environment

Klebsiella is not a focused concern of the regulatory environment.

References

1. Asimov I, Bassett DL, Beamer PR et al. (eds) (1966) Stedman's Medical Dictionary, 21st ed. Williams & Wilkins Co., Baltimore
2. Sonnenwirth AC (1973) The enteric bacilli and similar Gram-negative bacteria. In: Davis BD, Dulbecco R, Eisen HN, Ginsberg HS, Wood WB, McCarty M (eds) Microbiology including immunology and molecular genetics, 2nd ed. Harper & Row, New York, pp 769–771
3. Sahly H, Podschun R, Ullmann U (2000) *Klebsiella* infections in the immunocompromised host. Adv Exp Med Biol 479:237–249
4. Tsai WC, Stroeter RM, Zisman DA et al. (1997) Nitric oxide is required for effective innate immunity against *Klebsiella pneumoniae*. Infect Immun 65:1870–1875
5. Szostak MP, Hensel A, Eko FO et al. (1996) Bacterial ghosts: non-living candidate vaccines. J Biotech 44:161–170
6. Ratajczak HV, Thomas PT, House RV et al. (1995) Local versus systemic immunotoxicity of isobutyl nitrite following subchronic inhalation exposure of female B6C3F1 mice. Fund Appl Toxicol 27:177–184

KLH ELISA

▶ Plaque Versus ELISA Assays. Evaluation of Humoral Immune Responses to T-Dependent Antigens

Knock-In

A gene targeting approach in which the knockout construct used to remove a mouse gene carries a functional gene (usually not of mouse origin).

▶ Knockout, Genetic
▶ Transgenic Animals

Knockout Animal

An animal in which genetic code for a specific protein has been removed.

▶ Animal Models of Immunodeficiency
▶ Transgenic Animals
▶ Knockout, Genetic

Knockout, Genetic

JEANINE L. BUSSIERE · BRAD BOLON
Amgen Inc.
One Amgen Center Drive
Thousand Oaks, CA 91320-1799
USA

Synonyms

Genetically engineered mouse, gene-targeted mouse, knockout mouse, KO mouse, null mutant mouse, −/− mouse, targeted mutant mouse, tm mouse

Definition

Gene-targeted or "knockout" animals have been created to specifically lack an endogenous gene using molecular and cellular genetic engineering techniques (1). Homologous recombination is employed to replace the endogenous gene in an ▶ embryonic stem (ES) cell with engineered DNA. The DNA insertion may be a nonsense sequence that merely interrupts the endogenous gene, or it may contain a functional gene that encodes a different protein (a ▶ knock-in).

Characteristics

Conventional Knockout Technology

Targeting protocols employ homologous recombination between identical flanking sequences of nucleotides on:

- a targeting construct bearing an engineered DNA sequence
- the endogenous gene on a chromosome within ES cells.

Individual ES cells are cultured and then challenged with cytotoxic agents to remove those in which targeting was inaccurate. This selection is possible because one or more chemical-sensitive elements are located adjacent to the engineered sequence on the targeting construct; these elements promote death only in those ES cells in which the correct recombination event has not occurred. Surviving ES cells are cultured and injected into blastocysts, where they are incorporated at random into all the tissues of the developing embryo.

Progeny are born as ▶ chimeras (with tissues containing both normal and gene-targeted cells), grown to adulthood, and then bred to determine whether or not the targeted cells are present in the gonads and contributing to gamete production. Chimeric animals in which germline transmission occurs are used as the parental generation (founders) to breed ▶ homozygous knockout animals. This knockout technology currently is suitable only for certain strains of mice, as this species is the only one in which reliable ES cell lines have been produced.

Conditional Knockout Technology

Gene targeting can be limited to specific life stages and/or tissues by using a targeting construct that bears a site-specific recombinase (1). The prototype for this paradigm is the Cre/loxP system, in which the bacterial enzyme Cre excises any DNA located between two *loxP* sites (a short nucleotide sequence that does not occur in vertebrate DNA). This procedure requires the creation of two lines of genetically altered mice: a gene-targeted line in which the engineered sequence contains a functioning gene flanked by two *loxP* sites; and a transgenic line incorporating the *Cre* gene. The two parent lines of mice are normal, but crossing them results in progeny in which the *loxP*-flanked gene has been excised in cells that express the transgenic Cre protein. Gene inactivation is limited to a single tissue by placing the *Cre* transgene under the control of a tissue-specific gene promoter. Another form of conditional knockout mice utilizes chemically mediated inhibition of a particular gene product at the relevant stage of life. Conditional knockouts are especially useful for studying developmentally essential genes in adult mice where global knockout during gestation would result in embryonic lethality. In addition, this approach is more similar to the clinical situation where inhibition of a gene product occurs after birth (generally in adults) and avoids any adaptations or compensations that may occur in the animals by knock-ing out the gene product during embryonic development.

Transgenic Alternatives to Knockout Technology

Gene targeting techniques are quite time consuming, with production of knockout progeny often requiring a year or more. Other genetic engineering strategies have been developed to reduce or delete gene function rapidly and simply without resorting to actual removal of the endogenous gene (1). Three such techniques require production of transgenic mice, or animals in which DNA has been added rather than removed.

- First, the transgenic protein may act as a ▶ dominant negative inhibitor that overrides the activity of the endogenous protein without interacting with it.
- Second, the transgenic protein may bind and inactivate the endogenous protein.
- Third, a transgenic protein with cytotoxic activity may be used to kill specific populations of cells that normally harbor the endogenous gene.

Physiological Impact of Genetic Knockout

The physiological properties of the targeted gene and the replacement sequence will determine what the functional significance of the engineered null mutation will be to the knockout mouse. All targeting procedures disrupt the normal coding sequence of an endogenous gene, thereby preventing expression and function of its protein product (i.e. knockout). Insertion of an engineered sequence that contains a functional gene with constitutive activity (knock-in) may replace the role of the deleted endogenous gene, or yield a different functional or structural alteration. Both the genetic and phenotypic composition of knockout mice must be defined before useful information may be gathered regarding the mechanism, efficacy, and potential toxicity of the deleted gene product. The impact of null mutations and knock-in replacement genes is examined either in vivo or in likely target tissues in vitro. The consequences of genetic modification typically are investigated in young adult mice using various combinations of conventional anatomic, biochemical, clinical, and molecular methods. Morphologic assessment is considered the "gold standard" for phenotypic analysis. Specific endpoints that might be noted at the gross or microscopic level include alterations in the size, shape, color or location of organs, or the presence of aberrant elements (e.g. extra organs, tumors). Knockout and transgenic mice often are structurally normal even if functional abnormalities are apparent, while many engineered mice appear to lack both structural and functional defects. However, subtle ▶ phenotypes (functional and/or structural changes resulting from the genetic engineering event) sometimes may be unmasked using pharmacological challenges or other physiological stressors (2,3).

Preclinical Relevance

Knockout (and transgenic) mice are rapidly gaining acceptance as routine tools for mechanistic research and offer considerable promise for generating specific models of toxicological importance. Knockout mice have been used to assess drug specificity, to investigate mechanisms of toxicity, and to screen for mutagenic and carcinogenic activities of xenobiotics. Similarly, the impact of novel therapeutic candidates can be estimated in knockout mice; generation of viable and fertile animals with null mutations for a potential target protein implies that pharmacological inhibition of the molecule in vivo will elicit no major adverse effects. Furthermore, the apparent lack of an in vivo phenotype could be used in conjunction with substantial evidence of in vitro efficacy to support the selection of a likely no observable adverse effect level (NOAEL) for use in preclinical pharmacology and toxicology studies. Particular emphasis in future pharmacology and toxicology studies will be directed toward conditional knockout mice (to evaluate the impact of chemically-mediated inhibition of a particular gene product at the relevant stage of life) and "humanized" knock-in animals (in which the endogenous mouse gene is replaced with the homologous human gene to examine its role in disease or drug metabolism). Humanized mice are of particular importance as these animals can be employed to evaluate the efficacy and toxicity of human proteins that are not pharmacologically active in normal rodents or that induce a neutralizing antibody response that limits long-term exposure. It is important to remember that a "humanized" mouse is still a mouse, and that any phenotype, or absence thereof, in mice bearing a human gene knock-in models—but is not strictly analogous to—normal human biology (4). One particular criticism is that "humanized" mice manufacture one or a few human proteins of interest, but other proteins that interact with the human molecules are still of mouse origin. The physiological effect of human-mouse protein interactions may differ slightly—or substantially—from that of the normal human-human association. With respect to the immune system, the physiological functions and pathways for many genes important to normal immune function have been investigated using knockout (and transgenic) mice (5). Again, humanized mice are of particular importance in modeling the human immunologic response, as they have several advantages over conducting immunopharmacology and immunotoxicity studies in nonhuman primates (the only alternative if the human protein is not active in rodents). Rodent studies are simple, relatively inexpensive, and can include enough experimental subjects to achieve suitable statistical power. More importantly, immunotoxicity assays are well characterized in the mouse, in contrast to the nonhuman primate.

However, three caveats must be kept in mind when using genetically engineered mice for immunotoxicity assessment:

- the emphasis on morphologic assessment as the usual standard for phenotypic analysis means that the immune function of most genetically engineered mice is poorly characterized
- conclusions reached using a standard knockout mouse (in which the gene is missing throughout gestation and postnatal life) may not accurately reflect disease or pharmacological interventions in which genetic function is nullified only during adulthood (the most likely clinical scenario)
- most critically, the background strain on which the null mutation is carried (mice with different genetic backgrounds) respond very differently to immune stimuli.

The standard background of knockout mice is a mixture of C57BL/6 (the predominant component, derived from the blastocyst) and S129 (the major source of ES cells for gene targeting). Further, not all S129 ES cell lines are comparable, and knockout mice often are not bred to achieve genetic homogeneity on a suitable genetic background. Therefore, it is critical to first assess the immunopharmacologic response using standard immune assays in the background strain (C57BL/6) prior to testing knockout mice back-crossed to increase the C57BL/6 gene fraction, or to back-cross the null mutation onto mice with a genetic background relevant to immunotoxicity assessment (i.e. B6C3F1).

Relevance to Humans

The data generated from knockout and transgenic mice have been used to model immune responses based on the similarity between vertebrate immune systems. Such comparisons are made even more relevant by using "humanized" mice. Generally, the aim of such studies has been to discern molecular pathways and physiological functions, or to examine the efficacy of immunopharmacologic manipulations. Toxicity bioassays routinely are performed in knockout animals for some purposes (e.g. mutagenicity and carcinogenicity), but the relevance of such mice with respect to conventional immunotoxicity testing remains to be proven. With the increasing number of protein therapeutics on the market, these data become even more important to demonstrate that the knockout mice are a viable alternative to immunotoxicity testing in nonhuman primates, and are relevant to the findings seen in humans.

Regulatory Environment

Preclinical efficacy and safety studies, especially chronic studies, are notoriously difficult to perform when the candidate therapeutic agent is a human protein. Due to these difficulties, regulatory agencies are concerned with alternative means of assessing risk. These alternatives may include testing in nonhuman primates, homologous proteins in the appropriate animal species, or in assessing knockout or transgenic mice. It should be understood that these are all surrogates to testing the clinical candidate in humans, and that each of these options has its own set of caveats. However, "humanized" knockout and transgenic mice should provide a reasonable alternative, especially for immunotoxicity protocols in which the mouse response is well characterized.

References

1. Bolon B, Galbreath E, Sargent L, Weiss J (2000) Genetic engineering and molecular technology. In: Krinke G (ed) The Laboratory Rat. Academic Press, London, pp 603–634
2. Bolon B, Galbreath EJ (2002) Use of genetically engineered mice in drug discovery and development: Wielding Occam's razor to prune the product portfolio. Int J Toxicol 21:55–64
3. Doetschman T (1999) Interpretation of phenotype in genetically engineered mice. Lab Anim Sci 49:137–143
4. Liggitt HD, Reddington GM (1992) Transgenic animals in the evaluation of compound efficacy and toxicity: will they be as useful as they are novel? Xenobiotica 22:1043–1054
5. Ryffel B (1997) Impact of knockout mice in toxicology. Crit Rev Toxicol 27:135–154

KO Mouse

▶ Knockout, Genetic

Kupffer Cells

Specialised, macrophage-like cells in the liver. Kupffer cells phagocytise foreign particles, bacteria and old blood cells.

▶ Fish Immune System

L

Lactate Dehydrogenase (LDH) and/or Adenylate Kinase (AK) Leakage

Lactate dehydrogenase (LDH) and adenylate kinase (AK) are both enzymes which are essential for normal cell metabolism. Upon cell damage these proteins are released from the cytosol into the cell culture medium. There, the activity of these enzymes is detected with special measurement procedures and can finally be used as non-destructive parameters for the determination of cytotoxicity.
▶ Three-Dimensional Human Skin/Epidermal Models and Organotypic Human and Murine Skin Explant Systems

Lactoferrin

A non-heme β-globulin that acts as an iron-transporting protein.
▶ Respiratory Infections

Lamina propria

Layer of connective tissue between epithelium of the intestines and the underlying muscle layer of the epithelium—the muscularis mucosae. Site of lymphocyte extravasation and primary infections.
▶ Immunotoxic Agents into the Body, Entry of
▶ Mucosa-Associated Lymphoid Tissue

Langerhans Cells

Bone-marrow derived epidermal cells with a dendritic morphology expressing a CD1 marker in humans, and containing a cytoplasmic organelle called the Birbeck granule. They express class II MHC antigen and are the principal antigen-presenting cells of the skin, which emigrate to local lymph nodes to become dendritic cells. They are very active in presenting antigen to T cells.
▶ Skin, Contribution to Immunity
▶ Local Lymph Node Assay
▶ Local Lymph Node Assay (IMDS), Modifications

Large Granular Lymphocyte

A lymphocyte lineage characterized by large size and distinct cytoplasmic granules. This cell usually is thought to represent a natural killer cell.
▶ Lymphocytes

Laser Capture Microdissection (LCM)

A technique for isolating single cells from tissues using a laser, yielding a pure population of cells for the analysis of molecular function. LCM is especially useful for removing tumor cells from surrounding tissue.
▶ Polymerase Chain Reaction (PCR)

Lateral Line

A sensory organ of fish perceiving low frequency vibrations. The organ consists of a canal running down the side of the body that communicates to the outside via pored scales.
▶ Fish Immune System

LDA

▶ Limiting Dilution Analysis

Lectins

A family of carbohydrate-binding proteins found in animals and plants. Different types of lectins recognize specific carbohydrate ligands.
▶ Cell Adhesion Molecules
▶ Mitogen-Stimulated Lymphocyte Response

Leukemia

LEIGH ANN BURNS-NAAS
Pfizer Global Research & Development
10777 Science Center Dr.
San Diego, CA 92121
USA

Synonyms
Acute lymphocytic leukemia, acute myelogenous leukemia, chronic lymphocytic leukemia, chronic myelogenous leukemia, biphenotypic leukemia; prolymphocytic leukemia, hairy cell leukemia

Definition
Leukemia is a heterogeneous group of malignancies of the bone marrow and blood that affect individuals of all ages. As a result of hematopoetic dysregulation, these cancers can result in conditions such as ▶ anemia, ▶ thrombocytopenia, and ▶ neutropenia, and increase the risk for ▶ opportunistic infection in affected individuals. Though there are numerous types of leukemias, these disorders can be divided broadly into four major types:
- acute lymphocytic leukemia (ALL)
- acute myelogenous leukemia (AML)
- chronic lymphocytic leukemia (CLL)
- chronic myelogenous leukemia (CML).

ALL is the most common acute form in children, AML being the predominant acute form in adults. Chronic leukemias are uncommon in childhood and teenage years, but increase in incidence with increasing age. Biphenotypic leukemias have morphological characteristics of both myelogenous and lymphoblastic leukemias. Prolymphocytic and hairy cell leukemias are rare.

Characteristics
Acute leukemia is a rapidly progressing disease reflecting an imbalance between proliferation and differentiation that results in the accumulation of immature cells in the bone marrow and in the blood. Over time, these immature cells accumulate to an extent that can suppress the normal hematopoetic mechanisms, and as a result anemia, thrombocytopenia, and neutropenia can occur. Chronic leukemias represent an expanded population of cells that proliferate, but also can differentiate. Chronic leukemias generally progress more slowly and allow greater numbers of more mature, functional cells to be made. Leukemias are diagnosed via evaluation of the blood and often the bone marrow. In the acute forms of leukemia, red cell counts and platelet counts are decreased and there is an increase in the number of abnormal white cells (leukemic blasts) in the blood. This is confirmed by examination of the marrow, which almost always demonstrates the presence leukemia cells. ▶ Immunophenotyping and ▶ cytogenetic analyses can be used to confirm the disease. AML has several diagnostic subtypes defined by the identity and proportions of abnormal cells. This again reflects the heterogeneous nature of the disease. In CLL, there is also an increase in the number of white cells (lymphocytes) in the blood and bone marrow. These malignant cells are most often B lymphocytes (approximately 95% of CLL cases), but may also be T lymphocytes or natural killer (NK) cells. Low platelet counts and low red cell counts in the blood may be observed, but these cells are usually only slightly decreased in the early stage of the illness. Abnormal expression and regulation of genes (e.g. bcl2) that control normal cell death (apoptosis) appears to be malignant mechanism in the majority of CLL cases. Individuals with CLL may also have other immunological abnormalities, including autoimmune hemolytic anemia or immune-mediated thrombocytopenia.

Blood from patients with CML exhibits a few blasts and promyelocytes and a larger number of maturing and fully matured myelocytes and neutrophils. ▶ Cytogenetic analysis of bone marrow, and polymerase chain reaction (PCR) or fluorescence in situ hybridization (FISH) analysis of blood or bone marrow from an individual with CML would reveal the presence of an abnormal reciprocal translocation between pieces of chromosomes 9 and 22 (the Philadelphia chromosome). This chromosomal abnormality is characteristic of CML and results in expression of a chimeric oncogene, bcr-abl, that encodes a protein, tyrosine kinase controlling a variety of signaling pathways involved in cell proliferation.

There are other less common forms (or variants) of chronic leukemias. All result from a malignant transformation of a lymphocyte and the accumulation of these cells primarily in the marrow, blood, and the spleen. Large granular lymphocytic (LGL) leukemia is another type of chronic leukemia. The immunophenotype is either a T cell or an NK cell, but not a B cell. Unlike cells in other types of chronic lymphocytic leukemia, LGL leukemia is characterized by larger lymphocytes containing granules. Hairy cell leukemia is a chronic leukemia of lymphocytes, usually of a B cell. The malignant cells are larger than regular lymphocytes and have irregular, hair-like projections on them. These cells infiltrate the bone marrow and red pulp of the spleen. Prolymphocytic leukemia is a variant of CLL that displays larger than normal, mature-looking lymphocytes. Prolymphocytic leukemia differs from CLL by the presence of intense surface immunoglobulin staining, among other surface markers. The disease progresses more rapidly than the chronic form but more slowly than the acute form.

Preclinical Relevance

Alterations in ▶ cell-mediated immunity (CMI) are often considered to facilitate the development of neoplastic disease. This arm of the immune system is involved in controlling spontaneous tumors and infections among other things. The destruction of tumor cells can result from the cytolytic action of specific (cytotoxic) T lymphocytes (CTL), macrophages, and NK cells. These cells recognize specific antigens on tumor or virally infected cells and cause their death by one of several mechanisms considered here.

In cell-mediated cytotoxicity, the effector cell (CTL or NK) binds in a specific manner to the target cell. CTL recognize either foreign major histocompatability complex (MHC) class I on the surface of allogeneic cells, or antigen in association with self MHC class I (e.g. viral particles), while NK cell recognition of target cells involves the binding of the Fc portion of antigen-specific antibody coating a target cell to the NK cell via its Fc receptors. This NK cell mechanism of killing is also referred to as antibody-dependent cellular cytotoxicity (ADCC). Once the CTL or NK cells interact with the target cell, they undergo cytoplasmic reorientation so that cytolytic granules are oriented along the side of the effector that is bound to the target. The effector cell then releases the contents of these granules onto the target cell. The target cell may be damaged by the perforins or enzymatic contents of the cytolytic granules. In addition, the target is induced to undergo programmed cell death (apoptosis). Once it has degranulated, the CTL or NK cell can release the dying target and move on to kill other target cells.

The role of the macrophage in cell-mediated cytotoxicity involves its activation by T cell-derived cytokine (e.g. IFN-γ) and subsequent recognition of complement-coated target cells via complement receptors present on the surface of the macrophage. The result is enhanced phagocytic ability and the synthesis and release of hydrogen peroxide, nitric oxide, proteases, and tumor necrosis factor, all of which serve cytolytic functions. Macrophages may also kill tumor or infected cells via ADCC in a manner similar to that described for NK cells.

Drugs and non-drug chemicals and agents that alter cell-mediated immunity have the potential to cause an increased risk of opportunistic infections (e.g. viral, bacterial, parasitic) and development of neoplastic disease. By evaluating such things as the ability of the immune system to recognize and destroy tumor cells such as the P815 mastocytoma (used in the CTL assay) or YAC-1 cell (used in the NK assay) as well as evaluating proliferative ability to mitogens, cytokines, or allogeneic (foreign; non-self) cells, and evaluation of allograft rejection, immunotoxicity testing in rodents has identified many agents capable of causing suppression of cell-mediated immunity. A few of these agents are considered below.

Therapeutic agents have been developed that specifically inhibit several immune endpoints. These drugs are often used to treat symptoms associated with autoimmunity, in transplantation to prevent immune-mediated graft rejection, or to treat individuals with significant hypersensitivity responses. Cyclophosphamide is the prototypical member of a class of drug known as alkylating agents. Experimentally, cyclophosphamide is often used as a positive immunosuppressive control in immunotoxicology studies because it can suppress both humoral immunity and cell-mediated immune responses. Cell-mediated immune activities that are suppressed include the delayed hypersensitivity response (DTH), the cytotoxic T lymphocyte response (CTL), graft vs host (GVH) disease, and the mixed lymphocyte response (MLR). The immunosuppressive action of corticosteroids is also well known. Following binding to an intracellular receptor, these agents produce profound lymphoid cell depletion in rodent models and lymphopenia, associated with decreased monocytes and eosinophils and increased PMN, in primates and humans. Corticosteroids induce apoptosis and T cells are particularly sensitive. In general, corticosteroids suppress the generation of CTL responses, the MLR, NK activity, and general lymphoproliferation. A large range of cell-mediated immune reactivity is also reduced by azathioprine immunosuppressive treatment including the DHR, the MLR, and GVH disease. Although T cell functions are the primary targets for this drug, inhibition of NK function and macrophage activities has also been reported. Cyclosporin A is an important immunosuppressant that acts preferentially on T cells by inhibition of IL-2 gene transcription and subsequent inhibition of T cell proliferation.

In addition to therapeutic chemicals, an immunosuppressive effect on cell-mediated immunity by occupational and environmental chemicals has also been demonstrated. By far the most well characterized immunotoxic effects are those produced by benzene. In animal models, benzene induces anemia, lymphocytopenia, and hypoplastic bone marrow. In addition, it has been suggested that this myelotoxicity may be a result of altered differentiative capacity in bone marrow-derived lymphoid cells. Benzene exposure alters both humoral and cell-mediated immune parameters. The polycyclic aromatic hydrocarbons (PAHs) are a ubiquitous class of environmental contaminants that have been studied extensively regarding their immunotoxic potential. PAHs have been found to be potent immunosuppressants, having effects on humoral and cell-mediated immunity, as well as on host resistance to infection. A great deal of interest in recent years has been drawn to the untoward effects of cigarette smoking. Cigarette smoke contains literally hundreds of chemicals that the smoker inhales, many of which are known carcinogens, genotoxins, and/or immunosuppressants. Although the primary site of exposure of the immune system to cigarette smoke is the lung, other systemic immune parameters have been shown to be altered in smokers, including decreased serum immunoglobulin levels and lower NK cell activity.

There are numerous examples of drug and non-drug chemicals with the potential to alter immunocompetence in a manner that might influence susceptibility to spontaneous (primary or secondary) neoplasms and opportunistic infection, or reactivation of latent pathogens. This section has only considered a very few of these agents, but those which have limited or sufficient evidence for an association between suppression of immune function and predilection for leukemia. Clearly, though, other contributing factors (e.g. genotoxicity) cannot be excluded.

Relevance to Humans

Uncontrolled proliferation of specific lymphoid cells is the hallmark of the leukemias. Dysregulation of the mechanisms controlling proliferation and differentiation of these cells is likely to result from a series of transformations that affect chromosomes and the inappropriate expression of ▶ oncogenes and their abnormal proteins. There is evidence that supports a genetic predisposition to the development of some acute forms of leukemia. This evidence includes an increased incidence of leukemias in individuals with congenital disorders such as Down's syndrome, Fanconi's anemia, and ataxia telangiectasia, though a multistep process to malignancy is still likely to be in place. But predisposition resulting from congenital disorders cannot fully explain the overall incidence of leukemias observed. Other factors, such as therapeutic interventions and environmental influences that cause genetic damage and/or that suppress the immune system, are likely to play a role.

There is epidemiological evidence supporting an association between a variety of other agents and an increased risk for developing leukemia. Perhaps the most striking example involves the survivors of the atomic bomb explosions in Japan. In Hiroshima radiation victims were 30 times more likely to develop ALL, AML, or CLL than non-exposed individuals, and in Nagasaki the incidence of AML was even higher. Though these represent tremendous doses of radiation, there is some evidence that lower doses of ionizing radiation can also predispose individuals to the development of leukemia. Higher than expected incidences of AML have been also observed in patients receiving low doses of therapeutic radiation as well as individuals working at radium plants. While this observed increase in risk is probably due primarily to genetic damage from the radiation, there are indications that some forms of radiation are capable of causing varying degrees of systemic ▶ immunosuppression, particularly of cell-mediated immunity.

Altered immunocompetence has been associated with an overall increased risk for the development of leukemia. Individuals with acquired immunodeficiency syndrome (AIDS) have increased incidences of a variety of cancers, including leukemia, most likely as a result of the loss of the ability of the host to identify and eradicate neoplastic cells, particularly those infected with herpes simplex or Epstein-Barr viruses.

A similar effect has been observed in transplant recipients receiving chemotherapy with immunosuppressive drugs. AML is one of the most common ▶ secondary neoplasms after chemotherapy and radiation treatment for other cancers or after autologous transplantation. Typical drugs used in these treatments include alkylating agents and immunosuppressants. The alkylating agents have been clearly shown to suppress the immune system and sufficient evidence exists regarding their association with an increased risk of leukemia. The risk for developing AML appears to be related to cumulative doses of the alkylating agent(s) and may be enhanced by the addition of radiation therapy to the treatment regimen.

In addition to therapeutic interventions, exposure to some environmental chemicals has been associated with an increased risk for the development of leukemias. Among these, benzene (a widely used solvent) is probably the most studied and well recognized chemical leukemogenic agent. Benzene has been shown to be immunosuppressive and is associated with an increased incidence of AML in exposed individuals. Other industrial or agrochemicals also appear to confer an increased risk for developing acute leukemias including some pesticides, PAHs, ethylene oxides, and embalming fluids. Additionally, while there is a clear link between smoking and the development of neoplastic disease in the lung, it has been suggested that chronic cigarette smoking may increase the risk for developing AML. Of interest, cigarette smoke contains various levels of known immunosuppressive chemicals such as benzene and/or its metabolites, and the PAHs.

There is a clear association between suppression of immune function and an increased incidence of infectious and neoplastic disease in humans. Agents that produce immunotoxicity in animals have the potential to produce immune effects in the human population, and these effects may occur in the absence of observable disease. Of the agents described here that are associated with an increased risk for the development of leukemias, no direct and specific causal relationship between the development of cancer and the immunosuppressive action by these drugs and non-drug chemicals/agents has been demonstrated. However, a preponderance of epidemiological evidence exists showing that exposure to various immunotoxic chemicals is associated with increased risk for malignancies like leukemia and lymphoma that are also known to occur in immunocompromised patients. Thus, it is reasonable to conclude that alteration of immune function may contribute to the observed increase in risk.

Regulatory Environment

Because of the concerns regarding the potential for drugs and non-drug chemicals to cause any number of cancers, including leukemias, global regulatory bodies have established guidance and test guidelines for assessing this potential. These include specific assessment of carcinogenic potential (e.g. lifetime studies in rodents) as well as assessment of the mutagenic and/or clastogenic potential (short-term in vivo and/or in vitro tests) of chemicals. Though not directly related to genotoxicity and carcinogenicity assessment, guidelines regarding the assessment of immune status following repeated exposure to drug and non-drug chemicals have recently been and are continuing to be put into place. Of interest, not all immunotoxicology guidance/guidelines require the evaluation of cell-mediated immunity. Some examples of these guidances and test guidelines are provided in Table 1.

References

1. International Programme on Chemical Safety (1996) Health impact of selected immunotoxic agents. In: Environmental Health Criteria 180: Principles and Methods for Assessing Direct Immunotoxicity Associated with Exposure to Chemicals. World Health Organization, Geneva, pp 85–147
2. Leukemia & Lymphoma Society (2002) Public literature. The Leukemia and Lymphoma Society, White Plains
3. Luster MI, Simeonova P, Germolec DR, Portier C, Munson AE (1996) Relationships between chemical-induced immunotoxicity and carcinogenesis. Drug Info J 30:281–285
4. Miller KB, Grodman HM (2001) Leukemia. In: Lenhard RE Jr, Osteen RT, Gansler T (eds) Clinical Oncology. American Cancer Society, Blackwell Science, Malden, pp 527–551
5. Pitot HC, Dragan YP (2001) Chemical carcinogenesis. In: Klaassen CD (ed) Casarett and Doull's Toxicology: The Basic Science of Poisons, 6th ed. McGraw-Hill, New York, pp 280–286

Leukemia-Initiating Cells (L-IC)

The hematopoietic stem cells of the leukemia. Note that only a small subfraction of all leukemic cells can self-renew in an organism. In human leukemias, the incidence of the L-IC subfraction was found even lower than the incidence of normal hematopoietic stem cells occurring in bone marrow samples from healthy individuals.

▶ Hematopoietic Stem Cells

Leukocyte

White blood cell.

▶ CD Markers

Leukocyte Culture: Considerations for In Vitro Culture of T cells in Immunotoxicological Studies

MACIEJ TARKOWSKI
Nofer Institute of Occupational Medicine
Lodz
Poland

Synonyms

Characteristics

The the immune system comprises of innate and acquired mechanisms of defense against invading pathogens. Through the help of innate, non-antigen-specific mechanisms utilizing, for example, professional scavenging cells such as macrophages, it helps to clear mucosal tissues, to fight off the skin pathogens. The second line of defense consists of cells that through fetal–neonatal "education" can eliminate foreign for the host antigens and not react to those that are recognized as self. Both mechanisms of defense can be compromised by toxic effects of chemicals in occupational and non-occupational settings. The the immune system can be affected by chemicals to which we are exposed through inhalation, by food consumption or by skin contact. A growing number of allergies, development of autoimmune diseases, and

Leukemia. Table 1 Examples of regulatory guidance and test guidelines for assessment of immunotoxicity, genotoxicity, or carcinogenicity (where available, only harmonized guidance/guidelines are presented)

Regulatory body	Endpoint	Guidance/test guideline
US Food and Drug Administration (FDA)	Immunotoxicity	**Center for Food Safety and Nutrition (CFSAN)** Redbook 2000: Immunotoxicity studies **Center for Devices and Radiological Health (CDRH)** Immunotoxicity testing guidance **Center for Drug Evaluation and Research (CDER)** Immunotoxicology evaluation of investigational new drugs
	Genotoxicity	**Center for Food Safety and Nutrition (CFSAN)** Redbook 2000: Bacterial reverse mutation test In vitro mammalian chromosome aberration test In vitro mouse lymphoma TK$^{+/-}$ gene mutation assay In vivo mammalian erythrocyte micronucleus test **Center for Drug Evaluation and Research (CDER)** See ICH
	Carcinogenicity	**Center for Food Safety and Nutrition (CFSAN)** Carcinogenicity studies with rodents Combined chronic toxicity/carcinogenicity studies with rodents **Center for Drug Evaluation and Research (CDER)** See ICH
Committee for Proprietary Medicinal Products (CPMP)	Immunotoxicity	Note For Guidance on Repeated Dose Toxicity
	Genotoxicity	See ICH
	Carcinogenicity	See ICH
Japanese Ministry of Health, Labor, and Welfare (MHLW)	Immunotoxicity	Immunosuppression, Draft Guidance
	Genotoxicity	For Pharmaceuticals, see ICH
	Carcinogenicity	For Pharmaceuticals, see ICH
International Conference on Harmonization (ICH)	Genotoxicity	ICH S2: Genotoxicity Includes the standard battery of tests and guidance on specific aspects of testing
	Carcinogenicity	ICH S1: Carcinogenicity Includes the need for carcinogenicity studies, approaches to evaluating carcinogenic potential, and dose selection

compromised defense mechanisms to fight off infections could result from this exposure to toxicants. Depending on the exposure route differenT cells of the immune system can be involved, but antigen-specific reactions will always occupy dendritic cells and lymphocytes.

Recognizing the toxic effects of chemicals on the immune system is one of the main goals of toxicological assessment of the risk that they may pose for human health. Efforts to establish adequate tests to assess these immunotoxic effects are still ongoing. Those that are implemented, in general, test for the immunotoxic effects on innate, cell-mediated, or humoral-mediated defense mechanisms. They use different animal models and in vitro tests, as well as exposure-controlled human studies. One of the main interests for studying immunotoxic effects of chemicals is to recognize whether they can induce antigen-specific reactions or—as is the case for most of chemicals—hapten-specific immune responses, and what the consequences of these effects are.

As indicated earlier these responses are mediated by lymphocytes and antigen-presenting cells (APCs). The antigen-specific response of lymphocytes results in induction of their proliferation and differentiation. In particular, the analyses of proliferative responses

Leukemia. Table 1 Examples of regulatory guidance and test guidelines for assessment of immunotoxicity, genotoxicity, or carcinogenicity (where available, only harmonized guidance/guidelines are presented) (Continued)

Regulatory body	Endpoint	Guidance/test guideline
US Environmental Protection Agency (EPA)	Immunotoxicity	**Health Effects Test Guidelines** 870.7800 Immunotoxicity **Biochemicals Test Guidelines** 880.3550 Immunotoxicity 880.3800 Immune response
	Genotoxicity	**Health Effects Test Guidelines** 870.5100 Bacterial reverse mutation test 870.5140 Gene mutation in *Aspergillus nidulans* 870.5195 Mouse biochemical specific locus test 870.5200 Mouse visible specific locus test 870.5250 Gene mutation in *Neurospora crassa* 870.5275 Sex-linked recessive lethal test in *Drosophila melanogaster* 870.5300 In vitro mammalian cell gene mutation test 870.5375 In vitro mammalian chromosome aberration test 870.5380 Mammalian spermatogonial chromosomal aberration test 870.5385 Mammalian bone marrow chromosomal aberration test 870.5395 Mammalian erythrocyte micronucleus test 870.5450 Rodent dominant lethal assay 870.5460 Rodent heritable translocation assays 870.5500 Bacterial DNA damage or repair tests 870.5550 Unscheduled DNA synthesis in mammalian cells in culture 870.5575 Mitotic gene conversion in *Saccharomyces cerevisiae* 870.5900 In vitro sister chromatid exchange assay 870.5915 In vivo sister chromatid exchange assay
	Carcinogenicity	**Health Effects Test Guidelines**: 870.4200 Carcinogenicity 870.4300 Combined Chronic Toxicity/Carcinogenicity

have found applications in in vivo and in vitro immunotoxicologic tests. The in vivo ▶ local lymph node assay (LLNA) (1) uses this phenomenon to test for skin-sensitizing properties of chemicals, whereas in vitro proliferation tests were proven to help identify beryllium-specific responses of T cells isolated from bronchoalveolar lavage of beryllium-sensitized people (2).

Culture of cells in in vitro conditions is a technique used in many disciplines of research including immunotoxicology. Presently many different cell types of human or animal origin can be maintained in these conditions due to the recognition of their nutritional and other cell-specific requirements. Once the cells become available there are several aspects that have to be recognized before starting in vitro cell culture. The type of cells and the test for which they are going to be used will determine the specific culture conditions, thus the following must be considered:

- cell type nutritional requirements
- duration of cell culture
- presence of other cells (feeding cells) or/and soluble components.
- incubation requirements
- adequate culture vessels.

Two major divisions of cell types are used that determine specific culture conditions. There are cells growing in suspension, or in adherent conditions, which are reflected by the basic morphological and physiological characteristics of the cells. Lymphocytes grow in suspension, whereas cells such as keratinocytes, epithelial or endothelial cells grow as adherent cells. Cell culture can be short term or it may require a longer period of incubation and several passages. Some experiments like those assessing cell proliferation of lymphocytes

Leukemia. Table 1 Examples of regulatory guidance and test guidelines for assessment of immunotoxicity, genotoxicity, or carcinogenicity (where available, only harmonized guidance/guidelines are presented) (Continued)

Regulatory body	Endpoint	Guidance/test guideline
Organization for Economic Coordination and Development (OECD)	Genotoxicity	OECD 471 Bacterial reverse mutation test OECD 473 In vitro mammalian chromosomal aberration test OECD 474 Mammalian erythrocyte micronucleus test OECD 475 Mammalian bone marrow chromosomal aberration test OECD 476 In vitro mammalian cell gene mutation test OECD 477 Sex-linked recessive lethal test in *Drosophila melanogaster* OECD 478 Rodent dominant lethal test OECD 479 In vitro sister chromatid exchange assay in mammalian cells OECD 480 *Saccharomyces cerevisiae* gene mutation assay OECD 481 *Saccharomyces cerevisiae* miotic recombination assay OECD 482 DNA damage and repair, UDS in mammalian cells in vitro OECD 483 Mammalian spermatogonial chromosome aberration test OECD 484 Genetic toxicology: mouse spot test OECD 485 Genetic toxicology: mouse heritable translocation assay OECD 486 Unscheduled DNA synthesis test with mammalian liver cells in vivo OECD Draft In vitro Syrian hamster embryo (SHE) cell transformation assay
	Carcinogenicity	OECD 451 Carcinogenicity studies OECD 453 Combined chronic toxicity/carcinogenicity studies

can require isolation of peripheral blood mononuclear cells (PBMCs) and their in vitro stimulation with the antigen for 3–10 days, depending on the antigen. They can also be used to assess the production of cytokines in cultured supernatants. This direct use of cells in test differs from other prolonged cultures in which, first, cells of interest have to be grown to reach the appropriate cell number, confluence or differentiation profile. This is the case in most of the adherent cells which have to be grown to reach the appropriate cell number and confluence. The need for prolonged cultures is also needed when growing cells which are needed to differentiate, as in the case of establishing, for example, specific T cell lines, clones or differentiation of monocytes into dendritic cells.

Regardless of the cell type, for in vitro cultures specific conditions have to be met. One is to provide cells with culture media formulated the way it meet the cells optimal growth support. Following are the basic elements of the culture media.

Nutritional requirements of different cell types are fulfilled by culture media usually supplemented with serum and other specific components.

Culture media consist of basic elements such as balanced salts that provide adequate buffering conditions. There are culture media which utilize Hank's or Earle's salts depending whether the equilibration is required with the atmospheric or gas phase containing 5% or 10% of CO_2, respectively. These salt solutions provide conditions to grow the cells at a pH value that is appropriate for the cell type and to prevent pH fluctuations that may occur during the cell growth and/or metabolism. Usually, in addition to the culture media HEPES is added, which improves the control of pH over the range of 6.7–8.4. However, care must be taken because of its toxicity for different cell types. Adequate concentrations should be used, usually up to 25 mM. Additionally culture media are supplemented with essential amino acids, vitamins, and carbohydrates.

Depending on the cell type that is cultured they may be supplemented with different media.

Serum
Supplements include fetal bovine serum, newborn calf serum, horse serum, and human serum. However, different batches of serum are available from different vendors and they can contain proteins, growth factors, hormones, amino acids, lipids, vitamins, carbohydrates that may affect cell culture. Thus it is important to test several different sera from different sources before starting the cell culture, especially when no previous results on cell growth are available. In some cases of culturing human cells, to minimize any unwanted background measurements generated from calf serum, AB or autologous serum can be used. For some experiments, especially those using blood mononuclear cells, serum has to be heat inactivated to prevent complement-dependent cell lysis. In several cases serum-free media is used, for example for keratinocytes or hepatocytes that are often used in toxicological studies.

l-Glutamine
This amino acid belongs to the essential components of culture media. It may be utilized by some cells, especially those that quickly divide, in a short period of time. Thus adequate concentrations of it have to be used, otherwise the growth arrest may occur. Also, it is important to supplement culture media with fresh L-glutamine since it easily hydrolyzes and its by-products can even be toxic for cells.

Non-Essential Amino Acids
These are especially useful as supplements in culture media for long-term cell cultures.

Sodium Pyruvate
Sodium pyruvate is usually used when a higher energy source is required. It acts as an intermediate in sugar metabolism where it is converted to acetaldehyde and carbon dioxide.

Growth Factors
These include cell-specific factors for which growth of the cells can be highly dependent. They include endothelial cell growth factor, epidermal cell growth factor, nerve growth factor, and certain cytokines.

Antibiotics
These supplements include the mixture of streptomycin, penicillin and gentamicin. They are used to prevent infections with gram-positive or gram-negative bacteria, and *Mycoplasma*.

Attachment Factors
They are needed for surface adhesion and growth of some of the adherent cell types. Examples are fibronectin, laminin, and collagen.

Another important factor in cell growth is the use appropriate tissue vessels. The use of the particular type of the culture vessel will depend on the cell line, its growth extent and the type of the test that the cells will be used for. In the case of adherent types of cells, first they have to be grown to reach confluency before being tested. Their spread over the surface provides the means for their growth, and thus their increased numbers. This step usually is achieved by using flasks of different sizes. Once the cells reach confluence they can be transferred to the next flask for further growth (this is called the passage) or to other vessels, if used for the test.

For suspension of cells, the factor that determines the size of the vessel, or flask, is the concentration [of cells] in the culture medium and thus the volume.

No less important than the size is the material from which the vessels are made. Most cells are grown in polystyrnee built surfaces but some cells may need special conditions, for example addition of matrix proteins to support adhesion and growth of adherent cells, or specifically formulated surfaces for growth of neurons.

Cell growth may depend also on specific requirements of the biology of the cell. T cells are dependent on the presence of antigen and APCs for differentiation and proliferation. Thus for in vitro propagation of T cells, not only do adequate culture conditions have to be met but also the requirements that stem from the biology of these cells.

Animal or human studies on T cell reactions to toxicants frequently involve analysis of their proliferative responses. The T cell proliferation test is used as an indicator of the immune response of the host to the toxicant it was exposed to. In this case, PBMCs from people or animal cells (obtained by draining the lymph nodes at the site of exposure) are usually used. In some cases however, human cells of bronchoalveolar lavage are used, as in the case of sensitization to beryllium. This test uses the process of antigen-specific response of T cells and their subsequent proliferation to confirm the engagement of the host immune system in response to toxicant exposure. In this test isotope-labeled (H^3-labeled) thymidine is usually used which, during cell proliferation, incorporates into newly synthesized DNA. The in vitro analysis of proliferation of the cells indicated above does not require additional sources of APCs due to the fact that usually they are present in the cell preparations and they take 3–10 days depending on the antigen.

Another endpoint of T cell activity includes the measurement of the release of soluble mediators—cytokines—and the expression of surface receptors. However, to increase the sensitivity of the antigen-mediated reactions that could be detected in in vitro conditions it might be necessary to propagate antigen specific

T cells, to mimic an in vivo processes that lead to proliferation and differentiation of T cells.

T cell lines in humans can be derived from PBMCs. The chance of success increases with a high proportion of in vivo antigen-specific T cells. For example, higher numbers of pollen allergen-specific T cells will exist during the pollen season; also high numbers of tuberculin-purified protein derivative can be expected in patients immunized against tuberculosis. The culture and establishment of T cell lines specific for the antigen of interest can be achieved by their culture in appropriate conditions that provide cells with sufficient growth factors but also renewed sources of antigen and APCs. A general schematic presentation of this method is shown in Figure 1, and Figure 2 shows the preparation of APCs.

Although the source of APCs shown in Fig. 2 is B cells, some scientists prefer to use dendritic cells, which are differentiated from monocytes. These cells consist of a professional pool of APCs which can more effectively present antigen to T cells. Thus in some cases use of these cells rather than B cells may be needed, and may have more advantages.

The technique of preparing B cells shown in Figure 2 is the most widely used one. It uses the property of T cells to express the cluster determinant (CD) 2 receptor, which binds sheep red blood cells (SRBC). B cells do not possess this receptor and thus will stay in the interphase between the gradient and the buffer. To obtain pure populations of B cells it is necessary to incubate with SRBC twice. Because T cells are the ones whose activity is tested, B cells added to

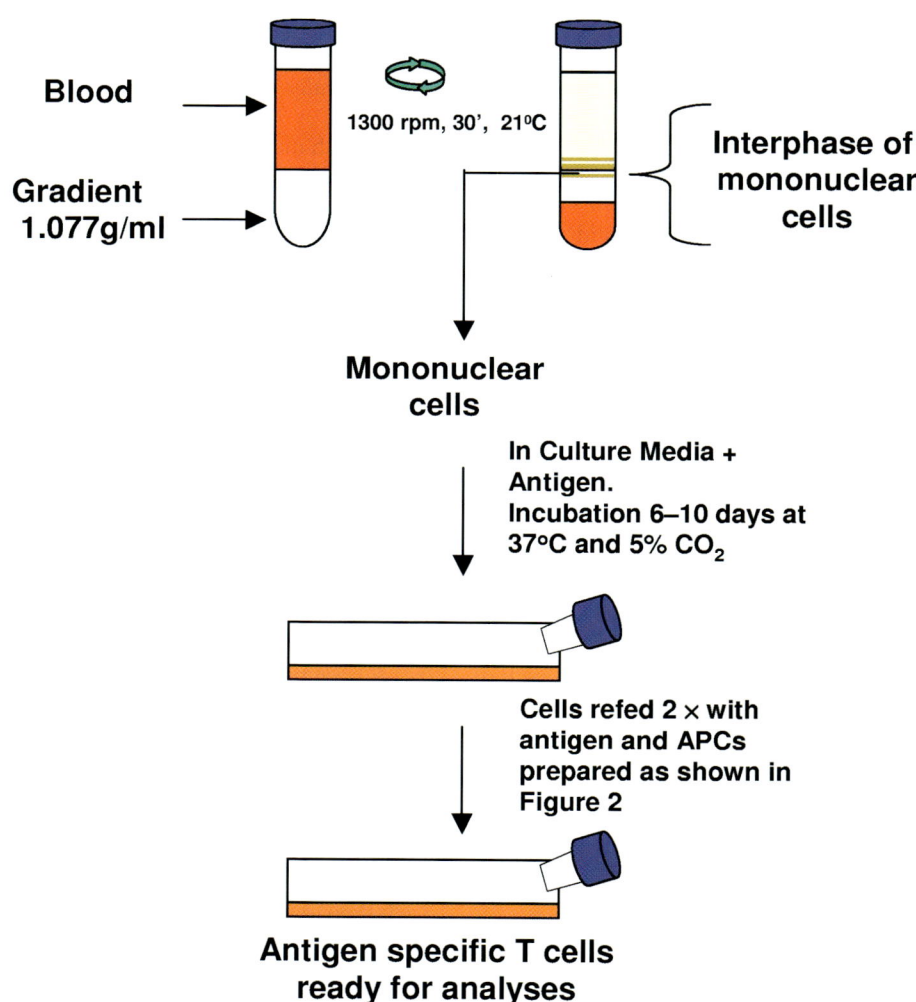

Leukocyte Culture: Considerations for In Vitro Culture of T cells in Immunotoxicological Studies.
Figure 1 Preparation of antigen-specific T cell lines.

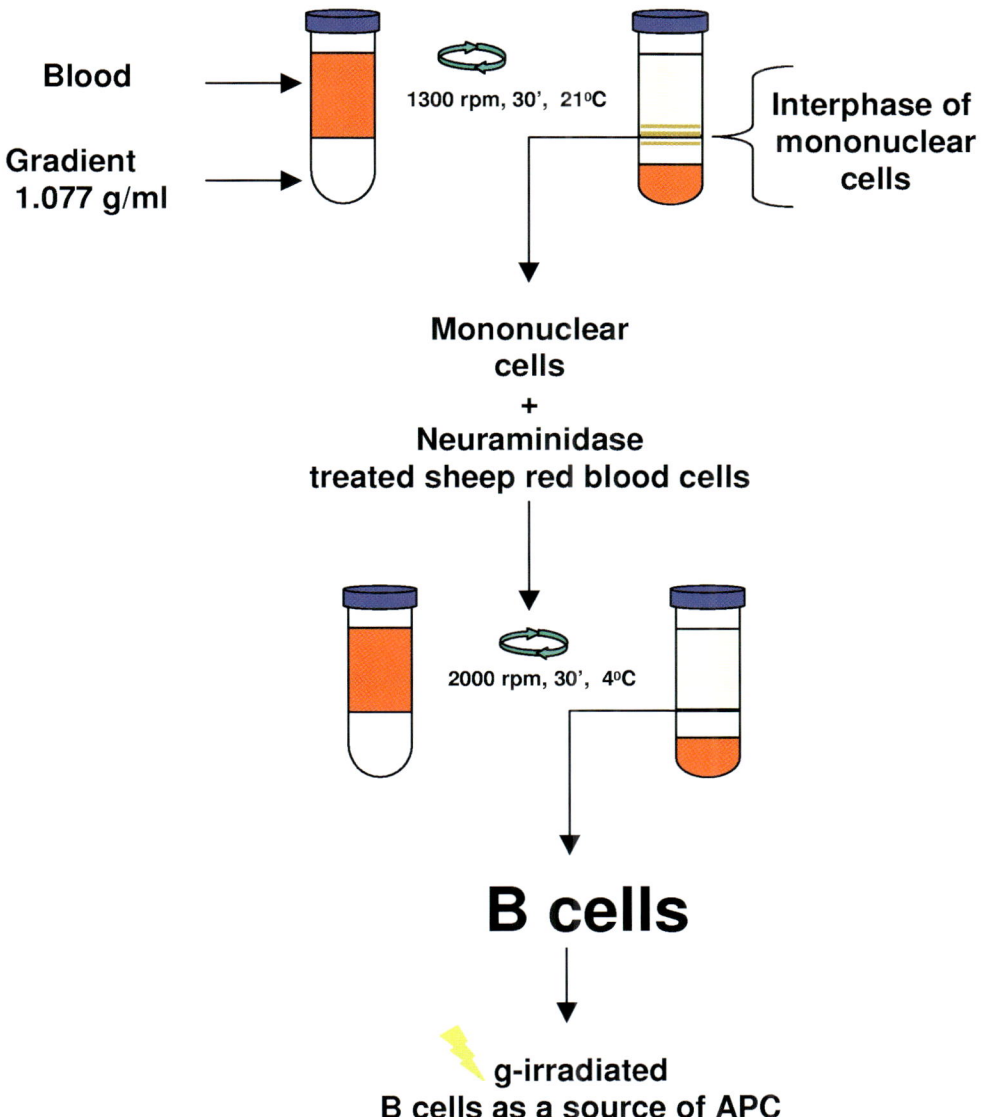

Leukocyte Culture: Considerations for In Vitro Culture of T cells in Immunotoxicological Studies.
Figure 2 Preparation of antigen-presenting cells.

the culture need to be inactive in terms of their own protein synthesis, which is usually inhibited by γ-irradiation.

Once the population of T cells is obtained they can be the source for the analysis of antigen-specific reactions. These can include cytokine production, expression of cell surface receptors, and intracellular signaling mechanisms. Cytokine analysis, especially of T helper (Th) type 2 cells or Th type 1, may provide important information about the properties of the antigen. Preferential increase of Th2 cytokines, such as interluekin(IL)-4 or IL-13, can indicate that antigen may induce a humoral type of immune response, and lead to allergic sensitization.

These cytokines provide signal for B cells to undergo immunoglobulin isotype switch to heavy-chain class E. Preferential production of IL-12 in cultures of mononuclear cells and interferon (IFN)-γ of T cells, on the other hand, can suggest that antigen may induce cell-mediated immune responses.

Also of Importance is the analysis of cell receptor expression. Analysis of the expression of chemokine receptors, ▶ adhesion molecules, the tumor necrosis factor receptor family, and cytokine receptors may provide important information about the activity of antigen-mediated stimulation of T cells. The expression of some of the receptors indicate important changes in cell migration, proliferation, inactivation, cytokine production, and others. Most of these can be measured by cytofluorometry, which can show qualitative and quantitative differences in receptor expression.

Pros and Cons

The study of T cell activity is one of the major fields in basic immunology. Rapid advances have been and still are being made in this field, making it easier to understand the immunological effects of toxic agents. Recognition of T cell reactions to the antigen of exposure may provide important information for risk assessment. The proliferation of T cells to toxic agents has already been established as a fundamental element in the detection of chemicals with skin sensitizing properties (1).

Further research may bring important information on the activity of these cells and respiratory sensitization or autoimmune reactions. Thus the technique of establishing antigen-specific T cells in an in-vitro system may provide researchers with an important tool for investigating immunotoxic effects, and may in the future create markers for easy identification of these actions.

Chemicals are usually haptens which by themselves can not induce an immune response—they require binding to host proteins. Antigen-specific T cells are difficult to establish in in-vitro conditions for such antigens; there is even evidence that the hapten induces a specific immune responses. Therefore, for now, high-molecular-weight antigens are being studied in the main. These studies, due to their complexity, will not be used as methods for measuring immunotoxic potentials of chemicals, but they may lead to finding easy determination markers.

References

1. Kimber I, Dearman RJ, Basketter DA, Ryan CA, Gerberick GF (2002) The local lymph node assay: past, present and future. Cont Dermat 47:315–328
2. Frome EL, Newman LS, Cragle DL, Colyer SP, Wambach PF (2003) Identification of an abnormal beryllium lymphocyte proliferation test. Toxicology 183:39–56

Leukocyte Differentiation Antigens

▶ CD Markers

Leukocyte Emigration

A highly selective and specific process involving the recruitment of leukocytes to a site of inflammation. The sequential events include rolling, activation, adhesion and transendolthelial cell migration of leukocytes.

▶ Inflammatory Reactions, Acute Versus Chronic

Leukocyte Function-Associated Antigen-3 (LFA-3)

Leukocyte function-associated antigen-3 (LFA-3) is a co-stimulatory molecule whose ligand is the CD2 molecule expressed on T lymphocytes. LFA-3 interactions co-stimulate cytokine secretion from T lymphocytes. It is also known as CD58.

▶ Interferon-γ

Leukocyte Margination and Adhesion

Specific events associated with leukocyte emigration that can be observed histologically at the site of an inflammatory response.

▶ Inflammatory Reactions, Acute Versus Chronic

Ligand Blotting

Ligand blotting is a modification of western blot analysis, in which protein receptors, following electrophoretic separation and transfer onto a blotting membrane, are detected by the specific binding of labeled ligands.

▶ Western Blot Analysis

Ligand Passer

Ligand passer is one of several hypotheses put forward to explain the pro-apoptotic influence of TNF-R2 in the absence of a death inducing signaling complex. It has been suggested that the high affinity and slow on/off kinetics of TNF-R2 for TNF-α may result in sequestration of biologically active TNF-α possible ligand passing to TNF-R1 thus facilitating TNF-R1-induced apoptosis.

▶ Tumor Necrosis Factor-α

Ligand Passing

A cytokine bound to a soluble receptor or to one type of a membrane-inserted cytokine receptor may leave the complex and bind to the signaling transmembrane cytokine receptor (i.e. from the p75 receptor for tumor necrosis factor to the p55 receptor).
▶ Cytokine Receptors

Ligand Traps

Ligand traps are recombinant molecules which contain the ligand binding domains of the α-chain and the β-chain of a heterodimeric cytokine receptor fused to the constant parts of a human immunoglobulin G molecule. Ligand traps are artificial binding molecules which bind their corresponding cytokine with high specificity and with a higher affinity than a homodimeric fusion protein containing only two α-chains of the cytokine receptor.
▶ Cytokine Receptors

Limiting Dilution Analysis

GEORG BRUNNER
Fachklinik Hornheide an der Universität Münster
Dorbaumstrasse 300
D-48157 Münster
Germany

Synonyms
LDA, limiting dilution analysis

Short Description
Limiting dilution analysis is a method in cell biology for estimating the frequency of a specific cell type in a complex cell mixture. This cell type is identified by its response to an activation signal, which can induce cell proliferation, differentiation, or the expression of specific cellular functions in the ▶ responder cells, such as ▶ cytotoxic activity or the release of cytokines or antibodies. To set up limiting dilution analysis, a dilution series of the cell mixture is prepared, and a large number of ▶ replicate cultures at each dilution is assayed for their response to the activation signal. The cultures are scored as responding or nonresponding. Mathematical analysis of the percentage of nonresponding cultures at each dilution allows the estimation of the ▶ responder cell frequency in the original cell population.

Characteristics
Limiting dilution analysis is a powerful technique to characterize quantitatively and qualitatively defined cellular subpopulations present in an unfractionated cell mixture, based on their functional response to specific stimulation. A number of different cell types can be analyzed responding to a variety of signals. The analysis is most commonly used to investigate T- and B-lymphocyte repertoires. It is particularly useful in evaluating the role of specific cell types in immune responses as well as in detecting responder cell types present at low frequency. Apart from applications of this technique in the analysis of lymphocyte repertoires, the quantitation of hematopoietic stem cells is commonly achieved by limiting dilution analysis in vivo or in vitro (1).

For a standard limiting dilution analysis, replicate cultures are set up at various dilutions of the cell population to be analyzed, in the presence and absence of a specific stimulus (e.g. antigen), and, if appropriate, in the presence of exogenously added cytokines (e.g. interleukin-2 or T cell growth factor) (2). The larger the number of replicates set up at each dilution, the more precise is the estimate of the responder frequency. Therefore, a minimum of 24 (up to 96) replica cultures is commonly used. The culture conditions should be optimal for the cell population analyzed such that the frequency of responders is the only limiting factor in the assay. This includes addition of an optimal concentration of cytokines as well as optimal time of incubation. Depending on the type of assay and the output signal measured, cultures are incubated for periods ranging between 3 and 18 days.

The most commonly used output signals are measurements of cytotoxic activity, cell proliferation, or cytokine production.

Analysis of cytotoxic activity is used in assays determining the frequency of cytotoxic T lymphocytes (CTLs), natural killer (NK) cells, lymphocyte-activated killer (LAK) cells, or antibody-dependent cellular cytotoxic (ADCC) cells. This effector function is measured via the release of ^{51}Cr from labeled ▶ target cells. The target cell types used depend on the specificities of the cell populations assayed. To confirm the specificity of cytotoxic analyses, multiple target cells can be used (▶ split-well analysis).

A less specific output signal is the measurement of cell proliferation. Proliferative cells are usually monitored by measuring the incorporation of tritiated thymidine into DNA. In this case, measurements need to be corrected for baseline proliferation in the cell population in the absence of specific stimulation as well as for proliferative activity due to cytokines added to the cultures (e.g. interleukin-2).

An alternative output signal is the production of specific proteins, e.g. lymphokines or antibodies, which can be measured in immunochemical assays such as the enzyme-linked immunosorbent assay (ELISA). Following incubation, the replicate cultures are scored as positive (responsive to stimulation) or negative (unresponsive). While positive cultures may arise from the presence of one or more responder cells, negative cultures unequivocally demonstrate the absence of responders. Therefore, the fraction of unresponsive cultures is determined for each dilution of the cell population assayed. These values are then used to estimate the frequency of responder cells (n) by plotting the log of the fraction of negative cultures (F) against the cell number in the respective replicate cultures (Fig. 1). The zero term of the Poisson equation ($F = e^{-n}$ where e is the base of the natural logarithm) predicts that an average of one responder cell per culture (n = 1) will result in a fraction of 0.37 (1/e) of the replicate cultures being negative (37%). Thus, the frequency of responder cells in the cell population assayed can be read directly from the graph (Figure 1). The most informative range of this analysis is between negative fractions 0.1–0.37 (3). Outside this range, it is advisable to extend the analysis to larger fractions of negative cultures.

Limiting dilution analysis of unfractionated cell populations may result in nonlinear kinetics. This is usually caused by the interaction of two subpopulations of responder cells, present at distinct cell frequencies, responding to the same stimulatory signal (4). The cell subpopulations do not behave independently but interact with each other and compete for cytokines. This exemplifies the versatility of limiting dilution analysis in monitoring, apart from cellular frequencies, complex cellular interactions in a biologically relevant setting.

The principle of limiting dilution of a cell population is also used preparatively to isolate single cells (▶ limiting dilution cell cloning). In this technique, cells are plated at very low cell densities (between 0.25 and 1 cells/culture), usually on top of a feeder cell layer. The output signal measured is cell proliferation, and positive cultures at the lowest cell density are likely to be derived from single cells. Since a small chance remains, according to Poisson statistics, that positive cultures arise from two cells rather than one, limiting dilution is usually repeated to verify clonality. Apart from the characterization of cell populations, limiting dilution analysis is also being used in molecular biology to quantitate DNA in complex biological samples. This is achieved by amplifying the target DNA sequence in the sample by ▶ limiting dilution polymerase chain reaction (PCR) (5). For this technique, PCR is optimized such that the amplification of endogenous DNA will take place in an all-or-nothing fashion, even at very low target DNA frequencies. By performing multiple replicate PCR reactions at serial dilutions of the sample, the frequency of the target DNA can be estimated from the fraction of negative reactions (no DNA product amplified) at each sample dilution using Poisson statistics as described above (see Figure 1).

Pros and Cons

Limiting dilution analysis allows the monitoring of a large number of variables in complex cell mixtures containing several types of responder cells. The cell population analyzed does not need to be fractionated, which enables cellular and molecular interactions to occur mimicking the in vivo situation.

The analysis is noninvasive and can easily be repeated multiple times. The results obtained with immune cells of healthy individuals are reproducible over a relatively long time period.

Reliability of the results is limited to a relatively narrow range of dilutions of the cell population, which is usually not known at the beginning of a new experimental series.

To obtain statistically significant results, relatively large numbers of replicate cultures (at least 24) are required for each dilution of the cell population.

The data obtained from limiting dilution analyses of unfractionated, mixed cell populations may deviate from linearity, which is usually due to the presence of more than one responder cell type in the sample. These responders interact with each other resulting in stimulatory and/or inhibitory effects, depending on the dilution of the sample (3).

Predictivity

Reliable information on cellular frequencies in mixed cell populations is only obtained in a relatively narrow range of dilutions. The most informative value is between multiplicities (responder frequency per well) of 2.5 and 1 (i.e. at dilutions of the cell population resulting in 10%–37% of the replicate cultures being unresponsive). At dilutions outside this range, reliability of limiting dilution analysis drops sharply. As a compromise, it is recommended to design the assays such that they cover lower (< 1) rather than higher multiplicities. The precision of limiting dilution assays correlates with the number of replicate cultures set up for each dilution of the cell population. A replicate number of 24 is minimally required. Larger replicate numbers will yield greater precision.

Limiting dilution analysis has been shown to be reproducible, yielding similar frequencies of antigen-responsive cells in humans over a time period of several months. Limiting dilution analysis can be used to determine the frequency of functionally corresponding cell types in both animal models and humans. However, the translation of data from animal models into the human system is limited by the complexity and

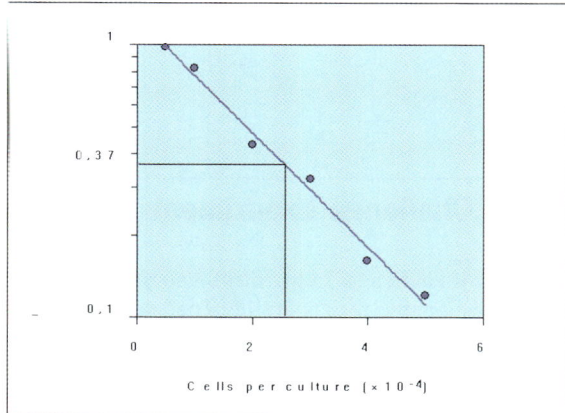

Limiting Dilution Analysis. Figure 1 Example of a graphical analysis of LDA data. The log of the fraction of negative cultures (F) is plotted against the cell number in the cultures. The zero term of the Poisson equation predicts that, at F=0.37, there is on average one responder cell per culture with the specificity tested. This allows the estimation of the responder frequency in the original cell mixture (in this example one responder cell in 2.5×10^4 cells).

species-specific properties of immune responses. This is exemplified by the T lymphocyte response to HIV-1 in an animal model for acute infection, which is much more pronounced than in humans (6).

Relevance to Humans

Limiting dilution analysis is of outstanding clinical importance, since it is the only technique allowing the quantitation of immune responses at the single-cell level. It is being used to assess T cell responses in humans, e.g. to different pathogens. The higher the frequency of antigen-specific T cells, the stronger the immune response will be following antigenic challenge. Such information is important for attempts to enhance the efficiency of vaccine design.

Limiting dilution analysis is also particularly useful in a variety of additional clinical settings involving the quantitation of immune or hematopoietic cells responding to a specific stimulus. Examples are the immunological monitoring of transplant recipients and cancer patients to assess their response to allo-MHC (major histocompatability complex) antigens of a particular organ or bone marrow donor and to specific antigens of the patient's tumor, respectively. In the case of organ or bone marrow transplantation, this is helpful in judging the patient's risk of rejecting the graft. Analysis of the immune response of cancer patients can be helpful in predicting and/or monitoring the success of therapeutic approaches involving the stimulation of the patient's immune system. The success of bone marrow transplantation significantly depends on the number of hematopoietic stem cells present in the graft. Limiting dilution analysis is widely used to determine the stem cell frequency in bone marrow grafts.

Limiting dilution PCR (polymerase chain reaction) has been used to assess the therapeutic success and the level of minimal residual disease in leukemia patients by quantifying leukemic lymphocytes in the blood. In addition, this technique has been evaluated for potential application in the detection of genetically modified organisms in food.

Regulatory Environment

Not applicable.

References

1. Sieburg HB, Cho RH, Müller-Sieburg CE (2002) Limiting dilution analysis for estimating the frequency of hematopoietic stem cells: Uncertainty and significance. Exp Hematol 30:1436–1443
2. Sharrock CEM, Kaminski E, Man S (1990) Limiting dilution analysis of human T cells: A useful clinical tool. Immunol Today 11:281–286
3. Fazekas de St Groth S (1982) The evaluation of limiting dilution assays. J Immunol Meth 49:R11–R23
4. Dozmorov I, Eisenbraun MD, Lefkovits I (2000) Limiting dilution analysis: From frequencies to cellular interactions. Immunol Today 21:15–18
5. Sykes PJ, Neoh SH, Brisco MJ, Hughes E, Condon J, Morley AA (1992) Quantitation of targets for PCR by use of limiting dilution. BioTechniques 13:444–449
6. Kent SJ, Corey L, Agy MB, Morton WR, McElrath MJ, Greenberg PD (1995) Cytotoxic and proliferative T cell responses in HIV-1-infected *Macaca nemestrina*. J Clin Invest 95:248–256

Limiting Dilution Cell Cloning

This method in cell biology is used to isolate single cells from complex cell mixtures.
▶ Limiting Dilution Analysis

Limiting Dilution Polymerase Chain Reaction

This method in molecular biology is an application of DNA amplification by standard polymerase chain reaction (PCR) for estimating the frequency of specific DNA sequences in complex biological samples.
▶ Limiting Dilution Analysis

Lipid Rafts

Lipid rafts are sphingolipid/cholesterol-enriched membrane domains found in all mammalian cell types. In unactivated lymphocytes they are small, but upon activation they cluster together to form large rafts. They are though to have a role in the initiation and organization of the signaling cascades, which lead to the activation of B cells and T cells by pre-assembling the signaling proteins in restricted areas of the membrane close to the receptor complexes. This clustering is believed to allow quick and efficient connection to signaling cascade upon receptor engagement.
▶ Signal Transduction During Lymphocyte Activation

Lipids

▶ Fatty Acids and the Immune System

Lipopolysaccharide (LPS)

Refers to lipid-containing polysaccharides found in the cell wall of Gram-negative bacteria. The lipopolysaccharide derived from *Escherichia coli* stimulates nonspecifically the proliferation of B cells of several species. It is widely used in immunology for studies involving lymphocyte proliferation. The lipid A component mediates toxicity and cytokine production whereas the polysaccharides are responsible for immunogenicity. Also called endotoxin. LPS can induce activation of proinflammatory cytokines including IL-1β, TNF-α and IL-6.
▶ Lymphocyte Proliferation
▶ Polyclonal Activators
▶ Interleukin-1β (IL-1β)
▶ Polyclonal Activators
▶ Cytokines
▶ Neonatal Immune Response

Listeria Monocytogenes

A facultative intracellular bacterium.
▶ Host Resistance Assays

Live-Death Discrimination

▶ Viability, Cell

Live Rate

▶ Viability, Cell

LLNA Challenge Experiment

An LLNA challenge experiment can be conducted when questionable results are obtained with an induction phase LLNA. It consists of two phases, the induction or sensitization phase, and the challenge or elicitation phase. Challenged groups are compared with the induction control in order to discover an allergy-relevant change in reactivity to the test chemical.
▶ Local Lymph Node Assay (IMDS), Modifications

Local Immune System

▶ Skin, Contribution to Immunity

Local Lymph Node Assay

IAN KIMBER · REBECCA J DEARMAN
Syngenta Central Toxicology Laboratory
Alderley Park, Macclesfield
Cheshire
SK10 4TJ
UK

Short Description

The murine local lymph node assay (LLNA) was developed initially as a method for the identification of chemicals that have the potential to cause skin sensitization and allergic contact dermatitis. Since its first description the LLNA has been the subject of detailed assessments in the context of national and international collaborative trials, and of extensive comparisons with both other predictive test methods and human data. The origins, development, evaluation and eventual validation of the LLNA have recently been described comprehensively elsewhere (1).
In contrast to guinea pig tests, the LLNA identifies contact allergens as a function of events induced during the induction phase of skin sensitization—specifically the stimulation of proliferative responses in draining ▶ lymph nodes.

Characteristics

The LLNA is based on an appreciation of the events

that characterize the induction phase of skin sensitization. Following topical exposure to a skin sensitizing chemical molecular and cellular processes are provoked that act in concert to elicit a cutaneous immune response. Among the key events are the mobilization of epidermal ▶ Langerhans cells (LC), and their migration from the skin to draining lymph nodes, via the ▶ afferent lymphatics. While in transit from the skin these cells undergo a functional maturation and localize in the paracortical region of the lymph nodes where they present antigen to responsive T lymphocytes. Antigen-activated T lymphocytes are stimulated to divide and differentiate, the former resulting in a selective clonal expansion of allergen-specific cells. This is the cellular basis of sensitization. If the now sensitized subject encounters the same allergen again—at the same site or a different skin site—then the expanded population of responsive T lymphocytes will mount an accelerated and more aggressive secondary immune response at the point of exposure. This will, in turn, provoke the dermal inflammatory reaction that is recognized clinically as allergic contact dermatitis (2). A unifying and mandatory characteristic of contact allergens is, therefore, the stimulation of T lymphocyte responses in lymph nodes draining the site of skin exposure, and this provides the mechanistic rationale for the LLNA.

Methodological descriptions of the LLNA are available elsewhere (3). Briefly, the standard assay is conducted as follows. Groups of mice (CBA strain) are exposed topically, on the dorsum of both ears, to various concentrations of the test chemical, or to an equal volume of the relevant vehicle alone. Treatment is performed daily for 3 consecutive days. Five days following the initiation of exposure mice receive an intravenous injection of [^{3}H] thymidine (^{3}H-TdR). Animals are killed 5 hours later and the draining (auricular) lymph nodes are excised. These are pooled for each experimental group, or alternatively are pooled on a per-animal basis. Single cell suspensions of lymph node cells (LNC) are prepared and incorporation of ^{3}H-TdR measured by β-scintillation counting. Interpretation of results is based upon derivation of a stimulation index (SI). For each concentration of test chemical an SI is calculated relative to the concurrent vehicle control. Skin sensitizers are defined as chemicals which—at one or more test concentrations—induce *at least* a 3-fold increase in LNC proliferation compared with the vehicle control (an SI of 3 or more) (see Figure 1).

There have been descriptions of proposed modifications to the standard LLNA—some relatively minor in scope, others more substantial (1). A consideration of these alternatives is beyond the scope of this essay (see ▶ Local Lymph Node Assay (IMDS), modifications). The standard assay as described above was developed specifically for the purposes of hazard identification, and the performance of the assay in this context is addressed below. However, the LLNA is now used also for the measurement of relative potency in the context of risk assessment. This application is predicated on an understanding that the vigour of T lymphocyte proliferation induced in draining lymph nodes correlates closely with the extent to which sensitization will be acquired. The thesis is, therefore, that activity in the LLNA will be informative also of potency. For this purpose an EC3 value is derived—the **E**stimated **C**oncentration (EC) of chemical required to induce an SI of exactly **3** (4).

Advantages and Disadvantages

In the context of hazard identification the LLNA offers a number of important advantages compared with guinea pig tests. Exposure is via the relevant route, there is no requirement for adjuvant, and the readout is objective and quantitative. Moreover, it is acknowledged that the LLNA also provides for significant animal welfare benefits with regard to both a reduction in the number of animals required and the trauma to which they are potentially subject.

The LLNA also permits accurate measurement of relative ▶ skin sensitization potency as a first step in the risk assessment process, in a way that was not usually possible with the standard guinea pig assays.

Predictivity

As indicated above the LLNA has been evaluated exhaustively and has been found to provide a robust and reliable method for the identification of skin sensitizing chemicals that can serve as a stand-alone alternative to guinea pig test methods. In common with all toxicity test methods the accuracy of the LLNA is not perfect. Thus it has been shown, for instance, that some—but by no means all—skin irritants may provoke low level responses. However, it has been demonstrated that the overall performance of the LLNA is at least comparable with (and in most instances superior to) methods such as the guinea pig maximization test and the occluded patch test of Buehler (5). The method has been endorsed in the USA by the Interagency Coordinating Committee on the Validation of Alternative Methods (ICCVAM) and in Europe by The European Centre for the Validation of Alternative Methods (ECVAM).

Relevance to Humans

Clearly an important—indeed critical—aspect of the evaluation and validation of the LLNA was establishing that the method is able to identify accurately chemicals that are known to cause skin sensitization in humans. As confirmed during validation, this is the case.

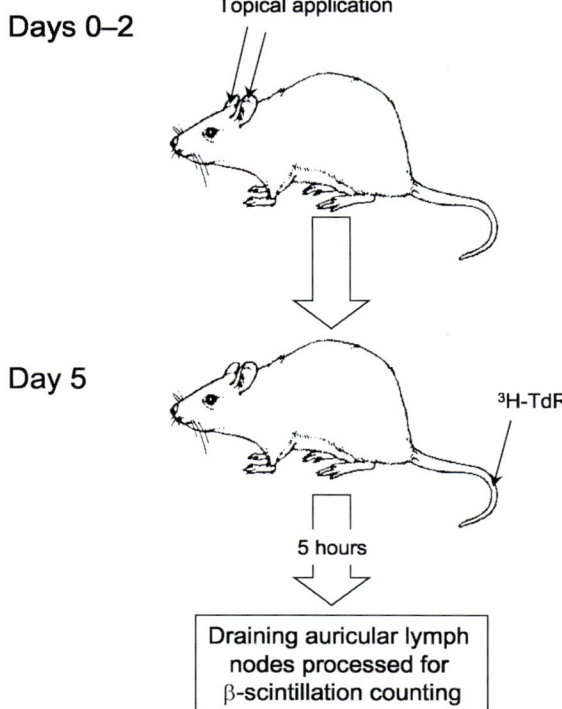

Local Lymph Node Assay. Figure 1 A schema of the standard local lymph node assay. Mice (CBA strain) receive three consecutive daily applications of 25 μl of various concentrations of the test material, or of the relevant vehicle alone, to the dorsum of both ears. Five days after the initiation of exposure, mice receive an intravenous injection, via the tail vein, of 20 μCi of ^{3}H-thymidine (^{3}HTdR) in 250 μl of phosphate-buffered saline. 5 hours later draining auricular lymph nodes are excised and pooled for each experimental group, or on a per-experimental-animal basis, and are processed for β-scintillation counting.

It has been demonstrated also that measurement of relative potency using the LLNA reflects clinical experience. Thus, derived EC3 values have been shown to correlate well with what is known of the potency with which chemical allergens cause skin sensitization in humans (6,7).

Regulatory Environment

In the period following endorsement of the method by ICCVAM (in 1999) the LLNA has been adopted by several regulatory agencies in the USA. The method has also now been incorporated into a new test guideline (number 429) entitled *Skin Sensitization: Local Lymph Node Assay* by the Organization for Economic Cooperation and Development (OECD). This was adopted formally in April 2002. In parallel, the European Union (EU) has prepared a new test guideline for the LLNA (B.42) entitled *Skin Sensitization: Local Lymph Node Assay* which has now been published in the Official Journal of the EU.

Concluding Comments

The LLNA provides an accurate method for the identification of chemicals that have the potential to cause skin sensitization. Moreover, the method is finding increasing application for the measurement of relative risk. There will undoubtedly be opportunities for further refinement of the LLNA that will enhance further characterization of hazards and risks associated with skin sensitization.

References

1. Kimber I, Dearman RJ, Basketter DA, Ryan CA, Gerberick GF (2002) The local lymph node assay: past, present and future. Cont Derm 47:315–328
2. Kimber I, Basketter DA, Gerberick GF, Dearman RJ (2002) Allergic contact dermatitis. Internat Immunopharmacol 2:201–211
3. Dearman RJ, Basketter DA, Kimber I (1999) Local lymph node assay: Use in hazard and risk assessment. J Appl Toxicol 19:299–306
4. Kimber I, Basketter DA (1997) Contact sensitization: A new approach to risk assessment. Human Ecolog Risk Assess 3:385–395
5. Gerberick GF, Ryan CA, Kimber I, Dearman RJ, Lea LJ, Basketter DA (2000) Local lymph node assay: Validation assessment for regulatory purposes. Amer J Cont Derm 11:3–18
6. Basketter DA, Blaikie L, Dearman RJ et al. (2000) Use of the local lymph node assay for the estimation of relative contact allergic potency. Cont Derm 42:344–348
7. Gerberick GF, Robinson MK, Ryan CA et al. (2001) Contact allergenic potency: correlation of human and local lymph node assay data. Amer J Cont Derm 12:156–161

Local Lymph Node Assay (IMDS), Modifications

Hans-Werner Vohr
PH-PD, Toxicology
Bayer HealthCare AG
Aprather Weg 18a
42096 Wuppertal
Germany

Peter Ulrich
Preclinical Safety, MUT2881.329
Novartis Pharma AG
Auhafenstrasse
CH-4132 Muttenz
Switzerland

Description

In 1998, after a decade of scientific evaluation and extensive interlaboratory validation, the local lymph node assay (see LLNA) was peer reviewed and was, in principle, endorsed as a stand-alone test by an independent panel of experts on behalf of the US Interagency Co-ordinating Committee on the Validation of Alternative Methods (ICCVAM) in 1999. However, improvements were suggested by the peer review panel concerning the discrimination of contact allergenic from irritation potential. Some irritating and phototoxic compounds were shown to cause a false-positive response, since they induced a vigorous cell proliferation in the ear-draining lymph nodes (1) or a marked LN ▶ hyperplasia (2,3). The occurrence of false-positive results is an obvious consequence of the tight physiological connection between inflammatory tissue processes and specific immune reactions, with the latter being more vigorous, when the first has reached a level sufficient to provide all adjuvant-like signals to the immune system. Thus, the proliferation of lymphocytes in the ear-draining lymph node is a mandatory, but not sufficient requirement for the induction of a contact allergic response (such as ▶ allergic contact dermatitis). A further reason for alternative LLNA protocols are the opinions of many researchers considering the use of radioactivity in the "classical" LLNA as not appropriate for various reasons.

Characteristics

Chemicals applied onto the skin and penetrating through the stratum corneum may induce two forms of skin reactions. First a compound specific immune reaction may be induced by binding to self-proteins, and second an acute inflammatory reaction by cytotoxic (irritating) properties of the chemical. The antigen-specific immune reaction is characterised by the development of memory T cells and B cells. Such memory cells are responsible for the elicitation of a pronounced skin reaction after re-exposure of low concentrations of the test chemical during the challenge phase. No such memory cells are induced in course of an acute skin reaction due to cell destruction (irritancy), although most of the skin sensitizing chemicals do express slight irritating properties in addition.

Additional endpoints were included in the test protocol with the aim to provide more information about the irritation potential of chemicals, which can be directly related to the reactivity in the LN. Homey et al (1998) (2) and Vohr et al. (2000) (4) reported that measurement of ear swelling in the LLNA can be used to assess the irritation potential of chemicals. Following a comparison of chemical-induced effects in the ear skin and the ear-draining LN, they concluded that LN hyperplasia in the absence of a significant primary ear irritation points to an allergic process, whereas the finding of LN hyperplasia together with ear irritation requires a more careful evaluation of the nature of the reactivity—(photo)allergy or true ▶ (photo)irritation. In a recent paper by Ulrich et al, the weight determination of circular biopsies of the ears has proven to be a useful marker for skin irritation (5).

The experimental set-up of the alternative LLNA is similar to the radioactive LLNA described by Kimber et al. (cf. related essay in this book), with the exception that the alternative approach does not require a 2-day resting phase between the 3-day treatment and killing the animals without being less sensitive. This study protocol was subject to an interlaboratory validation with the aim of establishing the measurement of LN hyperplasia and skin irritation as alternative endpoints (6). The interlaboratory validation thus used two modifications compared to the "classical" LLNA. These modifications measure lymph node weights and cell counts instead of radioactive labeling, and take the acute skin reaction (irritancy) into consideration for the judgment of the effects seen.

Irritancy

While measurement of cell counts (instead of incorporation of radioactive material) is already widespread (see below), measurement of the acute skin reaction too was introduced, as originally developed by Homey et al. (2). The reason for this modification was to flag up "false-positive" results due to skin irritation, as for example measured after sodium dodecyl sulfate (SDS) application. This modified LLNA was called the "integrated model for the discrimination of skin reactions" (IMDS).

Inclusion of the ability to detect skin irritancy in the model is one main advantage of the modified test. The irritating effect of chemicals can be taken into consideration by determining the acute skin reaction (increase in ear thickness and/or ear weight). Although a moderate irritation will also result in cell proliferation in the skin draining lymph node (false positive) a strong irritating, and therefore cytotoxic, reaction will induce more influx of lymph fluid, and by this an increase in lymph weight instead of cell proliferation (see also Table 1).

There is a dramatic difference in the cytokine pattern expressed during the induction phase of an immunological skin reaction, or after epidermal cell destruction by cytotoxicity. This cytokine balance, however, has an important impact on the skin reaction observed. During the onset of a specific (immunological) skin reaction the cytokines induced will cause migration of the ▶ Langerhans cells into the draining lymph nodes. In contrast to this mechanism the cytokines released during irritating/cytotoxic activation will attract leucocytes into the skin, reduce the tissue integrity at the inflammatory side, and increase the permeability for body fluids. In addition, the cytotoxic amounts of the test item reaching the draining lymph node can also result in cell destruction, and this process in some cases exceeds the cell proliferation/activation. Taken together, with increasing irritancy, decreased cell proliferation and relative increased lymph node weights will be observed.

There are several examples in the literature about such a "negative" influences on cell activation by increasing irritancy of a chemical or by pretreatment with sodium lauryl sulfate (SLS).

Cell Counts

In contrast to the consideration of the irritating potential in the LLNA, alternate endpoints (especially cell counting) have been described from the very beginning. All results published so far verify that there is no difference between determining proliferation in the draining lymph node on the basis of the cell counts or ^{3}H-thymidine incorporation. The sensitivity of both methods is comparable (Figure 1). It has to be clarified, however, that the "▶ positive level" or threshold value as defined for the radioactive method, that is, the EC3 value (see LLNA), is exclusively defined for the method and mouse strain used for LLNA. Such positive limits have to be calculated for each endpoint and strain of mice individually (see also below).

A modification of the assay which involves measuring cell counts instead of radioactive labeling provides not only comparable sensitivity but also has the advantage that the cell suspension can be further analyzed by different methods (flow cytometry, chemiluminescence, immunofluorescence) to gain an insight into mechanistic events.

By comparing the specific immune reaction induced by the test item in the draining lymph nodes (LN cell counts, LN weights) with the immediate unspecific acute skin reaction (ear swelling, ear weight) it is possible to discriminate the irritating potential from the sensitizing potential of the compound tested. International standards have successfully used this modification, and such modifications are also authorized in the Note of Guidance SWP/2145/00 of the CPMP (2001) and OECD Guideline No. 429.

International Catch-up Validation

The results of the interlaboratory validation of the modified LLNA in a working group of nine European laboratories have been published (6). This validation study has been conducted under the conditions recommended in the relevant OECD guidance, whereby the study was supervised by an independent coordinator, samples were blinded, and statistics were performed by an external independent statistician. This interna-

Local Lymph Node Assay (IMDS), Modifications. Table 1 Contact and irritant chemicals induce increases in acute ear reaction and in lymph node (LN) cell counts. On three consecutive days, five NMRI mice per group were topically treated with either 1% oxazolone, 1% croton oil, or vehicle on the dorsal surface of both ears. On day 3, ear thickness of both ears was measured, the local draining lymph nodes of the ears were removed, and individual lymph node cell counts were determined. Data indicate mean ear swelling or lymph node cell count indices which were defined as ratio of mean values from test groups to mean of the vehicle-treated control group. The maximum increases of stimulation indices were set to 2 for ear swelling and to 5 for cell counts, respectively.

	Acute ear Reaction [Index]	Lymph Node Cell Counts [Index]
Vehicle	1	1
1% Oxazolone	1.09* (≙9% increase)	4.13** (≙78% increase)
Vehicle	1	1
1% Croton oil	1.59** (≙59% increase)	2.58** (≙39% increase)

* $P < 0.05$.
** $P < 0.01$.

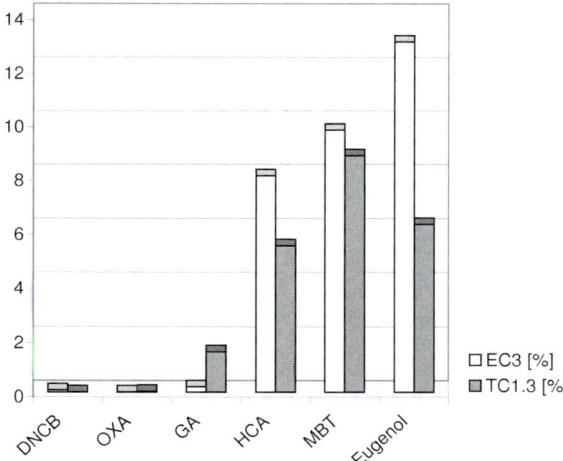

Local Lymph Node Assay (IMDS), Modifications. Figure 1 Comparison of threshold values obtained with (i) the radioactive version of the LLNA (white columns) and (ii) the alternative LLNA as described in the text. EC3: effective concentration, which causes a 3-fold increase in radioactivity in the LN preparations of chemical-treated groups in comparison to the vehicle control. This is the level considered to indicate a positive response in the radioactive LLNA. TC1.4: threshold concentration, which induces a 1.4-fold increase in the lymph node cellularity in comparison to the vehicle control.

ical concentrations, which cause mild to moderate LN hyperplasia for induction and a lower concentration for challenge. In cases of contact allergy, the response in the challenged group will be more vigorous than the induction control, which received the compound the first time. In addition, the clinical manifestation of contact allergy in terms of increased ear weights (or swelling) can be set into relation to the response in the LN (2, 4).

Potency

Of special interest is the fact that during recent years every effort has been made to rank sensitizing compounds according to their potency for inducing allergic skin reactions. Three or four categories of skin sensitizing compounds have been proposed, namely (weak), moderate, strong, and (extreme). In case of the LLNA the cut-offs for these categories is the so-called EC3 value (7). This "positive level" is the stimulation index of 3 based on radioactively labelling of the lymph node cells.

It is evident that compared to the cell counts much higher stimulation indices can be reached by the radioactive method. However, the individual standard deviation is also higher than for the non-radioactive method. Therefore, the "positive level" for the cell counts as determined on the basis of historical data as well as in the international validation study is 1.4

tional validation study using NMRI and BALB/c mice showed that the cell count index is a robust and reliable parameter for detecting skin sensitizing properties of test material. In addition, this parameter again proved to be as reliable as radioactive labelling. The data obtained for the acute skin reaction (ear weight, ear swelling) prevented all participating labs from classifying SDS as a sensitizing compound or overestimating the stimulation indices of test items with pronounced irritating properties.

The evaluation is based on the statistical comparison of the results of the test groups to the vehicle control using the Wilcoxon test. To this end, the parameters are measured for each individual animal. If statistically significant changes in lymph node parameters are obtained, these are related to the ear weights to account for the contribution of the primary skin irritation. This is quite another approach as used for the radioactive method, which calculates a ▸ stimulation index out of the radioactivity determined in group-wise pooled lymph nodes, without taking account of possible dermal irritation.

Pronounced irritation could cover a slight skin sensitizing property of the test item. For verification of such questionable results it is possible to perform a so-called ▸ LLNA challenge experiment using test chem-

Local Lymph Node Assay (IMDS), Modifications. Table 2

	LN	Ear
	0,001	0,001
	100	100
MBT	8,7500	100,0000
Eugenol	6,2000	100,0000
Isoeugenol	9,9000	100,0000
CinAld	5,4000	47,6200
OXA	0,0280	0,6250
DNCB	0,0105	0,3600
DNFB	0,0078	0,0560
GA	1,5000	0,5300
FA	24,0000	9,2000
CroOil	0,0240	0,0550
SDS	15,4000	15,0000
TCSA	0,0400	0,8150
TCSA/UVA	0,0260	0,2990
Acr	1,2220	100,0000
Acr/UVA	0,1175	0,0405
8-MOP/UVA	0,0078	0,0025

for NMRI and 1.55 for BALB/c, respectively. On the basis of these values a ranking of the sensitizing standards has been made which shows exactly the same order as described for the LLNA based on the radioactive labelling (Figure 1).

The interdependency of irritancy and sensitization for some international standard compounds is given in Figure 2. There are characteristic clusters of compounds which reveal a striking association of sensitizing potential and skin irritation. Chemicals with low or no skin irritation potential are at the same time weak sensitizers in the LLNA, and chemicals with a higher potential to irritate the skin bear a stronger sensitization potential. However, some irritants and photoirritants appear also in the cluster of chemicals with high LN activation potential.

Pros and Cons

The principal advantages of the LLNA in comparison to the classical GP assays have been discussed elsewhere (see LLNA). The addition of skin irritation as a further endpoint contributes to the larger specificity of this alternative LLNA. Some chemicals may be problematic when tested in an induction phase assay. In such cases, if there is no other information about the chemical, like the mode of action (▶ hapten theory) or the structure-activity relationship, a biphasic (secondary response) LLNA may be indicated, unless a certain degree of uncertainty in the positive classification of the chemical as a contact allergen is tolerable.

There are some technical limitations of the LLNA, which have to be considered. One problem is that the test item to be applied on the ear must be a solution or a fine suspension in a small volume. In addition, the test item must be prevented from dripping off the skin, and the vehicle should enhance the skin penetration of the test item because of the non-occlusive protocol. Therefore, the choice of vehicle is quite different to that for guinea pig assays, and aqueous formulations are not to be used without further modification of the protocol.

Predictivity

The predictivity of the modified LLNA (IMDS) is similar to that of the radioactive LLNA. However, its specificity could be improved by the introduction of skin irritation as an endpoint. Hence, the evaluation of both skin irritation and LN hyperplasia potentials of a chemical provides a better foundation for the distinction of contact allergenic activity from true skin irritation. Structure-activity relationships and a hypothesis on the mode of chemical action (▶ hapten theory) complete the evaluation and support the interpretation of LLNA results. In some cases the conduction of a biphasic LLNA consisting of an sensitization and an elicitation phase (secondary response), may be appropriate to bring final clarification.

Benzalkonium chloride (BKC) is a good example for the importance of measuring both the irritating properties as well as the sensitizing properties of a test compound. BKC is a strong irritant but with clear skin sensitizing properties. BKC produced a bell-shaped concentration-response curve for both the radioactive index (8) and the LN cell count index (Figure 3). In contrast to LN hyperplasia, the skin irritation expressed as ear weight index in the modified LLNA is increasing over a concentration range up to 10% (Figure 3). The conclusion from this pattern is that sensitization occurs at lower and less irritable concentrations of BKC, whereas at high concentrations the strong irritation counteracts the onset of a contact allergic response in the LN. This is exactly what is found in humans. Therefore, a recommendation exist to use BKC at concentrations below 1% in human patch tests to avoid non-specific irritant reactions.

Relevance to Humans

See also Predictivity. Potency assessment with LLNA is very enticing because of the objective nature of the endpoints. However, one has to consider that the strength of allergic response depends not only on the chemical features of the test compound, but also on the readiness to react of the individual's immune system. Furthermore, the choice of vehicle has a considerable influence on the strength of the response in an animal test. Thus, the conditions of human exposure should be carefully evaluated before extrapolating absolute animal data.

Regulatory Environment

- The LLNA had been accepted as a stand-alone test for skin sensitization first by the OECD (no. 429: Skin Sensitization: Local Lymph Node Assay by the Organisation for Economic Cooperation and Development). This was adopted formally in April 2002.
- In parallel, the European Union (EU) has prepared a new test guideline for the LLNA (B.42 Skin Sensitization: Local Lymph Node Assay) that has now been published in the Official Journal of the EU.
- In 2003 the US-EPA (Environmental Protection Agency) adopted its Skin Sensitization guideline to accept the lymph node assay as a stand alone test (OPPTS 870.2600).
- The LLNA is also described in the following guidelines of the EU: CPMP Note for guidance on non-clinical local tolerance testing of medicinal products. Adopted February 2001. CPMP/SWP/2145/00, and CPMP Note for guidance on photosafety testing. Adopted June 2002. CPMP/SWP/398/01.

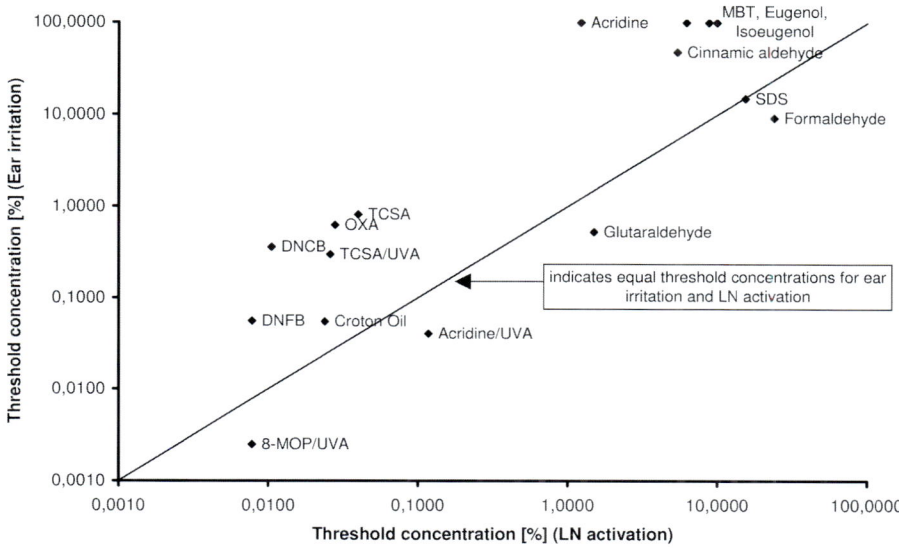

Local Lymph Node Assay (IMDS), Modifications. Figure 2 Relation of primary ear skin irritation and LN activation induced by chemicals. Threshold concentrations for primary LN activation and ear irritation were calculated for each chemical by applying curve-fitting algorithms to the concentration-response curves. The threshold indices for LN activation (1.3) derived from cell count data and ear irritation (1.1) assessed by weight measurement were approximated from the lowest applied concentrations of the chemicals leading to statistically significant responses. To support the definition of threshold concentrations a large set of historical data was included in the survey. Data taken from Ulrich et al. (5).

- The FDA provides also a guideline with a discussion of the LLNA: Guidance for Industry: Immunotoxicology evaluation of investigational new drugs. CDER 2002.

References

1. Scholes EW, Basketter DA, Lovell WW, Sarll AE, Pendlington RU (1992) The identification of photoallergic potential in the local lymph node assay. Photodermatol Photoimmunol Photomed 8:249–254
2. Homey B, von Schilling C, Blumel J et al. (1998) An integrated model for the differentiation of chemical-induced allergic and irritant skin reactions. Toxicol Appl Pharmacol 153:83–94
3. Ulrich P, Homey B, Vohr HW (1998) A modified murine local lymph node assay for the differentiation of contact photoallergy from phototoxicity by analysis of cytokine expression in skin-draining lymph node cells. Toxicology 125:149–168
4. Vohr HW, Blumel J, Blotz A, Homey B, Ahr HJ (2000) An intra-laboratory validation of the Integrated Model for the Differentiation of Skin Reactions (IMDS): Discrimination between (photo)allergic and (photo)irritant reactions in mice. Arch Toxicol 73 (10–11):501–509
5. Ulrich P, Streich J, Suter W (2001) Intra-laboratory validation of alternative endpoints in the murine local lymph node assay for the identification of contact allergic potential: Primary ear skin irritation and ear-draining lymph node hyperplasia induced by topical chemicals. Arch Toxicol 74:733–744
6. Ehling G, Hecht M, Heusener A, Huesler J, Gamer AO, v Loveren H, Maurer T, Riecke K, Ullmann L, Ulrich P, Vandebriel R, Vohr H-W (2004) An European Inter-Laboratory Validation of Alternative Endpoints of the Murine Loacl Lymph Node Assay. 1stROUND (and 2ndROUND) Toxicology, submitted
7. ECETOC (2003) Technical Report Number 87. Contact Sensitization: Classification According to Potency. Brussels: European Centre for Ecotoxicity and Toxicology of Chemicals
8. Gerberick GF, Cruse LW, Ryan CA (1999) Local lymph node assay: Differentiating allergic and irritant responses using flow cytometry. Methods 19:48–55

Local Lymph Node Assay (LLNA)

This uses a murine model to test for skin sensitization properties of chemicals.
▶ Leukocyte Culture: Considerations for In Vitro Culture of T cells in Immunotoxicological Studies
▶ Hapten and Carrier

Local Skin Immune Hyperreaction

▶ Contact Hypersensitivity

LPS

▶ Polyclonal Activators

LTT

▶ Lymphocyte Transformation Test

Lung Sensitization Test

▶ Animal Models for Respiratory Hypersensitivity

Lupoid Hepatitis

▶ Hepatitis, Autoimmune

Lymph

A clear, yellowish fluid found in intercellular spaces and in the lymphatic vessels.
▶ Lymph Nodes

Lymph Flow

▶ Lymph Transport and Lymphatic System

Lymph Gland

▶ Lymph Nodes

Lymph Node

An encapsulated and organisedcollection of immunologically competent lymphoid cells that receive lymph and antigen from local tissues via afferent lymphatics. Primary immune responses to antigen encountered in the surrounding tissues are induced in regional lymph nodes.
▶ Local Lymph Node Assay

Lymph Nodes

C FRIEKE KUPER
Toxicology and Applied Pharmacology
TNO Food and Nutrition Research
Zeist
The Netherlands

Synonyms
lymph node, lymph gland, nodus lymphaticus

Definition
Lymph nodes are secondary lymphoid organs (peripheral lymphoid organs) located in the course of ▶ lymphatic vessels. They filter the ▶ lymph during its passage from the tissues to the thoracic duct and initiate immune reactions.

Characteristics
General
Lymph nodes are found in mammals as rounded to bean-shaped lymphoid structures. They are surrounded by a collagenous capsule from which trabeculae enter into the node, and connected by lymph vessels. The basic anatomical features are the cortex/paracortex and medulla (1). The cortex/paracortex is the site of antigen encounter and initiation of immune reactions. The products of an immune response (activated cells, effector lymphocytes, inflammatory mediators) are generated in the medulla. The cortex consists of follicles, mainly just underneath the lymph node capsule, and interfollicular areas which extend to the medulla and are known as paracortex (Figure 1). The follicles contain mostly B cells, whereas T cells are the major lymphocyte population in the paracortex. The paracortex is easily distinguished by the presence of specific blood vessels, the so-called high endothelial venules. The arterial blood supply, entering the node at the medulla, ends in the cortex/paracortex as arteriolar capillaries. The capillary bed feeds venules which are lined with endothelial cells that become cuboidal ("high") when activated. Lymphocytes migrate through these high endothelial venules following adherence to the endothelium by specific receptor-ligand interactions. The specificity of these receptor-ligand interactions is different between lymph nodes draining the 'internal' organs and tissues and those draining secretory surfaces and mucosa-associated lymphoid tissues (MALT). This way, by using the same transport system (blood circulation), lymphocytes "belonging" to one of these systems can specifically access the internal or mucosal lymphoid nodes. After migration into the parenchyma, the cells move to their microenvironment in follicles or paracortical areas.

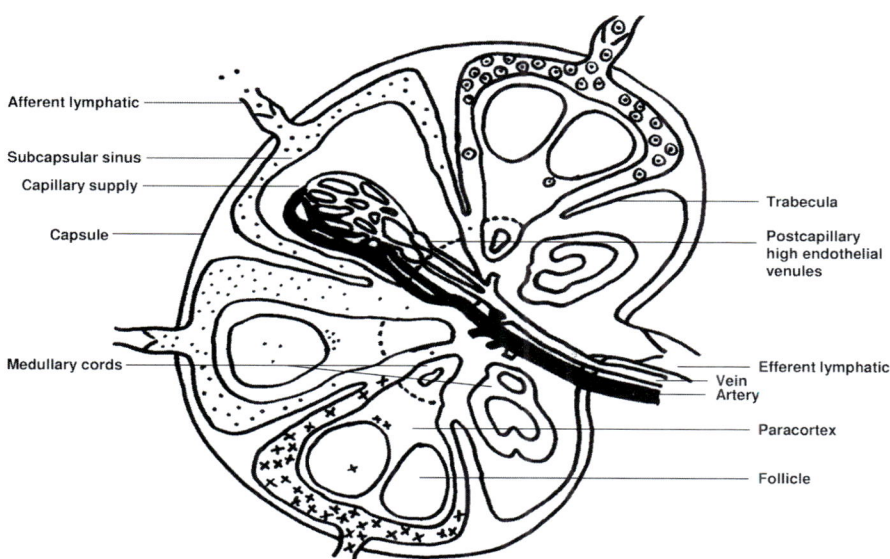

Lymph Nodes. Figure 1 Schematic presentation of lymph node. •, ×, and ⊙ represent antigenic substances that are drained from the respective tissues into the lymph node.

Reactions after Antigen Contact

The major route of entry for antigens and pathogens is by the afferent lymph flow, which ends in the subcapsular area in most animals (in pig the flow ends in the medulla). From there, antigens, either free or processed by macrophages ("veiled macrophages" in the afferent lymph), move to the paracortex. The subcapsular area is rich in macrophages (sinus macrophages) that can phagocytose free antigen. In the paracortex, antigens are presented to T lymphocytes for the initiation of the immune response. The main antigen-presenting cell population consists of interdigitating cells, which are a special type of cells closely related to Langerhans cells in the skin and veiled macrophages in the afferent lymph. These cells are often surrounded by T cells. The antigen-presenting cells express MHC (major histocompatability complex) class II antigens in high density, enabling the T cells to recognize the antigenic determinant with their α-β TCR (T cell receptor) and the polymorphic (self) MHC class II molecule. In this cellular interaction, the immune response starts with synthesis of cytokines. For B-cell activation, the process proceeds in the lymphoid follicle, which develops from a primary into a secondary follicle. The secondary follicle consists of a ▶ germinal center surrounded by a mantle. The activation and proliferation of B cells in the germinal center is accompanied by an isotype switch of the immunoglobulin class synthesized by the B cell. Following activation and clonal expansion in the germinal centers, B lymphocytes migrate to the medullary cords of the lymph nodes and either become plasma cells or exit the lymph nodes as memory B cells. Since antigen can remain in the follicular microenvironment for a long time, it causes a persistent activation of B cells and so contributes to immunological memory within the B lymphocyte compartment. After antigen disappearance, immunological memory in the B cells is short-lived and is taken over by the T cell population.

The major site of effector immune reactions is the medulla. Medullary cords contain macrophages and may contain activated effector cells, depending on the response initiated in the cortex. Among the effector cells are plasma cells and T cells, the latter bearing an α-β TCR that recognizes antigen in the context of the polymorphic determinant of MHC class I molecules. In addition, the medulla houses T cells, involved in a delayed type hypersensitivity response and synthesizing a variety of lymphokines. The reaction products leave the lymph nodes from the medulla via the efferent lymph and blood circulation for other sites in the body. For instance, plasma cell precursors home to the bone marrow, which supplies the major portion of intravascular immunoglobulins.

Growth and Involution

The lymph node architecture is a dynamic rather than a static feature. The unstimulated or resting lymph nodes, for example the popliteal or axillary lymph nodes, are very small, being a storage site for virgin T and B cells; most follicles are primary. After stimulation the organ increases in size in a relatively short time, showing high proliferative activity of lymphocytes and germinal center formation. After termination of the reaction or transfer of the reaction to the next draining node, it returns to its original size. Lymph nodes retain the ability to respond throughout life. An age-associated decrease in the number of germinal centres and overall inactive appearance of the node has been reported (2). Lymph nodes of aged animals can show the same degree of reaction as those observed in young rats upon local or systemic antigenic stimulation.

Preclinical Relevance
The lymph node is a lymphoid organ that drains specific organs and tissues via the interstitial tissue fluid and lymph. Direct immunotoxic effects of a compound may be manifested in specific lymph nodes whereas indirect effects, via the thymus and/or bone marrow or via systemic actions, may affect all lymph nodes more or less in an equal manner. For immunotoxicity assessment, lymph nodes can be divided into those that drain the site of exposure to the toxicants and those at some distance. This does not imply that the nondraining lymph nodes are not stimulated, because mucosa-draining lymph nodes almost always demonstrate some level of activation due to continuous exposure to antigenic substances in the air or in the food. For instance, the superior cervical (mandibular) lymph nodes in rodents contain many germinal centres in the cortex and plasma cells in the medullary cords, because they are continuously exposed to antigens via the oronasal mucosae, and the superior mesenteric lymph nodes house many macrophages, especially in the sinuses and paracortex due to continuous exposure to antigens via the intestines. The microscopic appearance of lymph nodes varies widely depending on the plane of sectioning. This should be kept in mind when evaluating these organs.

Relevance to Humans
The structure and function of lymph nodes varies only slightly between mammals. As with other lymphoid organs such as the thymus and spleen, the universality of the immune system observed in mammals and the results of toxicity studies observed so far indicate that data from laboratory animals can be extrapolated quite well to humans. Differences in immunotoxicity between laboratory animals and man appear to depend predominantly on differences in toxicokinetics and metabolism of toxicants, and these aspects should therefore be considered when immunotoxicity data from animals are extrapolated to humans.

Regulatory Environment
Current regulatory toxicity tests, e.g. for pharmaceuticals and industrial substances, include immune parameters but are still under development. Several guidelines require examination of both the lymph nodes that drain the exposure site as well as examination of nondraining lymph nodes (see Table 1 for lymph nodes and the area they drain). The relevance of weighing lymph nodes is well recognized in certain guidelines, especially those dealing with pharmaceuticals and those specifically designed for detection of immunomodulating substances (local lymph node assay and popliteal lymph node assay). The relevance and applicability of weighing the mesenteric lymph nodes in oral studies is still under debate.

References
1. Sainte-Marie G, Belisle C, Peng FS (1990) The deep cortex of the lymph node: Morphological variations and functional aspects. In: Grundmann E, Vollmer E (eds) Reaction Patterns of the Lymph Node, Part I. Cell Types and Functions. Springer Verlag, Berlin, pp 33–63
2. Losco P, Harleman H (1992) Normal development, growth and aging of the lymph node. In: Mohr U, Dungworth DL, Capen CC (eds) Pathobiology of the Aging Rat, Volume I. ILSI Press, Washington DC, pp 49–75
3. Tilney NL (1971) Patterns of lymphatic drainage in the adult laboratory rat. J Anat 109:369–383
4. Kuper CF, Schuurman H-J, Vos JG (1995) Pathology in immunotoxicology. In: Burleson GR, Dean JH, Munson AE (eds) Methods in Immunotoxicology, Volume I. Wiley-Liss, New York, pp 397–437

Lymph Transport and Lymphatic System

ANATOLIY A GASHEV · DAVID C ZAWIEJA
Department of Medical Physiology, College of Medicine, Cardiovascular Research Institute Division of Lymphatic Biology, Texas A&M University System Health Science Center
336 Reynolds Medical Building
College Station, TX 77843
USA

Synonyms
lymph flow, lymphodynamics

Definition
The lymphatic system is an organized network of vessels and nodes, which are present in most tissues and organs of the body. This system plays a vital role in fluid and macromolecular transport, lipid absorption, and immunity, and is involved in various pathologic processes of different origins. Effective functioning of the lymphatic system depends upon the central movement of lymph through its vascular and nodal network.

Lymph Nodes. Table 1 Drainage patterns of lymph nodes in the rat (adapted from (3,4))

Lymph node (group)	Area drained	Efferent drainage
Superficial cervical	Tongue, nasolabial lymphatic plexus	Posterior cervical nodes
Facial	Head; ventral aspect and sides of neck	Posterior cervical nodes
Internal jugular	Pharynx, larynx, proximal part of esophagus	Posterior cervical nodes
Posterior cervical	Superficial cervical, facial and internal jugular nodes, pharynx, larynx, proximal part of esophagus, NALT	Cervical duct
Brachial	Upper extremities, shoulders, chest	Axillary nodes
Axillary	Upper extremities, trunk, brachial nodes	Subclavian duct
Inguinal	Thigh, haunches, scrotum, lateral tail	Axillary nodes
Popliteal	Foot, hind leg	Lumbar, inguinal nodes
Gluteal	Tail	Caudal, lumbar, inguinal, popliteal nodes
Parathymic	Peritoneal cavity, liver, pericardium, thymus, lung	Mediastinal duct
Posterior mediastinal	Thoracic viscera, pleural space, pericardium, thymus	Mediastinal duct
Paravertebral	Diaphragm, thoracic viscera	Posterior mediastinal nodes
Caudal	Ventral tail, anus, rectum, gluteal nodes	Iliac nodes
Iliac	Pelvic viscera, popliteal, gluteal, caudal nodes	Renal nodes
Para-aortic	Pelvic viscera, popliteal, gluteal, caudal nodes	Renal nodes
Renal	Kidneys, suprarenal and lumbar lymphatics	Renal duct to cisterna chyli
External lumbar	Fat pad, psoas muscles, pelvic viscera	Lumbar lymphatics
Splenic	Splenic capsule and trabeculae	Posterior gastric nodes
Posterior gastric	Distal esophagus, stomach, pancreas, splenic node	Portal nodes
Portal	Liver, splenic, posterior and gastric nodes	Portal duct to cisterna chyli
Superior mesenteric	Small intestines, cecum, ascending and transverse colon, Peyer's patches	Superior mesenteric duct to cisterna
Inferior mesenteric	Descending and sigmoid colon	Inferior mesenteric duct to cisterna

Characteristics

In spite of growing efforts to discover the physiological mechanisms of lymph flow, there is no commonly accepted concept of the regulation of lymph flow. However, investigations over the last decades have presented many basic facts on lymph flow and its regulation. The lymphatic transport system moves fluid, macromolecules, and formed elements from within the interstitial spaces into the lymphatic capillaries in the form of lymph. To accomplish these tasks, the lymphatic vessels must act as both fluid pumps and pathways. Lymphatic vessels display genotypic and phenotypic characteristics of vascular, cardiac, and visceral myocytes, which are needed to fulfill the unique roles of the lymphatic system. Lymph formation in organs and tissues occurs due to the differences in interstitial and lymphatic capillary pressures (1). From the lymphatic capillaries, the lymph is propelled throughout a complicated network of lymphatic vessels and numerous lymph nodes toward to the lymphatic trunks and thoracic duct. Lymph enters the central veins from the lymphatic trunks and the thoracic duct which are functioning as the final outflow paths for the lymphatic circulatory pathways.

Lymphatic vessels are made up of chains of sequentially located chambers called ▶ lymphangions (2). A lymphangion is a morphological/functional unit of lymphatic vessels defined as the section of a lymphatic vessel between two adjacent lymphatic valves. Lymphangions have highly organized endothelial and smooth muscle layers in their walls. The presence of spontaneous oscillatory contractile activity has been observed in lymphangions from many regions of the body in many different species.

The spontaneous contractile activity of lymphangions is initiated by action potentials, apparently arising within cells in the smooth muscle layer. The phasic contractions of the lymphangions are termed the ▶ active lymph pump. It has been proven that this active pumping of the lymphatic vessels is a critically important mechanism for generating lymph flow. The phases of the lymphangion contractile activity that lead to the movement of lymph toward the venous circulation are similar to the phases of contractile activity of the heart chambers (3). There are also several passive forces that affect lymph flow. The term "passive" reflects the origin of these forces, which are not generated by active contractions of lymphatic muscle cells. The rate of lymph formation is the principal passive force that influences lymph flow in every lymphatic bed. Lymph formation depends on many factors and varies locally in connection with the level of physiological activity of the organ or tissue. The so-called ▶ passive lymph pump consists also of other passive lymph-driving forces: the contractions of skeletal muscles, fluctuations in central venous pressure, the influence of respiratory movements, pulsations of adjacent arteries, and influences of gravitational forces. Due to varying anatomical positions in the body, different lymphatic beds are affected differently by these forces. It is also known that lymphangions have inherent autoregulatory mechanisms, such as stretch-dependent and shear-dependent responses.

Two of the main physical factors that regulate lymph dynamics are ▶ intraluminal pressure and flow modulate active lymph pumping (4).

There are significant regional differences in the basic characteristics of lymphatic contractility. The regional variability in the active lymph pumps is likely predetermined by different hydrodynamic factors and by differences in regional outflow resistances in the lymphatics at their respective locations. The lymph pump has very complicated systems of extrinsic control, both neural and humoral (5), to match lymphatic pumping to the different physiological activities in the different parts of the body. Thus changes in the lymphatic contraction/relaxation state via the intrinsic/extrinsic pumping mechanisms or neural/humoral controllers can lead to changes in the lymph flow.

Preclinical Relevance

Damage to the transport capabilities of the lymphatic system causes lymphedema, which is associated with different pathologies, such as inflammation, invasions of parasites or bacteria, partial or full occlusions of lymphatic vessels and nodes after surgical manipulations or X-ray treatment of tumors. Despite of ongoing attempts to discover the mechanisms of and treatments for lymphatic-involved diseases, there are a many disputable or unknown issues regarding the physiology of lymph transport.

Relevance to Humans

There is strong evidence for the presence of spontaneous contractile activity in human lymphatic vessels. This evidence was obtained using many different techniques including visual observations, measurements of lymph pressure fluctuations in catheterized lymphatics, and some limited data from isolated lymphatic sections in vitro (6). This spontaneous contractile activity presumably results in changes in lymph pressure that are needed to produce lymph flow. The mechanism by which this pumping activity is regulated is still not well understood. We know that some factors can change lymph flow by modulating human lymphatic contractility (e.g. adrenergic and nonadrenergic regulation); however, our knowledge of the mechanisms which generate and regulate lymphatic pumping activity in humans is far from complete. In general, most of our understanding of the mechanisms responsible for controlling human lymphatic pump function

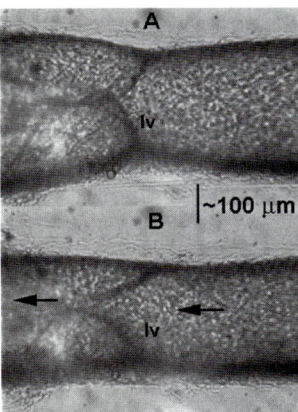

Lymph Transport and Lymphatic System.
Figure 1 Microphotographs illustrating the lymphatic contractile cycle. Two microphotographs were taken at the end of diastole (relaxation phase, A) and at the peak of contraction during systole (contraction phase, B) of isolated rat cervical lymphatic vessel. It is clear that contraction caused a decrease in lymphatic diameter. The increase in intralymphatic pressure due to the contraction led to the opening of the lymphatic valve (lv) and propulsion of fluid in the downstream direction (shown by arrows in area B).

is limited primarily to observational data and needs more investigation.

References

1. Schmid-Schonbein GW (1990) Microlymphatics and lymph flow. Physiol Rev 70:987–1028
2. Mislin H, Rathenow D (1962) Experimentelle Untersuchungen über die Bewegungskoordination der Lymphangione. Rev Suisse Zool 69:334–344 (in German)
3. McHale NG, Roddie IC (1976) The effect of transmural pressure on pumping activity in isolated bovine lymphatic vessels. J Physiol 261:255–269
4. Gashev AA (2002) Physiologic aspects of lymphatic contractile function: Current perspectives. Ann NY Acad Sc 979:178–187
5. von der Weid P (2001) Review article: Lymphatic vessel pumping and inflammation—the role of spontaneous constrictions and underlying electrical pacemaker potentials. Aliment Pharmacol Ther 15:1115–1129
6. Gashev AA, Zawieja DC (2001) Physiology of human lymphatic contractility: A historical perspective. Lymphology 34:124–134

Lymphadenopathy

This is a swelling of the lymph nodes, which can be classified as localized or generalized. Causes may range from benign infections to serious underlying illnesses, such as lymphoma, metastatic cancer and acquired immunodeficiency syndrome.

▶ Interferon-γ

Lymphangion

Morphological/functional unit of lymphatic vessels defined as the section of a lymphatic vessel between two adjacent lymphatic valves.

▶ Lymph Transport and Lymphatic System

Lymphatic

Vessel that collects fluid from interstitial spaces and lead it via lymph nodes to the thoracic duct and blood.

▶ Lymph Nodes

Lymphoblastic Lymphoma

▶ Lymphoma

Lymphocyte-Activated Killer (LAK) Cells

LAK cells are cytotoxic cells with a relatively broad target cell specificity and develop from peripheral blood lymphocytes upon stimulation with interleukin-2.

▶ Limiting Dilution Analysis

Lymphocyte-Activating Factor

▶ Interleukin-1β (IL-1β)

Lymphocyte Activation Test

▶ Lymphocyte Transformation Test

Lymphocyte Mitogenesis

▶ Lymphocyte Proliferation

Lymphocyte Proliferation

RENÉ CREVEL
Safety & Environmental Assurance Centre
Unilever Colworth
Sharnbrook, Bedford
MK44 1LQ
UK

Synonyms

lymphocyte transformation, lymphocyte mitogenesis, lymphocyte proliferative response

Definition

Lymphocyte proliferation is defined as the process whereby lymphocytes begin to synthesize DNA after cross-linking of their antigen receptor either following recognition of antigen or stimulation by a polyclonal activator (mitogen).

Characteristics.
Events in lymphocyte proliferation

Lymphocyte proliferation is a fundamental characteristic of the response of lymphocytes to antigenic stimulation. In physiological situations, contact between a lymphocyte and an antigen-presenting cell (APC) results in the formation of an immunological synapse. In this synapse, binding to the 3 T cell receptor (TCR) of the peptide–MHC complex carried by antigen-presenting cells, together with a costimulatory signal delivered by interaction between 3 ▶ CD28 and its ligand, initiates proliferation, while the synapse itself is stabilized by interactions between adhesion molecules and their ligands on the respective cells involved. The process can be short-circuited in vitro through the use of polyclonal activator (e.g. ▶ phytohemagglutinin, ▶ Concanavalin A (ConA) or anti-CD3 for T cells), which removes the requirements for antigen presenting cells and costimulation. These events result in a increase in the cytoplasmic concentration of the group of molecules known as NFκB, which in turn, upon translocating to the nucleus, initiates transcription of genes for IL-2 and its receptor, CD25. The increased production of interleukin-2, through an autocrine process, further stimulates proliferation. While cross-linking of the antigen receptor is a critical event in lymphocyte proliferation, other factors determine the outcome of the stimulatory process, in particular the signals delivered by different types of antigen-presenting cells. Only antigen-presentation by professional antigen-presenting cells, such as dendritic cells, is capable of providing the accessory signals needed to drive proliferation of T cells leading to an effective immune response to an antigen. Absence of some of those signals will lead instead to anergy or clonal deletion. The result of cell stimulation by antigen or mitogen is to shift the cell from the G0 (quiescent) phase of the cell cycle to the G1. At this particular phase of the cycle, regulation of progression through the cycle is governed by proteins which are involved in cell-cycle regulation in all cells (e.g. cyclins), so at this point control is no longer specific to cells of the immune system. Following activation and entry into G1, the lymphocyte will progress through the subsequent phases of the cycle—S phase (synthesis) where DNA synthesis takes place, G2 phase (gap 2), and M phase (mitosis), where cell division actually takes place, before returning to G1 (Fig. 1).

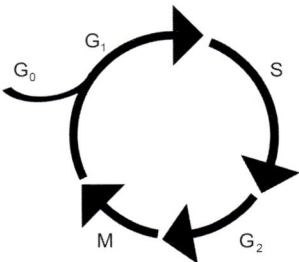

Lymphocyte Proliferation. Figure 1 The cell cycle.

Time course of lymphocyte proliferation

The initial events of lymphocyte proliferation, namely formation of the immunological synapse and the interaction between APCs and lymphocytes, take place over several hours, although there is debate over the actual time for which any one peptide–MHC complex needs to bind to the TCR. Once the initial event has taken place, proliferative activity peaks at different times depending on the species, the antigen, or mitogen concentration and—in vitro—the cell concentration. Typically rodent lymphocytes will show peak mitogenic responses after 48 hours, while antigenic responses require 72–96 hours. In man, the corresponding figures are 72 hours and 4–5 days, and sometimes longer.

Measurement of lymphocyte proliferation

Consideration of the events in cell proliferation shows that the process can be measured at several points. However, DNA synthesis is the parameter that is still most commonly used to measure lymphocyte proliferation. One of the reasons for preferring it to earlier phases is that it reflects a fundamental feature of the process, unlike some of the earlier events, and therefore can be argued to have greater predictive value. One of the most common ways of measuring DNA synthesis, and hence lymphocyte proliferation, is using radiolabeled nucleotides, typically tritiated thymidine. Lymphocytes are usually cultured in 96-well microplates, although other formats (60-well and 384-well) are also used, for a duration appropriate to the species and the stimulus used (mitogen or antigen). Typically, for in vitro lymphocyte cultures, tritiated thymidine is added up to 24 hours before termination of the culture and is incorporated into the dividing cells in proportion to their number in the culture. After the end of the incubation period with labeled nucleotide, the cells are harvested in such a way that the uptake of radioactivity by the cells can be measured. As part of the harvesting process, the cell cultures are aspirated onto glass fiber filter papers, or other suitable supports, which are then air dried and placed in vials containing a scintillation counting cocktail (when the radiolabel used is tritium). The disintegrations per minute (d.p.m.) reflect the intensity of proliferation in each culture. Results are expressed in a variety of ways, sometimes as d.p.m. corrected for background activity, and sometimes as a stimulation index (ratio of d.p.m. of stimulated cultures to d.p.m. of control cultures). A similar method is used to measure proliferation in vivo in tests such as the local lymph node assay, but the labeled nucleotide is injected intravenously, and the lymphocytes harvested from the relevant lymph nodes at an appropriate time after the injection. The cell cultures are directly added to the scintillation cocktail and radioactivity measured.

MTT (3-(4,5-dimethylthiazol-2-yl)2,5-diphenyl tetrazolium bromide) dye measurement relies on the incorporation of the dye by viable cells. In proliferating cultures, unstimulated cells tend to die off, thus the MTT method tends to measure survival as much as actual proliferation.

Flow cytometric methods for measurement of lymphocyte proliferation have been described. One uses the intracellular fluorescent dye carboxyfluorescein succinimidyl ester (CFSE). Indices of proliferation such as percentage of blastic transformation and mitotic activity correlated well with tritiated thymidine incorporation, after ▶ $CD3^+$ T cells were gated. Other variants of the technique include analysis based on a dye incorporated into the membrane of dividing cells, direct counting of cells, and analysis of the entry of cells into the S phase of the cells cycle.

Proliferative responses can be measured in different subpopulations of lymphocytes and in response to polyclonal lymphocyte activators (mitogens) as well as to specific antigen, if the cells are from an immunized donor. T cell mitogens include Concanavalin A (Con A) and phytohaemagglutinin (PHA), which can be used successfully with human cells, as well as those of several common laboratory species such as the mouse and rat. In contrast, B cells from different species appear to respond to different mitogens. While ▶ pokeweed mitogen (PWM) stimulates human cells, murine cells respond to lipopolysaccharide (LPS), while rat cells can be stimulated with ▶ *Salmonella typhimurium* mitogen (STM). The dose–response to Con A and PHA has been well investigated. It shows an optimum dose beyond which the response declines owing to toxicity of the mitogen to the cells (Fig. 2). This characteristic has implications for measurement of proliferation in immunotoxicity studies, inasmuch as measurement at a single dose may lead to misleading conclusions. Most antigens tend to produce responses which plateau, thus assessment can be limited to a dose which just gives a maximal response. The mixed leukocyte reaction (MLR) is also a form of proliferative assay, in which allogeneic cells are the stimulus, rather than a soluble antigen or mitogen.

Preclinical Relevance

Lymphocyte proliferation has been, and continues to be used extensively as one of the immunotoxicity endpoints. Recent studies incorporating it as an endpoint include an investigation of the Gulf War Syndrome, of the effects of sodium bromate, a byproduct of water disinfection by ozone, and of the effect of N,N-diethylaniline in mice. Its enduring popularity may be related in part to the fact that it does reflect a basic biological phenomenon in the immune system. Another advantage may be that direct comparisons can be made between the activity of animal cells used in toxicity studies and that of human lymphocytes exposed to the same test material. Lymphocyte proliferation measurements can be applied in several ways to investigate the potential immunotoxicity of test materials on immunocompetent cell populations. Probably the most common consists of administering the test compound in vivo, for instance in a standard repeat-dose study, and to examine the response to mitogens or specific antigen of suspensions of immunocompetent cells prepared from lymphoid organs removed at the end of the study. This type of study is relatively simple to perform in a laboratory familiar with cell culture techniques, but can be difficult to interpret in the absence of supplementary data on the composition of cell populations and other parameters of responsiveness, such as cytokine profiles. This difficulty is increased further by the observation that the response of cell populations from different lymphoid organs can be very different. For instance, in early studies in which carrageenan (a sulfated polysaccharide extracted from a type of seaweed) was administered orally to rats, the proliferative response of mesenteric lymph node cells was depressed, while that of cells from cervical lymph nodes was increased (1). Luster et al (2) examined the sensitivity and predictive ability of different measures, and combinations of measures of immune responsiveness, based on the extensive dataset generated in the B6C3F1 mice used in the US National Toxicology Program. They found that there was good concordance between lymphocyte proliferation to T cell mitogens and several other measures of immune responsiveness. However, they also found other combinations of tests to be more predictive and did not recommend the inclusion of proliferative responses in standard immunotoxicity test batteries, but this conclusion may depend on the exact experimental conditions. Lymphocyte proliferation in response to Con A and PHA was also included in the test battery used in the International IPCS CEU Collaborative Immunotoxicity Study

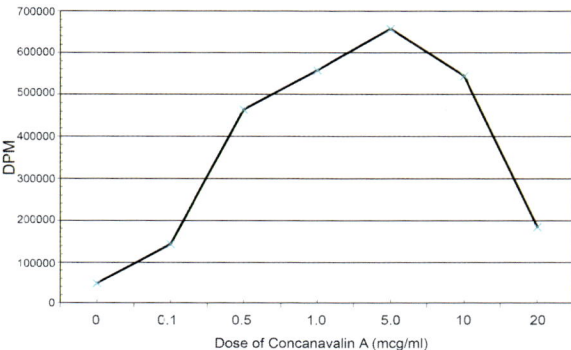

Lymphocyte Proliferation. Figure 2 Dose response of rat splenocytes to Concanavalin A.

(▶ ICICIS). This study found that lymphocyte proliferation was reasonably reproducible over several laboratories, if a standard protocol was used, and was of similar sensitivity to histopathologic methods for screening immunosuppressive immunotoxicants (Fig. 3) (3). Lymphocyte proliferation ex vivo as a measure of immune responsiveness also possesses some inherent weaknesses. The principal one is that it relies on the cells not recovering from the immunotoxic insult once they have been removed from the animal and placed in culture. Clearly this will depend on the nature of the immunotoxicant and therefore the effect that it has on cells in vivo. If, for instance, it is specifically toxic to certain subpopulations, or is able to prevent their activation, then this effect will be detectable ex vivo without the immunotoxicant in the cell culture, because, for instance, it will specifically alter the balance of cells in the lymphoid organ used. It is possible also to conceive of a situation where a test material produces an effect on lymphocytes which requires continued exposure to manifest itself. An example might be that of a change in membrane fluidity through incorporation into the membrane. Such an effect could rapidly wane once the cells had been placed in the culture environment.

An alternative way in which lymphocyte proliferation has been used is through exposure of immunocompetent cells to the putative toxicant in vitro. One advantage of this approach is that it permits a direct comparison between the responses of human cells and those of equivalent cells from experimental animals, exposed identically. It also permits the identification of transient effects such as those postulated in the preceding paragraph, although it is attended by other drawbacks. These include the potential for interactions with the mitogens or antigens themselves, which could result in false positives. An additional requirement is to ensure that the doses used are not cytotoxic per se. If immunotoxicity is due to a metabolite, rather than the parent compound, the test is likely to give a false negative result. A more serious difficulty is understanding how to extrapolate the results obtained to an estimate of effect in vivo. For this reason, the value of exposure in vitro is probably limited to screening assays and mechanistic studies.

Relevance to Humans

Lymphocyte proliferation is used as an index of immune function in many clinical applications. Specific instances include measurement of responsiveness to mitogens and specific antigens to identify possible immunodeficiency or hyporesponsiveness. One major application continues to be the monitoring of immune status in HIV/AIDS patients and indeed efforts in this area have been focused on standardization of the procedure to facilitate interlaboratory comparisons (4). In most instances, responses to mitogens rather than specific antigens are measured, thereby avoiding the problems associated with having to consider the patient's immunization history, as well as the need for longer culture periods. Responsiveness to specific antigens is arguably more sensitive and relevant, and its feasibility has been demonstrated, even in whole blood cultures.

Other clinical applications of lymphocyte proliferation include the diagnosis of chronic beryllium disease and of pulmonary fibrosis induced by exposure to feathers, monitoring of the course of cutaneous T cell lymphoma, and monitoring of the course of bee-venom allergen immunotherapy.

Lymphocyte proliferation is also used widely as a clinical research tool. A few illustrative examples include investigation of cellular immune responses to *Trypanosoma cruzi* proteins, of T cell function in tuberculosis patients, and of autoimmune responses of T cells in multiple sclerosis.

Lymphocyte proliferation has been used to monitor the effects of xenobiotics on the immune response of human subjects since the beginning of immunotoxicology. Well-known early studies include those underta-

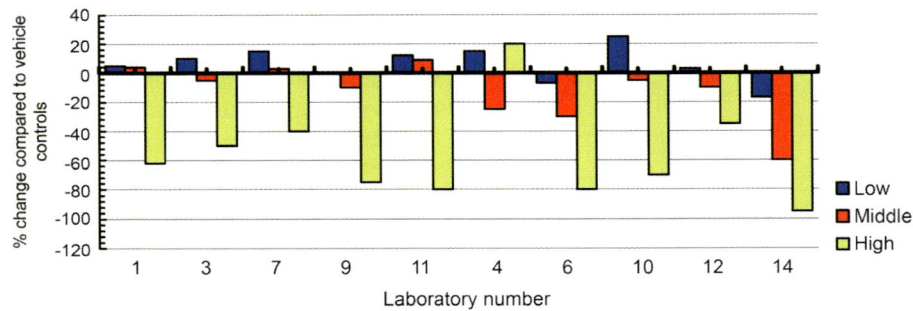

Lymphocyte Proliferation. Figure 3 ICICIS: proliferative response of rat spleen cells to concanavalin A in different laboratories (expressed as % change from control).

ken by Bekesi et al (5) on the populations exposed to polybrominated biphenyls (PBBs) in Wisconsin, as well as the investigations into the Yu-cheng (6) and Yu-sho episodes resulting from contamination of rice bran oil with PCBs (Fig. 4). This method has also been used as one of the immunological endpoints in an investigation of the immunomodulatory effects of conjugated linoleic acid isomers (7).

Regulatory Environment

Although widely used in clinical and preclinical trials for measuring disturbance of the activation of immunocompetent cells, lymphocyte proliferation assays are only mentioned in some immunotoxicity guidelines. However, lymphocyte proliferation features as an endpoint in the local lymph node assay (LLNA), a method developed to predict the potential of low-molecular-weight chemicals to induce allergic contact dermatitis (OECD 429). According to this guideline, the measurement of proliferation, using tritiated thymidine, is undertaken in vitro, even with radiolabeling taking place in vivo. On the other hand this guideline also mentions the possibility of measuring lymphocyte proliferation by other methods.

References

1. Bash JA, Vago JR (1980) Carrageenan-induced suppression of T lymphocyte proliferation in the rat in vivo suppression induced by oral administration. J Retic Soc 28:213–221
2. Luster MI, Portier C, Pait DG et al. (1992) Risk assessment in immunotoxicology. I. Sensitivity and predictability of immune tests. Fund Appl Toxicol 18:200–210
3. ICICIS Group Investigators (1998) Report of the validation study of assessment of direct immunotoxicity in the rat. Toxicology 125:183–201
4. Froebel KS, Pakker NG, Aiuti F et al. (1999) Standardization and quality assurance of lymphocyte proliferation assays for use in the assessment of immune function. European Concerted Action on Immunological and Virological Markers of HIV Disease Progression. J Immunol Methods 227:85–97
5. Bekesi JG, Roboz J, Fischblein A et al. (1985) Immunological, biochemical and clinical consequences of exposure to polybrominated biphenyls. In: JH Dean et al. (eds). Immunotoxicology and immunopharmacology. Raven Press, New York, pp 346–406
6. Lee T-P (1992) The toxic effects of polychlorinated biphenyls. In: Miller K, Turk JL, Nicklin S (eds). Principles and practice of immunotoxicology. Blackwell Scientific, London
7. Albers R, van der Wielen RP, Brink EJ, Hendriks HF, Dorovska-Taran VN, Mohedel C (2003) Effects of *cis*-9, *trans*-11 and *trans*-10, *cis*-12 conjugated linoleic acid (CLA) isomers on immune function in healthy men. Eur J Clin Nutr 57:595–560

Lymphocyte Proliferation Test

▶ Lymphocyte Transformation Test

Lymphocyte Proliferative Response

▶ Lymphocyte Proliferation

Lymphocyte Transformation

▶ Lymphocyte Proliferation

Lymphocyte Transformation Test

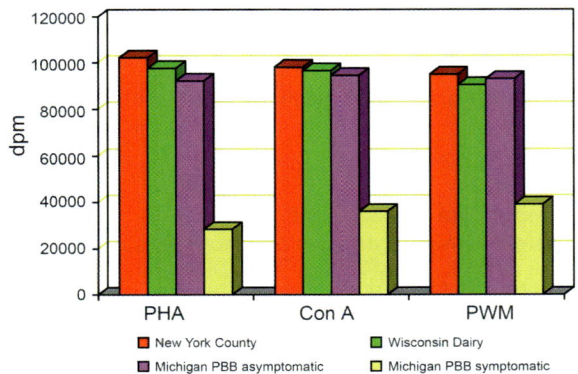

Lymphocyte Proliferation. Figure 4 Proliferative responses of individuals exposed to PBB. From Bekesi et al. (1985).

WERNER PICHLER
Klinik für Rheumatologie & klinische Immunologie/Allergologie
Inselspital-Universtät Bern
3010 Bern
Switzerland

Synonyms

lymphocyte proliferation test, lymphocyte activation test, LTT

Short Description

The aim of the lymphocyte transformation test is to determine, whether a patient has developed a T cell

Lymphocyte Transformation Test. Table 1 Diagnostic procedures in drug allergy – sensitivity and specificity (1)

	(Epicutaneous) skin tests	Lymphocyte transformation test
Retrospective:	Sensitivity: **64%**	Sensitivity 78/100 = **78%**
	Specificity: **85%**	Specificity 87/102 = **85%**
		(falsely+: mainly NSAID)
Prospective:	6/14 = **43%**	13/19 = **68%**

response against a certain drug. Such T cell responses are present in practically all types of allergic reactions (▶ type I–IV reactions), which means also in antibody-mediated reactions (1). For that purpose, peripheral blood mononuclear cells obtained after density gradient centrifugation (Ficoll-Hypaque) are cultured with different concentrations of the incriminated drug for 5–6 days (1–7). The drug concentration used in this in vitro assay has previously been determined by establishing toxic concentrations (e.g. by evaluating an inhibition on PHA-stimulations). The proliferation of the cells can be measured by different means: most frequently ^{3}H-thymidine incorporation is used, but evaluation of T cell activation by flow cytometry (e.g. CD69 upregulation on T cells, intracellular staining of newly synthesized cytokines) is also possible.

The test is interpreted as positive if the proliferation is at least two times higher than the control culture, which does not contain the drug but was performed in the same media. This is given as stimulation index (SI). However, some authors use even lower or higher cut-off values (SI 1.8–3.0). It is advisable to set the cut-off value for a novel drug by analyzing the reactivity to the drug in exposed, but not allergic, individuals (n = 20).

Characteristics

For many years it has been well known that certain drugs can induce the proliferation of some T cell clones bearing a specific T cell receptor. This proliferation is drug specific and dependent on the available T cells with a fitting T cell receptor repertoire, namely the ability of the drug to interact with a certain T cell receptor. These drug-specific T cells have been cloned (2–7). Both CD4 and CD8 cells react, but the majority of cells obtained in such proliferation assays are CD4 positive and express the αβ-T cell receptor. The T cells recognize the drug as a hapten, which means that it is bound to a certain carrier protein, or the drug can directly stimulate the T cell receptor, whereby interaction with the major histocompatability complex (MHC) is still required to completely activate the cell (8).

The drug-reactive T cells secrete high amounts of various cytokines, in particular the ▶ interleukin IL-5, which leads to ▶ eosinophilia—a typical hallmark of drug hypersensitivity. Measurement of IL-5 has been proposed as alternative read-out system. The type of T cell reaction corresponds to the clinical picture (8). Since the precursor frequency of drug-reactive T cells is often low, special care has to be given to optimal culture conditions. Too many macrophages may block proliferation by high prostaglandin PGE_2 secretion, and it is also advisable to perform the test both in autologous plasma as well as AB serum (20% of the media), since differences are not uncommon.

Pros and Cons

The pros of this test include the fact that it is easy to perform and can be rapidly adapted to new compounds. In addition:
- it is a test based on human cells and it is an in vitro test
- it can be performed with frozen cells
- high SI are clear indications of a sensitization.

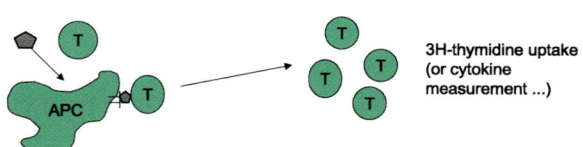

Lymphocyte Transformation Test.
Figure 1 Proliferation of T-cells in cell culture to the drug.
Interpretation:
SI > 3 (stimulation index=cpm plus drug/cpm minus drug) is indicatind a sensitization to β-lactam-antibiotics.
SI > 2 is indicating a sensitization to other drugs.

Lymphocyte Transformation Test.
Table 2 Lymphocyte transformation test-suitable drugs

- pure substance antibiotics
- antiepileptics
- antihypertensives
- NSAID
- contact allergens
- others, e.g. hydroxymethylcellulose
- almost all substances, if soluble, pure and not toxic

Lymphocyte Transformation Test. Table 3 Lymphocyte transformation test – positivity in drug allergic diseases

+	−
maculopapular exanthema	macular exanthema?
bullous exanthema	blood dyscrasia like ITP, hemolytic anaemia
drug fever	Guillain-Barré?
AGEP	TEN (− > +)
eosinophilia, DHS/DRESS	vasculitis (− > +)
hepatitis	
pancreatitis	
nephritis	
interstitial lung disease	
urticaria, anaphylaxis	

Lymphocyte Transformation Test. Table 4 Drug allergy tests for T cell reactions

+	−
in vitro regulatory and inflammatory T-cell reactions detectable	complicated, long lasting
not dependent on strong inflammatory response	not in acute reaction, results available long after reaction
positive with IgE mediated reactions	dependent on fresh cells (24–36 hours) standardization
sensitivity and specificity satisfactory	pure drugs necessary
open for new development	false reactions with tablets
open for research (cytokine determination,...)	„distant" to clinic
possible with frozen cells	TEN, vasculitis often negative metabolism(?)

The specificity of the test seems to be very good: quite a few studies show that positive results with drugs such as amoxicillin, lamotrigine, and carbamazepine, can only be obtained in sensitized individuals. However, non-steroidal anti-inflammatory drugs (NSAIDs) could reduce PGE_2 synthesis and thus enhance the proliferation. However, this pharmacological mechanism is not seen in all individuals.

Disadvantages of the use of this test include the following facts:

- it is dependent on patients' blood cells, which have been sensitized previously in vivo
- it cannot predict the immunogenicity of a drug, as non-sensitized individuals do not react
- some drugs are difficult to dissolve and need special solvents (e.g. DMSO) which must be tested as well
- this biological test is based on the in vitro proliferation of cells obtained from peripheral blood, so it is highly variable and it is problematic to set a correct cut-off point
- the test is highly dependent on optimal cell composition and serum supplementation of the media, and with time the presence of sensitized cells in the peripheral blood may go down
- the sensitivity and specificity are debatable, because some groups report a relatively high sensitivity for an allergy test (60%–70%) (1) and others do not, which might be because of the selection of different patients and of different drugs; a high sensitivity was reported for anti-epileptic-induced drug hypersensitivity reactions (> 95%) (6,7).

Predictivity

The test is based on the evaluation of already sensitized individuals. Therefore, it cannot be used in non-sensitised individuals where it is—by definition—negative. It could be used to follow people treated in phase 1–4 studies, by analyzing the proliferative response of exposed persons to the drug (± side effects).

Relevance to humans

There are not many tests available to indicate a drug hypersensitivity reaction and to pinpoint the relevant drug in drug hypersensitivity reactions (1). Therefore even a suboptimal test such as the LTT is considered to be useful. The growing recognition for a role of T cells in such hypersensitivity reactions makes the LTT even more attractive, as it is positive in a wide variety of human hypersensitivity reactions. However, as with other immunological tests, sensitization is not necessary associated with clinical symptoms.

References

1. Pichler WJ, Tilch J (2004) The lymphocyte transformation test in the diagnosis of drug hypersensitivity. Allergy 59:809–820
2. Nyfeler B, Pichler WJ (1997) Sensitivity and specificity of the lymphocyte transformation test to drugs. Clin Exp Allergy 27:175–181
3. Neukomm C, Yawalkar N, Helbling A, Pichler WJ (2001) T-cell reactions to drugs in distinct clinical manifestations of drug allergy. J Invest Allerg Clin Immunol 11:275–284
4. Maria VA, Victorino RM (1997) Diagnostic value of specific T-cell reactivity to drugs in 95 cases of drug-induced liver injury. Gut 41:534–540
5. Tsutsui H, Terano Y, Sakagami C, Hasegawa I, Mizoguchi Y, Morisawa S (1992) Drug-specific T-cells derived from patients with drug-induced allergic hepatitis. J Immunol 149:706–716
6. Naisbitt DJ, Britschgi M, Wong G et al. (2003) Hypersensitivity reactions to carbamazepine: characterization of the specificity, phenotype, and cytokine profile of drug-specific T cell clones. Mol Pharmacol 63:732–741
7. Naisbitt DJ, Farrell J, Wong G et al. (2003) Characterization of skin homing lamotrigine-specific t-cells from hypersensitive patients. J Allergy Clin Immunol 111:1393–1403
8. Pichler WJ (2003) Delayed drug hypersensitivity reactions. Ann Int Med 139:683–693

Lymphocytes

BRAD BOLON
GEMpath Inc.
2540 N 400 W
Cedar City, UT 84720-8400
USA

Definition

Lymphocytes are a subclass of leukocytes (white blood cells) that are the principal elements controlling the ▶ adaptive immune response, which develops in response to a particular antigenic stimulus. One special lymphocyte variety—the ▶ large granular lymphocyte—also plays a role in the antigen-independent ▶ innate immune response.

Characteristics

Lymphocyte Categories

In general, lymphocytes can be classified by either lineage (Figure 1) or function.

The three principal lineages of lymphocytes are B lymphocytes (B cells), T lymphocytes (T cells), and natural killer cells (NK cells). Activated B cells differentiate into ▶ plasma cells, which secrete the immunoglobulins (antibodies) that drive humoral immunity and serve to destroy extracellular pathogens and their products. Activated T lymphocytes power cell-mediated immunity, either by killing damaged cells directly—chiefly those expressing tumor antigens or products derived by intracellular pathogens (especially viruses)—or by regulating the activities of other immune effector cells (including B cells). Both B cell-mediated and T cell-mediated functions are controlled by antigen-specific receptors, which originate during the evolution of adaptive immunity to a specific antigen. In contrast, natural killer cells do not have antigen-specific receptors, but instead serve as components of the innate immune response that kill tumor cells or cells infected with intracellular pathogens (particularly viruses).

Functional distinctions also can be used to classify lymphocytes into various groups. Naive B and T cells are relatively inactive (or resting) until they encounter an appropriate target antigen, after which they proliferate and differentiate into antigen-specific effector lymphocytes and, in some cases, into immunological memory cells. The primary function of all B cell-derived effector and memory cells is antibody production. In contrast, different classes of activated T cells perform diverse tasks; cytotoxic T lymphocytes destroy diseased cells, while helper T lymphocytes and suppressor T lymphocytes activate or dampen, respectively, the operation of other leukocyte classes (particularly B lymphocytes and macrophages) as well as certain types of activated non-immune cells (e.g. fibroblasts).

These different lymphocyte classes can be identified by their CD surface markers. For example, in humans cytotoxic and suppressor T cells bear CD8, while helper T cells express CD4. Given populations of T cells may be divided further based on their specific tasks. Thus, CD4-positive T cells that activate macrophages and promote digestion of intracellular bacteria are said to facilitate the T helper type 1 (Th1) response, while those CD4-positive T cells that kill infected cells and direct the destruction of extracellular pathogens by activating B cells are said to promote the T helper type 2 (Th2) response. The Th1 and Th2 reactions are driven by divergent but overlapping groups of cytokines, which typically are secreted by T cells (i.e. ▶ lymphokines) and macrophages (monokines).

Lymphocyte Distribution

Lymphocytes are generated in the central or primary lymphoid organs, the bone marrow, and thymus. Lymphocytes first arise in bone marrow from ▶ pluripotent hematopoietic stem cells (HSCs), which give rise to other partially committed HSCs, including the common lymphoid progenitor cell. The lymphoid progenitor cells differentiate in the marrow into immature B cells and immature T cells. The B cells undergo further maturation in the marrow into naive B cells, hence the designation "B cell," for "bone marrow-de-

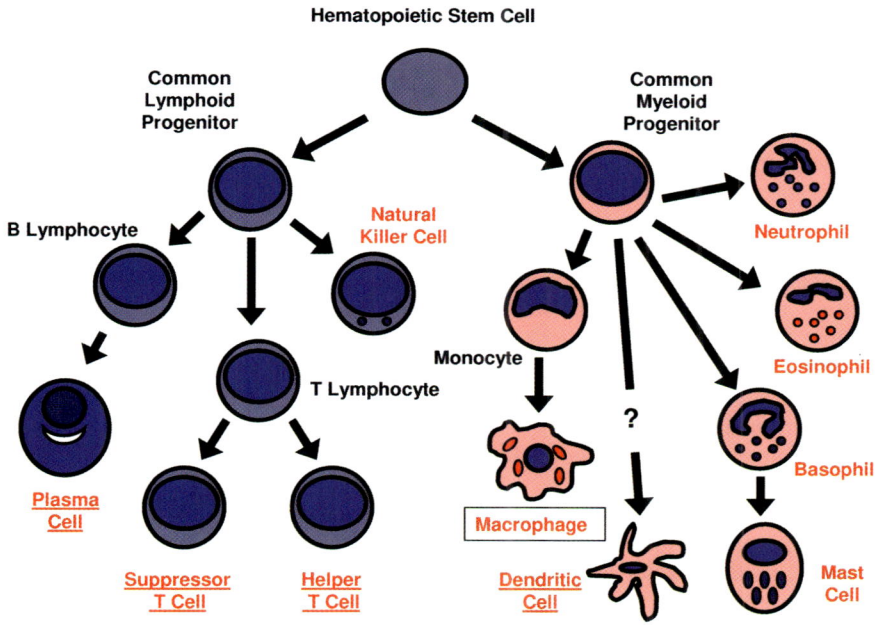

Lymphocytes. Figure 1 All leukocyte (white blood cell) lineages are derived from partially committed stem cells that arose from pluripotent hematopoietic stem cells. The common progenitor cells undergo further differentiation into various classes of lymphocytes (blue cells on the left) or myeloid cells (pink cells on the right). For each lineage, fully differentiated effector cells (labels in red) participate in either the adaptive immune response (labels underlined) or the innate immune responses, or both (labels in boxes). For detailed descriptions of the biology for the various leukocyte types, see the relevant entry.

rived." In contrast, immature T cells migrate to the thymus before maturing into naive T cells (thus the label "T cell" for "thymus-derived."). Naive B and T cells subsequently relocate to the peripheral or secondary lymphoid organs, where adaptive immune responses are initiated and resting lymphocytes are maintained in close proximity to nascent antigen-presenting cells (chiefly ▶ dendritic cells). These secondary lymphoid organs include the lymph nodes, various ▶ mucosa-associated lymphoid tissues (e.g. bronchi and gut), spleen, and tonsils. All lymphoid tissues contain myriad lymphocytes framed by numerous non-lymphoid cells and connective tissue fibers. The structures of secondary lymphoid organs are designed to filter circulating antigens from the blood or lymph, offer them to antigen-presenting cells (APCs), and then to facilitate the interaction between the APCs and naive lymphocytes in ▶ germinal centers. Exchanges between lymphocytes and the adjacent non-lymphoid cells (particularly antigen-presenting cells in the secondary lymphoid organs) direct later stages of lymphocyte development, lymphocyte activation, and long-term lymphocyte maintenance and retention. Lymphocytes also are common in many non-lymphoid organs. ▶ Ectopic lymphoid tissues may form in diseased tissue during some chronic inflammatory conditions. These structures consist of poorly circumscribed lymphocyte aggregates (chiefly T cells), with or without germinal centers, which are formed to provide a concentrated focus of antigen-specific effector cells. Lymphocytes also are found frequently in capillaries, particularly the winding sinusoids in organs that filter blood (e.g. kidney and liver) and in the submucosal connective tissues in organs with a direct connection to the outside environment (particularly the digestive, genitourinary, and respiratory tracts). In the blood, these lymphocytes typically are either naive (unstimulated) cells or memory cells (antigen-specific elements engaged in surveillance) and are merely passing through—rather than engaged in—a tissue-specific immune response. In contrast, lymphocytes in the submucosal connective tissue regulate the immune reaction to exogenous antigens that approach or penetrate the mucosa. A component of this response includes the movement of individual cells into the mucosa to serve as intraepithelial lymphocytes.

Lymphocyte Morphology

As shown in Figure 1, lymphocytes have a stereotypical appearance that has earned them (along with macrophages) the moniker ▶ mononuclear leukocyte (as apposed to ▶ granulocytes, which comprise the

▶ polymorphonuclear leukocytes). Prior to activation, unstimulated lymphocytes are small cells with few cytoplasmic organelles and relatively inactive nuclear chromatin. These features impart a unique aspect to both the B cell and T cell lineages: cells with a dark, round, dense nucleus surrounded by a thin rim of cytoplasm. Upon activation, T lymphocytes generally maintain their initial appearance. In contrast, activated B lymphocytes develop into plasma cells and assume a distinctive profile characterized by eccentric nucleus with a few radiating chromatin stripes (a so-called "cartwheel" nucleus) and a prominent perinuclear pale zone (the enlarged Golgi apparatus). Natural killer cells are large granular lymphocytes that are distinguished by their bigger overall size, enhanced quantity of cytoplasm, and prominent cytoplasmic granules.

Lymphocyte Physiology

Leukocytes—including large granular lymphocytes—engaged in the innate immune response bear several different cell surface receptors, each of which recognizes a different feature shared by many pathogens. In contrast, most lymphocytes efficiently participate in the adaptive immune response against a specific foreign antigen because each lymphocyte forms a unique antigen receptor during maturation by rearranging its complement of receptor genes. Collectively, the entire universe of T and B lymphocytes bears an enormous repertoire of receptors with highly diverse antigen-binding sites. When their specific antigen is presented to them, naive lymphocytes become activated, express many new cell surface antigens, and proliferate, thereby forming myriad clones that will seek to neutralize a single antigen. This process of clonal expansion to form an effector lymphocyte population takes several days upon the first exposure to an antigen (either by infection or vaccination). Most effector lymphocytes undergo apoptosis once their target antigen has been eliminated, but the persistence of memory cells after antigen clearance has occurred leads to a more rapid reaction upon subsequent exposures. Lymphocytes with receptors that recognize endogenous (self) antigens typically are deleted during development, thereby ensuring ▶ tolerance of the body to its own tissues and preventing the initiation of autoimmune disease.

Preclinical Relevance

Lymphocytes are found in all mammals. Thus, the working assumption is that preclinical evaluation of lymphocyte function in mammalian models will recapitulate the effects (including those endpoints relevant to immunotoxicology) that manifest in human beings. Typical endpoints used to evaluate lymphocyte physiology in animals and humans are a routine hematologic assessment (complete blood count, with white blood cell differential count and cytologic evaluation of blood smears) and various in vitro function tests, such as the lymphocyte transformation assay.

Immune function testing in animal models typically is performed to define the role of various lymphocyte classes in the pathogenesis of disease (i.e. the "toxicity" of lymphocytes to tissues) rather than to examine the toxicity of xenobiotics or other types of insults to lymphocytes. Hypotheses usually assess the molecular and cellular mechanisms responsible for the disease (1,2) and/or test potential therapies that might prevent or reduce the tissue destruction (2,3).

Lymphocyte numbers and/or activities in animals may be substantially reduced or markedly enhanced by spontaneous or engineered mutations. For example, many inherited rodent models of immunodeficiency have been reported to result, at least in part, from aberrant lymphocyte functions (4). Selected mouse models characterized by deficient lymphocytes include *bg/bg* (beige), which harbor defective cytotoxic T cells and NK cells; *nu/nu* (nude), which lack T cells; *ob/ob* (obese), which have less active T cells but hyperactive NK cells (on the C57BL/6J genetic background); *scid/scid* (severe combined immunodeficiency), which lack both B cells and T cells; and *xid/xid* (X-linked immunodeficiency), which lack B cells. The nude rat (*rnu/rnu*) has a T cell deficit equivalent to that in the *nu/nu* mouse (4). In contrast, lymphocyte numbers and functions are greatly elevated in MRL/Mp-*lpr/lpr* (lymphoproliferation) mice, which develop generalized hyperplasia of lymphocytes and plasma cells (4), and in animals with lymphoma, a malignancy in which the tumor cells arise from a lymphocytic lineage (typically B cells or T cells). A novel means of examining the impact of human lymphocytes in vivo using an animal test system is to transplant human cells (isolated from the ▶ buffy coat of a routine blood sample), lymphoid tissue fragments, or wedges of lymphocyte-rich diseased tissue (e.g. synovial biopsy from an arthritis patient) into an immunodeficient rodent (5).

Other models with reduced or exaggerated lymphocytic activity have been generated in immunocompetent animals. Rodent models of autoimmune diseases exhibit many clinical and morphologic features that resemble those of the corresponding human conditions. Potential lymphotoxic effects may be evaluated by administering chemicals, such as alkylating agents (e.g. cancer chemotherapies), diethylstilbestrol, glucocorticoids, and certain heavy metals, as well as by delivery of ionizing radiation (6). Preclinical examination of immunotoxicity testing typically is performed using such immunocompetent subjects, both because their intact immune systems more closely resemble that of humans and because the studies will be more cost-effective as they do not require the special husbandry practices needed to maintain animals with immunologic defects.

Relevance to Humans

Human disease counterparts have been discovered for many rodent models, especially those for autoimmune disease (1) and primary immunodeficiency (7). Autoimmune diseases in humans are known to be incited by certain xenobiotics, apparently as the result—at least in part—of enhanced function by B cells or helper T cells, or reduced control by suppressor T cells; the immune system in animals behaves in a similar fashion. Both animal and human immune systems are affected in a comparable manner by exposure to many immunosuppressive agents. The concordance between the response between the immune systems of immunocompetent animals and humans suggests that immunotoxicity data—including information with regard to potential lymphotoxic effects—will be correlated to a high degree across species.

Regulatory Environment

Due to the many structural and functional similarities between vertebrate immune systems, preclinical efficacy and safety studies in immunocompetent animals are routinely employed as surrogates to define potential lymphocytotoxic events that might be initiated in human beings exposed to xenobiotics. However, preclinical efficacy and safety studies are notoriously difficult to perform when the candidate therapeutic agent is a human protein, as many will not cross-react with elements in animal immune systems. Alternative means of assessing immunotoxic risk for such proteins include testing in non-human primates or investigating homologous molecules in the appropriate animal species. Evaluation of in vivo immunotoxicity toward human lymphocytes also might be examined using human tissue xenografts in one or more immunodeficient rodent models, but at present such assays are too technically difficult and expensive to perform on a routine basis. All these tests are surrogates to testing the clinical candidate in humans, and as such each of these options will have its own set of caveats. However, "humanized" immunodeficient mice should provide a reasonable alternative in the future, especially for immunotoxicity protocols in which the mouse response is well characterized.

References

1. Farine J-C (1997) Animal models in autoimmune disease in immunotoxicity assessment. Toxicology 119:29–35
2. van den Berg W (2000) What we learn from arthritis models to benefit arthritis patients. Baill Clin Rheumatol 14:599–616
3. Bendele A, McComb J, Gould T et al. (1999) Animal models of arthritis: Relevance to human disease. Toxicol Pathol 27:134–142
4. ILAR (Institute for Laboratory Animal Research of the National Research Council) Committee on Immunologically Compromised Rodents (2002) Hereditary immunodeficiencies. In: Immunodeficient rodents: A guide to their immunobiology, husbandry, and use. The National Academies Press, Washington DC, pp 36–139
5. Cleland LG, Fusco M, Proudman SM, Wing SJ, Spargo LDJ, Mayrhofer G (2001) Recruitment of mononuclear leucocytes to osteoarthritic human synovial xenografts in the ears of SCID mice. Immunol Cell Biol 79: 309–319
6. ILAR (Institute for Laboratory Animal Research of the National Research Council) Committee on Immunologically Compromised Rodents (2002) Induced immunodeficiencies. In: Immunodeficient rodents: A guide to their immunobiology, husbandry, and use. The National Academies Press, Washington DC, pp 140–147
7. NICHD (National Institute of Child Health and Human Development), National Institutes of Health, USA (2002) Primary immunodeficiency. http://www.nichd.nih.gov/publications/pubs/primaryimmunobooklet.htm (last accessed May 12, 2004)

Lymphocytes

Antigen-specific leukocytes such as B cells, T helper cells and cytotoxic T cells.

▶ Natural Killer Cells

Lymphodynamics

▶ Lymph Transport and Lymphatic System

Lymphoid Organ Compartments

Compartments within lymphoid organs/tissues feature one or more specific functions, and each houses lymphoid and non-lymphoid cells of different lineages and in different ratios.

▶ Histopathology of the Immune System, Enhanced

Lymphokine

A cytokine secreted by a lymphocyte to control the responses of various immune (lymphocytes, macrophages) and non-immune (chiefly cells of the fibroblast lineage) effector cells. Now replaced by the more general term cytokines.

▶ Cytokines
▶ Lymphocytes

Lymphoma

LEIGH ANN BURNS-NAAS
Pfizer Global Research & Development
10777 Science Center Dr.
San Diego, CA 92121
USA

Synonyms

Hodgkin's lymphoma, Hodgkin's disease, non-Hodgkin's lymphoma (NHL), Burkitt's lymphoma, lymphoblastic lymphoma, marginal zone lymphoma, anaplastic large cell lymphoma

Definition

Lymphomas are diverse neoplasms that begin by the malignant transformation, usually of a T cell, B cell, or natural killer (NK) cell, in the lymphatic system. These diseases result from damage (typically a mutation) to the cell's DNA that results in uncontrolled and excessive growth and confers a survival advantage on the malignant lymphocyte and the cells that are formed from its proliferation. Accumulation of these dividing cells results in the lymphadenopathy, often an initial indicator of the disease. While similar to leukemia in its general nature (malignant transformation of lymphocytes), lymphomas differ from leukemias in that they typically originate in the lymphatic system (e.g. lymph nodes) instead of the bone marrow.

Characteristics

Thirty or more subtypes of specific lymphomas or closely related lymphocytic leukemias have been categorized. Biopsies are usually required to make a definitive diagnosis of lymphoma. Tissue from the biopsy is examined microscopically to determine the pattern of the abnormalities and types of cells involved. Biopsied cells may be evaluated using immunophenotyping and cytogenetic analysis. ▶ Immunophenotyping may be used to provide additional diagnostic confirmation and to determine if the malignant cells are T cells, B cells, or NK cell types. ▶ Cytogenetic analysis is used to identify chromosomal abnormalities characteristic of certain of the lymphomas. Together, these diagnostics may help in the choice of drugs used for treatment.

Like the leukemias, non-Hodgkin lymphoma (NHL) is a diverse group of hematopoetic neoplasms, with the distinctions between types based on the characteristics of the neoplastic cells involved. Each histologic grouping is diagnosed and treated differently. Of interest, in many cases of diagnosed NHL, a translocation between certain sets of chromosomes is observed upon cytogenetic analysis. Several genes have been closed and sequenced at the breakpoints of these translocations (*bcl* genes). These genes encode specific disease-related proteins that alter the normal growth and proliferation of the cells. For example, one of the better-known genes is *bcl-2*, an ▶ oncogene whose overexpression can prevent normal programmed cell death. This gene is expressed in over 80% of cases of follicular NHL.

Over half of hematopoetic neoplasms are lymphomas with Hodgkin's disease representing about one-tenth of those diagnosed lymphomas. Hodgkin's disease is a unique form of lymphoma identified by the presence of a special cell type known as the Reed-Sternberg cell, a large multinucleated cell containing extensive eosinophilic cytoplasm and a large blue nucleoli. One of the important features of Hodgkin's disease is a decrease in the functional capacity of the immune system. Cell-mediated immune function, T cell-mediated immunity in particular, appears to be impaired and as a result, affected individuals with Hodgkin's disease are more susceptible to certain types of infection. The disease is more prevalent among adolescents and with diagnosis and proper treatment has a very high cure rate. Conversely, the incidence of NHL increases with increasing age and is slightly more prevalent in men than in women.

Marginal zone lymphomas include mucosal-associated lymphoid tissue (MALT) NHL (MALT-omas), monocytoid B-cell NHL, and primary splenic lymphomas. MALT-omas are extranodal neoplasms potentially involving one or several organ systems and may be associated with the presence of autoimmune disorders or *Helicobacter pylori* infection. Cells in spleen lymphomas are similar to other marginal zone lymphomas, however they have a distinguishing villous appearance that can be confused with hairy cell leukemia. Anaplastic large cell lymphomas have characteristic chromosomal abnormalities associated with overexpression of nucleoplasmin-anaplastic lymphoma kinase, as well as unique surface protein expression indicating the origin of the cell to be most likely from the T cell or null cell lineages.

Preclinical Relevance

While the etiology of the majority of lymphomas has yet to be determined, there are some associations that can be made (Table 1). Cancer patients undergoing treatment with cytotoxic drugs are more likely to develop secondary neoplasms in the years following therapy. Additionally, transplant patients on high doses of immunosuppressive drugs used to prevent graft rejection also appear to be at higher risk for developing blood cancers with the predominant form being NHL. A similar trend is noted in immunocompromised patients (e.g. HIV). One common thread in this increased susceptibility to lymphomas is a decrease in ▶ cell-mediated immunity in the susceptible individual.

The cell-mediated arm of the immune system is involved in controlling spontaneous tumors and infections, among other things. The destruction of tumor cells can result from the cytolytic action of specific (cytotoxic) T lymphocytes (CTL), macrophages, and NK cells. These cells recognize specific antigens on tumor or virally infected cells and cause their death by one of several mechanisms considered here.

In cell-mediated cytotoxicity, the effector cell (CTL or NK) binds in a specific manner to the target cell. CTL recognize either foreign major histocompatability complex (MHC) class I on the surface of allogeneic cells, or antigen in association with self MHC class I (e.g. viral particles), while NK cell recognition of target cells involves the binding of the Fc portion of antigen-specific antibody coating a target cell to the NK cell via its Fc receptors. This NK cell mechanism of killing is also referred to as antibody-dependent cellular cytotoxicity (ADCC). Once the CTL or NK cells interact with the target cell, they undergo cytoplasmic reorientation so that cytolytic granules are oriented along the side of the effector which is bound to the target. The effector cell then releases the contents of these granules onto the target cell. The target cell may be damaged by the perforins or enzymatic contents of the cytolytic granules. In addition, the target is induced to undergo programmed cell death (apoptosis). Once it has degranulated, the CTL or NK cell can release the dying target and move on to kill other target cells.

The role of the macrophage in cell-mediated cytotoxicity involves its activation by T cell-derived cytokine (e.g. IFN-γ) and subsequent recognition of complement-coated target cells via complement receptors present on the surface of the macrophage. The result is enhanced phagocytic ability, and the synthesis and release of hydrogen peroxide, nitric oxide, proteases, and tumor necrosis factor, all of which serve cytolytic functions. Macrophages may also kill tumor or infected cells via ADCC in a manner similar to that described for NK cells.

Drugs and agents that alter cell-mediated immunity have the potential to cause an increased risk of ▶ opportunistic infections (e.g. viral, bacterial, parasitic) and development of neoplastic disease. By evaluating such things as the ability of the immune system to recognize and destroy tumor cells such as the P815 mastocytoma (used in the CTL assay) or YAC-1 cell (used in the NK assay) as well as evaluating proliferative ability to mitogens, cytokines, or allogeneic (foreign; non self) cells, and evaluation of allograft rejection, immunotoxicity testing in rodents has identified many agents capable of causing suppression of cell-mediated immunity. A few of these agents are considered below.

Therapeutic agents have been developed that specifically inhibit several immune endpoints. These drugs are often used to treat symptoms associated with autoimmunity, in transplantation to prevent immune-mediated graft rejection, or to treat individuals with significant hypersensitivity responses. Cyclophosphamide is the prototypical member of a class of drug known as alkylating agents. It is often used as a positive immunosuppressive control in experimental immunotoxicology studies because it can suppress both humoral and cell-mediated immune responses. Cell-mediated immune activities that are suppressed include the delayed hypersensitivity response (DTH), the cytotoxic T lymphocyte response (CTL), graft-versus-host (GVH) disease, and the mixed lymphocyte response (MLR).

In addition to alkylating agents, other drugs used to intentionally suppress the immune system also alter cell-mediated immunity. The immunosuppressive action of corticosteroids is one example. Following binding to an intracellular receptor, these agents produce profound lymphoid cell depletion in rodent models and lymphopenia, associated with decreased monocytes and eosinophils and increased PMNs, in primates and humans. Corticosteroids induce apoptosis, and T cells are particularly sensitive. In general, corticosteroids suppress the generation of CTL responses,

Lymphoma. Table 1 Examples of agents associated with an increased risk for the development of primary or secondary lymphomas

Chemotherapeutic agents

Azathioprine
Cyclosporin A
FK-506
Rapamycin
OKT3 monoclonal antibody
Cyclophosphamide
Melphalan
Busulfan
Phenytoin
Methotrexate

Other agents

Infectious agents
   Reactivation of latent viruses (such as EBV)
   Helicobacter pylori
   Herpes virus 8
   Hepatitis C
   Human immunodeficiency virus / AIDS
Autoimmune diseases (such as rheumatoid arthritis)
Environmental chemicals (?)

the MLR, NK activity, and general lymphoproliferation. A large range of cell-mediated immune reactivity is also reduced by azathioprine immunosuppressive treatment including the DHR, the MLR, and GVH disease. Although T cell functions are the primary targets for this drug, inhibition of NK function and macrophage activities has also been reported. Cyclosporin A and FK-506 are important immunosuppressants that act preferentially on T cells by inhibition of IL-2 gene transcription and subsequent inhibition of T cell proliferation. Rapamycin is structurally related to FK-506, but inhibits T cell proliferation by blocking cell-cycle progression.

There are numerous examples of drug and non-drug chemicals that have the potential to alter immunocompetence in a manner that might influence one's susceptibility to spontaneous primary or spontaneous ▶ secondary neoplasms and opportunistic infection or reactivation of latent pathogens. This section has only considered a very few of these agents, but those which have limited or sufficient evidence for an association between suppression of immune function and predilection for lymphomas. Clearly, though, other contributing factors (e.g. genotoxicity) cannot be excluded.

Relevance to Humans

A definitive cause for most lymphomas has not been established, though exposure to some pathogens has been implicated. Infection with some viruses such as the Epstein-Barr virus (EBV), human immunodeficiency virus (HIV), and human T cell lymphocytotropic virus (HTLV) has been shown to be associated with an increased risk for the development of a variety of lymphomas including Burkitt's lymphoma, Hodgkin's disease, NHL, or T cell lymphoma. In fact, the incidence of lymphoma in HIV-infected persons has been 50–100 times the incidence rate expected in the uninfected population since the apid rise in HIV infection in the 1980s. Even bacterial infections such as *H. pylori* are associated with an enhanced susceptibility to MALT-oma. Although alterations in DNA still seem to be integral to the malignant transformation, the high frequency of infection may be a contributing factor to the development of disease.

Altered immunocompetence has been associated with an overall increased risk for the development of secondary neoplasms such as lymphomas. Individuals with acquired immunodeficiency syndrome (AIDS) have increased incidences of a variety of cancers, including lymphoma, most likely as a result of the loss of the ability of the host to identify and eradicate neoplastic cells, particularly those infected with pathogenic agents. A similar effect has been observed in transplant recipients and individuals with severe autoimmune disorders receiving chemotherapy with cytotoxic and/or immunosuppressive drugs. These drugs are known to alter cell-mediated immunity, an effect that can in effect increase the susceptibility to infectious agents such as viruses. In many cases of significant ▶ immunosuppression in people, there is a reactivation of latent viruses such as EBV that is associated with the development of secondary lymphomas. Of interest, there has been a suggestion that exposure to environmental chemicals such as pesticides or herbicides has contributed to the increased incidence of lymphoma. While some of these agents have been demonstrated to alter both ▶ humoral immunity and cell-mediated immunity, and some limited epidemiological studies suggests an increase in lymphoma in rural communities were farming is an important part of life, a clear association has yet to be definitively established.

There is a clear association between suppression of immune function and an increased incidence of infectious and neoplastic disease in humans. Agents that produce immunotoxicity in animals have the potential to produce immune effects in the human population, and these effects may occur in the absence of observable disease. Of the agents described here that are associated with an increased risk for the development of lymphomas, no specific causal relationship between the development of cancer and the immunosuppressive action by these drug or non-drug chemicals/agents has been clearly demonstrated. However, a preponderance of epidemiological evidence exists showing that exposure to various immunotoxic chemicals is associated with increased risk for malignancies (e.g. leukemia and lymphoma) that are also known to occur in immunocompromised patients. Thus, it is reasonable to conclude that alteration of immune function may contribute to the observed increase in risk.

Regulatory Environment

Because of the concerns regarding the potential for drugs and non-drug chemicals to cause any number of cancers, including lymphomas, global regulatory bodies have established guidance and test guidelines for assessing this potential. These include specific assessment of carcinogenic potential (e.g. lifetime studies in rodents) as well as assessment of the mutagenic and/or clastogenic potential (short-term in vivo and/or in vitro tests) of chemicals. Though not directly related to genotoxicity and carcinogenicity assessment, guidelines regarding the assessment of immune status following repeated exposure to drug and non-drug chemicals have recently been and are continuing to be put into place. Of interest, not all immunotoxicology guidance and guidelines require the evaluation of cell-mediated immunity. Some examples of these guidances and test guidelines are provided in Table 2.

▶ CD Markers

Lymphoma. Table 2 Examples of regulatory guidance and test guidelines for assessment of immunotoxicity, genotoxicity, or carcinogenicity*

Regulatory body	Endpoint	Guidance/test guideline
US FDA	Immunotoxicity	**Center for Food Safety and Nutrition (CFSAN)** Redbook 2000: Immunotoxicity Studies **Center for Devices and Radiological Health (CDRH)** Immunotoxicity Testing Guidance **Center for Drug Evaluation and Research (CDER)** Immunotoxicology Evaluation of Investigational New Drugs
	Genotoxicity	**Center for Food Safety and Nutrition (CFSAN)** Redbook 2000: Bacterial Reverse Mutation Test In Vitro Mammalian Chromosome Aberration Test In Vitro Mouse Lymphoma TK$^{+/-}$ Gene Mutation Assay In Vivo Mammalian Erythrocyte Micronucleus Test **Center for Drug Evaluation and Research (CDER)** See ICH
	Carcinogenicity	**Center for Food Safety and Nutrition (CFSAN)** Carcinogenicity Studies with Rodents Combined Chronic Toxicity/Carcinogenicity Studies with Rodents **Center for Drug Evaluation and Research (CDER)** See ICH
CPMP	Immunotoxicity	Note For Guidance on Repeated Dose Toxicity
	Genotoxicity	See ICH
	Carcinogenicity	See ICH
MHLW	Immunotoxicity	Immunosuppression, Draft Guidance
	Genotoxicity	For pharmaceuticals, see ICH
	Carcinogenicity	For pharmaceuticals, see ICH
ICH	Genotoxicity	**ICH S2: Genotoxicity** (includes the standard battery of tests and guidance on specific aspects of testing)
	Carcinogenicity	**ICH S1: Carcinogenicity** (includes the need for carcinogenicity studies, approaches to evaluating carcinogenic potential, and dose selection)
US EPA	Immunotoxicity	**Health Effects Test Guidelines** 870.7800 Immunotoxicity **Biochemicals Test Guidelines** 880.3550 Immunotoxicity 880.3800 Immune Response

References

1. International Programme on Chemical Safety (1996) Health impact of selected immunotoxic agents. In: Environmental Health Criteria 180: Principles and Methods for Assessing Direct Immunotoxicity Associated with Exposure to Chemicals. World Health Organization, Geneva, pp 85–147
2. Leukemia & Lymphoma Society (2002) Public literature. The Leukemia and Lymphoma Society, White Plains
3. Cheson BD (2001) Hodgkin's disease and the non-Hodgkin's lymphomas. In: Lenhard Jr RE, Osteen RT, Gansler T (eds) Clinical Oncology. American Cancer Society, Blackwell Science, Malden, pp 497–516
4. Luster MI, Simeonova P, Germolec DR, Portier C, Munson AE (1996) Relationships between chemical-induced immunotoxicity and carcinogenesis. Drug Info J 30:281
5. Pitot HC, Dragan YP (2001) Chemical carcinogenesis. In: Klaassen CD (ed) Casarett and Doull's Toxicology: The Basic Science of Poisons, 6th edn. McGraw-Hill, New York, pp 280–286
6. Rosenthal DS, Schnipper LE, McCaffrey RP, Andreson KC (2001) Multiple myeloma and other plasma cell dyscrasias. In: Lenhard Jr RE, Osteen RT, Gansler T (eds) Clinical Oncology. American Cancer Society, Blackwell Science, Malden, pp 517–525

Lymphoma. Table 2 Examples of regulatory guidance and test guidelines for assessment of immunotoxicity, genotoxicity, or carcinogenicity* (Continued)

Regulatory body	Endpoint	Guidance/test guideline
	Genotoxicity	**Health Effects Test Guidelines** 870.5100 Bacterial Reverse Mutation Test 870.5140 Gene Mutation in *Aspergillus nidulans* 870.5195 Mouse Biochemical Specific Locus Test 870.5200 Mouse Visible Specific Locus Test 870.5250 Gene Mutation in *Neurospora crassa* 870.5275 Sex-linked Recessive Lethal Test in *Drosophila melanogaster* 870.5300 In vitro Mammalian Cell Gene Mutation Test 870.5375 In vitro mammalian chromosome aberration test 870.5380 Mammalian spermatogonial chromosomal aberration test 870.5385 Mammalian bone marrow chromosomal aberraton test 870.5395 Mammalian erythrocyte micronucleus test 870.5450 Rodent dominant lethal assay 870.5460 Rodent heritable translocation assays 870.5500 Bacterial DNA damage or repair tests 870.5550 Unscheduled DNA synthesis in mammalian cells in culture 870.5575 Mitotic gene conversion in *Saccharomyces cerevisiae* 870.5900 In vitro sister chromatid exchange assay 870.5915 In vivo sister chromatid exchange assay
	Carcinogenicity	**Health Effects Test Guidelines** 870.4200 Carcinogenicity 870.4300 Combined chronic toxicity/carcinogenicity
OECD	Genotoxicity	OECD 471 Bacterial Reverse Mutation Test OECD 473 In vitro Mammalian Chromosomal Aberration Test OECD 474 Mammalian Erythrocyte Micronucleus Test OECD 475 Mammalian Bone Marrow Chromosomal Aberration Test OECD 476 In vitro Mammalian Cell Gene Mutation Test OECD 477 Sex-Linked Recessive Lethal Test in *Drosophila melanogaster* OECD 478 Rodent Dominant Lethal Test OECD 479 In vitro Sister Chromatid Exchange Assay in Mammalian Cells OECD 480 Saccharomyces cerevisiae, Gene Mutation Assay OECD 481 Saacharomyces cerevisiae, Miotic Recombination Assay OECD 482 DNA Damage and Repair, UDS in Mammalian Cells in vitro OECD 483 Mammalian Spermatogonial Chromosome Aberration Test OECD 484 Genetic Toxicology: Mouse Spot Test OECD 485 Genetic Toxicology: Mouse Heritable Translocation Assay OECD 486 Unscheduled DNA Synthesis Test with Mammalian Liver Cells in vivo OECD Draft In Vitro Syrian Hamster Embryo (SHE) Cell Transformation Assay

Lymphotoxin

▶ Tumor Necrosis Factor-α

Lytic Unit (LU)

A value assigned to a quantity of effector cells needed to kill a certain percentage of a predetermined number of target cells.

▶ Cytotoxicity Assays

M

M Cells

Specialized epithelial cells in mucosal epithelia overlying lymphoid structures in mucosae such as Peyer's patches and tonsils. M cells transport macromolecules and particles across the epithelium and are possibly involved in presentation to underlying lymphoid cells.
▶ Immunotoxic Agents into the Body, Entry of

MAbs

▶ Monoclonal Antibodies

Macrophage

A long-lived mononuclear phagocytic cell derived from monocytes in the blood that were produced from stem cells in the bone marrow. These cells have a powerful, although nonspecific, role in immune defense. These intensely phagocytic cells contain lysosomes and exert microbicidal action against microbes which they ingest. If monocytes enter a focus of inflammation they become the more activated inflammatory macrophage. Those provide signals that are critical to the development of an adaptive immune response.
▶ Trace Metals and the Immune System
▶ Therapeutic Cytokines, Immunotoxicological Evaluation of
▶ Opsonization and Phagocytosis
▶ Macrophage Activation

Macrophage Activation

JG LEWIS
Department of Pathology
Duke University Medical Center
Box 3712
Durham, NC 27710
USA

Synonyms

macrophage development, macrophage maturation, macrophage differentiation, alternative activation, type 2 activation
(Not all these terms are complete synonyms of general macrophage activation as defined here, but they are listed because many of the terms appear in older literature as synonyms, or in newer literature as subtypes of macrophage activation.)

Definition

All definitions of macrophage activation generally refer to morphological and functional changes in cells of the mononuclear phagocyte system (MPS) after they have fully differentiated into ▶ monocytes and have left the bone marrow and traveled to the tissues. Semantic debates have occurred over of what changes in mononuclear phagocytes are differentiation, maturation, or activation. For example, is the change from monocyte to macrophage that occurs in the tissues, differentiation, maturation, or activation? These definitions are not easily categorized to the satisfaction of all, but the present convention is that once a cell is a macrophage, all subsequent changes in morphology and function is mostly activation.

It is important to make a distinction between the strict definition of the term "macrophage activation" and the more general usage for a cell that is "activated" for a single function. Macrophage activation is used to identify ▶ macrophages in a defined stage of activation, and that stage of activation influences the ability of the cell to perform a large number of functions (see below). Conversely, the particular pattern of functions expressed by a macrophage defines its stage of activation.

progressed from a succinct well-defined functional system, to very complex mechanistic system, and is presently evolving into a branched system in which there may be at least three distinct but interrelated pathways of activation.

Preclinical Relevance

Macrophage activation as a concept is not at present part of any testing or regulation. Because of the pivotal role macrophages play in both innate and adaptive immunity and the numerous functions affected by the state of macrophage activation, any comprehensive investigation into the immunotoxicity of any environmental agent should evaluate this important indicator of immunological health.

Relevance to Humans

The differences between rodent and human mononuclear phagocytes offer a classic example of the problems of extrapolating animal data to the human situation. The expression of inducible nitric oxide synthetase (iNOS) and the subsequent secretion of nitric oxide (NO.) is a well-defined system in the mouse and this one function is believed by many to represent an excellent measure of full macrophage activation. Murine macrophages will secrete small amounts of NO. if exposed to relatively large amounts of LPS (μg/ml) but do not at physiologically relevant concentrations (ng/ml). Similarly, murine macrophages do not express iNOS or secrete NO. when exposed to even very large concentrations of IFN-γ. However, IFN-γ and LPS together are extremely synergistic and exposure of macrophages to 10 units/ml of IFN-γ lowers the amount of LPS required to get a full response for the expression of iNOS and secretion of NO. to the range of 1–10 ng/ml. In fact, IFN-γ makes murine macrophages so sensitive to LPS that any reagent purported to elicited NO. secretion in IFN-γ-primed macrophages must be rigorously shown to be free of LPS. The problem with human monocytes and macrophages is that they absolutely refuse to express iNOS or secrete any NO. in response to IFN-γ and LPS. This difference is made even more frustrating because we know human macrophages have the gene for iNOS and express it. iNOS and nitrosotyrosine—a byproduct of NO. secretion—are readily identified in macrophages surrounding granulomas in the lungs of TB patients, and nitrite (reduction product of NO.) can readily be detected in the plasma of patients harboring infections. Thus, we know human cells perform this important function but they do so in response to different (as yet unidentified) signals than the mouse. This glaring difference in human and murine macrophages means that any data on the effects of environmental agents on macrophages from rodent studies must be evaluated extremely carefully and confirmatory human studies are mandatory.

Several reports have linked macrophage activation to specific human diseases. It has been known since before the early 1920s that long-term chronic and granulomatous inflammation, in which activated macrophages serve as the end effector cells, are associated with carcinogenesis. This is particularly seen in patients with chronic tuberculosis in which tumors arise around granulomas, and in osteomyelitis where malignant squamous cell carcinomas, so called Marjolin's ulcers, arise at the sight of draining sinuses. There is data suggesting that abnormal activation of macrophages in the brain (the microglia) may be important in killing neurons in both Alzheimer's disease and AIDS dementia. Macrophages are known to be a reservoir for HIV and it is now known that the virus does not replicate until the macrophages become activated. Septic shock, which takes a huge toll in mortality in institutions that treat iatrogenically immunosuppressed patients undergoing organ transplantation or chemotherapy, is an end result of activation of macrophages and the release of TNF and IL-1. Interestingly, there is now a human clinical syndrome called macrophage activation syndrome (MAS). MAS is a potentially life-threatening process that occurs in the pediatric age group usually as a complication of the systemic onset of juvenile rheumatoid arthritis or its therapy.

Regulatory Environment

As mentioned above, macrophage activation as a specific indicator is not subject to regulation. Macrophage functions are included in many general screens for immunotoxicity but unfortunately the functions are usually relegated to a few innate immunity functions such as phagocytosis or chemotaxis. Given the above discussion it should be obvious that any observation on the effects of an environmental agent on the immune response, whether it is modulation of humoral, cellular, or innate immunity, may have its origin in the effects on mononuclear phagocytes. Therefore, an understanding of macrophage activation is important in any critical evaluation of the effects of environmental agents on the immune response.

References

1. Adams DO, Hamilton TA (1984) The cell biology of macrophage activation. Annual Rev Immunol 2:283–318
2. Ehrt S, Schnappinger D, Bekiranov S et al. (2001) Reprogramming of the macrophage transcriptome in response to interferon-gamma and *Mycobacterium tuberculosis*: signaling roles of nitric oxide synthase-2 and phagocyte oxidase. J Exp Med 194:1123–1140
3. Gordon S (2003) Alternative activation of macrophages. Nat Rev Immunol 3: 23–35
4. Mosser DM (2003) The many faces of macrophage activation. J Leukocyte Biol 73:209–212

M

M Cells

Specialized epithelial cells in mucosal epithelia overlying lymphoid structures in mucosae such as Peyer's patches and tonsils. M cells transport macromolecules and particles across the epithelium and are possibly involved in presentation to underlying lymphoid cells.
▶ Immunotoxic Agents into the Body, Entry of

MAbs

▶ Monoclonal Antibodies

Macrophage

A long-lived mononuclear phagocytic cell derived from monocytes in the blood that were produced from stem cells in the bone marrow. These cells have a powerful, although nonspecific, role in immune defense. These intensely phagocytic cells contain lysosomes and exert microbicidal action against microbes which they ingest. If monocytes enter a focus of inflammation they become the more activated inflammatory macrophage. Those provide signals that are critical to the development of an adaptive immune response.
▶ Trace Metals and the Immune System
▶ Therapeutic Cytokines, Immunotoxicological Evaluation of
▶ Opsonization and Phagocytosis
▶ Macrophage Activation

Macrophage Activation

JG LEWIS
Department of Pathology
Duke University Medical Center
Box 3712
Durham, NC 27710
USA

Synonyms

macrophage development, macrophage maturation, macrophage differentiation, alternative activation, type 2 activation
(Not all these terms are complete synonyms of general macrophage activation as defined here, but they are listed because many of the terms appear in older literature as synonyms, or in newer literature as subtypes of macrophage activation.)

Definition

All definitions of macrophage activation generally refer to morphological and functional changes in cells of the mononuclear phagocyte system (MPS) after they have fully differentiated into ▶ monocytes and have left the bone marrow and traveled to the tissues. Semantic debates have occurred over of what changes in mononuclear phagocytes are differentiation, maturation, or activation. For example, is the change from monocyte to macrophage that occurs in the tissues, differentiation, maturation, or activation? These definitions are not easily categorized to the satisfaction of all, but the present convention is that once a cell is a macrophage, all subsequent changes in morphology and function is mostly activation.

It is important to make a distinction between the strict definition of the term "macrophage activation" and the more general usage for a cell that is "activated" for a single function. Macrophage activation is used to identify ▶ macrophages in a defined stage of activation, and that stage of activation influences the ability of the cell to perform a large number of functions (see below). Conversely, the particular pattern of functions expressed by a macrophage defines its stage of activation.

The various stages of activation were largely defined in two experimental systems and it is critical that the system in which the terms are being used be known (see Characteristics of Activation below). The earliest references to activation of macrophages referred to activation for the killing of microbial organisms. It was observed early on that unactivated macrophages killed intracellular pathogens poorly, but ▶ activated macrophages had a greatly enhanced capacity for killing of microorganisms. With the advent of tumor immunology, a great amount of interest was placed on the activation of macrophages for killing tumor cells. A whole activation scheme was created for activation of macrophages for tumor-cell killing that differs from activation for microbial killing.

Characteristics

The classic system of macrophage activation in several precise well-defined stages was largely established in the murine peritoneal macrophage model of activation for tumor cell kill. In this model progression through various stages of activation generally correlated with enhanced functions that are associated with cellular killing and only those macrophages that can kill tumor cells in vitro were defined as fully activated (see Activation for Killing of Microorganisms below). In this model macrophages of defined stages of activation could be obtained in vivo or created in vitro (Table 1), thus:

- macrophages lavaged from the unmanipulated peritoneal cavity were termed ▶ resident macrophages (quiescent and inactivated)
- macrophages lavaged from the peritoneal cavity elicited by prior intraperitoneal injections of various sterile chemical irritants such as Brewer's thioglycolate or casein were termed ▶ inflammatory macrophages
- macrophages elicited by the injection of mimetics of microorganisms such as RNA or long-chain polymers were termed ▶ primed macrophages
- cells elicited by the injection of microorganisms such as mycobacteria were termed fully-activated macrophages.

In vitro, an inflammatory macrophage needs exposure to both an activating signal (interferon (IFN)-γ) and a triggering signal (bacterial lipopolysaccharide (LPS)) to become fully activated. A primed macrophage only requires exposure to LPS to become fully activated. An inflammatory macrophage exposed to IFN$_\gamma$ alone becomes a primed macrophage.

Confusing the issue of stages of activation were studies determining activation for killing of microorganisms. Resident macrophages do not kill tumor cells and kill microorganisms poorly. However, inflammatory macrophages have a greatly enhanced ability to kill microorganisms but do not kill tumor cells, and the same is true for primed cells. Therefore, inflammatory or primed cells as defined in the tumor cell system could be considered activated cells in terms of killing intracellular pathogens.

These simple functionalistic definitions and descriptions of macrophage activation have provided a basic scaffolding upon which has been erected an immense amount of information about how macrophages function as effector cells and how they interact with other cells and their environment. It was recognized early on that numerous functions not necessarily directly associated with killing tumor cells or microorganisms varied according to the stage of activation. The vast majority of cataloged functions increased in activity with increased activation state but some, such as arachidonic acid metabolism and secretion of prostaglandins and leukotrienes, actually decreased with activation. Table 2 lists some of these functions that vary with classical macrophage activation.

With the immense amount of information concerning cytokines, chemokines, cell-associated and soluble receptors for these molecules, and a vast array of other immunomodulating molecules, the simple functional model of macrophage activation has become an oversimplification. Microarray technology has demonstrated that the transcription of an immense number of genes (at least hundreds) are altered in macrophages as they progress along the pathways of activation. The identification of the specific signals that elicit various functions has also been complicated. For example, classically LPS was described as the signal that triggered IFN-γ primed cells, but we now know that LPS causes secretion of tumor necrosis factor (TNF), and TNF is a very potent trigger signal for macrophages. Thus, it is thought that LPS works indirectly and TNF is actually the proximate triggering signal. Adding to the complexity of the activation process are the observations that the presence of the many secreted cytokines varies the response and/or secretion of other cytokines that also affect activation. Thus, the old IFN-γ and LPS observations on function—while valid—are the tip of the iceberg, with the large complex interactions of numerous signals, cytokines, chemokines, and soluble and cell-associated receptors comprise the larger and more significant structure below the waterline.

Some interesting analyses and classifications have come out of this seemingly infinitely complex system. Macrophages exposed to IL-4 or glucocorticoids seem to be pushed into a different state of activation very unlike the classically activated macrophage and this has been termed the "alternatively activated" macrophage. These cells differ from classically activated macrophages in that they produce large quantities of arginase and thus fail to make any nitric oxide, and

Macrophage Activation. Table 1 Stages of activation for tumor cell killing of murine peritoneal macrophages

Stage of activation	In vivo	In vitro
Resident cells	Unmanipulated peritoneal cavity	N/A
Inflammatory	Injection of sterile irritants	N/A
Primed	Injection of polymers	Exposure of inflammatory macrophages to interferon-γ
Fully activated	Injection of microorganisms	Exposure of inflammatory macrophages to interferon-γ and lipopolysaccharide

Macrophage Activation. Table 2 Cellular functions that vary with stage of macrophage activation.

	Macrophage type			
	Resident	Inflammatory	Primed	Activated
Tumor cell kill	−	−	−	++++
Kill of microorganisms	±	++	+++	++++
Oxidative burst	−	+++	++++	++++
Phagocytosis	++	++++	++++	++++
Arachidonate metabolism	++++	++	+	+
Nitric oxide releases	−	−	−	++++
IL-1 secretion	−	−	−	++++
TNF secretion	−	−	±	++++
IL-12 secretion	−	−	+++	+
MCP-1	−	−	−	++++
MIP-1α	−	−	−	+++
MHC II expression	±	+	++++	+++

IL, interleukin; MCP, monocyte chemotactic protein; MHC, major histocompatibility complex; MIP, macrophage inflammatory protein; TNF, tumor necrosis factor.

they are poor at killing intracellular pathogens. Rather than supporting T cell proliferation that actually suppress it and they secrete large amounts of IL-10. It is thought that the alternative pathway of macrophage activation is a mechanism for both suppression of immune responses and a way to downregulate potentially damaging macrophage functions, such as oxidant release in places like the alveolar spaces where numerous macrophages exist and must deal with environmental exposures.

Other observations stemming from investigations into the relationships between the secretion of IL-12 by macrophages and the secretion of IFN-γ by T helper (Th) cells have spawned the classification of "type II" macrophage activation. In this model the classic activation pathway of IL-12 secretion by macrophages, IFN-γ secretion by T cells, and a Th1 adaptive immune response can be interrupted by the presence of various antigen antibody complexes. When macrophages bind these complexes to Fcγ-receptors and additionally bind ligands for Toll-like receptors, CD40, or CD44, they secrete large amounts of IL-10 instead of IL-12. This in turn induces T-cells to secrete IL-4 instead of IFNγ and leads to a Th2 adaptive immune response; hence the name type II activation. The large amount of secreted IL-10 suggests anti-inflammatory effects of these cells and indeed the presence of type II activated macrophages in mice significantly protects against LPS induced septic shock and death. Although alternatively activated macrophages and type II activated macrophages share responses to IL-4 and secretion of IL-10, type II activated macrophages are distinct from alternatively activated macrophages in that type II activated macrophages do not produce arginase and retain their ability to secrete cytokines shared with the classically activated macrophage such as TNF, IL-1, and IL-6.

In summary, the concept of macrophage activation has

progressed from a succinct well-defined functional system, to very complex mechanistic system, and is presently evolving into a branched system in which there may be at least three distinct but interrelated pathways of activation.

Preclinical Relevance

Macrophage activation as a concept is not at present part of any testing or regulation. Because of the pivotal role macrophages play in both innate and adaptive immunity and the numerous functions affected by the state of macrophage activation, any comprehensive investigation into the immunotoxicity of any environmental agent should evaluate this important indicator of immunological health.

Relevance to Humans

The differences between rodent and human mononuclear phagocytes offer a classic example of the problems of extrapolating animal data to the human situation. The expression of inducible nitric oxide synthetase (iNOS) and the subsequent secretion of nitric oxide (NO.) is a well-defined system in the mouse and this one function is believed by many to represent an excellent measure of full macrophage activation. Murine macrophages will secrete small amounts of NO. if exposed to relatively large amounts of LPS (μg/ml) but do not at physiologically relevant concentrations (ng/ml). Similarly, murine macrophages do not express iNOS or secrete NO. when exposed to even very large concentrations of IFN-γ. However, IFN-γ and LPS together are extremely synergistic and exposure of macrophages to 10 units/ml of IFN-γ lowers the amount of LPS required to get a full response for the expression of iNOS and secretion of NO. to the range of 1–10 ng/ml. In fact, IFN-γ makes murine macrophages so sensitive to LPS that any reagent purported to elicited NO. secretion in IFN-γ-primed macrophages must be rigorously shown to be free of LPS. The problem with human monocytes and macrophages is that they absolutely refuse to express iNOS or secrete any NO. in response to IFN-γ and LPS. This difference is made even more frustrating because we know human macrophages have the gene for iNOS and express it. iNOS and nitrosotyrosine—a byproduct of NO. secretion—are readily identified in macrophages surrounding granulomas in the lungs of TB patients, and nitrite (reduction product of NO.) can readily be detected in the plasma of patients harboring infections. Thus, we know human cells perform this important function but they do so in response to different (as yet unidentified) signals than the mouse. This glaring difference in human and murine macrophages means that any data on the effects of environmental agents on macrophages from rodent studies must be evaluated extremely carefully and confirmatory human studies are mandatory.

Several reports have linked macrophage activation to specific human diseases. It has been known since before the early 1920s that long-term chronic and granulomatous inflammation, in which activated macrophages serve as the end effector cells, are associated with carcinogenesis. This is particularly seen in patients with chronic tuberculosis in which tumors arise around granulomas, and in osteomyelitis where malignant squamous cell carcinomas, so called Marjolin's ulcers, arise at the sight of draining sinuses. There is data suggesting that abnormal activation of macrophages in the brain (the microglia) may be important in killing neurons in both Alzheimer's disease and AIDS dementia. Macrophages are known to be a reservoir for HIV and it is now known that the virus does not replicate until the macrophages become activated. Septic shock, which takes a huge toll in mortality in institutions that treat iatrogenically immunosuppressed patients undergoing organ transplantation or chemotherapy, is an end result of activation of macrophages and the release of TNF and IL-1. Interestingly, there is now a human clinical syndrome called macrophage activation syndrome (MAS). MAS is a potentially life-threatening process that occurs in the pediatric age group usually as a complication of the systemic onset of juvenile rheumatoid arthritis or its therapy.

Regulatory Environment

As mentioned above, macrophage activation as a specific indicator is not subject to regulation. Macrophage functions are included in many general screens for immunotoxicity but unfortunately the functions are usually relegated to a few innate immunity functions such as phagocytosis or chemotaxis. Given the above discussion it should be obvious that any observation on the effects of an environmental agent on the immune response, whether it is modulation of humoral, cellular, or innate immunity, may have its origin in the effects on mononuclear phagocytes. Therefore, an understanding of macrophage activation is important in any critical evaluation of the effects of environmental agents on the immune response.

References

1. Adams DO, Hamilton TA (1984) The cell biology of macrophage activation. Annual Rev Immunol 2:283–318
2. Ehrt S, Schnappinger D, Bekiranov S et al. (2001) Reprogramming of the macrophage transcriptome in response to interferon-gamma and *Mycobacterium tuberculosis*: signaling roles of nitric oxide synthase-2 and phagocyte oxidase. J Exp Med 194:1123–1140
3. Gordon S (2003) Alternative activation of macrophages. Nat Rev Immunol 3: 23–35
4. Mosser DM (2003) The many faces of macrophage activation. J Leukocyte Biol 73:209–212

5. Adams DO, Lewis JG, Dean JH (1988) Activation of mononuclear phagocytes by xenobiotics of environmental concern: analysis and host effects. In: Garder DE, Crapo JD, Massaro EJ (eds). Toxicology of the lung. New York: Raven Press; 354–373
6. Smits HA, Boven LA, Pereira CF, Verhoef J, Nottet HS (2000) Role of macrophage activation in the pathogenesis of Alzheimer's disease and human immunodeficiency virus type 1-associated dementia. Eur J Clin Invest 30:469–470
7. Lewis JG, Adams DO (1987) Inflammation, oxidative DNA damage, and carcinogenesis. Environ Hlth Persp 76:19–27
8. Ravelli A (2002) Macrophage activation syndrome. Curr Opin Rheumatol 14: 548–552

Macrophage Development

▶ Macrophage Activation

Macrophage Differentiation

▶ Macrophage Activation

Macrophage Inflammatory Protein-1

Macrophage inflammatory protein-1 (MIP-1) is an acidic protein with two variants designated MIP-1α (CCL3) and MIP-1β (CCL4), belonging to the C-C chemokine subgroup. Their biological activities are mediated by receptors that bind both factors. Both MIP-1α and MIP-1β are the major factors produced by macrophages following their stimulation with bacterial endotoxins. Both chemokines are involved in the activation of granulocytes and seem to be involved in acute neutrophilic inflammation. MIP-1α as well as MIP-1β stimulate the production of reactive oxygen species in neutrophils and the release of lysosomal enzymes. They also induce the synthesis of other proinflammatory cytokines such as IL-1, IL-6, and tumor necrosis factor (TNF) in fibroblasts and macrophages, and can induce the proliferation and activation of so-called chemokine activated killer cells. MIP-1α is also expressed by epidermal Langerhans cells. It is a potent T cell chemoattractant, shows chemotactic activity toward macrophages/monocytes and acts as a potent basophil agonist, inducing chemotaxis and the release of histamine and leukotrienes. MIP-1α also acts as an inhibitor of the proliferation of immature hematopoietic stem cells. MIP-1α is the primary stimulator of TNF secretion by macrophages, whereas MIP-1β antagonizes these inductive effects. MIP-1β is also expressed in monocytes and promotes adhesion of CD8$^+$ T cells to the vascular cell adhesion molecule VCAM-1. Recently, two novel C-C chemokines of the MIP-1 family were described. MIP-1γ (CCL9) was found in mice. It is expressed in contrast to any other chemokine constitutively by a wide variety of tissues. Intracerebroventricular injection of recombinant MIP-γ induces fever. It binds to the same high-affinity receptor on neutrophils as MIP-1α. Both factors appear to share common signaling pathways. MIP-1δ (CCL15) was originally isolated from a human fetal spleen cDNA library. It is expressed in T cells and B lymphocytes, in natural killer cells, monocytes, and monocyte-derived dendritic cells. In monocytes and dendritic cells the expression of MIP-1δ can be induced by other proinflammatory cytokines. It seems to be chemotactic for T cells and monocytes.

▶ Cancer and the Immune System

Macrophage Inflammatory Protein-1α (MIP-1α)

Macrophage inflammatory protein-1α (MIP-1α) is a CC chemokine also known as CCL3. Binds to CCR5, an HIV co-receptor.
▶ Chemokines
▶ Interferon-γ

Macrophage Inflammatory Protein-1β

Macrophage inflammatory protein(MIP)-1β is a chemoattractant and neutrophil-activating cytokine.
▶ Interleukin-1β (IL-1β)

Macrophage Inflammatory Protein 3 alpha (MIP-3α)

Macrophage inflammatory protein 3 alpha (MIP-3α) belongs to the group of CC(β) chemokines and binds to the chemokine receptor CCR-6.

▶ Three-Dimensional Human Skin/Epidermal Models and Organotypic Human and Murine Skin Explant Systems

Macrophage Maturation

▶ Macrophage Activation

Magnusson-Kligman Maximization Test

▶ Guinea Pig Assays for Sensitization Testing

Major Histocompatibility Complex Class I Antigen Presentation

▶ MHC Class I Antigen Presentation

Major Histocompatibility Complex Class II Antigen

▶ Antigen Presentation via MHC Class II Molecules

Major Histocompatibility Complex (MHC)

The major histocompatibility complex (MHC) is a cluster of highly polymorphic genes that encode membrane-associated glycoproteins called MHC molecules. In humans (HLA) and mice (H-2) there are two types MHC molecules, referred to as class I and class II. MHC class I molecules are expressed on all nucleated cells, and peptide-MHC class I complexes are recognized by $CD8^+$ T cells. MHC class II molecules are expressed on a small subset of cells, including dendritic cells, macrophages and activated B cells, and peptide–MHC class II complexes are recognized by $CD4^+$ T cells. Peptides derived from proteins in the cytoplasm or exocytic pathway are expressed with MHC class I molecules, whereas peptide fragments from proteins found outside the cell and degraded in the lysosomal-endosomal compartments are associated with MHC class II molecules. MHC molecules also provide the primary determinants responsible for the rapid rejection of grafts between individuals.

▶ Cytotoxic T Lymphocytes
▶ Graft-Versus-Host Reaction
▶ Autoimmune Disease, Animal Models
▶ Antinuclear Antibodies
▶ Autoantigens
▶ Systemic Autoimmunity
▶ Mixed Lymphocyte Reaction
▶ Cell-Mediated Lysis
▶ Helper T Lymphocytes
▶ Antibody-Dependent Cellular Cytotoxicity
▶ MHC Class I Antigen Presentation
▶ Antigen-Specific Cell Enrichment
▶ Polyclonal Activators

Major Histocompatibility Complex (MHC) Molecules

Heterodimeric membrane proteins, encoded by the large genetic locus identified as major histocompatibility complex (located on human chromosome 6 and mouse chromosome 17), that serve as peptide display molecules for recognition by T lymphocytes.

▶ Cytotoxicity Assays

MALT

Mucosa-associated lymphoid tissue (MALT) refers to secondary lymphoid tissue in the intestine (gut-associated lymphoid tissue, GALT), the respiratory tract (bronchus-associated lymphoid tissue, BALT; and nasal-associated lymphoid tissue, NALT), the genitourinary tract from the urethra to the ovaries and bladder or testes and bladder, the mammary gland, and the eye.

▶ Mucosa-Associated Lymphoid Tissue

Mannose-Binding Lectin (MBL)

A component of the complement system that interacts with carbohydrates, leading to the activation of serine proteases and the cleavage of the complement component C4.

▶ Complement, Classical Pathway/Alternative Pathway

Mannose-Binding Lectin Pathway

A pathway of the complement system that is independent of specific antibody and activated by foreign carbohydrates including mannose and N-acetylglucosamine. This pathway includes the complement

components MBL, MASP-1 and 2, C4, and C2, resulting in the formation of a C3 convertase to cleave C3.
▶ Complement, Classical Pathway/Alternative Pathway

MAP-Kinase

▶ Transforming Growth Factor β1 (TGF-β1)

Marginal Zone

The outer layer of the white pulp in the spleen and populated by intermediate-sized B lymphocytes, which have a major function in the T lymphocyte-independent antibody response.
▶ Spleen

Marginal Zone Lymphoma

▶ Lymphoma

Mast Cells

FRANK AM REDEGELD · MAURICE W VAN DER HEIJDEN
Dept Pharmacology and Patophysiology, Utrecht Institute of Pharmaceutical Sciences
University Utrecht
Sorbonnelaan 16
P.O. Box 80082
3508 TB Utrecht
The Netherlands

Synonyms
Mast cells, MC, connective tissue mast cells, CTMC, mucosal mast cells, MMC.

Definition
Mast cells are characterized by a granular, fattened appearance, which made Paul Ehrlich in 1875 describe them as "Mastzellen" (well-fed cells). Mast cells may be considered as the guards of connective tissue close to the veins, and of dermal and mucosal interfaces against intruders from the hostile environments of the skin, airways, and gut. Mast cells can initiate the inflammatory response to chemical compounds and pathogens by release of mediators from basophilic granules.

Characteristics
Localization
Mast cells and basophils are characterized mainly by their numerous cytoplasmic granules that stain metachromatically with basic dyes, such as toluidine blue or alcian blue/safranin. Mast cells do not circulate, but reside mostly in loose connective tissue compartments. Bone-marrow derived haematopoietic progenitor cells ($CD34^+$) enter the vascularized tissues, where they complete maturation to mast cells. Basophils are present in the blood and mature in the bone marrow before they enter the circulation.

Subpopulations
At least two subpopulations of mast cells have been identified, which havee distinct phenotypes. In humans, the population that is present at mucosal surfaces is characterized by the protease tryptase (MC_T). The predominant population that is found in connective tissues contains tryptase, chymase and carboxypeptidase (MC_{TC}). In rodents, the populations of mast cells that inhabit the mucosal surfaces and connective tissues are characterized by their mediator content, which is chondroitin sulfate in mucosal mast cells (MMC) and heparin in connective tissue mast cells (CTMC). In humans as well as in rodents, heterogeneity also exists for other mediators including cytokines. The connective tissue mast cell is regarded the most mature type. Maturation of mucosal mast cells to connective tissue type is stimulated by T helper type 2-derived cytokines that are released after an inflammatory stimulus.

Activation
Mast cells play a well-recognized and important role in immunoglobulin E (IgE) associated immune responses (acquired immunity) and allergic disorders. Cross-linking of IgE bound to its high-affinity receptor FcεRI is the classical example of mast cell activation. Cross-linking occurs when several FcεRI–IgE complexes at the same time recognise a single antigen (complex). The cross-linking event triggers a phosphorylation cascade that induces release of intracellular calcium from the ▶ endoplasmic reticulum, leading to degranulation. IgE that is bound to mast cells has a long half-life, enabling an immediate response after antigen encounter. Monomeric IgE bound to FcεRI has an anti-apoptotic effect on mast cells in the absence of cross-linking and causes enhanced cell-surface expression of FcεRI.

In addition to antigen-mediated cross-linking of IgE-sensitized mast cells, numerous other stimuli have activating properties or prevent apoptosis. Mast cells express an activating receptor for immunoglobulin G (IgG), which is FcγRI in human and FcγRIII in rodents, and an inhibitory IgG receptor (FcγRIIb). Stem cell factor (SCF), the ligand for the c-Kit receptor, is the main survival and growth factor for human and rodent mast cells. Adenosine can facilitate hypersensitivity reactions by potentiating mast cells via adenosine receptors (AR). Mast cells participate in the innate immune system by being activated directly by certain bacterial components. Alternatively, pathogens may activate mast cells indirectly via components of the complement system. Compounds from other immune cells, termed histamine release factors, are able to stimulate mast cells. Mast cells often reside around endings of sensory nerves in the vicinity of blood vessels and submucosal glands. The nerves are able to control the mast cells by releasing neuropeptides such as substance P. Proteases that are released from platelets during blood clotting, such as thrombin and coagulation factor Xa, are also known to activate mast cells. Other factors that are able to stimulate mast cell activation are shown in Figure 1. Table 1 summarizes activating and inhibiting receptors that are present on mast cells.

As was indicated previously, some stimuli lead to potentiation of mast cells, whereas others give activation (i.e. granule release). However, different stimuli may also give rise to differential release profiles; for example, human mast cells produce much more tumor necrosis factor (TNF)-α after cross-linking of FcγRI than cross-linking of FcγRI.

Secretion of Mediators

Upon activation, mast cells secrete a heterogeneous group of newly synthesized and preformed mediators. The preformed mediators, which are stored in secretory granules, include histamine and serotonin, proteases, hydrolases, proteoglycans, and inflammatory and chemotactic factors. The newly synthesized mediators include a number of cytokines and several lipid mediators (prostaglandins and leukotrienes).

Release of histamine, prostaglandins, leukotrienes, and other preformed or rapidly synthesized mediators leads to an increase in capillary permeability and contraction of smooth muscle (early-phase reaction). The slower release or synthesis of cytokines and chemokines from the activated mast cells results in attraction and migration of leukocytes and lymphocytes into the permeabilized tissues (late-phase reaction).

Preclinical Relevance

The function and dysfunction of mast cells have been associated with numerous disorders that involve hypersensitivity and inflammatory activities. At present, the main role for the mast cell is thought to be in the onset of inflammatory reactions. Leukocyte infiltration after mast cell activation leads to most clinical symptoms in contact sensitivity and ▶ delayed-type hypersensitivity (DTH) reactions, and this is probably also true for the IgE-mediated reactions in allergic patients. It should be noted that, in addition to their role in inflammatory reactions, mast cells are known to be able to phagocytose, to present antigen to T cells, and to act in non-immune issues such as tissue repair.

Mast Cells in Immediate-Type Hypersensitivity

Tissue mast cells sensitized with IgE antibodies are the main cause of immediate anaphylactic (type I) ▶ [immediate-type] hypersensitivity responses. The mast cells of allergic individuals are loaded with antigen-specific IgE. Cross-linking of the IgE by an allergen leads to mast cell degranulation. In the skin, degranu-

Type	Stimulus
Immunological stimuli	IgE, IgG, IgLC
Complement components	C3a, C4a, C5a
Cytokines	IL-1, IL-3, IL-8, SCF, TNF-α, IFN-γ, GM-CSF, NAP2, CTAP III
Chemokines	RANTES, MCP-1-4, MIP-1α
Neuropeptides	Substance P, somatostatin, CGRP, Neuropeptide Y, neurokinin-A
Other Peptides	Endothelin-1, bradykinin,
Calcium Ionophores	A23187, ionomycin
Basic compounds	Compound 48/80, poly-L-lysine/arginine, polymyxin B,
Proteases	Chymotrypsin, thrombin, factor Xa,
Bacterial compounds	Lipopolysaccharides (LPS), lectin, peptidoglycan
Other compounds	Opioids, smooth muscle relaxants, ATP, dextran, mellitin,
Physical factors	Physical stress (ultrasound, sunlight, beat, heat, cold, pressure, hypoxia, osmotic alterations)

Type	Mediator
Preformed	
	Histamine, serotonin
Proteases	Tryptase, chymase, cathepsin B, carboxypeptidase A, mouse mast cell proteases 1-6 (MMCP 1-6)
Proteoglycans	Heparin, chondroitin sulfate E
	β-hexaminidase
Newly synthesized	
	Adenosine
Lipid mediators	Platelet activating factor (PAF) Leukotrienes B4, C4, D4 Prostaglandin D2
Cytokines and chemokines	IL-1, IL-2, IL-3, IL-4, IL-5, IL-6, IL-8, IL-9, IL-13, IL-16, TNFα, IFNγ, GM-CSF, MCP-1, RANTES, MIPα, MIPβ

Mast Cells. Figure 1 Summary of mast cell stimuli (left side) and compounds that are released after stimulation (right side), grouped by their characteristics.

Mast Cells. Table 1 The main activating and inhibitory receptors on mast cells

Activating Receptors	Ligand
FcεRI	Immunoglobulin E
FcγRI (human)	Immunoglobulin G
FcγRIII (rodent)	Immunoglobulin G
c-Kit	Stem cell factor (SCF)
Adenosine receptor (AR)	Adenosine
Toll-like receptors (TLR) 2, 4, 6, 8	Bacterial components (lipopolysaccharides, lipoproteins, mannans, lectin, peptidoglycan)
Paired immunoglobulin-like receptor A (PIR-A)	?
gp49A	?
C3aR, C5aR (CD88)	C3a/C4a, C5a
Protease activated receptors (PAR) 1/ 3/ 4, 2	Thrombin, factor Xa/ trypsin/ tryptase
Neurokinin receptor-1 (NK-1)	Substance P
Inhibiting Receptors	**Ligand**
FcγRIIb	Immunoglobulin G
AR	Adenosine
PIR-B	?
Mast cell function-associated antigen (MAFA)	?
gp49B1	integrin αvβ3

lation is shown as an immediate wheal (edema from increased vascular permeability) and flare (from vasodilatation) reaction. In other tissues, such as ileum and trachea, mast cell mediators cause contraction of smooth muscle tissue and increased vasopermeability (e.g. leading to diarrhea).

Mast Cells in Contact Sensitivity and Delayed-Type Hypersensitivity

Contact sensitivity and delayed-hypersensitivity reactions are characterized by the activation of a T cell population, which is either directly cytotoxic, or activates macrophages (Th1) or eosinophils (Th2). T cell recruitment is probably mediated by the activation of mast cells, which secrete compounds that attract other inflammatory cells, enabling and to enable to enter the tissue at the site of inflammation.

Early after immunization with contact sensitizers, antigen-specific free immunoglobulin light chains are produced and released into circulation. These light chains apparently bind to mast cells in peripheral extravascular tissues. The sensitized mast cells release inflammatory mediators upon local antigen challenge, thereby initiating the T cell-dependent inflammatory response.

Relevance to Humans

Diseases that are characterized by the abnormal growth and accumulation of mast cells are grouped by the term mastocytosis. However, allergic and autoimmune disorders in which mast cells play a role are usually characterized by activation of the normal mast cell population.

Mast Cells in Allergic Disorders

Mast cells play a pivotal role in a number of immune diseases. The clinical manifestations of type I hypersensitivity reactions are generally termed ▶ atopy. Atopic diseases include the most common forms of asthma, rhinitis, and eczema. Once an allergic disease becomes chronic, the mast cell is no longer required for progression of the disease.

Testing for atopy is often done by the ▶ skin prick test, which involves injection of small amounts of allergen into the skin. The following wheal and flare response is an indication of allergy. Elevated serum levels of IgE are another well-recognized indication of atopy. Total and antigen-specific IgE levels can be measured by paper radioimmunosorbent test (PRIST) and radioallergosorbent test (RAST), respectively. Mast cell mediators, such as histamine or tryptase, can be detected in serum but are degraded much faster than IgE.

Mast Cells in Auto-immune Disorders

Mast cells have been proposed to play a role in the onset of several autoimmune diseases, including rheumatoid arthritis and multiple sclerosis. The importance of the mast cell in autoimmune disorders is not clear, although studies in patients and with animal models show that mast cells participate in the inflammatory process.

References

1. Benoist C, Mathis D (2002) Mast cells in autoimmune disease. Nature 420:875–878
2. Gurish MF, Boyce JA (2002) Mast cell growth, differentiation, and death. Clin Rev Allergy Immunol 22:107–118
3. Mekori YA, Metcalfe DD (2000) Mast cells in innate immunity. Immunol Rev 173:131–140
4. Redegeld FA, van der Heijden MW, Kool M et al. (2002) Immunoglobulin-free light chains elicit immediate hypersensitivity-like responses. Nature Med 8:694–701
5. Sharma BB, Apgar JR, Liu FT (2002) Mast cells (receptors, secretagogues, and signaling). Clin Rev Allergy Immunol 22:119–148

(Matrix) Metalloproteinase (MMP)

The group of (matrix) metalloproteinases consists of a set of enzymes (collagenases, gelatinases and stromelysins) which are essential in the field of in vivo tissue remodeling and wound healing. They catalyze the degradation of fibrillar collagen and other matrix proteins. These processes are known to be under the control of cytokines and chemokines. Furthermore, MMP-1, MMP-2 and especially MMP-9 are known to be involved in the processes leading to the directed migration of immunocompetent cells (monocytes, macrophages and Langerhans cells).

▶ Three-Dimensional Human Skin/Epidermal Models and Organotypic Human and Murine Skin Explant Systems

Maturation of the Immune Response

HUUB FJ SAVELKOUL
Cell Biology and Immunology Group
Wageningen University
Wageningen
The Netherlands

SCOTT B CAMERON · ANTHONY W CHOW
Division Infectious Diseases, Department of Medicine
University of British Columbia
Vancouver BC
Canada

Synonyms

T helper cell polarization, in vitro culture, intracellular cytokine staining

Definition

Properties of many diseases, particularly systemic autoimmune disease, strongly support the involvement of helper T lymphocytes (1,2). For example, pathogenic autoantibody responses generally are of high-affinity IgG class, after having undergone affinity maturation, which requires ▶ T helper cells. The protein antigens to which many autoantibodies are directed generally require T cell help. Many of the successful therapies, e.g. cyclosporin A, act primary on T cells. Besides roles as helpers, T cells may directly provoke cellular injury during inflammatory phases of the disease process. T cells, and in particular CD4$^+$ helper T cells produces effector molecules, called ▶ cytokines, upon activation. A multiplicity of cytokine abnormalities has been associated with various autoimmune and immune-mediated diseases. It is thus be-

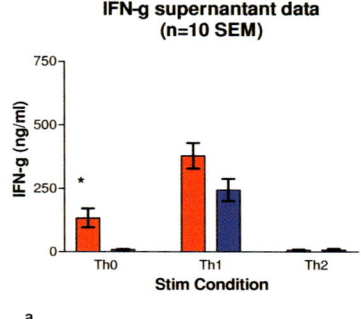
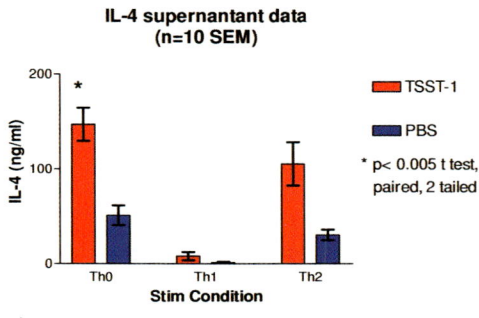

Maturation of the Immune Response. Figure 1 Absolute cytokine production of interferon-γ and interleukin-4. Absolute IFN-γ (1A) and IL-4 (1B) production by T helper cells isolated from mice treated repeatedly in vivo with phosphate-buffered saline (PBS) or TSST-1, as detected by ELISA (detection limit 7.5 pg/ml) under different in vitro polarizing conditions, including Th0, Th1 and Th2 conditions.

Metabolism, Role in Immunotoxicity. Table 1 Drug-metabolizing enzymes and their intracellular location

Reaction	Enzymes	Localization
Phase I		
Hydrolysis	carboxyesterase	microsomes, cytosol
Reduction	epoxide hydrolase	microsomes, cytosol
Oxidation	azo- and nitro-reduction	microflora, microsomes, cytosol
	quinone reduction	cytosol, microsomes
	reductive dehalogenation	microsomes
	alcohol dehydrogenase	cytosol
	monoamine oxidase	mitochondria
	prostaglandin H synthase	microsomes
	flavin monooxygenase	microsomes
	cytochrome P450	microsomes
Phase II		
Glucuronide conjugation	UDP-glucuronsyltransferase	microsomes
Sulfate conjugation	sulfotransferase	cytosol
Glutathione conjugation	glutathione S-transferase	cytosol, microsomes
Acylation	N-acetyltransferase	cytosol, mitochondria

Metabolism, Role in Immunotoxicity. Table 2 Human cytochrome P450s involved in xenobiotic activation

Isozymes	Xenobiotics
1A1	benzo[a]pyrene and polycyclic aromatic hydrocarbons
1A2	acetaminophen, 2-acetylaminofluorene, 2-naphthylamine, amino acid pyrolysis products (MeIQ, MeIQx, Glu-P-1, IQ, Trp-P-1)
2A6	dimethylnitrosamine, 4-(methylnitrosamino)-1-(3-pyridyl)-1-butanone (NNK)
2B6	5-aminochrysene, cyclophosphamide
2E1	acetaminophen, benzene, carbon tetrachloride, ethylene dibromide, ethyl carbamate, dimethylnitrosamine, vinyl chloride
3A4	acetaminophen, aflatoxin B1, benzo[a]pyrene-7,8-dihydrodiol, cyclophosphamide, sterigmatocystin, senecionine

tochrome P450 isozyme selectively without any toxicity in vivo at the dose for inhibition.

In vitro immunotoxicity testing with spleen cells has been used to detect immunotoxic chemicals, because spleen cells are able to respond to certain antigens added to cultures in an appropriate condition. However, a major pitfall is the limited xenobiotic-biotransforming activity of spleen cells. Therefore, it has been an issue to detect immunotoxic chemicals in vitro that require metabolic activation, such as dimethylnitrosamine, ethyl carbamate and cyclophosphamide. To solve this problem, coincubation of spleen cells with liver microsomes or primary hepatocyte cultures has been developed and used for the last 10 years or more. Spleen cells are coincubated with liver microsomes and test substances for 0.5 h or 1 h and separated for the culture with specific antigens, like sheep red blood cells (SRBC). Because liver microsomes do not contain phase II enzymes and because they are immunosuppressive when coincubated for a long time, the coincubation of spleen cells with primary hepatocyte cultures has subsequently been developed. The hepatocytes are anchorage-dependent and spleen cells are anchorage-independent, so the two cells can be separated very easily following coincubation. In addition, the primary cultures of hepatocytes maintain many differentiated functions of liver including phase I and phase II enzymes. Therefore, after coincubation with a test substance for up to 4 h, spleen cells are separated from hepatocytes for immunization with certain antigens in vitro to test variable immune functions. Although there are many things to be optimized, this hepatocyte-splenocyte coculture system has been shown to be an ideal method for detecting immunotoxic chemicals requiring metabolic activation.

Preclinical Relevance

As stated below there is no special guideline for studying the possible role of metabolic activation in chemical-induced immunosuppression. To investigate the role of metabolic activation in vivo, satellite groups should be assigned to induce or inhibit cytochrome P450 enzymes. If information on the specific isozymes of cytochrome P450 involved in the metabolism of test substance is available, specific inducers or inhibitors can be selected. In most in vivo studies, the T cell-dependent antibody response to sheep red blood cells has been used widely as an immunotoxic parameter. For in vitro studies, a coculture of spleen cells with hepatocytes is recommended. However, a skillful technique for isolation and culture of hepatocytes is required in this study. So the cell-free system using liver microsomes or S-9 fractions can be selected as a first choice to determine the role of metabolic activation in chemical-induced immunotoxicity.

Relevance to Humans

Before causing DNA damage many human carcinogens are activated by drug-metabolizing enzymes including cytochrome P450s; therefore, it is easy to speculate that certain immunotoxicants might require metabolic activation in humans. If there were information on the metabolic profiles of test substances in human subjects, or in human microsomal fractions, it would be clear whether or not the test substance requires metabolic activation for its immunotoxic effect.

Regulatory Environment

There is no regulatory guideline for determining a possible role of metabolism in chemical-induced immunotoxicity. No regulatory guidelines are under consideration at present.

References

1. Haggerty HG, Holsapple MP (1990) Role of metabolism in dimethylnitrosamine-induced immunosuppression: a review. Toxicology 63:1–23
2. Jeong TC, Jordan SD, Matulka RA, Stanulis ED, Kaminski EJ, Holsapple MP (1995) Role of metabolism by esterase and cytochrome P450 in cocaine-induced suppression of the antibody response. J Pharmacol Exp Therap 272:407–416
3. Kim DH, Yang KH, Johnson KW, Holsapple MP (1988) Role of the transfer of metabolites from hepatocytes to splenocytes in the suppression of in vitro antibody response by dimethylnitrosamine. Biochem Pharmacol 37:2765–2771
4. Tucker AN, Munson AE (1981) In vitro inhibition of the primary antibody response to sheep erythrocytes by cyclophosphamide. Toxicol Appl Pharmacol 59:617–619
5. Yang KH, Kim BS, Munson AE, Holsapple MP (1986) Immunosuppression induced by chemicals requiring metabolic activation in mixed cultures of rat hepatocytes and mouse splenocytes. Toxicol Appl Pharmacol 83:420–429

Metabolite

A product of xenobiotic biotransformation. In general, the metabolites are excreted easily when compared to the parent compounds. Sometimes, reactive metabolites can be produced by drug-metabolizing enzymes, including cytochrome P450s, causing immunotoxicity.

▶ Metabolism, Role in Immunotoxicity

Metals and Autoimmune Disease

DAVID A LAWRENCE
Laboratory of Clinical and Experimental
Endocrinology and Immunology
Wadsworth Center
Albany, NY 12201-0509
USA

Synonyms

Toxic metals (silver, Ag; gold, Au; cadmium, Cd; mercury, Hg; lead, Pb); essential trace metals (copper, Cu; iron, Fe; nickel, Ni; zinc, Zn), antibody-producing lymphocytes, B cells, helper T lymphocytes, CD4+ T helper cells, interleukins, cytokines, major histocompatibility complex antigens, MHC (mouse H-2, human HLA), antigen-presenting cells, APC, cells expressing MHC class II molecules.

Definition

Many metals such as copper, iron, and zinc are necessary, at relatively low doses, to maintain a healthy state, and they are referred to as essential trace elements. However, at higher doses these metals can be toxic. Metals such as silver, gold, cadmium, mercury, and lead have no known beneficial effects and have been collectively referred to as toxic heavy metals. If a metal damages the immune system, or causes an immune response that damages other tissues or organ systems, the metal is stated to be immunotoxic. An autoimmune disease occurs when the immune system develops reactivity to self constituents (self antigens), which causes damage to self cells, tissues, or organs, or disrupts normal functions by interfering with the activity of a self molecule, resulting in ill health or even eventual death. Immune responses to self antigens are not always bad; for example, the immune system can aid the clearance of senescent red blood cells, which is beneficial. This type of immunity to self does not result in disease and is referred to as autoimmunity. An antigen is a molecule that is recognized as being foreign to the immune system of an animal (including humans). When an antigen is capable of stimulating an immune response it is referred to as an immunogen. A metal is not an immunogen—a metal cannot induce an immune response unless it couples to a larger molecule. Although small molecules (haptens), such as metals, cannot induce an immune response (cannot be immunogenic) by themselves, specific responses can be elicited to a hapten that is conjugated to a carrier molecule, and once formed, antibodies can bind to the hapten itself.

Molecular Characteristics

Environmental agents are definitely involved in the development of autoimmune diseases, even though it is known that a person's genes have a significant influence on the development of autoimmune disease. This is apparent from evaluation of monozygotic twins, in whom autoimmune diseases develop in both twins less than 50% of the time. Thus, a person's genetics are only partially responsible for the onset of an autoimmune disease; the other influences come from environmental agents, including some metals.

Environmental influences on the development of autoimmune disease depend on the hosts' genetic make-up, their overall health, and their age and gender. Metals have been implicated in the development of autoimmune disease but, as with other environmental agents, the mechanisms responsible have not been fully delineated. In fact, with some metals (such as cadmium or lead) it is more likely that the metal exacerbates a disease that was previously subclinical (that is, the metals do not initiate the autoimmune disease). There is evidence to suggest that at least two metals, mercury (1) and gold (2), can induce an autoimmune disease. Nickel is well known for its ability to promote detrimental immune responses, apparent as allergic skin reactions to nickel bound to self molecules (contact dermatitis) (3), although it has not been shown to induce any autoimmune disease. Likewise, beryllium can induce a chronic inflammatory lung disease (berylliosis), but a self protein complexed with the metal may be needed to maintain the immunopathologic response.

The molecular characteristics associated with the etiology of an autoimmune disease need to be understood before a metal can be linked to its induction, and this requires research. Some metals may enhance the level of autoantibodies to autoantigens, but unless this causes a pathological condition it should be listed as an enhancer of autoimmunity and not as an inducer of autoimmune disease.

Like other toxicants, metals may be implicated indirectly in disease. A metal can increase oxidative stress resulting in cell damage. Excess of an essential metal such as iron can cause cytotoxicity by increasing the generation of reactive oxygen species. Metals such as mercury or tin can be even more toxic when present in their organic form (e.g. methylmercury or triethyltin). The state of the element also is critical in determining its toxicity. For example, the chromium and arsenic ions Cr^{3+} and As^{5+} are less toxic than Cr^{6+} and As^{3+}, respectively. In large part this is due to differences in cellular uptake of a particular metal depending on its valance.

The cytotoxicity that occurs after exposure to a toxic metal might generate release of greater than normal levels of an autoantigen or the process might cause denaturation of self molecules generating autoantigens. Both scenarios have been implicated in the breaking of tolerance (immune unresponsiveness) to a self molecule and development of immunity to the autoantigen, which could culminate in autoimmune disease (2,4). Alternatively, the stress induced by a metal could initiate the production of a stress response protein (for example, the proteins of the heat shock protein families such as HSP70 and HSP90) which are known to influence intracellular protein trafficking and processing as well as functions of antigen-presenting cells (APC).

Putative Interaction with the Immune System

With gold, silver, and mercury it is known that the background genetics of the animal are critical. For example, silver and mercury induce autoantibodies to nucleolar autoantigens only in $H-2^s$, $H-2^q$ and $H-2^f$ mice (5) although other strains also can generate autoantibodies if they have susceptibility genes, such as the *sle* loci of lupus-prone NZM mice or NZB/NZWF1 mice, which are $H-2^z$. Some mouse or rat strains develop autoimmune disease with no obvious environmental exposure—for example the NZM2410 strain has multiple *sle* loci, and both males and females develop systemic lupus erythematosus (SLE) early in life, whereas only female NZM2758 mice develop SLE later in life because they have fewer *sle* genes. Even though the *H-2* genes are known to influence lupus penetrance, both NZM2410 and NZM2758 have the same *H-2* genes. However, other strains (such as $H-2^s$ mice) do not develop disease unless exposed to a stressor, and their disease is usually less severe and/or is transient. Additionally, the sensitivity/susceptibility of $H-2^s$ mice to metal-induced autoimmune nephritis depends on more genes than just the $H-2^s$ gene complex (1).

Mercury enhances autoantibody levels to a number of autoantigens, but it appears that antibodies to fibrillarin are especially detrimental (1) and that they are possibly (in part) responsible for the ensuing autoimmune nephritis (renal pathology). Fibrillarin is a protein localized exclusively in the fibrillar region of the nucleolus. It is suggested that mercury alters the antigenic determinants of fibrillarin either by producing a mercury-containing determinant or uncovering a cryptic peptide, which stimulates production of helper T (Th) cells to altered self. These Th cells in turn stimulate B cells that have specificity to unaltered fibrillarin, which can maintain the disease in the absence of further mercury exposures. Such a model has been suggested for gold, in that gold bound to a protein can cause differential proteolysis and the newly formed "cryptic" peptides are foreign since they would not have been formed unless gold was present (6).

H-2 genes include the major histocompatability complex (MHC) class I and class II molecules of the mouse. The MHC class II molecules (I-A and I-E) influence the development and activation of the $CD4^+$ helpers. The I-A and I-E genes in this complex are known as the immune response genes, which are equivalent to the human leukocyte antigen(HLA)-DR, HLA-DP and HLA-DQ genes of humans. These MHC class II molecules pick up antigenic peptides within antigen-presenting cells (APC) and transport and present the peptides on the APC surface to the $CD4^+$ T helper cells. It is the final structural complex formed between the peptide and MHC class II molecules, which determines the ligand for the $CD4^+$T helper cell's antigen-specific receptor (TCR). As the $CD4^+$ (MHC class II responsive) and $CD8^+$ (MHC class I responsive) T cells develop within the thymus, many T cells with TCRs specific for self MHC class II or class I molecules and various self peptides are destroyed or rendered unresponsive. The $CD4^+$ T helper cells help B cells to proliferate and differentiate into efficient antibody-producing cells (plasma cells) as well as $CD8^+$ T cells to become active cytolytic T cells. Thus, $CD4^+$ T helper cells are important for all aspects of adaptive (antigen-specific) immunity, as apparent with acquired immunodeficiency syndrome (AIDS) when the $CD4^+$ T cells are selectively destroyed by human immunodeficiency virus (HIV). Since selection within the thymus as well as extrathymic mechanisms exist to maintain T cell tolerance to self molecules, various things must occur to account for T cell promotion of autoimmune diseases. Different mechanisms have been proposed to explain how an encounter with an environmental agent could be responsible for this immunopathologic process. A metal may bypass the lack of self-recognizing T cells. A metal may directly alter the structure of the MHC class II molecule; a metal may bind to a naturally forming self peptide, affecting the manner in which it associates with the MHC class II molecule; a metal may cause the formation of different peptides from self protein, described previously as cryptic peptides.

Alternatively, instead of altering the structure of the MHC-peptide complex a metal may enhance the signal(s) for T cell activation. This could occur by effects to APC or to T cells themselves. For example, a metal may stimulate through nonantigen-specific receptors in manners similar to antibodies to CD2 or CD3 molecules on the T cell surface. It also is known that a memory T cell requires less signal through its TCR than a naive T cell in order to become activated. This difference is due in part to the expression of accessory molecules, such as CD54, CD80, and CD86, on the surface of APC or the density of the MHC class II molecules presenting peptides, and metals may enhance the expression of these molecules or increase their affinities for their co-receptors on the T helper cells. Dendritic cells are thought to be the main type of APC presenting antigen to naive T helper (Th0) cells, which is likely because of their high density of MHC class II molecules.

Mercury and lead increase the expression of MHC class II molecules. Metal-induced changes in the expression of MHC class II molecules or accessory molecules aiding in the association of APC and Th0 cells and of extracellular or intracellular regulatory factors also could influence whether type 1 (Th1) or type 2 (Th2) CD4 helper T cells will preferentially develop from naive Th0 cells (see Figure 1). A change in the balance could affect the development of autoimmune responses. Gold, mercury and lead preferentially skew the development of Th0 cells toward Th2 cells. Prior to the development of significant mercury-induced nephritis, the host animals have signs of enhanced type 2 immunity, apparent mostly as increased levels of type 2 immunoglobulins (IgE and IgG_1) and ▶ cytokines (interleukins IL-4, IL-5, IL-10 and IL-13) in the mice or rats. However, at later times, increased levels of the type 1 cytokine interferon(IFN)-γ are observed, and have been suggested to further the immunopathology in the kidney. Some autoimmune diseases, such as rheumatoid arthritis and multiple sclerosis, are believed to be initiated mainly by type 1 immunity whereas the autoimmune disease SLE may begin with a predominance of type 2 immunity. Temporary imbalances between these regulatory T cells seems to be, at least, partially responsible for the immunopathologies associated with autoimmune diseases.

Loss of T cell tolerance to self antigen is usually responsible for development of an autoimmune disease. A metal also may bypass the need for a strong antigen-specific signal by enhancing the level of cytokines, soluble factors that can modulate immune reactivities. Cytokines include growth factors (T cell growth factor IL-2; B cell growth factor IL-5), differentiation factors (B cell differentiation factor IL-6), and even inhibitory factors (cytokine synthesis inhibitory factor IL-10). Overproduction of cytokines can be toxic, as observed with systemic inflammatory response syndrome (SIRS) and toxic shock syndrome. The ▶ proinflam-

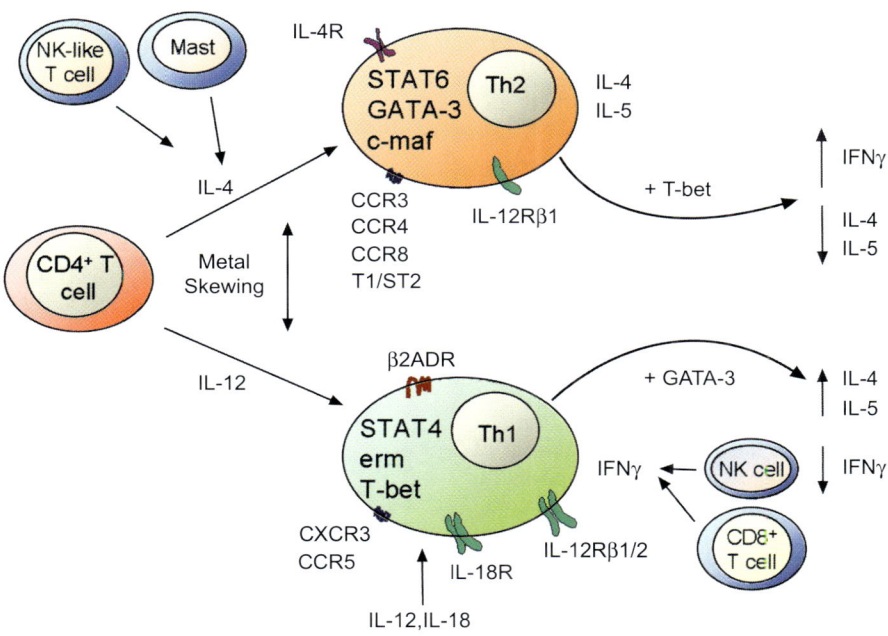

Metals and Autoimmune Disease. Figure 1 Metals can modify the development of Th1 (type 1 T helper CD4+ cells) and Th2 (type 2 T helper CD4+ cells) from naive Th0 cells. The Th1 and Th2 subsets differentially regulate cell-mediated and humoral immune responses and the activities of each other. Numerous differences in cytokines (e.g. interleukin IL-4 vs IL-12), neuropeptides (β2-andrenergic receptor (ARD)) and chemokines (o.g. CCR3 vs CXCR3) influence the development and activity of Th1 of Th2 cells. In ertain autoimmune diseases either Th1 or Th2 activity predominates.

matory cytokines (IL-1, IL-6 and TNF-α) are often implicated in these toxicities. The ability of lead to cause as much as a 1000-fold increase in mortality due to endotoxin has been shown to relate to excessive production of TNF-α. Under normal circumstances, when the proinflammatory levels get high the levels in the brain increase causing activation of the hypothalamus-pituitary-adrenal (HPA) axis. The HPA axis culminates in the production of glucocorticoids, which are immunosuppressive. However, in some autoimmune diseases, the animals have been shown to have a "blunted" HPA axis. Even although the IL-1, IL-6 and TNF-α concentration is high in the brain, the HPA axis is not activated and/or functional. As proposed in Figure 2 for lead, many metals are known to be capable of directly or indirectly increasing oxidative stress. Such effects could aid in elevation of the proinflammatory cytokines and concomitantly inhibit the HPA axis, resulting in the possible potentiation of an autoimmune disease. A scenario of this type indicates that a metal may affect autoimmune responses by effects on systems (nervous and endocrine systems) in addition to the immune system.

Relevance to Humans

The two best examples of a metal influencing autoimmune disease in humans are observed with mercury and gold. Despite observations made almost three decades ago that skin-whitening creams containing mercury can induce membranous nephropathy (or autoimmune nephritis), clinical reports documenting mercury-induced nephritis from such creams continue (7). Membranous nephropathy also has occurred from exposure to mercury in the fluorescent-tube recycling industry.

The following autoimmune diseases have been implicated with exposure to gold or mercury: Addison's disease, colitis, multiple sclerosis, thrombocytopenia and, as expected, glomerulonephritis. It has been suggested that some of the great artists (Rubens, Renoir, Dufy, and Klee) who used bright colors containing toxic heavy metals (like gold and mercury) developed rheumatoid arthritis and scleroderma (8).

Many of the suggested mechanisms for loss of tolerance to self molecules noted earlier have been implicated in human disease. An individual's genetic makeup is also involved; in any individual, a gene loci implicated in one autoimmune disease increases the ease by which an environmental agent can induce another autoimmune disease. For example, gold has been used for the treatment of rheumatoid arthritis, but in some of these patients the gold induces a second au-

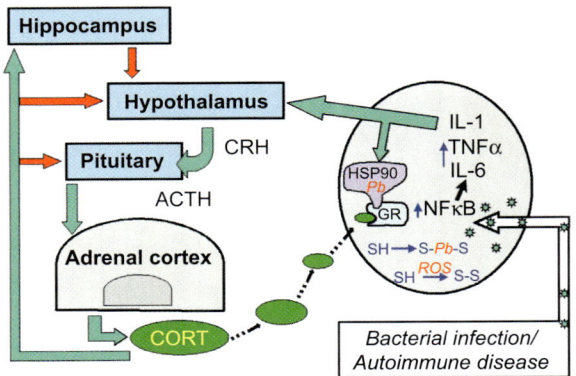

Metals and Autoimmune Disease. Figure 2 Metal interference with the hypothalamic-pituitary-adrenal (HPA) axis which can blunt hyperimmune reactivity (as occurs with autoimmune diseases) via glucocorticoid suppression of cytokine expression. Hyperreactivity of the immune system can lead to excessive levels of the proinflammatory cytokines interleukin(IL)-1, IL-6 and tumor necrosis factor(TNF)α, which in excess can be toxic. Infectious and autoimmune diseases can generate high levels of IL-1, IL-6 and TNFα, but the HPA-axis activity usually limits their overproduction. Metals elevate these cytokines by generation of reactive oxygen species (ROS) which concomitantly inhibit suppression of glucocorticoid feedback. CRH, corticotropin-releasing hormone; ACTH, corticotropin; CORT, cortisol (human); GR, glucocorticoid receptor.

toimmune disease, usually idiopathic thrombocytopenic purpura. Gold appears to induce the formation of autoantibodies to the autoantigen glycoprotein V (GPV) on platelets; however, gold is not bound to GPV in order for these autoantibodies to be reactive. As with the animal studies, humans with particular HLA (MHC class II) molecules appear to be more susceptible to metal-induced autoimmune diseases. Further, the metals inducing autoimmune diseases can modify the expression of cytokines that can affect immune regulation and type 1 versus type 2 immunity.

References

1. Nielsen JB, Hultman P (2002) Mercury-induced autoimmunity in mice. Environ Health Perspect 110 [Suppl 5]:877–881
2. Fournie GJ, Mas M, Cautain B et al. (2001) Induction of autoimmunity through bystander effects. Lessons from immunological disorders induced by heavy metals. J Autoimmun 16:319–326
3. Budinger L, Hertl M (2000) Immunologic mechanisms in hypersensitivity reactions to metal ions: an overview. Allergy 55:108–115
4. Weigle WO (1997) Advances in basic concepts of autoimmune disease. Clin Lab Med 17:329–340
5. Hansson M, Abedi-Valugerdi M (2003) Xenobiotic metal-induced autoimmunity: mercury and silver differentially induce antinucleolar autoantibody production in susceptible H-2s, H-2q and H-2f mice. Clin Exp Immunol 131:405–414
6. Griem P, Panthel K, Kalbacher H, Gleichmann E (1996) Alteration of a model antigen by Au(III) leads to T cell sensitization to cryptic peptides. Eur J Immunol 26:279–287
7. Soo YO, Chow KM, Lam CW et al. (2003) A whitened face woman with nephrotic syndrome. Am J Kidney Dis 41:250–253
8. Pedersen LM, Permin H (1988) Rheumatic disease, heavy-metal pigments, and the Great Masters. Lancet 1:1267–1269

3-methylcholanthrene, 3-MC, dibenz-a, h-anthracene, benz-a-anthracene, BA, BP-quinones

▶ Polycyclic Aromatic Hydrocarbons (PAH) and the Immune System

MHC

▶ Polyclonal Activators
▶ Autoimmune Disease, Animal Models

MHC Antigen Presentation

Antigen presentation is the way by which cells express antigen bound to MHC molecules on their cell surface for recognition by T lymphocytes. It can include antigen processing as well. This process decribes the degradation of proteins and the generation of peptides, which are suitable to be presented on MHC molecules on the cell surface.

▶ MHC Class I Antigen Presentation

MHC Class I Antigen Presentation

Hansjoerg Schild · Mark Schatz
Institute of Immunology
University of Mainz
Obere Zahlbacher 67
D-55131 Mainz
Germany

Synonyms

major histocompatibility complex class I antigen presentation

Definition

▶ MHC class I antigen presentation is the process by which antigen-presenting cells express foreign antigen complexed with major histocompatibility complex (MHC) class I molecules on their cell surface for recognition by lymphocytes. The antigens, which represent peptides, are generated from proteins of intracellular pathogens, such as viruses, intracellular bacteria, and, as a special case, tumors. The recognition of those presented peptides by specific CD8-expressing ▶ cytotoxic T lymphocytes (CTLs) triggers the killing of the antigen-presenting cell. In contrast to this, MHC class II antigen presentation is mainly responsible for the generation of peptides from proteins of extracellular pathogens and is essential for the activation of CD4-expressing T cells, which, for example, assist in the efficient production of antibodies by B cells. This is described in detail in another chapter.

Characteristics

Protein processing by the proteasome and its regulators

Almost all peptides presented by MHC class I molecules originate from the cytosol. Recent studies suggest that most of those peptides come from newly synthesized but defective proteins called defective ribosomal products (DRiPs).

The first step in antigen processing (the generation of presented peptides from an intact antigen) is mediated by the proteasome, the major endopeptidase and thus the main degrading enzyme in the cytoplasm. The so-called 20S core of the proteasome is a large, multicatalytic cylindrical complex, which consists of four stacked rings, each containing either seven α or seven β subunits. The rings are arranged in the order αββα, where the six proteolytic sites are located in the inner β subunits of the cylinder, three in each of the central rings. The proteins are broken down into peptides after their introduction into the hollow core. Most interestingly from an immunological point of view, is the fact that three of the constitutive subunits with catalytic activity can be exchanged in newly synthesized proteasomes by the interferon(IFN)-γ-inducible subunits low molecular-weight protein 2 (LMP2), LMP7, and MECL-1 which also have catalytic activity. This is especially remarkable, since IFN-γ is produced in response to viral infections. Interestingly, the genes for LMP2 and LMP7 are located in the MHC-locus, where many immunologically relevant genes are encoded. The proteasomes that do not have IFN-γ inducible subunits are called constitutive or standard proteasomes and are the common form in most cell types. In professional antigen-presenting cells and inflamed tissues the proteasomes with LMP2, LMP7 and MECL-1, termed immunoproteasomes, are dominant. It is well accepted, that either the constitutive or the immunoproteasome generate the exact C-terminus of most known CTL epitopes. Immunoproteasomes preferentially produce peptides with hydrophobic or basic C-terminal residues, which have a higher affinity for MHC class I molecules and are favoured by TAP (see below). However, there are a lot of epitopes known, that are generated solely by constitutive proteasomes. Not only do the inducible subunits alter the function of the proteasome, but also other molecules that can regulate its activity and specificity. One of them is the 19S particle, which together with the 20S core forms the 26S proteasome. In this complex, the 19S cap is responsible for the recognition of ubiquitinated proteins, which are due to their ubiquitination marked for degradation. Another interaction-partner is PA28 (11S regulator), which is just like the special immunoproteasomal subunits inducible by IFN-γ. PA28 is a hexameric ring consisting of two alternating subunits, PA28α and PA28β, which binds flat to an outside ring of the proteasome. The binding enhances the capacity of the proteasome to cleave proteins and thus to produce precursors of epitopes dramatically. PA28 does not only upregulate the activity but also changes the specificity of the proteasome.

Generation of the correct N-terminus of the epitope and transportation into the endoplasmic reticulum

As described earlier, the proteasome generates the correct C-terminus of an epitope but only rarely the correct N-terminus, which therefore usually leads to an aminoterminal extended version of the final epitope. To trim those peptides to the proper size, there is a need for aminopeptidases, which generate the right N-terminus. This can be done either in the cytosol or later on in the endoplasmic reticulum (ER). Candidates for the cytosol are the puromycin-sensitive aminopeptidase (PSA), leucine aminopeptidase (LAP) (inducible by IFN-γ), and bleomycin hydrolase (BH). Thimet oligopeptidase (TOP) also could play a role in the creation or destruction of certain epitopes.

Another interesting enzyme which fits in between the

proteasome and the aminopeptidases is the tripeptidyl peptidase II (TPP II). It cleaves off the first three N-terminal amino acids from a peptide at once. Furthermore, it can take over some of the functions of the proteasome and act as an endopeptidase to generate the correct C-terminus of at least some epitopes.

Either the final or the N-terminal extended epitope has to enter the ER to meet an MHC class I molecule. The translocation of the peptide from the cytosol into the endoplasmic reticulum is done by the transporter associated with antigen processing (TAP), which is located in the ER-membrane. TAP, as a member of the ATP-binding cassette (ABC) family of transporters, needs ATP for its translocation activity. The subunits of the heterodimer, TAP1 and TAP2, are encoded within the MHC itself and are upregulated upon exposure to IFN-γ. TAP binds peptides in the range of 8 to 13 residues, but can transport even longer ones with up to 40 residues. This makes it possible to transport either the final epitopes or longer precursor peptides into the endoplasmic reticulum (ER) for further trimming. Human TAP preferentially translocates peptides with a hydrophobic or a basic C-terminus. This preference is in agreement with the C-terminus of most peptides generated by proteasomes and with peptide binding to MHC class I molecules inside the ER. However, the binding affinity of the human TAP is not only influenced by the C-terminus. The three N-terminal residues also have a significant effect on the transport into the endoplasmic reticulum.

The question of how the peptides travel from the proteasome or the aminopeptidases to TAP can not be answered conclusively. One possibility is by diffusion; another is by peptide binding to cytosolic chaperones (e.g. heat shock proteins), which act like shuttles.

After the peptides are transported into the ER and before they bind to an empty MHC class I heterodimer, a large fraction still has to be trimmed at their N-terminus to have the correct length. A recently identified aminopeptidase, called ERAAP or ERAP1, seems to be mainly responsible for this action.

Structure and characteristics of the MHC class I molecule

Finally in the endoplasmic reticulum and with the correct length, the peptides get loaded onto the MHC class I molecules, which are heterodimers consisting of the membrane-spanning α-chain (43 kDa) and the non-covalently bound β_2-microglobulin (β_2m) (12 kDa). The α-chain can be subdivided into three domains: α1, α2, and α3. The first two domains together form the peptide-binding cleft, which is lined by two α-helices lying on a sheet of eight antiparallel β- strands. The main binding sites of the peptide, which is usually 8–10 amino acids long, are the amino- and the carboxy-terminus in addition to anchor residues, which positions and characteristics are defined by the MHC variant. The T cell receptor recognizes the whole complex by partially binding to exposed residues of the peptide and partially interacting with accessible elements of the two α-helices. In contrast to its counterpart MHC class II, the class I molecule is expressed on all nucleated cells. The highest expression is found on T cells, B cells, macrophages, neutrophils, and other antigen-presenting cells. In humans, the MHC molecule is usually termed HLA (human leukocyte antigen).

Assembly of the MHC class I complex and peptide loading

The loading of peptide on an MHC class I molecule in the ER is the final step of as MHC class I complex assembly. This process starts with the binding of the newly synthesized class I heavy chain to calnexin. After β_2m binds to the heavy chain, calnexin is exchanged for calreticulin. To form the MHC class I peptide-loading complex, a lot of other molecules become involved. One of them is tapasin, which acts like a bridge between the class I heterodimer and TAP. Tapasin is encoded within the MHC-locus and upregulated by IFN-γ. The major function of tapasin seems to bring the empty MHC class I molecule close to TAP and thereby increase the loading speed. Also, tapasin has an editing function. It catalyzes the exchange of low-affinity peptides bound to MHC class I molecules for high-affinity peptides. The usage of high affinity peptides leads to stable and therefore long-living MHC class I-peptide complexes on the cell surface later on. This again not only gives rise to a strong and sustained immune response but also prevents the exchange of intracellular against extracellular peptides after the MHC complex is transported to the cell surface. Another molecule of the MHC class I peptide-loading complex is ERp57, which interacts non-covalently with calreticulin and through an interchain disulfide bond with tapasin. ERp57 seems to be responsible for the correct disulfide bond formation in the class I heavy chain.

The exact time course of the MHC class I peptide-loading complex assembly is so far only speculation. Most likely tapasin and ERp57 bind to TAP with the help of calnexin. Following that, the empty MHC class I molecule interacts with this complex. Concurrently, calnexin is exchanged for calreticulin.

Quantitative analysis showed that the complex has a higher order. There are four tapasin molecules bound to each TAP heterodimer, whereas each tapasin interacts with one MHC class I molecule. In the complex there seems to be less calreticulin and ERp57 then stoichiometrically expected. However, it cannot be excluded that there are more molecules involved in the

assembly of the peptide-loading complex or are even part of it.
After binding the peptide, the MHC class I-peptide complexes dissociate from the loading complex, associate with calnexin, and stay in the endoplasmic reticulum for a little while. Then they exit the ER with the help of transport receptors through the Golgi apparatus to the plasma membrane, where they are presented to $CD8^+$ T cell.

Even on the way to the cell surface there is a possibility to process peptides. This is done by the protease furin, which resides in the *trans*-Golgi network. The importance of this alternative path, however, remains unclear.

In uninfected cells, peptides from degraded self proteins fill the peptide-binding clefts of MHC class I complexes, which then are presented at the cell surface but do not activate $CD8^+$ T cells. This constant turnover gives the body the opportunity to notice pathogens rapidly and elicit an immune response early in an infection.

Empty MHC class I molecules, which did not bind peptide, dissociate under secretion of β_2m and degradation of the heavy chain by the proteasome after retrograde translocation through the Sec61p complex from the ER into the cytosol. This control mechanism is necessary to prevent the appearance of empty MHC class I heterodimers on the cell surface, which could bind extracellular peptides and then lead to a false immune response.

To give a quantification of the antigen processing, it is estimated that around 2000 to 10 000 molecules of a protein have to be degraded to generate one cell surface class I-peptide complex.

Alternative MHC class I antigen-processing pathways

It should be mentioned that there are alternative MHC class I antigen-processing pathways which do not follow the general rule. In cross-priming, for example, macrophages and dendritic cells take up exogenous antigens under certain conditions, process them in a TAP-dependant and proteasome-dependent way, and present the epitopes on the cell surface in context with MHC class I molecules for the inspection by CTLs (1–3).

Relevance to Humans
MHC class I deficiency

A very clear picture of what happens if the MHC class I antigen presentation is absent can be seen in the so called MHC class I deficiency, also known as bare lymphocyte syndrome (BLS) type I. BLS type II is characterized by the absence of MHC class II molecules and type III by the absence of both class I and class II molecules. Because of this, the maturation of T cells in the thymus is equally affected by no positive selection during their development as are peripheral immune reactions due to the inability to present antigens. The medical manifestation of this syndrome in types II and III is an early onset of severe combined immunodeficiency (SCID). This differs from the class I deficiency, which is not associated with a par-

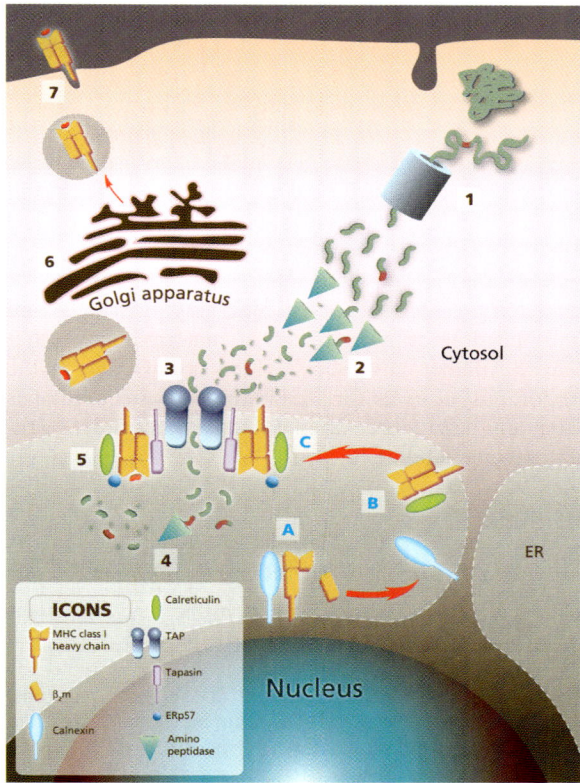

MHC Class I Antigen Presentation.
Figure 1 Processing and presentation of antigens by MHC class I molecules:
1) Break-down of proteins into smaller fragments by the proteasome in the cytosol; 2) Cleavage of N-terminal residues by cytosolic aminopeptidases; 3) Transport of peptides into the ER by TAP; 4) Further trimming of the N-terminus by endoplasmic aminopeptidases; 5) Loading of peptide onto the MHC class I molecule by the peptide loading complex; 6) Dissociation of the MHC class I/peptide complex from the loading complex and transportation through the Golgi apparatus to the cell surface; 7) Presentation of peptide on the cell surface.
Assembly of the MHC class I peptide-loading complex: A) Stabilization of the newly synthesized MHC class I heavy chain by calnexin; B) Binding of β2m to the heavy chain with the exchange of calnexin by calreticulin; C) Formation of the MHC class I peptide-loading complex consisting of TAP, tapasin, MHC class I heterodimer, calreticulin and Erp57.
Image work courtesy of Hazel Ang.

ticular disease during the first years of life, with the exception of chronic lung disease, which develops in late childhood. Systemic infections have not been described in HLA class I-deficient patients. Chronic bacterial infections, often beginning in the first decade of life, are restricted to the respiratory tract and extend from the upper to the lower airway. Bronchiectasis, emphysema, panbronchiolitis, and bronchial obstruction have been described, as well as a high frequency of nasal polyps and involvement of the nasal sinuses (4).

Escape mechanisms of viruses

The loss of MHC class I on the cell surface can also be initiated by the action of viruses. In viral infections a selective inhibition of different parts of the antigen-processing machinery by direct interference of some viral proteins with the antigen-processing pathway can be seen. This gives the pathogens the opportunity to escape the immune system, or at least persist in the body permanently. One of the viruses that has a whole set of such proteins is the human cytomegalovirus (hCMV). In CMV-infected cells, the MHC class I heavy chains can still assemble with β_2m but afterwards are degraded rapidly. Responsible for this effect is an array of genes called US2 through US11. The protein encoded by US3, for example, binds to the MHC class I heterodimer and retains it in the endoplasmic reticulum. US2 or US11 transport MHC class I heavy chains back into the cytosol for degradation by the proteasome. US6 binds the TAP complex from the ER-luminal side and inhibits peptide translocation. pp65, another hCMV protein, inhibits the generation of specific T cell epitopes. But CMV is not the only virus with such remarkable characteristics. The herpes simplex virus (HSV) can also drastically reduce the surface expression of MHC class I molecules. Responsible for this effect is ICP47, the product of the US12 gene, which seems to interfere with the translocation of peptides into the ER through direct interaction with TAP.

However, diminishing the MHC class I level on the cell surface does not prevent the cell from getting killed. In cases like this, natural killer (NK) cells can recognize the loss an MHC class I and trigger target cell death. To evade this fate, the human CMV contains a gene that encodes for an MHC homolog (UL18), which binds to β_2m, and acts as a camouflage MHC molecule on the cell surface.

The direct interference of viral proteins with the antigen-processing pathway is not always necessary for the virus to survive an immune response. This can also be done by escape mutations. Here, an amino acid exchange occurs, either within the epitope, which prevents the recognition of the infected cells by $CD8^+$ T cells, or outside of the epitope, which can prevent the generation of the peptide by interfering with its proper processing or its transport (5).

New vaccine development

Knowledge about the MHC class I antigen processing pathway can be applied to the possible development of new vaccines, which are not only more potent and less expensive, but also free of any side effects. This could be achieved by using the CTL epitope alone or fused to another protein, which would abrogate the need for an attenuated and possibly harmful virus. Furthermore, these vaccines could be produced by chemical means, by bacteria, by cow's milk, by banana plants, for example, which would not only make them inexpensive, but also available to a wider group of people. Additionally, the immune response could be influenced by adding immunostimulators or immunomodulators to the vaccine, by applying the epitope as a peptide, or encoded on RNA or DNA, by changing the method and location of its application, among many other ways. Another advantage of this knowledge is the opportunity it gives for predicting the correct epitopes from any newly discovered pathogen to create a working vaccine against it.

References

1. Van Kaer L (2002) Major histocompatibility complex class I-restricted antigen processing and presentation. Tiss Antigens 60:1–9
2. Pamer E, Cresswell P (1998) Mechanisms of MHC class I-restricted antigen processing. Annu Rev Immunol 16:323–358
3. Janeway CA, Travers P, Capra JD, Walport MJ (eds) (1999) Immunobiology: The immune system in health and disease, 4th ed. Garland, New York
4. de la Salle H, Donato L, Zimmer J et al. (1999) HLA class I deficiencies. In: Ochs HD, Smith CIE, Puck JM (eds) Primary immunodeficiency diseases: A molecular and genetic approach. Oxford University Press, New York, pp 181–188
5. Früh K, Ahn K, Peterson PA (1997) Inhibition of MHC class I antigen presentation by viral proteins. J Mol Med 75:18–27

MHC (Major Histocompatibility Complex) Class II Molecule

Cell surface molecules that participate in antigen presentation to $CD4^+$ T helper cells.

▶ Humoral Immunity

MHC Restriction

Due to the nature of the $\alpha\ \beta$ T cell receptor (TCR) the

characteristic of T lymphocytes is to recognize foreign peptides only when they are bound to a particular allelic form of a major histocompatibility complex (MHC) molecule. CD4⁺ T cells recognize peptides presented by MHC class II molecules, and CD8⁺ T cells recognize peptides presented by MHC class molecules. The exchange of the peptide as well as a different MHC molecule would abolish T cell activation.
▶ Superantigens
▶ Antigen Presentation via MHC Class II Molecules

MHC Tetramer

Soluble tetramers of major histocompatibility complex (MHC) class I or class II molecules are a biotechnological reagent produced in the laboratory to analyze the specificity of antigen receptors expressed by T cells. In nature T cell receptors (TCR) recognize peptide antigens bound to membrane MHC molecules. Quantitative molecular analysis of cell–cell interactions is intrinsically difficult, but MHC glycoproteins are integral membrane molecules that lose their function when dissociated from membranes. Complexes of four MHC molecules are stable in solution and associate to antigenic peptides. The tetramer–peptide complex binds to TCR in an antigen-specific manner, thus allowing enumeration of antigen-specific T cells in a polyclonal cell population. Tetramers can be tagged with fluorochromes for cytofluorometric analysis.
▶ Tumor, Immune Response to
▶ Enzyme-Linked Immunospot Assay

Microarray Technology

The simultaneous individual measurement of the mRNA expression level of thousands of genes in a given sample by means of hybridisation.
▶ Toxicogenomics (Microarray Technology)

Microenvironment of the Bone Marrow

This is the immediate area that surrounds the stromal cells. Bone marrow stromal cells produce an array of soluble growth factors, such as granulocyte-macrophage colony stimulating factor, that are necessary for the normal proliferation and differentiation of the cells of the hematopoietic system. The combined microenvironments of the bone marrow form a network of cells and vessels that facilitates the close association of hematopoietic stem cells with the stromal cells. This facilitates the proliferation and differentiation of these hematopoietic stem cells in response to these soluble growth factors. Immune cells are generated by hematopoiesis continuously during life.
▶ Developmental Immunotoxicology

Microfold Cells

Microfold cells, or M cells, are located on specialized epithelium in the intestinal tract and function to sample intestinal antigens.
▶ Mucosa-Associated Lymphoid Tissue

Microinjection

The insertion of some material, e.g., DNA, into a cell through a microelectrode using either hydrostatic pressure or a minute electric current.
▶ Transgenic Animals

Micronutrients

▶ Vitamins

Microparticles

These circulating vesicular particles are composed of lipids and shedded protein (complexes) derived from endothelial and circulating blood cells following exposure to mechanical stress, cytokines and other agonists or during apoptosis. Microparticles can be transferred between vascular and blood cells and platelets, which may thereby become enriched in specific proteins that are not associated with platelets per se.
▶ Blood Coagulation

Microsphere-Based Multiplex Assays

▶ Cytokine Assays

MIG

A monokine induced by interferon-γ which belongs to the group of CXC(α) chemokines and binds to the chemokine receptor CXCR-3.
▶ Three-Dimensional Human Skin/Epidermal Models and Organotypic Human and Murine Skin Explant Systems

Migration Inhibitory Factor (MIF)

A lymphokine which inhibits the migration of macrophages away from the site of interaction between lymphocytes and antigenss.
▶ Therapeutic Cytokines, Immunotoxicological Evaluation of

Migration of Neutrophils

▶ Chemotaxis of Neutrophils

Minipig

▶ Porcine Immune System

Mismatched or Matched Organ

The determination of the exact MHC haplotype expressed by a patient and by a possible organ donor allows the transplantation of organs that express (nearly) the same MHC alleles like found for the patient.
▶ Antigen Presentation via MHC Class II Molecules

Mitogen

A substance which causes cells, particularly lymphocytes, to undergo cell division.
▶ Mitogen-Stimulated Lymphocyte Response
▶ Primate Immune System (Nonhuman) and Environmental Contaminants

Mitogen-Activated Protein Kinase Cascade

Mitogen-activated protein(MAP) kinase cascades are signaling pathways responsible for transducing many of the intracellular effects of cell surface receptor activation. The signals are relayed via a cascade of kinases that phosphorylate each other. These cascades lead to the phosphorylation and activation of transcription factors.
▶ Signal Transduction During Lymphocyte Activation

Mitogen-Activated Protein Kinases (MAP Kinases)

The MAP kinase family of signaling pathways include ERK, JNK, and p38. These pathways are activated upon lymphocyte activation and cytokine receptor ligation and are critical for cell proliferation.
▶ Transforming Growth Factor β1; Control of T cell Responses to Antigens

Mitogen Assay

Immune function assay in which lymphocytes can be stimulated ex vivo with plant lectins, which results in cell activation and proliferation.
▶ Canine Immune System

Mitogen-Induced Lymphocyte Blastogenesis

▶ Mitogen-Stimulated Lymphocyte Response

Mitogen-Induced Lymphoproliferative Response

▶ Mitogen-Stimulated Lymphocyte Response

Mitogen Response

▶ Mitogen-Stimulated Lymphocyte Response

Mitogen-Stimulated Lymphocyte Proliferation Assay

▶ Mitogen-Stimulated Lymphocyte Response

Mitogen-Stimulated Lymphocyte Response

RALPH J SMIALOWICZ
Office of Research and Development, National Health & Environmental Effects Research Laboratory
US Environmental Protection Agency
Research Triangle Park, NC 27711
USA

This report has been reviewed by the Environmental Protection Agency's Office of Research and Development and approved for publication. Approval does not signify that the contents necessarily reflect the views and policies of the Agency, nor does mention of trade names or commercial products constitute endorsement or recommendation of use.

Synonyms
Mitogen response, mitogen-stimulated lymphocyte proliferation assay, mitogen-induced lymphoproliferative response, mitogen-induced lymphocyte blastogenesis

Short Description
Antigen-driven immune responses involve the activation and proliferation of lymphocytes. These steps are integral to in vivo cell-mediated and humoral immune responses in man and animals. The ▶ mitogen-stimulated lymphocyte response is an in vitro correlate of activation and proliferation of specifically sensitized lymphocytes by antigen in vivo. The mitogen-stimulated response has also been used to assess the immunotoxic potential of drugs and chemicals in humans and experimental animals. In vitro lymphocyte stimulation or transformation was first described by Nowell (1), who found that the addition of an extract from the red kidney bean *Phaseolus vulgaris* to cultures of human peripheral blood caused morphological changes in the small resting lymphocytes, which resulted in ▶ blastogenesis (i.e. the formation of large pyroninophilic lymphocytes with large nuclei and prominent nucleoli). ▶ Mitotic cells were observed in these leukocyte cultures after several days of culture. Several plant ▶ lectins, in addition to the *P. vulraris* extract phytohemagglutinin (PHA), which stimulates T lymphocytes, have been shown to induce blastogenesis and lymphocyte proliferation in vitro. These include:

- concanavalin A (Con A), which also stimulates T lymphocytes and which is derived from the jack bean *Canavalia enciformis*
- soybean agglutinin (SBA) and peanut agglutinin (PNA), which display species specificity
- pokeweed mitogen (PWM), isolated from the roots of *Phytolaca americana*, which consists of five mitogens Pa-1 to PA-5, all of which stimulate T lymphocytes. PWM Pa-1 also stimulates B lymphocytes to undergo mitosis and Ig secretion in the presence of macrophages and T lymphocytes.

In addition to plant lectins, a wide variety of chemically diverse agents have been demonstrated to be mitogenic for cultured lymphocytes, including bacterial products such as lipopolysaccharide (LPS) from gram-negative bacteria and purified protein derivative of ▶ tuberculin (PPD), both of which are B cell mitogens. Other B cell mitogens include sodium metaperiodate and dextran sulfate. Antibodies such as antilymphocyte serum and anti-α2- and β2-microglobulins are mitogenic, as is anti-CD3 which induces T cell activation. Certain metal ions (e.g. zinc, mercury and nickel), calcium ▶ ionophores (e.g. A23187), and certain proteolytic enzymes (e.g. the serine ▶ proteases trypsin and chymotrypsin) are also mitogenic. Since many mitogens preferentially bind to and activate only certain B and/or T lymphocyte subpopulations, they have been very useful in identifying defective lymphocyte populations in clinical as well as experimental situations (2).

Characteristics

The ability of mitogens to stimulate lymphocytes to mature and divide in vitro occurs in the absence of antigenic specificity of the lymphocyte receptor, and thus the designation "nonspecific" or "▶ polyclonal" mitogen stimulation. As a result, a larger number of lymphocytes are stimulated by mitogens than would occur following exposure to antigens, which typically induce cell proliferation in a smaller proportion of the lymphocyte population. In vitro mitogen nonspecific activation of lymphocytes results in myriad biochemical events culminating in DNA synthesis and cell division. These biochemical events have been most extensively studied using lectin mitogens and include membrane-related changes in the transport of monovalent cations such as:

- Na^+ and K^+, and the influx of the divalent cation Ca^{2+}
- phospholipid synthesis and turnover
- alterations in the intracellular concentrations of the nucleotides cyclic adenosine monophosphate

(cAMP) and cyclic guanosine monophosphate (cGMP)
- activation of protein kinase C
- gene activation.

Within hours of mitogen-induced lymphocyte activation, protein synthesis occurs followed by increased RNA synthesis, and finally, within 36–48 hours, DNA synthesis (3). While any of these biochemical events can be used as a marker of lymphocyte activation, DNA synthesis has been the most widely used end point. This has been measured by incorporation of ^{3}H- thymidine or the use of the tetrazolium salt 3-[4,5-dimethylthiazol-2-yl]-2,5-diphenyl tetrazolium bromide (MTT), which is cleaved by viable cells, to form a formazan dye that can be quantitated spectrophotometrically.

Pros and Cons

The mitogen-stimulated lymphocyte response is an in vitro assay that has utility in the identification of potential immunotoxicants in acute and chronic toxicity studies. This assay has been employed by a number of laboratories as part of the first tier or level of immune function assays in immunotoxicity testing. It is included at this level of immunotoxicity testing because it provides important information about the capability of T and B lymphocytes to proliferate—an essential event for most immune responses. Since this assay involves the in vitro stimulation of lymphocytes, and consequently does not require the use of animals specifically sensitized with antigens, it is easily applicable to immune function assessment in routine toxicity testing. In addition to its application for evaluating the immunotoxic potential of agents in animal studies, this assay also is useful for characterizing the direct action of an immunotoxicant, or its metabolites, upon addition to cultured lymphocytes. Unfortunately, mitogen-stimulated lymphoproliferation represents a polyclonal or global response of T and/or B lymphocytes which is nonspecific in nature. As such, it does not mirror an antigen-driven immune response. It is critical that the viability of the cells be determined so that the number of viable cells per microtiter well is equivalent across the cells obtained from all the animals being tested.

Predictivity

In the context of immunotoxicity testing, the mitogen-stimulated lymphocyte response represents one of the "first tier" tests for identifying agents which cause immunosuppression in rodent species. An interlaboratory study evaluated the sensitivity and predictability of a number of immune function assays by testing 40–46 chemicals. A predictive value of 0.67 was calculated for the T cell mitogen responses, while the B cell mitogen (LPS) response had a predictive value of 0.50. Based on these results, T cell mitogen responses were predictive of immunotoxicants and the B cell mitogen response was not (4).

Relevance to Humans

The mitogen-stimulated lymphocyte response has been used clinically to assess cellular immunity in patients suffering from immunodeficiency diseases, cancer, and autoimmunity, as well as in patients undergoing immunotherapy. It has also been shown to be useful for assessing lymphocyte function in ▶ asymptomatic AIDS patients (5).

Regulatory Environment

The mitogen-stimulated lymphocyte response is one of several rodent in vitro immune function tests employed to determine potential immunotoxicity of compounds. However, there are no immunotoxicity test guidelines which identify this assay as a required test.

References

1. Nowell PC (1960) Phytohemagglutinin: an initiator of mitosis in cultures of normal human leukocytes. Cancer Res 20:462–466
2. Smialowicz RJ (1995) In vitro lymphocyte proliferation assays: the mitogen-stimulated response and the mixed lymphocyte reaction in immunotoxicity testing. In: Burleson GR, Dean JH, Munson AE (eds) Methods in Immunotoxicology, Volume 1. John Wiley, New York, pp 197–210
3. Stites DP (1984) Clinical laboratory methods for detection of cellular immune function. In: Stites DP, Stobo JD, Fudenberg HH, Wells JV (eds) Basic and Clinical Immunology, 5th ed. Lange Medical Publications, St Louis, pp 353–372
4. Luster MI, Portier C, Pait DG et al. (1992) Risk assessment in Immunotoxicology. I. Sensitivity and predictability of immune tests. Fundam Appl Toxicol 18:200–210
5. Janossy G (1991) Immune parameters in HIV infection– A practical guide. Immunol Today 12:255–256

Mitogenic Stimulation

This is the process of stimulating cells to proliferate by exposure to mitogens, normally applied to T lymphocytes stimulated with concanavalin A (ConA) or phytohemagglutinin (PHA). Cytokine secretion is also stimulated by mitogens.

▶ Interferon-γ

Mitogens

▶ Polyclonal Activators

Mitomycin C

Acts as an alkylating agent by inhibiting DNA synthesis by crosslinking DNA to an extent proportional to its content of guanine and cytosine; its action is most prominent during the late G1 and early S phases of the cell cycle.

▶ Mixed Lymphocyte Reaction

Mitotic

Having to do with the presence of dividing or proliferating cells.

▶ Mitogen-Stimulated Lymphocyte Response

Mixed Leukocyte Culture

▶ Mixed Lymphocyte Reaction

Mixed Lymphocyte Reaction

RALPH J SMIALOWICZ
Office of Research and Development, National Health & Environmental Effects Research Laboratory
US Environmental Protection Agency
Research Triangle Park, NC 27711
USA

This report has been reviewed by the Environmental Protection Agency's Office of Research and Development and approved for publication. Approval does not signify that the contents necessarily reflect the views and policies of the Agency, nor does mention of trade names or commercial products constitute endorsement or recommendation of use.

Synonyms

mixed leukocyte reaction, mixed leukocyte culture, MLR

Short Description

The mixed lymphocyte reaction (MLR) is an in vitro assay in which leukocytes, from two genetically distinct individuals of the same species, are cocultured resulting in cell blast transformation, DNA synthesis and proliferation. Generation of the MLR occurs as a consequence of the incompatibility of the ▶ allogeneic determinants which are expressed on the surface of cell populations and which are encoded by the major histocompatibility complex (MHC) (1). The designations for the MHC in man, mice and rats are HLA, H-2 and RT1, respectively. There are different cell types, which stimulate naive ▶ alloreactive $CD4^+$ and $CD8^+$ T cells, with $CD4^+$ cells exhibiting higher anti-MHC responses. Dendritic cells (DC), which constitutively express MHC class I and MHC class II molecules, as well as intracellular adhesion molecule(ICAM)-1 and leukocyte function-associated antigen(LFA)-3, are potent antigen presenting cells (APC) which play a role in many immunological reactions including the MLR.

Characteristics

This assay can be performed either as a two-way MLR, in which cells from both donors will undergo blast transformation and proliferation, or as a one-way MLR, in which the cells from only one donor will proliferate because the cells from the other donor have been prevented from responding. For immunotoxicity testing, only the one-way MLR is employed (2). For example, H-2 locus ▶ histo-incompatible mice have been used for the one-way MLR assay in which C57BL/6J $H-2^b$ mice are the source of responder cells (i.e. proliferating cells) while CBA/J $H-2^k$ mice are the source of stimulator cells (i.e. cells prevented from proliferating). Stimulator cells are blocked/prevented from proliferating by treatment with either radiation or ▶ mitomycin C. The source of cells for the MLR include blood, lymph nodes, or spleen, the latter of which is predominately employed in immunotoxicity testing. The rat is an exception to the preponderant use of spleen cells in the MLR. Attempts to generate an MLR response using rat spleen cells have resulted in poor lymphocyte proliferation. It has been suggested that this weak response may be due to the presence of natural suppressor macrophages in the spleens of rats, and that inhibition of lymphocyte proliferation in the rat MLR is due to the direct effect of ▶ cytostatic products of oxidative L-arginine metabolism by these macrophages (3). Consequently, rat lymph node lymphocytes (e.g. Fischer 344 RT-1^{lv1} versus Wistar/Furth RT-1^u rats) are recommended over spleen cells for the rat MLR in immunotoxicity testing. The proliferative response in the MLR is measured primarily by ^{3}H-labeled thymidine uptake. An alternative non-radioactive cell proliferation assay employs the tetrazolium salt 3-[4,5-dimethylthiazol-2-yl]-2,5-diphenyl tetrazolium bromid (MTT), which is cleaved by viable cells, to form a formazan dye which can be quantitated spec-

trophotometrically. Each MLR test consists of four quadruplicate sets of wells prepared in a round bottomed 96-well microtiter plate. These four sets consist of the following:
- 1. "responder" cells only
- 2. "responder" and "stimulator" cells
- 3. "responder" cells and the T cell mitogen concanavalin A (Con A)
- 4. "stimulator" cells and Con A.

The last two sets serve as internal controls, while the MLR response is determined by subtracting set 1 from set 2.

Pros and Cons

The MLR has been employed in clinical settings to establish the compatibility of donors for bone marrow and living related renal ▶ allotransplantation. This is an antigen-driven response, which requires the participation of antigen-presenting cells. The MLR has utility in the identification of potential immunotoxicants in acute and chronic toxicity studies employing either mice or rats as the test species, with consideration of the caveats for the latter species indicated above. Unfortunately, the application of the MLR requires the use of additional animals; beyond those involved in immunotoxicity testing per se, to serve as the source of "stimulator" cells. This use of additional animals may be unacceptable under certain circumstances. It is also important to emphasize that the viability of "responder" cells be determined so that in preparing the MLR cultures the number of viable cells per microtiter well is equivalent across the cells obtained from all the animals being tested.

Predictivity

In the context of immunotoxicity testing, the MLR represents an in vitro antigen-specific lymphocyte activation/proliferation assay, which has been shown to identify agents, which block or inhibit lymphocyte DNA synthesis and cell proliferation. However, a study of the sensitivity and predictability of a number of immune assays, employed in an interlaboratory immunotoxicity test evaluation of 39 different compounds, indicated that the MLR provided poor predictability with a value of 0.56 (4). A similar evaluation of the predictability of the MLR in rats has not been undertaken.

Relevance to Humans

The MLR has been used clinically for histocompatibility typing of candidates and donors for organ and bone marrow transplants. It has also been employed in clinical diagnosis of immunodeficiency diseases.

Regulatory Environment

The MLR is one of several rodent in vitro immune function tests employed to determine potential immunotoxicity of compounds. However, there are no immunotoxicity test guidelines, which identify the MLR as a required test.

References

1. Bain B, Vas M, Lowenstein L (1964) The development of large immature mononuclear cells in mixed lymphocyte cultures. Blood 23:108–111
2. Smialowicz RJ (1995) In vitro lymphocyte proliferation assays: the mitogen-stimulated response and the mixed lymphocyte reaction in immunotoxicity testing. In: Burleson GR, Dean JH, Munson AE (eds) Methods in Immunotoxicology, Volume1. John Wiley and Sons, New York, pp 197–210
3. Hoffman RA, Langrehr JM, Billiar TR, Curran RD, Simmons RL (1990) Alloantigen-induced activation of rat splenocytes is regulated by the oxidative metabolism of L-arginine. J Immunol. 145:2220–2226
4. Luster MI, Portier C, Pait DG et al. (1992) Risk assessment in Immunotoxicology. I. Sensitivity and predictability of immune tests. Fund Appl Toxicol 18:200–210

Mixed Lymphocyte Response (MLR)

The activation of rodent or human T lymphocytes in culture by mixing untreated lymphocytes with lymphocytes from an allogeneic donor. In the most commonly used form, the one-way mixed lymphocyte response, the "stimulator" lymphocytes are inactivated by treating them with radiation or a DNA-binding agent to prevent proliferation so that only the proliferation of the "responder" lymphocytes will be detected.

▶ Polyclonal Activators
▶ Mixed Lymphocyte Reaction

MMC

▶ Mast Cells

Mobilization

The migration of hematopoietic stem cells and progenitor cells from the bone marrow (and spleen) into the blood, which occurs after treatment of an organism with hematopoietic growth factors or cytotoxic substances or a combination of both. Mobilization of hematopoietic stem cellsis transient, and may involve a

concomitant transient expansion of the hematopoietic stem cells pool in the bone marrow.
▶ Hematopoietic Stem Cells

Modeling

▶ Statistics in Immunotoxicology

Molecular Chaperones

These are proteins whose function is to bind and stabilize proteins at intermediate stages of folding, assembly, translocation across membranes and degradation. Many are heat shock proteins.
▶ Interferon-γ

Molecular Mimicry

ALAN EBRINGER · LUCY HUGHES · TAHA RASHID · CLYDE WILSON
Division of Life Sciences
King's College, University of London
150 Stamford Street
London
SE1 8WA
UK

Synonyms
molecular similarity, cross-reactivity, antigenic similarity, autoimmunity

Definition
Molecular mimicry or "molecular similarity" is an important concept in immunology and immunotoxicology, whereby external antigens that resemble some "self-antigens" will lead to an immune response which will cause tissue damage to the host, and this is called autoimmunity.

Immune responses are known to occur in all vertebrate animals and control the ability of such animals to deal with "external" or "non-self" biochemical agents. These external environmental factors, which evoke an "immune reaction" are generally known as "antigens". The response to an antigen through the production of an immune response is marked by cellular proliferation of lymphocytes, interaction with macrophages, production of plasma cells, and finally the appearance of specific molecules known as "antibodies" which can be demonstrated to bind to the original antigens or related structures.

External or "non-self" biochemical agents, which can act as antigens, can be either freely administered drugs, usually having some therapeutic effect, or molecules present in infectious agents such as viruses, bacteria, or parasites. The smallest size of an antigen is usually a peptide consisting of six amino acids or a carbohydrate molecule, which is approximately six monosaccharides in length. However smaller compounds can also have an immunological effect if they become attached to a larger molecule. The larger molecule is then known as the "carrier" whilst the smaller molecule is called the "hapten". Specific immune responses can then be evoked by the hapten-carrier molecule but once produced the hapten alone can produce pathological effects.

The classical example in immunotoxicology of such a deleterious effect is the problem of penicillin hypersensitivity, especially when applied to skin as a topical antibiotic. The penicillin molecule becomes covalently bound to skin surface proteins, which then act as hapten carriers and produce an immune response, usually in the IgE isotype. When the subject is exposed on a subsequent occasion to the penicillin molecule, an anaphylactic shock occurs which in some occasions could have a fatal outcome. Due to these problems, penicillin is no longer used as a topical antibiotic.

Characteristics
Autoimmune diseases are characterized by the presence of antibodies which bind to self-antigens and therefore are known as autoantibodies. In some diseases such autoantibodies can cause tissue damage. Many human diseases, such as systemic lupus erythematosus, juvenile diabetes, or rheumatoid arthritis are considered as examples of autoimmune diseases.

Two main theories have been proposed for the origin of such autoantibodies: either the immune system spontaneously starts producing tissue damaging immune cells or infection occurs by a microbiological agent, which possesses antigens exhibiting molecular similarity or "molecular mimicry" with some tissues of the host.

There is little evidence for the concept that autoimmune diseases occur as a result of some "lymphocyte mutation." Usually autoimmune diseases are characterized by relapses and remissions, strongly suggesting repeated exposure to an external antigen. If the "lymphocyte hypothesis" were to hold, one would expect continued clonal proliferation of cytotoxic cells leading eventually to target organ failure with death of the patient. The proposal that a lymphocyte mutates and starts attacking, say synovial tissues, to produce rheumatoid arthritis, suggests rather a neoplastic process but autoimmune diseases do not behave as tumors.

The "infection hypothesis" would appear to have greater merit in providing an explanation for the origin of autoimmune diseases. Following an infection by an external agent which carries biochemical structures showing molecular mimicry with "self-antigens" of the host, an immune response will occur with antibodies being produced against the invading microbe. A portion of these antibodies will bind to the "self-tissues" of the host which exhibit molecular mimicry and therefore these antibodies are acting as autoantibodies. When present in high concentrations, such autoantibodies can cause tissue damage and eventually lead to a disease involving the organ possessing structures resembling the invading pathogen. The cytotoxic effect of such antibacterial antibodies has been demonstrated in rheumatoid arthritis and ▶ ankylosing spondylitis using a simple sheep red cell assay (1).

The classical model of an autoimmune disease evoked by an infection is "rheumatic fever." The microbe *Streptococcus* possesses molecular sequences which resemble the human heart. Following an upper respiratory infection or tonsillitis, antistreptococcal antibodies bind to endocardial antigens and cause tissue damage. The patient then develops a cardiac murmur, fever, and muscle pains and is then said to suffer from "rheumatic fever". Since autoantibodies can cause tissue damage, this is an example of an autoimmune disease produced by an infection. Rheumatic fever has more or less disappeared in the western world over the last 50 years due to the widespread use of antibiotics such as penicillin.

Sydenham's chorea is another disease which involves molecular mimicry. Sydenham's chorea occurs in rheumatic fever patients who have a high titer of anti-*Streptococcal* antibodies. Some of these antibodies will bind to the basal ganglia of the brain, because similar antigens are present in the streptococci. The autoantibodies binding to the basal ganglia will produce ataxia and chorea. Rheumatic chorea is characterized by semi-purposive involuntary movements which are usually intensified by voluntary effort. The chorea usually wanes following treatment with high doses of antibiotics, such as penicillin and is an example of a neurological autoimmune disease evoked by an infection.

Autoantibodies can be evoked not only by microbial agents but also by therapeutic drugs. For instance patients suffering from hypertension have been treated in the past by the drug alpha-methyl-dopa, which binds to nor-adrenaline receptors but does not fire them. Autoantibodies against the patient's own red cells have been demonstrated using the direct ▶ Coomb's test and such patients developed an autoimmune haemolytic anaemia. However when the drug is stopped both the haemolytic anaemia and the positive Coombs test disappear without any further complications.

Many other drugs can act as haptens and give rise to autoimmune diseases. Some of the most frequently encountered therapeutic agents giving rise to autoimmune reactions are drugs to name a few, such as hydralazine, cephalothin, hydantoins, trimethadione and procaine amide, but there are many others. The general conclusion can be made that almost every drug can give rise to a hapten-carrier situation and therefore is capable through molecular mimicry of producing a deleterious autoimmune pathological response.

Putative interaction with the immune system
3 examples:

The interaction of molecular mimicry with the immune system in producing pathological consequences will be illustrated by 3 examples of autoimmune diseases evoked by exposure to microbiological agents possessing antigens crossreacting with self-tissues. The 3 examples involve the common diseases rheumatoid arthritis, ankylosing spondylitis and ▶ multiple sclerosis.

Rheumatoid Arthritis

Rheumatoid arthritis is an example of an autoimmune disease evoked by an infection. Over 95% of rheumatoid arthritis patients possess HLA-DR1/4 antigens while the frequency of these antigens in the general British population is about 35%. A particular amino acid sequence found in these HLA antigens is EQ(K/R)RAA and this exhibits molecular mimicry with the ESRRAL sequence found in *Proteus* haemolysin (2) (Fig. 1). Elevated levels of antibodies to the urinary pathogen *Proteus mirabilis* have been found in patients with rheumatoid arthritis from more than 14 different countries throughout the world (3).

When patients, usually middle aged or elderly women, are suffering from an upper urinary tract infection by *Proteus mirabilis* they will produce antibodies against all the antigens found in these microbes. Those antibodies targeted against *Proteus* haemolysin will also bind to HLA-DR1/4 antigens found in the synovial tissues and when the titer of antibodies is sufficiently high, it will activate the complement cascade and lead to a cytotoxic response involving tissue injury and synovial inflammation.

Another example of molecular mimicry in this disease involves the enzyme *Proteus* urease, which resembles a similar sequence found in type XI collagen (2). Hyaline cartilage is predominantly a component of the small joints of the hands and feet, and contains type XI collagen. This could explain why rheumatoid arthritis is essentially an arthritic disorder involving the small joints of the extremities. In presence of elevated levels of anti-*Proteus* urease antibodies, these target the antigens found in type XI collagen and cause cytotoxic damage especially in the small joints

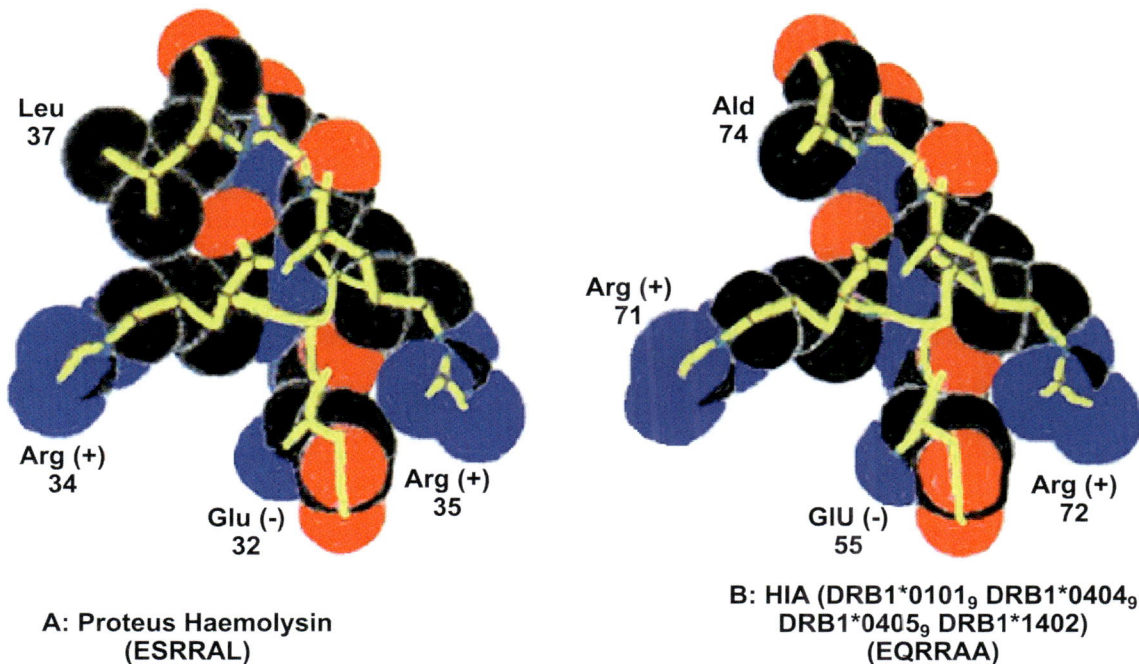

Molecular Mimicry. Figure 1 Molecular similarity between Proteus mirabilis hemolysin and HLA-DR1/4 amino acid sequence peptides. Modified from Wilson et al. (1995) with permission.

of the hands and feet, thereby providing a possible explanation for the anatomical distribution of the pathological lesions found in this disease.

A possible therapeutic implication of these results is that elimination of the external antigen, namely the *Proteus mirabilis* microbe from the upper urinary tract, by antibiotic or other means, might alleviate the severity of the arthritic episodes found in this disease. Clearly longitudinal studies are indicated to test this possibility.

Ankylosing Spondylitis

Ankylosing spondylitis is a chronic disease of the spine. It is also another example of an autoimmune disease evoked by an infection. The human leukocyte antigen HLA-B27 antigen is present in 96% of patients with ankylosing spondylitis but in only 8% of the general population. Again molecular mimicry is found to operate in this condition. A particular peptide sequence present in the HLA-B27 molecule is QTDRED which shows molecular mimicry with a similar sequence located in the nitrogenase enzyme found in the commensal bowel microbe *Klebsiella* (4) (Fig. 2). Another cross-reactive sequence showing molecular mimicry to HLA-B27 is found in the pullulanase enzyme of *Klebsiella*. The *Klebsiella* microbe is found in the large bowel, especially around the ileocecal junction, where it proliferates on monosaccha-

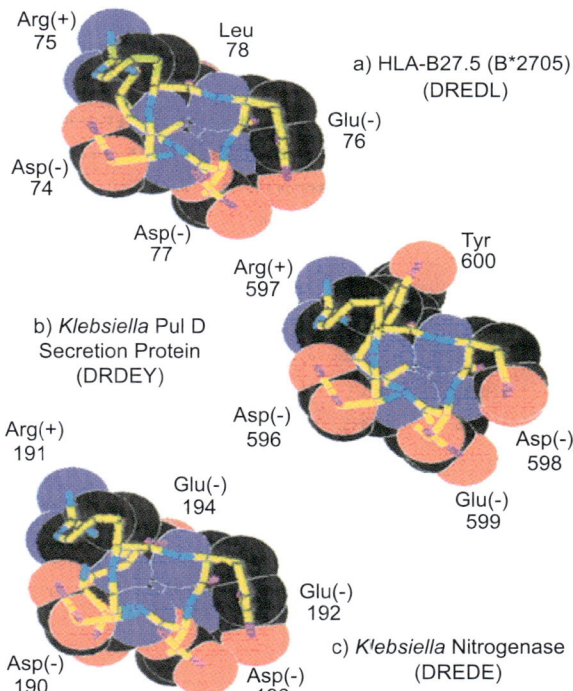

Molecular Mimicry. Figure 2 >Molecular similarity between *Klebsiella pneumoniae* pullulanase, nitrogenase and HLA-B27 amino acid sequence peptides. Taken from Fielder (1995) with permission.

ride and disaccharide substrates entering from the small intestine, which are derived from the consumption of starch-containing compounds such as bread, potatoes, cakes, and pasta. A "low-starch diet" appears to be of some benefit in patients with ankylosing spondylitis. Antibodies to *Klebsiella* microbes have been found to be present in ankylosing spondylitis patients from many different countries, such as Spain, the Netherlands, Japan, Germany and the UK. Furthermore the cytotoxic activity of such anti-*Klebsiella* antibodies has been demonstrated in sera obtained from ankylosing spondylitis patients and compared to the cytotoxicity of anti-*Proteus* antibodies present in patients with rheumatoid arthritis (1).

Multiple Sclerosis

The presence of clinical similarities between "experimental allergic encephalomyelitis" (EAE), especially in relation to hind-quarter paralysis and lower-limb ataxia in multiple sclerosis patients leads to the question whether some environmental agents might possess antigens resembling or cross-reacting with brain tissues.

EAE is considered as an animal model of multiple sclerosis. One of the main components in the central nervous system responsible for the production of EAE is a basic protein present in the white matter of the brain—myelin basic protein.

In 1970 a highly active peptide from myelin was identified which, when injected in microgram quantities into guinea pigs, would produce hind-leg paralysis, tremors, weight loss, and eventually death—features characteristic of EAE.

The hypothesis proposed was that in the environment there may be a microbe, which could possess proteins resembling brain tissues, similar to the situation of *Streptococcus* in rheumatic fever and *Proteus* in rheumatoid arthritis. Computer analysis of proteins in SwissProt database, revealed that the microbe *Acinetobacter* which is present in soil, on skin, in contaminated waters, and in fecal materials has such a sequence (5) (Fig. 3). The sequence is present in the molecule 4-carboxy-mucono-lactone decarboxylase of *Acinetobacter* and subsequently a similar sequence was found in gamma-carboxy-mucono-lactone decarboxylase of *Pseudomonas*. Both groups of microbes *Acinetobacter* and *Pseudomonas* belong to the same family of Gram-negative bacteria and share many antigens.

Multiple sclerosis patients were found to have elevated levels of antibodies to both *Acinetobacter* and *Pseudomonas* bacteria (6). The discovery that common environmental microbes such as *Acinetobacter* and *Pseudomonas* had sequences showing molecular mimicry with brain antigens suggested a possible mechanisms as to how patients with multiple sclerosis could have developed their disease. Over 50% of multiple sclerosis patients suffer from sinusitis and the sinuses are anatomical sites from which *Acinetobacter* and *Pseudomonas* bacteria can be readily isolated. Since IgG antibodies can cross the blood-brain barrier, such antibacterial antibodies could produce pathological lesions in brain tissues which eventually manifest themselves as the clinical features of multiple sclerosis.

Relevance to Humans

The concept of molecular mimicry has provided a better understanding of the toxicologic problems associated with the use of drugs which, through the hapten-carrier complex, produce immunological responses that may have pathological consequences for the patient.

The second important relevance of molecular mimicry to humans has been the identification of microbial agents which carry biochemical sequences resembling the "self-tissues" of the host.

These results in people with rheumatoid arthritis, ankylosing spondylitis, and multiple sclerosis open up entirely new therapeutic possibilities. The use of antibiotic therapy could decrease the severity of these diseases and may even avoid the permanent pathological sequelae of these conditions. After all, rheumatic fever and Sydenham's chorea have disappeared in the western world because of the early use of anti-*Streptococcus* therapy.

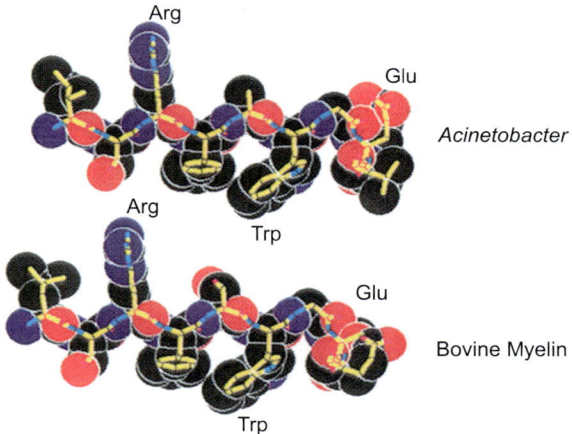

Molecular Mimicry. Figure 3 Molecular similarity between *Acinetobacter calcoaceticus* 4-carboxy-mucono-lactone decarboxylase and myelin basic protein amino acid sequence peptides. Modified from Ebringer et al. (1997) with permission.

References

1. Wilson C, Rashid T, Tiwana H et al. (2003) Cytotoxicity responses to peptide antigens in rheumatoid arthritis and ankylosing spondylitis. J Rheumatol 30:972–978

2. Wilson C, Ebringer A, Ahmadi K et al. (1995) Shared amino acid sequences between major histocompatibility complex class II glycoproteins, type XI collagen and *Proteus mirabilis* in rheumatoid arthritis. Ann Rheum Dis 54:216–220
3. Ebringer A, Rashid T, Wilson C (2003) Rheumatoid arthritis: proposal for the use of anti-microbial therapy in early cases. Scand J Rheumatol 32:2–11
4. Schwimmbeck PL, Yu DTH, Oldstone MBA (1987) Autoantibodies to HLA-B27 in the sera of HLA-B27 patients with ankylosing spondylitis and Reiter's syndrome. Molecular mimicry with *Klebsiella pneumoniae* as potential mechanism of autoimmune disease. J Exp Med 166:173–181
5. Ebringer A, Thorpe C, Pirt J, Wilson C, Cunningham P, Ettelaie C (1997) Bovine spongiform encephalopathy: is it an autoimmune disease due to bacteria showing molecular mimicry with brain antigens? Environ Health Perspect 105:1172–1174
6. Hughes LE, Bonell S, Natt RS et al. (2001) Antibody responses to *Acinetobacter sp.* and *Pseudomonas aeruginosa* in multiple sclerosis: Prospects for diagnosis using the myelin-*Acinetobacter*-neurofilament antibody index. Clin Diagn Lab Immunol 8:1181–1188

Molecular Similarity

▶ Molecular Mimicry

Monoclonal Antibodies

GEORGE TREACY · DAVID M KNIGHT
Centocor Inc.
200 Great Valley Parkway
Malvern, PA 19355
USA

Synonyms
Antibodies, mAbs.

Definition
Monoclonal antibodies (mAbs) are populations of identical, monospecific antibodies with defined specificity and affinity for a target ▶ antigen. These antibodies are produced by the daughter cells of a single antibody-producing lymphocyte, often using an immortal ▶ hybridoma cell line grown in vitro. Monoclonal antibodies can also be constructed synthetically and produced as engineered recombinant proteins.

Characteristics
Monoclonal antibody technology was originally developed using a mouse system in which antigen-specific B cells from immunized mice were fused to an immortal cell line to generate monoclonal hybridomas secreting fully murine mAbs (1). The technology has evolved to enable genetic modification of the murine antibody genes resulting in hybrid mAbs containing human sequences. Methods have also been developed to generate human mAbs using transgenic mice or molecular display technologies.

Monoclonal antibodies exhibit the same general structural and functional characteristics of antibodies, and can be of any ▶ antibody class (IgG, IgM, IgD, IgE, IgA) or any ▶ antibody isotype within a class. The most commonly encountered mAbs are of the IgG class, and are tetrameric proteins consisting of two identical heavy chains and two identical light chains (see Figure 1).

Each IgG molecule contains two antigen combining sites formed by the *N*-terminal regions of the heavy and light chains, which determine the antigen-recognition and binding properties of the monoclonal antibody. The specificity of a given monoclonal antibody is determined by the precise amino acid sequences of the heavy and light protein chains in the antigen-combining region; the sequences of this region are therefore somewhat variable among monoclonal antibodies of different specificities. The antigen-combining region is also known as the variable region because of this sequence variation. The remainder of the molecule is relatively constant in sequence among different mAbs, and functions to allow interactions with other immune system components. This region of the mAb determines the class and isotype of the mAb, and is known as the constant region, or ▶ Fc region. Recombinant DNA technologies can be used to manipulate mAb sequences to produce mAbs with improved properties such as higher affinity, increased functional activity, and reduced ▶ immunogenicity for in vivo ap-

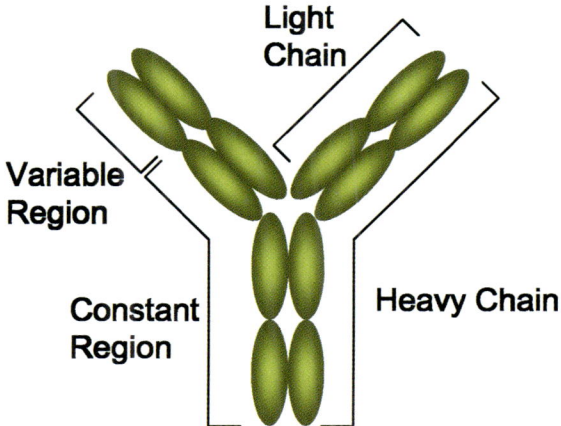

Monoclonal Antibodies. Figure 1 Schematic representation of a monoclonal IgG molecule.

plications (2). In addition to a high degree of specificity and affinity for the target antigen, mAbs also exhibit in vivo pharmacokinetic properties similar to naturally occurring antibodies.

Preclinical Relevance

The exquisite specificity and high affinity of monoclonal antibodies for their target antigens make them well suited for many applications. Monoclonal antibodies that block or neutralize the function of their targets can be useful to delineate biological pathways, and as therapeutics. Non-blocking mAbs that recognize their targets but do not inhibit their functions can be useful as assay and detection reagents. Highly specific and sensitive immunoassays utilizing mAbs are commonplace for the detection and quantification of a variety of types of molecules in biological and non-biological systems. Neutralizing mAbs have been used to great advantage both in vitro and in vivo to validate potential therapeutic targets by neutralizing the action of a specific target and analyzing the consequences. Target validation using mAbs in vivo is useful only for targets that are accessible to antibodies, i.e. cell surface or secreted molecules. Because of the high degree of specificity exhibited by mAbs, there is usually restricted cross-reactivity with target homologs from different species. For example, it is likely that a mAb that neutralizes a human protein target will not cross-react and neutralize the rodent homolog. For human antibody therapeutics, it is therefore often difficult or impossible to perform preclinical proof-of-concept efficacy studies in animals with the candidate mAb therapeutic. An additional "surrogate" or "analogous" mAb may have to be obtained that neutralizes the target in the desired species to facilitate preclinical studies. The mAb species cross-reactivity issue also impacts the ability to perform safety and toxicology studies for human antibody therapeutics, often restricting the species in which such studies can be performed to nonhuman primates.

Relevance to Humans

Monoclonal antibodies are well suited for diagnostic applications in clinical settings (3). They have been used to measure a variety of disease-related biomarkers in patient blood and other tissues and secretions as diagnostic and prognostic indicators of disease. Radioactively labeled mAbs have also been useful for diagnostic imaging of diseased tissue in patients. A variety of therapeutic mAbs are marketed (Table 1) and many others are in development. Marketed mAbs include engineered or modified mAbs, and in one case, a monoclonal antibody fragment. It was quickly recognized from early clinical experience with murine monoclonal antibodies that immune responses were likely to limit the therapeutic utility and increase safety risks inherent in these molecules. Several strategies have been developed to avoid these problems by increasing the percentage of human amino acid sequences in the therapeutic molecules under the reasonable assumption that immunogenicity can be reduced by mimicking natural human antibodies. Chimeric (murine variable regions fused to human constant regions) and humanized (murine hypervariable regions transplanted into human frameworks) antibodies are examples of the engineering of foreign proteins into a more human-like form. Of the 16 FDA-approved therapeutic monoclonal antibodies, 13 are either chimeric or humanized molecules, illustrating the success of this general approach. One of the advantages of mAbs as therapeutics is their favorable pharmacokinetics, which can lead to circulating half-lives of up to 3 weeks in humans. Such long clearance times may allow extended therapeutic effects with infrequent dosing.

Regulatory Environment

Regulatory guidelines for developing monoclonal antibodies (murine, human and engineered) for therapeutics and diagnostic use in humans have been issued by various regulatory agencies. In general, they are comparable among the European Union, Japan and United States. The approach for developing mAbs for human use is case-by-case based on science and relevance. The documents listed below are guidance documents and should not be considered regulatory requirements.

Relevant Guidelines

- ICH Harmonised Tripartite Guideline. S6 Preclinical Safety Evaluation of Biotechnology-Derived Pharmaceuticals, 1997 (http://www.ich.org.pdflCH/s6.pdf)
- US Food and Drug Administration (CBER). Points to Consider in the Manufacture and Testing of Monoclonal Antibody Products for Human Use, 1997 (http://www.fda.gov/cber/gdlns/ptc_mab.pdf)
- US Food and Drug Administration (CBER). Guidance for Industry: Monoclonal Antibodies Used as Reagents in Drug Manufacturing, 2001 (http://www.fda.gov/cber/gdlns/mab032901.pdf)
- European Medicines Evaluation Agency (EMEA) Guideline. Production and Quality Control of Monoclonal Antibodies, Directive 75/318/EEC, 1995 (http://www.q-one.com/guidance/emea.htm)

References

1. Kohler G, Milstein C (1975) Continuous cultures of fused cells secreting antibody of predefined specificity. Nature 256:495–497
2. Siegel D (2002) Recombinant monoclonal antibody technology. Transf Clin Biol 9:15–22

Monoclonal Antibodies. Table 1 Therapeutic monoclonal antibodies approved by the US Food and Drug Administration

Generic Name	Trade Name	Sponsor Company	Type	Approval Date
Muromonab-CD3	Orthoclone	Ortho Biotech	Murine	1986
Abciximab	ReoPro	Centocor	Chimeric Fab fragment	1994
Rituximab	Rituxan	Genentech	Chimeric	1997
Daclizumab	Zenapax	Hoffman-La Roche	Humanized	1997
Basiliximab	Simulect	Novartis	Chimeric	1998
Palivizumab	Synagis	MedImmune	Humanized	1998
Infliximab	Remicade	Centocor	Chimeric	1998
Trastuzumab	Herceptin	Genentech	Humanized	1998
Gemtuzumab Ozogamicin	Mylotarg	Wyeth-Ayerst	Humanized	2000
Alemtuzumab	Campath	Millennium/LEX	Humanized	2001
Ibritumomab Tiuxetan	Zevalin	IDEC	Murine	2002
Adalimumab	Humira	Abbott	Human	2002
Omalizumab	Xolair	Genetech	Humanized	2003
Efalizumab	Raptiva	Genetech	Humanized	2003
Bevacizumab	Avastin	Genetech	Humanized	2004
Cetuximab	Erbitux	Im Clone	Chimeric	2004

3. Khaw B (1999) Antibodies as delivery systems for diagnostic functions. Adv Drug Deliv Rev 37:63–80

Monoclonal Antibody (mAb)

Each individual B lymphocyte (and its descendants) produces exclusively antibodies with a unique specificity and affinity. By fusion with non-secreting myeloma cells they can be immortalized and thus provide an unlimited source of a single-specific—monoclonal—antibody.

▶ Humanized Monoclonal Antibodies
▶ Immunotoxicology of Biotechnology-Derived Pharmaceuticals
▶ Antibody-Dependent Cellular Cytotoxicity
▶ Antibodies, Antigenicity of
▶ Cytokine Inhibitors

Monocyte Chemoattractant Protein 1 (MCP-1)

Monocyte chemoattractant protein 1 (MCP-1) belongs to the group of CC (β) chemokines and binds to chemokine receptor CCR-2. It is also known as CCL2. Important in the migration of monocytic cells.

▶ Three-Dimensional Human Skin/Epidermal Models and Organotypic Human and Murine Skin Explant Systems
▶ Chemokines

Monocytes

Cells that develop from promonocytes in the bone marrow that enter the blood and circulate for a period of time before migrating into the tissues to become macrophages.

▶ Macrophage Activation

Monokine

Cytokine released from monocytes and macrophages. Now replaced by the general term cytokines.

▶ Cytokines

Mononuclear Cell Function

▶ Inflammatory Reactions, Acute Versus Chronic

Mononuclear Leukocyte

Immune cells that have nuclei with round (lymphocytes) or elliptical (macrophages) profiles.
▶ Lymphocytes

Mononuclear Phagocyte System (MPS)

A host-wide system of cells that are mononuclear (as opposed to polymorphonuclear), phagocytic, and have numerous other functions in both innate and adaptive immunity. They include blood monocytes, tissue macrophages, Kupffer cells, microglia, mesangial cells in the kidney, Langerhans cells in the skin, and dendritic cells throughout the body.
▶ Macrophage Activation

Monounsaturated Fatty Acids

These are fatty acids containing one double bond between a pair of adjacent carbon atoms.
▶ Fatty Acids and the Immune System

Morbilliform Eruptions

This form is probably the most usual manifestation of a drug eruption, usually occurring 1–2 weeks after beginning therapy. It is of note, however, that this eruption is often the initial presentation of more serious eruptions including Stevens-Johnson syndrome (SJS) and toxic epidermal necrolysis (TEN), and may evolve into generalized erythroderma with continued administration of the drugs. A viral exanthem is a matter for consideration in differential diagnosis, especially with respect to its early stage. This rash is usually symmetrical, consisting of erythematous macules and papules that may become confluent, and it often begins on the trunk or in areas of pressure or trauma.
▶ Drugs, Allergy to

Motif

A protein motif is a small structural domain comprised of a sequence of amino acids that can be found in a variety of proteins. Some motifs commonly found in proteins involved in intracellular signaling include, for example, ITAMs and SH2 and SH3 domains.
▶ Signal Transduction During Lymphocyte Activation

−/− Mouse

▶ Knockout, Genetic

Mouse Ear Swelling Test

SHAYNE COX GAD
Gad Consulting Service
102 Woodtrail Lane
Cary, NC 27511
USA

Synonyms
MEST.

Characteristics

The mouse ear swelling test (MEST) and variations on it were developed to overcome disadvantages inherent in traditional guinea pig-based tests (such as cost, length of test, a large amount of animal care space required, difficulty in assessing pigmented materials as antigens, and having a subjective endpoint).

Since Crowle formally proved that passive transfer of delayed-type contact hypersensitivity can be produced in the mouse in 1959, research immunologists have generated a wealth of information in attempts to understand the delayed-type hypersensitivity (DTH) re-

Test group Day 0	Control group Days 1, 3 and 5
1. Fur of abdomen is clipped	1. Fur of abdomen is clipped
2. ID Injection of FCA (Freund's complete adjuvant)	2. Topical application of substance or vehicle
3. Abdominal skin is tape stripped	3. Abdominal skin site is dried rapidly
4. Topical application of substance of vehicle	
5. Abdominal skin site is dried rapidly (electric dryer)	

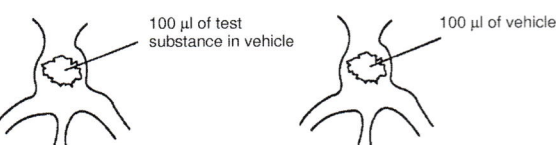

Mouse Ear Swelling Test. Figure 1 Details of mouse ear swelling test (MEST) challenge stage procedure.

sponse in this species. In particular, they have demonstrated that thymus-derived cells are necessary for inducing a DTH response. Also, the mouse has been used to investigate immunosuppressive properties of certain drugs, such as fluorinated steroids and corticosteroids. All of these have lead to the development of a formalized test procedure, the mouse ear swelling test (MEST), based on that the method as described by Gad et al. (1, 2) and Thorne (3) for evaluating test substances for their potential to cause dermal sensitization in mice. This method is shown diagrammatically in Figures 1–3. It evaluates contact sensitization by quantitatively measuring mouse ear thickness.

Procedure

- 6–8-week-old female mice (e.g. CF-1 or BALB/c) are used. They are observed for at least 1 week before the start of the study to detect any signs of illness. Any mouse displaying redness of either ear prior to the start of a test should be replaced.
- Mice, which have been randomly placed in cages upon arrival, are assigned to groups. Each test substance is investigated in a pretest group of at least 8 mice, a test group of at least 15 mice, and a control group of at least 10.
- If animals are not individually marked they should always be handled one at a time when each phase of this procedure is performed. The following procedure is conducted to prevent mixing animals during each phase (e.g. shaving, intradermal injections, tape-stripping, and dosing): all mice are removed from their original cage and placed in an empty cage for holding. One mouse is removed from the holding cage at a time, the phase activity is performed, and then the mouse is returned to its original cage. This step is repeated for each of the remaining mice in the holding cage.

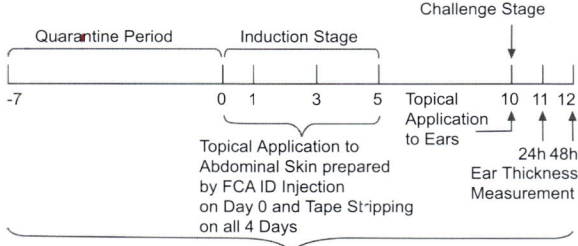

Mouse Ear Swelling Test. Figure 2 Optimal study design of the mouse ear swelling test (MEST).

Equipment

Before initiating a study the following materials must be on hand:
- Oditest Model D-1000 (Dyer) thickness gauge or similar
- small animal clipper (Oster) with a No. 40 blade
- microliter syringe
- glass tuberculin syringe
- Dermaclear (3M) tape or similar
- anesthesia facilities for mice
- some 30-gauge needles (25 mm)
- Freund's complete adjuvant (FCA).

Pretest
1 week before testing
- For 1 week prior to initiation of testing, animals are fed a diet enhanced with vitamin A acetate at 250 IU/g of feed.
- A dermal (abdomen and ear) irritation and toxicity probe study is conducted 1 week prior to the actual MEST in order to establish the maximum concentration of test substance that produces minimal irritation to the abdomen (belly) region after a single topical application of each of 4 days (if the substance does have potential to irritate skin) and to establish a concentration of test substance that is nonirritating to the ear after a single topical application. Also, dose levels of the test substance that produce systemic toxicity can be identified during the pretest (and subsequently avoided).
- The test substance is diluted, emulsified, or suspended in a suitable vehicle. A vehicle (such as acetone, 70% ethanol, 25% ethanol, or methyl ethyl ketone) is selected that is able to solubilize the test substance and be volatile.
- Two mice from the pretest group are used to test each concentration of test substance. As many as four concentrations can be evaluated. The mice used for belly irritation are also used for ear irritation testing. Levels that are irritating to one site (belly or ear) may not be to the other.

Day 0
- On day 0, the first day of the pretest, each animal is prepared by clipping the hair from the belly region using a small animal clipper with a size No. 40 blade.
- After clipping the belly, the outer layers of epidermis (stratum corneum) of each mouse are removed from the shaved belly region with a tacky transparent tape (25 mm) as Dermaclear. This procedure is referred to as "tape-stripping." It is not painful and no anesthetic is required. On day 0, the belly skin of each mouse is tape-stripped until the application region appears shiny. While an assistant supports the dorsal portion of the mouse, the tape is pressed

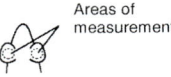

Mouse Ear Swelling Test. Figure 3 Details of the mouse ear swelling test (MEST) induction stage procedure.

firmly over the clipped belly region and quickly removed; this procedure is repeated as many times as needed until the skin appears glossy.
- After tape-stripping the belly, 100 μl of solvent containing test substance is applied to the belly region using a microliter pipette. At the same time the test substance is applied to the ventral surface (10 μl) and dorsal surface (10 μl) of the left ear of the mouse using a microliter pipette.

Day 1

- On day 1 (24 h after dosing the ears) the thickness of all probe animal ears is measured using an Oditest meter.
- Ether is used to anesthetize the mice in a fume hood while the ears are measured.
- When a mouse reaches the "surgical anesthesia" stage, it is removed from the jar and gently placed on the countertop of the fume hood, which is prepared with a protective lining.
- While supporting the mouse with one hand, the other hand is used to press the finger lever on the Oditest gauge in order to open the flat measurement contacts. One ear of the mouse is then inserted between the contacts until it is positioned with approximately 1–2 mm of the outer edge of the ear showing. After positioning the ear, the finger lever is released to allow the contacts to clamp onto the ear. The measurement is read from the gauge after the indicator needle is stabilized. If desired, one or two more measurements can be rapidly made to be certain of a reproducible reading.
- Once a reading is obtained, the contralateral ear is measured in the same manner. The animal's body is turned over in order to position the other ear for measurement.
- All measurements are recorded.

Days 1, 2, 3

- On subsequent days (days 1, 2, 3) the belly region is first tape-stripped until shiny, and then 100 μl of test substance is applied topically to the belly region using a microliter pipette.
- On day 4 (24 h after the last topical application) the belly skin of all animals is observed for dermal irritation and a description of the results is recorded.
- Any signs of systemic toxicity are observed on any of the pretest days should be noted.

Based on the results of the pretest data, a judgment is made as to which concentration will be used for topical induction applications to the belly and for topical challenge application to the ear. A minimal or mildly irritating concentration is preferable for induction so that the potential for achieving sensitization is maximal without harming or compromising the test animal. The highest nonirritating concentration identified is used for challenge application to maximize responsiveness while avoiding the possibility of having a confounding irritation response.

Main Study

Day 0

- Use 15 mice in the test and 10 mice in the control group. The belly of each mouse in the test and control groups is clipped free of hair.
- Immediately after clipping, two intradermal injections of Freund's complete adjuvant (FCA) emulsion are made at separate sites in the skin of the shaved belly (each site flanks the ventral midline). Approximately 20 μl of FCA emulsion is injected with a glass tuberculin syringe with a 30-gauge needle attached. Injections are performed in test and control mice.
- Following the intradermal injections, the belly skin of the test and control group animals is tape-stripped until the site gives a shiny appearance.
- After tape-stripping the belly, 100 μl of test substance (at a concentration determined by pretest) is topically applied to the belly skin of test group animals with a microliter pipette. Control animals receive a dose of 100 μl of vehicle. If greater certainty of identifying weak sensitizers is desired, an additional group of 15 mice is dosed with a concentration one-third of that identified in the pretest. This procedure avoids false-negative results by being in the downregulated response region (3).

Days 1, 3, 5

- The skin of the belly of test and control group animals is tape-stripped until shiny in appearance.
- After tape-stripping 100 μl of test substance is topically applied to the belly skin of test group animals and 100 μl of vehicle-solvent is topically applied to the control group animals.

Challenge Stage
Day 10
- Each test group mouse and all five control group mice are dosed with 20 µl of a concentration of test substance (determined by the data) on both the ventral and dorsal surfaces of the left ear. The contralateral (right) ears are dosed with 20 µl of 100% vehicle on the ventral and dorsal sides.

Day 11
- Ear thickness is measured 24 h after challenge dosing using the procedure as described on day 1.

Day 12
- Measurements of thickness are repeated 48 h after challenge dosing.

Rechallenge
Day 17
- If the test substance is judged to be a nonsensitizing agent after the first challenge application, or it causes dermal sensitization in only a few animals, or causes ear swelling that is weak or questionable, then a second and final challenge application should be performed on each test animal on day 17.
- The five control group mice from the first challenge are not rechallenged because they have been exposed to the test substance and are no longer true negative controls. The five remaining naive control animals (not used for the first challenge) are challenged for comparison with the test group.
- The procedure used for the first challenge application is used for the repeated challenge. Either the same concentration or a new concentration (higher or lower) of test substance may be used, depending on the results of the first challenge.

Days 18, 19
- Measurement of the thickness of both ears is performed on days 18 and 19 (24 h and 48 h after rechallenge, respectively). Measurement are recorded.

Interpretation of Results
Judgment concerning the presence or absence of sensitization is made for each animal. The judgment is based on the percentage difference (%) between test and control ears. A "positive" sensitization response is considered to have occurred if the test ear of one or more animals is at least 20% thicker than the control ear. This effect criterion is selected because it guarantees a level of false-positives of less than 1 in 1000 (2). The percentage of animals in a test group that is considered "positive" is then calculated and recorded as percentage responders.
Ear thickness measurements of the negative control group are used to identify any possible dermal irritation reactions, which would be interpreted as false-positive dermal sensitization responses.
In addition, the percentage of ear swelling is calculated for the test group. Measurements of the left (A) and right (B) ear thickness are added. Percentage ear swelling is calculated by the sum of A (test ear thickness) divided by the sum of B (control ear thickness) multiplied by 100, as in:

% Ear swelling = (A / B) × 100

When a second challenge application is performed, the data from both challenges are compared. If neither challenge procedure produces a positive sensitization reaction or both produce equivalent responses, the classification of the test substance is based on both challenge applications. If one challenge application (whether it is the first or second) produces a greater number of positive dermal reactions than the other, the classification of the test substance is based on the challenge with the most positive responses.
One or more unequivocally positive responses (20% or greater swelling compared to the control ear) in a group of 15 animals should be considered as positive result. A negative, equivocal, or single response indicates that a substance is not a moderate or strong sensitizer.

Pros and Cons
The MEST offers distinct advantages over the guinea pig dermal sensitization procedures:
- mice are far less expensive
- less vivarium space is required
- the duration of the test is shorter
- less test substance is used
- overall cost is significantly less
- the test is objective
- materials that stain the skin may be easily evaluated (several test materials are colored and difficult to evaluate by existing guinea pig methods)
- there is a low false-negative rate and no false-positive rate (if properly performed) (2,3)
- the test is more accurate for predicting relative hazard to humans (4).

There are, however, some disadvantages:
- the data base (though not small) is not as extensive as those for the guinea pig maximization test (GPMT) or Buehler tests
- fewer people have experience with this system
- although this test is very robust because it detects strong sensitizers and does not generate false positives, care must be taken with the technique; the study design is important for identifying moderate and weak sensitizers
- the test is included in both OECD and ICH/FDA immunotoxicity guidance but is not fully described.

Test System Enhancements

As with all other dermal sensitization procedures, increasing percutaneous absorption will increase the sensitivity. Factors that will enhance absorption (and techniques for achieving them) include the following:
- increased surface area of solids
- hydration of the area of skin exposed to chemical (by wetting of solids and using occlusive wrappings)
- irritation of induction application site (with 1% sodium lauryl sulfate in water)
- abrasion of application site (not to be done in combination with irritation of site)
- injection of test material (if possible)
- proper selection of solvent or suspending system
- removal of part of all of the "barrier layer" (stratum corneum) by tape-stripping
- increased number of induction applications; while not enhancing percutaneous absorption as such, mild stimulation of the immune system of test animals (e.g. by injecting FCA or another adjuvant alone, or FCA blended with the test material) increases responsiveness of the test system.

It is generally believed that using the highest possible concentration of test material (mildly irritating for induction; just below irritating for challenge) will guarantee the greatest possible sensitization response—and will therefore also serve to universally increase sensitivity. There are reports, however, that this is not true for all compounds, and that a design involving multiple doses (two or more concentrations) would increase sensitivity. Such designs, however, significantly increase costs.

Predictivity

As with all biologically based test systems, some aspects of the MEST do not always meet the needs of all specific situations and individuals, and some aspects could be improved. The procedures described above aim to provide a set of tools to deal with most variations in needs to perform the test.

Others have tried methods using incorporation of radioisotopes, in which radioactivity levels in the ear (the challenge site) are measured as a "more objective" or more sensitive indication of sensitization. But such methods have limitations: dealing with the radioactivity; and an inability to rechallenge animals. The method has not been shown uniformly to improve performance characteristics. A combination of the two techniques (ear thickness and isotope measurements) may, however, be powerful for detecting and characterizing weak sensitizers.

The original MEST design was not found uniformly to identify weak and moderate sensitizers, but the method presented here incorporates improvements that have been developed over several years, and these improvements significantly strengthen the test's performance.

Relevance to Humans

Most of the general problems associated with the currently used predictive test systems (for delayed contact sensitization) revolve around the question: What do the results mean in terms of hazard to people?

There are problems for a number of reasons, but two major problems arise from two facts. First, as a population humans exhibit greater variability in sensitivity than this animal test system; in trying to reduce this gap the current test systems do not give a true prediction of relative hazard in people—what portion of the human population will be sensitized, and how easily?. Second, what is evaluated in these models? It may be a mixture, such as a cosmetic. And if a chemical is found to be a sensitizer, we may be concerned about structurally related compounds that evoke a response in those already sensitized to the tested compound (that is, there may be cross-sensitization).

Once we have animal sensitization test data, we must relate these to potential hazards in humans. At one end of the scale, a negative finding does not guarantee that a material will not be a sensitizer in humans, although most investigators would expect it to be only a weak or mild sensitizer. On the other hand, it is not so easy to determine how significant it is if a material is found to be a strong or extreme sensitizer in each of these assays.

There are two options. The first involves human patch tests, performed in test groups that are large enough (100–200 people) and varied enough (e.g. different ages and skin types) to be representative of the population to be exposed. These results will indicate what to expect in humans, but the approach is expensive and has both ethical and liability concerns.

The second option is to use methodology that allows evaluation of potency in a human model. Such evaluations require dose-response testing, and a number of considerations should be taken into account.

Potency

A key issue in evaluating the potential hazard of agents inducing delayed contact hypersensitivity is being able to categorize their potency. As described earlier (4) a method for utilizing data from virtually any of the existing test systems, including the MEST, has previously been described. The method leads to calculation of a potency index (PI) which can be used to classify the potency of positive agents. A classification scheme based on the resulting index was devised, as follows:
- Class I PI > 4.0 Severe
- Class II 4.0 > PI ≥ 3.0 Strong
- Class III 3.0 > PI ≥ 2.0 Moderate

Mouse Ear Swelling Test. Table 1 Mouse Ear Swelling Test. Representative MEST test results

Chemical	MEST results Sensitized	Swelling	Class
Oxazolone	100	134	Severe
Toluene diisocyanate	100	142	Severe
Dinitrofluorobenzene (DNFB)	100	168	Severe
N, N-Dimethyl-p-nitrosoaniline	100	158	Severe
Picryl chloride	90	130	Severe
Dinitrochlorobenzene (DNCB)	80	130	Severe
p-Phenylenediamine	67	109	Strong
HMDI	67	139	Moderate
Glutaraldehyde	67	125	Strong
Dansyl chloride	60	124	Strong
Nickel sulfate	38	118	Moderate
Methyl methacrylate	44	118	Mild
Eugenol	42	119	Mild
Hexamethylenimine	40	106	Strong
Potassium dichromate	40	114	Moderate
Methyl ethyl ketoxamine	40	120	Moderate

Interestingly, ECETOC recently published a scheme which yields very similar results but is based on four categories (weak, moderate, strong and extreme). The ECETOC report (5) also contains a comparison between results obtained in local lymph node assays (LLNA) and maximization assays (GPMT).

- Class IV $2.0 > PI \geq 1.0$ Mild
- Class V $1.0 > PI \geq 0$ Weak or questionable

Values previously obtained from this classification scheme are also assigned below.
Interestingly, ECETOC (5) recently published a scheme yielding very similar results, which is based on four categories (weak, moderate, strong, and extreme). The ECETOC report also contains a comparison of results obtained in local lymph node assays (LLNA) and maximization assays (GPMT).

Cross-sensitization

A frequent situation is that one member of a structural series will evoke a positive response in individuals that have been sensitized; we call this broader response "cross-sensitization." This situation occurs because the structures of these materials complexed with a protein are not distinguished as different by the "educated" surveillance lymphocytes.
Any of the animal tests described here can be modified to see if cross-sensitization occurs among members of a series. The test is conducted with multiple groups of animals; those animals that are successfully sensitized are then rechallenged with other members of the class.

Mixtures

Mixtures become a particular problem in sensitization testing because frequently we are called upon first to evaluate a complex mixture in an animal test system; then, if it is found to be a sensitizer, we are called upon to determine which component is the cause of the positive response. If such a component can be identified, it is often possible to reformulate the mixture, thus serving the desired need without the problem component.
Such components can be identified by continued testing in a set of animals that have been previously sensitized to the mixture as a whole. Groups of positively sensitized animals are rechallenged with separate samples of different suspect components to identify which one evokes a positive response. The guinea pig methods offer an advantage here, in that multiple components may be simultaneously evaluated on different sites of the same animal.
The MEST provides a basic test paradigm which has a wide range of applications. Reviewing these in depth is beyond the scope of this chapter, but it has been put to a number of uses, and modified for predicting hazards to humans, thus:

- the basic design can also be used with rats and guinea pigs, or the ear challenge and swelling mea-

surement can be used in guinea pig assays for pigmented materials
- it can be used as a model for mechanisms of chemically induced dermal sensitization
- it can be used as a photosensitization model (6)
- it can be used to screen for suppression of T-cell modulated immune responses
- it provides a tool for studying tumor promotion mechanisms.

The Regulatory Environment

The MEST is currently listed as an accepted test system under both ICH and OECD guidance. While less commonly employed than the LLNA, GPMT, or Buehler test systems, data generated using the MEST are accepted by the FDA and under OECD.

References

1. Gad SC, Dunn BJ, Dobb DW (1985) Development of an alternative dermal sensitization test: Mouse Ear Swelling Test (MEST) In: Goldberg AM (ed) In Vitro Toxicology. Proceedings of 1984 Johns Hopkins Symposium, pp 539–551
2. Gad SC, Dunn BJ, Dobbs DW, Reilly C, Walsh RD (1986) Development and validation of an alternative dermal sensitization test. The mouse ear swelling test (MEST). Toxicol Appl Pharmacol 84:93–114
3. Thorne PS, Hawk C, Kaliszewski SD, Guiney PH (1991) The noninvasive mouse ear swelling assay: Refinements for detecting weak contact sensitizers. Fundam Appl Toxicol 17:790–806
4. Gad SC (1988) A scheme for the ranking and prediction of relative potencies of dermal sensitizers, based on data from several test systems. Appl Toxicol 8:301–312
5. ECETOC (2003) Contact Sensitization: Classification According to Potency. Technical Report 87 (04/03)
6. Neumann NJ, Homey B, Vohr HW, Lehmann P (2000) Methods for testing the phototoxicity and photosensitization of drugs. In: Kydonieus AF, Wille JJ (eds) Biocehmical Modulation of Skin Reactions. CRC Press, Boca Raton, pp 65–80

Mouse Immune System

▶ Rodent Immune System, Development of the

MSC

▶ Suppressor Cells

MSE

▶ Three-Dimensional Human Skin/Epidermal Models and Organotypic Human and Murine Skin Explant Systems

MTT Conversion (MTT Test)

The MTT test is commonly used for the determination of cell viability in ex vivo/in vitro test procedures. It is based on a redox reaction that takes place in every cell with detectable mitochondrial activity. In living cells —also in early apoptotic cells—the slightly yellow MTT reagent (3-(4,5-dimethylthiazole-2-yl)-2,5-diphenyl tetrazolium bromide) is converted to blue/violet formazan crystals. Cells that are hardly affected and dead cells lack the capacity to reduce the MTT reagent. Once extracted from the living cells by an organic solvent like isopropanol, the blue formazan solution can be determined by measurement of absorption (550–570 nm with 620 nm reference).

▶ Three-Dimensional Human Skin/Epidermal Models and Organotypic Human and Murine Skin Explant Systems

Mucociliary

The tracheobronchial region is lined with epithelial cells and is coated with mucus. The beating of cilia present in the tracheobronchial tree moves this mucus layer upwards, serving as an important clearance mechanism for deposited particles that are then carried to the oral cavity where they are swallowed or excreted.

▶ Respiratory Infections

Mucosa

The epithelial lining of body systems that communicates with the external environment (e.g. the digestive, genitourinary, and respiratory tracts).

▶ Lymphocytes

Mucosa-Associated Lymphoid Tissue

ROSANA SCHAFER · CHRISTOPHER CUFF
Dept. Microbiology, Immunology, and Cell Biology
West Virginia University
2095 Health Sciences North
P.O. Box 9177
Morgantown, WV 26506-9177
USA

Synonyms
Secretory immune system

Definition
Mucosal immunity protects non-keratinized or 'wet' epithelium in the gastrointestinal, respiratory, and genitourinary tracts, as well as the eye and mammary gland. Mucosal immunity includes innate humoral and cellular effectors, and both T cell mediated and B cell mediated adaptive immune responses. The hallmark feature of mucosal immunity is the predominant production of antigen-specific immunoglobulin A (IgA) following exposure of antigen at mucosal sites. Several other features of mucosal immunity distinguish it from systemic immunity including organization and distribution of lymphoid tissues and cells, activation states of inductive and effector sites, and the presence of unique populations of lymphocytes. Understanding unique properties of mucosal immunity contributes to rational immunotherapy for infectious and autoimmune diseases.

Characteristics
▶ Mucosa associated lymphoid tissue(MALT) includes secondary lymphoid tissues in the intestine (referred to as gut associated lymphoid tissue (GALT), the respiratory tract (▶ bronchus associated lymphoid tissue (BALT) and ▶ nasal associated lymphoid tissue (NALT)), the genitourinary tract (from urethra to the ovaries and bladder or testes and bladder), the mammary gland, and the eye. In general, mucosal lymphoid tissue is described as an inductive or effector tissue. Inductive sites include local lymph nodes or lymph node-like structures where antigen activation of specific lymphocytes is initiated. These activated lymphocytes differentiate and migrate to effector sites, which include the region just below the epithelial lining (the lamina propria) and within the epithelial lining. Both B and T cells generally populate the lamina propria, while the epithelium is usually highly enriched in T cells, and most commonly ▶ $CD8^+$ T cells.

Intestinal Lymphoid Tissue
The intestinal tract is the most well characterized of the various MALT sites. Lymphocytes in the gut are found in the ▶ Peyer's patches, the ▶ lamina propria, and the epithelium (1).

Peyer's Patches
Aggregates of lymphoid cells in organized structures in the intestine are referred to as Peyer's patches. These structures make up the inductive sites of the intestinal immune response where antigens first contact the immune system and mucosal immune responses are initiated. Peyer's patches are similar to lymph nodes in the periphery, but they differ in at least four important features:

- First, Peyer's patches are chronically stimulated with antigens that pass through the intestine and thus normally contain secondary lymphoid follicles characterized by the presence of germinal centers.
- Second, in contrast to peripheral lymph nodes, B cells—not T cells—are the predominant lymphocyte population.
- Third, Peyer's patches do not have afferent lymphatics that drain surrounding tissues. Rather, they are covered by specialized epithelium that contains microfold (M) cells that sample intestinal antigens.
- Finally, and perhaps most importantly, Peyer's patches are enriched in B cells that give rise to IgA-producing plasma cells following antigen stimulation.

The mechanism of preferential immunoglobulin class switching to IgA in the Peyer's patches is not entirely understood, but is likely an outcome of unique environmental factors such as production of transforming growth factor-β (▶ TGF-β), and perhaps specialized accessory cells that provide unique signals to antigen-activated B cells. Following antigen stimulation, lymphocytes leave the Peyer's patches via efferent lymphatics and migrate to the mesenteric lymph node. From there leukocytes migrate to the blood stream via the common thoracic duct and travel through the blood to effector sites in the intestine and other mucosal lymphoid tissues.

Additional aggregates of inductive lymphoid tissue appear to be present in the distal end of the gastrointestinal tract. Rectal immunization appears to induce significant mucosal immune responses, particularly in the genitourinary tract.

Lamina Propria
The lamina propria contains a variety of cells of hematopoietic origin including IgA-secreting plasma cells and both $CD4^+$ T helper cells and $CD8^+$ cytotoxic T cells. In addition, the lamina propria contains various accessory cell populations such as dendritic cells and macrophages. The lamina propria is considered an effector site of the GALT, where activated cells mi-

grate from the blood stream and mediate their effector function. Indeed, a hallmark of lamina propria lymphocytes is that they typically bear markers associated with cell activation such as ▶ CD45RO. IgA-producing plasma cells secrete dimeric IgA that is then bound by polymeric immunoglobulin receptor (pIgR) expressed by absorptive epithelial cells, and translocated into secretions where it protects mucosal surfaces, mainly by blocking adherence of pathogens or toxins, but may also act intracellularly during translocation. T helper and cytotoxic cells mediate typical functions such as cytokine production and cytotoxicity. However, in the healthy host their function appears to be tightly regulated to prevent deleterious response to antigens associated with normal flora or food antigens. It has been suggested that regulatory T cells that produce interleukin (IL)-10, TGF-b, or other regulatory cytokines, help control potentially pathologic responses.

Additional subpopulations of organized lymphoid aggregates called cryptopatches are found in the lamina propria. The function and significance of the cryptopatches is not well understood. It is thought that they contain lymphopoietic stem cells and thus contribute to the pool of intestinal lymphocytes.

Epithelium

Interdigitated between epithelial cells of many mucosal sites is a population of T cells described as ▶ intraepithelial lymphocytes (IELs). These are heterogeneous in terms of phenotype and function, and there is considerable variation among different species and anatomic locations. Although heterogeneous, 70%–90% of intestinal IELs are $CD8^+$, and most of them express the conventional heterodimeric form $CD8\alpha\beta$. However, a substantial fraction of intestinal IELs express the unconventional homodimeric $\alpha\alpha$ form of CD8. Most IELs are T cell receptor (TCR) $\alpha\beta^+$. There is also a significant fraction of $TCR\gamma\delta^+$ cells that variably increase in quantity from 2% to 10% in the duodenum, and up to 40% in the large intestine. While TCR $\alpha\beta^+$ IELs can express any of the CD8 and CD4 phenotypes, the most prevalent population is the conventional $CD8\alpha\beta^+$ $TCR\alpha\beta^+$ IELs. Of $TCR\gamma\delta^+$ IELs, up to approximately 60% express the homodimeric $CD8\alpha\alpha$, with the remainder being $CD4^-$ $CD8^-$.

The distinct phenotypes may correlate with ontogeny. Numerous studies undertaken in animals have suggested that $CD8\alpha\alpha^+$ $TCR\alpha\beta^+$ and $TCR\gamma\delta^+$ IELs may be derived from extrathymic or at least unconventional thymus-dependent maturation pathways. This is less clear in humans. Distinct phenotypes appear to correlate with function and specificity. IELs with the conventional $CD8\alpha\beta^+$ $TCR\alpha\beta^+$ phenotype contain conventional antigen-specific cytotoxic T cells that are major histocompatability complex (MHC) class I restricted and mediate effector cytotoxic function by perforin and Fas ligand (FasL) mechanisms. They also produce cytokines such as interferon(IFN)-γ, IL-8, and tumor necrosis factor-(TNF)-α and chemokines such as macrophage inflammatory protein(MIP)-1β. $CD8\alpha\alpha^+$ $TCR\alpha\beta^+$ IELs and $TCR\gamma\delta^+$ IELs also have the ability to mediate cytotoxicity and secrete IFN-γ, although both subsets have been shown to recognize alternative nonclassical MHC molecules on intestinal epithelial cells such as CD1d.

In addition to cytolysis, IELs influence epithelial barrier integrity. IELs have been shown to produce keratinocyte growth factor (KGF) that can aid in repair of damaged epithelium. Thus, different IEL subpopulations function to maintain normal epithelium growth homeostasis, nonspecifically prevent infection of the epithelium, specifically resolve infection and guard against reinfection, resolve inflammation, and repair epithelial damage.

Respiratory Lymphoid Tissue

The main compartments of the mucosal immune system of the respiratory tract are the NALT in the upper respiratory tract, and the BALT in the lower respiratory tract.

NALT

NALT is the initial site of defense against inhaled antigens and can lead to the generation of an immune response in the respiratory tract and a generalized response at other mucosal sites such as the gut (2). It is an important inductive site of the upper respiratory immune response that induces a predominant IgA antibody response after infection. In humans, ▶ Waldeyer's ring is a ring of lymphoid tissue that consists of the lingual tonsil, palatine tonsils, and nasopharyngeal tonsils (adenoids). NALT in rodents is composed of a paired organ that lies beneath the nasal epithelium of the nasal floor and is considered the functional equivalent of Waldeyer's ring. It has a well-organized structure similar to that of Peyer's patches described above. The nasal mucosa is drained by the superficial cervical lymph nodes that subsequently drain to the posterior cervical lymph nodes.

BALT

BALT has been best characterized in rabbits where it is comprised of organized lymphoid follicles and a specialized epithelium as described for GALT. However, the presence of BALT in other species such as rats and humans is variable as to its structure, organization, and number, therefore the exact function and role of BALT is not completely understood (3). Recent studies in rodents have suggested that BALT may not be an inductive site of a mucosal immune response in some species and that the predominant IgG response in the lower respiratory tract after infection is a result of the systemic immune response.

Genitourinary Tract

Organization and function of the mucosal immune system along the genitourinary tract is not well understood. There are significant anatomic differences among species, and the classic description of inductive and effector sites is probably less relevant at this site. There does not appear to be inductive lymphoid tissue in the genitourinary tract. Rather, intestinal immunization (particularly rectal immunization) appears to be most effective at inducing effector lymphocytes in the genitourinary tract. In human males, T cells and macrophages can be found associated with the epithelium that lines the male reproductive tract. Plasma cells are localized mainly to the penile urethra. In females, immune cells are relatively sparse, but T cells, dendritic cells and macrophages are found in the vaginal and uterine epithelium. Although few plasma cells are located in the noninflamed female reproductive tract, both IgG and secretory IgA is found in secretions.

Preclinical Relevance

Most potentially immunotoxic chemicals enter the body through mucosal surfaces and thus may affect mucosal immune responses. In particular, the respiratory tract and gastrointestinal tract are sites of exposure and absorption of immunotoxicants in the environment. However, with only a few exceptions, most assessments of immunotoxicity focus on analysis of peripheral lymphoid cells and tissues either from peripheral blood or spleen cells.

Relevance to Humans

Two issues of medical importance support the need for further study of mucosal immunity. First, many if not most pathogens invade through mucosal surfaces and thus mucosal immunity provides an important initial barrier to invasion. Therefore, development of vaccines that induce mucosal immunity is a high priority in vaccine research. Second, in many mucosal regions of the body such as the intestine, immunologic inductive and effector tissues are in close communication with a heavy antigenic load. How the mucosal immune system maintains a state of immunologic responsiveness to pathogenic molecules yet appears under normal circumstances to largely ignore nonpathogenic microbes and food antigens could provide approaches to treat autoimmune disease.

Acknowledgement

Research in the laboratory of CFC is supported by grants AI34544, RR16440, and T32-ES010953 from the National Institutes of Health. Research in the laboratory of RS is supported by grants ES07460 and T32-ES010953 from the National Institutes of Health.

References

1. Mowat AM, Viney JL (1995) The anatomical basis of intestinal immunity. Immunol Rev 156:146–166
2. Sminia T, Kraal G (1999) Nasal-asssociated lymphoid tissue. In: Ogra, P, Mestecky J, Lamm ME, Strober W, Bienenstock J, McGhee JR (eds) Mucosal Immunology. Academic Press, San Diego, pp 357–364
3. Pabst R (1992) Is BALT a major component of the human lung immune system? Immunol Today 13:119–122

Mucosal Mast Cells

▶ Mast Cells

Mucositis

▶ Oral Mucositis and Immunotoxicology

Multicomponent Enzyme Complexes

Multicomponent enzyme complexes are the backbone elements of the hemostasis system and show a composition principally consisting of several building blocks— enzyme, cofactor, substrate (pro-enzyme)— all bound and assembled on (activated platelet) surfaces. The cofactor in each complex determines its enzymatic efficiency and thereby contributes to the overall amplification of the system finally leading to thrombin generation.
▶ Blood Coagulation

Multiple Sclerosis

A potentially disabling neurologic disease due to damages in the myelin sheaths in the central nervous system characterized by various neurological features of disparate nature both in locations and time of appearance.
▶ Molecular Mimicry

Multiplex

Any number of technologies that detect or quantify several analytes simultaneously in a single sample.

▶ Cytokine Assays

Multiplicity

Multiplicity is the frequency of a specific cell type in a cell mixture.
▶ Limiting Dilution Analysis

Multipotential Stem Cell

The highest echelon stem cell of marrow, capable of spawning pluripotential precursors of all myeloid and lymphoid lineages.
▶ Colony-Forming Unit Assay: Methods and Implications

Multisystem Autoimmunity

▶ Systemic Autoimmunity

Murine Immune System

▶ Rodent Immune System, Development of the

Muteins

Mutation of proteins can alter their biological properties. The term mutein is mostly used for mutated cytokines which behave as receptor antagonists and thereby can inhibit cytokines.
▶ Cytokine Inhibitors

Myalgia

Muscle pain.
▶ Systemic Autoimmunity

Mycotoxins

Toxic compounds produced by certain fungi.

▶ Respiratory Infections

Myeloid Differentiation Factor 88

Myeloid differentiation factor(MyD)-88 is an adaptor molecule of the IL-1 signaling pathway.
▶ Interleukin-1β (IL-1β)

Myeloid Suppressor Cells

▶ Suppressor Cells

Myeloperoxidase

An enzyme found in neutrophils that catalyses oxidations by hydrogen peroxide. Myeloperoxidase activity promotes killing of microbes and tumor cells, inactivation of chemotactic factors, and cross-linking and iodination of proteins.
▶ Chemotaxis of Neutrophils

Myocardial Infarction

A type of heart attack caused by loss of blood supply to the heart due to complete occlusion (blockage) of a coronary artery, which leads to death of heart muscle as a result of lack of oxygen.
▶ Fatty Acids and the Immune System

Myocarditis

Inflammation of the heart muscle.
▶ Cardiac Disease, Autoimmune

Myosin

A large protein involved in muscle contraction.
▶ Cardiac Disease, Autoimmune

Naive Cell

A mature cell (especially a lymphocyte) that is capable of performing its full range of functions but which has yet to be stimulated to respond to a specific antigen.
▶ Lymphocytes

Naive T Cell

Mature T cell that has not been activated before.
▶ Antigen Presentation via MHC Class II Molecules

Nasal-Associated Lymphoid Tissue

Nasal-associated lymphoid tissue (NALT) refers to secondary lymphoid tissue in the respiratory tract.
▶ Mucosa-Associated Lymphoid Tissue

Natural Antibodies

▶ Autoantigens

Natural Antibody

An antibody produced without apparent antigenic stimulation.
▶ Autoantigens

Natural Killer Cell Assay

KARIN CEDERBRANT
Safety Assessment
Astra Zeneca
S-151 85 Södertälje
Sweden

Synonyms
Natural killer ^{51}Cr release assay, non-radioactive flow cytometric analysis of NK cell cytotoxicity.

Short Description
The described tests are used as in vitro or ex vivo assays to measure cytotoxic activity of effector cells —here, natural killer (NK) cells—from peripheral blood or spleen. In immunotoxicologic investigations these assays are intended for detection of possible alterations in NK cell function as an effect of exposure to drugs and chemicals.

Characteristics
Principle
The principle of the tests is based on co-cultivation of effector cells (E) and target cells (T) followed by determination of the proportion of dead target cells. The E:T ratios used in the assays normally vary between 25:1 and 200:1.

Target Cells and Mechanism of Killing
NK cells play a central role in the innate immune defence against intracellularly infected cells and tumours, without possessing any memory cell function based on prior antigen exposure. NK cells kill their targets either by the well known secretory/necrotic cytotoxic mechanism, associated with perforin-mediated killing when eliminating rare leukemic cell lines ex vivo, or by non-secretory/apoptotic mechanisms, associated with killing of solid tumour cells ex vivo. The described NK assays measure NK cell killing of the first type using a lymphoid leukemic cell line, YAC-1, as a target for rat or mice NK cells, or a myeloid leukemic cell line, K562, as a target for human NK cells. To obtain a high sensitivity of the assays target cells

need to be young and in the log phase of growth for efficient lysing.

Effector Cells

There are primarily two sources of effector cells: peripheral blood and spleen. The effector cells are prepared either as a mononuclear cell suspension by Ficoll separation of heparinized peripheral blood or by single-cell preparation of the spleen. For spleen cells, mechanical disaggregation is preferred since enzymatic treatment procedures might interfere with cell function. Effector cells should be prepared within 24 hours (preferably immediately) after sampling to minimize possible risks of secondary inhibitory effects on the cytolytic activity. It is recommended that effector cells are kept at room temperature, not being refrigerated or cryopreserved, because such treatment will cause loss in activity. All lymphoid cells in the peripheral blood or spleen preparations are referred to as "effectors" when establishing the different E : T ratios used in the assays. However, only approximately 5–15% of the lymphoid cells, in rats or humans, are NK cells by definition.

NK Cell Activity vs Cell Number

Determination of the NK cell proportion in the effector cell population can be performed by flow cytometric immunophenotyping using e.g. CD161 rat-specific antibodies, or CD16/CD56 human-specific antibodies. However, the NK phenotype is highly heterogenous and approximately 50 different subsets have been identified. All NK cells may not exhibit lytic activity against the target cell lines, thus a correlation between the total number of NK cells and measured cytotoxic activity is not always obvious. It should be noted that significant decreases in NK cell activity can be observed without any significant reduction in NK cell number (1). Therefore, phenotypic enumeration of NK cells and measurement of cytotoxic NK cell activity should be regarded as two separate parameters. Each laboratory should establish their own baseline values regarding normal cytotoxic activity, as this may vary between species, strains and compartment for isolation of effector cells.

51Cr Release Assay

The conventional ^{51}Cr release assay (2) is based on radiolabeling of target cells with ^{51}Cr. After 4 hours co-incubation with effector cells the amount of ^{51}Cr released into the supernatant is quantified in a gamma counter. The endpoint shows an indirect proportion of killed target cells given as percentage specific cytolysis and is calculated by the formula:
{(Release in experimental sample − spontaneous release) / (Total release − spontaneous release)} × 100
The method is limited by spontaneous leakage of ^{51}Cr from the target cells that increases with time. After 24 h the spontaneous release is > 50%. Approximately 1×10^6 effector cells are needed per sample using an E:T ratio of 100:1. Therefore, only splenic effectors will be numerous enough for rodent tests.

Non-Radioactive Assays

Current non-radioactive assays are generally based on flow cytometry. These methods have a good correlation with the ^{51}Cr release assay (2,3) and also have the advantage of being able to identify lytic events before they are detectable with the ^{51}Cr release assay. The target cells are labeled with a fluorescent dye to render discrimination between the populations of effector cells and target cells possible. To identify permeabilized target cells a fluorescent DNA stain, which labels only cells with compromised plasma membranes, is added. This technique enables a clear separation between live and dead target cells (Figure 1). The endpoint shows a direct proportion of killed target cells as percentage of specific cytolysis calculated by the formula:
{(% dead targets in the sample − % spontaneously dead targets) / (100 − % spontaneously dead targets)} × 100

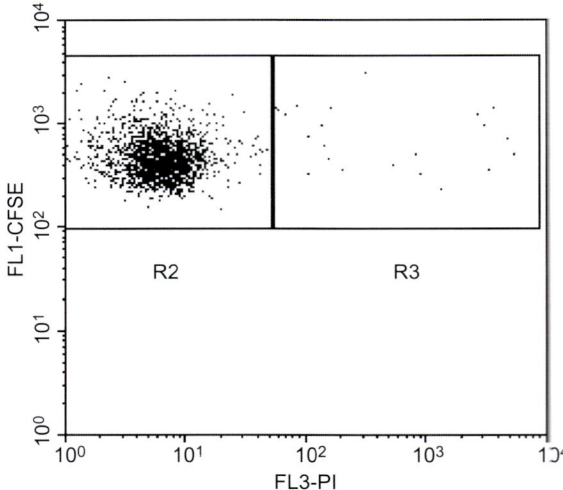

Natural Killer Cell Assay. Figure 1 Natural killer (NK) cell activity in Wistar rat peripheral blood measured by flow cytometric analysis of CFSE-labeled and PI-labeled YAC-1 targets. Results are shown as FL1 (CFSE)/FL3(PI) dot plots gated on target cells using CellQuest (Becton Dickinson, USA) analysis software. Viable targets are given in R2. Non-viable targets are given in R3. Cells were cultivated for 18 hours. *A*: Control without effector (E) cells shows 1.7% spontaneously dead target cells (T) in R3. *B*: An E:T ratio of 50:1 and addition of IL–2 shows an NK cell activity of 63.7% dead target cells in R3.

The choice of target cell labeling is crucial for this test to be successful. The target cell membranes are labeled with a primary fluorescent dye, e.g. F-18, FITC, PKH-2, PKH-26, or CFSE (5-(6)- carboxy-fluorescein succinimidyl ester). After co-incubation with effector cells, a secondary fluorescent DNA-binding dye (e.g. propidium iodide (PI) or TO-PRO-3 iodide) is added for discrimination of dead target cells. Leakage of the primary dye can sometimes contribute to cross-contamination of cells other than the targets as described with fluorescein or PKH-2. To avoid cross-contamination, CFSE is preferable as a primary dye. It is added to the unlabeled targets as non-fluorescent 5-(6)-carboxy-fluorescein diacetate, succinimidyl ester and will diffuse passively into the cells. Within the membrane of viable cells esterases remove the acetyl moieties leaving CFSE that binds to proteins and is well retained within the cell. CFSE is non-toxic, does not interfere with cell function, is suitable for long-term NK assays (16–20 h) which are needed to render reading of NK activity in certain animal species possible, and it has the fluorescent excitation and emission characteristics of fluorescein and is suitable for use in combination with the DNA binding dye, PI. CFSE and PI can be readily distinguished by flow cytometry, using the FL1- and FL3-channel respectively, omitting the need for compensation of spectral overlap. Targets and effectors are co-incubated for 18 hours (rat) or 4 hours (human) followed by flow cytometric analysis. The assay can be performed with or without addition of interleukin(IL)-2. Approximately 2.5×10^5 effector cells are needed per sample using an E : T ratio of 100 : 1. The low amount of effectors required indulges the use of peripheral blood in rodent studies as well as repeated testing of the same animal (3).

Another non-radioactive assay used for measuring NK cell activity is the lactate dehydrogenase (LDH) release assay. Since this assay cannot discriminate between LDH released from dead effector cells or dead target cells it has limited use, especially in long-term cultures.

Pros and Cons

One of the advantages of the flow cytometric assay compared to the classical ^{51}Cr release assay is the low amount of effector cells needed. This minimal requirement of effectors allows for the use of peripheral blood as an effector source and also for repeated testing of the same animal. A summary of advantages and disadvantages with the two methods are shown in Table 1.

Predictivity

Host resistance models are ultimate tests in risk assessment of consequences that a decrease in any immune parameter, such as NK cell activity, may have for resistance against infection or tumor cells. When changes in NK cell activity (^{51}Cr release), in mice, are compared to changes in host resistance the concordance is 73% (4). Consequently, NK cell function is regarded as a valuable predictor of experimentally induced immunotoxicity. Still, NK assays are generally more sensitive than host resistance assays. In mice, a decrease in NK activity of $\geq$ 50%–60% is required before a decrease in specific host resistance to tumors can be detected (1).

NK cell activity measured as described in this section covers only one NK-mediated pathway of cytotoxicity: secretory/necrotic killing by perforin/granzyme. Mechanisms for elimination of, for example, solid tumours are not covered. To make a complete determination of NK cell function per se, additional mechanism-related tests are needed (5). Also, for each chemical tested specific attention should be paid to which NK cell compartment is most suitable for isolation of effector cells (3). The test substance's route of exposure could be crucial considering this choice.

Relevance to Humans

Various drugs have shown to affect NK cell function in animal studies or after in vitro exposure of human cells (3). However, the clinical relevance of these data needs further evaluation. Impaired cytotoxic NK cell activity has been associated with acute virus infections, multiple sclerosis, AIDS, systemic lupus erythematosus, and leukemia. Evidences for a correlation between reduced NK cell activity and cancer in man are scarce but medium and high cytotoxic activity of peripheral blood is associated with reduced cancer risk, while low activity is associated with increased cancer risk.

Regulatory Environment

The European EMEA guidance on immunotoxicity testing of new drugs for human use (CPMP/SWP/1042/99; October 2000) encourages NK cell function as an immunotoxicity parameter in standard non-clinical repeat-dose toxicity studies. Testing of NK cell activity is suggested as a first-hand alternative to another functional assay (Primary antibody response to a T-cell dependent antigen) with the aim of avoiding the use of satellite animal groups required by the latter. It should be noted that none of these functional assays are meant to act as stand-alone tests and that the results always should be interpreted along with other toxicological findings and general health status of the animal. The FDA guidance for industry on immunotoxicology evaluation of investigational new drugs (October 2002) includes determination of NK function as a follow-up test if warranted based on findings in standard repeat-dose toxicology studies.

Natural Killer Cell Assay. Table 1 Advantages and disadvantages of ^{51}Cr vs flow cytometry (3)

Property		^{51}Cr	Flow Cytometry
Health and environment	The use of hazardous radioisotopes	Yes	No
Use in regulatory studies	Enough cells from peripheral blood	No	Yes
	Repeated testing of the same animal (rat)	No	Yes
Technical and mechanistic aspects	Target labeling affects cell function	Yes	No
	Suitable for long-term assays	No	Yes
	Results must be read within	27.7 days*	60 min
	Direct determination of dead/viable target cells	No	Yes
	Early detection of lytic events	No	Yes
	Spontaneous leakage of labeling from target cells	Yes	No**
	Same technique as for immunophenotyping	No	Yes
	Hands-on time	Shorter	Longer
Number of advantages		2	9

In summary, flow cytometry (3) is a suitable and advantageous alternative to the classical ^{51}Cr release assay.
* Half-life of ^{51}Cr
** When CFSE is used for target cell labeling

References

1. Wilson SD, McCay JA, Butterworth LF, Munson AE, White KL Jr (2001) Correlation of suppressed natural killer cell activity with altered host resistance models in B6C3F1 mice. Toxicol Appl Pharmacol 177:208–218
2. Friberg D, Bryant J, Whiteside TL (1996) Measurements of natural killer (NK) activity and NK cell quantification. Meth Enzymol 9:316–326
3. Marcusson-Ståhl M, Cederbrant K (2003) A flow cytometric NK-cytotoxicity assay adapted for use in rat repeated dose toxicity studies. Toxicology 193:269–279
4. Luster MI, Portier C, Pait DG et al. (1993) Risk assessment in immunotoxicology II: Relationships between immune and host resistance tests. Fund Appl Toxicol 21:71–82
5. Wahlberg BJ, Burholt DR, Kornblith P et al. (2001) Measurement of NK activity by the microcytotoxicity assay (MCA): a new application for an old assay. J Immunol Meth 253:69–81

Natural Killer Cells

DAVID SHEPHERD
Center for Environmental Health Sciences, Dept of Biomedical and Pharmaceutical Sciences
University of Montana
58 Skaggs Building, 32 Campus Drive
Missoula, MT 59812
USA

Synonyms

NK cells

Definition

Natural killer (NK) cells are lymphoid-derived cellular components of the ▶ innate immune system. These cells are part of the first line of defense that function to contain viral infections as well as the spread of tumor cells. The measurement of NK cell activity has recently become a critical assay used to identify potentially immunotoxic chemicals.

Characteristics

NK cells are derived from the bone marrow and comprise approximately 5%–15% of human peripheral blood ▶ lymphocytes (1). They lack surface expression of antigen-specific receptors, as well as many of the surface molecules such as CD3, CD4, CD8 or

CD19, that are typically found on B and T lymphocytes. NK cells contain large cytoplasmic granules that aid in the killing of pathogen-infected cells, tumor cells, and major histocompatibility complex (MHC) class 1 disparate cells. However, recognition and killing of target cells by NK cells is intricately regulated by a combination of activating and inhibitory receptors.

The activation of NK cells to kill infected cells is accomplished following the ligation of several distinct cell surface receptors (Table 1).

A heterogeneous family of NK cell-specific immunoglobulin-like molecules has recently been identified. They are known as natural cytotoxicity receptors, and they activate NK cells when bound to as yet undefined ligands. This receptor family includes NKp30, NKp44 and NKp46 (2,3). In addition, NKG2D (a member of the lectin superfamily) also activates NK cells following stimulation by the MHC class I polypeptide-related sequence A/B (MICA/MICB) and UL16-binding proteins (ULBPs). In another somewhat distinct mechanism, CD16 (an immunoglobulin G Fc receptor (FcγRIIIa) that is expressed on NK cells and phagocytic cells) can provide activating signals to NK cells following the cross-linking of bound IgG (2). The primary outcome of CD16-mediated NK cell activation is the generation of ▶ antibody-dependent cell-mediated cytotoxicity (ADCC). Alternatively, NK cells are activated by interferons (IFN-α and IFN-β) and cytokines such as the interleukins IL-2, IL-12, IL-15, IL-18 and tumor necrosis factor TNF-α (3). Stimulation of NK cells by these soluble mediators can increase their activity up to 100-fold as measured by the NK cell assay (1). Furthermore, activation by these cytokines leads to increased proliferation, trafficking, and ▶ cytotoxicity by NK cells, as well as inducing them to secrete copious amounts of IFN-γ—crucial in controlling various infections. NK cells can also secrete other effector cytokines including TNF, granulocyte macrophage colony stimulating factor (GM-CSF), IL-5, IL-10 and IL-13 (2).

To regulate the activation of NK cells and prevent the killing of normal host cells, a collection of MHC class I-specific inhibitory receptors exist (Table 1). In humans, NK cell inhibitory receptors include members of the killer-cell immunoglobulin-like receptor (KIR) family that bind to MHC class I molecules, and the CD94–NKG2A heterodimer that recognizes non-polymorphic human leukocyte antigen HLA-E molecules (2,3). In mice, the inhibitory receptors on NK cells that recognize non-self MHC class alleles are called Ly49 (1). Rodents also express the CD94–NKG2 receptor. Signaling via the NK inhibitory receptors suppresses the killing activity of NK cells and serves to protect healthy cells that express self MHC class I molecules.

Following the activation of NK cells (in the absence of inhibitory signals), tumor or virus-infected target cells can be killed by two separate mechanisms. The first mechanism of NK cell-mediated killing is identical to

Natural Killer Cells. Table 1 Receptors involved in the regulation of natural killer cell function

Receptor	Species	Ligand
Activating receptors		
NKG2D	Human	MICA/MICB, ULBP1, -2, -3
	Mouse	Rae-1, H60
NKp30	Human	Unknown
NKp44	Human	Unknown
NKp46	Human	Unknown
CD16	Human/mouse	IgG
Inhibitory receptors		
KIR2DL1	Human	HLA-C
KIR2DL2/3	Human	HLA-C
KIR3DL1	Human	HLA-B
KIR3DL2	Human	HLA-A
Ly49	Mouse	H-2K, H-2D
CD94-NKG2A	Human	HLA-E
	Mouse	Qa-1

HLA, human leukocyte antigen; KIR, killer cell immunoglobulin-like receptor; MHC, major histocompatibility complex; MICA/MICB, MHC class I polypeptide-related sequence A/B; ULBP, UL16 binding protein.

that used by cytotoxic T lymphocytes (CTL); cytotoxic granules are secreted onto the cell surface of bound target cells where they penetrate the membrane and induce apoptosis (1). The second mechanism also involves the induction of apoptosis in target cells although via a non-secretory pathway. This form of cytotoxicity results from the engagement of death receptor pathways such as CD178 (FasL) on the surface of NK cells and CD95 (Fas) on target cells (2). Other TNF family members including TNF and TNF-related apoptosis-inducing ligands (TRAIL) also mediate NK cell killing via their respective receptors on tumor cells.

Preclinical Relevance

NK cell activity is affected by a number of drugs and environmental chemicals. Although chemical-induced changes in NK cell activity have been demonstrated to possess a high concordance with altered host resistance, it remains to be determined if these alterations in immune function are physiologically relevant (5). The investigation of NK cells and NK cell activity following exposure to drugs and environmental chemicals is regulated by special guidelines. NK cell activity is commonly employed as an initial screening test for use in non-clinical, repeated-dose immunotoxicity studies.

Relevance to Humans

Because of the importance of the innate immune system in the early response to infection, loss of NK cells or their functions following exposure to drugs and xenobiotics could possibly lead to an increased susceptibility to disease. This point is illustrated in a number of known NK cell deficiencies and associated infectious susceptibilities that exist in humans (and also rodents). For examples, individuals harboring alterations in CD16 gene expression via several polymorphisms are at increased risk of developing frequent upper respiratory viral infections, recurrent bacterial infections, and even autoimmune disorders (4). In people suffering from functional NK cell deficiency (FNKD), NK cells are present but they lack one or more NK cell functions (4). These individuals are highly susceptible to several pathogens, including Epstein-Barr virus (EBV) and herpes simplex virus (HSV) due to a lack of NK cell cytotoxicity.

Another well-characterized immunodeficiency resulting from aberrations in NK cell cytotoxic effector mechanisms is Chediak-Higashi syndrome (CHS). NK cells in patients with CHS are defective in spontaneous cytotoxicity and ADCC, and possess abnormal morphology (4). The molecular pathogenesis of this disease involves mutations in the *LYST* gene which normally regulates lysosomal trafficking. CHS patients have recurrent respiratory bacterial infections, susceptibility to *Candida* and *Aspergillus*, and an associated lymphoproliferative syndrome. Interestingly, a natural mutation in the *LYST* (*Beige*) gene also exists in several mammalian species including mice, rats, and Aleutian minks, providing several experimental models of CHS (4).

Data collected from human studies looking at the effects of chemicals on NK cells and their functions correlates well with data generated from animals. Thus, experimental animal data on NK cells is useful for assessing the potential risk that immunotoxic chemicals may pose to humans.

Regulatory Environment

Recently, the European Agency for the Evaluation of Medicinal Products recommended that testing of NK cell activity be included in the initial immunotoxicity screening of medicines for human use (5). This test is to accompany other non-functional parameters of the immune system such as hematology, lymphoid organ weights, histopathology, bone marrow cellularity, and lymphocyte subset analysis. In addition, the US Food and Drug Administration (FDA) insists on the inclusion of NK cell activity testing for the safety evaluation of pharmaceuticals (5). However, this regulatory agency recommends that NK cell evaluation be considered a Tier II test, which is only to be performed if adverse effects are detected in non-functional Tier I assays as described above. NK cell activity is also currently being considered by the Japanese Ministry of Health, Labor and Welfare to be included as an assay in drug safety evaluation (5).

References

1. Janeway CA, Travers P, Walport M, Shlomchik M (2001) Immunobiology: The Immune System in Health and Disease, 5th edn. Induced innate responses to infection. Garland Publishing, New York, pp 82–85
2. Smyth MJ, Hayakawa Y, Takeda K, Yagita H (2002) New aspects of natural killer cell surveillance and therapy of cancer. Nature Rev Cancer 2:850–861
3. Moretta A (2002) Natural killer cells and dendritic cells: Rendezvous in abused tissues. Nature Rev Immunol 2:957–964
4. Orange JS (2002) Human natural killer cell deficiencies and susceptibility to infection. Microbes Infect 4:1545–1558
5. Cederbrant K, Marcusson-Stahl M, Condevaux F, Descotes J (2003) NK cell activity in immunotoxicity drug evaluation. Toxicology 185:241–250

Natural Killer ^{51}Cr Release Assay

▶ Natural Killer Cell Assay

Natural Killer (NK) Cell

Natural killer (NK) cells are non-T and non-B lymphocytes that can kill some types of tumor cells, virus-infected cells, and cells infected with other types of intracellular pathogens. NK cells have innate mechanisms for target cell recognition, as well as an antibody-dependent mechanism (ADCC). The innate mechanism is mediated by a family of stimulatory and inhibitory receptors that are encoded by the NK gene complex.
▶ Cell-Mediated Lysis
▶ Respiratory Infections
▶ Cytotoxicity Assays
▶ Flow Cytometry
▶ Dermatological Infections

Natural Killer (NK) Cell Assay

The chromium release assay involves the ex vivo killing of a suitable cell line which is labeled with radioactive chromium, typically using splenocytes or peripheral blood mononuclear cells from animals dosed with the test article.
▶ Canine Immune System
▶ Flow Cytometry
▶ Dermatological Infections

Negative Selection

The process in the thymus that leads to the elimination of unreactive or autoreactive thymocytes.
▶ Thymus: A Mediator of T Cell Development and Potential Target of Toxicological Agents

Neoantigen-Forming Chemicals

Chemicals that a capable of eliciting an immune response not necessarily have to act as haptens, but they may also alter normal self proteins, e.g. by modifying their expression, by oxidising single amino acid side chains, or by complex formation, in such a way that other peptides or larger amounts of certain peptides of the altered protein are presented by antigen-presenting cells. Since these peptides may not have been 'seen' by the immune system before (i.e. they are new antigens or neoantigens), no T-cell tolerance towards them is established and, therefore, an immune response against them may be initiated.

▶ Popliteal Lymph Node Assay, Secondary Reaction

Neonatal Immune Response

KENNETH S. LANDRETH · SARAH V. M. DODSON
Department of Microbiology, Immunology, and Cell Biology
West Virginia University Health Sciences Center
Morgantown, WV 26506
USA

Synonyms
Newborn immune function, neonatal tolerance

Definition
The immune responses of neonatal mammals, including man, remain immature at birth. Neonatal immune responses are characterized by failure to respond to polysaccharide antigens, delay in antibody production, absence of isotype switching, and unusual susceptibility to tolerance induction following antigen (Ag) exposure. This period of neonatal immunodeficiency persists for up to 6 weeks in rodents and several years in humans.

Characteristics
Architecture of the neonatal immune system. The immune system of mammals is formed embryonically in a progression of tissue microenvironments that develop sequentially. Prior to birth, precursors for ▶ immunocompetent cells are formed and expand in embryonic tissues, however, mature cell function can not be demonstrated prior to birth. Neonatal immunity differs from adult immunity in anatomical features and functional outcomes.

Neonatal antibody production. The spleen microenvironment in the newborn differs from that of adults. Splenic primary follicles of B lymphocytes are not observed until 2 weeks after birth in mice. Secondary follicles or germinal centers are detected in the spleen of mice until 4 weeks of age. The formation of splenic germinal centers is known to be essential for cell interactions necessary for T-dependent activation of B lymphocytes and subsequent antibody formation in response to T-dependent antigens (1). In spleens of human neonates, germinal centers are not detected until several months following birth.

B lymphocytes from neonates produce lesser amounts of antibodies than do cells from adult animals. Most importantly, neonates do not produce antibodies when challenged with polysaccharide antigens. This is has been particularly noted when mice were challenged with lipopolysaccharide-rich (LPS) antigens extracted from bacteria (2). It has also been determined that while stimulation of adult B cells by antigen binding to the B cell receptor (BCR) confers activation and cell proliferation, neonatal B cells often undergo apoptosis following antigen binding (3). This ease of ▶ tolerance induction in B cells may have importance in establishing tolerance to self-antigens early in life.

Neonatal B cells do not effectively antibody isotype switching from IgM to other Ig classes, including IgG and IgA for some time after birth. Because of neonatal B cells, IgA is not found in mucosal secretions until about 4 weeks of age in mice and at about 2 years in humans.

Transference of maternal antibody protection. Newborns acquire protective immunoglobulins (Ig) passively through the mother from placental transfer before birth and postnatal lactation. These antibodies protect newborns from foreign antigen exposure during the period of neonatal immunodeficiency; however, they also blunt active immunity in the neonate by eradicating antigen before it can stimulate newly formed B cells in the newborn. Maternal gammaglobulin antibodies have a half-life of approximately 90 days and decrease rapidly following cessation of breast feeding.

Neonatal cell mediate immune responses. CD4+ T helper cells and CD8+ cytotoxic T cells are present in secondary lymphoid tissues of neonates, however, cell mediated immune response are damped in neonates following engagement of the T cell receptor (TCR) (4). This observation is likely related to differences in cytokine secretion by these cells. Stimulated T cells from newborns produce less of the cytokines Interleukin-2 (IL-2), granulocyte-macrophage colony stimulating factor (GM-CSF), interferon-γ (INF-γ) and interleukin-5 (IL-5) when compared to cytokine production by adult T cells (3). This difference in cytokine production also results in reduced antibody responses to T cell dependent (TD) antigens and decreased immune responsiveness to antigen exposure in general (5). Some studies have also suggested that neonatal T cells require greater levels of co-stimulatory molecule interaction involving CD2, CD3, and CD28, in addition to TCR engagement than do adult T cells.

Preclinical Relevance

Studies of immune responsiveness to antigen challenge in neonatal mice and humans differ dramatically from that demonstrated for adults. Design of vaccination schedules for developing adaptive immunity to pathogenic organisms requires careful attention to this the period of neonatal immunodeficiency and the timing of maturation of the immune response to specific antigen challenges.

Relevance to Humans

Human infants do not respond to polysaccharide or T-dependent antigens some time after birth. This period of neonatal immunodeficiency has to be considered in both design of vaccination schedules and investigation of altered immune mechanisms related to immunodeficiency. It is also important to consider the role of maternal antibodies in responses to pathogenic organisms or vaccination to antigens during the period of neonatal immunity.

Regulatory Environment

There is considerable interest in detecting immunotoxicity following chemical exposure in animal models and in exposed human populations. Developing regulatory testing methods for immunotoxic compounds must consider both the unusual vulnerability and the altered responsiveness of the immune system during the perinatal period.

References

1. Astori M, Finke D, Karaoetian O, Acha-Orbea A (1999) Development of T-B cell collaboration in neonatal mice. Inter Immunol 11 (3):445–451
2. Muthukkumar S, Goldstein J, Stein K (2000) The ability of B cells and dendritic cells to present antigen increases during ontogeny. J Immunol 165:4803–4813
3. Marshall-Clarke S, Reen D, Tasker L, Hassan J (2000) Neonatal immunity: how well has it grown up? Immunol Today 21 (1):35–41

Neonatal Immune Response. Table 1 Maturation of the Immune Response

Immune Function		Rodent	Human
Antibody-Mediated Immune Responses	TI-1 Responses	at birth	at birth
	TI-2 Responses	1 month	1 year
	TD Isotype Switching	1 month	> 1 year
	TD Affinity Maturation	1 month	> 1 year
	TD Ab Heterogeneity	1 month	2 years
Cell-Mediated Immune Responses	Tolerance Induction	3 days	> 7 days
	Cytokine Production	1 month	> 1 month

4. Adkins B (1999) T-cell function in newborn mice and humans. Immunol Today 20 (7):330–335
5. Fadel S, Sarazotti M (2000) Cellular immune responses in neonates. Int Rev Immunol 19:173–193

Neonatal Tolerance

▶ Neonatal Immune Response

Neural Tube Defect

Defects where the vertebra or back of the skull fail to close, allowing spinal cord or brain—and associated tissues—to protrude. In humans such defects involving spinal cord and brain are also referred to as spina bifida and exencephaly, respectively.
▶ Birth Defects, Immune Protection Against

Neuroendocrine Response

▶ Stress and the Immune System

Neurons

The morphologic unit of the nervous system, consisting of the nerve cell body and its various processes, the dendrites and the axis cylinder process or nuraxon; the axis cylinder process of a nerve cell; a neuraxon.
▶ Serotonin

Neurotransmitter

A chemical which acts as a messenger between cells in the brain and nervous system; it transmits impulses across the gap from a neuron to another neuron, a muscle or a gland.
▶ Serotonin

Neutropenia

Neutropenia is a condition in which the normal concentration of neutrophils (also known as polymorphonuclear (PMN) cells) in the blood is decreased. A severe deficiency of neutrophils may enhance susceptibility to infectious disease.
▶ Leukemia
▶ Therapeutic Cytokines, Immunotoxicological Evaluation of
▶ Antiglobulin (Coombs) Test

Neutrophil

KATHLEEN RODGERS
Livingston Research
University of Southern California
1321 N. Mission Road
Los Angeles, CA 90033
USA

Synonyms
Polymorphonuclear neutrophil, granulocyte

Definitions
Neutrophils are terminally differentiated cells, rich in cytoplasmic granules, containing a lobulated chromatin-dense nucleus with no nucleolus. Four types of cytosolic granules have been characterized.

Characteristics
Neutrophils represent 50%–60% of the total circulating leukocytes in the nontraumatized human. In rodents, neutrophils represent a much smaller percentage of the circulating leukocytes. These cells constitute the first line of defense against infectious agents and other invading substances. Once an inflammatory response is initiated, neutrophils are often the first cells recruited to the site of infection or injury. Neutrophil microbiocidal processes consist of the formation of a combination of reactive oxygen (and possibly nitrogen) species and various hydrolytic enzymes and polypeptides.
Neutrophils mature in the bone marrow prior to being released into the circulation, where they spend only 4–10 hours before marginating and entering tissue pools, where they survive 1–2 days in the absence of survival factors such as granulocyte-colony stimulating factor (G-CSF). Senescent neutrophils are thought to undergo apoptosis prior to removal by macrophages. Apoptosis is a means to clear to neutrophil without release of their cytotoxic contents and may play a role in the termination of the acute phase of an inflammatory response. Cells of the circulating and marginating pools can exchange with each other.
Morphological maturation stages of neutrophils include myeloblasts, promyelocytes, myelocytes, metamyelocytes, bands, and finally segmented neutrophils.

Surface expression of various antigens during the stages of neutrophil development are well characterized. For example, CD16b, CD 35 and CD10 appear with neutrophilic maturation and CD49b and CD64 expression is down regulated during maturation (1).

Under normal circumstances, neutrophils are produced in the human bone marrow at the rate of 10^{11} cells per day (2). This process is controlled by two CSFs (G-CSF and granulocyte macrophage-CSF (GM-CSF)) that direct the production and differentiation of bone marrow progenitor cells. During states of stress (such as trauma) and infection, the rate of neutrophil differentiation can increase as much as 10-fold.

During an inflammatory response, chemotactic factors are generated which signal the recruitment of additional neutrophils to the site of injury and/or infection. Under normal conditions, neutrophils roll along microvascular walls via low-affinity interactions of selectins with endothelial carbohydrate ligands. During an inflammatory response, β_2-integrins and high affinity binding to intracellular adhesion molecules is activated on the activated endothelial cells signaling the first step in transmigration to the site of inflammation.

Preclinical Relevance

Neutrophils are the first line of defense in the control of infection and are intimately involved in the initiation of inflammatory responses in response to trauma. Alterations in neutrophil function can potentially affect host resistance to multiple invading agents. Further, increased neutrophil function or prolonged residence at the site of injury can result in tissue damage. Therefore, neutrophils have the potential to mediate or be involved in many of the immunopathological events contributing to disease.

Neutrophil Function in Infection Control

Polymorphonuclear neutrophils generate nonspecific immune responses capable of controlling bacterial invasion that also risk injuring or destroying normal, viable tissue. While many factors modify this response, the interaction between adhesion molecules on the vascular endothelium and ligands on circulating neutrophils lead to neutrophil attachment, priming, and activation. These interactions can affect the subsequent neutrophil function and resulting inflammatory responses. Various events, such as hemorrhagic shock, sepsis and tissue injury result in rapid upregulation and increased expression of adhesion molecules. Primed neutrophils are extremely sensitive to activating agents that render the cells capable of producing high levels of reactive oxygen metabolites (ROI). Together with phagocytosis and proteolytic enzymes, the generation of ROI by neutrophils is of central importance to the innate host defense to bacterial infection. In fact, chronic granulomatous disease is a rare genetic disorder characterized by severe, recurrent infections due to the inability of neutrophils and macrophages to mount an adequate respiratory burst to kill invading bacteria.

Neutrophil Recruitment after Trauma

Trauma, including major surgery, stimulates a cascade of events that mediate the inflammatory response. Activation of the complement system and of neutrophils is an early response to surgical trauma. Agents that are stimulated by surgical trauma and influence neutrophil production, apoptosis and function include interleukins IL-1, IL-6, IL-8, tumor necrosis factor (TNF), CSFs and bioactive lipids, such as platelet activating factor. In fact, several citations have reported a relationship between the degree of surgical trauma and the release of inflammatory mediators. Further, surgical stress (in particularly postoperatively) is associated with a marked increase in the level of circulating catecholamines. The α-adrenergic catecholamines markedly enhance neutrophil numbers.

Induction of anesthesia alone was shown to induce a slight increase in number of circulating neutrophils. With the initiation of surgery, the number of circulating neutrophils increased up to 4.5 times during the surgery itself. The amount and duration of increase may be related to the degree of trauma induced during the surgical procedure. This may be explained by the observation that major, but not minor, surgery correlates with a reduction in neutrophil apoptosis.

Peripheral blood neutrophil function has also been shown to be modulated by surgical trauma and induction of anesthesia. For example, it was shown that neutrophil chemotaxis was reduced up to 36% simply by the induction of anesthesia, but this inhibition was reversed by the surgical procedure. Further, neutrophil respiratory burst activity, enzyme content, microbiocidal killing and surface proteins have also been shown to be modulated by surgical trauma.

Relevance to Humans

In human beings, there are severe consequences to either a reduction in neutrophil number or function: increased incidence, severity and duration of bacterial and fungal infections, or increased neutrophil function (i.e. tissue damage and destruction leading to immunopathological changes).

Consequences of Neutropenia

Hematopoietic cells destined to become mature neutrophils move through three cellular compartments before they enter the blood. Neutropenia results from disorders of stem cells, defects in the processes of proliferation and differentiation and abnormalities in the distribution and turnover of blood cells (3,4). Neutropenia is defined as a decrease in the absolute neutrophil count (ANC) to below normal levels. The ANC can vary widely in healthy individuals due to exercise, emotional state and circadian rhythm. In general, the level of neutropenia is defined as mild, moderate and severe based upon the risk of pyogenic infections when that level of neutropenia is sustained over 2–3 months.

Neutropenic patients are usually infected by organisms of their endogenous flora, the resident bacteria of the mouth, oropharnyx, gastrointestinal tract and skin. Overall, gingivitis and mouth ulcerations are the most common problems resulting from neutropenia. In general, patients with severe chronic neutropenia have fewer serious infections than those with the same degree of neutropenia resulting from cancer chemotherapy. Chemotherapy-induced neutropenia broadly affects host-defense function, including the barrier role of the mucosal cells in the mouth and gastrointestinal tract.

Neutrophils in Inflammatory Disease States

Although the neutrophil response is designed to restrict the damage to the region surrounding the invading organism, collateral damage to surrounding tissues often occurs during the control of the invading organism. There is evidence from clinical studies that exaggerated recruitment and activation of neutrophils are linked to several inflammatory disorders including asthma, and chronic bronchitis (5). This may be due to the release of bioactive products that can actively contribute to the pathogenic process. Neutrophils can release cytokines that perpetuate neutrophil recruitment (e.g. tumor necrosis factor), along with proteolytic enzymes (e.g. elastase), bioactive lipids (e.g. prostaglandins) and ROI. While all of these products are necessary to the proper function of the neutrophil in the clearance of bacterial and fungal infections, they contribute to tissue damage when the number of activated neutrophils at a site is prolonged.

Regulatory Environment

Neutrophils are a granulocytic cell of hematopoietic origin that is central to the innate immune response to bacterial and fungal infection. Neutropenia results in an increased susceptibility to disorders they are meant to protect against; however, leukocytosis or prolonged neutrophilia can contribute to inflammatory disease pathogenesis.

Neutrophil numbers are evaluated in preclinical safety evaluation during hematological evaluation, and assessment of bone marrow. Further, immunopathological consequences of increase neutrophil number, residence or function is assessed during histopathological evaluation of multiple tissues. These studies are conducted as a routine in the evaluation of the safety of a new agent. Further evaluation of neutrophil function can occur during specialized tests to evaluate potential immunotoxicology. Initially, alterations in host resistance may indicate an effect on neutrophil number or function. If indicated, studies of neutrophil function may be conducted as a follow-up to alterations indicative of possible change in host resistance models.

References

1. Elghetany MT (2002) Surface antigen changes during normal neutrophilic development: a critical review. Blood Cell Mol Dis 28:260–274
2. Cannistra SA, Griffin JD (1988) Regulation of the production and function of granulocytes and monocytes. Semin Hematol 25:173–188
3. Dale DC, Guerry D, Wewerka JR, Bull J, Chusid M (1979) Cyclic neutropenia: a clinical review. Medicine 58:128–144
4. Pincus SH, Boxer LA, Stossel TP (1976) Chronic neutropenia in childhood. Analysis of 16 cases and a review of the literature. Amer J Med 61:849–861
5. Leirisalo-Repo M (1994) The present knowledge of inflammatory process and the inflammatory mediators. Pharmacol Toxicol 75 [Suppl 2]:1–3

Neutrophils

Short-lived bone-marrow-derived non replicating blood leukocytes, with a distinctive condensed chromatin (a nucleus of three to five lobes), that are specialized for phagocytosis and the killing of microbes.

▶ Opsonization and Phagocytosis
▶ Respiratory Infections

Newborn Immune Function

▶ Neonatal Immune Response

NF-kappa B (NFκB)

This is nuclear factor of kappa light-chain enhancer in B cells. It occurs in numerous cells and is activated by a range of stimuli. Both inappropriate activation and suppression are associated with adverse conditions, e.g. inflammatory processes, inappropriate immune cell development. After activation, NFκB translocates to the cell nucleus, where it binds to DNA and regulates transcription.

▶ Lymphocyte Proliferation

NHL

▶ Lymphoma

Niacin

Nicotinic acid; Vitamin B$_3$.
▶ Serotonin

Nitro-PAH

▶ Polycyclic Aromatic Hydrocarbons (PAHs) and the Immune System

Nitrophenyl-Chicken gamma Globulin

A T-dependent antigen prepared from chicken gamma globulin. Used primarily as a T-dependent antigen for ELISA determination.
▶ Plaque Versus ELISA Assays. Evaluation of Humoral Immune Responses to T-Dependent Antigens

NK Cells

▶ Natural Killer Cells

NK Gene Complex (NKC)

The NK gene complex (NKC) consists of a large family of cell surface receptors that act to either inhibit or stimulate the recognition and lysis of cells by natural killer (NK) cells.
▶ Cell-Mediated Lysis

NK Cell Killing

▶ Cell-Mediated Lysis

No Observable Adverse Effect Level (NoAEL)

The "no observable adverse effect level" is the xenobiotics dose at which no undesirable (toxic) effects are seen in animal studies.
▶ Knockout, Genetic

Nodus lymphaticus

▶ Lymph Nodes

Non-Caseating Granuloma

Non-caseating granuloma is a discrete nodule of multinucleated giant cells and lymphocytes that encapsulate persistent antigen to minimize tissue damage.
▶ Chronic Beryllium Disease

Non-Hodgkin's Lymphoma

▶ Lymphoma

Non-Obese Diabetic Mouse (NOD)

Murine model for Insulin-dependent diabetes mellitus. Genetic alteration of the specific MHC alleles in this model has helped to elucidate the role of gene polymorphisms in disease susceptibility.
▶ Autoimmune Disease, Animal Models

Non-Radioactive Flow Cytometric Analysis of NK Cell Cytotoxicity

▶ Natural Killer Cell Assay

Non-Steroidal Anti-Inflammatory Drugs (NSAIDs)

A class of compounds which inhibit the cyclooxygenase activity of prostaglandin H synthase (COX), thereby preventing prostaglandin synthesis and alleviating prostaglandin-induced symptoms such as pain, fever, and inflammation.

▶ Prostaglandins

Nonhuman Primates, Immunotoxicity Assessment of Pharmaceuticals in

WERNER FRINGS · GERHARD F WEINBAUER
Covance Laboratories GmbH
Kesselfeld 29
D-48163 Münster
Germany

Synonyms
Preclinical immunotoxicity evaluation in the nonhuman primate.

Definition
Assessment of immunotoxicity is a prerequisite for all new drugs in Europe. In the US only several classes of drugs require immunotoxicity evaluation. For conventional pharmaceuticals, rodent models commonly yield satisfactory results concerning effects on cells, organs, and functions of the immune system. The targeted design of more sophisticated drugs, especially biotechnology-derived substances, requires investigations in animal models that are more closely related to the human. Nonhuman primates are the species of choice to detect highly specific immunotoxic side effects, to avoid ▶ immunogenicity issues associated with rodent models and to discriminate toxicity from efficacy of immunomodulatory drugs (1).

Characteristics
Unlike immunotoxicity studies in rodents, no standardized tests and protocols are generally available for nonhuman primates as yet. However, the rodent tests can be transferred in principle. An exception is the host resistance assay. Since changes in the virulence of the challenge-pathogen can lead to variability in the assay, comparatively large group sizes are required which can pose a limitation upon nonhuman primate usage (see below). Specialized facilities are needed to maintain infected animals and separate them from the main colony. This is usually more demanding and expensive for nonhuman primates than for rodents. Using death as a test endpoint is obviated by ethical considerations.
The choice of species and assay should always depend on the precise question and/or test article. Cynomolgus monkeys and rhesus monkeys (*Macaca fascicularis* and *M. mulatta*) and the common marmoset (*Callithrix jacchus*) are the nonhuman primate species used most commonly in toxicology testing.
Due to the inherent properties of immunomodulatory substances (e.g. cross-reacting cytokines) immunotoxicity can show signs of either immunosuppression (potentially associated with reduced resistance to infections or cancer) or immunostimulation (potentially associated with autoimmunity or hypersensitivity). Therefore, immunotoxicity evaluation in nonhuman primates should not be confined to merely testing immunosuppression, as has been done until recently.
The design of (immuno)toxicity studies in nonhuman primates generally invokes specific considerations. For ethical and cost reasons the group sizes in these studies are usually kept comparatively small. This, and the fact that no inbred strains of nonhuman primates are available, can impair statistical power for detection of test article-related effects. Interindividual variation of parameters of interest can be higher for nonhuman primates than for humans or inbred rodents. Therefore, it is advisable to include individual comparisons of baseline (pre-dose) evaluation.
Evidently, more test article is needed for immunotoxicity studies in nonhuman primates as compared to rodents. If the amount of test article is limited, the use of the smaller marmoset monkey should be considered.

Preclinical Relevance
Preclinical relevance is generally the major justification for using nonhuman primates in toxicological studies. For the evaluation of biologics, monkeys should be chosen as test system for a variety of reasons:

- Closer homology for amino-acid sequence reduces the risk and probability for lack of biological activity. For example, the dose-limiting clinical toxicity of recombinant human interferon (IFN)-γ could only be reproduced in the nonhuman primate model.
- Closer homology for amino-acid sequence reduces the risk and probability of antibody formation and subsequent neutralizing antibody activity and absent or decreased exposure activity.
- Many antibodies for human immunological assays cross-react with nonhuman primate molecules and the same assay kits can be used for preclinical and clinical evaluation.

A rodent homolog or a transgenic rodent model represent the current alternatives for the use of nonhuman primate models. However, due to the interspecies differences in the immune systems and the multiple functions and feedback loops of immunomodulatory molecules, a species most closely resembling the human systems should always be used during evaluation of potential immunotoxicology.

Nonhuman Primates, Immunotoxicity Assessment of Pharmaceuticals in. Table 1 Methods in nonhuman primate immunotoxicology*

Function measured	Assay	Source
Structural integrity	Hematology/clinical chemistry	Ex vivo
	Histopathology	Necropsy
	Immunohistochemistry	Necropsy
	Flow cytometry (Immunophenotyping)	Ex vivo
B-cell function	Antibody production (e.g anti-KLH response)	In vivo/ex vivo
	Mitogenesis	Ex vivo
T-cell function	Delayed-type hypersensitivity	In vivo
	Cytokine analysis	Ex vivo/in vitro
	Mitogenesis or activation markers	Ex vivo/in vitro
Natural immunity	Natural killer cell function	Ex vivo
	Macrophage/neutrophil function	Ex vivo

KLH, keyhole limpet hemocyanin.
* Immunotoxicity tests available for nonhuman primates (modified from House & Thomas (5)). Other tests might as well be performed, since no standardized tests are available.

Relevance to Humans

When considering preclinical studies to evaluate the influence of biologics, immunomodulatory drugs or vaccines on the immune system, species most closely related to humans should be considered (2). Among the primates, hominoid monkeys (e.g. chimpanzees) are most closely related to humans with evolutionary separation $6–14 \times 10^6$ years ago. Ethical concerns, governmental impact and costs are the reasons for not using these animals in immunotoxicity studies. Old World monkeys (e.g. baboons or macaques) are at the second closest phylogenetic distance (25×10^6 years) followed by New World monkeys (e.g. marmosets, 40×10^6 years). Old World monkeys, as well as hominoid monkeys, for example, have major histocompatability complex (MHC) molecules which much resemble their human counterparts. Although the repertoire of MHC alleles might differ, the organization and expression pattern is comparable. New World monkeys in contrast show a condensed or smaller MHC as compared to humans. The strong relevance of Old World monkey models to humans is evident from the fact that the majority of the anti-human CD antibodies cross-react with the corresponding molecules in Old World monkeys but to a much lesser extent with New World monkeys. On the other hand, the marmoset immune system seems to share some—albeit not yet fully understood—similarities to humans. For example, the course of pertussis infection in marmosets mimics the infection of children more precisely than in Old World monkeys.

For the evaluation of reproductive or developmental immunotoxicity, macaques are the species of choice. There are close physiologic similarities with regard to the endocrine control of female and male gonadal functions, pregnancy, and prenatal development (3,4). The postnatal development of the immune system of macaques shares similarity to that of the developing human, for example in the levels of immunoglobulins and maturation of immune cells.

Regulatory Environment

Several guidelines for immunotoxicology evaluation have been issued. A new drug must be examined for immunotoxicity in the relevant species (for review see House and Thomas (5)). Nonhuman primates are frequently the relevant animal model for immuntoxicology evaluation of biologics.

Initially, descriptive parameters such as hematology, clinical chemistry, weight and microscopy of lymphoid organs and bone-marrow cellularity are required. The European Agency for the Evaluation of Medical Products (EMEA) also encourages immunophenotyping, and recommends a functional test for natural killer cell activity at that stage (CPMP/SWP/0142/99: Note for Guidance on Repeated Dose Toxicity: Appendix B, Guidance on Immunotoxicity, October 2000). Immunotoxicologic findings strongly suggest additional follow-up studies to investigate the nature and mechanism of the immunotoxic effects. The choice of assays for these studies should be scientifically motivated. The Center for Drug Evaluation and Research (CDER) within the US Food and Drug Administration recommends immunophenotyping and functional assays such as the natural killer cell activity or antibody response investigation (CDER: Guidance for Industry: Immunotoxicology Evaluation of Investigational New Drugs, October 2002).

References

1. Thomas PT (2002) Non-clinical evaluation of therapeutic cytokines: Immunotoxicologic issues. Toxicology 174:27–35
2. Kennedy RC, Shearer MH, Hildebrand W (1997) Non-human primate models to evaluate vaccine safety and immunogenicity. Vaccine 15:903–908
3. Hendrickx AG, Makori N, Peterson P (2000) Non-human primates: Their role in assessing developmental effects of immunomodulatory agents. Hum Exp Toxicol 19:219–225
4. Buse E, Habermann G, Osterburg I, Korte R, Weinbauer GF (2003) Reproductive/developmental toxicity and immunotoxicity assessment in the non-human primate model. Toxicology 185:221–227
5. House RV, Thomas PT (2002) Immunotoxicology: Fundamentals of preclinical assessment. In: Derelanke MJ, Hollinger MA (eds) Handbook of Toxicology, 2nd ed. CCR Press, Boca Raton, Florida

Nonparametric Statistics

A set of statistical techniques which make no assumptions regarding the underlying distribution of the data.
▶ Statistics in Immunotoxicology

Nonsteroidal Antiinflammatory Drugs

▶ Anti-inflammatory (Nonsteroidal) Drugs

Norepinephrine (Noradrenaline)

A catecholamine hormone secreted by the adrenal medulla and a neurotransmitter released by postganglionic nerve cells. It is released predominantly in response to hypotension and stress. It acts on α- and $\beta 1$-adrenergic receptors and is a powerful vasopressor.
▶ Stress and the Immune System

Northern

The transfer of RNA molecules from an agarose gel to a membrane by capillarity or an electric field. The immobilized RNA can be detected at high sensitivity by hybridization to a sequence specific probe.
▶ Southern and Northern Blotting

Nosocomial

Pertaining to or originating in a hospital.
▶ Streptococcus Infection and Immunity

NSAID-Activated Gene (NAG-1)

A divergent member of the transforming growth factor (TGF)-β superfamily which is induced by non-steroidal anti-inflammatory drugs (NSAIDs) and appears to play a role in the pro-apoptotic and anti-tumorigenic properties of NSAIDs.
▶ Prostaglandins

NSAIDs

▶ Anti-inflammatory (Nonsteroidal) Drugs

Nuclear Factor κB (NFκB)

Nuclear factor κB (NFκB) is a transcription factor central to a major signaling pathway induced by a plethora of stimuli including inflammatory agents such as TNF-α. NFκB is normally sequestered in the cytoplasm by interaction with inhibitor of κB proteins (IκB). Activation of the NFκB-inducing kinase cascade results in phosphorylation-dependent ubiquitination of IκB leading to proteolytic degradation and freeing of NFκB. NFκB then translocates to the nucleus via a nuclear localization signal sequence and binds to NFκB-response elements in the promoters of numerous genes involved in cell survival, proliferation, differentiation, and inflammation resulting in upregulation of gene expression.
▶ Tumor Necrosis Factor-α

Nucleic Acid Blotting

▶ Southern and Northern Blotting

Nucleic Acid Vaccines

▶ DNA Vaccines

Nude Mouse

This a genetically athymic mouse (e.g. devoid of T-cells and cell-mediated immune capability) which also carries a closely linked gene involved in hair production.
▶ Graft-Versus-Host Reaction

Null Mutant Mouse

▶ Knockout, Genetic

Nurse Cell

Epithelial cell in the outer cortex of the thymus; enclose multiple thymocytes and add to the thymocyte maturation process.
▶ Thymus

Nutrition and the Immune System

MICHELLE CAREY
NIEHS ND D2-01, Laboratory of Pulmonary Pathobiology
P.O.Box 12233
RTP, NC 27709
USA

Synonyms
Immunonutrition.

Definition
Many nutrients in the diet play important roles in maintaining optimal immune function. Nutrient deficiency is associated with an impaired immune response, particularly in cell-mediated immunity, phagocytic function, antibody response, and the complement system. Worldwide, malnutrition is the most common cause of immunodeficiency. Substances such as amino acids, nucleotides, probiotics and fatty acids can be added to standard nutritional support solutions, and the use of such formulations is known as immunonutrition.

Characteristics
There are three main sites in the immune system that can be targeted by specific nutrients: 1) the mucosal barrier; 2) cell mediated immunity; 3) local or systemic inflammation. The following substances are examples of nutrients that can be used to target one or more of these three important components of immunity.

Glutamine
Glutamine is an essential nutrient for immune cells both as a primary fuel and as a nitrogen donor for nucleotide precursor synthesis. Laboratory data have demonstrated numerous effects of glutamine on immune cells: macrophage phagocytosis in vitro declines when glutamine concentrations are lowered, glutamine supplementation significantly enhances phytohaemagglutinin (PHA)-stimulated lymphocyte proliferation and there is evidence of improved bactericidal function of neutrophils following glutamine supplementation in vitro. In addition, glutamine has been reported to restore mucosal immunoglobulin A, to enhance upper respiratory tract immunity, and enhance bacterial clearance in peritonitis.

Arginine
Arginine is considered a non-essential amino acid although endogenous supplies are reduced during trauma and sepsis. Importantly, arginine (via the arginine deaminase pathway) is a unique substrate for the production of the biological effector molecule nitric oxide and it is via this molecule that arginine is thought to mediate many of its immunomodulatory effects. Clinical evidence suggests that arginine enhances the depressed immune systems of patients suffering from injury, surgery, malnutrition or sepsis, by acting on cellular defense mechanisms. Arginine supplementation has many effects on immune cells such as enhanced lymphocyte and monocyte proliferation, enhanced T helper cell formation and activation of macrophage cytotoxicity.

Nucleotides
In the case of adequate protein intake, de novo synthesis is the main source of nucleotides with glutamine being the major nucleotide donor. During episodes of infection following trauma or injury, the demand for nucleotides is increased to facilitate the synthetic capacity of immune cells. Decreased nucleotide availability has many effects on immune cells such as impaired T cell function, weakened natural killer cell activity, suppressed lymphocyte proliferation, reduced phagocytosis, and impaired clearance of pathogens.

Probiotics
The gut flora are believed to confer immunological protection on the host by creating a barrier against pathogenic bacteria. Antibiotic use and disease can lead to disruption of this barrier leaving the host gut susceptible to pathogens. It is now believed that this barrier can be maintained by dietary supplements

called "probiotics" which are live "desirable" bacteria. In addition to creating a barrier effect, some probiotic bacteria produce proteins which inhibit the growth of pathogens or the probiotic bacteria themselves compete with pathogens for nutrients. There is also some evidence these beneficial bacteria can enhance gut immune responses: rat and mouse studies reveal that orally administered lactic acid bacteria increase numbers of T lymphocytes, $CD4^+$ cells and antibody secreting cells, enhance lymphocyte proliferation, natural killer activity, cytokine production and phagocytic activity in macrophages.

n-3 Polyunsaturated Fatty Acids

There is much interest in the anti-inflammatory effects of n-3 polyunsaturated fatty acids (PUFA). Dietary n-6 and n-3 PUFA modulate the lipid content of membrane phospholipids which in turn affects eicosanoid production. A diet rich in n-6 PUFA favors synthesis of eicosanoids derived from the arachidonic acid (AA) precursor, whereas a diet rich in n-3 PUFA shifts the balance of eicosanoids synthesized to favor those derived from docosahexaenoic acid (DHA) and eicosapentaenoic acid (EPA). DHA and EPA, which are found in fish oil, can decrease the production of the proinflammatory cytokines, decrease lymphocyte proliferation, and suppress autoimmune disease, although the exact mechanism is unclear.

Relevance to Humans

There are many clinical applications of immunonutrition. Various enteral formulas are available containing immune-modulating substances such as glutamine, arginine and n-3 PUFA, and the clinical benefits of such formulas have been shown in post-operative and critically ill patients. Clinical trials with these formulas show clear evidence for reduced incidence of infections, reduced duration of ventilation, and shortened hospital stays. Several trials have shown immune enhancing effects of dietary probiotic supplementation.

References

1. Andrews FJ, Griffiths RD (2002) Glutamine: essential for immune nutrition in the critically ill. Br J Nutr 87 [Suppl 1]:S3–S8
2. Calder PC, Kew S (2002) The immune system: a target for functional foods? Br J Nutr 88 [Suppl 2]:S165–S177
3. Chandra RK (2002) Nutrition and the immune system from birth to old age. Eur J Clin Nutr 56 [Suppl 3]:S73–S76
4. Suchner U, Kuhn KS, Furst P (2000) The scientific basis of immunonutrition. Proc Nutr Soc 59:553–563
5. Suchner U, Heyland DK, Peter K (2002) Immune-modulatory actions of arginine in the critically ill. Br J Nutr 87 [Suppl 1]:S121–S132

O

Obese Zucker rat

▶ Diabetes and Diabetes Combined with Hypertension. Experimental Models for

OECD

Organization for Economic Co-operation and Development, an international organization helping governments solve the economic, social and governance challenges of a globalized economy. Testing guidelines of the OECD recommend procedures for testing chemicals.
▶ Immunotoxicology

Oily Fish

Species of fish which store fat throughout their flesh (for example salmon, herring and mackerel)s.
▶ Fatty Acids and the Immune System

Oliguria

A urine volume insufficient to sustain life, usually less then 400 ml per 24 hours.
▶ Septic Shock

Oncogenes

Mutated genes that are the cause of a cancer. The normal gene is called the proto-oncogene. These genes are usually involved in the regulation of cell growth or survival or the intermediate steps in those processes.
▶ Lymphoma
▶ Leukemia

Opportunistic infection

Infections with bacteria, viruses, fungi, or protozoa to which individuals with a normal immune system are not usually susceptible. That is, infections that are caused by microbes that are not very infectious, but that can be so when the normal immune system is not functioning properly.
▶ Lymphoma
▶ Leukemia

Opsonin

A molecule that binds to antigen and phagocyte to enhance phagocytosis. C3b and C4b along with their degradation products are opsonins derived from the complement system. Immunoglobulins also function as opsonins.
▶ Complement, Classical Pathway/Alternative Pathway

Opsonins

All factors by which bacteria or other microorganism are altered by the attachment to the surface so that they are more readily and more efficiently engulfed by phagocytes, are collectively called Opsonins.
▶ Respiratory Infections

Opsonization

Phagocytosis of microorganisms such as bacteria can be enhanced by binding of antibodies, complement factors (mainly C3b), or blood plasma proteins, which are also collectively termed opsonins. These endogenous proteins cover a pathogen and thereby make it "visible" for sentinel cells (macrophages, dendritic cells, or neutrophilic granulocytes) which posses specific receptors for the opsonins.

▶ Immune Response
▶ Fish Immune System
▶ Complement Deficiencies
▶ Streptococcus Infection and Immunity

Opsonization and Phagocytosis

CHARLES J CZUPRYNSKI
Department of Pathological Sciences
University of Wisconsin
2015 Linden Drive W
Madison, WI 53706
USA

Synonyms

ingestion, uptake, internalization

Definition

Opsonization is the process by which a foreign particle, particularly a microbe, is coated with plasma proteins (opsonins) so as to facilitate the attachment and internalization of that particle by a professional phagocytic cell. In general, the process refers to coating of the microbe with immunoglobulin molecules (antibodies) that are specific for antigenic determinants on that organism, or with ▶ complement proteins (particularly C3b) deposited on the surface of the organism via either the classical or alternative activation pathways. The presence of these plasma proteins on the surface of the microbe facilitates their sequential interaction with ▶ immunoglobulin receptors (Fc receptors) or complement receptors (CR) on the phagocyte surface. These interactions result in encirclement of the particle by the cytoplasmic membrane of the phagocytic cell, until the particle is contained within a membrane-bound vacuole (phagosome) within the cell.

Characteristics

In general, the term phagocytosis refers to ingestion of microbes or other particles by professional phagocytic cells. These include granulocytes (principally ▶ neutrophils) and monocytes in the bloodstream, and mononuclear phagocytes (▶ macrophages) that are distributed throughout the various tissues of the body. Opsonization of a microbe by immunoglobulins (antibodies) reflects a specific immune response against antigenic epitopes on that microbe, as a result of natural infection or immunization, or a cross-reaction with antibodies against related antigens. Activation of the complement cascade on the surface of the microbe via the classical pathway, occurs following binding of certain immunoglobulin isotypes (usually IgG) to the surface of the microbe. This event leads to deposition of the C1 complex, which has C3 convertase activity and cleaves C3 to C3b on the surface of the microbe. The alternative pathway of activation occurs when a different C3 convertase (composed of C3 and factor B) forms on the microbial surface. Once the IgG or C3b are deposited on the microbial surface, they can interact with specific receptors (Fc and CR, respectively) on the cytoplasmic membrane of phagocytic cells. There are subtypes of both receptors present on various leukocyte populations, that may demonstrate either activating (immunoreceptor tyrosine-based activation motif or ITAM) or inhibitory (immunoreceptor tyrosine-based inhibition motif, or ITIM) activity. Other plasma proteins that have been reported to opsonize and facilitate ingestion of particles by phagocytic cells include fibronectin, fibrinogen, and C-reactive protein. The interactions amongst plasma proteins and phagocytic cells provides the critical first line of cellular defense in innate immunity against microbial infection.

Once it is internalized within the phagosome, the vacuole becomes acidified and the ingested microbe will be exposed to a variety of antimicrobial compounds. These include reactive oxygen intermediates and nitrogen intermediates that are produced in response to activation of ITAM-containing Fc receptors, and antimicrobial proteins and peptides (e.g. lysozyme, defen-

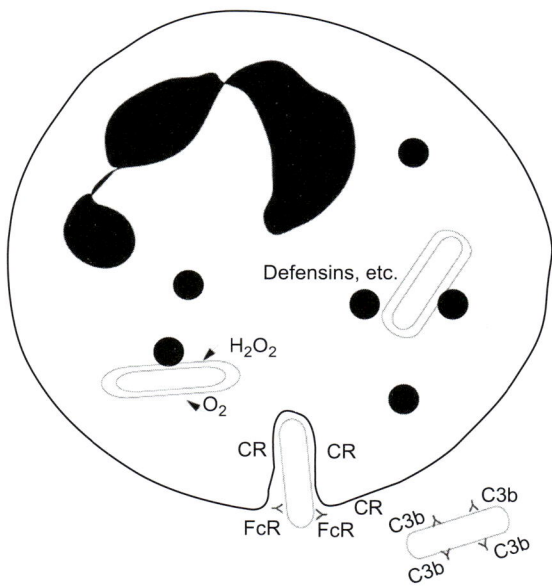

Opsonization and Phagocytosis. Figure 1 Simplified overview of how opsonization with IgG and C3b facilitates phagocytosis, phagolysosome formation, and release of reactive oxygen intermediates, defensins and other microbicidal peptides into the phagolysosome.

sins) that are present in preformed lysosomal granules in the phagocytic cell. Following granule fusion, the phagosome is referred to as a phagolysosome. Most microbes cannot survive in the hostile environment of the phagolysosome. Intracellular pathogens have evolved strategies to circumvent formation of this hostile environment (e.g. inhibition of phagosome acidification or phagolysosome formation), or resistance mechanisms to the toxic compounds contained within the phagolysosome (e.g. production of catalase or superoxide dismutase to scavenge reactive oxygen intermediates).

Preclinical Relevance
Experimental assessment of the effects of potential toxicants on innate immunity would include concerns about inhibition of the ability of the exposed individual to produce immunoglobulins that can opsonize pathogenic microbes, and the ability of their granulocytes and mononuclear phagocytes to ingest and kill the opsonized microorganisms. Various simple assays can be performed to assess phagocytosis of bacteria, yeast cells, or other particles. Opsonins (e.g. serum as a source of immunoglobulins and complement) might be added to facilitate evaluation of the phagocytic function of granulocytes and mononuclear phagocytes. Fluorescent beads (either opsonized or uncoated) can be used to evaluate phagocytosis by microscopy or flow cytometry.

Relevance to Humans
Phagocytic cells are essential for innate immunity against microbial infection. Individuals afflicted with genetically determined defects in their phagocytic cells often experience significant problems in controlling infectious agents. In some instances, the defect may be of sufficient severity to cause repeated severe infections or premature death. There is concern that toxicants in the environment might alter the ability of humans to produce, mobilize, and regulate the activity of these phagocytic cells, and by so doing render individuals susceptible to infectious diseases. An additional concern for inhaled particulate toxicants is that their ingestion by phagocytic cells in the lung might elicit release of inflammatory mediators (e.g. cytokines, eicosanoids) that could damage nearby cells, or attract inflammatory leukocytes that trigger hypersensitivity responses.

Regulatory Environment
Assessment of phagocytic cells is not universally required as part of the assessment of immune function. Nonetheless, it is frequently listed on tiers of immune function assays, and its assessment may be prudent or required if the compound in question is known to have an adverse effect on phagocytic cells, or on the production of immunoglobulins or other plasma proteins required for opsonization of microbes. Although assessment of opsonization is not frequently done in a regulatory situation, production of immunoglobulins, as assessed either by ELISA or a plaque-forming cell assay, is heavily relied upon for assessment of potential immunotoxicants. If a weak antibody response occurred following toxin exposure, it might indicate a potential for decreased resistance against extracellular pathogenic microbes that must be opsonized, ingested, and killed by phagocytic cells. The draft EPA Health Effects Test Guidelines (OPPTS 870.1350) for Acute Inhalation Toxicity with Histopathology include assessment of phagocytic activity by alveolar lavage macrophages using a fluorescent bead assay and microscopy.

References
1. Aderem AA, Underhill DM (1999) Mechanisms of phagocytosis in macrophages. Ann Rev Immunol 17:593–623
2. Janeway CA, Travers P, Walport M, Shlomchik M (2001) Immunobiology. Garland Publishing, New York, pp 24, 39–40, 49, 55–56, 371–373
3. Rosenberger CM, Finlay BB (2003) Phagocyte sabotage: disruption of macrophage signaling by bacterial pathogens. Nat Rev Mol Cell Biol 4:385–396

Oral Mucositis and Immunotoxicology

GARY J ROSENTHAL
Drug Development
RxKinetix Inc.
1172 Century Drive, Ste # 260
Louisville, CO 80027
USA

Synonyms
Mucositis, stomatitis, ulcerative mucositis, ulcerative stomatitis, radiation mucositis, oral ulcer

Definition
Oral mucositis is a frequent toxicological complication of high-dose chemotherapy as well as head and neck radiotherapy. This insidious condition manifests as inflammation of the moist mucosal lining the mouth and back of the throat and ranges from redness to severe ulceration over vast portions of the region. Symptoms of oral mucositis vary from local pain and discomfort to the inability to chew and/or swallow food or fluids, or to communicate.

Characteristics
Oral mucositis induced from either chemotherapy or

radiotherapy is characterized by painful and often incapacitating ulcerative lesions of the oropharyngeal mucosa (1). Chemotherapy-induced mucositis often presents as lesions involving the buccal and tongue mucosa, the soft palate, and the floor of the mouth. In contrast to this broad area of injury associated with chemotherapy, patients being treated for head and neck cancers with ionizing radiation manifest mucositis on those oral mucosal sites that lie in the direct path of radiation beam. The targeted oropharyngeal mucosa is lined by mucus membranes with a high mitotic index and is exceptionally sensitive to the antiproliferative effects of chemotherapy and radiotherapy.

At the tissue and cellular level, oral mucositis manifests initially as hypoplasia and destruction of superficial epithelial cells along with a lack of cell renewal. The subsequent erythematous areas proceed to desquamation and eventually ulcers covered by an exudate. From the host defense perspective, cancer treatment and resultant oral ulcers serve to weaken the defense system of the lining of the mouth leading to marked local infections. In addition, the potential for systemic infection due to opportunistic and acquired oral flora has been documented in cancer patients. This further complicates the already challenging health status of immunosuppressed patients where morbidity and mortality due to infection is of prime concern. While not considered life-threatening to the extent that chemotherapy-induced myelosuppression has historically been, oral mucositis is often identified by cancer patients as the single worst side effect of therapy. Symptoms may be so severe that they may limit a patient's ability to tolerate their chemotherapy or radiotherapy, resulting in delayed or shortened treatment and limited efficacy (2).

The early understanding of mucositis was of a disorder that simply resulted from non-specific toxicity of chemotherapy or radiotherapy against the basal epithelium. More recent research has broadened our understanding to suggest a more complex pathology with multifaceted interactions between connective tissue, endothelium and epithelium, myelosuppression and the oral microenviroment. A conceptual model for oral mucositis outlining the probable pathophysiology was published by Sonis in 1998 (3). In this model, mucositis is broken down into four phases (also see Figure 1):
- inflammatory/vascular phase
- epithelial phase
- ulcerative/bacteriological phase
- healing phase.

Inflammatory/Vascular Phase
In this early phase, chemotherapy or radiotherapy directly or indirectly induce events leading to local inflammatory events, including reactive oxygen-induced cell damage, NF-κB/early response gene activation, and proinflammatory cytokine induction, all of which serve as a foundation for local tissue damage and initiate the events leading to development of mucositis.

Epithelial Phase
Dividing cells of the epithelium begin to atrophy and cell renewal is diminished in the face of any continued antiproliferative cancer therapy. Inflammatory events serve to augment the negative effects of tissue destruction.

Ulcerative/Bacteriological Phase
This is generally considered to be the most symptomatic phase with ulcerative erosions of the mucosa and an altered opportunistic microbial microenviroment.

Healing Phase
This involves renewal of the epithelial cell population, re-establishment of microbial homeostasis and local immune function.

While oral mucositis is biologically complex and progresses as a continuum of these phases, the depiction outlined by Sonis in 1998 allows a focus on the characteristic and likely primary events of the disorder as it progresses from initiation to healing. Research conducted over the last 5 years (4,5) has served to add support for the model put forth by Sonis and elucidates in greater detail the cellular and subcellular events associated with oral mucositis. Considering the involvement of reactive oxygen species (ROS) in mediating other manifestations of chemotherapy or radiation toxicity, it seems likely that ROS play a role in the initiation and progression of mucosal injury (5).

Preclinical Relevance
Oral mucositis remains under extensive laboratory investigation. Animal models of mucositis have been developed in a variety of species using radiation alone or both chemotherapy and radiation protocols. Possessing a cheek pouch accessible to treatment and observation, the hamster has provided much of the currently available preclinical information, though rodents have also been used with some success. Critical understanding into the roles of mucosal immune dysregulation and wound healing are imperative areas of preclinical research that will improve prospects for effective prophylactic or treatment strategies.

Relevance to Humans
Myelosuppression was previously the major dose-limiting toxicity associated with of cancer therapies. With therapeutic advances in the 1980s relative to infection prevention and reduced myelosuppression via growth factors such as granulocyte colony stimulating factor

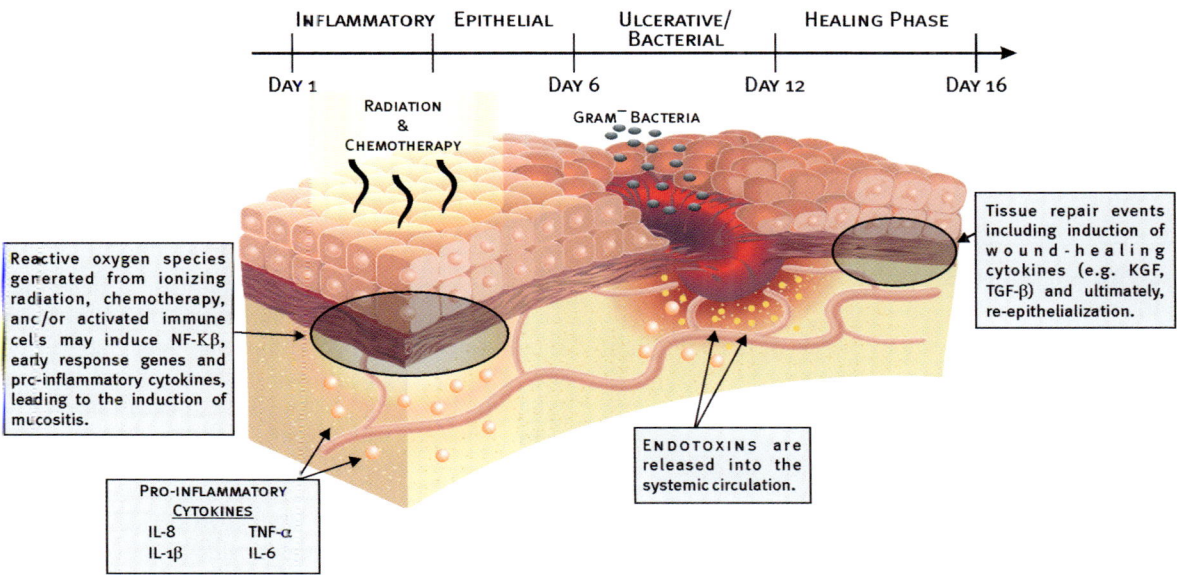

Oral Mucositis and Immunotoxicology. Figure 1 A conceptual model for oral mucositis outlining the probable pathophysiology.
Adapted from Sonis (1998).

(G-CSF), thrombopoietin and erythropoietin, non-hematologic toxicities have now become significant dose-limiting concerns. Of these non-hematologic toxicities, oral mucositis has emerged as one of the most problematic toxicities associated with current therapeutic regimens with direct impacts on cure rates and long-term survival (2).

Considering the frequency of oral mucositis, which is determined by the type of cancer therapy (e.g. approximately 40% in patients treated with systemic chemotherapy to nearly 100% of patients treated for head and neck cancer), this unmet clinical need remains an active area of clinical investigation.

Regulatory Environment

A plethora of approaches have undergone clinical assessment around the world with no single approach showing any consistent benefit. Currently no medication is approved by the Food and Drug Administration (FDA) to prevent or treat oral mucositis. While clinical trials investigating a variety of novel approaches continue, most patients and clinics manage symptoms with morphine or other narcotic analgesics, with mouth rinses, changes in diet, and cold liquids.

The FDA recognizes that oral mucositis is a serious illness that needs to be addressed expeditiously and as such has granted Fast Track designation to some candidate therapeutics in clinical development. Fast Track designation is intended to expedite the regulatory review and approval process for a product and claim that addresses a significant unmet medical need.

References

1. Berger AJ, Kilroy TJ (1997) Oral complications. In: Devita VT, Hellman S, Rosenberg SA (eds) Cancer Principles and Practice of Oncology, 5th ed. Lippincott Williams & Wilkins, Philadelphia, pp 2714–2725
2. Ohrn K, Sjoden P, Wahlin Y, Elf M (2001) Oral health and quality of life among patients with head and neck cancer or hematological malignancies. Supp Care Canc 9:528–538
3. Sonis ST (1998) Mucositis as a biological process: a new hypothesis for the development of chemotherapy-induced stomatotoxicity. Oral Oncol 34:39–43
4. Sonis ST, Scherer J, Phelan S et al. (2002) The gene expression sequence of radiated mucosa in an animal mucositis model. Cell Prolif 35 (Suppl 1):93–102
5. Blonder JM, Etter J, Samaniego A et al. (2001) Topical bioadhesive antioxidants reduce the severity of experimental radiation induced oral mucositis. Proc Amer Soc Clin Oncol 20:1606

Oral Ulcer

▶ Oral Mucositis and Immunotoxicology

Organ-Specific Autoimmunity

Autoimmune disease which affects a single organ in the body such as the pancreas in type 1 diabetes.
▶ Systemic Autoimmunity

Organogenesis

The formation of the various organs of the body from embryonic tissues during gestation.
▶ Developmental Immunotoxicology

Organotypic Murine or Human Skin Explant System

▶ Three-Dimensional Human Skin/Epidermal Models and Organotypic Human and Murine Skin Explant Systems

Orofacial Cleft

A birth defect in which the palate and/or associated structures fail to close along the midline during fetal development. Cleft palate is an orofacial cleft defect.
▶ Birth Defects, Immune Protection Against

Oryctolagus cuniculus

▶ Rabbit Immune System

Oxidative Stress

Oxidants released by phagocytes such as neutrophils or macrophages play an important role in early immune defense against pathogens. Free radicals are any atom that contain one or more orbital electrons with unpaired spin states. Some radical species are very reactive with other biomolecules (proteins, DNA) and others, like the normal triplet state of molecular oxygen, are relatively inert. Cells have multiple protective mechanisms against oxidative stress, e.g. protective agents like antioxidants to prevent cell damage.
▶ Rodents, Inbred Strains

Oxy-PAH

▶ Polycyclic Aromatic Hydrocarbons (PAHs) and the Immune System

p53

A tumor suppressor gene that codes for a protein of approximately 53 K molecular weight that controls cell death and cell cycling; this gene is mutated in many human cancers allowing transformed cells to escape cell death. p53 is often activated in response to DNA damage produced by genotoxic agents leading to inhibition of the cell cycle and induction of apoptosis.
▶ Polycyclic Aromatic Hydrocarbons (PAHs) and the Immune System

p53 Tumor Suppressor Protein

p53 tumor suppressor protein (p53) is a labile protein located in the nucleus. Agents which damage DNA induce p53 to become very stable by a post-translational mechanism, allowing its concentration in the nucleus to increase dramatically. The functional structure includes a strong transcriptional activation domain at the amino terminus, a central evolutionary highly conserved sequence-specific DNA binding domain and a tetramerization domain. p53 is a potent transcription factor and once activated it represses transcription of genes containing p53-binding sites (several of which are involved in stimulating cell growth) while stimulating expression of other genes involved in cell cycle control. The wild-type form of p53 exhibits strong anti-oncogenic properties. It arrests the cell cycle in response to DNA damage, thereby allowing DNA repair before the replication of the genome, and induces apoptosis if the damage to the cell is too severe. A loss of p53 function (oncogenic transformation) is therefore a key step in the neoplastic progression. It is also capable of strongly inhibiting transcription from many genes lacking p53-binding sites. Several oncogenic DNA viruses express viral gene products that associate with and inhibit the transcriptional activation function of p53. In cells, p53 can associate with a 90-kD protein, identified as the product of the *mdm-2* oncogene, which is amplified in some types of tumors. When bound to mdm-2, p53 can no longer function as an activator of transcription.
▶ Cancer and the Immune System

PAH

▶ Polycyclic Aromatic Hydrocarbons (PAHs) and the Immune System

Paramagnetic Cell Selection

▶ Antigen-Specific Cell Enrichment

Parturition

The act of giving birth to offspring.
▶ Developmental Immunotoxicology

Passive Cutaneous Anaphylaxis

▶ Assays for Antibody Production

Passive Immunotherapy

Immunotherapy based on the administration of preformed therapeutic agents of immunological origin, for example antibodies or cytokines. Passive immunotherapy entailing transfer of cells of the immune system is called "adoptive". In principle, passive immunotherapy does not depend on the immune system of the host, but in practice therapeutic efficacy frequently depends on the active participation of host's immune responses. For example the activity of monoclonal antibodies can depend on host's complement or cell-mediated cytotoxicity. *See also* Active immunotherapy.

▶ Tumor, Immune Response to

Passive Lymph Pump

Called also as an "extrinsic lymph pump." This term describes a sum of different forces which influence lymph flow but do not originate from intrinsic lymphatic contractions. Depending on the current conditions, these forces could sometimes generate pressure gradients in the lymphatic network supportive for central lymph flow. These passive lymph-driving forces include lymph formation, contractions of skeletal muscles, and fluctuations in central venous pressure, the influence of respiratory movements, pulsations of adjacent arteries, and influences of gravitational forces.
▶ Lymph Transport and Lymphatic System

Pathogenicity

The ability to produce a disease or morbid condition.
▶ Streptococcus Infection and Immunity

PBS

Phosphate-buffered saline.
▶ Maturation of the Immune Response

PCA

▶ Passive Cutaneous Anaphylaxis

PCDDs

▶ Dioxins and the Immune System

PCR

▶ Polymerase Chain Reaction (PCR)

PEF

Peak expiratory flow, a measurement of the rate of exhalation of air.
▶ Asthma

Pentavalent Vanadium

Pentavalent vanadium is the ionic form of vanadium when the maximal number (five) of the outer shell electrons (the two from 4s and all three from 3d) have been shed, thereby giving the atom an overall charge of +5.
▶ Vanadium and the Immune System

Peptide Regulatory Factors

▶ Cytokines

Peptides

▶ Immunotoxicology of Biotechnology-Derived Pharmaceuticals

Percent of Living Cells

▶ Viability, Cell

Perforin

A protein that, upon secretion, polymerizes to form membrane-spanning pores. Perforin monomers are synthesized by cytotoxic T lymphocytes (CTL) and natural killer (NK) cells and stored in cytoplasmic secretory granules.
▶ Cell-Mediated Lysis
▶ Cytotoxic T Lymphocytes
▶ Cancer and the Immune System

Peripheral Tolerance

Tolerance mechanisms taking place in the blood, spleen, lymph node and the mucosal immune system.

▶ Tolerance

Peyer's Patches

Lymphoid aggregates found in the wall of the small intestines which are mainly composed of B lymphocytes with many blasts organized in germinal centers. The predominant non-IgM isotype of B blasts in Peyer's patches is IgA, making these organs important players in mucosal immune responses.
▶ Immunotoxic Agents into the Body, Entry of
▶ Mucosa-Associated Lymphoid Tissue

PHA

▶ Polyclonal Activators

Phagocytic Cells

Cells that have the capability of ingesting bacteria, foreign material, and other cells.
▶ Respiratory Infections

Phagocytosis

Phagocytosis describes the ingestion of particles of more than 0.5 μm. This ability is largely restricted to specialized cells called phagocytes such as polymorphonuclear granulocytes or monocytes/macrophages, whereas pinocytosis (the uptake of small particles less than 0.5 μm) is a capacity of many cells. Phagocytosis depends on membrane receptors, including many receptors of the innate immune system or products of the adaptive immune system (e.g. antibodies). Phagocytosed biological materials including microbes are degraded and in most cases thereby inactivated. Phagocytosis by antigen-presenting cells is the prerequisite for the generation of small peptides, which are bound by MHC class II molecules and presented to T lymphocytes. Thus phagocytosis forms a link between innate and adaptive immunity.
▶ Immune Response
▶ Fish Immune System
▶ Mucosa-Associated Lymphoid Tissue
▶ Streptococcus Infection and Immunity
▶ Humoral Immunity

Phagocytosis Assay

Phagocytosis involves the engulfment of insoluble particles such as bacteria or erythrocytes resulting in their uptake into the cell cytoplasm where they would normally be degraded. Phagocytosis is commonly measured ex vivo using a microscopic or flow cytometry technique.
▶ Canine Immune System

Pharmacodynamics

A branch of pharmacology dealing with the interactions between drugs and living systems.
▶ Therapeutic Cytokines, Immunotoxicological Evaluation of

Pharmacokinetics

The study of bodily absorption, distribution, metabolism, and excretion of drugs. Developing the pharmacokinetic profile of drugs is an important component of setting safe and effective dose levels.
▶ Therapeutic Cytokines, Immunotoxicological Evaluation of

Phenotype

The combination of behavioral, physiological, and structural changes that occur in a genetically engineered animal. Phenotypes in knockout mice may be obvious, subtle, or not apparent, depending upon the presence of one or more compensatory genes (which can compensate for the engineered mutation) and the assays selected to assess the animals.
▶ Knockout, Genetic

Phosphatases

Phosphatases are enzymes responsible for the removal of phosphate groups from proteins. In many cases they are responsible for shutting down signaling pathways or preventing activation of signaling pathways in a cell.
▶ Signal Transduction During Lymphocyte Activation

Phosphorylative Balance

Phosphorylative balance is a descriptor of the status maintained by cellular proteins as a result of the activities of cellular phosphatases and kinases being regulated in a manner such that neither overwhelmingly predominates under normal conditions.
▶ Vanadium and the Immune System

Photoactivation

▶ Photoreactions

Photoallergic Contact Dermatitis

▶ Photoreactive Compounds

Photoallergy (Photoallergic Contact Dermatitis)

Photoallergy is an acquired immunological reactivity which does not occur after the first exposure of skin to certain chemicals (also after systemic administration) and subsequent exposure to light or UV radiation. It needs 1–2 weeks before a photoallergen-specific skin reactivity can be demonstrated after re-exposure and light or UV irradiation. It is a delayed hypersensitivity response manifested in the skin as eczema.
▶ Local Lymph Node Assay (IMDS), Modifications
▶ Photoreactions
▶ Photoreactive Compounds

Photodermatology

▶ Photoreactions

Photoirritation

Photoirritation is an acute toxic response elicited after the first exposure of skin to certain chemicals (also after systemic administration) and subsequent exposure to light or UV radiation.
▶ Local Lymph Node Assay (IMDS), Modifications

Photoreactions

THOMAS MAURER
Toxicology
Swissmedic
Erlachstrasse 8
CH-3000 Bern 9
Switzerland

Synonyms

photosensitization, photodermatology, photoactivation, photoallergy

Definition

The interactions of ▶ sunlight and skin may lead to beneficial effects such as vitamin D production, as well as adverse effects such as sunburn or cancer. Photoreactions may be due to direct effects of sunlight or due to the combination of an exogenous chemical and sunlight. The mechanisms underlying photoreactions are different and the immune system may or may not be involved. The sunlight effects are dependent on the wavelength of radiation and as a consequence on the penetration into the skin. Because of those influences it is clear that various types of reactions exist.

Characteristics

The solar spectrum is divided into the ultraviolet (UV) part, the visible part, and the infrared part. In the field of photodermatology, the ultraviolet region is the most important. In Table 1 ultraviolet is divided according to the "Commission Internationale d'Éclairage" and some characteristics for the three parts of ultraviolet radiation are given (1).
In many publications, the separation of UV-A and UV-B is made at 320 nm. In addition, the division between UV-B and UV-C is often made at 290 nm, due to the fact that shorter wavelengths are not present in terrestrial sunlight.
The laws of photochemistry are important in all photoreactions:
- Grotthus-Draper law: only radiation that is absorbed is capable of initiating a photochemical process.
- Bunsen-Roscoe law: the photochemical effect depends on dose (intensity × time) and not dose rate (intensity).
- Stark-Einstein law: each photon absorbed by a molecule activates one molecule in the primary step of a photochemical process.
- Planck's law: the energy of a photon is related to its wavelength.

A light-absorbing molecule, called chromophore, can

Photoreactions. Table 1 Ultraviolet

Light Type	% Solar Radiation on Earth	Wavelength	Depth of Penetration	Window Glass Penetration
UV-C	0%	280–100 nm	Epidermis	–
UV-B	1.7%	315–280 nm	Epidermis to papillary dermis	–
UV-A	6.3%	400–315 nm	Papillary to reticular dermis	+
Visible	92%	400–800 nm	Reticular dermis to subcutis	+

be present in the skin (such as DNA, proteins, lipoproteins, blood components, urocanic acid) or be exogenously applied (directly to the skin or be distributed to the skin after oral or parental administration). Molecules absorbing radiation get activated and can act directly by the oxidation of molecules, by radical or toxic photoproduct formation or by the formation of allergens based on the photoactivated binding of newly formed haptens with proteins.

Terms Often Used in Photodermatology

Photosensitization

This is often used as a general term for a chemical-induced reaction in the presence of ultraviolet or visible radiation including the terms phototoxicity and photoallergy.

Phototoxicity

Phototoxicity has been defined as an increased reactivity of the skin to UV radiation and/or visible light produced by a chemical agent on a non-immunological basis. It is an acute reaction which can be caused by a single exposure to a chemical and UV or visible light. In some publications the term photoirritation has been used specifically for phototoxic reactions after topical application of compounds. Phototoxic reactions may be induced in all persons exposed to a certain chemical and the appropriate dose of radiation.

Photoallergy

This is defined as an increased reactivity of the skin to UV radiation produced by a chemical agent on an immunological basis. The skin reaction does not occur after the first combined exposure to a chemical and light. An induction period of minimally 1 week or 2 weeks is required before skin reaction may be elicited. As in contact allergy, not all persons exposed to a combination of a certain chemical and UV radiation will react. The energy needed for the provocation of a photoallergic reaction is generally much lower than that needed for a phototoxic reaction.

Preclinical Relevance

Predictive testing of chemicals for their phototoxic or photoallergic potential has been performed for many years, even before official guidelines were available. However, these tests were only performed for certain classes of chemicals, such as optical brighteners, cosmetics, tetracyclines, non-steroidal anti-inflammatory drugs, etc. Testing for photomutagenicity or photocarcinogenicity has also been performed.

More details on the compounds involved in photoreactions are given in a separate chapter of this encyclopedia.

Relevance to Humans

Sunlight Effects Without Involvement of the Immune System

Acute and chronic effects in man are known due to direct exposure to sunlight. Acute effects, such as sunburn, occur in everyone if enough of the appropriate wavelength is absorbed. In the case of exposure to UV-B, a dose of 10–50 mJ/cm^2 can be sufficient to elicit a sunburn reaction. The erythema starts a few hours after exposure, leads to erythema and edema formation, and finishing with desquamation and long-lasting pigmentation.

Sunburn induced by the UV-A radiation range also occur. However, the dose needed lies in the range of 50–100 J/cm^2. The erythema starts earlier and the pigmentation induced is not of long duration.

The characteristic long-term clinical effects of sunlight exposure are wrinkling, atrophy, hyperpigmentation, actinic keratoses, and skin cancer formation (2,3).

Sunlight Effects with Involvement of the Immune System

The various forms of photoimmunologic effects in man are described in the monograph of Krutmann and Elmets (4). Photoreactions with an immunological mechanism are polymorphous light eruptions, chronic actinic dermatitis, lupus erythematosus, and solar urticaria.

The polymorphous light reaction is the most frequent

form of photoreaction. The reaction is mainly elicited by UV-A radiation.

Photourticaria is a rare type of photoreaction and often a chemical is additionally involved. The action spectrum varies and the reactions may be depend on UV-A, and visible light, as well as UV-B.

Sunlight may not only stimulate photoreactions in the skin in combination with allergens but also can suppress the immune system of the skin (5). Especially when the skin is exposed to UV-B, the function of the Langerhans cells may be inhibited and/or the number of cells in the skin is reduced. UV immunosuppressive effects may lead to the inhibition of contact allergic reactions and are involved in the promotion of skin cancer (4,5).

Sunlight Effects in Combination with Chemicals

Phototoxic reactions in man are much more frequent than photoallergic reactions. However, photoallergic reactions may lead to persistent light sensitivity and are, therefore, more relevant for the patients involved. Chemicals known to be involved in photoallergies are antipsychotic drugs, antihistamines, antidiabetics, diuretics, antibiotics, halogenated salicylanilides, UV filters, etc.

Photoallergic reactions are difficult to categorize according to the classification of Coombs and Gell. Due to the fact that they are seen 24 hours after photoprovocation, it is thought that most of the photoallergic reactions are of type IV. In most cases the photoallergic reactions are dependent on UV-A and the irradiation dose for the elicitation of a photoallergic reaction is generally much lower than that eliciting a phototoxic reaction or sunburn.

The combination of radiation and a chemical to induce a toxic effect has been used in recent years for therapy in dermatology and oncology (6). Compounds used for photodynamic therapy are hematoporphyrin derivative[s], aminolevulinic acid, benzoporphyrin derivative[s], or phthalocyanines. These chemicals are activated by wavelength of the low energetic visible part. This light part penetrates better, to deeper tissues, and should elicit phototoxic reactions only in the tissue where the chemical had been distributed but not in the normal tissue. With flexible lasers it is even possible to treat cancers in the body (e.g. cancers in the urinary bladder).

Regulatory Environment

For many years the only guideline including a list of methods to test for photoallergenic potential was the *Guideline for Toxicity Studies of Drugs Manual of Japan*. Eight methods were included in the list of possible methods without giving priority to any one of the eight methods.

Last year a new CPMP guideline of the European Agency for the Evaluation of Medicinal Products (EMEA) came into force: CPMP/SWP/398/01: *Note for Guidance on Photosafety Testing* (http://www.emea.eu.int). It is requested to test all new drugs, which absorb light between 290 nm and 700 nm and which are used topically or reach the skin or eyes following systemic exposure. The testing should include phototoxicity, photoallergy, photomutagenicity and photocarcinogenicity. The 3T3 NRU in vitro test is recommended for phototoxicity. This test has been validated and a draft OECD protocol is available; the final acceptance by the OECD is expected soon. For the other parts of phototesting general recommendations are made.

A guidance for industry paper on ▶ photosafety testing from the FDA (CDER) came into force in May 2003 (http://www.fda.gov/cder/guidance/index.htm). In the introduction, the following statement was made: "Use of the principles expressed in this guidance should reduce unnecessary testing while ensuring an appropriate assessment of photosafety". The guidance paper does not recommend specific methods. It mentions general aspects in photosafety testing which have to be considered and explains possible testing strategies that depend on the duration of use of a drug. Discussions on the influence of formulations on photoreaction are also included.

References

1. Diffey B-L (2002) What is light? Photodermatol Photoimmunol Photomed 18:68–74
2. Epstein JH (1999) Phototoxicity and photoallergy. Semin Cutan Med Surg 18:274–284
3. Berneburg M, Plettenberg H, Krutmann J (2000) Photoaging of human skin. Photodermatol Photoimmunol Photomed 16:239–244
4. Krutmann J, Elmets C-A (1995) Photoimmunology. Blackwell Science, New York
5. Meunier L (1999) Ultraviolet light and dendritic cells. Eur J Dermatol 9:269–275
6. Ceburkov O, Gollnick H (2000) Photodynamic therapy in dermatology. Eur J Dermatol 10:568–576

Photoreactive Compounds

Frank Gerberick
Human Safety Department
Procter & Gamble Company
P.O. Box 538707
Cincinnati, OH 45253-8707
USA

Synonyms

photoallergy, photoallergic contact dermatitis, contact photoallergy

Definition

▶ Photosensitivity is the broad term that is used to describe abnormal or adverse reactions to the sun or artificial light sources. These response may be phototoxic or photoallergic in nature. Photoallergy is a cell-mediated immunologic reaction to a chemical that has been made antigenic by the interaction with ultraviolet (UV) or visible radiation (1). This reaction is similar to allergic contact dermatitis, but differs in that chemicals require activation by light to elicit the response. Clinically, photoallergenic skin responses resemble phototoxic reactions but may be distinguished by the increased severity with repeat exposure and time course for eliciting a response. Further, photoallergenic responses are less frequent than phototoxic reactions. Histologically, the response is characterized by epidermal edema and vesicle formation and a dense perivascular infiltrate. The action spectrum for most photoallergic reactions is 310–400 nm, primarily UVA radiation. Photoallergy is prevented by protection from or avoiding light exposure. Many of the known human photoallergens are also phototoxic.

Molecular Characteristics

The common point of initiation for any biological response to light is absorption of photon energy by a ▶ chromophore. The probability of light absorption is dependent on the molecular structure of the chromophore and the wavelengths of light. The wavelengths of light of most concern are UV (100–400 nm) and, to a lesser extent visible radiation (400–760 nm). Because wavelengths below 290 nm are absorbed by the ozone layer and do not reach the surface of the earth, ▶ UVC radiation from sunlight is of little concern. It is important to keep in mind that the energy of UVR is inversely proportional to its wavelength. This relationship is meaningful when considering the probability and consequences of photoactivation.
Absorption of UV or visible photons results in electronically excited molecules; dissipation of this energy may result in an adverse phototoxic effect on biological systems. For a molecule to have a direct photobiological effect, it must absorb photons and dispense with this gain of energy in some manner. Otherwise, any impact of a xenobiotic on UVR-induced responses in the skin would be attributed to some secondary mechanism or response modifier, that is, changing the spectrum of light to which the skin is exposed. Thus, understanding the potential mechanism(s) of the interaction between UV and a chemical is critical when considering the phototoxicologic impact of a response (2). However, it is not possible at this time to predict the photoallergic potential of a compound from its molecular structure alone.

Putative Interaction with the Immune System

Photoallergy is a cell-mediated immunologic reaction to chemicals that histologically and mechanistically resembles allergic contact dermatitis (1). The critical factor differentiating these reactions is that the chemicals that produce photoallergy reactions require activation by UV radiation in order to induce and elicit the response. Two mechanisms have been postulated to explain formation of the photoproducts responsible for the induction of photoallergic reactions (2). In the first mechanism, a photoallergen in its excited state reacts with proteins to form an allergen. In the second, the excited state of the photoallergen is converted into a simple contact allergen that binds to protein. Following formation via either pathway, it is believed that the photoallergen is processed by epidermal Langerhans cells, which transport the processed antigenic determinant (photohapten) to regional lymph nodes. Therein, antigen-specific T helper cells recognize the antigen bound to Langerhans cells and are triggered to proliferate and promote the dissemination of effector and memory T cells that are able to elicit a cutaneous response upon subsequent encounter with the inducing antigen. In support, it has been shown that photoallergic contact dermatitis can be adoptively transferred with immune cells from animals with ▶ photosensitization to naive animals (3). Moreover, it has been demonstrated that photohapten-modified Langerhans cells are capable of stimulating lymphocytes from photoallergic animals in an antigen-specific manner (4).
Guinea pig and mouse models have been developed for predicting the photoallergic potential of photoreactive chemicals (5,6). Currently, there are no in vitro methods available for screening chemicals for photoallergy. However, an in vitro 3T3 Neutral Red Uptake assay has been developed for ▶ phototoxicity testing that has been proposed as being useful in tier testing for photoallergy (7).

Relevance to Humans

Photoallergy clinical responses characteristically range

from a simple erythema to a severe vesiculobullous eruption. The diagnosis of photoallergy is suspected by the clinical picture, including the character and distribution of the eruption (e.g. eruption is most notably in sun-exposed areas such as face and hands). Involvement extending beyond the exposed site frequently occurs. Similar to the identification of a contact allergen, the contact photoallergen is identified by photopatch testing by the dermatologist.

Over the years, photoallergic responses in humans have been clearly established with a number of compounds (Table 1). For example, antimicrobial agents, fragrances, plant derivatives, sunscreens, and some drugs have been reported to produce photoallergy in humans. Specifically, halogenated phenolic compounds, coumarins, musk ambrette, promethazine, chlorpromazine and p-aminobenzoic acid have been reported to be photoallergenic in humans (8).

References

1. Stephens TJ, Berstresser PR (1985) Fundamental concepts in photoimmunology and photoallergy. J Toxicol Cutan Ocular Toxicol 4:193–218
2. Kornhauser A, Wamer WG, Lambert LA (1996) Cellular and molecular events following ultraviolet irradiation of skin. In: Marzulli FN, Maibach HI (eds) Dermatotoxicology. Taylor & Francis. Washington DC, pp 189–220
3. Harber LC, Harris H, Baer RL (1966) Photoallergic contact dermatitis due to halogenated salicylanilides and related compounds. Arch Dermatol 94:225–230
4. Gerberick GF, Ryan CA, von Bargen EC, Stuard SB, Ridder GM (1991) Examination of tetrachlorosalicylicylanilide (TCSA) photoallergy using in vitro photohapten-modified Langerhans cell-enriched epidermal cells. J Invest Dermatol 97:210–218
5. Gerberick GF (1994) Predictive models for assessment of contact allergy. In: Dean JH, Luster MI, Munson AE (eds) Immunotoxicology and immunopharmacology. Raven Press, New York, pp 681–692
6. Ulrich P, Homey B, Vohr HW (1998) A modified murine local lymph node assay for differentiation of contact photoallergy from phototoxicity by analysis of cytokine expression in skin-draining lymph node cells. Toxicology 125:149–168
7. Spielmann H, Balls M, Dupuis J et al. (1998) The international EU/COLIPA in vitro phototoxicity validation study: Results of phase II (blind trial). Part 1: The 3T3 NRU Phototoxicity Test. Toxicol In Vitro 12:305–327
8. Kaidbey K (1991) The evaluation of photoallergic contact sensitizers in humans. In: Marzulli FN, Maibach HI (eds) Dermatotoxicology. Taylor & Francis, Washington DC, pp 595–605

Photoreactive Compounds. Table 1 Substances reported to produce photoallergic contact dermatitis in humans

Classes and Compounds	
Antimicrobial agents	3,3,4',5-Tetrachlorosalicylanilide
	3,4',5'-Tribromosalicylanilide
	3,4,4'-Tribromocarbanilide
	Hexachlorophene
	Bithionol
	Fentichlor
Fragrances	Musk ambrette
	6-Methylcoumarin
Plant derivatives	Balsam of Peru
	Wood mixture
	Lichen mixture
Sunscreens	p-Aminobenzoic acid (PABA)
	Octylbenzones
	Butyl methoxydibenzoylmethane
Drugs	Sulfanilamide
	Chlorpromazine
	Promethazine
	Enoxacin

Photosafety

Side effects due to a combination of parts of sunlight and a light absorbing chemical have different mechanisms (e.g. non-immunological or immunological based reactions). To cover all aspects of possible reactions, it is therefore necessary to test a chemical in various models. Guidelines for photosafety testing covers in general discussions on phototoxicity, photoallergy, photomutagenicity and photocarcinogenicity.

▶ Photoreactions

Photosensitivity

Used to describe a state of heightened reactivity to photons and may be due to photoallergy, phototoxicity or an unknown mechanism. This term is often used when the mechanism underlying the abnormal reaction is unknown.

▶ Photoreactive Compounds

Photosensitization

Thie term has two meanings:
- Light sensitivity produced after photons are absorbed by an exogenous chromophore or an abnormally large amount of an endogenous chromophore.
- Describes the physical process of increasing the (skin) sensitivity to the effects of light, mainly UVA- and UVB-absorbing substances (photosensitizer). When used to describe the reaction of skin to an exogenous chemical and UV or visible radiation, the term includes both photoirritation and photoallergic reactions.

▶ Local Lymph Node Assay (IMDS), Modifications
▶ Photoreactions
▶ Photoreactive Compounds

Phototoxicity

Although a general term for compounds able to induce photoreactions, i.e. photoirritation, photoallergy, photogenotoxicity, or photocarcinogenicity, it is, however, often used to substitute the term photoirritation, which broadly refers to photon-induced damage to cells or tissues. Thus, its use is usually restricted to damage that does not occur via an immune mechanism. Phototoxicity is used clinically to refer to the presence of morphological evidence of acute changes in the skin (e.g. erythema, edema, and scaling) following exposure to UV radiation with or without an exogenous photosensitizer.

▶ Photoreactive Compounds
▶ Local Lymph Node Assay (IMDS), Modifications

Physicochemical Properties

These are the many properties which can be used to describe a chemical, such as its formula, its melting point, pKa, lopP, molecular orbital data, electrophilicity and so on. Data on these properties is used to build quantitative structure–activity relationships.

▶ Chemical Structure and the Generation of an Allergic Reaction

Phytohemagglutin (PHA)

A plant lectin that can activate human and rodent T lymphocytes. It is a mitogen for T cells.

▶ Polyclonal Activators
▶ Nutrition and the Immune System

Pig

▶ Porcine Immune System

Pineal Gland

Corpus pineale; *epiphysis, epiphysis cerebri*, pineal body, ductless gland; a small endocrine gland in the brain, situated beneath the back part of the *corpus callosum*, which produces the hormones melatonin and serotonin.

▶ Serotonin

Pinocytic Uptake

A form of endocytosis whereby a small liquid droplet, a minute particle, and/or solute is ingested by a cell via a process that utilizes nonspecific membrane invagination and subsequent formation of endocytic vesicles containing the exogenous material.

▶ Chromium and the Immune System

Pituitary Gland

A small endocrine gland located at the base of the brain that regulates growth and metabolism. The gland is divided into the posterior and anterior pituitary, each producing its own unique hormones.

▶ Stress and the Immune System

Plaque Assay

▶ Plaque-Forming Cell Assays
▶ Plaque Versus ELISA Assays. Evaluation of Humoral Immune Responses to T-Dependent Antigens

Plaque-Forming Cell Assays

GREGORY LADICS
DuPont Haskell Laboratory
1090 Elkton Road
Newark, DE 19714
USA

Synonyms

Humoral immune assay, hemolytic plaque assay, plaque assay, antibody-forming cell, AFC, enzyme-linked immunospot, ▶ ELISPOT, sheep red blood cells, SRBC, enzyme linked immuno sorbant assay, ELISA, spot-forming cells, SFC

Short Description

The aim of such assays is to evaluate the ability of an individual to mount a humoral immuneresponse (i.e. antibody response) to a particular antigen, typically sheep red blood cells (SRBC). Following exposure to SRBCs or other ▶ T-cell dependent antigens, the generation of an antigen specific antibody response requires the cooperation and interaction of several immune cell types: antigen presenting cells (e.g. macrophages), T cells, and B cells. Thus, there are a number of steps in the process where alterations in the function of specific cells can impair the ability of B cells to produce antigen-specific antibody. Because of this complex interaction of cells, the quantification of the plaque-forming cell (PFC) response (i.e. the specific antibody-forming cell (AFC) SRBC response) was found to provide one of the best predictors of immunotoxicity in mice (1,2). The PFC response to SRBC uses immunocompetent cells from lymphoid organs, primarily the spleen. In addition, other antigens, known as T cell-independent antigens (e.g. TNP-lipopolysaccharide, requires B cells and macrophages or DNP-Ficoll, requires B cells only) can be used in the PFC assay to identify the primary immune cell type(s) targeted by a particular compound. These antigens bypass the need for T cells in eliciting production of antibody by B cells. Despite the antigen used, the PFC assay is based on SRBC, which include the antigen or conjugated hapten for the specific antibody being produced by B cells. A modification of the PFC assay that allows for the measurement of AFCs that produce antibody of different isotypes (immunoglobulins IgM, IgG, IgE, or IgA) is the ▶ ELISPOT assay. The ELISPOT assay is similar in methodology to the ELISA.

Characteristics

The hemolytic plaque assay was originally developed by Jerne and Nordin in 1963 as a means to measure the number of IgM antibody-forming cells specific to SRBC (3). In this method which has been subsequently modified, spleens from SRBC immunized animals are removed (for rodents, typically 4–5 days following immunization) and cells are then mixed with SRBC and complement in a semisolid media (e.g. agar). This mixture is then plated onto Petri dishes, covered with a glass coverslip, and incubated for 3 h at 37°C. During the incubation, B cells produce IgM antibody specific for SRBC, which then bind to SRBC membrane antigens and cause complement-mediated lysis of the SRBC and the subsequent formation of plaques (i.e. clear areas of hemolysis around each antibody-forming cell) which can then be counted visually. Data are usually expressed as IgM AFC (or PFC)/spleen or IgM AFC (or PFC)/million spleen cells. The secondary immune response (i.e. IgG AFC) can be evaluated using a minor modification of this same assay.

The PFC assay may also be conducted entirely in vitro with immunocompetent cells obtained from either chemically treated or naive animals using modifications of the Mishell-Dutton culture system (4). In the latter case, the immunocompetent cells are exposed to the test article (or metabolites if test article incubated with a metabolic activation system prior to in vitro exposure of cells) and SRBC for the first time in tissue culture. This approach also allows for separation-reconstitution studies with the cells involved in generating a primary humoral immune response (i.e. macrophages, B cells, and T cells) to identify the primary cell type(s) targeted by a test article.

The ELISPOT assay is a modification of the PFC assay that allows for the measurement of AFCs that produce antibody of different isotypes (IgM, IgG, IgE, or IgA). The ELISPOT assay is similar in methodology to the ELISA. Antigen is allowed to adhere to a solid support (e.g. plastic or nitrocellulose). A blocking solution is then added to bind any remaining protein-binding sites and subsequently decrease nonspecific binding of reagents. Immunocompetent cells are then added and during an incubation period at 37°C/5% CO_2 for 4–20 h, AFCs secrete antibody that binds to surrounding antigen. The AFCs detected by the ELISPOT assay are called spot-forming cells (SFC). To determine the isotype of SFC, an enzyme (e.g., alkaline phosphatase) linked anti-immunoglobulin antibody conjugate specific for different heavy chains is used. A substrate is then added and an insoluble product produced in areas where antibody is bound to antigen. Each spot produced by the insoluble product represents an AFC. The spots are counted under magnification (15–25×) and compared to negative controls (no cells or no antibody). Data are expressed as SFC/million cells or SFC/organ (e.g. small intestine or spleen).

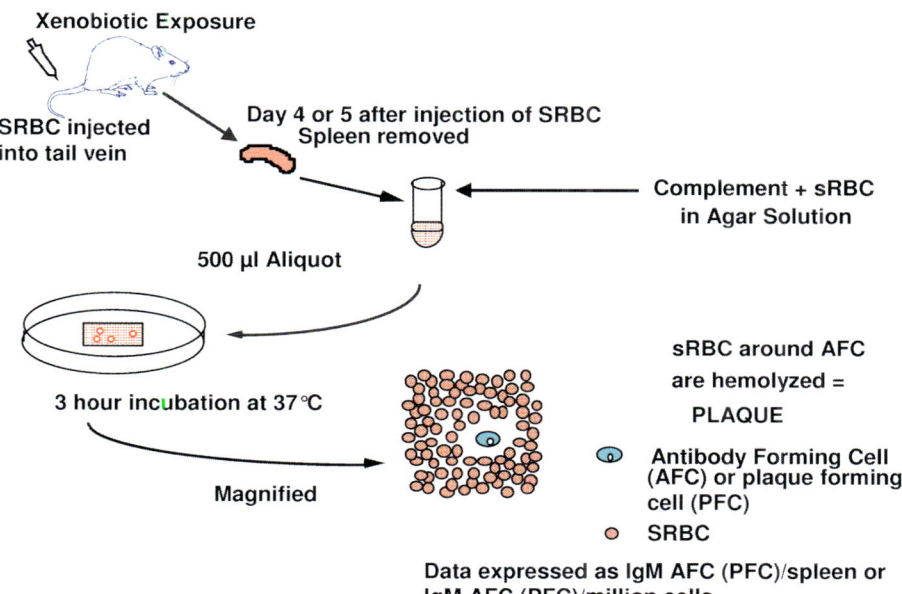

Plaque-Forming Cell Assays. Figure 1 Immunoglobulin M plaque-forming cell assay.

Pros and Cons

The PFC has some advantages.
- The quantification of the PFC response (i.e. the specific IgM antibody-forming cell response) was found to provide one of the best predictors of immunotoxicity in mice (1,2).
- The PFC assay is well-characterized and has been validated in a number of laboratories for its sensitivity and reproducibility and is recommended by the US National Toxicology Program to assess the humoral immune response.
- ELISPOT can measure isotypes and subclasses.
- The PFC assay can also be conducted entirely in vitro. This approach allows for separation-reconstitution studies to identify the primary cell type(s) targeted by a test article as well as a means to potentially distinguish between test article-induced direct or indirect (e.g. neuroendocrine alterations) effects on immunocompetent cells.

There are also some disadvantages.
- The PFC does not quantitate the amount of antibody produced, but rather the number of specific antibody-producing plasma cells in a particular tissue (e.g. spleen) and therefore does not account for antibody produced in other sites (e.g. bone marrow, lymph nodes).
- In addition, the PFC assay requires the sacrifice of the animal and thus does not allow for multiple samples to be taken from the same animal. As a result, a time course of humoral immune function cannot be conducted; a recovery period following test article administration cannot be evaluated; upon rechallenge with antigen a secondary IgG-mediated immune response cannot be measured.
- The PFC assay is expensive and very labor intensive.
- The PFC assay does not lend itself to be automated, although the ELISPOT assay, which can be conducted in nitrocellulose bottom 96-well plates, may be automated.

Predictivity

As indicated above quantification of the PFC response was found to provide one of the best predictors of immunotoxicity in mice (1). In a subsequent study, 51 chemicals were assessed in mice using a panel of immunotoxicologic assays. Of the assays evaluated, the highest correlations with immunotoxic potential for the chemicals were observed for the splenic IgM PFC response and cell-surface marker analysis (2). Therefore, the PFC response to SRBC may be the most sensitive immune parameter available to assess chemical-induced alterations to the immune system. The basis for this sensitivity, as previously discussed, stems from the fact that the generation of an antigen-specific antibody response requires the cooperation and interaction of several immune cell types: antigen presenting cells (e.g. macrophages), T cells, and B cells. Thus, there are a number of cells that a chemical can target to alter the ability of B cells to produce antigen-specific antibody.

With respect to ELISPOT, nothing is known up to now

about predictivity for the immunotoxic potential of a chemical.

Relevance to Humans

Due to the invasive nature (i.e. injection of antigen, acquisition of immunocompetent cells) of the PFC assay, predictive testing in humans is not conducted. However, limited predictive testing can be performed on human peripheral blood. Serum concentrations of each of the major immunoglobulin classes (IgM, IgG, IgA, and IgE) can be measured and natural immunity (i.e. antibody levels to ubiquitous antigens such as blood group A and B antigens, heterolysins, and antistreptolysin) can be assessed by ELISA. However, quantifying total immunoglobulin levels lacks the predictive value of assays that measure specific antibody responses following challenge with an antigen. Additionally, antibody responses following immunization to proteins (e.g. diphtheria, tetanus, poliomyelitis) and polysaccharides (e.g. pneumococcal, meningococcal) can be measured.

For the most part, the tests available for evaluating humoral immunity in humans only assess the secondary recall response, rather than a primary response to a new antigen. Primary immune responses, however, are a more sensitive measure of immune alteration compared to secondary responses (5). The clinical relevance of moderate or transient alterations in humoral immune function is also not known. Human data is limited to severe and longlasting immunosuppression resulting from therapeutic drug treatments. Furthermore, what human data are available are difficult to interpret due to the idiosyncrasies of the immune system. The age, sex, or genetic background of an individual, and a number of other factors such as stress, malnutrition, chronic infections, or neoplasia can effect a "normal" immune response. Risk assessment is further complicated due to a lack of human exposure data to xenobiotics in general. Additionally, a biologically significant change in immune function does not necessarily produce a clinical health effect until the patient encounters a stress or insult. Further problems arise when evaluating dose-response relationships due to the immune systems' reserve or redundant capacity.

Regulatory Environment

Relatively recently the regulatory agencies have begun to require the evaluation of the primary antibody response. In the USA, for example, in 1998 the Environmental Protection Agency EPA published guidelines requiring chemicals used as pesticides to undergo an evaluation of the primary humoral immune response to a T-dependent antigen (i.e. SRBC) using either the PFC or ELISA following the administration of a test article to mice and/or rats for 28 days. Testing the ability of pharmaceuticals to alter the antigen-specific antibody response is determined by a number of conditions. In the US, for example, the Food and Drug Agency (FDA) suggests considering follow-up studies to investigate mechanism(s) of immunotoxicity that may include evaluating the antibody response to a T-dependent antigen among other endpoints if:

- there is evidence of immunotoxicity in repeat-dose toxicology studies
- the test article or metabolites accumulate or are retained in reticuloendothelial tissues (i.e. there are pharmacokinetic effects)
- the test article is used for the treatment of HIV or a related disease
- there are effects suggestive of immunosuppression that occur in clinical trials.

In Europe, conventional pharmaceuticals (not biotechnology derived or vaccines) under CPMP guidance must undergo an initial 28-day screening study in which the primary humoral immune response to a T-dependent antigen (e.g. SRBC) is conducted if an analysis of lymphocyte subsets and natural killer cell activity are unavailable. Additional studies are conducted on a case-by-case basis, which consist of functional assays to further define immunological changes and may include a measure of the primary antibody if not evaluated in the initial screening study. The relevant guidelines are:

- FDA (CDER) Immunotoxicology Evaluation of Investigational New Drugs, 2002
- EPA OPPTS 870.7800 Immunotoxicity, 1998
- CPMP/SWP/2145/00 Note for Guidance on Non-Clinical Immunotoxicology Testing of Medicinal Products, 2001.

References

1. Luster MI, Munson AE, Thomas P et al. (1988) Development of a testing battery to assess chemical-induced immunotoxicity: National Toxicology Program's guidelines for immunotoxicity evaluation in mice. Fundam Appl Toxicol 10:2–19
2. Luster MI, Portier C, Pait DG et al. (1992) Risk assessment in immunotoxicology. I. Sensitivity and predictability of immune tests. Fundam Appl Toxicol 18:200–210
3. Jerne NK, Nordin AA (1963) Plaque formation in agar by single antibody producing cells. Science 140:405–412
4. Kawabata TT, White KL Jr (1987) Suppression of the in vitro humoral immune response of mouse splenocytes by benzo(a)pyrene and inhibition of benzo(a)pyrene-induced immunosuppression by α-naphthoflavone. Cancer Res 47:2317–2322
5. National Research Council. Biologic Markers in Immunotoxicology (1992) National Academy Press, Washington DC, pp 1–206

Plaque Versus ELISA Assays. Evaluation of Humoral Immune Responses to T-Dependent Antigens

Kimber L White
Dept. of Pharmacology & Toxicology
Virginia Commonwealth University
P.O.Box 980613
Richmond, VA 23298
USA

Synonyms
Plaque Assay, Antibody-Forming Cell Assay, SRBC ELISA, KLH ELISA, T-Dependent Antibody-Forming Cell Response, Evaluation of Humoral Immunity

Short Description
In other sections of this encyclopedia are descriptions on the Plaque assay and various ELISA assays used to evaluate the humoral immune response to T-dependent antigens. This essay will compare the Plaque assay to the ELISA assay with regards to the information obtained from each, focusing on the strengths and weaknesses of the various assays, proper data interpretation, and the state of validation and predictability obtained from the different assay techniques. It may be helpful for the reader to review the sections on the methodology of the assay in the encyclopedia to better appreciate the comparison of the two assays as they relate to evaluating humoral immune responses to T-dependent antigens.

Characteristics
The Plaque assay to the T-dependent antigen sheep erythrocytes (SRBC) has been the "gold standard" for evaluating potential immunotoxic effects of compounds on the immune system. While it is used primarily to evaluate effects on the humoral immunity, the complex nature of developing an antibody response to a T-dependent antigen requires multiple cell types, including antigen presenting cells, such as macrophages and dendritic cells, T cells including T-helper cells, specifically Th2 T-helper cells, and B cells that are capable of proliferating and differentiating into plasma cells capable of secreting antibody to the SRBC antigen. The Plaque assay has been validated in both mice and rats in national and international ring studies, and it has been shown to be the most predictive of the functional assays for identifying immunomodulatory compounds (see below) (1,2).

Several regulatory agencies have indicated that in evaluating responses to T-dependent antigens, assays other than the Plaque assay can be used. These agencies often site the work of Temple et al. (3) as the basis for these alternative assays. In the Temple paper, the effects of two potent immunosuppressive compounds, benzo(a)pyrene [▶ B(a)P] and cylophosphamide, were evaluated in both female B6C3F1 mice and Fischer 344 rats. Both the Plaque assay and the SRBC ELISA were conducted on the optimum day for each assay in the two rodent species. The results of these studies showed similar immunosuppressive results for both the Plaque assay and the SRBC ELISA in mice and rats. Due to the similar responses observed, the SRBC ELISA and, subsequently, ELISAs to other T-dependent antigens have been accepted by various regulatory agencies as appropriate for evaluating the potential immunotoxicity of drugs and chemicals.

While both the Plaque assay and ELISAs to T-dependent antigens are capable of evaluating the effects on the humoral immune response, one must be careful in the interpretation of the data, which are obtained from each of the assays. The key points of the data interpretation are shown below in Table 1.

The Plaque assay, conducted on splenocytes, measures only the production of antibody cells in the spleen. Accordingly, it only reflects effects that occur on spleen cell populations. The endpoint evaluated is not antibody levels, but the number of antibody-forming cells (AFC) also called plaque-forming cells (PFC) present in the spleen. Each of the AFCs is capable of producing one plaque, which is actually enumerated in the assay. In contrast, the ELISA to a T-dependent antigen measures the antibody levels in the serum (serum titer) and is either expressed in mass units, if a standard is available, or as a titer, which is routinely defined as the reciprocal of the dilution which meets the criteria of the laboratory's ELISA assay. Such criteria can include an "Endpoint Titer", which is the serum dilution of the test sample that does not differ from background control ± either two or three standard deviations of the background. Midpoint Titers are another approach for defining the concentration of antibody present in the serum. Often the Midpoint Titer is defined as the reciprocal of the dilution of serum obtained at an optical density of O.5, provided the value falls on the linear portion of the ELISA curve.

Regardless of how the serum titer is calculated, common to all approaches is the interpretation of data. Titers obtained from serum represent the antibody levels resulting from the production of antibody in the spleen, lymph nodes and the bone marrow. While most immunotoxicologist are familiar with the fact that spleen and lymph nodes are capable of producing antibody, the fact that the bone marrow is a major source of antibody production and the primary source for secondary antibody production is less well known. Accordingly, the two assays are actually measuring two different endpoints. In the case of the Plaque assay, the effect of the compound on the spleen only is being

Plaque Versus ELISA Assays. Evaluation of Humoral Immune Responses to T-Dependent Antigens.
Table 1 Immunological Interpretation of Plaque and ELISA Data

Assay	Antibody Production Site	Endpoint Evaluated
Plaque Assay	Spleen	Effects on Spleen Only
Serum Titers (ELISA)	Spleen, Lymph Nodes, and Bone Marrow	Holistic Evaluation of Antibody Production in the Animal

determined, while in the ELISA the effect of the compound on the spleen, lymph nodes and bone marrow is what is being evaluated. Thus, it is not surprising that compounds that affect either the spleen or bone marrow have the potential for producing different results in the two assays.

In studies conducted by the US National Toxicology Program (NTP), differential effects have been observed in the Plaque assay and the SRBC ELISA in the same animals treated with the test compound. For example, with ▶ AZT, a drug known to affect bone marrow, no effect was observed in the Plaque assay, while a statistically significant decrease in the serum titers, as determined by the SRBC ELISA, was observed. Similarly, with the drug thalidomide, which produces numerous changes in the spleen, a statistically significant enhanced response was observed in the Plaque assay, while no effect was observed in the serum antibody levels as determined by SRBC ELISA. Recently, Keyhole Limpet Hemocyanin (KLH), another T-dependent antigen, has been receiving considerable use in evaluating the effects on humoral immunity using ELISA technology to measure antibody levels to KLH in serum. KLH has several advantages over SRBC as a T-dependent antigen in conducting immunotoxicological evaluations. These are addressed in the next section. While the use of KLH is increasing, recent studies have suggested that it may be less sensitive than the Plaque assay (4). By increasing the amount of KLH antigen used to sensitize the test species, at least for mice and rats, sensitivity can be increased somewhat but it still is less than observed with the Plaque assay.

Since the Plaque assay and ELISA actually measure two different endpoints, the selection of what assay should be used depends in part on the question being asked. On occasion, regulatory agencies have requested evaluation of both the primary (IgM response following single sensitization) and secondary (IgG response following two sensitizations) responses. The characteristics of these responses are shown in Table 2.

In general the primary response, following a single injection of a T-dependent antigen, results predominately in the production of IgM. The primary response is very sensitive to modulation by compounds. In our studies we have never seen an effect on the IgG response, which did not also have an effect on the IgM response. In contrast to the primary IgM response, the secondary IgG response (following two injections of antigen) is extremely insensitive to modulation by compounds, if the primary response was allowed to develop intact. For example, B(a)P is a potent immunosuppressive compound that will significantly decrease the primary IgM response when administered before or during the time of antigen sensitization. However, if B(a)P is administered before or during the time of the second antigen administration, but after the primary response developed, B(a)P will not suppress the development of the secondary response. During the primary response high ▶ avidity antibody (IgM) is primarily produced. In the secondary response low ▶ avidity antibody (IgG) is the primary isotype formed to T-dependent antigens. Although IgM is a high avidity immunoglobulin, following a single injection the affinity of the antibody is usually very low. The high avidity and low affinity favor conducting a Plaque assay over an ELISA. The high avidity enhances interaction with SRBC in the Plaque assay and, even though the affinity is low, it is sufficient to bind the SRBC antigen and activate complement producing cell lysis. In contrast, a low affinity IgM antibody can be easily washed away in the multiple washing steps of an ELISA.

As indicated previously much of the secondary (IgG) antibody secretion after two sensitizations occurs in the bone marrow. Accordingly, evaluation of secondary response would favor using the ELISA approach over the Plaque assay. The high affinity antibody, which results following two injections as the antibody matures, also favors the use of ELISA over the Plaque assay for evaluating the secondary response.

Pros and Cons

Each assay used to evaluate the T-dependent antibody response, the Plaque assay and ELISA, have their strengths and weaknesses. The Plaque assay is well established, numerous ring studies have validated the assay in mice and rats, and it has been shown to be predictive for immunotoxicological effects of drugs and compounds. The assay is relatively simple and does not require expensive equipment. Furthermore,

Plaque Versus ELISA Assays. Evaluation of Humoral Immune Responses to T-Dependent Antigens. Table 2 Characteristics of Primary and Secondary Responses

Primary Response	Secondary Response
Sensitive to Modulation by Compounds	Insensitive to Modulation by Compounds
High Avidity Antibody (IgM)	Low Avidity Antibody (IgG)
Low Affinity Antibody	High Affinity Antibody
Favors Conducting	**Favors Conducting**
Plaque Assay	ELISA Assay

the assay can be conducted rapidly if multiple technicians are utilized. Data are obtained the same day the study is conducted. The data collected only consist of a small number of pages; i.e., for a 50-animal study two pages are generated. This facilitates data review, particularly when studies are conducted under Good Laboratory Practices (GLP). The major weakness with the assay is the reliance upon SRBC as the T-dependent antigen. Sheep must be screened to ensure they produce a good response and are not bled too frequently. On the other hand there are a few small companies providing SRBC commercially, e.g. Dr. Merk & Kollegen, Germany, or the Colorado Serum Company, Denver, USA. Another weakness of the assay is that the blood has a limited shelf life for sensitization. The ELISA assay, particularly the SRBC ELISA, has been shown to produce similar results as the Plaque assay when evaluated using potent immunosuppressive compounds. The SRBC ELISA, like all ELISAs, has the advantage in that the assay does not have to be conducted the same day animals are sacrificed. Serum or plasma can be collected and frozen away for evaluation at a future date. As with the Plaque assay the major weakness is related to the reliance on SRBC as the sensitizing T-dependent antigen as well as for preparing the SRBC membrane preparation needed for the ELISA. For the KLH ELISA, the KLH is available commercially which assures lot-to-lot consistency of the antigen. This is a major problem when SRBCs are used to prepare membrane preparation for use in the SRBC ELISA. More importantly, standards for both mice and rats of anti-KLH IgM and IgG antibodies are available which allow the data from KLH ELISA to be expressed in mass units, i.e., µg/ml of anti-KLH antibody.

Among the weaknesses of all ELISAs is the time required to properly analyze and evaluate the data. Furthermore, when conducting studies under GLP, a significant amount of ELISA data must be generated in order to ensure an adequate data trail. Unlike a 50-animal Plaque assay, which will generate two sheets of paper, the ELISA evaluation for the same 50 animals would generate two notebooks of data in order to meet the GLP requirements (see Figure 1). As addressed below, only limited validation of the ELISA to T-dependent antigens has been conducted.

Predictivity
Concordance

Studies conducted by the NTP in mice demonstrated that the Plaque assay was the most predictive of the functional immunotoxicological assays evaluated. The concordance with the Plaque assay alone was 78%. When a second assay was added, i.e., NK, the ▶ concordance increased to 94% and, with the plaque assay and two additional assays, a 100% concordance could be achieved with the proper selection of assays. As of this time the predictability of either the SRBC ELISA or the KLH ELISA has not been established in rodents or other species.

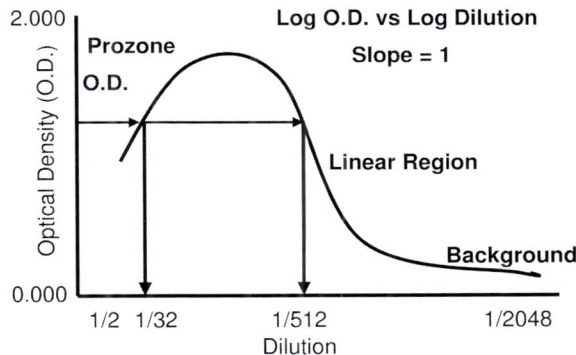

Plaque Versus ELISA Assays. Evaluation of Humoral Immune Responses to T-Dependent Antigens. Figure 1 ELISA Curve illustrating why multiple sample dilutions (Mulitpoint Analysis) are need when conducting ELISA. In the Prozone, area of high antibody concentration, the optical density (OD) actually increases as the sample is diluted due to antigen-antibody complex formation. In the linear region, the OD halves with each dilution when samples are serially diluted 1:2. Slope in the linear region is equal to 1. As the concentration of antibody decreases in diluted samples the OD approaches background. Using a single sample dilution (Single Point Analysis) can produce erroneous results.

Assay Validation

The Plaque assay has undergone numerous validation studies, at both the national and international level. Numerous studies have shown that in both mice and rats, the Plaque assay was capable of detecting effects of immunosuppressive compound in the various laboratories conducting the test. In these validation studies, as expected, reduced variability was observed in those assays, which had the greatest experience in conducting the Plaque assay. However, even those laboratories with little or no experience were capable of detecting effects of immunosuppressive compounds once they had the assay up and running. In contrast, only one validation study has been conducted using the SRBC ELISA (5). In this study, the Plaque assay was compared to the SRBC ELISA in both CD-1 mice and Sprague Dawley rats. The conclusion of this study was that the Plaque assay detected suppression at lower concentrations compared to the SRBC ELISA in both rats and mice. While numerous laboratories are using the KLH ELISA to assess humoral immune responses, as of this date, no validation study has been completed. Several validation studies for KLH are being discussed for both rodents and non-rodent species.

Relevance to Humans

Results from studies using the Plaque assay have been predictive of effects observed in clinical studies. However, the development of antibody levels in humans is evaluated by measuring antibody levels in the serum or plasma of the patient. This is routinely done using ELISA. Thus in terms of methodology, evaluating humoral immune responses in humans is more consistent with the evaluation of humoral immune responses in animals by ELISA than the Plaque assay, although a primary response by *in vitro* stimulation of PBC with SRBC would theoretically also be possible.

Regulatory Environment

In the past the Plaque assays were considered the "gold standard" for evaluating the effect of a test article on the humoral immune response. However, the US Environmental Protection Agency (EPA) initially and other regulatory agencies followed, in accepting the antibody response to a T-dependent antigen, evaluated using endpoints other than the Plaque assay. Among other endpoints considered acceptable was the SRBC ELISA, as well as ELISAs to other T-dependent antigens, including Keyhole Limpet Hemocyanin (KLH), Tetanus Toxoid, and ▶ nitrophenyl-chicken gamma globulin. The use of ▶ ELISPOT technology as a method of evaluating the humoral immune response has also been suggested by some regulatory agencies.

Relevant Guidelines

EPA OPPTS 880.3800 Biochemicals Test Guidelines, Immune Response, 1996
OECD 407, Guideline for 28-Day Repeated Dose Oral Toxicity Test in the Rat, 1998
The European Agency for the Evaluation of Medicinal Products. Evaluation of Medicines for Human Use, 1999

Draft Guidelines

FDA (CDER), Immunotoxicology Evaluation of Investigational New Drugs, Draft Guidance, 1002
ICH Guidelines on Immunotoxicology (in preparation)

References

1. Luster MI, Portier C, Pait DG, White KL Jr, Gennings C, Munson AE, Rosenthal GJ (1992) Risk assessment in immunotoxicology. I. Sensitivity and predictability of immune tests. Fund Appl Toxicol 18:200–210
2. Dayan AD, Kuper F, Madsen C, Smialowicz RJ, Smith E, Van Loveren H, Vos JG, White KL Jr (1998) Report of validation study of assessment of direct immunotoxicity in the rat. The ICICIS Group Investigators. Toxicology 125:183–201
3. Temple L, Kawabata TT, Munson AE, White KL Jr (1993) Comparison of ELISA and plaque-forming cell assay for measuring the humoral immune response to SRBC in animals treated with benzo(a)pyrene or cyclophosphamide. Fund Appl Toxicol 21:412–419
4. Shea JA, Peachee VL, White KL Jr (2003) Characterization of Keyhole Limpet Hemocyanin (KLH) as an alternative t-dependent antigen for ELISA immunotoxicological evaluations in mice. Toxicologists 72:504
5. Loveless SE, Ladics GS, Smith C, Holsapple MP, Woolhiser MR, Anderson PK, White KL Jr, Musgrove DL, Smialowicz RJ, Williams W (2002) Interlaboratory study of the primary antibody response to sheep red blood cells in outbred rodents. Toxicologist 66:1164

Plasma Cell

Terminally differentiated B lymphocyte that synthesizes and secretes antigen-specific immunoglobulin.
▶ Germinal Center
▶ B Lymphocytes
▶ Lymphocytes

Plasmodium

A parasite that causes malaria, which infects erythrocytes.
▶ Host Resistance Assays

Platelets

Little plates or plaques; specifically, blood platelets or thrombocytes; irregularly shaped disks, containing granules but no definite nucleus; about one-third to one-half the size of an erythrocyte, and containing no hemoglobin; called also Hayem's hematoblast, Zimmermann's corpuscles or particles; number from 200,000–800,000/cu mm of blood; fragments of the cytoplasm of older megakaryocytes.
▶ Serotonin
▶ Blood Coagulation

Plating Efficiency

The fraction of all cells seeded into a clonogenic semisolid culture assay which will give rise to colonies.
▶ Hematopoietic Stem Cells

PLN Index

The immune response going on in a lymph node is measured quantitatively using parameters, such as cell count, cell proliferation or expression of cell surface markers. The PLN index is calculated by dividing the value for the lymph node from the hind footpad treated with the test substance by the value for the lymph node from the other hind footpad treated with the vehicle (or left untreated). Similarly, the PLN index can be calculated from PLN parameters from a group of test chemical-treated animals and from a control group.
▶ Popliteal Lymph Node Assay, Secondary Reaction

PLNA

▶ Reporter Antigen Popliteal Lymph Node Assay
▶▶ Popliteal Lymph Node Assay
▶ Popliteal Lymph Node Assay, Secondary Reaction

Pluripotent

A state in which a primitive stem cell can differentiate into any of multiple daughter cell lineages. Partially committed stem cells can develop into only one or a few cell lineages.
▶ Lymphocytes

Pluripotential Stem Cell

A midechelon stem cell, ancestor of committed myeloid and lymphoid cell lines.
▶ Colony-Forming Unit Assay: Methods and Implications

Pneumococcal Disease Models

▶ Streptococcus Infection and Immunity

Pneumonia

Inflammation, or an acute infectious disease, of the lungs.
▶ Klebsiella, Infection and Immunity
▶ Respiratory Infections

Pokeweed Mitogen

A lectin (glycoprotein) extracted from the pokeweed (*Phytolacca Americana*). In immunology it is used to stimulate the division of lymphocytes, and can therefore be used to test their proliferative capacity. Unlike concanavalin A and phytohemagglutinin, it stimulates B cells and plasma cells, as well as T cells, and is therefore frequently used to investigate functions of those particular types of cell, including antibody production. *See also* Concanavalin A; Phytohemagglutinin.
▶ Lymphocyte Proliferation
▶ Polyclonal Activators

Polarization

Differentiation capacity.
▶ Maturation of the Immune Response

Polybrominated Biphenyls (PBBs)

PBBs and their chlorinated analogues (PCBs) are highly lipid soluble and environmentally persistent compounds which were used in a variety of applications. They were among some of the first materials investigated for potential immunotoxicity in both laboratory animals and man, following accidental exposure of populations.

▶ Lymphocyte Proliferation

Polychlorinated Biphenyls (PCBs) and the Immune System

JOHN L OLSEN
Stony Brook University Medical School
327 Sheep Pasture Road
Setauket, NY 11733
USA

Synonyms

chlorobiphenyl, Aroclor (USA), Clophen (Germany), Kanechlor (Japan), polychlorinated diphenyls

Definition

Polychlorinated biphenyls (PCBs) belong to the halogenated aromatic hydrocarbon family of chemicals that also includes dioxins and polychlorinated dibenzofurans (PCDFs). PCBs are a mixture of up to 209 individual chlorinated compounds, known as congeners, which are no longer produced in the USA but are among the most widespread of environmental pollutants. Electrical transformers, capacitors, lighting fixtures, hydraulic oils, paints, inks, and home appliances may contain PCBs if they were manufactured before 1977. Burning of some wastes can also release PCBs into the environment. Commercial PCB mixtures are known in the USA by the trade name Aroclor (Monsanto, St Louis MO, USA). There are no known natural sources of PCBs. Human exposure to PCBs occurs primarily through food contamination. Occupational exposure was reported to produce acute health effects as early as 1936. Indicators of exposure to PCBs in humans may include chloracne, altered hepatic enzyme levels, liver malignancy, decreased pulmonary function, susceptibility to lung infections, changes in the menstrual cycle, lower birth weights, eye irritation, immunosuppression, elevated neutrophil counts, and deficits in vision and intelligence measures.

Molecular Characteristics

A PCB has the formula $C_{12}H_{10-n}Cl_n$ with the general structure as shown in Figure 1.

PCBs are either oily liquids or solids that are colorless to light yellow. They are inert chemicals, resistant to biological and chemical degradation. Some PCBs can exist as a vapor and can travel long distances in the air to be deposited in areas where air currents are cooler than where the PCBs were released. In water, PCBs fix to organic particles and bottom sediments, though a fraction will remain undissolved. Aroclors are identified by a four-digit number. The first two digits indicate the number of carbon atoms; the last two digits refer to the percent weight of chlorine in the mixture. For example, Aroclor 1248 denotes 12 carbon atoms with 48% chlorine by weight. Aroclors 1232, 1248, 1260 and 1270 have an average number of chlorines per PCB molecule of 2, 4, 6, and 10, respectively. The pattern of chlorination of a congener determines its physical, chemical, and biological properties. For example, chlorines in the *meta* (carbon atoms 3, 3', 5 and 5') or *para* (carbon atoms 4 and 4') positions allow the PCB molecule to remain planar and to bind the aromatic hydrocarbon receptor (AhR), achieving effects similar to—but less potent than—2,3,7,8-tetrachlorodibenzo-p-dioxin (TCDD; dioxin). However, halogen substitution at the *ortho* positions (carbons 2 and 6) on one phenyl ring interfere with *ortho* halogens in the other phenyl ring (at carbon atoms 2' and 6') creating a non-planar molecule that cannot bind the AhR.

Coplanar dioxin-like PCBs, including 2,3,3',4,5,5'-hexachlorobiphenyl and 2,3,3',4,5'-pentachlorobiphenyl, are assumed to exert immunotoxic effects through an initial action of binding the cytosolic AhR, which is a member of the thyroid hormone/steroid superfamily of transcription factors. Over 400 environmental toxicants and endogenous compounds, including by-products of aspartate aminotransferase metabolism, bind the AhR (1). After ligand binding, AhR releases its chaperone heat shock proteins, moves into the nucleus and forms a dimer with the AhR nuclear translocator protein, Arnt. The basic region of AhR performs DNA binding and contains a nuclear localization sequence, whereas its helix-loop-helix and Per-Arnt-Sim (PAS) homology domains confer protein-binding capability. In the nucleus, the AhR/Arnt complex binds xenobiotic-responsive elements (XREs) in the genes of the so-called AhR battery involved in development, reproduction, toxicity response, oncogenesis, and oxidative stress, including phase I genes (p4501A1, 1A2, 1B1) and phase II genes (GSTα, NADPH-quinone-oxidoreductase, UDP-glucuronosyltransferase). Deregulation of retinoid and thyroid hormone homeostasis is an important consequence of induction of AhR battery genes. The drug-metabolizing enzymes of the battery carry out the destruction of their own inducing ligands or ligand metabolites. PCBs are metabolized mainly by mixed-function oxidases into a wide range of metabolites, but enzymatic activation is not necessary for PCB toxicity.

c-Src protein is functionally attached to the AhR and is activated by ligand binding in numerous animals, including rodents and humans. This activation is rapid, often potent at very low doses, and longlasting. Downstream targets of this activity include increased protein kinase C (PKC) activity in the thymus and upregulation of thyroid hormone receptor mRNA. p450 regulation is likely src-independent. The AhR also forms complexes with the retinoblastoma tumor suppressor in human MCF-7 cells, suggesting a role for the AhR in cell cycle regulation and G1 arrest.

Non-dioxin-like PCBs account for the bulk of the PCBs found in biological and environmental samples. Immunotoxic effects mediated independently of the AhR include:

- PCB metabolism to arene oxide intermediates that can alkylate cellular macromolecules to become toxic adducts
- congener-specific antagonistic or synergistic interactions between non-coplanar PCBs and dioxin
- some PCBs and PCB metabolites are thought to bind transport proteins in plasma, including transthyretin, thus altering the transport and plasma concentrations of endogenous molecules
- PCB-induced liver enzyme induction and metabolism of parent substrates (including medications or environmental pollutants) to daughter compounds that may have enhanced or diminished immunotoxic or therapeutic properties.

Regarding the second point, Aroclors 1248, 1254, 1260, and several congeners have been shown to antagonize the immunosuppressive effects of dioxin. Whether this antagonism is mediated by means of displacing dioxin from its binding site on the AhR or by actually counterbalancing the effects of dioxin through an alternate pathway is not clear at this point. The use of the term 'non-dioxin-like PCBs' is not necessarily practical. They are not a single class of chemicals and have multiple toxicities with discrete structure-activity relationships. Further congener-specific research may better classify these compounds and identify their specific targets in human tissues.

Putative Interaction with the Immune System

The potential for PCBs to be immunomodulatory has been the subject of extensive experimental investigations. Accumulating evidence indicates that the immune system is probably one of the most sensitive targets for PCB-induced toxicity.

Antibody Titers and Lymphocyte Populations

In mice, humoral suppression is dependent on a PCB congener's affinity for the AhR. In C57Bl/6 mice the coplanar, non-*ortho*-substituted 3,3',4,4'-tetrachlorobiphenyl (TCB) exhibits a high affinity for the AhR and causes severe humoral antibody suppression. But in the genetically different DBA/2D2 mice, for which TCB shows a lower binding affinity for the AhR, this humoral suppression does not occur. The essentially non-coplanar, di-*ortho*-substituted 2,2',5,5'-tetrachlorobiphenyl binds the AhR weakly in both animals and does not suppress humoral immunity in either strain. Several animal studies using Aroclor mixtures have shown that it is the more highly chlorinated PCBs (Aroclors 1260, 1254, 1248) that are responsible for diminished plaque-forming cell responses to sheep red blood cell challenge, decreased IgG titers to KLH antigen, and decreased spleen and lymph node cellularity. TCB and other AhR-binding PCBs cause thymus atrophy in animal models, specifically targeting immature thymocytes. Interestingly, PCBs given in less than immunotoxic levels to mice may reduce the effects of 2,3,7,8-TCDD-induced inhibition of the plaque-forming cell response to sheep erythrocytes.

Chronic exposure of rhesus monkeys to Aroclor 1254 was associated with decreased titers to SRBC; the calculated lowest observable adverse effect level (LOAEL) for this effect was 5 μg/kg/day. Rhesus monkeys exposed to Aroclor 1248 had reduced cortical and medullary areas in the thymus, inapparent germinal centers in lymph nodes and spleen, and hypocellular bone marrow. It is important to note that PCB suppression of humoral antibody responses can be both T cell-dependent (an SRBC response) and T cell-independent (an LPS response), indicating that PCBs can alter B cell differentiation into plasma cells independently of thymic alterations.

Data is limited and often conflicting regarding PCB-induced suppression of cell-mediated immunity in laboratory animals. Dogs exposed to Aroclor 1248 had a decreased delayed-type hypersensitivity reaction to intradermal injection of tetanus toxoid compared to controls. This, along with several in vitro experiments, indicates that a subpopulation of T-cells may be affected.

Effects of PCB exposure in humans were documented following ingestion of contaminated rice bran oil in Yusho, Japan in 1968, and of rice in Yu-Cheng, Taiwan in 1979). Contamination of the PCBs with PCDFs confounds the interpretation of these studies. Long-term studies of the more than 2000 people who were exposed during these events revealed increased mortality due to the food poisoning. In both of the studies,

Polychlorinated Biphenyls (PCBs) and the Immune System. Figure 1 General structure of polychlorinated biphenyls (PCBs).

normal IgG levels were seen, but IgM and IgA were decreased, returning to normal after 4 years. Reductions were seen in total, active, and helper T cells, as well as delayed-type hypersensitivity to tuberculin antigen.

The immunologic effects of in utero exposure have been studied in Yu-Cheng children born between 1978 and 1987 (2). Serum immunoglobulins and cell surface markers were not different compared to controls. It has been reported that children exposed to dioxin-like PCBs in breast milk have decreased numbers of CD8 cells. In a study of Dutch children monitored for prenatal and childhood exposures, PCBs were associated with lower antibody levels to mumps and measles at preschool age (3).

Infection and the Innate Immune Response

The maturation of the immune system is particularly vulnerable to PCBs. In the Yu-Cheng study (2) children exposed in utero exhibited immunosuppression, higher frequencies of bronchitis in the first 6 months of life, and higher frequencies of respiratory tract and ear infections in a 6-year follow-up. In the Dutch study (3) prenatal PCB exposure was associated with less shortness of breath with wheeze, and at 42 months of age PCB levels were associated with a higher prevalence of recurrent middle-ear infections and chicken pox, yet a lower prevalence of allergic reactions. It is therefore possible that exposure to dioxin-like PCBs in utero and during early childhood might lead to the recurrence of common infections that help to prevent the development of allergy. So far, though, proof of a causal relationship requires more evidence and improved reporting of confounding factors.

In experimental animals, mononuclear phagocytes are altered functionally by PCBs, showing decreased phagocytic activity and splenic clearance of bacteria and increased sensitivity to bacterial endotoxins. Interestingly, rhesus monkeys administered Aroclor 1254 have increased serum complement activity (CH_{50}). Mice have a decreased ability to clear *Listeriamonocytogenes* when given PCBs by acute peroral gavage or subacute parenteral injection, and this seems to be sex specific, as females are less sensitive to the effects of PCBs.

Studies in non-human primates, dogs, and human adults and stillborn babies exposed to PCBs show many common defects to structures that are important to the maintenance of innate immunity: modulation of skin and nail beds, eye exudate, inflammation of Meibomian glands, and neutrophil levels and function. Gastric hyperplasia, which could affect host mucosal immunity, has been found in monkeys but not in dogs. The natural killer (NK) cell is a cellular bridge between innate and adaptive immunity. In rat studies, NK cell activity was depressed by Aroclor 1254.

Paradoxically, Aroclor may be protective for tumor development in cancer-cell transplantation studies. Aroclor 1254 reportedly reduced mortality in mice following injection of Ehrlich's tumor cells. Rats transplanted with Walker 256 carcinosarcoma showed retarded tumor cell growth if also treated with Aroclor 1254.

In mice, many PCB congeners and methylsulfonyl metabolites bind the bronchial mucosa and are selectively retained in lung Clara cells by a PCB-binding protein that is homologous to human uteroglobin. The uteroglobin gene encodes a cytokine-like anti-inflammatory protein that has a potent-inhibitory activity against phospholipase A2. Lung tissues from PCB-poisoned Yusho patients harbored 16 different PCB methylsulfonyl PCBs. Clara cells are a key site of activities that are dependent on p450 enzymes that convert various substrates to reactive products. When compared to control mice, uteroglobin knockout mice fail to accumulate methylsulfonyl PCBs in lung. The role of uteroglobin in PCB-induced respiratory immunotoxicity has not been fully elucidated in humans.

Relevance to Humans

PCBs have been found in at least 500 of the 1600 National Priorities List sites identified by the US Environmental Protection Agency. As in other species, PCBs accumulate in humans in fatty tissues because of their lipophilic properties and resistance to degradation. Average levels in adipose tissue are 1 p.p.m. and 10 p.p.b. in blood, and have been declining in the population since the 1980s. In general, fish consumption is the major source of exposure to PCBs. The human health effects of exposure to PCBs in fish are a function of congener toxicity, PCB concentration in fish, and fish consumption rates. Some risk factors for having a high PCB body burden include older age, dermal and ingestion exposure, as well as complex metabolic factors including lactation, diet, and exercise activity. As a significant percentage of PCBs are liberated from drying soil, humans may be exposed via inhalation. There may be sex-specificity for some biological outcomes.

It has been reported that breastfed infants of mothers who ate high amounts of Great Lakes fish had a greater amount of microbial infections compared to control infants of mothers who did not eat fish. Studies of bacterial aerobic degradation of PCBs (removal of chlorines) in the environment indicate that this process decreased the risk of PCB-induced immunotoxicity, as measured by splenocyte proliferation assays.

PCBs elicit a wide spectrum of biochemical responses, some of which are similar to those caused by dioxin. Because dioxin-like compounds normally exist in biological and environmental samples as complex mixtures, the concept of toxic equivalency factors (TEF) has been developed to simplify risk assessment and regulatory management. Relative toxicities of dioxin-like compounds in relation to the reference compound, TCDD, were determined on the basis of results obtained in in vivo and in vitro studies. A dozen congeners in particular exhibit dioxin-like effects (Table 1). 3,3',4,4',5-pentachlorobiphenyl, 3,3',4,4',5,5'-hexachlorobiphenyl, 3,3',4,4'-tetrachlorobiphenyl have TEFs in humans of 0.1, 0.01, and 0.0001, respectively. Maximum dioxin-like activity is obtained when there are no *ortho*, two or more *meta*, and both *para* positions are occupied. A single chlorine in the *ortho* position weakens dioxin-like activity but increases AhR-independent toxicity of the compound. In population studies, TEFs should be calculated for PCBs, dioxins, and polychlorinated dibenzofurans. However, the antagonism between certain high-dose PCB congeners/mixtures and dioxin-like chemicals may confound models that assume TEFs can be additive.

Because PCBs cause immune dysregulation experimentally, it is plausible that they may alter cancer-cell surveillance. Consistent with animal studies, workers occupationally exposed to PCBs show increased deaths from several cancers including those of the liver, melanoma, thyroid, pancreas, and gastrointestinal and central nervous system neoplasms. However, it is very possible that these diseases affect PCB levels due to wasting and contraction of the body lipid compartment, so that retrospective studies need to be interpreted with caution. An intriguing prospective study reported a clear positive dose-response relation between a high level of background exposure to PCBs and development of non-Hodgkin lymphoma (5).

References

1. DeVito MJ, Birnbaum LS (1995) Dioxins: model chemicals for assessing receptor-mediated toxicity. Toxicology 102:115–123
2. Yu ML, Hsin JW, Hsu CC, Chan WC, Guo YL (1998) The immunologic evaluation of the Yucheng children. Chemosphere 37:1855–1865
3. Weisglas-Kuperus N, Patandin S, Berbers GA et al. (2000) Immunologic effects of background exposure to polychlorinated biphenyls and dioxins in Dutch preschool children. Environ Health Persp 108:1203–1207
4. Ahlborg UG, Becking GC, Birnbaum LS et al. (1994) Toxic equivalency factors for dioxin-like PCBs: Report on WHO-ECEH and IPCS consultation, December 1993. Chemosphere 28:1049–1067
5. Rothman N, Cantor KP, Blair A et al. (1997) A nested case-control study of non-Hodgkin lymphoma and serum organochlorine residues. Lancet 350: 240–244
6. Ahlborg UG, Becking GC, Birnbaum LS et al. (1994) Toxic equivalency factors for dioxin-like PCBs: Report on WHO-ECEH and IPCS consultation, December 1993. Chemosphere 28:1049–1067

Polychlorinated Dibenzodioxins

▶ Dioxins and the Immune System

Polychlorinated Biphenyls (PCBs) and the Immune System. Table 1 Polychlorinated biphenyl (PCB) congeners that resemble dioxin*

Congener number	IUPAC name
77	3,3',4,4'-tetrachlorobiphenyl
81	3,4,4',5-tetrachlorobiphenyl
105	2,3,3',4,4'-pentachlorobiphenyl
114	2,3,4,4',5-pentachlorobiphenyl
118	2,3',4,4',5-pentachlorobiphenyl
123	2,3',4,4',5'-pentachlorobiphenyl
126	3,3',4,4',5-pentachlorobiphenyl
156	2,3,3',4,4',5-hexachlorobiphenyl
157	2,3,3',4,4',5'-hexachlorobiphenyl
167	2,3',4,4',5,5'-hexachlorobiphenyl
169	3,3',4,4',5,5'-hexachlorobiphenyl
189	2,3,3',4,4',5,5'-heptachlorobiphenyl

*Congeners that show structural similarity to dioxin, bind to the AhR, induce dioxin-specific responses, and are persistent and accumulate in the food chain (Ahlborg UG, Becking GC, Birnbaum LS, et al (1994) Toxic equivalency factors for dioxin-like PCBs: Report on WHO–ECEH and IPCS consultation, December 1993. Chemosphere 28: 1049–67).

Polychlorinated Diphenyls

▶ Polychlorinated Biphenyls (PCBs) and the Immune System

Polyclonal

Describes the proliferation or products of different cells.

▶ Mitogen-Stimulated Lymphocyte Response

Polyclonal Activators

STEPHEN B PRUETT
Department of Cellular Biology and Anatomy
Louisiana State University
Health Sciences Center
Shreveport, Louisiana 71130
USA

Synonyms
Mitogens, polyclonal mitogens, PHA, phytohemagglutin, ConA, concanavalin A, LPS, lipopolysaccharide, SEB, streptococcal enterotoxin B, MLR, mixed lymphocyte response, pokeweed mitogen, PWM, major histocompatibility complex, MHC

Definition
Polyclonal activators are agents that activate a significant proportion of B or T lymphocytes (or both), inducing their proliferation and generally at least some of the functions that are induced by antigen-mediated cellular activation (e.g. cytokine production).

Characteristics
There are several classes of polyclonal activators, which differ in structure and mechanism of action. Several plant-derived lectins (which bind oligosaccharides on various cell surface proteins) activate T or B lymphocytes. For example, phytohemagglutin (PHA) and concanavalin A (ConA) activate T lymphocytes, and PWM activates T and B lymphocytes. These agents generally act by cross-linking surface molecules and initiating signaling events leading to cellular activation. Bacterial lipopolysaccharide (LPS) at high concentrations (usually greater than 10 µg/ml) polyclonally activates B lymphocytes, acting through Toll-like receptor 4 (1). Superantigens are agents that bind T cell receptors on a broad subset of T cells (e.g. all which express the Vb8 region on their T cell receptor) and also bind the MCH class II protein. This functions to link MHC II proteins on antigen-presenting cells and T cell receptors on T cells in a manner that allows T cell activation. Most superantigens are protein products of bacteria, and examples include staphylococcal enterotoxins, streptococcal pyrogenic toxins, and toxic-shock syndrome toxin (2). Staphylococcal enterotoxin B (SEB) is one of the most potent and most widely used superantigens. It could be argued that superantigens represent a more physiologically relevant method for activating lymphocytes, because the receptors that normally are involved in activation by antigen are involved in superantigen-induced activation. Antibodies specific for the antigen receptors or signal transduction components of T cells (TCR or CD3) or B cells (surface Immunoglobulin) can be used to activate these cells. In the case of B cells, anti-immunoglobulin antibodies plus interleukin(IL)-4 are sufficient to achieve substantial B cell activation. In the case of T cells, accessory cells (macrophages or dendritic cells) must be present, or the anti-CD3 antibody must be coated onto the surface of the culture plate to achieve full activation. These approaches are often regarded to be similar to antigen-induced activation of these cells, but it should be noted that there are likely several costimulatory signals induced during natural activation of cells by antigen, which are not induced by these stimuli. Finally, inactivated lymphocytes from allogeneic animals can be used as a polyclonal T cell activator. This is referred to as the mixed lymphocyte reaction or mixed lymphocyte response (MLR). It is based on the fact that a substantial percentage of T lymphocytes recognize and respond to major histocompatibility proteins found on allogeneic lymphocytes. If these "stimulator" cells are treated to prevent their reciprocal response (i.e. by radiation or DNA synthesis inhibitor), the response of the "responder" lymphocytes can be assessed by measuring their proliferation.

Preclinical Relevance
The preclinical relevance of polyclonal activators in immunotoxicity testing is indicated both by the rationale for their use and by experimental data. The rationale for evaluating responses to polyclonal activators is that the activation and proliferation of T cells and B cells are required for immune responses, and most of the cellular events and molecular components required for these processes are also required for cellular responses to polyclonal activators. Experimental results from studies sponsored by the National Toxicology Program demonstrate that the effects of a wide variety of drugs and chemicals on lymphocyte responses to polyclonal activators (PHA, ConA, and LPS) are strongly concordant with effects on other

important immunological end points, including T cell-dependent antibody responses (3).

Relevance to Humans

Clinical immunologists routinely use polyclonal activators to test T and B lymphocyte function in humans. Abnormal responses to these agents can be used in the diagnosis of hereditary or acquired immunodeficiency conditions.

Regulatory Environment

Cellular responses to polyclonal activators are not specifically required or recommended as immune function tests in immunotoxicity testing in Guidance documents from the US Environmental Protection Agency (4) or US Food and Drug Administration (5). However, they are included in the "extended studies" required by the European Agency for the Evaluation of Medicinal Products (EMEA) to characterize the effects of drugs that seem to have immunotoxic potential (6). In addition, they are often included in research studies, because they are simple, reproducible, and generally correlate well with other measures of immunotoxicity (3).

References

1. Kawai T, Adachi O, Ogawa T, Takeda K, Akira S (1999) Unresponsiveness of MyD88-deficient mice to endotoxin. Immunity 11:115–122
2. Kuby J (1994) Immunology. WH Freeman, New York, pp 218–219
3. Luster MI, Portier C, Pait DG et al. (1992) Risk assessment in immunotoxicology. I. Sensitivity and predictability of immune tests. Fundam Appl Toxicol 18:200–210
4. US EPA(1998) Health Effects Test Guidelines OPPTS 870.7800 Immunotoxicity. http://www.epa.gov/opptsfrs/OPPTS_Harmonized/870_Health_Effects_Test_Guidelines/Series/870-7800.pdf
5. US FDA (2002) Guidance for Industry. Immunotoxicology evaluation of investigational new drugs. http://www.fda.gov/cder/regulatory/applications/guidance.htm
6. European Agency for Evaluation of Medicinal Products (2000) Note for Guidance on repeated dose toxicity. http://www.emea.eu.int/pdfs/human/swp/104299en.pdf

Polyclonal Mitogens

▶ Polyclonal Activators

Polyclyclic Aromatic Hydrocarbons (PAHs) and the Immune System

SCOTT W BURCHIEL
College of Pharmacy Toxicology Program
University of New Mexico
Albuquerque, NM 87131-0001
USA

Synonyms

PAHs, benzo-a-pyrene, BaP, 7,12-dimethylbenz-a-anthracene, DMBA, 3-methylcholanthrene, 3-MC, dibenz-a, h-anthracene, benz-a-anthracene, BA, BP-quinones, BPQ, BP-7,8-diol, BP-7,8-diol, 9,10-epoxide, BPDE, anthracene, pyrene, benzo-e-pyrene, BeP, nitro-PAH, oxy-PAH

Definition

Polycyclic aromatic hydrocarbons are a class of diverse compounds that consist of three or more fused aromatic rings, primarily based on anthracene and pyrene. Environmentally, PAHs are formed during the combustion of fossil fuels and high temperature reactions of hydrocarbons, benzene and other aromatic compounds.

Molecular Characteristics

The biologic and toxicologic activities of PAHs are largely dependent on their ability to interact with AhR receptors present in many mammalian cells and tissues. In general, PAHs are semivolatile compounds that are quite lipophilic and exert both specific and non-specific membrane effects. Highly specific structure-activity relationships have been observed for the effects of PAHs on cells and tissues. Specific effects generally relate to the expression of AhR and the varying ability of individual PAHs to activate AhR-dependent gene promoter regions, referred to as dioxin response elements (DRE). The dioxin 2,3,7,8-tetrachlorodibenzo-p-dioxin (TCDD) is a pure AhR agonist that is approximately 100 times more potent than PAH in binding to AhR and activating DRE. TCDD and related agents are discussed elsewhere in this book.
PAHs are metabolized by cytochrome P450-dependent and -independent pathways. The P450-dependent pathways, most notably CYP1A1, CYP1A2 (liver only) and CYP1B1, are induced following binding of AhR and activation of DRE. In the case of BaP and other PAHs, these enzymes form oxidative metabolites, such as epoxides, which in the presence of epoxide hydrolase-1 (EPHX1) form dihydrodiols that can undergo another round of metabolism to form diol-epoxides such as benzo-a-pyrene(BaP)-diol-epoxide (BPDE). BPDE and several other BaP metabolites,

such as the BP-quinones (BPQ; discussed below) are strong electrophiles that trigger the activation of secondary metabolizing enzymes including glutathione-S-transferases (GST) and *N*-quinone oxidoreductase (NQO1) via interaction with electrophile response elements (EpRE) that are also referred to as antioxidant response elements (ARE) (1). Electrophilic BaP metabolites, most notably BPDE, covalently bind to DNA leading to genetic mutations responsible for tumor initiation. In addition, the formation of BPDE-DNA adducts leads to induction of the ▶ p53 pathway, which may trigger cell cycle changes and apoptotic pathways in target cells.

Several oxidative metabolites of PAHs, such as BP-quinones and DMBA-quinones are known to redox-cycle, leading to production of reactive oxygen species (ROS) (2). PAH-quinones are formed via P450-dependent and independent reactions, such as by peroxidases or by ultraviolet light (3). Redox-cycling results in the formation of ROS, including superoxide anion and hydrogen peroxide, which are known to activate lymphocytes. Redox-cycling requires reducing equivalents generally supplied by NADPH, and thus BPQ and related agents interact with the mitochondrial electron transport chain, leading to oxidative stress and ATP depletion. The immunotoxicologic effects of BPQ, other oxy-PAHs, and nitro-PAHs, have not been fully evaluated.

Thus, the biologic and toxicologic action of PAHs is a complicated interplay between the ability of specific PAHs to bind to endogenous AhR, the formation of oxidative and electrophilic metabolites, and the removal of reactive molecules via secondary metabolic processes (Fig. 1). The effects of these agents result from activation of both genotoxic and non-genotoxic pathways. Therefore, the toxicity of PAHs to target tissues is dependent upon the exposure of cells and tissues to circulating parent compounds and metabolites, their expression of AhR, and their propensity to form bioactive versus detoxified metabolites.

Putative Interaction with the Immune System

Carcinogenic PAHs have been found to suppress the immune system of animals (4,5). Initial studies showed that BaP (but not BeP), DMBA, and 3MC suppress humoral immunity, and later studies showed that many immune cells are targets of PAH action (as described below). The humoral immune response to T-dependent antigens is considered to be a sensitive indicator of immune suppression, although suppression of T cell and B cell proliferation has been observed at similar exposure levels. In general, there is a correlation between the carcinogenicity of a PAH and its immunotoxicity, which is likely due to requirements for AhR binding activity and metabolic activation pathways. The immunosuppression produced by BaP and DMBA has both similarities and differences. BaP is a moderate to strong AhR ligand, whereas DMBA is a weak AhR ligand. In addition (as discussed below) much of the immunotoxicity of BaP is due to P450 CYP1A1-dependent metabolism. CYP1A1 is expressed only at low levels in lymphocytes until AhR are activated. In the case of DMBA, the major metabolizing enzyme is P450 CYP1B1, which is expressed in many tissues constitutively.

Immune System Targets of PAH Toxicity
Bone Marrow
Both DMBA and BaP exert important effects on the bone marrow altering the formation of B cells. The mechanism of pre-B cell bone marrow suppression appears to be P450 CYP1B1-dependent and may be caused via pre-B cell apoptosis (6).

Thymus
Thymic atrophy is observed in mice treated with high concentrations (50–100 mg/kg) of DMBA. Interestingly, there does not appear to be a correlation be-

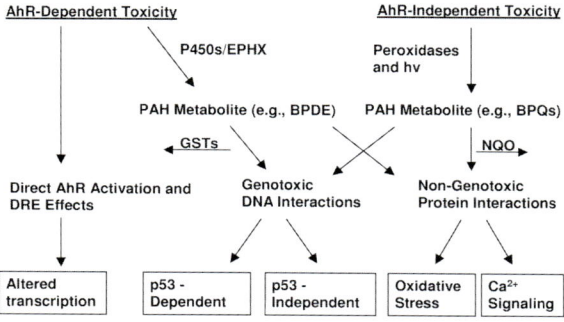

Polycyclic Aromatic Hydrocarbons (PAHs) and the Immune System. Figure 1 Potential molecular mechanisms of polycyclic aromatic hydrocarbon (PAH) immunotoxicity via Ah receptor(AhR)-dependent and AhR-independent pathways. Several AhR-dependent pathways produce immunotoxic metabolites of benzo-a-pyrenes (BaP) (such as BPDE) that appear to act via both p53-dependent (genotoxic) and p53-independent pathways. BPDE has also been shown to signal Ca^{2+}-dependent pathways, and BaP metabolites such as BP-quinones (BPQ) also signal via Ca^{2+} due to redox-cycling and mitochondrial oxidative stress. The immunotoxicity of BPQ has not been fully evaluated. Two secondary enzyme systems, glutathione-S-transferase (GST) and *N*-quinone oxidoreductase (NQO1), reduce the concentrations of PAH metabolites. Thus, the effects of PAHs such as BaP and DMBA involve complex interplay between bioactivating and inactivating pathways. The metabolites act via non-genotoxic (signaling and apoptotic pathways) and genotoxic pathways. (DRE: dioxin response elements.)

tween AhR phenotype and the ability of DMBA to decrease thymus weights (7).

Spleens

Many studies have been performed with spleens obtained from PAH-treated mice. In these studies both DMBA and BaP have been found to be immunosuppressive for humoral and cell-mediated immunity. DMBA was also found to be immunosuppressive when exposures occurred in vitro. In general, the most sensitive target cells for PAHs appear to be B and T helper cells for humoral immunity and cytotoxic T cells (CTL) for cell-mediated immunity, although macrophages and antigen-presenting cells (APC) have also been implicated. Splenic natural killer (NK) cells have also been shown to be suppressed by DMBA.

Peripheral Lymphoid Tissues

A few peripheral lymphoid tissues have been examined for the effects of PAHs following intratracheal or intragastric administration of BaP. It was found that BaP suppressed the antibody response of lung-draining lymph nodes in rats exposed to BaP intratracheally, and suppressed the response of murine mesenteric lymph nodes (MLN) and Peyer's patches following gavage.

Mechanisms of Immunosuppression by PAHs
AhR Ligands Suppress the Immune System

As mentioned above, many PAHs are moderate-to-strong AhR ligands. It has been previously noted that there is generally a positive correlation between the carcinogenicity and immunotoxicity of a PAH. This correlation likely exists because both carcinogenicity and immunotoxicity are largely dependent on AhR binding, increased P450 expression, and formation of bioactive metabolites. Direct AhR-dependent immunosuppression has been reported in both T and B cells for the pure AhR agonist, TCDD. Presumably these same pathways would be activated by BaP and other AhR ligands; however, a key difference is that the persistence of PAHs in the body is reduced due to metabolism.

PAH Metabolism Required for Immunotoxicity

P450 metabolism is very important for immunotoxicity of PAHs. Metabolism occurs in central (liver, lung) and peripheral tissues (lymph nodes, spleen, and bone marrow). The local metabolism of PAHs is complicated by the expression of AhR and constitutive levels of P450s such as CYP1B1. CYP1A1 is highly inducible in some white blood cells. The highest levels of P450 activity have been detected in monocytes and macrophages, and low levels have been detected in B and T cells. Most human and murine B and T cell lines used for in vitro modeling studies have little P450 activity.

Further support for the role of P450 metabolism in the immunotoxicity of PAHs has been obtained in studies that demonstrate that the Ah receptor antagonist and P450 CYP1A1/CYP1B1 inhibitor α-naphthoflavone (ANF) prevents the immunotoxicity of BaP and DMBA in murine spleens cells and human peripheral blood T cells (8). The CYP1A1 and CYP1B1 metabolite responsible for murine spleen cell and human T cell immunotoxicity due to BaP is likely BPDE.

The role of metabolism for DMBA bone marrow toxicity in mice has been established through the use of CYP1B1 null (knock-out) mice. DMBA-induced pre-B cell toxicity was nearly totally ablated in mice that did not express CYP1B1 (6).

Genotoxic Mechanisms of PAH Metabolite-Induced Immunotoxicity

The mechanisms whereby PAHs produce immunotoxicity should be divided into two general categories: genotoxic and non-genotoxic (Figure 1). Many PAHs are bioactivated to reactive metabolites that bind to DNA and exert genotoxicity. These compounds can be identified in an Ames assay. The mechanism whereby genotoxic chemicals produce immunotoxicity is likely p53-dependent. It is beyond the scope of this essay to discuss mechanisms of p53 signaling, but it is known that p53 plays a critical role in cell-cycle regulation and the induction of apoptosis. Bulky PAH-DNA adducts induce p53, which is turn inhibit cell cycling and induce apoptosis in many cells. DMBA has been found to produce immunotoxicity in bone marrow via the induction of apoptosis, so it is not surprising that some recent studies showed p53 knock-out mice are resistant to the bone-marrow suppressive effect of DMBA (9). It will be interesting to examine the p53-dependence for immunotoxicity produced by other PAHs, because it has been shown that DMBA induces apoptosis in several peripheral lymphoid organs (10). It is also known that many genotoxic chemicals that induce p53 are immunosuppressive. Therefore, p53 is probably an important pathway for immunotoxicity of numerous agents. It should be noted that in some studies good agreement has been found between the functional immunotoxicity observed and change in immunophenotypic markers, while in other studies there has been a lack of agreement. These differences probably relate to the amount of apoptosis and cell death that was produced by PAHs, as it is unlikely that functional changes would be detected simply by immunophenotypic analysis of spleen cells. Thus, immunophenotyping is not likely a stand-alone sensitive marker for immunotoxicity produced by many xenobiotics at non-cytotoxic concentrations.

Non-Genotoxic Mechanisms of PAH Immunotoxicity

Several studies have shown that PAHs can activate or interfere with lymphocyte signaling pathways in both murine and human B and T cells. Many xenobiotics produce biphasic effects on immune responses with low concentrations stimulating responses and high doses producing inhibition. Agents that mimic or alter signaling pathways may manifest these characteristics.

PAHs have been found to activate protein tyrosine kinase activity in lymphocytes leading to Ca^{2+}-dependent signaling in B and T cells in mice and humans. Two reviews have appeared on this topic (4,5). DMBA has been found to directly activate human T cells; however BaP metabolites are implicated in the activation of protein tyrosine kinases in human B and T cells. The consequences of altered B and T cell signaling may be a persistent suppression of immune response resulting from a tolerance-like mechanism on lymphocytes. One of the consequences of T helper cell tolerance would be the lack of production of interleukin-2, which is a key cytokine for both humoral and cell-mediated immunity. It has been found that DMBA does prevent the formation of IL-2 and that IL-2 can partially overcome immunosuppression.

Relevance to Humans

Humans are exposed to complex mixtures of PAHs via the air breathed (including significant quantities that are present in wood smoke, cigarette smoke, and various emissions) and via the diet. There is also occasional dermal exposure, usually associated with occupational exposures to tars, soots, and vapors. PAH exposures can be monitored using major urinary metabolites. Although there have been some reports, there is little epidemiologic evidence that PAHs produce immunosuppression under conditions of environmental or industrial exposure in humans. The concentrations of PAHs that are required to suppress humoral- and cell-mediated immunity in mice is extremely high, typically in the range of 10–50 mg/kg of BaP. These high concentrations have also been found to suppress humoral immunity in fish. In general, the concentrations of PAHs that are required to produce immunotoxicity are higher than those required to produce cancer. Nevertheless, it is clear that human white blood cells, and lymphocytes in particular, are sensitive to immunosuppression produced by BaP exposures in vitro. Cellular signaling pathways are also affected in human peripheral blood leukocytes (HPBL). In fact, in vitro data suggests that HPBL may be more sensitive than murine spleen cells in the in vitro suppression of T cell mitogenesis produced by BaP (8). Given that there are several genetic polymorphisms that may affect a person's susceptibility to PAHs, further research is necessary to determine who might be at risk for PAH immunosuppression.

References

1. Primiano T, Sutter TR, Kensler TW (1997) Antioxidant-inducible genes. Adv Pharmacol 38:293–328
2. Zhu H, Li Y, Trush MA (1995) Characterization of benzo[a]pyrene quinone-induced toxicity to primary cultured bone marrow stromal cells from DBA/2 mice: potential role of mitochondrial dysfunction. Toxicol Appl Pharmacol 130:108–120
3. Reed M, Monske M, Lauer F, Meserole S, Born J, Burchiel S (2003) Benzo[a]pyrene diones are produced by photochemical and enzymatic oxidation and induce concentration-dependent decreases in the proliferative state of human pulmonary epithelial cells. J Toxicol Environ Health 66:1189–1205
4. Davila DR, Davis DP, Campbell K, Cambier JC, Zigmond LA, Burchiel SW (1995) Role of alterations in Ca^{2+}-associated signaling pathways in the immunotoxicity of polycyclic aromatic hydrocarbons. J Toxicol Environ Health 45:101–126
5. Burchiel SW, Luster MI (2001) Signaling by environmental polycyclic aromatic hydrocarbons in human lymphocytes. Clin Immunol 98:2–10
6. Heidel SM, MacWilliams PS, Baird WM et al. (2000) Cytochrome P4501B1 mediates induction of bone marrow cytotoxicity and preleukemia cells in mice treated with 7,12-dimethylbenz[a]anthracene. Cancer Res 60:3454–3460
7. Thurmond LM, Lauer LD, House RV, Cook JC, Dean JH (1987) Immunosuppression following exposure to 7,12-dimethylbenz[a]anthracene (DMBA) in Ah-responsive and Ah-nonresponsive mice. Toxicol Appl Pharmacol 91:450–460
8. Davila DR, Romero DL, Burchiel SW (1996) Human T cells are highly sensitive to suppression of mitogenesis by polycyclic aromatic hydrocarbons and this effect is differentially reversed by alpha-naphthoflavone. Toxicol Appl Pharmacol 139:333–341
9. Page TJ, O'Brien S, Holston K, MacWilliams PS, Jefcoate CR, Czuprynski CJ (2003) 7,12-Dimethylbenz[a]anthracene-induced bone marrow toxicity is p53-dependent. Toxicol Sci 74:85–92
10. Burchiel SW, Davis DA, Ray SD, Archuleta MM, Thilsted JP, Corcoran GB (1992) DMBA-induced cytotoxicity in lymphoid and nonlymphoid organs of B6C3F1 mice: relation of cell death to target cell intracellular calcium and DNA damage. Toxicol Appl Pharmacol 113:126–132

Polymerase Chain Reaction (PCR)

JOHN L OLSEN
Stony Brook University Medical School
327 Sheep Pasture Road
Setauket, NY 11733
USA

Synonyms

PCR, competitive PCR, real-time PCR, real-time and quantitative PCR, RTqPCR, real-time reverse transcription PCR, semiquantitative PCR

Short Description

The purpose of the polymerase chain reaction (PCR) is to detect and characterize genes. PCR is an in vitro method for amplifying a DNA sequence using a heat-stable polymerase and two primers, one complementary to the (+) strand at one end of the sequence to be amplified and the other complementary to the (−) strand at the other end. The newly synthesized DNA strands then serve as templates for the same primers and successive rounds of primer annealing, strand elongation and dissociation produce a highly specific amplification of the sequence. PCR can be used in environmental monitoring assays to detect the existence or absence of a DNA sequence in a sample—for example, a gene specific for an infectious viral particle or bacterium. In semiquantitative PCR, an analysis of the intensity of the stained PCR end products separated by gel electrophoresis can give a rank-order assessment of the original copy amount of the gene among several samples. Real-time and quantitative PCR (RTqPCR) is a technique useful to determine cellular mRNA expression changes for genes regulated by toxic compounds, drugs, infectious agents, or biological processes. RTqPCR can also be used to directly detect and quantify the number of DNA gene copies in a sample, useful when monitoring bacterial load.

Characteristics

PCR kinetics was introduced in 1985 by Kary Mullis and takes advantage of DNA polymerase to amplify a specific fraction of the genome (1). In conventional PCR, nucleic acid is first extracted from cells or a biological matrix. Several commercial kits and purification columns are available to extract DNA or mRNA (Qbiogene; Promega; Qiagen). After nucleic acid is purified, its purity and quantity are measured using a spectrophotometer. If one is conducting a gene expression experiment and mRNA regulation is the target of study, reverse transcriptase is used to generate complementary DNA (cDNA) from mRNA. For eukaryotic cells, random hexamers, oligo d(T)s, or sequence-specific reverse primers may be used to generate cDNA during reverse transcription. cDNA can be used in the PCR reaction in a similar fashion to DNA. A requirement for amplifying a gene sequence is to know the nucleotides flanking the segment of DNA so that specific oligonucleotide primers can be designed and synthesized. A well designed primer should consist of about 18 to 30 bases and hybridize to its short sequence with insignificant hybridization to other sequences in the reaction tube. Two primers are included in the reaction tube, one for each of the complementary DNA strands. The complete target sequence spanning the region between the two primers is generally less than 600 base pairs.

PCR amplification of a DNA or cDNA sequence is carried out in a reaction tube with an optimized salt buffer. It is a three-step process, referred to as a cycle, that is repeated 30 to 40 times in an automated thermal cycler. The first step is denaturation of double-stranded DNA to single strands by heating the sample, usually to greater than 90 °C. The second step is annealing of the two primers to their targets on single-stranded DNA, achieved between 40 °C and 65 °C. Selecting the most favorable annealing temperature is dependent on the length and base sequence of the primers. The third step of the cycle is extension. Once the primers are annealed, the temperature is elevated to approximately 72 °C and the enzyme Taq DNA polymerase catalyses primer extension (beginning at the 3' end of the primer) with complementary nucleotides (dNTPs) in the 5' to 3' direction. Taq DNA polymerase is a recombinant heat-stable DNA polymerase derived from *Thermus aquaticus*. The thermal cycler carries out the reaction through 30 or more cycles of denaturing, annealing, and extension. In conventional PCR, an endpoint analysis is done by looking at the fluorescently stained PCR products separated by gel electrophoresis.

Fluorescence-based real-time and quantitative PCR was developed in the mid 1990s to better confirm, characterize, and quantify nucleic acids in different sample populations. The genetic material to be analyzed can be either DNA or cDNA produced following reverse transcription of a sample's mRNA. An oligonucleotide probe (▶ TaqMan; Roche, Mannheim, Germany) containing a reporter fluorescent dye on the 5' end and a quencher dye on the 3' end is synthesized. This probe is designed to hybridize to the gene of interest in a central region of the amplicon. In an intact probe the quencher produces Förster resonance energy transfer (FRET), which lessens the reporter dye's fluorescence. However, during the PCR reaction, when the polymerase replicates a template on which the Taqman probe is bound, the 5′ nuclease activity of the DNA polymerase cleaves the probe, separating the reporter and quencher dyes to bring about increased fluorescence of the reporter. RTqPCR detects accumulating levels of DNA indirectly by monitoring the increase in fluorescence of the probe (see Figure 1). Software calculates the number of cycles necessary to detect a signal, generally in the exponential phase of the PCR reaction, and can determine the starting level of the gene. The sample starting amount can be interpolated either by comparison with a serially diluted standard curve where the target gene copy number (in plasmid or gene fragment form) per well is known (absolute quantitation) or by comparison with a housekeeping gene, such as β-actin, amplified from the same sample (fold-change or ddCt quantitation). Careful consideration should be given when choosing internal reference housekeeping genes for data normalization, as a popular gene used for this purpose, GAPDH, plays a role in several cellular functions and its mRNA levels do not consistently remain constant.

Regularly used alternatives to the Taqman method directly measure DNA using a fluorescent DNA (▶ SYBR green) that hybridizes nonspecifically to double-stranded DNA or by using probes—known as molecular beacons and scorpions—that specifically hybridize with the gene of interest. The increase in fluorescence is measured real-time and analyzed similarly to the Taqman-based method described above.

When working with tissue samples, RNA template can be derived from homogenized biopsy material or, when it is preferable to study the RNA and/or protein of a specific cell type, cells are purified by flow cytometry or laser capture microdissection (LCM) of paraffin-embedded tissue sections mounted on a glass slide. RNA expression has been quantified even after immunohistochemical staining of tissue sections using LCM, opening up the possibility of studying any archived tissue specimen. This combination of LCM and RTqPCR is particularly useful for studying the genetic regulation of tissue infiltration by tumor cells.

Pros and Cons

RTqPCR has several advantages over other methods of RNA quantitation. The assays are highly specific and sensitive to genetic regulation, and are easier and cheaper to perform than traditional hybridization techniques, such as northern blotting. The PCR reaction is carried out in a 96-well plate, making it a high-throughput method to test numerous conditions in the same experiment. The starting amount of RNA can be very small (nanograms) and single cells or tens of cells can be studied. SYBR green has the advantage of being applicable to any PCR product, doing away with the need to design specific probes for each gene target, but at the same time increasing the possibility of falsely positive amplification. The RT and PCR reactions may be done in the same tube (multiplex) but because of competition for reagents in the well, this protocol is less sensitive than the two-tube method and the results of these experiments should not be directly compared to two-tube protocols. Careful consideration should be given when choosing an RNA purification method, as different manufacturers' kits may result in RNA preparations that yield dissimilar amounts of cDNA template. Probes and primers made by different companies, or different lots from the same manufacturer may have different chemistries, requiring end-users to verify these reagents with cDNA standards.

Some other factors should be considered when performing RTqPCR. Dispensing the low-volume solutions can become tedious, and the cost of robotic preparation is out of reach for most laboratories. Also, though most protocols call for DNase treatment of RNA preparations, many researchers hesitate to use DNase because of the possible residual RNases that might degrade their sample. In order to protect against contamination, significant laboratory space and dedicated equipment is necessary to physically separate the different stages of the PCR process. The potential presence of inhibitors of PCR in the matrix of biological samples is another limitation.

PCR detection methods for pathogens offer the capacity of replacing immunoassays (e.g. ELISA) and cell culture. Viral samples can be characterized immediately without the need for culturing. PCR can detect

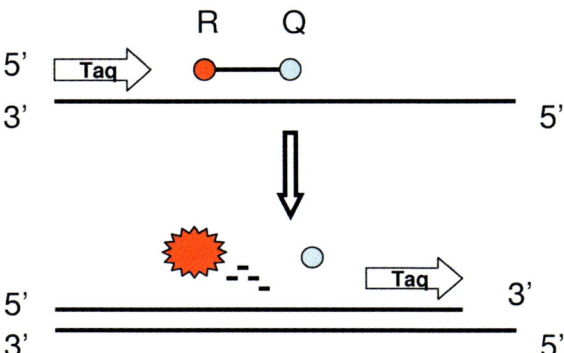

Polymerase Chain Reaction (PCR). Figure 1 In the PCR reaction the polymerase replicates a template on which the Taqman probe is bound. The 5′ nuclease activity of the DNA polymerase cleaves the probe, separating the reporter and quencher dyes to bring about increased fluorescence of the reporter. RTqPCR detects accumulating levels of DNA indirectly by monitoring the increase in fluorescence of the probe.

pathogens that cannot be cultured in the lab. When considering the analysis of viruses, RTqPCR cannot determine directly viral infectivity and may not detect the new nucleotide sequences of emerging viruses. Useful technologies to enrich for viruses in sample material include using host cell receptors or monoclonal antibodies to capture virus particles for further genetic analysis. Microbiologists have also developed PCR-based ▶ DNA fingerprinting methods to distinguish between various sources of fecal bacteria.

Allele-specific discrimination is accomplished by using two different Taqman probes, each with a unique fluorescent dye attached. These assays help to genotype individuals and have been used to analyze gene polymorphisms that contribute to the susceptibility to toxic insult, including cytochromes and glutathione S-transferases. Genetic variants of cytokine receptors have also been characterized in human populations using RTqPCR ▶ allelic discrimination technology. Most assays discriminate between two alleles of single nucleotide polymorphisms (SNPs). An initial mutation screening study to identify different variants of a gene is performed, using direct sequencing, PCR-single strand conformation polymorphism (PCR-SSCP), or denaturing high performance liquid chromatography (D-HPLC). Once polymorphisms have been identified, allele-specific probes are designed for both alleles A and B (assuming a bi-allelic system for the sake of illustration, though the method is capable of discriminating an unlimited number of alleles). Binding of the probe for allele A to allele B DNA is suboptimal because of the mismatch within the Taqman probe and the target sequence. For a bi-allelic system, an unknown sample of DNA will be identified as having present allele A, B, or both A and B. These genotypes (AA, AB, and BB) are designated based on real-time PCR growth curves and can be validated by direct DNA sequencing of PCR products separated by agarose gel electrophoresis.

Predictivity

One of the major advantages of RTqPCR is its ability to detect very rare events (one copy of a specific gene/10^7 cells). This high sensitivity makes the assay vulnerable to false-positives caused by DNA or RNA contamination. Also, computer quantitation of gene copy number can be sensitive to operator-dependent effects. Results can be produced with insufficient knowledge of the statistical analyses being performed by the software that comes bundled with the RTqPCR machines, and some authors are not forthcoming in describing their entire statistical analysis and normalization protocols. To make results meaningfully predictive of human health outcomes, investigators should ensure validation of reagents and RNA samples and move towards standardized and universal methods and reagents.

Relevance to Humans

The RTqPCR assay has become the gold standard and most commonly used mRNA quantification method for monitoring transcription in vitro and for observing the effects of receptor signaling. The majority of the RTqPCR immunotoxicology experiments up to this point have been performed either using animal models or using human or animal cell lines. The goals of these and human in vivo experiments are to define biomarkers (any plausible and measurable biological endpoint relevant to the toxicity of a compound), or ▶ surrogates (statistically validated endpoints with medical importance) of clinical illness, and to develop therapies for these processes. For example, Spink and coworkers used RTqPCR to investigate the effects of estrogen on aryl hydrocarbon responsiveness in dioxin-treated human breast cancer cells (2). RTqPCR can be used to validate in a quantitative manner the findings of microarrays, which measure genome-wide responses to toxic compounds but also generate overwhelming data sets that are cumbersome for rapid toxicology screens. Morgan and coworkers used cDNA microarrays to explore the effects of a diverse group of toxicants including ouabain and hydrogen peroxide in a human hepatocellular carcinoma cell line (3). Their array data found a subset of seven genes that can be used in an RTqPCR platform to test compounds for cellular oxidative stress.

Real-time quantitative PCR is applied extensively in the study of cancer and the response of cancer cells to drugs. It has been used to monitor the toxicology and dissemination of adenoviral vectors synthesized to treat cancers. It can also effectively detect small numbers of blood-borne tumor cells and lymphangiogenesis, but is in general still in the developmental stage of accurate cancer diagnosis. PCR-based assays can identify specific chromosome aberrations, such as have been shown to be associated with benzene exposure. Biernaux et al. used nested RT-PCR to detect the BCR-ABL fusion in RNA from the t(9;22) (q34; q11) translocation (4). Appropriately designed primer pairs and probes can preferentially amplify mutant genomic DNA and single nucleotide mutations even in the presence of a 1000-fold excess of wild-type DNA.

Several important factors should be taken into account when designing and interpreting PCR-based toxicology experiments. First, because of its high sensitivity, RTqPCR can detect small changes in genetic regulation. However, RTqPCR does not give information about RNA stability, protein production, or biological activity of proteins, so that the presence of transcriptional changes should be verified by protein analysis. Second, given the variability of expression of genes in normal subjects ('natural variation') and the presence of highly polymorphic genes such as CYP2D6, which has at least 70 allelic variants with a range of biological activities and a high degree of interethnic variability ('population-specific variation'), it is advised to have large sample sizes and background information about the normal allelic variation of the gene under consideration. Third, the extent of tissue sampling within an organ should be sufficient to accommodate for any local differences in genetic distribution. One lung lobe may have a different blood perfusion and airflow than another, so care should be taken to consistently harvest similar biological material.

Regulatory Environment
While PCR detection methods clearly play an important role in hazard identification, risk assessment and contaminant regulation, a major problem associated with its use in the regulatory context is how to properly interpret positive and negative results. Regulatory agencies decide whether detection-only of pathogens can be used for risk assessment or if that data must be coupled with evidence of infectivity. While positive findings of a genetic mutation in a tissue sample indicate the presence of that altered gene, the medical implication is not always clear. In Biernaux's study, the BCR-ABL translocation was found in 40% of normal healthy subjects, complicating the health risk assessment (5). Negative results need to be interpreted in light of the manner of sample collection, processing and storage methods that might degrade genetic material, and the detection limit of the assay.

Lack of standardization of PCR methods is a major concern for developing health-impact assessments. Quality assurance and quality control measures, sample preparation protocols, and internal controls need to be tested in many laboratories and standardized in order to move beyond the developmental stage. Many PCR primers are constructed by different laboratories and this may confound pathogen occurrence monitoring. PCR methods are rapidly changing, so a performance-based approach with benchmark values is needed to know if the data being generated is helping to address specific health problems.

As gene array-based approaches remain semiquantitative, RTqPCR assays are still the most common method to quantitate gene expression patterns in different populations. Microarrays produce a comprehensive picture of the genomic-wide response to toxic compounds. Microarray techniques are most appropriate for individual compound studies rather than for screening of multiple chemicals. Real-time PCR is designed to monitor single endpoints or small sets of endpoints among several samples. RTqPCR is therefore the method of choice to verify and quantify microarray results or resolve questions when discrepancies appear in direction of gene change on microarray plates produced by different companies. Microarray and RTqPCR methodologies are used together to develop molecular signatures for classes of toxicants with distinct actions on specific sets of genes. These molecular signatures are valuable when predicting the toxicity of other compounds that have incomplete toxicological information.

Until the advent of RTqPCR and cDNA microarrays, molecular toxicology studies relied almost exclusively on rodent models and in vitro human systems to identify cellular changes that may be important in the human population. Those toxicological techniques required complicated methodologies to deal with dose extrapolation, interspecies extrapolations and interindividual differences. With the integration of RTqPCR into immunotoxicology risk assessment, epidemiological biomarkers can now be accurately quantified in order to resolve some of the uncertainties involved with previous approaches.

References
1. Mullis KB (1990) The unusual origin of the polymerase chain reaction. Sci Am 262:56–61
2. Spink DC, Katz BH, Hussain MM, Pentecost BT, Cao Z, Spink BC (2003) Estrogen regulates Ah responsiveness in MCF-7 breast cancer cells. Carcinogenesis 24:1941–1950
3. Morgan KT, Ni H, Brown HR, Yoon L et al. (2002) Application of cDNA microarray technology to in vitro toxicology and the selection of genes for a real-time RT-PCR-based screen for oxidative stress in Hep-G2 cells. Toxicol Pathol 30:435–451
4. Biernaux C, Loos M, Sels A, Huez G, Stryckmans P (1995) Detection of major bcr-abl gene expression at a very low level in blood cells of some healthy individuals. Blood 86:3118–3122

Polymeric Immunoglobulin Receptor (pIgR)

Polymeric immunoglobulin receptor (pIgR) is a specialized receptor that binds IgA and translocates it across epithelium into secretions.

▶ Mucosa-Associated Lymphoid Tissue

Polymorphism

Polymorphism is a term often used to represent any variation of a character or gene within a population. A more precise definition is the existence of a character in two or more variant forms in a population, where the least common form is present in more than 1% of

individuals. Consequently, a polymorphic character is neither very rare nor very common.
▶ AB0 Blood Group System

Polymorphonuclear Leukocyte

This is a collective term for granulocytes (basophils, eosinophils, neutrophils) derived from the bone marrow and characterized by a multilobed nucleus. These phagocytic cells migrate from the circulation to the site of inflammation in response to chemotactic signals.
▶ Streptococcus Infection and Immunity
▶ Lymphocytes

Polymorphonuclear Neutrophil

▶ Neutrophil

Polynucleotide Vaccines

▶ DNA Vaccines

Polyunsaturated Fatty Acids

These are fatty acids which contain more than one double bond.
▶ Fatty Acids and the Immune System

Popliteal Lymph Node Assay

RAYMOND PIETERS
Head Immunotoxicology
Institute for Risk Assessment Sciences (IRAS)
Yalelaan 2
P.O. Box 80.176
3508 TD Utrecht
The Netherlands

Synonyms
PLNA.

Short Description
The popliteal lymph node assay (PLNA) is in use to test the immunostimulatory capacity of low molecular weight compounds, in particular pharmaceuticals. Briefly, compounds are injected subcutaneously into the hind footpad of either mice or rats and 6–8 days later the draining popliteal lymph node (PLN) is excised and compared with popliteal lymph nodes from vehicle-treated animals. Comparison is basically made between the weight or cellularity of the nodes; the ratio of these parameters from compound-treated over vehicle-treated PLNs (designated the PLN index) indicates whether or not the compound is immunostimulatory. Immunostimulation is regarded as one of the possible prerequisites for pharmaceuticals and other compounds to activate and sensitize T cells and to induce systemic allergy or autoimmune-like derangements.

Characteristics
In the late 1970s and early 1890s, Gleichmann and colleagues realized that graft-versus-host (GvH) reactions might be the basis for pathogenetic mechanisms behind the development of drug-induced allergy and autoimmunity (1). To substantiate this hypothesis they studied effects of the hypersensitizing drug diphenylhydantoin in the PLNA, which was already in use to study GvH reactions (2). Kammüller and colleagues (3) further explored the PLNA as a method to screen for the immunostimulatory potential of chemicals by comparing series of structural homologues of drugs (e.g. hydantoin, zimeldine) and by studying of the immunological changes in the draining PLN induced by a variety of immunostimulatory chemicals (among these also contact sensitizers). Originally conducted in mice, Descotes and colleagues (4) used the rat as species to perform the PLNA and they found that the rat responds equally well to a series of chemicals (Note: some rats may have more than one popliteal lymph node per paw).

Usually, experimental groups contain 4–6 animals per group. The contralateral PLN can be used as internal control, but because in some cases these nodes also show signs of immunostimulation separate control groups are recommended.

The volume of the solution injected (with a 24 G needle) in the paw of mice is usually 50 µl in the case of aqueous solutions (both saline and PBS have been used) and 10 µl (alone or mixed with 40 µl saline) in the case of DMSO (for hydrophobic compounds). In the rat up to 100 µl can be used as injection volume. Specific buffers are used in case a compound is less soluble or not soluble at physiological pH.

The dose of the chemicals injected is mostly about 1 mg per mouse or up to 5 mg per rat, but for new chemicals this should be carefully evaluated as the dose may strongly vary and depends on the toxicity of the chemical ($HgCl_2$ for instance is used in a dose of only 50 µg per animal). Injection is either from toe-to-heel or from heel-to-toe which does not seem to make any difference.

The size of the lymph node (by weight, cell number of even diameter) is the most frequently used parameter for lymph node activation because it is easy to detect. However, this is not a very informative parameter and certainly does not distinguish sensitizers from non-sensitizing irritant chemicals. For this, immunologically more relevant parameters need to be included. Immunohistology, for instance, can be used to detect the presence of germinal centres, which are known to be T cell-dependent. Cytokine production by isolated PLN cells after mitogen-specific or hapten carrier-specific restimulation, either combined or not combined with flow cytometrical analyses of the cells, can also be used to provide information on T cell-dependency and also on the type of immune response that is elicited by the compound.

Pros and Cons

The primary PLNA is a straightforward, objective and cheap assay that allows fast screening of the immunostimulatory potential of compounds. Concentration-effect relationships of series of structural homologues, for instance of a newly synthesized pharmaceutical compound, can be easily tested in the PLNA, and by using immunologically more sophisticated read-out parameters the PLNA also offers a means to investigate initiating mechanisms of chemically induced adverse immune reactivity.

An important disadvantage of the unmodified simple PLNA, with only the PLN size as parameter, is that strong irritant chemicals are false positive in the PLNA. This is probably due to the fact that these chemicals induce too much local damage and consequently elicit a strong inflammatory response. However, mere irritants can be excluded as non-sensitizers by using more sophisticated variations of the PLNA (secondary PLNA, adoptive transfer PLNA, RA-PLNA). In particular, pro-haptens that require metabolic conversion (most well known in this respect is the anti-arrhythmic drug procainamide) are false negative in the PLNA. However, some of these false negative compounds can be detected by including a metabolizing system (addition of S9 mix or oxidizing polymorphonuclear cells) in the injection solution. Compounds that do not induce adverse immune reactions by immunostimulation, but for instance by interfering with immunoregulatory mechanisms, may also turn out to be false negative in the PLNA. Important to note is also that different strains of mice respond to a different degree to some chemicals, indicating genetic dependency of the PLNA response.

A major limitation of the PLNA with regard to risk assessment is that the exposure route (subcutaneous exposure) is not the usual route of exposure for drugs. In addition, subcutaneous paw injection limits the volume that can be injected, so some poorly soluble chemicals are not fully dissolved and need to be injected as suspensions. This may cause aspecific activation of macrophages and elicitation of inflammatory responses, so it is generally regarded as undesirable, although on the other hand aspecific activation by poorly soluble compounds may also occur in the gastrointestinal tract in case of those chemicals.

In all, it has to be realized that essentially the PLNA is a screening assay to assess a chemical's potential to stimulate the immune system. When extended with immunologically relevant parameters the PLNA can also indicate whether a chemical has the potential to sensitize the immune system, that is it may cause allergy or autoimmune-like phenomena in susceptible individuals under certain predisposing conditions. As a research tool the PLNA has without doubt contributed considerably to the fundamental knowledge about drug-induced immunostimulation.

It is important to note that over the past 20–25 years more than 130 chemicals (including drugs and homologues, as well as other chemicals) have been tested in the PLNA (5), yet the assay has never been formally validated. But available data of chemicals that have been tested in various laboratories have been compared, good reproducibility and similar predictivity, with only a few false-negative or false-positive compounds.

Predictivity

When using the PLNA as a screening test one should of course be aware of false-positive (e.g. strong irritants) and false-negative (e.g. pro-haptens) compounds. A false-negative compound that is indeed a pro-hapten can be dealt with by using a metabolic system as mentioned above. Strong irritants that induce severe inflammatory responses and necrosis at the side of injection should not be tested, or should be tested at a dose that has less aspecific effect. The use of immunologically relevant parameters or modifications of the PLNA that allow detection of anamnestic immune responses may provide more information in the case of false-positive compounds and identify such chemicals.

When this is taken into account, the correlation between PLN enlargement and the ability of a chemical to cause systemic autoimmune disorders is good to very good, considering that of the 130 or so chemicals tested only some 10 false-negative and false-positive chemicals could not be explained (3,5).

Notably, also known allergens (e.g. contact or respiratory allergens) are positive in the PLNA. This stresses that the PLNA may indicate the immunosensitizing potential of compounds, but it does not provide information about the kind of adverse immune effect that can be expected in humans. For that far more information is needed on the vast array of predisposing, immunoregulatory factors, as well as environmental factors, that influence the development of a sensitized immune system to actual clinical symptoms.

Relevance to Humans

The PLNA should be regarded as a test to screen for the possible hazard of a compound to cause immunosensitization. As such, the assay can be regarded as a first step to evaluate whether the compound has also the potential to stimulate and sensitize the immune system via the relevant (mostly oral) route of exposure. Unfortunately, models to assess the immunosensitizing potential of chemicals, in particular drugs, via the okal route are lacking. Conceivably, for the time being, assessment of the immunogenic potential of a chemical may be assessed by combining limited animal data (e.g. PLNA data) with information about dose-response relationships and predisposing conditions.

Regulatory Environment

The PLNA has been successfully used as a screening test for drug-induced adverse immune effects for over two decades, but the test has never been formally validated. Because of its many advantages as a simple screening assay and because of its added value to other local lymph node tests such as the LLNA it is considered important to validate the PLNA. Attempts to do this are ongoing in the framework of the International Life Sciences Institute-Health and Enviromental Sciences Institute (ILSI-HESI) Immunotoxicology Technical Commitee (ITC).

References

1. Gleichmann E, Pal ST, Rolink AG, Radasziewicz T, Gleichmann H (1984) Graft versus host reactions: clues to the etiopathology of a spectrum of immunological diseases. Immunol Today 5:324–332
2. Gleichmann H (1981) Studies on the mechanism of drug sensitization: T-cell-dependent popliteal lymph node reaction to diphenylhydantoin. Clin Immunol Immunopathol. 18:203–211
3. Kammüller ME, Thomas C, De Bakker JM, Bloksma N, Seinen W (1989) The popliteal lymph node assay in mice to screen for the immune dysregulating potential of chemicals-a preliminary study. Int J Immunopharmacol 11:293–300
4. Verdier F, Virat M, Descotes J (1990) Applicability of the popliteal lymph node assay in the Brown Norway rat. Immunopharmacol Immunotoxicol 12:669–677
5. Pieters R, Albers R (1999) Screening tests for autoimmune-related immunotoxicity. Env Health Persp 107:673–677

Popliteal Lymph Node Assay, Secondary Reaction

PETER GRIEM
Toxikology/Regulatory Affairs
Clariant GmbH
Am Unisys-Park 1
D–65843 Sulzbach
Germany

Synonyms

PLNA, secondary PLNA, adoptive transfer PLNA

Short Description

The secondary PLNA (reviewed in (1–3)) aims at determining a secondary T lymphocyte response to a low-molecular-weight chemical by injecting the chemical subcutaneously into one hind footpad of already sensitized in mice or rats. The administered dose has to be small enough not to elicit a primary immune response in nonsensitized animals, i.e. it should not suffice for stimulation of naïve T lymphocytes. The immune response against the injected chemical or antigen is quantified by analyzing one or more parameters of the draining popliteal lymph node (PLN), such as PLN weight, PLN cell number, cell surface marker expression on PLN cells or PLN cell proliferation measured, for example, as ^{3}H-methyl thymidine incorporation.

Characteristics

In the direct PLNA, a single injection of the test chemical is administered subcutaneously into one hind footpad. The contralateral footpad serves as an internal control and is usually injected with the vehicle (test solution without test chemical) or left untreated. After injection the test chemical is transported via the afferent lymphatics to the nearest ▶ draining lymph node, the PLN, where a primary immune response may take place. After 5–8 days, the left and right PLN are isolated and analyzed. PLN weight and/or other PLN parameters like cell count, cell proliferation and expression of cell surface markers are determined. Results are usually expressed as a PLN index which is the ratio of values obtained from the experimental and control sides. A primary PLN response normally peaks between days 4 and 10 and than returns to preinjection level within 3–5 weeks, unless a persistent material, such as silica particles, is injected.

While the primary PLNA is capable of distinguishing immunostimulating chemicals from immunologically inactive chemicals, this assay fails to distinguish haptens or ▶ neoantigen-forming chemicals from mere inflammatory irritants, unless more sophisticated analyses are performed, such as determination of cytokine production in PLN cells and/or expression of activation markers on the cell surface. If, however, a chemical elicits a specific T cell response part of the activated T lymphocytes (also called primed T cells) will differentiate into memory T cells. The latter can be detected by their capacity to mount a secondary immune response, which is characterized by faster kinetics and lower ▸ elicitation doses than those required for a primary response (i.e. T cell priming during sensitization). In order to test for a secondary response, a secondary PLNA can be performed in one of two ways (1–3).

Secondary PLNA in sensitized animals

Sensitization of naive animals (usually mice) can be accomplished by treating the animals as for the direct PLNA. Upon complete resolution of the primary PLN response (which is shown by satellite groups analyzed at different time intervals after injection) the animals are challenged in the same paw with a lower dose of the same chemical, which is substimulatory in the direct PLNA. After 4–6 days assessment of the specific secondary response is determined as ▸ PLN index by measuring the same parameters as in the direct PLNA. Results are expressed as PLN indices (mean and standard deviation of groups of at least five animals) and statistical tests, such as Student's *t* test or ANOVA, are used to determine significant differences between treatment and control groups.

Alternatively, the animals can be sensitized not via the hind footpad, but via a different route that is more relevant for human exposure, e.g. intranasally, orally, intramuscular or intravenously (see Table 1 for examples). The treatment during this sensitization phase can be extended to the time period desired. The animals are then challenged by injection of the test chemical into the hind footpad and the PLN response is evaluated as above.

Adoptive Transfer PLNA

This test measures secondary T lymphocyte responses from a chemically-treated donor animal following their transfer to a syngeneic (genetically identical) recipient animal. The T cell donor animals can be treated with the test chemical under conditions of exposure as to dose, route, frequency, and duration that mimic the human situation (see Table 1 for examples). After the treatment period, spleen cells, total splenic T lymphocytes, or a selected T cell subpopulation of the donor animals are transferred by subcutaneous injection into one hind footpad of recipient animals, using T cells from unexposed or vehicle-treated animals as a negative control. One day after the transfer, recipient animals are challenged at the same site by subcutaneous injection of the test chemical at a sub-stimulatory dose. The secondary PLN response is assessed 3–6 days after the challenge using the same read-out parameters as described above.

A positive response in the secondary PLNA indicates that exposure of the (donor) animals had resulted in T-cell sensitization and subsequent generation of memory T cells. The adoptive transfer PLNA provides direct evidence for the T cell dependence of the response to the test chemical. Table 1 provides a few examples of secondary and adoptive transfer PLNAs published in the literature.

Pros and Cons

The secondary PLNA allows determination in vivo of secondary T cell responses which sometimes cannot be measured in vitro, e.g. in the lymphocyte proliferation assay. The secondary PLNA could be performed on satellite groups of standard subacute and subchronic toxicity studies to identify a possible sensitization hazard of the test chemical.

Chemicals that cause hypersensitivity or autoimmunity through activation of specific T cells can be distinguished from chemicals that cause 'unspecific' or 'polyclonal' immunostimulation, such as hexachlorobenzene.

An interesting aspect of the secondary PLNA is that for challenge not only can the chemical used for sensitization be employed, but also a metabolite of the chemical that is presumed to be the ultimate ▸ hapten of the chemical. This metabolite may be a pure, synthesized chemical, it may be contained in a metabolite-generating enzyme solution or cell homogenate, or it may be contained in a homogenate of cells isolated from animals after nontreatment. In this way, the secondary PLNA can be used to demonstrate sensitization against a reactive intermediate metabolite of a chemical (see Table 1 for examples).

Performing the PLNA requires some technical skill and expertise. Especially the adoptive transfer PLNA is rather laborious and requires a comparatively large number of animals. Using a sensitization route that mimics human exposure may require repeated treatment of animals over weeks or months.

Predictivity

The predictivity of the secondary PLNA has not been formally evaluated. It has often been employed to investigate mechanisms of action of chemicals known to cause hypersensitivity or autoimmunity in humans. In general, involvement of chemical- or metabolite-specific T lymphocytes could be shown in cases where the direct PLNA or other sensitization tests gave false-negative results.

Relevance to Humans

While the secondary PLNA indicates the presence of

Popliteal Lymph Node Assay, Secondary Reaction. Table 1 Examples of chemicals studied in secondary popliteal lymph node assays

Chemical [reference]	Use, adverse effects in humans, immunogenicity	Treatment of animals during sensitization phase	Secondary PLNA method and test chemical used	Outcome of PLNA
p-Benzoquinone (BQ) [5]	Chemical (e.g. in photographic developers) causing allergic contact dermatitis. Positive in direct PLNA	C57BL/6 mice Single subcutaneous injection of 100 nmol BQ into hind footpad in a direct PLNA	Secondary PLNA performed in sensitized animals after primary PLN reaction had subsided completely (13 weeks). Test chemicals were 0.1 nmol BQ or 0.1 nmol benzene	No reaction occurred in mice that had received only solvent during priming; a specific secondary reaction occurred in BQ-primed mice to BQ and to benzene, indicating that enough benzene was metabolized locally to BQ to elicit a secondary T cell reaction
Gold(I) disodium thiomalate (Au(I)TM) [6]	Antirheumatoid drug. Induces dermatitis, hypergammaglobulinemia, and/or immune glomerulonephritis in high percentage of patients after prolonged treatment. Negative in direct PLNA	C57BL/6 mice Weekly intramuscular injections of 22.5 mg/kg Au(I)TM for 6–12 weeks	Adoptive transfer of splenic T cells (MACS removal of B cells). Test chemicals were Au(I)TM, gold(III) tetrachloride, homogenated peritoneal phagocytes from Au(I)TM-treated mice. Homogenated peritoneal phagocytes incubated with Au(I)TM in vitro	Transferred T cells from Au(I)TM-treated mice showed positive response to gold(III), not to Au(I)TM. They also reacted to phagocytes from Au(I)TM-treated mice and to phagocytes incubated with Au(I)TM in vitro, indicating that T cells were sensitized to gold (III), a reactive metabolite of gold(I) formed in phagocytes
Mercuric chloride ($HgCl_2$) [7]	Potential to induce autoimmunity in humans unclear. Induces antinuclear and antinucleolar autoantibodies in susceptible mouse and rat strains. Positive in direct PLNA	B10.S mice Subcutaneous injections of 0.5 mg/kg $HgCl_2$ three times a week for 1 or 8 weeks	Adoptive transfer of splenic T cells (MACS removal of B cells). Test chemicals were $HgCl_2$, nuclei from spleen cells of $HgCl_2$-treated or buffer-treated animals, and isolated nuclear protein fibrillarin, either preincubated with $HgCl_2$ or buffer	Transferred T cells from mice treated with $HgCl_2$ for 1 or 8 weeks showed positive responses to $HgCl_2$ and also to nuclei of $HgCl_2$-treated mice and to $HgCl_2$-treated fibrillarin. In addition, after 8 weeks of $HgCl_2$ treatment, T cells showed responses to untreated nuclei and fibrillarin, indicating a shift from mercury-specific to autoimmune T cells over time

T cell sensitization against the test chemical, it does not indicate whether this sensitization would cause (clinically relevant) disease. For linking of chemical exposure to induction of disease and for establishing dose-response relationships more sophisticated animal models are required. For some autoimmunogenic chemicals, e.g. mercuric chloride and gold(I) salts, prolonged treatment with the test chemical allowed detec-

Popliteal Lymph Node Assay, Secondary Reaction. Table 1 Examples of chemicals studied in secondary popliteal lymph node assays (Continued)

Chemical [reference]	Use, adverse effects in humans, immunogenicity	Treatment of animals during sensitization phase	Secondary PLNA method and test chemical used	Outcome of PLNA
Procainamide (PA) [8]	Antiarrhythmic drug Induces lupus-like syndrome in high percentage of patients after prolonged treatment Negative in direct PLNA	A/J strain (slow acetylator) and C57BL/6 (fast acetylator) mice Subcutaneous injections of 8 and 16 µmol PA, respectively, three times a week for 16 weeks Additional weekly intraperitoneal injections of 600 ng PMA in some groups of C57BL/6 mice	Adoptive transfer of splenic T cells (MACS removal of B cells) Test chemicals were PA, N-hydroxy-PA, N-acetyl-PA, and homogenated peritoneal phagocytes from PA-treated mice	Transferred T cells from A/J mice showed positive response to N-hydroxy-PA and to phagocytes from PA-treated mice, but not to PA or N-acetyl-PA T cells from C57BL/6 mice treated with PA +PMA, but not T cells from mice treated only with PA, reacted to N-hydroxy-PA and to phagocytes from PA +PMA-treated mice, but not to PA, N-acetyl-PA or phagocytes from mice treated only with PA, indicating that T cells were sensitized to N-hydroxy-PA, a reactive metabolite of PA While slow acetylator mice formed enough metabolite for sensitization, fast acetylator mice were only sensitized after additional stimulation of phagocytes

MACS, magnetic-activated cell sorting; PLNA, popliteal lymph node assay; PMA, phorbol myristate acetate.

tion of antinuclear autoantibodies and immune glomerulonephritis. Secondary PLNAs can help elucidating the mechanisms underlying allergic and autoimmune reactions to chemicals (especially drugs because of the relatively high exposure dose).

Regulatory Environment

While validated skin sensitization tests are available, there is currently no validated or widely applied standard toxicity test for the identification of compounds with the potential to induce systemic allergic or autoimmune reactions. Validation of the direct PLNA is currently considered (4). For regulatory purposes, results from secondary PLNA may currently only be used as supplementary (mechanistic) information.

References

1. Pieters R, Ezendam J, Bleumink R, Bol M, Nierkens S (2002) Predictive testing for autoimmunity. Toxicol Lett 127:83–91
2. Pieters R, Albers R (1999) Screening tests for autoimmune-related immunotoxicity. Environ Health Perspect 107 (Suppl 5):673–677
3. Goebel C, Griem P, Sachs B, Bloksma N, Gleichmann E (1996) The popliteal lymph node assay in mice: screening of drugs and other chemicals for immunotoxic hazard. Inflamm Res 45 (Suppl 2):S85–90
4. Vial T, Carleer J, Legrain B, Verdier F, Descotes J (1997) The popliteal lymph node assay: results of a preliminary interlaboratory validation study. Toxicology 122:213–218
5. Ewens S, Wulferink M, Goebel C, Gleichmann E (1999) T cell-dependent immune reactions to reactive benzene metabolites in mice. Arch Toxicol 73:159–167
6. Goebel C, Kubicka-Muranyi M, Tonn T, Gonzalez J, Gleichmann E (1995) Phagocytes render chemicals

immunogenic: oxidation of gold(I) to the T cell-sensitizing gold(III) metabolite generated by mononuclear phagocytes. Arch Toxicol 69:450–459
7. Kubicka-Muranyi M, Kremer J, Rottmann N et al. (1996) Murine systemic autoimmune disease induced by mercuric chloride: T helper cells reacting to self proteins. Int Arch Allergy Immunol 109:11–20
8. Kubicka-Muranyi M, Goebels R, Goebel C, Uetrecht J, Gleichmann E (1993) T lymphocytes ignore procainamide, but respond to its reactive metabolites in peritoneal cells: demonstration by the adoptive transfer popliteal lymph node assay. Toxicol Appl Pharmacol 122:88–94

Popliteal Plymph Node Assay (PLNA)

▶ Reporter Antigen Popliteal Lymph Node Assay

Population Studies

▶ Epidemiological Investigations

Porcine Immune System

Ricki M Helm
Arkansas Children's Hospital Research Institute,
Arkansas Children's Nutrition Center
University of Arkansas for Medical Sciences
1120 Marshall Street
Little Rock, AR 72202
USA

Synonyms
swine, minipig, pig

Short Description
The physiological and immunological similarities of swine and humans have become important features in large-animal models for biomedical research as evidenced in both veterinary and human literature citations (1). Swine are suitable for studies of developmental immunology, xenotransplantation, wound healing, immunization schemes, allergy and human asthma. The physiological relevance of the swine as an intended target species directly affects the laboratory "proof of concept" that can be translated to successful clinical treatments in the lieu of human trials. The different strains of pigs and miniature pigs available, crucial natural disease models that occur in outbred populations, combined with the rapidly growing swine immune reagent repertoire, will provide cost-effective studies that will reveal the importance of the swine as valid animal model systems that will more closely extrapolate to the human situation under investigation.

Characteristics
For comparative studies on immunoglobulins, Butler and Howard (2) summarized the immunoglublulins and Fc receptors from the Comparative Immunoglobulin Workshop. Interestingly, the sequences of $C\mu$, $C\alpha$, $C\gamma$, $C\kappa$, $C\lambda$ and Vh of swine are most similar to their human counterparts. Based upon flow cytometric analysis of T cells, $\gamma\delta$ T cells, B cells, myeloid cells, activation/maturation markers, and CD45-specific antibodies, the Third International Swine CD Workshop listed a total of 38 pig leukocyte CD/SWC determinants for pig leukocytes (3).

Data from CD3, CD4, CD8 immunophenotyping suggests the porcine $\alpha\beta$ thymocytes require 15 days to fully differentiate, while $\gamma\delta$ thymocytes differentiate in less than 3 days and migrate asynchronously from the thymus to the periphery. A summary of the flow cytometry and immunohistochemistry data, which identified 38 monoclonal antibodies to cluster group ligands, was presented in the second-round analysis of the B-cell sections at the Third International Swine CD Workshop (4).

Cloned pig cytokines include tumor necrosis factor (TNF)- α, interleukins IL-1α, IL-1β, IL-2, IL-4, IL-6, IL-8, IL-12, IL-15, tumor growth factor(TGF)-β, granulocyte-colony stimulating factor (G-CSF), monocyte chemoattractant protein-1, and the recent addition of IL-5. Stimulation of porcine peripheral blood mononuclear cells (PBMC) with CpG oligodeoxynucleotide (ODN) activates and upregulates proinflammatory cytokines IL-6, TNF-α and the T-helper 1-associated cytokine IL-12. The addition of swine IL-3 and c-kit ligand (KL) to fresh cultures of human and swine bone marrow cells enhanced swine hematopoietic chimersim.

A Th1-biased immune response—activation/stimulation/secretion of swine PBMC results in proinflammatory cytokines IL-6, TNF-α and IL-12—that correlates well with human PBMC responses. Direct skin tests and passive cutaneous anaphylaxis serum results in the swine closely mimic antigen/allergen responses with respect to delayed and immediate hypersensitivity responses characterized in human biological assessments.

Similar gastrointestinal mucosal architecture, maturation and immune responses in swine will provide a significant understanding to the immune mechanisms, antigen handling, and innate immune defenses and disease prevention that could be applied to human neonates. For example, neonatal piglets sensitized to food allergens followed by oral challenge mimic the physical and immunologic characteristics of food allergy in humans (5). The model should prove to be useful in investigations to determine IgE-mediated mechanisms, immunotherapeutic intervention strategies, and provide a predictive model for assessing novel proteins as potential allergens. A caveat is that commercial anti-swine IgE is still not available and heat-inactivated passive cutaneous anaphylaxis reactions are still needed to indicate IgE-mediated diseases. In related investigations, genetically modified maize (Bt-maize) was shown to be substantially equivalent (a characteristic required by regulatory agencies) to parental maize with respect to nutrition. Porcine models of wound healing and therapeutic options are offering significant contributions to treatment of human burn victims. Airway hyperreactivity, eosinophil infiltration, treatments with agonists on airway mechanics and tryptase inhibitors are providing mechanistic applications to the pathophysiology of asthma.

Because of the growing shortage of human organs and tissues for transplantation, the swine has become a major candidate for xenogenic organ donation. Problems with transmission of zoonoses, donor organ anatomy and function are still areas of concern; however, immunological rejections can be over come with remarkable progress being made in organ transplantation. With respect to immunological organization and anatomy, the porcine larynx is similar to that of the human larynx, suggesting a clearcut case for laryngeal transplants. The mechanical properties, antithrombogeniety and tissue compatibility of decellularized, heparinized carotid swine arteries may also be suitable for patients in need of grafts. The gene for hyperacute rejection was the first knockout gene identified in swine when organs from swine were transferred to primates. Recently, genetic manipulation of porcine B7 molecules, such as in the CD80- knockout swine or the soluble CD80 transgenic swine, may potentially provide the basis for therapeutic strategies to regulate the human response to graft organs. Continued improvements in gene knockout swine models, homologous recombination of fetal somatic cells, and nuclear transfer in swine suggest that specific modifications made to the swine genome may be extrapolated, in time, to humans.

Depending upon the balance between effector and regulatory function, and disturbances in this balance in swine and humans, the similarities of fetal placentation during gestation and the development of immunocompetence during ontogeny of the neonate, swine provide distinct advantages for studies related to human diseases. Differences of extrathymic $CD4^+CD8^+$ double-positive T cells, porcine blood T cells with high proportions of $\gamma\delta$ T cells and features of lymphoid tissue structure/lymphocyte transport, B cell differentiation that follows different pattern, maternal/fetal and neonatal interaction that differ from other experimental animals, the swine is by far the more realistic animal model that will allow a better extrapolation of disease models to the human.

Pros and Cons

The swine has a similar immune physiology to that of the human and its use as an animal model will more directly allow an extrapolation that can be translated to clinical outcomes. Similarities include fetal placentation, duration of gestation and development of immunocompetence during ontogeny of the neonate. Swine offer natural disease models that can be used to analyze host-pathogen interactions, infectious agents, xenotransplants and allergy that occur in outbred populations. Sequence similarity of MHC and immunoglobulins should provide distinct advantages in xenotransplantation and humanization of porcine proteins in transgenic systems. Intensive work has advanced the use of swine as large animal models with respect to immunological reagents. Cost, convenience, ease of handling, experience, and the available research tools —monoclonal antibodies, gene probes, knockouts— are significant problems; however, the benefits and advances in swine immunological reagents are fast becoming a reality.

Predictivity

The neonatal swine model of food allergy is being investigated as a model to predict allergenicity of genetically modified foods (5). In this model, novel proteins introduced into crops destined for the food marketplace are being systematically compared to a profile of known food allergens and food tolerant sources for IgE-mediated immune mechanisms. In other models, the risk of environmental hazards for developmental immunotoxicology studies in large animals is becoming prominent. With reference to transcutaneous immunizations, epidermal thickness, relative proportions of stratum epithelium and uncornified epidermis, and the distribution of Langerhans cells and their dendrites, swine appear to be more appropriate models than mice for extrapolation to the human system.

Relevance to Humans

Notable is the development of transgenic pigs for studies of organ transplantation. Because of its size, availability and limited risk of zoonosis, use of the pig as a donor for xenotransplantation investigations has been reached as a consensus in transplantation medicine.

The neonatal swine food allergy model closely resembles gastrointestinal food allergy in humans and should offer significant contributions into mechanisms of IgE-mediated allergy and therapeutic options for treatment of allergy.

Skin parameters of epidermal thickness, proportion of the epidermal layers and the distribution of Langerhans cells in swine appear to be appropriate models for skin testing, transcutaneous immunization studies, and human wound healing studies.

Investigations such as these provide a clinical basis for future studies using pigs as large animal models; however, as in any animal model, extrapolation to human studies must be critically evaluated.

Regulatory Environment

When proposing to use swine as an animal model in the United States, investigators must stringently adhere to the guidelines established by two significant agencies:
- Association or Assessment and Accreditation of Laboratory Animal Care or similar agency
- the respective Institutional Care and Use Committee reviews.

Similar agencies are in place worldwide to provide the investigator with the appropriate guidelines in the use of animals for research purposes. These agencies safeguard both the animal and the investigator from improper use of any animal in laboratory investigations.

References

1. Tumbleson ME, Schook LB (eds) (1996) Advances in Swine in Biomedical Research, Volumes 1,2. Plenum Press, New York
2. Butler JE, Howard C (2002) Summary of the Comparative Immunoglobulin Workshop (CIgW) on immunoglobulins and Fc receptors. Vet Immunol Immunopathol 87:481–484
3. Haverson K et al. (2001) Overview of the Third International Workshop on Swine Leukocyte Differentiation Antigens. Vet Immunol Immunopathol 80:5–32
4. Boersma WJ, Zwarat RJ, Sinkora J, Rehakova Z, Haverson K, Bianchi AT (2001) Summary of workshop findings for porcine B-cell markers. Vet Immunol Immunopathol 80:63–78
5. Helm RM, Furuta GT, Stanley JS et al. (2002) A neonatal swine model for peanut allergy. J Allergy Clin Immunol 109:136–142

Positive Level

This threshold level has to be exceeded—independent of statistical significance—to classify a reaction as positive. In case of lymph nodes assays these levels are empiric values generated from historical control values.
▶ Local Lymph Node Assay (IMDS), Modifications

Positive Selection

The process in the thymus of selecting thymocytes which recognize self peptides presented by self major histocompatibility complex proteins.
▶ Thymus: A Mediator of T Cell Development and Potential Target of Toxicological Agents
▶ Antigen-Specific Cell Enrichment

Preclinical Immunotoxicity Evaluation in the Nonhuman Primate

▶ Nonhuman Primates, Immunotoxicity Assessment of Pharmaceuticals in

Preclinical Safety Assessment

Studies that are performed in vitro or in vivo in animal models to establish the pharmacokinetics, metabolic profile, tolerability, and safety of a drug candidate formulation. These studies are performed to set safe dose levels prior to initial human clinical trials.
▶ Therapeutic Cytokines, Immunotoxicological Evaluation of

Prednisone

A synthetic steroid with antiinflammatory and immunosuppressive activity.
▶ Cyclosporin A

Prenatal Immunotoxicology

▶ Developmental Immunotoxicology

Prevention of Infection

▶ Attenuated Organisms as Vaccines

Primary Antibody Response

The immune response that is induced by initial exposure to an antigen. It is mediated largely by IgM anti-

body and develops more slowly and to a lesser extent than a secondary response. Antibody responses to the antigen are measured as an indication of immune competence.
▶ B Lymphocytes
▶ Assays for Antibody Production
▶ Canine Immune System

Primary Humoral Immune Response

▶ Assays for Antibody Production

Primary Immune Response

The immune response that is induced by initial exposure to an antigen, which activates naive lymphocytes. It is mediated largely by IgM antibody and sensitized T cells. It develops more slowly and to a lesser extent than a secondary immune response.
▶ Flow Cytometry

Primary Lymphoid Organs

Organs in which lymphocyte precursors mature into antigenically committed, immunocompetent cells. In mammals, the bone marrow and thymus are the primary lymphoid organs in which B cell and T cell maturation occur, respectively.
▶ Flow Cytometry

Primate Immune System (Nonhuman) and Environmental Contaminants

HELEN TRYPHONAS
Toxicology Research Division
Food Directorate, Health Products and Food Branch
Frederick G Banting Research Center, Tunney's Pasture
Ont. K1A 0L2 Ottawa
Canada

Synonyms
Symian, ape

Definition
The subject of chemically induced alterations on structural and functional components of the immune system is rapidly becoming a major part of research in toxicology. Such research is required to support the evaluation of potentially adverse health effects of chemicals. While a few examples of direct exposure of humans to potentially harmful chemicals exist, the majority of data supporting evaluation is currently generated in experimental animal models such as rodents and canine species. However, it is known that considerable differences exist in the structure and function of the immune systems of human and experimental animal models. Therefore, it is recommended that the extrapolation of data from experimental animals to human be made with caution. To maximize the degree of relevance of experimental data to the human situation, scientists in established nonhuman primate centers continue to develop and validate methodologies in several monkey species. The need for the development of a nonhuman primate model in immunotoxicology is becoming increasingly important in safety evaluations as new and potentially immunotoxic pharmacologic agents and biotechnology products are added to the market.

Characteristics
The use of nonhuman primate species in immunotoxicology offers distinct advantages over rodent species. These are especially useful in situations where established guidelines for toxicity testing of chemicals require, in addition to a rodent model, the use of a nonrodent experimental animal model (1). The advantages include the following:
1. Phylogenetic proximity to man: As a result of their phylogenetic proximity to man nonhuman primates share a number of physiologic, metabolic and behavioural similarities with humans. For example, the ovarian cycles of the nonhuman primate are similar to those of humans. Absorption, biotransformation and excretion of several drugs and chemicals are similar in monkeys and man. In view of the evolution of the brain, which is the last stage before man, social behaviour in monkeys is similar in many respects to that of humans.
2. The anatomy and function of the immune system of nonhuman primates is similar in many respects to that of humans: This allows the application of reagents available for use in human immunology to be used to study corresponding immune parameters in monkeys. For example, cross-reactivity between mouse antihuman monoclonal antibodies (mAbs) and monkey leukocyte surface antigens has been demonstrated for several human mAbs using the whole blood lysis technique in two-color, fluorescein isothiocyanate (FITC) and phycoerythrin (PE), flow cytometric analysis (1). Reference values have been established for infants and adult *Macaque fascicularis* (cynomolgus), the *Macaque mulatta* (rhe-

sus), the *Macacca nemestrina* (pig-tailed macaque), the *Cercocebus atys* (sooty mangabeys), the *Callithrix jacchus* (marmoset) and the *Pan troglodytes* (chimpanzees) monkeys.

In general, members of the New World monkeys (such as the marmosets) and of the Old World monkeys (including the cynomolgus) are preferred over other species such as the *Papio* (baboons) and rhesus monkeys. Firstly, they are easy to breed. This allows the establishment and maintenance of a large colony thus avoiding the problems associated with wild-caught monkeys. Secondly, they are relatively small in size with short pregnancy periods thus making breeding and husbandry relatively easy and, thirdly, they are easy to hand tame. This minimizes the need of restraint methods in administering the experimental dose or in performing other procedures such as blood or milk collection and fat biopsies.

The use of nonhuman primates as experimental animal models is not without drawbacks. These, although not unsurmountable, include the potential for anthropozoonoses especially to B virus infection, and the increased cost over rodent and canine species for procuring and maintaining especially large monkeys such as the rhesus monkey or baboons.

Furthermore, ethical issues raised against the use of animals in research are also applicable to the use of nonhuman primates in immunotoxicology. Such ethical issues often impose restrictions on the numbers of monkeys used in research. This can present a problem especially in cases where multiple doses of an agent must be studied and in cases where there is a need to increase the number of monkeys/test dose so as to enhance the power of statistics.

Preclinical Relevance
Immune Parameters Available for Use in Nonhuman Primates

A number of immunologic parameters have been developed and validated for application to nonhuman primates. In addition to hematologic profiles (total white blood cell counts and differentials) and immunohistopathology techniques, there are a number of other assays which are typically used in a clinical laboratory and have been successfully applied to monkeys (2). They are grouped as follows:

- Assays to study effects on humoral immunity
  - total serum immunoglobulin (IgG, IgA and IgM) levels using the enzyme-linked immunosorbent assay (ELISA)
  - challenge with specific antigens: sheep red blood cells (SRBC), tetanus toxoid (tt), pneumococcal (pneu) antigens and determination of antigen-specific antibody levels in serum using the hemmaglutination and ELISA techniques
- Assays to study effects on cell-mediated immunity (CMI)
  - lymphocyte transformation (^{3}H-thymidine incorporation) (LT) in response to the ▸ mitogens phytohemagglutinin-P (PHA-P), concanavalin A (Con A) and poke weed mitogen (PWM) or specific antigens such as tetanus toxoid
  - mixed lymphocyte cultures using allogeneic lymphocytes: delayed-type hypersensitivity (DTH) using either dinitrochlorobenzene (DNCB) as the sensitizing agent or the multitest kit (Multitest-CMI) which contains a group of ▸ recall antigens [*Candida albicans, Trichophyton mentagrophytes, Proteus mirabilis,* tuberculin purified protein derivative (PPD), streptococcus group C, diphtheria, tetanus toxoid, and a glycerine control
- Assays to study effects on nonspecific immunity
  - monocyte function (activation, phagocytosis and respiratory burst activity) using latex particles, SRBC in flow cytometric techniques and activating agents such as phorbol myristate acetate or zymosan
  - natural killer cell activity using the cell line K562 as target cells in a 4-h chromium-51 release assay
- Assays used for mechanistic studies
  - cell surface marker analysis (CSMA) using cross-reacting mouse anti-human monoclonal antibodies or monkey-specific monoclonal antibodies when available
  - serum complement levels
  - cytokine levels—basal and in lectin-activated cultures
  - hydrocortisone levels.

The majority of these assays have been developed or adapted for use in the cynomolgus, rhesus, the pig-tailed macaque and the marmoset monkeys. In comparison, fewer assays have been developed or adapted for use in the baboons, squirrels and chimpanzee monkeys. While validation of these assays across laboratories is an issue that needs to be addressed, assays for CMI, challenge with foreign antigens, NK cell assay and CSMA have been reproduced in several laboratories. Of these assays, challenge with foreign antigens, CSMA and NK have repeatedly proven to be the most sensitive for detecting chemical-induced immunotoxicity.

Examples of Immunotoxicity Studies in which Nonhuman Primates were Used

Nonhuman primates have been used extensively in studies designed to investigate the potential immunotoxic effects of chemicals particularly those which are of environmental concern such as polychlorinated biphenyls (PCBs) (2), 2,3,7,8-TCDD (dioxins) (3) and toxaphene (4). Following is a brief discussion of data available for each of these chemicals.

PCBs

The majority of the immune parameters listed above have been applied to studies concerning the potential immunotoxic effects of the PCB mixture known as Aroclor 1254, in adult and infant rhesus monkeys. This chronic, multidose, two-generation toxicity/reproductive/immunotoxicity study generated a great deal of data on effects of PCBs not only in adults but also in infant monkeys. Table 1 lists the number of parameters investigated and their outcome. A large number of immune parameters were affected by PCB treatment (2). In particular the response to SRBC antigens was significantly affected at levels of Aroclor 1254 as low as 5 μg/kg body weight/day.

TCDD

Immunotoxicity data regarding the effects of TCDD in nonhuman primates are scarce. Changes in the $CD4^+ : CD8^+$ cell ratio similar to that observed in the rhesus PCB studies have been reported in rhesus monkeys exposed to 5 or 25 p.p.t. of TCDD in the diet (3). These changes were not associated with T-cell function as measured by the lymphocyte transformation assay in response to mitogens, alloantigens, or xenoantigens. Also, NK activity and antibody production following immunization with tetanus toxoid were not affected by treatment. Offspring of the exposed dams had increased levels of antitetanus antibodies which correlated with TCDD levels in tissues. Changes in T-cell subsets characterized by a decrease in $CD4^+$ and an increase in $CD8^+$ cells were also reported in marmosets treated with 10 ng/kg TCDD (1).

Toxaphene

Recent studies have shown that the pesticide toxaphene, a complex mixture of chlorinated bornanes with more than 13 000 individual isomers, is also immunotoxic in cynomolgus monkeys (Table 2). In this study young adult female monkeys, with ten monkeys per group, were administered doses of 0.00, 0.1, 0.4 or 0.8 mg/kg body weight/day for 75 weeks while five male monkeys/group were administered toxaphene at a dose of 0.8 or 0.0 mg/kg body weight/day (4). A striking feature of this study was the statistically significant reduction in antibody titers in response to immunization with SRBC and tetanus toxoid antigens without any significant effect on the antibody response to pneumococcal antigens indicating that the T-cell dependent humoral immune response was compromised. The effect was highly significant at the 0.8 mg/kg dose. Studies on the infants of the same monkeys indicated that there were no effects on the humoral immune response but there was a statistically highly significant increase in the $CD4^+$ cell number with a concurrent highly significant decrease in the $CD8^+$ cell population, suggesting that the regulatory cells of the immune system were affected by treatment.

Relevance to Humans

The relevance of immunotoxicity data generated in monkeys in relation to the human population remains unresolved. However, the available data in humans accidentally or occupationally exposed to various agents of environmental concern strongly indicate that the human immune system is a target for chemical-induced immunotoxic effects (5). Examples of these include the populations exposed to PCBs, polychlorodibenzofurans (PCDFs) and quaterphenyls (PCQ) via contaminated rice oil (The Yusho and Yu-Cheng episodes), humans consuming fatty fish species from the Baltic sea, and studies on the Inuit (Northern Quebec) populations consuming large amounts of fish fat (5). Studies in newborn and children exposed in utero to ambient levels of PCBs and dioxins suggest that this population of humans may be particularly sensitive to the immunotoxic effects of environmental chemicals. This is due to the fact that certain chemicals including the PCBs are known to cross the human placenta and to be secreted in large amounts in the mother's milk. Examples of such studies include women working in a capacitor manufacturing factory in the Shiga Prefecture, Japan, studies on fish-eating populations from the Great Lakes, and the Netherlands studies (5). All these studies report effects on several parameters of the immune system. Many of the affected parameters are similar to those for which effects were shown in nonhuman primates exposed to PCBs or dioxins (2).

Regulatory Environment

The ultimate purpose of conducting immunotoxicity studies is to enable the regulatory agencies to determine "safe" levels of unwanted chemicals in the environment and in the food chain. To facilitate this process, several countries have issued guidelines for immunotoxicity testing in rodents and the majority of the studies performed during the last decade followed these guidelines. Although no such guidelines exist for nonhuman primate models, several of the assays used in monkeys correspond to those listed in the proposed guidelines for rodents making the process of across species comparisons possible (5).

Studies in experimental animals such as guinea pigs, rabbits and rodents have been helpful in identifying a No Observed Adverse Effect Level (NOAEL) for several of the chemicals which are of environmental concern. The best example of such studies would be those

Primate Immune System (Nonhuman) and Environmental Contaminants. Table 1 PCB-induced immunotoxic effects in adult and infant rhesus monkeys (modified from (2))

Parameter	Adults 22 weeks	55 weeks	Infants
Antibody titers to:			
SRBC primary IgM and IgG at 23 months	9		9
SRBC secondary IgM and IgG at 55 months		9	ND
Pneumococcal antigens		NS	ND
Cell surface markers:			
T lymphocytes	NS	9	ND
T helper/inducer cells (Th/i)	9	NS	ND
T suppressor/cytotoxic cells (Ts/c)	8	NS	ND
Th/I :Ts/c ratio	9	NS	ND
B lymphocytes	NS	NS	ND
Lymphocyte transformation			
³H-thymidine incorporation			
Phytohemagglutinin	NS	9	NS
Concanavalin A	NS	9	NS
Poke weed mitogen	NS	NS	NS
Mixed lymphocyte culture	NS	NS	ND
Total serum IgG, IgM and IgA	NS	ND	ND
Monocyte function			
Stimulant: Zymosan:			
Peak reading	ND	9	ND
Time to peak reading	ND	NS	ND
Stimulant: phorbol myristate acetate:			
Peak reading	ND	9	ND
Time to peak reading	ND	8	ND
Interleukin-1	ND	9	ND
Tumor necrosis factor	ND	NS	ND
Natural killer cell activity	ND	8	NS
Serum complement (CH$_{50}$) activity	ND	8	ND
Thymosin alpha-1	ND	8	
Thymosin beta-4	ND	8	ND
Interferon levels (Con A-stimulated peripheral blood)	ND	8	ND
Serum hydrocortisone	8	ND	ND

Ig, immunoglobulin; SRBC, sheep red blood cells
NS: not significantly different from control $P \geq 0.05$; ND: not done
* Statistically significant increase at $P \leq 0.05$.
** Statistically significant decrease at $P \leq 0.05$.

that were performed using several of the commercially available PCBs. The calculated NOAELs in these animals were high in comparison to those calculated from similar data generated in monkeys (2). This was attributed to the higher rate of PCBs eliminated in mice and rats compared to the rate observed in monkeys. Increased sensitivity of monkeys has been reported not only for commercially available PCBs but also for a mixture of congeners representative of those commonly found in human milk. The PCB immunotoxicity data in monkeys has been used extensively by regulatory and advisory agencies. The Agency for Toxic Substances and Disease Registry (ATSDR) has derived a Minimal Risk Level (MRL) of 0.02 µg/kg/day for chronic-duration oral exposure to PCBs. The chronic oral MRL is based on a Lowest Observed Adverse Effect Level (LOAEL) of 0.005 mg/kg/day for immunological effects in adult monkeys that were evaluated after 23 and 55 months of exposure to Aroclor 1254 (6). Typically these calculations apply an

Primate Immune System (Nonhuman) and Environmental Contaminants. Table 2 Immunotoxic effects of toxaphene in adult cynomolgus monkeys

Parameter	Females	Males	Infants
Response to: sheep red blood cells (SRBC)			
Primary IgM	9	9	NS
Primary IgG	9	NS	NS
Secondary IgM	9	ND	NS
Secondary IgG	NS	ND	NS
Response to:			
Tetanus toxoid IgG	9	ND	ND
pneumococcus IgG	NS	ND	ND
T-lymphocyte subsets			
CD4	NS	NS	8
CD8	NS	NS	9
CD4/CD8 ratio	NS	NS	9
B lymphocyte numbers (absolute)	9	NS	NS
Lymphocyte transformation in response to mitogens	NS	ND	ND
Natural killer cell activity	NS	ND	ND
Natural killer cell numbers	NS	NS	8
Delayed-type hypersensitivity (DTH) to DNCB	NS	ND	ND
Serum hydrocortisone levels	NS	ND	ND

Adult monkey data compiled from reference (4). Infant data compiled from a manuscript in preparation.

uncertainty factor of 300 (10 for extrapolating from a LOAEL to a NOAEL, 3 for extrapolating from monkeys to humans, and 10 for compensating for the observed variability among humans). Similarly, the US Environmental Protection Agency (EPA) has calculated an oral reference dose (RfD) of 0.02 µg/kg/day for Aroclor 1254 (IRIS 2000) based on the evaluation of dermal/ocular and immunologic effects in monkeys, and an oral RfD of 0.07 µg/kg/day based on reduced birth weight in monkeys (6).

An MRL of 0.03 µg/kg/day has been derived for intermediate-duration oral exposure to PCBs. The intermediate oral MRL is based on a LOAEL of 0.0075 mg/kg/day for neurobehavioral alterations in infant monkeys that were exposed to a PCB congener mixture representing 80% of the congeners typically found in human breast milk. This MRL was also supported by immunologic effects reported for the same monkeys (6).

References

1. Neubert R, Helge H, Neubert D (1996) Nonhuman primates as models for evaluating substance-induced changes in the immune system with relevance for man. In:.Smialowicz RJ, Holsapple MP (eds) Experimental Immunotoxicology. CRC Press, Boca Raton, pp 63–117
2. Tryphonas H, Feeley M (2001) Polychlorinated biphenyl-induced immunomodulation and human health effects. In: Robertson LW, Hansen LG (eds) PCBs. Recent advances in environmental toxicology and health effects. The University Press of Kentucky, pp 194–209
3. Hong R, Taylor K, Abonour R (1989) Immune abnormalities associated with chronic TCDD exposure in Rhesus. Chemosphere 18:313–321
4. Tryphonas H, Arnold DL, Bryce F et al. (2001) Effects of toxaphene on the immune system of cynomolgus (*Macaca fascicularis*) monkeys. Food Chem Toxicol 39:947–958
5. Tryphonas H (2001) Approaches to detecting immunotoxic effects of environmental contaminants in humans. Environ Health Persp 109 [Suppl 6]:877–884
6. US DoH. (2000) Toxicological Profile for Polychlorinated Biphenyls (Update). U.S. Department of Health & Human Service Agency for Toxic Substances and Disease Registry, November 2000

Primed Macrophages

Inflammatory macrophages exposed to interferon-γ or elicited with certain polymers such as poly-IC become primed what have further enhanced functions associated with killing.

▶ Macrophage Activation

Procainamide

Procainamide (4-amino-N-(2-(diethylamino)ethyl)-benzamide) is used to treat cardiac arrhythmia (abnormal heart rate).
▶ Systemic Autoimmunity

Progesterone

Progesterone is a steroid hormone produced in the ovaries. Biological activity of progesterone is mediated by interaction with the progesterone steroid hormone receptor.
▶ Steroid Hormones and their Effect on the Immune System

Programmed Cell Death

▶ Apoptosis

Proinflammatory Cytokine

A cytokine that supports an inflammatory response by stimulating leukocytes to an enhanced activity towards microbial agents or other antigenic compounds. Usually associated with induction of other inflammatory mediators, e.g. prostaglandins or leukotrienes. Principal proinflammatory cytokines are TNF-α and IL-1 or IL-6. These cytokines activate multiple inflammatory response pathways, including the lymphocyte response in delayed type IV hypersensitivity.
▶ Cytokines
▶ Metals and Autoimmune Disease
▶ Chronic Beryllium Disease

Prolymphocytic Leukemia

▶ Leukemia

Promoter

A regulatory region usually located 5' of a gene's coding region which regulates where (i.e. which cell types) and when genes are expressed.
▶ B Lymphocytes

Pronucleus

Haploid nucleus resulting from meiosis. In animals the female pronucleus is the nucleus of the unfertilized ovum.
▶ Transgenic Animals

Prostaglandins

TINA SALI
NIEHS Mail Drop E4–09, Laboratory of Molecular Carcinogenesis
111 Alexander Drive
P.O.Box 12233
Research Triangle Park, NC 27709
USA

Synonyms
▶ prostanoids, ▶ eicosanoids, cyclooxygenase, COX

Definition
Prostaglandins are lipid metabolites that induce a diverse spectrum of biological effects. They are potent modulators of signaling pathways that regulate pain, inflammation, cell proliferation, transformation, angiogenesis, metastasis, and apoptosis. Production of prostaglandins from *cis*-unsaturated fatty acids, such as ▶ arachidonic acid, requires the activity of the cyclooxygenase (COX) enzymes, of which there are two isozymes (COX-1 and COX-2). Non-steroidal anti-inflammatory drugs (NSAIDs), such as aspirin, prevent prostaglandin production through the inhibition of COX.

Characteristics
Prostaglandins are secondary messengers that are produced in response to a variety of stimuli. They act as short-lived, local hormones in order to alter activities in the cells in which they are synthesized, as well as in adjoining cells. Prostaglandins thereby regulate numerous pathways which are important both for normal cellular function as well as the body's response to cell injury and disease. For example, prostaglandins have been shown to stimulate inflammation and induce endothelial cell growth, as well as relax and contract various types of smooth muscle, modulate synaptic transmission, regulate blood flow to certain organs, control ion and water transport in the kidneys, and induce sleep. Alterations in prostaglandin metabolism have been implicated in the pathogenesis of hypertension, asthma, inflammation, pain, fever, swelling, redness, headache, and the formation of ulcers and tu-

mors. Failure to regulate prostaglandin production can also enhance tumor spread and metastasis.

On the molecular level, prostaglandins are polyunsaturated fatty acids which contain a cyclopentane ring and two alkyl side chains. They are synthesized from C-20 *cis*-unsaturated fatty acids: icosatrienoic acid (1-series), eicosatetraenoic acid, which is also known as arachidonic acid, (2-series), and eicosapentainoic acid (3-series). Letters A–J designate the nature and position of substituents on the cyclopentane ring, as well as the presence and position of double bonds within the ring. Numerical subscripts (1, 2, or 3) indicate the number of double bonds in the alkyl side chains, which is a direct reflection of the fatty acid precursor from which the prostaglandin is synthesized. The prostaglandin (PG) family includes PGA, PGB, PGC, PGD, PGE, and PGF as well as the PGG and PGH intermediates, PGI (prostacyclin) and PGJ. The PGG and PGH intermediates are the direct products of COX and they are quickly converted to other prostaglandins by a second class of enzymes known as the synthases. PGA, PGB, and PGC are believed not to occur naturally, but to be produced only artificially during extraction procedures.

Arachidonic acid is the most abundant fatty acid precursor in most mammals, including humans, making the 2-series prostaglandins predominant in these organisms. The first step in the metabolism of prostaglandins from arachidonic acid is the release of arachidonic acid from membrane glycerophospholipids by phospholipase A_2 (PLA_2) (Figure 1). COX (also known as prostaglandin H synthase) has both cyclooxygenase and peroxidase activities. The cyclooxygenase activity of COX first converts arachidonic acid into PGG_2, which is followed by the subsequent conversion of PGG_2 to PGH_2 by the peroxidase activity of COX. The PGH_2 intermediate can then be further metabolized into other prostaglandins (PGD_2, PGE_2, $PGF_{2\alpha}$), prostacyclin (PGI_2), or thromboxane (TXA_2) by the corresponding synthases. Thromboxane is formed from PGH_2 by thromboxane synthase, but it is not formally considered a prostaglandin due to the insertion of oxygen in the cyclopentane ring. Thromboxane is a vasoconstrictor that induces platelet aggregation, while prostacyclin serves the opposite function of being a vasodilator that reduces platelet aggregation. It is therefore believed that the balance of these two lipid metabolites is important for maintaining vascular homeostasis. While PGD_2 promotes sleep, $PGF_{2\alpha}$ is important for many aspects of reproduction. The primary metabolite of the arachidonic acid pathway, however, is PGE_2. Among other functions, PGE_2 is an important mediator of the inflammatory response and the major prostaglandin found in colonic polyps. It is formed from PGH_2 by prostaglandin E synthases (PGES) and its formation appears to be coordinated by the co-regulation of PGES and COX.

Once produced, prostaglandins activate cell-signaling cascades which are linked to prostaglandin-specific G protein-coupled receptors. A variety of receptors have been identified which bind thromboxane, PGD_2, PGE_2, $PGF_{2\alpha}$, and PGI_2 as ligands (Figure 1). These receptors show selective ligand-binding specificity, and vary in terms of their abundance and tissue distribution. The effect elicited by any given prostaglandin is therefore dependent upon the location in the body where it is synthesized, as well as the presence and number of receptors to which it may bind.

The COX-1 and COX-2 isozymes are heme proteins which are highly homologous in terms of their amino acid sequence (61% identity), size and overall structure. COX-1 and COX-2 are encoded by separate genes and differ in their tissue distribution and subcellular localization. COX-1 is constitutively expressed and is responsible for various housekeeping duties such as maintaining normal gastric and kidney function. Production of COX-2, however, is rapidly induced by a variety of inflammatory stimuli, tumor promoters, and growth factors, and it is the isoform found mainly in cancer cells. It is believed that increased activity of the arachidonic acid pathway due

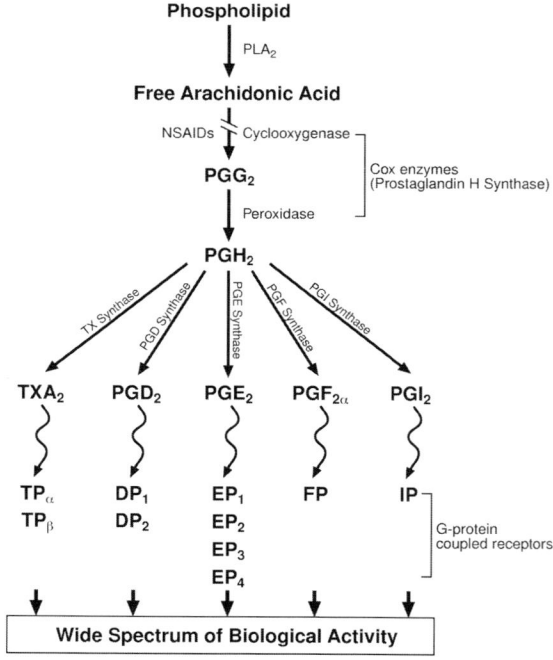

Prostaglandins. Figure 1 The synthesis of prostaglandins and thromboxane from arachidonic acid is shown, as well as the G protein-coupled receptors to which they may bind.

to chronic irritation or inflammation results in an increased risk of cancer. In support of this, both in vitro and in vivo data suggest that COX-1 and COX-2 may play a role in human cancer of the prostate, colon, and breast and other cancers. Excessive prostaglandin production due to overexpression of COX-2 is a commonly observed feature of neoplastic transformation, and COX-2 expression also appears to be involved in angiogenesis. A widespread role for COX-2 in carcinogenesis is evidenced by increased levels of COX-2 having been found in biopsies from tumors of the breast, stomach, lung, esophagus and skin. Furthermore, expression of COX-2 is increased by phorbol ester treatment, which is linked to tumor promotion, while COX-2 inhibitors suppress tumor proliferation and stimulate apoptosis in human colon cancer cells. Dramatic reductions in tumor size and number were observed upon independent knockout of the COX-1 and COX-2 genes in APC(adenomatous polyposis coli)-deficient mice—like human FAP (familial adenomatous polyposis) patients, these spontaneously develop a large number of intestinal tumors. These results suggest a role for both isoforms of COX in cancer.

NSAIDs reduce pain, inflammation, and other prostaglandin-induced symptoms by competitive inhibition of the cyclooxygenase activity of COX. A variety of drugs are currently available which vary in terms of their potency and toxicity profiles. Three classes of NSAIDs exist:

- non-selective COX inhibitors such as aspirin, ibuprofen, and naproxen
- drugs which selectively inhibit COX-1 such as the experimental drug SC-560
- COX-2-specific inhibitors like celecoxib and rofecoxib.

Selective COX inhibition can be attributed, at least in part, to positional amino acid differences between the two isozymes. Superposition of the 3-D structures of COX-1 and COX-2 isoforms reveals two non-conserved residues in the cyclooxygenase binding site which are responsible for a difference in the size and potential ligand interactions for the two active sites: Ile^{523} and His^{513} in COX-1, which correspond to Val^{523} and Arg^{513} in COX-2. However, molecular modeling and mutagenesis experiments confirm the importance of additional, conserved residues in ligand selectivity as well. In recent years, there has been great interest in the development of drugs which specifically inhibit COX-2. This is because of the important role COX-1 plays in the constitutive expression of prostaglandins for normal cellular function. For example, prostaglandins in the gastrointestinal tract act to protect the stomach and intestinal mucosa from inflammation, inhibit gastric acid secretion, and promote the healing of gastric and duodenal ulcers. The inhibition of COX-1 by NSAIDs therefore often results in unwanted side effects, such as stomach ulcers. These unwanted side effects can often be avoided by taking NSAIDs which specifically target the inducible COX-2 enzyme. On the other hand, studies in mice show that COX-1 specific inhibitors hold promise for delaying the onset of premature labor during pregnancy, without interfering with closure of the ductus arteriosus (a fetal blood vessel that joins the aorta and the pulmonary artery)—a side effect which can occur with the use of non-selective COX inhibitors.

Studies reveal a 40–50% decrease in colorectal cancer mortality due to the use of NSAIDs, as well as a decrease in the number and size of colorectal polyps in human and mouse models. The use of NSAIDs has also been linked to chemoprevention of breast and lung cancer, though to lesser extents. While there is data demonstrating this anti-tumorigenic activity is due to COX inhibition, other data suggest NSAIDs have COX-independent effects as well. For example, NSAIDs have been shown to induce cultured human colorectal cancer cells devoid of COX to undergo apoptosis. In addition, the amount of COX inhibitor required to induce apoptosis is usually greater than that needed to inhibit COX. One possible mechanism for these COX-independent effects involves an additional role for NSAIDs in the regulation of gene expression. One gene recently found to be greatly upregulated by NSAID treatment is *NAG-1* (NSAID-activated gene). *NAG-1* is a divergent member of the transforming growth factor-beta (TGF-β) superfamily, and expression of *NAG-1* has been shown to result in apoptosis and anti-tumorigenic activities in several model systems. The induction of *NAG-1* expression has been observed with a variety of NSAIDs and in several cancer cell lines including colorectal, lung, breast, and prostate. *NAG-1* induction occurs independently of COX inhibition and is not affected by prostaglandins. Therefore, *NAG-1* appears to be a common link between NSAIDs and their pro-apoptotic activity.

Preclinical Relevance

The regulation of prostaglandin metabolism has potential clinical relevance for the endocrine, reproductive, nervous, digestive, respiratory, cardiovascular, and renal systems. Daily intake of a low dose of aspirin may reduce the risk of heart attack by inhibiting COX-1 present in platelets (inhibition of COX-1 reduces the formation of TXA_2, thereby preventing the platelets from clumping). The use of NSAIDs is also potentially relevant to the prevention of cancer, though this effect cannot be exclusively attributed to the inhibition of prostaglandin production by COX. The development of COX-specific inhibitors shows promise for prevention of premature labor (COX-1-specific) and treatment of rheumatoid arthritis and osteoarthritis (COX-

2-specific). COX inhibitors are also potentially useful in the treatment of a variety of ailments which include cardiovascular disease, osteoporosis, cataract formation, diabetes and Alzheimer's disease.

Relevance to Humans

Prostaglandins have been found in nearly all human tissues and fluids examined, though they usually occur only in minute amounts and are degraded soon after synthesis. This nearly ubiquitous production of short-lived prostaglandins at low concentrations underscores both their biological importance and their potency. Prostaglandins and related lipids regulate a wide range of biological actions and alterations in their production are linked to a number of important human diseases and cancer. A number of drugs are currently on the market which are related to prostaglandins, the most notable of which are the NSAIDs.

Regulatory Environment

The only relevant regulations governing prostaglandin and NSAID studies are the standard regulations for human clinical trials and animal experiments.

References

1. Hsi LC, Eling TE (2002) Carcinogenesis involving cyclooxygenase and lipoxygenase. In: Harris RE (ed) COX-2 Blockade in Cancer Prevention and Therapy. Humana Press, Totowa, pp 245–255
2. Narumiya S, Sugimoto Y, Ushikubi F (1999) Prostanoid receptors: structures, properties, and functions. Physiol Rev 79:1193–1226
3. Filizola M, Perez JJ, Palomer A, Mauleon D (1997) Comparative molecular modeling study of the three-dimensional structures of prostaglandin endoperoxide H2 synthase 1 and 2 (COX-1 and COX-2). J Molec Graphics Model 15:290–300
4. Steele VE, Hawk ET, Viner JE, Lubet RA (2003) Mechanisms and applications of non-steroidal anti-inflammatory Drugs in the chemoprevention of cancer. Mutat Res 523–524:137–144
5. Baek SJ, Kim KS, Nixon JB, Wilson, LC, Eling TE (2001) Cyclooxygenase inhibitors regulate the expression of a TGF-β superfamily member that has proapoptotic and anti-tumorigenic activities. Molec Pharmacol 59:901–908

Prostaglandins

A subgroup of eicosanoids (lipid metabolites) that are produced by the enzymatic activity of cyclooxygenase.
▶ Prostaglandins

Prostanoids

A subgroup of eicosanoids (lipid metabolites) consisting of the prostaglandins and thromboxanes.
▶ Prostaglandins

Proteases

Enzymes that catalyze the breakdown of peptide bonds.
▶ Mitogen-Stimulated Lymphocyte Response
▶ Respiratory Infections

Protein Blotting

Protein blotting is a synonym for western blot analysis.
▶ Western Blot Analysis

Protein Kinases

Protein kinases are enzymes that alter the activity or confirmation of other proteins by adding a phosphate group to specific tyrosine, serine or threonine residues.
▶ Signal Transduction During Lymphocyte Activation

Protein-Modifying Compound

▶ Hapten and Carrier

Proteins

▶ Immunotoxicology of Biotechnology-Derived Pharmaceuticals

Pseudoallergic Reaction

A reaction that resembles allergy but is not due to the interaction of antigen with specific antibody.
▶ Complement, Classical Pathway/Alternative Pathway
▶ Complement and Allergy

Pseudoallergy

Symptoms similar to an IgE-mediated allergy (urticaria, angioedema, anaphylaxis), but without sensitization of the individual. Frequently caused by direct mast cell stimulation and subsequent degranulation. Typically caused by prostaglandin synthesis inhibitors, muscle relaxants, or contrast media.
▶ IgE-Mediated Allergies

Psoriasis

Psoriasis vulgaris is a frequently occurring skin disease in fair-skinned individuals. Disease starts often in the second decade of life triggered by infections like angina or measles. Pathogenesis involves hyperproliferation of keratinocytes, and accelerated migration of keratinocytes from the basal to the horny layer in 4 days (normal 28 days). Symptoms are intense desquamation of the skin and itching.
▶ Cyclosporin A

PUFA

▶ Fatty Acids and the Immune System

Pulmonary Hypersensitivity

▶ Exposure Route and Respiratory Hypersensitivity

Pulmonary Infections

▶ Respiratory Infections

PWM

▶ Polyclonal Activators

Pyrene

▶ Polycyclic Aromatic Hydrocarbons (PAHs) and the Immune System

Pyrogen

Pyrogens are compounds that induce fever. Exogenous pyrogens are bacterial constituents such as lipopolysaccharide (LPS). Endogenous pyrogens are proinflammatory cytokines such as TNF-α or IL-1.
▶ Cytokines

Q

Quantitative Analysis

▶ Statistics in Immunotoxicology

Quantitative Structure–Activity Relationships (QSARs)

Quantitative structure–activity relationships (QSARs) relate physicochemical properties of haptens to their relative ability to cause allergy in such a way that the potency of the allergen can be predicted. SARs omit the quantitative element of this, simply indicating in binary fashion whether or not a chemical has the capacity to behave as an allergen.

▶ Chemical Structure and the Generation of an Allergic Reaction

Quenching

Quenching is the inhibition or elimination of one process by another process. The stimulated emission of a laser oscillator can be quenched by a pulse of radiation of the same frequency traversing the oscillator in a different direction. This pulse induces the excited ions to emit radiation in a direction apart from the oscillating mode, and hence the oscillation is decreased.

▶ Viability, Cell

R

RA

▶ Reporter Antigen Popliteal Lymph Node Assay

RA-PLNA

▶ Reporter Antigen Popliteal Lymph Node Assay

Rabbit

▶ Rabbit Immune System

Rabbit Immune System

ROSE G MAGE
Molecular Immunogenetics Section Laboratory of Immunology
NIAID Building 10 11 N 311, MSC 1892 NIH, 10 Center Drive
Bethesda, MD
USA

Synonyms
Rabbit, *Oryctolagus cuniculus*.

Definition
The rabbit immune system consists of the organs, tissues, cells, and molecules that interact to contribute to specific responses to foreign antigens, infectious agents, or—in autoimmune conditions—to self antigens. Included among the important molecules are the genes and gene products that are necessary for the development and proper functioning of the immune system including antigen-specific T cell and B cell receptors and B cell-secreted immunoglobulins. The innate immune system that constitutes the first line of defense will not be considered here, although it is now clearly recognized as important for early recognition of "infectious non self" and initiation of events that induce adaptive immune responses. Rather, special emphasis will be placed here on the unique characteristics of the rabbit ▶ humoral immune system. The cells and molecules that contribute to ▶ cell-mediated immunity, though less well characterized than those of mice and humans, appear to be comparable.

Characteristics
An extensive review of the rabbit immune system has been published previously (1). Many key ideas about the immune system were first developed through studies of the rabbit model. The rabbit is rich in genetic variants (▶ allotypes) that provided markers used for documenting allelic exclusion, *cis* expression of linked genes, and germline recombination within the heavy chain locus. Figure 1 is a stick diagram of a rabbit immunoglobulin G molecule, showing some structural features and the locations of just a few of the many markers that distinguish inheritable sequence differences of heavy and light chains.

Although a normal IgG molecule would have two identical light chains, in this illustration the upper light chain depicts an unusual inter-domain disulfide bond that connects the variable and constant domains of most kappa chains of $C_\kappa 1$ type (allotypes b4, b5, b6, and some light chains from b9 type). The lower

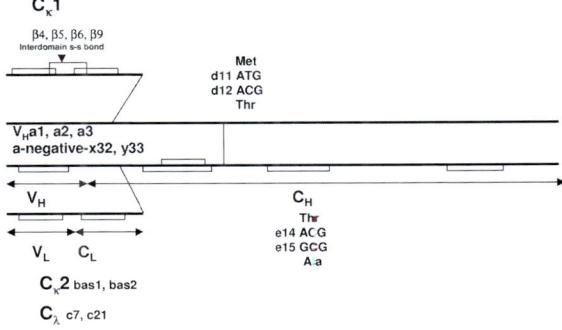

Rabbit Immune System. Figure 1 Stick diagram of rabbit immunoglobulin G showing locations of disulfide bonds and some allelic types (allotypes) of heavy and light chains.

light chain is drawn with only the intra-domain disulfide bonds typically found in other species, as well as in rabbit kappa 2 and lambda light chains. I have speculated that the great stability and long "shelf life" of rabbit antibodies may in part result from stabilization of kappa 1 light chain structures by the unusual inter-domain disulfide bond. Conversely, the Cys at position 80 in most rabbit V_κ genes presents a problem in generating recombinant chimeric Fab molecules with human C_κ because the Cys 80 in V_κ leads to an unpaired thiol group. Compared to b4 rabbits, those of the rare b9 and mutant *bas* types express a higher proportion of V_κ that lack the Cys 80. When rabbits of these types were immunized and recombinant rabbit-human Fab generated by phage display, yields of distinct and specific high affinity Fab increased (2).

A summary of some of the rabbit kappa and lambda genetic types and the organization of the kappa and lambda light chain loci is shown in Figure 2. The rabbit has an unusual duplication of the kappa light chain locus ($C_\kappa 1$ and $C_\kappa 2$ in Figure 2). The allelic forms of the $C_\kappa 1$ genes b4, b5, b6 and b9 differ by multiple amino acids in their constant regions and seem to have somewhat different sets of associated V_κ genes. There are more than 100 different V_κ genes but they are not fully mapped and sequenced. Some of V_κ genes and $C_\kappa 2$ are located about 2 Mb away from $C_\kappa 1$ in the duplicated rabbit kappa locus. In wild-type rabbits, kappa 1 light chains are the major expressed ▸ isotype along with 10–30% of lambda light chains. However, in the mutant *Basilea* strain (*bas*) which has a defective acceptor site for splicing J_κ to C_κ in mRNA for kappa 1 light chains, there is elevated expression of both kappa 2 and lambda light chains. The allelic forms of kappa 2 chains are the result of a single amino acid replacement change in the $C_\kappa 2$ sequences.

Figure 3 shows a schematic diagram of the heavy chain locus with V_H (the first few of more than 100 V_H genes are shown), D_H and J_H genes, and the genes that encode the constant regions of IgM, IgG, IgE and IgA (μ, γ, ε and α). Rabbits are again unusual in having only one γ gene but 13 α genes. A gene encoding a rabbit homologue of IgD has not been identified in the region downstream of rabbit μ where the δ gene is found in some species. Perhaps most unusual of all are the inherited forms of heavy chain variable regions that are detectable using anti-allotype antisera raised by immunization of rabbits of one type with IgG of another type. The reason why allelic forms of rabbit heavy chain variable regions are detectable became clear when it was found that in most rabbit B lymphocytes, the first gene in the locus, $V_H 1$ is rearranged; the different allelic forms have amino acid differences encoded by the $V_H 1$ alleles in framework regions 1 and 3. This $V_H 1$ gene can rearrange to one of several D_H genes and one of three functional J_H genes, to form $V_H D_H J_H$. As shown in Figure 3, $V_H 1$ is usually rearranged. In a mutant strain named *Alicia* (*ali*), the $V_H 1 a2$ gene is deleted and the first gene that is functional, $V_H 4$, is frequently found rearranged along with a few other upstream genes.

Rabbit H- and L-Chain Diversity is Generated by Rearrangements, Somatic Hypermutation and Gene Conversion

Today the rabbit remains a major source of polyclonal antibodies found in catalogs of commercial suppliers. There are some unique characteristics of the immune system of rabbits that contribute to their special capability to produce diverse highly specific high-affinity polyclonal antibodies. These include use of both gene conversion (GC) and somatic hypermutation (SHM) to alter the sequences of rearranged antibody heavy and light chain genes; selection of favorable amino acid replacements during clonal expansion of antigen-specific B lymphocytes in ▸ germinal centers; great diver-

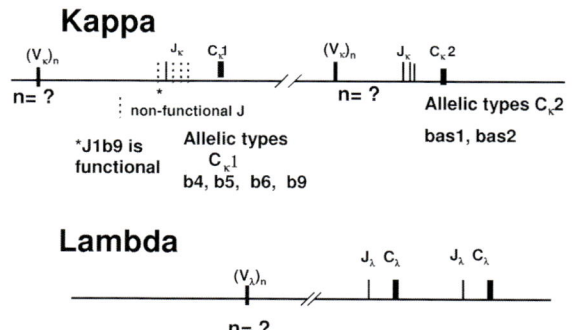

Rabbit Immune System. Figure 2 Diagram (not to scale) of the rabbit light chain kappa and lambda loci.

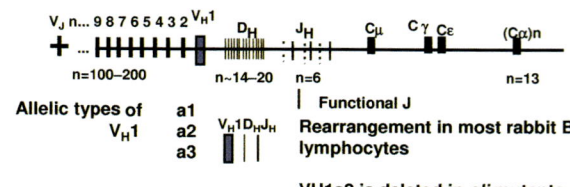

Rabbit Immune System. Figure 3 Diagram (not to scale) of the rabbit heavy chain locus.

sity of kappa light chain variable region genes; and unusual germline V_κ-encoded variability of the length of complementarity-determining region 3 (LCDR3). Compensation for limited heavy chain $V_H D_H J_H$ by diverse light chain $V_\kappa J_\kappa$ occurs even before the start of somatic diversification processes. Despite this, gene conversion further diversifies rearranged $V_\kappa J_\kappa$ both in the appendix of young non-immunized rabbits and in the spleens and lymph nodes of immunized rabbits.

The rabbit $V_H a$ allotypes are encoded by the 3' $V_H 1$ gene which rearranges in most B cells. Some diversity is generated by the choice of one of several D_H and J_H genes. Even before diversification by GC and SHM, there is diversity generated at the sites of V_H to D_H and D_H to J_H DNA recombination by insertions and deletions of bases at the sites of joining. At the points of joining, the additions and deletions of bases that occur lead to great variability in the sequences of the heavy chain third complementarity determining region (HCDR3). In most rabbit B cells, only one chromosome of the allelic pair undergoes complete rearrangement, and the order of arrangement may also be V_H to D_H followed by $V_H D_H$ to J_H; this differs from the order D_H to J_H followed by V_H to $D_H J_H$ shown in most textbooks. The sequence of the rearranged $V_H D_H J_H$ gene is further diversified by gene-conversion-like changes. Sequence blocks that vary in length are acquired from upstream donor V_H genes. This was first described as the mechanism of V_H-gene diversification in the chicken, where it occurs in specialized gut-associated lymphoid tissue (GALT), the bursa of Fabricius of embryos and young chicks, and later in life in splenic germinal centers. In young rabbits, these changes also take place in specialized GALT sites such as the appendix (3,4), and in older rabbits in germinal centers of spleens and lymph nodes in response to foreign antigens. Comparisons of the chicken bursa and rabbit appendix were first published in the 1960s and suggested that the rabbit appendix might be a homologue of the chicken bursa, based on similarities in follicle development and the finding that neonatal thymectomy of rabbits had no effect on appendix development. The independence of appendix cell development from the thymus, as well as remarkable histological resemblance, suggested that rabbit appendix may be a central lymphoid tissue analogous to chicken bursa. Subsequently it was shown that removal of appendix and Peyer's patches resulted in severe depletion of B cells and blunted immune responses. Once gene conversion was discovered to contribute to sequence diversification in both chicken bursa and rabbit appendix, it was also shown that removal of rabbit GALT structures limited—but did not eliminate—diversification of rearranged heavy chain sequences. There are similarities and differences between development and diversification of B cells in the two species, some of which are summarized in Table 1.

Rabbit Central and Peripheral B cell Development and V Gene Diversification

In rabbit appendix, development of the primary pre-immune antibody repertoire requires endogenous gut flora. The gut flora may primarily provide B cell survival and proliferation signals, either directly or indirectly through interactions that activate the innate immune system. The rearranged $V_H D_H J_H$ and $V_L J_L$ in appendix B cells diversify by gene conversion and somatic hypermutation but the receptors may not be specific for a provoking antigen (4). The sequences of rearranged heavy and light chains within a single expanding clone are strikingly diverse in CDR3. This led to the view that cells that diversify within individual clones in appendix may not develop receptors specific for a single antigenic epitope; the clonal diversification contrasts with the response to specific antigens in peripheral lymphoid tissues such as the spleen, lymph nodes, and Peyer's patches where germinal centers develop. There, B cells also diversify rearranged heavy and light chain sequences by somatic hypermutation and gene conversion. This antigen-driven diversification leads to increased affinity of the receptors on some B cells. Selection for cells with good affinity for the immunizing antigen occurs via interactions with antigen on the surface of specialized follicular dendritic cells (FDC). The cells with high affinity may process antigen picked up from FDC and present processed antigen to germinal center T cells which then release stimuli for proliferation, class switching and development into plasma cells or memory B cells. Gene conversion and somatic mutation may also decrease the affinity of antigen receptors. Cells with decreased affinity may die by apoptosis or possibly undergo further rounds of mutation and selection. We have also speculated that in adults peripheral germinal centers may have a secondary role comparable to the role of appendix in young rabbits. For example if some cells with decreased affinity survive and exit as antigen-responsive cells, the germinal centers could be a source of new repertoire in adults.

Rabbit Leukocyte Markers, Chemokines and Cytokines

Tables of rabbit leukocyte antigens, T cell receptors and associated proteins and accessory molecules involved in signaling, leukocyte and endothelial adhesion molecules and some chemotactic molecules described in rabbits for which probes and/or monoclonal antibodies are available can be found in reference (1). Data on cytokines and chemokines summarized at the time of this publication were limited. Although some progress has been made in this area (5), no commercial kits are available for rabbit. Some reagents specific for

Rabbit Immune System. Table 1 Similarities and differences between chicken bursa and rabbit appendix B cell development

Chicken Bursa	Rabbit Appendix
VDJ and VLJL rearrangements in spleen, yolk sac Generally, rearrangement on only one chromosome	VDJ and VLJL rearrangements in bone marrow, fetal omentum, fetal liver, young spleen Generally, rearrangement on only one chromosome.
Migration to embryonic bursa	Migration to newborn appendix
Rapid B cell expansion in bursal follicles even before exposure to exogenous (foreign) antigens Endogenous stimuli not characterized Diversification by gene conversion and SHM pre-and post-hatching to develop preimmune repertoire	Rapid B cell expansion requires presence of gut flora (exogenous) There may be some effects of endogenous stimuli such as CD5 Diversification by gene conversion and somatic hypermutation (SHM) after about 2 weeks of age to develop preimmune repertoire
Emigration from bursa to periphery Bursa involutes by sexual maturity	Emigration from appendix to periphery Appendix changes in appearance and possibly function but does not involute
Emigrants represent the chicken's preimmune repertoire Further diversification by gene conversion and hypermutation occurs in germinal centers of spleen after immunization	Emigrants thought to represent rabbit's preimmune repertoire However, the diversification seen in spleen after immunization suggests some new B cells may also seed adult spleen and initiate germinal centers

human markers (produced in species other than rabbit) cross-react with homologous rabbit proteins.

Preclinical Relevance

The special characteristics of the rabbit immune system that lead to high affinity and specificity of antibodies described above make the rabbit a major source of polyclonal antibodies used in diagnostics and immunopathology. Although rabbits have been used in toxicology for tests of eye irritation potential (Draize rabbit eye irritancy test), as well as for tests of dermal toxicity, many members of the scientific community and animal welfare organizations have criticized the tests as subjective and inhumane. In the United States, the validation status of in vitro screening assays for ocular irritation is currently being evaluated.

Relevance to Humans

Rabbit models for diseases of immunological relevance include various infectious diseases such as anthrax, syphilis, tuberculosis, virus-induced papilloma, and HTLV1: a rabbit model of hemolytic disease of newborns; complement deficiencies; and a variety of autoimmune diseases. Rabbits have been used as the starting source of potential humanized therapeutic monoclonal antibodies because they produce highly specific antibodies with high affinities (2). A polyclonal rabbit anti-human thymocyte globulin (Thymoglobulin), approved by the United States Food and Drug Administration in December 1998 (http://www.sangstat.com), is widely used for the treatment of renal transplant acute rejection, in conjunction with concomitant immunosuppression. However, such a therapeutic cannot be used to treat patients who are not immunosuppressed because they would mount immune responses to the foreign rabbit immunoglobulin. Attempts are currently under way to genetically engineer rabbits that will produce therapeutic human polyclonals (http://www.polyclonals.com). The technology for production of rabbit monoclonal antibodies has also developed to the point that highly specific high-affinity rabbit monoclonal antibodies may find use in drug discovery, diagnostics and possibly as the starting point for development of humanized therapeutic monoclonals (http://www.epitomics.com/technology/tech.html).

References

1. Mage RG (1998) Immunology of lagomorphs. In: Pastoret PP, Bazin H, Griebel HP, Govaerts H (eds) Handbook of Vertebrate Immunology. Academic Press, London, pp 223–260
2. Popkov M, Mage RG, Alexander CB, Thundivalappil S, Barbas CF, Rader C (2003) Rabbit immune repertoires as sources for therapeutic monoclonal antibodies: the impact of kappa allotype-correlated variation in cysteine content on antibody libraries selected by phage display. J Molec Biol 325:325–335
3. Pospisil R, Mage RG (1998) Rabbit appendix: A site of development and selection of the B cell repertoire. In: Kelsoe G, Flajnik M (eds) Current Topics in Microbiology and Immunology, Vol 229: Somatic Diversification

of Immune Responses. Springer-Verlag, Heidelberg, pp 59–70
4. Seghal D, Obiakor H, Mage RG (2002) Distinct clonal Ig diversification patterns in young appendix compared to antigen-specific splenic clones. J Immunol 168:5424–5433
5. Perkins HD, van Leeuwen BH, Hardy CM, Kerr PJ (1999) The complete cDNA sequences of IL-2, IL-4, IL-6 and IL-10 from the European rabbit (*Oryctolagus cuniculus*). Cytokine 12:555–565

Radiation Mucositis

▶ Oral Mucositis and Immunotoxicology

Radioimmunoassay (RIA)

An immunoassay that is based on the use of radioactivity (e.g. ^{125}Iodine-labeled antigens) to generate counts per minute upon the binding of a radiolabeled antigen with its antibody.
▶ Immunoassays

Randomized Complete Blocks Design

An experimental design utilizing several homogeneous groups of subjects. There are as many subjects in a block as there are treatment conditions, and within each block subjects are randomly assigned to treatment conditions.
▶ Statistics in Immunotoxicology

RANTES

RANTES (regulated on activation normal T cell expressed and secreted; CCL5) is a member of the C-C subgroup of chemokines. RANTES is secreted by circulating T cells and is chemotactic for T cells, eosinophils, and basophils and plays an active role in recruiting leukocytes into inflammatory sites. It increases the adherence of monocytes to endothelial cells, selectively supports the migration of leukocytes, and causes the release of histamines. RANTES binds to CCR5 which is an HIV co-receptor.
▶ Cancer and the Immune System
▶ Chemokines
▶ Interferon-γ

Ras

Ras is a small-molecular-weight G protein responsible for regulating the MAP kinase cascades, which lead to activation of transcription factors.
▶ Signal Transduction During Lymphocyte Activation

Rat Immune System

▶ Rodent Immune System, Development of the

Reaction

▶ Delayed-Type Hypersensitivity

Reactive Oxygen Intermediate (ROI)

Products, like hydrogen peroxide and superoxide anion, of the oxidative burst that occurs in neutrophils, macrophages, and other cells in response to phagocytosis or other forms of receptor stimulation. These reactive intermediates can be released into the phagosome, where they can attack ingested microbes, or are secreted outside the cell where they might attack extracellular pathogens, or contribute to inflammation and local tissue damage.
▶ Opsonization and Phagocytosis
▶ Antibody-Dependent Cellular Cytotoxicity

Real-Time and Quantitative PCR

▶ Polymerase Chain Reaction (PCR)

Real-Time Polymerase Chain Reaction

A system that detects and quantifies gene expression or concentration of a pathogen. PCR product is monitored cycle-by-cycle by combining thermal cycling, fluorescence detection, and application-specific software.
▶ Polymerase Chain Reaction (PCR)

Real-Time Reverse Transcription PCR

▶ Polymerase Chain Reaction (PCR)

Rearrangement

During B cell development in the bone marrow a rearrangement of the genomic DNA takes place. The gene encoding the variable domain of the light chain is generated by the stepwise recombination of two gene elements, the V_L gene and the J_L gene. The gene encoding the variable domain of the heavy chain is generated by the stepwise recombination of three gene elements, the V_H gene, the D_H element and the J_H gene.

▶ B Cell Maturation and Immunological Memory

Recall Antigens

Antigens, usually of microbial origin, such as tetanus toxoid or pneumococcal antigens, to which the organism has been previously exposed to and to which the organism has developed a memory capacity.

▶ Primate Immune System (Nonhuman) and Environmental Contaminants

Receptor Shedding

Some transmembrane cytokine receptors can be released from the surface by proteolytic cleavage through ektoproteases. Receptor shedding has two effects: it rapidly deprives the target cell of functional receptors on the cell surface and thus interrupts or terminates cytokine signaling. It also provides soluble cytokine receptors which may have agonist properties, e.g. by protecting the circulating cytokine from proteolytic degradation, or may have antagonistic effects by scavenging and neutralizing cytokines.

▶ Cytokine Receptors

Receptors for Mediators of the Immune System

▶ Cytokine Receptors

Recombinant

▶ Transgenic Animals

Recombinant Antibodies

Antibody molecules produced in prokaryotic and eukaryotic cells in culture or whole animals and plants using genetic engineering methods.

▶ Antibodies, Antigenicity of

Reconstructed Human Skin/Epidermis

▶ Three-Dimensional Human Skin/Epidermal Models and Organotypic Human and Murine Skin Explant Systems

Red Pulp

Part of the spleen comprising venous sinuses filled with blood and splenic cords. Main function is phagocytosis of particulate material and removal of aged erythrocytes from blood. In some species, the red pulp is a site of hematopoiesis.

▶ Spleen

Regenerative Anemia

Anemia characterized by the presence of increased reticulocyte count or increased polychromasia, indicative of adequate bone-marrow response.

▶ Antiglobulin (Coombs) Test

Regression Analysis

A statistical technique in which the relationship between the dependent variable and an independent variable or variables is fit using linear or nonlinear equations. Often used for deriving a prediction equation.

▶ Statistics in Immunotoxicology

Regulated on Activation, T Cell Expressed and Secreted (RANTES)

RANTES is a chemokine involved in intracellular signaling including stimulation of G protein-coupled receptor activity, and tyrosine phosphorylation of multiple proteins.

▶ Interleukin-1β (IL-1β)

Regulatory Cells

Specialized populations cells that modulate the function of other immune cells to prevent uncontrolled or prolonged responses.

▶ Autoimmunity, Autoimmune Diseases

Regulatory Environment

▶ Regulatory Guidance in Immunotoxicology

Regulatory Guidance in Immunotoxicology

ROBERT V HOUSE
DynPort Vaccine Company LLC
64 Thomas Johnson Drive
Frederick, MD 21702
USA

Synonyms
Regulatory environment, guidelines in immunotoxicology.

Definition
From its inception in the late 1970s, immunotoxicology has developed from an essentially academic discipline to an important tool for assessing the risk of human exposure to various xenobiotics. From its early days, immunotoxicology has been virtually synonymous with immunosuppression; this is perhaps due to the dual influences of early assays used to assess immunotoxicity and the more immediately obvious sequelae of decreased host resistance in comparison to, for example, autoimmunity. However, it is increasingly recognized that any perturbation of the immune response from its tightly regulated normal range can have serious adverse consequences on health. In recognition of this, most of the regulatory guidance specific for immunotoxicology emphasizes individual evaluation of an agent based on prior information and its expected/intended molecular mechanism of action. In this review regulatory guidance is divided into generalized chemical class, with the understanding that overlap is inevitable.

Characteristics
Industrial and Environmental Chemicals
Some of the earliest codified immunotoxicology test guidelines were developed to augment toxicological assessment of chemicals with some of the greatest potential for large-scale human exposure, namely pesticides. In 1996 the Office of Prevention, Pesticides and Toxic Substances (OPPTS) of the US Environmental Protection Agency (EPA) published guidelines entitled *Biochemicals Test Guidelines: OPPTS 880.3550 Immunotoxicity* (1), which described the preferred study design for evaluating potential immunotoxicity in biochemical pest control agents. The panel of tests included in this guideline is exceptionally thorough, including standard toxicology tests as well as many of the standard functional tests being employed at that time, including both humoral and cell-mediated immune function (the exceptions being primarily cytokine quantification and flow cytometry). Although this document explains the "how" of testing, it is lean on the "why". To address this deficiency, a second document was published concurrently, entitled *Biochemicals Test Guidelines: OPPTS 880.3800 Immune Response* (2). This companion document provides a good rationale for why pesticides must be tested for immunotoxicity, together with more detailed explanations for testing strategies, and additional details on advanced (mechanistic) tests including host resistance and bone marrow function.

Whereas immunotoxicity evaluation encompassed by the *880* series of guidelines would arguably detect any type of immunotoxicity, its breadth would probably render it tremendously expensive and time consuming. In 1998, the EPA followed up with *Health Effects Test Guidelines: OPPTS 870.7800 Immunotoxicity* (3) which described immunotoxicology testing for non-biochemical agents that would be regulated by EPA. This document provides descriptions of both why and how, with a far more abbreviated panel of testing to be performed. While the *880* series of immunotoxicology guidelines are probably excessive, the testing approach mandated by *870.7800* has stood up well in intervening years and reflects the more limited, case-by-case approach currently favored. Most notably, the functional assessment is pared down to T-dependent antibody formation (plaque assay), natural killer (NK) cell function, and quantitation of T cells and B cells; this combination is derived from the early work of Luster and colleagues which demonstrates the greatest predictivity of known immunotoxicants using these three assays. This study design described in this document is amenable for testing a wide range of industrial and environmental chemicals.

In Europe, the Organisation for European Cooperation and Development (OECD) regulates testing of chemicals for toxicity. The OECD *Guideline 407* entitled *Repeated Dose 28-day Oral Toxicity Study in Rodents* (4), while not specific for immunotoxicology, includes a variety of toxicological endpoints that can provide early evidence of immune system alterations. Missing, however, are any functional assays to directly measure any immune deficit. Although meetings have been held to suggest the addition of functional assays (e.g. Immunology Work Group Meeting, 11–12 December 1996), at present the *407* guideline does not include such assays.

Food Additives

After industrial and environmental chemicals, food additives may have the greatest potential for human exposure. In the USA these chemicals are regulated by the Food and Drug Administration's (FDA) Center for Food Safety and Applied Nutrition. In March 1993 the FDA published the *Draft Redbook II*, which recommended safety testing practices for food additives. This document contained an extensive description of immunotoxicology testing; although *Redbook II* was never finalized, the approach was described in some detail in a number of publications (5,6). In general, the *Redbook* guidelines resembled the "tier" approach that was used with such success in the early development and qualification studies performed under the aegis of the National Toxicology Program. However, *Redbook* emphasized a step-wise approach, beginning with "retrospective level I" (expanded) studies utilizing data obtained in standard toxicology testing as an initial indicator of potential immunomodulation. Subsequent stages included enhanced (expanded) level I, level II, and enhanced (expanded) level II testing designs. This approach was very much case-by-case, with each level predicated on positive findings in its predecessor.

In 2001, the FDA began offering an electronic version of *Redbook*, entitled *Toxicological Principles for the Safety of Food Ingredients (Redbook 2000)* (7). As of the writing of this review, the guidelines for immunotoxicity studies exist only in outline form in *Redbook 2000*.

Pharmaceuticals

In the USA, safety testing of small molecule pharmaceuticals is the purview of the US FDA Center for Drug Evaluation and Research (FDA CDER). In October of 2002, the CDER released a long-awaited document entitled *Guidance for Industry: Immunotoxicology Evaluation of Investigational New Drugs* (8).

This document is arguably the most comprehensive of any published guidance, describing a diversity of adverse events including immunosuppression, immunogenicity, hypersensitivity, autoimmunity, and adverse immunostimulation. The document describes each of these types of immunotoxicity (more accurately, immunomodulation) in detail, and provides not only approaches but also suggests methodology for evaluating each type. Like the document produced by the Committee for Proprietary Medicinal Products (CMPM) (as described below) the FDA CDER guidance advocates the use of information derived from standard repeat-dose toxicity studies to provide early evidence of immunotoxicity, with subsequent evaluations to be rationally designed to use a minimum of animals and resources while deriving the maximum amount of information. Subsequent to the publication of the FDA CDER the primary purpose of this particular document was to describe an overall approach to safety testing of pharmaceuticals, it was important as the first guidance document mandating specific immunotoxicology screening for pharmaceuticals. An appendix of this document describes a staged evaluation, emphasizing that information gained in standard toxicology evaluation can be useful as a primary indicator for immunotoxicity. Functional tests may be incorporated to gain additional information, first as an initial screen and then progressing to extended studies as indicated. The choice of assays to be used includes combinations of functional tests known to be predictive of immunotoxicity, as described in the early National Toxicology Program publications.

As the first published document requiring immunotoxicology evaluation, *CPMP/SWP/1042/99* predictably was met with a combination of resistance and confusion. Much of this was allayed in a workshop held in Noordwijk in the Netherlands in November of 2001, sponsored by the Drug Information Association (DIA). At this meeting the intent of the guideline was clarified. A summary of the workshop has been published (11).

A second CPMP document that includes reference to immunotoxicity assessment is *Note for Guidance on the Quality, Preclinical and Clinical Aspects of Gene Transfer Medicinal Products (CPMP/BWP/3088/99)* (12) currently in draft form. This document recognizes the possibility of adverse immunological events as a consequence of gene transfer therapy, although it makes no specific recommendations for testing.

Japanese regulatory agencies have been cautious in promulgating immunotoxicology guidelines. In 1999, the Japanese Pharmaceutical Manufacturers Association (JPMA) published two documents, *International Trends in Immunotoxicity Studies of Medicinal Products* (13) and *Survey on Antigenicity and Immunotoxicity Studies of Medicinal Products* (14). These comprehensive documents provided a survey of immunotoxicologic methods and study designs in use in Japan and elsewhere, without advocating or requiring any studies per se.

At the DIA meeting in Noordwijk (11) a representative from the Japanese Pharmaceutical Manufacturers' Associated presented an *Interim Draft Guidance for Immunotoxicity Testing* (15), which describes the current thinking on such testing. As of the preparation of this review, this draft guidance document has not been published and is not readily available for review. Thus, as of 2004, there are no published Japanese guidance documents specifically regulating immunotoxicology evaluation.

Biologicals

Biologicals (for the purposes of this review defined as therapeutics derived by biotechnology) present a unique challenge for immunotoxicity assessment for two primary reasons. First, many of these agents (e.g. cytokines and other immunomodulatory molecules) are intended to modulate therapeutically the immune response. Therefore, it can be difficult to differentiate between the agent's efficacy and a truly adverse reaction. Second, because many of these agents are proteins or peptides, their introduction into a host often triggers an immune response directed against the molecule itself. This can lead to alterations in pharmacodynamics, or to other adverse reactions. Thus, development of appropriate guidance on testing these agents is problematic. One approach is promulgated by the International Conference on Harmonisation (ICH) in the document *Preclinical Safety Evaluation of Biotechnology-Derived Pharmaceuticals S6* (16). This document includes sections on immunogenicity (as described above) as well as a brief mention of immunotoxicity studies. In short, the *S6* document recognizes the inappropriateness of a structured tier approach, opting instead for careful design of screening studies, followed by mechanistic studies to clarify any potential evidence of immunotoxicity. Specific techniques and approaches are not described in the *S6* document.

Safety evaluation of biological drugs is regulated in the USA by the FDA Center for Biologics Evaluation and Research (CBER). To date, the CBER has not promulgated any written guidance on immunotoxicology. The reason for this lack of written guidance is the extreme diversity of biological therapeutics, which makes it difficult to design a standardized testing approach. Rather, the approach of the CBER to addressing potential immunotoxicology has always been case-by-case, generally following suggestions provided in the *ICH S6* document.

Currently there are institutional changes underway within FDA that would put therapeutic proteins now regulated by CBER under the regulatory authority of CDER; therefore the CDER guidance document could apply to these products

Vaccines

Along with certain biologicals, vaccines present a challenge for immunotoxicological evaluation since they are specifically designed to induce an immune response—a situation deemed undesirable (or potentially so) for most of the other agents described in this review. Since methodology is well established to evaluate the desirable immunomodulation produced by vaccine, the concern of regulatory agencies is the propensity of these therapeutics to produced undesired or deleterious effects on the immune system.

European regulation of vaccines is described in *Note for Guidance on Preclinical Pharmacological and Toxicological Testing of Vaccines* (17) by the CPMP. In this document, immunotoxicology is to be considered during toxicology testing. In particular, vaccines should be considered for their immunological effect on toxicity, such as antibody complex formation, release of cytokines, induction of hypersensitivity reactions (either directly or indirectly), and association with autoimmunity. No specifics are described for methods or approaches; rather, each vaccine is to be evaluated on a case-by-case basis.

The FDA CBER is tasked with regulating vaccines in the US. One of the primary documents describing vaccine studies is Guidance for Industry for the Evaluation of Combination Vaccines for Preventable Diseases: Production, Testing and Clinical Studies (U.S. Department of Health and Human Services, 1997). Animal immunogenicity is covered in detail in the document, although immunotoxicity is not specified as an area of concern. On the other hand, the CBER's Considerations for Reproductive Toxicity Studies for Preventive Vaccines for Infectious Disease Indications (18). Although this is intended primarily to assess effects of vaccination on reproductive function (including generalized toxicity such as fetal malformations), it acknowledges the potential immunological reactions resulting from the vaccination process to exert unintended consequences. No specific guidance is provided on methods or approaches to be used in this evaluation.

Devices and Radiological Agents

It has been recognized by the FDA that immunotoxicity may result not only from chemical or biological agents that dynamically interact with the physiology of humans (such as small molecule drugs or biologicals), but also from medical devices that contact the body externally (via skin or mucosa), or internally (implantable devices), or by external communication to the blood or tissue.

Thus, FDA Center for Devices and Radiological Health published the guidance entitled *Guidance for Industry and FDA Reviewers: Immunotoxicology Testing Guidance* (19) in May 1999 that addresses testing for medical devices. This guidance is based on *General Program Memorandum G95-1,* an FDA-modified version of *International Standard ISO-10993, Biological Evaluation of Medical Devices—Part 1: Evaluation and Testing. Immunotoxicology Testing Guidance* provides detailed guidance for determining when immunotoxicity testing should be performed (including a flowchart and numerous tables), but does not provide details on which methods should be employed, or for overall study design.

Some additional details on the use of this guidance were published by Anderson and Langone in 1999 (20). This manuscript, similar to the guidance, provides little information on which specific assays to use. It is, however, a useful adjunct to the guidance document.

American Society for Testing and Materials

The American Society for Testing and Materials (ASTM) is a not-for-profit organization promoting the development of voluntary standards for materials, products, systems and services. ASTM develops documents that serve as a basis for manufacturing, procurement, and regulatory activities. Since the ASTM standards are voluntary, they are included in this review only for the sake of completeness. The two relevant documents are F1905-98 (*Standard Practice for Selecting Tests for Determining the Propensity of Materials to Cause Immunotoxicity*) (21) and F1906-98 (*Standard Practice for Evaluation of Immune Responses in Biocompatibility Testing Using ELISA Tests, Lymphocyte Proliferation, and Cell Migration*) (22).

Hypersensitivity

Although much attention is paid to immunosuppression (low immune response) in the majority of guidance documents, it is hypersensitivity (hyperactive immune response) that is the most common type of immunomodulation resulting from exposure to xenobiotics. Due to the acknowledged frequency of this occurrence, as well as the multiplicity of testing methods that have been developed, a complete coverage of this condition will not be included here. However, one method for assessing hypersensitivity has taken priority in assessing contact hypersensitivity, namely the murine local lymph node assay (LLNA). Detailed explanations of this assay and its use are covered in the OECD *429* guideline, entitled *Skin Sensitisation: Local Lymph Node Assay* (23); the US EPA document *OPPTS 870.2600 Skin Sensitization* (24), and the ASTM document *Standard Practice for Evaluation of Delayed Contact Hypersensitivity Using the Murine Local Lymph Node Assay (LLNA)* (25).

Regulatory Environment

The extended bibliography below includes guidelines and guideline drafts which are to be considered for the special aspects of immunotoxicologic screenings mentioned here.

Acknowledgement

This article was prepared under the Immunotoxicology Workgroup supported by the EPA Office of Research and Development (National Center for Environmental Assessment), the EPA Office of Children's Health Protection, National Institute of Environmental Health Sciences (National Toxicology Program) and National Institute for Occupational Safety and Health (Health Effects Laboratory Division). Members of the workgroup not included as authors are Laura Blanciforti (NIOSH), David Chen (EPA/OCPH), Dori Germolec (NIEHS, NTP), Michael Kashon (NIOSH), Marquea King (EPA/ORD/NCEA), Robert Luebke (EPA/ORD/HERL) Michael Luster (NIOSH) Christine Parks (NIEHS) and Yung Yang (EPA, OPPTS). Special thanks to Bob Sonawane (EPA/ORD/NCEA) for helping to organize this effort.

References

1. Biochemicals Test Guidelines: OPPTS 880.3550 Immunotoxicity. United States Environmental Protection Agency, February 1996
2. Biochemicals Test Guidelines: OPPTS 880.3800 Immune Response. United States Environmental Protection Agency, February 1996
3. Health Effects Test Guidelines: OPPTS 870.7800 Immunotoxicity. United States Environmental Protection Agency, August 1998
4. OECD Guideline for the Testing of Chemicals 407: Repeated Dose 28-day Oral Toxicity Study in Rodents. Adopted 27 July 1995
5. Hinton DM (1995) Immunotoxicity testing applied to direct food and colour additives: US FDA 'Redbook II' Guidelines. Hum Exp Toxicol 14:143–145
6. Hinton DM (2000) US FDA "Redbook II" immunotoxicity testing guidelines and research in immunotoxicity evaluation of food chemicals and new food proteins. Toxicol Pathol 28:467–478
7. Toxicological Principles for the Safety of Food Ingredients: Redbook 2000. Draft. Food and Drug Administration
8. Guidance for Industry: Immunotoxicology Evaluation of Investigational New Drugs. US Department of Health and Human Services, Food and Drug Administration Center for Drug Evaluation and Research (CDER). October 2002
9. Hastings KL (2002) Implications of the new FDA/CDER Immunotoxicology guidance for drugs. Int Immunopharmacol 2:1613–1618

10. Committee for Proprietary Medicinal Products (CPMP). Note for Guidance on Repeated Dose Toxicity (CPMP/SWP/1042/99). October 2000
11. Putman E, van Loveren H, Bode G et al. (2002) Assessment of the immunotoxic potential of human pharmaceuticals: a workshop report. Drug Info J 36: 417–427
12. Committee for Proprietary Medicinal Products (CPMP). Note for Guidance on the Quality, Preclinical and Clinical Aspects of Gene Transfer Medicinal Products (CPMP/BWP/3088/99). Draft version
13. International Trends in Immunotoxicity Studies of Medicinal Products. JPMA Drug Evaluation Committee Fundamental Research Group, Data 92. April 1999
14. Survey on Antigenicity and Immunotoxicity Studies of Medicinal Products. JPMA Drug Evaluation Committee Fundamental Research Group, Data 93. April 1999
15. Interim Draft Guidance for Immunotoxicity Testing. MHLW/JPMA, 2001 (unpublished)
16. ICH Topic S6: Preclinical Safety Evaluation of Biotechnology Derived Pharmaceuticals (CPMP/ICH/302/95). March 1998
17. Committee for Proprietary Medicinal Products (CPMP). Note for Guidance on Preclinical Pharmacological and Toxicological Testing of Vaccines (CPMP/SWP/4654/95). June 1998
18. Guidance for Industry: Considerations for Reproductive Toxicity Studies for Preventive Vaccines for Infectious Disease Indications. US Department of Health and Human Services, Food and Drug Administration Center for Biologics Evaluation and Research. Draft version, August 2000
19. Guidance for Industry and FDA Reviewers: Immunotoxicology Testing Guidance. US Department of Health and Human Services, Food and Drug Administration Center for Devices and Radiological Health. 6 May 1999
20. Anderson JM, Langone JJ (1999) Issues and perspectives on the biocompatibility and immunotoxicity evaluation of implanted controlled release systems. J Control Rel 57:107–113
21. American Society for Testing and Materials: Standard Practice for Selecting Tests for Determining the Propensity of Materials to Cause Immunotoxicity. F1905–1998
22. American Society for Testing and Materials: Standard Practice for Evaluation of Immune Responses in Biocompatibility Testing Using ELISA Tests, Lymphocyte Proliferation, and Cell Migration. F1906–1998
23. OECD Guideline for the Testing of Chemicals 429: Skin Sensitisation: Local Lymph Node Assay. Adopted 24 April 2002
24. Health Effects Test Guidelines: OPPTS 870.2600 Skin Sensitization. US Environmental Protection Agency, March 2003
25. American Society for Testing and Materials: Standard Practice for Evaluation of Delayed Contact Hypersensitivity Using the Murine Local Lymph Node Assay (LLNA). F 2148–21401

Regulatory T Cells

A specific T cell subset controlling the response of other T cells by cell-cell contact and secretion of cytokines.

▶ Tolerance
▶ Suppressor Cells
▶ Hapten and Carrier

Relative Risk

STEPHEN B PRUETT
Department of Cellular Biology and Anatomy
Louisiana State University
Health Sciences Center
Shreveport, Louisiana 71130
USA

Synonyms

None (in specific situations odds ratios can be numerically similar to relative risk but they are calculated differently).

Definition

Relative risk is the probability of an outcome in individuals exposed to a particular factor or condition divided by the probability of that outcome in individuals not exposed (1).

Characteristics

The characteristics of relative risk can best be understood by considering an example, as in Table 1.
The relative risk of cancer for persons exposed to this toxicant is $(25 \div 5025) \div (5 \div 5005) = 0.004975 \div 0.000999 = 4.98$. Thus, the risk of developing cancer is 4.98 times greater for the group exposed to the toxicant than for the nonexposed control group. Statistical analysis using a chi-square test or Fisher's exact test is done to determine the statistical significance of this difference and to determine the confidence intervals. In this example, the P value is < 0.005 and the 95% confidence interval is 1.9–13.0. The fact that this interval excludes 1.0 (the value expected if there is no difference in risk between the exposed and unexposed groups) can also be used to demonstrate that the relative risk noted in this case is significant.

Relative risk analysis is most often used with prospective studies (either cohort studies or randomized clinical trials). It is not suitable for case-control studies because these involve selection of cases on the basis of outcome, not exposure (1). The odds ratio can be

Relative Risk. Table 1 An example of the characteristics of relative risk

	Cancer	No cancer	Total
Toxicant exposure	25	5 000	5 025
No exposure	5	5 000	5 005
Total exposure	30	10 000	10 030

used for case-control studies. Interpretation of the biological relevance of relative risk data depends in part on an awareness of the distinction between relative risk and absolute risk. Increases in relative risk are more meaningful when the underlying absolute risks are relatively large than when they are small. Thus, a relative risk of 2.0 for an exposed population may cause relatively little concern if the frequency of the adverse outcome in the control population is 1:10 000 000, whereas more concern would be raised if the frequency of the adverse outcome is 1:100 in the control population.

Preclinical Relevance

Relative risk values could be calculated for most preclinical toxicology studies. However, this is usually not done, because other methods (e.g. analysis of variance with a post hoc test to compare means of multiple groups) are generally more appropriate and have greater statistical power.

Relevance to Humans

Relative risk is commonly used in epidemiological studies to determine if exposure to toxicants is significantly associated with adverse health effects or changes in values for standard clinical tests (2).

Regulatory Environment

Relative risk analyses from human studies can be useful in risk assessment, particularly when toxicant exposure is associated with increased relative risk for a detrimental outcome. However, it should be noted that population sizes in published studies are generally not sufficient to permit small (but potentially meaningful) increases in risk to be demonstrated. For example, an increase from 2 cases of a particular outcome per 100 000 people to 4 cases per 100 000 people represents a relative risk of 2.0, but the 95% confidence interval is 0.37–10.9. Therefore, this two-fold change in risk is not statistically significant. An additional difficulty in using relative risk from human studies in the risk assessment process is that exposure assessment has generally received little attention, so that it is not usually possible to quantify the amount of exposure required to produce an adverse effect. Thus, extrapolation from animal studies (with uncertainty factors added for cross-species extrapolation and sensitive human populations) is still required for most toxicants in the risk assessment and regulatory process.

References

1. Rosner B (2000) Fundamental Biostatistics. Duxbury Thompson Learning, Pacific Grove, CA, pp 54–58
2. Immune Deficiency Foundation. The Clinical Presentation of the Primary Immunodeficiency Diseases. A Primer for Physicians. The Laboratory Diagnosis of Immunodeficiency. http://www.primaryimmune.org

RELISPOT

▶ Enzyme-Linked Immunospot Assay

Repeated Measures Design

An experimental design in which each subject is measured at multiple time points. Measures over time within a subject will be correlated with one another and it is necessary to incorporate this into the analysis.
▶ Statistics in Immunotoxicology

Replicate Cultures

For many assays in cell biology (e.g. **limiting dilution analysis**), parallel cultures are set up under identical conditions to increase the precision of quantitative measurements.
▶ Limiting Dilution Analysis

Reporter Antigen

▶ Reporter Antigen Popliteal Lymph Node Assay

Reporter Antigen Popliteal Lymph Node Assay

RAYMOND PIETERS
Head Immunotoxicology
Institute for Risk Assessment Sciences (IRAS)
Yalelaan 2
P.O. Box 80.176
3508 TD Utrecht
The Netherlands

Synonyms
Popliteal lymph node assay, PLNA, reporter antigen, RA, RA-PLNA, ELISPOT

Short Description
Abbreviations
RA-PLNA:
RA-PLNA=reporter antigen popliteal lymph node assay
TNP-OVA=trinitrophenyl-ovalbumin
TNP-Ficoll=trinitrophenyl-Ficoll
ASC=antibody secreting cells

The reporter antigen-popliteal lymph node assay (RA-PLNA) is a modification of the PLNA to determine compound-induced specific antibody responses (i.e. number of antibody secreting cells (ASC) by ELISPOT) to selected bystander antigens (1). This approach allows assessment of the nature and type of the immune response induced by chemicals in a straightforward and simple manner. The T-cell independent type 2 antigen TNP-Ficoll, which is susceptible to noncognate T cell help, is used as a reporter antigen (RA) that indicates or reports whether a chemical can induce neoantigen specific T cell help. By using the T cell-dependent antigen TNP-OVA (in a nonsensitizing concentration) as RA one can assess whether a chemical has adjuvant or sensitizing potential. By using the RA approach, characteristics of the chemically induced immune response (T cell dependency, adjuvant potential, type of immune response) can be determined without the need to know the neoantigens that are elicited by a compound and without the need to isolate or synthesize these neoantigens for assessment of anamnestic immune reactivity. So, as for the primary PLNA, the RA-PLNA allows fast screening of compounds for immunopotentiating effect, but in addition it enables discrimination between immunosensitizing and mere adjuvant or irritant potential of compounds.

Characteristics
The RA-PLNA differs from the PLNA in two ways: the RA are injected together with the chemical under investigation; and the TNP-specific antibody secretion is determined in addition to cell numbers of the draining lymph nodes. Briefly, chemicals and RA (fixed final dose of 10 μg per mouse) are mixed in solution and injected into the hind footpad of a mouse. After 7 days the PLN is excised and used to prepare single-cell suspensions. PLN cells are counted and amounts of TNP-specific ASC in PLN suspensions are determined by ELISPOT.

TNP-OVA and TNP-Ficoll are chosen because of their specific immunogenic properties. TNP-Ficoll is a T cell-independent type 2 antigen, that cannot be recognized by T cells, but that is very well capable of triggering B cells to produce immunoglobulin(Ig)M. Once triggered by TNP-Ficoll, B cells become susceptible to noncognate T cell help, meaning that such B cells will also produce switched isotypes such as IgG_1 when soluble T cell help is available. Thus an IgG_1 response to TNP-Ficoll indicates that T cells are activated (and possibly sensitized) and that these T cells recognize neoantigens induced by the chemical. These can be hapten-carrier complexes or other neoantigens (e.g. previously cryptic epitopes or hidden autoantigens).

TNP-OVA is a protein antigen that can be recognized by T cells as well as B cells. The dose of TNP-OVA (10 μg per animal) that is used in the RA-PLNA is unable to elicit a specific immune response by itself. For a measurable immune response to TNP-OVA, extra or adjuvant signals are necessary. These adjuvant signals can be provided by any chemical that has some irritant or proinflammatory effect. In other words, a specific response against TNP-OVA indicates that the chemical is at least able to elicit an adjuvant signal. A specific antibody response against TNP-OVA does not, however, exclude sensitizing potential. But if an IgG_1 response is elicited against TNP-OVA—but not against TNP-Ficoll—it is highly probable that the compound itself is not a sensitizer but that it has mere adjuvant potential. A chemical that does not eli-

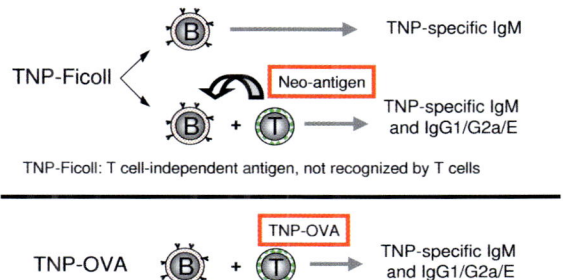

Reporter Antigen Popliteal Lymph Node Assay. Figure 1 Reporter antigens.

cit an IgG response against any of the two RA can be considered as non-immunopotentiating. So, by combining the outcomes of the antibody (IgG_1 or other switched isotypes) responses against TNP-Ficoll and TNP-OVA one can assess whether a chemical is a sensitizer (IgG_1 to TNP-Ficoll and TNP-OVA) or a mere adjuvant (IgG_1 to TNP-OVA but not to TNP-Ficoll), or is unable to elicit an immune response at all (no IgG_1 to any of the RA) (see Figure 1).

Interestingly, by changing the isotype specificity of the detection antibody in the ELISPOT assay (i.e. by using anti-IgG_{2a}, anti-IgG_{2b} and anti-IgE in addition to anti-IgM and anti-IgG_1) it is possible to determine the type (type 1 vs type 2) of the immune response. This can be done particularly well with TNP-OVA. This RA is also suitable to follow the immune response over a longer period of time (at least 4–5 weeks) and the type of memory response can be evaluated again without the need to know the relevant antigen induced by chemical exposure (2).

Recent studies have been published in which the RA-PLNA with TNP-Ficoll was combined with oral exposures to the drug diclofenac (3). Twenty days after single oral exposure to diclofenac, TNP-specific IgG_1 responses were observed in the PLN upon footpad injection of a subsensitizing dose of diclofenac together with TNP-Ficoll. Hence, it appears possible to detect compound-specific anamnestic responses by using TNP-Ficoll which is susceptible to non-cognate neoantigen specific T cell help.

It is important to note that coinjection of TNP-OVA or TNP-Ficoll did not appear to interfere with the type of immune response raised by the chemical.

Pros and Cons

The evident advantage of the RA-PLNA over the primary PLNA is that by the use of RA nonsensitizers can in principle be distinguished from nonsensitizing irritants. Moreover, the RA-PLNA provides a more robust indicator of immunostimulation (for chemicals the cell number varies from $1-2 \times 10^6$ to around 10^7 cells per PLN, whereas the number of IgG_1 AFC varies from $0-10/10^6$ to around $1300/10^6$ PLN cells) and immunologically relevant information can be obtained in particular with respect to the type of immune response that is elicited. The kinetics of the immune response to TNP-OVA initiated by a certain chemical can be easily followed over a certain period of time. This allows easy verification of the adjuvant potential of a chemical by detecting memory responses to the RA.

Apart from this, the RA-PLNA has pros and cons resembling those of the primary PLNA: it is a simple, straightforward, objective, and cheap test (despite the fact that it requires detection of ASC by ELISPOT) that allows fast screening of the immunostimulatory potential of compounds. Moreover, the outcome of the response depends likewise on the genetic background of the strain of mice used. Although false-positive chemicals (if they are so by irritancy) can be distinguished, false-negative pro-haptens remain undetectable without a metabolizing system.

The RA-PLNA is also limited by the irrelevance of the route of exposure. Interestingly, however, the RA technique can be used in combination with exposure to compounds via the oral route.

Predictivity

The RA-PLNA was developed in 1996 and has been used primarily for fundamental research into immunomodulating capacity of low-molecular-weight chemicals. Hence, the number of compounds tested in the RA-PLNA is limited to around 20–30. Although not formally evaluated or validated, based on a comparison between two independent laboratories, the chemicals tested in the two laboratories showed similar outcomes (1,2,4). The RA-PLNA seems to be more robust then the primary PLNA (because of the use of an immunological parameter) so the selectivity of the assay may be improved by the use of RA.

Relevance to Humans

As for the primary PLNA, the RA-PLNA should be regarded as a screening test for the possibility of a compound to cause immunosensitization, and as a first step to evaluate whether the compound has also potential to stimulate the immune system via the relevant route of exposure. The relevance of the outcomes of the RA-PLNA might be higher because it gives substantially more information about the possible effect of the chemical exposure.

Regulatory Environment

The RA-PLNA was developed only 5–6 years ago and was mainly used to perform mechanistic studies. The RA-PLNA is regarded as a modification of the PLNA that has improved predictivity; it is therefore included in the ILSI-HESI initiative to be evaluated as a predictive test.

References

1. Albers R, Broeders A, van der Pijl A, Seinen W, Pieters R (1997) The use of reporter antigens in the popliteal lymph node assay to assess immunomodulation by chemicals. Toxicol Appl Pharmacol 143:102–109
2. Albers R, de Heer C, Bol M, Bleumink R, Seinen W, Pieters R (1998) Selective immunomodulation by the autoimmunity-inducing xenobiotics streptozotocin and HgCl2. Eur J Immunol 28:1233–12342
3. Gutting BW, Updyke LW, Amacher DE (2002) BALB/c mice orally pretreated with diclofenac have augmented and accelerated PLNA responses to diclofenac. Toxicology 172:217–230

4. Gutting BW, Schomaker SJ, Kaplan AH, Amacher DE (1999) A comparison of the direct and reporter antigen popliteal lymph node assay for the detection of immunomodulation by low molecular weight compounds. Toxicol Sci 51:71–79

Resident Macrophages

Monocytes that migrate into normal tissues downregulate many activities and become resident macrophages which have reduced phagocytic and killing capacities but enhance signaling ability.

▶ Macrophage Activation

Respiratory Allergy Assay

▶ Animal Models for Respiratory Hypersensitivity

Respiratory Burst

The activation of oxidative metabolism of neutrophils, which is manifested by the production of highly reactive oxygen species, such as superoxide, hydroxyl radical and hydrogen peroxide. Respiratory burst is based on the activation of a multicomponent enzyme, NADPH-oxidase, in the neutrophil plasma membrane in response to various activators and during phagocytosis.

▶ Chemotaxis of Neutrophils

Respiratory Hypersensitivity Test

▶ Animal Models for Respiratory Hypersensitivity

Respiratory Infections

DONALD E GARDNER · SUSAN C GARDNER
Inhalation Toxicology Associates Inc.
P.O. Box 97605
Raleigh, NC 27624
USA

Synonyms

Bioaerosols, pulmonary infections, pneumonia, influenza, bronchitis, common cold, SARS, airborne contagion, epidemics, consumption, air pollution, host defense systems

Short Description

Humanity has always been vulnerable to microbes that cause respiratory disease. Presently respiratory infection is the sixth leading cause of death in the USA—a situation which may intensify in coming years. Being aware of the health risk of respiratory infections has become more critical for five reasons:
- the selection of resistant microbial flora due to the multitude of antimicrobial drugs currently available
- the rapid international transport of microbes due to population migration
- ▶ immunodeficiency diseases
- increased mean life expectancy
- the development of (formerly unavailable) surgical and systemic therapies for treating diseases.

The severity and risk of infection varies with the virulence, ▶ antigenicity (▶ immunogenicity) and viability of the invading organism, the number of viable organisms at the target site, their ability to damage host tissue by the production of toxins, and the function of the individual's normal microbial defenses. Microorganisms are highly versatile and widely distributed, occurring nearly everywhere in the environment. They are capable of replicating themselves or merely surviving in habitats that are extremely diverse and hostile. ▶ Ambient air that contains living organisms such as viruses, bacteria, fungi, protozoa, and algae (as well as products of their metabolism or their decomposition, such as toxins) is referred to as ▶ bioaerosols. When bioaerosols are inhaled and deposited in the respiratory tract, a normal host defense system exists to maintain health, and when this system is impaired, an individual's risk of respiratory disease is increased.

Characteristics

In the natural environment, healthy people exist in equilibrium with microorganisms. Microbes can be classified as pathogens, opportunists, or nonpathogens. Opportunistic microbes are organisms that normally are not capable of causing disease in a healthy immunocompetent person, but can cause disease in those with impaired host defense. The respiratory system is a most vulnerable target for such infectious agents because it is directly exposed to the external environment and has nearly four times the total surface area ($70m^2$) as the combined total surface areas of the gastrointestinal tract and the skin. Although microbial uptake through ingestion and through the skin is generally intermittent, inhalation provides a continuous means of exposure. Thus, for airborne biological agents the respiratory system is a major route of entry into the body.

An important distinction must be made between infection and disease. Infection implies that a microorganism has taken up residence in a host and is multiplying within that host—perhaps with no outward signs of disease. Thus it is possible to be infected with an agent, yet not have the disease (although disease may develop at a latter time). In contrast, those who appear "sick" are said to have a "disease".

Humans are the primary host for many microorganisms. After the microorganism has been inhaled and deposited in the respiratory tract, there are a number of elaborate defense mechanisms by which the respiratory system can protect this large surface area, including anatomical barriers (complicated shape of the nasal passages), physical clearance mechanisms (sneezing, coughing and ciliary activity), local antibody production (mainly immunoglobulin A, the preferred bacterial ▶ opsonin), production of ▶ interferons, ▶ proteases, ▶ antiproteases, antioxidants, ▶ fibronectin, ▶ lactoferrin, phosholipids, ▶ phagocytic cells (alveolar macrophages and ▶ neutrophils) and immune effector cells (lymphocytes and ▶ natural killer cells). Together these defenses maintain the integrity of the respiratory system. Despite this protective system there are a number of ways that the inhaled invading organism can circumvent these defenses and cause disease either from deficiency in one or more of the mechanisms of defense or from exacerbation of virulence of the microorganisms such that the host defense system is overcome. Known factors which result in a weakened defense system include:

- reduced physical removal of inhaled microorganism by the ▶ mucociliary mechanisms
- dysfunction of the macrophages
- alteration of the acellular lining material of the deep lung (▶ surfactant) or the mucus in the upper airways
- presence of edema fluid or inflammatory exudates in the airways
- pulmonary immunosuppression
- influence of environmental pollutants
- stress
- preexisting disease
- attenuated cytotoxic T cell function in malnourished children, alcoholics and elderly individuals
- immunocompromised individuals such as HIV-positive patients or transplant patients receiving immunosuppressive therapy (1).

Environmental Factors Influencing Infectious Disease

The presence of microbes in humans may be considered as the normal state, and the process of disease is a disturbance of the equilibrium between the host, the parasite, and the environment. The process of respiratory infection and the subsequent disease involves the interaction of a host, a microbe and the environment. Thus, it becomes necessary that in addition to considering the virulence of the biological agent and the susceptibility of the host, attention must be given to a variety of environmental and physiological factors that might influence the course and severity of the disease. A person exposed to a combination of stresses, such as those of a physical or chemical nature, may be more susceptible to certain biological agents and thus may be at a greater risk of contacting a disease. A variety of gaseous and particulate airborne pollutants may adversely affect the normal functioning of the host's respiratory defenses, which increases susceptibility to pulmonary infections. These include numerous inhaled metals (e.g. cadmium, lead, vanadium, nickel, manganese), gaseous pollutants (e.g. ozone, nitrogen dioxide, sulfur dioxide, phosgene, benzene, toluene, HCHO), particles (e.g. sulfuric acid), and complex mixtures (e.g. auto exhaust, tobacco smoke, wood smoke, fly ash) (1,2).

Mechanisms by which these environmental factors could exacerbate the incidence and severity of infectious respiratory disease in individuals with normal immune function includes the following:

- enhancing deposition
- interfering with clearance and ▶ bactericidal activity
- initiating inflammation leaving damaged epithelial tissue vulnerable to microbial invasion
- by a mechanism that enhances delivery or infectivity of viruses, fungi and bacteria.

Relevance to Humans

The human relevance of understanding the source, transmission, pathogenesis, and the need for testing of bioaerosols is obvious. This has been clearly demonstrated with the recent epidemic of the new ▶ coronavirus that is most likely the cause of severe acute respiratory syndrome (SARS). Within 4 weeks of the first appearance of SARS, the microorganism infected people around the world. Success in controlling such epidemics can be contributed to the worldwide global network of laboratories that pool resources to collaborate on such outbreaks.

The 21st century faces a significant health threat from of the use of biological agents as terrorist weapons. The potential to cause large numbers of serious casualties among military forces and civilians provides an excellent reminder to medical planners of the limits of medicine. Biological weapons include any organism or toxin found in nature that can be used to incapacitate, kill, or impede people. Examples of human respiratory diseases associated with biological agents include anthrax, meningitis, plague, tularemia, brucellosis, smallpox, viral encephalitis and hemorrhagic fever. A variety of microorganisms are capable of producing toxins such as ▶ aflatoxins, botulism, ▶ ricin, staphyloccal enterotoxin, ▶ mycotoxins and perfringen toxins. Toxin inhalation may cause acute illness with fever, sweating, muscle aches, ▶ rhinitis, and asthma. Currently it is estimated that about 17 countries are suspected of having offensive biological warfare programs, making the use of biological agents on military and civilian populations a greater threat than ever. Biological agents are easy to acquire, to synthesize, and to use. Only small quantities of microorganisms are needed to cause respiratory disease in people in metropolitan areas, so it is relatively easy to conceal, transport and disseminate them. These agents are difficult to detect or protect against since they are invisible, odorless, and tasteless, and their dispersal can be performed silently.

Assessment of Risk

All microorganisms and chemical agents have the potential to be harmful under certain conditions of exposure. Examples of microorganisms causing respiratory infection or sensitization when inhaled are presented in Table 1.

The important issue is not just of toxicity but also of risk. All humans accept some degree of risk in their daily activities. However, it is important to determine the probability that such an exposure will cause an adverse effect under actual conditions of human exposure. The National Academy of Science/National Research Council provides a structured approach to the process that has been widely used for assessing the risk of health effects resulting from exposure to chemical agents. There has been interest in developing similar risk assessment models to evaluate the likelihood of adverse human effects from exposure to infectious microorganisms. While the methods presently in use for chemicals are not directly appropriate for assessing risk from exposure to airborne pathogens, they do provide a conceptual framework for developing a similar process. Issues that are unique include assessment of the pathogen-host interaction, consideration of secondary spread, the possibility of short-term and long-term protective immunity, and assessment of the conditions that might allow microorganisms to propagate. Thus, the development of a process for pathogenic risk assessment is complex and should be expected to consist of several interrelated components, which are conceptually distinct steps.

Regulatory Environment

US government agencies with regulatory responsibilities, including the Environmenta Protection Agency (EPA) and the Food and Drug Administration (FDA), have recommendations for guidelines for immunotoxicity testing strategies (3,4). These are discussed in greater detail in other chapters. Such testing is intended to provide information on changes in the functioning of the immune system, which might occur as a result of exposure to a variety of agents. In the past, a battery of different assays, structured in a multitiered approach, has been used to assess immunotoxicity. Some of these assays included models for detecting respiratory susceptibility to bacterial and viral exposure.

Respiratory susceptibility to infectious agents can best be measured if the immune system is asked to perform its normal functions (that is to defend against infectious agents) and to correlate changes in various host resistance animal models with changes in specific immune parameters. Since an increase in the incidence and/or severity of infection has been consistently identified as one of the hallmark indicators of immune malfunction, a great deal of research has been conducted to design and characterize such host resistance models (5). These studies have consistently indicated that changes in specific respiratory immune functional parameters are associated with the changes in host resistance models (making the host more susceptible to pulmonary infection). While these animal model systems are sensitive indicators for examining the enhancement of microbial infection of the lungs from exposure to airborne agents, it is now generally accepted that such host resistance models are not feasible choices as initial predictors of immunotoxicity because of their high complexity and cost. It is thought, therefore, that these models are best positioned in the second tier of a testing strategy.

Control and Prevention

The detection, prevention and management of airborne respiratory infectious disease depend on preventing the exposure to the biological agents. The ultimate aim must be to quickly identify the causative agent and to establish reliable approaches for prevention and control. A well-designed, well-implemented surveillance program can detect unusual clusters of respiratory disease, document an outbreak, estimate the magnitude of the problem, and identify factors responsible for its emergence. It is important to eliminate the reservoir of the microorganisms, to interrupt the transmission of the infection, and to increase the resistance of the individual to the microorganism. Personal hygiene and cleanliness in the living environment may lessen the spread of the disease, as well as using personal protective equipment in living and work places, proper disposal of waste (especially those suspected of microbial contamination), and the proper design and construction of buildings to avoid buildup of fungal growth. Special care is necessary in the case of immunocompromised and particularly sensitive or susceptible individuals. Drug-resistant bacterial, viral, and pro-

Respiratory Infections. Table 1 Airborne microorganisms causing infection or sensitization

Bacterial disease	Causative Organism	Primary Reservoir
Pneumonia	*Streptococcus pneumoniae*	Humans
Nosocomial pneumonia	*Klebsiella pneumoniae*	Humans
Pneumonia	*Haemophilus influenzae*	Humans
Walking pneumonia	*Mycoplasma pneumoniae*	Humans
Q fever	*Coxiella burnetii*	Animals
Ornithosis, psittacosis	*Chlamydia psittaci*	Birds
Brucellosis*	*Brucella melitensis*	Animals
Legionnaires disease	*Legionella pneumophila*	Water
Tuberculosis	*Mycobacterium tuberculosis*	Humans
Hypersensitivity pneumonitis	*Thermoactinomyces*	Water, soil, compost
Diphtheria	*Corynebacterium diphtheriae*	Humans
Pertussis (whooping cough)	*Bordetella pertussis*	Humans
Inhaled anthrax	*Bacillus anthracis*	Animals
Bubonic plague*	*Yersinia pestis*	Animals
Tularemia	*Francisella tulaensis*	Animals
Viral diseases		
Influenza	Influenza A, B and C viruses	Humans
Severe acute respiratory syndrome (SARS)	Coronaviruses	Humans, animals
Croup	Parainfluenza viruses	Humans
Bronchiolitis pneumonia	Respiratory syncytial virus	Humans
Mumps*	Mumps virus	Humans
Measles	Rubella virus	Humans
Common cold	Rhinoviruses, coronaviruses, parainfluenza	Humans
Chicken pox*	Varicella zoster virus	Humans
Small pox*	Variola virus	Humans
Fungal diseases		
Asthma, rhinitis	*Alternaria*	Outdoor air, dead plants
Asthma, rhinitis	*Cladosporium*	Outdoor air, dead plants
Asthma, rhinitis	*Penicillium*	Damp organic material
Pulmonary aspergillosis	*Aspergillus*	Soil, compost
Coccidioidomycosis	*Coccidioides immitis*	Soil of arid regions
Histoplasmosis	*Histoplasma capsulatum*	Animals, soil, feathers
Protozoan/ algal diseases		
Hypersensitivity pneumonitis	Protozoa	Water reservoirs
Asthma, rhinitis	Algae	Water reservoirs

* Disease transmitted via respiratory tract, but signs of infection and disease are seen elsewhere in body.

tozoan pathogens pose a serious and growing problem for all people, regardless of their age, gender, or socio-economic background.

Microorganisms that are resistant to antibiotics cause the vast majority of infections that people acquire in hospitals. Drug resistance is accumulating and accelerating, thereby reducing the ability of drugs to combat infectious disease. While the emergence of microbial resistance cannot be stopped, the National Academy of Science/Institute of Medicine has addressed the urgency of this problem at the recent Forum on Emerging Infections. A number of specific suggestions were identified that will aid better understanding of microbial resistance, mitigating its impact on human health, thus transforming this growing threat into a manageable problem.

References

1. Gardner DE (2001) Bioaerosols and disease. In: Bingham E, Cohrssen B, Powell CH (eds) Patty's Industrial Hygiene and Toxicology, Volume 1, 5th ed. Wiley, New York, pp 679–711
2. Cohen MD, Zelikoff JT, Schlesinger RB (eds) (2000) Pulmonary Immunotoxicology, Kluwer, Norwell
3. US Food and Drug Administration (2002) Immunotoxicology Evaluation of Investigational New Drugs. FDA October 2002, pp 1–35

4. Environmental Protection Agency (1998) Health Effects Test Guidelines. OPPTS 870.7800 Immunotoxicity. EPA 712-C-96-351 October 1998, pp 1–11
5. Conn CA, Green FH, Nikula KJ (2000) Animal models of pulmonary infection in the compromised host. Inhal Toxicol 12:783–827

Responder Cell

Specific cell types present in complex cell mixtures can be identified based on their functional response to specific stimulation.
▶ Limiting Dilution Analysis

Responder Cell Frequency

In **limiting dilution analysis**, the frequency of a specific cell type in a cell mixture is estimated based on its functional response to specific stimulation.
▶ Limiting Dilution Analysis

Reverse Enzyme-Linked Immunospot Assay

RELISPOT is a synonym for ELISPOT.
▶ Enzyme-Linked Immunospot Assay

Reye's Syndrome

Fatal, fulminating hepatitis with cerebral edema.
▶ Anti-inflammatory (Nonsteroidal) Drugs

Rheumatic Fever

Late complication of dermatological or pharyngeal infection. Some serologic subtypes of β-hemolytic streptococci lead to the production of antibodies against the bacterial cell wall protein (M protein). Some of these antibodies cross-react with myocardial sarcolemmal proteins, leading to carditis.
▶ Dermatological Infections

Rheumatoid Arthritis and Related Autoimmune Diseases, Animal Models

JEANNE M SOOS
Immunologic Toxicology, Preclinical Safety Assessment
GlaxoSmithKline R&D
709 Swedeland Road
P.O.Box 97605
King of Prussia, PA 19406-0939
USA

Synonyms
Arthritis models, autoimmune models

Short Description
Animal models of rheumatoid arthritis are experimental models of induced joint and digit inflammation that can be utilized to investigate the mechanisms contributing to arthritic inflammation, to investigate potential therapies for rheumatoid arthritis, and to assess the potential for substances to either induce or downregulate autoimmune inflammatory responses. Autoimmune arthritis can be induced by injection or immunization by several different types of antigens in multiple susceptible strains of mice, rats, and higher animals (1).

Characteristics
Each of the arthritis models described below is characterized by induction of disease through immunization with either a self antigen or injection with a mixed antigen preparation. Assessment of arthritis in animal models is achieved through visual inspection of the front and hind paws. Inflammation can occur in the ankle and throughout the digits depending on the severity of disease. The scoring system for assessing the severity of inflammation is presented in Table 1. The scores for each joint are added and thus the maximum score for an individual animal is 16.

Rheumatoid Arthritis and Related Autoimmune Diseases, Animal Models. Table 1 Scoring system for visual evaluation of experimental arthritis

Score	Pathology on Visual Inspection
0	Normal size and structure of paw
1	Swelling observed in a single digit
2	Swelling observed in more than one digit
3	Swelling observed in the joint
4	Complete distention and swelling of digits and joint

Additional methods for assessing development of disease in models of arthritis include the measurement of paw thickness using a caliper, mercury or water plethysmography, histopathology of the joint, and radiographic evaluation of the joint.

Collagen-Induced Arthritis (CIA)

Collagen-induced arthritis can be induced in a variety of rodent strains as well as nonhuman primates. Type II collagen, the specific self antigen used for disease induction, is a major component of cartilage. Similar to human rheumatoid arthritis, susceptibility to collagen-induced arthritis is linked to the expression of certain major histocompatability complex (MHC) class II molecules (2). Disease is monophasic and can result in full resolution of disease or ankylosis of the joint. Pathogenesis of disease in this model involves both T cell and B cell responses.

Adjuvant Arthritis (AA)

Adjuvant arthritis can be induced in a variety of rat strains with the Lewis rat most commonly used. Rats are immunized with heat-killed *Mycobacterium tuberculosis,* strain H37Ra, in incomplete Freund's adjuvant and evaluated over time for development of disease. The disease tends to be monophasic and often results in irreversible joint ankylosis. Disease can also be induced by adoptive transfer of mycobacteria-specific T cells or lymph node cells of immunized rats.

Streptococcal Cell Wall Arthritis (SCW)

Streptococcal cell wall arthritis is inducible in a wide variety of rat strains and, depending on the strain, the susceptibility to disease development and severity of disease can vary. Streptococcal cell wall group A peptidoglycan-polysaccharide polymer preparation isolated from *Streptococcus pyogenes* cell walls is injected intraperitoneally for development of disease. The course of the disease is characterized by acute onset during the 48 h after injection followed by a chronic phase.

Measurement of Cellular Responses and Cytokines in Arthritis Models

Humoral responses can be evaluated by measuring antibody titers to the antigens used for induction of disease in the models described above. T cell responses can be measured by cellular proliferation assays as well as by the generation of antigen-specific T cell lines and clones. Cytokine analysis can be a valuable method for evaluating the mechanisms for modulation of disease. A list of the characteristics of each of these models is presented in Table 2.

Pros and Cons

There are some advantages of the individual models of arthritis. The use of a specific self antigen, such as type II collagen in the CIA model, allows for the dissection of multiple cellular inflammatory mechanisms by the methods described above. Induction of disease in the SCW arthritis model illustrates molecular mimicry and allows for study of the initial pathogenesis and potential triggers of the autoimmune response (3). Also the SCW is a chronic model of disease that more closely reflects the course of rheumatoid arthritis in humans.

Disadvantages also exist for these models. The course of disease is limited only to an acute phase, with no observations of a longer, chronic phase in the collagen-induced and adjuvant arthritis models. Further, for both the collagen-induced and SCW models, great attention must be paid to the quality and purity of the antigen preparation used to induce disease.

Predictivity

These models are the methods of choice for initial pharmacological studies, having good predictive value. However they may not be fully predictive of what may be observed in humans. For example, blockade of an inflammatory cytokine was shown to be therapeutic in animal models of autoimmunity (4,5) while blockade of that same cytokine induced autoimmunity in clinical studies (6,7). Immunological mechanisms leading to autoimmunity in humans are more complicated and less well understood than the prevention of autoimmunity (i.e. by blockade of T helper type 1 cytokines) in these animal models. For immunotoxicology, these models of inflammation provide a valuable means for evaluating potential for proinflammatory activities of new drugs and substances.

Relevance to Humans

Animal models of arthritis serve as a means for evaluating new therapies for human rheumatoid arthritis.

Regulatory Environment

The field of autoimmunity and the use of autoimmune models such as rheumatoid arthritis do not require regulation, but the value of these models is recognized for understanding the mechanisms of immunotoxicologic potential and for risk assessment by the FDA in the guidance for Immunotoxicology Evaluation of Investigational New Drugs.

Relevant Guidance

FDA (CDER) Immunotoxicology Evaluation of Investigational New Drugs, 2002

Rheumatoid Arthritis and Related Autoimmune Diseases, Animal Models. Table 2 Characteristics of the most widely employed animal models of experimental rheumatoid arthritis

Animal Models	Species/Strains	Antigen for Immunization	Disease Course	Induction of Disease
Collagen-induced arthritis	Mouse: primarily DBA/1, strains of the I-Aq and I-Ar class II allleles Rat: variety of strains from multiple class II Rt alleles	Chicken or bovine type II collagen	Monophasic	Mouse: days 21–35 Rat: high responder days 8–10 Rat: low responder days 30–60
Adjuvant arthritis	Lewis rat	Heat-killed *Mycobacterium tuberculosis* H37Ra	Monophasic	Days 10–17
Streptococcal cell wall arthritis	Lewis rat primarily Multiple rat strains with varying susceptibility	Streptococcal cell wall group A peptidoglycan-polysaccharide polymers	Biphasic	Acute phase: 48 hours Chronic phase: days 10–21 with continued disease for months

References

1. Coligan JE, Kruisbeek AM, Margulies DH, Shevach EM, Stober W (eds) (1991) Current Protocols in Immunology. John Wiley and Sons, New York
2. Wooley PH, Luthra HS, Stuart JM, Cavid CS (1981) Type II collagen-induced arthritis in mice. I: Major histocompatibility complex linkage and antibody correlates. J Exp Med 154:688–700
3. Taylor JE, Ross DA, Goodacre JA (1994) Group A streptococcal antigens and superantigens in the pathogenesis of autoimmune arthritis. Eur J Clin Invest 24:511–521
4. Mori L, Iselin S, de Libero G, Lesslauer W (1996) Attenuation of collagen-induced arthritis in 55-kDa TNF receptor type 1 (TNFR1)-IgG1-treated and TNFR1-deficient mice. J Immunol 157:3178–3182
5. Korner H, Goodsall AL, Lemckert FA et al. (1995) Unimpaired autoreactive T-cell traffic within the central nervous system during tumor necrosis factor receptor-mediated inhibition of experimental autoimmune encephalomyelitis. Proc Natl Acad Sci USA 92:11066–11070
6. Sicotte NL, Voskuhl RR (2001) Onset of multiple sclerosis associated with anti-TNF therapy. Neurology 57:1885–1888
7. Mohan N, Edwards ET, Cupps TR et al. (2001) Demyelination occurring during anti-tumor necrosis factor alpha therapy for inflammatory arthritides. Arthritis Rheumatol 44:2862–2869

Rheumatoid Arthritis (RA)

A common inflammatory disease caused by an autoimmune reaction directed to joints. It is mostly mediated by humoral immune reactions and immune complex deposits. So-called rheumatoid factors, i.e. autoantibodies against the constant region (Fc) of antibodies, have been described for this disease for the first time. This chronic potentially disabling arthritic condition with a female predominance, is characterized by peripheral symmetric polyarthritis with or without other associated systemic involvements, occurring predominantly in individuals who are HLA-DR4/DR1 positive.

▶ Hypersensitivity Reactions
▶ Systemic Autoimmunity
▶ Fatty Acids and the Immune System
▶ Complement Deficiencies
▶ Molecular Mimicry
▶ Cyclosporin A

Rhinitis

Inflammation of the nasal mucus membrane, marked by sneezing, lacrimation and watery mucus.
▶ Respiratory Infections

Ribonucleic Acid (RNA)

A biomolecule that has an informational, structural, and enzymic role. The structure is of ribose units joined in the 3' and 5' positions through a phosphodiester linkage with a purine or pyrimidine base attached to the 1' position. (RNA = ribonucleic acid).
▶ Southern and Northern Blotting

Ricin

A highly toxic lectin and hemagglutin occurring in seeds of castor beans. Used as a chemical warfare agent.
▶ Respiratory Infections

Risk Assessment

▶ Statistics in Immunotoxicology

Rodent Immune System, Development of the

KENNETH S. LANDRETH
Department of Microbiology, Immunology, and Cell Biology
West Virginia University Health Sciences Center
Morgantown, WV 26506
USA

Synonyms
Murine immune system, mouse immune system, rat immune system

Definition
Our understanding of cells and tissues that make up the immune system of vertebrates came from seminal studies of avian species. However, rodents have become the animal model of choice for studies of immunocompetence, and most of our knowledge of the development of the immune system has come from studies using mice or rats. The rodent immune system develops through a set of critical windows of vulnerability during embryonic and adult life that can be used for evaluating effects of environmental exposure to potentially toxic compounds (1). Most of the information in this review comes from studies of mice and, in all cases tested, closely parallels that found in other rodents, including rats.

Characteristics
Embryonic development. Embryonic development of the immune system in rodents initiates with formation of multipotential ▶ hematopoietic stem cells (HSC) in intraembryonic spanchnoplure surrounding the heart and in association with endothelial cells of the extraembryonic yolk sac (2). Hematopoietic stem cells undergo a temporal migration from intra-embryonic mesenchyme to fetal liver, fetal spleen, and ultimately to final residence in bone marrow and thymus: organs which continue to produce immunocompetent cells throughout life (Figure 1).

HSC first appear in embryonic mice during the 7th day of gestation. These cells develop in both intraembryonic splanchnopleuric mesenchyme surrounding the heart, a tissue often identified as the aorto-gonadomesonephric region (AGM), and in extraembryonic blood islands of the yolk sac (3). However, HSC in these two embryonic tissues differ dramatically in their developmental fate. Intraembryonic stem cells, but not those that arise in the yolk sac, contribute to sustained blood cell development and functional immune-responsive cells in postnatal rodents. The standard experimental assay for HSC has been the in vivo spleen colony forming cell (CFU-S) assay. This assay relies on the unique ability of rodent stem cells to migrate to the spleen after adoptive transfer and to initiate formation of macroscopic colonies of hematopoietic cells that are clonally derived. More recent in vitro assays for multipotential hematopoietic cells have been developed and enumerate cells capable of forming colonies in a semisolid matrix that contain multiple blood cell lineages. The initial period of stem cell formation for the hematopoietic system culminates as newly developed stem cells migrate to the embryonic liver and spleen.

At approximately day 10 of gestation in mice, HSC relocate from the AGM to the developing fetal liver. In fetal liver, HSC develop into more differentiated and lineage-restricted stem cells. These lineage-restricted stem cells further differentiate in this tissue site to form more mature progenitor cells which retain the ability to proliferate but are more restricted in their developmental potential. Lineage-restricted stem cells are operationally defined by their proliferation and/or differentiation in response to specific hematopoietic cytokines.

Progenitor cells are routinely enumerated using in vitrocolony forming unit (CFU) cell assays. The availability of recombinant cytokines and use of these assays have been invaluable in enumerating specific progenitor cells in embryonic tissues that are otherwise indistinguishable by morphologic analysis. These assays are central to any study of direct effects of toxic compounds on hematopoietic tissues and blood cell formation in the developing embryo or in postnatal animals.

Fetal Liver. Fetal liver continues to be the principal hemato-lymphopoietic organ until near the end of gestation, however, few morphologically or functionally identifiable mature leukocytes are found in the embryo until near the time of birth (4). B lymphocyte production in the rodent fetal liver has been well characterized and serves as a prototype of fetal development of the immune system (5). Cells with immunoglobulin (Ig) gene rearrangements are first found in the liver on gestational day 11 and increase rapidly to easily detectable numbers by gestational day 13. Cells with cell surface Ig (B lymphocytes) are not detected in fetal liver until day 18 of gestation, and remain at low frequency until birth. Numbers of hematopoietic cells decline in the fetal liver as the bone marrow assumes primary hematopoietic function at gestational day 18. HSC and lineage-restricted lymphoid progenitor cells are found in the developing spleen on gesta-

Development of the Human and Rodent Immune Systems

Human →	week 8–10	week 10–16	week 16–birth	birth–1 year	1–18 years
Rodent →	day 7–9	day 9–16	day 13–birth	birth–day 30	day 30–60
	Initiation of Hematopoiesis	Migration of Stem Cells; Expansion of Progenitor Cells	Colonization of Bone Marrow and Thymus	Maturation to Immune Competence	Establishment of Immune Memory

Conception — Birth — Sexual Maturity

Rodent Immune System, Development of the. Figure 1 Organs which continue to produce immunocompetent cells throughout life.
Adapted from: Dietert RR et al. (2000) Environ Health Perspect 108 [Suppl 3]:483–490

tional day 13, approximately the same time they are found in the fetal liver and remain detectable in that tissue until a few weeks after birth. Unlike the bone marrow, lymphopoiesis rapidly wanes in the spleen after birth and can not be demonstrated in adult mice.

Thymus. Organogenesis of the thymus initiates from the 3rd and 4th pharyngeal pouches in mice on gestational day 11. The thymus is immediately colonized by immigrant HSC, which are detectable on day 11 of gestation. The thymus continues to be a source of lymphoid cells (T lymphocytes) in postnatal rodents until somewhat after sexual maturity when the thymus regress and ceases function.

Lymph nodes. Lymph nodes form in the developing embryo by endothelial budding of the venous circulatory system, a process that initiates on gestational day 10.5 in mice. The formation of these organs is dependent on interaction of immature lymphoid cells with developing endothelial cells. Peyer's patches develop from clusters of cells on the proximal end of the intestine on gestational day 15.5 and nasopharyngeal lymphoid tissues form after birth in mice.

Bone marrow. Long bones of the embryo mineralize, and the central marrow cavity is excavated to create a marrow cavity on gestational day 17.5 in mice. This new site is immediately populated by hematopoietic cells derived from AGM (*not yolk sac*) (2). This population of cells establishes the hematopoietic bone marrow which serves as a reserve of HSC, blood cell development, and production of immune-responsive cells for the remainder of postnatal life.

Relevance to humans

Observations made on development of human lymphoid tissues and cells suggest that the temporal sequence of events described here for rodents is largely duplicated in human development (Figure 1). However, there are notable differences between the gestational appearance of immune function in rodents and humans. In general, immune responsive cells appear relatively earlier in gestation in human embryos, and tissues of the immune system are more mature at birth than in rodents. Of particular importance, cytokines that stimulate lymphoid cell proliferation and differentiation are conserved between humans and rodent species and, in many cases, retain sufficient homology to be active on cells from either species.

Regulatory Environment

Rodents are the preferred model for studies of immunology and immunotoxicology and most regulatory testing is carried out on these species. Despite the fact that we know significantly more about the development of the immune system in mice, rats continue to be a preferred model for immunotoxicology testing for historical rather than scientific reasons.

References

1. Dietert RR, Etzel RA, Chen D, Halonen M, Holladay S, Jarabek AM, Landreth K, Pedan D, Pinkerton K, Smialowicz RJ, Zoetis T (2000) Workshop to identify critical windows of exposure for children's health: immune and respiratory systems work group summary. Environ Health Perspect 108 [Suppl 3]:483–490
2. Cumano A, Godin I (2001) Pluripotent hematopoietic stem cell development during embryogenesis. Curr Opin Immunol 13:166–171
3. Metcalf D, Moore MAS (1971) Haemopoietic Cells. In: Neuberger A, Tatum EL (eds) Frontiers of Biology, Vol. 24. North-Holland, Amsterdam, pp 172–271
4. Landreth KS (2002) Critical windows in development of the rodent immune system. Hum Exp Toxicol 21:493–498
5. Landreth KS (1993) B lymphocyte development as a developmental process. In: Cooper EL, Nisbet-Brown E (eds) Developmental Immunology. Oxford University Press, New York, pp 238–273

Rodents, Inbred Strains

INA HAGELSCHUER
PH-R ZfV, Geb. 516
Bayer HealthCare AG
Aprather Weg 18
D-42096 Wuppertal
Germany

Synonyms
Inbred strains, inbreds, genetically defined rodents

Definition
General items
Most of the current definitions of inbred strains are established from rats and mice. These rodents have short-generation intervals and sufficient numbers of offspring to allow the application of close inbreeding. In contrast, only a few strains being inbred by definition (see below) are available from hamsters or guinea pigs.

Inbred strain
A strain shall be regarded as inbred when it has been mated brother × sister (hereafter called b×s) for 20 or more consecutive generations. Parent × offspring matings may substitute b×s matings provided that, in the case of consecutive parent × offspring matings, the mating in each case is to the younger of the two parents. Inbred strains are designated by three to five capital letters (for example the mouse strain CBA or the rat strain BN). Before creating new designations relevant databases have to be checked in order to avoid duplications. Present information on nomenclature rules can be drawn from the Jax Mice Catalogue (1).

Substrain
Inbred strains are divided into substrains when there are known or assumed genetic differences due to residual heterozygosity during colony set-up, mutation, or genetic contamination. These reasons are likely when:
- the strains are separated before inbreeding generation F40
- the present strain has been bred separately for 50 or more generations
- genetic differences have been proven.

Designation of a substrain
Parental strain or symbol for differentiation (examples of symbols):
- numbers, e.g. DBA/1 or DBA/2
- laboratory codes, e.g. A/J for Jackson Laboratory
- a combination of both if more than one substrain is developed in the same laboratory, e.g. CF/1Ztm, CF/2Ztm.CF/4Ztm for Central Laboratory Animal Facility, Medical University Hannover.

Coisogenic strain
Inbred strains are coisogenic to each other if they differ in only one allelic character. This difference can be only caused by a mutation and subsequent fixation of the mutated allele.

Designation of a coisogenic strain
Symbol of the background strain followed by the differentiating allele in italic letters, separated by a hyphen, e.g. C.B-Igh-1^b/IcrTac-$Prkdc^{scid}$ (formerly C.B-17-$Prkdc^{scid}$).
This strain is an example of a mutation within a congenic strain.
- symbol: $Prkdc^{scid}$
- symbol name: severe combined immune deficiency (the mutation scid)
- gene name: protein kinase, DNA-activated catalytic polypeptide (Prkdc)
- symbol description: mutation in the gene encoding the catalytic subunit of DNA-activated protein kinase, Prkd. Arose in the C.B-17 congenic strain.

Congenic strain
A congenic strain is developed by transfer of a chromosomal segment (consisting of the differentiating locus and flanking genes) from a donor strain to another strain (inbred background or recipient strain). The genetic background of the initial crosses have to be purified by at least 10 backcrosses with the recipient strain. Afterwards the original inbred strain and the congenic strain (recipient strain) should only differ in this introduced segment. A congenic strain should be mated b×s after having finished the backcross process (see Fig. 1).
The designation of a congenic strain contains the symbols or abbreviations of the recipient and donor strain, separated by a dot. The transferred gene segment is added in italic letters separated by a hyphen.
Thus, for C.B-Igh-1^b/IcrTac (formerly.C.B-17)
- donor strain: C57BL/Ka strain (B)
- background strain: BALB/c, (C)
- symbol name: immunglobulin H-1 ^{b}
- gene name: heavy chain allele (Igh-1^b)
- [C.B-17 = BALB/c.C57BL/Ka-Igh-1^b] (number 17 originated from the backcross number 17).

For B10.129P-$H12^b$/(6M)SnJ
- donor strain: 129P3/J
- background strain: C57BL/10n
- symbol name: histocompatibility 12b
- gene name: histocompatibility 12.

F1-hybrid

The cross of two inbred strains leads to an F1 hybrid. All F1 animals of the same parental strains are genetically identical. They are heterozygous at all loci having different alleles in the parental strains. F1 hybrids share the advantages of inbred strains. In addition they are robust against environmental influences resulting in low quantitative character variability. An important limitation is the discontinuation of breeding. The designation of a F1-hybrid consists at first of the symbol of the female parent followed by that of the male parent. Short or full symbols may be used. For example, B6C3F1 is the result of a female C57BL/6 mouse crossed to a C3H male mouse.

Characteristics

Advantages of inbred strains

Inbred animals of the same strain can be regarded as identical twins and can be reproduced unlimited.

Long-term genetic stability

Inbred strains can be assumed to be genetically constant for a long period of time, supposed they are correctly bred and genetically monitored. Under these conditions background data on strain characteristics may be comparable for many years.

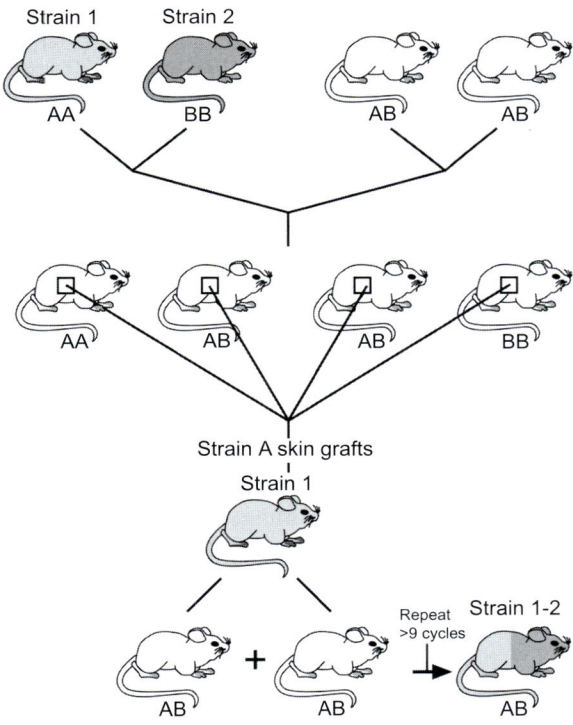

Rodents, Inbred Strains. Figure 1 Congenic strains. From: Cruse JM, Lewis RE (2004) Atlas of Immunology. CRC Press, Boca Raton.

Isogenicity

All individuals of a strain are isogenic. Thus skin grafts and tumors may be transplanted within an inbred strain without immunological rejection. Genetic data can be accumulated within the same strains. Of special interest for immunotoxicologists are data on major histocompatibility complex (MHC) (Fig. 2).

Homozygosity

Inbreds are homozygous at virtually all loci, and thus will breed true within the strain.

International distribution

Most of the used inbred strains have an international distribution, thus the genetic results of research can be easily compared.

Identification/monitoring

Inbreds can be identified by their strain-specific genetic profiles consisting of DNA polymorphisms, immunologic markers (e.g. MHC haplotypes in the mouse; RT1 haplotypes in the rat) and biochemical markers. The regular monitoring of genetic profiles can reduce the risk of unnoticed genetical contamination. The strain monitoring has to be performed of course by competent laboratories.

Uniformity

Genetic variation is reduced to nearly zero within strains by constant inbreeding. The use of inbred strains enable a much better standardization of the test conditions. Application of inbreds can improve the quality, the repeatability and comparability of results. The extent of analysis may be reduced or kept at a minimum by selecting an appropriate genetic model. Such a principle is in line with animal welfare legislation calling for reduction and refinement of animal experiments.

Individuality

Each inbred strain represents an unique genotype. This genotype can lead to a phenotype of biomedical interest (for example, inbreds with high or low tumor incidence, or high or low disease resistance).

Limitations of inbred strains

Individuality

Each inbred strain represents only one genotype. To extrapolate experimental results different inbred strains have to be considered, ideally including also their F1 hybrids.

Isogenicity and uniformity

The variability of quantitative characters within a single strain is exclusively due to non-genetic factors, like environmental or methodological factors.

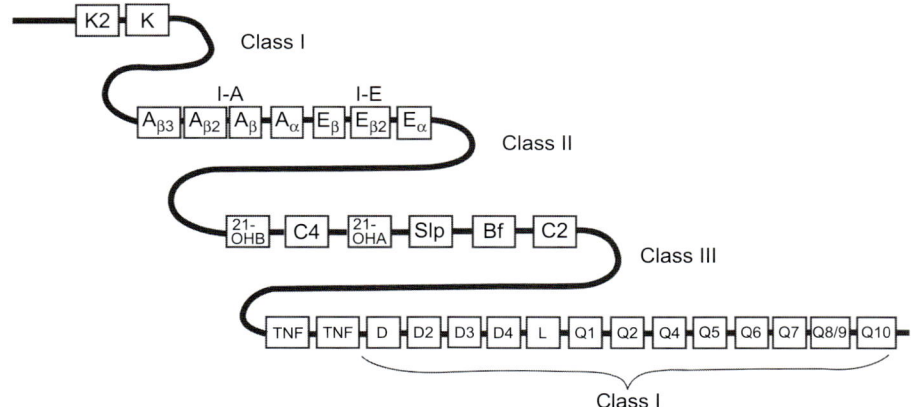

Rodents, Inbred Strains. Figure 2 H-2 histocompatibility system is the major histocompatibility complex in the mouse. From: Cruse JM, Lewis RE (2004) Atlas of Immunology. CRC Press, Boca Raton.

Examples of commonly used inbred strains in the field of immunotoxicology

Brown Norway rat (BN) (RT-1 haplotype n)

The BN rat is characterized by high basal level of serum IgE which can be induced massively by immunization with proteins. The inducibility of the IgE response is currently under evaluation as an indicator for allergic asthma and for chemically induced autoimmunity. As a reason for this high responsiveness the reduced antioxidant levels in the BAL cells is considered, resulting in reduced ability to adapt to ▶ oxidative stress induced by allergen induction.

In 2002 Vohr et al reported on a model that uses the reactions of regional lymph node as indicators for the induction of respiratory allergy.

The BN rat was also found to be a suitable model in a modified local lymph node assay (LLNA), self-limiting increase of IgE, reduced endogenous antioxidant level in BALF, and food allergy test.

B6C3F1 and C57BL/6 (h-2 haplotype b/k and b)

Historically both strains have been used extensively in the National Toxicology Program to set up an immunotoxicologic database which consists of a vast variety of tests of over 50 compounds (for a review article see Luster et al 1994). Both strains gave acceptable results in comparison study of mouse strains in a LLNA.

BALB/c (h-2 haplotype d)

This strain was used in the first LLNA. Its feasibility has been confirmed recently by Woolhiser et al in 2000, and Hüsler et al in 2003. BALB/c mice are IgE high responders and they are occasionally used for investigations of the regulation and induction of this antibody subclass. Because of this they have also been the object of intensive investigations for establishing short-term models for the differentiation between respiratory and (skin) contact allergy.

CBA/Ca and CBA/J (h-2 haplotype k)

The strain CBA/Ca was recommended by Kimber and Weisenberger in 1989 for the murine LLNA as a result of comparison of four murine strains. Young adult male or female mice of CBA/Ca (or CBA/J strain in the USA) are also recommended for the LLNA in the OECD guideline 429, 2002. The Fischer rat (F344-*RT-1 haplotype lv1*), Lewis rat (LEW-*RT-1 haplotype l*), and A/J-mouse *(h-2 haplotype a)* are also widely used in different immunotoxicity studies.

Preclinical Relevance

Inbred strains have made an essential contribution to biomedical research. Much progress in research, especially in the field of immunology, has followed from the development of inbred strains. Use of inbred strains enables much better standardization of the conditions. Inbreds help to improve the quality and reproduction of the results obtainable. A comparability of the results is given. Selecting an appropriate genetic model reduces the analysis needed, and may even keep it to a minimum. These points are in line with the animal welfare legislation calling for reduction and refinement of animal experiments.

On the other hand, the increased variability of immune responses of non-inbred strains (so-called outbred stocks) does reflect the human situation in a more realistic way. Results obtained by the use of outbred animals may thus have a better impact on risk assessment. Therefore, there is still debate about which of the strains mentioned are to be used for immunotoxicologic investigations. The various inbred rat and mouse strains show differences in their immunological reactions. Many differences occur especially in the field of sensitivity (from high-responders to non-responders). Other 'normal' immunoreactions can also differ, like the plaque-forming cell assay (PFCA) against sheep erythrocytes (SRBC). Undesired hyperreactions or autoimmune reactions may also depend on the genetic background, as well as host-resistance analysis. So it is essential that a rat or mouse strain must be checked for its immunological reactivity before using it for immunotoxicologic examinations. In addition, positive controls have to be established.

Relevance to Humans

Inbred strains serve as models of human diseases in various disciplines like immunobiology, transplantation medicine, autoimmunity, and oncology. Inbred mice and rats of one strain can be seen as identical twins.

The rat can also be used as an experimental model for nutrition research, with a reliable correlation of approximately 0.98 between rat and human in the digestibility of nutrients.

Research applications of the rat seems to be dominated by assigning function to the complete genomic sequence, particularly with respect to those regions involved in common human diseases. All in all, the rat offers the best 'functionally' characterized mammalian model system.

Regulatory Environment

With the exception of the above-mentioned OECD guideline 429 there are no recommendations for the use of special inbred or outbred strains in other immunotoxicity guidelines.

References

1. Jax Mice Catalogue (2001) http://www.informatics.jax.org/menus/strain_menu.shtml
2. Baker HJ, Lindsey JR, Weisbroth SH (1979) The laboratory rat, Vol. 1: Biology and diseases. Academic Press, New York
3. Festing M (1993) International index of laboratory animals, 6th ed. The British Library, London
4. Foster HL, Small JD, Fox JG (1981) The mouse in biomedical research, Volume I: History, genetics, and wild mice. Academic Press, New York
5. Hedrich HH (1990) Genetic monitoring of inbred strains of rats. Gustav Fischer Verlag Stuttgart, New York
6. Heinecke H (1989) Angewandte Versuchtierkunde. VEB Gustav Fischer Verlag, Jena
7. Klein J (1986) Natural history of the major histocompatibility comple. John Wiley and Sons, New York
8. Krinke GJ (2000) The laboratory rat. Academic Press, New York

Rosetting Techniques

These techniques are used to identify or isolate particles or cells bound by indicator cells or erythrocytes. For example, mixing sheep erythrocytes with human blood cells results in rosetting of human T cells surrounding the sheep erythrocytes. This erythrocyte (E) rosette was the first technique in separating T from B cells. The receptor responsible for this binding is the CD2 molecule on T cells. The rosettes are named after the central particle, i.e. E-rosette (see above), erythrocyte antibody (EA) rosette, or erythrocyte antibody complement (EAC) rosette.

▶ Rosetting Techniques

RT1.B, RT1.D (Rat)

▶ Antigen Presentation via MHC Class II Molecules

RTqPCR

▶ Polymerase Chain Reaction (PCR)

S

Safety Assessments

▶ DNA Vaccines

SAg

▶ Superantigens

Salmonella, Assessment of Infection Risk

WIM H DE JONG · ROB DE JONGE ·
JOHAN GARSSEN · KATSUHISA TAKUMI · ARIE
H HAVELAAR
Laboratory for Toxicology, Pathology and Genetics
National Institute for Public Health and the
Environment (RIVM)
PO Box 1
3720 BA Bilthoven
The Netherlands

Synonyms

Salmonella enterica, Salmonella enterica serovar Enteritidis, Salmonella Enteritidis, salmonellosis, Salmonella food poisoning, Salmonella food-borne disease, gastroenteritis

Definition

Salmonella spp. can cause severe enteric infections. Especially the typhoid (*S. typhi*) and paratyphoid bacteria result in septic conditions with typhoid-like fever and nausea, vomiting, abdominal cramps, and diarrhea. Nontyphoid salmonellae generally induce less severe symptoms restricted to the gastrointestinal tract—designated ▶ salmonellosis. Infection usually occurs by food poisoning of which raw meats and poultry, or poultry products (eggs), are common infection sources.

Characteristics
General Aspects

Contamination of food and water by pathogenic microorganisms such as *Salmonella* spp. is a risk for the population, especially for specific subgroups with a much higher risk profile. Microbiological risk assessment is an emerging tool for the evaluation of the microbiological safety of food and water. In this process, hazardous microorganisms are identified and exposure of the consumer to these organisms is estimated by a combination of observational data and mathematical modeling. The health risk arising from exposure is then estimated by use of a dose-response model, that gives a quantitative description for the relationship between the exposure to a certain number of pathogens (the dose) and the probability of an effect, such as infection or disease (1).

Experimental salmonellosis in rodents has been extensively studied and detailed information is available on the host-pathogen interaction from in vivo as well as in vitro experiments. Whereas in humans nontyphoid salmonellae induce self-limiting gastroenteritis with only occasional bacteremia, systemic disease develops in rats and mice. After oral exposure of rats to salmonellae the intestine is rapidly colonized, and salmonellae can be detected in the small intestine and cecum within 2 h. Intestinal colonization is concentrated in the distal ileum and cecum, and may be detected by fecal excretion. Invasion is quickly established via the M cells in Peyer's patches, causing salmonellae to be detectable in mesenteric lymph nodes within 8 h (2,3). Systemic infection results in high numbers of pathogens present in spleen and liver, and is associated with an increase in organ weights.

A major line of defence against invasion of salmonellae is uptake by macrophages. Later, these macrophages play a crucial role in the induction of T cell-dependent immunity that can be detected by the development of a delayed-type hypersensitivity reaction (3–5). However, virulent salmonellae have adapted to survive and can grow inside macrophages and may induce cell death by apoptosis (6). Polymorphonuclear leucocytes (mainly neutrophils) play a key role in the clearance of *Salmonella* infections (7). Despite the available insight in the pathogenesis and host-response, the published literature is of limited use for dose-response modeling of food-borne and waterborne pathogens. Usually only high doses were administered, often inoculated via the intraperitoneal or intravenous route, instead of orally. Most papers only considered a few endpoints to detect infection, and it is not known whether colonization of the intestine and invasion occur in a dose-dependent manner.

Animal Model

Young adult Wistar Unilever male rats age 6–8 weeks were infected by gavage with various doses of *Salmonella enterica* serovar Enteritidis 97–198, a patient's isolate origin RIVM, or for control with a nonpathogenic nalidixic acid-resistant *Escherichia coli* WG5 (3). In order to promote the survival of *Salmonella* so that they could reach the target sites for colonization, the standard model included overnight fasting of the animals and neutralization of gastric acid by suspending the inoculum in a sodium bicarbonate solution. This method also allows the investigation of the colonization potency of low doses of *Salmonella*. The dose-response for bacterial counts in feces is shown in Figure 1, and colonization of the spleen is presented in Table 1

Although in general no significant indications for overt clinical disease were noticed, bacterial counts in mesenteric lymph nodes and spleen indicated systemic infection. This was reflected by the time-related increase in the number of neutrophilic granulocytes (Figure 2). Indicative for the induction of a systemic immune response was the delayed-type hypersensitivity reaction after challenge of the animals with heat killed *Salmonella* antigens (Figure 3).

Mathematical Modeling

Dose-infection models as recommended by WHO and FAO (8) are based on several assumptions:
- the single hit hypothesis: each inoculated organism has a (possibly very small but not zero) probability to cause infection; each surviving organism grows to produce a clone of cells
- the hypothesis of independent action: the mean probability per inoculated pathogen to cause a

Salmonella, Assessment of Infection Risk. Table 1 Colonization of the spleen after oral administration of various doses of *Salmonella* Enteritidis

Dose (cfu/animal)	Animals exposed	Animals infected	% Infected
10	4	0	0
150	4	1	25
300	5	2	40
910	4	2	50
18 000 +	4	4	100

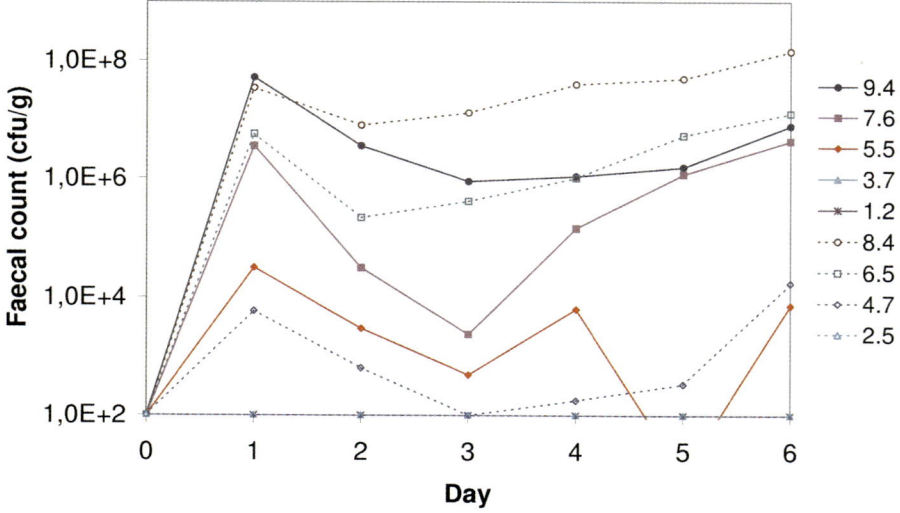

Salmonella, Assessment of Infection Risk. Figure 1

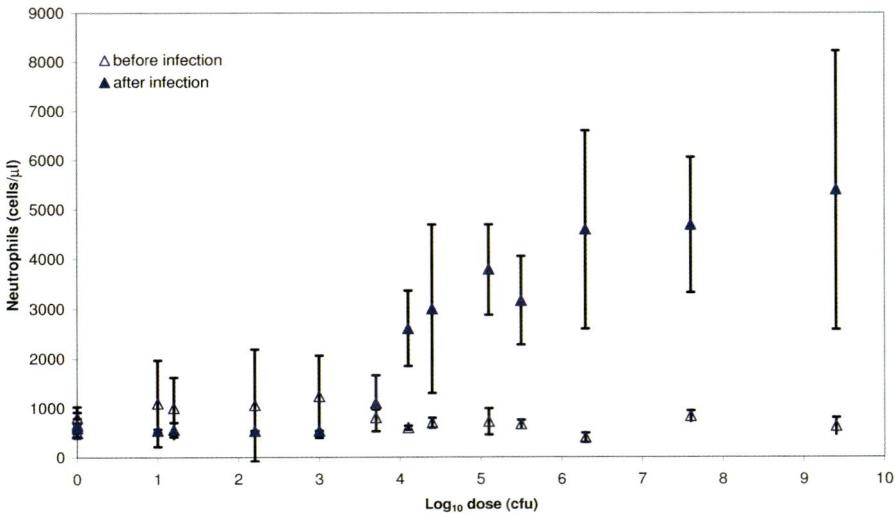

Salmonella, Assessment of Infection Risk. Figure 2

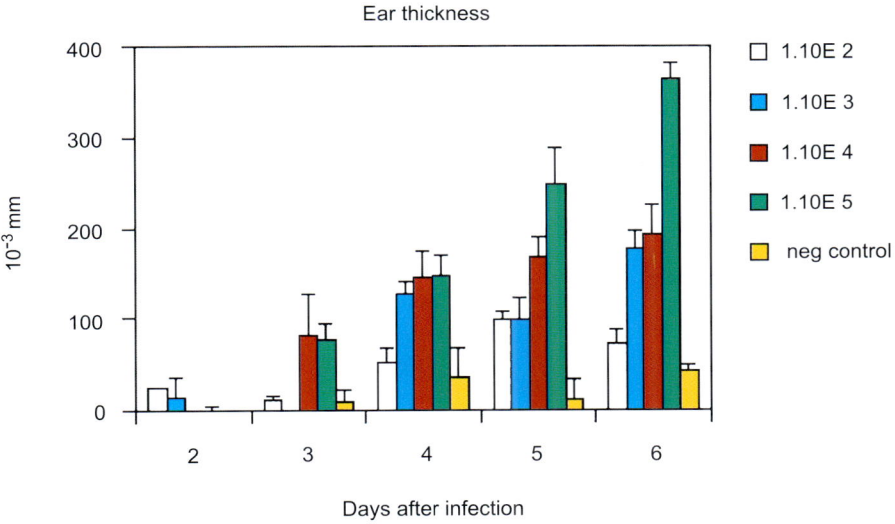

Salmonella, Assessment of Infection Risk. Figure 3

(symptomatic or fatal) infection is independent of their number
- microorganisms behave as discrete particles and cannot be divided in units smaller than one
- microorganisms are randomly distributed in the inoculum; this assumption is made for mathematical convenience but is not necessary; models for non-random distributions have been described (9).

These assumptions led to the family of single-hit models that are routinely used to describe microbial dose-response relationships. The best known models are the exponential model, the Beta-Poisson model and the hypergeometric model (10). Using this model, dose-response parameters can be estimated. *Salmonella* infection occurred at doses as low as 500 colony-forming units (cfu) and occasionally even lower. Marked histopathological alterations were mainly observed in the large intestine at doses above 10E6 cfu.

When prepared under laboratory conditions, repeatedly using the same kind of *Salmonella* preparations originating from the same deep-frozen stock, it appears reasonable to assume that the probability of any single organism to establish itself in the animal and cause infection is constant. Indeed, data for infection of the spleen could be fitted with the exponential model and indicated that on average, 1 per 860 cells in the inoculum infected the spleen (Figure 4). In contrast, the reproducibility of detecting infection by microbiological examination of mesenteric lymph nodes was poor, both with respect to the fraction of positive animals and the mean counts in positive animals.

Immune Responses to Salmonella

Rats are able to mount a vigorous cellular immune response in relation to *Salmonella* infection. There is a consistent dose-response relationship for most leucocyte subsets (3). The only exception was a lack of response by eosinophils, which was not unexpected because these cells are mainly related to parasitic infections. The strong increase in monocytes on day 5 can be interpreted as a consequence of the innate, nonspecific immune response in which tissue macrophages are recruited from a pool of blood monocytes to engulf the invading salmonellae. Initial depletion of the monocyte pool in blood by migration into tissues is overcompensated by increased production in the bone marrow (11). Survival and growth in macrophages was found to be a major virulence mechanism for *Salmonella* (7). Macrophage death by apoptosis is a host defence mechanism, which leaves the bacteria susceptible to subsequent phagocytosis and killing by neutrophils, of which the number increased significantly in the blood of infected animals. Later, the nonspecific immune response is succeeded by a specific response, which is manifest by a strong increase in lymphocytes on day 11.

Th1-Th2 Balance Influences Immune Response to Salmonella

T helper 1-T helper 2 balance can be skewed. Dose-response studies in two different rat strains skewed towards either a T helper-type 1 (Th1) directed immune response (Lewis rat) or a Th2-type directed immune response (Brown Norway rat) showed that the probability of infection per single *Salmonella* Enteritidis cell was approximately 100 times higher in Brown Norway rats than in Lewis rats (12). The probability of infection per *Salmonella* Enteritidis cell was 1:25 000 for Lewis rats, 1:800 for Wistar Unilever rats, and 1:200 for Brown Norway rats (Figure 5). So, the Th2 animals (Brown Norway rats) were more prone to infection than Th1 responding animals (Lewis rats); the Brown Norway rats were even more sensitive to infection compared to the originally used Wistar Unilever rats (12). The Wistar Unilever rat showed a kind of intermediate response between the more Th1 and Th2 type of responses shown by the Lewis and Brown Norway rats, respectively. It is plausible to attribute this difference in sensitivity to the fact that Lewis rats are Th1 prone and Brown Norway rats are Th2 prone. However, it should be taken into account that host factors other than the Th1-Th2 balance might be involved also in the sensitivity for *Salmonella* Enteritidis infection.

The delayed-type hypersensitivity (DTH) reaction was also found to be a sensitive and reproducible parameter to detect an immune response to *Salmonella* infection. Positive reactions were observed at challenge doses above 100 cfu, which is equivalent to doses that lead to colonization of the spleen. Not surprisingly, Lewis rats showed a more vigorous DTH re-

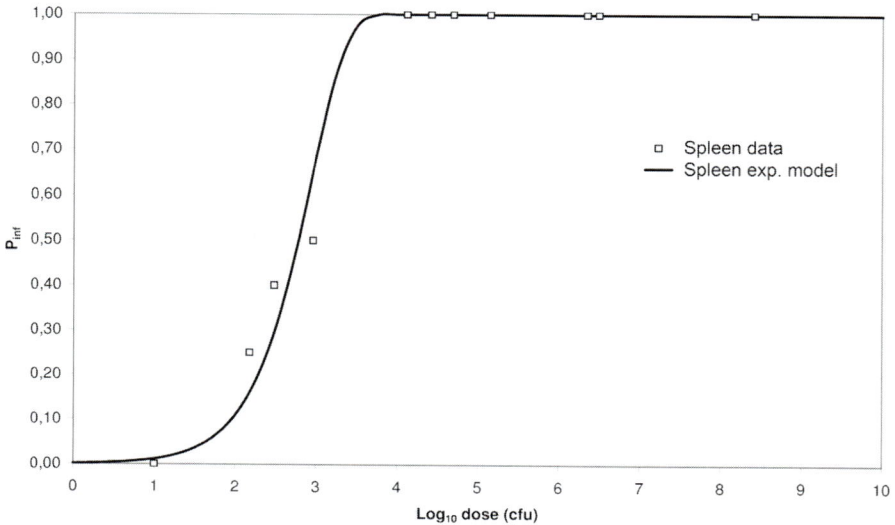

Salmonella, Assessment of Infection Risk. Figure 4

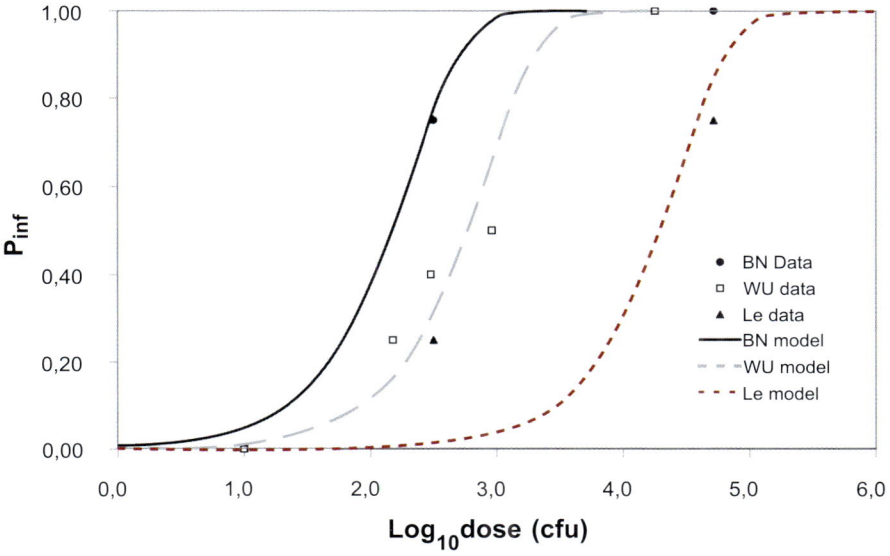

Salmonella, Assessment of Infection Risk. Figure 5

sponse than the Brown Norway rats (12). Cellular immune responses were more pronounced in Lewis rats but antibody responses were higher in the Brown Norway rats, thus indicating the influence of the Th1-Th2 balance on infection with *Salmonella*.

In both rat strains the neutrophilic response in the blood remained a very sensitive read out system for infection. Although initial infection levels are more prone to occur in the Th2 background (Brown Norway rats), when infected in animals prone to a Th1-type response (Lewis), systemic infection was more intense (12).

Preclinical relevance

The dose-response relationship of exposure by intragastric gavage of adult, male WU rats to ▸ *Salmonella Enteritidis* is well reproducible. In all experiments, the animals were shedding the bacteria in their feces after exposure to intermediate to high doses ($> 10^4$ cfu). At lower doses, no fecal excretion was detected within the 6-day period of observation but salmonellae could be isolated from the spleen and mesenteric lymph nodes. These findings are in accordance with the fact that *Salmonella* Enteritidis is highly invasive in rodents and that the intestinal tract may not be the major site of multiplication. Pathological results partly confirm these observations, indicating that the gastrointestinal tract (although portal of entry) shows relatively little abnormalities in animals that succumb to severe systemic illness after oral inoculation with very high doses of *Salmonella* Enteritidis. However, at lower, nonlethal doses lesions typical for gastroenteritis were observed in the ileum, cecum and proximal colon (3). No evidence of clinical illness was associated with these histological abnormalities.

Relevance to Humans

Dose-response models can be based on observational or experimental data. Observational data (usually from food-borne or water-borne outbreaks) have the advantage that they are based on actual situations, but generally provide limited information as the dose may be unknown (rare undetectable contamination, source unknown, uncontrolled storage conditions, sample unavailable) and the size of the exposed population is often not known. Experimental data, either from human volunteers or animal studies, have the advantage that they are obtained under well-controlled conditions and can therefore be subjected to mathematical analysis.

Limitations for experiments in human volunteers are that they are performed in healthy volunteers, and that the dose range and pathogenicity of the microorganism investigated are restricted to an infection resulting in a mild, self-limiting disease only. In animal models the dose-response relationship for infection can be assessed at a much broader range. Such models should enable risk assessors to evaluate the effect of single factors related to the host, pathogen and food matrix, and to make inferences about dose-response relations in humans.

Typical questions are: what is the effect of factors such as age, immunological status, nonspecific barriers (e.g. gastric acid, innate immunity) on the susceptibility of the host, and factors such as bacterial adaptation or protection by fatty foods on the infectivity of the pathogen.

As a next step kinetic models of the infection process can be developed, that describe the dynamics of the host-pathogen interaction in the alimentary tract, for which the animal models should provide insight in important mechanisms and should provide parameter estimates (13).

References

1. WHO (1999) Risk Assessment of Microbiological Hazards in Foods. Geneva, World Health Organization
2. Naughton PJ, Grant G, Spencer RJ, Bardocz S, Pusztai A (1996) A rat model of infection by *Salmonella typhimurium* or *Salm.* Enteritidis. J Appl Bacteriol 81:651–656
3. Havelaar AH, Garssen J, Takumi K et al. (2001) A rat model for dose response relationships of *Salmonella* Enteritidis infection. J Appl Microbiol 91:442–452
4. Collins FM, Mackaness GB (1968) Delayed hypersensitivity and Arthus reactivity in relation to host resistance in *Salmonella*-infected mice. J Immunol 101:830–845
5. Takumi K, Garssen J, Havelaar A (2002) A quantitative model for neutrophil response and delayed-type hypersensitivity reaction in rats orally inoculated with various doses of *Salmonella* Enteritidis. Int Immunol 14:111–119
6. Jones BD, Falkow S (1996) Salmonellosis: host immune responses and bacterial virulence determinants. Ann Rev Immunol 14:533–561
7. Vassiloyanakopoulos AP, Okamoto S, Fierer J (1998) The crucial role of polymorphonuclear leukocytes in resistance to *Salmonelladublin* infections in genetically susceptible and resistant mice. Proc Natl Acad Sci USA 95:7676–7681
8. WHO (still in press?) Guidelines for Hazard Characterization of Pathogens in Water and Food. Geneva, World Health Organization
9. Haas CN (2002) Conditional dose-response relationships for microorganisms: development and application. Risk Analysis 22:455–463
10. Teunis PFM, Havelaar AH (2000) The Beta Poisson dose-response model is not a single hit model. Risk Analysis 20:513–520
11. Volkman A, Collins FM (1974) The cytokinetics of monocytosis in acute *Salmonella* infection in the rat. J Exper Med 139:264–277
12. Havelaar AH, Garssen J, Takumi K et al. (2003) Intraspecies variability in the dose-response relationship for *Salmonella* Enteritidis, associated with genetic differences in cellular immune response. Submitted for publication
13. Takumi K, De Jonge R, Havelaar A (2000) Modeling inactivation of *Escherichia coli* by low pH: application to passage through the stomach of young and elderly people. J Appl Microbiol 89:935–943

Salmonella enterica

▶ Salmonella, Assessment of Infection Risk

Salmonella enterica serovar Enteritidis

▶ Salmonella, Assessment of Infection Risk

Salmonella Enteritidis

▶ Salmonella, Assessment of Infection Risk

Salmonella Enteritidis Rat Model

Animal model for estimation of dose response relationship after *Salmonella* infection. Parameters for infection include spleen counts, neutrophilic granulocyte response, and delayed-type hypersensitivity reaction. Low dosages already can induce systemic infection as indicated by spleen colonization after oral infection.

▶ Salmonella, Assessment of Infection Risk

Salmonella Food-Borne Disease

▶ Salmonella, Assessment of Infection Risk

Salmonella Food Poisoning

▶ Salmonella, Assessment of Infection Risk

Salmonella typhimurium Mitogen

This is a substance which stimulates non-specifically the proliferation of rat B cells, but does not cause them to differentiate. It is used in immunology to investigate the function of rat B cells.

▶ Lymphocyte Proliferation

Salmonellosis

Gastroenteric infection due to *Salmonella* spp., usually caused by consumption of inadequately heated contaminated food such as (raw) meats, poultry or poultry products. Symptoms after infection can include fever, severe nausea, vomiting, abdominal cramps and diarrhea.

▶ Salmonella, Assessment of Infection Risk

SAPS

▶ Cytokine Polymorphisms and Immunotoxicology

SARS

▶ Respiratory Infections

Saturated Fatty Acids

These are fatty acids which contain no double bonds.
▶ Fatty Acids and the Immune System

SCID Mouse

Severe combined immunodeficient mouse.
▶ Animal Models of Immunodeficiency

Scleroderma

Hardening of the skin. Systemic scleroderma has hardening of skin as well as involvement of other organs, especially the lungs, esophagus, kidneys and heart.
▶ Systemic Autoimmunity

SDS-PAGE

Sodium dodecylsulfate (SDS)–polyacrylamide gel electrophoresis (PAGE) separates proteins in an electric field based on their molecular size.
▶ Western Blot Analysis

SEB

▶ Polyclonal Activators

Secondary Antibody Response

The immune response that is induced following a second exposure to antigen and mediated by memory cells and largely by IgG antibody, which allows for a more rapid and stronger response than the primary response.
▶ B Lymphocytes
▶ Assays for Antibody Production

Secondary Cytokines

In immune responses like an inflammation master cytokines (e.g. interleukin-1, tumor necrosis factor α) are rapidly released which organize the immune reaction by stimulating tissue cells and leukocytes to produce further mediators of the immune system like Interleukin-6 or chemokines. These cytokine-induced cytokines are called secondary cytokines.
▶ Cytokine Receptors

Secondary Humoral Immune Response

▶ Assays for Antibody Production

Secondary Immune Response

▶ Memory, Immunological

Secondary Lymphoid Organs

Organs and tissues in which mature immunocompetent lymphocytes encounter trapped antigens and are activated into effector cells. In mammals, the lymph nodes, spleen, and mucosal-associated lymphoid tissue (MALT) like Peyer's plaques constitute the secondary lymphoid organs.
▶ Flow Cytometry
▶ Antigen Presentation via MHC Class II Molecules

Secondary Neoplasms

Secondary neoplasms are cancers that arise in an individual as a result of previous chemotherapy and/or

radiation therapy. These new cancers may surface months or years after the initial treatment for a variety of cancers.
▶ Leukemia
▶ Lymphoma

Secondary PLNA

▶ Popliteal Lymph Node Assay, Secondary Reaction

Secondary Prevention

Strategies employed to prevent reoccurrence of a life-threatening medical event.
▶ Fatty Acids and the Immune System

Secretory Immune System

▶ Mucosa-Associated Lymphoid Tissue

Secretory Immunoglobulin A

Secretory immunoglobulin A (sIgA) is a complex of two IgA molecules joined by a additional J-chain and a secretory component (Sc). Sc is made by epithelial cells of intestines and lungs as part of an IgA receptor and is involved in the transportation of IgA across the epithelial cells to mucosal surfaces where it can react with pathogens.
▶ Immunotoxic Agents into the Body, Entry of

Selection

▶ Cell Separation Techniques

Self Antigen

Many structures of a healthy organism could be recognized by the adaptive immune system leading to an autoimmune response. A number of mechanisms retain tolerance of the immune system against these structures.
▶ Antigen Presentation via MHC Class II Molecules
▶ Autoantigens

Self-Renewal

The ability of a hematopoietic stem cell to produce at least one other stem cell after cell division, and thus to maintain the stem cell status.
▶ Bone Marrow and Hematopoiesis
▶ Hematopoietic Stem Cells

Semiquantitative PCR

▶ Polymerase Chain Reaction (PCR)

Sensitization

Induction of IgE antibodies by an allergen. In general also used for the induction of allergic reactions.
▶ Food Allergy
▶ Hapten and Carrier

Sensitizer

A chemical that causes a substantial proportion of exposed people or animals to develop an allergic reaction in normal tissue after repeated exposure to the chemical.
▶ Three-Dimensional Human Skin/Epidermal Models and Organotypic Human and Murine Skin Explant Systems

Septic Shock

JUTTA LIEBAU
Fachklinik Hornheide
Dorbaumstrasse 300
D-48157 Münster
Germany

Synonyms
bacteremic shock, endotoxin shock, septicemic shock

Definition
Septic shock is a clinical syndrome of acute circulatory failure resulting from acute invasion of the bloodstream by microorganisms or their toxic products (1). In adults, it is characterized by persistent arterial hypotension unexplained by other causes. Hypotension is

defined by a systolic arterial pressure less than 90 mmHg, mean arterial pressure less than 60 mmHg or a reduction in systolic pressure of more than 40 mmHg. In children, septic shock is defined as ▶ tachycardia with signs of decreased organ perfusion including decreased peripheral pulses compared to central pulses, altered alertness, flash capillary refill or capillary refill longer than 2 seconds, mottled or cool extremities, or decreased urine output (2).

Terms with close relation to "septic shock" are sepsis, severe sepsis, and SIRS (systemic inflammatory response syndrome). SIRS is present if patients have more than one of the following findings: body temperature of more than 38.0° C or less than 36.0° C, heart rate of more than 90/min, respiratory rate of more than 20/min or PaCO2 lower than 32 mmHg, white blood cell count of more than 12 000 cells/μl or less than 4000 cells/μl. In this concept "sepsis" is defined as SIRS plus infection, "severe sepsis" as sepsis with organ dysfunction, hypoperfusion, or hypotension, and "septic shock" as sepsis with arterial hypotension despite adequate fluid resuscitation (2,3).

In septic shock, there are signs of inadequate organ perfusion. Common symptoms are fever, chills, tachycardia, ▶ tachypnea, and altered mental state. Circulatory insufficiency with low systemic vascular resistance and decreased myocardial function causes diffuse cell and tissue injury and organ failure (1).

Different scores related to sepsis have been used. PIRO is a new classification scheme for sepsis (where the acronym stands for: predisposition, insult (infection), response, organ dysfunction) that has recently been proposed, which is now being tested (2).

Characteristics

Septic shock occurs if pathogens and/or their toxins from a septic focus enter the blood and the tissues. Microbial factors important in septic shock include, among others, polysaccharides (LPS) from Gram-negative bacteria, enzymes, and exotoxins of Gram-positive bacteria. A variety of mediators are active in the pathogenesis of septic shock, among them active metabolites of the complement system (with impaired function of polymorphonuclear cells (PMNs), the coagulation system and factors released from stimulated cells like tumor necrosis factor (TNF-α) and interleukins IL-1, IL-6, and others. These important mediators are produced by LPS-stimulated monocytes and macrophages. There are proinflammatory cytokines (IL-1, TNF-α and IL-6) and antiinflammatory cytokines (IL-4, IL-10) which modulate the proinflammatory immune response. Cytokines induce secondary mediators which cause damage to cell function and structure. Activated tissue macrophages and monocytes are the cellular promotors of inflammation (1,4,5).

Septic shock is caused by:
- Gram-negative bacteria (60%–70%) such as *Escherichia coli*, *Klebsiella*, *Enterobacter*, *Proteus*, *Pseudomonas*, *Serratia*)
- Gram-positive bacteria (20%–40%) including staphylococci, pneumococci, and streptococci
- opportunistic fungi (2%–3%)
- (rarely) other agents like *Mycobacteria*, protozoans (*Plasmodium falciparum*) or viruses such as dengue fever.

Gram-negative bacteremia is associated with shock in 40% of patients (1).

Patients with septic shock suffer of hemodynamic instability with decreased systemic vascular resistance caused predominantly by high levels of nitric oxide. Initially an increased cardiac output is seen associated with ▶ hypovolemia and low blood pressure. About 50% of the patients suffer of impaired myocardial function (septic cardiomyopathy) in the course of the disease. About the same amount of all patients experience some form of end-organ damage caused by altered distribution of cardiac output, impaired microcirculation, and capillary leak syndrome associated with deterioration of the complement system. Complications include renal and liver failure, adult respiratory distress syndrome (ARDS) with respiratory failure, and disseminated intravascular coagulation (1).

MODS (multiple organ dysfunction syndrome) means failure or dysfunction of two or more organs and is responsible for the high mortality of sepsis.

Clinical manifestations of septic shock are fever of more than 38.3° C (some patients have hypothermia less than 36.0° C), chills, tachycardia, tachypnea of more than 30 breaths per minute, ▶ encephalopathy with mental state changes, hypotension, ▶ acrocyanosis, and gastrointestinal manifestations.

Laboratory data show leukocytosis of more than 12,000 cells/μl or leukopenia of fewer than 4000 cells/μl, elevated C reactive protein of more than 2 standard deviations over normal value, toxic anemia, low angiotensin III levels, pathologic urine analysis, and elevated lactate levels (1,2).

Hemodynamic parameters are arterial hypotension (see definition above), cardiac index of more than 3.5 L/min/m^2 and mixed venous saturation of more than 70%.

Organ dysfunction parameters are arterial hypoxemia, acute ▶ oliguria, hyperbilirubinemia, thrombocytopenia, ileus, creatinine increase of more than 0.5 mg/dL and coagulation abnormalities (2).

Blood cultures are positive in about 60% of patients. Cultures of wounds, urine, and tracheal secret may be positive (1,2).

Basic treatment must be started rapidly and consists of removal of the source of infection (e.g. wounds, catheters), support of respiration, hemodynamic support (e.g. monitored volume replacement, noradrenaline (norepinephrine)), parenteral nutrition, treatment of acidosis, and bactericide antibiotics. New treatment concepts include administration of hydrocortisone, activated protein C and intensive insulin therapy. Monitoring in an intensive care unit is necessary (4).

Preclinical Relevance
In industrialized countries most cases of septic shock are seen in hospitals. Measures to reduce the risk of septicemia in the population include proper vaccination (i.e. for pneumococci and meningococci), proper surgical wound treatment, and calculated antibiotic treatment of community-acquired infections like pneumonia and pyelonephritis (1).

Relevance to Humans
The overall mortality of septic shock is 50%–80%. The prognosis of septic shock is worse in patients with rapidly fatal diseases (4). Mortality of patients with Gram-positive bacteremia is higher than with Gram-negative bacteremia (5). Severe sepsis is the most common cause of death in noncoronary critical care units (more than 200 000 patients annually in the USA) (2).

About 50% of cases of septic shock are seen in already hospitalized patients (nosocomial infection) (5). Most patients with septic shock are elderly, chronically ill, or suffer of underlying diseases or procedures that make them susceptible to bloodstream invasion.

Patients at risk for septic shock include those with cancer (cytostatic therapy), malnutrition, chronic infection, renal failure, diabetes mellitus, immunosuppression, polytrauma, cardiac shock, burns, and organ perforation. Thus, for all invasive medical procedures, the general condition of the patient and his chronic diseases must be taken into account to allow a reasonable risk-benefit calculation. This concerns not only operations but also invasive diagnostic procedures, intravenous lines, and catheters (1).

In the last decades, mortality of septic shock has remained high in spite of multiple improvements in supportive therapy (4). Prevention of shock requires repeated physical examination, swabs, blood cultures, and intensive use of imaging techniques for the detection and treatment of the septic focus. Early and precise antibiotic therapy for endangered patients with infections is strongly encouraged (1,4,5).

Regulatory Environment
The national vaccination guidelines are:
- Control and Prevention of Meningococcal Disease: Recommendations of the Advisory Committee on Immunization Practices (ACIP) (2001). MMWR Recommendations and Reports
- The International Sepsis Forum (2001) Guidelines for the management of severe sepsis and septic shock. Intens Care Med 27[Suppl 1]:51–134

References
1. Braunwald E, Fauci A, Kasper DL, Hanser SL, Longo DL, Jameson JL (eds) (2001) Harrison's principles of internal medicine. McGraw Hill, New York
2. Levy MM, Fink MP, Marshall JC et al. (2003) International Sepsis Definitions Conference 2001 SCCM/ESICM/ACCP/ATS/SIS. Intens Care Med 29:530–538
3. American College of Chest Physicians/Society of Critical Care Medicine (ACCP/SCCM) Consensus Conference Committee (1992) Definitions for sepsis and organ failure and guidelines for the use of innovative therapies in sepsis. Crit Care Med 20:864–874
4. Eckart J, Forst H, Burchardi H (2002) Intensivmedizin. Ecomed, Landsberg
5. van Aken H, Reinhart K, Zimpfer M (2000) Intensivmedizin. Thieme, Stuttgart

Septicemia

A systemic disease associated with the presence and persistence of pathogenic microorganisms or their toxins in the blood.

▶ Streptococcus Infection and Immunity

Septicemic Shock

▶ Septic Shock

SEREX

Serological expression cloning (SEREX) of tumor antigens is a molecular cloning technique based on the screening of cDNA expression libraries from tumor specimens using sera of cancer patients. Cloning strategies usually aim at tumor antigens recognized by high-titer antibodies of the IgG class, thus requiring T cell recognition and help for immunoglobulin class switch. For this reason SEREX frequently picks up genes coding for antigens recognized both by antibodies and by T cells.

▶ Tumor, Immune Response to

Serine Protease Inhibitors

Circulating plasma proteins—including antithrombin or protein C inhibitor, which serve as pseudo-substrates for coagulation proteases such as thrombin, factor Xa or activated protein C—form covalent inactive complexes with the enzyme and thereby block its action. Heparin, or other glycosaminoglycans to which these inhibitors and the proteases can bind simultaneously, catalyse the enzyme inhibition several fold and thereby are effective anticoagulants.

▶ Blood Coagulation

Serotonin

HELEN V RATAJCZAK
Boehringer Ingelheim Pharmaceuticals
900 Ridgebury Road
Ridgefield, CT 06877
USA

Synonyms
serotonin, 5-hydroxytryptamine, 5-HT, enteramine

Definition
Serotonin (5-hydroxytryptamine, 5-HT) is classed as a ▶ hormone and has been found to be the most diverse physiological substance in the body. Serotonin is a ▶ neurotransmitter in the central nervous system, regulating functions such as sleep, mood, and appetite. Serotonin influences production of other hormones and interacts with the immune system.

Serotonin was first isolated from enterochromaffin cells originating from the gastric and intestinal mucosa by Ersparmer and Vialli in 1937. It was characterized by its ability to cause smooth muscle contraction and was named enteramine. About the same time Rapport isolated a substance from serum that caused vasoconstriction, and named it serotonin. After purification, structure elucidation and chemical synthesis, enteramine was found to be identical to serotonin (1).

Characteristics
The synthesis of serotonin takes place primarily in gastric and intestinal mucosa and in the ▶ pineal gland. It is synthesized from the essential amino acid ▶ tryptophan which is also a precursor of the vitamin ▶ niacin (nicotinamide) and of another hormone, ▶ melatonin. Under normal conditions biosynthesis accounts for only 2% of ingested tryptophan, leading to a daily production of about 10 mg serotonin. The major part of tryptophan is utilized for protein synthesis (1). Tryptophan and niacin are found in milk, beef, whole eggs, salt pork, wheat flour, corn, lean meats, poultry, fish, peanuts, organ meats, and brewer's yeast, with lower amounts in beans, peas, other legumes, most nuts, whole grains, and enriched cereals (2).

In addition to the ▶ gastric mucosa and the pineal gland, synthesis of serotonin occurs in the brain, spinal cord, bronchi, thyroid, pancreas and thymus. Circulating serotonin does not enter the brain by crossing the blood-brain barrier. Depots of serotonin in mammals are the enterochromaffin cells of the gastrointestinal tract (accounting for approximately 80% of total body serotonin), serotonergic ▶ neurons of the brain, the pineal gland, and ▶ platelets. Serotonin can be released from cells by stimulation with acetylcholine, noradrenergic nerve stimulation, increased intraluminal pressure and a decline of intestinal pH (1).

- Serotonin is a potent ▶ vasoactive amine. In the circulation it is almost entirely confined to platelets, thereby rendered functionally inactive. Clearance mechanisms have evolved to decrease plasma serotonin concentrations: platelets possess an active serotonin uptake system
- the liver catabolizes serotonin
- pulmonary endothelial cells take up serotonin
- specific macromolecules bind free serotonin.

Circulating plasma serotonin is taken up by platelets mainly by an active transport mechanism. Platelet serotonin content is elevated in people with serotonin-secreting carcinoid tumors and during long-term serotonin ingestion. Platelet serotonin half-life is about 4.2 days, which approximates to that of platelets. In platelets, serotonin is stored in dense granules. The platelet membrane contains two types of serotonin binding sites. One site mediates uptake, and the other causes platelet aggregation. In addition to the active uptake, a passive uptake process occurs at high extracellular serotonin concentrations and is proportional to serotonin levels (1).

There is interaction between the central nervous and the immune systems. Serotonin binding sites have been demonstrated on lymphocytes, eosinophils, and macrophages. Serotonin and its receptors are present in the immune system in three types of molecular structures:

- guanine nucleotide-binding protein-coupled receptors
- ligand-gated ion channels
- transporters.

Conversely, lymphokines and their respective receptors are present in the nervous and neuroendocrine systems. Another major pathway of interaction between the two systems is direct neural connections through the innervation of lymphoid organs. Serotonin is synthesized in epithelial cells, peptigergic neurons, and in leukocytes themselves in lymphoid organs, and is released in active form in the periphery (1).

Serotonin. Figure 1

Preclinical Relevance

The pharmacology of serotonin is particularly complex, with several receptor subtypes mediating responses at the different affinities identified. Brain serotonin influences the immune response via the hypothalamic-pituitary axis. An increase of brain serotonin has been shown to be immunosuppressive (3).
In peripheral blood, serotonin is localized primarily in platelets and is released at sites of injury. It has been shown to affect immune function both in vivo and in vitro.
In vivo studies include complex effects of serotonin on:
- lymphocyte subpopulation numbers
- peripheral leukocyte function
- suppression of immune response (decreased immunoglobulin IgM and IgG plaque-forming responses to sheep red blood cells in mice), and
- a permissive role in delayed-type hypersensitivity.

In vitro effects include:
- a 50% decrease in mitogen-induced lymphocyte proliferation and nearly complete inhibition of the production of interferon(IFN)-γ and other lymphokines
- suppression of IFN-γ-induced Ia expression by macrophages and phagocytes
- release of certain lymphokines and a polymorphonuclear cell chemotactic factor,
- up to 50% augmentation of natural killer (NK) cell cytotoxicity.

The serotonin receptor subtype 5-HT$_{1a}$ mediates the effect of serotonin on the activation of NK cells by monocytes and participates in the release of adrenocorticotropic hormone (ACTH) from the hypothalamus or pituitary. Serotonin has also been shown to affect potassium channels in a transformed lymphocyte cell line. Because the 5-HT$_{1a}$ receptor subtype is a target for newly developed anxiolytic medications, these drugs may have effects on immune function (4). An inverse relationship exists between brain serotonin levels and antibody production: reduction of brain serotonin levels stimulates antibody production (during the primary immune response), increases longevity, and delays onset of tumor growth. Interactions between tumor growth and the pineal are likely the consequence of nutrient and metabolic changes in tissues supporting the tumor cells, which in turn are due to an altered endocrine balance. The inhibitory effects of 5-hydroxytryptophan (5-HTP) can be reversed with exogenous luteinizing hormone, follicle-stimulating hormone, and ACTH, all of which are influenced by serotonergic pathways. In contrast, injection of 5-hydroxy tryptophane—the immediate precursor of serotonin—has been found to suppress the immune response, with increased latent period of antibody formation and decreased intensity of both the primary and secondary immune responses (2).

Relevance to Humans
Physiology

Serotonin is involved in a variety of physiological processes, including smooth muscle contraction, blood pressure regulation and both peripheral and central nervous system neurotransmission. In the central nervous system serotonin acts as a neurotransmitter-neuromodulator that is implicated in sleep pattern regulation, appetite control, sexual activity, aggression and drive. Central nervous system serotonin exerts its actions in concert with other neurotransmitters. In the periphery serotonin acts as a vasoconstrictor and proaggregator when released from aggregating platelets, as a neurotransmitter in the enteric plexuses of the gut and as an autocrine hormone when released from enterochromaffin cells from the gut, pancreas and elsewhere (1).

A basic knowledge of immune neuroendocrine interactions may be important to understanding certain disease mechanisms. Abnormalities of serotonin-related processes give rise to various pathological conditions; aberrations in central nervous system function are implicated in anorexia, anxiety, depression and schizophrenia, whereas degeneration of serotonergic neurons has been noted in Alzheimer disease and Parkinson disease, peripheral aberrations in drug-induced emesis, hypertension, migraine, genesis of cardiac arrhythmias, Raynaud's disease, fibrotic syndromes and some symptoms of the carcinoid syndrome.

The quantitatively most pronounced aberration in serotonin production and metabolism is in people with carcinoid tumors. Midgut carcinoid tumors produce and secrete serotonin. Carcinoid patients may convert as much as 60% of dietary tryptophan to serotonin. Long-term augmentation of the serotonin biosynthetic pathway may result in serious reduction of the free-tryptophan body pool, causing niacin deficiency and subsequent development of pellagra-like symptoms (1).

Platelet serotonin content is age-dependent—but not gender-dependent. Platelet serotonin concentration in elderly subjects is significantly lower than in adults and children and significantly higher than in newborns. There is no significant variation in platelet serotonin content over a period of 24 hours, nor is there any difference in levels in different seasons of the year. Human in vivo concentration is 168 ± 13 ng/ml whole blood and 341 ng/10^9 platelets. In contrast, both free and total plasma tryptophan (the serotonin precursor) have a circadian rhythm, with maximum values observed in the afternoon and minimum values at night (1).

Platelet serotonin and other related compounds are increased in the presence of serotonin-producing carcinoid tumors. Platelet serotonin is a more sensitive marker for increased serotonin production by carcinoid tumors than urinary 5-hydroxyindoleacetic acid (5-HIAA). In cases with high serotonin secretion rate, platelet serotonin reaches a maximum at approximately 50 nmol/10^9 platelets, whereas urinary 5-HIAA does not. Other neuroendocrine tumors and celiac disease give moderately increased platelet and plasma serotonin content, urinary serotonin and 5-HIAA excretion. Increased concentrations of platelet-poor plasma serotonin have been found in several disease states such as preeclampsia and type I diabetes. Menstrual cycle dependency was found, with higher periovulatory and premenstrual concentration of serotonin found in platelet-poor plasma (1).

Significantly reduced platelet serotonin can be found in subjects using selective serotonin reuptake inhibitors. Plasma tryptophan levels are dependent on dietary intake and have been found reduced in malabsorption syndromes, in several psychiatric disease states, and in carcinoid disease. Cerebrospinal fluid (CSF) levels of indoles are also dependent on dietary intake of tryptophan and are reduced in several neurodegenerative and psychiatric disease states (1).

Blood platelets play a major role in normal hemostasis and in the formation of occlusive thrombotic disorders. Acquired platelet dysfunction after coronary clot formation likely affects short-term and long-term outcomes in patients after acute coronary events. Therefore, inhibiting platelet function is an important therapeutic goal in patients with acute coronary artery disease (5).

Clinical depression has been identified as an independent risk factor for increased mortality in patients after an acute myocardial infarction, with increased platelet activity suggested as the mechanism for this adverse association. The prevalence of depression after an acute myocardial infarction is 15%–20%. Studies suggest that serotonin not only has a role in the neuropathophysiology of depression but also promotes thrombogenesis directly by enhancing platelet aggregation. Depressed patients exhibited 41% more platelet activation and higher procoagulant properties than healthy controls (5).

Excessive transcardiac accumulation of serotonin has been demonstrated in patients when chronic stable angina is converted to unstable coronary syndromes. Serotonin has been shown to be an important mediator of intermittent coronary obstruction caused by platelet aggregation and dynamic vasoconstriction. Recent clinical evidence showed there were reduced restenosis rates in people after angioplasty treated with selective serotonin reuptake inhibitors (5).

Regulatory Environment
Serotonin is not a focused concern of the regulatory environment.

References
1. Kema IP, deVries EGE, Muskiet FAJ (2000) Clinical chemistry of serotonin and metabolites. J Chromatog B 747:33–48
2. Mahan LK, Arlin M (eds) (1992) Krause's Food, Nutrition and Diet Therapy, 8th edn. WB Saunders, Philadelphia
3. Ader R (ed) (1981) Psychoneuroimmunology. Academic Press, New York
4. Plotnikoff N, Murgo A, Faith R, Wybran J (eds) (1991) Stress and Immunity. CRC Press, London
5. Nair GV, Gurbel PA, O'Connor CM, Gattis WA, Murugesan SR, Serebruany VL (1999) Depression, coronary events, platelet inhibition, and serotonin reuptake inhibitors. Am J Cardiol 84:321–323

Serum Sickness

Serum sickness is caused by i.v. injection of foreign proteins (antigen) which then lead to formation of immune complexes (antibody-antigen complexes). This type III hyperreaction can be interpreted as a generalized Arthus reaction. It was first described in 1905 by Piquet and Schick. The host's immunological hyperreaction is characterized by rash, fever, lymphadenopathy, athralgia, and/or nephritis.
▶ Serum Sickness

Seveso-Dioxin

▶ Dioxins and the Immune System

Seveso-Poison

▶ Dioxins and the Immune System

SFC

▶ Plaque-Forming Cell Assays

Shared Tumor Antigens

Tumor antigens shared by many tumors of various histologic types. Opposed to unique tumor antigens that are ideally present in single individual tumors without cross-reactivity with other tumors of the same or of other histotypes.
▶ Tumor, Immune Response to

Sheep Red Blood Cell Receptor

These bind to the receptor CD2 on T cells.
▶ Leukocyte Culture: Considerations for In Vitro Culture of T cells in Immunotoxicological Studies

Sheep Red Blood Cells (SRBC)

The most common antigen used by immunotoxicologists to induce a T cell-dependent antibody response.
▶ Glucocorticoids
▶ Immunoassays
▶ Plaque-forming Cell Assay

Shingles

Caused by reactivation of latent varicella virus in sensory root ganglia in patients previously infected with chickenpox. The lesions are vesicular on erythematous base and usually appear along the line of one or two dermatomes.
▶ Dermatological Infections

Signal Transduction During Lymphocyte Activation

KATHLEEN M BRUNDAGE
Dept of Microbiology, Immunology and Cell Biology
West Virginia University
P.O.Box 9177
Morgantown, WV 26506-9177
USA

Synonyms
Signaling through antigen receptors.

Definition
Signal transduction is by definition the conversion of a signal from one form to another. For lymphocytes, signal transduction begins at the plasma membrane and is initiated by the binding of antigen to the

T cell receptor (TCR) or the B cell receptor (BCR). As a result of this binding the activation of several signaling cascades occurs, resulting in the propagation and expansion of the initial signal. For lymphocytes, ultimately the response to extracellular signals is the induction of a new gene transcription pattern.

Characteristics

Activation of signal transduction pathways within lymphocytes occurs when the antigen specific receptors (TCR for T cells and BCR for B cells) bind their specific antigen. Receptor engagement results in the clustering and reorganization of the plasma membrane into large ▶ lipid rafts. It is hypothesized that these lipid rafts allow for the quick and efficient activation of signaling cascades upon antigen binding by the receptor. In T cells, these rafts are enriched with hyperphosphorylated CD3ζ chain and most of the T cell receptor complex (TCR complex)-associated phosphorylated and activated Zap70. In B cells, these rafts are enriched with B cell receptor complexes (BCR complexes) containing phosphorylated immunoglobulin (Ig)α and Lyn.

Signaling Through the TCR

Engagement of the TCR by peptide-MHC (major histocompatability complex) presented on antigen-presenting cells (APCs) results in the hyperphosphorylation of ▶ ITAM motifs (▶ immunoreceptor tyrosine-based activation motifs) on the CD3ε, γ and δ chains by the ▶ Src family kinase members Lck and Fyn (Figure 1). Phosphorylated ITAMs recruit the tyrosine kinase Zap70 to the receptor complex. Lck activates Zap70 by phosphorylation. Activated Zap70 recruits and activates two ▶ adaptors, LAT (linker of activation in T cells) and Slp76. These two adaptors recruit other proteins that bind through their SH2 domains. One of the proteins recruited in this manner is the Tec kinase Itk. Itk is phosphorylated by Lck and phosphorylates PLCγ (phospholipase C gamma) thereby activating PLCγ. PLCγ is responsible for the cleavage of the plasma membrane phospholipid phosphatidylinositol bisphosphate (PIP2) into inositol trisphosphate (IP3) and diacylglycerol (DAG). IP3 binds its receptor on the endoplasmic reticulum resulting in the release of calcium ions (Ca^{2+}) from intracellular stores. Release of the intracellular Ca^{2+} stores triggers the

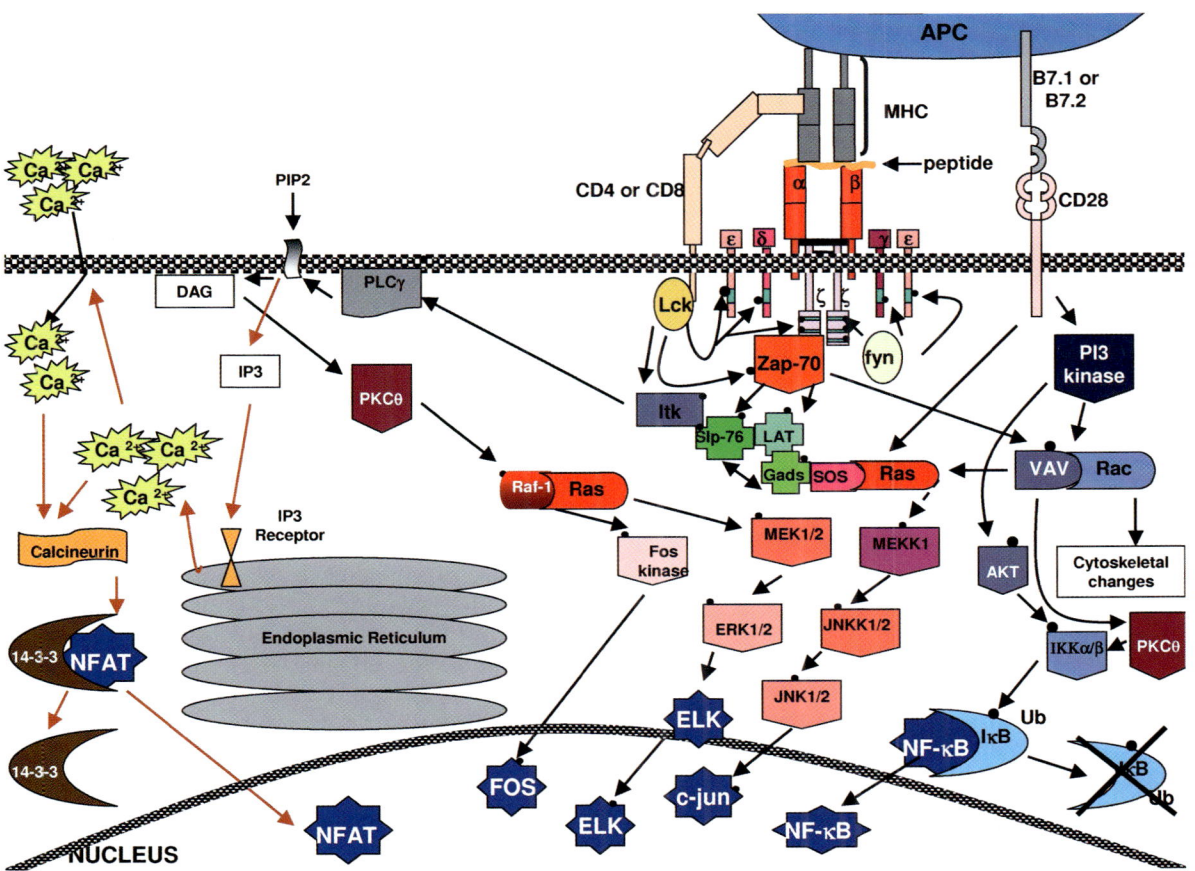

Signal Transduction During Lymphocyte Activation. Figure 1 Signaling through the T cell receptor.

opening of calcium channels in the plasma membrane, allowing more Ca^{2+} into the cell. This increase in intracellular Ca^{2+} activates the calcium-dependent serine/threonine phosphatase calcineurin. Dephosphorylation of the ▶ transcription factor NFAT (nuclear factor of activated T cells) by calcineurin releases NFAT from its cytoplasmic binding partner 14-3-3. Once released from 14-3-3, NFAT translocates into the nucleus where it binds to its target DNA binding sites located in the promoters of some genes, including cytokines such as the interleukin (IL)-2.

TCR complex → Zap70 → Itk → PLCγ → IP3
↘ DAG

IP3 → Ca^{2+} release → NFAT

The other product of PIP2 cleavage is DAG. DAG remains associated with the plasma membrane and activates members of the protein kinase C (PKC) family in particular PKCθ. PKCθ is a serine/threonine ▶ protein kinase that is responsible for phosphorylating the serine/threonine kinase Raf-1, which interacts with the small G protein ▶ Ras. The Ras-Raf-1 complex is responsible for activating several ▶ MAP kinase cascade pathways. These pathways eventually result in the activation of several ▶ transcription factors including c-fos, Elk and c-jun.

DAG → PKCθ → Ras → MAP kinase cascades
↘c-fos ↘ELK ↘c-jun

In T-cells there are at least two pathways that can activate Ras. One of these pathways involves the adaptor protein LAT, which as discussed above, is activated by Zap70. Activated LAT binds to the adaptor Gads, which in turn, interacts with and activates the ▶ guanine exchange factor (GEF) SOS. Once activated SOS activates Ras, which as previously discussed, goes on to activate the ▶ MAP kinase pathways. Ras is also activated when the co-stimulatory molecule CD28, which is expressed on the surface of T cells, interacts with its ligand B7.1 or B7.2 (CD80) expressed on APCs. In addition to activating Ras, CD28 engagement also activates phosphatidylinositol 3-OH kinase (PI-3 kinase). PI-3 kinase can activate several other proteins including the adaptor VAV. Besides being an adaptor, VAV is also a GEF similar to SOS. In addition, VAV can also be activated by Zap70. VAV activates Rac, a small G protein, and this results in the cytoskeletal rearrangements necessary for T cell activation. VAV also activates PKCθ leading to the activation of Raf-1 and the activation of the MAP kinase pathways as described above. In addition PKCθ directly activates the Iκ kinases (IKK) α/β. These kinases are also activated by AKT, another kinase that is activated by PI-3 kinase. Activated IKK phosphorylates IκB, the inhibitory protein responsible for retaining the transcription factor NF-κB in the cytoplasm. For NF-κB to be released from IκB, IκB has to be phosphorylated, ubiquitinated and degraded. Once released from IκB, NF-κB translocates into the nucleus where it binds to its target DNA binding site located in the promoters of many genes.

CD28 → Ras → MAP kinase cascade → c-fos, ELK and c-jun
CD28 → PI3 kinase → VAV → Rac → cytoskeleton changes
↘AKT → IκB → NF-κB

Signaling Through the BCR

The initial steps in B cell activation involve the cross-linking of the cell surface immunoglobulin receptor (BCR), by antigen. Prior to antigen binding, the ▶ Src family kinases Blk and Lyn are only weakly associated with the unphosphorylated ITAMs of the invariant Igα and Igβ chains of the BCR complex. Upon crosslinking and clustering of the BCR, Blk and Lyn phosphorylate the ITAMs on the Igα and Igβ chains (Figure 2). The phosphorylated ITAMs recruit the tyrosine kinase Syk, which is activated by phosphorylation by other Syk proteins as well as by Blk and Lyn. Activated Syk phosphorylates the adaptor BLNK (B-cell linker adaptor protein) also known as Slp-65. BLNK activates the Tec family kinase Btk. Btk acts in a manner similar to Itk in T-cells, activating PLCγ the protein responsible for cleaving PIP2 into DAG and IP3. As previously described, DAG and IP3 are responsible for the activation of RAS, which activates the MAP kinase signaling cascade, and the calcium-dependent pathways that lead to the activation of NFAT. BLNK also interacts with the adaptor Shc, which binds the adaptor Grb2. The Shc-Grb2 complex binds the GEF SOS, which in turn activates ▶ Ras as previously described. In B cells, Ras is also activated when the B cell co-receptor CD19 binds its ligand. CD19-induced Ras activation is through activation of VAV, which is activated by PI-3 kinase as previously described for T cells. Activated VAV activates the small G protein Rac which results in cytoskeletal rearrangements. As previously described for T cell activation, PI-3 kinase will also activate the kinase AKT as part of the signaling cascade, resulting in the activation of the transcription factor NF-κB.

Signaling Through Co-Receptors

In addition to the signaling pathways activated by TCR and BCR complexes described above, there are other co-stimulatory receptors on the surface of T cells and B cells that have a role in cell activation. These other receptors include CD19 (previously described), CD21 (complement 2 receptor) and CD81 (TAPA-1)

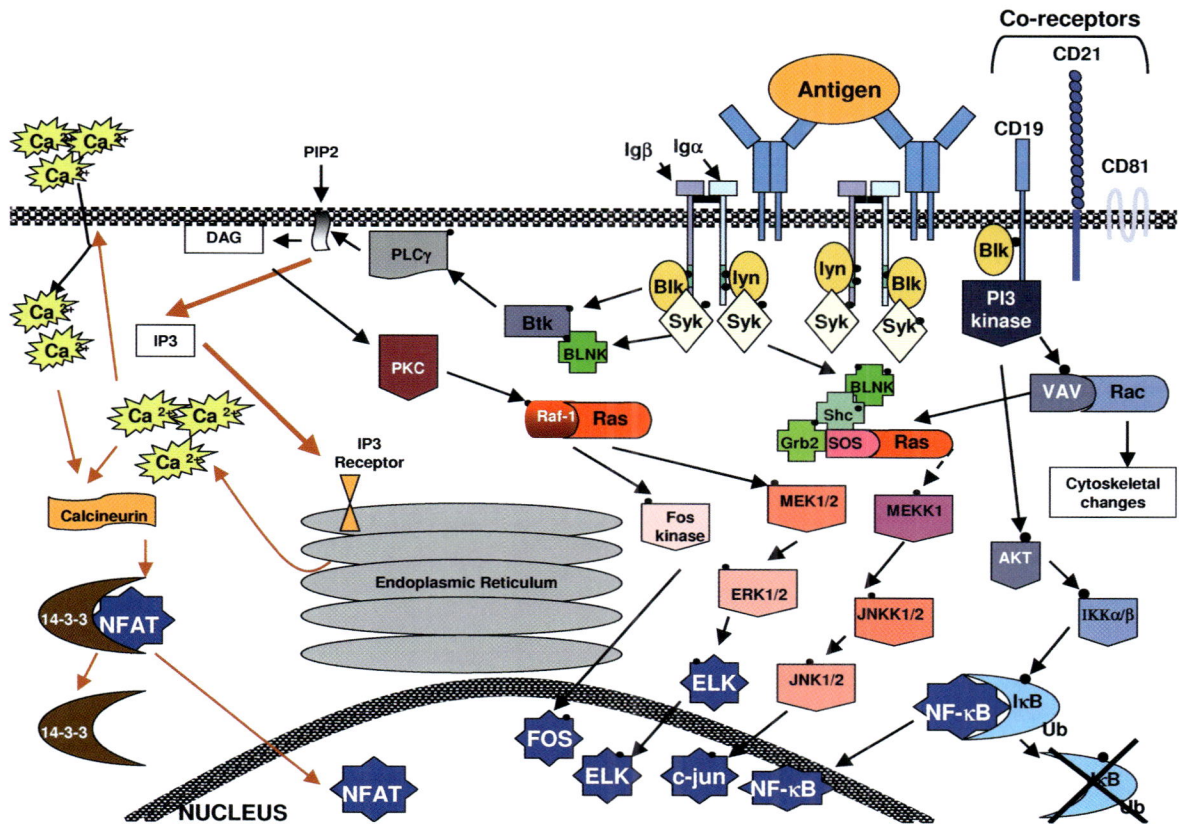

Signal Transduction During Lymphocyte Activation. Figure 2

on B cells, and CD28 (previously described) and CD40L on T cells, as well as cytokine receptors and adhesion molecules. Thus, the activation of a lymphocyte is a complex process involving the interplay of many proteins and the activation of many interconnected signaling cascades to expand and propagate the initial signals. As stated previously, the ultimate goal of the activation of these different signaling cascades is the initiation of new gene expression patterns. Besides activation signaling cascades there are also inhibitory signaling cascades. For lymphocytes inhibitory signals usually block the response by raising the threshold at which signal transduction can occur. On B cells, receptors such as Fc-receptor gamma IIB-1 (binds Fc portion of IgG), CD22 and PIR-B (paired immunoglobulin-like receptor) are involved in inhibiting activation. On T cells, receptors such as CTLA-4 (cytotoxic T-lymphocyte-associated) and KIRs (killer inhibitory receptors) inhibit the activation of T cells. A common ▶ motif found in the cytoplasmic tail of many inhibitory receptors is the ITIMs (immunoreceptor tyrosine-based inhibitory motifs). This motif, upon phosphorylation, recruits the inhibitory ▶ phosphatases SHP-1 (Src homology 2 (SH2) domain-containing protein tyrosine ▶ phosphatase-1), SHP-2 (src homology 2 (SH2) domain-containing protein tyrosine ▶ phosphatase-2) and SHIP (SH2-containing inositol phosphatase). These phosphatases are responsible for the in activation of many protein kinases including Btk and Itk as well as IP3 (an initiator of the calcium activation pathway).

A universal mechanism for the activation of signaling pathways is phosphorylation by protein kinases. Once these pathways have been activated they must be inactivated in order to prevent unregulated growth. In many cases to inactivate these pathways, phosphatases which can be activated by protein kinases, dephosphorylation the protein kinases thereby inactivating them. In addition, some transcription factors initiate transcription of their own inhibitors. For example NF-κB initiates transcription of its inhibitor IκB, the protein responsible for retaining NF-κB in an inactive state in the cytoplasm. The activation of lymphocytes is a tightly regulated process. During the lifespan of an individual, the lymphocytes and the immune system must maintain a delicate balance between both the activation and inhibitory pathways in order to function properly and protect an individual from infection or the development of autoimmune diseases.

BCR complex → Syk → Btk → PLCγ → IP3 and DAG
IP3 → Ca^{2+} release → NFAT
DAG → PKC → Ras → MAP kinase cascades → c-fos, ELK and c-jun
CD19 → PI3 kinase → VAV → Ras → MAP kinase cascades → c-fos, ELK and c-jun
↘ VAV → Rac → cytoskeleton changes
↘ AKT → IκB → NF-κB

Preclinical Relevance
The investigations into the effect of chemicals on the signaling pathways that lead to lymphocyte activation are not regulated under special guidelines. However, there are several examples in the literature of chemicals (e.g. cannabinol and the herbicide propanil) which have been shown to alter the signaling pathways that lead to the activation of lymphocytes.

Relevance to Humans
Any chemical or compound that inhibits the ability of lymphocytes to respond to antigenic stimulation and become activated will ultimately have an affect on the ability of an individual to fight an infection. Paradoxically, a chemical or compound which enhances the immune response and the ability of lymphocytes to become activated can also be detrimental. Individuals who are predisposed to develop autoimmune disease or who have a pre-existing autoimmune disease an enhanced immune response can increase both the potential to develop an autoimmune disease and/or the severity of the disease.

Regulatory Environment
At this time there are no specific guidelines for determining effects of chemicals on the signaling pathways that lead to the activation of lymphocytes. However there are National Toxicology Program's guidelines for Immunotoxicology evaluation in mice which recommend examining the effect of a chemical on lymphocyte blastogenesis in response to a mitogen and a mixed lymphocyte response assay to allogenic lymphocytes (1). These assays measure the proliferative and activation capacity of lymphocytes. In order for proliferation or activation to occur a lymphocyte must turn on new or alter gene transcription. It is clear that alteration of these responses by chemical exposure could be the result of alteration in the activation of one or more of the signaling cascades described here.

References
1. Luster MI, Munson AE, Thomas PT et al. (1988) Development of a testing battery to assess chemical-induced Immunotoxicology: National Toxicology Program's guidelines for immunotoxicity evaluation in mice. Fund Appl Toxicol 10:2–19
2. Lewis RS (2001) Calcium signaling mechanisms in T lymphocytes. Ann Rev Immunol 19:497–521
3. Gauld SB, Dal Porto J, Cambier JC (2002) B cell antigen receptor signaling: roles in cell development and disease. Science 296:1641–1642
4. Dong C, Davis RJ, Flavell RA (2002) MAP kinases in the immune response. Ann Rev Immunol 20:55–72
5. Jordan MS, Singer AL, Koretzky GA (2003) Adaptors as central mediators of signal transduction in immune cells. Nature Immunol 4:110–116

Signaling Through Antigen Receptors

▶ Signal Transduction During Lymphocyte Activation

Single Amino Acid Polymorphisms

▶ Cytokine Polymorphisms and Immunotoxicology

Single Base Transition

▶ Cytokine Polymorphisms and Immunotoxicology

Single Nucleotide Polymorphisms

▶ Cytokine Polymorphisms and Immunotoxicology

Sjögren Syndrome

A chronic systematic inflammatory disorder characterized by dryness of mucus membranes.
▶ Systemic Autoimmunity

Skin, Contribution to Immunity

Emanuela Corsini
Department of Pharmacological Sciences
University of Milan
Via Balzaretti 9
20133 Milan
Italy

Synonyms
Local immune system

Definitions

Atopy is the clinical manifestation of type I hypersensitivity reactions including eczema, asthma and rhinitis. General predisposition toward development of IgE-mediated hypersensitivity reactions toward common environmental antigens. ▸ Atopic dermatitis is a chronic, itching, inflammation of the skin in atopic individuals. Contact hypersensitivity is a delayed inflammatory reaction on the skin seen in type IV hypersensitivity, resulting from allergic sensitization. Dermatitis is an inflammatory skin disease showing redness, swelling, infiltration, scaling and sometimes vescicles and blisters. Percutaneous absorption represents the passage of compounds across the skin. The composition and structure of the stratum corneum determine the pathways for diffusion as well as the solubility and diffusivity of compounds within the skin. The skin is a physical barrier between an organism and its environment. It can be divided into four different regions: the stratum corneum, the viable epidermis, the ▸ dermis and the hypodermis. Skin irritation is a form of skin inflammation induced by primary contact with chemicals and is thought not to be mediated by lymphocytes. Stratum corneum is the outermost layer of the skin, the primary barrier to percutaneous absorption. The organization of the stratum corneum can be viewed as a brick wall with the bricks representing the corneocytes and the mortar representing the intercellular lipids. It contains approximately 15% water, 70% protein and 15% lipid.

Characteristics

The skin represents the interface between the environment and internal organs and it is essential for survival. The skin protects the body against external insults and prevents water loss. The skin participates directly in thermal, electrolyte, hormonal, metabolic, and immune regulation. Skin is constantly exposed to many antigens. Rather than merely repelling them, the skin has a rapid response immune system of its own to ward off invasion by infectious agents, and to react to noxious chemicals applied to the skin.

Skin Structure

In order to understand the defensive capacities of the skin, it is important to define its structural basis. The skin consists of two major components: the outer epidermis and the underlying dermis, which are separated by a basal membrane. The epidermis is a multilayered epithelium composed of several different cell types. **Keratinocytes** (KC) are the major epidermal cell type, and represent about 95% of epidermal cell mass. They are responsible for the biochemical and physical integrity of the skin via the formation of the stratum corneum. **Langerhans cells** (LC) comprise the second most prominet cell type in the epidermis. LC are bone-marrow derived dendritic cells and represent only 2%–5% of epidermal cell population. ▸ Melanocytes represent 3% of epidermal cells and by generating melanin they protect the skin against ultraviolet radiation. The blood supply to the epidermis is guaranteed by the capillaries located in the rete ridges at the dermal-epidermal junction. The dermis is separated from underlying tissues by a layer of adipocytes (hypodermis). The dermis makes up approximately 90% of the skin and has mainly a supportive function. In addition, epidermal appendages (hair follicles, sebaceous glands and eccrine glands) span the ▸ epidermis and are embedded in the dermis.

Skin Immune System

Besides its barrier function, the skin has been recognized as an immunologically active tissue. The immune system, associated with mucosal surfaces, evolved mechanisms discriminating between harmless antigens and commensal microorganisms and dangerous pathogens (1). LC are the principal antigen-presenting cells in the skin, and the KC act as signal transducers, converting nonspecific exogenous stimuli into the production of cytokines, chemotactic factors and adhesion molecules. Cells in the dermis and epidermis, including dermal **dendritic cells** (DC), epidermal LC, melanocytes, and migrating lymphocytes, are all important in skin immune reaction and are know to produce a great variety of cytokines.

The skin innate immune response is rapid, provides the initial line of defense against microorganisms, is antigen nonspecific, and lacks immunological memory. Cellular constituents of the skin innate immune system include keratinocytes, DC, macrophages, natural killer cells and neutrophils. The innate immune system is able to recognize the conserved pathogenic patterns on microbes by pattern recognition receptors, such as the Toll-like receptor and others. Adaptive immunity, in contrast, is delayed in time, is antigen specific, and involves a recall response. A proper balance between innate and adaptive immunity is essential, because if the innate response is inadequate septic shock may result from a cutaneous bacterial infection, or the exaggerated innate response may yield a chronic inflammatory response.

Epidermal Cytokines

The major mechanisms used by epidermal cells to participate in immune and inflammatory skin reactions are the production of cytokines and responses to cytokines (2). Within the epidermis, the KC are the major source of cytokines, along with the LC and melanocytes (3). Epidermal cells can produce constitutive or following activation an arsenal of cytokines, strongly supporting the idea that the skin functions as an immune organ and that an important role of the skin is to provide an immune barrier between the external environment and internal tissues. Table 1 gives a list of the cytokines produced by epidermal cells.

The histopathological pattern of nearly every inflammatory skin disease can be accounted for by the appropriate cytokine or combination of cytokines (4). It is important to remember that multiple mechanisms and cell types may be involved in the induction of skin toxic responses. Determining the source, the kinetics of production, and the regulation of inflammatory mediators in the skin will be of value in predicting the various toxicities arising from exposure to environmental agents. Differences in skin toxic responses may be the result of early differences in the epidermal response to cell injury, and in the production of inflammatory signals.

Contribution of Keratinocytes to Skin Immunity

In the last two decades it has become clear that the KC play an important role in the initiation and perpetuation of skin inflammatory and immunological reactions. While resting KC produce some cytokines constitutively, a variety of environmental stimuli, such as tumor promoters, ultraviolet light and chemical agents, induce epidermal KC to release inflammatory cytokines (IL-1, TNF-α), chemotactic cytokines (IL-8, IP-10), growth promoting cytokines (IL-6, IL-7, IL-15, GM-CSF, TGF-α) and cytokines regulating humoral versus cellular immunity (IL-10, IL-12, IL-18) (3). Cytokine production by KC can affect migration of inflammatory cells, can have systemic effects on the immune system, can influence KC proliferation and differentiation processes; and regulate the production of other cytokines.

Of all the cytokines produced by KC, only IL-1α, IL-1β and TNF-α activate a sufficient number of effector mechanisms to independently trigger cutaneous inflammation (5). Unstimulated KC contain large amount —in biological terms—of preformed and biologically active IL-1α, in addition to inactive pro-IL-1β. Damage to the KC releases IL-1α, which essentially is a primary event in skin defense. IL-1α stimulates further release of IL-1α and the production and release of other cytokines such as IL-8, IL-6, and GM-CSF. Thus, by cytokine cascades and networks, an inflammatory response can be rapidly generated. In this scenario, KC act as proinflammatory signal transducers, responding to nonspecific external stimuli with the production of inflammatory cytokines, adhesion molecules, and chemotactic factors, preparing the dermal stroma for specific immunological activity.

On the other hand, in the skin, TNF-α is stored in dermal mast cells, but following stimulation it may be produced by KC and LC. Antibody to TNF-α abolishes many inflammatory skin reactions, including allergic and ▶ irritant contact dermatitis (6). An important mechanism by which TNF-α influences the development of an inflammatory reaction is induction of the expression of cutaneous and endothelial adhesion molecules.

Contribution of Langerhans cells to Skin Immunity

DC and LC are unique cell types, in that they function in both innate and specific immunity, depending on their state of maturation and local microenvironmental conditions. It has been recently reported that LC have the potential to reduce inflammatory responses in skin (reviewed by Nickoloff (7)). There is evidence that LC in normal skin may actually function as antiinflammatory agents to counter balance the proinflammatory tendencies of keratinocytes. When LC are decreased in number or function within the epidermis an enhanced inflammatory reaction in skin is observed.

LC are intraepidermal antigen-presenting cells whose dendrites intercalate between KC. In normal skin, immature LC monitor the environment as sentinel cells, on guard to detect foreign intruders. An antigen when absorbed through the epidemis will likely encounter a

Skin, Contribution to Immunity. Table 1 Cytokines expressed by epidermal cells

Epidermal Cells	Cytokine (Constitutive or Inducible Expression)
Keratinocytes	IL-1α, IL-1β, IL-1RA, IL-3 (mouse), IL-6, IL-7, IL-8 (human), IL-10, IL-12, IL-15, IL-18, IL-20, TNF-α, G-CSF, GM-CSF, M-CSF, GRO, MIP-2 (mouse), IP-10, RANTES, MCP-1, TGF-α, TGF-β
Langerhans cells	IL-1α, IL-1β, IL-6, IL-15, IL-18, TNF-α, Gro, MIP-2, MIP-1α, TGF-β
Melanocytes	IL-1α, IL-1β, IL-6, IL-7, IL-8, IL-10, IL-12, IL-24, TNF-α, G-CSF, GM-CSF, M-CSF, GRO, MIP-2 (mouse), RANTES, MCP-1, TGF-α, TGF-β

G-CSF=granulocyte-colony stimulating factor; GRO=Growth Related Oncogene, IL=interleukin; MCP=monocyte chemotactic protein; MIP=macrophage inflammatory protein; RANTES=regulated on activation, T-cell expressed and secreted; TGF=transforming growth factor; TNF=tumor necrosis factor.

LC. The antigen is then captured and processed by LC. Activated LC move out of the epidermis into the dermis, and into the regional lymphatic system, eventually finding their way to the regional draining lymph node. In the lymph node, LC differentiate into mature dendritic cells and present antigen to specific T lymphocytes, using major histocompatibility complex (MHC) class II molecules to hold the processed antigen in place. Adhesion molecules on both the antigen-presenting cell (i.e. B7) and the T cell (i.e. CD28) ensure appropriate contact and costimulation. Following appropriate stimulus, a clone of T cells is produced with the ability to react to the antigen which caused their production. The activation and clonal expansion of allergen-reactive T cells is the pivotal event in the acquisition of skin sensitization.

If the antigen persists or if the antigen is re-encountered, again LC in the epidermis take up the antigen and migrate into the dermis. In theory, there is contact with specific antigen-responsive T cells in the dermis. This is because clones of lymphocytes stimulated by skin-encountered antigens "home" to the skin using special receptors on the cell surface that attach only the skin microvasculature. Antigen is presented to these antigen specific T cells. The lymphocytes divide and release a large amount of proinflammatory cytokines that mediate the ensuing inflammatory response. These T cell products also activate KC, which in turn produce cytokines. Inflammation eventually eliminates the antigen, either by killing the viable organisms, washing out the antigen with edema fluid, or engulfing particulates in activated macrophages, neutrophils, or other phagocytic cells.

At least in experimental animals it has been shown that there are two major populations of DC in afferent lymph draining the skin that differ in their capacity to stimulate CD4 and CD8 T cells. While expression of the costimulatory molecules CD80, CD86 and CD40 appear similar for both DC populations, differences in expression of cytokine transcripts have been shown: the $CD11a^+/SIRPalpha^-/CD26^+$ population synthesizes more IL-12, whilst the $CD11a^-/SIRPalpha^+$ population produces more IL-10 and IL-1α. This is likely to affect the bias of the immune response following presentation of antigen to T cells by one DC subpopulation or the other. It is proposed that $SIRPalpha^-$ DC would promote a T helper type 1(Th1)-biased response. No definitive data are available relatively to the lymphoid or myeloid origin of the two DC populations: $SIRPalpha^-$ cells are propably myeloid, while $SIRP^+$ cells are probably lymphoid (9).

Preclinical Relevance
Whereas irritant contact dermatitis is a form of skin inflammation induced by primary contact with chemicals and is thought not to be mediated by lymphocytes, ▶ allergic contact dermatitis represents a lymphocyte-mediated delayed-type hypersensitivity reaction that requires previous sensitization by the same chemicals. The biochemical mechanisms involved in skin irritation are complex and not fully understood. Different skin irritants can trigger different inflammatory process. In addition to destroying tissue directly, chemicals can alter cell functions and/or trigger the release, formation, or activation of autocoids, such as histamine, arachidonic acid metabolites, kinins, complement, reactive oxygen species and cytokines. Substances that are keratin solvents, dehydrating agents, oxidizing or reducing agents, and others, may be irritants. Due to the eterogenicity of skin irritants, there is no reliable method for assessing irritancy based on chemical structure. Virtually any chemical substance may be an irritant under conditions of exposure that predispose to the occurrence of an irritant response. The biological processes necessary for producing hypersensitivity are grouped into two phases: an induction phase and an elicitation phase. Induction has been referred to as the afferent phase, the initial exposure through clonal expansion and the release of memory cells. Elicitation, the efferent phase, consists of local recognition of the antigen by the memory cells, the release of cytokines, and the activity of inflammatory mediators which are generated locally and produce the dermatitis.

Relevance to Humans
The two most frequent manifestations of skin toxicity are irritant contact dermatitis and allergic contact dermatitis. Hypersensitivity reactions to drugs and industrial chemicals are relatively common in man. Depending on the country, dermatoses account for 20%–70% of all occupational diseases, and 20%–90% of these are irritant contact dermatitis; most of the remaining contact dermatitis is allergic contact dermatitis.

Hypersensitivity reactions are often considered to be increased at such a rate to become a major health problem in relation to environmental chemical exposure. Contact hypersensitivity is characterized by an eczematous reaction at the point of contact with an allergen. It develops normally in two temporally distinguished phases: the induction phase and the elicitation phase. Exposure to allergen will induce in susceptible individuals the immune response necessary for sensitization (the induction phase). Sensitization takes 10–14 days in humans. Re-exposure to the same antigen will result in elicitation of the inflammatory reaction after a characteristic delay of usually 12–48 h (the elicitation phase). The nature of the immune responses induced by chemical allergens is essentially no different from that which characterizes protective immunity. Allergic contact dermatitis is a multifactorial disease, the onset of which depend on the nature of the chemical, concentration, type of exposure, age, sex, genetic susceptibility, and not genetic idiosyncrasies.

Cutaneous antigens are generally of low molecular weight (hapten). It is clear that some allergens (urushiol, dinitrocholorobenzene, oxazolone, diphenylcyclopropenone) are very potent sensitizers and appear able to induce sensitization in all normal people, whereas some others (metallic salts of nickel, cobalt and chromium, and a huge variety of organic compounds) are weak antigens and appear to sensitize only susceptible individuals.

Regulatory Environment

As a wider varieties of xenobiotics can cause cutaneous damage, it is requested by legislative agencies to characterize this potential when developing, registering, or certifying new materials (*see also* "Contact Hypersensitivity").

Skin Irritation Testing in Animals

Predictive animal tests are routinely employed to assess topical irritation responses, such as acute or cumulative irritation, corrosion or photoirritation. Despite multitudinous objections, the test most widely used for prediction potential skin irritation is the Draize test (10) and its modifications, in which the tested material is topically applied on abraded or intact skin of the albino rabbit, clipped free of hair, under an occluded patch for 4–24 h. Other species have been used as well, such as albino guinea pigs, hairless mice, swine and beagle dogs, but albino rabbits are still used for most skin irritation testing, with the recognition that rabbits are more responsive than other species (including humans) to mild or moderate irritant insults. Recently, alternative in vitro tests have been validated and accepted by regulatory authorities in the EU and OECD member countries to characterize the corrosive and phototoxic potential of xenobiotics.

Skin Allergy Testing in Animals

Several standardized predictive tests in guinea pigs (cf. "Guinea Pig Assays for Sensitization Testing") and some tests in mice use the response in the efferent phase as an indication of immune reactivity to the chemical. Some newer predictive assays in mice involve stimulation of lymphocytes in local draining lymph nodes during the afferent phase as the endpoint (cf. "Local Lymph Nodes Assay (LLNA)" and "IMDS").

References

1. Tlaskalova-Hogenova H, Tuckova L, Lodinova-Zadnikova R et al. (2002) Mucosal immunity: its role in defense and allergy. Int Arch Allergy Immunol 128:77–89
2. Corsini E, Galli CL (2000) Epidermal cytokines in experimental contact dermatitis. Toxicology 142:203–211
3. William IR, Kupper TS (1996) Immunity at the surface: homeostatic mechanisms of the skin immune system. Life Sci 58:1485–1507
4. Luster MI, Wilmer JL, Germolec D et al. (1995) Role of keratinocyte-derived cytokines in chemical toxicity. Toxicol Lett 82/83:471–476
5. Kupper TS (1990) Immune and inflammatory process in cutaneous tissues. Mechanisms and speculations. J Clin Invest 86:1783–1789
6. Piguet PF, Grau GE, Houser C, Vassalli P (1991) Tumor necrosis factor is a critical mediators in hapten-induced irritant and contact hypersensitivity reactions. J Exp Med 173:673–679
7. Nickoloff BJ (2002) Cutaneous dendritic cells in the crossfire between innate and adaptive immunity. J Dermatol Sci 29:159–165
8. Kimber I, Dearman RJ (2002) Allergic contact dermatitis: the cellular effectors. Contact Dermat 46:1–5
9. Howard CJ, Hope JC, Stephens SA, Gliddon DR, Brooke GP (2002) Co-stimulation and modulation of the ensuing immune response. Vet Immunol Immunopath 87:123–130
10. Draize JH, Woodward G, Calvery HO (1944) Methods for the study of irritation and toxicity of substances applied topically to the skin and mucous membranes. J Pharm Exp Therap 82:377–389

Skin Prick Test

The skin prick test is the most common test for atopy. It involves injection of small amounts of allergen into the skin, leading to an immediate wheal-and-flare reaction in allergic individuals.

▶ Mast Cells
▶ Food Allergy

Skin Sensitization Assay

▶ Guinea Pig Assays for Sensitization Testing

Skin Sensitization Potency

The inherent potency with which a contact allergen will induce skin sensitization. Activity is considered usually as a function of the amount of chemical required for the acquisition of sensitization.

▶ Local Lymph Node Assay

SLE

▶ Autoimmune Disease, Animal Models

Sleeping Sickness

Sleeping Sickness is a disease in humans caused by the protozoan parasites *Trypanosoma brucei rhodesiense* and *T.brucei gambiense* which are transmitted by tsetse flies (*Glossinidae*). Sleeping Sickness occurs in sub-Saharan Africa. The disease is untreated 100% fatal.

▶ Trypanosomes, Infection and Immunity

Slp 76 (SH2 Domain-Containing Leukocyte Protein of 76 kDa)

Slp 76 is an adaptor protein found in T cells. It is phosphorylated by the tyrosine kinase Zap70 and is involved in the activation of the pathway that leads to activation of PLCγ and the release of intracellular Ca^{2+} as well as the activation of the Ras pathway.

▶ Signal Transduction During Lymphocyte Activation

Smad

Members of the Smad family of signaling proteins serve as substrates for the TGF-β receptor type I kinase, and function to transduce signals from the membrane into the nucleus and regulate ligand-specific gene transcription. A genetic screen in *Drosophila* designed to identify mutations that would modify the *dpp* mutant phenotype isolated a gene that was named *Mothers against Dpp* (*mad*). A genetic screen in *Cunninghamella elegans* designed to identify components of TGF-β1 isolated a gene that was named *sma*. It was later shown that Sma proteins are homologous to the Mad proteins, and thus the *Drosophila* and *C. elegans* nomenclature was combined to Smads.

▶ Transforming Growth Factor β1; Control of T cell Responses to Antigens

Small Secreted Cytokines

▶ Chemokines

SNPs

▶ Cytokine Polymorphisms and Immunotoxicology

Somatic Hypermutation

Somatic hypermutations occur during the immune response to thymus-dependent (TD) antigens and are inserted in the variable regions of antibody H chains and L chains. Somatic hypermutations cause an increase in the affinity of antibodies during the primary as well as following immune responses and are thus responsible for the phenomenon of immune maturation.

▶ Idiotype Network

Southern

The transfer of DNA molecules from an agarose gel to a membrane by capillarity or an electric field. The

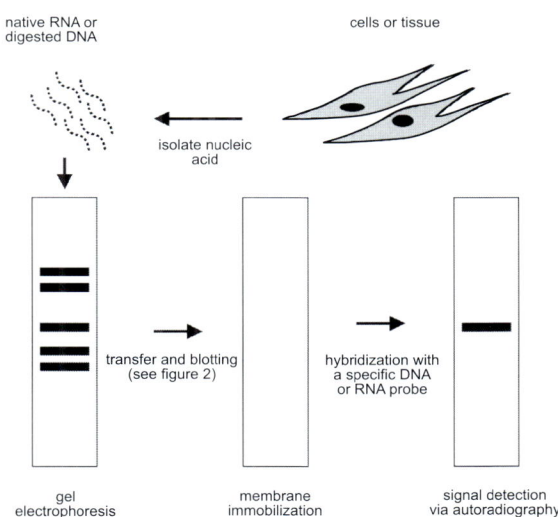

Southern and Northern Blotting. Figure 1 The southern and northern blotting techniques. Enzymatically digested DNA or denatured RNA is fractionated by gel electrophoresis. After transfer and fixation of the DNA fragments or RNA to a membrane support, the membrane is incubated with a labeled DNA or RNA probe that is specific for the target of interest. Because the probe hybridizes to only the target fragment/molecule, a specific band is visualized following detection.

immobilized DNA can be detected at high sensitivity by hybridization to a sequence specific probe.
▶ Southern and Northern Blotting

Southern and Northern Blotting

KEVIN TROUBA
NIEHS Mail Drop C1-04, Envionmental Immunology Laboratory
111 Alexander Drive
P.O.Box 12233
Research Triangle Park, NC 27709
USA

Synonyms
DNA and RNA blotting, nucleic acid blotting.

Short Description
Nucleic acid detection can be performed using a technique known as ▶ blotting, which is the transfer of DNA (deoxyribonucleic acid) or RNA (ribonucleic acid) onto a membrane support. When this procedure is used to examine DNA or RNA it is referred to as Southern or northern blotting, respectively.

Characteristics
Southern blotting is the transfer of DNA from an agarose gel to a membrane support (1). Since its inception, Southern blotting has been used to determine gene copy number, detect restriction fragment length polymorphisms, determine modifications to DNA (e.g. methylation), and detect the presence of genetically inserted transgenes. It also is used to study RNA splicing and antibody and T cell receptor formation, and has aided in the identification of gene rearrangements associated with a variety of human genetic disorders and cancers.

Northern blotting is the transfer of RNA from an agarose gel to a membrane similar to that used in Southern blotting (2). Northern blotting is used to evaluate ▶ gene expression in a manner that is both qualitative (in which tissues or cells, or under which physiological conditions a gene is expressed) and quantitative (level of gene expression). This method also is used to detect the expression of foreign genes in transgenic organisms and is a useful adjunct to complementary DNA (cDNA) cloning because mRNA molecular weights can be compared with those of cloned DNAs. The basic principles of Southern and northern blotting are similar; however, there are a few notable exceptions. RNA is more labile than DNA and is very sensitive to degradation by enzymes called RNases. RNA degradation reduces the quality of data and the ability to quantify message expression. RNA molecules often form secondary structures that significantly alter their mobility, requiring gel ▶ electrophoresis to be performed under denaturing conditions (in the presence of glyoxal or formaldehyde). With Southern blotting, ▶ genomic DNA is digested with restriction enzyme(s) to produce small fragments that can be easily fractionated, whereas RNA is analyzed as an intact molecule. Genomic DNA also does not require many of the special handling precautions necessary with RNA.

The general schemes used in Southern and northern blotting are depicted in Figures 1 and 2. DNA or RNA is isolated from cells or tissues and then fractionated in a solid support (gel) in an electric field (electrophoresis). Following size separation, the DNA or RNA is transferred onto a membrane (generally nylon or nitrocellulose) and immobilized by exposure to ultraviolet radiation or heat.

Membrane-bound DNA or RNA undergoes ▶ hybridization to a sequence homologous, labeled nucleic acid probe (usually prepared using cDNA or RNA). Detection of the probe depends on the labeling method employed. For example, with ^{32}P-labeled probes the signal is detected and quantified using X-ray film (autoradiography). Probes are generated by several methods, including random-priming, nick-translation, and polymerase chain reaction (PCR). Sequences with only partial homology (e.g. species-specific cDNA or genomic DNA fragments) are used to prepare probes as well.

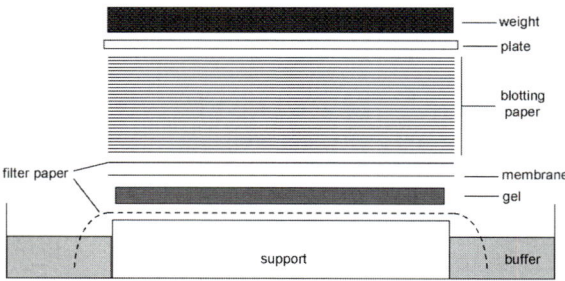

Southern and Northern Blotting. Figure 2 Capillary transfer of DNA or RNA from agarose gels. Buffer is drawn from a reservoir via a paper wick and passes through the gel into a blotting material (e.g. paper towels, blotting paper). The DNA or RNA is eluted from the gel by the flow of buffer and is deposited onto a membrane. A weight applied to the top of the blotting apparatus ensures a tight junction between the layers of material used in the transfer system.

Pros and Cons
Northern blotting remains the standard for detection and quantification of mRNA despite the development

of more sensitive molecular techniques such as the reverse transcription-polymerase chain reaction (RT-PCR). It presents several advantages over newer techniques because it is a useful method for determining mRNA size and for identifying alternatively spliced transcripts and multigene family members. One advantage of Southern blotting is that it can be used to examine stretches of DNA such as the multikilobase restriction fragments produced by some restriction enzymes. For both Southern and northern blotting, the analysis of different genes via membrane stripping and reprobing also is an advantage over traditional PCR gel analysis techniques.

A disadvantage of northern blotting is that it is the least sensitive of newer RNA analysis methods like PCR, gene microarrays, and ribonuclease protection assays (RPA). One disadvantage of Southern blotting is that it usually requires a relatively large quantity of high quality DNA. For this reason, PCR techniques are becoming widely applied to tasks for which Southern blotting was traditionally used (e.g. DNA fingerprinting and identification) because much less DNA is required and, in many cases, PCR techniques can be performed using partially degraded DNA. The use of radioisotopes in Southern and northern blotting also is considered a disadvantage from a health perspective, and the short half-life of phosphorus ^{32}P (14 days) necessitates that the probe be prepared and used in an expeditious manner. Although it can be an advantage over traditional PCR gel analysis, reprobing of a Southern or northern membrane usually requires removal of the initial probe before hybridizing with a second probe. This process can be time consuming and can reduce subsequent DNA and RNA detection, a distinct disadvantage.

Predictivity

Southern and northern blotting is used to better understand the molecular basis of disease. For example, northern blotting, which displays a high degree of correlation with RT-PCR, is used to monitor target gene expression or to identify biomarkers in isolated tissues and organs that results from a potential toxicological event or challenge. Genes of toxicological interest include proinflammatory cytokines, growth factors, cytochrome P450s, glutathione-S-transferases, cyclin dependent kinases, and cell cycle inhibitors.

Relevance to Humans

DNA and RNA underlie many aspects of human health, both in function and dysfunction. Southern and northern blotting will continue to aid in the identification of genetic events involved in neoplastic and non-neoplastic diseases, inherited disorders, genetic susceptibility, and predisposition to multigenic diseases. The accurate diagnosis and classification of acquired infectious diseases and neoplastic disease, as well as their prediction, prognosis, and therapeutic monitoring, will be aided by these technologies.

Obtaining a detailed picture of how genes and other DNA sequences function together and interact with environmental factors ultimately will lead to the discovery of pathways involved in normal processes and in disease. Such knowledge will have a profound impact on the way disorders are diagnosed, treated, and prevented, and will bring about changes in clinical and public health practices. The characterization of (southern) and expression profiling (northern) of genes in normal and diseased human tissues is an important first step to understanding the potential role of genes in human disease.

Regulatory Environment

Southern and northern blotting technology is applicable to any study where the detection, characterization, or quantification of a nucleic acid in biological material is required. These methods are often used in basic research, in the analysis and diagnosis of human inherited disease, as well as in a forensic capacity.

References

1. Southern EM (1975) Detection of specific sequences among DNA fragments separated by gel electrophoresis. J Mol Biol 98:503–517
2. Alwine JC, Kemp DJ, Stark GR (1977) Method for detection of specific RNAs in agarose gels by transfer to diazobenzyloxymethyl paper and hybridization with DNA probes. Proc Natl Acad Sci USA 74:5350–5354

Spherocytes

Abnormal erythrocyte characterized by its smaller, round size, and darker color without central pallor. It is seen on Romanofski-stained peripheral blood smears of patients with immune-mediated hemolytic anemia (IMHA).

▶ Antiglobulin (Coombs) Test
▶ Hemolytic Anemia, Autoimmune

Spleen

C Frieke Kuper
Toxicology and Applied Pharmacology
TNO Food and Nutrition Research
Zeist
The Netherlands

Synonyms

Definition
A large vascular, lymphoid organ in the upper part of the abdominal cavity of vertebrates near the stomach, with two main compartments designated ▶ red pulp and ▶ white pulp. Its has two main functions: as a blood-filtering organ (clearance of particulates, aged or defective erythrocytes and platelets) and in the production of immune responses against blood-borne antigens. In some species like the dog, the spleen functions as a storage organ of red blood cells.

Characteristics
General
The spleen is a large blood-filtering and blood-forming organ, consisting of two main compartments that are named the red pulp and the white pulp. The white pulp is visible grossly on cross-section of the spleen as white patches within the deep red (red pulp) organ, and represents the lymphoid compartment of the spleen. The spleen contains about a quarter of the body's total lymphocyte population; during lymphocyte recirculation more cells pass through the spleen than through all the lymph nodes. The main immunological function of the spleen is to guard the blood compartment of the body. It does so, among other ways, by generating T cell-independent antibody responses to bacterial polysaccharides and by exerting enormous phagocytic power. The antibody response may not be completely T cell-independent in all cases, since T cell-derived factors enhance the response to some of these antigens. Antibodies generated are mainly of IgM class. The ▶ marginal zone plays a key role in these antibody responses and retains B lymphocyte memory.

Red Pulp
The red pulp consists of venous sinuses lined by reticuloendothelial cells and splenic cords (cords of Billroth), which contain macrophages, lymphocytes, and plasma cells. The red pulp is also rich in natural killer cells. Macrophages perform a major function in blood cell clearance (for example, of old red blood cells) and in phagocytosis, especially of nonopsonized particles. This high volume blood-filtering function is made possible by two factors: the direct contact, unobstructed by blood-vessel walls, between phagocytic cells and blood-borne particles and the large blood supply. The phagocytic function is especially important in the case of intravascular pathogenic microorganisms, before formation of antibody and subsequent opsonization occur (early bacteremia). The splenic macrophages together with the hepatic phagocytic system synthesize the majority of complement components involved in the classical complement cascade. In the case of systemic septicemia, the red pulp increases and contains large proportions of (immature) granulocytes.

In the fetus, the spleen is an important site of hematopoiesis. This function is maintained in certain animal species (rats and mice) throughout adulthood, whereas in others extramedullary hematopoiesis is resumed only in certain diseases.

White Pulp
The white pulp is composed of lymphocytes, macrophages and plasma cells within a mesenchymal reticular network. Description of the histology of the spleen varies according to the animal species (1). Generally, the white pulp consists of a central arteriole surrounded by the periarteriolar lymphocyte sheath (PALS), the inner part being a T lymphocyte area (Figure 1).

The outer PALS contains B lymphocytes and plasma cells. Adjacent follicles contain mainly B cells. Around the PALS and follicles is a zone containing B cells and called the marginal zone; this region is easily distinguished, especially in rats, but is less prominent in dogs, and may even be absent in the human spleen (2). The PALS has a microenvironment and passenger leukocyte content similar to that of the lymph node paracortex. The follicles are comparable in structure and function with those of lymph follicles. In contrast, the marginal zone has a microenvironment that is unique to the spleen. Histologically, the B cells at this site are of medium size; they are larger and paler than B cells in primary follicles and the follicular mantle of secondary follicles. In addition, they do not show the morphology of centrocytes or centroblasts found in germinal centers, and the phenotypic expression indicates that the marginal zone B cells are a separate population of B cells. Also there are characteristic macrophages, namely the marginal zone macrophages, which are dispersed throughout the marginal zone, and the marginal metallophilic macrophages, which border the marginal sinus. Hematogenically spreading infectious pathogens are controlled initially by being trapped to marginal zone macrophages, a process that is enhanced by binding the pathogens to natural antibodies (3). Subsequently, marginal zone B cells proliferate, B cell foci are formed and IgM antibodies against the pathogens are produced.

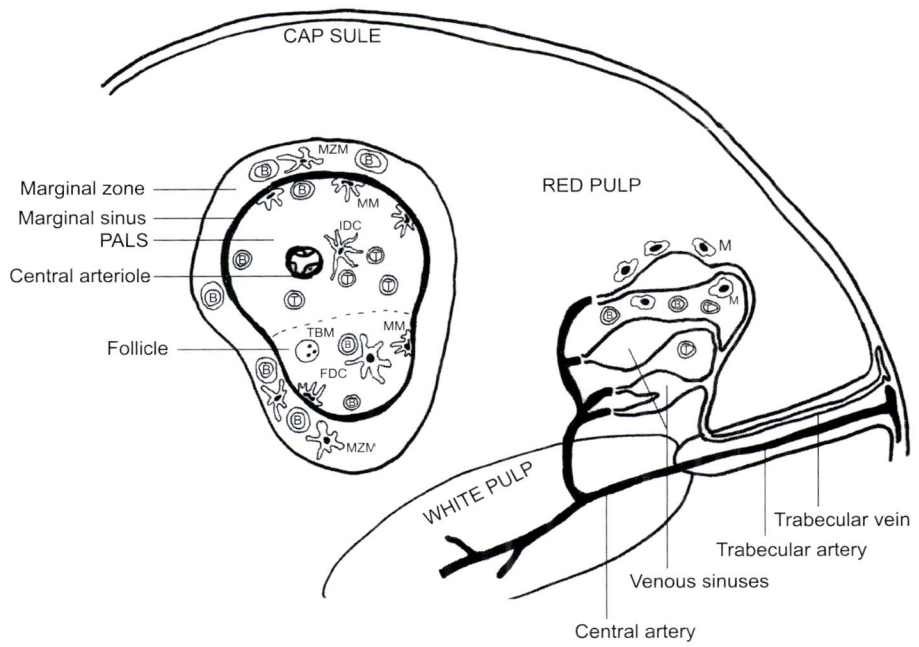

Spleen. Figure 1 Schematic presentation of the spleen compartments. Marginal zone, sinus, periarterial lymphocyte sheath (PALS) and follicle are depicted in a cross-section of the white pulp. The white pulp contains marginal zone macrophages (MZM), marginal metallophillic macrophages (MM), tingible body macrophages (TBM), interdigitating dendritic cells (IDC), follicular dendritic cells (FDC), B lymphocytes (B) and T lymphocytes (T). The splenic cords (cords of Billroth) in the red pulp contain predominantly macrophages (M), and several lymphocytes (T and B).

The white pulp reaches peak development in rats, as in man, at puberty, which coincides with peak thymus function, followed by a gradual involution with age. A loss of functional splenic lymphocytes does occur with age, however, with a corresponding increase in the number of reticular cells and macrophages. It is probable therefore that—at least in rodents—the spleen (unlike the thymus) continues to increase in weight with age, maintaining a fairly consistent ratio of organ weight to body weight (4).

Preclinical Relevance
The spleen has both lymphoid and nonlymphoid functions. Primary follicles are characteristic of germ-free rodents and are frequently found in the spleen of laboratory rodents, reflecting the relative protection of this lymphoid organ from exogenous antigen. Immunotoxicity that is manifested in the splenic lymphoid compartment can be a direct effect of a compound or occurs indirectly via effects on thymus and/or bone marrow. The reaction to antigenic substances involves germinal centre development. The enormous phagocytic system is affected by macromolecules and other "reticuloendothelial system expanders" (5) and by increased hemosiderin accumulation due to enhanced erythrocyte degradation. Hemodynamic derangements and various types of anemia lead to changes in the spleen, in the red pulp as well as in the white pulp.

Relevance to Humans
Spleen histology varies somewhat between laboratory animal species and between some laboratory species and humans, the major difference being the distinction of the boundaries between PALS, the follicles and marginal zone, and the prominence of the marginal zone. In the human spleen, the white pulp architecture and the distribution of lymphocyte subpopulations is comparable to those in the rat and mouse, although the marginal zone is less prominent or may even be absent. Also splenic function is comparable between experimental animals and humans, as is exemplified in humans by splenectomy, after which reduced nonspecific phagocytosis of nonopsonized particles, lowered serum IgM levels, and increased susceptibility to infections by encapsulated bacteria have been documented. Hemopoiesis is common in adult rats, but in humans only present under pathological conditions. Differences in immunotoxicity between laboratory animals and humans appear to depend predominantly on differences in toxicokinetics and metabolism of substances, and these aspects should therefore be con-

sidered when results from animal studies are extrapolated to humans.

Regulatory Environment
Most guidelines require that the spleen is weighed and examined histopathologically. Moreover, functional tests as required by guidelines often include the measurement of antibody responses in spleen cells, and analysis of surface markers on immune cells using blood and spleen cells.

References
1. Zapata AG, Cooper EL (1990) The Immune System: Comparative Histophysiology. John Wiley, Chichester
2. Young B, Heath JW (2000) Immune system. In: Functional Histology. Churchill Livingstone, Edinburgh, pp 193–222
3. Ochsenbein AF, Zinkernagel RM (2000) Natural antibodies and complement link innate and acquired immunity. Immunol Today 21:624–630
4. Losco P (1992) Normal development, growth, and aging of the spleen. In: Mohr U, Dungworth DL, Capen CC (eds) Pathobiology of the Aging Rat, Volume I. ILSI Press, Washington, DC, pp 75–95
5. Gopinath C, Prentice DE, Lewis DJ (eds) (1987) The lymphoid system. In: Atlas of Experimental Toxicological Pathology. MTP Press, Lancaster, pp 122–137

Split-Plot or Split-Unit Design

An experimental design structure in which there are variance components which must be estimated between subjects as well as within subjects.
▶ Statistics in Immunotoxicology

Split-Well Analysis

To confirm the specificity of cytotoxic analyses, multiple target cells are used.
▶ Limiting Dilution Analysis

Spot-Forming Cells

▶ Plaque-Forming Cell Assays

SRBC ELISA

▶ Plaque Versus ELISA Assays. Evaluation of Humoral Immune Responses to T-Dependent Antigens

Src Family Kinases

A group of protein kinases anchored to the cell membrane by a lipid moiety attached to their amino terminal region. Src family kinases are responsible for phosphorylating other proteins. They are activated by phosphorylation of their tyrosine residue in the kinase domain and inhibited by phosphorylation of the tyrosine residue in their carboxy terminus.
▶ Signal Transduction During Lymphocyte Activation

Src Homology 2 (SH2) Domain

A protein domain found in many signaling proteins that binds to phosphotyrosine residues on other proteins. SH2 domains are used to recruit intracellular signaling molecules to the activated receptors on lymphocytes.
▶ Signal Transduction During Lymphocyte Activation

Src Homology 3 (SH3) Domain

A protein domain found in many signaling proteins. SH3 domains bind proline-rich regions in other proteins recruiting these other proteins to the signaling complexes (see Adaptors).
▶ Signal Transduction During Lymphocyte Activation

Statistics in Immunotoxicology

MICHAEL L KASHON
Biostatistics Branch
National Institute for Occupational Safety and Health
Morgantown, WV 26505
USA

Synonyms
Quantitative analysis, experimental design, modeling, risk assessment

Definition
Statistical methodologies are powerful and indispensable tools in risk assessment and the experimental analysis of the immune system. As such, they are utilized at all levels including hazard identification, dose-response assessment, exposure assessment, and risk characterization. Statistical methods are utilized to

summarize, model, and draw inferences based on empirically collected data.

Characteristics

A detailed description of the variety of statistical techniques that are utilized in immunotoxicology and toxicology in general is well beyond the scope of this essay. A well written and practical guide for the appropriate use of various statistics in the field of toxicology is presented by Gad and Weil (1). This essay will emphasize the inseparable link between sound experimental design and valid statistical tests, and give an overview of some of the fundamental statistical tests used in immunotoxicology experiments. In all experiments, the process in which data are collected, and the nature of the data (that is, continuous, discrete, categorical) will dictate the type of statistical analysis that is performed. It is critical that effective communication between statisticians and research scientists occur at the time of study design to ensure that the experimental design is appropriate and that the data are collected in a manner that allows the hypothesis under investigation to be answered using the appropriate statistical methods.

Experimental Design Issues

The process in which the data are collected, irrespective of the specific laboratory techniques used, is a direct reference to the experimental design. Experimental design can be parceled into two parts including the treatment structure and the design structure.
- Treatment structure appropriately belongs in the realm of the scientist and reflects such things as the choice of treatment groups, dosing regimens, time courses, and the selection of appropriate control groups.
- The design structure is in the realm of the statistician and reflects issues including randomization of experimental units, appropriate degree of replication for sufficient power, prevention of confounding, and generally striving to obtain the maximum amount of information from a finite pool of resources. Some common design structures include the completely randomized design, ▶ randomized complete blocks design, split-plot or split-unit designs, ▶ crossover designs, and ▶ repeated measures designs.

It is important to note that for a given treatment structure, any number of design structures are appropriate, and the selection of a particular design structure is determined by factors not limited to the treatment structure.

Analysis Issues

With the exception of the completely randomized design (the simplest design structure) all of these experimental designs fall into the category of analytical procedures known as mixed models, in that they have more than one random variance component that must be estimated.

The simplest of these is the randomized complete blocks design in which groups of experimental subjects, representing one from each treatment combination, are processed simultaneously for many or all parts of the experiment. The number of these groups or blocks determines the sample size. This design is particularly useful when labor intensive laboratory procedures limit the number of samples that can be processed simultaneously. Ultimately, any variation which occurs from block to block is accounted for in the analysis and removed from the error term that is utilized for the statistical test. This reduction in the magnitude of the experimental error term increases the likelihood of finding significant treatment effects. Mixed models can be complex and the analyses require that the covariance structure of the random effects (blocking factors, correlation structure of repeated measures) be fit prior to estimating the fixed effects of the experiment (dose, sex, time). It is often the case that an investigator has utilized one of the above designs without realizing it and proceeds with data analysis as though it were a completely randomized design. Unfortunately the incorrect—usually inflated—error term is utilized in tests of significance, causing a type II error and significant treatment effects are missed. Some very powerful mixed model statistical procedures are available in SAS (2) and the analysis of such data should be performed by those who are familiar with the program to ensure proper coding. Types of data subject to statistical analysis include numerical, ordinal and nominal. Numerical data can be either continuous as in the case of organ weights, or discrete as in the case of tumor incidence. Ordinal data are essentially ranks and might include a scale indicating the severity of histopathological changes. Finally, nominal data are categories and might include the presence or absence of a particular attribute.

Data generated by many immune function assays are continuous numerical data such as organ weights, or burden of infectious agent in host-resistance assays. Other types include discrete count data such as the number of antibody-producing cells which are in such high numbers that they can be treated as though they are continuous numerical data. Many of these variables follow a normal or Gaussian distribution and can be analyzed using normality based statistical methods, such as ▶ analysis of variance, ▶ analysis of covariance, and ▶ regression analysis.

However, many biological parameters follow a lognormal distribution and thus may violate the assumptions of the above-mentioned statistical tests, particularly the assumption regarding homogeneity of variance (thus rendering inaccurate P values). Traditionally, these data have been analyzed using normality based statistical tests following a transformation of the data which serves to reign in extreme values and make the variances comparable between groups. Another analytical approach which allows for the original data to be analyzed in the case of heterogeneous error variance is to model each group variance separately. This can be performed using the SAS procedure Proc Mixed (2,3). Additional options include using nonparametric statistical methods which make no assumptions regarding the underlying distribution of the data (4). These tests can often be as robust and as powerful as their parametric counterparts.

All of the normality based statistical analyses described above fall into the category of linear models. Linear models assume that the response variable is a linear combination of the independent variables, and that the model is linear in all of its parameters. There are many cases in which linear models are inadequate to describe the data appropriately. When this occurs, one can use nonlinear models which do not force the data to fit a given linear structure when in fact a linear model is inadequate.

A good example of the use of nonlinear models is in the case of the local lymph node assay (LLNA). This assay is often used in hazard identification to determine the sensitizing potential of a given compound. Typically, if application of a chemical to the skin results in a three-fold or greater increase in cell proliferation in the local draining lymph nodes, then the chemical is marked as a potential sensitizer. Estimation of the concentration which yields a three-fold increase in cell proliferation (EC3) was initially performed by interpolation between two doses; one above the point in which a three-fold stimulation relative to control animals occurs, and one below. Nonlinear regression models allow for a better fit of the available dose-response relationship, and allow for more precise estimates of the EC3 (5). Additional analysis of these data utilizing ▶ bootstrap resampling methods allows for confidence intervals to be calculated around this point estimate and thus yields an uncertainty margin for sensitizing concentrations of a given chemical.

Many host-resistance assays generate nominal data, such as whether the animal survives a specified length of time or not. These data are not amenable to the statistical analyses under linear or nonlinear models. The data are often analyzed in a contingency table using a Chi-squared statistic, or exact probabilities are generated using Fisher's exact test to assess significance of the treatments. If dose-response studies are utilized in the host-resistance assay, tests for trend can be performed using the Cochran-Armitage linear trend test. It is also feasible to record the length of time taken for an animal to show clinical signs of disease. When time to an event is the outcome of an experiment, then survival analysis is the appropriate statistical analysis to use.

Screening procedures for evaluating the potential immunotoxicity of compounds will utilize dose-response experiments and often attempt to determine at which dose a compound is toxic. Answers to these questions require that multiple comparisons between treatment groups be made, and how these are made should be determined prior to the experiment. It is common practice to compare all treatment groups with a single control group; this results in a series of statistical tests which are not independent of one another, and thus influences the overall error rate, specifically increasing the likelihood of a type I error. Adjustments such as Dunnett's test will make appropriate changes to the critical value of the t test and reduce the family-wise error rate.

When pair-wise comparisons between all treatment groups are desired, several useful procedures for adjusting the overall error rate can be found in nearly all textbooks on experimental design. Often the effect of a compound on immune function is strongly suspected or known to occur in a specific direction (it increases or decreases). When this is the case, an increase in the power of the statistical test can be achieved by using an ordered alternative hypothesis. There are parametric as well as nonparametric versions of these tests which derive order-restricted means and calculate tests of significance using the new means.

Immunotoxicity screening studies are often performed using a tiered testing approach and the assessment of multiple endpoints. In this regard, tests which assess cell-mediated immunity, humoral-mediated immunity, and various immunopathologic parameters are often performed such that the likelihood of determining whether a compound is immunotoxic is increased at the expense of knowledge regarding specificity of the immune defect. Positive indications of toxicity in the first tier lead to more detailed tests of immune function in a second tier, including host-resistance assays. The result of this approach is that the identification of immunotoxic chemicals, and the organ systems affected, will not be determined from a single immune function test and reliance on its associated statistical analysis. This strategy of utilizing converging lines of evidence to conclude that a given compound is immunotoxic appropriately places the emphasis on proper experimental design.

In the end, so long as an experiment is designed and carried out appropriately, a legitimate statistical model can be fit and the effects of a given compound can be accurately assessed. However, there is often little that can be done statistically to salvage a poorly designed or poorly performed experiment.

Regulatory Environment

While specific guidelines for immunotoxic testing of compounds have been published by the Environmental Protection Agency (EPA), the Food and Drug Administration (FDA), and the (EMEA), there are no specific guidelines regarding the use of statistical methods in immunotoxicological studies. There is however, a section from FDA/CFSAN with statistical considerations for toxicity studies examining food ingredients. Further, there are clear statistical guidelines put forth by the FDA with respect to statistical methods used in clinical trials which would cover investigations of immunotoxic potential of compounds for consumption or medicinal purposes.

References

1. Gad SC, Weil CS (1994) Statistics for toxicologists. In: Hayes AW (ed) Principles and Methods of Toxicology, 3rd edn. Raven Press, New York, pp 221–274
2. Littell RC, Milliken GA, Stroup WW, Wolfinger RD (1996) SAS System for Mixed Models. SAS Institute, Cary
3. Milliken GA, Johnson DE (2002) Analysis of Messy Data, Volume III: Analysis of Covariance. Chapman and Hall/CRC, New York
4. Hollander M, Wolfe DA (1999) Nonparametric Statistical Methods, 2nd ed. John Wiley, New York
5. van Och FMM, Slob W, de Jong WH, Vandebriel RJ, van Loveren H (2000) A quantitative method for assessing the sensitizing potency of low molecular weight chemicals using a local lymph node assay: employment of a regression method that includes determination of the uncertainty margins. Toxicology 146:49–59

Stem Cells

A class of cells having two generative capabilities: replication and formation of more differentiated offspring.

▶ Colony-Forming Unit Assay: Methods and Implications

Stereochemistry

The spatial relationship of atoms and groups in a molecule of a substance and its effect(s) on the properties of the substance.

▶ Chromium and the Immune System

Sterilizing Immunity

▶ Attenuated Organisms as Vaccines

Steroid Hormones

Steroid hormones are lipophilic compounds derived from cholesterol metabolism, which typically mediate their biological activity by binding to a specific intracellular cytosolic receptor.

▶ Steroid Hormones and their Effect on the Immune System

Steroid Hormones and their Effect on the Immune System

MICHAEL LAIOSA
NIAID/NIH
Center Dr., Room 111
Bethesda, MD 20892
USA

Synonyms

steroid hormones, gonadal hormones, estrogen, testosterone, androgens, glucocorticoids, glucortisol, environmental estrogens, endocrine disrupting chemicals

Definition

Steroid hormones, which include ▶ glucocorticoids and the sex hormones ▶ estrogen, ▶ progesterone, and testosterone, possess immunomodulatory roles that include effects on T and B cell development, lymphoid organ size, lymphocyte cell death, immune function, and susceptibility to autoimmune disease. ▶ Steroid hormones are lipophilic compounds derived from cholesterol metabolism, which typically mediate their biological activity by binding to an intracellular cytosolic receptor. Hormone binding to its receptor causes the receptor to translocate to the nucleus where it forms a homodimer with another ligand-activated receptor. The steroid receptor homodimer will then bind to specific DNA sequences in steroid responsive genes modulating their transcription (Figure 1) (1–4). Biological specificity of steroid hormone activity is dependent on interaction between the hormone and its specific receptor—glucocorticoid receptor (GR), progesterone receptor (PR), estrogen receptor (ER), and ▶ androgen receptor (AR). All four ▶ steroid receptors have been found in immune cells and tissues, making the immune system an important target for steroid activity (1–4).

Environmental estrogens or ▶ endocrine disrupters are chemicals found throughout the environment that have estrogenic activity and include compounds in plastics such as bisphenol-A, and phthalates, in detergents and surfactants such as octylphenol and nonylphenol, in pesticides such as methoxychlor, dichlorodiphenyl-trichloroethane (DDT), hexachlorobenzene, and dieldrin, in industrial chemicals such as polychlorinated biphenyls (PCBs) and 2,3,7,8,tetrachlorodibenzo-p-dioxin (TCDD), and in natural plant estrogens (genistein and coumesterol) (5). The presence of steroid receptors in immune tissues and cells make the immune system a potential target of environmental estrogens because of the ability of these molecules to bind to steroid receptors, mimic hormones, antagonize hormones, alter hormonal binding to receptors, and/or alter metabolism of natural or endogenous hormones (3–5).

Characteristics

Glucocorticoids

Glucocorticoids mediate their biological activity by binding to the intracellular cytosolic glucocorticoid receptor (GR). The GR exists in the cytosol in a complex with heat shock proteins and immunophilin proteins. Receptor binding by ligand leads to the GR translocating to the nucleus where it forms a homodimer and binds to specific DNA sequences referred to as glucocorticoid responsive elements (GREs). GR binding to GREs may enhance or inhibit transcription of the particular gene (1). Additional biological activity of glucocorticoids is conferred through cross-talk between the ligand-activated GR and other transcription factors including AP-1, nuclear factor NFκB, CREB, STAT3 and STAT5 (1).

One of the most profound effects of glucocorticoids on the immune system is the rapid thymic atrophy that occurs within hours after exposure to a pharmacological dose of the hormone. Moreover, thymic atrophy has been observed following periods of psychological, emotional, or physical stress when glucocorticoid levels are abnormally elevated. Glucocorticoid-induced thymic atrophy has been attributed to selective loss of $CD4^+CD8^+$ thymocytes due to apoptosis or programmed cell death (1). Moreover, the apoptosis in $CD4^+CD8^+$ thymocytes induced by glucocorticoids is due to the low levels of the antiapoptotic protein Bcl-2 found in these cells compared with other thymocytes and T-cells (1).

Although $CD4^+CD8^+$ thymocytes are acutely sensitive to glucocorticoid-induced apoptosis during drug administration or periods of stress, peripheral T-cells are not induced to die. Nevertheless, it has been long appreciated that glucocorticoids have potent immunosuppressive effects on peripheral T-cells and have been used extensively for treating a variety of autoimmune and inflammatory diseases. Recent studies have demonstrated that glucocorticoids mediate their immunosuppressive effects by inhibiting the production of T-cell growth factors and cytokines, including interleukins IL-1, IL-2, IL-3, IL-4, IL-6, IL-7, IL-10, IL-13, granulocyte-macrophage colony-stimulating factor (GM-CSF), the tumor necrosis factor TNF-α, and the interferon IFN-γ (1). Inhibition of cytokine pro-

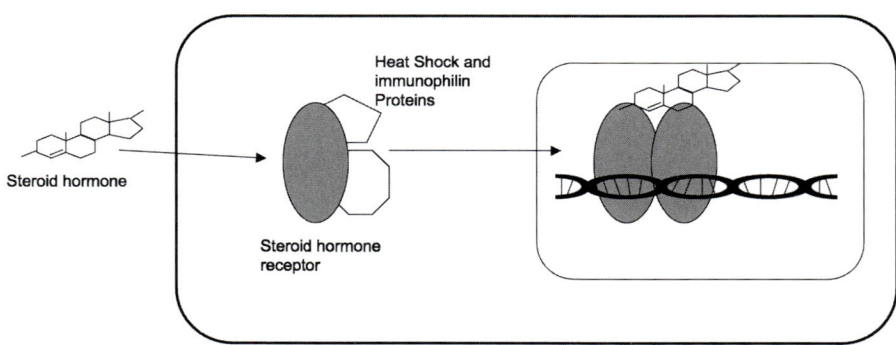

Steroid Hormones and their Effect on the Immune System. Figure 1 General biological response to steroid hormones: Steroids are lipophillic compounds that freely pass through the cellular membrane where they bind to a specific steroid hormone receptor. Steroid receptors are typically localized in the cytosol in an inactive state, complexed with heat shock and immunophilin proteins (GR, PR, AR), but may be found in the nucleus (ER). Once activated by binding to their specific ligand, steroid receptors translocate to the nucleus and form homodimers which then modulate expression of genes containing response elements specific to the ligand—steroid receptor homodimer. In some instances, additional biological responses may be made by cross talk between the activated steroid receptor and other nuclear transcription factors (GR can interact with AP-1, NF-kB, CREB, STAT3 and STAT5 transcription factors).

duction appears to be at the level of gene expression such that the ligand activated GR may form protein-protein interactions with other transcription factors thereby inhibiting DNA binding and function. One such factor, AP-1, has been implicated in the GR-mediated inhibition of IL-2 transcription (1).

Androgens

Biological activity of androgens is mediated through their interaction with the intracellular androgen receptor. As with the glucocorticoid receptor, the AR is a ligand-activated transcription factor that is involved in modulation of gene expression. Castration and hormone replacement studies have revealed androgenic effects on both B- and T-cell development (3). In numerous different organisms including cattle and rodents significant thymic enlargement has been observed following male castration. Moreover, thymic enlargement following castration was even seen in aged males (3). Using mice that possess a spontaneous mutation in their AR and thus are androgen unresponsive, it has been shown that the effect of thymus enlargement following castration is AR dependent (3).
The AR protein has been found in all thymocyte subsets with the highest expression found in the immature $CD8^+CD3^{lo}$ thymocyte subset (3). AR expression has also been found in the cortical and medullary stromal elements of the thymus (3). To determine the cellular target of androgen activity, a set of in vivo experiments utilizing both AR defective, and C57Bl/6 mice were used to generate hematopoietic chimeras such that the ▶ stromal cells expressed the normal AR and the thymocytes expressed the mutant AR. In these AR chimeric mice it was found that the thymus size would only enlarge following castration (and shrink with androgen replacement) when a functional AR was present in the stromal tissue (3). These data indicate that AR expression in thymic stromal cells is necessary for mediating androgen-induced thymic atrophy. It has been proposed that the mechanism for androgen-induced thymic atrophy may be due to accelerated thymocyte apoptosis mediated by the thymic stromal cells (3). Alternatively, the AR may increase transcription of cytokines such as the transforming growth factor TGF-β, which would inhibit T-cell development (3).
B-cell development in the bone marrow is also affected by androgens with significant expansion of immature B-cells occurring following male castration. Furthermore, androgen replacement reverses the B-cell expansion in the bone marrow, but has no affect on spleen size or B-cell number. The probable explanation for why B-cells in the bone marrow are androgen sensitive, but splenic B-cells are not, comes from the finding that the AR is expressed only in immature B lymphocytes but not in mature splenic B-cells (3).

Progesterones

Unlike AR expression, which has been found in both lymphocytes and stromal cells, the progesterone receptor (PR) has thus far only been found in stromal cells in the thymus (4). Nevertheless, at doses of progesterone comparable to the levels present during pregnancy, progesterone has been shown to cause thymic atrophy, especially when administered simultaneously with estrogen (4). Moreover, thymic atrophy following simultaneous administration of both estrogen and progesterone was shown to depend on the presence of the PR in the thymic stroma. These data indicate that progesterone inhibits T-cell development indirectly by affecting the stromal cells (4).

Estrogens

Classical estrogen signaling is mediated by the estrogen receptors ERα and ERβ. The inactive receptors are localized to the nucleus where they remain in a complex with other proteins until ligand activation results in the ER binding to and activating genes that possess estrogen-responsive elements in their promoters. There is high homology between the DNA-binding domains of ERα and ERβ and, therefore, it has been hypothesized that functional specificity is likely to be conferred by the differential tissue expression observed between the two receptors (2).
The list of immunological endpoints that are implicated as targets of estrogenic activity is considerable and diverse. Estrogen has been shown to inhibit both B- and T-cell development (5). Additionally, estrogen has effects on B and T lymphocytes in the lymph node and spleen (5). Both lymphocytes and non-lymphoid stromal cells have been shown to express estrogen receptors, indicating there may be multiple cellular targets of estrogen in the immune system (5). Estrogen also is suspected to play an important role in autoimmune diseases, especially systemic lupus erythematosus (SLE), because of the disproportionate number of women who are affected by this disorder of the immune system (5-7). B-cells producing immunoglobulins directed against self antigens, anti-DNA antibodies, anti-actin, and anti-cardiolipin antibodies have all been found to be elevated in mice exposed to estrogen (5). Plasma B-cell numbers in the spleen (mature B-cells) were found to be nearly 10 times higher in these same mice (5). Moreover, splenic B-cells from estrogen-treated mice had heightened expression of the anti-apoptotic protein Bcl-2 and were consequently more resistant to apoptosis following activation (5).
Immunological defects have also been observed following exposure to the broad class of chemicals collectively referred to as environmental estrogens or endocrine disruptors. Bottle-nosed dolphins found in the Gulf of Mexico experienced decreased proliferative responses to T-cell mitogens correlating with elevated

DDT and PCB levels (5). A similar inhibition of splenocyte and peripheral blood leukocyte T-cell proliferation was observed in DDT-exposed and PCB-exposed Beluga whales (5). Endocrine disruptors have also been shown to inhibit lymphocyte development in chicks where embryos exposed to PCB 126 during incubation had decreased numbers of B- and T-cells (5). Furthermore, male rats exposed to the pesticide methoxychlor during perinatal and prepubertal developmental periods had decreased antibody production and thymic weights (5).

In addition to the effects of endocrine disruptors on B- and T-lymphocytes, macrophages and antigen-presenting cells have also been shown to be affected by endocrine disruptors. For example, bisphenol-A has been shown to inhibit plastic adherence of rat peritoneal macrophages. The adherence deficiency observed in vitro is therefore likely to affect the process of phagocytosis, which is dependent on macrophage adherence to surfaces and particles (5).

Preclinical Relevance

One of the most striking aspects of steroid hormones is their influence on the gender bias of autoimmune diseases, whereby women are afflicted far more and often with much greater severity than men (6,7). Approximately 5% of the population in the USA has been diagnosed with an autoimmune disorder and, depending on which specific disorder one is analyzing, 60%–95% of the afflicted patients are women (6,7). Specifically, 60%–70% of people diagnosed with multiple sclerosis (MS) and rheumatoid arthritis (RA) are women (7). Women make up 85%–95% of all cases of thyroid disease, SLE, and Sjögren syndrome (7). Evidence of the influence of steroid hormones comes in part from observations made during pregnancy when hormonal fluctuations, notably estrogen and progesterone, are the greatest. The severity of MS and RA is lessened during the third trimester of pregnancy when estrogen and progesterone levels are highest. However, flare-ups in the disease state are often observed when hormone levels drop after birth (7). In contrast to these diseases, SLE seems to worsen during pregnancy. The differences in disease states have been attributed to direct hormonal effects on T-cell-dependent responses known as T-helper type 1 (Th1) and T-helper type 2 (Th2) responses. Th2-dependent responses, which enhance antibody production, appear to be favored during pregnancy (7).

In addition to the direct affects of sex steroids on the

Steroid Hormones and their Effect on the Immune System. Table 1 Effects of steroid hormones.

Steroid hormone	Receptor	Developmental effects	Peripheral immune effects
Glucocorticoids Cortisol Corticosterone	GR	Thymic atrophy Apoptosis of $CD4^+CD8^+$ thymocytes	Immunosuppressive Inhibits T-cell growth factors IL-1, IL-2, IL-3, IL-4, IL-6, IL-7, IL-10, IL-13, GM-CSF, TNFα, IFNγ
Progesterone	PR	Thymic atrophy	Unknown
Androgens Dihydrotestosterone	AR	Thymus size and bone marrow B-cell number increase following castration and decrease with testosterone administration	No effect on splenic B-cells due to loss of AR expression as B-cells mature
Estrogens 17β-estradiol	ER	Both T-cell and B-cell development are inhibited	Women tend to have higher incidence and severity of autoimmune diseases T-helper type 2-dependent responses are enhanced during pregnancy Animal models exposed to elevated estrogen have higher titers of auto-antibodies, and plasma B-cell numbers are increased
Environment disrupting chemicals (EDCs)	Various	Thymic atrophy Inhibition of B-cell and T-cell development	Decreased proliferative responses to T-cell mitogens Decreased antibody production Decreased phagocytosis of macrophages Regulation of cytokines and chemokines has been shown in vitro to be affected by various EDCs

Steroid Hormones and their Effect on the Immune System. Figure 2 Chemical Structure of four principle steroid hormones, corticosterone, 17β-estradiol, progesterone and dihydrotestosterone.

immune system, which make women more susceptible to autoimmune diseases, additional indirect neural-endocrine-immune interactions may also influence the immune response (7). In a potential feedback loop involving glucocorticoids, sex hormones are known to modulate the hypothalamic-pituitary-adrenal axis, possibly affecting stress responses. Interestingly, females tend to have higher circulating concentrations of the glucocorticoid hormone corticosterone. Furthermore, glucocorticoids can suppress sex hormone production and action (7).

Relevance to Humans

The prevalence of autoimmune disease in women is significantly higher than for men. However, the precise mechanism by which different hormones may contribute to the type and severity of disease is still unknown. Moreover, studies in animal models suggest there are likely to be genetic differences, which make certain individuals and/or populations more susceptible to environmental factors (including steroid hormones), which may affect disease onset and presentation (7).

In addition to the effects of steroids on autoimmune disease, a growing number of studies on humans accidentally exposed to a variety of endocrine disruptors have revealed numerous alterations to the immune system. Men who ingested PCB-contaminated fish have decreased activity of natural killer cells compared to unexposed individuals (5). Workers exposed to pesticides have enhanced macrophage activity and reduced antibodies (5). Additionally, workers who have been chronically exposed to pesticides have a higher ratio of $CD4^+$ to $CD8^+$ lymphocytes (5).

Diethylstilbestrol (DES) is a model endocrine disruptor, originally used under the mistaken assumption that it helped to prevent miscarriages. It has been shown to have negative health effects in both the mothers and children exposed prenatally. Exposure has been associated with an increased risk for clear cell adenocarcinoma (a rare vaginal/cervical cancer) in prenatally exposed daughters, as well as with significant increases in autoimmune diseases (5).

Finally, a number of endocrine disruptors have been shown to cause changes in cytokine and chemokine regulation in vitro using human cell lines and human estrogen receptor constructs (5). However, far more research is needed to assess the mechanism of action of this large class of compounds, the effects of these compounds in vivo, and dosage ranges where deleterious effects may be seen.

Regulatory Environment

The understanding that there are compounds in the environment possessing properties that can modulate hormonal pathways and responses has captivated worldwide attention. As a result, several government-sponsored agencies and committees have initiated research goals, guidelines, and rules for identifying endocrine disrupting chemicals (EDCs), and determining the risk of EDCs to the health of wildlife and human populations. Specifically, in 1995 US Environmental Protection Agency held two international workshops to identify research needs for risk assessment of EDCs. The workshops suggested the potential effects of these chemicals on reproductive, neurological, and immunological tissues, as well as carcinogenesis, were especially important to understand and expand research (8). In 1996 the Food Quality Protection Act and the Safe Drinking Water Act were amended by US Congress to include testing of use of pesticides in food and drinking water contaminants, for estrogenicity and other hormonal activity (8). In 1998 the Endocrine Disruptor Screening and Testing Advisory Committee (EDSTAC), under the auspices of the EPA, defined 'other hormonal activity' to be androgens and compounds affecting thyroid function (8). EDSTAC also recommended an extensive process for prioritizing, screening, and testing the nearly 70 000 chemicals regulated by the EPA for endocrine disrupting activity. The screening battery involves a combination of in vitro and in vivo assays involving several different taxa (8).

In 1998 the EPA-sponsored Research Plan for Endocrine Disruptors posed a number of questions to stimulate research priorities. Some of the questions were:

- What effects are occurring in populations?
- What are the chemical classes and their potencies?

- What are the dose-response characteristics in the low-dose region?
- Do our testing guidelines adequately evaluate potential endocrine-mediated effects?
- What extrapolation tools are needed?
- What are the effects of exposure to multiple EDCs and will a toxic equivalency factor approach be feasible?
- How and to what degree are human and wildlife populations exposed to EDCs?
- What are the major sources and environmental fates of EDCs?
- How can unreasonable risks be managed?

Current progress on each of these questions can be found in the excellent review by Daston et al (8). More recently, the Global Assessment of the State of the Science of Endocrine Disruptors was published in 2002 by the World Health Organization, United Nations Environmental Program, and International Labor Organization. This report concluded that there is enough evidence from the known effects of endogenous and exogenous hormones to suggest EDCs are capable of harming certain human biological functions, including reproductive and developing systems. The report goes on to emphasize that the possible effects on human populations is a cause for concern and is an area of high research priority (9).

References

1. Ashwell JD, Lu FW, Vacchio MS (2000) Glucocorticoids in T cell development and function. Annu Rev Immunol 18:309–345
2. Hall JM, Couse JF, Korach KS (2001) The multifaceted mechanisms of estradiol and estrogen receptor signaling. J Biol Chem 276:36869–36872
3. Olsen NJ, Kovacs WJ (2001) Effects of androgens on T and B lymphocyte development. Immunol Res 23:281–288
4. Tibbetts TA, DeMayo F, Rich S, Conneely OM, O'Malley BW (1999) Progesterone receptors in the thymus are required for thymic involution during pregnancy and for normal fertility. Proc Natl Acad Sci USA 96:12021–12026
5. Ahmed SA (2000) The immune system as a potential target for environmental estrogens (endocrine disrupters): a new emerging field. Toxicology 150:191–206
6. Vidaver R (2002) Molecular and clinical evidence of the role of estrogen in lupus. Trends Immunol 23:229–230
7. Whitacre CC (2001) Sex differences in autoimmune disease. Nat Immunol 2:777–780
8. Daston GP, Cook JC, Kavlock RJ (2003) Uncertainties for endocrine disrupters: our view on progress. Toxicol Sci 74:245–252
9. Damstra T (2003) Endocrine disrupters: the need for a refocused vision. Toxicol Sci 74:231–232

Steroid Receptors

Steroid receptors are typically localized in the cytoplasm in an inactive state until activated by binding to their specific steroid hormone ligand: glucocorticoids with glucocorticoid receptors, estrogen with estrogen receptors, testosterone with androgen receptors, and progesterone with progesterone receptors. When the hormone binds to its receptor it causes the receptor to translocate to the nucleus, where it forms a homodimer with another ligand-activated receptor. The steroid receptor homodimer will then bind to specific DNA sequences in steroid-responsive genes modulating their transcription.

▶ Steroid Hormones and their Effect on the Immune System

Stevens–Johnson Syndrome (SJS)

This syndrome is characterized by widely distributed erythematous or purpuric macules and papules, with fever and involvement of two or more mucus membrane surfaces. Although the term is frequently used as a synonym for erythema multiforme major, this syndrome is more severe, with target-like lesions extending to the trunk and face, and it is though to be drug-induced. According to the criteria proposed by Roujeau et al (1), this syndrome is defined as cases with limited areas of epidermal detachment (< 10%).

▶ Drugs, Allergy to

Stimulation Index

For each concentration of test chemical a stimulation index (SI) is calculated relative to the concurrent vehicle control. By definition the SI of the control (vehicle) group is set to 1. The interpretation of results is based upon derivation of these SI, thus an SI of 2 means a doubling of the values measured for the controls.

▶ Local Lymph Node Assay (IMDS), Modifications

Stomatitis

▶ Oral Mucositis and Immunotoxicology

Streptococcal Enterotoxin B (SEB)

A protein toxin produced by *Streptococcus* that acts as a superantigen for mouse and human T lymphocytes.
▶ Polyclonal Activators

Streptococcus Infection and Immunity

S GAYLEN BRADLEY
College of Medicine/Research Affairs
Penn State University
500 University Drive
Hershey, PA 17033
USA

Synonyms
host resistance assays, pneumococcal disease models, in vivo immunotoxicology testing

Definition
Streptococci are spherical Gram-positive bacteria that lack the enzyme catalase and grow as ▶ facultative anaerobes. They are members of the so-called "lactic acid" bacteria because they share with *Lactobacillus* the ability to grown in high concentrations of sodium chloride and produce lactic acid as a primary endproduct of carbohydrate metabolism. Streptococci are widely distributed in nature with respect to host species and site of colonization on or in a host. They are members of the normal microbial flora of the mouth and intestinal tract of humans, but the genus also includes agents of serious human disease, in part by invasive processes and in part by autoimmune-like processes. *Streptococcus* is a heterogeneous genus, some members of which produce polysaccharide capsules, while others produce a variety of extracellular ▶ virulence factors.

Characteristics
Streptococci are important agents in human infectious disease throughout the lifespan, and therefore drugs that alter host resistance may have serious consequences. *Streptococcus pyogenes* may invade the skin leading to edema or enter the lymphatics and ultimately the blood stream after trauma or surgical wounds, or establish local infections in the throat that may spread to the middle ear and the meninges. In the course of streptococcal ▶ bacteremia, the bacteria may attach to heart valves, leading to fatal acute endocarditis. Even members of the normal streptococcal flora may enter the circulatory system from the oral cavity, intestine, or urinary track, leading to subacute endocarditis. In streptococcal toxic shock syndrome, the cell wall protein antigens act as superantigens, stimulating T cells to release ▶ cytokines that mediate shock and tissue injury. Several weeks after a streptococcal infection, especially a sore throat, the host may react to bacterial cell membrane antigens that cross-react with human heart tissue antigens, leading to rheumatic fever—the most serious sequela of *Streptococcus* infection. Alternatively, antibody complexes involving *Streptococcus* may attach to the glomerular basement membrane, leading to blood in the urine and acute glomerulonephritis. The organism multiplies in the tissues, and the pneumococcal cell wall activates procoagulation activity at the surface of the endothelial cell, leading to release of fibrin into the alveoli (1).

Enterococcus faecalis is a member of the normal intestinal floral, but it is among the more common agents causing ▶ nosocomial infections, especially in intensive care units of hospitals. The prevalence and seriousness in the hospital setting reflects the high level of resistance of *E faecalis* to most antibiotics.

The streptococci have a wide range of virulence factors and factors to protect the bacterium from host defenses. Anticapsular antibody to one of the many polysaccharide capsular antigenic types does not afford protection against infection by *S. pneumoniae* with a different capsular type. In *S. pyogenes,* the cell surface M protein protects the bacterium from ▶ phagocytosis by ▶ polymorphonuclear leukocytes. M protein preparations contain antigenic determinants eliciting antibodies that cross-react with human cardiac sarcolemma. The streptococcal cell wall contains many components contributing to the organism's virulence, including recruitment of leukocytes into the lung and subarachnoid space, enhancement of permeability of cerebral endothelia and alveolar epithelia, induction of cytokine production, initiation of the procoagulation cascade, and stimulation of platelet-activating factor production (2). Lipoteichoic acid is a cell wall constituent that facilitates colonization of host structures by forming a complex with the M protein of the bacterium and fibronectin on epithelial cells of the host. C-reactive protein, a serum protein that binds the lipoteichoic acid of *S. pneumoniae,* has bactericidal activity and blocks the adherence of the bacterium to the platelet- activating factor receptor on the epithelial surface (3). Streptokinase is an enzyme that converts human plasminogen into plasmin, an activated enzyme that digests fibrin. Hyaluronidase is an enzyme that degrades connective tissue thereby allowing *S. pyogenes* to spread more rapidly. Streptococci produce several hemolysins, enzymes that lyse erythrocytes. Two well-characterized hemolysins are streptolysin O and streptolysin S. *S. pneumoniae* does not produce a battery of toxins and enzymes but relies on its polysaccharide capsule to avoid host defenses.

The immune system is a complex interactive network of cells and humoral products that protects the intact animal from infectious agents and other foreign matter, and removes certain damaged or markedly altered host cells (4). Numerous arms of the immune system are available to the host in its defense against *S. pneumoniae* infections; and the sum of the integrated attack on the invader is more effective than the sum of the parts. ▶ Complement, for example, enhances recruitment of polymorphonuclear leukocytes to infected sites. The first line of defense against *S. pneumoniae* consists of phagocytosis and killing of the bacteria after ▶ opsonization by complement C3 (5,6). A successful host defense must not only stop the growth of the invading bacterium, but must also foster repair of damage to host tissue. During the recovery process, polymorphonuclear leukocytes recruited to the infected lung adhere to the pulmonary endothelium leading to the release of proteases that can digest the fibrotic deposits (1).

Preclinical Relevance

The clinical relevance of ▶ immunotoxicity detected in preclinical assays should be interpreted in conjunction with the pharmacologic dose, the reversibility of the immune alteration, and effects on other target organs. It is generally recognized that decreases in serum immunoglobulin are an insensitive indicator of immune perturbation and more reliable indicators must be used to detect immune alternation reliably. Several in vitro assay have proven to be useful indicators of immune impairment; for example, the T-dependent antibody response to sheep erythrocytes as detected in a hemolytic plaque assay. There is an excellent correlation between in vitro assays and host resistance assay in that impaired host resistance is invariably linked to a depressed in vitro response (7). On rare occasions statistically significant alterations have been measured in vitro, but altered host resistance has not been detected, documenting the important role of compensatory mechanisms in the complex immune network (4). The *S. pneumoniae* infectious model has been found to be a useful, reproducible, and informative tool for detecting altered host defense subsequent to exposure to a candidate drug (8). By considering timing of morbidity and mortality as well as final results, insight into the mechanism of immune impairment can be predicted. Signs of early morbidity and mortality usually indicate a diminution of complement activity. Morbidity and mortality that develop 2–4 days after challenge are indicative of an altered response of the polymorphonuclear leukocytes. Effects on this cell type may reflect impaired migration, phagocytosis, and/or nitric oxide production. Differential expression occurring beyond the fifth day after challenge indicates an effect on antibody production. The *S. pneumoniae* model is one of the most sensitive host resistance assays that has been used in evaluating immunotoxicity of drug candidates and industrial and environmental chemicals (9). One of the reasons for its predictability is that it is one of the least complicated host resistance models, and one that measures the capability of the host to mount a T-independent antibody response to the polysaccharide capsule, the functional capacity of granulocytic and monocytic cells, and serum complement activity.

There are practical and social limitations to the use of animal models as a screening tool to detect immunotoxicity. These assays use a large number of mice and are expensive to conduct. The mice require special facilities to isolate the infected animals and require special facilities to grow and manipulate the pathogen (9,10). Pathogenic strains of streptococci do not grow well on most bacteriologic media unless enriched with blood or tissue fluid, and may require an atmosphere containing 10% carbon dioxide. Accordingly, most streptococcal models rely on death as the endpoint rather than counts of colony-forming units in infected tissues and organs.

Relevance to Humans

An intact immune system is critical for good health. Modest changes in immune function, although reproducible and statistically significant, do not necessarily indicate immunotoxicity that precludes further development of a drug candidate. Experimental animal models are the best surrogates for detecting the harmful effects of drugs and chemicals and for detecting their mechanisms of action. Although the mouse is not usually considered the species of choice for toxicologic studies, the mouse immune system is the model of choice for immunotoxicologic evaluations because it is well characterized and most critical reagents are available. Animal models provide the means whereby dose-response and mechanistic studies can be performed to provide a basis for hazard evaluation as related to human risk (10).

S. pneumoniae is the causative agent of streptococcal pneumonia and bacterial meningitis. It accounts for the majority of bacterial pneumonia, in part because so many people carry strains capable of causing disease. At least 1 million children die of pneumococcal disease every year, mostly in developing countries. *S. pneumoniae* is the most common cause of bacterial meningitis in the USA (2). In the US and other developed countries, morbidity and mortality from pneumococcal diseases occur most often among the elderly population. People whose spleens have been damaged or removed, and individuals with sickle cell disease are also at increased risk to *S. pneumoniae* infection.

Numerous studies have found that host resistance models are the best available means to illustrate a link between immunosuppression and clinical manifestations of disease endpoints (5,10).

Regulatory Environment

The effects of new drugs on the immune system should be assessed before the drug is approved for clinical trials. Evidence of immunotoxicity may be detected during standard nonclinical toxicology evaluations, but additional studies directed specifically to immunotoxicologic assessment are often warranted. Any follow up investigations might include in vitro immune function or in vivo host resistance assays (10). Guidelines are given by FDA (CDER), Guidance for Industry: Immunotoxicology Evaluation of Investigational New Drugs. October 2002, p 35. (http://www.fda.gov/cder/guidance/index.htm)

References

1. Wang E, Simard, M, Ouellet N, Bergeron, Y, Beauchamp D, Bergeron MG (2002) Pathogenesis of pneumococcal pneumonia in cyclophosphamide-induced leukopenia in mice. Infect Immun 70:4226–4238
2. Tuomanen EI, Austrian R, Masure HR (1995) Pathogenesis of pneumococcal infection. N Engl J Med 332:1280–1284
3. Gould JM, Weiser JN (2002) The inhibitory effect of C-reactive protein on bacterial phosphorylcholine-platelet activating factor receptor mediated adherence is blocked by surfactant. J Infect Dis 186:361–371
4. Keil D, Luebke RW, Pruett SB (2001) Quantifying the relationships between multiple immunological parameters and host resistance: Probing the limits of reductionism. J Immunol 167:4543–4552
5. Bradley SG (1995) Streptococcus host resistance model. In: Burleson GR, Dean JH, Munson AE (eds) Methods in immunotoxicology, volume 2. New York: Wiley & Sons; 159–168
6. Bradley SG, Munson AE, McCay JA et al. (1995) Subchronic 10-day immunotoxicity of polydimethyl siloxane (silicone) fluid, gel and elastomer and polyurethane disks in female B6C3F1 mice. Drug Chem Toxicol 17:175–220
7. Luster MI, Portier C, Pait, DG et al. (1993) Risk assessment in immunotoxicology. II. Relationships between immune and host resistance tests. Fund Appl Toxicol 21:71–82
8. Bradley SG (1995) Introduction to animal models in immunotoxicology: Host resistance. In: Burleson GR, Dean JH, Munson AE (eds) Methods in immunotoxicology, volume 2. Wiley & Sons, New York, pp 135–141
9. Bradley SG (1985) Immunologic mechanisms of host resistance to bacteria and parasites. In Dean J, Luster MI, Munson AE, Amos (eds) Immunotoxicology and immunopharmacology. Raven Press, New York, pp 45–53
10. Talmadge D (Chair: Subcommittee on Immunotoxicology) (1992) Animal models for use in detecting immunotoxic potential and determining mechanisms of action. In: Biologic markers in immunotoxicology. National Academy of Science USA, Washington DC, pp 83–98

Streptococcus pneumoniae

A bacterium that causes respiratory tract infection.
▶ Host Resistance Assays

Streptokinase

Streptococcus pyogenes produces the enzyme streptokinase which activates the plasminogen activator. Plasminogen activator catalyses the conversion of plasminogen in plasmin. Plasmin degrades the fibrin layer around pathogens.
▶ Dermatological Infections

Streptozotocin-Induced Diabetes Mellitus, STZ-Induced Diabetic Rat

▶ Diabetes and Diabetes Combined with Hypertension, Experimental Models for

Streptozotocin-Induced Spontaneously Hypertensive Rat,

▶ Diabetes and Diabetes Combined with Hypertension, Experimental Models for

Stress

The sum of the biological reactions to any adverse stimulus, physical, mental, or emotional, internal or external, that tends to disturb the organism's homeostasis.
▶ Stress and the Immune System

Stress and the Immune System

HELEN G. HAGGERTY
Bristol-Myers Squibb Co.
6000 Thompson Road
East Syracuse, NY 13057
USA

Synonyms

neuroendocrine response

Definition

▶ Stress is any natural or experimentally contrived circumstance that poses an actual or perceived threat to an animal's well being. The ▶ stressor can be psychological, physical, environmental, immunological, or chemical. When an animal encounters a stressor, its "fight-or-flight" response is activated. Systems needed to deal with an immediate threat are upregulated, while those that are not necessary, such as the digestive, reproductive, and immune system are downregulated. In response to a stress, the central nervous system modulates the immune system by a complex network of signals that result in the bidirectional interaction of the nervous, endocrine, and immune systems. The hypothalamic-pituitary-adrenal (HPA) axis and the sympathetic-adrenal medullary (SAM) axis are activated, triggering a cascade of events that results in the down regulation of immune function. In addition, the activation of the immune system due to an infection, allergic response, or inflammation, can also cause an increase in the activity of the HPA axis, which in turn leads to the dampening of that immune response.

Characteristics

Stress-induced activation of the HPA axis and the SAM axis stimulates the release of corticotropin-releasing hormone (CRH) from the ▶ hypothalamus. CRH promotes the release of adrenocorticotropic hormone (ACTH) from the ▶ pituitary gland, which in turn stimulates the release of ▶ glucocorticoids (corticosterone in rats and mice and cortisol in humans) from the adrenal cortex. Glucocorticoids have been shown to have potent immunosuppressive as well as antiinflammatory and antiallergic properties. By inhibiting the function and trafficking of many immune cells such as lymphocytes, macrophages, monocytes, and neutrophils, glucocorticoids can downregulate the immune response and keep it from becoming excessive. Glucocorticoids inhibit the release of a number of inflammatory mediators and cytokines, as well as diminish the effect of these mediators on their target tissue. Circulating glucocorticoids also exert a negative feedback on the HPA axis downregulating its activity.

CRH can also stimulate the production and release of ▶ norepinephrine (noradrenaline) and ▶ epinephrine (adrenaline) from the adrenal medulla. Lymphoid and myeloid cells have been shown to exhibit β-2 adrenergic receptors that allow them to respond to signals from the SAM axis. Generally, these signals have been found to downregulate the immune response. Furthermore, the primary and secondary lymphoid tissues are innervated by noradrenergic and sympathetic nervous system, which release neurotransmitters that can subsequently affect the immune system.

Activated immune cells can also release a number of mediators, such as proinflammatory cytokines, that can then influence the HPA axis and nervous system. For example, immune cells can be stimulated to release proinflammatory cytokines such as IL-1β, which can directly stimulate the release of CRH from the hypothalmus, initiating the cascade of events described.

How stress affects the immune system can also be dependent on the degree of stress and its duration, acute or chronic. Mild stress for a short period of time can have no effect on the immune system or

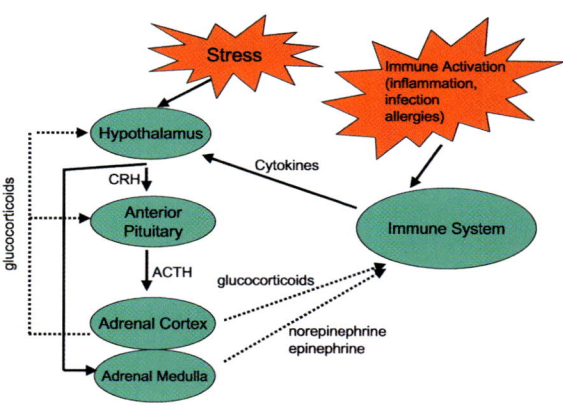

Stress and the Immune System. Figure 1 In response to a stress or immune activation, the hypothalamic-pituitary-adrenal (HPA) axis and the sympathetic-adrenal medullary (SAM) axis trigger a cascade of events that results in the downregulation of immune function. Stimulation of the hypothalamus by either stress or immune activation via cytokines leads to the release of corticotropin-releasing hormone (CRH). CRH stimulates the release of adrenocorticotropic hormone (ACTH) from the pituitary gland, which promotes the release of glucocorticoids. Circulating glucocorticoids dampen the immune response and inhibit further release of ACTH. CRH also stimulates the release of epinephrine (adrenaline) and norepinephrine (noradrenaline) which can also down modulate immune responses.

may increase the activity for some immune parameters. However, intense or long-term stress has been found to significantly lower the immune response. It is hypothesized that following an initial stress response, resting immune cells become activated to deal with that insult. However, if that stress response is not removed, the cells become tolerant and no longer respond to an activating signal.

Preclinical Relevance

There are a number of preclinical models that have been developed to study stress such as restraint, heat or cold, electric shock, auditory stimulation, and crowding or isolation. All these stresses have been shown to activate the HPA axis and effect various immune parameters, such as B cell and T cell proliferation, cytokine production, antibody production, natural killer cell cytotoxicity, and chemotaxis of monocytes and neutrophils. Therefore, when assessing the immunotoxicity of compounds in preclinical studies, it is important that these types of conditions, along with stress due to the handling of laboratory animals during experimental testing, are carefully controlled so that they do not complicate the interpretation of study data assessing immune parameters.

Many chemicals and pharmaceuticals can induce a neuroendocrine stress response that can be immunosuppressive. In addition, the toxicological testing of chemicals and pharmaceuticals in animals is often conducted at high doses at or near the maximum tolerated dose (MTD). The significant toxicity exhibited at these high doses can often be a source of stress for the animal inducing the activation of the HPA axis. Thus, when a change is observed on an immune parameter at toxic doses, it can often be difficult to discern whether it is due to a direct effect of the chemical on the immune system or a stress-induced immunological change. In some cases, adrenal hypertrophy and/or thymic depletion can be observed suggesting a stress-mediated effect. Therefore, when studying the effects of chemicals on the immune system, doses that are not overtly toxic should be tested so that a stress response does not lead to the erroneous identification of a test compound as an immunotoxicant.

Corticosterone can account for a significant portion of the suppression of several immunological parameters in rodents treated with potent chemical stressors. A linear model has been developed which demonstrates a clear relationship between the cumulative plasma corticosterone exposure and suppression of several immunological parameters. It is hoped that this model may allow the determination of the contribution of stress to an immunosuppressive response by a chemical by measuring a single neuroendocrine mediator.

Relevance to Humans

In humans, chronic and repeated stress has been shown to impair immune function to the extent that it can impact human health. Psychological factors that can cause stress have been found to be associated with increased susceptibility to and/or progression of a variety of pathophysiologic processes which involve the immune system such as infections, tumors, allergies, and autoimmune diseases. For example, in a healthy person the immune system plays an important role in keeping viruses under control. Many viruses are present in healthy people, but remain latent due to the host's immune defenses. If the cellular immune response becomes compromised, viruses can become reactivated and lead to active infection. Studies conducted in individuals undergoing psychological stress, such as medical students during exams or caregivers of people with Alzheimer's disease, have demonstrated a reduction in cellular immune responses and an increase in antiviral antibody titers, suggesting viral reactivation. They also demonstrate a reduced response to viral vaccines, as well as a higher incidence of respiratory infections.

Regulatory Environment

The effect of stress on the immune system is addressed in the FDA guidance document on Immunotoxicity Testing of Investigational New Drugs. The FDA recognizes the role stress can play in contributing to an immunosuppressive response and recommends that it be carefully controlled. Even when there is a potential indirect mechanism, such as stress, for alterations in immune parameters, it is recommended that the pattern of that response be carefully evaluated to determine if immune function studies would be warranted. One approach that is cited for determining the contribution of stress in an immunosuppressive response is the quantitation of stress-related blood hormones, such as corticosterone, and comparison with systemic drug exposure.

References

1. Black PH (1994) Central nervous system-immune system interactions: Psychoneuroendocrinology of stress and its immune consequences. Antimicrob Agents Chemother 38:1–6
2. Pruett SB, Collier S, Wu WJ, Fan R (1999) Quantitative relationship between the suppression of selected immunological parameters and the area under the corticosterone concentration versus time curve in B6C3F1 mice subjected to exogenous corticosterone or to restraint stress. Toxicol Sci 49:272–280
3. US Food and Drug Administration (2002) Guidance for Industry: Immunotoxicology evaluation of investigational new drugs. FDA, Rockville
4. Whitnall MH (1993) Regulation of the hypothalamic corticotropin-releasing hormone neurosecretory system. Prog Neurobiol 40:573–629

5. Yang EV, Glaser R (2002) Stress-induced immunomodulation and the implications for health. Int Pharmacol 2:315–324

Stressor

A stimuli that elicits a stress reaction.
▶ Stress and the Immune System

Stroma

Matrix of adherent multilayered cellular complex that provides the microenvironment for hematopoiesis. Composed of endothelium, fibroblasts, reticular cells, fat cells, and macrophages. Furnishes lodging and issues short-range growth factors.
▶ Colony-Forming Unit Assay: Methods and Implications

Stromal Cell-Derived Factor-1 (SDF-1)

Stromal cell-derived factor-1 (SDF-1) is a CC chemokine also known as CCL12 which is important in the migration of lymphocytes, hemopoietic stem cells and other types of cells. It binds to CXCR4, an HIV co-receptor.
▶ Chemokines

Stromal Cells

Stromal cells are epithelial cells present in immune tissues (thymus, lymph node, spleen, bone marrow) that express numerous growth factors, ligands, and receptors, and which facilitate lymphocyte development, maturation, apoptosis, and immune responses.
▶ Steroid Hormones and their Effect on the Immune System

Structural Alert

A chemical substructure commonly associated with reactivity and thus with the ability of a substance to behave as an allergen.
▶ Chemical Structure and the Generation of an Allergic Reaction

Structure Activity Relationships

▶ Chemical Structure and the Generation of an Allergic Reaction

STZ-SHR Rat

▶ Diabetes and Diabetes Combined with Hypertension, Experimental Models for

Sunlight

Sunlight includes various parts—the ultraviolet (UV) part, the visible part and the infrared (IR) part. Most important for the induction and elicitation of reactions due to a combination of radiation and a light absorbing chemical is the UV-A part (315–400 nm) of the solar spectrum. The UV-A part of sunlight in general does not induce sunburn reactions. Sunburn reactions are mainly induced by the UV-B part (280–315 nm).
▶ Photoreactions

Superantigens

THOMAS HERRMANN
Institute for Virology and Immunobiology
University of Würzburg
Versbacher Strasse 7
D-97078 Würzburg
Germany

Synonyms
sAg, superantigens.

Definition
Proteins that bind and activate most or all T cells that express a particular set of T-cell antigen receptors (TCR) β chain variable genes (Vβ). In the broader sense, superantigens are lymphocyte-activating molecules binding specifically to V gene-encoded parts of antigen receptors.

Characteristics
The term superantigen (SAg) has been coined by Kappler and Marrack (1) as an operational definition of various T-cell activating substances with specificity for T cell antigen receptors comprising certain superantigen-specific ▶ Vβ. This means that many or all T cells

expressing these Vβ respond to a superantigen, independent of their original antigen-specificity and ▶ MHC restriction: e.g. the bacterial superantigen *Staphylococcus enterotoxin* B (SEB) activates Vβ7- and Vβ8-positive mouse T cells while SEA, for example, activates Vβ 3-, 10-, 11- and Vβ12-positive mouse T cells. The Vβ specificity, the limited number of Vβ genes in mouse (up to 25) and man (about 60) together with the reactivity of several Vβ for one superantigen results in a high frequency of superantigen-specific T cells. This frequency can reach nearly half of all primary T cells compared with less than 1 in 10 000 responding to a typical peptide antigen. Table 1 gives a comparison of the key features of antigens and superantigens, and Figure 1 shows a highly schematic view of the binding of superantigens to TCR and MHC molecules.

Various methods are used to determine the Vβ specificity of a T cell response. In vitro, cell cultures of unprimed T cells are performed with or without superantigen and subsequently Vβ−specific activation is determined. This can be done by parallel measurement of early activation markers such as CD69 and Vβ expression or by testing changes in the Vβ frequency after several days of cell culture. An alternative method is the analysis of activation of T cell clones or T cell hybridomas expressing known Vβ. In vivo, superantigens can be administered by various routes and cause a transient increase in the frequency of cells expressing superantigen-specific Vβ, which often is followed by apoptosis or anergy of the superantigen-activated cells.

Superantigens cannot be defined by structural criteria, therefore it is extremely important to ensure origin and purity of a T-cell activating agent before claiming its superantigenicity. Particularly, this holds true for products of *Staphylococcus aureus* and *Streptococcus pyogenes*. Both bacteria produce a wide range of superantigens (Table 2) and (cross)-contamination has led to erroneous claims of superantigenic features. An example of the practical importance is the originally reported activation of mouse Vβ 3 T cells by SEB, which is due to the minute contamination from SEA found in some commercial preparations. Another instructive example is the Vβ-specific stimulation of human T cells by Epstein-Barr (EB) virus-infected cells, which is not caused by an EBV product, but by a protein of the human endogenous retrovirus K (HERV-K), which becomes transactivated after EBV infection.

It also should be pointed out that changes in the Vβ distribution in a T cell pool found after antigenic challenge or in medical conditions do not necessarily result from action of superantigens. Indeed, a nonrandom Vβ usage of TCR comprising certain Vβ has been documented also for the immune response to various peptide antigens.

The superantigens produced by the gram-positive bacteria *S. aureus* and *S. pyogenes* are characterized in some detail. They show little sequence similarity to each other but share a carboxy domain, which is structurally similar to the immunoglobulin-binding motifs of streptococcal proteins G and L and an amino-terminal domain with an oligosaccharide oligonucleotide (OB)-binding fold, which can be found in the B subunits of cholera and pertussis toxin. Based on sequence similarity, they can be further divided into five groups as listed in Table 2. Table 2 also lists superantigens of gram-negative bacteria and viruses which are less well characterized with respect to structure, MHC and TCR-binding (2–5).

The most intensively studied viral superantigens are those encoded by germline-integrated proviruses (*mtv*) of the retrovirus mouse mammary tumor virus (MMTV). They are found in all laboratory mouse strains and also in wild mice. Thymic expression of mtv superantigen induces a Vβ specific deletion of

Superantigens. Table 1 Comparison of superantigens (SAgs) and peptide antigens

	Superantigens	Peptide Antigens
Genes encoding TCR contact sites	BV	AV, AJ, BV, BJ, DJ, BJ
TCR structures interacting with (S)Ags	CDR1, 2, HV4 of the β-chain and framework region	CDR1, 2, 3
Frequency of specific primary T cells	1/2–1/100	1/10000–1/100000
Presenting molecule	MHC class II	MHC class I and class II
Processing required for presentation	No (with exception of mtv-SAgs)	Yes
Size of presented molecule	15–30 kDa	1–3 kDa

HV4, hypervariable loop 4; MHC, major histocompatability complex; TCR; T cell receptor.

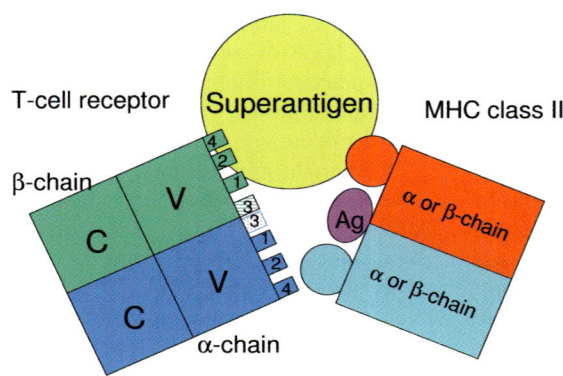

Superantigens. Figure 1 Schematic diagram of the complex of T cell receptor, superantigen (Sag) and MHC class II. Numbered boxes in the T cell receptor show CDR1-3 and hypervariable loop 4 (HV4). Note that some SAg contact also the antigenic peptide and/ or bind both chains of the MHC class II molecule.

mtv-SAg-specific thymocytes. The mtv superantigen expressed by peripheral B cells can lead to a massive stimulation in mixed leukocyte cultures which was originally used to define the so-called "minor lymphocyte stimulatory locus" (Mls).

Like antigens, superantigens need to be presented. Efficient binding has been only reported for MHC class II molecules and, in contrast to peptide antigens, bacterial superantigens are presented as intact molecules. MHC class II binding varies between superantigens (Table 2) and can involve the α chain, the β chain, or both chains of the MHC class II molecule. Exact binding sites of the TCR vary for different types of superantigen but, consistent with the Vβ specificity, contacts are mostly found in the BV-encoded domain, especially the complementarity determining regions 1 and 2, the framework regions, and the fourth hypervariable loop (Figure 1; Table 1).

Due to the unique topology of the TCR-Sag-MHC II complex, the original MHC restriction of the T cells plays only a minor (or no) role in superantigen recognition. Consequently—although to a variable extent—CD4 as well as CD8 cells respond to the superantigens presented by MHC class II molecules. Furthermore, presentation across the species barrier occurs. It can even improve the T cell response if the presenting xenogenic MHC class II molecule binds superantigen better than the autologous restriction element.

MHC class II binding of superantigens varies not only between species but also between MHC class II isotypes (in human leukocyte antigens(HLA)-DR, - DP, - DQ or in mouse H2A, H2E) and their alleles. So far, most superantigens have been isolated from human pathogens. Often they bind much better to human than to mouse or rat MHC class II molecules, which may be because of an adaptation of the superantigen producing pathogen to the host, and this to some extent explains species-specific differences in the dose response.

The biological function of most superantigens is unclear. Bacterial superantigens induce a massive antigen-independent activation of the immune system. This activation can cause a flooding of the organism with cytokines and, subsequently, disease symptoms. It can also interfere with an antigen-specific immune response and thus may be advantageous for the invading microorganism. In the case of MMTV, superantigen-induced lymphocyte activation is needed for virus replication and the replication cycle is disturbed if superantigen-specific T cells are missing. As mentioned above, this can happen as a consequence of intrathymic mtv-SAg-induced deletion, meaning that expression of a superantigen by an endogenous retrovirus may protect from infection by an exogenous retrovirus.

Preclinical Relevance

The Vβ specificity of superantigen-mediated T cell activation, together with the possibility to track superantigen reactive cells with Vβ-specific monoclonal antibodies, has made superantigen a widely used tool in T cell immunology. Moreover, superantigen-induced T cell activation interferes with antigen-induced immune responses, so that both exacerbation and suppression have been reported for animal models of autoimmunity (especially experimental autoimmune encephalomyelitis) as well as for animal models of allergy and asthma.

The contribution of some of the staphylococcal enterotoxins (SE) and toxic shock syndrome toxin 1 (TSST-1) to the etiology of the toxic shock syndrome is undisputed, but not faithfully reflected by the commonly used mouse models. In these models, SEB in conjunction with hepatotoxic agents such as D-galactosamine induces a toxic shock syndrome-like disease, but notably this disease shows a tumor necrosis factor (TNF)α–induced lethal hepatocellular apoptosis, which has not been reported for patients with clinical toxic shock syndrome.

Relevance to Humans

Many superantigens are potent toxins—but some aspects of their toxicity like the food poising caused by SE—are independent of the T cell activating properties. Nevertheless, there is a clear correlation between the massive superantigen-induced cytokine release and various forms of toxic shock syndrome, scarlet fever and illness caused by systemic *Yersinia pseudotuberculosis* infection. In other medical conditions (such as atopic dermatitis, psoriasis, Kawasaki syndrome-like

Superantigens. Table 2 Vβ specificity of superantigens (SAgs) produced by different organisms (grouped according to reference (4)).

Organism and Superantigen		BV (Vβ) Specificity in Humans	Binding to MHC Chain	Associated Disease(s)	Group
Staphylococcus aureus	TSST-1[5]	2	Class II α	Toxic shock syndrome	I
	SEB[2]	3, 12, 13.2, 14, 15, 17, 20[5]		Toxic shock syndrome (preferentially), food poisoning[3]	II
	SEC1[2]	12, 13.2			
	SEC2[2]	12, 13.2, 14, 15, 17, 20			
	SEC3[2]	5, 12, 13.2		?	
	SEG	3, 12, 13a, 14	?	?	
Streptococcus pyogenes	SSA[5]	1, 3, 15	?	?	
	SPE-A[5]	2, 12, 14, 15	?	Toxic shock syndrome, scarlet fever	
Staphylococcus aureus	SEA[5]	1.1, 5.3, 6.3, 6.4, 7.3, 7.4, 9.1, 18	Class II α + β	Food poisoning (preferentially)[3], toxic shock syndrome	III
	SED	5, 12	Class II α + β	?	
	SEE	5.1, 6.3, 6.4, 6.9, 8.1, 18	Class II α + β		
	SEG	3, 12, 13a, 14	?		
	SEH[5,4]				
Streptococcus pyogenes	SPE-C[5]	1, 2, 5.1, 10	β	Toxic shock syndrome, scarlet fever	IV
	SPE-G	2.1, 6.9, 12.3	?	?	
	SPE-J	2.1	?	?	
	SMEZ-1	2.1, 4.1, 7.3, 8.1	β	?	
	SMEZ-2	4.1, 8.1	?	?	
	SPE-H	2.1, 7.3, 9.1	?	?	
	SPE-I	18.1, 9.1, 22a	?	?	

Superantigens. Table 2 Vβ specificity of superantigens (SAgs) produced by different organisms (grouped according to reference (4). (Continued)

Organism and Superantigen		BV (Vβ) Specificity in Humans	Binding to MHC Chain	Associated Disease(s)	Group
Staphylococcus aureus	SEI	1, 5, 6b, 9, 23	?	?	V
	SEK	5.1, 5.2, 6.7	?	?	
	ETA	2	?	?	
	ETB	2	?	Scalded skin syndrome[3]	
Mycoplasma arthritidis MAS/MAM		8, 175	Class I α + β	Arthritis in rodents; unclear association in humans[3]	
Yersinia pseudotuberculosis YPM		3, 9, 13	?	Systemic disease after Yersinia infection	
Human endogenous retrovirus K HERK[6]		7, 13	?	Juvenile diabetes type I?	
Mouse mammary tumor virus (MMTV)		Various types stimulate different sets of Vβ	?	SAg required for replication cycle of virus	
Rabies virus nucleocapsid[8]		8	?	Rabies[3]	
Urtica dioica agglutinin UDA		Specific for 8.3 mouse T cells	Class I and II	?	

EBV, Epstein-Barr virus; ETA, exfoliate exotoxin A; MAS/MAM, Mycoplasma arthritidis SAg/mitogen; MHC, major histocompatability complex; SEB, Staphylococcus enterotoxin B; SMEZ, Streptococcus pyogenes mitogenic toxin; SPE-A Streptococcus pyrogenic exotoxin A; SSA Streptococcus SAg; TSST, toxic shock syndrome toxin; YPM, Yersinia pseudotuberculosis mitogen.

[1] Vβ specificity of staphylococcal and streptococcal products according to reference (5).
[2] 3-D crystal structure for substance or of substances complexed with TCR and/or MHC available.
[3] No (or unclear) contribution of superantigenic activity to disease symptoms.
[4] Appears to be an "untypical" SAg, which binds to MHC II β chains and activates only human Vα 10 T cells
[5] Also designated as 19.
[6] Transactivated during EBV infection.
[7] Exclusively found in mus spp.
[8] So far tested only by one laboratory.

illness) and various autoimmune diseases the role of superantigens remains to be clarified.

The analysis of the human T cell response to superantigens can be complicated by neutralizing antibodies found in many human sera. Also, differential binding of Vβ-specific monoclonal antibodies to Vβ alleles can lead to misinterpretation in the analysis of a superantigen response.

Regulatory environment
No specific regulation for superantigens as such.

References
1. White J, Herman A, Pulle AM, Kubo R, Kappler JW, Marrack P (1989) The Vβ specific superantigen staphylococcal enterotoxin B: Stimulation of mature T cells and clonal deletion in neonatal mice. Cell 56:27–35
2. Acha-Orbea H, MacDonald HR (1995) Superantigens of mouse mammary tumor virus. Ann Rev Immunol 13:459–486
3. Sundberg EJ, Li Y, Mariuzza RA (2002) So many ways of getting in the way: Diversity in the molecular architecture of superantigen-dependent T cell signaling complexes. Curr Opin Immunol 14:36–44
4. McCormik JK, Yarwood JM, Schlievert PM (2001) Toxic shock syndrome and bacterial superantigens: An update. Ann Rev Microbiol 55:77–104
5. Alouf JE, Müller-Alouf H (2003) Staphylococcal and streptococcal superantigens: Molecular, biological and clinical aspects. Int J Med Microbiol 292:429–440

Suppressor Cells

NANCY I KERKVLIET
Oregon State University
Dept. Environmental and Molecular Toxicology
Corvallis, OR 97331
USA

Synonyms
CD4; CD25 regulatory T cells, T_R1 cells, myeloid suppressor cells, MSC

Definition
Cells that function to actively suppress the activity of other cells within the context of an immune response are called suppressor cells. Non-specific as well as antigen-specific suppressor cells have been described. Together, they comprise a basic homeostatic mechanism within the immune system that has been recognized for more than 25 years. Suppressor cell activity is appreciated for its role in the maintenance of tolerance to self-antigens, and is maligned for its ability to prevent effective immunological responsiveness to tumors and certain other pathogens. Although the concept of suppressor cells fell into disfavor for several years, following an inability to clone the cells or to precisely characterize the mechanisms of their activity, new evidence for a population of natural suppressor T cells, now referred to as T regulatory cells, has reinvigorated the field (1). At the same time, significant advances have been made in the characterization of non-lymphoid natural suppressor cells derived from the myeloid lineage (2).

Characteristics
T Regulatory Cells
$CD4^+$ T regulatory cells have been characterized phenotypically by their natural expression of CD25 (the alpha chain of the interleukin-2 receptor). In mice, these cells also express high levels of CD62L and low levels of CD45RB, distinguishing them from antigen-activated T cells that transiently express CD25 but downregulate the expression of CD62L and upregulate expression of CD45RB. In some models, regulatory T cells also express CD152 (CTLA-4), a signaling molecule that inhibits T cell activation. In rats, T regulatory cells express the $CD4^+CD45RC^-$ phenotype. In humans, $CD4^+CD25^+$ T cells with regulatory cell functions have been isolated from peripheral blood.

$CD4^+$ T regulatory cells have a low proliferative capacity in vitro, which is dependent on IL-2. When repetitively stimulated in the presence of IL-10 a subpopulation of $CD4^+$ T cells develop (designated T_R1 cells) that secrete high amounts of IL-10 as well as transforming growth factor-β (TGF-β), and suppress the immune response of other T cells in vitro and in vivo. In other models, IL-4 production is also associated with suppressive activity.

Activation of $CD4^+$ T regulatory cells is T cell receptor (TCR)-dependent. However, once activated, these cells inhibit proliferation of other T cells in an antigen-non-specific manner. The inhibitory effect of T regulatory cells induces the target cells to undergo cell-cycle arrest at the G0/G1 stage.

$CD4^+$ T regulatory cells have been shown to inhibit autoimmune-related pathology in several experimental disease models including experimental allergic encephalitis (EAE), diabetes, inflammatory bowel disease, as well as the immunopathology associated with neonatal thymectomy (for primary references see reference (1)). Depletion of $CD4^+CD25^+$ T cells can also enhance immune responses to foreign antigens associated with microbial pathogens and to tumor antigens. Furthermore, expansion of $CD4^+CD25^+$ T cells or augmentation of their activity can suppress allograft rejection. Thus, the activity of natural $CD4^+CD25^+$ T regulatory cells not only influences the occurrence of autoimmune disease but can also suppress the response of T cells to exogenous antigens.

Myeloid Suppressor Cells

Myeloid suppressor cells (MSC) morphologically resemble granulocyte-monocyte progenitor cells, expressing the granulocyte-monocyte markers CD11b (Mac-1) and Gr-1 (Ly-6G), and appear to be similar to the "natural suppressor (NS) cells" described more than two decades ago. When cultured, MSC lost their expression of Gr-1 but retained high expression of CD11b. They also stained positive for F4/80 and CD31. The cells did not express class II, B220, CD3, or CD16, whereas a subpopulation of the cells expressed a low level of CD11c, DEC 205, and/or CD86. Depending on the cytokines present during differentiation in vitro, suppressive (IL-4) or activating (IL-4 + GM-CSF) functions could be induced, suggesting a common myeloid progenitor.

As summarized in Mazzoni et al. (2), numerous investigators have shown that MSC accumulate in the spleens and lymph nodes of tumor-bearing mice, where they inhibit antibody production, generation of cytotoxic T lymphocytes (CTL), and lymphocyte proliferative responses. MSC activity correlated with the release of GM-CSF by the tumor cells and was simulated by administration of GM-CSF to normal mice. In addition to tumor-bearing states, MSC also accumulate in the spleens of mice undergoing intense immune stimulation—such as that associated with graft-versus-host disease or treatment with superantigens—and may function to restrain otherwise overwhelming immune responses. Several soluble mediators, including inhibitory cytokines (TGF-β and IL-10), prostaglandins, reactive oxygen intermediates, hydrogen peroxide, and nitric oxide (NO) have been implicated in the suppressive mechanism of MSC, of which NO appears to be particularly important. Production of peroxynitrite ($ONOO^-$), a potent oxidant that inhibits T cell activation and induces T cell apoptosis, has been shown to mediate the suppressive effects of MSC in some studies.

Preclinical Relevance

Suppressor cells can profoundly influence the outcome of an immune response. If suppressor cell activity is selectively augmented by exposure to an exogenous chemical, the subsequent immune response will be suppressed. Depending on the immune context in which suppressor cell activity is induced, the suppression may be beneficial or deleterious. In the 1980s, suppressor cell activity was invoked to explain numerous situations of immune suppression following chemical exposure—however, definitive characterization of the suppressor cells remained elusive. The possibility that 2,3,7,8-tetrachlorodibenzo-p-dioxin and other immunotoxic AhR ligands act through induction of suppressor cell activity was debated for many years and remains unresolved. The enhanced susceptibility of mice to tumor growth following exposure to ultraviolet (UV) light was also attributed to the induction of suppressor cell activity. Similarly, induction of suppressor cells by UV light was associated with suppressed systemic immune responsiveness to contact sensitizers.

Loss of suppressor cell activity induced by exposure to an immunotoxicant could also lead to uncontrolled immune responses, resulting in development of immune-mediated pathologies and exacerbation of autoimmune disease. For example, treatment of mice with cyclophosphamide (CY) induced the onset of diabetes in male and female non-obese diabetic mice at an age when spontaneous diabetes rarely occurs. Since spleen cells from CY-treated diabetic mice were capable of transferring the disease to irradiated non-diabetic recipients, the loss of suppressor cells by CY was concluded. CY was also reported to act as a selective toxicant toward suppressor cells, resulting in a potentiated allergic contact dermatitis response. Loss of suppressor cells has also been investigated for a role in mercury-induced and D-penicillamine-induced autoimmunity. Recently, mercury-induced autoimmunity was shown to be exacerbated by treatment of the mice with a reagent that blocked CTLA-4; similar treatment induced disease in mice genetically resistant to the disorder. These results suggest that the activity of T regulatory cells may be important in this model.

Relevance to Humans

Numerous disease states may benefit from therapeutic manipulation of suppressor cell function, ranging from induction of effective antitumor immunity to prevention and treatment of autoimmune diseases and transplant rejection. On the other hand, inadvertent changes in suppressor cell function by exposure to immunotoxicants could induce or exacerbate the disease process. As our understanding of suppressor cell immunology progresses, and as therapeutic reagents that target suppressor cells are developed, monitoring suppressor cell activity may become a routine measurement for assessing immunologic status, and regulatory requirements for their assessment may logically ensue.

Regulatory Environment

There are no regulations that require the analysis of suppressor cells in immunotoxicity assessments.

References

1. Maloy KJ, Powrie F (2001) Regulatory T cells in the control of immune pathology. Nature Immunol 2:816–821
2. Bronte V, Serafini P, Mazzoni A, Segal DM, Zanovello P (2003) L-arginine metabolism in myeloid cells controls T lymphocyte functions. Trends Immunol 24:302–306

3. Mazzoni A, Bronte V, Visintin A et al. (2002) Myeloid suppressor lines inhibit T cell responses by an NO-dependent mechanism. J Immunol 168:689–695

Surface Plasmon Resonance (SPR)

A biophysical method for detecting binding of ligands, such as antibodies, to immobilized antigens on a sensor chip surface in real-time. In SPR-based assays the signals (resonance units) are generated when the binding of a molecule to the biosensor surface causes a change in refractive index (i.e. light reflection from a conducting film at the interface between two media).
▶ Immunoassays

Surfactant

Surface-acting agents that form a monomolecular layer over pulmonary alveolar surfaces, to stabilize alveolar volume by reducing surface tension and protecting against risk of collapse (atelectasis).
▶ Respiratory Infections

Surrogate

A statistically-validated health outcome measure which is not the true object of screening or diagnosis but which is intended to substitute for the final health outcome of interest. Surrogates are usually chosen because they can be measured before the primary health outcome.
▶ Polymerase Chain Reaction (PCR)

Sweetbread (When Used as Food)

▶ Thymus

Swine

▶ Porcine Immune System

SYBR Green

In real-time PCR detection, this dye binds to double-stranded DNA and upon excitation emits light. There is no need to design a probe for the gene target being analyzed. The dye cannot distinguish between specific and non-specific DNA products.
▶ Polymerase Chain Reaction (PCR)

Symian

▶ Primate Immune System (Nonhuman) and Environmental Contaminants

Syngeneic

The genetic relationship between individuals of a species in an inbred population is designated as syngeneic. In humans syngeneic refers to the fact that within an individual chromosomes are identical.
▶ Idiotype Network
▶ Graft-Versus-Host Reaction

Systemic Anaphylaxis

An allergic reaction to systemically administered antigen which in serious cases (life-threatening) is known as anaphylatic shock. It is characterized by circulatory collapse and suffocation due to tracheal swelling. It is mediated by antigen-specific IgE antibodies binding to mast cells of connective tissue throughout the body.
▶ Hypersensitivity Reactions

Systemic Autoimmunity

K MICHAEL POLLARD
Department of Molecular and Experimental Medicine
The Scripps Research Institute
10550 North Torrey Pines Road
La Jolla, CA 92037
USA

Synonyms
Autoimmunity, multisystem autoimmunity, systemic autoimmunity, connective tissue diseases

Definition
An exact definition of ▶ autoimmunity is difficult because the etiology of the majority of autoimmune diseases is unknown. There is also confusion as to what

constitutes an autoimmune disease. In general terms the immune system (of healthy individuals) is tolerant of the tissue constituents of its host, but intolerant of non-host (or "non-self") matter such as invading viruses, bacteria or other pathogens. In autoimmunity the immune system mounts responses against tissue constituents of the host organism. This response is most commonly referred to as recognition of "self", as opposed to "non-self" (e.g. viruses, bacteria). The "self/non-self" paradigm is argued to be the basis for understanding how the immune system determines whether or not to respond to a suspected challenge. Considerable debate exists as to the definition of "self" and "non-self" with both "infection" and "danger" being suggested as possible discriminators. Irrespective of the final outcome of these ongoing discussions it is clear that autoimmunity constitutes a significant disruption in the mechanisms that regulate the immune systems' ability to discriminate at the molecular level. An autoimmune disease may be described in one of two ways. The presence of autoantibodies is deemed by some to be sufficient indication that the immune system has failed to regulate "self/non-self" discrimination. Others argue that it is important to determine if the presence of autoimmunity is causally related to disease, and not the byproduct of disease or simply an innocent bystander. Clearly the most rigorous definition is that autoimmunity (e.g. autoantibodies) is the cause of disease, however this can be difficult to establish, particularly in human patients.

Autoimmune diseases consist of two types:
- organ-specific, in which the autoimmune response targets a particular organ system
- systemic autoimmunity, which targets multiple organ systems of the body.

Systemic autoimmunity comprises a group of disorders best characterized by systemic lupus erythematosus (SLE), rheumatoid arthritis (RA), ▶ Sjögren syndrome (SS), and ▶ scleroderma. Systemic autoimmune diseases are complex clinical entities that may involve all of the organ systems of the body so that two patients without overlapping clinical findings may still be classified as having the same disease. Individual systemic autoimmune diseases are identified according to clinical features (see below).

Characteristics

Systemic autoimmune diseases are disorders of unknown etiology. They are thought to be influenced by genetics, gender, and the environment. Systemic autoimmune diseases are classified clinically by their respective spectrum of signs and symptoms (1). There is no individual diagnostic test which identifies any of these disorders and their complexity makes diagnosis particularly difficult. For disorders such as SLE and scleroderma patients are classified according to a list of criteria that have been drawn up for each disorder after clinical evaluation of the signs and symptoms of selected patients. Certain clinical features may be more common in one type of disease than another. For example RA is most commonly associated with an inflammatory polyarthritis of the small joints, and the presence of rheumatoid factor. In contrast the arthritis of SLE is non-deforming and non-erosive. SLE patients commonly have photosensitivity, serositis, lymphopenia, renal disease, and antibodies against double-stranded (ds) DNA. Scleroderma includes disorders ranging from the generalized form of diffuse cutaneous systemic sclerosis to the far less severe limited cutaneous form, also known as ▶ CREST (calcinosis, Raynaud's phenomenon, esophageal involvement, sclerodactyly, and telangiectasia). Skin-thickening is a common feature of the forms of scleroderma, as is ▶ fibrosis and ▶ vasculopathy.

A characteristic feature of autoimmune diseases is the presence of autoantibodies. In ▶ organ-specific autoimmunity autoantibodies react with the particular organ or tissue associated with the disease (e.g. anti-acetylcholine receptor autoantibodies in myasthenia gravis). In systemic autoimmunity autoantibodies react with common cellular components that appear to bear little resemblance to the underlying clinical picture (e.g. anti-dsDNA autoantibodies in SLE). In both types of autoimmune disease autoantibody specificities can serve to *aid* diagnosis. Autoantibody specificities may occur at very different frequencies in a variety of diseases, and the resulting profile of specificity versus frequency can be indicative of a particular clinical syndrome.

With regard to the relationship between systemic autoimmunity and immunotoxicology it is important to note that numerous studies have shown that interactions between certain xenobiotics (chemicals and therapeutics) and the immune system can result in autoimmunity. The most common cases resemble SLE, or scleroderma. Drug-induced SLE is most commonly associated with drug treatment as the induced clinical signs and symptoms are temporally related to starting a drug, and generally abate with discontinuation. However, disease induction can require months to years of exposure, and symptoms such as autoantibodies may persist for many months after cessation of exposure. Approximately 40 medications are associated with the appearance of systemic autoimmune features. In some cases drug-induced SLE appears to be mediated by reactive drug metabolites generated from the ingested medication by neutrophil-mediated oxidative transformation. The two most frequently described are ▶ procainamide and ▶ hydralazine. When injected into the ▶ thymus of mice, the reactive metabolite of procainamide—procainamide hydroxyla-

T Helper 1 Cytokines (Th1)

Cytokines such as IFN-α and IL-2, are autocrine and paracrine signaling molecules produced by CD4$^+$ T cells in response to MHC class II antigen stimulation, and stimulate growth and activation of immunocytes and other inflammatory cells.
▶ Chronic Beryllium Disease

T Helper 1 (Th1) Cells

CD4 T cells that produce cytokines such as interferon (IFN)γ, and interleukin IL-2 but not IL-4 and IL-5. By this they direct cellular immune responses.
▶ Helper T lymphocytes
▶ Flow Cytometry

T Helper 1 (Th1) Response

A reaction mediated by CD4-positive T cells that serves to activate macrophages and promote digestion of intracellular bacteria. Th1 cells secrete cytokines such as interferon-γ (which activates macrophages) and lymphotoxin-α (which activates macrophages, inhibits B lymphocytes, and is directly cytotoxic to some cells).
▶ Lymphocytes
▶ Chronic Beryllium Disease

T Helper 1–T Helper 2 Balance

Balance in immune response after contact with antigen, especially important for response to allergens where T helper activity drives immune response either to cellular (delayed-type hypersensitivity, Th1) or antibody (IgE, Th2) mediated allergic reaction. Rats prone to a T helper 1 reaction (Lewis rats) showed more resistance to *Salmonella* infection compared to rats prone to a T helper 2 reaction (Brown Norway rats).
▶ Salmonella, Assessment of Infection Risk

T Helper 2 Cells

▶ Helper T lymphocytes

T Helper 2 (Th2) Cells

CD4 T cells that produce cytokines such as interleukins IL-4 and IL-5 but not IL-2 and interferon IFNγ. By this they direct humoral immune responses.
▶ Helper T lymphocytes
▶ Flow Cytometry

T Helper 2 (Th2) Response

A reaction mediated by CD4-positive T cells that kill infected cells and direct the destruction of extracellular pathogens by activating B cells. Th2 cells secrete cytokines such as the interleukins IL-4 and IL-5 (which activate B lymphocytes) and IL-10 (which inhibits macrophage activation).
▶ Lymphocytes

T Helper Cell

CD4$^+$ helper cell subgroups that are defined by a different pattern of cytokine release. The Th1 subgroup produces a cytokine profile to induce inflammation and cell-mediated immunity. The Th2 subgroup produces a cytokine profile to induce antibody synthesis. Both subgroups act antagonistically to each other to secure an enhanced, but balanced immune response.
▶ Cytokines
▶ Maturation of the Immune Response
▶ Leukocyte Culture: Considerations for In Vitro Culture of T cells in Immunotoxicological Studies
▶ Food Allergy

T Helper Cell Polarization

▶ Maturation of the Immune Response

T Helper Lymphocyte

▶ Trace Metals and the Immune System

T Lymphocyte

White blood cell with characteristic appearance, cell-surface markers, and function. They undergo differen-

tiation in the thymus. T lymphocytes control most aspects of the immune response, and are involved directly in attack on virus-infected cells and aberrant cells, such as malignant cells and cells originating from a different individual (as in a transplanted organ).
▶ CD Markers
▶ Canine Immune System
▶ Delayed Type Hypersensitivity

3T3 Neutral Red Uptake (NRU) Test

The in vitro 3T3 neutral red uptake (NRU) phototoxicity test was developed and validated in a joint EU/COLIPA project (1992–97). The aim was to establish a valid in vitro alternative to the available in vivo tests. The parameter for the detection of cell viability and for measuring the total activity of a cell population is based on the uptake of the vital dye neutral red into cellular lysosomes of living murine BALB/c 3T3 fibroblasts.
▶ Three-Dimensional Human Skin/Epidermal Models and Organotypic Human and Murine Skin Explant Systems

T Regulatory Cells (T$_{regs}$)

T regulatory or suppressor T cells play important roles in the regulation of immune responses and mediate a dominant immunologic tolerance. The mechanisms by which naturally occurring T$_{regs}$ are able to suppress CD4$^+$ and CD8$^+$ T cell proliferation are not yet known. The CD4$^+$CD25$^+$ T$_{regs}$ represent a subset of suppressor T cells and have been shown to play a critical role in the prevention of organ-specific autoimmunity and allograft rejection.
▶ Transforming Growth Factor β1; Control of T cell Responses to Antigens

T Suppressor Lymphocyte

▶ Trace Metals and the Immune System

Tachycardia

Heart rate above 100 beats per minute.
▶ Septic Shock

Tachypnea

Increased number of breaths per minute.
▶ Septic Shock

TAPA-1 (Target of an Antiproliferative Antibody-1)

TAPA-1 (CD81) is a 26 kDa surface protein expressed on the surface of B cells as well as T cells. It binds several different integrins and is believed to be involved in activation, cell adhesion and migrations.
▶ Signal Transduction During Lymphocyte Activation

Taqman

A DNA probe (labeled with a fluorescent reporter dye and a fluorescent quencher) used to detect specific sequences in PCR products. When amplification occurs the Taqman probe is degraded by the 5' exonuclease activity of Taq DNA polymerase, thus separating the quencher from the reporter. The increase of reporter dye fluorescence is used to determine the presence of specific gene sequences.
▶ Polymerase Chain Reaction (PCR)

Target Cell

The cytotoxic activity of immune cells is targeted towards specific cell types, which vary depending on the cytotoxic cell type involved.
▶ Limiting Dilution Analysis

Target Cell Killing

▶ Cell-Mediated Lysis

Targeted Mutant Mouse

▶ Knockout, Genetic

TbAT1 Genes

Trypanosomes are unable to synthesize purines de novo and rely on nucleoside transporters. The *Trypanosoma brucei* adenosine transporter 1 (TbAT1), also described as the trypanosomal P2-transporter, enables adenosine uptake. In addition, it confers susceptibility to antitrypanosomal drugs such as arsenicals. Various point mutations have been identified in the *TbAT1* gene of resistant trypanosomes.

▶ Trypanosomes, Infection and Immunity

TCDD

▶ Dioxins and the Immune System

Telomeres

Telomeres are the physical ends of chromosomes. They are specialized nucleoprotein complexes that have important functions, primarily in the protection, replication, and stabilization of the chromosome ends. In most organisms telomeres contain repeated simple DNA sequences composed of a G-rich strand and a C-rich strand (called terminal repeats). These terminal repeats are highly conserved—in fact all vertebrates appear to have the same simple sequence repeat (up to 2000 times) in telomeres (TTAGGG)n. After each cell division, telomere shortening takes place. Telomere length is therefore indicative for the numbers of divisions a cell has been through. Critically short telomeres trigger replicative senescence and cell cycle arrest. The innate immune system provides the first line of defence against many microorganisms and is essential for the control of common bacterial infections. It comprises macrophages, neutrophils, and natural killer cells. These cells of the innate immune response play also a pivotal role in the initiation of a subsequent adaptive immune response.

▶ Aging and the Immune System

TEQ/TEF

The complex nature of polychlorinated dibenzo-*p*-dioxins, dibenzofurans and biphenyls, which are usually generated together in occupational or environmental exposure complicates the risk evaluation for humans. This is a concept introduced to facilitate risk assessment and regulatroy control of exposure to these mixtures. 2,3,7,8-TCDD has been assigned a toxic equivalency factor (TEF) of 1.0. TEF values for individual congeners of dioxins, furans, and biphenyls in combination with their concentration can be used to calculate the total TCDD toxic equivalents concentration (TEQs) contriubted by all dioxin-like congeners in the mixture using appropriate equations. Compounds are included in the scheme and assigned a TEF if they show structural relationships to PCDD or PCDF, bind to the aryl hydrocarbon receptor, elicit aryl hydrocarbon receptor mediated biochemical and toxic responses, and persist and accumulate in the food chain.

▶ Dioxins and the Immune System

Teratogen

Any substance or exposure that causes birth defects.

▶ Birth Defects, Immune Protection Against

Testosterone

▶ Steroid Hormones and their Effect on the Immune System

Tests for Autoimmunity

RAYMOND PIETERS
Head Immunotoxicology
Institute for Risk Assessment Sciences (IRAS)
Yalelaan 2
P.O. Box 80.176
3508 TD Utrecht
The Netherlands

Short Description

A considerable number of chemicals, including many drugs, are capable of inducing autoimmune-like diseases in man (1–3).
Autoimmunogenic chemicals rarely induce similar clinical adverse effects in test animals and are hardly ever identified in general toxicity testing. Hence, autoimmune-like symptoms often become apparent only after introduction to the market. In combination with the fact that these symptoms can induce very serious or life-threatening conditions, the autoimmunogenicity of chemicals, and drugs in particular, poses a huge problem to certain sectors of society—patients, clinicians, pharmaceutical companies, and governmental

agencies. Conceivably, there is an urgent need for screening tests to identify such chemicals.

The main reason for the inability to assess a chemical's potential to cause autoimmune-like diseases is that the underlying mechanism is very complex, and can involve the interplay of many predisposing factors. An important factor is the genetic make-up, the major histocompatability complex (MHC) haplotype, non-MHC regulatory genes, metabolic polymorphisms, and gender; but many environmental factors (such as ongoing infections and food ingredients) are also known to co-influence autoimmune phenomena (Figure 1).

The complexity of the etiology may be the reason that only a few drug-using patients develop autoimmune-like derangements, but also explain why symptoms suddenly appear after a long period of symptom-free drug usage.

A set of rational criteria to establish autoimmune etiology of diseases in man was postulated in 1962 and is reviewed in (4). One of the criteria requires the presence of circulating antibodies or cell-mediated autoimmunity. Others require that the corresponding autoantigen should be identified, and—more importantly —that the disease can be reproduced by passive transfer of that antibody or the self-reacting cells or by immunization with the self antigen.

Today, it is realized that autoreactivity (both autoreactive B as well as T cells) is a normal and necessary property of a healthy immune system and that only few self-reactive autoantibodies or autoreactive lymphocytes may be considered pathogenic (i.e. directed against a pathologically relevant autoantigen and capable of causing tissue damage and reproducing disease in experimental animals). The frequency of these pathogenic autoreactive antibodies or lymphocytes may be significantly higher in diseased compared to control population (2,3).

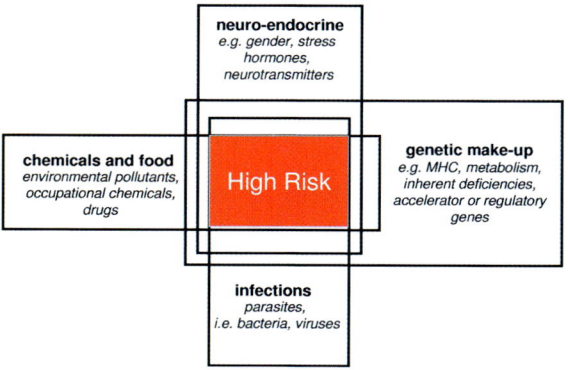

Tests for Autoimmunity. Figure 1 Representation of risk factors that are possibly involved in development of autoimmune derangements. Adapted from (7).

Although autoreactivity is a healthy phenomenon and changes in autoimmune-linked parameters do not necessarily result in an autoimmune disease, it is important to note that changes in such parameters may be used to flag a chemical as *possibly* autoimmunogenic.

Characteristics

At present no clearly defined screening tests for autoimmunity in animals exist. The popliteal lymph node assay (PLNA) is a simple straightforward local lymph node assay that may be useful to screen for initial immunostimulating capacity of chemicals. But this assay can only be regarded as a first screening test for immunosensitizing potential and to indicate that a chemical *might* induce autoimmune-linked symptoms. It is preferable for screening tests for autoimmunity to use relevant exposure routes and demonstrate systemic changes in parameters indicative of autoimmune-linked responses. Diagnosis of autoimmune-linked diseases in test animals, like rats or mice, may be based on a combination of general well-being, routine clinical tests and (immuno)histology. Clinical investigations should include general hematology (e.g. to check for anemias) as well as tests for complement activity, or acute-phase proteins, and for erythrocyte sedimentation. Liver and renal impairment should be monitored biochemically (3). Morphologically, a wide range of organs should be checked for indications of inflammation, overt apoptosis (in the thymus in particular). Peripheral immunologic organs should be checked for indications of activation (e.g. hyperplasia or formation of germinal centers) (5).

Morphological indications of tissue inflammation, activated immune organs or immunomodulation (e.g. thymus atrophy) should be followed up by more thorough investigations into alterations of autoimmune parameters (5). Because development of actual autoimmune disease depends on a complex interplay of (non)inherent factors (see Figure 1), relevant changes in any of the animals should be considered as an alert to pursue further investigations. This is certainly the case in outbred animals which are used for evaluation of toxicity, but also in inbred animals which are also not always 100% responsive.

The initial focus in follow-up studies should be on detection of autoantibodies, which can be directed against a wide spectrum of autoantigens (2). In case the target autoantigen is not yet known, and particularly for screening purposes, the indirect immunofluorescence (IF) technique may be useful. The immunofluorescence technique, which is also used in the clinic, has been used successfully in animal studies. Briefly, cryosections or isolated cells grown on microscopic slides (for instance HepG2 tumor cells for antinuclear antibodies (ANA) or freshly isolated granulocytes for antineutrophil cytoplasmic antibodies

(ANCA)) are an incubated with serum suspected to contain autoantibodies followed by a incubation with fluorochrome-labeled second-step antibody. Interestingly, the immunofluorescence technique can be applied to cryosections of a range of relevant organs (such as kidney, thyroid, liver, skin, adrenals, and sex organs), although false-positive staining (perhaps as a result of antibody binding to Fc receptors) particularly in inflamed tissue has to be taken into account. When the specificity is known, autoantibodies can be detected by various other techniques (most notably enzyme-linked immunoassay, ELISA). In many cases, it may suffice to perform an ELISA for ANA. Compound-specific lymphocyte transformation tests (LTT) can be used in cases of drug allergy or chemical exposure (see for instance Schnyder et al 2000) but detection of autoreactive T cells in case of chemicals is much more difficult. This is mainly due to the fact that the relevant autoantigen (chemically altered or previously cryptic epitopes) is hardly ever known, and also because specific autoreactive T cells are relatively scarce even in clinical situations. A solution would be to immortalize selected self-reactive T cell clones, but this is not easy to incorporate in a general testing model for autoimmunity.

Pros and Cons

All of these methods may at best provide circumstantial evidence for autoimmune effects and/or etiology in animals. The advantage of these methods is that they can all be used in animal toxicity studies without interference with the study per se, and only ask for more extensive analyses of samples (blood, serum and organs) that are already to be isolated at dissection. But, importantly, as the immunological effects depend greatly on genetic make-up, autoimmune effects may be easily missed when small groups of outbred test animals are used. So animal tests for autoimmunity (including the parameters discussed here) should be done with larger test groups and should be performed over relatively long periods of exposure (> 90 days). Importantly, adverse effects which are indicative of autoimmunity—even if they occur in only one animal—should already be taken as an alert to execute follow-up studies with inbred animal strains, such as the frequently used high-IgE-responding Brown Norway (BN) rat. To date, only a limited number of compounds (e.g. $HgCl_2$, gold salts, D-penicillamine, nevirapine, hexachlorobenzene) have been shown to induce autoimmune-like phenomena in this rat strain.

Relevance to Humans

Chemical-induced autoimmune effects detected in animals can be predictive for the human situation. However, as in humans the prevalence of autoimmune effects will be low in (outbred) animals as well. Studies with particularly sensitive rat strains, for instance, such as the BN rat, may identify much better the hazard of autoimmunogenic potential of a chemical. Notably, such a sensitive rat has to be regarded as a representative of very susceptible humans.

Regulatory Environment

At present guidelines for detection of autoimmunogenic capacity do not exist. It should be realized that none of the present animal models—including the BN rat model—is capable of detecting autoimmunogenic potential of a wide range of different chemicals. The popliteal lymph node assay (PLNA) is an animal model that may indicate whether a chemical is immunostimulatory. Immunostimulation may result in sensitization of the immune system and is considered one of the prerequisites for inducing autoimmunity.

A number of the parameters proposed here, however, could be easily or are already incorporated in existing guidelines. For instance, the OECD guideline 407 includes the hematology, clinical biochemistry and pathology of a series of organs. But without further analyses of (auto)antibody levels, larger test groups of inbred animals and long exposure periods (> 90 days) a chemical's potential to induce autoimmunity will hardly ever be detected in these toxicity studies.

So, future research to design predictive protocols and screening models is greatly needed. This could be initiated by thorough research into the relevance of the above-mentioned parameters in repeated-dose studies over a relatively long period with inbred strains of rats (e.g. BN and Lewis strains) as well as mice (e.g. SJL and C3H/He strains), but also in outbred animals that are normally used in toxicity studies. Such studies should first be performed in a limited number of well-equipped laboratories, and should be followed by more extensive ring studies.

References

1. D'Cruz D (2000) Autoimmune diseases associated with drugs, chemicals and environmental factors. Toxicol Letters 112–113:421–432
2. Verdier F, Patriarca C, Descotes J (1997) Autoantibodies in conventional toxicity testing. Toxicology 119:51–58
3. D'Cruz D (2002) Testing for autoimmunity in humans. Toxicol Letters, 127:93–100
4. Shoenfeld Y, Isenberg D (eds) (1990) The Mosaic of Autoimmunity, Factors Associated with Autoimmune Disease. Introduction. Research Monographs in Immunology, Volume 12. Elsevier, Amsterdam
5. Frieke Kuper C, Schuurman H-J, Bos-Kuijpers M, Bloksma N (2000), Predictive testing for pathogenic autoimmunity: the morphological approach. Toxicol Letters 112–113:433–442
6. Schnyder B, Burkhart C, Schnyder-Frutig K et al. (2000) Recognition of sulphamethoxazole and its reactive

metabolites by drug-specific CD4⁺ T cells from allergic individuals. J Immunol 164:6647–6654

7. Kammüller ME, Bloksma, N, Seinen W (1989) Immune disregulation induced by drugs and chemicals. In: Kammüller ME, Bloksma N, Seinen W (eds) Autoimmunity and Toxicology. Elsevier, Amsterdam, pp 3–25

2,3,7,8-tetrachlorodibenzo-p-dioxin

▶ Dioxins and the Immune System

Tetravalent Vanadium

Tetravalent vanadium is the ionic form of vanadium when four outer shell electrons (that is, two from 4s and two from 3d orbitals) have been shed, thereby giving the atom an overall charge of +4.

▶ Vanadium and the Immune System

TGF-β1

▶ Transforming Growth Factor β1; Control of T cell Responses to Antigens

Th1/Th2 Balance

An important mechanism in the immune regulation involves homeostasis between the T helper 1 (Th1) and T helper 2 (Th2) activity of CD4⁺ T helper cells expressing different cytokine patterns. T helper cells showing Th1 activity are more prone to induce a cell-mediated immunity whilst T helper cells obtaining Th2 activity are more prone to induce a humoral-type immune response. T helper cells showing either Th1-type or Th2-type reactivity are exclusively characterized by differences in cytokine expression. Briefly, Th1 reactivity is predominantly connected to interferon (IFN)-γ, IL-2, and IL-12 secretion. In contrast Th2 cells express mainly IL-4, but also IL-5, IL-6, IL-10 and IL-13. The Th1/Th2 balance is integrated in the immune regulation in a dynamic and reversible manner, depending also on kinetics and dose–response of the immune response.

▶ Cancer and the Immune System

Three-Dimensional Human Skin/Epidermal Models and Organotypic Human and Murine Skin Explant Systems

HANS-WERNER VOHR · ECKHART HEISLER
PH-PD, Toxicology
Bayer HealthCare AG
Aprather Weg 18
D-42096 Wuppertal
Germany

Synonyms

human skin recombinants, reconstructed human skin/epidermis, 3-D human skin/epidermal equivalents, in vitro engineered skin/epidermal substitutes, artificial skin/epidermis, organotypic murine or human skin explant system, MSE, HSE, hOSEC

Definition

Human full-thickness skin models and reconstituted epidermal equivalents are in vitro-engineered tissue cultures that provide a three-dimensional architecture which is biochemically, morphologically and functionally comparable to human epidermal tissue/skin in vivo. Organotypic skin explant systems are based on ex vivo skin removed from humans or mice and subsequently cultured in toto. All the models were shown to be useful in screening for topically applied irritating, ▶ corrosive or photocytotoxic compounds. Results from experiments with systemically applied compounds have already been published with such models, too. Furthermore, in recent studies it was demonstrated that 3-D skin models also provide the capacity to further characterize and screen for substances with a sensitizing potential.

Characteristics

Reconstructed Human Epidermal Models

Reconstructed human epidermal models are built up from proliferating, differentiating and cornifying keratinocytes which are airlift-cultured on a porous polymeric membrane. The design of the cell culture conditions (air-liquid interphase and medium/ingredients) drives the cells to differentiate and form a three-dimensional (3-D) epidermal multilayer with a functional and stratified surface. Most of the key structural elements of native epidermis like ▶ keratins, ▶ transglutaminase and lipid composition that characterize the status of keratinocyte differentiation are present in 3-D human epidermal equivalents.

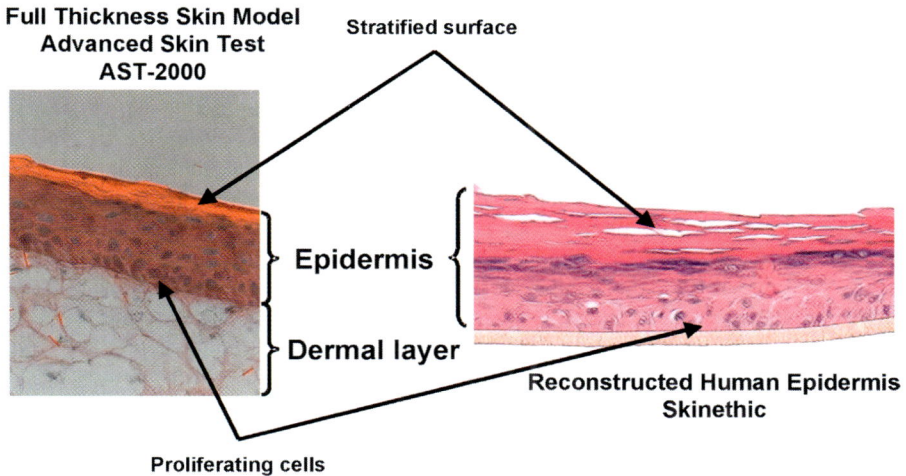

Three-Dimensional Human Skin/Epidermal Models and Organotypic Human and Murine Skin Explant Systems. Figure 1 Two different reconstructed tissues: H&E stained sections of untreated reconstructed human full thickness skin model (Advanced Cell Systems, AST-2000) and epidermal model (Skinethic RHE). Both models are comercially available. (Picture of AST-2000 by kind permission of Advanced Cell Systems, St. Katharinen, Germany).

Full-Thickness Human Skin Substitutes

Full-thickness human skin substitutes additionally provide a dermal layer that usually consists of a collagen matrix which is populated by living fibroblasts. In an early state of research the use of de-epidermized human dermis as the backbone of full-thickness skin equivalents has been discussed as well. In comparison to single-cell culture systems, the most predominant feature of these artificial skin models is the existence of a physiological and functional barrier (the stratum corneum) that regulates percutaneous absorption/penetration of compounds as well as transepidermal water loss. Although the barrier functions of artificial skin models are different from the situation in vivo, the results from studies evaluating the penetration properties of various reference test compounds have shown a good correlation to in vivo data.

Organotypic Skin Explant Systems

Organotypic skin explant systems from human, rats or —to a lesser extent—mice have also been established for evaluating percutaneous absorption and penetration. However, in comparison to reconstructed skin models the explant cultures naturally provide a physiological cell composition and micro-architecture including immunocompetent cells (e.g. Langerhans cells).

For toxicological and immunotoxicologic research, both topical treatment (application of compounds to the dry stratum corneum) as well as systemic-like treatment (application of substances directly into the cell culture medium) are possible using 3-D skin models. Hazard identification is based on the measurement of decreased cell viability and changes in tissue morphology after treatment (histological examination; see below). In recent studies it was also shown that topical and systemic-like treatment of 3-D skin models with hazardous compounds often results in induced expression and/or release of immunomodulating proteins (cytokines, chemokines, matrix metalloproteinases, growth factors, and other parameters which are involved in a variety of biochemical pathways; see below). The determination of these parameters gives a detailed overview of the cell status which can additionally confirm the results from viability testing and histological examinations (multiple endpoint analysis; MEA).

Screening for Irreversible Cutaneous Toxicity (Corrosion)

Screening for Acute Irritation

Both artificial skin models and organotypic skin explant systems are suitable for screening for dermal irritation induced by topically applied irritating or photo-irritating compounds and formulations. Most likely in this situation is that in vivo substances with a strong ▶ irritant potential provoke severe destruction of the reconstructed or explanted tissues and affect the integrity of residential cells. The use of these systems to test chemicals, compounds, or formulations according to their irritant properties depends on the measurement of cell viability after topical treatment with compounds and additional time-related incubation. Cytotoxic and photocytotoxic effects cause a significant decrease in cell viability. For this reason, determination of cell viability is essential for the assessment of compound biocompatibility using artificial skin models or organotypic skin explant systems. However, in most of the published test protocols MTT conversion is used as a single endpoint parameter for the determination of cell viability and consequently the degree of cytotoxicity caused by irritation and photoirritation. Recently the identification of more specific parameters allows a multiple endpoint analysis (cell viability, histological examination, release of IL-1α;).

Expression of Immunomodulating Proteins and Screening for Dermal Sensitization

Both irritation and sensitization of the skin are related to the expression and release of immunomodulating proteins such as cytokines, chemokines and cell surface proteins, especially within the epidermis. The local immune system of the skin in vivo is based on the interactions between epidermal keratinocytes, epidermal Langerhans cells, and dermal fibroblasts. Once activated by antigen uptake and processing, Langerhans cells undergo morphological changes and start to migrate to the local draining lymph nodes. There T cells become activated upon successful antigen presentation. In cases of cutaneous irritation causing epidermal cell damage, keratinocytes release a cocktail of proinflammatory proteins from their intracellular reservoirs. This finally results in a non-specific activation of the skin's immune system (see also▶ contact hypersensitivity section).

Considerable efforts have been made to integrate Langerhans cells into reconstructed human skin models. However, there is still no complex in vitro system available that provides functional antigen-presenting cells in the epidermis or dermis. Nevertheless, keratinocytes are also thought to be involved in the initial steps of irritation and sensitization. Topical treatment of artificial skin models with irritating compounds leads especially to the release of interleukins IL-1α and IL-8 by keratinocytes. Furthermore, the subsequent analysis of cell culture supernatants by different ELISA techniques (enzyme-linked immunoassay) additionally show an induced release of different chemokines and cytokines as shown in Table 1.

The profile of released proteins depends on the kind of model used for the experiments. In comparison to reconstructed epidermal models, full-thickness skin models provide a set of parameters that are related to the interaction between epidermal keratinocytes and dermal fibroblasts. In recent studies carried out with sensitizing substances the ratio between IL-1α and IL-8 release after topical treatment with the compounds revealed promising results that suggest that reconstructed human skin models are capable of discriminating ▶ sensitizers from compounds with an exclusively irritant potential. Other studies identified promising parameters (increased release of the chemokines monocyte chemoattractant protein 1 (MCP-1) and interferon-inducible protein (IP-10) from a human full-thickness skin model AST-2000 after treatment with the standard sensitizer (oxazolone) that certainly can contribute to a successful discrimination between sensitizers and irritants in vitro.

From an immunological point of view, however, it is of prime importance for sensitization testing to analyze parameters (▶ MIG, Langerin, TARC, etc.) that are characteristic for the cross-talk between keratinocytes, fibroblasts and antigen-presenting cells in their natural setting. For this reason, research on skin sensitization (screening, mechanistic) is particularly focussed on the use of organotypic skin explant systems as well as the development of skin recombinants that incorporate functional antigen-presenting cells.

Pros and Cons
Experimental Strategies

Methods used in in vitro dermal toxicology are often based on single-cell culture systems, which in turn are built up from either freshly isolated primary cells derived from cosmetic surgery, foreskins or well-established cell lines. Methods for cytotoxicity and photocytotoxicity testing, like the 3T3 neutral red uptake (NRU) test, have been successfully validated. However, test principles based on single-cell cultures are subject to some limitations due to their lack of a physiological barrier. For this reason they are usually restricted to soluble substances and therefore fail when it comes to testing hydrophobic compounds or formulations. Furthermore, the concentrations of compounds inducing irritation in single-cell cultures are significantly lower than those determined in in vivo experiments. Due to the absence of a stratified surface, false positive results may also occur, because substances may be classified as (photo)cytotoxic by 3T3 NRU although they are physicochemically unable to pass through the physiological barrier (the stratum corneum).

By using 3-D skin models it is possible to overcome these problems, and they offer a promising test system for topical and systemic-like compound administration. Furthermore, artificial skin models and organotypic skin explant systems may be suitable for screening for sensitizing properties of compounds in vitro. With respect to this last point research is still in progress, but a convincing system may be available in the near future.

Test Principles

As already mentioned, MTT testing is often used as a single-endpoint parameter for predicting the irritant potentials of substances, although cytotoxicity is not

Three-Dimensional Human Skin/Epidermal Models and Organotypic Human and Murine Skin Explant Systems. Table 1 Expression/release of immunomodulating proteins from 3-D skin models

Parameter	Expression/Release	Release Inducible
Interleukin-1α	++ (a,b,c)	Yes
Interleukin -1β	+ (b,c)	Slightly
Interleukin -6	+++ (+a,b,c)	Yes
Interleukin -8	+++ (+a,b,c)	Yes
Tumor necrosis factor-α	+ ((a),b,c)	Slightly
Monocyte chemoattractant protein MCP-1	+++ (+b,c)	Yes
MIG	+ (c)	Slightly
Interferon-inducible protein IP-10	+ (b,c)	Slightly
Macrophage inflammatory factor MIP-3α	+ (c)	Slightly
Matrix metalloproteinase MMP-3	++ (+b,c)	Slightly
Matrix metalloproteinase MMP-9	++ (b,c)	Yes

a, epidermal model; b, full thickness skin model; c, organotypic skin explant system; +, low level; ++, medium level; +++, high level; (+), high background.

a sufficient stand-alone parameter for predicting cutaneous irritation. In vitro testing associated with MTT conversion is always subject to some limitations, because the test principle is based on a chemical redox reaction which may also run without any participation of living cells. This may lead to false positive results. Another problem with MTT, especially concerning 3-D skin models, was observed when test results were compared to histological examinations of reconstructed skin models after compound treatment. Due to cellular activity, formazan crystals were found to be formed especially in the cells from the basal layer. For this reason, it is not possible to detect undesired compound-related effects on cells from the stratum spinosum or stratum granulosum by MTT (see Figure 2). Other test principles for the determination of cell viability are based on the quantitative analysis of enzymes from the cytosol of cells. When cells lose their integrity through damage to the plasma membranes, the leakage of these proteins can be recorded and quantified by bioluminometric or other optical enzymatic test systems.

In this context the measurement of lactate dehydrogenase (LDH) and/or adenylate kinase leakage is often discussed as a defined parameter for the analysis of substance-related cytotoxic effects on in vitro cell systems.

Finally, the induced release of proinflammatory mediators like IL-1α additionally serves as a good parameter for the characterization of skin irritation, because IL-1α was found to be released from cells which are influenced by irritating chemicals. Although MTT is a reliable and valid parameter for the analysis of cell viability, the results should be supplemented and verified additionally by multiple endpoints, such as expression and release of proinflammatory mediators, decrease of the barrier function determined by transepidermal water loss (TEWEL) and/or evaluation of morphological changes (histologic examination).

Comparison to In Vivo Test Principles

The replacement of in vivo methods for corrosivity and irritancy according to Draize by in vitro reconstructed skin models is often discussed, especially from an ethical point of view. In addition, the use of 3-D skin models is less time-consuming than in vivo

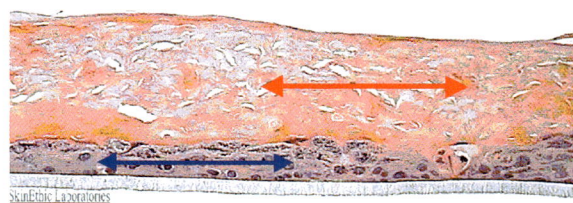

Three-Dimensional Human Skin/Epidermal Models and Organotypic Human and Murine Skin Explant Systems. Figure 2 H&E stained section of Skinethic RHE after treatment with 0,4% SDS and 24 hours of incubation (5% CO2, 37°C, max hum.) The area marked with the red arrow shows massive destruction of cells in the upper epidermal layers. However, the basal layer (blue arrow) is not affected. Here, MTT test gave false negative results. Although cell viability was correctly determined, the integrety of the cells in the upper epidermal layer was hardly affected. This effect however, was undetectable by MTT alone (By kind permisson of SkinEthic Laboratories, Nice, France).

testing, and if the costs for animal health and care are taken together, reconstructed skin models are less cost-effective than animal testing, too. However, a ranking between strong irritation and weak or mild irritation based on experimental results from testing with reconstructed skin/epidermis still seems to be questionable. The establishment of in vitro test methods for sensitization is not that easy, although protein fingerprinting of cells from organotypic skin explant systems and reconstructed epidermal/skin models revealed promising results that contribute to the in vivo situation. In recent studies it was shown that the expression and release of immunomodulating proteins (Table 1) serve as good parameters for the characterization of compounds with sensitizing properties. However, the use of these parameters as criteria for predicting sensitization has not been validated so far. For this reason, guinea pig assays like those described by Buehler or Magnusson and Kligman are still the most reliable methods for sensitization testing, even though they are based on visible subjective parameters like the formation of erythema. In this context, another valid method is LLNA/IMDS (local lymph node assay/integrated model for the differentiation of (chemical)-induced skin reactions) which characterizes sensitizing compounds with the help of cellular parameters, but is still based on animal treatment.

Predictivity

Irritation of the skin caused by exposure of individuals to different kinds of hazardous compounds or formulations is the most common non-specific immune reaction observed in human skin. In vivo (animal) test principles according to the methods of Draize are frequently used for the identification of substances with irritant potential. For several reasons, however, these test methods are questionable. The analysis of substances according to Draize testing is mainly based on the evaluation and scoring of macroscopic parameters such as overcasting of the rabbit eye cornea or redness of the skin after treatment with the compounds being tested. As far as this point is concerned, it has recently been shown that the choice of endpoints for the assessment of acute skin irritation according to international standards (methods according to Draize) may lead to misclassification of substances. Furthermore, the transfer of established data from animal testing to the human situation in vivo is still controversially discussed. For this reason the human patch test was established. This ideally meets the requirements, but patch testing in human is restricted to weak or moderate irritating compounds. These pragmatic disadvantages of in vivo animal and human testing for skin corrosion or acute skin irritation are furthermore accompanied by the discussion of the ethical justification of animal testing in toxicological research. With the use of reconstructed tissue models it is possible to overcome most of the problems described above. From multiple endpoint analysis (see Characteristics) reliable parameters are available that are simple to determine, while the output is more stringent than visual evaluation of results.

Furthermore, artificial skin models were proven to be reproducible in intra- and interlaboratory multicenter studies. As mentioned above, the predictivity of reconstructed tissue models is limited. In comparison to human in vivo skin the different physiological barrier function of the reconstructed stratified surface may cause problems because the risk of false positive results cannot be totally excluded. In addition, distinguishing between weak and moderate irritating compounds is sometimes not easy. However, research is focussing on new parameters that could help to solve these problems. Despite this early state of affairs, it is possible to state that human reconstructed tissue models exhibit acceptable predictivity in screening for corrosive compounds (sensitivity and specificity > 80%). Although the validation and catch-up validation studies for acute irritancy of topical applied formulation and/or raw materials are still in progress, a high correlation of sensitivity has already been estimated by the relevant ECVAM Task Force.

Another main topic of interest concerns alternative in vitro models for skin sensitization. At present, no reconstructed tissue model is available that meets the guideline criteria for adequate screening. However, considerable efforts have been made to search for parameters (cytokines, chemokines) which specifically characterize the complexity of the processes leading to skin sensitization (skin penetration, formation of protein-hapten complexes, antigen uptake and processing, migration of LC to the local draining lymph nodes, presentation of antigen to T cell populations in the draining lymph nodes). In the light of this complexity, the use of organotypic skin explant systems seems to be very promising, because they provide the same micro-architecture and the same cell composition as in vivo skin and are therefore potent tools for mechanistic studies.

Relevance to Humans

The test results from animal testing for irritancy and corrosion according to Draize are controversially discussed among toxicologists. In cases of acute irritation these test methods have never been validated and they principally depend on a collection of cross-connected empirical clinical and preclinical data. For this reason, the use of reconstructed human tissues is of particularly great value, because the cells used for these skin constructs are of human origin. Although some differences in the characteristic barrier function have been described, the experimental design closely matches the human situation in vivo.

Unfortunately, screening for sensitization in vitro is even more complex because artificial tissue structures are necessary which must in addition provide immunorelevant cross-talk activities. For hazard identification, on the other hand, fingerprinting of proteins released from 3-D in vitro skin models has already been evaluated and some of these parameters were shown to hold key positions in immunological pathways (IL-8, MCP-1, IL-1α, IL-6, etc.). These may therefore help to screen for compounds with a sensitizing potential in vitro. As mentioned earlier, human skin explant systems in particular are believed to be very suitable models for further characterization of immunorelevant parameters. In the heat of discussion about testing for sensitization, one should keep in mind that in vivo animal testing (guinea pig assays or the (modified) local lymph node assay) or human patch testing, as well as all possible *in vitro* models which are going to be established and validated in the future, are not capable of taking all parameters influencing the induction of skin sensitization into account (individual parameters such as genotype, age, sex, side of contact/penetration and of course the overall condition of the skin).

Regulatory Environment
Skin Irritation/Corrosion

The international standards for skin irritation and corrosion are still based on in vivo test principles according to the methods of Draize et al. (1944). However, the considerable efforts of organizations like ECVAM, ICCVAM, COLIPA, the Steering Committee on Alternatives to Animal Testing (SCAAT) have had a favorable and lasting influence on the establishment of in vitro test methods of international guidelines. The use of several in vitro human skin models for skin corrosion was validated by ECVAM in 2000. For acute skin irritation, however, a first prevalidation study failed but the process of improving the use of reconstructed tissue models especially in this field of toxicologic research is strictly ongoing.

Guidelines for Determination of Substance-Induced Skin Corrosion
- OECD Guideline 402: Acute Dermal Tox.
- OECD Guideline 404/405: Acute Dermal/Ocular Tox Irritation and Corrosion
- OECD Guideline 410: Repeated Dose Dermal Tox.
- OECD Guideline 430: In Vitro Skin Corrosion—Rat TER (Trans Epidermal Resistance) Test
- OECD Guideline 431: In Vitro Skin Corrosion—Human Skin Models
- Annex V of Directive 67/548/EEC (1997)
- US Code of Federal Regulations (1991)

Sensitization

Up to now no in vitro screening model has been available to correctly predict exclusively sensitizing properties of compounds. From an immunological point of view this is not surprising because of the lack of antigen-presenting cells in most of the reconstructed human tissues. However, the induced release of immunomodulating proteins indicates promising parameters for successful discrimination between irritating and sensitizing substances. As long as none of the reconstructed or organotypic models match the criteria for a successful prevalidation study, immunotoxicologic research must rely on guinea pig test principles according to Bühler, Magnusson and Kligmann, or on a refined test assay like the LLNA or the integrated model for the differentiation of (chemical)-induced skin reactions (IMDS).

Guidelines for Determination of Substance-Induced Sensitization
- OECD Guideline 406: Skin Sensitization (1992)
- OECD Guideline 429: LLNA (2002)
- U.S. EPA-OPPTS Harmonized Test Guideline 870.2600 on Skin Sensitization (1998)
- FDA (CDER) (Draft) Immunotoxicology Evaluation of Investigational New Drugs (2001)
- CPMP/SWP/398/01 (Draft) Note for Guidance on Photosafety Testing (2001) (as modified LLNA)

References
1. Botham PA, Earl LK, Fentem JH, Roguet R, van de Sandt JJM (1998) Alternative methods for skin irritation testing: the current status. ECVAM Skin Irritation Task Force Report 1. ATLA 26:195–211
2. Zuang V et al. (2002) Follow-up to the ECVAM prevalidation study on in vitro tests for acute skin irritation. ECVAM Skin Irritation Task Force Report 2. ATLA 30:109–129
3. Spielmann H et al. (2003) Report of the Second SkinEthic Workshop: In Vitro Reconstructed Human Tissue Models in Applied Pharmacology and Toxicology Testing, Nice, France
4. Coquette A, Berna N, Vandenbosch A, Rosdy M, De Wever B, Poumay Y (2003) Analysis of interleukin-1 alpha (IL-1 alpha) and interleukin-8 (IL-8) expression and release in in vitro reconstructed human epidermis for the prediction of in vivo skin irritation and/or sensitization. Toxicol In Vitro 17:311–321
5. Heisler E, Ahr HJ, Vohr HW (2001) Local immune reactions in vitro: Skin models for the discrimination between irritation and sensitization. Exp Clin Immunobiol 204:1–2

Three Rs

Reduction (fewer animals), refinement (less severe

procedures), and replacement (in-vitro alternatives) of animal experiments, first proposed by Russel and Burch in 1959.
▶ Canine Immune System

Thrombin

Thrombin is a multifunctional serine protease that has procoagulant activities when diffusable in the blood stream. But it loses this ability and initiates a potent anticoagulant pathway when bound to its endothelial cell receptor thrombomodulin, thereby mediating generation of the anticoagulant enzyme-activated protein C. The cellular activities of thrombin on platelets, endothelial or smooth muscle cells are mediated through G protein-coupled protease-activated receptors (PAR) that are initially cleaved by thrombin before a newly generated peptide motif of the receptor can serve as an internal tethered ligand for initiation of cell signaling.
▶ Blood Coagulation

Thrombocytopenia

Thrombocytopenia is a condition in which the normal concentration of platelets (thrombocytes) in the blood is decreased. A significant shortage of platelets can result in bruising and easy bleeding.
▶ Leukemia
▶ Antiglobulin (Coombs) Test

Thrombocytopenic Purpura

A rare autoimmune disorder characterized by a shortage of platelets, leading to bruising and spontaneous bleeding. Approximately half of the cases are idiopathic (unknown cause). Other cases are caused by drugs, infections or autoimmune disorders such as lupus erythematosus.
▶ Interferon-γ

Thymic Hypoplasia

An immunodeficiency that selectively affects the T lymphocyte limb of the immune response. There is lymphopenia with diminished T cell numbers.
▶ Trace Metals and the Immune System

Thymocyte Development

▶ Thymus: A Mediator of T Cell Development and Potential Target of Toxicological Agents

Thymocyte Education

▶ Thymus: A Mediator of T Cell Development and Potential Target of Toxicological Agents

Thymocyte Selection

▶ Thymus: A Mediator of T Cell Development and Potential Target of Toxicological Agents

Thymus

C Frieke Kuper
Toxicology and Applied Pharmacology
TNO Food and Nutrition Research
Zeist
The Netherlands

Synonyms
Thymus, thymus gland, sweetbread (when used as food)

Definition
The thymus is a primary lymphoid organ in vertebrates; in mammals it is located in the cranioventral mediastinum and lower part of the neck. The prime functions of the thymus in mammals are the development of immunocompetent T lymphocytes from bone-marrow-derived stem cells, the proliferation of mature naive T cells to supply the circulating lymphocyte pool and peripheral tissues and the development of immunological self-tolerance. The thymus elaborates a number of soluble factors (thymic hormones) which regulate several immune processes, including intrathymic and post-thymic T-cell maturation, and neuroendocrine processes such as the synthesis of neuroendocrine hormones by the central nervous system.

Characteristics
Anatomy and Histology
The thymus is located in the cranioventral mediastinum and lower part of the neck, whereas small islands of thymic tissue may be present near the thyroid and

parathyroid glands. In young animals it is roughly pyramid-shaped with its base located ventrally. The gland consists of two lobes, fused in the midline by connective tissue. The two thymic lobes are enclosed by a fibrous capsule from which septa traverse into the organ, dividing it into lobules. The lobules have basically the same architecture, with a subcapsular area, a ▶ cortex, corticomedullary junction and a ▶ medulla. The cortex is easily recognizable in hematoxylin and eosin(H&E)-stained sections by its high density of thymocytes (immature lymphocytes) and therefore darker appearance when compared with the less densely populated medulla. The framework of the thymus is formed of epithelial reticular cells in which the bone-marrow-derived lymphoid (thymocytes/lymphocytes) and non-lymphoid cells (macrophages, dendritic cells) are packed. The vast majority of lymphocytes are T cells, but accumulations of B cells do occur. Epithelial aggregates with centrally located cell debris, the so-called ▶ Hassall's bodies, are a characteristic feature in the medulla.

The different thymic compartments are associated with different T cell maturation processes, namely early (cortical) maturation and late (medullary) maturation, which in turn are associated with differences in the marker expression and cytology of epithelial cells, lymphocytes, macrophages and interdigitating cells (Figure 1).

Moreover, the capacity of epithelial cells to synthesize thymic hormones differs, the major site of hormone synthesis being the medullary epithelium (1). A characteristic and unexplained microenvironment is formed by the cortical and medullary areas which are devoid of epithelial cells but full of thymocytes, the so-called epithelial-free areas or EFAs (2). The function of these EFAs is unknown, although medullary EFAs may be associated with autoimmune diabetes. Foci of myelopoiesis are found in the connective tissue septa, within the lymphoid tissue at the outer rim of the lobules, and at the corticomedullary zone. Hemoglobin-containing cells can be found among the myelocytic series in the interlobular septa, at the outer rim of the lobules. In the medulla no erythroid precursors have been observed. Blood vessels enter the lobules via the interlobular trabeculae/septa and branch at the corticomedullary area to supply the cortex and medulla. Postcapillary venules in the corticomedullary region have a specialized cuboidal epithelium similar to that of the high-endothelial venules of the lymph node, which allows passage of lymphocytes into and out of the thymus. Sheaths of connective tissue and an epithelial cell layer with its basement membrane are found around the blood vessels. The space between the epithelial basement membrane and the vessel lining is often quite broad around the corticomedullary vessels and is called the perivascular space. This space may contain all kind of blood cells and most often contain fine lymphatics. Nerves course along the blood vasculature.

During ontogeny, hematopoietic progenitor cells migrate into the thymic epithelial primordium between days 11–13 of fetal life in mice. Small lymphocytes can be found in the thymic primordium at about day 14 (mouse) or day 15 (rat) of fetal life. The thymus is fully developed, meaning a cortex and medulla can be distinguished, at day 17 of fetal life in the mouse and by days 19–21 in rats, and the organ grows considerably immediately after birth. This growth is caused by the immense postnatal antigen stimulation; at that time large numbers of mature T cells are demanded. The thymus starts to involute after adulthood is reached. With age, the two thymic lobes diverge caudally and in old animals are almost completely separated; the thymus is then restricted to the area cranially to the aortic arch. The number of lymphocytes decrease, especially in the outer cortex. Although areas with different lymphocyte density, suggesting the presence of cortex and medulla, are often present in advanced age, the general arrangement of the cortex enclosing the medulla is not strictly maintained. This gives the thymus an irregular appearance. The expanding perivascular connective tissue meshwork and increasing perivascular lymphocyte accumulations may further disturb the normal pattern. The septa and capsule harbor increasing numbers of adipose cells, which eventually invade the thymic parenchyma.

In addition to the expansion of the connective tissue component, epithelial cords and tubules are large and numerous in the old thymus and the epithelial Hassall's bodies become relatively more prominent though in absolute numbers they decrease. Adrenergic innervation of the gland is maintained in old animals. Thymic involution may be related to changes in the hormonal status of the individual; circulating thymic hormone is reduced to very low levels in adults. The consequences of age-related involution are obvious: the emigration of lymphocytes from the thymus shows a dramatic decrease. Apparently, the persistent generation of new antigen-recognition repertoire in the T cell population of adults is not needed. Instead, the body can defend itself using the established repertoire and extra-thymic self renewal of the T cells. Pregnancy in rodents results in radical, but reversible changes. After an initial rise in thymic weight in early pregnancy, involution starts with lymphocyte cell death in the cortex. In wild populations, cyclical enlargement and regression is documented. For instance, most birds showed an involuted thymus at the time of mating and laying, whereas on subsequent egg incubation the thymus size is increased.

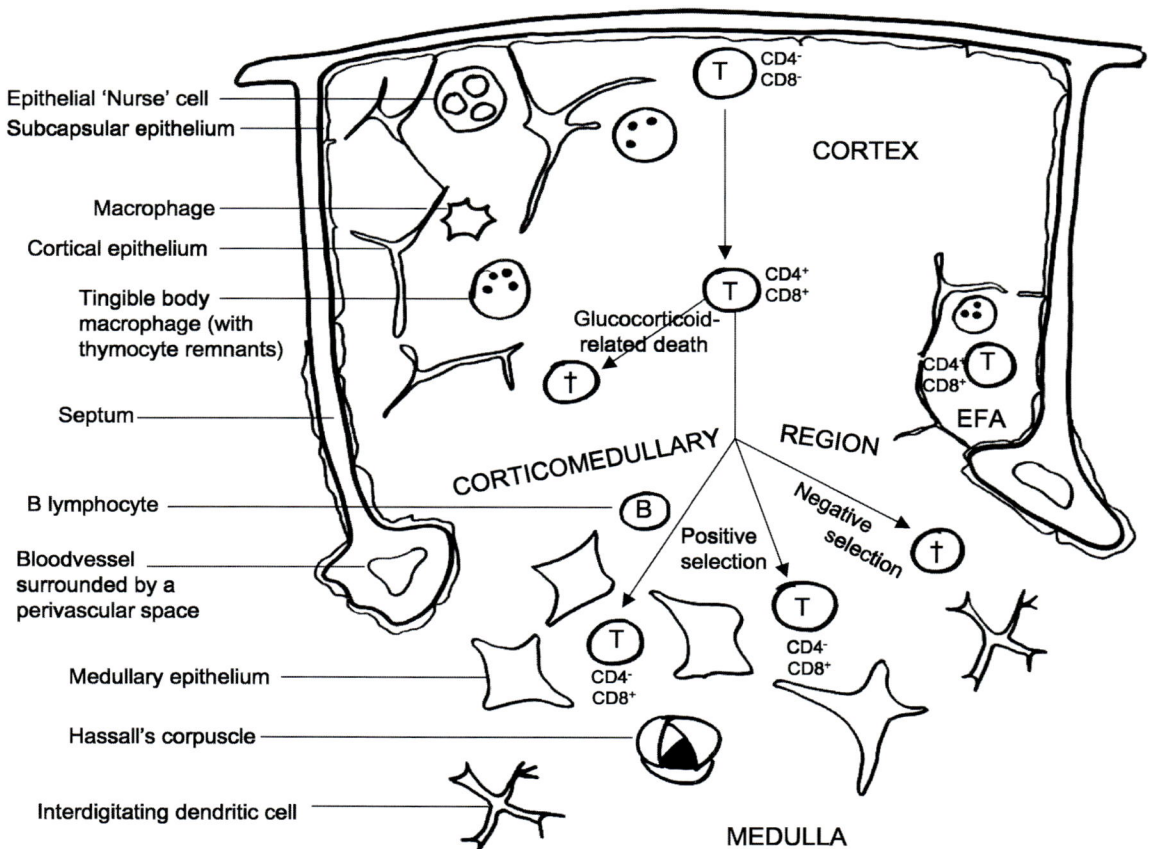

Thymus. Figure 1 Schematic presentation of a thymus lobule with cortex, corticomedullary region, medulla and an epithelial-free area (EFA). In the lobule, a simple overview of thymocyte maturation is presented: round cells representing T lymphocytes (T) with their membrane markers CD4 and/or CD8

T-Cell Maturation

T cells reside in the thymus during their maturation from progenitor cells to immunocompetent T cells. The process of T-cell maturation includes a number of steps which are associated with location in different microenvironments (3). (See Figure 1).

The immature cells, which enter the lobules by the blood vasculature at the corticomedullary junction, first move to the outer subcapsular cortex, where they appear as large lymphoblasts. They then pass through the cortex where the cells become small lymphocytes with scanty cytoplasm. Finally, the cells move to the medulla, where they appear as medium-sized lymphocytes. These translocational stages in development can be monitored by the immunologic phenotype: cells change from $CD4^-CD8^-$ (double negative) at a very immature stage into a $CD4^+CD8^+$ (double positive) phenotype, which is characteristic for almost all lymphocytes in the cortex. In the medulla, T cells have the phenotype of relatively mature cells, with distinct $CD4^+CD8^-$ (about 70%) and $CD4^+CD8^+$ (about 30%) populations. This phenotypic change is accompanied by a crucial aspect of intrathymic T-cell maturation: the genesis of the T cell receptor (TCR) consisting of the alpha-beta heterodimer (4). The DNA genomic organization encoding these chains is in germ-line configuration, with a variety of gene segments encoding the variable part of the receptor molecule. Before transcription and translation into TCR becomes possible, combinations have to be made of gene segments encoding the variable and constant parts of the TCR.

This process of gene rearrangement requires the thymus microenvironment. The cell can synthesize the receptor after completion of this gene rearrangement. The receptor is then expressed on the cell membrane with the CD3 molecule, which acts as the transmembrane signal-transducing molecule after TCR stimulation. Even when the TCR has not yet been synthesized, this CD3 molecule is already present in the cytoplasm of the cell. T cells at this stage of maturation can be recognized by cytoplasmic staining with CD3 reagents.

TCR gene rearrangement is similar to the rearrangement of genes encoding immunoglobulin heavy and light chains that takes place in the bone marrow microenvironment. However, after surface expression of the TCR, the cell undergoes a process unique to T cells, namely, specific selection on the basis of recognition specificity. First, the cell is examined for its affinity for its own major histocompatibility complex (MHC; self restriction). T cells with an intermediate affinity for self MHC peptides are allowed to expand (positive selection). Secondly, T cells with a high affinity for self MHC are deleted (negative selection). In this way, the random pool of antigen-recognition specificities of T cells is adapted to the host's situation. The T cell repertoire in germ-line configuration cannot be fully expressed but is influenced by the individual's own MHC haplotype.

It is generally accepted that the epithelial microenvironment of the thymic cortex plays a major role in positive selection. This microenvironment expresses MHC class I and class II products and morphologically (at electron microscopic level) shows close interactions with lymphocytes. This close interaction is reflected by the complete inclusion of lymphocytes inside the epithelial cytoplasm (thymic ▶ nurse cell). Negative selection has been ascribed to either the epithelial compartment or the medullary dendritic cells. The cortex can be considered a primary or central lymphoid organ because of its antigen-free microenvironment. In contrast, antigens can move relatively freely into the medulla and encounter antigen-presenting dendritic cells as well as antigen-reactive T cells. Thus the medulla has properties of a secondary lymphoid organ.

Preclinical Relevance

The dynamics of the thymus with ongoing reactions of cell proliferation and differentiation, and gene amplification, transcription and translation makes it highly susceptible to toxic insults. Compounds that interfere with these processes are often immunotoxic. Therefore, a decrease of thymus weight in preclinical studies is often a first indicator of toxic action of a xenobiotic agent on the immune system, although some compounds, like cyclosporine, profoundly alter thymic histophysiology, without apparent effect on thymus weight. The dynamic nature of the immune system provides it with great regenerative capacity: the original architecture of the thymus is restored rapidly following involution induced, for example, by irradiation, or treatment with glucocorticosteroids or organotin compounds.

Thymus in aged or immunocompromised animals may hardly be visible. For histology adipose and connective tissues from the cranioventral mediastinum, which contains thymic tissue, should then be collected. The thymus is also very susceptible to acute (glucocorticoid-related) stress (5). It is conceivable that with age the thymus becomes less sensitive to toxic insults and that toxic effects on the thymus with age have less functional importance, because of age-related thymic involution. However, the components that constitute the various thymic compartments are still present in healthy old animals, as was shown by reconstitution studies. Therefore, a decreased sensitivity to toxic compounds may not be a general property of the involuted thymus in aged animals.

Relevance to Humans

The use of data obtained in laboratory animal species for man presents difficulties when species differ in organ anatomy and histophysiology and sensitivity. The thymus is present in all vertebrates, possibly with few exceptions, and there are only a few structural differences between the species (6). Anatomical differences relate to thymus location and number of thymic lobes, the prominence of epithelial aggregates with centrally located cell debris, the so-called Hassall's bodies, and the presence of B cell follicles. During the third month of gestation the thymic primordium becomes colonized by marrow-derived stem cells. When these stem cells are indeed thymocyte precursor cells, their migration into the thymic primordium at that time is considerably earlier—relative to gestation time—in humans than in mice or rats. Differences in immunotoxicity between laboratory animals and man appear to depend predominantly on differences in toxicokinetics and metabolism of substances. Moreover, the interindividual differences and the age-related intraindividual variations are probably more marked than interspecies differences. It should be emphasized that the "normal" architecture of the thymus, as known from textbooks, can be expected only between the late gestational period and young adulthood, and before pregnancy.

The universality of the immune system observed in mammals and the data obtained so far indicate that data from laboratory animals can be extrapolated quite well to humans.

Regulatory Environment

Regulatory toxicity testing, which uses immune parameters, is still under development. This applies to pharmaceuticals and industrial substances as well. Nevertheless, most guidelines recognize the importance of the thymus. For instance, the European Union guidelines on repeated-dose toxicity testing with pharmaceuticals require the macroscopic and microscopic examination of the spleen, thymus, and some lymph nodes with respect to the immune system.

Moreover, a multilaboratory, 28-day oral toxicity study (OECD guideline 407) with the model immunotoxicants azathioprine and cyclosporine demonstrated that the most consistent effects were observed in the thymus (7).

References

1. Dabrowski MP, Dabrowski-Bernstein BK (1990) Immunoregulatory Role of the Thymus. CRC Press, Boca Raton
2. Bruijntjes JP, Kuper CF, Robinson J, Schuurman H-J (1993) Epithelium-free area in the thymic cortex of rats. Dev Immunol 3:113–122
3. Van Ewijk W (1991) T-cell differentiation is influenced by thymic microenvironments. Ann Rev Immunol 9:591–615
4. Werlen G, Hausmann B, Naeher D, Palmer E (2003) Signaling life and death in the thymus: Timing is everything. Science 299:1859–1863
5. Godfrey DI, Purton JF, Boyd RL, Cole TJ (2000) Stress-free T-cell development: glucocorticoids are not obligatory. Immunol Today 21:606–611
6. Zapata AG, Cooper EL (1990) The immune system: comparative histophysiology. In: The Thymus. John Wiley, Chichester, pp 104–150
7. International Collaborative Immunotoxicity Study (ICICIS) Group Investigators (1998) Report of validation study of assessment of direct immunotoxicity in the rat. Toxicology 125:183–210

Thymus: A Mediator of T Cell Development and Potential Target of Toxicological Agents

MICHAEL LAIOSA
NIAID/NIH
Bethesda, MO 20897
USA

ALLEN SILVERSTONE
Upstate Medical University
166 Irving Ave.
Syracuse, NY 13210
USA

Synonyms

T-cell development, thymocyte development, T-cell selection, thymocyte selection, thymocyte education, positive selection, negative selection, thymus, thymus atrophy, thymus involution

Definition

T-cell development is the process by which hematopoietic progenitor cells from the bone marrow home to the thymus and undergo a complex process of differentiation, proliferation and selection to become mature T-cells that will emigrate from the thymus to peripheral lymphoid organs such as the spleen and lymph nodes. Additional maturation and differentiation into T-helper (Th) type 1 and Th type 2 subsets occur in the periphery and are discussed elsewhere.

Characteristics

The ▶ thymus is the central organ for T-cell development in the body, and the principle function of the thymus is to regulate T-cell recognition of self antigens presented by the body to insure that useless or self-reactive T-cells do not mature. T-cell development is characterized by progenitor cells that originate in the fetal liver or bone marrow and enter the thymus through the blood stream (1). The thymocytes then undergo a highly regulated process of differentiation, proliferation, selection, and maturation to become T-cells. The stages of murine thymocyte differentiation can be distinguished by differentially expressed surface molecules stained with fluorochrome-labeled antibodies and detected using flow cytometry. The thymocyte subpopulation that appears earliest is identified by expression of the lymphoid homing receptor CD44 and cKit, the receptor for the stem cell factor, ($CD44^+CD25^-$, DN1) (1). Subsequently, the high affinity interleukin receptor IL-2α (CD25) and the heat stable antigen (HSA, CD24) are upregulated and the proliferation rate of this population also increases (DN2) (1). Following expression of CD25, CD44 is down modulated leading to the next stage of differentiation, $CD44^-CD25^{hi}$ (DN3) (1). In the DN3 population, the $\alpha\beta$ and $\gamma\delta$ T-cell antigen receptor (TCR) lineages begin to diverge as recombination activating gene products 1 and 2 (RAG1, RAG2) begin somatic gene rearrangement of the TCR β locus (1). Successful rearrangement and surface expression of a functional TCR β chain in a complex with the pre-Tα protein results in a burst of proliferation and the gradual reduction of CD25 expression on the cell surface (DN4) (1). Subsequent to successful expression of TCR β, rearrangement of TCR α begins and the CD8 and CD4 molecules are expressed on the cell surface(1). It has been calculated that it takes 3–4 days for a DN3 cell to differentiate into the DP stage of T-cell development (1).

Once TCRα rearrangement is complete, the $CD4^+CD8^+$ double positive (DP) thymocytes begin a rigorous selection process by engaging their $\alpha\beta$ TCR with complexes of self peptides bound to major histocompatibility complex (MHC) class I and II proteins (1), expressed by epithelial, myeloid, and dendritic antigen presenting cells (APCs) in the cortex of the thymus (2). The TCR-MHC interaction leads to one of three possible outcomes depending on the nature of the interaction. TCRs with no or weak affinity for MHC will die by neglect. In comparison, potentially self-reactive TCRs with too high or strong affinity for the peptide MHC complex undergo negative selection. Only TCRs with the appropriate affinity for peptide MHC complexes will undergo maturation, CD4 (class II MHC) or CD8 (class I MHC) lineage commitment and ▶ positive selection (1).

The signal transduction that results in positive selection begins with phosphorylation of the intracellular portion of the TCRζ chain by the src kinase Lck. Phosphorylation of TCRζ results in the subsequent recruitment of Zap70, which becomes activated and phosphorylates the linker of activated T-cells (LAT). The phosphorylated LAT acts as a docking complex, which recruits and activates a number of molecules involved in TCR signal transduction and calcium ion (Ca^{2+}) flux (3). The generation of a Ca^{2+} flux has been shown to depend on phospholipase C γ (PLCγ), which generates inositol-3-phosphate (IP3) and diacylglycerol (DAG) (1). IP3 is responsible for the increase in intracellular Ca^{2+} and leads to the activation of the calcineurin pathway and the NFAT family of transcription factors (1). In contrast DAG is involved in activating protein kinase C (PKC) family members and can be a mediator in activation of the Ras pathway. In DP thymocytes it is thought that DAG activates the guanine nucleotide exchange factor RasGRP1 leading to activation of the extracellular signal-related kinase (ERK) (1). ERK activation in thymocytes undergoing positive selection is thought to be involved in activating the early growth response-1 (EgR-1) nuclear transcription factor (1). The positively selected DP thymocytes then upregulate Bcl-2 and mature to become either class II restricted ($CD4^+$; T helper) or class I restricted ($CD8^+$; T cytotoxic) single positive thymocytes. Additional selection occurs in the medulla of the thymus before final maturation and emigration of the SP T-cells into the periphery (1).

Although ▶ negative selection results in a profoundly different outcome (cell death rather than maturation) many of the signaling pathways utilized are the same or similar. Most current data on thymocyte selection favor a model where the affinity between a TCR and self peptide-MHC complexes determines whether a thymocyte will be positively selected or deleted. High affinity interactions with TCR and self peptide-MHC may activate additional signaling pathways such as the Jnk pathway, which ultimately lead to apoptosis (1). In comparison, TCRs with weak or no affinity for self peptide-MHC complexes will die by neglect in the thymus within 1–3 days (1). Only thymocytes possessing the appropriate affinity and duration of binding between a TCR and self peptide-MHC complexes can be positively selected (1).

A number of toxicological agents have been identified which can interrupt or inhibit various stages of T-cell development, which ultimately leads to atrophy of the thymus. Agents that have been shown to cause thymic atrophy in vivo include corticosteroids, estrogens and estrogen-like compounds, polychlorinated biphenyls (PCBs), and polychlorinated dibenzodioxins and dibenzofurans (PCDD and PCDF). Representative agents that are known to induce thymic atrophy and possible mechanisms by which they can induce atrophy are listed in Table 1 (4,5).

Evidence of thymic atrophy after toxicant exposure has a relatively strong correlation to predicting if an agent will be immunotoxic as defined by classic immunotoxicity assays such as delayed-type hypersensitivity (DTH), and the sheep red blood cell (SRBC) challenge assay (6). However, linking immunotoxicant-induced defects in thymic development to deficiencies in a functional response has been a major obstacle in the field of immunotoxicology. Relating thymic atrophy to alterations of functional responses have suffered from a lack of data and agreement on the type of assays, kinetics, and dosing protocols to be used.

Relevance to Humans

The thymus has been shown to be essential for development of T-dependent immune responses. Indeed, patients with the rare ▶ DiGeorge syndrome who lack a thymus present with a severe immunodeficiency associated with a complete lack of T-cells. The DiGeorge T-deficiency can be completely restored by the transplantation of an allogeneic thymus graft (7). The essential role for the thymus in T-cell development has been further appreciated in recent clinical studies. These studies show that despite the longstanding observation of ▶ thymus atrophy with increasing age, the adult thymus is fully capable of producing and selecting new T-cells following periods of systemic T-cell depletion. Following chemotherapy, production of new thymic-derived naive T-cells has been observed (7). Additionally, infection with HIV has been shown to cause a dramatic thymic pathology characterized by thymic atrophy and a block in T-cell development at the $CD3^-CD4^-CD8^-$ stage of development. However, thymopoieis can be restored in some HIV patients undergoing highly active antiretroviral therapy (▶ HAART) (7). Finally, evidence of TCR gene rearrangement in recent thymus emigrants has been observed in normal adults of at least 60 years of age (7). These data strongly support an active and dynamic role for the thymus organ in mediating new T-cell development throughout an individual's life.

The effect of ▶ immunotoxicants as mediators of thymic atrophy in humans has been controversial and difficult to assess for some time. The lack of consensus on whether a particular toxicant can cause thymic at-

Thymus: A Mediator of T Cell Development and Potential Target of Toxicological Agents. Table 1 Agents known to cause thymic atrophy and mechanism of atrophy induction[1]

Agent	Mechanism
Androgens	Loss of DP thymocytes; mediated by androgen receptor
Cisplatin	Apoptosis in proliferating thymocytes
Cyclosporin A	Prevents Ca^{++} mobilization; inhibits positive selection; delayed negative selection
Dexamethasone (and other corticosteroids)	Apoptosis in DP thymocytes
Dibutyl and tributyltin	Possible apoptosis; inhibition of proliferation of DN thymocytes
Diethylstilbestrol (DES), estradiol, estrogens and estrogen-like chemicals	No evidence of apoptosis, possible effects on progenitors and cell cycle; estrogen receptor-mediated
Ethylene glycol monomethyl ether	Reduction in DP thymocytes, but no evidence of apoptosis Reduction of lymphocyte progenitor capacity
Ethanol	Apoptosis; increase in $CD4^+$ mature cells, loss of $CD25^+$ DN cells; evidence of Ca^{++} increase and protein kinase C activation
Malnutrition, vitamin deficiency	Increase of glucocorticoid levels; apoptosis of DP thymocytes
2,3,7,8, tetrachlorodibenzo-*p*-dioxin	No evidence of apoptosis in vivo; inhibition of bone marrow progenitors; inhibition of cell proliferation in thymic DN cells; all effects mediated by the aryl hydrocarbon receptor
T-2 toxin and other mycotoxins	Elimination of putative lymphocyte progenitor cells in fetal liver; no evidence of apoptosis induction.

[1] Adapted from Luster et al. (4) and Silverstone (5)

rophy is due in part to the obvious ethical considerations with human studies. Moreover, the vast majority of immunotoxicity assays that have been developed are in rodent models that possess inherent flaws when attempting to determine dose, pharmacokinetic, and risk assessment comparison models to humans. The challenges of relating risk assessment models to humans should be overcome in the future as immunotoxicologists begin to develop nonhuman primate models, novel in vitro models and comparative ▶ toxicogenomic studies to fill in the gaps in knowledge about particular toxicants as related to T-cell development and immunotoxicity (6,8,9).

Regulatory Environment

Regulatory agencies in the USA have recently started to stress the importance of understanding how immunotoxicants affect the developing immune system in children. The need to understand the effects of immunotoxicants in children is particularly important because of the possibility that during the period when the immune system is most actively developing, it may be especially sensitive to the effects of an immunotoxicant. Moreover, immunotoxicant exposure in children may lead to more severe effects and/or a higher risk for long-term deleterious outcomes when compared to doses determined for adults (9). Although there are currently limited data comparing adult and child responses to immunotoxicants on the developing immune system, several possibilities for differences exist. An immunotoxicant may affect the developing immune system of a child but not an adult. Furthermore, an immunotoxicant may affect the developing immune system of a child at a lower dose than in an adult (9).

In an attempt to get the full picture about childhood exposure to immunotoxicants and the effect of exposure on the developing immune system of children, several EPA sponsored workshops have listed the need for expanding exposure studies in very young animals as a high priority. These workshops include the EPA sponsored workshop on endocrine disruptors held in 1995, and the EPA sponsored workshop by the Risk Science Institute of the International Life Sciences Institute held in 1996. More recently, the EPA added a recommendation to the two-generation reproductive study (OPPTS 870-3800), stating: for F1 and F2 weanlings that are examined macroscopically, the following organs should be weighed for one randomly selected pup per sex per litter: brain, spleen and thymus (9). The recommendation to use thymus and spleen weights was made because numerous studies have concluded that thymic and splenic weight may be immunotoxicant predictors (6).

In 2001 the EPA created a developmental immunotox-

icology working group. The mission of the this group is to determine:
- the state of science to support the creation of a guideline for developmental immunotoxicology
- what should be included in such a guideline
- how this guideline would be validated
- when a developmental immunotoxicology guideline would be used (9).

Lastly, in 2003, the National Institute of Environmental Health Sciences (NIEHS) and National Institute for Occupational Safety and Health (NIOSH) cosponsored a consensus workshop on methods to evaluate developmental immunotoxicity. This workshop made several recommendations for immunotoxicant screening assays as well as assays that needed further validation and assays for research development (4). The recommended screening assays for developmental immunotoxicants were the primary antibody response (T-dependent), delayed-type hypersensitivity response, complete blood count (CBC), and weights of thymus, spleen and lymph nodes. Assays that require additional validation include phenotypic analyses, macrophage function and natural killer cell activity. Finally, stem cell functional assays were listed as assays that require additional research and development (4).

References

1. Starr TK, Jameson SC, Hogquist KA (2003) Positive and negative selection of T cells. Ann Rev Immunol 21:139–176
2. Anderson G, Jenkinson EJ (2001) Lymphostromal interactions in thymic development and function. Nat Rev Immunol 1:31–40
3. Germain RN, Stefanova I (1999) The dynamics of T cell receptor signaling: complex orchestration and the key roles of tempo and cooperation. Ann Rev Immunol 17:467–522
4. Luster MI, Dean JH, Germolec DR (2003) Consensus workshop on methods to evaluate developmental immunotoxicity. Environ Health Persp 111:579–583
5. Silverstone AE (1997) T cell development. In: Sipes G, McQueen CA, Gandolfi AJ (eds) Comprehensive Toxicology, 1st edn. Elsevier Science, New York, pp 39 ff
6. Holladay SD, Blaylock BL (2002) The mouse as a model for developmental immunotoxicology. Hum Exp Toxicol 21:525–531
7. Spits H (2002) Development of alpha-beta T cells in the human thymus. Nat Rev Immunol 2:760–772
8. Buse E, Habermann G, Osterburg I, Korte R, Weinbauer GF (2003) Reproductive/developmental toxicity and immunotoxicity assessment in the nonhuman primate model. Toxicology 185:221–227
9. Holsapple MP (2003) Developmental immunotoxicity testing: a review. Toxicology 185:193–203

Thymus Atrophy

Loss of thymocyte weight and cellularity after exposure to an immunotoxicant.
▶ Thymus: A Mediator of T-Cell Development and Potential Target of Toxicological Agents

Thymus-Dependent Antigen

Thymus-dependent antigens (TD) are protein antigens which only can induced an antibody response with the help of thymus-derived T helper cells. This T cell help is also essential for the class switch observed during TD immune responses.
▶ Idiotype Network

Thymus Gland

The thymus is a primary lymphoid organ, the site of T-cell development. It is situated in the anterior superior mediastenum, behind the breastbone. The organ, in particular its epithelial cells and connective tissue provide the microenvironment wherein thymocytes proliferate, rearrange their T-cell receptor genes, and undergo positive and negative selection. The thymus slowly atrophies after puberty, but can become fully functional again in clinical situations like radiation therapy and stem cell transplantation.
▶ Thymus
▶ Dioxins and the Immune System
▶ Thymus: A Mediator of T Cell Development and Potential Target of Toxicological Agents
▶ Systemic Autoimmunity

Thymus Involution

▶ Thymus: A Mediator of T-Cell Development and Potential Target of Toxicological Agents

Tight Junctions

An intercellular junctional structure, typically found in epithelia and endothelia. In the tight junction the two membranes of neighboring cells are brought into close proximity through binding of specific transmembrane proteins. This results in a selectivity barrier that seals the apical lumen from the basolateral intercellular

space and also establishes cellular polarity by preventing membrane-linked molecules from freely diffusing between the apical and the basolateral cell surface.
▶ Cell Adhesion Molecules

Time-Resolve Fluorometry

An instrumental design to collect emission at a certain time interval after the pulsed excitation and to improve the detection sensitivity by means of a temporal rejection of background.
▶ Cytotoxicity Assays

Tissue Factor

This cellular receptor for factor VII/VIIa is constitutively expressed on cells of the media and adventitia of the vessel wall. When it is exposed to plasma clotting factors at sites of vascular injury it serves as a potent (extrinsic) cofactor for the activation of factor X. Tissue factor is also associated with platelets and microparticles and is responsible for intravascular activation of blood clotting in the absence of tissue damage.
▶ Blood Coagulation

Tm Mouse

▶ Knockout, Genetic

TNF-α

▶ Tumor Necrosis Factor-α

Tolerance

ANKE KRETZ-ROMMEL
Principal Scientist
Alexion Antibody Technologies
Suite A, 3958 Sorronto Valley Rd
San Diego, CA 92121
USA

Synonyms
Immunological unresponsiveness

Definition
The primary function of the immune system is to protect the host from foreign materials while at the same time ensuring that no attack against self proteins occurs. Immunological tolerance is the absence of immunological responsiveness to specific antigens, encompassing unresponsiveness to self antigens, but also tolerance to therapeutics such as antibodies, recombinant proteins and conventional drugs. Breakdown of immune tolerance is defined by the appearance of T-cells or antibodies to self antigen or the therapeutic entity. The result may be autoimmune disease or allergic or anaphylactic reactions. Furthermore, an immune response to a drug may reduce its efficacy.

Immune tolerance is an active process at both the B cell and T-cell level, involving processes taking place in central lymphoid organs (thymus and bone marrow) and peripheral lymphoid organs (blood, spleen, lymph node, mucosal immune system). The underlying mechanisms are subject to a continuous debate involving clonal deletion, anergy, regulatory T cells and regulatory dendritic cells. In this chapter these concepts will be outlined with reference to drugs affecting various tolerance mechanisms, and the interested reader is referred to more in depths reviews.

Characteristics of T Cell Tolerance
Central Mechanisms
T cells develop in the thymus. Recombination of gene segments creates the two chains that make up the T cell receptor (TCR) resulting in a large repertoire of receptor specificities. To ensure the export to the periphery of T cells that recognize peptides in the context of self major histocompatiblity complex (MHC), but do not strongly react to self antigens, the cells have to undergo positive and negative selection processes as outlined in Figure 1.

Selection is a rigorous process that results in the death of approximately 95% of T cells. T cells first have to undergo positive selection on self peptide presented in the context of self MHC. Successful signaling through the TCR has been suggested to raise the threshold of activation of these T cells possibly through the production of negative regulators (1). If the T cells still can be activated in a subsequent encounter of self peptide presented by MHC the T cell will undergo clonal deletion by apoptosis, a process termed negative selection. This leaves only T cells to be exported to the periphery with a threshold of activation that can not normally achieved by self peptides. Interference with negative selection in the thymus has been proposed as a mechanism for the induction of autoimmunity. TCDD and cyclosporine have been evoked to affect both positive and negative selection processes. The reactive metabolite of the antiarrhythmic procainamide hydroxylamine (PAHA) has been shown to interfere

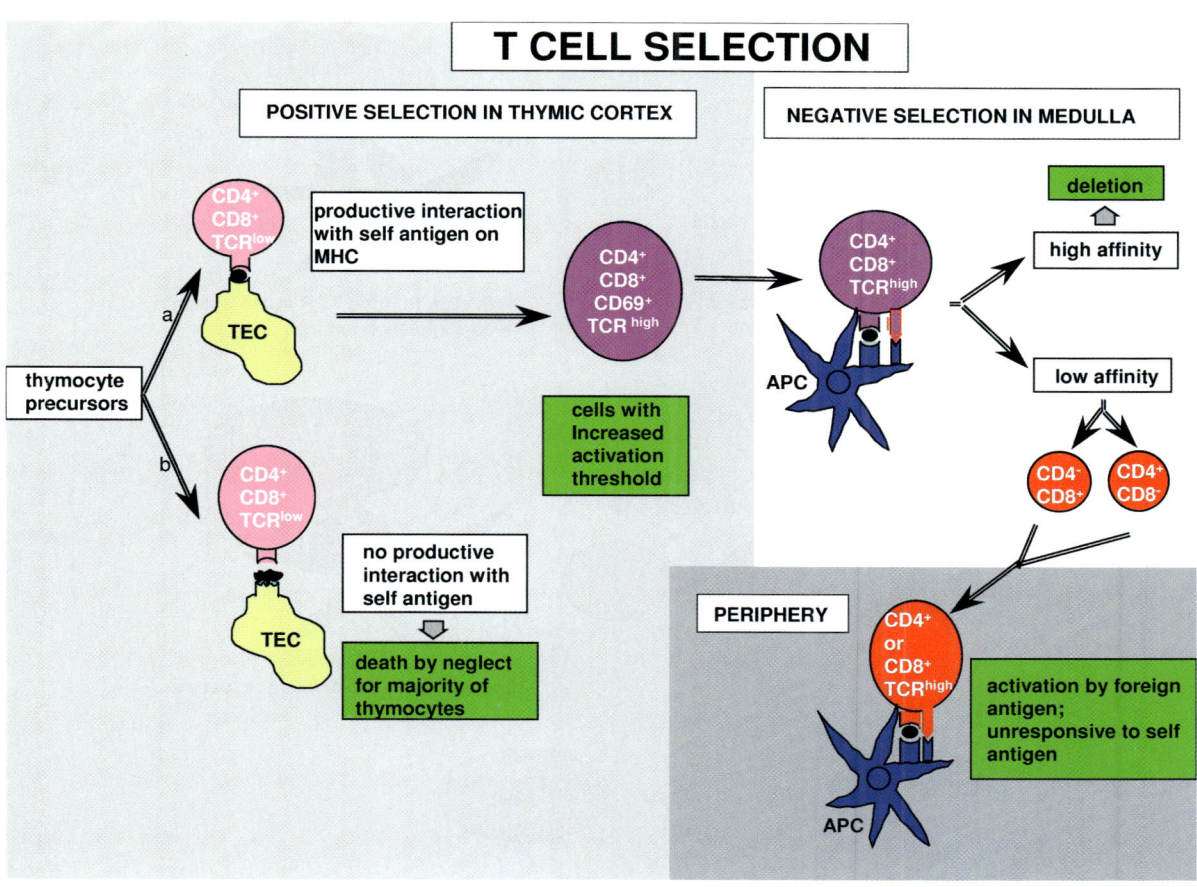

Tolerance. Figure 1 Central tolerance mechanisms. After migration from the bone marrow to the thymus, T cells first undergo selection on self peptides presented by thymic epithelial cells. Cells productively interacting with the presented peptide proceed to negative selection resulting in deletion of cells with high affinity for self peptide. TEC=thymic epithelial cell; APC=antigen presenting cell.

with positive selection in the thymus, resulting in the export to the periphery of autoreactive T cells and autoantibody production similar to that observed in patients with drug-induced lupus.

Peripheral Mechanisms

T cells leaving the thymus still might respond to self antigens if the antigens are present in such high concentration that they can bind to "weak" receptors or if they did not encounter the self peptide in the thymus which might be the case for certain tissue-specific antigens. A number of peripheral mechanisms can control these potentially self-reactive cells (2) as summarized in Figure 2.

Lack of Costimulation

Activation of T cells not only requires interaction of the TCR with peptide presented by MHC on antigen-presenting cells (APC), but also a second signal (costimulation). Among the most important of these costimulatory molecules are members of the B7 family, interacting with CD28 on the T cell. Ligation of CD28 by either B7-1 or B7-2 lowers the threshold of TCR signaling needed to induce T-cell activation and increases the effect of that signal by promoting T cell expansion and proliferation. Recently, additional members of the B7-CD28 family involved in the development or maintenance of immune tolerance have been identified such as ICOS which is expressed by activated T cells. Ligation of ICOSL by ICOS prolongs T cell activation. If a T cell receives a signal through the TCR in the absence of costimulation, cells are unresponsive to subsequent stimulation by the peptide in context of MHC in the presence of costimulation—a process termed anergy. While this phenomenon has only been demonstrated in vitro, it recently has been recognized that naive T cells (T cells that have not been stimulated before) in the periphery require frequent interaction with peptide presented by MHC in order to survive. This has been suggested to

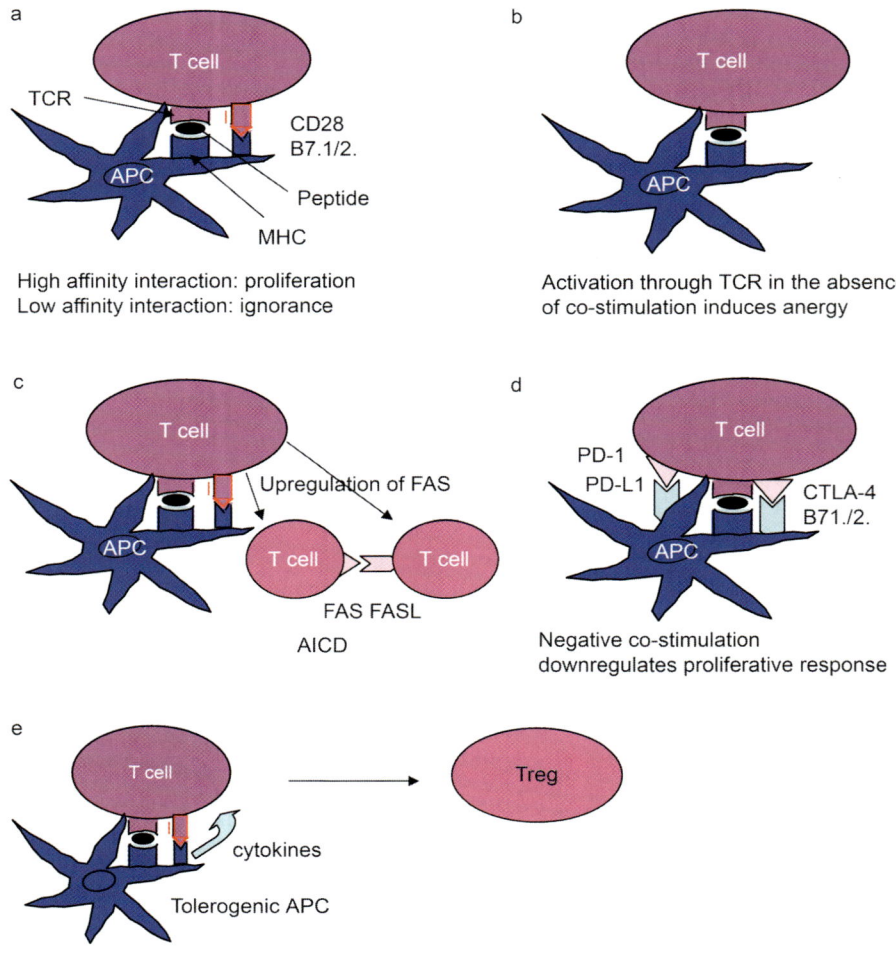

Tolerance. Figure 2 Peripheral tolerance mechanisms. A: Only T cells with high affinity for the antigen presented by antigen presenting cells (APC) will proliferate. Since thymic emigrants have been tuned to have a threshold of activation generally above that achieved by most self peptides, T cell interaction with self peptide presented by major histocompatability complex (MHC) does not result in proliferation. B: If a T cell sees antigen in the context of MHC in the absence of costimulatory signals, anergy can be induced. The T cell is subsequently unresponsive to challenge with the cognate antigen by APCs, even in the presence of costimulatory molecules. C: Death molecules such as fatty acid synthetase (FAS) and tumor necrosis factor (TNF) get upregulated in the course of a T cell response to limit proliferation and cytokine production. T cells involved in the response to antigen will undergo activation-induced cell death (AICD) by apoptosis. D: A number of immunoreceptors downregulate the T cell response. Some of them are upregulated during the T cell response to limit it, and some of them are constitutively expressed on tissues to prevent damage by T cells. E: Tolerogenic dendritic cells can induce T_{reg} which control the response of other T cells.

be an important mechanism of peripheral tolerance, maintaining a high activation threshold of T cells which can only be overcome by foreign antigen. Drugs could potentially provide a "danger" signal to the immune system resulting in upregulation of costimulatory molecules and activation of self-reactive T cells. However, clearly not all drugs inducing cell stress or cell death result in an activation of the immune system. Evidence is emerging though that some compounds can alter dendritic cells resulting in upregulation of the costimulatory molecule CD86 or provoking migration of dendritic cells by upregulating the CCR7 receptor.

Failure to Encounter Self Antigens (Immune privilege)

Under normal conditions, the cells in nonlymphoid

organs throughout the body are not in contact with T cells and are thus sequestered from the immune system. This lowers the probability of a low affinity self-reactive T cell encountering a specific self antigen. Only in the presence of "danger" signals such as provided by bacteria can T cells enter non-lymphoid organs. Certain tissues are particularly protected from the entry of T cells, such as the interior of the eye, brain and testes. Constitutive expression of immunosuppressive receptors and cytokines ensures protection of these organs from immune-mediated damage.

Receipt of Death Signals

An important mechanism of maintaining immune homeostasis is the downregulation of the immune response after activation. Activation of antigen-presenting cells by bacteria or viruses results in upregulation of costimulatory molecules and production of proinflammatory cytokines such as tumor necrosis factor (TNF)-α. Persistence of an inflammatory environment increases the risk of activating T cells by self peptides by providing costimulation to these cells. Also, there is a risk of cross-reactivity of T cells activated by pathogens with self antigen, because activated T cells require less costimulation. Therefore, most activated T cells ultimately undergo a process of programmed death or apoptosis. Apoptosis of activated cells (activation-induced cell death, AICD) occurs by cytokine withdrawal and by induction through fatty acid synthetase (FAS) and TNF-α. FAS acts on FAS ligand expressed on activated T cells. These cells therefore can kill themselves as well as activated B cells and macrophages. Also, expression of other receptors mediating immune suppression play an important role in downregulating the immune response as discussed in the following section.

Immunosuppressive Receptors

The immune response can be terminated by upregulation of the T cell surface molecule CTLA-4. While CTLA-4 is present at very low levels on resting T cells, it is markedly upregulated after T cell activation. Similar to the positive costimulatory molecule CD28, CTLA-4 binds to B7.1 and B7.2. Due to its substantially higher affinity for these molecules, CTLA-4 outcompetes CD28, thereby transducing inhibitory signals to the activated T cell. More inhibitory molecules have recently been identified. PD-1 is expressed on activated T cells, B cells and myeloid cells, and engagement by its ligands PD-L1 and PDL-2 inhibits T cell proliferation and cytokine production. Expression of immunosuppressive receptors on nonlymphoid organs is another safeguard mechanism against self attack of T cells.

Regulatory Cells and Cytokine Milieu

A minor population of T cells known as regulatory T cells (T_{reg}) suppresses the proliferative response and production of inflammatory cytokines of other T cells. They may constitute a specialized T cell subset to reduce the activity of autoreactive T cells. T_{reg} constitutively express CTLA-4 and secrete transforming growth factor(TGF)-β and interleukin(IL)-10. Mechanisms of action are still under debate, but they seem to require direct cell-cell contact.

In addition to regulatory T cells, dendritic cells and macrophages play a major role in immune tolerance. The functional activities of dendritic cells are mainly dependent on their state of activation and differentiation. Terminally differentiated mature dendritic cells can efficiently induce the development of T effector cells, whereas immature dendritic cells are involved in maintenance of peripheral tolerance. The means by which immature dendritic cells maintain peripheral tolerance are not entirely clear, however, their functions include the induction of anergic T cells, T cells with regulatory properties as well as the generation of T cells that secrete immunomodulatory cytokines. Depending on the cytokines produced by the macrophage/dendritic cell, the immune response can be steered towards a Th1 or Th2 response. Th1 cells produce IFN-γ, IL-2 and TNF-α and regulate classical delayed (type IV) hypersensitivity. Th2 cells secrete IL-4, IL-5, IL-6 and IL-10 and participate in immediate (type I) hypersensitivity reactions and B cell antibody-mediated immunity. The effect of drugs on cytokine production and the importance of the cytokine milieu resulting in drug-induced autoimmunity are being studied extensively.

Characteristics: B cell Tolerance

Similar to T cells, B cells are constantly being tolerized to self antigens. For a thorough discussion of B cell tolerance the reader might refer to Jacquemin et al. (3).

Central Mechanisms

B cells mature and undergo selection on self peptides in the bone marrow. A large population of B cells with different specificities is created by genetic recombination within the immunoglobulin locus generating a broad range of heavy- and light-chain sequences that rearrange to form a B cell receptor (BCR). If the immature B cell encounters extracellular antigen capable of crosslinking its BCR, a signal is created that will block further development of this autoreactive cell. The B cell will initiate the receptor editing process to produce BCR with new antigen specificities. If it cannot alter its BCR effectively, the immature B cell will be deleted by apoptosis. Some autoreactive

B cells escape deletion and enter the peripheral circulation in an anergic state.

Peripheral Mechanisms

After recognition and uptake of antigen in the periphery, these partially activated B cells migrate through the lymphoid tissue. If an activated B cell encounters a T cell that has been activated by the same antigen, antibodies against that antigen are produced. B cells cannot respond to most antigens without receiving help from T helper cells. Therefore, ensuring self tolerance of T cells is an important mechanism of keeping B cells from producing autoantibodies. However, drugs affecting B cell tolerance can ultimately result in autoimmunity when the individual has other predisposing factors, as might be the case for pristane.

Additional Mechanism for Drugs to Break Immune Tolerance

The most common hypothesis of how drugs result in immune stimulation is the formation of drug-protein conjugates by reactive drug metabolites with self antigens. The resulting haptens might be recognized as foreign by the immune system. Although formation of haptens has been demonstrated for a number of drugs associated with idiosyncratic immune adverse reactions (e.g. phenytoin, carbamazepine, halothane, tielinic acid, procainamide and diclofenac) these adducts are not a predictive factor for adverse immune reactions indicating that additional factors are required to induce the immune response. It has been demonstrated that binding of halothane to CF3CO proteins mimics very closely the structure of the E2 subunit proteins of the 2-oxoacid dehydrogenase complexes and protein X—autoantigens associated with halothane hepatitis. Furthermore, binding of drugs to protein can alter their cleavage and presentation after cell death. Exposure of macrophages to mercuric chloride has been show to alter fibrillarin processing, resulting in the appearance of self epitopes not normally encountered by the immune system.

In addition to covalent drug binding to proteins, non-covalent interactions of drugs such as sulfamethoxazole with MHC-peptide complexes have been implicated in immunological adverse reactions.

While disruption of immune tolerance by classical chemical drugs leaves many unanswered questions, immune responses after administration of bioengineered drugs is far more straightforward. The importance of antibodies in therapeutics gains increasing recognition. Often, these antibodies are of mouse origin and certain residues are recognized as foreign by the human immune system. Engineering methods known as "humanization" and pegylation decrease the risk of an immune response against the therapeutic.

Preclinical Relevance

Adverse drug reactions affecting immune tolerance are difficult to address in the preclinical setting. However, a number of assays have been developed to address the potential of drugs to sensitize the immune system, such as the popliteal lymph node assay that assesses the effects of drugs on macrophages, or assays looking for altered cytokine profiles. Few animal models demonstrating chemically induced autoimmunity are available, but are specific for the compound used.

As far as immunogenicity of biotherapeutics is concerned, some animal models have proved to be useful. For example, transgenic mice were developed to produce and secrete human tissue plasminogen activator to which they developed immune tolerance. These mice were capable of producing antibodies to a form of human tissue plasminogen activator that had been modified by a single amino acid substitution. Furthermore, nonhuman primates have been used successfully in predicting the relative immunogenicity of different forms of human growth hormone. Also, computer modeling methods are used in predicting the immunogenicity of proteins.

Relevance to Humans

Adverse drug reactions account for 2%–5% of all hospital admissions, a portion of which is based on immune-mediated reactions. With more than 80 recombinant proteins in clinical use and more than 400 therapeutic antibodies in clinical trials, immune tolerance to these proteins is a major issue and predicting immunogenicity is crucial (4).

Regulatory Environment

Regulatory issues for drug-induced autoimmunity and allergy are covered in their respective chapters. For clinical trials of recombinant proteins, patients are screened for the development of antidrug antibodies.

References

1. Grossman Z, Singer A (1996) Tuning of activation thresholds explains flexibility in the selection and development of T cells in the thymus. Proc Natl Acad Sci USA 93:14747–14752
2. Sharpe AH, Freeman GJ (2002) The B7-CD28 superfamily. Nat Rev Immunol 2:116–126
3. Jacquemin MG, Vanzieleghem B, Saint-Remy JM (2001) Mechanisms of B-cell tolerance. Adv Exp Med Biol 489:99–108
4. Pendley C, Schantz A, Wagner C (2003) Immunogenicity of therapeutic monoclonal antibodies. Curr Opin Mol Ther 2:172–179

Tolerance and the Immune System

Unresponsiveness to antigenic stimulation that is either mediated by genetics, or is acquired by special conditions of antigenic exposure. The immune system has established several mechanisms that prevent immune reactions against self antigens. Of central importance is the tolerance of the immune regulatory helper T cells. Activation of helper T cells can be controlled by tolerance induction in the thymus, by sequestration of antigens in immune privileged sites (brain, testis, cornea) and by active suppression of immune responses by regulatory T cells.
▶ Antigen Presentation via MHC Class II Molecules
▶ Graft-Versus-Host Reaction
▶ Autoantigens
▶ Autoimmune Disease, Animal Models
▶ Antinuclear Antibodies
▶ Lymphocytes
▶ Transforming Growth Factor β1; Control of T cell Responses to Antigens

Toll-Like Receptors

A family of receptors expressed by cells of the innate immune system and directed against conserved structures present on many micro-organisms. Ten members of this receptor family are present in humans (e.g. TLR4 specific for lipopolysaccharide; TLR2 for peptidoglycan; TLR5 for flagellin). They are named after the *Drosophila* protein Toll which is involved in the antibacterial defense of the fruit fly.
▶ B Cell Maturation and Immunological Memory

Toxic Epidermal Necrolysis (TEN)

Toxic epidermal necrolysis (TEN) represents the most serious extreme of the febrile mucocutaneous syndrome in which there is a full-thickness sloughing of the epidermis. According to the criteria, TEN is defined as detachment affecting about 30% of the body surface area. Stevens-Johnson syndrome is similar to TEN in terms of the histopathology and the responsible drugs, indicating that these two conditions are part of the same spectrum. Fas-Fas L interactions appear to be involved in the epidermal necrolysis.
▶ Drugs, Allergy to

Toxic Oil Syndrome (TOS)

An illness associated with the ingestion of adulterated rapeseed oil in Spain in 1981. The most distinctive lesion is a non-necrotizing vasculitis involving different types and sizes of blood vessels in every organ.
▶ Systemic Autoimmunity

Toxicogenetics

The genetic basis for individual differences in susceptibility to toxicity, with single nucleotide polymorphisms (SNPs) being the prime source of variability in the genome.
▶ Toxicogenomics (Microarray Technology)

Toxicogenomic Studies

Studies in toxicology which screen for global changes in gene expression following exposure to a toxicological agent.
▶ Thymus: A Mediator of T Cell Development and Potential Target of Toxicological Agents

Toxicogenomics

The measurement of altered gene expression upon exposure to a compound or drug, thereby identifying the toxicant and characterising its mechanism of action.
▶ Toxicogenomics (Microarray Technology)

Toxicogenomics (Microarray Technology)

ROB J VANDEBRIEL
Laboratory for Toxicology, Pathology and Genetics
National Institute for Public Health and the Environment
3720 BA Bilthoven
The Netherlands

Synonyms

Gene profiling, expression profiling, global gene expression analysis

Definition

▶ Microarray technology is the simultaneous individual measurement of the mRNA expression level of thousands of genes in a given sample by means of hybridization. ▶ Toxicogenomics is the measurement of altered gene expression upon exposure to a compound or drug, thereby identifying the toxicant and characterizing its mechanism of action.

Characteristics

Although individual differences exist, the basic principle of microarray technology is the same for different platforms (1). The term platforms means types of arrays or array suppliers, in the latter case combined with dedicated hardware and software. First, per gene, a single probe or a few different probes are generated, using either polymerase chain reaction(PCR)-amplified complementary DNA (cDNA), or synthetic DNA segments (oligonucleotides or oligos) devised on the basis of these cDNA sequences. Usually they are spotted onto a glass surface in a regular array. This process is called spotting or arraying, and requires dedicated machinery. Some companies manufacture oligos in situ, either using photolithography (Affymetrix) or chemical coupling (Agilent).

Several options exist to obtain arrays:

- ready-made arrays (e.g. Affymetrix, Agilent)
- custom-made arrays (e.g. Affymetrix, Agilent)
- in-house spotting of a PCR-amplified clone collection (e.g. Invitrogen) or of an oligo collection (MWG, Operon, Sigma).

Other manufacturers of ready-made arrays include Operon, MWG, and Phase-1, but this list is by no means exhaustive.

A clone collection is a collection of bacteria, each containing a plasmid consisting of a different cDNA insert. Care has to be taken that the individual clones indeed contain the correct insert; verifying clone sets by sequencing the inserts is not uncommon. Second, RNA or mRNA is isolated from cells or tissues and cDNA is synthesized. This cDNA is labeled using a fluorescent label, either during or after synthesis. The labeled cDNAs are then hybridized to the array. The Affymetrix platform uses a single labeled cDNA (Cy3) per hybridization, whereas other platforms rely on two labeled cDNAs (Cy3 and Cy5; most often test and control). The array is then read using a scanner (with fitted laser(s) that measures for each spot the fluorescence intensity. These data are then transferred to a personal computer. This process is outlined in Figure 1.

During and after this process a number of controls have to be performed to assure that the results obtained are correct. For the arrays these controls include the shape of the spots and the amount of DNA spotted (e.g. by hybridization of labeled random hexamers). After hybridization these controls include a similar average staining intensity over the entire array, and plotting the intensity ratio of both labels against the intensity of the label for the control sample. This ratio should be independent of the intensity for most of the genes interrogated. To exclude artifacts caused by differential incorporation of the two labels into the cDNAs a dye swab is useful. If replicate samples are tested, statistics can be performed. Ratios of test vs control of > 2 are generally considered significant. If several time points, dose groups, or organs are analyzed, more advanced statistics can be done, such as cluster analysis and/or principal component analysis (2). To this end several algorithms have been written, most of them being freely available on the internet. Commercial software packages have the advantage of easier data handling, compared to the tedious process of uploading data-sets to algorithms on the web (see Baxevanis and Francis Ouelette for a primer on the subject) (3).

The number of genes to be analyzed is of interest. Obviously, for mechanistic studies as well as for seeding databases that are ultimately aimed at identifying toxic profiles of compounds, the number of genes should be maximal, nowadays meaning virtually all genes. With statistics aiding in the process of gene selection, signatures of toxicity (such as peroxisome proliferators) or pathology (such as liver necrosis) may eventually be addressed by interrogating a small number of genes.

Preclinical Relevance

A first important issue of toxicogenomics is to establish specific types of toxicity, or even compounds on the basis of signature expression profiles. A proof-of-principle approach to obtain such signature profiles proved to be successful (4,5). A first step towards preclinical relevance is to obtain a database consisting of gene profiles for a range of model compounds.

Since studies aimed at seeding such a database are usually divided between different laboratories and the outcome has to be useful also for laboratories outside the study group, care has to be taken that results from these laboratories can be compared, or used back and forth. With the current state of technology, various methodologies and platforms exist for assessing gene expression, making it difficult to compare and compile data across laboratories.

An important initiative in this respect is the "minimum information about a microarray experiment" (MIAME) document (6), produced by the microarray gene expression database (MGED) society (http://www.mged.org). This set of guidelines is in the process of extension for toxicogenomics (MIAME/Tox), aiming to define the core that is common to most

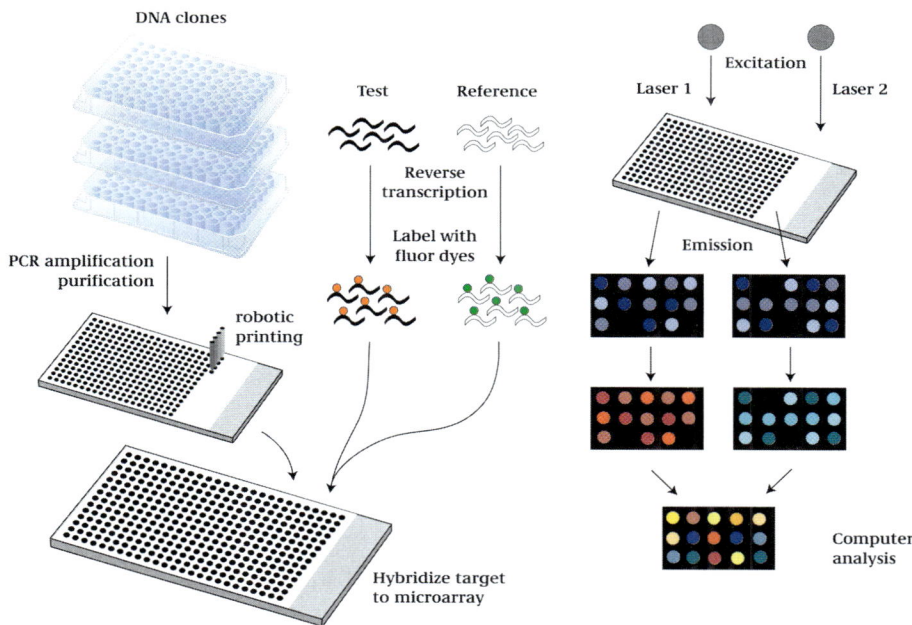

Toxicogenomics (Microarray Technology). Figure 1 Schematic illustration of microarray analysis. In this particular example PCR amplified cDNAs are dotted.

toxicogenomic experiments. The major objective of MIAME/Tox is to guide the development of toxicogenomics databases and data management software. The draft document can be found at http://hesi.ilsi.org. Efforts to build international public toxicogenomics databases are underway at the National Center for Toxicogenomics, National Institute of Environmental Health Sciences, USA (http://www.niehs.nih.gov/nct) and at the EMBL European Bioinformatics Institute (http://www.ebi.ac.uk/microarray/index.html) in conjunction with the International Life Sciences Institute Health and Environmental Sciences Institute (http://hesi.ilsi.org). This database will be made public in late 2003 or early 2004.

A provisional conclusion from experiments conducted so far is that multiple sources of variability exist, including expected sources of biological variability, isolation and labeling of mRNA samples, hardware and software settings, microarray lot numbers and gene coverage, and annotation. Nevertheless, the gene expression profiles relating to biological pathways are robust enough to allow insight into mechanism, strong information on topographic specificity is provided, dose-dependent changes are observed, and concerns of over sensitivity may be unfounded (http://hesi.ilsi.org).

Relevance to Humans

A second important issue of toxicogenomics is the genetic basis for individual differences in susceptibility to toxicity. Much of the variability in the genome stems from single nucleotide polymorphisms or SNPs, that occur roughly every 1000 nucleotides. A map describing over 1.4 million SNPs (7) is available (http://snp.cshl.org). The next step is then to find an association of a particular SNP and a disease trait. Generally, two approaches can be taken to find such associations: one is a candidate gene approach, where genes in key biochemical pathways are investigated for SNPs, and in the second approach SNPs and thereby target genes are identified by whole genome approaches. Mixed approaches can of course also be taken. An example of a successful candidate gene approach is the SNP mapping of the hypersensitivity response (HSR) to the drug abacavir. Over 100 SNPs were tested on the basis of candidate genes. Polymorphisms from two of the candidate genes (tumor necrosis factor(TNF)-α and human leukocyte antigen (HLA)-B57) were found to be highly associated with the hypersensitivity response to abacavir (8).

Similar to gene profiling, creating a database that describes associations between SNPs and disease is an important goal. Using high-density SNP mapping it should be feasible to study the genetic basis for several common diseases simultaneously. For drug adverse effects this will surely be more difficult since only few patients with a certain drug prescribed will show adverse effects.

A recent development comes from the finding that the human genome can be parsed into haplotype blocks, being regions over which there is little evidence for historical recombination and within which only a few haplotypes are observed (9). Markers for these haplotype blocks are now available, which makes it possible to identify the genetic control of responses to toxicants without the necessity to identify the specific SNP responsible.

Regulatory Environment

Regulations that rely on genomics are not yet in place but there is little doubt that within the next 5–10 years gene expression profiles will be used for safety as well as efficacy assessment. This requires a firm database of expression profiles that can be directly related to well characterized toxicological and pathological endpoints.

Second, risk assessment has traditionally been performed across whole populations with widely varying responses. The goal is that by genetically identifying sensitive subpopulations, the accuracy of risk assessment can be improved. Possibly, this may eventually lead to personalized risk profiles.

References

1. Duggan DJ, Bittner M, Chen Y, Meltzer P, Trent JM (1999) Expression profiling using cDNA microarrays. Nature Genet 21S:10–14
2. Eisen MB, Spellman PT, Brown PO, Botstein D (1998) Cluster analysis and display of genome-wide expression analysis. Proc Natl Acad Sci USA 95:14863–14868
3. Baxevanis AD, Francis Ouelette BF (eds) (2001) Bioinformatics. John Wiley & Sons, New York
4. Hamadeh HK, Bushel PB, Jayadev S et al. (2002) Gene expression analysis reveals chemical-specific profiles. Tox Sci 67:219–231
5. Hamadeh HK, Bushel PB, Jayadev S et al. (2002) Prediction of compound signature using high density gene expression profiling. Tox Sci 67:232–240
6. Brazma A, Hingamp P, Quackenbush J et al. (2001) Minimum information about a microarray experiment (MIAME)—toward standards for microarray data. Nature Genet 29:365–371
7. Sachidanandam R, Weissman D, Schmidt SC et al. (2001) A map of human genome sequence variation containing 1.42 million single nucleotide polymorphisms. Nature 409:928–933
8. Roses AD (2002) Genome-based pharmacogenetics and the pharmaceutical industry. Nature Rev Drug Disc 1:541–549
9. Gabriel SB, Schaffner SF, Nguyen H et al. (2002) The structure of haplotype blocks in the human genome. Science 296:2225–2229

T_R1 Cells

▶ Suppressor Cells

Trace Metals

Those metals commonly found in minute amounts in the organism.
▶ Trace Metals and the Immune System

Trace Metals and the Immune System

JUDITH T ZELIKOFF
Depart. of Environmental Medicine
New York University School of Medicine
57 Old Forge Road
Tuxedo, NY 10987-5007
USA

Synonyms

$CD4^+$, T helper lymphocyte, $CD4^+/CD8^-$, T helper lymphocyte, $CD8^+$, T suppressor lymphocyte, COPD, chronic obstructive pulmonary disease, asthma, bronchitis, emphysema.

Definition

Trace metals are normally present in minute quantities in the body. Many of them are also transition elements, essential for life due to their ability to control metabolic and signaling functions, such as zinc (Zn), manganese (Mn), and copper (copper). However, these same essential metals can also be toxic because of their ability to evade established controls for cellular uptake, transport, and compartmentalization. Aluminum (Al) is a toxic trace element, unavoidable by the general population because of its widespread environmental distribution. The immunotoxicity of trace metals other than Al, copper, Mn, and zinc can be found in a number of review articles (1–3).

Molecular Characteristics
Aluminum

Aluminum is the third most prevalent element in the Earth's crust. It is an A-type metal, or hard acid, that strongly prefers oxygen-donor ligands; hydroxide, citrate, phosphate, and nucleoside phosphate groups are probably the most important low-molecular-mass bioligands for the predominant trivalent cation (Al^{3+}). It also binds readily to the two high-affinity iron-binding sites of the serum transport protein, ▶ transferrin (TF). There is a wide variation in the ability of different ligands to solubilize and transport the Al^{3+} ion to critical target sites.

Copper

Copper is a Group II (or IB) element, the third most

abundant transition metal found in living things. It exists in one of two stable oxidation states: as cuprous (Cu^{1+}) and cupric (Cu^{2+}) ions. Consequently, its biological chemistry is dominated by participation in redox reactions. Copper is necessary in the diet for iron utilization and as a cofactor in enzymes associated with oxidative metabolism. It is transported in serum bound initially to albumin and later more firmly to α-ceruloplasmin where it is exchanged in the cupric form; normal copper serum level is 120–145 µg/l. At elevated levels, copper is toxic to cells, presumably by binding indiscriminately to thiol moieties or by catalyzing a Fenton-type reaction to produce reactive hydroxyl radicals. Binding of copper by biological ligands such as small peptides, large proteins, and enzymes is required to minimize potential deleterious effects. Most stored copper is usually bound to metal-lotheinein (MT), a ubiquitous class of proteins that is well suited to the role of metal sequestration.

Manganese

Manganese is the only Group VIIB element commonly found in biological environments. Although the inorganic chemistry of manganese displays a range of stable oxidation states, its biological chemistry is dominated by the divalent form (Mn^{2+}). Because Mn^{2+} is very similar in size and charge density to magnesium (Mg^{2+}) and Zn^{2+} and also prefers to assume thetrahedryl and octahedral geometric structures, Mn^{2+} can replace Mg^{2+} in the enzyme pyruvate carboxylase and Zn^{2+} in superoxide dismutase (SOD) with only negligible effects on enzyme activities.

Zinc

Zinc is found in large quantities in the vertebrate body (second only to iron); it is the first member of Group IIB elements and forms stable complexes with sulfur, phosphate, and carbon atoms. Biological complexes contain zinc only in the divalent oxidation state (Zn^{2+}). Since Zn^{2+} is the only stable oxidation state of the metal, it does not play a redox-active role in biological processes. However, Zn^{2+} can actively participate in enzymatic reactions as a Lewis acid or as a structural cofactor. Zinc is part of, or a cofactor for,
such enzymes as carbonic anhydrase, carboxypeptidase, SOD, lactate dehydrogenase, phosphatase, and glutamate dehydrogenase. Zinc also displays a structural role in biological systems, as exemplified by its role in maintaining the integrity of zinc finger transcription factors that bind to DNA and regulate the transcription of genetic information.

Relevance to Humans
Aluminum

While some daily exposure to aluminum is unavoidable, inhalation by the general population is usually considered negligible (i.e. 0.14 mg aluminum dust per day). However, smelters, miners, welders and other workers involved in various metal industries are often acutely exposed to localized atmospheres containing 2–4 mg/m^3 of aluminum, resulting in time-weighted-average (TWA) intakes of > 23 mg per 8-hour shift. Increases in pneumonia, bronchitis, asthma, pneumoconiosis, lung cancers, and pulmonary fibrosis have been described in occupationally exposed workers. In addition, there is little doubt that aluminum can cause encephalopathy, osteopathies, and anemia in kidney dialysis patients. Although early studies set 100 µg/l plasma as the level of aluminum below which neurotoxicity failed to occur, recent studies have demonstrated subtle neurocognitive and/ or psychomotor effects, as well as EEG abnormalities in dialysis patients expressed at levels well below this limit. Infants are a particularly susceptible subgroup for aluminum toxicity partly due to their rapidly growing and immature brain and skeleton and their developing blood-brain barrier; preterm infants are generally recognized to be at risk for aluminum loading due to their immature kidney function. While the reference range for blood aluminum levels in healthy individuals is < 10 µg/l, studies in infants have demonstrated plasma aluminum levels > 50 µg/l after oral intake of aluminum-containing antacids.

Copper

As an ▶ essential element, copper promotes iron absorption from the gastrointestinal system, it is involved in the transport of iron from tissues into plasma, it helps maintain myelin in the nervous system, it is necessary for hemoglobin synthesis, and it is important in the formation of bone and brain tissue. Apart from occupational exposure, daily copper intake averages ∼ 0.02 mg. The fine balance required for copper in humans is evident in genetically inherited inborn errors of copper metabolism. For example, in Wilson's disease there is failure to excrete copper from the liver to the bile, resulting in copper overload in the liver, brain, kidneys, and cornea; and in Menkes disease, which is characterized by severe copper deficiency due to an error in copper transport from the intestines. Copper, usually in the form of cuprous oxide and cupric hydroxide (which converts to cupric oxide), is generally encountered in high concentrations in the air of metallurgical processing plants, iron and steel mills, and around coal-burning power plants. In contrast to airborne copper concentrations in rural/suburban areas that average 0.01–0.26 µg/m^3, particulate copper levels in workplace sites be 50–900 µg/m^3. Inhalation of such levels can result in an immunologically-based condition called "copper fever".

Manganese

Manganese, an essential trace element for all living organisms, is necessary for bone formation, cholesterol and fatty acids synthesis, and as a dissociable cofactor for several enzymes including SOD. Despite its essentialness, the toxic effects of manganese are well known, particularly those associated with the nervous system (4). Manganese is widely employed in many industries: in alloy steel manufacture for deoxidation and to promote hardenability; in the electric industry for production of dry cells; in the chemical industry, where they are used as oxidants, for the manufacture of fertilizers, paints and varnishes, and in the production of glass and glazes (3). Apart from the direct release of manganese into the air by several types of mining industries and alloy and steel production facilities, manganese is introduced into the ambient environment by the combustion of manganese-containing fossil fuels (used as anti-knock additives and combustion improvers). Manganese (whose toxicity in many cases depends upon compound solubility) has been found at measurable levels in the majority of suspended particulate matter (including coal flyash) in urban environments. While air levels of manganese in many metropolitan areas containing steel or alloy plants can range from 0.5–3.3 $\mu g/m^3$, the majority have levels ≤ 0.1 $\mu g/m^3$; average air levels in the absence of any contributing point sources are in the range 0.03–0.07 $\mu g/m^3$. Alternatively, occupational airborne levels of manganese are usually in the range $1-\geq 100$ $\mu g/m^3$ (although levels as high as 1 mg/m^3 have been measured); workplace permissible exposure limits (PEL) of 300 (TWA) and 500 $\mu g/m^3$ have been recommended by the World Health Organization and OSHA (Occupation Safety and Health Association), respectively.

Zinc

Zinc is ubiquitous in the environment and present in most foodstuffs, water, and air. It is a nutritionally essential element that serves as a cofactor for more than 70 metalloenzymes. Daily dietary intake of zinc is usually 12–15 mg/day and ~ 20%–30% of ingested zinc is absorbed; zinc deficiency results in a wide spectrum of clinical effects depending upon age, stage of development and deficiencies of related metals (i.e. zinc deficiency can exacerbate impaired copper nutrition and exacerbate cadmium and lead toxicity). Airborne concentrations of zinc are usually < 1 $\mu g/m^3$, with the majority of zinc being derived from automobile exhaust, soil erosion, and local commercial, industrial or construction activities. In urban areas, atmospheric zinc concentrations are in the range 0.02–0.50 $\mu g/m^3$; rural air contains 0.01–0.06 $\mu g/m^3$. Because zinc also contaminates certain workplace environments, national guidelines of 1.0 mg, 5–10 mg, and 0.1 mg/m^3 have been established for soluble zinc, insoluble zinc oxide, and carcinogenic zinc chromate, respectively.

Putative Interactions with the Immune System
Aluminum

Although limited in number, immunotoxicologic studies using a variety of animal models have demonstrated that injection of soluble aluminum compounds increases mononuclear cell mitotic index; injection of the metal or insoluble aluminum agents alter monocyte/macrophage numbers and immune function (1). Dietary exposure of rodents to soluble aluminum reduces cytokine production, T helper (Th) and T suppressor (Ts) cell numbers, and host resistance to *Listeria monocytogenes* infection. Repeated inhalation exposure of rabbits and hamsters to soluble aluminum increases lung immune cell numbers; similar effects were not seen in aluminum-exposed workers. Effects of inhaled aluminum on host resistance are inconsistent, showing decreased resistance to subsequent bacterial challenge in some studies and no effect in others. Differences between the studies are thought to be due to intratracheal versus inhalation exposure routes. *In vitro* studies employing soluble aluminum salts demonstrate a range of effects on immune cells derived from a variety of animal species including humans. For example, aluminum chloride treatment of rat alveolar macrophage reduced reactive oxygen intermediate production.

Copper

Much the same as for manganese and zinc, studies of copper immunotoxicity are complicated by the fact that copper is essential to maintenance of immunocompetence and, thus, most immunotoxicity occurs as a result of copper insufficiency. Splenomegaly and thymic atrophy are consistent findings in copper-deficient mice. Alterations in antibody response and B-lymphocyte function are also well documented with experimentally induced copper deficiency. In contrast, serum antibody levels in humans with nutritional or genetic copper-deficiency are reported to be normal. B-lymphocytes are increased in number in copper-deficient animals, but they respond poorly to mitogen stimulation. Although the effects of copper deficiency on T-lymphocyte populations are well characterized, the overall effect on cell-mediated immunity is unclear. Except in female rats, ▸ CD4$^+$ and ▸ CD8$^+$ subsets are decreased in the peripheral blood and spleen of copper-deficient rodents. Though no gender effect has been observed in mice, the immune system of male rats appears more susceptible to copper deficiency than that of females. Clinical studies involving healthy men on low copper diets fail to support the animal studies with respect to circulating T-lymphocytes and CD4$^+$ and CD8$^+$ subsets. While effects of copper-deficiency on innate immunity are inconclusive in animal studies, reduced neutrophil numbers and functionality are well defined in clinical studies. The most consistent immune defect associated with copper deficiency in epidemiologic, clinical, and toxicological studies is impaired host resistance due primarily to suppressed antibody-mediated responses and phagocyte antimicrobicidal activities (2).

Manganese

While relatively few studies have investigated the effects of manganese on the immune system, immunotoxicity appears dependent (like many of the other metals discussed herein) upon compound solubility. Immune responses of the lung appear particularly sensitive to the immunomodulating effects of manganese. Inhalation of insoluble manganese compounds reduces the ability of the lungs to resist and clear subsequent bacterial/viral infections and exacerbates ongoing viral infections (3). Inhalation studies examining the effects of soluble manganese reveal little effect on lung immune cell-related functionality. In contrast, studies wherein rabbit alveolar ▶ macrophages were exposed *in vitro* to manganese chloride demonstrated decreased cell viability and number, increased incidence of cell lysis, and reduced phagocytic activity.

Zinc

Zinc deficiency (like copper) can impair humoral and cell-mediated host immunocompetence. Zinc-deficient children and laboratory animals consistently present with ▶ thymic hypoplasia; oral administration of zinc supplements appear to reverse this effect. In zinc-deficient animals, secondary antibody responses to T-dependent antigens are suppressed in conjunction with accelerated thymic hypoplasia and a decreased number of CD4$^+$ cells. Administration of zinc to immunosuppressed human populations appears to increase numbers of CD4$^+$ and CD8$^+$ thymocytes which, in turn, give rise to increased numbers of Th-cells important for the activation of cytotoxic T- and B-lymphocytes. In contrast, zinc suppresses concanavalin A-induced T lymphocyte proliferation by *in vitro*-exposed human immune cells (2) and compromises pulmonary host resistance against bacterial infection (5); suppressive effects of inhaled zinc on pulmonary antimicrobial activity are most likely due to zinc-induced reductions in macrophage phagocytic activity. While the main immunological effect of occupational zinc exposure is metal fume fever, inhalation of particulate zinc by occupationally exposed workers also alters pyrogenic, chemotactic, and anti-inflammatory cytokines. While zinc appears to play an important regulatory role in membrane-associated events of certain nonspecific immune cell types, effects of zinc on innate immunity are conflicting.

References
1. Zelikoff JT, Cohen MD (1997) Metal immunotoxicology. In: Massaro EJ (ed) Handbook of Human Toxicology. CRC Press, New York, pp 811–852
2. Omara FO, Brousseau P, Blakley BR, Fournier M (1998) Iron, zinc, and copper. In: Zelikoff JT, Thomas PT (eds) Immunotoxicology of Environmental and Occupational Metals. Taylor and Francis, London, pp 231–262
3. Cohen MD (2000) Other metals: aluminum, copper, manganese, selenium, vanadium, and zinc. In: Cohen M, Zelikoff JT, Schlesinger RB (eds) Pulmonary Immunotoxicology. Kluwer, Boston, pp 267–299
4. Inoue N, Makita Y (1996) Neurological aspects of human exposures to manganese. In: Chang LW (ed) Toxicology of Metals. CRC Lewis, New York, pp 415–421
5. Zelikoff JT, Chen LC, Cohen MD et al. (2003) Effects of inhaled ambient particulate matter (PM) on pulmonary anti-microbial immune defense. Inhal Toxicol 15:101–120

Trans-Signaling

The soluble Interleukin-6 receptor α-chain binds interleukin-6 and can then interact with the transmembrane receptor subunit glycoprotein 130 (gp130) and induce signal transduction. Thus a cell lacking an endogenous binding subunit of the interleukin-6 receptor can respond to Interleukin-6 in the presence of the soluble receptor-derived from distant producer cells, hence trans-signaling.

▶ Cytokine Receptors

Transcription Factors

Proteins (enzymes) that bind to regulatory sequences (response elements) in the promoter region of a gene, forming a complex to which RNA polymerase binds. The process of transcription converts the genetic information contained in DNA into an RNA message for synthesis of a specific protein.

▶ Glucocorticoids
▶ Signal Transduction During Lymphocyte Activation

Transendothelial Migration

This is the exit of circulating leukocytes from blood into tissue by means of traversing the microvascular endothelium. This process involves loose interactions of blood leukocytes with the luminal side of blood vessels (mostly mediated by selectins), and this is followed by firm adhesion and leukocyte transmigration. This latter step critically depends on the rapid and transient modulation of integrin function, which itself is controlled by chemokine receptor signaling. Only those leukocytes firmly arrest that bear the appropriate set of chemokine receptors and, therefore, the chemokines present on the luminal side of microvessels are viewed as key controllers of leukocyte extravasation.

▶ Immune Cells, Recruitment and Localization of

Transferrin

A protein that combines with and competes for iron with bacteria.

▶ Trace Metals and the Immune System

Transferrin Receptor

These are cell membrane receptors for transferrin. They play a role in iron uptake by the cell, and are highly expressed in proliferating cells.

▶ Interferon-γ

Transforming Growth Factor β1; Control of T Cell Responses to Antigens

SUSAN C MCKARNS
Laboratory of Cellular and Molecular Immunology
NIAID/NIH
Building 4, Room 111, MSC 0420, 4 Center Drive
Bethesda, MD 20892
USA

Synonyms
TGF-β1(the nomenclature is used worldwide with the number designating the isoform)

Definition
The transforming growth factor-β (TGF-β) superfamily consists of more than 40 structurally related secreted proteins (1). Three members (TGF-β1, 2, 3) are expressed in mammals; despite a 70%–76% sequence homology, these isoforms have expression pattern and functional differences. Whereas, TGF-β2 and TGF-β3 are important for cellular differentiation, development, and embryogenesis, the effects of TGF-β1 are predominantly—albeit not exclusively—immunologic. Lymphoid cells selectively produce TGF-β1. The name 'transforming' is something of a misnomer because this factor is not always associated with oncogenesis. TGF-β1 is possibly the most pleotropic of all the ▶ cytokines and growth factors, and its activity is cell-type- and context-dependent. The ability of TGF-β1 to *suppress* cell growth distinguishes it from most other cytokines/growth factors. Mechanistically, it converts receptor ligation at the cell surface into an enzymatic signaling cascade within the cell to change the level of expression of target genes. In this manner, it is able to target a vast array of immune cell lineages to modulate their ability to proliferate, differentiate, survive, perform effector functions, and migrate to sites of antigen presentation and/or inflammation. These events are vital to the initiation, progression, and resolution of inflammatory responses. Dysregulated expression or function of TGF-β1 is implicated in autoimmune disease, chronic inflammation, and tumor progression. The driving force behind TGF-β1 seems to be maintaining homeostasis of controlled immune responses, and it achieves its goal by orchestrating a network of intracellular signaling crosstalk that enables cells to rapidly respond to changes in their environment.

Characteristics
Cellular sources
TGF-β1 expression is present at the four-cell embryo stage and persists, in most tissues, during morphogenesis and into adulthood. Most all mature cells have been shown to produce this factor. Likewise, nearly all cell types have functional TGF-β1 receptors. Although controversial, it has been postulated that the primary effector function of a small cohort of regulatory T cells (e.g. Th3 and CD4$^+$CD25$^+$) is to secrete TGF-β1.

Regulation of activity
It is well established that the bioavailability and activity of TGF-β1 are influenced by the environment (Table 2), and it generally is accepted that some of

Transforming Growth Factor β1; Control of T Cell Responses to Antigens. Table 1 GenBank accession numbers for transforming growth factor-β1 (TGF-β1)

Species	Accession numbers (partial listing only)	
	Gene	Protein
Human (*Homo sapiens*)	J04431, J05114	PO1137
Mouse (*Mus musculus*)	AH003562	P04202

these changes increase human susceptibility to immunologic-related diseases. TGF-β1 predominantly is secreted as a biologically inert complex consisting of mature TGF-β1, latency associate protein (LAP), and latent TGF-β1-binding protein (LTBP). Prior to binding to TGF-β receptors, the latent complex must be cleaved into the 25 kDa active TGF-β1 homodimer.

TGF-β1 signaling

The predominant mechanism by which TGF-β1 elicits its activity is through modulation of gene transcription. TGF-β1 mediates the association of transmembrane type II (TβRII) and type I (TβRI) receptors. Ligand binding propagates signaling through phosphorylation of multiple effector proteins. The only known direct TGF-β1 signaling effectors are a class of structurally similar ▶ Smad proteins (2). Once the ligand has bound to the serine-threonine kinase receptor a signaling complex is formed, leading to the phosphorylation of Smad 2 and Smad 3 and their subsequent trafficking to the nucleus, where they bind well-defined Smad response elements and function as transcriptional modulators to regulate transcription of TGF-β1 target genes (Fig. 1).

Smads can also positively regulate gene expression by recruiting coactivators such as CBP/p300 or negatively by forming complexes with histone deacetylases (HDACs) or corepressors (such as c-ski and SnoN) which themselves associate with HDACs. One key negative regulation of Smad signaling is the expression of the inhibitory Smad, Smad7. Smad7 blocks TGF-β signaling by competing with R-Smads for association with TβRI, or by targeting receptors for ubiquitin-mediated degradation. Potent inducers of Smad7 are interferon-γ (IFN-γ), tumor necrosis factor (TNF)-α, and interleukins IL-1β, and IL-7. Induction of Smad7 represents an important regulatory interplay between TGF-β1 and cytokines in immune cell function. It is noteworthy that type TβRI and TβRII receptors distinguish themselves from other cytokine/growth factor receptors by their specificity for serine/threonine, rather than tyrosine kinase, activity.

Smad-independent signaling pathways also regulate TGF-β1 signaling. For instance, TGF-β1 activates mitogen-activated protein (MAP) kinases including the extracellular regulated kinases (ERKs), c-Jun N-terminal kinases (JNKs) and p38 kinases. TGF-β1 has also been shown to activate Rho-like GTPases and phosphatidulinostiol-3-kinase (P13K) and signal through protein phosphatase 2A (PP2A). One key point of cross-talk among signaling intermediates is MAP kinase activation that occurs downstream of growth factors, integrins, and chemokine receptors. In most cases, activated ▶ MAP kinases promote the actions of TGF-β1 to enhance cell migration. Thus, activation of growth factor receptor and the pattern of cytokine/chemokine signaling have a tremendous impact on the response of cells to TGF-β1 (Fig. 1). Finally, TGF-β1

Transforming Growth Factor β1; Control of T Cell Responses to Antigens. Table 2 Factors that modulate bioactivity of transforming growth factor-β1 (TGF-β1)

Enhance TGF-β synthesis and secretion

Liver hepatotoxicants (carbon tetrachloride, acetaminophen, alcohol)

Tissue injury (liver, renal, and lung)

Hypoxia

Stress

Viral infection

Parasitic infection

Steroid hormones (retinoids, vitamin D, and tamoxifen)

Activate extracellular latent TGF-β1

Mannose 6-phosphate/insulin-like growth factor 2 receptor (M6P/IGF2R)

Transglutaminase

Plasmin/plasminogen activator

Apoptotic T cells

Reactive oxygen species

αvβ6 Integrin receptor

Suppress activation of extracellular latent TGF-β1

α2-Macroglobulin

Decorin

Endoglobin

Mucosal mast cell protease (MMCP)

Antagonize TGF-β1 signaling

Cytokines: tumor necrosis factor-α, interferon-γ, interleukins IL-1β, IL-6, IL-2

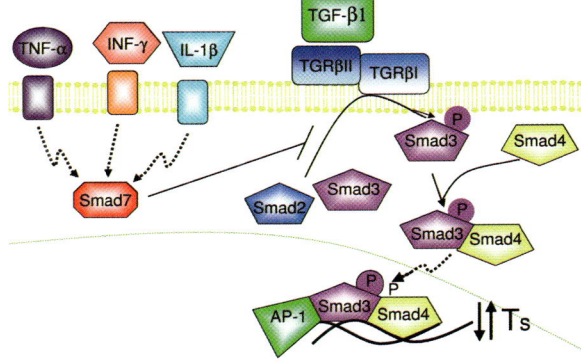

Transforming Growth Factor β1; Control of T Cell Responses to Antigens. Figure 1

signaling can also modulate protein stability. For example, it enhances degradation of TβRI.

Immunological activities

A loss-of-function mutation in TGF-β1 results in the rapid onset of lethal multiorgan inflammation and autoimmune phenotype. These transgenic mouse models clearly establish the critical role of this factor in maintaining immune homeostasis for the prevention of disease and chronic inflammation (Fig. 2).

A role for TGF-β1 has been implicated, in several different mouse models, including ▶ tolerance, and particularly in mucosal immunity. While these studies clearly demonstrate the onset of inflammation in the absence of TGF-β1 signaling, more recent data suggest that the role for TGF-β1 in controlling T cell homeostasis may be restricted to preventing inappropriate responses to self- or environmental antigens, rather than regulating T cell responses to low-avidity self-ligands (3). In addition to its immunosuppressive and anti-inflammatory properties, TGF-β1 is capable of promoting inflammation (4). For example, at the early stages of inflammation, it enhances lymphoid, neutrophil, monocyte, and macrophage migration, presumably to enhance the localization of these cells at the site inflammation. Probably TGF3 also prolongs the inflammation associated with numerous autoimmune disorders by actively sequestering activated T cells at the site of inflammation. TGF-β1 also exerts numerous suppressive effects on T and B lymphoid effector and antigen-presenting cells and many of these effects are summarized in Table 3.

Preclinical Relevance
Implications for disease

Dysregulated expression of TGF-β1 or response of immune cells to TGF-β1 signaling have been implicated in the pathogenesis of many human diseases, including hypersensitivity reactions such as asthma and food allergies, as well as autoimmune disorders, including encephalomyelitis, arthritis, systemic lupus erythematosus, and allograft rejection. While it is commonly accepted that both environmental and genetic factors contribute to the incidence of these immune responses, it remains unclear why some individuals are susceptible to these disorders while others are not. Appropriate levels of TGF-β1 have been shown to be essential for maintaining immunologic balance, to prevent the pathogenesis of hypersensitivity reactions/chronic inflammation and autoimmune disorders. Perhaps a better mechanistic understanding of how it modulates cellular and molecular pathways will provide important insights that will enhance our understanding of susceptibility to these diseases. Outlined below are three prevalent immune disorders that occur in response to common environmental exposure; which, TGF-β1 is key to regulation of the ensuing pathological immune responses.

Asthma

The development of asthma in response to environmental antigens affects up to 20% of the population in developed countries. Asthma is a chronic inflammatory disease of the airways that is characterized by mononuclear infiltration, eosinophil degranulation, and bronchoconstriction. TGF-β1 is constitutively expressed by airway epithelial cells, eosinophils, T lymphocytes, macrophages, and fibroblasts, and stored in the extracellular matrix of the airways. Rodent models of asthma suggest that it mediates both anti-inflammatory and profibrotic effects (5). Prior to allergen exposure, it is thought to play a critical protective role against the onset of asthma by suppressing airway inflammation and hyper-responsiveness through the suppression of T lymphocytes, dendritic cells, eosinophils, mast cells, and IgE production. Notably, mononuclear cell infiltration into the lungs is prevalent in TGF-β1 null mice. Additionally, TGF-β1 may further suppress airway CD4$^+$ T cell allergen exposure by enhancing activity of T regulatory cells. However, repeated long injury is accompanied by a profound TGF-β1-mediated recruitment of fibroblasts into the airways, a progressive deposition of extracellular matrix, and subsequent fibrosis and bronchoconstriction.

Food allergy

Food allergy is characterized by an adverse hypersensitive response to food consumption. A normal healthy gastrointestinal immune response must discriminate between harmful pathogens and harmless dietary antigens and commensal bacterial flora. The mucosal immune system has generated two adaptive immune responses to meet this challenge: induction of a local secretory IgA response, which is propagated in the absence of a measurable systemic immune response, to clear potentially dangerous antigens; and induction of oral tolerance, a state of non-responsiveness or

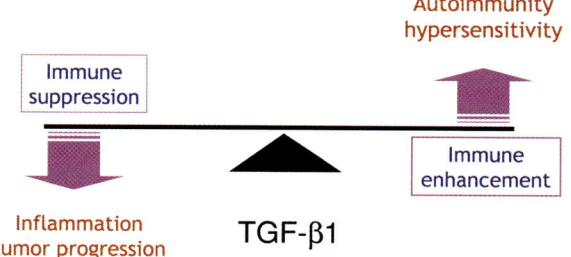

Transforming Growth Factor β1; Control of T Cell Responses to Antigens. Figure 2

Transforming Growth Factor β1; Control of T Cell Responses to Antigens. Table 3 Biological activities of transforming growth factor-β1 on the immune system

Parameter	TGF-β1-mediated effect
T lymphocytes	
TCR-induced CD4$^+$ and CD8$^+$ T cell proliferation	Suppress; memory CD4$^+$ resistant to G1 cell cycle arrest
IL-2-induced CD4$^+$ and CD8$^+$ T cell proliferation	Suppress, dependent upon IL-2 concentration
Th1 differentiation	Suppress, but dependent upon strength of T cell stimulation; inhibits T-bet and INF-γ expression
Th2 differentiation	Suppress; inhibits GATA-3 and IL-4 expression
Th1 effector function	Suppress; inhibits INF-γ and IL-2 production; inhibits IL-12 signaling
Th2 effector function	Suppress; inhibits IL-4 and IL-5 production
CD8$^+$ cytotoxic T cell effector function	Suppress
CD4$^+$ and CD8$^+$ migration/adhesion	Enhance; increases CXCR4 and α4β7 expression
CD4$^+$CD25$^+$ regulatory T cell function	Enhance; increases Foxp3, GITR, CD103, CTLA-4 expression
IL-12 signaling	Suppress; downregulates IL-12 receptor β2 chain
T cell apoptosis	Suppresse or enhance, dependent on microenvironment
B lymphocytes	
Proliferation	Suppress
Effector function	Enhance IgA and IgG2b and suppressed most other isotypes
Antigen-presenting cells	
MHC class I and class II molecules	Suppress
Monocytes and macrophages	
Monocyte chemotaxis	Enhance or suppress
Macrophage chemotaxis	Suppress
Neutrophils	
Neutrophil chemotaxis	Enhance or suppress
Chemokine and receptor expression	Suppress or enhance: chemokine, receptor, and cell type-dependent

CTLA=cytotoxic T-lymphocyte-associated protein 4; GITR= glucocorticoid induced TNF receptor; Ig=immunoglobulin; IL=interleukin; INF=interferon; MHC=major histocompatibility complex; TCR=T cell receptor; TGF=transforming growth factor; Th=T helper cell.

hypo-responsiveness which minimizes unnecessary immune reactions against harmless antigens. A failure to induce or an inability to maintain oral tolerance may leads to a food allergy. TGF-β1 is abundant throughout the mucosa and has been shown in several experimental models to play a profound role in maintaining oral tolerance (5). It is well documented that a population of TGF-β1-secreting T helper cells is generated when low doses of antigen are consumed. These TGF-β1-secreting regulatory T cells also produce various amounts of IL-4 and/or IL-10. Secreted TGF-β1 suppresses T cell proliferation and promotes class switching of B cell IgA isotypes to modulate the adaptive immune response. Moreover, it also enhances the preservation of the epithelial barrier between environmental antigens in the gut flora and T lymphocytes in the mucosa to add yet another level of regulatory control over adaptive T cell responses.

Childhood food allergies have been associated with a reduction in the number of mucosal TGF-β1-producing lymphocytes. Aberrant levels of mucosal TGF-β1 and associated dysregulated responses to the normal gut flora have also been implicated in the pathogenesis of inflammatory bowel disease. Collectively, these data implicate a role for TGF-β1 to maintain oral tolerance in humans.

Autoimmunity

Discordance in incidence of autoimmune disease in monozygotic twins demonstrates a role for environmental exposure in regulating immune homeostasis. Although numerous environmental factors have been implicated, the underlying mechanisms remain relatively undefined. The autoimmune phenotype of the TGF-β1 knockout mouse, characterized by circulating antinuclear antibodies and glomerular deposit s of immune complexes, probably best defines the role of this factor in the disease process. 100% of TGF-β1 knockout mice succumb to a massive multiorgan inflammation involving the heart, lung, liver, gut, salivary glands, eyes, brains, and other tissues. The inflammatory infiltrates are predominantly perivascular and vary from neutrophilic in the stomach to lymphocytic in the brain. In agreement, systemic administration of exogenous TGF-β1 or adoptive transfer of TGF-β1 –producing T cells protect against autoimmune diseases in several experimental models, including diabetes, encephalomyelitis, inflammatory bowel disease, arthritis, systemic lupus erythematosus, and allograft rejection. Targeted deletion of TGF-β signaling in T cells alone has been demonstrated to be sufficient to induce an autoimmune phenotype. It remains to be determined whether other non-T cell TGF-β1-producing cells (e.g. macrophages) contribute to the disease process as well. The precise mechanisms of actions underlying the ability of this factor to regulate autoimmune disorders remains speculative. Recent evidence implicates a significant role for regulatory T cells. $CD4^+$ $CD25^+$ regulatory T cells, also called suppressor T cells, can be delineated into two subsets of $CD4^+$ $CD25^+$ T cells with inherent activity to suppress autoreactive T cells: these are 'natural' regulatory $CD4^+$ $CD25^+$ T cells that emerge from the thymus, and adaptive regulatory $CD4^+$ $CD25^+$ T cells that are induced in the periphery. TGF-β1 has been shown to positively regulate both subsets, and TGF-β1-mediated expansion of $CD4^+$ $CD25^+$ T cells protects against autoimmune diabetes (7). However, in view of the diversity of pathology associated with autoimmune disorders, it is highly likely that TGF-β1 also utilizes other critical mechanisms of action. For instance, modulation of Th2/Th1 cytokine balance, cell survival, migration, effector function, and Th3-mediated tolerance represent likely alternative mechanistic routes (8).

Relevance to Humans

The phenotype of the TGF-β1 mouse resembles human SLE, Sjögren syndrome, graft-versus-host disease, and polymyositis, suggesting that TGF-β1 may play a similar regulatory role in human immunologic disorders. The levels of TGF-β1in serum and of its mRNA in tissue can be measured and have been used as diagnostic or prognostic markers for other human diseases. For example, high levels of the factor in RNA in tissues are associated with gastric cancer. High serum levels also correlate with the development of fibrosis in patients with breast cancer who have received radiation therapy. Understanding the mechanisms of action of environment-induced immune disorders in experimental models will potentiate the development of better predictive risk assessment assays to prevent disease as well as more specific therapeutic regimens aimed at increasing effectiveness and diminishing deleterious side effects.

Regulatory Environment

Although interaction of chemicals with cytokines or chemokines may have an important impact on the function and regulation of the immune system they are not regulated by any specific immunotoxicity guideline. The cytokine network is mentioned in different guidelines or guideline drafts but exclusively in connection with extended 'case-by-case' investigations.

The regulation of the immune system is complex, and identification of the mechanism, of action of chemical-induced immune toxicity is critical for the understanding of the disease process. The regulatory cytokine TGF-β1 may be of special interest for such investigations. A better understanding of the disease process will provide the basis for the development of more sensitive and predictable assays for risk assessment. Chemical-induced immunotoxicity may be indirectly mediated via the soluble potent immune modulator, TGF-β1. TGF-β1 may be a useful biomarker of chemical-induced and/or environmental-induced immunotoxicity. A critical challenge is to determine the appropriate therapeutic level of active TGF-β1 or signaling pathways that positively influence cell responsiveness to ameliorate disease while minimizing deleterious side effects.

References

1. Flanders CF, Roberts AB (2000) TGF-β. In: Oppenheim JJ, Feldman M, Durum SK, Hirano T, Vilcek J, Nicola NA (eds) Cytokine reference: A compendium of cytokines and other mediators of host defense. Academic Press, New York, pp 719–746
2. Shi Y, Massague J (2003) Mechanisms of TGF-β signaling from cell membrane to the nucleus. Cell 13;113:685–700
3. Gorelik L, Flavell RA (2002) Transforming growth factor-β in T-cell biology. Nat Rev Immunol 2:46–53
4. McCartney-Francis NL, Frazier-Jessen M, Wahl SM (1998) TGF-β: A balancing act. Int Rev Immunol 16:553–580
5. Duvernelle C, Freund V, Frossard N (2003) Transforming growth factor-β and its role in asthma. Pulmon Pharmacol Ther 16:181–196

6. Weiner HL (2001) Oral tolerance: immune mechanisms and the generation of Th3-type TGF-β-secreting regulatory cells. Microbes Infect 11:947–954
7. Peng Y, Laouar Y, Li MO, Green EA, Flavell RA (2004) TGF-β regulates in vivo expansion of Foxp3-expressing $CD4^+CD25^+$ regulatory T cells responsible for protection against diabetes. Proc Natl Acad Sci USA 101:4572–4577
8. Prud'homme GJ, Piccirillo CA (2000) The inhibitory effects of transforming growth factor-beta-1 (TGF-β1) in autoimmune diseases. J Autoimmun 1:23–24

Transforming Growth Factor β1 (TGF-β1)

TGF-β1 is the prototype for a superfamily of secreted proteins that control many aspects of growth and development. It was named transforming growth factor because upon its discovery it was shown to induce a transformed or tumor cell phenotype in normal cells. TGF-β1 is now known to regulate a diverse array of cellular functions unrelated to cell transformation. Within the immune system, TGF-β1 is critical for cell growth, differentiation, effector cell function, survival, and migration.

▶ Transforming Growth Factor β1; Control of T cell Responses to Antigens
▶ Mucosa-Associated Lymphoid Tissue

Transgenic Animals

PETER J BUGELSKI
Experimental Pathology
Centocor, Inc.
R-4-2, 200 Great Valley, Parkway
Malvern, PA 19355
USA

Synonyms
Knock-out, knock-in, genetically modified, recombinant

Definition
Animals whose genome has been modified using recombinant DNA technology so as to have a foreign gene expressed (knock-in) or a native gene suppressed (knock-out) in a heritable fashion. A number of transgenic species have been created: mice, rats, pigs, goats, cattle, sheep and fish. Currently, the vast majority of transgenic animals are mice. Transgenic mice have applications in numerous areas of biomedical research (e.g. neurologic, inflammatory, autoimmune, neoplastic and cardiovascular disease), in immunology, mutagenesis and carcinogenesis research, and in novel target evaluation and drug discovery. In immunotoxicology, although the potential usefulness of transgenic mice is widely recognized, practical application has been limited.

Characteristics
Generation of Transgenic Animals (1)
There are two principal methods by which transgenic mice are created: ▶ microinjection of genetic material into the ▶ pronucleus of a fertilized ova; or gene transfection of embryonic stem cells (ES cells) cells followed by injection of the transgenic cells into a blastocyst. In either case, the resulting transgenic embryo is implanted into a recipient female prepared for pregnancy. The principal difference between the techniques is that, when successful, microinjection of the pronucleus results in homozygous offspring, while transfection of ES cells results in chimerae (see ▶ chimera) that must be selectively bred to yield homozygous animals.

There are two principal types of transgenic events, those with one or more random insertions of the transgene and those where ▶ homologous recombination results in targeted insertion. Either type of transgenic event can result in a knock-in (KI) that will express the coding sequences of the transgene. Although random insertion will by definition result in a mutation in the recipient genome, as most DNA is noncoding these mutations are generally silent. Homologous recombination is used to selectively disrupt expression of the homologous gene, resulting in a knock-out (KO). This requires design of the inserted DNA so as to contain sequences homologous to the desired host species' gene.

A third type of transgenic animal—the knock-in-knock-out (KI-KO) mouse—has also been created. If the inserted DNA has a sequence homologous to a murine gene and also codes for a foreign protein, a KI-KO mouse can be created in a single step. Alternatively, these two types of transgenic events can be combined in a two-step process to result in a KI-KO strain.

Transgene Expression
In some cases, expression of the transgene by the host is not important. For example in mutagenesis research the endpoint can be a mutation in the transgene that will be expressed and detected ex vivo. In most cases however, expression of the transgene is desired. Insertion of multiple copies of the gene and linkage to a potent promoter (e.g. simian virus 40 promoter, will likely ensure widespread and high-level ▶ gene expression. Depending on the experiments to be conducted, however, it may be important that the site(s)

of expression, the magnitude of expression and timing of expression be controlled. This can be accomplished, but is no means guaranteed by selecting the gene promoter sequences included in the transgene. Techniques are now reasonably well established for controlling the sites and magnitude of expression and ▶ "conditional" gene expression in transgenic animals is an active field of research.

Preclinical Relevance
Transgenes
Transgenic mice have been created which express a wide and ever increasing range of genes (as of June 2003 the database maintained by BioMed Net lists 2300 transgenic mice (2)). These genes include reporter constructs (e.g. β-galactosidase or green fluorescent protein (GFP)) and viruses (e.g. hepatitis C), and a wide variety of human proteins.

Of greatest relevance to immunotoxicology is the expression of human cytokines, cell surface markers and immunoglobulins. Mice transgenic for mutant reporter genes have application in genotoxicity, and mice transgenic for mutant oncogenes have application as short-term replacements for 2-year cancer bioassays.

In efforts to facilitate xenotransplantation, transgenic pigs have been created that express a small number of human genes. There have also been reports on a model of colitis in rats expressing a human major histocompatibility antigen (HLA)-B27.

Function
Expression, however, is not sufficient for the transgenic strain to have preclinical relevance. The transgene gene product (i.e. the protein) must be functionally active in the transgenic animal. In the case of immunotoxicology, this will generally require that the human protein binds, and will result in signal transduction in its respective murine receptor (e.g. CCR and FcR or binding proteins such as major histocompatability complex (MHC) class II).

Genetic Background
Two strains of inbred mice are widely used for creating transgenic mice; 129 and C57 black. Once a transgenic strain has been created however, it may be possible to "move" the transgene into an alternate genetic background (e.g. BALB/c, by selective breeding. This can be of critical importance in application of transgenic mice in immuntoxicology where the genetic background of the mice can have a significant impact on the experiment (e.g. delayed-type hypersensitivity, transplantation, immunogenicity and host defense against infection or neoplasia).

Fecundity
Fecundity can also be an important factor in determining the success of application of transgenic mice in toxicology. Many strains of transgenic mice show low fecundity as determined by fertility and number of offspring. As we must have sufficient numbers of animals for study, low fecundity can have a serious impact on our ability to conduct a given experiment. Simple animal husbandry (e.g. selection of proven breeders) may be sufficient to solve this issue. Maintaining the breeding colony as heterozygotes may be required. However, as the offspring of heterozygotes will be a mix of transgenic and nontransgenic, the offspring must be genotyped or phenotyped prior to enrolment in studies.

Application of Transgenic Rodents in Immunotoxicology
Transgenic mice have been used extensively for studying the immune system. As of June 2003 the National Institutes of Health Medline lists over 2800 papers describing the use of transgenic mice to study immunology. Obviously, there are far too many examples to list here. However, applications of direct relevance to immunotoxicology are much rarer. Some selected examples are listed in Table 1.

Relevance to Humans
As with any animal system, the relevance of transgenic animals to humans is somewhat limited. Some of the factors which lead to this limited relevance are listed in Table 2. One must also keep in mind that in most cases while the transgenic animal may be transgenic for one human protein (and therefore immunotolerant to that human protein) it will likely not be inherently tolerant to any administered human therapeutic protein. With these caveats in mind, a priori, transgenic mice should have a much relevance to humans as any murine system.

Regulatory Environment
Use of transgenic animals for demonstrating pharmacologic activity and safety are gaining increasing acceptance by regulatory authorities. They are specifically dealt with in the following guidance documents:
- FDA Guidance for Industry. Clinical Development Programs for Drugs, Devices, and Biological Products for the Treatment of Rheumatoid Arthritis (RA) http://www.fda.gov/cder/guidance/1208fnl.pdf
- FDA Guidance for Industry. Immunotoxicology Evaluation of Investigational New Drugs. http://www.fda.gov/cder/guidance/4945fnl.doc
- ICH Guidance for Industry. S1B Testing for Carcinogenicity of Pharmaceuticals http://www.fda.gov/cder/guidance/1854fnl.pdf

▶ Animal Models of Immunodeficiency

Transgenic Animals. Table 1 Examples of application of transgenic rodents in immunotoxicology (KI, knock-in; KO, knock-out)

Transgenic system	Application	Reference
Various cytokine KI and KO mice	Drug hypersensitivity	3
TNF-α receptor KO	Mechanism of toluene diisocyanate asthma	4
Human CD4 KI-murine CD KO	General and immunotoxicity of a chimeric antihuman CD4 monoclonal antibody	5
Human CD4 KI-murine CD KO	Embryo–fetal and immunotoxicologic development study of a chimeric antihuman CD4 monoclonal antibody	6
Human growth hormone KI rats	Immunogenicity	7
Human interferon-α KI	Breaking immune tolerance to interferon-α	8
Human carcinoembryonic antigen KI mice	Safety of human carcinoembryonic antigen tumor vaccine	9

Transgenic Animals. Table 2 Examples of sources of limitation of the relevance of transgenic mice to human immunotoxicity

Physiology	
Kinetics	Generally more rapid clearance of xenobiotics and therapeutic proteins in mice
Metabolism	Differences between murine and human P450 usage and inducibility, substrate specificity and metabolite profile
Immunology	
Ontogeny	Differences in timing of cytogenesis, histogenesis and organogenesis of the immune system
Receptors	Differences in binding affinity and signal transduction of human proteins for murine receptors and binding proteins
Immunogenicity and tolerance	Differences between human and murine antigen processing and MHC restrictions
T cells	Differences in T helper 1 and 2 usage and switching
B cells	Differences in immunoglobulin class switching
Macrophages	Differences in Fc receptor utilization

References

1. Hofker MH, Van Deursen J (eds) (2002) Transgenic Mouse: Methods and Protocols. Methods Molecular Biology, Vol. 209. Humana Press, Clifton NJ
2. BioMed Net. http://www.biomednet.com/db/mkmd (accessed June 2003)
3. Moser R, Quesniaux V, Ryffel B (2001) Use of transgenic animals to investigate drug hypersensitivity. Toxicology 158:75–83
4. Matheson JM, Lemus R, Lange RW, Karol MH, Luster MI (2002) Role of tumor necrosis factor in toluene diisocyanate asthma. Am J Respir Cell Mol Biol 27:396–405
5. Bugelski PJ, Herzyk DJ, Rehm S et al. (2000) Preclinical development of keliximab, a Primatized anti-CD4 monoclonal antibody, in human CD4 transgenic mice: characterization of the model and safety studies. Hum Exp Toxicol 19:230–243
6. Herzyk DJ, Bugelski PJ, Hart TK, Wier PJ (2002) Practical aspects of including functional endpoints in developmental toxicity studies. Case study: immune function in HuCD4 transgenic mice exposed to anti-CD4 MAb in utero. Hum Exp Toxicol 21:507–512
7. Takahashi R, Ueda M (2001) The milk protein promoter is a useful tool for developing a rat with tolerance to a human protein. Transgenic Res 10:571–575
8. Braun A, Kwee L, Labow MA, Alsenz J (1997) Protein aggregates seem to play a key role among the parameters influencing the antigenicity of interferon alpha (IFN-alpha) in normal and transgenic mice. Pharm Res 14:1472–1478
9. Francini G, Scardino A, Kosmatopoulos K et al. (2002) High-affinity HLA-A(*)02.01 peptides from parathyroid hormone-related protein generate in vitro and in vivo antitumor CTL response without autoimmune side effects. J Immunol 169:4340–4849

Transgenic Mouse

Transgenic mice are genetically engineered mice that over-express foreign DNA and are typically referred to as transgenic, while those in which foreign DNA has replaced an endogenous gene are termed gene targeted (or knockout). In the strict sense, however, both these procedures yield a transgenic mouse (i.e. one with added genetic material).
▶ Knockout, Genetic

Transglutaminase

The epidermal keratinocyte transglutaminase I is a calcium-dependent enzyme that plays a central role in keratinocyte cornification. It catalyzes the cross-linking between glutamine and lysine residues of isopeptides at the inner surface of keratinocyte cell membranes, which is an essential step for the stabilization of their cornified cell envelope (CCE).
▶ Three-Dimensional Human Skin/Epidermal Models and Organotypic Human and Murine Skin Explant Systems

Transition Element

Elements that occupy the middle portions (the d-block) of the periodic table, have valence electrons in two or more shells instead of only one, and are characterized in most cases by variable oxidation states and magnetic properties.
▶ Chromium and the Immune System
▶ Vanadium and the Immune System

Transporter Associated with Antigen Processing (TAP)

TAP is composed of two subunits, TAP1 and TAP2. This heterodimer, which belongs to the ABC (ATP-binding cassette) transporter family is responsible for the shuttling of peptides from the cytosol into the lumen of the endoplasmic reticulum.
▶ MHC Class I Antigen Presentation

Trichinella spiralis

A helminthic parasite, invading the gut mucosa and residing as larvae in striated muscle tissues.
▶ Host Resistance Assays

Triglycerides

Tricglycerides are molecules that consist of a glycerol backbone esterified to three fatty acids.
▶ Fatty Acids and the Immune System

Trivalent Chromium

The ionic form of chromium when three outer shell electrons (one from 4s and two from 3d orbitals) have been shed, thereby giving the atom an overall charge of +3.
▶ Chromium and the Immune System

Trypanosomes, Infection and Immunity

RONALD KAMINSKY
Centre de Recherche Santé Animale
Novartis
CH-1566 St-Aubin
Switzerland

Synonyms
hemoflagellates

Definition
Trypanosomes are protozoan parasites of the family of *Trypanosomatidae*, belonging to the order of *Kinetoplastida* of the class of *Zoomastigopohora*.
Three species are pathogenic to man—*Trypanosoma brucei gambiense* and *T brucei rhodesiense* cause African human sleeping sickness in sub-Saharan Africa, while *T. cruzi* causes Chagas disease in South America (Table 1).

Characteristics
Characteristics of the parasites
The prominent morphological feature of the unicellular protozoan parasites is the kinetoplast, an organelle which contains about 15% of the cells DNA. The kinetoplast can be visualized by Giemsa staining or flu-

Trypanosomes, Infection and Immunity. Table 1 Characteristics of human pathogenic trypanosomes

Trypanosoma species	Disease	Transmission vector	Mode of transmission	Animal reservoirs	Geographic distribution
T brucei gambiense	Sleeping sickness	Tsetse flies (Glossina spp.)	Bite	Mainly dogs, pigs, and certain game animals	West and Central Africa
T. brucei rhodesiense	Sleeping sickness	Tsetse flies (Glossina spp.)	Bite	All major domestic animals and various game animals	East Africa
T. cruzi	Chagas disease	Reduviid bugs (Triatoma spp., Rhodnius spp., Panstrongylus spp.)	Contamination by bug feces	Domestic (dogs, cats, guinea-pigs), rodents and wild animals (opossums etc.)	Southern and Central America

orescent dies like DAPI. Movement of trypanosomes is via a flagellum which originates at the basal body near the kinetoplast and which is attached to the body of the parasite by an undulating membrane.

The African trypanosomes are extracellular parasites (16–30 μm long) which move within the blood (hence their designation as hemoflagellates) or within the cerebral spinal fluid. T. cruzi occurs in man in both as extracellular and intracellular form. After introduction into the blood T. cruzi invades various cell types including macrophages and muscle cells. The intracellular form (3 μm in diameter) is much smaller than the extracellular form and does not posses a flagellum, but still contains the kinetoplast.

All bloodstream forms of African trypanosomes are coated with variable surface glycoproteins (VSGs). The VSGs are anchored through a glycosyl phosphatidyl inositol lipid to the body of the parasite. These highly immunogenic VSGs have, at any one point of time, the same structure resulting in a specific variant antigen type (VAT). However, the VSGs are periodically removed and replaced with the result that the parasite population bearing one VSG are killed by an antibody response and are replaced by a new population with another variant antigen type. This ▶ antigenic variation is a mechanism that plays a key role in the escape of trypanosomes from total destruction by the immune response of their mammalian hosts (1).

Cyclical transmission

African trypanosomes are cyclically transmitted by tsetse flies (various Glossina species) (Table 1). After a fly has taken a blood meal from an infected host, the trypanosomes undergo various changes and multiplication within the fly. They finally mature in the salivary glands of the ▶ tsetse fly, to infectious metacyclic forms which are transmitted to a naive host.

The American T. cruzi is transmitted cyclically by 'kissing' bugs, the family of Reduviidae (Table 1), not by direct inoculation when the vector is feeding but by contamination through parasites in feces. Tri-

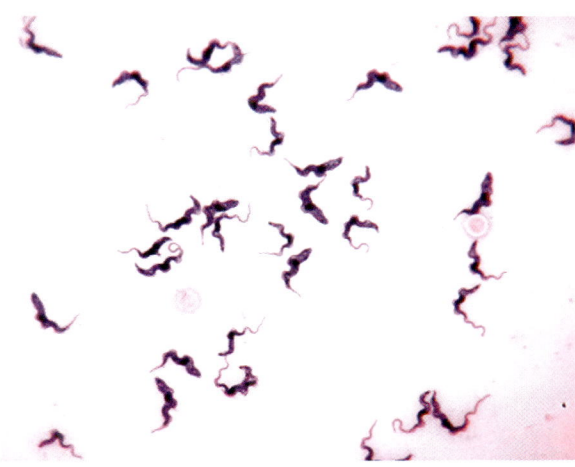

Trypanosomes, Infection and Immunity.
Figure 1 Trypanosoma brucei brucei bloodstream forms.

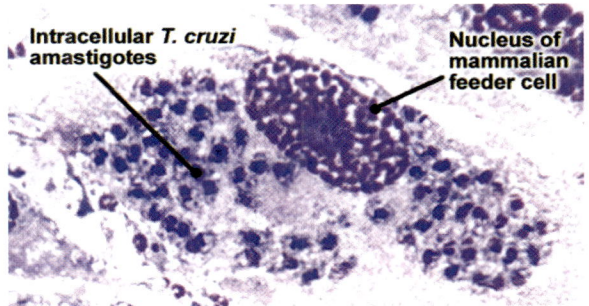

Trypanosomes, Infection and Immunity.
Figure 2 Trypanosoma cruzi: in vitro cultured amastigote forms in mammalian feeder cells.

atoma infestans is the major transmitting species, but various others species including *Triatoma* spp., *Rhodnius* spp., and *Panstrongylus* spp. are capable of transmitting *T. cruzi*.

Characteristics of the diseases

▶ Sleeping sickness is 100% fatal if left untreated. There are two disease stages for human African trypanosomiasis. A chancre, a primary lesion at the site of the bite, it is not observed frequently. The first stage, when trypanosomes are in the blood, is characterized by non-specific symptoms, such as fever, severe headache, joint or muscle aches. The second stage of the disease (also called late stage) starts with the invasion of the central nervous system by trypanosomes, which cross the blood-brain barrier 3–6 month post infectionem. It is in the late stage of the disease that the characteristic symptoms of sleeping sickness occur, such as sleep disturbances, alteration of mental state, muscle tone disorders, abnormal movements, and sensory and coordination disorders, up to a final general apathy.

The acute form of Chagas disease is characterized by general malaise with a variety of clinical manifestations. Symptoms can be very mild and atypical. At the site of entry of *T. cruzi* a local inflammation called a chagoma may develop; this is known as a Romana sign if it occurs at the eyelid. The acute form is followed by a period of an indeterminate form without any clinical symptoms. It is estimated that 20%–50% of persons with the indeterminate form of the infections will suffer from cardiac, digestive, or neurological damage 10–20 years after infection (2).

Preclinical Relevance

African and American trypanosomes can be manipulated in vitro and in various animal models. *T. brucei gambiense* appears to be the most difficult species for laboratory work. In vitro assays and in vivo models are being used to identify new active compounds, non-variant vaccine targets, and to monitor drug resistance. Some forms play an important role in host resistance models for immunotoxicity screenings.

Relevance to Humans
Infection

Sixty Million people in 36 countries of sub-Saharan Africa live at risk of acquiring sleeping sickness. In 1999 around 45 000 cases were reported, but the number of people thought to have the disease at any one time is between 300 000 and 500 000. Chagas disease affects 16–18 million people, and about 100 million (25% of the population of Latin America) are at risk of acquiring Chagas disease. Due to the chronic character (indeterminate stage) of Chagas disease, transmission occurs not only via insect vectors, but also by congenital transmission, and from transfusions with contaminated blood, and organ transplantations.

Immunity

Due to antigenic variation of the African trypanosomes, which can express approximately 1000 different variant antigen types (1), immunity to the parasites develops only to specific VATs but does not provide protection against infection.

Treatment

Two drugs, pentamidine and suramin, are used in the first stage of sleeping sickness prior to CNS involvement. The first-line treatment for late-stage cases, when trypanosomes are established in the CNS, is the arsenic-based drug melarsoprol (3). The drug has been in use since 1949. However, up to 5% of treated patients may die because of lethal encephalopathy due to the drug. Recently a new treatment schedule (4) was designed, but the number of patients with encephalopathy syndromes was the same as before. Nevertheless, the new 10-day schedule is a useful alternative to the present standard 26-day treatment schedule. Eflornithine (DFMO) is used mainly as a back-up in instances of melarsoprol-refractory *T. brucei gambiense*. Its efficacy against East African sleeping sickness is limited due to an innate lack of susceptibility of *T. brucei rhodesiense* based on higher ornithine decarboxylase turnover.

The unsatisfactory treatment situation for sleeping sickness is hampered further by the occurrence of melarsoprol-resistant trypanosomes (5) in several regions of sub-Saharan Africa. A molecular mechanism in the resistant isolates was identified: the majority of individual resistant isolates from geographically distant localities contained the same set of point mutations in their ▶ *TbAT1* genes (6), which codes for an adenosine transporter (7).

The drug of choice for treatment of Chagas disease is nifurtimox, with benznidazole as a back-up. However, these drugs are associated with side effects (2). Nifurtimox and benznidazoles were introduced at the beginning of the 1970s. Treatment success varies according to the phase of Chagas disease, the period of treatment, and the dose, the age, and geographical origin of the patients. Good results have been achieved in the acute phase, in recent chronic infection, and congenital infections. However, there is still controversy about their use in chronic cases (2).

Regulatory Environment

At present there is only one new antitrypanosomal drug on clinical trial in Africa—the diamidine derivative DB289. Identification of novel compounds and their development to drugs is pursued by various private-public initiatives. Registration of new drugs might be facilitated when these drugs are classified as orphan drugs. As mentioned above, trypanosomes are indirectly regulated by different immunotoxicology guidelines by the recommendation for infection models using these parasites in host-resistance assays. More detailed information is given in the relevant entries in this book.

References

1. Borst P (2002) Antigenic variation and allelic exclusion. Cell 109:5–8
2. Coura JR, de Castro SL (2002) A critical review on Chagas disease chemotherapy. Mem Inst Oswaldo Cruz 97:3–24
3. Legros D, Ollivier G, Gastellu-Etchegorry M et al. (2002) Treatment of human African trypanosomiasis—present situation and needs for research and development. Lancet Infect Dis 2:437–440
4. Burri C, Nkunku S, Merolle A, Smith T, Blum J, Brun R (2000) Efficacy of new, concise schedule for melarsoprol in treatment of sleeping sickness caused by *Trypanonosoma brucei gambiense*: a randomized trial. Lancet 355:1419–1425
5. Kaminsky R, Mäser P (2000) Drug resistance in African trypanosomes. Curr Opin Anti-infect Invest Drugs 2:76–82
6. Matovu E, Geiser F, Schneider V et al. (2001) Genetic variants of the TbAT1 adenosine transporter from African trypanosomes in relapse infections following melarsoprol therapy. Molec Biochem Parasitol 117:71–81
7. Mäser P, Sütterlin C, Kralli A, Kaminsky R (1999) A nucleoside transporter from *Trypansoma brucei* involved in drug resistance. Science 285:242–244

Tryptophan

α-Amino-β-indole-propionic acid; a component of proteins; it is chromogenic, producing a violet color with chlorine or bromine solution.
▶ Serotonin

Tsetse Fly

Tsetse flies (*Glossinidae;* more than 30 species) are sub-Saharan bloodsucking flies (*Diptera*). The females do not lay eggs but give birth to living larvae. Both sexes feed on the blood of humans, livestock, and wild animals. Tsetse flies transmit human and animal pathogenic trypanosomes. Ingested trypanosomes of an infested host undergo a development cycle in the tsetse fly to mature to metacyclic forms which are infective for the next host.
▶ Trypanosomes, Infection and Immunity

TSK

An acronym for tight skin which is associated with thickened skin and fibrosis due to mutations in the fibrillin gene.
▶ Systemic Autoimmunity

Tuberculin

Mixture of antigens obtained from the culture of *Mycobacterium tuberculosis*.
▶ Mitogen-Stimulated Lymphocyte Response

Tuberculin-Type Reaction

A classical example of a delayed-type hypersensitivity (DTH) is the tuberculin-type reaction. In sensitized individuals, it is induced by an intradermal injection of tuberculin, an extract of *Mycobacterium tubercolosis*. This particular example of DTH was first described by R. Koch. He who observed that patients with tubercolosis reacted with fever and shock after the subcutaneous injection of tuberculin. Typically, the T cell mediated local immune reaction appears one or two days after the application.

The tuberculin test, however, is not an allergic reaction. It is a diagnostic proof for the previous infection with *M. tubercolosis* and also other pathogens such as *M. leprae* or *Leishmania tropica*.
▶ Delayed-Type Hypersensitivity

Tumor Antigen

Any molecule leading to immune recognition of tumor. A generic term that encompasses tumor-specific antigens, antigens shared by normal and neoplastic cells, and specificities recognized by xenogeneic antibodies (e.g. human molecules bound by mouse monoclonal antibodies) that can be non-antigenic in the species of origin.
▶ Tumor, Immune Response to

Tumor-Associated Antigens

Tumor-associated antigens (TAA) are tumor-specific proteins that can be recognized by immune effector cells of the host. To date, a variety of TAA are known. These are derivatives of either (i) physiological self-antigens or tissue specific differentiation antigens that are dramatically overexpressed by tumor cells in comparison to other cells, (ii) mutated self-proteins or specific oncogenic antigens inappropriately expressed by tumor cells, or (iii) those derived from virally encoded antigens. The recognition pattern induced by TAA allows the immune system to distinguish the transformed neoplastic cells from surrounding normal tissue cells and triggers the immune cascade against them.

▶ Cancer and the Immune System

Tumor, Immune Response to

PIER-LUIGI LOLLINI
Cancer Research Section
Department of Experimental Pathalogy, University of Bologna
Viale Filopanti 22
I-40126 Bologna
Italy

Synonyms
Immune response to cancer, anti-tumor immunity.

Definition
The immune system of the host responds to tumor growth as it does to infectious agents, with specific (e.g. T cells and antibodies) and non-specific (e.g. natural killer cells and cytokines) effector and regulatory mechanisms. The immune response reduces the number of tumors arising in the host, but is no longer effective against established tumors. Tumor immunotherapy is the attempt to elicit a therapeutic immune response in cancer patients.

Characteristics
The immune response against tumors was formally demonstrated in the late 1940s and early 1950s using transplantable tumors induced with chemical carcinogens or retroviruses in inbred mice (1). The experiments showed that mice vaccinated with a given tumor reject a subsequent challenge with the same tumor (immune memory), but fail to reject an unrelated tumor (specificity).

The immunization-challenge system was extensively used to characterize the effector and regulatory mechanisms of the immune response against tumors using two strategies:
- cellular and molecular analysis of local and systemic components elicited by immunization and/or involved in rejection
- use of mice with selective immune deficiencies of genetic origin (spontaneous mutation or genetically modified mice) or induced by exogenous treatments like monoclonal antibodies or drugs.

Specific immune responses against tumors are mainly due to T cells. Cytotoxic T cells (CTL) expressing the CD8 surface molecule are the final effectors capable of tumor cell lysis. Helper T cells (Th) expressing CD4 play a fundamental positive or negative regulatory role. Tumor immunologists tend to downplay the importance of B cells, antibodies and complement because solid tumors are resistant to complement-mediated cytotoxicity (tumor cells express complement inhibitors like CD55 and CD59) and in immunization-challenge systems B cells can even favor tumor growth ("enhancement").

Most cells of the innate (also called natural or non-adaptive) immune system directly affect tumor growth, and are required for the generation of T cell immunity. Professional and non-professional phagocytes destroy tumor cells and generate antigenic material that is subsequently picked up by antigen-presenting cells (APC) like dendritic cells, that are indispensable to activate T cell responses. Natural killer (NK) cells can kill tumor cells in tissues and in the bloodstream, thus are important in the control of systemic metastatic spread. In the course of the immune response many cytokines released by various cell types have regulatory and effector activities. Interferons IFN-α, IFN-β, and IFN-γ and tumor necrosis factors TNF-α and TNF-β, in addition to their roles as internal mediators of the immune system, directly inhibit tumor cell proliferation, trigger apoptosis, and induce the secretion of anti-angiogenic chemokines like MIG and IP-10 (2).

Immune Surveillance
The ▶ immune surveillance hypothesis, originally proposed in the late 1950s, postulates that the immune system protects the host not only from infectious agents, but also from tumor onset (1). Two predictions can be derived from the theory:
- tumors that grow despite the immune system have found a way to escape surveillance, thus must be poorly immunogenic
- tumor incidence should be higher in immunodepressed than in immunocompetent individuals.

The low immunogenicity of spontaneous (as opposed to carcinogen-induced or viral-induced) tumors in mice was easily verified, and is also a property of human tumors. Demonstration of the second prediction has been more controversial, because the degree and duration of immunodepression in experimental systems and in human conditions is highly variable and rarely complete. Only recently, with the advent of knockout mice, has it been clearly demonstrated that aging immunodepressed mice develop significantly more tumors than immunocompetent mice (1). Tumors arising in such immunodepressed mice are more immunogenic than tumors of immunocompetent mice, thus providing a further demonstration of the hypothesis. In long-term immunodepressed adult humans (e. g. transplant recipients or HIV-infected patients) the incidence of virus-induced tumors (such as Kaposi sarcoma or cervical carcinoma) is increased, but many other tumor types display an incidence similar to that of the immunocompetent population.

Tumor Antigens

The search for ▶ tumor antigens in human tumors was conducted for many years by means of antisera and monoclonal antibodies obtained after immunization of rodents with human cells or tissues. This endeavor led to the discovery of a wealth of molecules expressed by human tumors that are recognized by xenogeneic antibodies. However some molecules detected by rodent antibodies display little or no antigenicity in the human species, or data on recognition by the human immune system are not available. Application of the term "tumor antigens" to molecules that are not recognized as such in the species of origin is inappropriate, whereas "tumor markers" is more appropriate. Even though the immunological import of tumor markers is dubious, they have a great clinical relevance in tumor diagnosis, prognosis, and follow-up. Some examples are lactate dehydrogenase (used to monitor treatment of testicular cancer, Ewing's sarcoma and other human tumors), neuron-specific enolase (neuroblastoma and small cell lung cancer), and DU-PAN-2 (pancreatic carcinoma).

To distinguish "true" tumor antigens, that can induce a specific immune response in the species of origin, leading to tumor rejection, the terms "tumor rejection antigens" or "tumor specific transplantation antigens" are sometimes used. Having clearly established the distinction between tumor markers and tumor antigens, here we will simply use the latter term. Molecular cloning of tumor antigens became possible in the 1980s thanks to technologies based on T cell recognition under syngeneic or autologous conditions. Identification of tumor antigens and measure of specific responses are currently based on T cell clones with helper or cytotoxic activity in vitro, identification of peptides bound to major histocompatibility complex (MHC) molecules on the surface of tumor cells, molecular cloning of T cell receptor (TCR) genes from ▶ tumor-infiltrating lymphocytes (TIL), soluble ▶ MHC tetramers produced in vitro and bound to synthetic peptides, and screening of DNA libraries with patient's sera (▶ SEREX) (2). It can be noted that SEREX is antibody-based, however it makes use of high affinity human IgGs that derive from a Th-induced immunoglobulin class switch, thus SEREX can be viewed as a T-B hybrid technology.

The main groups of tumor antigens are shown in Table 1 (3). One major fact is that most tumor antigens are not tumor specific. The protein expressed by tumor cells, and the antigenic peptides derived from it are identical to those of normal cells, thus leading to the conclusion that the immune response to tumors is actually an autoimmune response. Experimental and clinical proofs of the autoimmune nature of anti-tumor immune responses were obtained in melanoma-bearing individuals, who develop autoimmune vitiligo as a consequence of vaccination with tumor antigens (2). The autologous nature of many tumor antigens is one of the reasons why tumors are poorly immunogenic, suggesting that a break of immune tolerance is a prerequisite to an effective anti-tumor immune response. In a few cases immune tolerance does not operate, either because normal cells expressing the antigen are in immunologically privileged sites (e.g. ▶ cancer-testis antigens), or because the antigen is involved in a physiological network of immune responses (idiotypes of T and B cell neoplasms). The only truly ▶ tumor-specific antigens are those that derive from mutations of oncogenes (RAS, CDK4) or tumor suppressor genes (p53), from chimeric proteins encoded by chromosomal translocations (BCR-ABL), or from tumor-specific alternative splicing (MUC-1, possibly HER-2). Experimental evidence shows that, even when tumor antigens are not shared by normal cells and are tumor-specific, spontaneous immune responses are quite low and ineffective in the tumor-bearing host.

Low Immunogenicity of Tumors

A complete understanding of the reasons why tumors are poorly immunogenic is of paramount importance to devise immunotherapeutic strategies to induce a protective response (1). Basically tumors are tolerated by the immune system because their antigenic profile is almost identical to that of normal cells. In addition, genetic instability of tumor cells generates a large array of phenotypes that can escape immune recognition using a variety of passive and active strategies. Down-regulation of antigen expression is an obvious alternative that has been incompletely investigated. The most common defect in human tumors (80%–90% of all solid tumors) is a partial down-regulation of MHC class I molecules required for peptide binding and T cell recognition (4). Active strategies may include the induction of regulatory (i.e. suppressive) cells of myeloid ($CD11b^+/Gr1^+$) or lymphoid ($CD4^+/CD25^+$) origin, the secretion of suppressive cytokines like TGF-β or IL-10, or the expression of pro-apopto-

Tumor, Immune Response to. Table 1 Examples of tumor antigens

Tumor antigen group	Examples*
Cancer-testis antigens	MAGE-A1–A12, B1–B4, C1, C2 BAGE GAGE-1–8 NY-ESO-1
Differentiation or lineage-specific tumor antigens	gp100 Melan-A (MART-1) Prostate specific antigen (PSA) Tyrosinase Tyrosinase-related proteins (TRP)
Shared tumor antigens	Carcinoembryonic antigen (CEA) HER-2/neu MUC-1 Telomerase catalytic unit (TERT)
Mutated antigens	RAS β-catenin Cyclin-dependent kinase 4 (CDK4) MUM-1 p53
Fusion proteins	BCR-ABL PML-RARα PAX3-FKHR SYT-SSX1/2 EWS-WT1, EWS-FLI1

* Complete listing is given in reference 3.

tic surface molecules (1). On top of all immune regulations, an expanding tumor could overcome the immune response by sheer cell kinetics.

From Tumor Immunology to Immunotherapy

Spontaneous immune responses are incapable of eradicating established tumors (spontaneous regression has been rarely described in human melanoma and renal cell carcinoma). Preclinical evidence demonstrates that the immune response, if properly activated, can cure tumors. Analogous conclusions can be drawn from some successful clinical approaches. A convincing clinical example is the ability of allogeneic T cell transplants to reduce the risk of leukemic relapse by 30%–40%, a phenomenon known as "graft versus leukemia" (GvL). The main strategies to induce a therapeutic immune response in human patients are based on the administration of preformed immunologic "drugs" (▶ passive immunotherapy) or of therapeutic vaccines (▶ active immunotherapy).

Passive immunotherapy is currently the most successful way to target human tumors (2). A small number of monoclonal antibodies with significant activity against human tumors emerged from clinical trials and is approved for clinical use. The best known examples are trastuzumab (Herceptin), a humanized monoclonal antibody against HER-2 for breast cancer, and rituximab, a monoclonal antibody against CD20 for non-Hodgkin's lymphoma (NHL). Several other monoclonal antibodies against similar, or different, target antigens are being developed. It is interesting to note that the therapeutic activity of monoclonal antibodies is only partly mediated by classical immune functions such as complement-mediated cytotoxicity and ▶ antibody-dependent cell-mediated cytotoxicity▶ (ADCC). Therapeutic effect is also attributable to the activity of monoclonal antibodies as "receptor antagonists", through the inhibition of receptor dimerization and signaling, and the induction of receptor internalization and degradation.

Immunotherapy with cytokines (2) received a considerable attention throughout the 1980s and 1990s. The major clinical drawback has been the high toxicity of cytokines. Immune cytokines physiologically reach high *local* concentrations, but high *systemic* dosages are usually associated with severe toxicity. Toxicity hampered the clinical development of promising molecules such as IL-2, TNF-α, and IL-12. IFN-α is a good example of a cytokine that can be administered systemically at active dosages to cancer patients with tolerable toxicity. IFN-α initially showed therapeutic activity against hairy cell leukemia, and has significantly prolonged survival in chronic myeloid leukemia (CML). It is also used for some solid tumors, such as melanoma, with a significantly lower activity than against hematologic malignancies. As previously noted for therapeutic monoclonal antibodies, IFN-α owes its anti-tumor activity to a combination of immune and non-immune effects. The latter include inhibition of tumor cell proliferation, induction of cell differentiation, and inhibition of neo-angiogenesis.

Molecular definition of tumor antigens prompted a large number of vaccination trials based on a variety of immunological approaches to ▶ cancer vaccines (2). One possibility is to vaccinate with the DNA encoding a tumor antigen that is picked up and translated by host cells (DNA vaccination). Alternatively, vaccines are made of whole cells, recombinant proteins or synthetic peptides admixed with adjuvants. Dendritic cell-based vaccines exploit the pivotal role of antigen presentation in the generation of T cell responses. Dendritic cells cultured in vitro are fed ("pulsed") with tumor antigens and then injected in vivo. Promising results were obtained in small phase I/II clinical trials, but definitive evidence of a marked clinical benefit from therapeutic cancer vaccines is still lacking.

Preclinical Relevance

Study of the immune response to tumors was largely conducted in preclinical model systems. The results were mostly confirmed by human studies, when possible, thus it is generally assumed that preclinical evidence and features of the immune response to tumors apply to human and clinical situations.

One area that requires caution is the toxicity of cytokines endowed with anti-tumor activity. Because of species specificity, human cytokines are inactive, or partially active in rodents, therefore mouse cytokines must be used for mouse studies. This situation is quite different from the development and testing of conventional anticancer drugs, in which the same molecule is used in preclinical and in clinical studies. Some cytokines like TNF-α that display a potent anti-tumor activity in mice are clinically useless because in humans the maximum tolerated dose is much lower than the effective dose (2).

A major stumbling block for cancer vaccines, as well as for other forms of tumor immunotherapy, is the fact that most clinical trials recruit advanced patients that are heavily immunosuppressed and poorly responsive to vaccines, whereas preclinical data clearly show that the ideal use of vaccines would be for cancer prevention in healthy individuals at risk, or for adjuvant therapy against micrometastatic foci, rather than for therapy of bulky and advanced lesions (5). In conclusion it is conceivable that active immunotherapy will demonstrate its anti-tumor potential only when clinical studies will follow the path clearly marked by preclinical data.

Relevance to Humans

Cancer patients treated with monoclonal antibodies respond to therapy only if the tumor expresses high levels of the target antigen, thus demonstration of high antigen levels in tumor lesions is a prerequisite for therapy. Some indications are available for specific antigens (e.g. HER-2), for which clinical benefit of antibody therapy at intermediate antigen levels is dubious.

Assessment of the anti-tumor immune response in cancer patients is not routinely performed outside clinical trials of immunotherapy. Most immune tests applied in patients receiving immunotherapy are not standardized, and in some instances are of questionable value. For example it is not clear if tests performed on peripheral blood lymphocytes (the easiest sampling route) correlate with the immune response at tumor sites. Correlation of positive or negative clinical results with the immune status of patients and with modifications of the immune response induced by immunotherapy is a major open issue (2).

Regulatory Environment

Preclinical data are required to design clinical trials, but analysis of human immune responses to tumors is confined to in vitro systems, thus there is no need for guidelines concerning animal testing. Within clinical trials, study of the immune response is highly dependent on treatment (e.g. type of vaccine or cytokines used) and in most instances there are no gold standards or guidelines pertinent to immune testing of cancer patients for what concerns antibody responses, cytokine release or T cell cytotoxicity against autologous or non-autologous tumor cells. Skin tests (cf. delayed-type hypersensitivity) are used to detect responses elicited by anti-tumor vaccines.

Immunotherapy trials are being conducted with a wide range of approaches that span practically all classes of therapeutic agents and therapies. The range of adverse or unwanted effects that can affect patients is correspondingly wide. In addition to general toxicity of the therapeutic agent, and to the presence of contaminants in the preparation, a specific type of potential adverse effect of cancer immunotherapy is the induction of autoimmunity. In practice all treatments aimed at inducing an immune response against tumor antigens shared by normal cells involve an autoimmune response. It must be underlined that regulatory requirements for prophylactic vaccines to be administered to healthy individuals in the general population are quite different from those applying to cancer patients. In fact some relatively mild forms of autoimmunity induced by immunotherapy in cancer patients, such as vitiligo in melanoma, are regarded as surrogate markers of anti-tumor immune response (2).

In various instances a single therapeutic rationale can be implemented with different treatment modalities that depend on different regulatory environments. For example, to enhance tumor antigen recognition cytokines can be administered systemically to the patient (drug therapy), or tumor cells can be genetically modified to secrete the cytokine (gene therapy), or be fused with autologous or allogeneic dendritic cells (adoptive cell therapy). A complete listing of all the guidances, guidelines, and regulations pertaining to each and every immunotherapeutic approach goes beyond the scope of this article. The reader is referred to the web sites of the regulatory bodies for comprehensive listings and full text of documents.

Regulatory Bodies and Agencies

- European Agency for the Evaluation of Medicinal Products (EMEA) http://www.emea.eu.int
- European portal to the pharmaceutical regulatory sector (EudraPORTAL) http://www.eudra.org
- US Food and Drug Administration (FDA) http://www.fda.gov
- FDA Center for Biologics Evaluation and Research (CBER) http://www.fda.gov/cber
- FDA Center for Drug Evaluation and Research (CDER) http://www.fda.gov/cder
- Japanese Ministry for Health, Labour and Welfare http://www.mhlw.go.jp
- Organization for Economic Co-operation and Development (OECD) http://www.oecd.org
- International Conference on Harmonisation of Technical Requirements for Registration of Pharmaceuticals for Human Use (ICH) http://www.ich.org

References

1. Pardoll D (2003) Does the immune system see tumors as foreign or self? Ann Rev Immunol 21:807–839
2. Rosenberg SA (2000) Principles and practice of the biologic therapy of cancer. Lippincott Williams & Wilkins, Philadelphia
3. Renkvist N, Castelli C, Robbins PF, Parmiani G (2001) A listing of human tumor antigens recognized by T cells. Cancer Immunology Immunotherapy 50:3–15
4. Garrido F, Algarra I (2001) MHC antigens and tumor escape from immune surveillance. Adv Cancer Res 83:117–158
5. Lollini PL, Forni G (2002) Antitumor vaccines: is it possible to prevent a tumor? Cancer Immunol Immunother 51:409–416

Tumor Immunology

▶ Cancer and the Immune System

Tumor-Infiltrating Lymphocytes

Lymphocytes (usually T cells) isolated from tumor specimens. Tumor-infiltrating lymphocytes (TILs) can be cultured in vitro to analyze their functional and molecular features, and can be also injected in vivo for therapeutic purposes.

▶ Tumor, Immune Response to

Tumor Necrosis Factor (TNF)

TNF-α and TNF-β lymphotoxin are produced by macrophages and T lymphocytes. First described as cytotoxins for tumor cells, later as important cytokines for the inflammatory response, cooperation with other leukocytes, induction of fever and interference with fat metabolism; therefore TNF is also named cachectin.

▶ Cytokines

Tumor Necrosis Factor-α

VICTOR J JOHNSON
Toxicology and Molecular Biology Branch
National Institute for Occupational Safety and Health
1095 Willowdale Road
Morgantown, WV 26505
USA

Synonyms

tumor necrosis factor-α, lymphotoxin, cachectin, TNF-α

Definition

Tumor necrosis factor-α (TNF-α) is a pleiotropic proinflammatory cytokine that mediates key roles in homeostasis, cell growth and proliferation, tissue damage, repair and chronic diseases. TNF-α production is induced by a plethora of stimuli including bacterial products, oxidative stress, other cytokines, and general tissue damage. As such, this cytokine has a central role in orchestrating many injury and disease states including immunotoxicity.

Molecular Characteristics and Mediators of Signaling

Human TNF-α is synthesized as a 26-kDa pro cytokine destined for expression on the plasma membrane. Proteolytic processing by members of the matrix metalloproteinase family of enzymes results in the extracellular release of the mature soluble 17-kDa form of TNF-α. Both the membrane-bound and soluble forms are biologically active and play important roles in overlapping and distinct signaling processes. Signaling is achieved through ligation of two structurally distinct receptor subtypes, TNF-receptor 1 (TNF-R1) and TNF-R2. TNF-R1 is constitutively expressed on most nucleated cells whereas TNF-R2 has a more restricted expression, mainly on cells of the immune system and is inducible. A schematic of the major signaling pathways and molecular mediators is shown in Figure 1.

Membrane and soluble TNF-α form trimers that induce the trimerization of the TNF- receptor upon binding. Activation of the receptor initiates the formation of unique signaling complexes that are distinct for each receptor subtype. A ▶ death-inducing signaling complex is formed at the intracellular domain of TNF-R1 involving the recruitment and binding of a number of accessory proteins, including TNF-R1-associated death-domain-containing factor (TRADD), Fas-associated death-domain-containing protein (FADD) and TNF-R-associated factor-2 (TRAF2). Binding is achieved through mutual death domains (DD) present on TNF-R1 and the accessory proteins, and the DD sequence is unique to the intracellular portion of TNF-R1. The resulting complex recruits

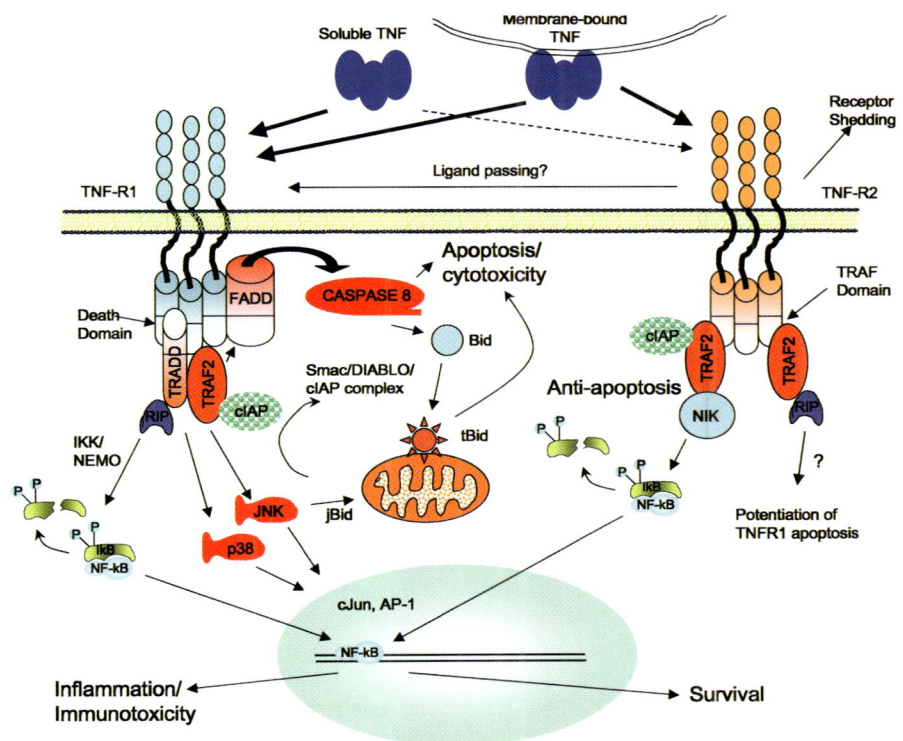

Tumor Necrosis Factor-α. Figure 1 Death and survival pathways in tumor necrosis factor (TNF)-α signaling. TNF-α exerts its biological effects through ligation of two distinct receptors, TNF-R1 and TNF-R2. Activation of TNF-R1 results in the formation of a death-inducing signaling complex, consisting of TNF-R1 intracellular DD, TRADD, FADD and TRAF2. This complex recruits other intracellular signaling molecules that activate pathways culminating in cell death (caspase and JNK pathways) and cell survival (NFκB activation pathway). Additionally, activation of MAP kinases and NFκB can result in upregulation of genes involved in inflammation, including TNF-α itself. On the other hand, activation of TNF-R2 leads to the formation of a signaling complex via mutual TRAF domains. This complex is known to lead to NFκB activation and anti-apoptotic signaling. Increasing evidence suggests a role for TNF-R2 in apoptosis, possibly through potentiation of TNF-R1 pro-apoptotic signaling. Overall, these signaling pathways can contribute to immunotoxicity by directly inducing cell death and tissue damage, initiating and contributing to inflammation, and/or altering the proliferative capacity of cells and tissues.

other proteins with enzymatic activity culminating in the induction of several major signaling pathways, including the caspase pathway, mitogen-activated protein (MAP) kinase pathways and pathways that lead to nuclear factor κB (NFκB) activation. The cell death and tumor regression properties of TNF-α are attributed to TNF-R1-mediated activation of caspases and c-Jun NH2-terminal kinase (JNK). JNK activation has been shown to cleave the Bcl-2 interacting domain (Bid) resulting in jBid translocation to the mitochondria. This induces the release of Smac/DIABLO from the mitochondria, which then sequesters inhibitor of apoptosis (cIAP) proteins leading to FADD-induced activation of caspase 8. Caspase 8 activates effector caspases 3 and 7 leading to the cleavage of several intracellular proteins and apoptotic cell death. Apoptosis is not the predominant outcome in most cell types in vivo and can be prevented through the activation of the NFκB survival pathway. Activation of NFκB results from the coordinated action of receptor interacting protein (RIP), NFκB inducing kinase (NIK) and inhibitor of κB (IκB) kinases. The outcome is phosphorylation-dependent ubiquitination and degradation of IκB, which results in the release of NFκB into the cytoplasm which then translocates to the nucleus via a nuclear localization sequence. Heterodimers and homodimers of the NFκB/Rel family drive the transcription of many survival and inflammation genes containing NFκB response elements. Therefore, this pathway functions to prevent cell death in normal healthy cells and upregulates the production of proteins involved in inflammatory processes.

The intracellular domain of TNF-R2 lacks the DD and instead has TRAF domains responsible for the recruitment of TRAF1, TRAF2, and TRAF3. This complex recruits cIAP and NIK both responsible for NFκB-mediated anti-apoptotic signaling. However, several investigators have reported that TNF-R2 may be important in the regulation and potentiation of TNF-R1-induced apoptosis. Several mechanisms have been proposed, including TNF-R2, acting as a high-affinity trap/▶ ligand passer and TNF-R2-induced upregulation of endogenous TNF-α production, both leading to autocrine and paracrine activation of TNF-R1-mediated apoptosis. A dual role in cell survival and death has been shown such that in the absence of TNF-R1 signaling, TNF-R2 promotes not only proliferation of naïve T lymphocytes but also apoptosis in activated $CD8^+$ T lymphocytes. Additionally, the affinity of TNF-R2 for membrane-bound TNF-α is much greater than that for the soluble form (affinities are equal for TNF-R1), suggesting that TNF-R2 is important in direct cell to cell regulation and localized immune responses. TNF-R2 can also be shed from the plasma membrane resulting in soluble TNF-R and downregulation of TNF-α signaling.

Relevance to Immunotoxicity

TNF-α is produced by many cell types including immune cells (macrophages, monocytes, dendritic cells, T lymphocytes, B lymphocytes), endothelial cells, epithelial cells, and fibroblasts following activation by appropriate stimuli. Therefore, this cytokine can be envisioned to play a role in immunotoxicity in many target organs. Indeed, TNF-α, through its ability to influence inflammatory processes, has been demonstrated in response to immuntoxicants targeting the liver, kidney, lung, muscle, eye, skin and brain, to name a few sites. Gene knockout of TNF-α, its receptors, or neutralizing-antibody studies have been used to investigate the role of TNF-α in immunotoxicity. For example, mice deficient in TNF-R1 and TNF-R2 show reduced lung inflammation and cytokine changes in response to toluene diisocyanante, an occupational asthmogen. Removal of TNF- signaling almost completely abrogated inflammation and fibrosis in the liver following treatment with the known immunotoxicant, carbon tertrachloride. TNF-α plays an important role in the immunotoxicity of many mycotoxins such as vomitoxin and fumonisin B_1. Kidney TNF-α levels increase in adriamycin-induced nephropathy and may play a direct role in the associated proteinuria. Immediate production of TNF-α in the skin is evident following exposure to agents that cause allergic and irritant contact dermatitis and is positively correlated with the inflammatory response. Significantly, mice deficient in TNF-Rs show reduced development of contact dermatitis. In addition to its role in chemical-mediated immunotoxicity, TNF-α also mediates critical events in the pathogenesis of iodiopathic diseases involving the immune system, including bacterial infection and sepsis, chronic inflammatory lung diseases, cancer, and autoimmunity.

The vastness of TNF-α involvement in chemical and idiopathic immunotoxicity stems from its central role in many cytokine/chemokine networks. TNF-α signaling (see Figure 1) can modulate the expression of other important mediators of inflammation required for the recruitment and activation of effector cells (macrophages, lymphocytes, neutrophils, eosinophils) that can contribute to tissue injury. Therefore, enhanced synthesis and release of TNF-α following immunotoxicant exposure or during disease can initiate and exacerbate acute and chronic inflammation—known contributors to tissue injury, repair and remodeling (fibrosis).

Relevance to Humans

Examining the association between genetics and disease prevalence and disease severity provides valuable insight into the mechanisms of disease. As such, greater than 1000 scientific studies have been conducted investigating the association between polymorphisms in the TNF family and diverse human diseases. Genetic variation in TNF-α has been associated with chemical toxicities and idiopathic diseases of the immune system and diseases with immune involvement. Examples include occupational lung diseases, such as silicosis and coal workers' pneumoconiosis, chemotherapy-induced pulmonary fibrosis, adverse drug reactions, response to hepatitis B vaccination, asthma, diabetes, and rheumatoid arthritis. The strong association with human disease has prompted research into potential therapies related to inhibition of TNF-α signaling. To date, several US FDA approved protein-based injectable inhibitors that block TNF-TNF-R interactions have been used to successfully treat human disease, including rheumatoid arthritis and juvenile chronic arthritis. Ongoing clinical trials show promise for these therapeutics in other diseases like psoriasis, psoriatic arthritis, ankylosing spondylitis, and Crohn's disease. Second-generation small-molecule inhibitors of TNF are now undergoing clinical trials and function through blocking specific mediators in the TNF signaling cascade. Caution must be exercised in the use of these treatments as side effects including potential exacerbation of congestive heart failure, activation of latent tuberculosis infection, development of antinuclear antibodies, and systemic lupus erythematosis have been reported. Nevertheless, TNF-α is a pinnacle cytokine in acute and chronic inflammatory disease and toxicity and represents a promising therapeutic target.

References

1. Luster MI, Simeonova PP, Gallucci R, Matheson J (1999) Tumor necrosis factor alpha and toxicology. Crit Rev Toxicol 29:491–511
2. Palladino MA, Bahjat FR, Theodorakis EA, Moldawer LL (2003) Anti-TNF-α therapies: the next generation. Nature Rev Drug Discov 2:736–746
3. Gupta S (2002) A decision between life and death during TNF-α-induced signaling. J Clin Immunol 22:185–194
4. Chen G, Goeddel DV (2002) TNF-R1 signaling: a beautiful pathway. Science 31:1634–1635
5. Lui Z-G (2004) Adding facets to TNF signaling: the JNK angle. Mol Cell 12:795–796
6. Schook L, Laskin D (eds) (1994) Xenobiotics and Inflammation. Academic Press, San Diego CA

Tumor Necrosis Factor Receptor-Associated Factor-6

TRAF-6 induces multiple signals from TOLL-like receptors that sense infection.

▶ Interleukin-1β (IL-1β)

Tumor-Specific Antigen

Antigen expressed by tumor cells, but not by normal cells. Truly specific tumor antigens are generated by oncogenic genetic lesions, such as mutations in oncogenes and tumor suppressor genes, or chromosomal rearrangements leading to the synthesis of fusion proteins. The idiotype of T and B cell receptor in lymphoid malignancies is considered a tumor-specific antigen (but there might be non-neoplastic clones sharing the same idiotype). Cancer-testis antigens are also considered tumor specific because male germ line cells (the only normal cell type sharing such antigens) lack MHC expression, thus cannot present the antigen to the immune system.

▶ Tumor, Immune Response to

Type I Error

A decision error in which a true null hypothesis is incorrectly rejected.

▶ Statistics in Immunotoxicology

Type I Reactions According to Gell and Coombs

▶ IgE-Mediated Allergies

Type I–IV Reactions

Gell and Coombs described antibody and T cell-mediated reactions with distinct clinical pathology and underlying pathomechanism. The type IV reactions can be subdivided in type IVa–IVd reactions, which reflect the involvement of distinct effector cells.

▶ Lymphocyte Transformation Test
▶ Hypersensitivity Reactions

Type 1 or Type 2 T Cell Responses

Subset of T lymphocytes called T helper (Th) cells can respond to different stimuli by secreting different cytokine patterns. Two well characterized patterns are categorized as type 1 (Th1) and type 2 (Th2) responses. Type 1 responses promote inflammation and cell mediated immunity primarily through the production of IFN-γ. Type 2 responses promote allergies

and antibody mediated responses primarily through production of IL-4 and IL-5. A type 1 response is antagonistic to a type 2 response and vice versa.
▶ Cytokine Assays

Type II Activation

▶ Macrophage Activation

Type II Error

A decision error in which a false null hypothesis is not rejected.
▶ Statistics in Immunotoxicology

Type II Interferon

▶ Interferon-γ

U

Ulcerative Mucositis

▶ Oral Mucositis and Immunotoxicology

Ulcerative Stomatitis

▶ Oral Mucositis and Immunotoxicology

Upper Respiratory Infection

Invasion by living pathogenic microorganisms of the higher part of the lungs, where conditions are favorable to their growth and from where their toxins may gain access to, and act injuriously upon, the tissues.
▶ Klebsiella, Infection and Immunity

Uptake

▶ Opsonization and Phagocytosis

Urinary ract Infection

Invasion by living pathogenic microorganisms of the urine-producing tract, where conditions are favorable to their growth and from where their toxins may gain access to, and act injuriously upon, the tissues.
▶ Klebsiella, Infection and Immunity

Urticaria

Urticaria is characterized by pruritic, erythematous wheals, usually resolving within 24 hours. Urticaria may develop as part of cutaneous manifestations of systemic anaphylaxis including the pulmonary, circulatory, and gastrointestinal systems. Although allergic IgE-mediated reactions to drugs are generally thought to be frequent causes of drug-induced urticaria, systemic anaphylaxis or urticaria can also be caused by drugs or biological agents that are responsible for direct mediator release from mast cells or basophils. Urticaria heals without skin lesions.
▶ Drugs, Allergy to
▶ Anti-inflammatory (Nonsteroidal) Drugs
▶ IgE-Mediated Allergies

US Clean Air Act

Federal law that regulates air emissions from area, stationary, and mobile sources and authorizes the US Environmental Protection Agency to establish National Ambient Air Quality Standards to protect public health and the environment.
▶ Asthma

UVA Radiation

320–400 nm range. Also known as "long-wave ultraviolet" or "near UV" (because it is adjacent to visible), or "black light." Subdivided into UVA II (320–340 nm) and UVA I (340–400 nm). In Europe, frequently defined as 315–400 nm, based on CIE denotation.
▶ Photoreactive Compounds

UVC Radiation

200–290 nm range. Also called "germicidal radiation" or "shortwave UV". Not found in sunlight at the surface of the earth because it is filtered by ozone and water vapor.
▶ Photoreactive Compounds

Vβ

See T cell antigen receptor.
▶ Superantigens

Valency

The combining power of a atom with respect to its ability to gain, lose, or share electrons in its outer orbitals/shells.
▶ Chromium and the Immune System

Vanadate

Vanadate is a pentavalent monomer of vanadium oxide that can exist either as the meta- or ortho- form depending on the number of oxygen ligands (meta- if n=3; ortho- if n=4) about the vanadium atom.
▶ Vanadium and the Immune System

Vanadium and the Immune System

MITCHELL D COHEN
Department of Environmental Medicine
New York University School of Medicine
57 Old Forge Road
Tuxedo, NY 10987
USA

Definition

Although Andres del Rio was first to "discover" vanadium in 1801 (he named it erythronium), he later came to believe that he had only rediscovered lead chromate. Credit for its true discovery in 1831 went to Nils Sefstrom who, using iron ore, was the first to isolate an oxide of a new metal that he termed vanadium in honor of the Norse goddess of love and beauty, Vanadis.

Characteristics

Vanadium (V), a group VB ▶ transition element, can exist in multiple valences (0, +2, +3, +4, +5) in both anionic and cationic forms. Although the tetravalent and pentavalent forms are the most stable, discrete ions of each do not exist in nature. Most commonly, these ions are bound to oxygen as negatively-charged polymeric oxyanions that readily complex with polarizable ligands such as S or P. In nature, ▶ pentavalent vanadium is most often encountered in the form of vanadium pentoxide (V_2O_5), though ferrovanadium, vanadium carbide, and various forms of ▶ vanadates also exist. Colloidal V_2O_5 can liberate vanadate (VO_3^- and VO_4) agents by loss of water, and the resulting monomeric vanadate ions can be further converted to higher polymeric forms (Figure 1), akin to how chromate ions link during olation. These conversions, and therefore the distribution, of vanadium species in solution depend on pH and vanadium concentration. As a rule, as vanadate unit numbers in the polymer increase, overall toxicity declines; however, even large polymers like ▶ decavanadate can give rise to toxicities.

Putative Interaction with the Immune System
Putative Non-Immune System Interactions and Toxicities of Vanadium Agents

While vanadium has been shown to be a mutagen and a clastogen in numerous mammalian and prokaryote systems, little is known regarding carcinogenic/mutagenic effects of vanadium agents in humans and animal models; in addition, only a few studies regarding its teratogenic/embryotoxic effects exist. Following life-long feeding of rodents with ▶ tetravalent vanadium, there was inconclusive evidence for carcinogenicity. This is likely the result of the low level of gastrointestinal uptake, as is the case with many carcinogenic metals that do not display carcinogenic potentials. In contrast, studies using rodents inhaling V_2O_5 indicated a dose-related increase in incidence of pulmonary/sinonasal epithelial hyperplasia and metaplasia. Epidemiological studies noted that acute and/or chronic exposure to moderate-to-high levels of V_2O_5 or vanadate in dusts/fumes resulted in increased localized fibrotic foci and lung weights, and an enhanced incidence of lung cancer initiated by other agents.

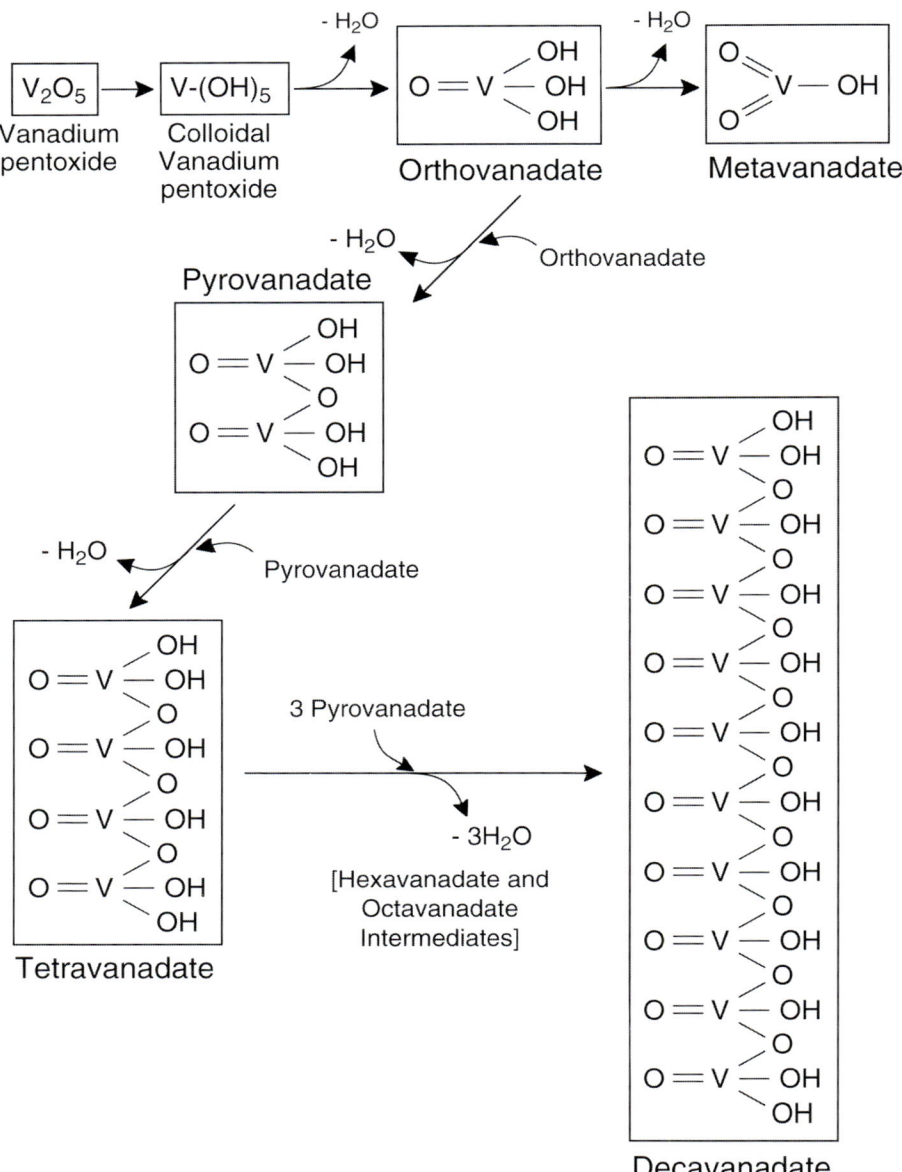

Vanadium and the Immune System. Figure 1 Formation of vanadates and higher polymers from vanadium pentoxide.

Apparently, vanadium might not act as a direct carcinogen, but rather it exerts secondary (▶ immunosuppressive) toxicities in hosts, allowing initiated cancers to progress to neoplasms. It should be noted that certain vanadium compounds have also been shown to act as anticarcinogens. When given to rodents bearing Erlich ascites or liver tumors, vanadocene displays cancerostatic activity. In addition, certain vanadium compounds in the form of dietary supplements have been shown to block cancer induction by other known carcinogenic agents. While the mechanisms of the anticarcinogenic activity in these studies are not clear, it should be noted that the effect observed with vanadocene is not unique; similar results have been obtained using metallocene complexes with other transition elements.

Putative Immune System Interactions and Immunotoxicities of Vanadium Agents

In immune system cells (as with all cell types), vanadate ions are able to enter the cytoplasm through the channels utilized by phosphate and chromate anions;

insoluble forms of vanadium enter through pinocytic uptake. Once in the cell (Figure 2), vanadate is rapidly acted upon by cellular reductants (such as NAD(P)H, glutathione, ascorbate, catechols) and converted to tetravalent vanadyl species.

Unlike other toxic metal oxyanions (such as chromate), vanadyl can then either be bound to proteins *or* can be readily oxidized back to the vanadate ion. As it is thought that only pentavalent vanadium can exit the cell, this represents a means of detoxification. However, shuttling back and forth between oxidation states also represents a means for retoxification. Not only are levels of cellular reductants depleted during shuttling events, but pentavalent vanadium gives rise to its own toxic effects, including generation of reactive oxygen species, inhibition of enzymes in nuclear and cytoplasmic processes, and alterations in the ▶ phosphorylative balance of several proteins secondary to inhibition of cellular phosphatases.

As either vanadate or pentoxide, vanadium has been shown to alter immunological responses in humans and experimental animals since the early 1900s. Its immunosuppressive effects became apparent long after its initial pharmacological use as an immune enhancer. Workers exposed to atmospheric vanadium had increased occurrences of coughing spells, tuberculosis, and respiratory tract irritation. Postmortem examinations of these workers revealed extensive lung damage; the primary cause of death was most often respiratory failure secondary to bacterial infection. Later studies demonstrated that acute exposure to high, and/or chronic exposure to moderate, levels of vanadium-bearing dust or fumes resulted in higher incidence of several pulmonary diseases, including asthma, rhinitis, pharyngitis, pneumonia, and bronchitis. Detailed cytologic studies with cells from these exposed workers noted vanadium-induced disturbances in neutrophil and plasma cell numbers, immunoglobulin production, and lymphocyte mitogenic responsiveness.

Vanadium-induced changes in human immunological function are reproduced in animal models. Subchronic and acute exposures of rodents to pentavalent vanadium agents have been shown to alter

- mitogen-induced lymphoproliferation
- alveolar/peritoneal macrophage phagocytosis and lysosomal enzyme activity or release
- host resistance to bacterial endotoxin (LPS) and intact microorganisms
- lung immune cell populations
- *in situ* induction of interferon-γ (IFN-γ) and interleukin(IL)-6 by polyinosinic-polycytidilic acid
- mast cell histamine release.

Exposure also produced pathological alterations in immune system organs (Peyer's patches, thymus, and spleen). Studies with macrophage cell lines or mice exposed to NH_4VO_3 noted decreased surface levels of Fc-receptors for immunoglobulin binding and diminished production and activity of tumor necrosis factor(TNF)-α and IL-1α. The latter study also showed that in the absence of exogenous stimuli, vanadium-exposed macrophages released significantly greater amounts of inflammatory prostaglandin E_2 than did untreated controls.

Studies of host resistance to infection, by *Listeria monocytogenes* after acute/subchronic vanadate exposure reported that resident peritoneal and alveolar macrophage function, and consequently cell-mediated immunity, were adversely affected. At the sites of infection, bacterial numbers increased rapidly, but no increase in macrophage or neutrophil numbers occurred. Macrophages recovered from vanadate-treated mice displayed decreased capacities to phagocytoze opsonized *Listeria* or to kill those few organisms ingested. These defects were thought attributable to vanadium-induced disturbance in cell superoxide anion formation, ▶ glutathione redox cycle activity, and hexose-monophosphate shunt activation, events critical to maintaining energy for phagocytosis/intracellular killing.

While precise mechanisms underlying the immunomodulatory effects of vanadium are not yet clear, in vivo and in vitro studies have begun to yield information to enable hypothetical mechanisms to be proposed. In macrophages (as well as other cell types) vanadate ions have been shown to:

- disrupt microtubule and microfilament structural integrity
- induce alterations in local pH due to vanadate polyanion formation
- modify lysosomal enzyme release and activity
- alter secretory vesicle fusion to lysosomes
- disrupt cell protein metabolism at both the level of synthesis and catabolism
- modulate both the inducibility and magnitude of reactive oxygen intermediate formation/release.

Though these structural/biochemical changes may directly contribute to immunomodulation, they may have underlying roles in a reduced ability of vanadium-exposed cells to interact with, and respond to, signaling agents during an immune response. Disrupted endocytic delivery of surface receptor-ligand complexes to lysozomes, subsequent complex dissociation, and receptor recycling/*de novo* receptor synthesis, can diminish the magnitude of macrophage cytokine-induced and/or antigen-induced responses. Along these lines, studies have indicated that macrophage priming by T-lymphocyte-derived IFNγ was adversely affected by vanadium exposure. Activation of cellular protein kinase A (PKA) or protein kinase C (PKC) and in-

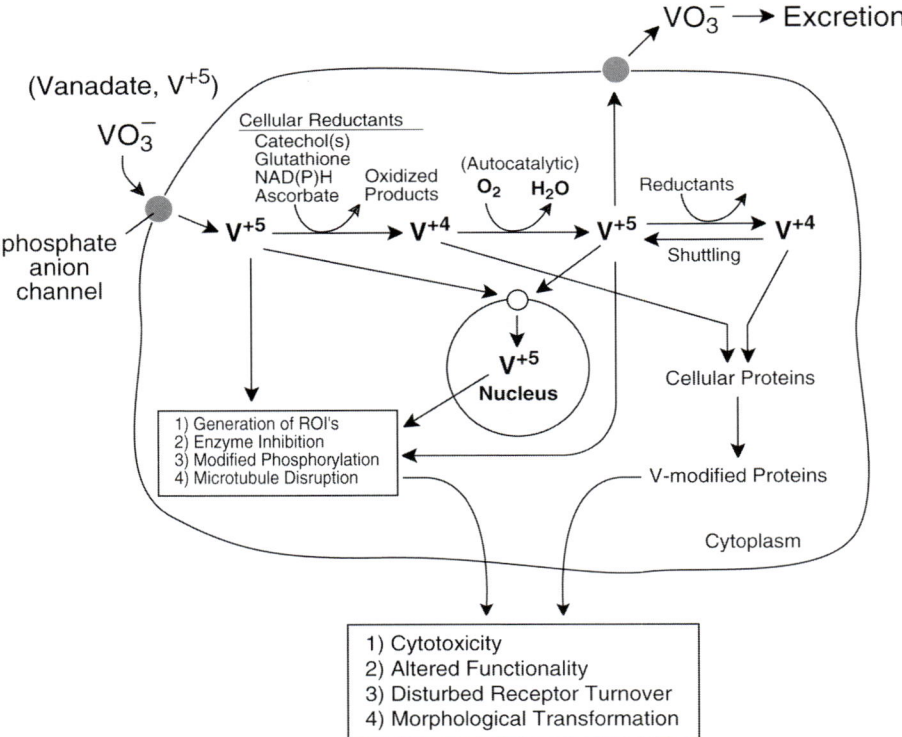

Vanadium and the Immune System. Figure 2 Metabolism of vanadate after entry into cell and select toxic manifestations that may result.

creases in intracellular calcium concentrations, events that can result in a downregulation of IFNγ receptor (IFNγR) expression, have been observed in macrophages harvested from mice, as well as in J774 murine macrophage cultures treated with vanadium. In mouse WEHI-3 macrophages, both the levels of two classes of surface IFNγR and their binding affinities for IFNγ were greatly modified by vanadium treatment. In these cells, IFNγ-inducible responses (including, enhanced calcium ion influx, MHC class II antigen expression, and zymosan-inducible reactive oxygen intermediate formation) were diminished secondary to the changes in IFNγR expression/binding activity.

Though these indirect mechanisms may clearly be a means by which decreased expression of surface receptors occurs on vanadium-exposed macrophages, it is also believed that vanadium may directly modify receptor proteins themselves (via interactions with amino acid R-groups in various regions of the proteins). It has also been hypothesized that modified receptor responses may be due to induced changes in cellular protein phosphorylation/dephosphorylation states secondary to modifications in the activities of cell phosphatases. Prolonged phosphorylation of receptor proteins (or cytokine-induced second messengers) might induce false states of cell activation and, as a result, cytokine receptor modification in the exposed cells. Similarly, prolonged phosphorylation of proteins could also lead to bypass of normal signal transduction pathways and subsequent activation of cytokine DNA response elements that, in turn, lead to downregulated cytokine receptor expression/function.

In view of the evidence to indicate that vanadium compounds are immunomodulating (primarily at the level of the macrophage; see Figure 3) and the increased concern regarding potential exposures of human populations, vanadium has now been included as a US EPA Superfund target inorganic chemical.

Relevance to Humans

Vanadium is one of the more ubiquitous trace metals in the environment. Since clays and shales can contain > 300 ppm, coals up to 1% vanadium (by weight), and petroleum oils 100–1400 ppm depending on site of recovery, fossil fuel combustion is the most identifiable source for delivery of vanadium-bearing particles into the atmosphere. Ambient air levels of vanadium vary from 0.02% by weight in soil-derived aerosols, to 0.02%–0.20% in automobile-derived fumes and 0.54%–0.82% in oil combustion-generated aerosols, depending on the region under study. Typical rural vanadium levels are 0.25–75 ng/m^3 while urban set-

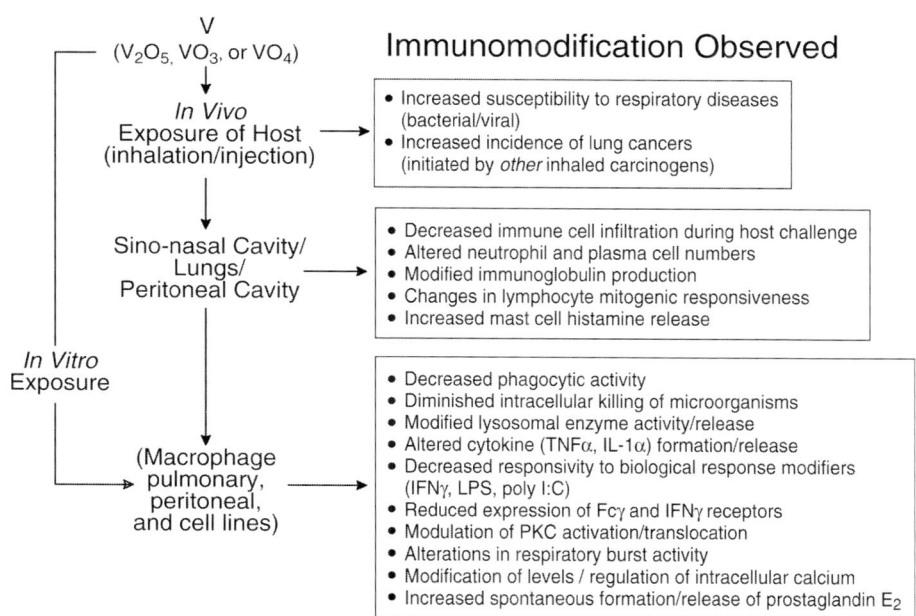

Vanadium and the Immune System. Figure 3 Immunomodulatory events known to occur following *in vivo* or *in vitro* exposure to pentavalent vanadium agents.

tings are usually higher (60–300 ng/m^3); on average, ambient vanadium concentrations in cities are often several µg per m^3. Seasonal variations (winter urban air vanadium levels are 6-fold greater than summer levels) arise from increased combustion of vanadium-bearing oils, shales, and coals for heat and electricity. At these levels ($\approx$ 50 ng/m^3) and based on experimental inhalation studies it is estimated that $\approx$ 1 µg vanadium enters the average adult human lung each day. Clearance of vanadium from the lungs depends on solubility of the agent inhaled. With insoluble V_2O_5 or more soluble vanadates the initial clearance is fairly rapid, with $\approx$ 40% of both chemical classes cleared within 1 h of intake. However, significant amounts of the cleared material can enter the systemic circulation and give rise to absorption levels of 50%–85% of an inhaled dose (depending on agent solubility). After 24 h the two forms diverge in ability to be cleared, with the insoluble form persisting longer. Thus, total clearance of vanadium is never achieved, with 1%–3% of the original dose persisting as long as 65 days or more. As a result, lung vanadium burden can increase with length of time spent in contaminated environments.

Exposure to vanadium also readily occurs via oral ingestion. Levels of vanadium are higher in freshwater than in seawater (0.3–200 vs 1–3 µg/l, respectively), due mostly to saline-induced precipitation of vanadium ions. Municipal water concentrations are usually < 10 µg/l, so drinking water is not considered an important source. Food represents the primary source of noninhaled vanadium intake in both humans and animals. As vanadium levels in most natural foodstuffs are only several parts per billion, daily human dietary intake is estimated to be from 0.01–2 mg/day. Unlike copper, lead, and tin, whose increased presence in consumable products arises from purposeful supplementation or from product-induced container leaching, the major contaminating source of foodstuffs by vanadium is soil. After oral intake of vanadium-contaminated water, soil, or foodstuff (or by swallowing vanadium-bearing sputum), absorption from the gastrointestinal tract is low (average 0.1%–1%), irrespective of the parent compound. There appears to be greater intestinal uptake of vanadium in younger animals than in adults, possibly due to greater nonselective permeability of the immature intestinal barrier. Oddly, exposure via noninhalational routes (e.g. *per os*) can still give rise to increased lung vanadium burdens and toxicities.

Irrespective of route of entry, vanadium that does enter circulation is preferentially distributed to the kidney, liver, blood, and bone. Though each have their own clearance mechanisms and kinetics, it appears that the site for long-term vanadium retention is bone. Because bone acts as a repository for vanadium, possible effects on hematological endpoints, including immune cell development and function, are great.

Though it has been suggested that vanadium is an essential element for chickens and rats, its essentialness in humans and most other animals is still not clear. For now, it appears that only in certain plants (e.g. marine algae), bacteria (e.g. nitrogen-fixing *Azotobacter*), fungi (e.g. *Amanita* species) and a few lower life forms (e.g. tunicate ascidians *A. nigra*, *A. ceratodes* and the fan worm *Psedopotamilla occelata*) does vanadium have some demonstrable function in the host's biochemical life processes.

References

1. Cohen MD (1998) Vanadium. In: Zelikoff J, Thomas P (eds) Immunotoxicology of Environmental and Occupational Metals. Taylor and Francis, London, pp 207–229
2. Cohen MD (2000) Other metals: aluminum, copper, manganese, selenium, vanadium, and zinc. In: Cohen MD, Zelikoff J, Schlesinger RB (eds), Pulmonary Immunotoxicology. Kluwer Academic, Norwel, pp 267–300
3. Zelikoff J, Cohen MD (1997) Metal immunotoxicology. In: Massaro EJ (ed) CRC Handbook on Human Toxicology, CRC Press, Boca Raton, pp 811–852
4. ATSDR (1991) Toxicological Profile for Vanadium and Compounds. US Public Health Service, Atlanta
5. Cohen MD (2004) Pulmonary immunotoxicology of select metals: Aluminum, arsene, cadmium, chromium, copper, manganese, nickel, vanadium, and zinc. Journal of Immunotoxicology I (I):39–69

Variable Region (V Region)

The segments of immunoglobulin heavy or light chains that vary in sequence in chains of the same allotype and isotype, i.e. responsible for antigen specificity.

▶ B Lymphocytes
▶ Rabbit Immune System

Variable Surface Glycoprotein (VSG)

Surface coats (synonym 'glycocalyx') are present in pathogenic protozoan and helmintic endoparasites. The body of parasites is covered with variable surface glycoproteins (VSGs), which are anchored through a glycosyl phosphatidyl inositol lipid (GPI-anchor). VSGs are highly immunogenic. Regular changing of the VSG results in different variant antigen types (VATs), an immune escape process of the parasite called antigenic variation.

▶ Trypanosomes, Infection and Immunity

Variance, Analysis of

A set of procedures where the total variability from a set of observations in an experiment is partitioned into those that account for the systematic or treatment effect, and those that account for chance or random factors.

▶ Statistics in Immunotoxicology

Variant Antigen Type (VAT)

The highly immunogenic variable surface glycoprotein (VSGs) are at any one point of time of identical structure in individual trypanosomes and within the majority of a population in a host resulting in a specific variant antigen type. Only very few individual parasites express different VATs which are selected for when the immune response of the host will eliminate all trypanosomes covered by the major VAT.

▶ Trypanosomes, Infection and Immunity

Vasculitis

An inflammatory reaction of a blood vessel or a lymph vessel. Vasculitis may occur in many different sites. Tissue damage starts by complement activation by immune complexes sticking at these sites.

▶ Hypersensitivity Reactions
▶ Systemic Autoimmunity

Vasculopathy

Any disease of blood vessels.
▶ Systemic Autoimmunity

Vasoactive Amine

A molecule released from mast cells, basophils, and platelets that induce contraction of endothelium and smooth muscle (examples: histamine and 5-hydroxytryptamine).

▶ Serotonin

VDJ

Rearrangement and joining of the V (variable), D (diversity), and J (joining) gene segments is responsible for the vast potential repertoire of different heavy chain variable regions in the B cell receptor.
▶ B Lymphocytes

VDJ Region

The genetic code for immunoglobulin variable domain J.
▶ Animal Models of Immunodeficiency

Viability, Cell

ANDREA ENGEL
BD Biosciences
Life Science Research
Tullastr. 8–12
D-69126 Heidelberg
Germany

Synonyms
Percent of living cells, live rate, live-death discrimination.

Short Description
Cellular viability indicates the proportion of live cells in a population. The methods for determination of cell viability are based on different parameters. These parameters are the membrane integrity, the physiological status e.g. enzyme activity, othe electrochemical gradient between intact cell compartments or the capacity of proliferation. A more indirect, exclusive parameter is the measurement of cell proliferation by labeling of the dividing DNA molecules in viable cells.

Characteristics
Measurable characteristics of cell viability, are:
- the integrity of the cell membrane
- the physiological status of the cell
- electrochemical gradients
- the capacity of proliferation.

These provide the basis for several types of assays for cell viability (summarized in Table 1), which can be monitored by colorimetric detection, fluorescence detection (microscopy or flow cytometry) or radioisometric detection. This chapter summarizes the methods for discrimination of living, damaged, and dead cells. Methods for monitoring apoptosis are covered in another section.

Membrane Integrity
Proof of the membrane integrity can be evaluated using a number of dyes that specifically label dead or damaged cells. All these probes accumulate in dead or damaged cells and do not enter viable ones. The use of fluorescent DNA binding probes such as propidium iodide (PI), 7-amino-actinomycin D (7-AAD) and TO-PRO-3 without thiazol orange in a flow cytometric method is well established. This method is simple, and the dead and damaged cells are clearly distinguishable from the viable ones.

7-AAD and TO-PRO-3 allow somewhat more flexibility in combination with fluorochromes other than PI. These methods are applicable to bacteria, mammalian cells, protozoa and yeasts (2,4,5,12). Identification of living cells is afforded by thiazolorange (TO) or several of its derivatives, such as the SYBR probes, YOYO-1 or TOTO-1 (molecular probes). Combination of 'live cell probes' with 'dead cell probes' such as TO or SYBR with PI or 7-AAD thus allows definition of viable, injured and dead cells (2). An example of a flow cytometric analysis is shown in Figure 1. The use of the TO derivatives YO-PRO-1 and TO-PRO-1 has advantages because they do not affect the proliferative capacity of the viable cells. In addition, these probes show a higher DNA affinity and are also valid for the detection of apoptotic cells. An example for dye combination is the use of ethidium bromide (EB) with acridine orange (AO). EB penetrates only dead or injured cells due to the loss of their membrane integrity and stains these cells red. Viable cells appear green caused by AO.

For evaluation using fluorescence microscopy, several different fluorescent probes are available (4). For example, the combination of AO and EB is widespread. A very fast and uncomplicated means of live–dead discrimination is afforded using light microscopy. This direct observation is primarily used before further cell culture. Morphologic changes can be determined as the simplest criterion. Trypan blue is a commonly used dye (6). It is actively excluded by viable cells with intact cell membranes. Nonviable cells retain the dye and are stained. Another dye, erythrosin B is used in a similar way, but is not that widespread in use.

The detection of intracellular enzymes like DNase or trypsin in the cellular environment can also be used to indicate the existence of dead cells. Also the penetration of cytoplasmic markers (like antitubulin, anticytokeratin-specific antibodies) can prove the presence of damaged plasma membranes (4).

Viability, Cell. Table 1 Summary of probes and detection systems which can be used for the determination of viable and non-viable cells

	Analyte	Cells detected	Detection system
Membrane integrity	PI, 7-AAD, EB, TO-PRO-3	Nonviable	Fluorescence (e.g. FCM, IF)
	TO, SYBR derivatives, YOYO-1, TOTO-1, YO-PRO-1, TO-PRO-1	All (viable and nonviable)	Fluorescence (e.g. FCM, IF)
	Escape of intracellular enzymes	Nonviable	Colorimetric (e.g. ELISA)
	Penetration of cytoplasmic markers	Nonviable	Fluorescence (e.g. IF, FCM)
Enzyme activity	Fluorescein diacetate, BCECF, calcein AM, dihydroethidium, MTT, XTT, Wst-1, Wst-8	Viable	Fluorescence (e.g. FCM, ELISA) Luminescence
Electrochemical gradient	Rhodamine 123, DiOC$_3$	Viable	Fluorescence (e.g. FCM, IF) Colorimetric (e.g. ELISA)
	Oxol	Nonviable	Fluorescence (e.g. FCM, IF) Colorimetric (e.g. ELISA)
Proliferation	^{3}H-thymidine	Viable, proliferating	Radioactive (liquid scintillation counter)
	Bromodeoxyuridine	Viable, proliferating	Fluorescence (e.g. FCM, IF) Colorimetric (e.g. ELISA)

7-AAD, 7-amino-actinomycin D; BCECF, 2',7'-bis-(2-carboxyethyl)-5-(and-6)- carboxyfluorescein; EB, ethidium bromide; ELISA, enzyme-linked immunosorbent assay; FCM, flow cytometry; IF, immunofluorescence microscopy; PI, propidium iodide; SYBR, SYBR (molecular probes); TO, thiazol orange; TOTO, TOTO-3 iodide (molecular probes); Wst-1, 4-[3-(4-iodophenyl)-2-(4-nitrophenyl)-2H-5-tetrazolio]-1,3-benzene disulfonate; Wst-8, 2-(2-methoxy-4-nitrophenyl)-3-(4-nitrophenyl)-5-(2,4-disulfophenyl)-2H-tetrazolium; YOYO-1, YOYO-1 iodide (molecular probes).

Physiological Status

Viability can be assessed by verification of a specific cell function, e.g. an enzyme activity. In most cases, this measurement is also dependent on the membrane integrity, in so much as it influences the availability of, or retention of, the probe. Here one uses a lipid-soluble probe that readily crosses the membrane, and which is nonfluorescent, e.g. fluorescein diacetate (FDA). The activity of cellular ▶ esterases in viable cells converts the probe to a highly fluorescent form (in this case, free fluorescein). Viable cells retain the fluorochrome and are strongly fluorescent; nonviable ones are dim or nonfluorescent. FDA is used with bacteria, protozoa, phytoplankton, plant cells, yeasts and mammalian cells (4). One variant of this dye, 2',7'-bis-(2-carboxyethyl)-5-(and-6)- carboxyfluorescein (BCECF) (Molecular Probes), is excluded from the cells in an energy-dependent manner and the retention in the viable cells is therefore optimized. Carboxyfluorescein and calcein AM work in the same way. The long retention time and the small pH sensitivity of calcein AM are favorable. Dihydroethidium is taken up by viable cells and cleaved by esterases to an ethidium monomer, which binds to cellular DNA and causes their red fluorescence. Dead cells remain unstained. Using this dye the viable ▶ intracellular parasites can also be identified by flow cytometry (4). The choice of the optimal probes depends upon the cell system. Differences both in uptake efficiency and in the retention time should be taken into consideration.

Metabolic activity can also be evaluated using different water-soluble tetrazolium salts (e.g. 3-(4,5-dimethylthiazol-2-yl)-2,5-diphenyltetrazolium bromide (MTT); sodium 3´-[1-(phenylaminocarbonyl)-3,4-tetrazolium]-bis (4-methoxy-6-nitro) benzene sulfonic acid hydrate (XTT); 4-[3-(4-iodophenyl)-2-(4-nitro-

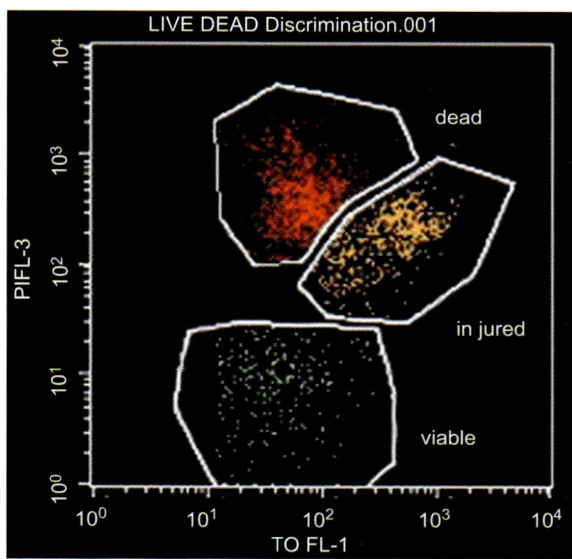

Viability, Cell. Figure 1 A bacterial sample was stained using the BD Cell Viability Kit (BD Biosciences) with thiazolorange (TO) and propidium iodide (PI) and analyzed on a BD FACSCalibur. All cells are TO-positive. Viable cells are only TO-positive; injured ones TO-positive and weakly PI-positive; dead cells are positive for both markers (2).

phenyl)-2H-5-tetrazolio]-1,3-benzene disulfonate (Wst-1); 2-(2-methoxy-4-nitrophenyl)-3-(4-nitrophenyl)-5-(2,4-disulfophenyl)-2H-tetrazolium (Wst-8)). These are cleaved by the metabolic activity of mitochondria into water-soluble colored formazanes. The experiments are easily and quickly performed in a microtiter plate format. The measurement system is, however, strongly influenced by the assigned cell number and the cell culture environment (e.g. pH, D-glucose concentration) (7).

Electrochemical Gradients

Viable cells maintain characteristic electrochemical gradients across their plasma membrane. The maintenance of the pH and other ion gradients is often affected by the loss of viability. Probes typically used to measure electrochemical gradients are lipophilic and charged. They thus concentrate in particular subcellular compartments, depending on the relative membrane potential. Rhodamine-123 (a cationic lipophilic probe) accumulates in mitochondria. Cells with active mitochondria are stained bright green, but dead or damaged cells remain unstained because of the loss of the ion gradient of their mitochondrial membrane. Rhodamine-123 is used to assess cell viability for bacteria, yeast and mammalian cells (4). When using rhodamine-123 and some other comparable probes, one should be aware of the fact that certain cell types actively exclude these by a glycoprotein pump, thus leading to weaker staining. An additional technical tip: in the presence of glutathione, cells may have hyperpolarized mitochondria, which causes unspecific staining of dead or damaged cells. The presence of aliphatic side-chains supports the probe uptake and its retention time in the cell. One example is the cyanine probe DiOC6. It accumulates more strongly in the mitochondria and works in the same way as rhodamine-123. All the dyes mentioned here could be combined with DNA labeling probes, which detect nonviable cells (e.g. PI or EB combined with rhodamine-123).

Lipophilic anionic probes stain nonviable cells. One example, oxol, has been used for determinations in bacteria and protozoa (4, 12). The loss of the negative potential in comparison to the environment leads to the accumulation of oxol in dead cells. Some caution in interpretation is required, in light of a publication reporting PI-negative and oxol-positive bacteria (8).

Proliferation

Proliferation is another indication of cell viability. The ability to form colonies in vitro, after plating at low density, is one of the oldest, but lengthiest procedures. More often, a direct measurement of new DNA synthesis is used as a criterion of cell growth. Here, analogues of the DNA base deoxythymidine, e.g. radioactive ^{3}H-thymidine or BrdU, are added to the culture and are incorporated into the DNA in place of thymidine in cells undergoing DNA synthesis (13, 14). The level of ^{3}H-thymidine incorporation is determined in a liquid scintillation counter. The incorporation of bromodeoxyuridine (BrdU) is detected with an anti-BrdU antibody. The incorporation can be visualized either in a cellular immunoassay (ELISA) or on a single cell basis by fluorescence microscopy or flow cytometry. The most appropriate method for measuring cellular viability depends on the test conditions and the question to be addressed. Many considerations that are relevant to the choice have been discussed above. In addition, safety considerations may require that the cells have to be fixed prior to analysis. The methods described thus far are not applicable to fixed cells. Several modifications of the same types of measurement have, however, proven to be suitable for use when subsequent fixation is required. For example, ethidium monoacid (EMA) is positively charged and labels dead cells by DNA intercalation. Exposure of the samples to visible light causes cross-linking of the probe with the DNA so that the surplus dye can be washed out before cell fixation. A combination of PI and Hoechst is also convenient. PI incubation prior to fixation labels the dead cells. After ethanol fixation, cells are counterstained with Hoechst. PI, which is only present in the dead cells, ▶ quenches the Hoechst fluorescence and labels these cells red. Styryl-751

(LDS-751) penetrates into viable and nonviable cells. Dead and damaged cells are substantially more intensely labeled than living ones. Cross-linking of the DNA and the dye is not necessary.

Relevance to Humans

The regulation of cellular viability is an important criterion in the evaluation of in vitro and in vivo experiments. The determination of cellular viability and growth is critical, for example, in the evaluation of the effect of cytostatics or antibiotics, in drug-screening for the development of therapeutics, in the search for optimal fermentation conditions for protein secretion or bioproduction, or the evaluation of mutagenic, carcinogenic and cytotoxic characteristics of different substances.

Regulatory Environment

Various guidelines and draft guidelines for cell viability testing are available.
- TGA (Therapeutic Goods Administration) Guidelines for sterility testing of therapeutic goods, 2002
- EPA (Environmental Protection Agency) Health effects test guidelines OPPTS 870.5300: Detection of gene mutations in somatic cells in culture
- EPA (Environmental Protection Agency) Health effects test guidelines
- OPPTS 870.5375; In vitro mammalian cytogenetics
- Draft guidelines on skin corrosivity
- OECD Guideline 431, March 2001
- OECD Guideline 404, June 2001
- Draft guidelines on phototoxicity
- OECD Draft Guideline 432, March 2003 Thanks to Dr. Pamela Schu-Werner for her assistance with corrections in expression and spelling in English.

References

1. Alvarez-Barrientos AJ, Cynton R, Nombela C, Sanchet-Perez M (2000) Applications of flow cytometry to clinical microbiology. Clin Microbiol Rev 13:167–195
2. Alsharif R, Godfrey W (2002) Bacterial detection and live/dead discrimination by flow cytometry. Microbial cytometry application note. BD Biosciences Immunocytometry Systems, San Jose
3. Alsharif R, Tapia M, Godfrey W, Wanalund J, Nagar M (2002) Bacterial disinfectant efficacy using flow cytometry. Microbial cytometry application note. BD Biosciences Immunocytometry Systems, San Jose
4. Current protocols in flow cytometry, UNIT 9.2 Assesment of cell viability
5. Graham JK (2001) Assessment of sperm quality: a flow cytometric approach. Anim Reprod Sci 68:239–247
6. Johnson JE (1995) Methods for studying cell death and viability in primary neuronal cultures. Methods Cellular Biology 46:243–276
7. Mosmann TR (1983) Rapid colorimetric assay for cellular growth and survival: application to proliferation and cytotoxicity assays. J Immunol Methods 65 (1–2):55–63
8. Nebe-von Caron G, Badley RA (1996) Bacterial characterization by flow cytometry. In: Al-Rubeal M, Emery AN, eds. Flow cytometry applications in cell culture. Marcel Dekker, New York, pp 257–290
9. Nebe-von Caron G, Stephens PJ, Badley RA (1999) Bacterial detection and differentiation by cytometry and fluorescent probes. Proc Royal Microbiol Soc 34:321–327
10. Nebe-von Caron G, Stephens PJ, Hewitt CJ, Powell JR, Badley RA (2000) Analysis of bacterial function by multi-colour fluorescence flow cytometry and single cell sorting. J Microbiol Meth 42:97–114
11. Shapiro HN (2000) Microbial analysis at the single-cell level: tasks and techniques. J Microbiol Meth 42:3–16
12. Sims PJ, Waggoner AS, Wang CH, Hoffman JF (1974) Studies on the mechanism by which cyanine dyes measure membrane potential in red blood cells and phosphatidylcholine vesicles. Biochemistry 13 (16):3315–3330
13. Steel GG (1977) In: Growth kinetics of tumours. Clarendon Press, Oxford
14. Takagi S et al. (1993) Detection of 5-bromo-2-deoxyuridine (BrdUrd) incorporation with monoclonal anti-BrdUrd antibody after deoxyribonuclease treatment. Cytometry 14:640
15. Thornton R, Godfrey W, Gilmour L, Alsharif R (2002) Evaluation of yeast viability and concentration during wine fermentation using flow cytometry. Microbial Cytometry Application Note. BD Biosciences, Immunocytometry Systems, San Jose

Vimentin

Belongs to the class III intermediate filaments, together with desmin and glial fibrillary acidic protein. Extracellular matrix protein (~ 54 kDa molecular mass) produced by fibroblasts in conective tissue with a function in maintaining the cell shape.
▶ Immunotoxic Agents into the Body, Entry of

Virulence

The degree of pathogenicity of a microorganism as indicated by invasiveness and mortality.
▶ Streptococcus Infection and Immunity

Vitamins

RENÉ CREVEL
Safety & Environmental Assurance Centre
Unilever Colworth
Sharnbrook, Bedford
K44 1LQ UK
UK

Synonyms

micronutrients, essential trace nutrients

Definition

Vitamins constitute a heterogeneous range of micronutrients which are essential to the maintenance of good health and, indeed, life itself in many cases. With some exceptions, they cannot be synthesized by the organism and must therefore be supplied by the diet. Vitamins are commonly grouped into fat-soluble and water-soluble types. Fat-soluble vitamins include vitamin A (retinol and its derivatives), vitamins D_2 and D_3 (ergocalciferol, calciferol), vitamin E (α-tocopherol), and vitamins K_1 and K_2 (phylloquinone, farnoquinone). Water-soluble vitamins include those of the B complex: vitamins B_1 (thiamine), B_2 (riboflavin), B_3 (niacin), B_6 (pyridoxine), B_{12} (cyanocobalamin, pantothenic acid, folic acid), and vitamin C (L-ascorbic acid).

Characteristics

Given their varied and crucial biological roles, vitamins might be expected to modulate the activity of the

Vitamins. Figure 1 a

Vitamins. Figure 1 b

Vitamins. Figure 1 c

immune system, and its functional correlates, including inflammation, resistance to infection, and tumor progression. Studies investigating such effects have, however, largely concentrated on a few vitamins. These include vitamin A and retinoids, vitamin D, vitamin E and vitamin C. Only a few studies, and in some cases none, have been undertaken into the effects of the other vitamins on the immune system. This section will summarize postulated activities and, where identified, mechanisms for each vitamin.

Vitamin A

Vitamin A is involved in the development of virtually all cells, with actions at a fundamental level (inhibition of potassium currents, PKC-associated signal transduction). In cells of the immune system, as well as in others, these actions are mediated through two types of receptor, the retinoic acid receptor (RAR), and the retinoid X receptor (RXR). These receptors belong to a family of nuclear receptors which also includes the vitamin D receptor (VDR). Vitamin A and retinoids drive the immune system towards a T helper type 2 cell (Th2) phenotype, inhibiting the development of the Th1 phenotype, an activity mediated through the RXR. Conversely, vitamin A deficiency leads to a Th1 phenotype, manifested by increased interferon-γ secretion and a downregulation of antigen-presenting cell activity, as well as reduced secretion of interleukin (IL-5). This was well illustrated by the development of a Th1 response to the parasite *Trichinella spiralis* instead of the normal Th2 response in vitamin A-deficient mice, accompanied by reduced antibody responses. The RAR is also involved in immune modulation by vitamin A, according to studies in mouse Th1 cell clones which revealed that retinoic acid inhibits IFN-γ production through the CD28 costimulatory pathway.

Retinol also acts as an important cofactor in T cell activation, apparently influencing the G0 to G1 transition and inducing a marked increase of proliferation of human peripheral blood mononuclear cells (PBMC). However, this effect requires special culture conditions to demonstrate, explaining the apparent contradiction with early findings that appeared to show no effect in standard lymphocyte proliferation experiments. The effect depends on the lymphoid organ from which the cells originate: $CD4^+$ and $CD8^+$ thymic cells, but only $CD4^+$ peripheral T cells respond.

The effects of vitamin A extend to the innate immune system. Vitamin A acetate was found to inhibit multiplication of tubercle bacilli in macrophages in vitro.

Vitamin D

The active form of Vitamin D is $1\alpha(OH)_2D_3$, which is produced by cytochrome P450-dependent metabolism of calciferol and ergocalciferol, and is also known as calcitriol. It was initially identified through its effect on calcium metabolism and bone formation, and is not a vitamin in the true sense, as it can be produced endogenously by the action of sunlight on the skin. More recent studies, including the discovery of the ubiquitous vitamin D receptor (VDR), have revealed that it plays a much more wide-ranging role, of which immunomodulatory effects form part.

Vitamin D modulates the function of both lymphocytes and accessory cells, including antigen-presenting cells, interacting with the cells through the VDR. Early studies showed that it downregulated activation of T helper/inducer T cells, but not suppressor T cells or B cells. Subsequent work found that it also modulated cytokine production in the PBMCs of both man and experimental animals, and altered monokine/cytokine production in vivo, particularly reducing secretion of tumor necrosis factor TNF-α. More recent findings indicate that the effect on lymphocytes is principally a skewing of the response away from a Th1 type towards Th2, with IFN-γ, IL-12, and TNF-α being particularly affected by suppression, the effect reflecting an enhancement of Th2 manifestations rather than suppression of those of Th1. Key recent findings suggest that the immunomodulatory effects of vitamin D may be mediated mainly through their influence on accessory cells, rather than directly on lymphocytes. Vitamin D inhibits maturation of dendritic cells and monocytes, resulting in low level expression of key markers of maturation (such as surface major histocompatibility complex (MHC) [class] II [antigens], CD40, CD25, and IL-12, but is not cytotoxic or cytostatic except at very high doses. Interestingly, IL-4 is reported to reverse the effect of vitamin D on monocyte differentiation. Vitamin D also enhances antibody- and complement-mediated phagocytosis.

Vitamin E

Vitamin E is a naturally occurring antioxidant, which is found in several isomeric forms, the most active of which is α-tocopherol. In vivo, it counteracts free radical damage and is particularly important to immune cells owing to the higher concentration of polyunsaturated fatty acids (PUFAs) in their membranes. Vitamin E is postulated to reduce prostaglandin E_2 (PGE_2) production, possibly through its action on cyclooxygenase 2 (COX-2), the expression of which is reduced in vitamin E-loaded macrophages from old mice to the level normally found in the macrophages of young mice.

Immune cells exposed to higher levels of vitamin E and tested in vitro secrete greater amounts of the cytokines IL-2, IFN, IFN-γ, while production of IL-6, IL-1, and TNF-α (proinflammatory cytokines) is reduced. Vitamin E has also been shown to inhibit programmed cell death due to T cell activation, through inhibition of CD95 ligand expression.

Vitamin K

Vitamin K (phylloquinone) plays a role in blood clotting, controlling the production of several coagulation factors. The main symptoms of deficiency are the consequences of coagulation defects (e.g. easy bruising, impaired clotting). The effects of vitamin K on the immune system have not been systematically examined, and only anecdotal results are available, without any exploration of mechanisms.

Vitamin B1

Vitamin B_1 (thiamine) is closely involved in intracellular metabolism, in particular the glycolytic pathway and Krebs cycle.

Vitamin B2

Vitamin B_2 (riboflavin) acts as an essential coenzyme in many oxidation-reduction reactions involved in carbohydrate metabolism. It is thus an important component in the maintenance of oxidant status. One of the signs of deficiency is cutaneous lesions.

Vitamin B3

Vitamin B_3 (niacin, nicotinic acid) also acts as a coenzyme in oxidation-reduction reactions, and deficiency results in severe consequences in many organ systems.

Vitamin B6

Vitamin B_6 (pyridoxine) participates in protein metabolism as the coenzyme in many enzyme systems. It is also involved in fat metabolism and in energy transformation in several tissues. Direct addition of vitamin B_6 to lymphocyte cultures indicates that it plays a role in the induction of serine hydroxymethyl transferase, an enzyme which is induced during mitogenic stimulation. Deficiency at the cellular level would therefore be expected to reduce any proliferation-dependent responses. One report suggests vitamin B_6 can bind to the CD4 cell surface receptor, but this has not been further explored.

Vitamin B12

Vitamin B_{12} was identified through its role in the prevention of pernicious anemia. Deficiency is also associated with neurological impairment. The biochemical defect appears to be in the conversion of deoxyuridylate to thymidylate, and also implicates folic acid. Early studies on lymphocytes from individuals with pernicious anemia showed reduced proliferative activity in response to mitogens.

Vitamin C

Vitamin C (L-ascorbic acid) was first identified for its anti-scorbutic activity, but numerous studies since have demonstrated its strong antioxidant activity. It influences a wide range of processes dependent on its oxidative-reduction properties. Specifically it is involved in the synthesis of collagen, carnitine and neurotransmitters, as well as in cholesterol metabolism. A study in which cultured cell lines were preloaded with dehydroascorbic acid, showed that it inhibited TNF-α activation of NFκB activation, and concluded that vitamin C can influence inflammatory, neoplastic, and apoptotic processes.

Preclinical Relevance
Vitamin A

Most animal studies on the effects of vitamin A on the immune system have focused on the effects of deficiency, which remains a problem for a significant proportion of the world's population. These studies have largely confirmed those mechanistic studies which indicated that deficiency produced a shift to a Th1 phenotype. An experimental model of infection-induced (*Staphylococcus aureus*) arthritis thus showed enhanced T cell responses, but not antibody B cell responses in the vitamin A-deficient rats. The disease itself was longer-lasting and more severe in those rats, and measures of innate immunity (complement activity and phagocytosis) were also depressed. In line with the Th1 bias, reduced antibody responses to T cell-dependent antigens, but not to T-independent ones such as LPS were noted in rats deficient in vitamin A, together with reduced natural killer cell and neutrophil activity. As might be expected, vitamin A deficiency affects secondary antibody responses as well as primary ones, as illustrated by the reduced response of human PBMC from tetanus toxoid immune donors in a vitamin A-deficient SCID mouse model.

In the mouse, deficiency leads to increased production of IFN-γ and IL-12, while supplementation with vitamin A led to the development of a Th2 profile with increased levels of interleukins IL-4, IL-5 and IL-10. Effect on the number of T cells and B cells was relatively modest. Other studies revealed that vitamin A deficiency reduced secondary (IgG) responses more than primary (IgM) ones, because of impaired clonal expansion of B cells, rather than reduced antibody production per cell.

Large excesses of vitamin A resulted in increased phagocytic activity in Kuppfer cells from rats, as well as increased prostaglandin E_2 and TNF-α secretion from those cells as well as peripheral blood mononuclear cells. Mice supplemented with vitamin A produced delayed-type hypersensitivity responses to *Mycobacterium bovis* immunization, whereas unsupplemented mice did not—even though their diet was adequate in vitamin A.

The effects of vitamin A have also been investigated in other species. Chicks showed reduced T lymphocyte proliferative responses with low vitamin A intakes and enhanced ones with high vitamin A intakes. Vitamin A supplementation of the diet of Holstein cows after the end of lactation led to transient increases in concanavalin A-stimulated lymphocyte proliferation. However the effects of supplementation post-partum and during lactation were more difficult to interpret, with both suppression and stimulation being observed under different conditions.

Vitamin D

The effects of Vitamin D on immune responses in experimental animals are consistent with findings from in vitro experiments, namely reduced immune activation—much of it attributable to reduced antigen presentation. Early experiments demonstrated a reduced antibody response to keyhole limpet hemocyanin (KLH), as well an attenuated delayed-type hypersensitivity response to 2,4-dinitrochlorobenzene (DNCB). Treatment of female mice with dendritic cells exposed to Vitamin D prolonged the survival of syngeneic male skin grafts, and reduced the clearance of injected male splenocytes. Consistent with those results, lymph nodes from VDR knockout mice were significantly larger than nodes from wild type mice, illustrating how vitamin D modulates immune activation.

Supplementation with vitamin D also demonstrates clinically significant effects, attenuating or completely suppressing the disease in animal models of allergic encephalomyelitis (EAE), type 1 diabetes, and transplant rejection.

Vitamin E

Deficiency of vitamin E in various species of animals (sheep, pigs, dogs, chicken) reduces a range of commonly assessed measures of immune function, including mitogen responses (T and B cells), IL-2 production, natural killer cell (NK) activity, antibody titer, plaque-forming cell response to sheep red cell immunization, and phagocytosis by neutrophils and, with some exceptions, macrophages. Vitamin E deficiency can also increase the virulence of Coxsackievirus B3 in mice, rendering an avirulent strain pathogenic. However, vitamin E deficiency can sometimes protect—SCID mice with good vitamin E status, but that were unable to mount an adaptive immune response, succumbed to infection with *Plasmodium yoelli*, whereas their vitamin E-deficient counterparts survived.

Vitamin E supplementation increases or restores various measures of immune responsiveness in aged experimental animals, mainly mice and rats—delayed type hypersensitivity (DTH) responses, mitogen responsiveness, IL-2 production. Rats fed a diet high in vitamin E diet showed increased numbers of $CD4^+$ cells in thymus, and an increased ratio of $CD4^+$ to $CD8^+$ (helper/inducer to suppressor/cytotoxic cells).

Functional measures of improved immune function following vitamin E supplementation have been demonstrated through reductions in mortality from, or increased resistance to, experimental infections in mice, and increased resistance to pathogens in chicken, sheep, and pigs. For instance, the influenza virus titers of mice supplemented with high levels of vitamin E were reduced to the same levels as those of young mice. In a murine AIDS model supplementation with 15 times the normal intake of vitamin E restored immune function, including mitogen responsiveness, NK activity, cytokine secretion pattern, and reduction of evidence of hyperactivity. The mechanism was postulated as antioxidant activity protecting against programmed cell death and viral replication.

Vitamin K

In early studies, water-soluble derivatives of vitamin K act as adjuvants for antibody production in mice to a soluble protein antigen, although they failed to boost the induction of delayed-type hypersensitivity. Dietary studies revealed a protective effect of vitamin K deficiency in tuberculosis, although blood levels were not measured to verify the deficient state. In chicken, vitamin K deficiency increased the mortality from infection with the parasite *Eimeria tenella*, and reduced the dose of parasite required. In rats, vitamin K and synthetic analogues are reported to increase the immune response, although no details are given.

Vitamin B1
A thiamine-deficient diet had little effect on the resistance of rats to a normally avirulent *Corynebacterium* infection, but severe debilitation might have masked any subtle signs. Another study showed no difference in survival after *Mycobacterium tuberculosis* inoculation between a diet deficient in the B-vitamin complex and a standard laboratory animal diet, while an excess of the complex decreased survival. However, the data do not permit a distinction between the individual components of the B-vitamin complex. In mice and guinea-pigs, thiamin deficiency induced by injection of an antagonist, resulted in thymus atrophy and inhibition of T cell-mediated responses.

Vitamin B2
Vitamin B_2 enhanced host resistance in mice to a variety of microorganisms, possibly through stimulation of innate immunity. At levels above those required to alleviate deficiency, it also protected against endotoxin and exotoxin shock, as well as infection by Gram-positive and Gram-negative bacteria.

Vitamin B3
Only one study examined the effect of vitamin B_3 on the immune system. It found enhanced mitogen-driven proliferation, as well as increased anti-sheep erythrocyte antibody levels, but a reduced DTH response to the hapten trinitrochlorobenzene.

Vitamin B6
The effects of vitamin B_6 on the immune status of mice and rats are consistent with reported in vitro effects on proliferative responses. Most of these effects have been investigated in the context of antagonism of vitamin B_6 activity by the ammonia caramel contaminant 2-acetyl-4(5)-(1,2,3,4-tetrathydroxybutyl)imidazole (THI). Principal findings from the most comprehensive studies include severe lymphopaenia, affecting particularly the T helper/inducer cell population, with accompanying decreases in mitogen-driven proliferative responses, as well as T-dependent antibody responses, and host resistance. One study confirms the effects on thymic lymphocytes, but reports increased proliferative activity of thymocytes (not splenic lymphocytes as in other studies) consistent with a reduction of the proportion of immature thymocytes. Vitamin B_6 deficiency induced by administration of an antagonist (4-deoxypyrindine) reduced the inflammatory and antibody responses to *Trichinella spiralis* in mice. A contrasting report relates that vitamin B_6 deficiency increased the IgE response to the soluble antigen dinitrophenyl-ovalbumin.

Vitamin B12
Limited data exist on the effect of vitamin B_{12} on immune parameters in animals. Deficiency produced an increased ratio of CD4 to CD8 lymphocytes in mice, with a higher proportion of the $CD4^+$ cells secreting IL-4 than IFN-γ. Consistent with these observations, serum IgE levels were increased, while IgG and IgM levels were reduced, as was the C3 component of complement. Similar findings were observed in rats, although observations were limited to CD4 to CD8 ratios and IgG, IgM and C3 levels. Very early host resistance studies in rodents failed to demonstrate any adverse effects of vitamin B_{12} deficiency, but this may have been due to difficulties in inducing sufficiently profound deficiency.

Vitamin C
Few animal models exist of vitamin C deficiency, because guinea-pigs, fruit-eating bats, and primates are the only species, apart from man, that do not synthesize vitamin C. Furthermore, in studies of deficiency, effects on the immune system are potentially very difficult to distinguish from the overall adverse effects on health.

Supplementation of the diets of mice and weanling pigs with substantial amounts of vitamin did not alter measures of immune function, such as lymphocyte proliferation, DTH responses, or antibody production.

Relevance to Humans
Vitamin A
Vitamin A deficiency remains a major public health problem in many countries, with impaired immune functioning well documented as one of the major effects, both in clinical trials and epidemiologic observations. Those studies reveal that the effects of vitamin A deficiency in people are largely consistent with findings in experimental animal and in in vitro studies. One study demonstrated underlying abnormalities in the T cell populations of vitamin A-deficient children, compared to those supplemented with vitamin A who had with higher CD4 to CD8 (helper/inducer to suppressor/cytotoxic) cell ratios, higher proportions of CD4 cells, and lower proportions of CD8 45RO cells. A study in children confirmed a Th1 cytokine pattern in vitamin A deficiency, accompanied by an increased number of NK cells. Interestingly, mortality was not related to vitamin A levels, suggesting that the benefits of the Th1 shift in this instance (increased activity against intracellular pathogens, viruses) may have outweighed the deleterious ones. However, in general, studies of vitamin A deficiency in HIV-in-

fected populations have shown that low serum vitamin A levels are associated with increased mortality, more rapid disease progression, and increased maternal-fetal transmission of disease. Also, because of the wide-ranging effects of retinol, deficiency can impair immunity by 'non-specific' mechanisms involving other cell types, such as inadequate repair of damaged mucosal surfaces, as well as by reducing the activity of cells of the innate immune system. A low serum provitamin A and carotenoid level was associated with an increased risk for heterosexual HIV acquisition in patients with sexually transmitted diseases, in a study in India.

In contrast to some of the equivocal effects on infections cleared handled by Th1 mechanisms, vitamin A deficiency clearly impairs immunity where Th2 responses are critical. Thus, children who suffered from vitamin A deficiency and received vitamin A supplements produced increased antibody responses to tetanus toxoid and measles. Similarly, supplementation with vitamin A reduced malarial febrile episodes, as well as spleen enlargement and parasite load, in a population exposed to the malarial parasite, although there was no consistent effect on the proportion infected or anemia in that population.

Vitamin D
Studies in man have shown that experimental findings in vitro and in experimental animals have clinical application. Thus HIV-infected patients have reduced levels of $1\alpha(OH)_2D_3$, and their chronic immune activation can be attenuated by administration of this compound. Vitamin D_3 analogues are also used in the treatment of psoriasis.

Epidemiological data also indicate that some autoimmune diseases (insulin-dependent diabetes mellitus, rheumatoid arthritis) are more common in regions of vitamin D deficiency, although a causal relationship remains to be demonstrated.

Vitamin E
Severe deficiency of vitamin E is generally rare, and has not been described to the same degree as for other vitamins. Nevertheless, low serum vitamin E levels have been associated with an increase in oxidative stress in HIV-infected individuals, and early studies showed vitamin E supplementation (twice normal intake) reduced AIDS progression, with beneficial effects on several immune parameters.

Aside from AIDS sufferers, many studies of vitamin E supplementation in elderly humans have been undertaken, largely to test the hypothesis derived from animal studies that vitamin E could retard age-associated immune function. Measures of immune function such as DTH, mitogen responses, and antibody response to vaccines (hepatitis B, tetanus toxoid) were used to evaluate the effects. The results generally suggest an improvement in immune function and reduction of inflammation, but are not sufficient to define an appropriate level of supplementation. Some epidemiological data support the view that vitamin E supplementation can restore immune function in the elderly or counteract its decline, but many of the studies are confounded by other factors, such as concurrent intakes of other vitamins, the health status of participants, and assessment of vitamin E status. Some intervention studies suggest that while low-dose supplementation boosts immune responses, very high doses may reduce it.

Vitamin K
No studies were found of the effect of vitamin K on the immune system in humans.

Vitamin B1
Vitamin B_1 (thiamine) intakes above the recommended daily allowance (RDA) are associated with improved survival of HIV-infected individuals, and slow progression to AIDS. Vitamin B_1 deficiency is associated with parasitemia, but the studies do not demonstrate any relationship between immune function and vitamin B_1 deficiency. More interestingly, an intervention study showed that administration of B-complex vitamins restored immune function (DTH, mitogen response) in surgical cancer patients.

Vitamin B2
An epidemiological study failed to identify a relationship between vitamin B_2 (riboflavin) status and parasitemia with *Plasmodium falciparum*.

Vitamin B3
Several studies have considered the effects of vitamin B_3 on human measures of immunocompetence, but always in combination with other vitamins. Thus supplementation reduced re-infection with some parasites in one study, while in another it improved weight gain in HIV-infected pregnant women. However, these outcomes only suggest a beneficial effect on immunity, in the absence of measures of immune function. In another study, niacin intakes were positively correlated with IgG levels in endurance-trained athletes. Interestingly, supplementation with very large doses of B vitamins, including niacin, in an elderly population was associated with reduced circulating lymphocyte counts, and no evidence of improved immune function.

Vitamin B6
Human studies indicate that vitamin B_6 affects lymphocyte maturation and differentiation, with reduced DTH and antibody responses. Thus, in HIV-infected individuals at an early stage of infection, reduced

vitamin B_6 levels were associated with reduced immune function, manifested by decreased mitogen responsiveness and NK cell activity. No relationship was found with lymphocyte population profiles or serum immunoglobulin levels. On the other hand, evidence does not indicate that supplementation well above normal requirements benefits immune responses.

Vitamin B12

Clinical or epidemiological studies on the effects of vitamin B_{12} on the immune system are limited. Consistent with the findings in rats and mice, a small human study reported increased CD4 to CD8 ratios against a reduction in total lymphocyte numbers, together with reduced NK cell activity. However serum immunoglobulin levels were unchanged, as were mitogen-driven proliferative responses. Vitamin B_{12} administration restored the altered parameters. In a prospective cohort study, HIV-infected individuals with low serum vitamin B_{12} concentrations developed AIDS faster than those with adequate serum levels. Another study found an association between low serum levels of vitamin B_{12} and infection with *Helicobacter pylori* in healthy individuals.

Vitamin C

Because of the severity of profound vitamin C deficiency, studies have been limited to a few investigations of the effects of moderate deficiency. Lymphocyte proliferation responses were not affected, although reductions in DTH responses, measured by skin testing, were observed.

Supplementation with vitamin C has been studied much more extensively, particularly in relation to the hypothesis of its beneficial activity against the common cold. Taken together these studies indicate that vitamin C supplementation does not reduce incidence, but reduces the severity and duration of the common cold. No mechanism has been identified to date, and there remain questions over the most appropriate dose. The studies also show that vitamin C may also protect against lower respiratory tract infections.

Other studies have investigated whether vitamin C could protect against the increased susceptibility to infection following intense exertion. Results are inconclusive. Some studies show a reduction in the incidence of upper respiratory tract infections (URTI) after racing, but a recent very carefully controlled study concluded that vitamin C did not alter post-race manifestations of oxidative stress or immunity. Incidence of post-race URTI was not reported, which may indicate that any beneficial effect of vitamin C may be mediated by non-immune mechanisms.

Another study of supplementation revealed a transient increase in NK cell activity and a reduction in proportion of cells undergoing apoptosis.

Regulatory Environment

There is no regulatory application of the effects of vitamins on the immune system.

References

1. Calder PC, Kew S (2002) The immune system: a target for functional foods? Br J Nutr 88 [Suppl 2]:S165–177
2. Erickson KL, Medina EA, Hubbard NE (2000) Micronutrients and innate immunity. J Infect Dis 182 [Suppl 1]: S5–10
3. Han SN, Meydani SN (2000) Antioxidants, cytokines, and influenza infection in aged mice and elderly humans. J Infect Dis 182 [Suppl 1]:S74–80
4. Lin R, White JH (2004) The pleiotropic actions of vitamin D. Bioessays 26:21–28
5. Meydani SN, Beharka AA (2001) Vitamin E and immune response in the aged. Bibl Nutr Diet 55:148–158
6. Stephensen CB (2001) Vitamin A, infection and immune function. Ann Rev Nutr 21:167–192

W

Waldeyer's Ring

A ring of lymphoid tissue in humans that is composed of the lingual tonsil, palatine tonsils and the nasopharyngeal tonsils.
▶ Mucosa-Associated Lymphoid Tissue

Warm Autoantibodies

Autoantibodies that react as well, or more strongly, at 37 °C than lower temperatures.
▶ Hemolytic Anemia, Autoimmune

Western Blot Analysis

GEORG BRUNNER
Fachklinik Hornheide an der Universität Münster
Dorbaumstrasse 300
D-48157 Münster
Germany

Synonyms

▶ Immunoblotting, ▶ electroblotting, protein blotting.

Short Description

Western blotting is a protein analysis technique which combines protein separation in an electric field with immunochemical methods of protein detection. It is used to identify, characterize, quantify, and/or isolate proteins or antibodies present in complex biological samples. First, proteins are separated according to their molecular size and/or net charge by electrophoretic migration through three-dimensional polyacrylamide gels. This is followed by the electrophoretic transfer of the proteins out of the gel onto a two-dimensional, protein-binding membrane support. Finally, proteins bound to the membrane are visualized by the binding of labeled, protein-specific antibodies or ligands.

Characteristics

Analogous to the membrane blotting techniques for DNA (Southern blotting) or RNA (Northern blotting), electrophoretic blotting and analysis of proteins on a membrane support was developed and termed Western blotting (1,2). Western blot analysis allows the identification of proteins in complex mixtures as well as their characterization, i.e. determination of quantity, molecular size, net charge, epitopes for antibody recognition, or ligand binding sites. It is also a particularly versatile technique to identify the presence, quantity, and specificity of antibodies, e.g. polyclonal antibodies present in serum samples.

The sensitivity of protein detection is in the range of 10–100 fmoles (1–10 ng of protein). The sensitivity is mainly determined by the specificity (signal-to-noise ratio) and sensitivity of the detection system, but also depends on the protein loading capacity of electrophoresis, the efficiency of protein transfer, and the concentration of the proteins of interest in the sample.

Western blotting involves a number of different steps (Figure 1), such as the separation of the proteins by electrophoretic techniques (I), the electrophoretic transfer of the separated proteins onto a membrane support (II), the blocking of free protein binding sites on the membrane (III), and the detection of the proteins of interest using specific antibodies or ligands (IV).

(I) Proteins in complex biological samples such as cell or tissue extracts are diluted into sodium dodecylsulfate(SDS)-containing gel electrophoresis sample buffer. Alternatively, cell or tissue samples can be extracted directly in sample buffer followed by solubilization of the proteins by boiling.

Electrophoresis is usually performed using discontinuous standard Tris/glycine SDS-polyacrylamide (▶ SDS-PAGE) gels (3), separating the proteins present in the sample based on their relative molecular size. However, several other techniques are possible and have been used to increase the resolution and versatility of electrophoretic protein separation. Thus, polyacrylamide gradient gels enhance the resolution of proteins in crude cell or tissue lysates over a much wider size range than standard gels of uniform concentration. Proteomics analysis requires maximum protein resolution, and complex mixtures of proteins

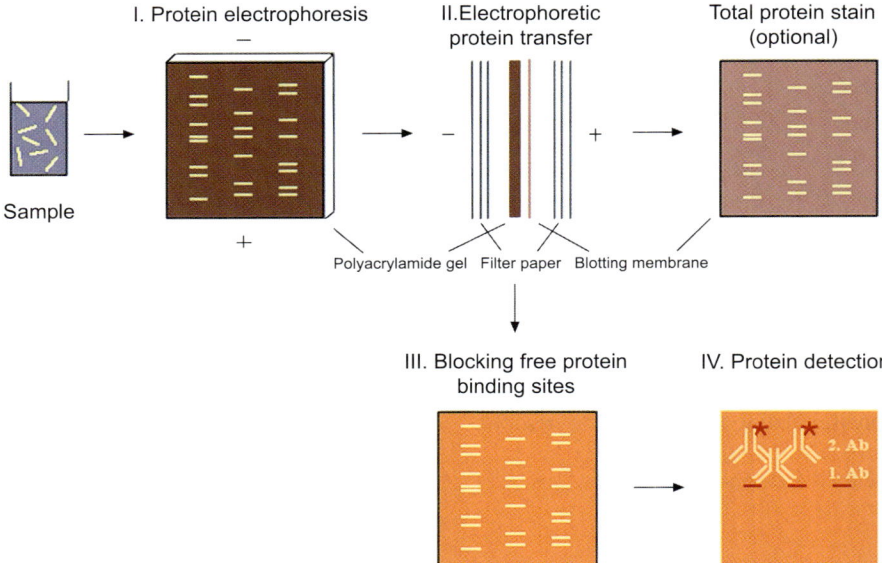

Western Blot Analysis. Figure 1 Western blot analysis.

are therefore subjected to ▶ 2-D gel electrophoresis which can easily resolve 1500 different proteins. Proteins are separated, in the first dimension, by isoelectric focusing according to their isoelectric point, followed by separation, in the second dimension, by SDS-PAGE on the basis of their molecular size.

(II) Following electrophoretic separation, proteins are transferred out of the 3-D polyacrylamide gel matrix onto a 2-D membrane support in order to render them accessible for antibody recognition or ligand binding (3). Whereas DNA and RNA are efficiently blotted by capillary diffusion, electrophoretic elution is the preferred technique for ▶ protein blotting. The membrane is placed in direct contact with the polyacrylamide gel in an electric field, which induces the migration of the SDS-loaded, negatively charged proteins out of the gel and onto the membrane. Alternatively, particularly when ultra-thin gels are being used, proteins can be transferred simply by diffusion (diffusion blotting). Protein transfer by diffusion is less efficient as compared to electrophoretic transfer, but efficiency can be significantly increased by raising the temperature to 70°C.

Typical ▶ blotting membrane materials are nitrocellulose, polyvinylidene difluoride (PVDF), and nylon, which vary in their physical properties and protein binding characteristics and capacity.

▶ Electrophoretic protein transfer can be performed by immersing the gel-membrane sandwich in a blotting chamber in an appropriate buffer to produce the desired electric field strength (wet transfer). Alternatively, protein elution can be achieved by placing the sandwich, between filter paper soaked with the respective transfer buffer, in direct contact with two plate electrodes (semi-dry transfer). While the latter procedure is considerably faster than the wet transfer, it is not very efficient for larger proteins (> 100 kDa) due to the limited time of transfer and electric field strength that can be applied.

The efficiency of the protein transfer depends on the pore size of the polyacrylamide gel and the efficiency of the protein binding to the blotting membrane. While small proteins (< 20 kDa) easily migrate out of the gel, their binding to the membrane is often inefficient and requires a membrane with a small pore size as well as methanol in the transfer buffer. Methanol, however, shrinks the gel, thus decreasing the pore size of the gel and further impairing protein transfer. On the other hand, large linear molecules, such as proteoglycans, are often difficult to transfer and require high electrophoretic field strength and cooling as well as extended transfer time periods (24–48 hours). Addition of SDS to the buffer enhances the transfer of proteins out of the gel but can interfere with protein binding to the membrane support.

Following transfer, the protein pattern blotted onto the membrane can be visualized, on a duplicate blot run in parallel, using protein stains such as india ink (irreversible) or Ponceau S (reversible).

(III) Following protein transfer, free protein-binding sites on the membrane support are blocked by incubation with a concentrated protein solution such as bovine serum albumin, non-fat dried milk, or casein, or with a detergent such as Tween 20. This prevents nonspecific interactions of the detection system with the membrane support during subsequent incubation steps.

(IV) Proteins of interest are most commonly detected via the recognition of antigenic epitopes by specific antibodies. This requires the epitopes to be resistant to denaturation. Polyclonal antisera are more likely to contain antibodies reacting with denatured antigens than monoclonal antibodies. To control for non-specific antibody-protein interactions, pre-immune or irrelevant antibodies of the same species are applied on a duplicate blot run in parallel. Alternatively, protein receptors blotted onto the membrane can be detected via the binding of specific ligands.

Binding to defined protein bands is visualized either by using directly labeled primary antibodies or ligands, or by incubating the membrane with unlabeled primary followed by labeled secondary antibodies. Antibodies are either radioactively labeled with ^{125}iodine (detection by radiography) or conjugated to enzymes such as alkaline phosphatase or horseradish peroxidase (detection by enzyme assay using chromogenic or chemoluminogenic substrates). The sensitivity of detection can be enhanced by labeling the secondary antibodies with biotin, followed by incubation with labeled streptavidin, thus amplifying the detection system.

Western blotting analysis can be combined with other techniques to enhance the sensitivity of protein detection or to purify antibodies. To enhance the sensitivity (i.e. the signal-to-noise ratio) the proteins of interest can be concentrated and enriched prior to electrophoresis. This is most commonly done by immunoprecipitation using specific antibodies bound to protein A-agarose beads. To affinity-purify antigen-specific antibodies, the respective protein bands can be excised, following incubation with a polyclonal antiserum, and bound antibodies are eluted from the membrane pieces at an acidic pH.

Pros and Cons

Western blot analysis combines the high resolution power of protein electrophoresis with the sensitivity and specificity of antibody-antigen or receptor-ligand interactions.

Sample preparation does not require protein purification, since crude biological samples can be subjected directly to SDS-PAGE. In addition, problems of protein insolubility or instability are overcome by extracting or diluting cells or tissue samples in SDS-containing electrophoresis sample buffer, followed by a boiling step which inactivates protease activities potentially present in the sample.

Samples can be selectively labeled for western blot analysis in order to allow for the specific analysis of defined protein fractions, for example by biotinylation of cell surface proteins prior to extraction.

Protein detection following electrophoretic transfer can be modified in order to visualize proteins bound to the membrane based on their functional activity, instead of via the recognition by specific antibodies. For example, blotted receptors can be detected by the specific binding of labeled ligands.

Proteins are usually denatured during sample preparation, electrophoresis, or electrophoretic transfer, which can interfere with their detection on the membrane and does not allow further functional characterization of the blotted proteins.

Although the signal obtained in western blot analysis correlates with the protein amount present on the membrane, at least over a certain range of protein concentration, detection is only semi-quantitative. Protein transfer efficiency out of the SDS-PAGE gel as well as protein binding to the membrane support are influenced by the size, the conformation, and other physical properties of the proteins and the membrane. In addition, the linear relationship between protein amount and signal obtained depends on the linearity of the detection system and is limited by the protein binding capacity of the membrane.

Detection of minor proteins in complex biological samples can be difficult due to non-specific hydrophobic or electrostatic protein-protein interactions of antibodies or ligands with other proteins present in the sample. In this case, partial purification of the proteins is required, for example by immunoprecipitation, protein chromatography, of cell/tissue fractionation prior to electrophoresis.

Predictivity

Western blotting allows the fast and sensitive qualitative and semi-quantitative analysis of complex protein mixtures as well as the identification of antigen-specific antibodies in serum. However, the results strongly depend on the specificity and sensitivity of the detection system as well as on the antigenic and physical properties of the proteins under investigation. Therefore, the preservation of antigenic epitopes or ligand binding sites, as well as the efficiency of protein transfer, need to be verified, and the detection system requires standardization to allow quantification of the results. Proteolytic cleavage of proteins in the sample prior to electrophoresis, resulting in protein fragments still containing the epitope for antibody recognition, can complicate the interpretation of the results.

Relevance to Humans

Western blot analysis is widely used as a tool for the analysis of specific proteins as well as for broader proteomic approaches (particularly in combination with 2-D electrophoretic techniques), allowing the qualitative and quantitative analysis of complex biological samples, e.g. crude cell and tissue extracts or plasma or serum samples. It can be used for protein screening, quantification, as well as for antibody isolation.

Western blotting has important clinical applications, like the confirmatory test for human immunodeficiency virus type 1 (HIV-1) infection. Viral proteins are separated electrophoretically, transferred onto a membrane support, and incubated with patient sera. Binding of serum antibodies to defined viral antigens is then visualized using Western blot analysis.

Regulatory Environment

Not applicable.

References

1. Burnette WN (1981) Western blotting: Electrophoretic transfer of proteins from sodium dodecylsulfate-polyacrylamide gels to unmodified nitrocellulose and radiographic detection with antibody and radioiodinated protein A. Anal Biochem 112:195–203
2. Towbin H, Staehelin T, Gordon J (1979) Electrophoretic transfer of proteins from polyacrylamide gels to nitrocellulose sheets: Procedure and some applications. Proc Natl Acad Sci USA 76:4350–4354
3. Laemmli UK (1970) Cleavage of the structural proteins during the assembly of the head of bacteriophage T4. Nature 227:680–685

White Pulp

Part of the spleen around central arterioles where lymphoid cells reside. Comprises three major compartments: the periarteriolar lymphocyte sheath or PALS, follicles and marinal zone.

▶ Spleen

WHO

World Health Organization. The WHO has responsibility for regulating vaccines.

▶ Immunotoxicology

X

Xenogeneic

This term describes the genetic relationship between individuals of different species, with respect to antigenicity.
▶ Idiotype Network
▶ Graft-Versus-Host Reaction

XRE-Element (Xenobiotic Response Element)

Conserved short DNA sequence in the promoter regions of many genes, XRE-elements are the binding site for the ligand complexed aryl hydrocarbon receptor. Elimination of the XRE-element from a promoter abrogates the inducibility of the respective gene by dioxins or other aryl hydrocarbon receptor agonists. Another name for the XRE is dioxin-responsive element, or DRE.
▶ Dioxins and the Immune System

XSCID

X-linked severe immunodeficiency syndrome resulting in the loss of B and T lymphocyte function because of the dysfunction of multiple interleukin receptors which share a common protein chain.
▶ Canine Immune System

YAC

A murine lymphoma cell line that is highly sensitive to rodent NK-mediated lysis.
▶ Cytotoxicity Assays

Z

Zap70 (Zeta-Associated Protein of 70 kDa)

Zap70 is a tyrosine kinase expressed in T cells that is activated through phosphorylation by Lck. It is involved in enhancing and propagating the activation signal from the TCR complex at the cell surface. It is responsible for phosphorylating and thus activating the adaptor proteins LAT and Slp 76.

▶ Signal Transduction During Lymphocyte Activation

Zucker Rat

▶ Diabetes and Diabetes Combined with Hypertension, Experimental Models for

Appendix

List of Entries

Essays are shown in bold

- 3'
- 5'
- **AB0 Blood Group System**
  GEOFF DANIELS, MARCELA CONTRERAS
- AB0 Histo-Blood Group System
- Abscess
- Acquired Immunity
- Acrocyanosis
- Activated Macrophages
- Activation-Induced Cell Death (AICD)
- Activator Surface
- Active Immunotherapy
- Active Lymph Pump
- Acute Graft-Versus-Host Disease
- Acute Inflammation
- Acute Lymphocytic Leukemia
- Acute Myelogenous Leukemia
- Adaptive Immune Response
- Adaptive Immunity
- Adaptors
- ADCC
- Adherens Junctions
- Adhesion Molecules
- Adoptive Transfer PLNA
- Adrenocorticotropic Hormone (ACTH)
- Adult Respiratory Distress Syndrome (ARDS)
- Advanced or Extended Histopathology
- Afferent Lymphatics
- Affinity Maturation of the Immune Response
- Aflatoxins
- Agglutination
- **Aging and the Immune System**
  ANNA MARIA WOLF
- Ah Receptor (AhR)
- Air Pollution
- Airborne Contagion
- Alexin
- Allelic Discrimination
- Allergen
- Allergen Hypothesis
- Allergic Contact Dermatitis
- Allergic Reactions
- Allergic Reactions to Drugs
- Allergic Rhinitis (Hay Fever)
- Allergy
- Alloantigens
- Allogeneic
- Allogeneic Determinants
- Alloreaction
- Alloreactive
- Allotransplantation
- Allotype
- Allotypic Epitopes
- Alternative Activation
- Alternative Pathway
- Ambient Air
- Amnestic (or Recall) Immune Response
- ANA
- Anaphylactic Shock (Anaphylaxis)
- Anaphylatoxin
- Anaphylaxis
- Anaplastic Large Cell Lymphoma
- Androgen
- Anemia
- Anemia Associated with Immune Response
- Anergic
- Anergy
- Angioedema
- Angiogenesis/Angiostasis
- **Animal Models for Respiratory Hypersensitivity**
  JÜRGEN PAULUHN

▶ **Animal Models of Immunodeficiency**
  KENNETH L HASTINGS, SHUKAL BALA
▶ Ankylosing Spondylitis
▶ Anterior (Head) Kidney
▶ Anthracene
▶ Anti-Cancer Antibodies
▶ Anti-DNA Antibodies
▶ Anti-Double Stranded (ds) DNA Antibodies
▶ Anti-Histone Antibodies
▶ Anti-Inflammatory Antibodies
▶ Anti-Inflammatory Cytokine
▶ **Anti-Inflammatory (Nonsteroidal) Drugs**
  PETIA P SIMEONOVA
▶ Anti-Single Stranded (ss) DNA Antibodies
▶ Anti-Tumor Immunity
▶ **Antibodies, Antigenicity of**
  EUGEN KOREN
▶ Antibody
▶ Antibody Class
▶ Antibody-Dependent Cell-Mediated Cytotoxicity (ADCC)
▶ Antibody-Dependent Cellular Cytotoxic (ADCC) Cells
▶ **Antibody-Dependent Cellular Cytotoxicity**
  JORGE GEFFNER
▶ Antibody-Dependent Cytotoxicity
▶ Antibody-Forming Cell
▶ Antibody-Forming Cell Assay
▶ Antibody Forming Cell Response
▶ Antibody Fragments
▶ Antibody Isotype
▶ Antibody Response to Therapeutic and Diagnostic Antibodies
▶ Anticytokines
▶ Antigen
▶ Antigen-Antibody Binding Assay
▶ Antigen-Dependent B Cell Development
▶ **Antigen Presentation via MHC Class II Molecules**
  FRANK STRAUBE
▶ Antigen-Presenting Cell (APC)
▶ **Antigen-Specific Cell Enrichment**
  KGC SMITH
▶ Antigenic Similarity
▶ Antigenic Variation
▶ Antigenicity

▶ Antigens, T Dependent and Independent
▶ **Antiglobulin (Coombs) Test**
  ANNE PROVENCHER BOLLIGER
▶ Antihistamines
▶ **Antinuclear Antibodies**
  MICHAEL HOLSAPPLE
▶ Antioxidant (Levels)
▶ Antiprotease
▶ Aorto-Gonadomesonephros Region (AGM)
▶ Ape
▶ **Apoptosis**
  SHIGEKAZU NAGATA
▶ Arachidonic Acid
▶ ARNT (Aryl Hydrocarbon Receptor Nuclear Translocator)
▶ Aroclor
▶ Arthritis Models
▶ Artificial Determinant
▶ Artificial Skin/Epidermis
▶ Aryl Hydrocarbon Receptor
▶ Aryl Hydrocarbon Receptor Nuclear Translocator (ARNT)
▶ Aspirin
▶ Aspirin-Like Drugs
▶ **Assays for Antibody Production**
  GREGORY LADICS
▶ **Asthma**
  MERYL KAROL
▶ Asthma Models
▶ Asymptomatic
▶ Atopic Allergy
▶ Atopic Dermatitis
▶ Atopy
▶ Attenuated Bacilli
▶ **Attenuated Organisms as Vaccines**
  KOERT J STITTELAAR
▶ **Autoantibodies, Tests for**
  JAN GMC DAMOISEAUX, JAN WILLEM COHEN TERVAERT
▶ Autoantibody
▶ Autoantibody Detection
▶ **Autoantigens**
  MICHAEL HOLSAPPLE
▶ Autoimmune Chronic Active Hepatitis
▶ Autoimmune Chronic Hepatitis
▶ Autoimmune Disease

▶ **Autoimmune Disease, Animal Models**
   DORI GERMOLEC
▶ Autoimmune Disorders
▶ Autoimmune Heart Disease
▶ Autoimmune Models
▶ Autoimmune Reaction
▶ Autoimmunity
▶ **Autoimmunity, Autoimmune Diseases**
   NOEL R ROSE
▶ Autologous
▶ Autoreactive Cells
▶ Avidity
▶ Azathioprine
▶ Azo Dyes
▶ AZT
▶ B Cell
▶ B Cell Antigen Receptor (BCR)
▶ **B Cell Maturation and Immunological Memory**
   CLAUDIA BEREK
▶ B Cell Receptor Complex
▶ B Cell Receptor Modification
▶ **B Lymphocytes**
   NORBERT E KAMINSKI, COURTNEY EW SULENTIC
▶ B Memory Cell
▶ B7.1 and B7.2
▶ Bacteremia
▶ Bacteremic Shock
▶ Bactericidal
▶ B(a)P
▶ Bcl-2 Interacting Domain (Bid)
▶ *Beige* Mouse
▶ Benzo-E-Pyrene
▶ Berylliosis
▶ Beryllium Disease
▶ Beryllium-Stimulated [or Beryllium-Specific Peripheral Blood?] Lymphocyte Proliferation Test (BeLPT)
▶ *bg/nu/xid* Mouse
▶ Bioaerosols
▶ Biologic-Response Modifiers
▶ Biologics
▶ Biotherapeutics
▶ Biotransformation
▶ Biphenotypic Leukemia

▶ **Birth Defects, Immune Protection Against**
   STEVEN HOLLADAY
▶ Blastogenesis
▶ Blood Cell Formation
▶ Blood Clotting
▶ **Blood Coagulation**
   KLAUS T PREISSNER
▶ Blood Coagulation and Fibrinolysis
▶ Blood Group System
▶ Blotting
▶ Blotting Membrane
▶ **Bone Marrow and Hematopoiesis**
   REINHARD HENSCHLER
▶ Bootstrap
▶ BPDE
▶ BP-7,8-diol
▶ BPQ
▶ Bronchitis
▶ Bronchus-Associated Lymphoid Tissue
▶ Buehler Test
▶ Buffy Coat
▶ Burkitt's Lymphoma
▶ C-Reactive Protein
▶ C3 Convertase
▶ C5 Convertase
▶ Cachectin
▶ CAMs
▶ **Cancer and the Immune System**
   JÖRG BLÜMEL
▶ Cancer Immunoediting
▶ Cancer Immunosurveillance
▶ Cancer-Testis Antigens
▶ Cancer Vaccine
▶ **Canine Immune System**
   MARK WING
▶ **Carcinogenesis**
   I BERNARD WEINSTEIN
▶ **Cardiac Disease, Autoimmune**
   NOEL R ROSE
▶ Cardiac Output (CO)
▶ Cardiomyopathy
▶ Carrier
▶ CAS Number 17646-01-6-
▶ Caspase
▶ CD (Cluster of Differentiation)
▶ **CD Markers**

H Zola, B Swart
- CD Molecule
- CD3
- CD4
- CD4⁺/CD8⁻
- CD4⁺ T Cells
- CD8
- CD28
- CD40 Ligand
- CD45RO
- **Cell Adhesion Molecules**
  Kris Vleminckx
- Cell Adhesion Receptors
- Cell-Based Bioassays
- Cell-Mediated Allergic
- Cell-Mediated Hypersensitivity, Type IV Immune Reaction
- Cell-Mediated Immunity
- **Cell-Mediated Lysis**
  B Paige Lawrence
- **Cell Separation Techniques**
  Mario Assenmacher
- Cell Sorting
- Cellular Immune Reactions
- Central Tolerance
- Centroblast
- Centrocyte
- CFA
- CFU-Meg
- Chagas Disease
- Chemical Allergen
- **Chemical Structure and the Generation of an Allergic Reaction**
  David Basketter
- Chemoattractants
- Chemokine
- Chemokine Receptor Antagonists
- Chemokine Receptors
- **Chemokines**
  Rafael Fernandez-Botran
- Chemotactic Cytokines
- Chemotaxis
- **Chemotaxis of Neutrophils**
  Lasse Leino
- Chimera
- Chip Array
- Chlorobiphenyl
- $C_{62}H_{111}N_{11}O_{12}$
- Chromate
- **Chromium and the Immune System**
  Mitchell D. Cohen
- Chromium Release Assay
- Chromophore
- **Chronic Beryllium Disease**
  Sally S Tinkle, Ainsley Weston
- Chronic Graft-Versus-Host Disease
- Chronic Inflammation
- Chronic Inflammatory Disease
- Chronic Lymphocytic Leukemia
- Chronic Myelogenous Leukemia
- Chronic Obstructive Pulmonary Disease
- Classical Pathway
- Clonal Deletion
- Clonal Expansion
- Clonotypic Antibodies, T Cell
- Clophen
- Cluster Determinant (CD)
- Cluster of Differentiation (CD)
- Co-Cultured
- Co-Stimulation
- Cold Agglutinins Disease
- Cold Autoantibodies
- Colony Forming Unit (CFU)
- **Colony-Forming Unit Assay: Methods and Implications**
  Hava Karsenty Avraham, Byeong-Chel Lee, Shalom Avraham
- Committed Precursor of Megakaryocytes
- Common Chain
- Common Cold
- Competitive PCR
- Complement
- **Complement and Allergy**
  Jean F Regal
- Complement Cascade
- **Complement Deficiencies**
  Michelle Carey
- Complement Fixation Test
- Complement Fragments
- **Complement System**
  Jean F Regal
- Complementarity-Determining Region

- (CDR)
- ▶ Complementary DNA (cDNA)
- ▶ Complete Freund's Adjuvant
- ▶ Completely Randomized Design
- ▶ ConA
- ▶ Concanavalin A (ConA)
- ▶ Concordance
- ▶ Conditional Gene Expression
- ▶ Conditioning
- ▶ Connective Tissue Diseases
- ▶ Connective Tissue Mast Cells
- ▶ Constant Regions (C Regions)
- ▶ Consumption
- ▶ Contact Dermatitis
- ▶ **Contact Hypersensitivity**
  SHAYNE COX GAD
- ▶ Contact Inhibition
- ▶ Contact Photoallergy
- ▶ Coombs Test
- ▶ COPD
- ▶ Coreceptor Competition
- ▶ Coreceptor of the TCR
- ▶ Cornifying/Cornification
- ▶ Coronavirus
- ▶ Corrosive
- ▶ Cortex
- ▶ Corticosteroid-binding globulin (CBG)
- ▶ Corticotrophin-Releasing Hormone
- ▶ Corticotropin-Releasing Hormone/Factor
- ▶ Covariance, Analysis of
- ▶ COX
- ▶ COX Inhibitors
- ▶ Coxsackievirus
- ▶ CpG Motifs
- ▶ CPMP
- ▶ CREST
- ▶ Criteria Air Pollutants
- ▶ Cross-Reactivity
- ▶ Crossover Design
- ▶ Crossreactivity
- ▶ Cryopreservation of Immune Cells
- ▶ CTL
- ▶ CTL Activity
- ▶ CTMC
- ▶ Cutaneous Anaphylaxis, Passive (PCA)
- ▶ Cyclooxygenase (COX)

- ▶ **Cyclosporin A**
  P ULRICH
- ▶ Cytochrome P450s
- ▶ Cytogenetics
- ▶ Cytokeratin
- ▶ **Cytokine Assays**
  CURTIS C. MAIER
- ▶ **Cytokine Inhibitors**
  KLAUS RESCH
- ▶ Cytokine Network
- ▶ **Cytokine Polymorphisms and Immunotoxicology**
  BERRAN YUCESOY, VICTOR J JOHNSON, MICHAEL I LUSTER
- ▶ Cytokine Receptor
- ▶ Cytokine Receptor Complexes
- ▶ **Cytokine Receptors**
  MICHAEL U. MARTIN
- ▶ **Cytokines**
  MARIANNE NAIN, DIETHARD GEMSA
- ▶ Cytomegalovirus
- ▶ Cytoskeleton
- ▶ Cytostatic
- ▶ Cytotoxic Activity
- ▶ Cytotoxic T Cell
- ▶ Cytotoxic T Lymphocyte (CTL) Assay
- ▶ **Cytotoxic T Lymphocytes**
  B PAIGE LAWRENCE
- ▶ Cytotoxic T Lymphocytes (CTL)
- ▶ Cytotoxicity
- ▶ **Cytotoxicity Assays**
  ELIZABETH R GORE
- ▶ 2-D Gel Electrophoresis
- ▶ 3-D Human Skin/Epidermal Equivalents
- ▶ DAT
- ▶ DBPCFC
- ▶ Death-Inducing Signaling Complex
- ▶ Decavanadate
- ▶ Decoy Receptor
- ▶ Defensin
- ▶ **Delayed-Type Hypersensitivity**
  HANS-WERNER VOHR
- ▶ Delayed-Type Hypersensitivity (DTH)
- ▶ Delayed Type IV Hypersensitivity
- ▶ Dendritic Cell
- ▶ Deoxyribonucleic Acid (DNA)

- ▶ Depletion
- ▶ **Dermatological Infections**
  S. Hanneken, Norbert J. Neumann
- ▶ Dermis
- ▶ Desaturation
- ▶ Desmosomes
- ▶ **Developmental Immunotoxicology**
  John Barnett
- ▶ Dexamethasone
- ▶ DHA
- ▶ **Diabetes and Diabetes Combined with Hypertension, Experimental Models for**
  PA van Zwieten
- ▶ Diethylstilbestrol (DES)
- ▶ Differentiation
- ▶ DiGeorge Syndrome
- ▶ Dilated Cardiomyopathy
- ▶ 7,12-dimethylbenz-a-anthracene
- ▶ DIOC18$_3$
- ▶ Dioxin Response Elements (DRE)
- ▶ Dioxins
- ▶ **Dioxins and the Immune System**
  Charlotte Esser
- ▶ Direct Antiglobulin Test
- ▶ Direct Coombs test
- ▶ Disseminated Intravascular Coagulation (DIC)
- ▶ DMBA
- ▶ DNA and RNA Blotting
- ▶ DNA-Derived Products
- ▶ DNA Fingerprinting
- ▶ **DNA Vaccines**
  Deborah L Novicki
- ▶ Docosahexanoic Acid
- ▶ Dog
- ▶ Dolichos Biflorus
- ▶ Dominant Negative
- ▶ Dominant Peptides
- ▶ Draining Lymph Node
- ▶ Drug Allergies
- ▶ Drug Hypersensitivity Syndrome
- ▶ Drug-Induced Hypersensitivity
- ▶ Drug-Induced Hypersensitivity syndrome (DIHS)
- ▶ **Drugs, Allergy to**
  Tetsuo Shiohara
- ▶ DTH
- ▶ Duffy Antigen
- ▶ ECSA
- ▶ Ectopic Lymphoid Tissue
- ▶ Eczema Herpeticum
- ▶ Edema
- ▶ Effector Cells
- ▶ Eicosanoids
- ▶ Eicosapentanoic Acid
- ▶ Electroblotting
- ▶ Electrochemoluminescent Immunoassay (ECLIA)
- ▶ Electrophoresis
- ▶ Electrophoretic Protein Transfer
- ▶ Elicitation Dose
- ▶ ELISA
- ▶ ELISPOT
- ▶ ELISPOT Assay
- ▶ ELR
- ▶ Embryonic Stem (ES) cell
- ▶ Emphysema
- ▶ *Encapsulatus*
- ▶ Encephalopathy
- ▶ Endocrine Disrupters
- ▶ Endocrine Disrupting Chemicals
- ▶ Endocytosis
- ▶ Endoplasmic Reticulum
- ▶ Endotoxin
- ▶ Endotoxin Shock
- ▶ Enhanced, Extended or Advanced Histology/Histopathology
- ▶ Enhanced Histological Assessment
- ▶ Enhancer
- ▶ Enrichment
- ▶ Enteramine
- ▶ Enterocytes
- ▶ Enterotoxin
- ▶ Environmental Estrogens
- ▶ Environmental Protection Agency (EPA)
- ▶ Enzyme-Linked Immunosorbent Assay
- ▶ Enzyme-Linked Immunosorbent Assay (ELISA)
- ▶ Enzyme-Linked Immunospot
- ▶ **Enzyme-Linked Immunospot Assay (ELISPOT)**
  Gernot Geginat

- Eosinophilia
- Eosinophilia–Myalgia Syndrome (EMS)
- Eotaxin
- EPA
- Epidemics
- **Epidemiological Investigations**
  ANDREW HALL
- Epidermal Cells
- Epidermis
- Epinephrine (Adrenaline)
- Epitope
- 9,10-epoxide
- Epstein–Barr Virus (EBV)
- Erythema Multiforme (EM)
- Erythrasma
- Erythroderma
- Erythroid Colony Stimulating Activity
- Erythropoiesis Stimulating Factor
- **Erythropoietin**
  RACHEL R. HIGGINS, YAACOV BEN-DAVID
- ESF
- Essential Elements
- Essential Fatty Acids
- Essential Trace Nutrients
- Esterases
- Estrogen
- Europium
- Evaluation of Humoral Immunity
- Evans Syndrome
- Ex vivo
- Experimental Design
- **Exposure Route and Respiratory Hypersensitivity**
  B JEAN MEADE, KIMBERLY J FAIRLEY
- Expression Profiling
- Extracellular Matrix
- Extravasation
- Extravascular Hemolysis
- Extrinsic Control, Neural and Humoral
- FACS
- Facultative Anaerobes
- Fas
- Fats
- **Fatty Acids and the Immune System**
  PARVEENN YAQOOB
- FBS

- FC
- Fc Region
- FcR
- FcRγ Chain
- FDA
- Fetal Bovine Serum (FBS)
- FEV
- Fibronectin
- Fibrosis
- Field Studies
- **Fish Immune System**
  BETTINA HITZFELD
- Fixed Drug Eruptions (FDE)
- Flow Cytometry
- **Flow Cytometry Technique**
  DANIELLE ROMAN
- Fluorescence Activated Cell Sorter
- Fluorescence-Activated Cell Sorting
- Fluoroimmunoassay (FIA)
- Follicular Dendritic Cell (FDC)
- Food Allergies
- **Food Allergy**
  AC KNULST
- Food and Drug Administration (FDA)
- Formylated Peptides
- Framework Regions (FR)
- Freund's Complete Adjuvant
- Fusion Proteins
- G Proteins
- Gamma Interferon Activation Sites (GAS)
- Gastric Mucosa
- Gastroenteritis
- Gene Conversion (GC)
- Gene Expression
- Gene Expression Analysis
- Gene Profiling
- Gene-Targeted Mouse
- Genetic Polymorphism
- Genetic Predisposition
- Genetic Susceptibility
- Genetic Vaccines
- Genetically Defined Rodents
- Genetically Engineered Mouse
- Genetically Modified
- Genomic DNA
- Germ Center

- **Germinal Center**
  C FRIEKE KUPER
- Germinal Center Reaction
- Germinal Centers
- Global Gene Expression Analysis
- Glomerulonephritis
- Glucocorticoid Receptors
- **Glucocorticoids**
  BOB LUEBKE
- Glucocorticoids and Stress
- Glucocorticosteroids (Glucocorticoids)
- Glucortisol
- Glucose Tolerance Factor (GTF)
- Glutathione Redox Cycle
- Glycosyltransferase
- Goblet Cells
- Gonadal Hormones
- Goodpasture's Syndrome
- Graft-Versus-Host Disease
- **Graft-Versus-Host Reaction**
  MICHAEL HOLSAPPLE
- Gram-Negative Bacteria
- Granulocyte
- Granulocyte-Macrophage Colony-Stimulating Factor (GM-CSF)
- Granuloma
- Granulopoiesis
- Granzymes
- Green Fluorescent Protein (GFP)
- Guanine-Exchange Factor
- Guidelines in Immunotoxicology
- **Guinea Pig Assays for Sensitization Testing**
  HANS-WERNER VOHR
- Gut-Associated Lymphoid Tissue (GALT)
- H Antigen
- H-Chain and L-Chain
- H-2IA
- H-2IE (Mouse)
- HAART
- Haemopoiesis
- Hairy Cell Leukemia
- Haplotype
- Hapten
- **Hapten and Carrier**
  HANS ULRICH WELTZIEN
- Hapten Theory
- Hassall's Bodies
- Helper T Cells (Th Cells)
- **Helper T Lymphocytes**
  ARATI KAMATH
- Hematopoiesis
- Hematopoietic Growth Factors (HGF)
- Hematopoietic Progenitor Cell (HPC)
- **Hematopoietic Stem Cells**
  REINHARD HENSCHLER
- Hematopoietic Stem Cells (HSC)
- Hemoflagellates
- Hemoglobinuria, Paroxysmal Nocturnal
- **Hemolytic Anemia, Autoimmune**
  ANNE PROVENCHER BOLLIGER
- Hemolytic Disease of the Newborn (HDN)
- Hemolytic Plaque Assay
- Hemolytic Transfusion Reaction
- Hemopoiesis
- Hemostasis
- HEPA Filtration
- **Hepatitis, Autoimmune**
  NEIL R. PUMFORD PH.D, KATHLEEN M. GILBERT PH.D
- Hereditary Spherocytosis
- Heterophilic Adhesion
- Heterotypic Adhesion
- Hexacoordinate
- Hexavalent Chromium
- Hexose–Monophosphate Shunt
- Histocompatibility
- Histoincompatible
- Histones
- **Histopathology of the Immune System, Enhanced**
  C FRIEKE KUPER
- HIV Co-Receptors
- HLA-DP
- HLA-DQ
- HLA-DR (Muman)
- Hodgkin's Disease
- Hodgkin's Lymphoma
- Homeostatic Control
- Homogeneity/Heterogeneity of Variance
- Homologous Recombination
- Homophilic Adhesion
- Homotypic Adhesion

- Homozygous
- Hormone
- HOSEC
- Host Defense Systems
- Host Resistance
- **Host Resistance Assays**
  H Van Loveren, H-W Vohr
- HSC
- HSE
- 5-HT
- HuCD4 Mouse
- Human Leukocyte Antigens (HLA)
- Human Pathogen
- Human Skin Recombinants
- **Humanized Monoclonal Antibodies**
  Joel B Cornacoff, Jill Giles-Komar
- Humanized Monoclonal Antibody
- Humanized Mouse
- Humoral Autoreactivity Assays
- Humoral Immune Assay
- Humoral Immune Function
- Humoral Immune Response
- Humoral Immune System
- **Humoral Immunity**
  Courtney EW Sulentic, Norbert E Kaminski
- Humoral-Mediated Immunity
- Hu-SCID Mouse
- Hybridization
- Hybridoma
- Hydralazine
- 5-hydroxytryptamine
- Hygiene Hypothesis
- Hyperchimeric
- Hypergammaglobulinemia
- Hyperplasia
- Hyperreactions
- **Hypersensitivity Reactions**
  Hans-Werner Vohr
- Hypersensitivity Reactions to Drugs
- Hypothalamic-Pituitary-Adrenal (HPA) Axis
- Hypothalamus
- Hypovolemia
- Hypoxia-Inducible Factor HIF1-β
- ICH
- ICICIS
- ICS
- Idiopathic Chronic Active Hepatitis
- Idiopeptides
- Idiosyncratic Drug Reactions
- Idiosyncratic Reaction
- Idiotype
- **Idiotype Network**
  Hilmar Lemke
- Idiotypic Cross-Reactivity
- Idiotypic Epitopes
- IFN-γ
- Ig V-Region
- IgE
- **IgE-Mediated Allergies**
  Werner Pichler
- IMHA
- Immediate-Type Hypersensitivity
- Immediate-Type or Delayed-Type Hypersensitivity
- **Immune Cells, Recruitment and Localization of**
  Bernhard Moser, Mariagrazia Uguccioni
- Immune Competence
- Immune Complexes and Complement Activation
- Immune Interferon
- Immune-Mediated Heart Disease
- Immune-Mediated Hemolytic Anemia (IMHA)
- **Immune Response**
  Michael U Martin, Klaus Resch
- Immune Response to Cancer
- Immune Surveillance
- Immunity
- Immunization
- Immunoabsorbent
- Immunoassay
- **Immunoassays**
  Danuta J Herzyk
- Immunoblotting
- Immunochemical-Based Assays
- Immunocompetence
- Immunocompetent
- Immunodeficiency
- Immunodeficient Animal

- ▶ Immunogenicity
- ▶ Immunoglobulin
- ▶ Immunoglobulin Class Switching
- ▶ **Immunoglobulin. Subclasses and Functions**
  DAVID SHEPHERD
- ▶ Immunohistochemical Staining
- ▶ Immunological Memory
- ▶ Immunological Synapse
- ▶ Immunological Unresponsiveness
- ▶ Immunomodulation
- ▶ Immunonutrition
- ▶ Immunopathology
- ▶ Immunopharmacology
- ▶ Immunophenotyping
- ▶ Immunopoiesis
- ▶ Immunopotentiation
- ▶ Immunoreceptor Tyrosine-Based Activation Motif
- ▶ Immunosenescence
- ▶ Immunosuppression
- ▶ Immunosuppressive
- ▶ Immunoteratology
- ▶ Immunotox
- ▶ **Immunotoxic Agents into the Body, Entry of**
  GEORG KRAAL, JANNEKE N SAMSOM
- ▶ Immunotoxic Intermediates
- ▶ Immunotoxicant
- ▶ Immunotoxicity
- ▶ **Immunotoxicological Evaluation of Therapeutic Cytokines**
  PETER T THOMAS
- ▶ **Immunotoxicology**
  DENNIS K FLAHERTY
- ▶ Immunotoxicology, Definition of
- ▶ **Immunotoxicology of Biotechnology-Derived Pharmaceuticals**
  GARY J. ROSENTHAL
- ▶ Immunotransmitters
- ▶ Impetigo Contagiosa
- ▶ In utero Immunotoxicology
- ▶ In vitro Culture
- ▶ In vitro Engineered Skin/Epidermal Substitutes
- ▶ In vivo Immunotoxicology Testing
- ▶ Inbred Strains
- ▶ Inbreds
- ▶ Indirect Coombs Test
- ▶ Indirect Immunotoxicity
- ▶ Infection Models
- ▶ Infectious Agents Models
- ▶ Inflammation
- ▶ Inflammatory Cytokines
- ▶ Inflammatory Heart Disease
- ▶ Inflammatory Macrophages
- ▶ **Inflammatory Reactions, Acute Versus Chronic**
  MICHAEL I LUSTER
- ▶ Influenza
- ▶ Influenza Virus
- ▶ Ingestion
- ▶ Innate Immune Response
- ▶ Innate Immune System
- ▶ Innate Immunity
- ▶ Intercellular Adhesion Molecule-1 (ICAM-1)
- ▶ **Interferon-γ**
  RAFAEL FERNANDEZ-BOTRAN
- ▶ Interferon (IFN)
- ▶ Interferon-Inducible Protein 10 (IP-10)
- ▶ Interferon-γ Receptor (IFNγR)
- ▶ Interferons
- ▶ Interleukin (IL)
- ▶ Interleukin-1 Receptor Accessory Protein
- ▶ Interleukin-1 Receptor Antagonist
- ▶ Interleukin-1 Receptor Associated Kinase I
- ▶ Interleukin-1 Receptor Associated Kinase II
- ▶ **Interleukin-1β (IL-1β)**
  DOROTHY COLAGIOVANNI
- ▶ Interleukin-1F2 (IL-1F2)
- ▶ Interleukin-4 (IL-4)
- ▶ Interleukin-5 (IL-5)
- ▶ Interleukin-8 (IL-8)
- ▶ Interleukin-12 (IL-12)
- ▶ Interleukin-18 (IL-18)
- ▶ Internalization
- ▶ **International Collaborative Studies on the Detection of Immunotoxicity**
  ANTHONY D DAYAN, HENK VAN LOVEREN, RALPH J SMIALOWICZ, IAN KIMBER
- ▶ Intracellular Adhesion Molecule
- ▶ Intracellular Cytokine Staining (ICS)
- ▶ Intracellular Parasites
- ▶ Intracellular Staining by Flow Cytometry

- Intraepithelial Lymphocytes
- Intraluminal Pressure and Flow
- Intravascular Hemolysis
- Ionophores
- IP-10
- Irritant
- Irritant Contact Dermatitis
- Isolation
- Isotype
- ITAM
- ITIM
- JAKS
- Janus Kinase (JAK)
- Jun NH2-Terminal Kinase
- Juvenile Periodontitis
- K-562
- Kanechlor
- Keratinocytes
- Keratins
- Keyhole Limpet Hemocyanin (KLH)
- **Klebsiella, Infection and Immunity**
  HELEN V RATAJCZAK
- KLH ELISA
- Knock-In
- Knockout Animal
- **Knockout, Genetic**
  JEANINE L. BUSSIERE, BRAD BOLON
- KO Mouse
- Kupffer Cells
- Lactate Dehydrogenase (LDH) and/or Adenylate Kinase (AK) Leakage
- Lactoferrin
- Lamina propria
- Langerhans Cells
- Large Granular Lymphocyte
- Laser Capture Microdissection (LCM)
- Lateral Line
- LDA
- Lectins
- **Leukemia**
  LEIGH ANN BURNS-NAAS
- Leukemia-Initiating Cells (L-IC)
- Leukocyte
- **Leukocyte Culture: Considerations for In Vitro Culture of T cells in Immunotoxicological Studies**
  MACIEJ TARKOWSKI
- Leukocyte Differentiation Antigens
- Leukocyte Emigration
- Leukocyte Function-Associated Antigen-3 (LFA-3)
- Leukocyte Margination and Adhesion
- Ligand Blotting
- Ligand Passer
- Ligand Passing
- Ligand Traps
- **Limiting Dilution Analysis**
  GEORG BRUNNER
- Limiting Dilution Cell Cloning
- Limiting Dilution Polymerase Chain Reaction
- Lipid Rafts
- Lipids
- Lipopolysaccharide (LPS)
- *Listeria Monocytogenes*
- Live-Death Discrimination
- Live Rate
- LLNA Challenge Experiment
- Local Immune System
- **Local Lymph Node Assay**
  IAN KIMBER, REBECCA J DEARMAN
- **Local Lymph Node Assay (IMDS), Modifications**
  PETER ULRICH, HANS-WERNER VOHR
- Local Lymph Node Assay (LLNA)
- Local Skin Immune Hyperreaction
- LPS
- LTT
- Lung Sensitization Test
- Lupoid Hepatitis
- Lymph
- Lymph Flow
- Lymph Gland
- Lymph Node
- **Lymph Nodes**
  C FRIEKE KUPER
- **Lymph Transport and Lymphatic System**
  ANATOLIY A GASHEV, DAVID C ZAWIEJA
- Lymphadenopathy
- Lymphangion
- Lymphatic
- Lymphoblastic Lymphoma
- Lymphocyte-Activated Killer (LAK) Cells

- Lymphocyte-Activating Factor
- Lymphocyte Activation Test
- Lymphocyte Mitogenesis
- **Lymphocyte Proliferation**
  RENÉ CREVEL
- Lymphocyte Proliferation Test
- Lymphocyte Proliferative Response
- Lymphocyte Transformation
- **Lymphocyte Transformation Test**
  WERNER PICHLER
- **Lymphocytes**
  BRAD BOLON
- Lymphocytes
- Lymphodynamics
- Lymphoid Organ Compartments
- Lymphokine
- **Lymphoma**
  LEIGH ANN BURNS-NAAS
- Lymphotoxin
- Lytic Unit (LU)
- M Cells
- MAbs
- Macrophage
- **Macrophage Activation**
  JG LEWIS
- Macrophage Development
- Macrophage Differentiation
- Macrophage Inflammatory Protein-1
- Macrophage Inflammatory Protein-1α (MIP-1α)
- Macrophage Inflammatory Protein-1β
- Macrophage Inflammatory Protein 3 alpha (MIP-3α)
- Macrophage Maturation
- Magnusson-Kligman Maximization Test
- Major Histocompatibility Complex Class I Antigen Presentation
- Major Histocompatibility Complex Class II Antigen
- Major Histocompatibility Complex (MHC)
- Major Histocompatibility Complex (MHC) Molecules
- MALT
- Mannose-Binding Lectin (MBL)
- Mannose-Binding Lectin Pathway
- MAP-Kinase
- Marginal Zone
- Marginal Zone Lymphoma
- **Mast Cells**
  FRANK AM REDEGELD, MAURICE W VAN DER HEIJDEN
- (Matrix) Metalloproteinase (MMP)
- **Maturation of the Immune Response**
  HUUB FJ SAVELKOUL, SCOTT B CAMERON, ANTHONY W CHOW
- MC
- Medulla
- Megakaryocyte Colony-Forming Unit (CFU-Meg)
- Melanocytes
- Melanomacrophages
- Melatonin
- Membrane Attack Complex (MAC)
- Memory Cells
- **Memory, Immunological**
  WILLIAM LEE
- Memory T Cells
- MEST
- Metabolic Activation
- **Metabolism, Role in Immunotoxicity**
  TAE CHEON JEONG
- Metabolite
- **Metals and Autoimmune Disease**
  DAVID A LAWRENCE
- 3-methylcholanthrene, 3-MC, dibenz-a, h-anthracene, benz-a-anthracene, BA, BP-quinones
- MHC
- MHC Antigen Presentation
- **MHC Class I Antigen Presentation**
  HANSJOERG SCHILD, MARK SCHATZ
- MHC (Major Histocompatibility Complex) Class II Molecule
- MHC Restriction
- MHC Tetramer
- Microarray Technology
- Microenvironment of the Bone Marrow
- Microfold Cells
- Microinjection
- Micronutrients
- Microparticles
- Microsphere-Based Multiplex Assays

- MIG
- Migration Inhibitory Factor (MIF)
- Migration of Neutrophils
- Minipig
- Mismatched or Matched Organ
- Mitogen
- Mitogen-Activated Protein Kinase Cascade
- Mitogen-Activated Protein Kinases (MAP Kinases)
- Mitogen Assay
- Mitogen-Induced Lymphocyte Blastogenesis
- Mitogen-Induced Lymphoproliferative Response
- Mitogen Response
- Mitogen-Stimulated Lymphocyte Proliferation Assay
- **Mitogen-Stimulated Lymphocyte Response**
  RALPH J SMIALOWICZ
- Mitogenic Stimulation
- Mitogens
- Mitomycin C
- Mitotic
- Mixed Leukocyte Culture
- **Mixed Lymphocyte Reaction**
  RALPH J SMIALOWICZ
- Mixed Lymphocyte Response (MLR)
- MMC
- Mobilization
- Modeling
- Molecular Chaperones
- **Molecular Mimicry**
  ALAN EBRINGER, LUCY HUGHES, TAHA RASHID, CLYDE WILSON
- Molecular Similarity
- **Monoclonal Antibodies**
  GEORGE TREACY, DAVID M KNIGHT
- Monoclonal Antibody (mAb)
- Monocyte Chemoattractant Protein 1 (MCP-1)
- Monocytes
- Monokine
- Mononuclear Cell Function
- Mononuclear Leukocyte
- Mononuclear Phagocyte System (MPS)
- Monounsaturated Fatty Acids
- Morbilliform Eruptions
- Motif
- –/– Mouse
- **Mouse Ear Swelling Test**
  SHAYNE COX GAD
- Mouse Immune System
- MSC
- MSE
- MTT Conversion (MTT Test)
- Mucociliary
- Mucosa
- **Mucosa-Associated Lymphoid Tissue**
  ROSANA SCHAFER, CHRISTOPHER CUFF
- Mucosal Mast Cells
- Mucositis
- Multicomponent Enzyme Complexes
- Multiple Sclerosis
- Multiplex
- Multiplicity
- Multipotential Stem Cell
- Multisystem Autoimmunity
- Murine Immune System
- Muteins
- Myalgia
- Mycotoxins
- Myeloid Differentiation Factor 88
- Myeloid Suppressor Cells
- Myeloperoxidase
- Myocardial Infarction
- Myocarditis
- Myosin
- Naive Cell
- Naive T Cell
- Nasal-Associated Lymphoid Tissue
- Natural Antibodies
- Natural Antibody
- **Natural Killer Cell Assay**
  KARIN CEDERBRANT
- **Natural Killer Cells**
  DAVID SHEPHERD
- Natural Killer ^{51}Cr Release Assay
- Natural Killer (NK) Cell
- Natural Killer (NK) Cell Assay
- Negative Selection
- Neoantigen-Forming Chemicals
- **Neonatal Immune Response**
  KENNETH S. LANDRETH, SARAH V.

M. Dodson
- Neonatal Tolerance
- Neural Tube Defect
- Neuroendocrine Response
- Neurons
- Neurotransmitter
- Neutropenia
- **Neutrophil**
  Kathleen Rodgers
- Neutrophils
- Newborn Immune Function
- NF-kappa B (NFκB)
- NHL
- Niacin
- Nitro-PAH
- Nitrophenyl-Chicken gamma Globulin
- NK Cells
- NK Gene Complex (NKC)
- NK Cell Killing
- No Observable Adverse Effect Level (NoAEL)
- Nodus lymphaticus
- Non-Caseating Granuloma
- Non-Hodgkin's Lymphoma
- Non-Obese Diabetic Mouse (NOD)
- Non-Radioactive Flow Cytometric Analysis of NK Cell Cytotoxicity
- Non-Steroidal Anti-Inflammatory Drugs (NSAIDs)
- Nonhuman Primates, Immunotoxicity Assessment of Pharmaceuticals in
  Werner Frings, Gerhard F Weinbauer
- Nonparametric Statistics
- Nonsteroidal Antiinflammatory Drugs
- Norepinephrine (Noradrenaline)
- Northern
- Nosocomial
- NSAID-Activated Gene (NAG-1)
- NSAIDs
- Nuclear Factor κB (NFκB)
- Nucleic Acid Blotting
- Nucleic Acid Vaccines
- Nude Mouse
- Null Mutant Mouse
- Nurse Cell
- **Nutrition and the Immune System**
  Michelle Carey

- Obese Zucker rat
- OECD
- Oily Fish
- Oliguria
- Oncogenes
- Opportunistic infection
- Opsonin
- Opsonins
- Opsonization
- **Opsonization and Phagocytosis**
  Charles J Czuprynski
- **Oral Mucositis and Immunotoxicology**
  Gary J. Rosenthal
- Oral Ulcer
- Organ-Specific Autoimmunity
- Organogenesis
- Organotypic Murine or Human Skin Explant System
- Orofacial Cleft
- *Oryctolagus cuniculus*
- Oxidative Stress
- Oxy-PAH
- p53
- p53 Tumor Suppressor Protein
- PAH
- Paramagnetic Cell Selection
- Parturition
- Passive Cutaneous Anaphylaxis
- Passive Immunotherapy
- Passive Lymph Pump
- Pathogenicity
- PBS
- PCA
- PCDDs
- PCR
- PEF
- Pentavalent Vanadium
- Peptide Regulatory Factors
- Peptides
- Percent of Living Cells
- Perforin
- Peripheral Tolerance
- Peyer's Patches
- PHA
- Phagocytic Cells
- Phagocytosis

- Phagocytosis Assay
- Pharmacodynamics
- Pharmacokinetics
- Phenotype
- Phosphatases
- Phosphorylative Balance
- Photoactivation
- Photoallergic Contact Dermatitis
- Photoallergy (Photoallergic Contact Dermatitis)
- Photodermatology
- Photoirritation
- **Photoreactions**
  Thomas Maurer
- **Photoreactive Compounds**
  Frank Gerberick
- Photosafety
- Photosensitivity
- Photosensitization
- Phototoxicity
- Physicochemical Properties
- Phytohemagglutin (PHA)
- Pig
- Pineal Gland
- Pinocytic Uptake
- Pituitary Gland
- Plaque Assay
- **Plaque-Forming Cell Assays**
  Gregory Ladics
- **Plaque Versus ELISA Assays. Evaluation of Humoral Immune Responses to T-Dependent Antigens**
  Kimber L. White
- Plasma Cell
- *Plasmodium*
- Platelets
- Plating Efficiency
- PLN Index
- PLNA
- Pluripotent
- Pluripotential Stem Cell
- Pneumococcal Disease Models
- Pneumonia
- Pokeweed Mitogen
- Polarization
- Polybrominated Biphenyls (PBBs)
- **Polychlorinated Biphenyls (PCBs) and the Immune System**
  John L Olsen
- Polychlorinated Dibenzodioxins
- Polychlorinated Diphenyls
- Polyclonal
- **Polyclonal Activators**
  Stephen B Pruett
- Polyclonal Mitogens
- **Polycyclic Aromatic Hydrocarbons (PAHs) and the Immune System**
  Scott W Burchiel
- **Polymerase Chain Reaction (PCR)**
  John L Olsen
- Polymeric Immunoglobulin Receptor (pIgR)
- Polymorphism
- Polymorphonuclear Leukocyte
- Polymorphonuclear Neutrophil
- Polynucleotide Vaccines
- Polyunsaturated Fatty Acids
- **Popliteal Lymph Node Assay**
  Raymond Pieters
- **Popliteal Lymph Node Assay, Secondary Reaction**
  Peter Griem
- Popliteal Plymph Node Assay (PLNA)
- Population Studies
- **Porcine Immune System**
  Ricki M Helm
- Positive Level
- Positive Selection
- Preclinical Immunotoxicity Evaluation in the Nonhuman Primate
- Preclinical Safety Assessment
- Prednisone
- Prenatal Immunotoxicology
- Prevention of Infection
- Primary Antibody Response
- Primary Humoral Immune Response
- Primary Immune Response
- Primary Lymphoid Organs
- **Primate Immune System (Nonhuman) and Environmental Contaminants**
  Helen Tryphonas
- Primed Macrophages
- Procainamide

- Progesterone
- Programmed Cell Death
- Proinflammatory Cytokine
- Prolymphocytic Leukemia
- Promoter
- Pronucleus
- **Prostaglandins**
  TINA SALI
- Prostaglandins
- Prostanoids
- Proteases
- Protein Blotting
- Protein Kinases
- Protein-Modifying Compound
- Proteins
- Pseudoallergic Reaction
- Pseudoallergy
- Psoriasis
- PUFA
- Pulmonary Hypersensitivity
- Pulmonary Infections
- PWM
- Pyrene
- Pyrogen
- Quantitative Analysis
- Quantitative Structure–Activity Relationship (QSARs)
- Quenching
- RA
- RA-PLNA
- Rabbit
- **Rabbit Immune System**
  ROSE G MAGE
- Radiation Mucositis
- Radioimmunoassay (RIA)
- Randomized Complete Blocks Design
- RANTES
- Ras
- Rat Immune System
- Reaction
- Reactive Oxygen Intermediate (ROI)
- Real-Time and Quantitative PCR
- Real-Time Polymerase Chain Reaction
- Real-Time Reverse Transcription PCR
- Rearrangement
- Recall Antigens
- Receptor Shedding
- Receptors for Mediators of the Immune System
- Recombinant
- Recombinant Antibodies
- Reconstructed Human Skin/Epidermis
- Red Pulp
- Regenerative Anemia
- Regression Analysis
- Regulated on Activation, T Cell Expressed and Secreted (RANTES)
- Regulatory Cells
- Regulatory Environment
- **Regulatory Guidance in Immunotoxicology**
  ROBERT V HOUSE
- Regulatory T Cells
- **Relative Risk**
  STEPHEN B PRUETT
- RELISPOT
- Repeated Measures Design
- Replicate Cultures
- Reporter Antigen
- **Reporter Antigen Popliteal Lymph Node Assay**
  RAYMOND PIETERS
- Resident Macrophages
- Respiratory Allergy Assay
- Respiratory Burst
- Respiratory Hypersensitivity Test
- **Respiratory Infections**
  DONALD E GARDNER, SUSAN C GARDNER
- Responder Cell
- Responder Cell Frequency
- Reverse Enzyme-Linked Immunospot Assay
- Reye's Syndrome
- Rheumatic Fever
- **Rheumatoid Arthritis and Related Autoimmune Diseases, Animal Models**
  JEANNE M SOOS
- Rheumatoid Arthritis (RA)
- Rhinitis
- Ribonucleic Acid (RNA)
- Ricin
- Risk Assessment
- **Rodent Immune System, Development of the**

KENNETH S. LANDRETH
▶ **Rodents, Inbred Strains**
INA HAGELSCHUER
▶ Rosetting Techniques
▶ RT1.B, RT1.D (Rat)
▶ RTqPCR
▶ Safety Assessments
▶ SAg
▶ **Salmonella, Assessment of Infection Risk**
WIM H DE JONG, ROB DE JONGE,
JOHAN GARSSEN, KATSUHISA TAKUMI, ARIE
H HAVELAAR
▶ *Salmonella enterica*
▶ *Salmonella enterica serovar* Enteritidis
▶ *Salmonella* Enteritidis
▶ *Salmonella* Enteritidis Rat Model
▶ *Salmonella* Food-Borne Disease
▶ *Salmonella* Food Poisoning
▶ *Salmonella typhimurium*
▶ Salmonellosis
▶ SAPS
▶ SARS
▶ Saturated Fatty Acids
▶ SCID Mouse
▶ Scleroderma
▶ SDS-PAGE
▶ SEB
▶ Secondary Antibody Response
▶ Secondary Cytokines
▶ Secondary Humoral Immune Response
▶ Secondary Immune Response
▶ Secondary Lymphoid Organs
▶ Secondary Neoplasms
▶ Secondary PLNA
▶ Secondary Prevention
▶ Secretory Immune System
▶ Secretory Immunoglobulin A
▶ Selection
▶ Self Antigen
▶ Self-Renewal
▶ Semiquantitative PCR
▶ Sensitization
▶ Sensitizer
▶ **Septic Shock**
JUTTA LIEBAU
▶ Septicemia

▶ Septicemic Shock
▶ SEREX
▶ Serine Protease Inhibitors
▶ **Serotonin**
HELEN V RATAJCZAK
▶ Serum Sickness
▶ Seveso-Dioxin
▶ Seveso-Poison
▶ SFC
▶ Shared Tumor Antigens
▶ Sheep Red Blood Cell Receptor
▶ Sheep Red Blood Cells (SRBC)
▶ Shingles
▶ **Signal Transduction During Lymphocyte Activation**
KATHLEEN M BRUNDAGE
▶ Signaling Through Antigen Receptors
▶ Single Amino Acid Polymorphisms
▶ Single Base Transition
▶ Single Nucleotide Polymorphisms
▶ Sjögren Syndrome
▶ **Skin, Contribution to Immunity**
EMANUELA CORSINI
▶ Skin Prick Test
▶ Skin Sensitization Assay
▶ Skin Sensitization Potency
▶ SLE
▶ Sleeping Sickness
▶ Slp 76 (SH2 Domain-Containing Leukocyte Protein of 76 kDa)
▶ Smad
▶ Small Secreted Cytokines
▶ SNPs
▶ Somatic Hypermutation
▶ Southern
▶ **Southern and Northern Blotting**
KEVIN TROUBA
▶ Spherocytes
▶ **Spleen**
C FRIEKE KUPER
▶ Split-Plot or Split-Unit Design
▶ Split-Well Analysis
▶ Spot-Forming Cells
▶ SRBC ELISA
▶ Src Family Kinases
▶ Src Homology 2 (SH2) Domain

- Src Homology 3 (SH3) Domain
- **Statistics in Immunotoxicology**
  MICHAEL L KASHON
- Stem Cells
- Stereochemistry
- Sterilizing Immunity
- Steroid Hormones
- **Steroid Hormones and their Effect on the Immune System**
  MICHAEL LAIOSA
- Steroid Receptors
- Stevens–Johnson Syndrome (SJS)
- Stimulation Index
- Stomatitis
- Streptococcal Enterotoxin B (SEB)
- **Streptococcus Infection and Immunity**
  S GAYLEN BRADLEY
- *Streptococcus pneumoniae*
- Streptokinase
- Streptozotocin-Induced Diabetes Mellitus, STZ-Induced Diabetic Rat
- Streptozotocin-Induced Spontaneously Hypertensive Rat,
- Stress
- **Stress and the Immune System**
  HELEN G. HAGGERTY
- Stressor
- Stroma
- Stromal Cell-Derived Factor-1 (SDF-1)
- Stromal Cells
- Structural Alert
- Structure Activity Relationships
- STZ-SHR Rat
- Sunlight
- **Superantigens**
  THOMAS HERRMANN
- **Suppressor Cells**
  NANCY I KERKVLIET
- Surface Plasmon Resonance (SPR)
- Surfactant
- Surrogate
- Sweetbread (When Used as Food)
- Swine
- SYBR Green
- Symian
- Syngeneic
- Systemic Anaphylaxis
- **Systemic Autoimmunity**
  K MICHAEL POLLARD
- Systemic Lupus Erythematosus (SLE)
- Systemic Vascular Resistance (SVR)
- T Cell
- T Cell Antigen Receptor (TCR)
- T Cell Antigen-Specific Receptor
- T Cell-Dependent Antibody Response
- T Cell-Dependent Antigen
- T Cell-Independent Antigen
- T Cell Oligoclonality
- T Cell Receptor (TCR)
- T Cell Receptor (TCR) Complex
- T Cell Selection
- $\gamma\delta$T Cells
- T-Dependent Antibody-Forming Cell Response
- T Helper 1 Cells
- T Helper 1 Cytokines (Th1)
- T Helper 1 (Th1) Cells
- T Helper 1 (Th1) Response
- T Helper 1–T Helper 2 Balance
- T Helper 2 Cells
- T Helper 2 (Th2) Cells
- T Helper 2 (Th2) Response
- T Helper Cell
- T Helper Cell Polarization
- T Helper Lymphocyte
- T Lymphocyte
- 3T3 Neutral Red Uptake (NRU) Test
- T Regulatory Cells (T_{regs})
- T Suppressor Lymphocyte
- Tachycardia
- Tachypnea
- TAPA-1 (Target of an Antiproliferative Antibody-1)
- Taqman
- Target Cell
- Target Cell Killing
- Targeted Mutant Mouse
- TbAT1 Genes
- TCDD
- Telomeres
- TEQ/TEF
- Teratogen

- Testosterone
- **Tests for Autoimmunity**
  Raymond Pieters
- 2,3,7,8-tetrachlorodibenzo-p-dioxin
- Tetravalent Vanadium
- TGF-β1
- Th1/Th2 Balance
- **Three-Dimensional Human Skin/Epidermal Models and Organotypic Human and Murine Skin Explant Systems**
  Hans-Werner Vohr, Eckhart Heisler
- Three Rs
- Thrombin
- Thrombocytopenia
- Thrombocytopenic Purpura
- Thymic Hypoplasia
- Thymocyte Development
- Thymocyte Education
- Thymocyte Selection
- **Thymus**
  C Frieke Kuper
- **Thymus: A Mediator of T Cell Development and Potential Target of Toxicological Agents**
  Michael Laiosa, Allen Silverstone
- Thymus Atrophy
- Thymus-Dependent Antigen
- Thymus Gland
- Thymus Involution
- Tight Junctions
- Time-Resolve Fluorometry
- Tissue Factor
- Tm Mouse
- TNF-α
- **Tolerance**
  Anke Kretz-Rommel
- Tolerance and the Immune System
- Toll-Like Receptors
- Toxic Epidermal Necrolysis (TEN)
- Toxic Oil Syndrome (TOS)
- Toxicogenetics
- Toxicogenomic Studies
- Toxicogenomics
- **Toxicogenomics (Microarray Technology)**
  Rob J Vandebriel
- T
- Trace Metals

- **Trace Metals and the Immune System**
  Judith T Zelikoff
- Trans-Signaling
- Transcription Factors
- Transendothelial Migration
- Transferrin
- Transferrin Receptor
- **Transforming Growth Factor β1; Control of T Cell Responses to Antigens**
  Susan C McKarns
- Transforming Growth Factor β1 (TGF-β1)
- **Transgenic Animals**
  Peter J Bugelski
- Transgenic Mouse
- Transglutaminase
- Transition Element
- Transporter Associated with Antigen Processing (TAP)
- *Trichinella spiralis*
- Triglycerides
- Trivalent Chromium
- **Trypanosomes, Infection and Immunity**
  Ronald Kaminsky
- Tryptophan
- Tsetse Fly
- TSK
- Tuberculin
- Tuberculin-Type Reaction
- Tumor Antigen
- Tumor-Associated Antigens
- **Tumor, Immune Response to**
  Pier-Luigi Lollini
- Tumor Immunology
- Tumor-Infiltrating Lymphocytes
- Tumor Necrosis Factor (TNF)
- **Tumor Necrosis Factor-α**
  Victor J Johnson
- Tumor Necrosis Factor Receptor-Associated Factor-6
- Tumor-Specific Antigen
- Type I Error
- Type I Reactions According to Gell and Coombs
- Type I–IV Reactions
- Type 1 or Type 2 T Cell Responses
- Type II Activation

- ▶ Type II Error
- ▶ Type II Interferon
- ▶ Ulcerative Mucositis
- ▶ Ulcerative Stomatitis
- ▶ Upper Respiratory Infection
- ▶ Uptake
- ▶ Urinary ract Infection
- ▶ Urticaria
- ▶ US Clean Air Act
- ▶ UVA Radiation
- ▶ UVC Radiation
- ▶ Vβ
- ▶ Valency
- ▶ Vanadate
- ▶ **Vanadium and the Immune System**
  MITCHELL D. COHEN
- ▶ Variable Region (V Region)
- ▶ Variable Surface Glycoprotein (VSG)
- ▶ Variance, Analysis of
- ▶ Variant Antigen Type (VAT)
- ▶ Vasculitis
- ▶ Vasculopathy
- ▶ Vasoactive Amine
- ▶ VDJ
- ▶ VDJ Region
- ▶ **Viability, Cell**
  ANDREA ENGEL
- ▶ Vimentin
- ▶ Virulence
- ▶ **Vitamins**
  RENÉ CREVEL
- ▶ Waldeyer's Ring
- ▶ Warm Autoantibodies
- ▶ **Western Blot Analysis**
  GEORG BRUNNER
- ▶ White Pulp
- ▶ WHO
- ▶ Xenogeneic
- ▶ XRE-Element (Xenobiotic Response Element)
- ▶ XSCID
- ▶ YAC
- ▶ Zap70 (Zeta-Associated Protein of 70 kDa)
- ▶ Zucker Rat